THE INTERNATIONAL SERIES OF MONOGRAPHS ON CHEMISTRY

THE INTERNATIONAL SERIES OF MONOGRAPHS ON CHEMISTRY

1. J. D. Lambert: *Vibrational and rotational relaxation in gases*
2. N. G. Parsonage and L. A. K. Staveley: *Disorder in crystals*
3. G. C. Maitland, M. Rigby, E. B. Smith, and W. A. Wakeman: *Intermolecular forces: their origin and determination*
4. W. G. Richards, H. P. Trivedi, and D. L. Cooper: *Spin-orbit coupling in molecules*
5. C. F. Cullis and M. M. Hirschler: *The combustion of organic polymers*
6. R. T. Bailey, A. M. North, and R. A. Pethrick: *Molecular motion in high polymers*
7. Atta-ur-Rahman and A. Basha: *Biosynthesis of indole alkaloids*
8. J. S. Rowlinson and B. Widom: *Molecular theory of capillarity*
9. C. G. Gray and K. E. Gubbins: *Theory of molecular fluids. Volume 1: Fundamentals*
10. C. G. Gray, K.E. Gubbins, and C. G. Joslin: *Theory of molecular fluids. Volume 2: Applications*
11. S. Wilson: *Electron correlations in molecules*
12. E. Haslam: *Metabolites and metabolism: a commentary on secondary metabolism*
13. G. R. Fleming: *Chemical applications of ultrafast spectroscopy*
14. R. R. Ernst, G. Bodenhausen, and A. Wokaun: *Principles of nuclear magnetic resonance in one and two dimensions*
15. M. Goldman: *Quantum description of high-resolution NMR in liquids*
16. R. G. Parr and W. Yang: *Density-functional theory of atoms and molecules*
17. J. C. Vickerman, A. Brown, and N. M. Reed (editors): *Secondary ion mass spectrometry: principles and applications*
18. F. R. McCourt, J. Beenakker, W. E. Köhler, and I. Kuščer: *Nonequilibrium phenomena in polyatomic gases. Volume 1: Dilute gases*
19. F. R. McCourt, J. Beenakker, W. E. Köhler, and I. Kuščer: *Nonequilibrium phenomena in polyatomic gases. Volume 2: Cross-sections, scattering and rarefied gases*
20. T. Mukaiyama: *Challenges in synthetic organic chemistry*
21. P. Gray and S. K. Scott: *Chemical oscillations and instabilities: non-linear chemical kinetics*
22. R. F. W. Bader: *Atoms in molecules: a quantum theory*
23. J. H. Jones: *The chemical synthesis of peptides*
24. S. K. Scott: *Chemical chaos*
25. M. S. Child: *Semiclassical mechanics with molecular applications*
26. D. T. Sawyer: *Oxygen chemistry*
27. P. A. Cox: *Transition metal oxides: an introduction to their electronic structure and properties*
28. B. R. Brown: *The organic chemistry of aliphatic nitrogen compounds*
29. Y. Yamaguchi, Y. Osamura, J. D. Goddard, and H. F. Schaeffer: *A new dimension to quantum chemistry: analytic derivative methods in* ab initio *molecular electronic structure theory*
30. P. W. Fowler and D. E. Manolopoulos: *An atlas of fullerenes*
31. T. Baer and W. L. Hase: *Unimolecular reaction dynamics*
32. A. J. Stone: *The theory of intermolecular forces*
33. L. Bruch, M. Cole, and E. Zaremba: *Physical absorption: forces and phenomena*

Theory of Molecular Fluids

Volume 2: Applications

C.G. GRAY, K.E. GUBBINS
and
C.G. JOSLIN

OXFORD
UNIVERSITY PRESS

OXFORD
UNIVERSITY PRESS

Great Clarendon Street, Oxford OX2 6DP

Oxford University Press is a department of the University of Oxford.
It furthers the University's objective of excellence in research, scholarship,
and education by publishing worldwide in

Oxford New York

Auckland Cape Town Dar es Salaam Hong Kong Karachi
Kuala Lumpur Madrid Melbourne Mexico City Nairobi
New Delhi Shanghai Taipei Toronto

With offices in

Argentina Austria Brazil Chile Czech Republic France Greece
Guatemala Hungary Italy Japan Poland Portugal Singapore
South Korea Switzerland Thailand Turkey Ukraine Vietnam

Oxford is a registered trade mark of Oxford University Press
in the UK and in certain other countries

Published in the United States
by Oxford University Press Inc., New York

First published 2011

British Library Cataloguing in Publication Data
Data available

Library of Congress Cataloging in Publication Data
Data available

Typeset by SPI Publisher Services, Pondicherry, India
Printed and bound by
CPI Group (UK) Ltd, Croydon, CR0 4YY

ISBN 978–0–19–855621–3

1 3 5 7 9 10 8 6 4 2

PREFACE

The aim of this book is to present an account of the equilibrium statistical mechanics of fluids of nonspherical molecules, especially liquids, including theories and applications to physical property behaviour. Since several excellent texts exist on the theory of simple liquids (e.g. J.-P. Hansen and I.R. McDonald, *Theory of Simple Liquids*, 3rd edition, Academic Press, San Diego, 2006), having central forces between atoms, we focus in particular on the effects arising from the acentricity in the intermolecular forces. In our previous volume (C.G. Gray and K.E. Gubbins, *Molecular Theory of Fluids*, Volume 1, *Fundamentals*, Clarendon Press, Oxford, 1984) we gave an account of the general theory, including basic statistical mechanics, intermolecular forces, and perturbation and integral equation theory. In this volume we consider the application of theory to particular classes of physical properties. We focus on equilibrium properties; time-dependent properties are discussed briefly in a few sections. Much of our aim is to present applications and tests of the theory against molecular simulations and experimental results. However, we also introduce new theory as needed for specific classes of properties, for example, the theory of hard nonspherical particles (§ 6.12), Wertheim's theory of associating liquids (§§ 6.13 and 7.9), conformal solution theories for simple mixtures (§ 7.7), Kirkwood–Buff theory of mixtures (§§ 7.2–7.6), density functional theory for inhomogeneous fluids (§§ 8.4, 8.6, 8.7), scattering theory (§ 9.1), and the theory of the structure factor (Chapter 9), dielectric properties (Chapter 10), and equilibrium spectral properties (Chapter 11). We have tried to make the book self-contained (the mathematical apparatus needed is given in Appendices A, B, and C of Volume 1, as well as in appendices of some chapters in the present volume, e.g. Appendices 6A and 8A), with a few exceptions where a full account of a theory would have lengthened the book considerably. The book is aimed at senior undergraduates, beginning graduate students, and research workers in chemistry, physics, life sciences, and engineering. We assume an undergraduate level knowledge of thermodynamics, statistical mechanics, electromagnetic theory, matrices, and vector analysis. A few sections require basic quantum mechanics.

Molecular simulation methods, particularly Monte Carlo and molecular dynamics methods for numerically solving the equations of statistical mechanics, have risen greatly in importance over the past few decades, thanks to increasing computer power and more efficient algorithms. They play an important role in testing analytic theories, validating intermolecular force models, testing the limits of macroscopic concepts at the nano scale, and determining properties that cannot yet be measured experimentally. We make extensive use of the results of such simulations, which make it possible to test theories without the added uncertainties associated with the intermolecular force fields. However, we do not describe these numerical simulation methods in detail here, since several excellent monographs already exist (e.g. D. Frenkel and B. Smit, *Understanding Molecular Simulation: From Algorithms to Applications*, 2nd edition, Academic Press, San Diego, 2002; M.P. Allen and D.J. Tildesley, *Computer Simulation of Liquids*, Clarendon Press, Oxford, 1987).

This book has had a long gestation period, more than thirty years. We had imagined that this might qualify as some sort of record in procrastination, but have been assured by one of the Delegates of Oxford University Press that 'thirty years is nothing'. Moreover, we are informed by the Press that the procrastination record is held by an Oxford don who produced his manuscript only after a period exceeding 70 years, an achievement with which we cannot hope to compete.

We are indebted to many friends and colleagues for helpful discussions, reading parts of the text, and for help in preparing the manuscript and the figures. In particular we benefited from helpful discussions with J.S. Rowlinson, B. Widom, R. Evans, K. Kaneko, M. Schoen, S.H.L. Klapp, M. Śliwinska-Bartkowiak, R. Raab, J.-Z. Wu, E. Santiso, and J.D. Moore; we are grateful to G. Jackson for reading Chapters 6 and 7, and for helpful comments on them; to G.H. Findenegg for comments on parts of Chapter 8; to D.E. Sullivan for reading and commenting on Chapter 8; and to A.D. Buckingham for reading an early draft of Chapter 10. We thank Demetra Dentes, June McKoy, Connie York, Erik Santiso, Joshua Moore, and Brenda Law for their assistance with the manuscript preparation, and Joshua Moore and Pauline Gubbins for their expert help in preparing the figures. We are also grateful for support of our research by the National Science Foundation, Department of Energy Basic Sciences, American Chemical Society, and the Natural Sciences and Engineering Research Council of Canada. Finally, it is a pleasure to thank Sir John Rowlinson for his help and encouragement in planning and carrying out this project, and the Oxford University Press for their patience and courtesy during the years these volumes were being written.

Guelph, Raleigh, and Lake Oswego C.G.G.
February 2011 K.E.G.
C.G.J.

TO
VIRGINIA, PAULINE AND TRISTANIA

VOLUME 1: FUNDAMENTALS

CONTENTS

VOLUME 2: APPLICATIONS

NOTE ON UNITS

In dealing with molecular properties and intermolecular forces and their effect on fluid properties we are concerned almost entirely with electrostatics; electromagnetic forces and macroscopic circuit problems play only a very minor role. We have therefore found it convenient to use mainly the electrostatic system of units rather than the rationalized Système International (SI) in our treatment of these topics (particularly in Chapter 2 and Appendices C and D of Volume 1, and in Chapters 8, 10 and 11 of this volume). In Chapter 9 we use the SI system in two sections (the ones on X-ray and electron scattering). In Chapter 11 we use the Gaussian system (which derives naturally from the esu system) when dealing with light absorption and scattering. This makes for a more compact notation, since the factors of $4\pi\varepsilon_0$ that occur in the SI system are avoided. In Appendix D.3 of Volume 1 the relations among the electrostatic (esu), SI, and atomic units (au) systems are discussed in some detail.

6

THERMODYNAMIC PROPERTIES OF PURE FLUIDS

> [Statistical thermodynamics] differs from classical thermodynamics in that the thermodynamic functions for the assemblies and phases with which we deal are not left unspecified, or to be derived solely from measurements, but are always constructed a priori by the application to particular molecular models of the fundamental theorems of statistical mechanics with which we start.
>
> Ralph Fowler and E. A. Guggenheim, Introduction to *Statistical Thermodynamics* (1949)

In this chapter we consider the thermodynamic properties of a pure isotropic, homogeneous fluid in the classical, rigid molecule approximation (quantum corrections are briefly discussed in § 6.9). In the canonical ensemble the fundamental relation is that for the configurational free energy (the configurational part, in the classical limit, of the total free energy), given by Eq. (3.11) of Vol. 1,

$$A_{\mathrm{c}} = -kT \ln Q_{\mathrm{c}}, \tag{6.1}$$

where Q_{c} is the configurational partition function, given by (3.82):

$$Q_c = \frac{1}{N!\Omega^N} \int \mathrm{d}\mathbf{r}^N \mathrm{d}\omega^N \exp[-\beta \mathcal{U}(\mathbf{r}^N \omega^N)] \equiv \frac{1}{N!\Omega^N} Z_c. \tag{6.2}$$

Here $\beta = 1/kT$, $\Omega = 4\pi$ or $8\pi^2$ for linear and nonlinear molecules, respectively, and Z_{c} is the configurational phase integral†. Expressions for the other thermodynamic properties in terms of the partition function can be derived from (6.1) using the usual relations of classical thermodynamics, e.g.

$$p = -\left(\frac{\partial A_c}{\partial V}\right)_{TN}, \tag{6.3}$$

$$U_{\mathrm{c}} = \left(\frac{\partial(\beta A_{\mathrm{c}})}{\partial \beta}\right)_{VN}, \tag{6.4}$$

† Although the total partition function, Q_{cl}, of eq. (3.77) of Vol. 1 is dimensionless, as it must be, the factored parts of the partition function do have dimensions. Thus, Q_{c} has dimensions of volume to the Nth power, and this leads to the logarithm of a dimensional quantity in Eq. (6.1). The final result for Q_{cl} is, however, dimensionally consistent. Similar points arise later and we do not comment on every one.

$$C_{\mathrm{vc}} = \left(\frac{\partial U_{\mathrm{c}}}{\partial T}\right)_{VN}. \tag{6.5}$$

From (6.1) and (6.3)–(6.5) it is easy to write down the relations for the pressure p, configurational energy U_{c}, and configurational heat capacity C_{vc} in terms of the partition function Q_c. If, in addition to the classical and rigid-molecule approximations we assume the potentials to be pairwise additive (Eq. (1.13)), then we can also relate these properties to the pair correlation function $g(\mathbf{r}\omega_1\omega_2)$. These relations are derived in §§6.1–6.4 below. Alternative expressions for the pressure and chemical potential are provided by the compressibility relation (§ 6.3), and test particle expressions (also referred to as the potential distribution theorem, § 6.5), respectively. These expressions do not depend on either the rigid molecule or pairwise additivity approximations.

The energy, pressure, compressibility, free energy, and chemical potential equations ((6.9a), (6.15), (6.16), (6.24), (6.35), (6.37), and (6.49)) provide alternative routes to the thermodynamic properties. Four of these can be used to check the consistency of any theory of $g(\mathbf{r}\omega_1\omega_2)$ by comparing results obtained by the four methods; these should be identical if the theory is exact, so that the degree of inconsistency gives a rough measure of the inexactness of the theory. Consistency is not a sufficient condition for exactness, of course; thus if we use a perturbation expansion for $g(\mathbf{r}\omega_1\omega_2)$ identical results will be obtained by each of the four routes regardless of the order at which the series is truncated.

In addition to the general expressions for p, U, χ, and A in terms of the partition function, intermolecular potential, and pair correlation function, it is often useful to express these properties in terms of the spherical harmonic expansion coefficients for the potential and the pair correlation function. Moreover, if the pair potential is of the site–site type some of these properties can be related to the simpler site–site correlation functions, rather than the full pair correlation function (in the case of χ such a relation exists irrespective of the type of potential). These expressions are therefore given in §§6.6 and 6.7. In §6.8 we note a rigorous inequality that exists for the free energy of a molecular fluid. Quantum corrections are briefly discussed in §6.9, and virial coefficients are considered in §6.10. Comparisons between theory and experiment are given in §6.11. The scaled particle approach for fluids of hard convex molecules, and its empirical extension to non-convex molecules, is described in §6.12, and theories for associating fluids are discussed in § 6.13.

6.1 The energy equation

The relation between the configurational energy and the partition function is obtained from (6.4) and (6.1) as

$$U_{\mathrm{c}} = -\left(\frac{\partial \ln Z_c}{\partial \beta}\right)_{VN}.$$

On performing the differentiation, we get:

$$U_c = \int d\mathbf{r}^N d\omega^N \mathcal{U} \frac{\exp[-\beta\mathcal{U}]}{Z_c} \equiv \langle \mathcal{U} \rangle. \tag{6.6}$$

Using the pairwise additivity approximation and noting that the sum of the $u(ij) \equiv u(\mathbf{r}_{ij}\omega_i\omega_j)$ contains $\frac{1}{2}N(N-1)$ terms, each of which gives the same result on averaging, we have $U_c = \frac{1}{2}N(N-1)\langle u(12) \rangle$, or

$$\begin{aligned} U_c &= \frac{1}{2}\int d\mathbf{r}_1 d\mathbf{r}_2 d\omega_1 d\omega_2 u(12) \left\{ N(N-1) \int d\mathbf{r}^{N-2} d\omega^{N-2} \frac{\exp[-\beta\mathcal{U}]}{Z_c} \right\} \\ &= \frac{1}{2}\rho^2 \int d\mathbf{r}_1 d\mathbf{r}_2 \langle u(\mathbf{r}_{12}\omega_1\omega_2) g(\mathbf{r}_{12}\omega_1\omega_2) \rangle_{\omega_1\omega_2}, \end{aligned} \tag{6.7}$$

where (3.105) has been used in carrying out the last step. This integral can be further simplified by noting that the integrand is a function of r_{12} only, $F(r_{12}) \equiv \langle u(12)g(12) \rangle_{\omega_1\omega_2}$. On changing integration variables from $(\mathbf{r}_1, \mathbf{r}_2)$ to $(\mathbf{r}_1, \mathbf{r}_{12})$, where $\mathbf{r}_{12} = \mathbf{r}_2 - \mathbf{r}_1$, we have

$$\begin{aligned} \int\int d\mathbf{r}_1 d\mathbf{r}_2 F(r_{12}) &= \int\int d\mathbf{r}_1 d\mathbf{r}_{12} \left| \frac{\partial \mathbf{r}_2}{\partial \mathbf{r}_{12}} \right| F(r_{12}) = \int\int d\mathbf{r}_1 d\mathbf{r}_{12} F(r_{12}) \\ &= V \int d\mathbf{r}_{12} F(r_{12}). \end{aligned} \tag{6.8a}$$

Expressing $d\mathbf{r}_{12}$ in polar coordinates, $d\mathbf{r}_{12} = r_{12}^2 dr_{12} d\omega_{12}$, and carrying out the integrations over the angles $\omega_{12} = \theta_{12}\phi_{12}$ gives

$$\begin{aligned} \int\int d\mathbf{r}_1 d\mathbf{r}_2 F(r_{12}) &= V \int_0^\infty dr_{12} r_{12}^2 F(r_{12}) \int_0^\pi d\theta_{12} \sin\theta_{12} \int_0^{2\pi} d\phi_{12} \\ &= 4\pi V \int_0^\infty dr_{12} r_{12}^2 F(r_{12}). \end{aligned} \tag{6.8b}$$

From (6.7) and (6.8b) we obtain the *energy equation*,

$$U_c = 2\pi\rho N \int_0^\infty dr r^2 \langle u(\mathbf{r}\omega_1\omega_2) g(\mathbf{r}\omega_1\omega_2) \rangle_{\omega_1\omega_2}, \tag{6.9a}$$

which provides a route to the thermodynamic properties if $g(\mathbf{r}\omega_1\omega_2)$ is known. For example, it can be used to derive an equation for the pressure by first using $U_c = \partial(\beta A_c)/\partial\beta$ to derive an expression for A_c, followed by the use of $p = -(\partial A_c/\partial V)$.

For brevity (6.9a) is often written (since the integrand is independent of ω)

$$U_c = \frac{1}{2}\rho N \int \mathrm{d}\mathbf{r} \langle u(\mathbf{r}\omega_1\omega_2) g(\mathbf{r}\omega_1\omega_2)\rangle_{\omega_1\omega_2}. \tag{6.9b}$$

Similar relations hold for the pressure, free energy, and other equations given in subsequent sections.

6.2 The pressure equation

From (6.1) and (6.3) the pressure is given by

$$p = p_c = kT\left(\frac{\partial \ln Z_c}{\partial V}\right)_{TN}, \tag{6.10}$$

or

$$p = \frac{kT}{Z_c}\frac{\partial}{\partial V}\int_0^V \mathrm{d}\mathbf{r}^N \int \mathrm{d}\omega^N \exp[-\beta\mathcal{U}(\mathbf{r}^N\omega^N)]. \tag{6.11}$$

The evaluation of the volume derivative is complicated by the fact that V appears as the upper integration limit for the integrations over the $\mathbf{r}_i$. Such integrals have already been considered in §3.4.1. The differentiation can be performed by a change of integration variables suggested by H. S. Green,[1]

$$\begin{aligned} \mathbf{r}_i &= V^{1/3}\mathbf{r}'_i \\ \mathrm{d}\mathbf{r}_i &= V\mathrm{d}\mathbf{r}'_i \end{aligned} \qquad i = 1, 2 \ldots, N \tag{6.12}$$

so that (6.11) gives

$$p = \rho kT - V^N \int_0^1 d\mathbf{r}'^N \int d\omega^N \frac{\partial\mathcal{U}(\mathbf{r}'^N\omega^N)}{\partial V}\frac{\exp[-\beta\mathcal{U}(\mathbf{r}'^N\omega^N)]}{Z_c} \tag{6.13}$$

Assuming pairwise additivity of the potentials, we have

$$\frac{\partial\mathcal{U}}{\partial V} = \sum_{i\neq j}\frac{\partial u(ij)}{\mathrm{d}\mathbf{r}_i}\cdot\frac{\partial\mathbf{r}_i}{\partial V} = \frac{1}{6V}\sum_{i\neq j} r_{ij}\frac{\partial u(ij)}{\partial r_{ij}}. \tag{6.14}$$

If we substitute (6.14) into (6.13), note that each of the $N(N-1)$ terms in the ij sum gives the same result on averaging, and transform back to the original $\mathbf{r}_i$ variables, we get $p = \rho kT - N(N-1)/6V\langle r_{12}u'(12)\rangle$, or

$$p = \rho kT - \frac{1}{6V}\int \mathrm{d}\mathbf{r}_1\mathrm{d}\mathbf{r}_2\mathrm{d}\omega_1\mathrm{d}\omega_2 r_{12}u'(12)$$
$$\times\left\{N(N-1)\int \mathrm{d}\mathbf{r}^{N-2}\mathrm{d}\omega^{N-2}\frac{\exp[-\beta\mathcal{U}(\mathbf{r}^N\omega^N)]}{Z_c}\right\}$$

or

$$
\begin{aligned}
p &= \rho kT - \frac{\rho^2}{6}\int d\mathbf{r}\,r\langle u'(\mathbf{r}\omega_1\omega_2)g(\mathbf{r}\omega_1\omega_2)\rangle_{\omega_1\omega_2} \\
&= \rho kT - \frac{2\pi}{3}\rho^2\int_0^\infty dr\,r^3\langle u'(\mathbf{r}\omega_1\omega_2)g(\mathbf{r}\omega_1\omega_2)\rangle_{\omega_1\omega_2},
\end{aligned}
\tag{6.15}
$$

where $u' = \partial u/\partial r$, and in the last step we have used (6.8a). Equation (6.15) is the *pressure equation*. The pressure equation provides a second route to the thermodynamic properties. Like the energy equation, (6.9a), it assumes rigid molecules and pairwise additivity. The pressure equation can also be derived from the virial theorem (see Eq. (E.14) of Vol. 1).

We note that in regions of r for which $\partial u/\partial r > 0$ (attractive region) the contribution of the intermolecular potential term on the right-hand side of (6.15) will result in a decrease in pressure, while the reverse is true where $\partial u/\partial r < 0$ (repulsive region).

6.3 The compressibility equation

The compressibility equation provides a third route to the thermodynamic properties, and has been written down in Eq. (3.113). It is most easily derived[2] by differentiating the grand canonical expression for the first order distribution function $f(\mathbf{r}_1\omega_1)$ with respect to the chemical potential. Such derivatives have been considered in §§3.4.1 and 3.4.2, and (3.238) gives the general result of such a differentiation for the distribution function of order h. Putting $h = 1$ in (3.238) gives

$$
\begin{aligned}
\rho kT\chi &= 1 + \rho\int d\mathbf{r}[\langle g(\mathbf{r}\omega_1\omega_2)\rangle_{\omega_1\omega_2} - 1] \\
&= 1 + \rho\int d\mathbf{r}[g(r) - 1],
\end{aligned}
\tag{6.16}
$$

which is the *compressibility equation*. Here $\chi = \rho^{-1}(\partial\rho/\partial p)_T$ is the isothermal compressibility and $g(r)$ is the centres pair correlation function. This expression does not rest on any assumptions concerning the form of the intermolecular potential energy (e.g. pairwise additivity or rigid molecules), and is therefore more general than the pressure equation. It holds for either isotropic or anisotropic fluids (e.g. liquid crystals), provided that they are homogeneous. Equation (6.16) relates χ to the centres correlation function, in contrast to the energy and pressure equations, where the full pair correlation function $g(\mathbf{r}\omega_1\omega_2)$ is needed.

As shown in Appendix 3E, it is also possible to express χ in terms of the direct correlation function, $c(\mathbf{r}\omega_1\omega_2)$, or the centres direct correlation function $c(r) = \langle c(\mathbf{r}\omega_1\omega_2)\rangle_{\omega_1\omega_2}$,

$$\rho kT\chi = [1 - \rho \int d\mathbf{r}\langle c(\mathbf{r}\omega_1\omega_2)\rangle_{\omega_1\omega_2}]^{-1}, \tag{6.17}$$

which holds for an isotropic fluid.

6.4 The free energy equation

In order to relate the free energy to the pair correlation function $g(\mathbf{r}\omega_1\omega_2)$ we use the coupling constant, or generalized charging process, approach.[3] The intermolecular potential energy is written

$$\mathcal{U}_\lambda(\mathbf{r}^N\omega^N) = \mathcal{U}_0(\mathbf{r}^N\omega^N) + \lambda\mathcal{U}_p(\mathbf{r}^N\omega^N), \tag{6.18}$$

so that $\lambda = 0$ corresponds to the reference system and $\lambda = 1$ to the real system. The configurational free energy $A_{c\lambda}$ for this system is

$$A_{c\lambda} = -kT\ln Q_{c\lambda}, \tag{6.19}$$

where

$$Q_{c\lambda} = \frac{1}{N!\Omega^N}\int d\mathbf{r}^N d\omega^N \exp[-\beta(\mathcal{U}_0 + \lambda\mathcal{U}_p)]. \tag{6.20}$$

The difference $A - A_0$ in free energies between the real and reference systems is given by

$$A - A_0 = \int_0^1 d\lambda \frac{\partial A_{c\lambda}}{\partial \lambda}. \tag{6.21}$$

Combining (6.19) and (6.20) and differentiating gives

$$\begin{aligned}\frac{\partial A_{c\lambda}}{\partial\lambda} &= \int d\mathbf{r}^N d\omega^N \mathcal{U}_p \frac{\exp[-\beta\mathcal{U}_\lambda]}{Z_{c\lambda}} \\ &\equiv \langle\mathcal{U}_p\rangle_\lambda,\end{aligned} \tag{6.22}$$

where $\langle\cdots\rangle_\lambda$ indicates an ensemble average for a system with potential $\mathcal{U}_0 + \lambda\mathcal{U}_p$. For pairwise additive $\mathcal{U}_p$, (6.22) becomes (cf. the derivation of (6.9a))

$$\langle\mathcal{U}_p\rangle_\lambda = \frac{1}{2}\rho N \int_0^\infty d\mathbf{r}\langle u_p(\mathbf{r}\omega_1\omega_2)g_\lambda(\mathbf{r}\omega_1\omega_2)\rangle_{\omega_1\omega_2}, \tag{6.23}$$

where g_λ is the pair correlation function corresponding to the system with pair potential $u_0 + \lambda u_p$. From (6.21), (6.22), and (6.23) we therefore have

$$A - A_0 = \frac{1}{2}\rho N \int d\mathbf{r} \int_0^1 d\lambda \langle u_p(\mathbf{r}\omega_1\omega_2)g_\lambda(\mathbf{r}\omega_1\omega_2)\rangle_{\omega_1\omega_2}, \tag{6.24}$$

which is the *free energy equation*. It provides a fourth route from the pair correlation function to the thermodynamic properties. For example, using $p = -(\partial A/\partial V)$ and (6.24) yields a fourth expression for the pressure, in addition to those obtained from the energy, pressure, and compressibility equations. Equation (6.24) can be used in computer simulation studies to calculate free energy changes.[4] The Monte Carlo values of $(A - A_0)$ given in Tables 4.1 and 4.2 were calculated in this way, as were the MSA and GMF values of $(A - A_0)$ obtained in §§5.4.4 and 5.4.9.

6.5 The test particle expressions for thermodynamic properties

6.5.1 *The Widom method*[5]

It is possible to relate the configurational thermodynamic properties to the interaction of a single molecule in the fluid with all of the other surrounding molecules.[6] We call this single, specified molecule the 'test particle'. These expressions are useful in both theoretical (see, for example, §6.12) and computer simulation (see later in this section) studies, and provide an alternative formalism to the expressions obtained in the previous four sections.

We first derive the expression for the chemical potential. Since we shall want to use these expressions for mixtures it is convenient to carry out the derivation for a mixture in which there are N_A molecules of component A, N_B of component $B, \cdots$, and N_R of component R; the total number of molecules is $N = \sum_\beta N_\beta$. The configurational part of the chemical potential $\mu_{\alpha c}$ of component α is given by

$$\mu_{\alpha c} = \left(\frac{\partial A_c}{\partial N_\alpha}\right)_{TVN'} = -kT\left(\frac{\partial \ln Q_c}{\partial N_\alpha}\right)_{TVN'}, \tag{6.25}$$

where N' means all $N_{\beta \neq \alpha}$, and Q_c for mixtures is (cf. (3.82))

$$Q_c = \frac{1}{\prod_\beta N_\beta! \Omega_\beta^{N_\beta}} \int dx^N \exp[-\beta \mathcal{U}_N(x^N)], \tag{6.26}$$

where $x^N = x_1 x_2 \cdots x_N$ with $x_i = \mathbf{r}_i \omega_i$, and $dx_i = d\mathbf{r}_i d\omega_i$. Since the N_β are large we can approximate (6.25) by

$$\mu_{\alpha c} \approx -kT \ln \frac{Q_c(N_\alpha)}{Q_c(N_\alpha - 1)} = -kT \ln\left[\frac{1}{N_\alpha \Omega_\alpha} \frac{Z_c(N_\alpha)}{Z_c(N_\alpha - 1)}\right], \tag{6.27}$$

where $Q_c(N_\alpha)$ is given by (6.26) and $Q_c(N_\alpha - 1)$ is the corresponding partition function for the system of $N - 1$ molecules (one α molecule removed),

$$Q_c(N_\alpha - 1) = \frac{1}{(N_\alpha - 1)! \Omega_\alpha^{N_\alpha - 1} \prod_{\beta \neq \alpha} N_\beta! \Omega_\beta^{N_\beta}}$$

$$\int dx_2 dx_3 \cdots dx_N \exp[-\beta \mathcal{U}_{N-1}(x_2 \cdots x_N)]. \tag{6.28}$$

Here the product is over all components except α, $\mathcal{U}_{N-1}$ is the intermolecular potential energy for the $(N-1)$ molecule system, and we label the missing α molecule as molecule 1. The quantities $Z_c(N_\alpha)$ and $Z_c(N_{\alpha-1})$ are the corresponding phase integrals given by (3.53). We now introduce $\mathcal{U}_{t\alpha}(x_1; x^{N-1})$, the intermolecular potential energy due to the interaction of the test particle of component α (molecule 1) with all other molecules in the system, by writing

$$\mathcal{U}_N(x_1x_2\cdots x_N) = \mathcal{U}_{t\alpha}(x_1; x_2\cdots x_N) + \mathcal{U}_{N-1}(x_2x_3\cdots x_N). \tag{6.29}$$

The potentials $\mathcal{U}_{t\alpha}$ and $\mathcal{U}_{N-1}$ can be expressed in terms of the coordinates of the molecules relative to those of the test particle, $x_{21}x_{31}\cdots x_{N1}$, where $x_{j1} = \mathbf{r}_{j1}\omega_{j1}$ denotes positional $(\mathbf{r}_{j1} = \mathbf{r}_j - \mathbf{r}_1)^\dagger$ and orientational ω_{j1} coordinates of molecule j relative to molecule 1; these coordinates are then independent of the coordinate x_1. We can therefore write $Z_c(N_\alpha)$ as

$$\begin{aligned}
Z_c(N_\alpha) &= \int \mathrm{d}x_1\mathrm{d}x_2\cdots\mathrm{d}x_N \exp[-\beta\mathcal{U}_{N-1}(x_2\cdots x_N)]\exp[-\beta\mathcal{U}_{t\alpha}(x_2\cdots x_N)] \\
&= \int \mathrm{d}x_1\mathrm{d}x_{21}\mathrm{d}x_{31}\cdots\mathrm{d}x_{N1}\exp[-\beta\mathcal{U}_{N-1}(x_{21}\cdots x_{N1})] \\
&\quad \times \exp[-\beta\mathcal{U}_{t\alpha}(x_{21}\cdots x_{N1})] \\
&= V\Omega_\alpha \int \mathrm{d}x_{21}\mathrm{d}x_{31}\cdots\mathrm{d}x_{N1}\exp[-\beta\mathcal{U}_{N-1}(x_{21}\cdots x_{N1})] \\
&\quad \times \exp[-\beta\mathcal{U}_{t\alpha}(x_{21}\cdots x_{N1})] \\
&= V\Omega_\alpha Z_c(N_\alpha - 1)\langle\exp[-\beta\mathcal{U}_{t\alpha}(x_{21}\cdots x_{N1})]\rangle_{N-1},
\end{aligned} \tag{6.30}$$

where $\langle\cdots\rangle_{N-1}$ means an ensemble average over the $(N-1)$ molecule system with intermolecular potential energy $\mathcal{U}_{N-1}(x_2\cdots x_N)$. From (6.27) and (6.30) we have[7,8]

$$\mu_{\alpha c} = -kT\ln\left[\frac{1}{\rho_\alpha}\langle\exp[-\beta\mathcal{U}_{t\alpha}]\rangle_{N-1}\right], \tag{6.31}$$

where $\rho_\alpha = N_\alpha/V$ is the number density of α molecules. The quantity $\langle\exp[-\beta\mathcal{U}_{t\alpha}]\rangle$ can be interpreted as follows. The test particle is placed with some fixed coordinates $x_1 = \mathbf{r}_1\omega_1$ in the fluid of $N-1$ molecules, and $\exp[-\beta\mathcal{U}_{t\alpha}]$ is then obtained as a function of the coordinates of all of the $N-1$ molecules; a Boltzmann-weighted average over the coordinates $x_2\cdots x_N$ is then taken. Alternatively, we could imagine fixing the $N-1$ molecules and moving the test particle freely among them; an unweighted average would then be taken over all possible positions and orientations of the test particle. In either case the test particle acts as a fictitious or 'ghost' particle, since it has no influence on the other $N-1$ molecules.

† We use here the convention for $\mathbf{r}_{j1}$ opposite to that used in Eq. (6.8a).

It is often more convenient to work with residual thermodynamic properties rather than configurational ones. The residual chemical potential $\mu_{\alpha r}$ is given by $\mu_{\alpha r} = \mu_\alpha - \mu_\alpha^{\text{id}} = \mu_{\alpha c} - \mu_{\alpha c}^{\text{id}}$, where the superscript id signifies the ideal gas value at the same temperature, density, and composition as the real mixture. For an ideal gas $\mathcal{U}_N = 0$ and $Q_c^{\text{id}} = V^N / \prod_\beta N_\beta!$ from (6.26), so that, using Stirling's approximation for $\ln N_\beta!$,

$$\ln N_\beta! = N_\beta \ln N_\beta - N_\beta \qquad (N_\beta \text{ large}) \tag{6.32}$$

we have

$$\mu_{\alpha c}^{\text{id}} = -kT\left(\frac{\partial \ln Q_c^{\text{id}}}{\partial N_\alpha}\right)_{TVN'} = kT \ln \rho_\alpha, \tag{6.33}$$

so that[8]

$$\mu_{\alpha r} = \mu_{\alpha c} - kT \ln \rho_\alpha. \tag{6.34}$$

From this result and (6.31) we have

$$\mu_{\alpha r} = -kT \ln \langle \exp[-\beta U_{t\alpha}]\rangle_{N-1}. \tag{6.35}$$

An alternative but closely related expression for $\mu_{\alpha r}$ can be obtained by noting that

$$\begin{aligned}
\langle \exp[-\beta U_{t\alpha}]\rangle_{N-1} &= \frac{\int \mathrm{d}x_2 \mathrm{d}x_3 \cdots \mathrm{d}x_N \exp[-\beta \mathcal{U}_{N-1}] \exp[-\beta \mathcal{U}_{t\alpha}]}{\int \mathrm{d}x_2 \mathrm{d}x_3 \cdots \mathrm{d}x_N \exp[-\beta \mathcal{U}_{N-1}]} \\
&= \frac{\int \mathrm{d}x_2 \cdots \mathrm{d}x_N \exp[-\beta \mathcal{U}_N]}{\int \mathrm{d}x_2 \cdots \mathrm{d}x_N \exp[+\beta \mathcal{U}_{t\alpha}] \exp[-\beta \mathcal{U}_N]} \\
&= \langle \exp[\beta \mathcal{U}_{t\alpha}]\rangle_N^{-1},
\end{aligned} \tag{6.36}$$

where $\langle \cdots \rangle_N$ is an ensemble average over the N molecule system for which the intermolecular potential energy is $\mathcal{U}_N$. From (6.35) and (6.36) we have[9]

$$\mu_{\alpha r} = kT \ln \langle \exp[\beta \mathcal{U}_{t\alpha}]\rangle_N. \tag{6.37}$$

Note that in (6.37) the average is taken over a system in which the test particle *does* interact with the other $N - 1$ molecules, i.e. the test particle is not a 'ghost'. We also note that both (6.35) and (6.37) are quite general. They do not rely on the assumption of pairwise additivity, nor on the rigid molecule approximation provided we suitably interpret the ensemble average, and are not restricted to uniform fluids.[7] They provide yet another route (the μ-route) to the thermodynamics; note that μ_α is *not* expressed in terms of $g(12)$, however.[10]

Both (6.35) and (6.37) have been used in computer simulation studies to obtain the chemical potential.[11,12] In using (6.35) it is common to introduce a fixed lattice of test particles having various orientations, and to obtain the interaction energy of the test particles with the real molecules; the average value of $\exp[-\beta \mathcal{U}_{t\alpha}]$ is then found. The major contribution to this average is from configurations in which the interaction energy is attractive ($\mathcal{U}_{t\alpha}$ negative), and the method works well for low and moderate densities. At high densities (much above $\rho\sigma^3 = 0.5$ for Lennard-Jones (LJ) molecules[12]) the test particles overlap real molecules for the majority of configurations, leading to predominantly repulsive configurations; at such densities the usual Monte Carlo and molecular dynamics procedures do not effectively sample the regions of phase space most important in determining $\mu_{\alpha r}$, and the statistical errors become large. One way of overcoming this problem is to bias the sampling so as to effectively cover the desired regions of phase space.[12] Similar problems are encountered in using (6.37) in simulations. In that case the dominant contributions to the average of $\exp[\beta\mathcal{U}_{t\alpha}]$ are from configurations in which $\mathcal{U}_{t\alpha}$ is repulsive (i.e. positive); since the test particle is a real molecule moving under the influence of forces from its neighbours, it tends to spend much of its time in attractive configurations, again leading to large statistical uncertainties in $\mu_{\alpha r}$ as calculated from the simulation. The situation can be illustrated schematically by rewriting (6.35) and (6.37) in the forms (for a pure fluid):

$$\langle \exp[-\beta\mathcal{U}_t]\rangle_{N-1} = \int_{-\infty}^{\infty} f(\mathcal{U}_t)\exp[-\beta\mathcal{U}_t]d\mathcal{U}_t \tag{6.38}$$

$$\langle \exp[-\beta\mathcal{U}_t]\rangle_{N} = \int_{-\infty}^{\infty} g(\mathcal{U}_t)\exp[\beta\mathcal{U}_t]d\mathcal{U}_t, \tag{6.39}$$

where $f(\mathcal{U}_t)$ is the probability density for the test particle energy $\mathcal{U}_t$ in the $N-1$ molecule system, and $g(\mathcal{U}_t)$ is the corresponding probability density in the N molecule system. As seen in Fig. 6.1, at high densities it is not enough to know $f(\mathcal{U}_t)$ or $g(\mathcal{U}_t)$ in the region of the maximum, since there are substantial contributions to the integrand for regions of $\mathcal{U}_t$ where f or g is small.

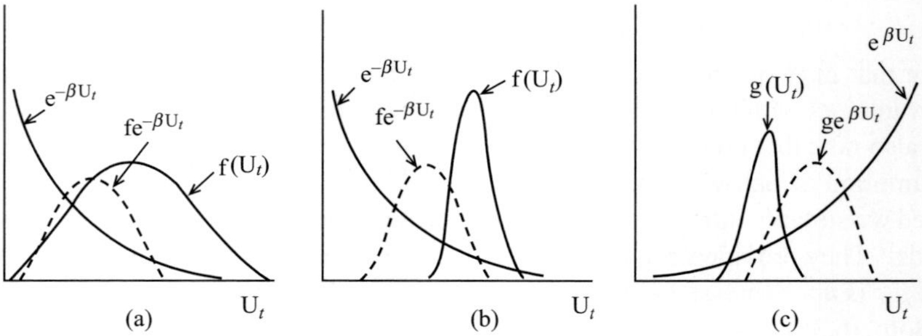

FIG. 6.1 Probability density distributions needed in the test particle method for a typical liquid state condition (schematic): (a) moderate density, (b) and (c) high density. The function f is a broad distribution and extends to large $\mathcal{U}_t$ values; the width of this distribution is compressed here for convenience of plotting.

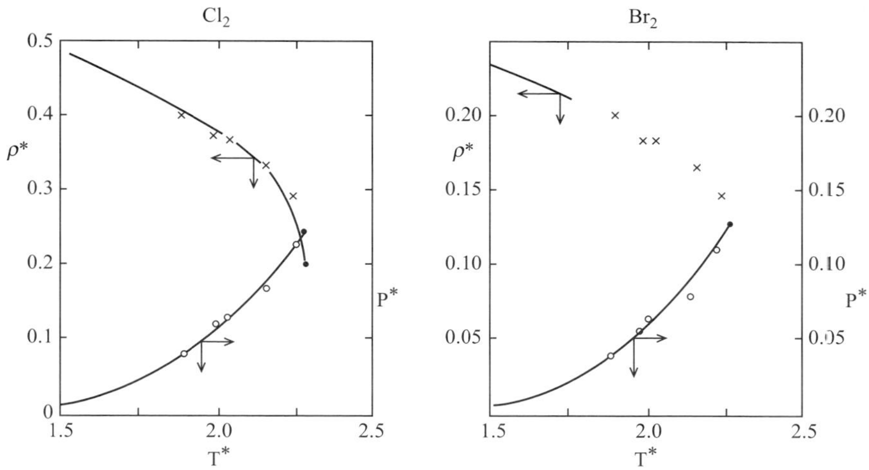

FIG. 6.2 Comparison of experiment (lines) and Monte Carlo results (points) for the coexisting liquid density ($\rho^* = \rho\sigma^3$) and vapour pressure ($p^* = p\sigma^3/\varepsilon$) for diatomic Lennard-Jones fluids with bond-length $\ell^* = \ell/\sigma = 0.630$. For Cl_2 the site–site parameters $\varepsilon/k = 185$ K, $\sigma = 3.538$ Å while for Br_2 $\varepsilon/k = 257.2$ K, $\sigma = 3.538$ Å. These parameters gave the best fit to the available coexistence data. Units of P* are the same in both figures. (From, Powles,[11] based on simulation data of Romano and Singer.[11])

An example of the use of such computer simulation results for μ_r is shown in Fig. 6.2 for two-centre LJ models of chlorine and bromine. In this case μ_r is calculated from (6.35), and the vapour pressure and coexisting liquid density are calculated by solving $\mu_l = \mu_g$ and $p_l = p_g = p$. The gas-phase properties μ_g and p_g are obtained from the virial expansion. Since the conventional Monte Carlo sampling procedure is used it is only possible to obtain the coexistence properties for the liquid at rather high temperatures (low densities).

It should be noted that for fluids of hard bodies $\mathcal{U}_{t\alpha}$ can only take the values 0 or ∞. The average of $\exp(-\beta\mathcal{U}_{t\alpha})$ that appears in (6.35) is still well behaved under these conditions (see §6.12 for applications to hard bodies). However, the average of $\exp(\beta\mathcal{U}_{t\alpha})$ in (6.37) is indeterminate, so that this expression is not useful for such fluids.

It is possible to derive equations for other properties in terms of $\mathcal{U}_{t\alpha}(1) = \mathcal{U}_{t\alpha}(x_1; x_2 \cdots x_N)$ by methods that closely follow that used above for the chemical potential.[6] For example, the configurational energy is $U_c = \langle \mathcal{U}_N \rangle_N$. If the intermolecular potential energy is pairwise additive we can write

$$\mathcal{U}_N = \frac{1}{2} \sum_{\alpha} \sum_{i=1}^{N_\alpha} \mathcal{U}_{t\alpha}(i), \tag{6.40}$$

and it is then easy to show that[13]

$$U_c = \frac{1}{2} V \sum_{\alpha} N_\alpha \Omega_\alpha \frac{Z_c(N_\alpha - 1)}{Z_c(N_\alpha)} \langle \mathcal{U}_{t\alpha}(1) \exp[-\beta \mathcal{U}_{t\alpha}(1)] \rangle_{N-1}$$

or, using (6.27), (6.33), and (6.35),

$$U_c = \frac{1}{2} \sum_{\alpha} N_\alpha \langle \exp[-\beta \mathcal{U}_{t\alpha}(1)] \rangle_{N-1}^{-1} \langle \mathcal{U}_{t\alpha}(1) \exp[-\beta \mathcal{U}_{t\alpha}(1)] \rangle_{N-1}. \tag{6.41}$$

By similar arguments, we can derive the following equation for the pressure, starting from (6.13)and (6.14):

$$p = \rho kT + \frac{1}{6} \sum_{\alpha\beta} \rho_\alpha \langle \exp[-\beta \mathcal{U}_{t\alpha}(1)] \rangle_{N-1}^{-1} \langle \Phi_{\alpha\beta}(1) \exp[-\beta \mathcal{U}_{t\alpha}(1)] \rangle_{N-1}, \tag{6.42}$$

where

$$\Phi_{\alpha\beta}(1) = -\sum_{j=2}^{N_\beta} r_{1j} \frac{\partial u_{\alpha\beta}(1j)}{\partial r_{1j}}. \tag{6.43}$$

Equation (6.42) has been used to determine pressures for hard core systems,[14] including the tensorial components of the pressure for inhomogeneous systems,[15] and for continuous potentials.[16]

The surface tension can also be expressed in terms of a test particle equation,[17] and the pair correlation function $g_{\alpha\beta}(\mathbf{r}\omega_1\omega_2)$ can be expressed in terms of $\mathcal{U}^{(2)}_{t\alpha\beta}(x_1 x_2; x_3 \cdots x_N)$, the intermolecular potential energy of interaction of two test particles of species α and β with all of the remaining molecules of the system.[6,7]

6.5.2 *The Kirkwood method*

Starting from (6.27) it is possible to derive a further expression for the chemical potential, due originally to Kirkwood,[3,18] that involves the pair correlation function $g(12) \equiv g(\mathbf{r}\omega_1\omega_2)$. In this approach molecule 1 of species α is coupled to the remaining $N-1$ molecules by a coupling parameter λ. Equation (6.29) is now replaced by[19]

$$\mathcal{U}_N(x^N; \lambda) = \lambda \mathcal{U}_{t\alpha}(x_1; x^{N-1}) + \mathcal{U}_{N-1}(x^{N-1}), \tag{6.44}$$

where $0 \leq \lambda \leq 1$ and x^{N-1} represents the configuration of all the other $(N-1)$ molecules. Thus $\lambda = 0$ and $\lambda = 1$ correspond to the $N-1$ and N molecule systems, respectively. We have $Z_c(\lambda = 1) = Z_c(N_\alpha)$ and $Z_c(\lambda = 0) = V\Omega_\alpha Z_c(N_\alpha - 1)$, so that from (6.27) and (6.34) we obtain

$$\mu_{\alpha r} = -kT \ln \left[\frac{Z_c(\lambda = 1)}{Z_c(\lambda = 0)} \right] = -kT \int_0^1 \frac{\partial \ln Z_c(\lambda)}{\partial \lambda} d\lambda, \tag{6.45}$$

where

$$Z_c(\lambda) = \int dx^N \exp[-\beta \mathcal{U}_N(x^N; \lambda)]. \tag{6.46}$$

From (6.44)–(6.46) we get

$$\mu_{\alpha r} = \int_0^1 d\lambda \int dx^N \mathcal{U}_{t\alpha}(x_1; x^{N-1}) \exp[-\beta \mathcal{U}_N(x^N; \lambda)]/Z_c(\lambda). \tag{6.47}$$

We now introduce the approximation of pairwise additivity. We label the molecules of species A as $1, 2, \cdots, N_A$, those of B as $1, 2, \cdots, N_B$, etc., so that

$$\mathcal{U}_{t\alpha} = \sum_\beta \sum_j^{N_\beta} u_{\alpha\beta}(1j). \tag{6.48}$$

Substituting this in (6.47), noting that the sum over j contains $(N_\beta - \delta_{\alpha\beta})$ terms (each giving the same result on integration), and using the definition of $g_{\alpha\beta}(12)$ given in (3.252), we get

$$\mu_{\alpha \mathrm{r}} = \sum_\beta \rho_\beta \int_0^1 d\lambda \int d\mathbf{r}_{12} \langle u_{\alpha\beta}(12) g_{\alpha\beta}(12; \lambda) \rangle_{\omega_1\omega_2}, \tag{6.49}$$

which is the Kirkwood result. Here $g_{\alpha\beta}(12; \lambda)$ is the pair correlation function for β molecules around a central $\alpha 1$ molecule when the $\alpha 1$ molecule is coupled to the extent λ. This equation is less general than (6.35) and (6.37) of the previous subsection, since it is based on the assumptions of pairwise additivity and rigid molecules. However, it is a useful additional route to the chemical potential, and has been widely used in the theory of solutions.[20]

6.6 Thermodynamic properties in terms of spherical harmonic expansion coefficients

The energy, pressure, compressibility, and free energy equations each involve integrals over functions of intermolecular distances and orientations. It is often convenient to expand these functions in spherical harmonics, and thus obtain an expression for the property in terms of the harmonic coefficients of $g(\mathbf{r}\omega_1\omega_2)$ and $u(\mathbf{r}\omega_1\omega_2)$. The property harmonic expansions are most readily derived using the generalized Parseval theorem; this is derived in Appendix B of Vol. 1. We consider here only the case of linear molecules, and employ space-fixed axes. The corresponding expressions for nonlinear molecules, and for the intermolecular **r**-frame, or for the space-fixed and **k**-frame expansions in **k**-space, can be obtained by using the appropriate form of the generalized Parseval theorem given in Appendix B.

For linear molecules and space-fixed axes, the generalized Parseval theorem is

$$\int d\mathbf{r}\langle A(\mathbf{r}\omega_1\omega_2)B(\mathbf{r}\omega_1\omega_2)\rangle_{\omega_1\omega_2} = (4\pi)^{-3}\sum_{\ell_1\ell_2\ell}(2\ell+1)\int d\mathbf{r}A(\ell_1\ell_2\ell;r)^*B(\ell_1\ell_2\ell;r), \tag{6.50}$$

where $A(\mathbf{r}\omega_1\omega_2)$ and $B(\mathbf{r}\omega_1\omega_2)$ are real and $A(\ell_1\ell_2\ell;r)$ is the space-fixed harmonic coefficient defined by

$$A(\mathbf{r}\omega_1\omega_2) = \sum_{\ell_1\ell_2\ell}\sum_{m_1m_2m} A(\ell_1\ell_2\ell;r)C(\ell_1\ell_2\ell;m_1m_2m) \times Y_{\ell_1m_1}(\omega_1)Y_{\ell_2m_2}(\omega_2)Y_{\ell m}(\omega)^*, \tag{6.51}$$

where $\omega_i \equiv \theta_i\phi_i$ is the direction of the symmetry axis of molecule i, $\omega \equiv \theta\phi$ is the direction of $\mathbf{r}$, and C is a Clebsch-Gordan coefficient (App. A).

We can now apply (6.50) to the energy equation (6.9b) by taking $A = u$ and $B = g$, so that

$$U_c = \frac{1}{2(4\pi)^3}\rho N\sum_{\ell_1\ell_2\ell}(2\ell+1)\int d\mathbf{r}u(\ell_1\ell_2\ell;r)g(\ell_1\ell_2\ell;r), \tag{6.52}$$

where we have made use of the fact that $u(\ell_1\ell_2\ell;r)$ and $g(\ell_1\ell_2\ell;r)$ are real (see §§ 2.3 and 3.2.1) and depend only on $r = |\mathbf{r}|$, and $\int d\mathbf{r}(\ldots)$ here can be replaced by $4\pi\int r^2dr(\ldots)$. We note that the spherical harmonic expansion has reduced the original sevenfold integral to the sum of one-dimensional integrals in (6.52). Similarly, applying (6.50) to (6.15), (6.16), (6.17), and (6.24) gives

$$p = \rho kT - \frac{1}{6(4\pi)^3}\rho^2\sum_{\ell_1\ell_2\ell}(2\ell+1)\int d\mathbf{r}ru'(\ell_1\ell_2\ell;r)g(\ell_1\ell_2\ell;r) \tag{6.53}$$

$$\rho kT\chi = 1 + (4\pi)^{-3/2}\rho\int d\mathbf{r}h(000;r) \tag{6.54}$$

$$= [1 - (4\pi)^{-3/2}\rho\int d\mathbf{r}c(000;r)]^{-1} \tag{6.55}$$

$$A - A_0 = \frac{1}{2(4\pi)^3}\rho N\sum_{\ell_1\ell_2\ell}(2\ell+1)\int d\mathbf{r}\int_0^1 d\lambda u_p(\ell_1\ell_2\ell;r)g_\lambda(\ell_1\ell_2\ell;r), \tag{6.56}$$

where $u'(\ell_1\ell_2\ell;r) = du(\ell_1\ell_2\ell;r)/dr$ and $h(000;r) = g(000;r) - (4\pi)^{3/2}$. In deriving (6.54) and (6.55) we note that $B = 1$, so that $B(\ell_1\ell_2\ell;r) = (4\pi)^{3/2}\delta_{\ell_10}\delta_{\ell_20}\delta_{\ell 0}$, and only the (000) harmonic of B is nonvanishing.

It is also possible to write down the corresponding expansions for U_c, p, χ, and A in terms of the intermolecular-frame harmonic coefficients, by using the appropriate form of the generalized Parseval theorem, (B.85), in place of (6.50).

Thus, for the configurational energy we obtain[21]

$$U_{\mathrm{c}} = \frac{1}{2(4\pi)^2}\rho N \sum_{\ell_1\ell_2 m} \int d\mathbf{r} u(\ell_1\ell_2 m; r) g(\ell_1\ell_2 m; r), \tag{6.57}$$

where $A(\ell_1\ell_2 m; r)$ is defined by

$$A(\mathbf{r}\omega_1'\omega_2') = \sum_{\ell_1\ell_2 m} A(\ell_1\ell_2 m; r) Y_{\ell_1 m}\left(\omega_1'\right) Y_{\ell_2 \underline{m}}\left(\omega_2'\right), \tag{6.58}$$

where $\omega_i{}'$ refers to the **r**-frame and $\underline{m} \equiv -m$. The space-fixed and **r**-frame coefficients of A are related by (3.147) and (3.148).

The intermolecular harmonics have been extensively used in computer simulation studies.[22] However, if the intermolecular potential consists of a small number of harmonic terms, the space-fixed harmonic expansion gives a simpler form for the properties. For example, if the anisotropic potential is $u_a = u_{QQ}$, the quadrupole–quadrupole interaction, then there is only one nonvanishing $u(\ell_1\ell_2\ell; r)$ harmonic, namely $u(224; r)$, (see (2.177) and (2.24)):

$$u(224; r) = (4\pi)^{3/2} \frac{(70)^{1/2}}{15} \frac{Q^2}{r^5}. \tag{6.59}$$

Thus, the expression (6.52) for the anisotropic contribution to U_{c}, for example, reduces to a single term,

$$U_{\mathrm{c}}^{\mathrm{a}} = \frac{9}{2(4\pi)^3}\rho N \int d\mathbf{r} u(224; r) g(224; r), \tag{6.60}$$

so that only one space-fixed harmonic of g, namely $g(224; r)$, need be known. In the intermolecular-frame expansion, (6.57), for the same intermolecular potential one would need *five* harmonics $u(\ell_1\ell_2 m; r)$ and $g(\ell_1\ell_2 m; r)$, viz. $\ell_1\ell_2 m = 22\underline{2}, 22\underline{1}, 220, 221, 222$. Also, since $ru' = -5u$ for the u_{QQ} interaction we see that $p^a = \frac{5}{3}\rho U_c^a / N$ for this case.

The convergence of the spherical harmonic expansion using the intermolecular frame has been studied[23] for homonuclear diatomic molecules for an atom–atom LJ potential model together with molecular dynamics simulation results for the $g(\ell_1\ell_2 m; r)$. It is found that for moderate elongations, $\ell^* < 0.45$ (with $\ell^* = \ell/\sigma$ where ℓ is the bond length and σ is the atom diameter), the series converges quite rapidly, but that the convergence quickly gets poorer as ℓ^* increases above this value. Typical results are shown in Table 6.1. The convergence has been found[22] to be rapid for central LJ + quadrupole–quadrupole (QQ) potentials. For potentials that are quite *hard*, and *elongated*, the convergence worsens considerably.[22,24] Of course, if the potential consists of only a few harmonic terms (e.g. LJ + QQ) speed of convergence is irrelevant for many properties, such as the energy, pressure, and G_2, which involve only particular low order harmonics.

Table 6.1

Convergence of the spherical harmonic expansion for homonuclear diatomic molecules using an atom–atom Lennard-Jones model

ℓ^*	T^*	ρ^*	Q^*	Number of harmonics†	U^*/N Series	U^*/N Exact	p^* Series	p^* Exact
0.425	1.543	0.50	0	4	−13.06	−13.11	−2.41	−2.34
0.425	1.543	0.50	0	8	−13.18	−13.11	−2.84	−2.34
0.5471	1.29	0.56	0.50	25	−16.74	−14.45	−1.11	−0.61
0.5471	1.29	0.52	1.00	25	−13.18	−14.13	0.08	−1.38
0.5471	2.20	0.52	2.00	25	−13.07	−16.91	0.63	−0.67

*Reduced quantities are $\ell^* = \ell/\sigma$; $T^* = kT/\varepsilon$; $\rho^* = \rho\sigma^3$; $Q^* = Q/(\varepsilon\sigma^5)^{1/2}$; $U^* = U/\varepsilon$; $p^* = p\sigma^3/\varepsilon$, where σ and ε are the atom–atom LJ parameters. Here ℓ is bond length.

† Four terms include coefficients $\ell_1\ell_2m$ = (000), (200), (020) and (220); eight terms include in addition to these (400), (040), (420), and (240); twenty-five terms include the additional harmonics (440), (600), (620), (260), (640), (460), (660), (800), (820), (280), (840), (480), (860), (680), (880), (10,0,0), and (0,10,0). From ref. 23.

6.7 Thermodynamic properties in terms of the site–site correlation functions

In this section we derive relations for the thermodynamic properties U_c, p, χ, and A_c in terms of site–site correlation functions. For the compressibility such relations can be derived for an arbitrary intermolecular potential, and are as general as the compressibility equation itself.† The energy, pressure, and free energy equations can only be transformed to relations involving the $g_{\alpha\beta}$ correlation functions if the intermolecular potential is a sum of site–site potentials, as in (2.5).

6.7.1 Compressibility

We first consider the compressibility, χ. We have seen in § 3.1.6 that $h_{\alpha\beta}(k)$, the Fourier transform of $h_{\alpha\beta}(r) = g_{\alpha\beta}(r) - 1$, is given by

$$h_{\alpha\beta}(k) = \langle h(\mathbf{k}\omega_1\omega_2)\exp[-i\mathbf{k}\cdot\mathbf{r}_{ci\alpha}]\exp[i\mathbf{k}\cdot\mathbf{r}_{c2\beta}]\rangle_{\omega_1\omega_2}, \tag{6.61}$$

where $h(\mathbf{k}\omega_1\omega_2)$ is the transform of $h(\mathbf{r}\omega_1\omega_2)$ and $\mathbf{r}_{ci\alpha}$ is the position of site α with respect to the molecular centre of molecule i. Taking the limit $k \to 0$ in this relation gives (we use the notation $\tilde{f}(k)$ in place of $f(k)$ when there is a possibility of ambiguity, as in (6.62))

$$\tilde{h}_{\alpha\beta}(0) = \langle\tilde{h}(0\omega_1\omega_2)\rangle_{\omega_1\omega_2}, \tag{6.62}$$

and combining this result with (3.113) gives the compressibility equation

† Other properties rigorously related to the site–site correlation functions $g_{\alpha\beta}$, regardless of the form of the potential, are the structure factor, dielectric constant (for nonpolarizable molecules), and the angular correlation parameter G_2 relevant for the Kerr constant and depolarized light scattering (see Chapters 9, 10, and 11).

$$\rho kT\chi = 1 + \rho\tilde{h}_{\alpha\beta}(0) = 1 + \rho \int d\mathbf{r} h_{\alpha\beta}(r) \tag{6.63}$$

or

$$\rho kT\chi = 1 + \rho \int d\mathbf{r}[g_{\alpha\beta}(r) - 1]. \tag{6.64}$$

Hence χ can be obtained from any one of the site–site correlation functions $g_{\alpha\beta}$ (equivalently, χ is unaffected by the choice of molecular centres used for $g(r)$ in (6.16), as expected physically).

It is possible to derive a similar expression for χ in terms of the reference interaction site model (RISM) site–site direct correlation functions $c_{\alpha\beta}$, as follows. The (matrix) OZ relation between $h_{\alpha\beta}(k)$ and $c_{\alpha\beta}(k)$ is (5.238),

$$\mathbf{h} = \boldsymbol{\omega}\mathbf{c}\boldsymbol{\omega} + \rho\boldsymbol{\omega}\mathbf{c}\mathbf{h}, \tag{6.65}$$

where $\boldsymbol{\omega}(k)$ is the matrix with elements $\omega_{\alpha\alpha'}(k) = \sin(k\ell_{\alpha\alpha'})/k\ell_{\alpha\alpha'}$. In the limit $k \to 0$, $\boldsymbol{\omega}$ approaches the matrix with elements $\omega_{\alpha\alpha'} = 1$ for all α, α'.

Hence from (6.65) we have:

$$\begin{aligned} h_{\alpha\beta} &= \sum_{\alpha'\beta'} \omega_{\alpha\alpha'}c_{\alpha'\beta'}\omega_{\beta'\beta} + \rho \sum_{\substack{\alpha'\ \alpha'' \\ \beta'\ \beta''}} \omega_{\alpha\alpha'}c_{\alpha'\alpha}\omega_{\alpha''\beta''}c_{\beta''\beta}\omega_{\beta'\beta} + \cdots \\ &\overset{\mathrm{k}\to 0}{\longrightarrow} \sum_{\alpha'\beta'} c_{\alpha'\beta'} + \rho \sum_{\substack{\alpha'\ \alpha'' \\ \beta'\ \beta''}} c_{\alpha'\alpha''}c_{\beta''\beta'} + \cdots \\ &\equiv \mathrm{C} + \rho\mathrm{C}^2 + \cdots, \end{aligned} \tag{6.66}$$

where

$$C = \sum_{\alpha\beta} \tilde{c}_{\alpha\beta}(0) \tag{6.67}$$

and $\tilde{c}_{\alpha\beta}(0) \equiv c_{\alpha\beta}(k = 0)$. Hence we have

$$\begin{aligned} 1 + \rho\tilde{h}_{\alpha\beta}(0) &= 1 + \rho C + (\rho C)^2 + \cdots \\ &= (1 - \rho C)^{-1} \\ &= \left[1 - \rho \sum_{\alpha\beta} \tilde{c}_{\alpha\beta}(0)\right]^{-1} \end{aligned} \tag{6.68}$$

Thus the compressibility equation can be written

$$\rho kT\chi = \left[1 - \rho \sum_{\alpha\beta} \tilde{c}_{\alpha\beta}(0)\right]^{-1} \tag{6.69}$$

or

$$\rho kT\chi = \left[1 - \rho \sum_{\alpha\beta} \int \mathrm{d}\mathbf{r} c_{\alpha\beta}(r)\right]^{-1}. \tag{6.70}$$

The number of $\tilde{c}_{\alpha\beta}$ in the sum over $\alpha\beta$ sites in (6.69) and (6.70) corresponds to the number of $\tilde{h}_{\alpha\beta}$ in (6.65). In the event that only one site A is chosen in each molecule, e.g. at the molecular centre, (6.70) reduces to

$$\rho kT\chi = \left[1 - \rho \int \mathrm{d}\mathbf{r} c_{AA}(r)\right]^{-1},$$

as one expects from the atomic liquid case, (5.20). It should be stressed that $c_{AA}(r)$ defined by the OZ equation (6.65) is not equal to $c(r)$, where $c(r) \equiv \langle c(\mathbf{r}\omega_1\omega_2)\rangle_{\omega_1\omega_2}$ is the centres direct correlation function defined by (3.122), although their $\mathbf{r}$-integrals are equal. In contrast, for this case $h_{AA}(r)$ *is* equal to $\langle h(\mathbf{r}\omega_1\omega_2)\rangle_{\omega_1\omega_2}$; see also §§ 3.1.5 and 5.5.2).

For homonuclear diatomic AA molecules with two sites per molecule, (6.64) and (6.70) reduce to

$$\begin{aligned} \rho kT\chi &= 1 + \rho \int \mathrm{d}\mathbf{r}[g_{AA}(r) - 1] \\ &= \left[1 - 4\rho \int \mathrm{d}\mathbf{r} c_{AA}(r)\right]^{-1}, \end{aligned} \tag{6.71}$$

which can also be derived directly from (5.251).

6.7.2 *Internal energy*[25–27]

We now turn to the properties U, A, and p for the particular case of a site–site pair potential, (2.5). The energy equation, (6.9a), becomes

$$U_c = \frac{1}{2}\rho^2 \frac{1}{\Omega^2} \sum_{\alpha\beta} \int \mathrm{d}\mathbf{r}_1 \mathrm{d}\mathbf{r}_2 \mathrm{d}\omega_1 \mathrm{d}\omega_2 u_{\alpha\beta}(r_{\alpha\beta}) g(\mathbf{r}\omega_1\omega_2), \tag{6.72}$$

where $\Omega = \int \mathrm{d}\omega$ and $r_{\alpha\beta}$ is the distance between site α in molecule 1 and site β in molecule 2. The integral in this equation can be written more simply in terms of the site–site correlation function $g_{\alpha\beta}(r_{\alpha\beta})$, which is proportional to the probability density that sites α and β (on different molecules) are at a distance $r_{\alpha\beta}$ apart, irrespective of the molecular orientations (see § 3.1.6). To carry out this simplification we transform the integration variables in (6.72) from $(\mathbf{r}_1\mathbf{r}_2\omega_1\omega_2)$ to

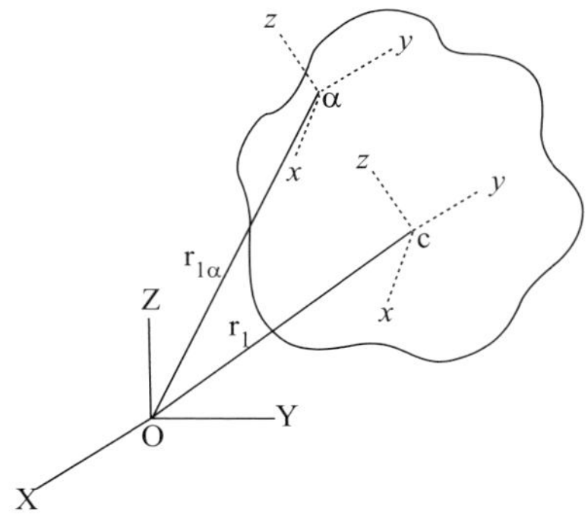

FIG. 6.3 Vectors $\mathbf{r}_1$ and $\mathbf{r}_{1\alpha}$ locate the centre and site α of molecule 1, respectively; XYZ are the space-fixed axes, and *xyz* are the body-fixed axes (kept parallel in the transformation of centres).

$(\mathbf{r}_{1\alpha}\mathbf{r}_{2\beta}\omega_1\omega_2)$, where $\mathbf{r}_i$ is the position of the molecular centre and $\mathbf{r}_{i\alpha}$ is the position of site α in molecule i (see Fig. 6.3); $\omega_i = \phi_i\theta_i\chi_i$ (or $\theta_i\phi_i$ in the case of linear molecules) is the molecular orientation of the body-fixed axes *xyz* with respect to space-fixed axes *XYZ*, and is unchanged in this transformation. The probability of finding molecules 1 and 2 in the element of configuration space $(\mathrm{d}\mathbf{r}_1\mathrm{d}\mathbf{r}_2\mathrm{d}\omega_1\mathrm{d}\omega_2)$ at $(\mathbf{r}_1\mathbf{r}_2\omega_1\omega_2)$ is $P(\mathbf{r}_1\mathbf{r}_2\omega_1\omega_2)\mathrm{d}\mathbf{r}_1\mathrm{d}\mathbf{r}_2\mathrm{d}\omega_1\mathrm{d}\omega_2$, where $P(\mathbf{r}_1\mathbf{r}_2\omega_1\omega_2)$ is the specific pair distribution function (see § 3.1.3). This probability is equal to that of finding the variables $(\mathbf{r}_{1\alpha}\mathbf{r}_{2\beta}\omega_1\omega_2)$ in their corresponding ranges, i.e.

$$P(\mathbf{r}_1\mathbf{r}_2\omega_1\omega_2)\mathrm{d}\mathbf{r}_1\mathrm{d}\mathbf{r}_2\mathrm{d}\omega_1\mathrm{d}\omega_2 = P(\mathbf{r}_{1\alpha}\mathbf{r}_{2\beta}\omega_1\omega_2)\mathrm{d}\mathbf{r}_{1\alpha}\mathrm{d}\mathbf{r}_{2\beta}\mathrm{d}\omega_1\mathrm{d}\omega_2. \tag{6.73}$$

Converting these specific distribution functions to (generic) correlation functions g (see (3.91) and (3.105)) we have, for the isotropic, homogeneous case,

$$g(\mathbf{r}\omega_1\omega_2)\mathrm{d}\mathbf{r}_1\mathrm{d}\mathbf{r}_2\mathrm{d}\omega_1\mathrm{d}\omega_2 = g(\mathbf{r}_{\alpha\beta}\omega_1\omega_2)\mathrm{d}\mathbf{r}_{1\alpha}\mathrm{d}\mathbf{r}_{2\beta}\mathrm{d}\omega_1\mathrm{d}\omega_2, \tag{6.74}$$

where $g(\mathbf{r}\omega_1\omega_2)$ is the usual angular pair correlation function and $g(\mathbf{r}_{\alpha\beta}\omega_1\omega_2)$ is proportional to the probability density that a pair of molecules are at orientations ω_1 and ω_2, with site–site separation $\mathbf{r}_{\alpha\beta} = \mathbf{r}_{2\beta} - \mathbf{r}_{1\alpha}$. The site–site correlation function is related to $g(\mathbf{r}_{\alpha\beta}\omega_1\omega_2)$ by (cf. (3.126))

$$g_{\alpha\beta}(r_{\alpha\beta}) = \langle g(\mathbf{r}_{\alpha\beta}\omega_1\omega_2)\rangle_{\omega_1\omega_2}. \tag{6.75}$$

Since (6.74) is true for each $\alpha\beta$ term in the sum in (6.72) we can substitute (6.74) in (6.72); note that $u_{\alpha\beta}(r_{\alpha\beta})$ is independent of ω_1 and ω_2, and use (6.75) to get

$$U_\mathrm{c} = \frac{1}{2}\rho^2 \sum_{\alpha\beta} \int \mathrm{d}\mathbf{r}_{1\alpha}\mathrm{d}\mathbf{r}_{2\beta}u_{\alpha\beta}(r_{\alpha\beta})g_{\alpha\beta}(r_{\alpha\beta}). \tag{6.76}$$

Changing variables from $\mathbf{r}_{1\alpha}\mathbf{r}_{2\beta}$ to $\mathbf{r}_{1\alpha}\mathbf{r}_{\alpha\beta}$, integrating over $\mathbf{r}_{1\alpha}$, and using polar coordinates for $\mathbf{r}_{\alpha\beta}$ (see (6.8a) and (6.8b)) gives

$$U_c = 2\pi N\rho \sum_{\alpha\beta} \int_0^\infty \mathrm{d}r_{\alpha\beta} r_{\alpha\beta}^2 u_{\alpha\beta}(r_{\alpha\beta}) g_{\alpha\beta}(r_{\alpha\beta}). \tag{6.77}$$

Thus a knowledge of the site–site correlation functions is sufficient to calculate the configurational energy, provided that the pair potential is of the site–site form. For the special case of homonuclear diatomic (AA) molecules, (6.77) becomes

$$U_c = 2N\rho \int \mathrm{d}\mathbf{r} u_{AA}(r) g_{AA}(r). \tag{6.78}$$

For the case of atomic liquids, having one site per molecule, (6.77) reduces to the standard result

$$U_c = \frac{1}{2} N\rho \int \mathrm{d}\mathbf{r} u(r) g(r). \tag{6.79}$$

6.7.3 *Pressure*[25–28]

For a site–site potential the pressure equation, (6.15), becomes

$$p = \rho kT - \frac{1}{6}\frac{\rho^2}{V\Omega^2} \sum_{\alpha\beta} \int \mathrm{d}\mathbf{r}_1 \mathrm{d}\mathbf{r}_2 \mathrm{d}\omega_1 \mathrm{d}\omega_2 r_{12} \frac{\partial u_{\alpha\beta}(r_{\alpha\beta})}{\partial r_{12}} g(\mathbf{r}_{12}\omega_1\omega_2)$$

or, using (6.74)

$$p = \rho kT - \frac{1}{6}\frac{\rho^2}{V\Omega^2} \sum_{\alpha\beta} \int \mathrm{d}\mathbf{r}_{1\alpha} \mathrm{d}\mathbf{r}_{2\beta} \mathrm{d}\omega_1 \mathrm{d}\omega_2 r_{12} \frac{\partial u_{\alpha\beta}(r_{\alpha\beta})}{\partial r_{12}} g(\mathbf{r}_{\alpha\beta}\omega_1\omega_2). \tag{6.80}$$

The derivative $\partial u_{\alpha\beta}(r_{\alpha\beta})/\partial r_{12}$ is at fixed molecular orientations and fixed ω_{12}, and is given by

$$\left(\frac{\partial u_{\alpha\beta}(r_{\alpha\beta})}{\partial r_{12}}\right)_{\omega_1\omega_2\omega_{12}} = \left(\frac{\partial r_{\alpha\beta}}{\partial r_{12}}\right)_{\omega_1\omega_2\omega_{12}} \frac{\mathrm{d}u_{\alpha\beta}(r_{\alpha\beta})}{\mathrm{d}r_{\alpha\beta}}. \tag{6.81}$$

To evaluate $(\partial r_{\alpha\beta}/dr_{12})$ we note that (see Fig. 3.8)

$$\mathbf{r}_{\alpha\beta} = \mathbf{r}_{12} + \mathbf{r}_{c2\beta} - \mathbf{r}_{c1\alpha} = \mathbf{r}_{12} + \mathbf{r}_{c\alpha\beta}, \tag{6.82}$$

where $\mathbf{r}_{c\alpha\beta} = \mathbf{r}_{c2\beta} - \mathbf{r}_{c1\alpha}$. Thus we have

$$r_{\alpha\beta}^2 = \mathbf{r}_{\alpha\beta} \cdot \mathbf{r}_{\alpha\beta} = r_{12}^2 + r_{c\alpha\beta}^2 + 2r_{12}\hat{\mathbf{r}}_{12} \cdot \mathbf{r}_{c\alpha\beta}, \tag{6.83}$$

where $\hat{\mathbf{r}}_{12} = \mathbf{r}_{12}/r_{12}$ is the unit vector in the $\mathbf{r}_{12}$ direction. Differentiating (6.83), noting that $r_{c\alpha\beta}$ and $\mathbf{r}_{c\alpha\beta}$ are fixed (since ω_1 and ω_2 are fixed), we have

$$2r_{\alpha\beta}\frac{\partial r_{\alpha\beta}}{\partial r_{12}} = 2r_{12} + 2\hat{\mathbf{r}}_{12} \cdot \mathbf{r}_{c\alpha\beta} = 2\left(r_{12}^2 + \mathbf{r}_{12} \cdot \mathbf{r}_{c\alpha\beta}\right)/r_{12}.$$

Thus, rearranging and using (6.82), we have

$$\begin{aligned}\frac{\partial r_{\alpha\beta}}{\partial r_{12}} &= (\mathbf{r}_{12}\cdot\mathbf{r}_{12}+\mathbf{r}_{12}\cdot\mathbf{r}_{c\alpha\beta})/r_{\alpha\beta}r_{12}\\ &= \mathbf{r}_{12}\cdot(\mathbf{r}_{12}+\mathbf{r}_{c\alpha\beta})/r_{\alpha\beta}r_{12}\\ &= \hat{\mathbf{r}}_{12}\cdot\hat{\mathbf{r}}_{\alpha\beta} = \cos\gamma_{\alpha\beta}\end{aligned} \tag{6.84}$$

where $\hat{\mathbf{r}}_{12} = \mathbf{r}_{12}/r_{12}$ and $\hat{\mathbf{r}}_{\alpha\beta} = \mathbf{r}_{\alpha\beta}/r_{\alpha\beta}$ are the unit vectors along $\mathbf{r}_{12}$ and $\mathbf{r}_{\alpha\beta}$, respectively, and $\cos\gamma_{\alpha\beta} = \hat{\mathbf{r}}_{12}\cdot\hat{\mathbf{r}}_{\alpha\beta}$ (i.e. $\gamma_{\alpha\beta}$ is the angle between $\mathbf{r}_{12}$ and $\mathbf{r}_{\alpha\beta}$). Substituting (6.81) and (6.84) into (6.80), and using (6.75), we get

$$p = \rho kT - \frac{1}{6}\frac{\rho^2}{V}\sum_{\alpha\beta}\int \mathrm{d}\mathbf{r}_{1\alpha}\mathrm{d}\mathbf{r}_{2\beta}\langle r_{12}\cos\gamma_{\alpha\beta}\rangle_{r_{\alpha\beta}}\frac{\mathrm{d}u_{\alpha\beta}(r_{\alpha\beta})}{\mathrm{d}r_{\alpha\beta}}g_{\alpha\beta}(r_{\alpha\beta}), \tag{6.85}$$

where $\langle\cdots\rangle_{r_{\alpha\beta}}$ denotes a weighted average over the orientations keeping the site–site distance $r_{\alpha\beta}$ fixed, i.e.

$$\langle r_{12}\cos\gamma_{\alpha\beta}\rangle_{r_{\alpha\beta}} \equiv \frac{\int \mathrm{d}\omega_1\mathrm{d}\omega_2 r_{12}\cos\gamma_{\alpha\beta}g(\mathbf{r}_{\alpha\beta}\omega_1\omega_2)}{\int \mathrm{d}\omega_1\mathrm{d}\omega_2 g(\mathbf{r}_{\alpha\beta}\omega_1\omega_2)}. \tag{6.86}$$

Changing variables in (6.85) from $\mathbf{r}_{1\alpha}\mathbf{r}_{2\beta}$ to $\mathbf{r}_{1\alpha}\mathbf{r}_{\alpha\beta}$, integrating over $\mathbf{r}_{1\alpha}$, and using polar coordinates for $\mathbf{r}_{\alpha\beta}$ gives

$$p = \rho kT - \frac{2\pi}{3}\rho^2\sum_{\alpha\beta}\int_0^\infty \mathrm{d}r_{\alpha\beta}r_{\alpha\beta}^2\langle r_{12}\cos\gamma_{\alpha\beta}\rangle_{r_{\alpha\beta}}u'_{\alpha\beta}(r_{\alpha\beta})g_{\alpha\beta}(r_{\alpha\beta}), \tag{6.87}$$

where $u_{\alpha\beta}'(r_{\alpha\beta}) = \mathrm{d}u_{\alpha\beta}(r_{\alpha\beta})/\mathrm{d}r_{\alpha\beta}$. This is the pressure equation for a site–site potential. Equation (6.87) is formally of the site–site type, but since $\langle r_{12}\cos\gamma_{\alpha\beta}\rangle_{r_{\alpha\beta}}$ requires the complete $g(\mathbf{r}\omega_1\omega_2)$ for its evaluation, it is not completely analogous to the relations (6.64) and (6.77) and (6.98) for χ, U_c, and A_c. Thus p *cannot* be expressed directly in terms of $g_{\alpha\beta}$ even for a site–site potential.[29] For an atomic liquid, with one site per atom, (6.87) reduces to the usual result, (5.23). For diatomic AA molecules (6.87) becomes

$$p = \rho kT - \frac{8\pi}{3}\rho^2\int_0^\infty \mathrm{d}r_{AA}r_{AA}^2\langle r_{12}\cos\gamma_{AA}\rangle_{r_{AA}}u_{AA}{}'(r_{AA})g_{AA}(r_{AA}). \tag{6.88}$$

We now consider the special case where the molecules are composed of hard spheres rigidly joined together (with sphere α of diameter $\sigma_{\alpha\alpha}$, etc.). The site–site potential is therefore

$$\begin{aligned}u_{\alpha\beta}(r_{\alpha\beta}) &= \infty \qquad r_{\alpha\beta}\le\sigma_{\alpha\beta}\\ &= 0 \qquad r_{\alpha\beta}>\sigma_{\alpha\beta},\end{aligned} \tag{6.89}$$

where $\sigma_{\alpha\beta} = \frac{1}{2}(\sigma_{\alpha\alpha} + \sigma_{\beta\beta})$. It follows that

$$\begin{aligned} \exp[-\beta u_{\alpha\beta}(r_{\alpha\beta})] &= 0 \qquad r_{\alpha\beta} \leq \sigma_{\alpha\beta} \\ &= 1 \qquad r_{\alpha\beta} > \sigma_{\alpha\beta} \end{aligned} \tag{6.90}$$

and

$$\frac{\mathrm{d}}{\mathrm{d}r_{\alpha\beta}} \exp[-\beta u_{\alpha\beta}(r_{\alpha\beta})] = \delta(r_{\alpha\beta} - \sigma_{\alpha\beta}), \tag{6.91}$$

where $\delta(r_{\alpha\beta} - \sigma_{\alpha\beta})$ is the Dirac delta function. Equation (6.91) follows because $\exp[-\beta u_{\alpha\beta}(r_{\alpha\beta})] \equiv \theta(r_{\alpha\beta})$ is the unit step function. The relation $\mathrm{d}\theta(x)/\mathrm{d}x = \delta(x)$ is verified by noting that the defining conditions ($\delta(x) = 0$ at $x \neq 0$, $\delta(x) = \infty$ at $x = 0$, and $\int \mathrm{d}x\delta(x) = 1$) for the δ-function are satisfied (see Appendix B).

It follows from (6.91) that

$$-\beta \frac{\mathrm{d}u_{\alpha\beta}(r_{\alpha\beta})}{\mathrm{d}r_{\alpha\beta}} \exp[-\beta u_{\alpha\beta}(r_{\alpha\beta})] = \delta(r_{\alpha\beta} - \sigma_{\alpha\beta})$$

or

$$\frac{\mathrm{d}u_{\alpha\beta}(r_{\alpha\beta})}{\mathrm{d}r_{\alpha\beta}} = -\frac{1}{\beta} \exp[\beta u_{\alpha\beta}(r_{\alpha\beta})]\delta(r_{\alpha\beta} - \sigma_{\alpha\beta}). \tag{6.92}$$

Substituting this result into (6.85), and using the fact that $y_{\alpha\beta}(r_{\alpha\beta}) = \exp[\beta u_{\alpha\beta}(r_{\alpha\beta})]g_{\alpha\beta}(r_{\alpha\beta})$ is continuous, gives

$$p = \rho kT + \frac{2\pi}{3}\rho^2 kT \sum_{\alpha\beta} \sigma^2_{\alpha\beta} \langle r_{12} \ \cos\gamma_{\alpha\beta}\rangle_{r_{\alpha\beta}=\sigma_{\alpha\beta}} g_{\alpha\beta}\left(\sigma^+_{\alpha\beta}\right), \tag{6.93}$$

where $g_{\alpha\beta}\left(\sigma^+_{\alpha\beta}\right)$ is the value of the site–site correlation function at contact.

For a hard sphere atomic fluid we have $\langle r_{12} \cos\gamma_{\alpha\beta}\rangle_{r_{\alpha\beta}=\sigma} = \sigma$, and (6.93) reduces to the usual result, (5.24). For a fluid of hard homonuclear diatomic *AA* molecules, each of the terms in the summation over α and β in (6.93) is the same, and the pressure equation becomes

$$p = \rho kT + \frac{8\pi}{3}\rho^2 kT\sigma^2 \langle r_{12}\cos\gamma\rangle_{r_{AA}=\sigma} g_{AA}(\sigma^+). \tag{6.94}$$

Using (6.82) and (6.84) we find that

$$\begin{aligned} r_{12}\cos\gamma|_{r_{AA}=\sigma} &= \left.\frac{\mathbf{r}_{AA}\cdot(\mathbf{r}_{AA}+\mathbf{r}_{c1A}-\mathbf{r}_{c2A})}{r_{AA}}\right|_{r_{AA}=\sigma} \\ &= \sigma\left[1 + \frac{\ell^*}{2}\left(\cos\theta'_{1A} - \cos\theta'_{2A}\right)\right], \end{aligned} \tag{6.95}$$

where $\ell^* = \ell/\sigma$, $\ell = 2r_{c1A} = 2r_{c2A}$ being the site–site distance, and $\theta_{1A}{}'$ and $\theta_{2A}{}'$ are the polar angles corresponding to $\mathbf{r}_{c1A}$ and $\mathbf{r}_{c2A}$ with Z chosen to lie along $\mathbf{r}_{AA'}$ (see Fig. 6.4).

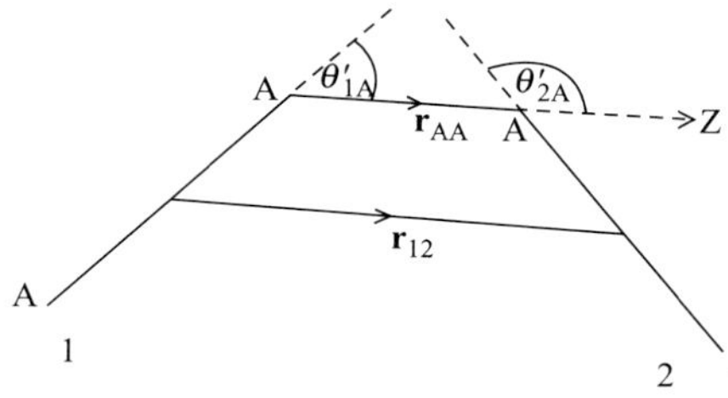

FIG. 6.4 Coordinates for two hard AA molecules.

6.7.4 *Free energy*[28,30]

In § 6.4 we have shown that (see (6.21), (6.22))

$$A - A_0 = \int_0^1 d\lambda \langle \mathcal{U}_p \rangle_\lambda \tag{6.96}$$

where A is the Helmholtz free energy for the full system with potential energy $\mathcal{U}(\mathbf{r}^N\omega^N)$, A_0 is the value for a reference system with potential energy, $\mathcal{U}_0(\mathbf{r}^N\omega^N)$, $\mathcal{U}_p \equiv \mathcal{U} - \mathcal{U}_0$ is the perturbing energy, and $\langle \ldots \rangle_\lambda$ is an average over configurations with Boltzmann weight corresponding to $\mathcal{U}_\lambda = \mathcal{U}_0 + \lambda \mathcal{U}_p$. If $\mathcal{U}_p$ is pairwise additive, and the intermolecular pair potentials $u_p(ij)$ are of site–site form, then from (6.77) we have

$$\langle \mathcal{U}_p \rangle_\lambda = 2\pi N\rho \sum_{\alpha\beta} \int_0^\infty dr r^2 u^p_{\alpha\beta}(r) g^\lambda_{\alpha\beta}(r) \tag{6.97}$$

and substituting this result in (6.96) gives

$$\begin{aligned} A - A_0 &= 2\pi N\rho \sum_{\alpha\beta} \int_0^1 d\lambda \int_0^\infty dr r^2 u^p_{\alpha\beta}(r) g^\lambda_{\alpha\beta}(r) \\ &= \frac{1}{2} N\rho \sum_{\alpha\beta} \int_0^1 d\lambda \int d\mathbf{r} u^p_{\alpha\beta}(r) g^\lambda_{\alpha\beta}(r), \end{aligned} \tag{6.98}$$

where $g^\lambda_{\alpha\beta}(r)$ is the site–site pair correlation function for the system with potential energy $\mathcal{U}_\lambda = \mathcal{U}_0 + \lambda \mathcal{U}_p$. If the reference system is the ideal gas, then $u^\mathrm{p}_{\alpha\beta}$ is replaced by the full site–site potential $u_{\alpha\beta}$ in the above expressions. For diatomic AA molecules (6.98) becomes

$$A - A_0 = 2N\rho \int_0^1 \mathrm{d}\lambda \int \mathrm{d}\mathbf{r} u^{\mathrm{p}}_{AA}(r) g^{\lambda}_{AA}(r). \tag{6.99}$$

The above method of introducing the charging parameter λ is not suitable for fluids composed of fused hard sphere molecules. For such fluids it is more convenient to use a charging parameter that scales both[28] the site–site diameters $\sigma_{\alpha\beta}$ and the intramolecular centre–site distances $\mathbf{r}_{c\alpha}$. Thus molecules characterized by the value λ of this parameter have site–site diameters $\sigma^{\lambda}_{\alpha\alpha}$ and intramolecular distances $\mathbf{r}^{\lambda}_{c\alpha}$ defined by

$$\begin{aligned} \sigma^{\lambda}_{\alpha\beta} &= \lambda\sigma_{\alpha\beta} \\ \mathbf{r}^{\lambda}_{c\alpha} &= \lambda\mathbf{r}_{c\alpha}, \end{aligned} \tag{6.100}$$

so that $\lambda = 0$ corresponds to the ideal gas of point noninteracting molecules. By a derivation analogous to that of (6.24) in § 6.4 we find that

$$\begin{aligned} A - A_0 &= \int_0^1 \mathrm{d}\lambda \left\langle \frac{\partial \mathcal{U}_{\lambda}}{\partial \lambda} \right\rangle_{\lambda} \\ &= \frac{1}{2}N\rho \sum_{\alpha\beta} \int_0^1 \mathrm{d}\lambda \int \mathrm{d}\mathbf{r}_{12} \left\langle \frac{\partial u^{\lambda}_{\alpha\beta}\left(r^{\lambda}_{\alpha\beta}\right)}{\partial \lambda} g^{\lambda}(\mathbf{r}_{12}\omega_1\omega_2) \right\rangle_{\omega_1\omega_2}, \end{aligned} \tag{6.101}$$

where A_0 is the Helmholtz free energy of the ideal gas of point particles at the same temperature, density, and number of molecules as the fluid of fused hard spheres. We note that (6.101) cannot be reduced to a form involving site–site correlation functions, since the term $\partial u^{\lambda}_{\alpha\beta}\left(r^{\lambda}_{\alpha\beta}\right)/\partial\lambda$ involves the molecular orientations ω_1 and ω_2 (see below).

We must now evaluate $\partial u^{\lambda}_{\alpha\beta}/\partial\lambda$. The potential $u^{\lambda}_{\alpha\beta}$ depends on λ through both $\sigma^{\lambda}_{\alpha\beta}$ and $r^{\lambda}_{\alpha\beta}$, since (see (6.82))

$$\mathbf{r}^{\lambda}_{\alpha\beta} = \mathbf{r}_{12} + \lambda\mathbf{r}_{c\alpha\beta}, \tag{6.102}$$

where $\mathbf{r}_{c\alpha\beta} = \mathbf{r}_{c2\beta} - \mathbf{r}_{c1\alpha}$. Thus we have

$$\frac{\partial u^{\lambda}_{\alpha\beta}}{\partial\lambda} = \frac{\partial u^{\lambda}_{\alpha\beta}}{\partial\sigma^{\lambda}_{\alpha\beta}}\frac{\partial\sigma^{\lambda}_{\alpha\beta}}{\partial\lambda} + \frac{\partial u^{\lambda}_{\alpha\beta}}{\partial r^{\lambda}_{\alpha\beta}}\frac{\partial r^{\lambda}_{\alpha\beta}}{\partial\lambda}. \tag{6.103}$$

From (6.92) we get

$$\frac{\partial u^{\lambda}_{\alpha\beta}}{\partial\sigma^{\lambda}_{\alpha\beta}} = -\frac{\partial u^{\lambda}_{\alpha\beta}}{\partial r^{\lambda}_{\alpha\beta}} = \frac{1}{\beta}\exp\left[\beta u^{\lambda}_{\alpha\beta}\left(r^{\lambda}_{\alpha\beta}\right)\right]\delta\left(r^{\lambda}_{\alpha\beta} - \sigma^{\lambda}_{\alpha\beta}\right). \tag{6.104}$$

Similarly $\partial u^{\lambda}_{\alpha\beta}/\partial r^{\lambda}_{\alpha\beta}$ is given by (6.92). To obtain $\partial r^{\lambda}_{\alpha\beta}/\partial\lambda$ we use (6.83),

$$\begin{aligned} 2r^{\lambda}_{\alpha\beta}\frac{\partial r^{\lambda}_{\alpha\beta}}{\partial\lambda} &= 2r^{\lambda}_{c\alpha\beta}\frac{\partial r^{\lambda}_{c\alpha\beta}}{\partial\lambda} + 2\mathbf{r}_{12}\cdot\frac{\partial \mathbf{r}_{c\alpha\beta}}{\partial\lambda} \\ &= 2\mathbf{r}^{\lambda}_{\alpha\beta}\cdot\mathbf{r}_{c\alpha\beta}, \end{aligned}$$

where (6.82) and (6.100) have been used. Thus we have

$$\frac{\partial r^{\lambda}_{\alpha\beta}}{\partial\lambda} = \hat{\mathbf{r}}^{\lambda}_{\alpha\beta}\cdot\mathbf{r}_{c\alpha\beta}, \tag{6.105}$$

where $\hat{\mathbf{r}}_{\alpha\beta} = \mathbf{r}_{\alpha\beta}/r_{\alpha\beta}$. Substituting these results into (6.103), noting that $\partial\sigma^{\lambda}_{\alpha\beta}/\partial\lambda = \sigma_{\alpha\beta}$ from (6.100), and substituting the resulting form for $\partial u^{\lambda}_{\alpha\beta}/\partial\lambda$ into (6.101) gives

$$\begin{aligned} A - A_0 = {} & \frac{2\pi N\rho}{\beta}\sum_{\alpha\beta}\int_0^{\sigma_{\alpha\beta}} \mathrm{d}\sigma^{\lambda}_{\alpha\beta}\sigma^{\lambda^2}_{\alpha\beta}g^{\lambda}_{\alpha\beta}\left(\sigma^{\lambda+}_{\alpha\beta}\right) \\ & - \frac{2\pi N\rho}{\beta}\sum_{\alpha\beta}\frac{1}{\sigma_{\alpha\beta}}\int_0^{\sigma_{\alpha\beta}} \mathrm{d}\sigma^{\lambda}_{\alpha\beta}\int_0^{\infty} dr r^2\left\langle\delta\left(r^{\lambda}_{\alpha\beta} - \sigma^{\lambda}_{\alpha\beta}\right)g^{\lambda}(\mathbf{r}\omega_1\omega_2)\left[\hat{\mathbf{r}}^{\lambda}_{\alpha\beta}\cdot\mathbf{r}_{c\alpha\beta}\right]\right\rangle_{\omega_1\omega_2}. \end{aligned} \tag{6.106}$$

We note that the first term on the right-hand side of this expression can be written in terms of $g^{\lambda}_{\alpha\beta}$, because the term $\left(\partial u^{\lambda}_{\alpha\beta}/\partial\sigma^{\lambda}_{\alpha\beta}\right)\left(\partial\sigma^{\lambda}_{\alpha\beta}/\partial\lambda\right)$ in (6.103) is independent of ω_1, ω_2; the same is not true of the second term on the right in (6.106) since $\mathbf{r}_{c\alpha\beta}$ depends on ω_1 and ω_2. In (6.106) both $g^{\lambda}_{\alpha\beta}$ and $g^{\lambda}(\mathbf{r}\omega_1\omega_2)$ refer to the fluid with potential $\mathcal{U}_{\lambda}$. By using the identity $p = -\partial A/\partial V$, together with the virial theorem, it is possible to derive[28] the equation of state (6.93) from (6.106).

An alternative expression for the free energy can be obtained[30] for fused hard sphere fluids by using a charging parameter λ' that scales only the site diameters $\sigma_{\alpha\beta}$, so that (6.100) is replaced by

$$\sigma^{\lambda'}_{\alpha\beta} = \lambda'\sigma_{\alpha\beta} \tag{6.107}$$

and the centre–site distances $r_{c\alpha}$ (and the angles between the $\mathbf{r}_{c\alpha}$) remain unchanged during the charging process and equal to the real values. In this case $\lambda' = 0$ corresponds to an ideal gas composed of fused hard sphere molecules with infinitesimally small spheres. This gas, whose Helmholtz free energy is written $A_0{}' \equiv A(\lambda' = 0)$, differs from that used as the reference state in (6.106) in that the molecules possess rotational kinetic energy. By retracing the steps in deriving (6.106) it is clear that in this case the free energy will be given by

$$A - A_0' = \frac{2\pi N\rho}{\beta} \sum_{\alpha\beta} \int_0^{\sigma_{\alpha\beta}} d\sigma_{\alpha\beta}^{\lambda'} \sigma_{\alpha\beta}^{\lambda'^2} g_{\alpha\beta}^{\lambda'} \left(\sigma_{\alpha\beta}^{\lambda'+}\right). \tag{6.108}$$

It should be noted that $g_{\alpha\beta}^{\lambda'}\left(\sigma_{\alpha\beta}^{\lambda'+}\right)$ in (6.108) is the site–site correlation function for molecules with variable site diameters $\sigma_{\alpha\beta}^{\lambda'}$, but with the fixed real values of the $\mathbf{r}_{c\alpha}$; in (6.106), by contrast, $g_{\alpha\beta}^{\lambda}\left(\sigma_{\alpha\beta}^{\lambda+}\right)$ is the site–site correlation function when both the $\sigma_{\alpha\beta}^{\lambda}$ and $\mathbf{r}_{c\alpha}^{\lambda}$ are variable. Equations (6.106) and (6.108) must give identical results if the exact correlation functions are used. If an approximate theory is used for the correlation functions they provide alternative routes to the thermodynamic properties which may give different results.

For a fluid of hard homonuclear diatomic *AA* molecules, (6.108) becomes

$$A - A_0' = \frac{8\pi N\rho}{\beta} \int_0^{\sigma} d\sigma^{\lambda'} \sigma^{\lambda'^2} g_{AA}^{\lambda'}(\sigma^{\lambda'}+). \tag{6.109}$$

6.8 A rigorous inequality for the free energy[31–41]

In this section we discuss an application of the classical Gibbs inequality[31] for the configurational Helmholtz free energy. The quantum version of the inequality (due to Bogoliubov) can also be derived;[32–35] the result then applies to the entire free energy, rather than just the configurational part.

If we write the total intermolecular potential energy as the sum of a reference and a perturbation part, $\mathcal{U}_\lambda = \mathcal{U}_0 + \lambda\mathcal{U}_\mathrm{p}$, then the free energy difference between the real and reference systems is given by (6.96). Moreover, we have

$$\left(\frac{\partial\langle\mathcal{U}_p\rangle_\lambda}{\partial\lambda}\right)_{NV\beta} = -\beta\langle(\mathcal{U}_p - \langle\mathcal{U}_p\rangle_\lambda)^2\rangle_\lambda, \tag{6.110}$$

which is the analogue of (3.249) in the canonical ensemble. The right-hand side of (6.110) must be negative or zero. It follows that $\langle\mathcal{U}_\mathrm{p}\rangle_\lambda$ is a decreasing function of λ, or a constant. Hence the maximum and minimum values of the integrand in (6.96) are $\langle\mathcal{U}_\mathrm{p}\rangle_{\lambda=0}$ and $\langle\mathcal{U}_\mathrm{p}\rangle_{\lambda=1}$, respectively. Multiplying these by the integration interval in (6.96) (unity), we arrive at bounds for the integral itself,

$$\langle\mathcal{U}_\mathrm{p}\rangle \le A - A_0 \le \langle\mathcal{U}_\mathrm{p}\rangle_0, \tag{6.111}$$

where $\langle\mathcal{U}_\mathrm{p}\rangle \equiv \langle\mathcal{U}_\mathrm{p}\rangle_{\lambda=1}$. The lower bound $\langle\mathcal{U}_\mathrm{p}\rangle$ in (6.111) is probably of little computational value, since it requires knowledge of the real fluid distribution functions. The upper bound

$$A - A_0 \le \langle\mathcal{U}_\mathrm{p}\rangle_0 \tag{6.112}$$

is valid for angle-independent or dependent $\mathcal{U}_p$, and is referred to as the *Gibbs-Bogoliubov inequality*.[32,33,36] Atomic fluid applications are discussed by Münster.[36] When $\langle \mathcal{U}_p \rangle_0$ is nonzero it can be used[39,40] as a basis for a variational determination of $A - A_0$.

If the fluid is molecular and the Pople reference system is chosen (§ 4.5), then $\langle \mathcal{U}_p \rangle_0 = 0$ and

$$A \leq A_0 \tag{6.113}$$

i.e. the anisotropic perturbation rigorously lowers the free energy.[41] The inequality (6.113), which holds to all orders and for arbitrarily large Pople-type $\mathcal{U}_p$, is manifestly satisfied for multipolar perturbations in second order (see (4.49)). For this case we also have, from (6.111), that $\langle \mathcal{U}_p \rangle \leq 0$; this is expected, since the Boltzmann averaging places more weight on the negative parts of $\mathcal{U}_p$.

The quantity $\langle \mathcal{U}_p \rangle_0$ equals the first-order perturbation term A_1 (see (4.5)), so that A_1 gives a rigorous upper bound for the free energy difference $A - A_0$.

It is clear from (6.110) that (for $\beta \neq 0$) the equalities in (6.111) hold if and only if $\mathcal{U}_p$ is a constant, equal to its average value for all configurations. If $\langle \mathcal{U}_p \rangle_0 = 0$ the only possible value is $\mathcal{U}_p = 0$. For $\beta \to 0$ (infinite temperature) the situation is more subtle;[41] we can have $A - A_0 = 0$ or $A - A_0 = -\infty$, depending on the nature of the potential.

General inequalities for the entropy or internal energy do not exist; it is possible that such inequalities can be derived for special potentials.[41]

6.9 Quantum corrections[42]

When quantum corrections are small, as is expected for most liquids except He and H_2, (see § 1.2.2 and Table 1.1 of Vol. 1), their effect on the thermodynamic properties can be treated by expanding the partition function in powers of $\hbar^2$. Details of this expansion are given in Appendix 3D of Vol. 1, and by Powles and Rickayzen.[42] The corresponding expansion for the Helmholtz free energy, $A = -kT \ln Q$, gives the following results for molecules of the symmetries indicated:

Linear ($I_x = I_y \equiv I$):

$$A - A_{c\ell} = \frac{\hbar^2 N}{24(kT)^2}\left[\frac{\langle F^2 \rangle}{m} + \frac{\langle \tau^2 \rangle}{I}\right] - \frac{\hbar^2 N}{6I} + O(\hbar^4) \tag{6.114}$$

Spherical top ($I_x = I_y = I_z \equiv I$):

$$A - A_{c\ell} = \frac{\hbar^2 N}{24(kT)^2}\left[\frac{\langle F^2 \rangle}{m} + \frac{\langle \tau^2 \rangle}{I}\right] - \frac{\hbar^2 N}{8I} + O(\hbar^4) \tag{6.115}$$

Symmetric top ($I_x = I_y \equiv I_\perp,\ I_z \equiv I_{||}$):

$$A - A_{c\ell} = \frac{\hbar^2 N}{24(kT)^2}\left[\frac{\langle F^2\rangle}{m} + \frac{\left\langle \tau_{||}^2 \right\rangle}{I_{||}} + \frac{\left\langle \tau_\perp^2 \right\rangle}{I_\perp}\right] - \frac{\hbar^2 N}{24 I_\perp}\left(4 - \frac{I_{||}}{I_\perp}\right) + O(\hbar^4) \tag{6.116}$$

Asymmetric top ($I_x,\ I_y,\ I_z$ not equal):

$$A - A_{c\ell} = \frac{\hbar^2 N}{24(kT)^2}\left[\frac{\langle F^2\rangle}{m} + \sum_{\alpha=x,y,z}\frac{\left\langle \tau_\alpha^2 \right\rangle}{I_\alpha}\right] - \frac{\hbar^2 N}{24}\sum_{\text{cyclic}}\left(\frac{2}{I_x} - \frac{I_x}{I_y I_z}\right) + O(\hbar^4), \tag{6.117}$$

where A is the quantal and $A_{c\ell}$ the classical value of A, and $\langle \ldots \rangle$ is a *classical* configurational average. Each of these expressions contains three terms of order $\hbar^2$; the first, involving $\langle F^2\rangle$, arises from translational (diffraction) effects, the second involves $\langle \tau^2\rangle$ from rotational potential energy effects, and a final term (e.g. $\hbar^2 N/6I$ in (6.114)) arises from rotational kinetic energy effects (see § 1.2.2 for a qualitative discussion). Here $\langle F^2\rangle$ and $\langle \tau^2\rangle$ are the classical mean squared force and mean squared torque (about the centre of mass) on a molecule, respectively, and can be obtained experimentally from the isotope separation factor (see § 1.2.2) or from spectral moments (infrared, Raman, or neutron), as discussed in Chapter 11. These quantities can also be calculated theoretically or by computer simulation (see Chapter 11). The quantities m and I_α are the molecular mass and principal $\alpha\alpha$-component of the moment of inertia, respectively, and in (6.117)

$$\sum_{cyclic}\left(\frac{2}{I_x} - \frac{I_x}{I_y I_z}\right) = \frac{2}{I_x} + \frac{2}{I_y} + \frac{2}{I_z} - \frac{I_x}{I_y I_z} - \frac{I_y}{I_x I_z} - \frac{I_z}{I_x I_y}. \tag{6.118}$$

Also, τ_α is the principal α component of $\boldsymbol{\tau}$, $\left\langle \tau_{||}^2 \right\rangle \equiv \langle \tau_z^2\rangle$, and $\langle \tau_\perp^2\rangle = \langle \tau_x^2\rangle + \left\langle \tau_y^2 \right\rangle$ is the total mean squared torque perpendicular to the symmetry axis z.

For linear molecules, (6.114) can be rearranged to the dimensionless form

$$\frac{A - A_{c\ell}}{\hbar^2 N/24 m\sigma^2} = \frac{1}{T^{*2}}\left(\langle F^{*2}\rangle + \frac{\langle \tau^{*2}\rangle}{I^*}\right) - \frac{4}{I^*} \tag{6.119a}$$

or, alternatively,

$$\frac{A - A_{c\ell}}{NkT} = \frac{1}{96\pi^2}\frac{\Lambda^{*2}}{T^{*3}}\left(\langle F^{*2}\rangle + \frac{\langle \tau^{*2}\rangle}{I^*}\right) - \frac{\theta_r}{3T}, \tag{6.119b}$$

where $F^{*2} = F^2/(\varepsilon/\sigma)^2$, $\tau^{*2} = \tau^2/\varepsilon^2$, and $I^* = I/m\sigma^2$ are the reduced quantities, with ε and σ being the usual energy and distance parameters in the isotropic part of the potential, and m the molecular mass; the quantities $\Lambda^* = h/2\pi\sigma(m\varepsilon)^{1/2}$ and $\theta_r = \hbar^2/2Ik$ are the dimensionless de Broglie wavelength and the characteristic rotational temperature, respectively, introduced in § 1.2.2. The magnitude of these corrections for liquids with dipole–dipole and

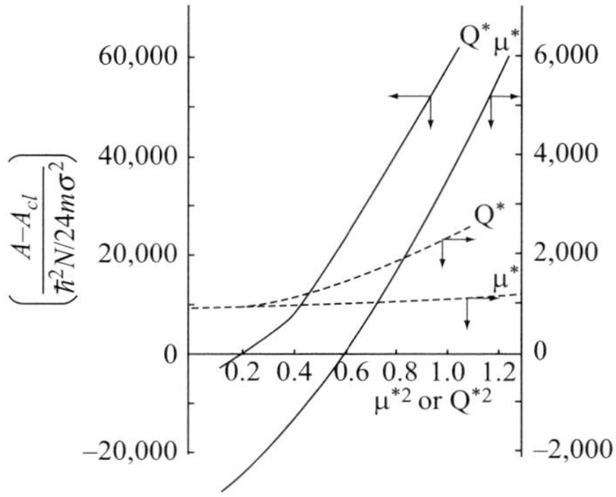

FIG. 6.5 Quantum correction to the free energy for a liquid of linear molecules with pair potential $u_{LJ} + u_a$, where u_{LJ} is the Lennard-Jones potential and u_a is either the dipole–dipole (curves labelled μ^*) or quadrupole–quadrupole (curves labelled Q^*) term; the state condition is $\rho^* = 0.800$, $T^* = 0.719$. Solid and dashed curves are for $I^* = I/m\sigma^2 = 0.001$ and 0.07, respectively.

Table 6.2

Quantum corrections to configurational energy and pressure for orthobaric liquids

Liquid	T(K)	$\frac{U-U_{c\ell}}{U_{c\ell}} \times 100$	$\frac{p-p_{c\ell}}{\rho kT}$
N_2	73	−5	0.8
F_2	59	−1.7	0.26
Cl_2	250	−0.7	0.07
HCl	188	−2	0.16
$CH_4^\dagger$	207	−4.4	0.09

† This state point is not orthobaric. The density is 24.9 mol ℓ^{-1}. From ref. 42.

quadrupole–quadrupole forces is shown in Fig. 6.5, based on Monte Carlo simulation values[43,44] of $\langle F^{*2}\rangle$ and $\langle \tau^{*2}\rangle$. Values of I^* for most linear molecules lie in the range 0.001 (HI) to 0.07 (I_2), and results for these two extremes are shown in Fig. 6.5. The relative magnitudes of the three terms in (6.119a) are strongly dependent on the value of I^* and on the strength of the anisotropic forces. Thus, when $I^* \sim 0.001$ it is clear that the rotational quantum corrections are the dominant ones unless the electrostatic forces are very weak. Conversely, for large I^*, e.g. $I^* \sim 0.07$, the translational correction dominates even for relatively strong electrostatic forces.

Rough estimates of the quantum corrections to the configurational energy and pressure have been made for several liquids by Powles and Rickayzen,[42] and are shown in Table 6.2. These calculations are based on computer simulation values

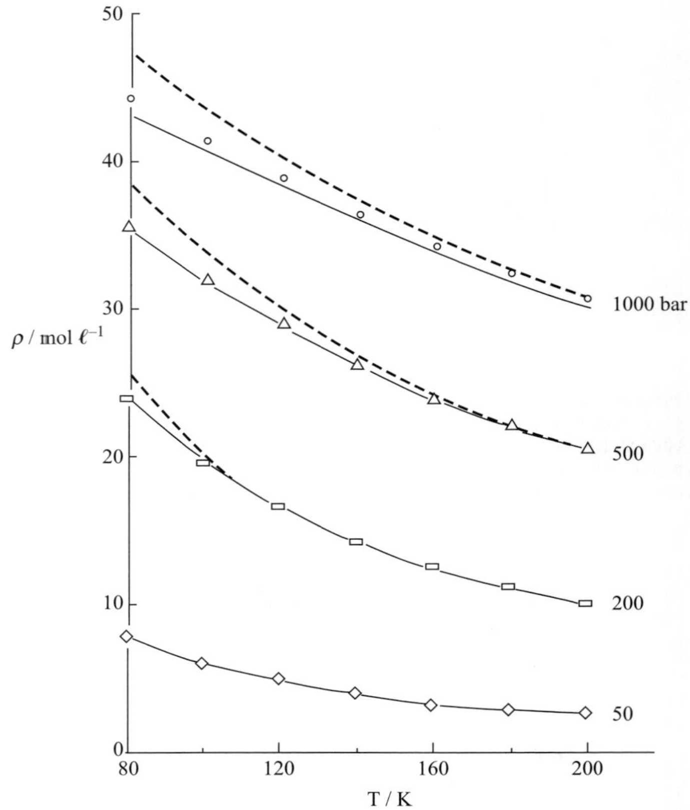

FIG. 6.6 Density of hydrogen from experiment[48] (points), theory neglecting quantum effects (- - -), and theory including the $O(\hbar^2)$ quantum correction (—) as given by (6.114). (From ref. 45.)

of $\langle \tau^2 \rangle$ and $\langle F^2 \rangle$ for site–site potential models; for nitrogen the quadrupole–quadrupole potential was also included. The ratio of the rotational to the translational potential energy terms in (6.119a), i.e. $\langle \tau^2 \rangle m / \langle F^2 \rangle I$, is about 0.6 for N_2 and F_2, and about 4 for HCl. The results show that the corrections to U are small, but not negligible. For the pressures the quantum corrections are larger. Furthermore, the quantum corrections to the pressure may be a large fraction of the total pressure for liquids near their triple points, since the total pressure is then small. For orthobaric liquid N_2 at 73 K, for example, $p \approx 1$ bar, whereas $p - p_{c\ell} = 140$ bar. For compressed fluid CH_4 at 207 K, 24.9 mol ℓ^{-1} (well away from the triple point), on the other hand, $p \approx 818$ bar, while $p - p_{c\ell} =$ 69 bar.

For hydrogen the quantum corrections are particularly large, so that the $\hbar^2$ term alone in (6.114) is not sufficient at low temperatures, and a full quantal

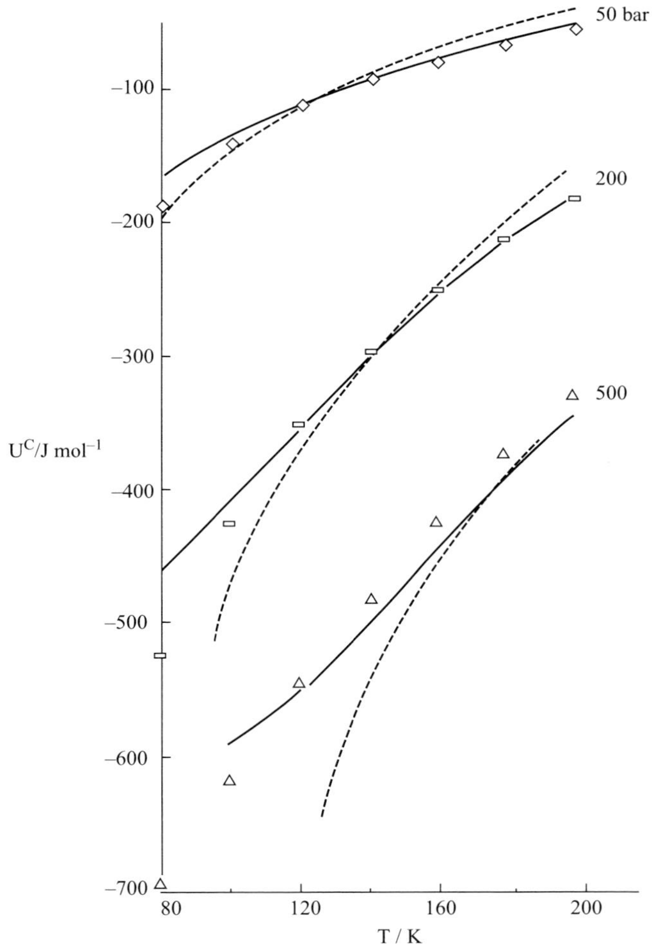

FIG. 6.7 Configurational internal energy of hydrogen. Key is as described in legend to Fig. 6.6. (From ref. 45.)

treatment becomes necessary.[45,46] Nevertheless, the $\hbar^2$ correction significantly improves the accuracy of calculated thermodynamic data for the temperature range 100–200 K.[45] Figures 6.6 and 6.7 show such results, compared to a purely classical calculation, for equation of state data. The calculations are based on the pair potential model

$$u(12) = u_{\mathrm{LJ}}(r) + u_{\mathrm{QQ}}(224) + u_{\mathrm{dis}}(202 + 022 + 224) + u_{\mathrm{ov}}(202 + 022), \tag{6.120}$$

where $u_{LJ}(r)$ is the Lennard-Jones model of (2.2), $u_{QQ}(224)$ is the quadrupole–quadrupole potential given by (2.185), $u_{dis}(202 + 022 + 224)$ is the anisotropic dispersion potential as given by the $\ell_1\ell_2\ell = 202$, 022, and 224 terms (see (2.223) and also (2.228) and (2.232)), and $u_{ov}(202 + 022)$ is the anisotropic overlap potential as given by the $\ell_1\ell_2\ell = 202$ and 022 terms (see (2.247), (2.248)). Calculations are based on the Padé approximant of (4.47), with the anisotropic potential parameters $Q = 0.65 \times 10^{-26}$ esu, $\kappa = 0.125$ (see (D.8)), and $\delta = 0.1$ (values of Q and κ are taken from sources listed in Appendix D of Vol. 1). Values of δ, ε, and σ are obtained by matching theory with experimental density and configurational energy data at higher temperatures, where quantum effects are small. It is seen from Figs. 6.6 and 6.7 that the quantum corrections increase at lower temperatures and higher pressures, as expected. Quantum corrections are appreciable at temperatures below 160 K, particularly at the higher pressures.

Quantum corrections to the critical constants have been calculated for LJ fluids.[47] The quantum corrections cause the critical temperature, pressure, and density to decrease.

6.10 Virial coefficients

The second virial coefficient is given by (3.268) and (3.269) as

$$B_2 = -2\pi \int_0^\infty \langle \exp[-\beta u(\mathbf{r}\omega_1\omega_2) - 1] \rangle_{\omega_1\omega_2} r^2 \mathrm{d}r \tag{6.121a}$$

$$= -\frac{2\pi}{3}\beta \int_0^\infty \langle u'(\mathbf{r}\omega_1\omega_2) \exp[-\beta u(\mathbf{r}\omega_1\omega_2)] \rangle_{\omega_1\omega_2} r^3 \mathrm{d}r, \tag{6.121b}$$

where $u' = \partial u/\partial r$. These expressions are based on the assumptions that (a) the molecules are 'rigid' (see § 1.2.1), (b) quantum effects can be neglected (§ 1.2.2), and (c) the angle-averaged pair potential, $\langle u(\mathbf{r}\omega_1\omega_2)\rangle_{\omega_1\omega_2}$, decreases faster than r^{-3} for large r. Assumption (a) can lead to errors for molecules that have internal rotations or are otherwise flexible, while (b) may not hold for light gases at low temperature. Nevertheless, (a) and (b) lead to negligible errors for most small molecules that do not contain hydrogen. Condition (c) is necessary in order that the integrals appearing in (3.268) and (3.269) be independent of the shape and size of the system, and is required if the thermodynamic properties are to exist. Condition (c) will hold for all cases considered in this book; note that (c) is satisfied for a potential of the form $u_0(r) + u_{\mu\mu}$, where u_0 is shorter-ranged than r^{-3} and $u_{\mu\mu}$ is the dipole–dipole potential (which varies as r^{-3}), since $\langle u_{\mu\mu}\rangle_{\omega_1\omega_2} = 0$.

Qualitatively $B_2(T)$ behaves as shown in Fig. 6.8 as a function of temperature T. At low temperature the attractive forces dominate molecular behaviour. On

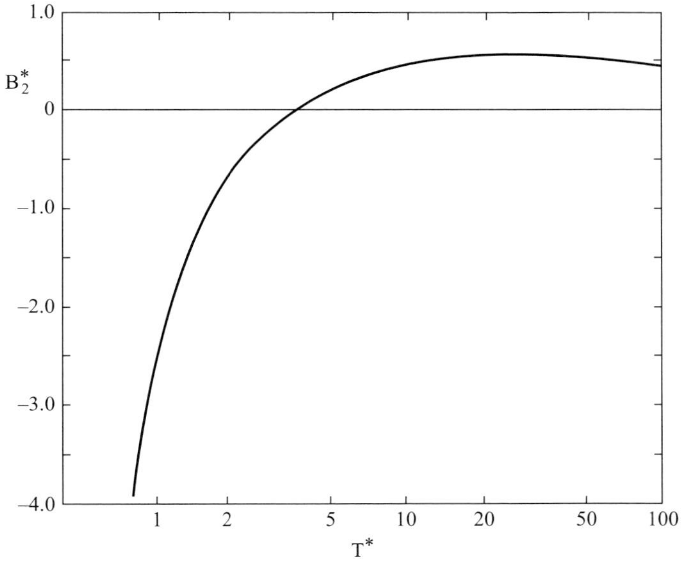

FIG. 6.8 General form of the reduced second virial coefficient, $B_2^* = B_2/b_o$ with $b_o = 2\pi\sigma^3/3$, as a function of reduced temperature, $T^* = kT/\varepsilon$. The curve shown is for a classical LJ gas.

average the molecules spend more time in each others 'spheres of influence' than they would in the absence of interaction. Since the molecules then hit the walls less frequently, p is *reduced* below the ideal gas or kinetic value ρkT, so that B_2 is *negative*. At high temperatures the opposite effect occurs; the molecules bounce off each other, due to the repulsive forces, hit the walls more frequently, and hence *increase* p over ρkT, giving a *positive* B_2. The rapid decrease with temperature of $B_2(T)$ at very low temperatures is caused by more and more (exponentially more) molecules becoming mutually trapped in bound states. The slow decrease with temperature of $B_2(T)$ at very high temperatures is due to the slowly decreasing effective diameter of the molecules; as the collisions become more energetic, the molecules penetrate further into the core region of the potential.[49]

6.10.1 B_2 and the pair potential

For the majority of cases, where the above conditions of rigidity and classical behaviour are fulfilled, (6.121a) shows that measurements of B_2 provide a valuable source of information about the pair potential. Complications due to the influence of three-body potentials or the need for an approximate theory, always a concern in interpreting liquid state data, are not present here. Information about the pair potential can be extracted from B_2 data either by inversion of the data, or by comparing calculated B_2 values with the data for various model potentials. The first procedure is the most desirable, and is possible in principle for atomic gases.

Thus, if the molecules are spherical, and the pair potential has a monotonically repulsive portion $u_+(r_+)$ for $r \equiv r_+ \leq r_m$, and a monotonically attractive portion $u_-(r_-)$ for $r \equiv r_- > r_m$ (cf. Fig. 2.4), where r_m is the separation at the potential minimum, it is possible to rearrange (6.121b) to the form[50,51]

$$\frac{3}{2\pi}\frac{1}{\beta}\exp[-\beta\varepsilon]B_2(\beta) = \int_0^\infty \Delta(\phi)\exp[-\beta\phi]\mathrm{d}\phi = \mathcal{L}[\Delta(\phi)], \tag{6.122}$$

where $\phi \equiv u + \varepsilon$ is the pair potential referred to the minimum of the well as zero, ε is the well depth, $\mathcal{L}[\ ]$ is the Laplace transform, and:

$$\begin{aligned}\Delta(\phi) &= r_+^3(\phi) - r_-^3(\phi) \qquad & \phi \leq \varepsilon \\ &= r_+^3(\phi) \qquad & \phi > \varepsilon\end{aligned} \tag{6.123}$$

Thus, if accurate B_2 data are available over a range of temperatures, it should be possible to invert the Laplace transform in (6.122) to obtain $\Delta(\phi)$, i.e. the potential itself for the purely repulsive region ($\phi > \varepsilon$), and the well-width function $(r_+^3 - r_-^3)$ for $\phi \leq \varepsilon$. Such a procedure has been successfully used to obtain the repulsive part of the potential for helium;[52] for this case data are available over a very wide range of temperatures and the attractive forces are weak. This inversion procedure has proved difficult for other atomic gases, since the inverse transform is highly sensitive to inaccuracies in the data, and because in many cases data are not available over a sufficient temperature range.[53] Less formal methods of inversion based on semiempirical techniques have been devised, and have proved successful for atomic gases.[51,53,54]

For *molecular* gases it is not possible to write (6.121a) or (3.272) as a Laplace transform that could be used to obtain information about $u(12)$, so that a formal inversion procedure does not exist in this case.[55] The use of B_2 data to investigate potentials has therefore been by the second of the methods mentioned earlier, i.e. comparisons of experiment with calculated B_2 values using various models. Unfortunately, it is found that B_2 is only sensitive to the form of the intermolecular potential energy function when data are available over a wide temperature range. When data are available over only a limited range B_2 is relatively insensitive to the form of the potential (in particular the anisotropic part), so that a variety of models often give an equally good fit to the data. Thus, for nitrogen and carbon dioxide it is found that the Kihara, LJ atom–atom (two sites), and Gaussian overlap potential models, with or without the addition of a quadrupole–quadrupole term, all give an equally good description of the available second virial coefficient data, although they do not work equally well for other properties (e.g. heat of sublimation, crystal lattice parameter).[56] This situation is largely due to the insensitivity of the temperature-dependent pseudopotential $u_0(r)$, defined by (3.273), to the nature of the anisotropic intermolecular forces; thus both multipolar and nonspherical shape forces tend to have a similar effect on $u_0(r)$.[57] Indeed, this insensitivity is

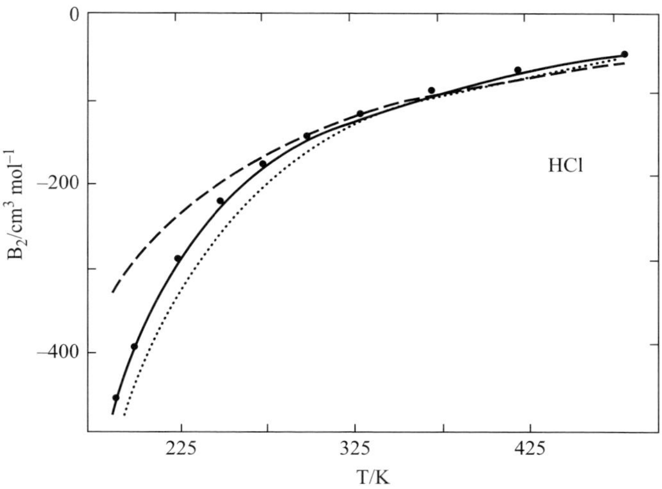

FIG. 6.9 Second virial coefficient of HCl: • experiment;[57] – – – site–site Lennard-Jones (SSLJ) fitted to liquid; —— SSLJ + multipole terms fitted to B_2; .... second model fitted to liquid. (From ref. 63.)

the basis for simple and widely used corresponding states methods for predicting second virial coefficients.[58]

Second virial coefficient data have often been used to estimate values for anisotropic potential parameters, by comparison of the data with calculations for a particular model; multipole moments and anisotropic overlap parameters have often been obtained in this way, by using a generalized Stockmayer potential model.[59] Multipole moments so obtained are often unreliable because of uncertainties in the potential model used (see Appendix D of Volume 1). A more stringent test of the potential model is to compare theory and experiment for both the pressure and the dielectric second virial coefficients[60] (see § 10.6, and also the collision-induced absorption virial coefficients, Chapter 11), and where possible the mean squared torque for the gas[61] (see Chapter 11).

Although second virial data alone are of limited value in testing potential models, a model that fits other data should, of course, also fit the virial data if it is the correct pair potential. A test of this sort on two potential functions whose parameters were obtained by comparing computer simulation results with experimental thermodynamic data for liquid hydrogen chloride is shown in Fig. 6.9. Either of these potential functions can be made to fit the B_2 data if the site–site parameters $\varepsilon_{\alpha\beta}$ and $\sigma_{\alpha\beta}$ are suitably adjusted (here α, β are H or Cl); the solid line shows the result for the second potential, which includes multipole terms. When the parameters are adjusted to fit liquid data, however, the potentials no longer fit the virial data. The potential including multipole terms appears to be the better of the two. The failure of the liquid potentials to fit the gas data could be due to errors in the pair potential, the neglect of multibody potential terms in treating the liquid, or a combination of these two effects.

6.10.2 Calculations for simple model potentials

There have been many calculations of B_2, and in some cases higher virial coefficients, for a wide variety of simple model potentials of the types discussed in § 2.1; the earlier work has been reviewed elsewhere.[51,64,65] These have included calculations for the generalized Stockmayer model, site–site LJ models with and without point charges, the Kihara model, Gaussian Overlap model, etc. (These models are discussed in §2.1 of Vol. 1.) The addition of electrostatic forces causes B_2 to decrease at any given temperature. This is because such forces are, on the average, attractive, and cause a decrease in the frequency of molecules hitting the wall, and hence a decrease in the pressure.

Fluids of hard nonspherical molecules are of particular interest since they are frequently used as reference fluids in perturbation theories for both isotropic and anisotropic (e.g. liquid crystal) fluids. For a fluid of hard spheres of diameter σ it is easy to show from (6.121a) that

$$B_2 = 4v_{\mathrm{m}}, \tag{6.124}$$

where $v_{\mathrm{m}} = \pi\sigma^3/6$ is the volume of a sphere. For a fluid of hard nonspherical molecules, each of volume v_{m}, it is convenient to write

$$B_2 = 4v_{\mathrm{m}}f_B, \tag{6.125}$$

where f_B is a factor whose difference from unity indicates the degree of departure from hard sphere behaviour.

The simplest molecules to treat are convex ones, i.e. ones in which any line segment whose endpoints are inside the body lies wholly within the body (see Appendix 6A); examples of such bodies are spherocylinders and ellipsoids (a nonconvex example is two fused hard spheres). For such molecules f_B is always greater than unity, so that convex nonspherical molecules always have second virial coefficients larger than those for spherical molecules of the same volume. The physical reason for $f_B > 1$ is that, as the molecule rotates, the effective volume swept out is greater than that of the molecule, thus tending to increase B_2. Isihara[66] has derived the following general expression for f_B for the convex molecule case (see Appendix 6A.5 for derivation):

$$f_B = 1 + \frac{1}{4}(\gamma - 3), \tag{6.126}$$

where $\gamma = r_{\mathrm{m}}s_{\mathrm{m}}/v_{\mathrm{m}}$ is the *shape factor*, and r_{m} and s_{m} are the mean radius of curvature and the surface area of the molecule, respectively (see Appendix 6A for an account of the geometry of hard convex bodies). From (6.125) and (6.126) we have, for hard convex molecules

$$B_2 = v_{\mathrm{m}}(1 + \gamma). \tag{6.127}$$

We note that γ, and hence B_2, is independent of temperature as in the case of hard spheres. For spheres of radius a we have $r_{\rm m} = a$, $s_{\rm m} = 4\pi a^2$, $v_{\rm m} = 4\pi a^3/3$, $\gamma = 3$, and $f_B = 1$. Formulae for $r_{\rm m}$, $s_{\rm m}$, $v_{\rm m}$, and f_B for other shapes have been worked out;[65,66] some of these are given in Table 6A.1.

For general nonconvex molecules there is no simple expression analogous to (6.127). For homonuclear (AA) hard sphere diatomic molecules (dumbbells), Isihara has derived an exact analytic expression for B_2.[67] This result has been generalized to heteronuclear hard diatomics (AB), unlike pair diatomics (AB/CD), and symmetrical linear triatomics (ACA) by Wertheim.[68] Numerical calculations of B_2 have been made for other fused hard sphere molecules, and some of these results are shown in Fig. 6.10 together with results for spherocylinders. It should be noted that for hard spheres $B_2/4v_m = 1$ so that $f_B > 1$ for all of the nonspherical shapes shown; f_B is greatest for hard diatomics. The results for homonuclear diatomics and spherocylinders are closely similar for moderate ℓ^* values, but become significantly different for $\ell^* > 1$.

Higher virial coefficients up to B_5 have been calculated for a variety of convex and non-convex molecular shapes; these calculations have been the subject of reviews.[69–71] Two of the most widely studied shapes are spherocylinders and dumbbells (see Fig. 6.11).

If the virial coefficients up to B_5 are known, for example, it is possible to form various Padé approximants to the virial series (3.267); such approximants provide a means of estimating contributions from higher order terms in the series (see Appendix B.6, vol. 1, for a discussion of Padé approximants). The 2/2 Padé has been widely studied and reproduces the series up to the B_5 term; it is (cf. (B.104), (B.108), (B.109))

$$\frac{p}{\rho kT} = \frac{1 + p_1\rho + p_2\rho^2}{1 + q_1\rho + q_2\rho^2} \tag{6.128}$$

where

$$p_1 = B_2 + q_1, \qquad p_2 = B_3 + B_2 q_1 + q_2$$

$$q_1 = \frac{B_3 B_4 - B_2 B_5}{B_2 B_4 - B_3^2}, \qquad q_2 = \frac{B_3 B_5 - B_4^2}{B_2 B_4 - B_3^2}.$$

Equation (6.128) is found to be superior at high density to the series truncated at B_5, for many convex and non-convex shapes. A typical result[72,73] is shown in Fig. 6.12.

The virial coefficient calculations described above are essentially exact, within the limitations of the numerical procedures used. Calculations for fluids of hard molecules based on approximate theories are described in § 6.12 (see Table 6.7).

Direct numerical evaluation of the virial coefficients have been carried out by the Monte Carlo integration method[74] and by nonproduct integration techniques,

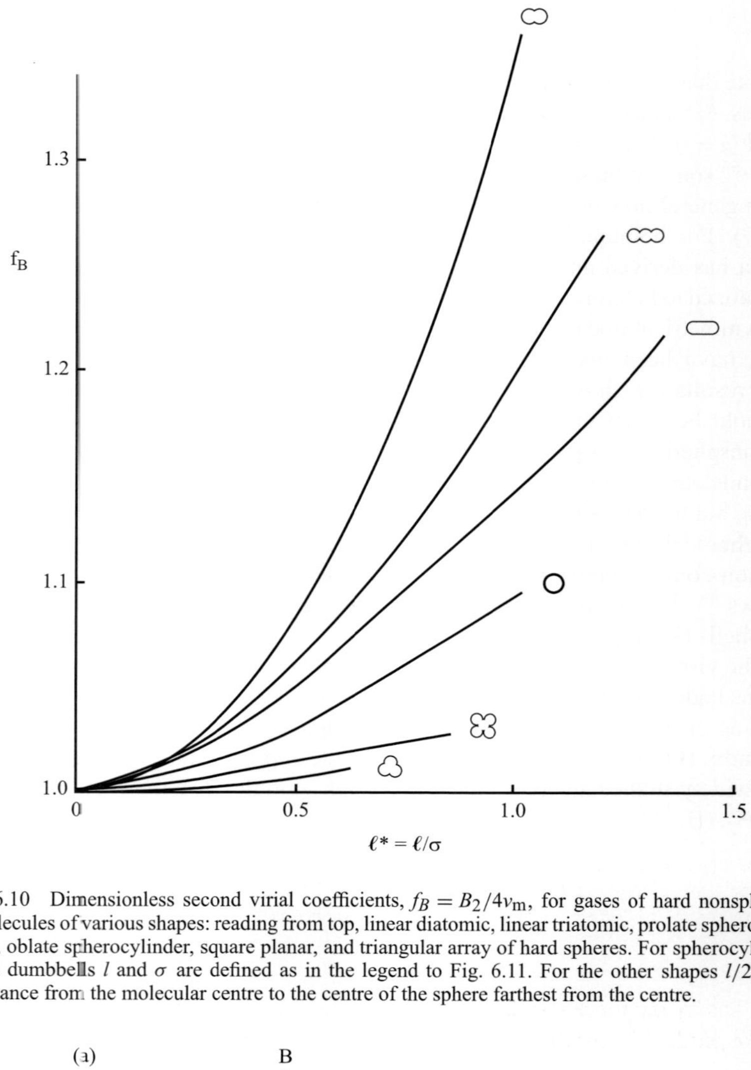

FIG. 6.10 Dimensionless second virial coefficients, $f_B = B_2/4v_m$, for gases of hard nonspherical molecules of various shapes: reading from top, linear diatomic, linear triatomic, prolate spherocylinder, oblate spherocylinder, square planar, and triangular array of hard spheres. For spherocylinders and dumbbells l and σ are defined as in the legend to Fig. 6.11. For the other shapes $l/2$ is the distance from the molecular centre to the centre of the sphere farthest from the centre.

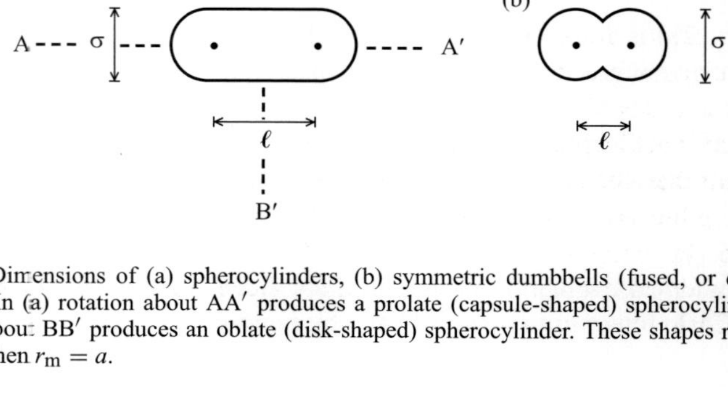

FIG. 6.11 Dimensions of (a) spherocylinders, (b) symmetric dumbbells (fused, or diatomic hard spheres). In (a) rotation about AA′ produces a prolate (capsule-shaped) spherocylinder, whereas rotation about BB′ produces an oblate (disk-shaped) spherocylinder. These shapes reduce to hard spheres when $r_m = a$.

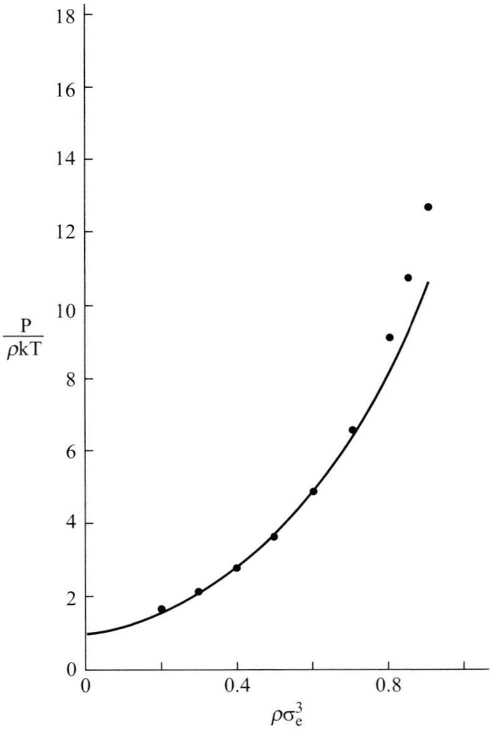

FIG. 6.12 The pressure for hard dumbbells of bond length $l^* = l/\sigma = 0.6$ from Monte Carlo simulation[68] (points) and from the 2/2 Padé approximant[67] formed from the virial coefficients up to B_5 (Eq. (6.128)). Here σ_e is the diameter of a hard sphere whose volume equals that of the dumbell, i.e. $v_m = \pi\sigma_e^3/6$. (see Eq. (5.245) of Vol. 1)

as well as by the more standard numerical integration methods. The Monte Carlo and nonproduct integration methods are most useful for nonlinear molecules.

6.10.3 Quantum corrections

The $O(\hbar^2)$ quantum correction to the classical expression (6.121a) for the second virial coefficient is given by (3.280)–(3.282) for linear and spherical top molecules. The quantities $\nabla^2 u(12)$ and $\nabla_\omega^2 u(12)$ that appear in (3.281) and (3.282) are given explicitly in Chapter 11. The calculation of the $O(\hbar^2)$ correction to B_2 is no more difficult than the calculation of B_2 itself. The correction is generally small, but not negligible, for many molecular gases at temperatures in the liquid range. For hydrogen the correction is large below 200 K. Table 6.3 shows typical results for models of H_2 and HCl.[75, 76] These calculations are based on the intermolecular potential models:

Table 6.3

*Quantum corrections to the second virial coefficient for models of H_2 and HCl**

Gas	T(K)	$B_{2c\ell}$(cm^3 mol^{-1})	B_t(cm^3 mol^{-1})	B_r(cm^3 mol^{-1})	B_2 (theory) (cm^3 mol^{-1})	B_2(expt) (cm^3 mol^{-1})
H_2[75]	123	−3.06	4.23	1.00	2.17	2.56
	173	5.68	2.58	0.46	8.72	9.14
	223	9.49	1.76	0.28	11.53	12.08
	323	13.15	1.13	0.15	14.43	15.14
	423	15.22	0.64	0.08	15.94	15.68
HCl[76]	150	−997.8	12.56	28.29	−956.9	—
	170	−646.1	5.98	12.22	−627.9	—
	190	−462.2	3.34	6.27	−452.6	−456.0
	250	−229.5	1.02	1.55	−226.9	−221.0
	330	−122.8	0.40	0.49	−121.9	−116.0

* Here $B_{2c\ell}$ is the classical value given by (6.121a), $B_t = \hbar^2 \langle F^2 \rangle_o / 24(kT)^3 m$ is the $O(\hbar^2)$ translational quantum correction, and $B_r = \hbar^2 \langle \tau^2 \rangle_o / 24(kT)^3 I$ is the $O(\hbar^2)$ rotational quantum correction; see eqns. (3.280)–(3.282). From refs. 75 and 76.

$$H_2: \quad u(12) = -\frac{a}{r^6} + \frac{b}{r^{12}} + \frac{c}{r^{12}}\left(\cos^2\theta_1' + \cos^2\theta_2'\right) \tag{6.129}$$

$$\mathrm{HCl}: \quad u(12) = \sum_{\alpha\beta} u_{\alpha\beta}^{\mathrm{LJ}}(r_{\alpha\beta}) + u_{\mu\mu} + u_{\mu Q} + u_{Q\mu} + u_{QQ}, \tag{6.130}$$

where *a, b, c* are constants[75] (fitted to the quantum mechanical calculations[77] of u), θ_i' refers to the intermolecular axis, $u_{\alpha\beta}^{\mathrm{LJ}}$ is the site–site Lennard-Jones potential for a two-site model, the sites being at or near the H and Cl nuclei, and $u_{\mu\mu}$, $u_{\mu Q}$, $u_{Q\mu}$, and u_{QQ} are the usual dipole–dipole, dipole–quadrupole, etc. terms; the parameters $\varepsilon_{\alpha\beta}$ and $\sigma_{\alpha\beta}$ in $u_{\alpha\beta}^{\mathrm{LJ}}$ were adjusted to best fit the virial data. From Table 6.3 it can be seen that the quantum corrections for H_2 are appreciable even at 423 K, and are large (10% or more) below 300 K. The $O(\hbar^4)$ rotational quantum correction for H_2 has also been calculated;[75] this correction is of opposite sign to the $O(\hbar^2)$ term (B_r in Table 6.3), and is appreciable at the lower temperatures, amounting to 40% of the $O(\hbar^2)$ term at 123 K. Thus the $\hbar^2$ expansion breaks down at temperatures much below this. The standard methods (e.g. Padé approximants—see Appendix B) used to accelerate series convergence have been applied to atomic fluids[78] (e.g. He with LJ potential). For very low temperatures, one of the more sophisticated semi-classical methods, or the full quantal calculation, must be used.[79] For HCl the quantum correction amounts to about 10 cm^3 mol^{-1} or 2% at the boiling point (188 K), and about 27 cm^3 mol^{-1} or 3.5% at the triple point (160 K). For H_2 the translational quantum correction is the dominant one, whereas the rotational correction is the larger in the case of HCl.

Similar calculations have been carried out for other molecules.[80,81] For water and ammonia it has been shown[81] that the quantum corrections may be as large as 10% even at 300 K, and that the rotational correction term dominates, as expected due to the presence of small moment of inertia components.

6.11 Perturbation theory results

Perturbation theory (PT) is usually both accurate and convenient for thermodynamic properties, and this approach has been widely used in preference to integral equation methods.[82] Comparisons of PT with computer simulation results have been made in Chapter 4. In this section comparisons with experimental data are discussed, but these have been more limited. This is in part because the calculations are tedious for some theories,[83] and in part because of our ignorance of the form of the intermolecular potentials. Some potential parameters must usually be fitted to experimental thermodynamic data, and it is then difficult to discriminate between different potential models unless data are available over a wide range of state conditions.

6.11.1 The u-expansion

The form of PT that has been most widely used is that due originally to Barker and Pople, and described in § 4.5 of Vol. 1. The Padé approximant to the first few terms of this series, (4.47),

$$A = A_0 + A_2/(1 - A_3/A_2) \tag{6.131}$$

is particularly useful, since it gives quite an accurate account of the effects of electrostatic forces on thermodynamic properties (see Figs. 4.8, 4.9, and Tables 4.1 to 4.3). By expanding the pair potentials that occur in A_2 and A_3 in ordinary or generalized spherical harmonics it is possible to carry out the averages over orientation in the perturbation terms; this yields general expressions for A_2 and A_3 that are simple to use; explicit expressions are available for these terms,[84–86] and useful expressions are available[87,88] for the reference fluid integrals that appear in these terms.

A comparison of (6.131) with experimental data for orthobaric (saturated) liquid CO_2 is shown in Fig. 6.13. The intermolecular pair potential used for these calculations was a generalized Stockmayer model (§ 2.1.1),

$$\begin{aligned} u(12) = u_0^{(n,6)}(r) + u_{QQ}(224) + u_{ov}(202 + 022) \\ + u_{dis}(202 + 022 + 224) \end{aligned} \tag{6.132}$$

where $u_0^{(n,6)}$ is the $(n, 6)$ potential model,

$$u_0^{(n,6)}(r) = \frac{n}{n-6}\left(\frac{n}{6}\right)^{6/(n-6)} \varepsilon\left[\left(\frac{\sigma}{r}\right)^n - \left(\frac{\sigma}{r}\right)\right]^6, \tag{6.133}$$

which reduces to the usual LJ model when $n = 12$. In (6.132) u_{QQ} is the quadrupole–quadrupole term, Eq. (2.184) of Vol. 1, $u_{\text{ov}}(202 + 022)$ is the anisotropic overlap potential modelled as the leading (202,022) harmonic terms and given by (2.248), and $u_{\text{dis}}(202 + 022 + 224)$ is the anisotropic dispersion potential as given by the 202, 022, and 224 harmonic terms (see (2.228), (2.230), and (2.232)). The following values of the anisotropic potential parameters were

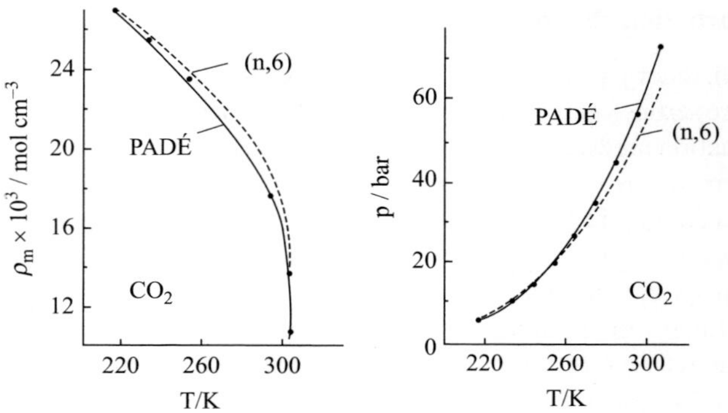

FIG. 6.13 The molar density ($\rho_m = \rho/N_A$, where N_A is Avogadro's number) and the vapour pressure for CO_2 along the saturation line from experiment[90] (points), Eqs. (6.131) and (6.132) (solid lines), and from the $(n, 6)$ model, (6.133) alone, (dashed line). (Adapted from ref. 89.)

used: $Q = -4.30x10^{-26}$ esu (from induced birefringence measurements on the gas—see Appendix D.), $\gamma/\alpha = 0.771$ (from gas phase data), and $\delta = 0.05$ (rough empirical estimate). The three isotropic potential parameters n, ε, and σ appearing in (6.133) were fitted to the experimental values of ρ and p for the orthobaric liquid; this gave $n = 16$, $\varepsilon/k = 254.6$ K, $\sigma = 3.736$ Å. The dashed line shown in Fig. 6.13 is the result obtained using the isotropic $(n, 6)$ model alone, again adjusting n, ε, and σ to the data. The anisotropic potential model of (6.132) clearly gives a better description of the data with the same number of adjustable parameters. A comparison of the Padé with data for dense fluid CO_2 is shown in Table 6.4.

Similar comparisons of theory and experiment have been made for a variety of small molecules (e.g. refs. 84, 91–99); some typical results for coexisting gas and liquid properties are summarized in Table 6.5. In these calculations the intermolecular potential model for C_2H_2, C_2H_4, and C_2H_6 was the same as for CO_2, i.e. (6.132), that for HCl and HBr consisted of an $(n, 6)$ term and multipole terms up to quadrupole–quadrupole, and that for CH_4 and CF_4 consisted of an $(n, 6)$ plus an octopole–octopole term.

The isothermal compressibility χ is related to the centres pair correlation function $g(r)$ by (6.16). In the Pople expansion for $g(r)$ the first-order term vanishes, (see (4.55)), so that to first order we have

$$g(r) \approx g_0(r). \tag{6.134}$$

Computer simulation results show that for liquids the electrostatic forces have little effect on $g(r)$—see Fig. 4.12—so that for such cases (6.134) is quite a good approximation. This further suggests the approximation, based on (6.16) and (6.134), that

Table 6.4

Comparison of thermodynamic properties of dense fluid CO_2 calculated from the Padé approximant of (6.131) with experiment[90]*

		P(bar)		$-U_c$(J mol^{-1})		$-S_r$(J mol^{-1} K^{-1})	
T(K)	ρ_m(mol ℓ^{-1})	Expt.	Calc.	Expt.	Calc.	Expt.	Calc.
315.50	14.6120	101	114	6887	6922	12.49	12.89
	19.0938	203	223	9002	8842	17.78	17.51
	20.8073	304	347	9781	9650	20.32	19.13
	22.6085	507	569	10711	10520	22.77	22.34
	25.1095	1013	1106	11754	11730	26.39	26.42
343.15	5.7611	101	102	3335	2970	5.71	5.06
	15.0674	203	210	6665	6874	11.46	12.59
	17.9902	304	314	8004	8123	15.82	15.55
	20.6372	507	539	9312	9327	18.79	18.76
	23.7128	1013	1072	10717	10750	22.93	23.15
373.15	4.3561	101	101	2350	2157	3.57	3.51
	11.1326	203	208	5091	5073	8.25	8.70
	15.1477	304	314	6685	6694	12.41	12.07
	18.7009	507	530	8254	8213	15.80	15.72
	22.4079	1013	1067	9980	9858	20.28	20.39
403.15	3.6713	101	101	1800	1741	2.62	2.73
	8.5677	203	203	3913	3856	6.13	6.31
	12.6724	304	310	5525	5502	9.79	9.44
	16.8683	507	511	7273	7141	13.18	13.04
	21.1758	1013	1061	9260	9053	17.98	18.08

* Here U_c is configurational internal energy and $S_r = S(NTV) - S_{id}(NTV)$ is the residual entropy; S_{id} is the ideal gas value at the same N, T, V as the real fluid. From ref. 89.

Table 6.5

*Mean deviation between experiment and calculations based on Eq. (6.131) for coexisting liquid (L) and vapour (G) phases**

	ρ^G		ρ^L		p	
	$\Delta\rho^G$(mol ℓ^{-1})	(%)	$\Delta\rho^L$(mol ℓ^{-1})	(%)	Δp(bar)	(%)
Xe	0.016	1.31	0.048	0.27	0.092	0.80
HCl	—	—	0.150	0.53	0.023	8.07
HBr	—	—	—	—	0.010	1.71
CO_2	0.063	2.57	0.150	0.67	0.180	1.06
C_2H_2	0.048	3.91	0.094	0.52	0.055	0.85
C_2H_4	0.063	4.40	0.046	0.28	0.070	0.56
C_2H_6	0.023	0.92	0.160	0.90	0.048	0.50
CH_4	0.031	2.29	0.074	0.33	0.110	1.35
CF_4	—	—	0.034	0.18	0.011	5.49

* Here $\Delta X \equiv \sum_{i=1}^{n} \left| X_i^{\text{theory}} - X_i^{\text{expt}} \right| / n$, where n is the number of data points for the various temperatures; the percentage deviation listed is $\sum_i \left| X_i^{\text{theory}} - X_i^{\text{expt}} \right| \times 100/nX_i^{\text{expt}}$ (critical point is excluded). From ref. 84.

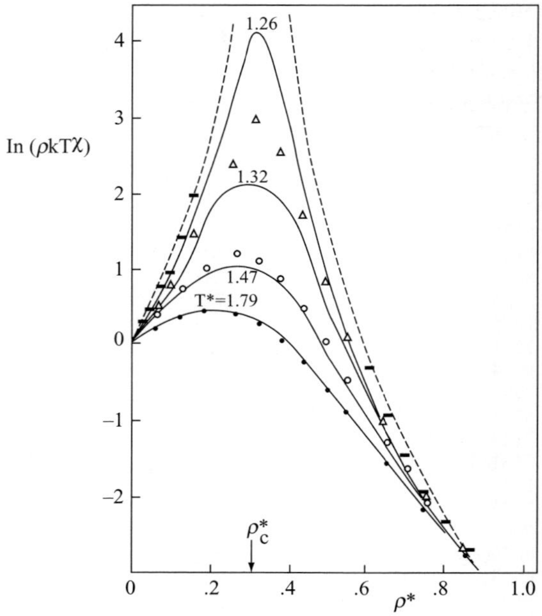

FIG. 6.14 Comparison of the function $\rho kT\chi$ for carbon dioxide and argon. Solid lines are isotherms for argon at the $T^* = kT/\varepsilon$ values shown. Dashed lines give values along the saturation line. Points are experimental data for carbon dioxide: filled rectangles, 298 K ($T^* = 1.26$); triangles, 312 K ($T^* = 1.32$); open circles, 348 K ($T^* = 1.47$); filled circles, 423 K ($T^* = 1.79$). Potential parameters are $\varepsilon_{CO_2}/\varepsilon_{Ar} = 1.96$, $\sigma_{CO_2}/\sigma_{Ar} = 1.087$, with $\varepsilon_{Ar} = 120.7$ K, $\sigma_{Ar} = 3.434$ Å. (From ref. 100.)

$$\chi \approx \chi_0 \tag{6.135}$$

for dense liquids. Equation (6.135) is found[84, 100] to give good results for liquids at densities $\rho > 2\rho_c$, and also for supercritical fluids except in the density range $0.25 < \rho^* < 0.55$. For these cases the higher order contributions to χ are small compared to χ_0. A test of (6.135) for CO_2 is shown in Fig. 6.14. Equation (6.135) has been used as a basis for engineering correlations and prediction methods for thermodynamic properties.[101]

Many of the existing calculations have been for molecules that either are linear or possess some other high degree of symmetry (CH_4, NH_3, etc.). As a result the multipolar forces usually involve only a single multipole moment for each ℓ value (see § 2.4.3), and similar simplifications occur for other types of anisotropic intermolecular forces. For molecules of more general shape the nonaxial nature of the intermolecular forces can have a significant effect on the thermodynamic properties. The Padé approximant (6.131) has been used to study the effect of a nonaxial quadrupole moment on the thermodynamic properties.[93] For such multipole forces A_2 and A_3 are given by (4.38) and (4.39), respectively. The expressions for these terms in the general nonaxial case differ from the axial expressions in that 'the' quadrupole moment Q appearing in the axial case is replaced by new quantities $\langle Q\rangle$, $\langle\mu^2 Q\rangle/\mu^2$, and $\langle\langle Q\rangle\rangle$, the replacement to be used depending on the perturbation term being considered. Thus in A_{2A} and A_{3E} we must replace the axial Q by $\langle Q\rangle$, while in A_{3A} we must replace $\mu^2 Q$ by $\langle\mu^2 Q\rangle$ and Q^3 by $\langle\langle Q\rangle\rangle^3$. These new quantities are rotationally invariant,[102] and are defined by[93] (see also § 2.4.3)

$$\begin{aligned}\langle Q\rangle^2 &= \tfrac{1}{3}(Q_{xx}-Q_{yy})^2+Q_{zz}^2\\ &= \tfrac{2}{3}\left(Q_{xx}^2+Q_{yy}^2+Q_{zz}^2\right)\\ &= \tfrac{4\pi}{5}\left(Q_{20}^2+Q_{22}^2\right)\end{aligned} \tag{6.136}$$

$$\begin{aligned}\langle\mu^2 Q\rangle &= \mu_x^2 Q_{xx}+\mu_y^2 Q_{yy}+\mu_z^2 Q_{zz}\\ &= \left(\tfrac{8\pi}{3}\right)\left(\tfrac{\pi}{5}\right)^{1/2}\left[Q_{20}\left(Q_{10}^2-|Q_{11}|^2\right)+\left(\tfrac{3}{2}\right)^{1/2}Q_{22}\left(Q_{11}^2+Q_{1\underline{1}}^2\right)\right]\end{aligned} \tag{6.137}$$

$$\begin{aligned}\langle\langle Q\rangle\rangle^3 &= Q_{zz}^3-Q_{zz}(Q_{xx}-Q_{yy})^2\\ &= \tfrac{4}{3}\left(Q_{xx}^3+Q_{yy}^3+Q_{zz}^3\right)\\ &= 4Q_{xx}Q_{yy}Q_{zz}\\ &= \left(\tfrac{8\pi}{5}\right)\left(\tfrac{\pi}{5}\right)^{1/2}\left[Q_{20}^3-6Q_{20}Q_{22}^2\right],\end{aligned} \tag{6.138}$$

where the $Q_{\alpha\alpha}$ and Q_{2n} are the body-fixed principal components in the quadrupole principal axes, with $Q_{\alpha\alpha}$ and Q_{2n} denoting Cartesian and spherical components, respectively. For axial molecules we have $Q_{xx}=Q_{yy}$ and $Q_{22}=0$ so that $\langle Q\rangle^2=Q^2$, $\langle\mu^2 Q\rangle/\mu^2=Q$, and $\langle\langle Q\rangle\rangle^3=Q^3$, where $Q=Q_{zz}=(4\pi/5)^{1/2}Q_{20}$ is 'the' quadrupole moment. The expression for A_{2A} for nonaxial molecules is identical to that for the axial case, but with Q replaced by $\langle Q\rangle$. If the quadrupole forces are small the perturbation series can be terminated at the A_2 term, and the calculations carried out with a single 'effective axial quadrupole moment' $\langle Q\rangle$. In the 'effective axial approximation' it is assumed that one can use the axial expressions for both A_2 and A_3 with the axial Q replaced by $\langle Q\rangle$. This will only be a good approximation if either A_3 is small, or if $\langle\mu^2 Q\rangle/\mu^2\langle Q\rangle\approx 1$ and $\langle\langle Q\rangle\rangle^3/\langle Q\rangle^3\approx 1$. Values of these ratios, and of the moments themselves, are shown in Table 6.6 for several nonaxial molecules.[103–105] It is seen that for ethylene (with $\mu=0$) $\langle\langle Q\rangle\rangle^3/\langle Q\rangle^3=-1$, since this ratio appears squared in A_{3A} the effective axial approximation will be very good.[106] For most other cases it will be very poor, however. Thus for water most of the terms contributing to A_{3A} would be in error by more than a factor of 100. Thus the correct non-axial treatment should be used in general.

Perturbation theory can also be used to study the influence of various types of anisotropic intermolecular forces on the critical properties T_c, p_c, and ρ_c, and

Table 6.6

*Dipole, quadrupole, and effective moments for some nonaxial molecules**

Molecule	Principal axes	μ_x, μ_y, μ_z (10^{-18} esu)	Q_{xx}, Q_{yy}, Q_{zz} (10^{-26} esu)	$\langle Q \rangle$ (10^{-26} esu)	$\langle \mu^2 Q \rangle / \mu^2$ (10^{-26} esu)	$\langle\langle Q \rangle\rangle$ (10^{-26} esu)	$\frac{\langle \mu^2 Q \rangle}{\mu^2 \langle Q \rangle}$	$\frac{\langle\langle Q \rangle\rangle^3}{\langle Q \rangle^3}$
(i) Non-polar								
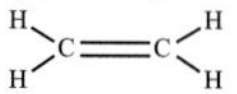 Ethylene	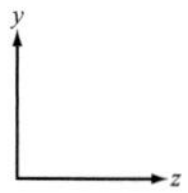	0	−4.09 +2.09 +2.0	4.09	—	−4.09	—	−1.0
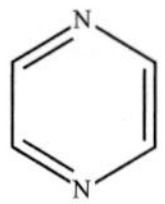 Pyrazine	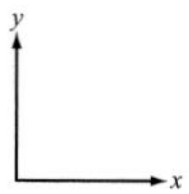	0	+14.83 −12.89 −1.94	16.1	—	11.4	—	0.36
(ii) C_{2v} polar								
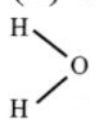 Water	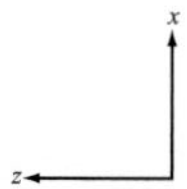	0 0 1.85	+2.63 −2.50 −0.13	2.96	−0.13	1.51	−0.044	0.13
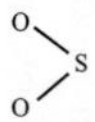 Sulphur dioxide	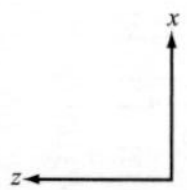	0 0 −1.63	−5.3 +4.0 +1.3	5.52	1.30	−4.79	0.24	−0.65
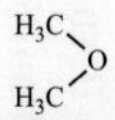 Dimethyl Ether	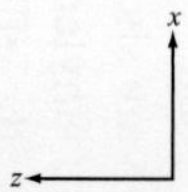	0 0 1.30	+3.3 −1.3 −2.0	3.32	−1.99	3.25	−0.60	0.93

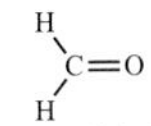 Formaldehyde	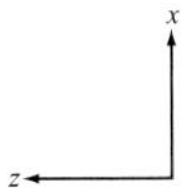	0 0 2.33	+0.2 −0.1 −0.1	0.20	−0.10	0.20	−0.50	1.00
 Pyridine	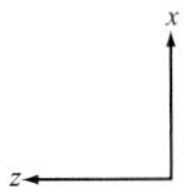	0 0 2.19	+9.7 −6.2 −3.5	9.82	−3.51	9.44	−0.36	0.89
(iii) Other polar								
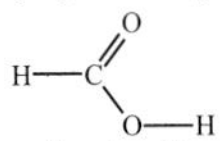 Formic acid	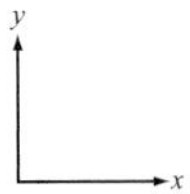	— 0.26 1.39 0.0 $\mu = 1.415$	+5.2 −5.3 +0.1	6.06	−4.93	−2.23	−0.81	−0.050
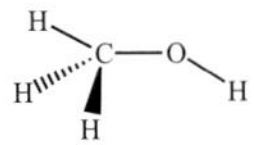 Methanol	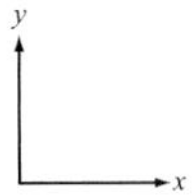	0.98 1.38 0.0 $\mu = 1.70$	+0.45 +0.62 −1.07	1.07	0.56	−1.06	0.52	−0.97

Most μ and Q values are from Appendix D of Vol. 1. The origin of the principal axes is to be taken at the centre of mass.

* The signs given for $\langle\langle Q\rangle\rangle$ are those that yield the correct sign for $\langle\langle Q\rangle\rangle^3$. The quantity μ listed for molecules in section (iii) of the table is the magnitude of the dipole moment. From ref. 93.

on the vapour–liquid coexistence properties. If the Stockmayer potential is used, i.e. $u = u_{LJ} + u_{\mu\mu}$, where u_{LJ} is the spherical LJ and $u_{\mu\mu}$ is the dipole–dipole potential, and if the Pople series is terminated at A_2, it is easy to relate A_2 to thermodynamic properties of the reference fluid, and the resulting equations take a particularly simple form.[107, 108] The reduced critical temperature, pressure, and density are obtained by solving

$$\left(\frac{\partial p^*}{\partial \rho^*}\right)_{T^*} = 0 \qquad \left(\frac{\partial^2 p^*}{\partial \rho^{*2}}\right)_{T^*} = 0 \tag{6.139}$$

where $T^* = kT/\varepsilon$, $p^* = p\sigma^3/\varepsilon$, and $\rho^* = \rho\sigma^3$. To second order it is found that[107]

$$\begin{aligned} T_c{}^* &= T_{c0}{}^* + \frac{1}{6}\mu^{*4} \\ p_c{}^* &= p_{c0}{}^* + \frac{5}{24}\mu^{*4} \\ \rho_c{}^* &= \rho_{c0}{}^* + \frac{1}{24}\mu^{*4}, \end{aligned} \tag{6.140}$$

which predicts that the critical temperature, pressure, and density are all increased by the addition of the dipolar forces; here $\mu^* = \mu/(\varepsilon\sigma^3)^{1/2}$. At a given temperature, the addition of the dipole–dipole interaction causes the vapour pressure and coexisting vapour density to decrease, while the coexisting liquid density increases. These results are as expected on physical grounds, since the dipolar forces are, on the average, attractive. These general trends are, for the most part, confirmed by calculations based on the Padé of (6.131), and are in agreement with experiment.

6.11.2 Nonspherical molecule reference perturbation theory

The Padé approximant, (6.131), gives a good account of the influence of multipolar electrostatic forces, but cannot be expected to describe nonspherical shape effects accurately. This is because the perturbation in the anisotropic overlap potential will be very large for certain orientations, regardless of the spherical reference potential that is chosen. The principal advantage of perturbation treatments based on reference fluids of nonspherical molecules is that the nonspherical shape of the molecules is built into the reference system. Such treatments can be expected to give good results when the primary source of acentric interaction is the nonspherical overlap force. When both nonspherical shape and electrostatic interactions are important, the choice of a nonspherical reference potential is difficult, and rapid convergence is less likely.

Nonspherical reference perturbation theories can be divided into two types (§ 4.8). In the first, the full pair potential is divided into a repulsive and an attractive part for a given pair of molecular orientations, according to Eqs. (4.89) and (4.90). In the resulting first-order theory, (4.88), the reference fluid free energy A_0 is usually obtained from a blip function expansion about A_d, the free energy for a fluid of hard nonspherical molecules; the latter is calculated from an equation of state for hard body fluids (see § 6.12). The reference fluid pair correlation function, $g_0(\mathbf{r}\omega_1\omega_2)$, is usually obtained from a zeroth order expansion about a fluid of spherical molecules, e.g. the f_y expansion described in § 4.6. In the second type of nonspherical reference perturbation theory, an interaction site–site potential is used, and the site–site potential is divided into repulsive and attractive parts as shown in Eqs. (4.100) and (4.101). The first-order theory is then given[109] by (4.102). In this equation both A_0 and g_0 are related to the corresponding quantities for a fluid of hard interaction site molecules through a blip function expansion. A_0 is then calculated from one of the hard body equations of state (§ 6.12), and g_0 is obtained from the RISM theory. Such an approach works best for dense fluids, where the structure is largely determined by the repulsive forces, but convergence is poorer at intermediate and low densities, where the attractive forces have an important influence on the structure. Monson and co-workers[110,111] have proposed a cluster expansion perturbation theory for site–site potential fluids, in which the long-range attractive dispersion and electrostatic potential effects are treated as a perturbation, and are calculated using a partially resummed cluster expansion. In this theory some contributions to the structure and free energy from these attractive forces are summed to infinite order in the perturbation. Since the theory includes the effects of long-range forces on the fluid structure, it has the ability to describe fluids of nonspherical molecules at low to intermediate density, where the first-order theories give poor results (see, e.g., Figs. 4.3 and 4.4), and may also be capable of describing fluids in which both anisotropic dispersion and electrostatic forces are important. The theory has been used to predict thermodynamic properties and vapour–liquid coexistence curves for a number of simple fluids (N_2, Cl_2, F_2, CO_2, C_2H_6).

Extensive comparisons of the first type of theory, based upon division of the full potential, with experimental data have been made. Among these are studies of simple inorganic fluids such as $N_2, O_2, F_2, Cl_2, Br_2, CO, CO_2, CS_2$;[110,112–115] of symmetrical polyatomic inorganics including CCl_4, CF_4, and SF_6;[116,117] and of organic fluids, including CH_4,[114] C_2H_6,[110,118] propane,[119,120] cyclopropane,[119] ethylene,[115] neopentane,[116] and benzene.[121,122] An example of a comparison of the theory with experimental data for benzene is shown in Fig. 6.15. The agreement is good for the Helmholtz energy, and also for the vapour pressure and saturated gas density, but less satisfactory for the saturated liquid density. For benzene a 6-site LJ model was used, each site representing one of the CH groups; electrostatic forces were not included. The second type of theory, in which the site–site potentials are divided, has been successfully used to predict the thermodynamic properties of fluid fluorine, bromine, and chlorine.[123]

In addition to these tests for intermolecular potentials of a general form, comparisons with experiment have been made[126–130] using the perturbation theory of convex molecule fluids (§ 4.8) for a few liquids, including N_2, CO_2, O_3, CH_4, C_2H_2, C_2H_4, C_2H_6, n-C_3H_8, cyclopropane, *n*-butane, isobutane, 2-butyne,

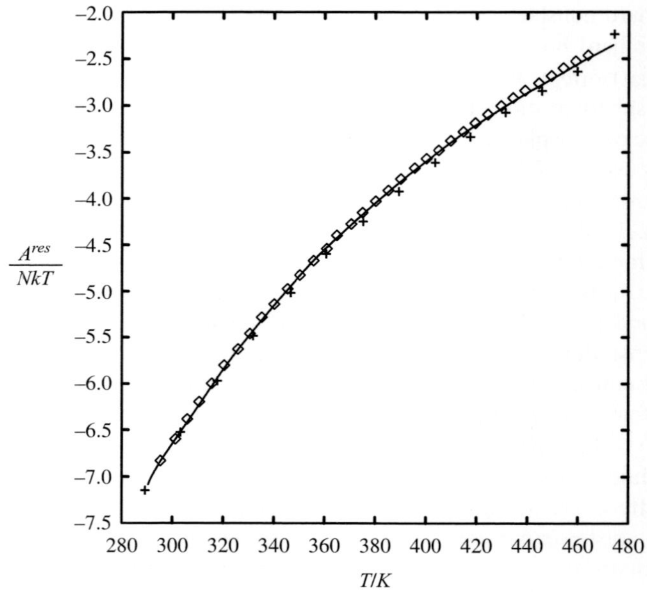

FIG. 6.15 Residual Helmholtz energy (excess over ideal gas value) for benzene. The full line is experimental data from the equation of Goodwin,[124] diamond points are experimental data of Kratzke *et al.*,[125] and + points are perturbation theory results for the 6CLJ model. (From ref. 122.)

propane, $C(CH_3)_4$, benzene, and *n*-alkanes up to C_{16}. A Kihara potential with a nonspherical core was used in these calculations. Some typical comparisons with experiment are shown in Figs. 6.16 and 6.17 for several hydrocarbons. Agreement is generally good, except in the critical region where errors due to the mean field treatment are apparent.

Nonspherical reference perturbation theory has also been used as a basis for approximations of the van der Waals type (see § 4.11); such approximations often provide useful semiempirical equations of state.[133]

There have been few comparisons of the second type of theory, based on division of the site–site potential, with experiment for pure fluids. Applications to mixtures have been more extensive, and are discussed in § 7.8.

6.12 Hard nonspherical molecules: scaled particle and related theories

Fluids composed of hard bodies exhibit several simplifying features. The configurational partition function and pair correlation functions are independent of temperature. Moreover the thermodynamic properties can be expressed in terms of the angular pair correlation function at contact, as shown, for example, in (6.93) for the case of fused hard spheres; it is not necessary to know $g(\mathbf{r}\omega_1\omega_2)$

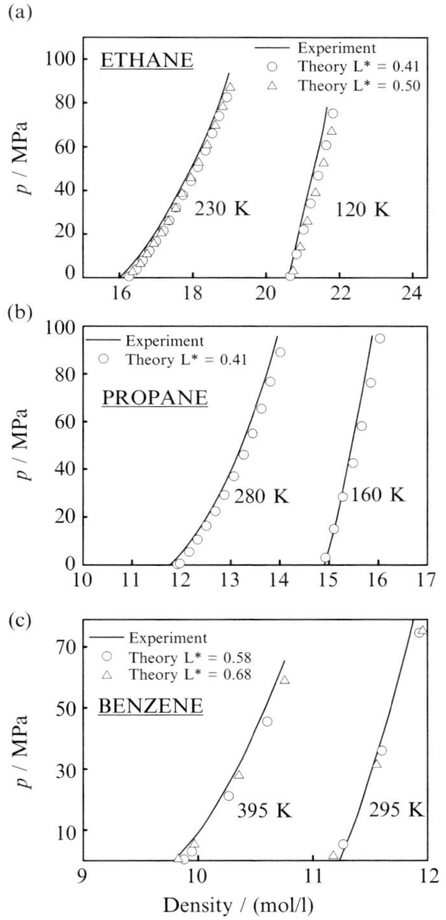

FIG. 6.16 Pressure as a function of density for two selected isotherms for several hydrocarbons. Lines are experimental data, points are from convex body perturbation theory. L* is length parameter in the Kihara potential. (From ref. 131.)

for all separations **r**. Also we need not concern ourselves with nonadditive potentials. Since there are no attractive forces there is no gas–liquid transition. However, other, entropy-driven, fluid–fluid phase transitions occur, including transitions between isotropic fluid and various (nematic, smectic, columnar, cholesteric, etc.) liquid crystal phases;[134–138] for mixtures of hard convex bodies, transitions can occur between two isotropic fluid phases of different composition.[139]

The theory of hard-body fluids is useful for investigating molecular size and shape effects, including transitions involving liquid crystal phases,[134,140] and in thermodynamic perturbation theory, where such fluids often provide a con-

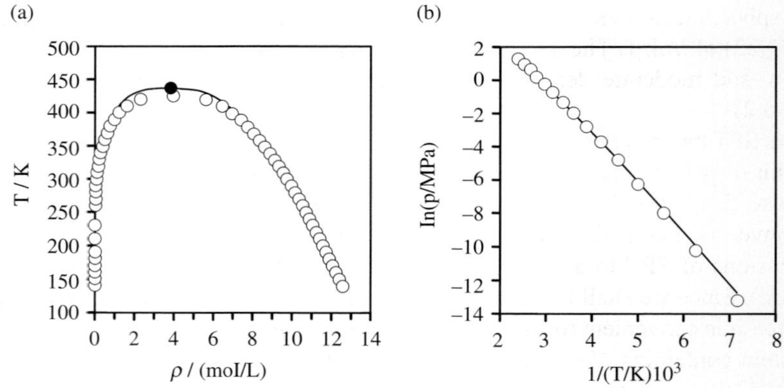

FIG. 6.17 (a) Vapour–liquid coexistence curve, and (b) vapour pressure for *n*-butane from experiment (points) and convex body perturbation theory (lines). *n*-Butane is modeled as the *trans* conformer. (From ref. 132.)

venient reference system. A dilute solution of large hard spheres in a solvent of small hard spheres forms an important model of colloidal suspensions (see, e.g. Cao *et al.*,[141] Boublik,[142] and Viduna and Smith[143]). We focus here on scaled particle and related theories, which yield the thermodynamic properties. For hard bodies composed of bonded hard spheres, the associating fluid theory of Wertheim also gives good results and is discussed in § 6.13. Theories for the structure are less well developed, and are discussed[144] in Chapters 4 and 5. The thermodynamic behaviour of hard-body fluids has been the subject of several reviews.[134, 145–147]

6.12.1 *Scaled particle theory (SPT)*

The reversible work of adding a point particle to a fluid of hard nonspherical bodies can be calculated exactly, as shown below. This work is nonzero because the point particle creates an associated 'excluded volume', from which the centres of the surrounding nonspherical bodies are excluded. In scaled particle theory (SPT) one estimates the free energy change due to the 'scaling up' of the point particle to the full-sized nonspherical body. This estimate is approximate, but the expression used is consistent with the known exact results for the point particle and for the large (macroscopic) nonspherical body limit. The scaled particle expression is thus a simple (but, it turns out, quite accurate) interpolation formula for the change in free energy (and hence in the partition function) on adding a molecule. SPT is less general than perturbation or integral equation theory, since it does not yield the molecular correlation functions.[148] It is primarily useful in that it yields the thermodynamic properties.

The theory was first developed by Reiss *et al.*[149] for hard-sphere fluids.[150] The equation of state obtained by the theory is identical to that obtained for hard spheres from Percus–Yevick (PY) theory[151] using the compressibility route, e.g. (5.22) of Vol. 1. The agreement with computer simulation results is excellent at low and moderate densities, and is quite good even at very high densities (Fig. 5.2).

The SPT has been extended to fluids of hard convex bodies by Gibbons[152] and it is this derivation that we give here.[153, 154] The restriction to convex bodies arises because the convexity greatly simplifies the geometry involved; the geometry of convex bodies is discussed in Appendix 6A. We consider an isotropic fluid (extensions of SPT to anisotropic fluids are briefly discussed at the end of this section). Since we shall be interested in using this theory for mixtures in the next chapter, it is convenient to carry out the derivation for mixtures here. We consider a system containing $N-1$ molecules, of which N_a molecules are of component a, N_b molecules of component $b, \ldots, N_r$ molecules of component r, with the total number of molecules being $N-1=\sum_\alpha N_\alpha$. The molecules are hard convex bodies; molecules of species a, b, etc. possess characteristic radii R_a, R_b, etc. For spheres R_α is the radius; for prolate spherocylinders of fixed length/radius ratio it is the radius of the cylinder, etc. (see Table 6A.1). When we later scale up a molecule, we shall do so at *constant shape*, so that the molecules can be assumed to have single characteristic dimensions R_α.

Suppose we now add to this solvent system at fixed V another (solute) convex body B having characteristic radius R_B. The configurational partition function is then (cf. (6.26))[155, 156]

$$Q_c(N) = \frac{1}{\Omega_B \prod\limits_\alpha N_\alpha! \Omega_\alpha^{N_\alpha}} \int dx^{N-1} dx_B \exp[-\beta \mathcal{U}_{\mathrm{N}}(x^{N-1}; x_B)] \tag{6.141}$$

where $x^{N-1} = \mathbf{r}^{N-1}\omega^{N-1}$ are the coordinates of the solvent molecules, and $x_B = \mathbf{r}_B\omega_B$ those of the solute. Here B is taken to be a distinct species, different from the other species present in the solvent, and the product over α includes solvent species only.

Consider a fixed configuration of solvent molecules $1, 2, \cdots, N-1$, and suppose that none of these molecules touch each other. We wish to carry out the integration over $\mathbf{r}_B\omega_B$, and so calculate the change in Q_c due to the addition of molecule B. If B touches any of the other $N-1$ molecules then $\mathcal{U}_N = \infty$, the exponential term in (6.141) vanishes, and such configurations do not contribute to $Q_c(N)$. If B does not touch any other molecule $\mathcal{U}_N = 0$ and the exponential term is unity. The problem then is to determine the volume of space over which the centre of B can move without B touching one of the other molecules. Consider one of the $N-1$ fixed molecules, A say. Surrounding A there will be an 'excluded volume' $v_{A+B} = v_{A+B}(\omega_A, \omega_B)$, which the centre of B cannot enter (Fig. 6.18). The centre of A, O_A, and the orientations ω_A and ω_B are each fixed. The excluded volume is found by moving B around A, keeping B in contact with A and ω_B fixed. The excluded volume is that enclosed by the dashed curve in Fig. 6.18.

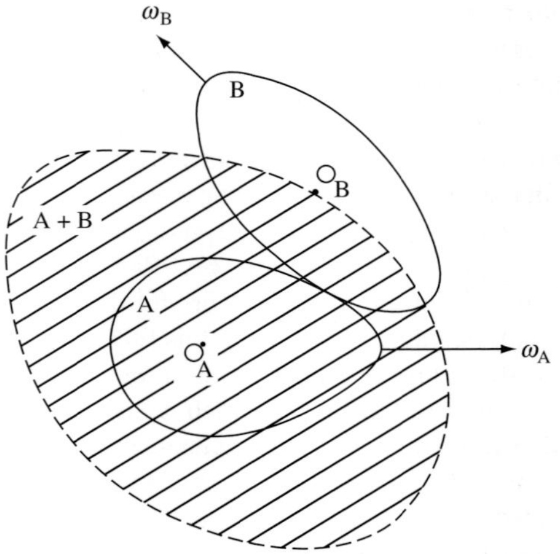

FIG. 6.18 The shaded region represents the excluded volume v_{A+B} due to body A. This is defined by the surface formed from the locus of O_B as B moves with fixed orientation ω_B over the surface of A, which is itself fixed in position and orientation.

Provided O_B does not lie within this region the pair potential energy is zero. There will be similar excluded volumes surrounding each of the original N molecules. Although it is easy to calculate the excluded volume due to a single molecule A, it is unfortunately not possible to calculate the net excluded volume V_{ex} for a dense fluid, since the excluded volumes for the various individual molecules overlap in a complicated way.

Suppose, however, that the characteristic radius of body B is less than or equal to 0,[157] $R_B \leq 0$; i.e. the centre of body B is able to touch the surface of, or penetrate body A. The excluded volume v_{A+B} would then lie within the confines of body A, and it would not be possible for the excluded volumes of neighbouring A molecules to overlap. In such cases the total excluded volume is simply the sum of the excluded volumes for each of the $N-1$ original molecules, i.e. $V_{ex} = \sum_\alpha \sum_i v_{i\alpha+B}$, where $v_{i\alpha+B}$ is the volume around the centre of molecule i of component α from which the centre of B is excluded. From (6.141) we find in this case

$$\frac{Q_c(N)}{Q_c(N-1)} = \frac{1}{\Omega_B}\int d\omega_B \int_{V_f} d\mathbf{r}_B$$
$$= V\left[1 - \frac{1}{V}\sum_{\alpha=a}^{r}\sum_{i=1}^{N_\alpha}\langle v_{i\alpha+B}\rangle_{\omega_B}\right], \qquad R_B \leq 0, \tag{6.142}$$

where $Q_c(N-1)$ is the configurational partition function for the solvent system and $V_f = V - V_{ex}$ is the free volume over which the centre of B can move without

B touching any of the remaining molecules. The term $\langle v_{i\alpha+B}\rangle_{\omega_B}$ is the unweighted average of $v_{i\alpha+B}$ over all possible orientations of B, and is independent of ω_A and of which one of the N_α molecules of species α is considered. Thus the sum over i in (6.142) can be replaced by $N_\alpha \langle v_{\alpha+B}\rangle_{\omega_B}$. The residual chemical potential of the solute B molecule can now be obtained from (6.142) together with (6.27) and (6.34) as[156] (note that in (6.34) $\rho_B = 1/V$)

$$\mu_{Br} = -kT \ln\left[1 - \sum_\alpha \rho_\alpha \langle v_{\alpha+B}\rangle_{\omega_B}\right], \qquad R_B \leq 0. \tag{6.143}$$

Here $\rho_\alpha = N_\alpha/V$ is the number density of α molecules and $\langle v_{\alpha+B}\rangle_{\omega_B}$ is the average excluded volume for the B molecule near an α molecule. We note from (6.143) that the addition of a particle B with characteristic radius $R_B \leq 0$ requires work, since its presence at $\mathbf{r}_B$ creates an excluded volume which the centres of other molecules cannot enter.

Equation (6.143) applies only for $R_B \leq 0$. A second exact relation for μ_{Br} can be obtained when R_B is very large, by using thermodynamic arguments.[149,150] The introduction of body B can be viewed as the formation of a cavity in the fluid. For R_B macroscopically large the leading term in μ_{Br}, i.e. the work of forming the cavity, will be a pressure–volume term,[158]

$$\mu_{Br} \to p v_{mB}, \qquad R_B \text{ large} \tag{6.144}$$

where p is the pressure and v_{mB} is the volume of molecule B. The work will also contain terms arising from the formation of the surface of the cavity, which will be important for somewhat smaller cavities; these terms will be proportional to the surface area and characteristic radius of the cavity (the latter term is the leading correction for the dependence of the surface tension on the curvature of the cavity surface).[159] The mean radius of curvature r_m, surface area s_m, and volume v_m (see Appendix 6A for definitions of these quantities) can conveniently be written in terms of the characteristic radius R_B as

$$\begin{aligned} r_{mB} &= a_B R_B \\ s_{mB} &= b_B R_B^2 \\ v_{mB} &= c_B R_B^3, \end{aligned} \tag{6.145}$$

where a_B, b_B, and c_B are constants that depend only on the molecular shape (see Table 6A.1). From the thermodynamic arguments given above and (6.145) it is clear that for large R_B the expression for the work of introducing body B is cubic in R_B. Moreover, when $R_B \to \infty$, the pv_{mB} term of (6.144) will be the dominant one.

The approximation of SPT consists of writing μ_{Br} as a cubic polynomial in R_B for all $R_B > 0$. Thus, we assume

$$\mu_{Br}(R_B) = \mu_{Br}^{(0)} + \mu_{Br}^{(1)} R_B + \mu_{Br}^{(2)} R_B^2 + \mu_{Br}^{(3)} R_B^3, \quad R_B > 0. \tag{6.146}$$

The first three terms on the right-hand side of this equation are regarded as the initial terms in a Taylor series expansion of μ_{Br} about $R_B = 0$, since it is known[149, 150] that for hard spheres μ_{Br} and its first and second derivatives are finite at $R_B = 0$. For pure hard spheres the third derivative of μ_{Br} is discontinuous at $R_B = 0$, the fifth derivative is discontinuous at

$$R_B = \sigma\left(\frac{1}{\sqrt{3}} - \frac{1}{2}\right),$$

where σ is the diameter of the solvent sphere, etc..[149, 150, 160] Since the discontinuities involve only rather high derivatives, it seems reasonable to represent μ_{Br} by the smooth polynomial (6.146). We assume that these results hold also in the nonspherical body case. The coefficient $\mu_{Br}^{(3)}$ is therefore approximated by setting the cubic term in (6.146) equal to the macroscopic limiting value given by (6.144). Thus we have

$$\mu_{Br}(R_B) = \mu_{Br}^{(0)} + \mu_{Br}^{(1)} R_B + \mu_{Br}^{(2)} R_B^2 + p v_{mB}, \quad R_B > 0. \tag{6.147}$$

The coefficients $\mu_{\mathrm{Br}}^{(0)}$, $\mu_{\mathrm{Br}}^{(1)}$, and $\mu_{\mathrm{Br}}^{(2)}$ can now be determined from (6.143). We first note that the average excluded volume $\langle v_{\alpha+B}\rangle_{\omega B}$ is given in terms of the fundamental measures of the bodies α and B by (6A.22) as

$$\langle v_{\alpha+B}\rangle_{\omega B} = v_{\mathrm{m}\alpha} + v_{\mathrm{m}B} + r_{\mathrm{m}B} s_{\mathrm{m}\alpha} + r_{\mathrm{m}\alpha} s_{\mathrm{m}B}. \tag{6.148}$$

It is at this point that the restriction to convex bodies occurs.

From this expression and (6.145) we find that

$$\begin{aligned} \langle v_{\alpha+B}\rangle_{\omega B}\big|_{R_B=0} &= v_{\mathrm{m}\alpha} \\ \frac{\partial}{\partial R_B}\langle v_{\alpha+B}\rangle_{\omega B}\bigg|_{R_B=0} &= a_B s_{\mathrm{m}\alpha} \\ \frac{\partial^2}{\partial R_B^2}\langle v_{\alpha+B}\rangle_{\omega B}\bigg|_{R_B=0} &= 2 b_B r_{\mathrm{m}\alpha}. \end{aligned} \tag{6.149}$$

Using these results and (6.143) gives

$$\mu_{Br}^{(0)} = -kT\ln\left[1 - \sum_\alpha \rho_\alpha v_{\mathrm{m}\alpha}\right] \tag{6.150}$$

$$\mu_{Br}^{(1)} = kTa_B\left(\sum_\alpha \rho_\alpha s_{\mathrm{m}\alpha}\right)\left(1 - \sum_\alpha \rho_\alpha v_{\mathrm{m}\alpha}\right)^{-1} \tag{6.151}$$

$$\begin{aligned} \mu_{Br}^{(2)} &= \tfrac{1}{2}kTa_B^2\left(\sum_\alpha \rho_\alpha s_{\mathrm{m}\alpha}\right)^2\left(1 - \sum_\alpha \rho_\alpha v_{\mathrm{m}\alpha}\right)^{-2} \\ &\quad + kTb_B\left(\sum_\alpha \rho_\alpha r_{\mathrm{m}\alpha}\right)\left(1 - \sum_\alpha \rho_\alpha v_{\mathrm{m}\alpha}\right)^{-1}. \end{aligned} \tag{6.152}$$

Substituting (6.150)-(6.152) in (6.147) gives

$$\mu_{Br} = -kT\ln(1-\rho v_{mx}) + \rho kT(r_{\mathrm{m}B}s_{\mathrm{m}x} + s_{\mathrm{m}B}r_{\mathrm{m}x})/(1-\rho v_{\mathrm{m}x}) + \frac{1}{2}\rho^2 kT(r_{\mathrm{m}B}s_{\mathrm{m}x})^2/(1-\rho v_{\mathrm{m}x})^2 + pv_{\mathrm{m}B}, \tag{6.153}$$

where $\rho = N/V$ is the total number density and $r_{\mathrm{m}x}$, $s_{\mathrm{m}x}$, and $v_{\mathrm{m}x}$ are mole fraction averages of the mean radius of curvature, surface area, and volume, respectively,

$$r_{\mathrm{m}x} = \sum_\alpha x_\alpha r_{\mathrm{m}\alpha} \tag{6.154}$$

$$s_{\mathrm{m}x} = \sum_\alpha x_\alpha s_{\mathrm{m}\alpha} \tag{6.155}$$

$$v_{\mathrm{m}x} = \sum_\alpha x_\alpha v_{\mathrm{m}\alpha} \tag{6.156}$$

and $x_\alpha = N_\alpha/N = \rho_\alpha/\rho$.

We now consider the case where the solute molecule B is one of the solvent components, α. Equation (6.153), with B replaced by α, gives the residual part of the chemical potential. The ideal gas chemical potential, μ_α^{id}, is readily obtained from the expressions (cf. (3.21) and (3.77)-(3.82) of Vol. 1),

$$\mu_\alpha^{\mathrm{id}} = \left(\frac{\partial A^{id}}{\partial N_\alpha}\right)_{TVN'} = -kT\left(\frac{\partial \ln Q^{\mathrm{id}}}{\partial N_\alpha}\right)_{TVN'} \tag{6.157}$$

$$Q^{\mathrm{id}} = V^N \prod_\gamma \Lambda_{t\gamma}^{-3N_\gamma}\Lambda_{r\gamma}^{-N_\gamma}(N_\gamma!)^{-1} q_{qu\gamma}^{N_\gamma}, \tag{6.158}$$

where $N' =$ all N_γ for $\gamma \neq \alpha$; thus, with the aid of Stirling's approximation, $\ln N! = N\ln N - N$, we get

$$\mu_\alpha^{\mathrm{id}} = kT\ln\left(\frac{\rho_\alpha \Lambda_{t\alpha}^3 \Lambda_{r\alpha}}{q_{qu\alpha}}\right). \tag{6.159}$$

From (6.153) and (6.159), the chemical potential of any component α in the mixture is given by

$$\mu_\alpha = kT\ln\left(\frac{\rho_\alpha \Lambda_{t\alpha}^3 \Lambda_{r\alpha}}{q_{qu\alpha}}\right) - kT\ln(1-\rho v_{\mathrm{m}x}) + \rho kT(r_{\mathrm{m}\alpha}s_{\mathrm{m}x} + s_{\mathrm{m}\alpha}r_{\mathrm{m}x})/(1-\rho v_{\mathrm{m}x}) + \frac{1}{2}\rho^2 kT(r_{\mathrm{m}\alpha}s_{\mathrm{m}x})^2/(1-\rho v_{\mathrm{m}x})^2 + pv_{\mathrm{m}\alpha}. \tag{6.160}$$

This is not yet an explicit expression for μ_α in terms of the densities and temperature because p occurs on the right-hand side. To obtain an expression for the pressure we make use of the Gibbs-Duhem equation,[161]

$$-SdT + Vdp = \sum_{\alpha} N_\alpha d\mu_\alpha. \tag{6.161}$$

For fixed T this gives a relation between p and the μ_α,

$$\left(\frac{\partial p}{\partial \rho_\beta}\right)_{T\rho'} = \sum_{\alpha} \rho_\alpha \left(\frac{\partial \mu_\alpha}{\partial \rho_\beta}\right)_{T\rho'}, \tag{6.162}$$

where ρ' is all $\rho_\gamma \neq \rho_\beta$. Carrying out the differentiation of (6.160), substituting the result in (6.162), and solving for $\partial p/\partial \rho_\beta$, gives

$$\begin{aligned}\frac{\partial p}{\partial \rho_B} = \frac{kT}{(1-\rho v_{\mathrm{m}x})} \big[& 1 + \rho(v_{\mathrm{m}\beta} + s_{\mathrm{m}\beta} r_{\mathrm{m}x} + r_{\mathrm{m}\beta} s_{\mathrm{m}x})/(1-\rho v_{\mathrm{m}x}) \\ & + \rho^2 \left(2 v_{\mathrm{m}\beta} r_{\mathrm{m}x} s_{\mathrm{m}x} + s_{\mathrm{m}\beta} s_{\mathrm{m}x} \left(r_{\mathrm{m}}^2\right)_x\right)/(1-\rho v_{\mathrm{m}x})^2 \\ & + \rho^3 v_{\mathrm{m}\beta} s_{\mathrm{m}x}^2 \left(r_{\mathrm{m}}^2\right)_x/(1-\rho v_{\mathrm{m}x})^3 \big],\end{aligned} \tag{6.163}$$

where

$$\left(r_{\mathrm{m}}^2\right)_x = \sum_{\alpha} x_\alpha r_{\mathrm{m}\alpha}^2. \tag{6.164}$$

If our representation of μ_{Br} in (6.147) is valid, p in (6.160) should be the corresponding approximation to the pressure; at fixed T changes in the latter must be expressible as an exact differential in terms of changes in the densities ρ_B, i.e.

$$\mathrm{d}p = \sum_{\beta} \left(\frac{\partial p}{\partial \rho_\beta}\right) \mathrm{d}\rho_\beta. \tag{6.165}$$

The condition for dp to be an exact differential is

$$\frac{\partial}{\partial \rho_\alpha} \frac{\partial p}{\partial \rho_\beta} = \frac{\partial}{\partial \rho_\beta} \frac{\partial p}{\partial \rho_\alpha}. \tag{6.166}$$

Carrying out the differentiation of (6.163) shows that this condition does not hold for a mixture of bodies of different shapes, but that it does hold for mixtures of the same shape. Thus for mixtures of different shapes (6.163) is not the differential coefficient of any function. This rather surprising result suggests an error in the form of (6.147) for such mixtures, possibly through the identification of p on the right-hand side of the equation with the pressure, or because of a discontinuity in one of the other terms.

For mixtures of molecules of the same shape, i.e. mixtures in which the constants a, b, and c in (6.145) are the same for all components, it is easily shown that (6.166) does hold. The solution of (6.163) is then

$$\frac{p}{\rho kT} = \frac{1}{1-\eta} + \frac{\gamma\eta}{(1-\eta)^2} + \frac{\left(r_{\mathrm{m}}^2\right)_x \gamma^2 \eta^2}{3 r_{\mathrm{m}x}^2 (1-\eta)^3} \quad (SPT), \tag{6.167}$$

where

$$\eta = \rho v_{mx}, \quad \gamma = r_{mx} s_{mx}/v_{mx} \tag{6.168}$$

are the reduced density (packing fraction) and shape factor for the mixture, respectively.

Equation (6.167) can be verified by carrying out the differentiation with respect to ρ_β and using the fact that for molecules of the same shape,

$$s_{m\beta} s_{mx} \left(r_m^2\right)_x = r_{m\beta}^2 s_{mx}^2, \tag{6.169}$$

which follows from the use of (6.145) with *a, b, c* the same for all species.

For the special case of a mixture of hard spheres of diameters $\sigma_a, \sigma_b, \cdots, \sigma_r$ (6.167) is identical to the PY pressure using the compressibility route—see (5.22) of Vol. 1 and the curve labelled PY(χ) in Fig. 5.2. In this case

$$\rho v_{mx} = \frac{\pi}{6} \sum_\alpha \rho_\alpha \sigma_\alpha^3$$

(the mixture generalization of the η of (5.18)),

$$s_{mx} = \pi \sum_\alpha x_\alpha \sigma_\alpha^2, r_{mx} = \frac{1}{2} \sum_\alpha x_\alpha \sigma_\alpha, \qquad \text{and } \left(r_m^2\right)_x = \frac{1}{4} \sum_\alpha x_\alpha \sigma_\alpha^2.$$

For a pure fluid of convex bodies (6.167) reduces to[162]

$$\frac{p}{\rho kT} = \frac{1}{1-\eta} + \frac{\gamma\eta}{(1-\eta)^2} + \frac{\gamma^2\eta^2}{3(1-\eta)^3} \quad \text{(SPT)}, \tag{6.170}$$

where $\eta = \rho v_m$ and $\gamma = r_m s_m / v_m$ are the reduced density and shape factor for the pure fluid. For spheres $\gamma = 3$ and (6.170) reduces to the PY(χ) equation of state (5.22),

$$\frac{p}{\rho kT} = \frac{1+\eta+\eta^2}{(1-\eta)^3}. \tag{6.171}$$

We can now substitute the expression for the pressure, (6.167), into (6.160) to obtain a closed form for the chemical potential. This gives

$$\begin{aligned}
\beta\mu_\alpha = {} & \ln\left(\frac{\rho_\alpha \Lambda_{t\alpha}^3 \Lambda_{r\alpha}}{q_{qu\alpha}}\right) - \ln(1-\eta) \\
& + \gamma[(r_{m\alpha}/r_{mx}) + (s_{m\alpha}/s_{mx}) + (v_{m\alpha}/r_{mx}s_{mx})]\eta/(1-\eta) \\
& + \frac{1}{2}\gamma^2\left[\left(r_{m\alpha}^2/r_{mx}^2\right) + 2(v_{m\alpha}/r_{mx}s_{mx})\right]\eta^2/(1-\eta)^2 \\
& + \gamma^2\left[v_{m\alpha}\left(r_m^2\right)_x / v_{mx} r_{mx}^2\right]\eta^3/3(1-\eta)^3. \quad \text{(SPT)}
\end{aligned} \tag{6.172}$$

For pure fluids this becomes

$$\beta\mu = \ell\text{n}\left(\frac{\rho\Lambda_t^3\Lambda_r}{q_{qu}}\right) - \ell\text{n}(1-\eta) + \frac{\eta(2\gamma+1)}{(1-\eta)} + \frac{\eta^2(\gamma^2+2\gamma)}{2(1-\eta)^2} + \frac{\eta^3\gamma^2}{3(1-\eta)^3}. \quad \text{(SPT)} \tag{6.173}$$

The equation for the residual Helmholtz free energy, $A_r = A(N\rho T) - A_{id}(N\rho T)$, where N represents $N_a, N_b, \cdots, N_r$ and $A_{\text{id}}(N\rho T)$ is the value for the ideal gas at the same composition, density, and temperature as the real fluid, can also be obtained from the expression for the pressure, by using the thermodynamic relation

$$p = -\left(\frac{\partial A}{\partial V}\right)_{NT}. \tag{6.174}$$

Thus by integrating (6.174) we get

$$\begin{aligned}\beta A_r/N &= -\frac{1}{N}\int_{\infty}^{V}\left(\frac{p}{kT}-\rho\right)\text{d}V \\ &= \int_0^{\rho}\left(\frac{p}{\rho kT}-1\right)\frac{\text{d}\rho}{\rho} = \int_0^{\eta}\left(\frac{p}{\rho kT}-1\right)\frac{\text{d}\eta}{\eta},\end{aligned} \tag{6.175}$$

where we have used $p_{\text{id}} = \rho kT$ and $\text{d}\rho = -(\rho^2/N)\text{d}V$. Using (6.167) in (6.175), and evaluating the standard integrals[163] involved gives

$$\beta A_r/N = -\ln(1-\eta) + \frac{\gamma\eta}{1-\eta} + \frac{\left(r_{\text{m}}^2\right)_x\gamma^2\eta^2}{6r_{\text{m}x}^2(1-\eta)^2} \quad \text{(SPT)}. \tag{6.176}$$

For pure fluids this becomes

$$\beta A_r/N = -\ln(1-\eta) + \frac{\gamma\eta}{(1-\eta)} + \frac{\gamma^2\eta^2}{6(1-\eta)^2} \quad \text{(SPT)}. \tag{6.177}$$

SPT has been tested against computer simulation for particles of various shapes. Table 6.7 shows such a test for the virial coefficients of prolate spherocylinders against molecular simulation results.[164] For molecules of arbitrary convex shape the SPT equation of state yields the following expression for the reduced virial coefficients, $B_n^* = B_n/v_{\text{m}}^{n-1}$, for $n > 2$,

$$B_n^* = 1 + (n-1)\gamma + \frac{(n-1)!}{3!(n-3)!}\gamma^2 \quad (\text{SPT}, n > 2), \tag{6.178}$$

and for prolate spherocylinders we have

$$\gamma = 3(\ell^*+1)(\ell^*+2)/(3\ell^*+2), \tag{6.179}$$

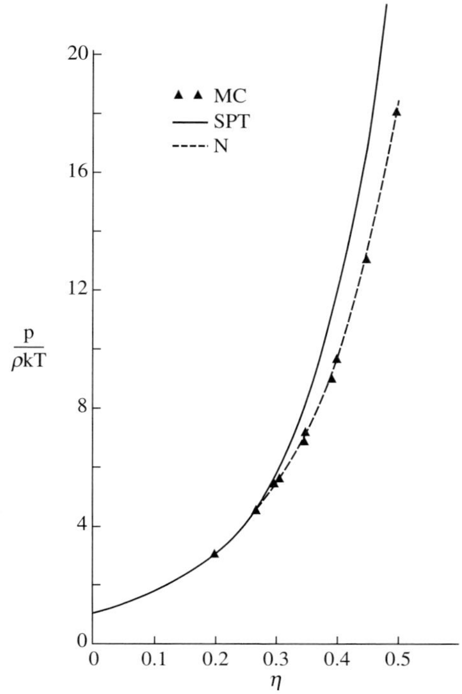

FIG. 6.19 The pressure as a function of packing fraction, $\eta = \rho v_m$, for prolate spherocylinders of length $\ell^* = \ell/\sigma = 2$ from Monte Carlo simulation (points); SPT, (6.170) (solid line); and the Nezbeda (N) equation of state, (6.180) (dashed line). (Adapted from ref. 166.)

where $\ell^* = \ell/\sigma$ is the reduced length of the spherocylinder, excluding the hemispherical caps (Fig. 6.11). For B_2, SPT yields the exact result, (6.127). SPT gives the exact value of B_3 for hard spheres, and is quite accurate for spherocylinders for $\ell^* \leq 3$; it is increasingly poor for larger ℓ^*, however, as seen from Table 6.7. For B_4 and B_5 the SPT gives poor results for all but the shortest bond lengths; at $\ell^* = 5$ it predicts values that are too large by more than a factor of 2. A comparison of the SPT values of B_3, B_4, and B_5 with Monte Carlo calculations for hard ellipsoids has also been made,[165] with results similar to those shown in Table 6.7.

Comparisons between SPT and Monte Carlo simulations for $p/\rho kT$ for prolate spherocylinders at higher densities[166, 167] are shown in Table 6.8 and Fig. 6.19. The SPT agrees well with the simulation data at the lower densities, but gives values that are too high at higher densities. This defect becomes more noticeable as ℓ^* increases.

In addition to the applications to isotropic fluids discussed here, modified forms of the SPT have been used to study the nematic-isotropic phase transition in liquid crystals. The liquid crystal is treated as a mixture, in which each species

Table 6.7

*Virial coefficients for hard prolate spherocylinders**

	$B_2/4v_m$	$B_3/(4v_m)^2$				$B_4/(4v_m)^3$				$B_5/(4v_m)^4$		
$\ell^* = \ell/\sigma$		MC	SPT	N	B	MC	SPT	N	B	MC	SPT	N
0	1.000	0.625	0.625	0.625	0.625	0.287	0.297	0.281	0.281	0.111	0.121	0.109
1.0	1.150	0.771	0.783	0.778	0.783	0.352	0.387	0.346	0.346	0.125	0.161	0.131
2.0	1.375	1.013	1.047	1.016	1.047	0.438	0.543	0.438	0.449	0.144	0.232	0.159
3.0	1.614	1.277	1.360	1.193	1.360	0.498	0.736	0.529	0.568	0.155	0.322	0.182
4.0	1.857	1.558	1.727	1.459	1.727	0.517	0.963	0.615	0.700	0.156	0.427	0.197
5.0	2.103	1.855	2.133	1.748	2.133	0.494	1.221	0.696	0.842	0.246	0.549	0.206

* MC = Monte Carlo values.[164] These are believed to have the following accuracies: 0.2% for B_3, 1% for B_4, 4% for B_5 (B_2 is calculated exactly using (6.127)). SPT = scaled particle theory, (6.178); N = Nezbeda′s equation, (6.190); B = Boublik′s equations, (6.189) and (6.187).

correspond to a particular molecular orientation. We do not enter into a discussion of these applications here, but refer the reader to the substantial literature that exists.[168–172]

SPT has also been used to calculate properties of real fluids[173] by substituting experimental density values in the equations for the particular temperature and pressure of interest. Most calculations have been based on hard spheres, and have included calculations for thermal expansion coefficients, compressibilities, surface tensions, latent heats of vapourization, specific heats, entropies, and solubilities of gases in liquids.

6.12.2 *Other equations of state*

SPT gives a good, though not exact, description of the equation of state for fluids of convex bodies. Attempts to improve this equation have been of two types: (a) the development of more accurate equations for hard convex bodies, and (b) the development of equations for nonconvex bodies, particularly fused hard spheres.

Hard convex bodies

We first consider improvements in the equations for hard convex bodies. For hard spheres we have already seen (§ 5.1) that the addition of an extra term to the SPT equation, (5.22), results in virtually exact agreement with computer simulation data (Fig. 5.2). The resulting equation is that of Carnahan and Starling (CS), (5.27). The CS equation was originally derived[174] by an examination of the virial expansion. It was noted that the first six reduced virial coefficients could be approximated by integers, and an examination of these led to a general formula for the nth coefficient B_n. This enabled the virial series to be summed, giving (5.27). A similar approach is possible for hard convex bodies.[175,176] In this case the virial coefficients depend on the shape factor of the molecules. An examination of the data for B_3 and B_4 for prolate spherocylinders suggests simple expressions for these quantities,[175] and a simple approximate form for the general coefficient B_n. The virial series can then be summed to give the Nezbeda equation,[175]

$$\frac{p}{\rho kT} = \frac{1}{1-\eta} + \frac{\gamma\eta}{(1-\eta)^2} + \frac{\left(\frac{1}{9}\gamma^2 + \frac{4}{3}\gamma - 2\right)\eta^2 - \frac{1}{3}\gamma\left(\frac{5}{3}\gamma - 4\right)\eta^3}{(1-\eta)^3} \quad (N). \tag{6.180}$$

It is readily verified that this reduces to the CS equation, (5.27), in the cases of hard spheres ($\gamma = 3$). The mixture generalization of this equation is due to Pavlicek, Nezbeda, and Boublik,[177]

$$\begin{aligned}\frac{p}{\rho kT} &= \frac{1}{1-\eta} + \frac{\gamma\eta}{(1-\eta)^2} \\ &+ \frac{1}{9(1-\eta)^3}\left\{\left[3\left(\left(r_{\mathrm{m}}^2\right)_x / r_{\mathrm{m}x}^2\right) + 12(u_{\mathrm{m}x}/s_{\mathrm{m}x}) - 18\left(u_{\mathrm{m}x}w_{\mathrm{m}x}/r_{\mathrm{m}x}s_{\mathrm{m}x}\right) - 2\right]\gamma^2\eta^2\right. \\ &\left. + \left[\left(r_m^2\right)_x (12u_{\mathrm{m}x} - 5s_{\mathrm{m}x})/r_{\mathrm{m}x}^2 s_{\mathrm{m}x}\right]\gamma^2\eta^3\right\} \text{ (PNB)},\end{aligned} \tag{6.181}$$

where η and γ are defined in (6.168), and

$$u_{\rm mx} = \sum_{\alpha} x_\alpha v_{\rm m\alpha}/r_{\rm m\alpha} \tag{6.182}$$

$$w_{\rm mx} = \sum_{\alpha} x_\alpha v_{\rm m\alpha}/s_{\rm m\alpha}. \tag{6.183}$$

Several variations on this equation have been proposed.[178–180] Equation (6.180) is of the form

$$\frac{p}{\rho kT} = \frac{1}{1-\eta} + \frac{C_2\eta}{(1-\eta)^2} + \frac{C_3\eta^2}{(1-\eta)^3} + \frac{C_4\eta^3}{(1-\eta)^3} (NPL). \tag{6.184}$$

If we require that C_2, C_3, C_4 yield the correct second, third, and fourth virial coefficients we find that

$$C_2 = B_2{}^* - 1,\ C_3 = B_3{}^* - 2B_2{}^* + 1,\ C_4 = B_4{}^* - 3\,(B_3{}^* - B_2{}^*) - 1\,, \tag{6.185}$$

where $B_n{}^* = B_n/v_{\rm m}^{n-1}$. One approach is to use the exact virial coefficients but this is usually inconvenient since B_3 and B_4 require a tedious Monte Carlo calculation and have only been evaluated for a few cases. Boublik[179] has suggested the following simple approximations for $B_3{}^*$ and $B_4{}^*$ for convex bodies,

$$B_3{}^* = 1 + 2\gamma + \frac{1}{3}\gamma^2 \tag{6.186}$$

$$B_4{}^* = 1 + \frac{14}{3}\gamma + \frac{1}{3}\gamma^2. \tag{6.187}$$

When these expressions, together with the exact expression (6.127) for $B_2{}^*$, $B_2{}^* = 1 + \gamma$, are used in (6.184), the following expression is obtained:[179]

$$\frac{p}{\rho kT} = \frac{1}{1-\eta} + \frac{\gamma\eta}{(1-\eta)^2} + \frac{\frac{1}{3}\gamma^2\eta^2 - \frac{1}{3}\gamma(2\gamma-5)\eta^3}{(1-\eta)^3} \quad (B). \tag{6.188}$$

The mixture generalization of this equation is[179]

$$\frac{p}{\rho kT} = \frac{1}{1-\eta} + \frac{\gamma\eta}{(1-\eta)^2} + \frac{\left[(r_{\rm m}^2)_x/r_{\rm mx}^2\right]\gamma^2(1-2\eta)\eta^2 + 5\gamma\eta^3}{3(1-\eta)^3} \quad (B). \tag{6.189}$$

We note that all of these hard convex body equations—(6.180), (6.181), (6.184), (6.188), and (6.189)—differ from the scaled particle theory expression (6.167) only in the $(1-\eta)^{-3}$ terms. The residual free energy corresponding to each of these equations for the pressure can be obtained by using the appropriate expression in (6.175).

These equations have been tested against computer simulation results for fluids of prolate spherocylinders. Results for the virial coefficients are shown in Table 6.7 for the N and B equations. The expression for the virial coefficients corresponding to the N equation, (6.180), is

$$B_n{}^* = 1 + (n-1)\gamma + \frac{1}{2}(n-1)(n-2)\left(\frac{1}{9}\gamma^2 + \frac{4}{3}\gamma - 2\right)$$
$$- \frac{1}{6}(n-2)(n-3)\gamma\left(\frac{5}{3}\gamma - 4\right) \quad (N), \tag{6.190}$$

while those for the B equation are given by (6.186) and (6.187). For B_3 the N equation gives quite good results for all ℓ^*, and is almost exact for $\ell^* \leq 2$. The B equation (6.186) for B_3 is identical to the SPT expression, (6.178). For B_4 and B_5 the N equation is superior to the others, although the B expression for B_4 is superior to that of SPT. The N equation gives quite good results for B_4 when $\ell^* < 3$ and for B_5 when $\ell^* < 2$. However, it fails to describe the behaviour of B_4 and B_5 for the larger ℓ^* values.

Figure 6.19 and Table 6.8 show comparisons of these equations of state with Monte Carlo data for prolate spherocylinders at higher densities. The N, NPL, and B equations are all a substantial improvement over the SPT equation. The N equation appears to be marginally better than the others, and is almost exact for spherocylinders of $\ell^* \leq 2$ for $\eta \leq 0.5$.

The colloidal limit

Binary hard sphere mixtures in which the dilute solute is a sphere of very large diameter compared to the small solvent molecules (the so-called 'colloidal limit') are of particular interest for the study of colloidal systems. The Boublik–Mansoori–Carnahan–Starling–Leland (BMCSL) equation of state[181]

Table 6.8

*The compressibility factor $p/\rho kT$ of hard prolate spherocylinders from Monte Carlo calculations and from various equations of state**

		$p/\rho kT$				
ℓ^*	η	SPT	N	NPL	B	MC
1	0.2	2.71	2.67	2.67	2.67	2.65 ± 0.02
	0.3	4.77	4.56	4.58	4.56	4.48 ± 0.07
	0.4	8.87	8.10	8.22	8.08	8.20 ± 0.10
	0.5	17.84	15.28	15.76	15.20	15.20 ± 0.20
2	0.2	3.18	3.06	3.06	3.09	3.07 ± 0.03
	0.3	5.96	5.40	5.41	5.48	5.40 ± 0.10
	0.35	8.28	7.23	7.24	7.34	7.17 ± 0.11
	0.4	11.67	9.74	9.75	9.89	9.60 ± 0.10
	0.45	16.73	13.24	13.27	13.44	13.00 ± 0.16
	0.5	24.50	18.25	18.30	18.50	18.00 ± 0.40

* Here $\ell^* = \ell/\sigma$; $\eta = \rho v_m$; SPT, N, and B refer to Eqs. (6.170), (6.180), and (6.188), respectively; and NPL refers to (6.184) and (6.185). From refs. 166 and 167.

provides a good description of hard sphere mixtures for cases where the sizes of the two molecular species are not very different, and reduces to the accurate Carnahan–Starling equation for pure hard sphere fluids. However, it is not accurate in the colloidal limit. Several modifications of the BMCSL equation, and of the corresponding equations for the radial distribution functions at contact, have been proposed and offer greater accuracy for the colloidal limit.[141–143, 182, 183]

Nonconvex bodies

We now consider briefly the attempts to extend scaled particle and related equations to nonconvex bodies. One approach is to use the NPL equations,[178] (6.184) and (6.185), or some similar expressions,[180] but this requires a knowledge of the virial coefficients B_2, B_3, and B_4, which are not always readily available. Another method is to carry through the SPT derivation for nonconvex bodies. The difficulty is then the evaluation of the excluded volume $\langle v_{A+B}\rangle_\omega$. Nezbeda[25] has used this approach, using an approximate expression for the excluded volume for dumbbells. The resulting equation of state for dumbbells gives good agreement with computer simulation results for low elongations and densities, but predicts $p/\rho kT$ values that are too high when η or ℓ^* are large. An alternative approach, and the one most widely used, is to use one of the convex body equations given earlier together with some semiempirical prescription for obtaining γ for the nonconvex body. One such prescription[184] is to use (6.125) for B_2, from which we have

$$\gamma = \frac{B_2}{v_m} - 1 = 4f_B - 1, \tag{6.191}$$

where f_B is the B_2-shape factor introduced in (6.126). Although this holds only for convex bodies, it can be used to define a semiempirical shape parameter γ for nonconvex bodies, and this parameter can then be used in any of the convex body equations of state given above. An alternative approach[185] is to calculate $\gamma = r_m s_m / v_m$ using the s_m and v_m appropriate to the nonconvex body, and calculating r_m for a corresponding convex body. For dumbbells (fused hard spheres of equal diameter), for example, a prolate spherocylinder of the same ℓ and σ as the dumbbell is used. The shape factor for dumbbells using this method is given by:

$$\gamma = 3(1+\ell^*)(2+\ell^*)/(2+3\ell^* - \ell^{*3}), \quad \ell^* \leq 1. \tag{6.192}$$

These two methods for evaluating γ for non-convex bodies give similar results[186] as shown in Table 6.9. The Nezbeda equation is seen to give values of $p/\rho kT$ that are too low at the higher densities, while the NPL equation gives values that are too high.

Tildesley and Streett[187] have fitted Monte Carlo data for hard dumbells with bond lengths in the range $0 \leq \ell^* \leq 1$ and densities $0 \leq \rho\sigma_e^3 \leq 0.9$ ($0 \leq \eta \leq 0.47124$), where σ_e is the diameter of the sphere having a volume equal to that of the dumbbell (see Eq. (5.245)), to an equation of state of the form

Table 6.9

*The compressibilty factor $p/\rho kT$ of hard dumbells (homonuclear fused hard spheres) for $\ell^* = \ell/\sigma = 0.6$ from Monte Carlo simulations and from various equations of state**

	$p/\rho kT$					
η	N†	N‡	NPL	PT_f	TS	MC
0.10472	1.63	1.63	1.63		1.64	1.63
0.20944	2.74	2.75	2.78	2.82	2.79	2.78
0.31416	4.80	4.80	4.97	5.01	4.94	4.95
0.41888	8.81	8.82	9.44	9.49	9.21	9.23
0.47124	12.25	12.27	13.41	13.21	12.90	12.87

* Here $\eta = \rho v_m$; N = Nezbeda equation, (6.180); NPL = equations (6.184) and (6.185); TS = Tildesley–Streett equation, (6.193); PT_f = *f*-expansion form of perturbation theory (see § 4.6); and MC = Monte Carlo values.[187]
† With γ calculated from (6.191).
‡ With γ calculated from (6.192).

$$\frac{p}{\rho kT} = \frac{1 + (1 + a_1\ell^* + a_3\ell^{*3})\eta + (1 + b_1\ell^* + b_3\ell^{*3})\eta^2 - (1 + c_1\ell^* + c_3\ell^{*3})\eta^3}{(1-\eta)^3} \quad (TS), \tag{6.193}$$

where $a_1 = 0.37836$, $a_3 = 1.07860$, $b_1 = 1.30376$, $b_3 = 1.80010$, $c_1 = 2.39803$, $c_3 = 0.35700$. This fits the $p/\rho kT$ values with an average deviation of 0.4% and a maximum deviation of 1.1%; the accuracy of the Monte Carlo values is estimated to be within 2%. This is substantially more accurate than the other semiempirical equations of state (see Table 6.9).

Other theoretical equations of state for nonconvex bodies include the *f*-expansion form of perturbation theory (see § 4.6), the RISM theory (§ 5.5.1), and various equations derived from Wertheim's theory of associating fluids (see § 6.13 and refs. 188–190). Some of the *f*-expansion results[191] are included in Table 6.9, and are in moderately good agreement with the MC values. The RISM results are known to be poor for the thermodynamic properties[192] (see, e.g. Fig. 5.16).

6.13 Associating fluids

6.13.1 Introduction

The perturbation theories used in § 6.11 provide a satisfactory approach for fluids composed of fairly small, simple molecules—those that interact with dispersion, overlap, and weak or moderately strong electrostatic and induction forces. Many hydrocarbons, simple inorganic fluids (N_2, O_2, CO, CO_2, N_2O, etc.), and fluids of small organic molecules (e.g. methyl fluoride, toluene) can be successfully treated by these methods. However, many other fluids, and particularly mixtures (see § 7.9), do not fall within these classes; examples include fluids where association (strong attraction, e.g. hydrogen bonding, charge transfer complexing, ion or dipole pairing, and other types of complexing) occurs between molecules of the same or different species, where charged species (electrolytes, plasmas, fused salts), liquid crystals, or chain molecules are present. In this section we consider theories for associating fluids, particularly liquids. Included in this class are fluids such as water, ammonia, various other inorganic molecules (e.g. HCl, HF), alcohols, carboxylic acids, amines, alkyl halides, nitro-compounds, ketones, aldehydes, amphiphiles, and some liquid crystals; all of these undergo H-bonding. The intermolecular forces involved in such fluids are stronger than those for simple fluids, usually by an order of magnitude or more,[193] as shown in Table 6.10; however, they are weaker than those that characterize true chemical bonds. Such intermolecular forces have often been referred to in the chemical literature as 'chemical' or 'specific' forces. We shall avoid such terminology here, since we wish to emphasize a point of view that is physical rather than chemical.

Table 6.10

Approximate energies of intermolecular potentials and interatomic bonds

	Bond	Energy (kJ/mol)
Simple Fluids	Ar ··· Ar	0.89
	CH_4 ··· CH_4	0.6
	CO_2 ··· CO_2	0.9
	C_6H_6 ··· C_6H_6	1.6
Associating fluids*	NH_3 ··· NH_3	15.9
	HF ··· HF	21.1
	H_2O ··· H_2O	22.3
	s-TNB ··· C_6H_6	26
	HCl ··· CH_3CHO	28
	H_2O ··· NH_3	31
	s-TNB ··· mesitylene	40
	H_2O ··· Li	57.6
	s-TNB ··· hexamethylbenzene	77
	BH_3 ··· BH_3	164
	Mg_2Sb ··· Mg_2Sb	168
	LiH ··· LiH	199
	$TiNi_3$ ··· $TiNi_3$	246
Chemical bonds	C–C	347
	N–H	389
	C–H	414
	O–H	465
	C=C	611
	C=O	737

* Some energies are from quantum mechanical calculations.[194] Bond energies for charge transfer complexes between s-trinitrobenzene (TNB) and aromatic molecules are from Thomson and de Maine.[195] Energies for intermetallic compounds are estimates from heats of formation based on the ideal chemical theory.[196] Based on ref. 193.

From Table 6.10 it is clear that there is a continuous distribution of intermolecular energies, so that there is no sharp dividing line between 'simple' and 'associating' fluids. In fact, the theory to be described below can be applied to simple fluids also. For the purposes of this section a precise definition is not necessary. A useful, if somewhat arbitrary, definition[197] of an associating fluid is one in which associating complexes (consisting of two or more molecules in some characteristic configuration) occur, and exist for lifetimes that are long compared to the period of a typical bond vibration (usually about 10^{-13} seconds). In such fluids these complexes are sufficiently long-lived to give rise to experimental evidence (e.g. new absorption peaks in the infrared region) for the presence of new molecular assemblies or species. Such a definition is particularly useful for gases and dilute associating species in solutions. For liquid phases it is harder to determine the structure of associated complexes, since experimental data are more difficult to interpret. More precise definitions for association are needed in molecular simulations, and are usually specific to the particular system considered.

Association has a large effect on many physical properties.[198–204] It leads to unexpectedly high melting and boiling points, and large enthalpies of vapourization, dielectric constants, and viscosities. Water, which can form up to four hydrogen bonds with neighbouring water molecules, is isoelectronic with neon, so that one expects the dispersion and overlap intermolecular forces for the two fluids to be similar; however, their properties are completely different. For example, the enthalpy of vapourization of water at its normal boiling point is about 9.72 kcal/mol (40.65 kJ/mol), as opposed to 0.41 kcal/mol (1.71 kJ/mol) for neon, while the critical temperatures are 647.1 and 44.4 K, respectively. These large differences in thermodynamic properties are primarily due to the strong H-bonding that occurs in the case of water. More direct experimental methods for studying association include ultrasonic and dielectric relaxation measurements, NMR chemical shifts, and a variety of spectroscopic techniques, such as infrared absorption spectra. Ultrasonic and dielectric relaxation measurements, and NMR chemical shifts, can be used to study the kinetics of disassociation of associated complexes, and so give information on the lifetime of such complexes. Infrared spectral measurements can probe changes in vibration frequencies of bonds that result from association. Such changes occur because H-bonding and other kinds of association result in strong constraints on the molecules involved, leading to changes in bond vibrations. It is generally observed that X-H stretching modes are shifted towards lower frequencies on hydrogen bonding. An example is shown in Fig. 6.20 for dilute acetone/methanol mixtures in a carbon tetrachloride solvent.[205] In these mixtures the concentration of methanol is kept constant and dilute to avoid self-association of methanol. As the acetone concentration is increased, H-bonding between acetone and methanol increases, and this leads to a new, broad peak for OH vibration for those methanol molecules that are H-bonded. The area under this peak provides a measure of the fraction of molecules that are H-bonded.

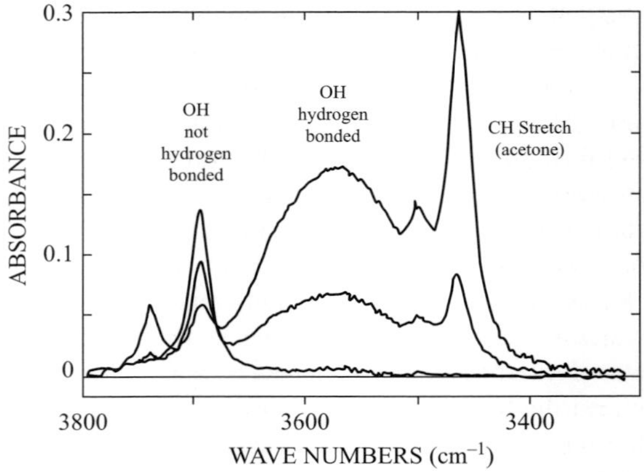

FIG. 6.20 Infrared spectra for dilute acetone/methanol mixtures in a CCl_4 solvent. Methanol concentration is kept fixed and dilute (to avoid self-association), and acetone concentration is increased in the three spectra. Reprinted with permission from ref. 205(i). Copyright 1989, Taylor & Francis, Ltd.

Theoretical approaches to associating fluids have been of two general kinds, which we refer to as *chemical* and *physical* theories, respectively. In the chemical theory, due originally to Dolezalek,[206] associated molecular complexes are regarded as new and distinct chemical species, arising from a chemical reaction (or series of reactions) between the original species present. Unusual thermodynamic behaviour is attributed to the change in composition of the fluid. The chemical theory is best illustrated by a simple example, the strong association of acetic acid to form dimers:

$$2\mathrm{CHCOOH} \leftrightarrow \mathrm{CHC}^{\mathrm{O\cdots HO}}_{\mathrm{OH\cdots O}}\mathrm{CCH} \tag{6.194}$$

The equilibrium condition for this reaction is

$$K = \frac{a_{(\mathrm{CH_3COOH})_2}}{a^2_{\mathrm{CH_3COOH}}}, \tag{6.195}$$

where $a_\alpha \equiv f_\alpha / f_\alpha^0$ represents the activity (see §§ 7.4 and 7.5) of the monomer or dimer, f_α and f_α^0 are the fugacity of component α in the mixture and in some pure liquid standard state, respectively, and $K = \exp(-\Delta G^0/kT)$ is the usual chemical equilibrium constant, with ΔG^0 the standard Gibbs energy change in the forward ($\rightarrow$) reaction (6.194). In the original chemical theory of Dolezalek,[206] it was assumed that the resulting mixture was either an ideal mixture or an ideal gas. For these cases (6.195) reduces to

$$K = \frac{\hat{\rho}_{(\mathrm{CH_3COOH})_2}}{\hat{\rho}^2_{(\mathrm{CH_3COOH})}}, \tag{6.196}$$

where $\hat{\rho}_\alpha = \rho_\alpha/\rho_\alpha^0$ and ρ_α and ρ_α^0 are the number density of species α and standard state value, respectively. If the mixture is assumed to be an ideal gas, these densities can be replaced by partial pressures, using the ideal gas equation. If the equilibrium constant K is available, from direct experimental measurement of the equilibrium system (e.g. by infrared spectroscopy) or from statistical mechanical calculations for the ideal gas based on spectroscopic data, (6.196) and the known total density of CH_3COOH molecules can be used to predict the equilibrium concentrations of monomer and dimer. Once these are known, gas–liquid equilibrium conditions can also be calculated, most simply by assuming an ideal liquid mixture and an ideal gas phase. More sophisticated treatments do not make either the ideal liquid mixture or ideal gas approximations. In those cases the calculations must be based on the rigorous equilibrium equation ((6.195) in the example described here), and an equation of state or a group contribution method is used to relate the activities to the concentrations of the various species. Approaches of this type have been reviewed by Prausnitz *et al.*[202] and by Müller and Gubbins.[193]

The case of dimerization of acetic acid is a particularly simple one, since only one simple reaction occurs. In many cases the situation is much more complex. In the case of alkanols, for example, H-bonding leads not only to dimers, but to bonded chains of higher mers of various lengths. Similarly, water forms a complex tetrahedrally bonded structure in the liquid phase. For such cases there are numerous ‘reactions’, and it is often not at all obvious what these are. Thus, a major problem with the chemical approach is that each association step leading to equilibrium must be known in advance. A second difficulty is that the reaction equilibrium constants are unknown in general, and may have to be treated as adjustable parameters; moreover, these constants vary with temperature, requiring further fitting of parameters. Thus, chemical theories are of limited predictive value.

In the physical theories, the thermodynamic properties are related to intermolecular forces in the usual way via statistical mechanics. Thus, associating fluids are treated on the same footing as simple fluids. The intermolecular potential models used must take into account the forces that produce association. A fundamental approach of this kind has several advantages over the chemical theory. It is not necessary to know what chemical reactions are involved, nor is it necessary to know equilibrium constants. Since there is a well-defined Hamiltonian for the system, approximations made in the theory can be tested against molecular simulations in the usual way. Such tests can be used to choose between different theoretical routes, and to refine the theory, before making comparisons with experiment. The principal challenge in such a physical approach is the specification of intermolecular potential models that are sufficiently realistic to describe the association, while at the same time being tractable.

Early physical theories relied on lattice models.[207–210] Such theories can be solved to give qualitative features of the thermodynamics and phase equilibrium behaviour. However, placing molecules on lattice sites gives a poor description of the fluid structure and underestimates the entropy, so that quantitative agreement with experimental data cannot be achieved. Most recent theories have been continuum in nature. These have taken the form of cluster expansions, renormalized perturbation expansions, and integral equation theories.[211–219] It was quickly found that fugacity expansions converged faster than density expansions for associating fluids. These ideas were extended by Wertheim,[220] who introduced a multi-density formalism, in which the expansion was expressed in terms of the densities of the various bonding states of the monomers.

Since the Wertheim theory has been particularly successful and is easy to use, we devote the remainder of this section to a description of the theory and some of its applications to real associating fluids. This theory and its applications have been the subject of reviews.[221–224]

6.13.2 Wertheim's theory

We consider a fluid of associating molecules interacting only with pair potentials. These potentials are assumed to have a core potential that includes a strongly repulsive component (nonspherical, in general), and a sum of interactions between association sites. We consider a mixture of components $A, B, \cdots, \alpha, \cdots \beta, \cdots R$, since we will use these equations also in Chapter 7 for mixtures. Association sites in each molecule are labeled $a, b, \cdots, i, \cdots j, \cdots$. Thus the pair potential between molecules 1 of species α and 2 of species β can be written

$$u_{\alpha\beta}(12) = u_{\alpha\beta,0}(12) + u_{\alpha\beta,1}(12) = u_{\alpha\beta,0}(12) + \sum_i \sum_j u^{ij}_{\alpha\beta}(r_{ij}) \qquad (6.197)$$

where $u_{\alpha\beta,0}$ is the reference (core) potential and $u^{ij}_{\alpha\beta}$ is the short-range, attractive potential between site i in the molecule of species α and site j in the molecule of β. It is the short-range, attractive potentials that give rise to the association. Here 12, as usual, represents $(\mathbf{r}_{12}, \omega_1, \omega_2)$. Wertheim takes the core potential to be purely repulsive, but this restriction is not a necessary one, and some later authors take the core potential to include dispersion interactions.

Wertheim's theory[220] makes use of graphical expansions (§ 3.6 of Vol. 1) for $\ln \Xi$ and for ρ_α, the log of the grand partition function and the number density of species α. Since the full graphical treatment is lengthy and relies on a detailed knowledge of graph manipulation,[225,226] we give only a brief outline of the theory here. The reader is referred to the original papers[220] for full details; the extension to mixtures is given by Joslin *et al.*[227] and by Chapman.[228] Following the discussion of Wertheim's graphical expansion theory, we give a simpler and more heuristic derivation due to Joslin *et al.*[227]

In order to introduce the geometry of the association interaction from the start, and to simplify graph cancellation due to steric effects, Wertheim departs from the usual expansion in the singlet densities, ρ_α, of § 3.6. These departures are of two kinds: (a) the use of the expansion in terms of *activities*, $\hat{z}_\alpha$, in place of number densities, and (b) classification of the resulting graphs into classes belonging to the densities of monomers, molecules bonded at association site 1, molecules bonded at association site 2, and so on. Here $\hat{z}_\alpha$ (with dimensions of inverse volume) is the mixture generalization of (8.3),[229] i.e.

$$\hat{z}_\alpha = \frac{\exp(\beta\mu_\alpha)q_{qu,\alpha}}{\Lambda_{t,\alpha}^3 \Lambda_{r,\alpha} \Omega_\alpha}. \tag{6.198}$$

Use of the $\hat{z}$-expansion enables graph cancellation due to steric incompatibility to be carried out much more efficiently, while classification of graphs into classes corresponding to different densities leads to a strong similarity to the Mayer theory expansion based on density alone, which in turns opens up the possibility of making use of the existing theory developed for correlation functions and direct correlation functions. The Wertheim theory rests on the assumption of three kinds of approximate steric constraints[230] on possible bonding between molecules (see Fig. 6.21):

1. The repulsive cores of any two molecules bonded together at sites i and j prevent a third molecule from bonding at either of these sites.
2. A site on a given molecule cannot bond simultaneously to two sites on another molecule. This will hold provided that the angle subtended at the molecule's centre by the two sites is large enough that the interaction ranges of the three sites cannot simultaneously overlap (see Fig. 6.21b).
3. Multiple bonding between two molecules is not allowed (Figure 6.21c).

These approximations are physically reasonable as far as hydrogen bonding interactions are concerned, and enable many of the more complex graphs to be

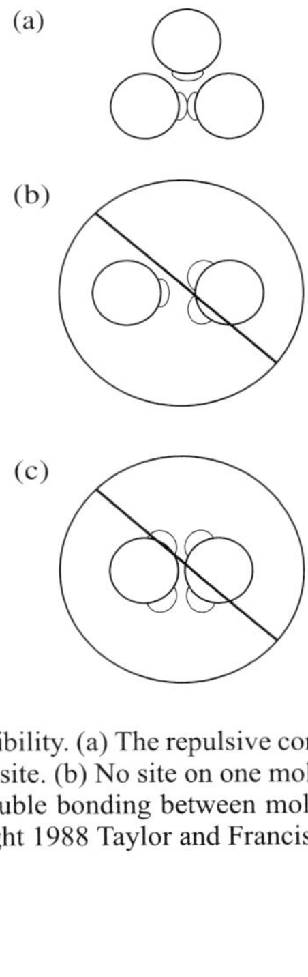

FIG. 6.21 Types of steric incompatibility. (a) The repulsive cores of the molecules prevent more than two molecules bonding at a given site. (b) No site on one molecule can bond simultaneously to two sites on another molecule. (c) Double bonding between molecules is not allowed. Reprinted with permission from ref. 235. Copyright 1988 Taylor and Francis Ltd.

neglected. Following elimination of these graphs, Wertheim carries out a topological reduction of the remaining graphs to obtain the final result. The resulting Helmholtz energy can be expressed in a first order thermodynamic perturbation theory form, TPT1,

$$\frac{A - A_0}{NkT} = \sum_\alpha x_\alpha \left[\sum_i \left(\ln X_\alpha^i - \frac{X_\alpha^i}{2} \right) + \frac{1}{2} n_\alpha \right], \tag{6.199}$$

where $x_\alpha = N_\alpha / N$ is the mole fraction of component α, X_α^i is the fraction of α molecules *not* bonded at site *i*, and n_α is the number of bonding sites on a molecule of species α; the sum is over all components, $\alpha = 1, 2, \cdots, m$. An expression for the mole fractions not bonded at a given site *i*, X_α^i, is obtained by noting that, at equilibrium in the canonical ensemble, the Helmholtz energy must be a minimum.[227] This minimization gives the implicit equation for X_α^i

$$X_\alpha^i = \left[1 + \rho \sum_\beta \sum_j x_\beta X_\beta^j \Delta_{\alpha\beta}^{ij} \right]^{-1}, \tag{6.200}$$

where $\rho = N/V$ is the total number density and $\Delta_{\alpha\beta}^{ij}$ is given by

$$\Delta_{\alpha\beta}^{ij} = \int \left\langle g_{\alpha\beta,0}(12) f_{\alpha\beta}^{ij}(12) \right\rangle_{\omega_1\omega_2} \mathbf{dr}_{12}, \tag{6.201}$$

where $g_{\alpha\beta,0}(12)$ is the reference fluid pair correlation function and $f_{\alpha\beta}^{ij}(12) = \exp\left[-u_{\alpha\beta}^{ij}(12)/kT \right] - 1$ is the Mayer *f*-function for the attractive site–site part of the intermolecular potential.

Expressions for other properties can be readily obtained from (6.199) by the use of standard thermodynamic identities. Of particular importance is the pressure $p = -(\partial A/\partial V)$, or compressibility factor, $Z = p/\rho kT$,

$$\frac{p}{\rho kT} = \rho \left(\frac{\partial (A/NkT)}{\partial \rho} \right)_{\{N_\alpha\},T}, \tag{6.202}$$

where $\{N_\alpha\}$ means the set of numbers of molecules of each component. Applying (6.202) to (6.199) gives

$$\frac{p}{\rho kT} = \left(\frac{p}{\rho kT} \right)_0 + \sum_\alpha \rho_\alpha \left[\sum_i \left(\frac{\partial X_\alpha^i}{\partial \rho} \right)_{\{N_\alpha\},T} \left(\frac{1}{X_\alpha^i} - \frac{1}{2} \right) \right]. \tag{6.203}$$

Heuristic derivation

Although the full graphical expansion method for deriving Wertheim's theory is rigorous, it is also involved. Several authors[227,228,231,232] have presented shorter and more heuristic derivations. We present here the derivation of Joslin *et al.*[227] In order to keep the notation simple, we treat the case where the molecules have only one association site. This means we will find only dimers, and no trimers, etc. The extension to the multiple site case is straightforward. The total Helmholtz energy is written as the sum of an ideal gas and a residual part (see also § 6.5.1),

$$A = A_{\mathrm{id}} + A_{\mathrm{r}}, \tag{6.204}$$

where (see (3.11)) $A_{\mathrm{id}} = -kT \ln Q_{\mathrm{id}}$ and Q_{id} is the partition function for the ideal gas at the temperature, density, and composition of interest, and is given by (see (3.18), (3.21), (3.77)–(3.81), and (3.250), together with (6.32))

$$\frac{A_{\mathrm{id}}}{VkT} = \sum_{\alpha} [\rho_\alpha \ln(\rho_\alpha \Lambda_\alpha) - \rho_\alpha], \tag{6.205}$$

where $\Lambda_\alpha \equiv (\Lambda_{\mathrm{t}\alpha}^3 \Lambda_{\mathrm{r}\alpha}/q_{\mathrm{qu},\alpha})$ and $\Lambda_{\mathrm{t}\alpha}$, $\Lambda_{\mathrm{r}\alpha}$, and $q_{\mathrm{qu},\alpha}$ are the translational, rotational, and quantal parts of the molecular partition function; the former are given by (3.79) and (3.81), and q_{qu} contains any other parts of the internal molecular partition function (vibration, electronic, etc.). The residual contribution to the Helmholtz energy, A_{r}, derives solely from the interactions between molecules. In conventional perturbation theory the first-order term for the Helmholtz energy is given by the mixture generalization of (4.19),

$$A_1 = \frac{1}{2} \sum_{\alpha} \sum_{\beta} \rho_\alpha \rho_\beta \int \mathrm{d}\mathbf{r}_1 \mathrm{d}\mathbf{r}_2 \left\langle g_{\alpha\beta,0}(12) \left(\frac{\partial u_{\alpha\beta,\lambda}(12)}{\partial \lambda} \right)_{\lambda=0} \right\rangle_{\omega_1 \omega_2}, \tag{6.206}$$

where $u_{\alpha\beta,\lambda}(12)$ is a pair potential defined so that it becomes the full potential, $u_{\alpha\beta}$, when $\lambda = 1$ and the reference fluid potential, $u_{\alpha\beta,0}$, when $\lambda = 0$. In the f-expansion (§ 4.6) $u_{\alpha\beta,\lambda}(12) = u_{\alpha\beta,0}(12) - kT \ln[1 + \lambda f_{\alpha\beta,1}(12)]$, so that $(\partial u_{\alpha\beta,\lambda}(12)/\partial\lambda)_{\lambda=0} = -kTf_{\alpha\beta,1}(12)$. Here $f_{\alpha\beta,1}(12) = \exp(-u_{\alpha\beta,1}(12)/kT) - 1$ is the Mayer f-function for the attractive (perturbation) part of the pair potential, $u_{\alpha\beta,1} = u_{\alpha\beta} - u_{\alpha\beta,0}$. Using these results in (6.206) gives the first-order contribution to the Helmholtz energy as[233]

$$\frac{(A_{\mathrm{r}} - A_{\mathrm{r}0})}{VkT} = \frac{(A_{\mathrm{c}} - A_{\mathrm{c}0})}{VkT} = -\frac{1}{2} \sum_{\alpha} \sum_{\beta} \rho_\alpha \rho_\beta \int \mathrm{d}\mathbf{r}_{12} \langle g_{\alpha\beta,0}(12) f_{\alpha\beta,1}(12) \rangle_{\omega_1 \omega_2}. \tag{6.207}$$

Equation (6.207) neglects higher terms in the series, of relative order $\rho \int \mathrm{d}\mathbf{r}_{12} \langle f_{\alpha\beta,1}(12) \rangle_{\omega_1\,\omega_2}$. For strongly associating fluids, the attractive potential $u_{\alpha\beta,1}(12)$ will be large and negative for orientations corresponding to the molecular association, so that $f_{\alpha\beta,1}(12)$ will be large and positive. Consequently, the conventional perturbation theory approach fails for strongly associated liquids.

As a way around this difficulty, Wertheim[220] proposed treating monomers, dimers, etc. as distinct entities, so that the total number density of species α can be written as

$$\rho_\alpha = \rho_{\mathrm{m}\alpha} + \rho_{\mathrm{d}\alpha}, \tag{6.208}$$

where $\rho_{\mathrm{m}\alpha}$ is the density of monomers and $\rho_{\mathrm{d}\alpha}$ is the density of α molecules present as dimers (we are considering the case where only dimers can form). For weak association $\rho_\alpha \approx \rho_{\mathrm{m}\alpha}$, but when the association is strong $\rho_{\mathrm{m}\alpha} << \rho_\alpha$. This suggests a renormalization of the splitting of A into two parts, and of the series for $A_{\rm r}$. For strong association it is profitable to write

$$A = A'_{\rm id} + A'_{\rm r}, \tag{6.209}$$

where $A_{\rm id}{}'$ and $A_{\rm r}{}'$ are renormalized quantities given by

$$\frac{A'_{\rm id}}{VkT} = \sum_{\alpha} \left[\rho_\alpha \ln \left(\rho_{{\rm m}\alpha} \Lambda_\alpha^3 \right) - \rho_{{\rm m}\alpha} \right] \tag{6.210}$$

$$\frac{A'_{\rm r} - A'_{\rm r0}}{VkT} = -\frac{1}{2} \sum_{\alpha} \sum_{\beta} \rho_{{\rm m}\alpha} \rho_{{\rm m}\beta} \int {\rm d}\mathbf{r}_{12} \langle g_{\alpha\beta,0}(12) f_{\alpha\beta,1}(12) \rangle_{\omega_1\omega_2}. \tag{6.211}$$

Comparison of (6.207) and (6.211) shows that in (6.211) we are including only the configurational interactions between *monomers* at density $\rho_{{\rm m}\alpha}$. Thus, (6.211) neglects higher terms of relative order $\rho_{{\rm m}\alpha} \int {\rm d}\mathbf{r}_{12} \langle f_{\alpha\beta,1}(12) \rangle_{\omega_1\omega_2}$, i.e. substantially smaller than the neglected terms in (6.207) since $\rho_{{\rm m}\alpha} << \rho_\alpha$ for strong association. We note, however, that the contributions to the free energy from the densities of bonded molecules are not neglected; rather, they are implicitly summed in the expression for A'_{id} in Eq. (6.210).

To determine $\rho_{{\rm m}\alpha}$, the number density of monomers of species α, we note that at equilibrium in the canonical ensemble the Helmholtz energy must be a minimum, i.e. $(\partial A/\partial \rho_{{\rm m}\alpha})_{\rho_\beta} = 0$. Applying this condition to equations (6.209)-(6.211) gives

$$\rho_\alpha = \rho_{{\rm m}\alpha} + \rho_{{\rm m}\alpha} \sum_{\beta} \rho_{{\rm m}\beta} \int {\rm d}\mathbf{r}_{12} \langle g_{\alpha\beta,0}(12) f_{\alpha\beta,1}(12) \rangle_{\omega_1\omega_2}. \tag{6.212}$$

Equations (6.209)-(6.212) can be combined to give a simpler form for the free energy. We first note that for the reference system in which bonding interactions are absent $\rho_{{\rm m}\alpha} = \rho_\alpha$, and

$$\frac{A'_{id,0}}{VkT} = \sum_{\alpha} [\rho_\alpha \ln(\rho_\alpha \Lambda_\alpha) - \rho_\alpha] \tag{6.213}$$

and from this and (6.210) we have

$$\frac{A'_{\rm id} - A'_{\rm id,0}}{VkT} = \sum_{\alpha} \rho_\alpha [\ln X_\alpha - X_\alpha + 1], \tag{6.214}$$

where $X_\alpha = \rho_{{\rm m}\alpha}/\rho_\alpha$ is the fraction of α molecules that are not bonded. From (6.211) and (6.214),

$$\frac{A - A_0}{VkT} = \sum_{\alpha} \rho_\alpha [\ln X_\alpha - X_\alpha + 1] - \frac{1}{2} \sum_{\alpha} \sum_{\beta} \rho_{m\alpha} \rho_{m\beta} \Delta_{\alpha\beta} \qquad (6.215)$$

with

$$\Delta_{\alpha\beta} = \int \langle g_{\alpha\beta,0}(12) f_{\alpha\beta,1}(12) \rangle_{\omega_1\omega_2} \mathbf{dr}_{12}. \qquad (6.216)$$

From (6.212) the last term on the right in (6.215) equals $-\left(\frac{1}{2}\right) \sum_\alpha \rho_\alpha (1 - X_\alpha)$, so that (6.215) becomes

$$\frac{A - A_0}{VkT} = \sum_{\alpha} \rho_\alpha \left[\ln X_\alpha - \frac{1}{2} X_\alpha + \frac{1}{2} \right], \qquad (6.217)$$

which is (6.199) for the case of molecules having a single bonding site. Moreover, by dividing Eq. (6.212) by ρ_α we obtain the implicit expression for the fraction of α molecules not bonded as

$$X_\alpha = \left[1 + \sum_\beta \rho_\beta X_\beta \Delta_{\alpha\beta} \right]^{-1}, \qquad (6.218)$$

which is (6.200) for the single site case.

For a dilute gas in which association occurs it is possible to derive equations analogous to those derived above without making any approximations.[193, 202, 221] These equations are the same as those given above, but with the reference pair correlation function replaced by its low density limit, (3.261),

$$g_{\alpha\beta,0}(12) = \exp[-u_{\alpha\beta,0}(12)/kT]. \qquad (6.219)$$

Thus, the Wertheim theory gives the correct low density limit.

Test against molecular simulation results

Tests of the Wertheim theory against molecular simulations show good agreement, even for strongly associating systems.[193, 227, 234–237] A variety of intermolecular potential models have been used. In early studies the reference potential in (6.197) was taken to be a hard sphere model, but later workers have often used a LJ spherical interaction as the reference; the attractive bonding interactions have been represented by off-centre point charge[227, 236] or square well[234, 235, 237] sites. In the case of point charge sites, the Coulomb interaction between sites on different molecules was set to 0 beyond some short cutoff distance. Thus these two representations of associating sites, point charge and square well, produce similar intermolecular pair potentials. Several kinds of square well interactions have been used, of which two cases are illustrated in Fig. 6.22; in both cases the larger sphere represents either a hard sphere or a LJ core, and the geometry of the square well sites is chosen to prevent bonding of more than one molecule per site. In model (a) spherical square well sites of diameter σ_b are placed at a distance l_b from the centre of the sphere, and the

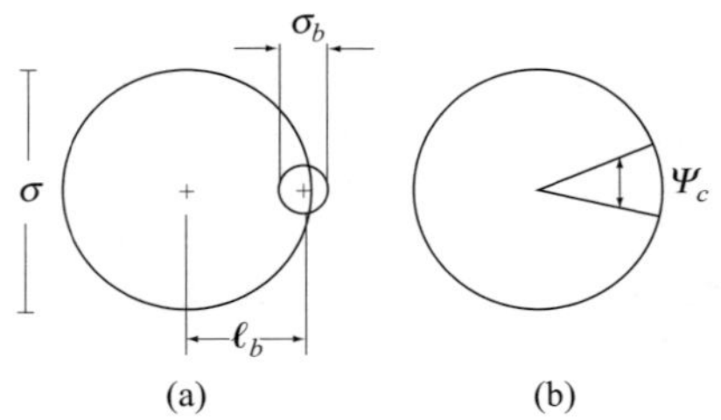

FIG. 6.22 Two square well models for describing associating fluids. (a) Spherical bonding site of diameter σ_b at distance ℓ_b from the centre of the molecule; (b) conical bonding site with cone angle ψ_c. (Adapted from ref. 237.)

site–site bonding potential between site i in molecule 1 and site j in molecule 2 is given by

$$\begin{aligned} u_1^{ij}(r_{ij}) &= -\varepsilon_{ij}^{\text{bond}} \text{ if } r_{ij} \leq \sigma_b \\ &= 0 \qquad\quad \text{if } r_{ij} > \sigma_b \end{aligned} \tag{6.220}$$

where r_{ij} is the distance between site i in molecule 1 and site j in molecule 2. With this model one can restrict the bonding to dimer formation by requiring that $(\sigma - 2l_b) < \sigma_b \leq \left[\sqrt{\left(\sigma^2 - \sqrt{3}\sigma l_b + l_b^2\right)} - l_b\right]$, where σ is the diameter of the molecular core. The second kind of square well site model, Fig. 6.22(b), has conical sites that have their origin at the centre of the core of the reference potential, and have a spherical cap that coincides with the surface of the core sphere. The potential is characterized by the angle ψ_c made by projecting the cone onto a plane (as shown in Fig. 6.22(b)). For this model the bonding potential between sites on molecules 1 and 2 is given by

$$\begin{aligned} u_1^{ij}(\mathbf{r}_{12}\omega_1\omega_2) &= -\varepsilon_{ij}^{\text{bond}} \text{ if } |\mathbf{r}_{12}| \leq r_c \text{ and } \left|\theta_i'\right| \leq \psi_{ic} \text{ and } \left|\pi - \theta_j'\right| \leq \psi_{jc} \\ &= 0 \qquad\qquad\qquad\qquad\qquad \text{otherwise,} \end{aligned} \tag{6.221}$$

where $\mathbf{r}_{12}$ is the vector from the centre of molecule 1 to that of 2 and θ' is the polar angle in the intermolecular frame, relative to $\mathbf{r}_{12}$ (see Fig. 2.9 of Vol. 1). In this model the two molecules will only bond at sites i and j if the molecular centres are closer than some cutoff distance r_c, and are oriented correctly. In most cases the two angular cutoffs are taken to be the same, i.e. $\psi_{ic} = \psi_{jc}$. Bonding can be restricted to dimers by ensuring that $\sigma < r_c \leq \sigma/(2\sin\psi_c)$. The second model, while being somewhat less physically realistic than the first, has the advantage that molecular simulations require less time, since it is not necessary to calculate the site–site distances, r_{ij}.

Comparisons of the Wertheim theory with Monte Carlo simulations for the case where the reference core potential is hard spheres, and conical square well sites are used to represent the association sites, are shown in Figs. 6.23 and 6.24. In each case figure (a) shows the results for a single bonding site of strength $\varepsilon^{\text{bond}} = 7kT$, while figure (b) shows results for molecules with two bonding sites of strength $\varepsilon^{\text{bond}} = 5kT$ placed diametrically on opposite sides of the centre of the core. In the theoretical calculations the hard sphere pair correlation function $g_0(r)$ of Verlet and Weis[238] was used in evaluating the integral Δ of Eq. (6.201) or (6.216). The Verlet–Weis function matches simulation results for $g_0(r)$ closely. For the case of square well bonding sites approximate analytic expressions[235] are also available for Δ. From Figs. 6.23 and 6.24 it is seen that the theory is in essentially quantitative agreement with the simulation results for these properties; equally good results are obtained for the configurational energies. The monomer mole fraction (Fig. 6.24) falls as density increases, as expected.

The theory has also been applied to cases where the reference potential is the LJ model.[237,239–245] The reference fluid pair correlation function appearing in Δ is now a complicated function of temperature and density, rather than of density alone. Ideally this should be obtained from simulations on the LJ fluid. Many workers have used approximate forms for $g_0(r)$, particularly the WCA approximation, which is appropriate for dense fluids (see § 4.1 and Figs. 4.1 and 4.2),

$$g_0(r) = g_{\text{WCA}}(r), \tag{6.222}$$

where $g_{\text{WCA}}(r)$ is the pair correlation function for a fluid interacting with the Weeks–Chandler–Andersen[246] cut and shifted potential,

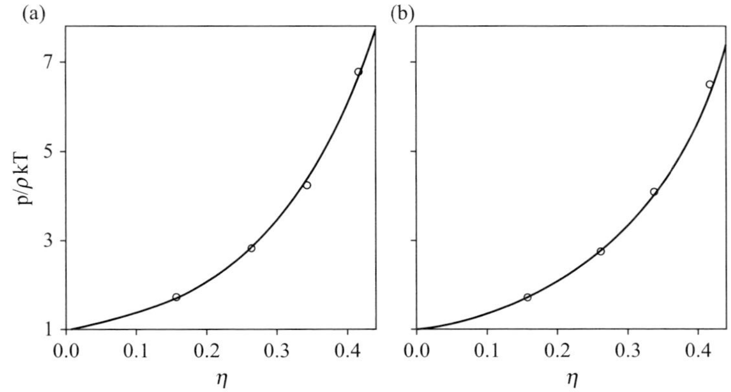

FIG. 6.23 The compressibility factor for a fluid of molecules modeled as hard spheres of diameter σ with conical square well association sites, as a function of reduced density $\eta = (\pi/6)\rho\sigma^3$: (a) one bonding site of strength $\varepsilon^{\text{bond}} = 7kT$, with $\psi_c = 27°$, $r_c = 1.05\sigma$; (b) two bonding sites, each of strength $\varepsilon^{\text{bond}} = 5kT$, with $\psi_c = 27°$, $r_c = 1.10\sigma$. Points are NPT Monte Carlo simulation results; line is Wertheim theory. (From ref. 235.)

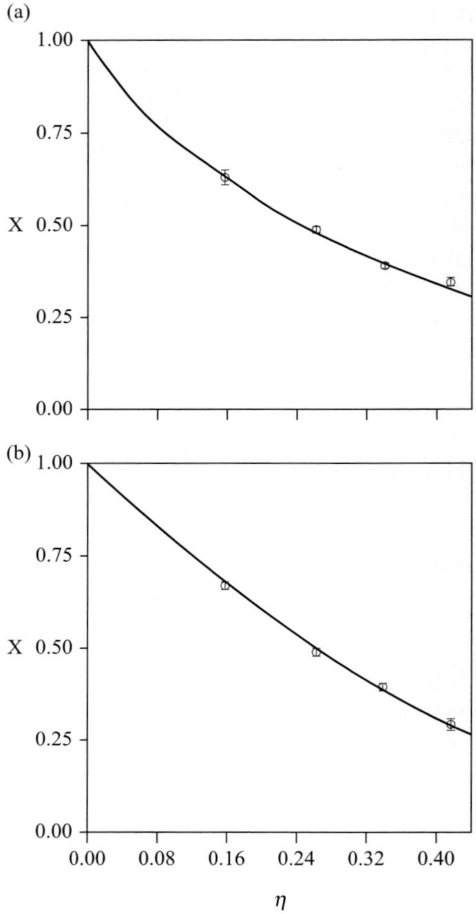

FIG. 6.24 The mole fraction of monomers X from *NPT* Monte Carlo simulations (points) and from Wertheim theory (lines). Key as described in the legend to Fig. 6.23. (From ref. 235.)

$$\begin{aligned} u_{\mathrm{WCA}}(r) &= u_{\mathrm{LJ}}(r) + \varepsilon_{\mathrm{LJ}} \quad && \text{if } r < r_{\mathrm{m}} \\ &= 0 && \text{if } r > r_{\mathrm{m}}, \end{aligned} \tag{6.223}$$

where $\varepsilon_{\mathrm{LJ}}$ is the well depth of the LJ potential and r_{m} is the separation corresponding to the potential minimum. However, considerable care is needed when using approximate $g_0(r)$ functions, since the integral Δ is sensitive to small errors in this function.[247] Comparisons of the Wertheim theory results using a LJ reference are in good agreement with simulation results, as shown for a fluid with one conical square well site in Figs. 6.25 and 6.26.

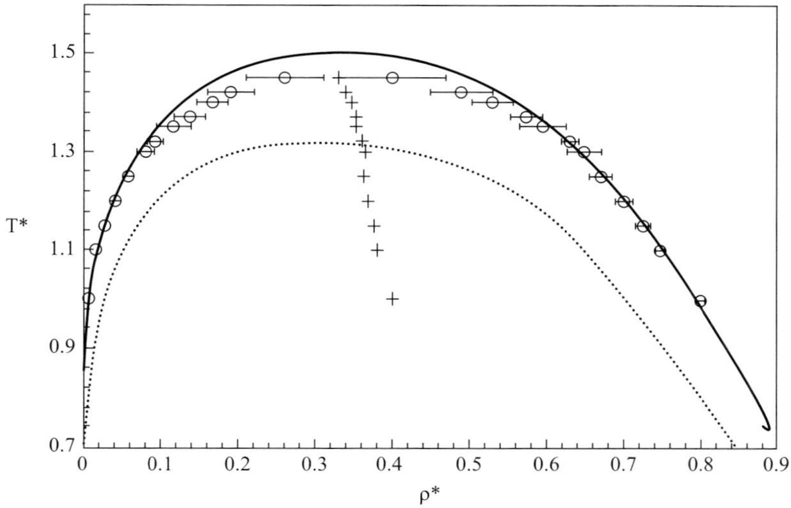

FIG. 6.25 Saturated liquid and vapour densities for a pure fluid in which the molecules consist of a spherical LJ core plus an off-centre conical square well site, with $\varepsilon^{\mathrm{bond}}/\varepsilon_{\mathrm{LJ}} = 10$, $r_c = \sigma$, and $\psi_c = 27°$. Circles are Gibbs ensemble Monte Carlo results, the + symbols show the line of rectilinear diameters ($\rho_m = (\rho_l + \rho_v)/2$), and the solid line is Wertheim's theory using for $g_0(r)$ a weighted average of $g_{\mathrm{WCA}}(r)$ and the low density result of Eq. (6.219). The dotted line is the phase diagram for the pure LJ fluid. Reprinted with permission from ref. 237. Copyright 1992 Taylor and Francis Ltd.

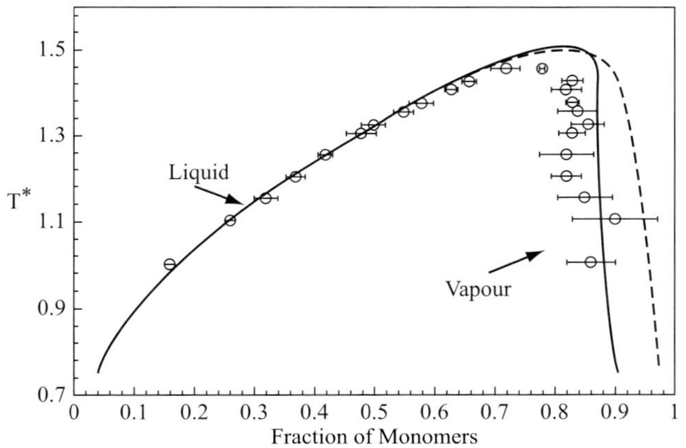

FIG. 6.26 The equilibrium fraction of monomers in the liquid and vapour phases for the same fluid as shown in Fig. 6.25. Circles are results of Gibbs ensemble MC simulation, and the solid line is the Wertheim theory result using the same approximate form for $g_0(r)$ as in the previous figure. The dashed line is the result from Wertheim's theory using the WCA approximation for $g_0(r)$ for all densities. Reprinted with permission from ref. 237. Copyright 1992 Taylor and Francis Ltd.

6.13.3 *Chain molecules*

If we take the limit where the bonding strength becomes large we obtain a fluid of chain molecules. If the unassociated molecules have only one site this produces a fluid of dimers. The resulting equation of state is in excellent agreement with molecular simulation results for both the thermodynamic properties for tangent hard sphere dimers[248] and for Lennard-Jones dimers.[249] For monomer units having two or more sites, longer chains can form. For finite bonding strengths, there will be a distribution of chain lengths when the system is in equilibrium, with some average length $\bar{m}$, where m is the number of monomer units in the chain. This distribution, and also the average chain length, will vary with bonding strength.

The ability of the theory to describe chain molecules was noted by Wertheim,[250] who made comparisons between calculations based on the theory with MC calculations by Dickman and Hall[251] for tangent hard sphere chains (also referred to as 'pearl necklace' chains) of length $m = 4, 8, 16$. In the simulations the chains were all of fixed length, whereas the theory was for a fluid with a distribution of chain lengths. In making the comparison, Wertheim made calculations for a fluid in which the chains had the same average length as those used in the simulations. Quite good agreement between theory and simulation was obtained, but discrepancies became larger as chain length increased, and were quite noticeable for chains of $m = 16$.

Chapman *et al.*[252] derived an expression for the free energy for a mixture of chain molecules in which the chain lengths were specified, thus making comparisons with simulation results easier. They did this by considering a mixture of spherical monomer molecules with bonding sites in the limit where the fraction of molecules in which certain sites are not bonded is zero; the stoichiometric ratio of spheres having different numbers of sites is chosen to produce chains of the desired lengths. Consider a M-component mixture of spherical monomers, each of which is in general labeled as a different component. Thus $M = N$, the total number of spheres in the most general case. These individual spheres will be bonded together to form R chains of $m_1, m_2, \cdots, m_J, \cdots, m_R$ spheres, so that $N = \sum_{J=1}^{R} m_J$. By restricting that sphere 1 only bonds to sphere 2, that sphere 2 only bonds to spheres 1 and 3, ... and that sphere m_J only bonds to sphere m_{J-1}, we form chains of m_J spheres (Fig. 6.27). Thus, for chain J there are $m_J - 2$ spheres having two bonding sites, and two end spheres having only one

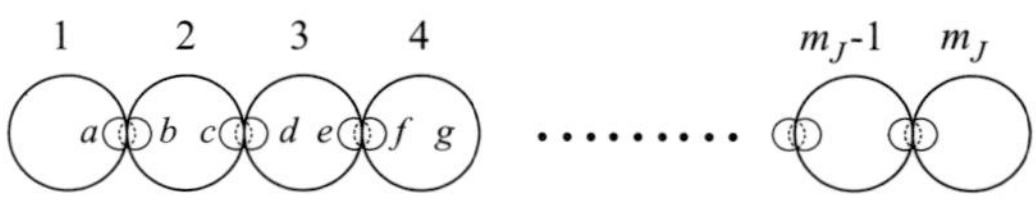

FIG. 6.27 Each chain J is formed by bonding m_J spheres together. Sphere 1 bonds only to sphere 2, sphere 2 only bonds to 1 and 3, etc. Since the first order Wertheim theory does not take into account the angle between any two bonding sites on a single sphere, or specify where the sites are located, it cannot distinguish between different configurations or isomers of a chain.

site. Referring to Fig. 6.27, we note that a given interaction site, for example e on monomer 3 in the chain, is only allowed to bond to one particular site, f on monomer 4, and not to any other site. Thus, $u_{34}^{ef} \neq 0$ (similarly u_{12}^{ab}, u_{23}^{cd}, etc. are non-zero), but all other u_{34}^{ij}, with $ij \neq ef$, are set equal to 0. It follows from (6.201) that $\Delta_{34}^{ef} \neq 0$, but all other $(ij \neq ef)\Delta_{34}^{ij}$ terms vanish.

Total bonding will be imposed on the system by setting the fraction of spheres not bonded at the appropriate sites to 0. We note that the spheres in the chain may have other bonding sites, not involved in chain formation, but these are not allowed to interfere with the formation of the chain.

The change in the compressibility factor due to bonding to form chains can now be obtained from (6.203). This can be rewritten as a sum over the chains we want to form, i.e.

$$\left(\frac{p}{\rho kT}\right)_{\text{chain}} = \left(\frac{p}{\rho kT}\right) - \left(\frac{p}{\rho kT}\right)_0 = \sum_{J=1}^{R} \rho_{chain}^J \left[\sum_{\beta=1}^{m_J} \sum_i \left(\frac{\partial X_\beta^i}{\partial \rho}\right)_{\{N_\alpha\},T} \left(\frac{1}{X_\beta^i} - \frac{1}{2}\right)\right], \tag{6.224}$$

where $\rho_{\text{chain}}^J = N_{\text{chain}}^J / V$ is the number density of chains of type J, while ρ remains the number density of the monomer spheres, N/V. The first sum is over each chain $J = 1, 2, \cdots, R$. The sum over $\beta = 1, 2, \cdots, m_J$ is over all the spheres that make up chain J, and that over i is over all binding sites for a given sphere. As earlier, X_β^i is the fraction of a given site i on sphere β that is not bonded.

If we now consider the mole fraction of species (monomer) α that is not bonded at i, we see from (6.200) and Fig. 6.27 that only one term under the summations in (6.200) is non-vanishing. For example, for site b on mer 2 in Fig. 6.27,

$$X_2^b = \frac{1}{1 + \rho_1 X_1^a \Delta_{21}^{ba}}$$

since all $u_{2\beta}^{bj}$ vanish except for $j = a$, $\beta = 1$. We now assume that only homonuclear chains are formed; i.e. all monomer spheres in a given chain are identical (except that the end spheres have only one site). Then, for a given chain J, since each site can only bond with one type of site that is of the same kind, $X_\beta^i = X_J$, i.e. the fraction of sites not bonded is the same for all sites on chain J. We can now solve for X_J using (6.200). Noting that the number of chains of type J is the same as the number of a given sphere β that makes up the chain, $N_{\text{chain}}^J = N_\beta$, this gives

$$1 - X_J - \rho_{\text{chain}}^J X_J^2 \Delta_{JJ} = 0. \tag{6.225}$$

Taking the derivative of this equation with respect to the number density ρ, and noting that $\left(\partial \rho_{\text{chain}}^J / \partial \rho\right) = \rho_{\text{chain}}^J / \rho$, gives

$$\frac{\partial X_J}{\partial \rho}\left(\frac{1}{X_J} - \frac{1}{2}\right) = -\frac{1}{2} X_J^2 \left[\frac{\rho_{\text{chain}}^J \Delta_{JJ}}{\rho} + \rho_{\text{chain}}^J \left(\frac{\partial \Delta_{JJ}}{\partial \rho}\right)\right]. \tag{6.226}$$

Substitution of this result in (6.224) gives

$$\left(\frac{p}{\rho kT}\right)_{\text{chain}} = -\frac{1}{2}\sum_{J=1}^{R}\rho_{\text{chain}}^{J}\left\{\sum_{\beta=1}^{m_J}\sum_{i}X_J^2\left[\frac{\rho_{\text{chain}}^{J}\Delta_{JJ}}{\rho}+\rho_{\text{chain}}^{J}\left(\frac{\partial\Delta_{JJ}}{\partial\rho}\right)_{\{N_\alpha\},T}\right]\right\}. \tag{6.227}$$

The inner sums over all bonding sites (β and i) for each monomer on the left side of this equation can be evaluated immediately. The two end spheres in chain J each have only one site, while the remaining $(m_J - 2)$ spheres each have two sites. Thus these two summations give a factor of $2(m_J - 2) + 2 = 2(m_J - 1)$ for each chain J. Thus (6.227) becomes

$$\left(\frac{p}{\rho kT}\right)_{\text{chain}} = -\sum\nolimits_{J=1}^{R}\rho_{\text{chain}}^{J}(m_J-1)X_J^2\Delta_{JJ}\left[\frac{\rho_{\text{chain}}^{J}}{\rho}+\rho_{\text{chain}}^{J}\left(\frac{\partial\ln\Delta_{JJ}}{\partial\rho}\right)_{\{N_\alpha\},T}\right]. \tag{6.228}$$

We now take the limit of complete bonding between chain-forming sites, i.e. the bonding strength tends to infinity and $X_J = 0$. We see from (6.200) that in this limit Δ_{JJ} tends to infinity, but from (6.225) the product $X_J^2\Delta_{JJ}$ remains finite, $X_J^2\Delta_{JJ} = (1 - X_J)/\rho_{\text{chain}}^{J} \to 1/\rho_{\text{chain}}^{J}$, so that (6.228) becomes

$$\left(\frac{p}{\rho kT}\right)_{\text{chain}} = -\sum\nolimits_{J=1}^{R}\frac{\rho_{\text{chain}}^{J}}{\rho}(m_J-1)\left[1+\rho\left(\frac{\partial\ln\Delta_{JJ}}{\partial\rho}\right)_{\{N_\alpha\},T}\right]. \tag{6.229}$$

Equation (6.229) is the final expression for the contribution to the compressibility factor due to chain formation.

Hard chains

If we consider the special case of chains formed from hard spheres at contact, and if we restrict the bonding potential to be infinitely attractive only at hard sphere contact, then from (B.67) we find $\Delta_{JJ} = g_{JJ}^{\text{hs}}(\sigma_{JJ})$, where $g_{JJ}^{\text{hs}}(\sigma_{JJ})$ is the contact value of the pair correlation function for the fluid of hard spheres of diameter σ_{JJ} in a hard sphere mixture, and σ_{JJ} is the diameter of the spheres making up chain J; thus (6.229) becomes

$$\left(\frac{p}{\rho kT}\right)_{\text{chain}} = -\sum_{J=1}^{R}\frac{\rho_{\text{chain}}^{J}}{\rho}(m_J-1)\left[1+\rho\left(\frac{\partial\ln g_{JJ}^{\text{hs}}(\sigma_{JJ})}{\partial\rho}\right)_{\{N_\alpha\},T}\right]. \tag{6.230}$$

For a hard sphere mixture the contact value of the pair correlation function is given to good accuracy by the equation of Mansoori *et al.*[253,254]

$$g_{JJ}^{hs}(\sigma_{JJ}) = \frac{1}{1-\zeta_3}+\frac{3\sigma_{JJ}\zeta_2}{2(1-\zeta_3)^2}+\frac{\sigma_{JJ}^2\zeta_2^2}{2(1-\zeta_3)^3}, \tag{6.231}$$

where

$$\zeta_\ell = \frac{\pi}{6}\sum_{J=1}^{R} \rho_{\text{chain}}^{J} m_J \sigma_{JJ}^{\ell}. \qquad (6.232)$$

The corresponding equation for the compressibility factor for the reference mixture of hard spheres is[253,254]

$$\left(\frac{p}{\rho kT}\right)_{\text{hs}} = \frac{6}{\pi\rho}\left[\frac{\zeta_0}{1-\zeta_3} + \frac{3\zeta_1\zeta_3}{(1-\zeta_3)^2} + \frac{3\zeta_2^3}{(1-\zeta_3)^3} - \frac{\zeta_3\zeta_2^3}{(1-\zeta_3)^3}\right] \qquad (6.233)$$

and the full equation of state for the mixture of hard sphere chains is given by the sum of (6.230) and (6.233). For a pure fluid of homonuclear hard sphere chains these equations simplify to give

$$\left(\frac{p}{\rho kT}\right) = \left(\frac{p}{\rho kT}\right)_{\text{hs}} + \left(\frac{p}{\rho kT}\right)_{\text{chain}}, \qquad (6.234)$$

where

$$\left(\frac{p}{\rho kT}\right)_{\text{hs}} = \frac{1+\eta+\eta^2-\eta^3}{(1-\eta)^3}, \qquad (6.235)$$

where $\eta = \pi\rho\sigma^3/6$ is the reduced density. This equation of state for hard spheres was first proposed by Carnahan and Starling.[255] The chain contribution for chains of length m is obtained from (6.235) as

$$\left(\frac{p}{\rho kT}\right)_{\text{chain}} = -\frac{m-1}{m}\left[1 + \rho\left(\frac{\partial \ln g^{hs}(\sigma)}{\partial\rho}\right)_{\{N_\alpha\},T}\right]. \qquad (6.236)$$

The residual part of the compressibility factor for chain formation is found by subtracting the ideal gas value for chain contribution, given for $\rho \to 0$ by

$$\left(\frac{p}{\rho kT}\right)_{\text{chain}}^{\text{id}} = -\left(\frac{m-1}{m}\right), \qquad (6.237)$$

so that the residual compressibility factor is

$$\left(\frac{p}{\rho kT}\right)_{\text{chain}}^{\text{res}} = \left(\frac{p}{\rho kT}\right)_{\text{chain}} - \left(\frac{p}{\rho kT}\right)_{\text{chain}}^{\text{id}} = -\frac{m-1}{m}\rho\left(\frac{\partial \ln g^{\text{hs}}(\sigma)}{\partial\rho}\right). \qquad (6.238)$$

The Helmholtz free energy is related to the compressibility factor, $Z = (p/\rho kT)$, by the identity

$$\left(\frac{\partial(A/NkT)}{\partial\rho}\right)_{\{N_\alpha\},T} = \frac{1}{\rho}Z, \qquad (6.239)$$

so that from (6.239) and (6.238) the residual Helmholtz energy is

$$\left(\frac{A}{NkT}\right)^{\text{res}}_{\text{chain}} = -\frac{m-1}{m}\int_0^{\rho}\frac{\partial \ln g^{\text{hs}}(\sigma)}{\partial \rho}d\rho = -\frac{m-1}{m}\ln g^{\text{hs}}(\sigma), \qquad (6.240)$$

where we note that at $\rho = 0$ the pair correlation function at contact is $g^{\text{hs}}(\sigma) = \exp[-u^{\text{hs}}(\sigma)/kT] = 1$. Since $N_{\text{chain}} = N/m$, this result can also be written

$$\left(\frac{A}{N_{\text{chain}}kT}\right)^{\text{res}}_{\text{chain}} = -(m-1)\ln g^{\text{hs}}(\sigma). \qquad (6.241)$$

For a mixture this becomes[252,256] (see (6.229))

$$\left(\frac{A}{N_{\text{chain}}kT}\right)^{\text{res}}_{\text{chain}} = -\sum_{J=1}^{R} x_J(m_J-1)\ln g^{\text{hs}}_{JJ}(\sigma_{JJ}). \qquad (6.242)$$

The contact value of the pair correlation function, $g(\sigma)$, is given by (5.24) as

$$g^{\text{hs}}(\sigma) = \frac{Z^{\text{hs}}-1}{4\eta}, \qquad (6.243)$$

where Z^{hs} is the compressibility factor for hard spheres, and is given to a good approximation by (6.235). From (6.243) together with (6.235) we find that

$$1 + \rho\frac{\partial \ln g^{\text{hs}}(\sigma)}{\partial \rho} = \frac{1+\eta-\eta^2/2}{(1-\eta)(1-\eta/2)}. \qquad (6.244)$$

Thus, from (6.234)–(6.236) and (6.244), the hard chain equation of state becomes[257]

$$\left(\frac{p}{\rho kT}\right) = \frac{1+\eta+\eta^2-\eta^3}{(1-\eta)^3} - \frac{m-1}{m}\frac{1+\eta-\eta^2/2}{(1-\eta)(1-\eta/2)}, \qquad (6.245)$$

where m is the number of spheres in a chain and η is the reduced density of the spheres.

An alternative approach to chain formation

An alternative derivation of the expression for the free energy of chain formation is possible, and is useful in giving insight into the approximations involved in the derivation, as well as in later extensions of the theory. It is possible to relate the Helmholtz energy of chain formation to the so-called y-function, or cavity function (see (3.98) of Vol.1).[218,258] We first consider a system of N spherical molecules. For a pair of molecules the two-body cavity function is given by (3.99) as

$$y(\mathbf{r}_1,\mathbf{r}_2) \equiv e^{u(\mathbf{r}_1\mathbf{r}_2)/kT}g(\mathbf{r}_1\mathbf{r}_2) = \frac{V^2\int d\mathbf{r}_3 d\mathbf{r}_4\ldots d\mathbf{r}_N \exp[-\mathcal{U}(\mathbf{r}^{N-2})/kT]}{Z_c(N)}, \qquad (6.246)$$

where $Z_c(N)$ is the N-body configuration integral of Eq. (6.2). Here $\mathcal{U}(\mathbf{r}^{N-2})$ is $\mathcal{U}(\mathbf{r}^N)$ exclusive of $u(\mathbf{r}_1\mathbf{r}_2)$, i.e. $\mathcal{U}(\mathbf{r}^{N-2}) \equiv \mathcal{U}(\mathbf{r}^{N-2};\mathbf{r}_1\mathbf{r}_2)$, the latter indicating that molecules $3, 4, \ldots$ interact with 1 and 2. We first consider the free energy required to form a dimer in the dense fluid, over and above the energy to form it in vacuum. Consider a system of N spherical monomers in which two monomers, 1 and 2, form a dimer; we will show that the free energy change is simply related to the two-body cavity function. The total number of molecules is now $N-1$, i.e. one dimer ($1D$) and $N-2$ monomers $(3, 4, \cdots, N)$. The configuration integral is

$$Z_c(N-2, 1D) = \int \mathrm{d}\mathbf{r}_{1D}\mathrm{d}\omega_{1D}\mathrm{d}\mathbf{r}_3\mathrm{d}\mathbf{r}_4 \ldots \mathrm{d}\mathbf{r}_N \exp[-\mathcal{U}(\mathbf{r}^{N-2,1D})/kT], \quad (6.247)$$

where $\mathbf{r}_{1D}$ and ω_{1D} are the position of the centre of mass, and orientation, respectively, of the dimer. If we first integrate over $\mathbf{r}_3, \mathbf{r}_4, \ldots, \mathbf{r}_\mathrm{N}$ in (6.247), by symmetry the result is independent of the position ($\mathbf{r}_{1D}$) and orientation (ω_{1D}) of the dimer. The integrals over $\mathbf{r}_{1D}$ and ω_{1D} thus will give a trivial factor of $V\Omega$, where $\Omega = 4\pi$. Thus from (6.247) and (6.246) we have

$$Z_\mathrm{c}(N-2, 1D) = \frac{\Omega}{V} Z_\mathrm{c}(N) y(\mathbf{r}_1, \mathbf{r}_2). \quad (6.248)$$

For a fluid of N monomers, the Helmholtz energy is given by Eq. (3.11) as $A = -kT \ln Q$, with Q given by (3.77)–(3.79) and (3.82), while for the ideal gas

$$A^\mathrm{id}(N) = -kT \ln Q^\mathrm{id} = -kT \ln \left(\frac{\Lambda_t^{-3N} V^N}{N!} \right), \quad (6.249)$$

so that, from (3.11) and (3.77)–(3.79), (3.82), the residual free energy for N monomers is

$$A^\mathrm{res}(N) = A(N) - A^\mathrm{id}(N) = -kT \ln \left(\frac{Z_c(N)}{V^N} \right). \quad (6.250)$$

For a fluid of 1 dimer and $N-2$ monomers the corresponding equations are (see (3.80))

$$A^\mathrm{id}(N-2, 1D) = -kT \ln \left(\frac{\Lambda_t^{-3(N-1)} \Lambda_r^{-1} V^{N-1}}{(N-2)!} \right) \quad (6.251)$$

$$A^\mathrm{res}(N-2, 1D) = -kT \ln \left(\frac{Z_c(N-2, 1D)}{V^{(N-1)}\Omega} \right). \quad (6.252)$$

From (6.250) and (6.252) the free energy change due to formation of one dimer in a sea of monomers is

$$A^\mathrm{res}_\mathrm{1dimer} = A^\mathrm{res}(N-2, 1D) - A^\mathrm{res}(N) = -kT \ln \left(\frac{Z_\mathrm{c}(N-2, 1D) V}{\Omega Z_\mathrm{c}(N)} \right)$$

or, using (6.248),

$$A^\mathrm{res}_\mathrm{1dimer} = -kT \ln y(\mathbf{r}_1, \mathbf{r}_2), \quad (6.253)$$

where $y(\mathbf{r}_1, \mathbf{r}_2)$ is the two-body cavity function for the monomer solvent. Equation (6.253) is an exact result for the residual free energy of formation of a single dimer at infinite dilution in a monomer solvent, and appears to have been first derived by Ben-Naim.[258] (The total free energy of formation, inclusive of the formation energy in vacuum, is $-kT\ln y(r_{12}) + u(r_{12})$. This is the potential of mean force.) For the special case of a dimer composed of two hard spheres at contact, in a sea of hard sphere monomers of diameter σ, this becomes

$$A^{\text{hs, res}}_{1\text{dimer}} = -kT\ln g^{\text{hs}}(\sigma). \tag{6.254}$$

In general, we are interested in the free energy change when all monomers form dimers. In order to make this additional step, we assume that the change in free energy on dimerization is proportional to the number of dimers formed, N_{dimer}. Thus, we assume that the influence of a nearby dimer on the dimerization process is the same as that of two monomers. With this approximation, the free energy of complete dimerization, $A^{\text{res}}_{\text{dimer}}$, is obtained from (6.253) and (6.254) as

$$A^{\text{res}}_{\text{dimer}} = -kTN_{\text{dimer}}\ln y(\mathbf{r}_1, \mathbf{r}_2) \tag{6.255}$$

and

$$A^{\text{hs, res}}_{\text{dimer}} = -kTN_{\text{dimer}}\ln g^{\text{hs}}(\sigma). \tag{6.256}$$

Equation (6.256) has been shown[259,260] to provide a highly accurate equation of state for hard dimer fluids. The above derivation is readily extended to the case of heteronuclear dimers.[260]

The derivation of (6.253) above, for the free energy change on forming a single dimer from two monomers, is easily generalized to the case of forming a chain of m monomers in a sea of monomer solvent. The derivation parallels that given above,[218,258,261] with the result

$$A^{\text{res}}_{1\text{chain}} = -kT\ln y(\mathbf{r}_1, \mathbf{r}_2, \mathbf{r}_3, \ldots, \mathbf{r}_m), \tag{6.257}$$

where $y(\mathbf{r}_1, \mathbf{r}_2, \mathbf{r}_3, \ldots, \mathbf{r}_m)$ is the m-body cavity function and $\mathbf{r}_i$ is the location of monomer i in the chain. For the case of a chain of hard spheres in contact (also called tangent hard sphere chains, or 'pearl necklace' chains) this becomes

$$A^{\text{hs,res}}_{1\text{chain}} = -kT\ln g^{\text{hs}}(\sigma_{12}, \sigma_{23}, \sigma_{34}, \ldots \sigma_{(m-1)m}), \tag{6.258}$$

where σ_{ij} is the bond length connecting the centres of neighbouring mers i and j. Equations (6.257) and (6.258) are exact results for the formation of a single m-mer at infinite dilution. To obtain the free energy change for polymerization of all of the monomers to form m-mers, we again assume that the change in Helmholtz energy is proportional to the number of chains formed, N_{chain}, so that (6.257) and (6.258) lead to

$$A^{\text{res}}_{\text{chain}} = -kTN_{\text{chain}}\ln y(\mathbf{r}_1, \mathbf{r}_2, \mathbf{r}_3, \ldots, \mathbf{r}_m) \tag{6.259}$$

$$A^{\text{hs,res}}_{\text{chain}} = -kTN_{\text{chain}}\ln g^{\text{hs}}(\sigma_{12}, \sigma_{23}, \sigma_{34}, \ldots, \sigma_{(m-1)m}). \tag{6.260}$$

The m-body correlation functions in these equations are, in general, unknown, so that we must resort to some further approximation to evaluate them. The so-called *linear approximation*[226,261,262] is usually used for this purpose, which approximates the m-body correlation function as a product of two-body ones for adjacent mers, i.e.

$$y(\mathbf{r}_1, \mathbf{r}_2, \mathbf{r}_3, \ldots \mathbf{r}_m) = y(r_{12})y(r_{23})y(r_{34})\ldots y(r_{(m-1)m}) \tag{6.261}$$

and

$$g^{\mathrm{hs}}(\sigma_{12}, \sigma_{23}, \sigma_{34}, \ldots \sigma_{(m-1)m}) = g^{\mathrm{hs}}(\sigma_{12})g^{hs}(\sigma_{23})g^{hs}(\sigma_{34})\ldots g^{\mathrm{hs}}(\sigma_{(m-1)m}). \tag{6.262}$$

This approximation has been shown to give good agreement with molecular simulation results for chain molecules, particularly for nearly linear arrangements of the mers in the chain. We note that for a chain of m mers there are $(m-1)$ sphere–sphere contacts. For a branched chain of m mers there are also $(m-1)$ sphere–sphere contacts; for ethane treated as bonded spheres representing the C and H atoms, for example, $m = 8$ and there are 7 sphere–sphere contacts. Thus, for homonuclear hard sphere chains (6.262) becomes

$$g^{\mathrm{hs}}(\sigma_{12}, \sigma_{23}, \sigma_{34}, \ldots \sigma_{(m-1)m}) = g^{\mathrm{hs}}(\sigma)^{m-1}. \tag{6.263}$$

Using this result in (6.260) gives

$$\left(\frac{A}{N_{\mathrm{chain}}kT}\right)^{\mathrm{res}}_{\mathrm{chain}} = -(m-1)\ln g^{\mathrm{hs}}(\sigma), \tag{6.264}$$

which is the Wertheim TPT1 result for hard chains given in Eq. (6.241).

Comparison with molecular simulation results

It should be noted that although Eqs. (6.230)–(6.245) are for hard chains, Eq. (6.229) holds for any reference potential, u_0. The theory is subject to the approximations noted earlier. In particular, any dependence of the thermodynamic properties on the angles between the vectors from the molecular centre to the various association sites is neglected. When the bonding site angles are small enough for the spherical cores to prevent simultaneous bonding at the sites, the theory fails. Thus, the steric self-hindrance of the chain structures is neglected. This approximation in the theory leads to the same thermodynamic properties for different isomers of a given molecule; for example, n-pentane and neo-pentane will have identical equations of state. The theory also predicts that the thermodynamic properties of mixtures of chains in which all the hard spheres have the same diameter depend only on the average chain length.[250] This is called the *principle of congruence*, and was first applied to alkane mixtures by Brønsted and Koefoed[263,264] in 1946. A comparison of the theory with Monte Carlo simulations for flexible hard sphere chains[252] is shown in Fig. 6.28. For $m = 2$ excellent agreement is obtained at all densities. Agreement remains good for $m = 4$, although the theory overestimates the compressibility factor slightly in

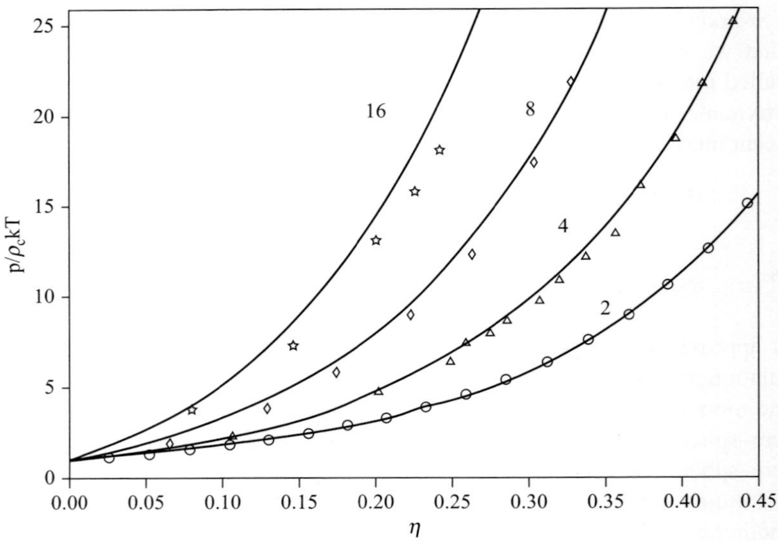

FIG. 6.28 The compressibility factor for 'pearl necklace' chains of hard spheres of $m = 2$, 4, 8, and 16 monomers. Here ρ_c is the number density of chain molecules. Lines are Eq. (6.245), points are Monte Carlo simulation results from Tildesley and Streett[248] for $m = 2$ and from Dickman and Hall[251] for other values of m. Reprinted with permission from ref. 252. Copyright 1988 Taylor and Francis Ltd.

the intermediate density range. Deviations become progressively larger for longer chains, presumably because the theory does not account for steric self-hindrance, which becomes important for longer chains. Other calculations for homonuclear tangent hard sphere chains have been reported,[265,266] for chain lengths up to 201 monomers,[267] and with similar results. The theory can also be readily extended to heteronuclear tangent hard sphere chains.[260,268–270]

Calculations for chains composed of spherical particles that possess attractive, as well as repulsive, interactions have been made, including square well[271–273] and Lennard-Jones[274–278] chains. The only added complication in such calculations is the need for accurate values of the pair correlation function for soft and sticky spheres. Johnson *et al.*[275] fitted simulation data for $g_0(r)$ for the LJ fluid as a function of density and temperature to an empirical function, and reported calculations for LJ chains up to 100 monomers in length. Some of their results are shown in Fig. 6.29 for 100 monomer chains. As for the hard chains, the theory gives pressures that are somewhat too high. The internal energies are given accurately at moderate and high densities. The theory does poorly at low densities, predicting energies and pressures that are too high. In this low density regime many of the chains coil up, and the main contribution is from intramolecular, rather than intermolecular, interactions. These intramolecular interactions are not accounted for in the present form of the theory (extension to include such interactions are discussed in the next section). Excluding the

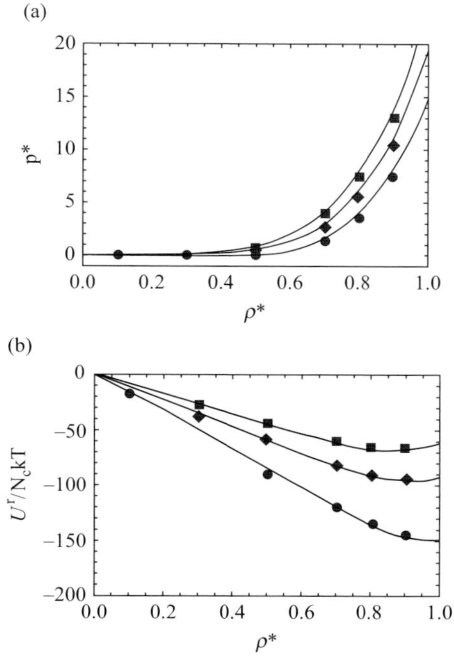

FIG. 6.29 (a) Reduced pressure, $p^* = p\sigma^3/\varepsilon$ and (b) reduced residual internal energy for Lennard-Jones (LJ) tangent 100-mers as a function of reduced monomer density, $\rho^* = \rho\sigma^3$. Points are molecular dynamics simulation results, lines are theory. Lines and symbols are for $T^* = kT/\varepsilon = 5.0$, 4.0, and 3.0 from top to bottom, respectively. Here ε and σ are the LJ parameters for the spheres; the bonding distance between centres of neighbouring spheres is taken to be σ. (From ref. 275.)

very low density region, the average absolute deviations between simulation and experiment are 17.2% for the pressure and 2.0% for the residual internal energy.

Associating chains

The above expressions can be readily extended to cases where the chains have additional association sites.[252] The expression for the compressibility factor is then of the form

$$\left(\frac{p}{\rho kT}\right) = \left(\frac{p}{\rho kT}\right)_0 + \left(\frac{p}{\rho kT}\right)_{\text{chain}} + \left(\frac{p}{\rho kT}\right)_{\text{bond}}, \tag{6.265}$$

where the first term on the right-hand side is the contribution for the reference system of non-bonded monomers (e.g. hard spheres or LJ spheres), the second term is the contribution for chain formation, given by (6.229), and the third term is the additional contribution due to bonding between association sites on the spheres making up the chain. This last term is given by (6.203), which can be rewritten as a sum over chains J (cf. (6.224)),

$$\left(\frac{p}{\rho kT}\right)_{\text{bond}} = \sum_{J=1}^{R} \rho_{\text{chain}}^{J} \left[\sum_{i} \left(\frac{\partial X_J^i}{\partial \rho}\right)_{\{N_\alpha\},T} \left(\frac{1}{X_J^i} - \frac{1}{2}\right) \right], \tag{6.266}$$

where the sum over J is over all chains, and that over i is for all bonding sites in chain J. The fraction of site i in chain J that is not bonded is given by (6.200) as

$$X_J^i = \left[1 + \sum_{J'=1}^{R} \sum_{j} \rho_{\text{chain}}^{J'} X_{J'}^{j} \Delta_{JJ'}^{ij} \right]^{-1}, \tag{6.267}$$

where the sum over j is over all sites on chain J', and $\Delta_{JJ'}^{ij}$ is given by (6.201) as

$$\Delta_{JJ'}^{ij} = \int \left\langle g_{JJ',0}(12) f_{JJ'}^{ij}(12) \right\rangle_{\omega_1\omega_2} d\mathbf{r}_{12}, \tag{6.268}$$

where $g_{JJ',0}$ (12) is the sphere–sphere correlation function between the sphere on chain J and the sphere on chain J', which is approximated as the radial distribution function for a mixture of non-bonded spheres. This approximation neglects effects due to steric self-hindrance between spheres in the chain. Nevertheless, this approximation is expected to be a good one at higher densities. As before, the theory does not account for any dependence on the position of the site or on the angles between different sites on a given molecule. Bonding at different sites on a given molecule are treated as independent events.

For a pure chain fluid equations (6.266)–(6.268) simplify to

$$\left(\frac{p}{\rho kT}\right)_{\text{bond}} = \rho_{\text{chain}} \sum\nolimits_i \left(\frac{\partial X^i}{\partial \rho}\right)_{N,T} \left(\frac{1}{X^i} - \frac{1}{2}\right) \tag{6.269}$$

$$X^i = \left[1 + \rho_{\text{chain}} \sum\nolimits_j X^j \Delta^{ij} \right]^{-1} \tag{6.270}$$

$$\Delta^{ij} = \int \langle g_0(12) f^{ij}(12) \rangle_{\omega_1\omega_2} d\mathbf{r}_{12}, \tag{6.271}$$

where the summations over i and j in (6.269) and (6.270) are over all sites in a given chain molecule and X^i is the fraction of chain molecules not bonded at site i. In general, the integral in (6.268) and (6.271) must be evaluated numerically, although approximate analytical forms have been proposed.[252]

Several studies of the influence of association sites on the behaviour of fluids of chain molecules have been reported.[252,278–280] Chapman *et al.*[252] carried out calculations of vapour-liquid coexistence curves for fluids of chain molecules with one or two association sites on each chain. The association sites were modeled using square well interactions, Eq. (6.220). They used a reference system of hard chains, and extended (6.265) to include a mean field term, since otherwise a fluid of non-associating chains does not exhibit gas-liquid equilibria,

$$\left(\frac{p}{\rho kT}\right) = \left(\frac{p}{\rho kT}\right)_0 + \left(\frac{p}{\rho kT}\right)_{mf} + \left(\frac{p}{\rho kT}\right)_{chain} + \left(\frac{p}{\rho kT}\right)_{bond} \tag{6.272}$$

where

$$\left(\frac{p}{\rho kT}\right)_{\mathrm{mf}} = -\frac{\varepsilon_{mf}\eta}{kT}, \tag{6.273}$$

where $\varepsilon_{\mathrm{mf}}$ is a measure of the strength of the attractive dispersion and other forces. This parameter should not be confused with the Lennard-Jones ε for fluid–fluid interactions; rather it represents the integral over the attractive (dispersion and other) potential for all separations (see (4.157) of Gray and Gubbins[281]). Vapour–liquid coexistence curves are shown in Fig. 6.30 for tangent hard sphere chains

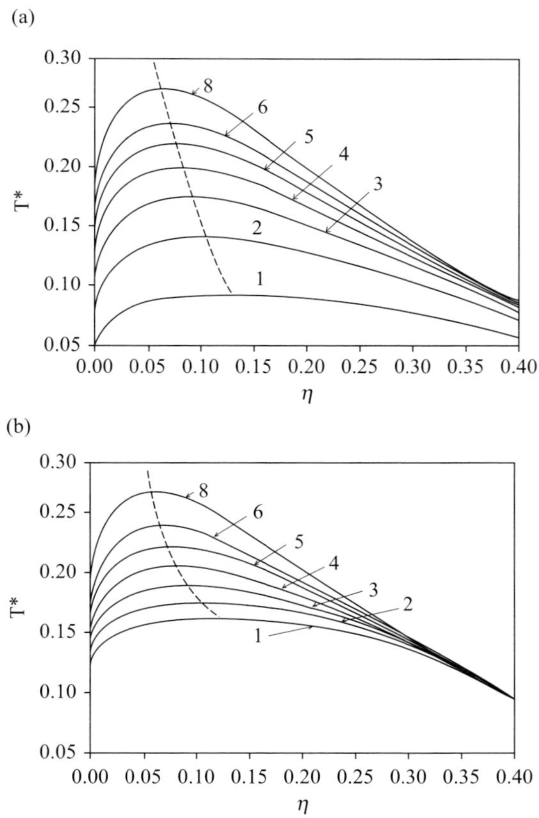

FIG. 6.30 Coexisting vapour and liquid densities, η, as a function of reduced temperature, $T^* = kT/\varepsilon$, for a fluid of chain molecues for different values of m, the number of hard spheres making up the chain: (a) no association sites; (b) two association sites on each chain molecule. The strength of the association sites is $\varepsilon_{\mathrm{HB}}/\varepsilon_{\mathrm{MF}} = 1.0$. The locus of the critical point for different values of m is shown by the dashed curve. Reprinted with permission from ref. 252. Copyright 1988 Taylor and Francis Ltd.

without association sites, Fig. 6.30(a), and with two association sites per chain of strength $\varepsilon_{\mathrm{bond}}/\varepsilon_{\mathrm{mf}} = 1.0$, Fig. 6.30(b); chain lengths up to $m = 8$ are shown. The association site strength chosen is typical of values for small H-bonding molecules such as acetic acid.[235] The effect of increasing chain length is to increase the critical temperature and decrease the critical density, as is seen experimentally for homologous series such as the *n*-alkanes. When association sites are added to the molecules the main effect is to further stabilize the liquid phase, so that the coexistence region is enlarged and extends to higher temperatures. The critical temperature is increased. This effect is greatest for the shortest chain molecules, $m = 1$, and is reduced as the chains become longer. For the longest chain studied, $m = 8$, the coexistence curve is almost the same as that for non-associating chains. In these longer chains the chain size becomes the dominant effect. The influence of association sites on these longer chains would be greater for stronger association sites, or for more association sites per chain. The mole fraction of non-bonded chains in the coexisting vapour and liquid phases are shown in Fig. 6.31. As expected, the degree of bonding decreases with increasing chain length, and the curves are shifted to the right for the largest values of m. The presence of two or more sites per chain makes it possible for a series of chain molecules to bond together, a chain of chains. This is not possible when there is only one association site per chain, so that the behaviour of one-site chains is fundamentally different from that of chains with more than one site.[252] Other studies have examined the effect of varying the strength of the association sites for chains of 4-mers,[279] and the effect of association sites on heteronuclear chains.[278]

Jog and Chapman[280] adopted a different approach to obtain an equation of state for fluids of dipolar chain molecules. Rather than applying Wertheim's theory to

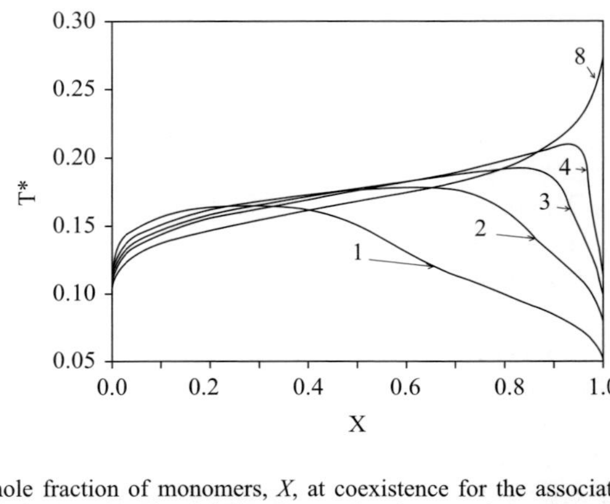

FIG. 6.31 The mole fraction of monomers, X, at coexistence for the associating chains with two bonding sites of Fig. 6.30(b). The right-hand part of each curve shows the monomer mole fraction for the vapour, while the left-hand part of the curve shows the corresponding values for the coexisting liquid phase. Reprinted with permission from ref. 252. Copyright 1988 Tayloy and Francis Ltd.

a fluid of hard spheres in the limit when all spheres are bonded at the contact sites, as above, they applied Wertheim theory to a mixture of nonpolar and dipolar hard spheres to produce a fluid of chains with alternate hard spheres and dipolar spheres. The contribution to chain formation, previously given by (6.236), now involves two terms, one for hard sphere interactions and one for dipolar hard sphere interactions. The latter involves the angular pair correlation function for dipolar hard spheres. This was approximated by the hard sphere pair correlation function.

Overlapping spheres

The above equations have been derived for chains of tangent spheres, i.e. the centres of the monomers making up the chain are separated by a distance σ (hard sphere or LJ diameter). For real molecules there is considerable overlap of the spheres forming the chain as a result of covalent bonding. It would clearly be desirable to extend the theoretical equations to describe chains of overlapping spheres. An approximate method for achieving this has been suggested,[282,283] based on equating the expressions for the second virial coefficient from the Wertheim theory to the exact expression for hard convex bodies, (6.127). The second virial coefficient for a fluid of hard chains composed of tangent spheres is, from the Wertheim theory,

$$B_2 = \frac{1}{2} v_m (3m + 5). \tag{6.274}$$

Comparison of this expression with (6.127) gives an expression for the effective number of tangent hard spheres, m_e, corresponding to a chain of overlapping spheres,

$$m_e = \frac{2}{3}\gamma - 1, \tag{6.275}$$

where $\gamma = r_m s_m / v_m$ is the molecular shape factor. Although the molecular surface area, s_m, and volume, v_m, are defined for nonconvex bodies, the mean radius of curvature, r_m, is not—it is rigorously defined only for convex bodies (§ 6.12). However, Boublik and Nezbeda[284] have shown that this quantity can be effectively calculated by replacing the nonconvex body by an equivalent convex one, which is obtained by figuratively wrapping the molecule in a taut membrane. This approach has been described in § 6.12.2, and is shown to give good results (see Table 6.9). This enables the shape factor, and hence the effective number of tangent spheres, to be calculated using (6.275); the resulting value of m_e is then used in the above equation of state, (6.245), to obtain a description of a fluid of chains of overlapping spheres. The predictions of this method have been tested against molecular simulation results[285] for a variety of model fluids in which the molecules are composed of between 2 and 6 overlapping hard spheres.[282] The agreement is generally excellent, as seen in Fig. 6.32. The accuracy of the method is not known for longer chains of overlapping spheres.

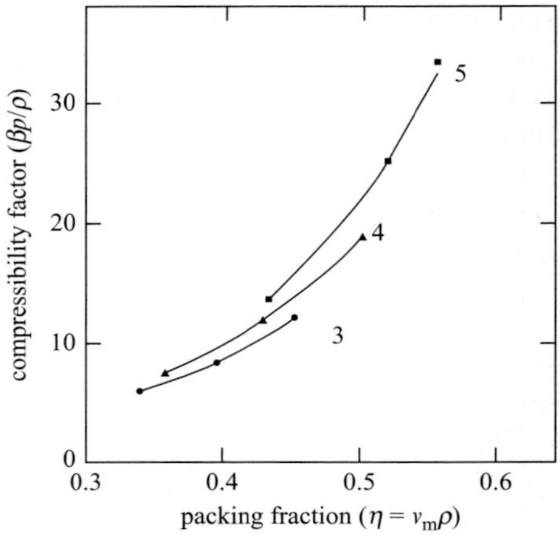

FIG. 6.32 The compressibility factor vs reduced density for chains of overlapping hard spheres from molecular dynamics simulation[285] (points) and from (6.245) (lines), with m replaced by m_e as given by (6.275). Results are shown for linear homonuclear overlapping hard sphere chains of 3, 4, and 5 spheres, each with bond lengths of $\ell/\sigma = 0.3979$. Reprinted with permission from ref. 282. Copyright 1990 American Chemical Society.

Hard dimer chains

A further modification to the Wertheim theory of chains described above is to replace the hard sphere reference system by one of hard dimers.[286–291] One first bonds two hard spheres to form hard dimers, using the Wertheim theory to obtain the compressibility factor. The theory is then used to obtain the change in the compressibility factor when these dimers are further bonded to form longer chains. This results in an expression for Z that involves the contact value of the hard dimer site–site pair correlation function, $g_{HD}(\sigma)$, in addition to the hard sphere function, $g_{HS}(\sigma)$. The resulting expression is known as the statistical associating fluid theory-dimer (SAFT-D) theory and is in somewhat better agreement with simulation results for longer hard chain fluids.

6.13.4 *Extensions to Wertheim's theory*

The theory described in §§ 6.13.2 and 6.13.3 is the first-order thermodynamic perturbation theory form of Wertheim's theory, or TPT1. While TPT1 is quite successful, because of the approximations made, the theory suffers from a number of limitations, including:

(a) In its original form the theory does not account for either dispersion or long-range electrostatic interactions.
(b) While straight chain and tree-like structures are accounted for, ring-like structures are not considered.

(c) When multiple sites are present on a single molecule, the relative positions of the sites has no effect on the properties. For example, for a triatomic molecule there is no dependence on the bond angle. For chain molecules, various isomers (e.g. straight chain and branched chain) will have the same properties.
(d) *Intra*molecular interactions are not accounted for.

These limitations can be overcome to a large extent by suitable extensions, and some of these are briefly described below.

Including dispersion and electrostatic interactions

Dispersion interactions can be included by choosing the reference potential in Eq. (6.197) to include dispersion. Alternatively, the effects of dispersion can be incorporated through an additional perturbation term in the free energy or pressure equations. This has already been done in a crude fashion in (6.272) and (6.273) by the inclusion of a mean field term. Jackson and coworkers[292–294] have included a variable range attractive potential $-\varepsilon\phi(r;\lambda)$ as a perturbation potential; here ϕ is a function representing some attractive potential, such as a square well, Sutherland, or Yukawa model, and λ is a parameter that determines the range of the potential. Comparisons with molecular simulation results show good agreement for potentials of various ranges. The influence of multipolar interactions can be included in a similar fashion,[295] by using the u-expansion perturbation theory (§ 4.5), and particularly the Padé approximant form of Eq. (4.47). This approach gives good results when compared against molecular simulations,[296] and improved agreement with experimental data for polar fluids such as water, alkanols, and sulphur dioxide.[295–297]

Ring structures

Wertheim's TPT1 theory can account for straight and branched chain structure formation, but not for ring formation from mers. This is because the graphs corresponding to rings are omitted in the theory. However, an equation that accounts for ring formation is needed to describe fluids of ring-like molecules (cyclopentane, cyclohexane, benzene, benzene derivatives, etc), and for molecules that form ring-like structures due to H-bonding or other association forces. Examples of the latter include acetic acid, which forms cyclic dimers in the gas phase (Eq. (6.194)), and hydrogen fluoride, which forms ring aggregates having 4 to 8 monomers in the gas phase.[298] Several authors[231,261,299–305] have extended the TPT1 theory to account for ring formation. The simplest approach to including ring formation is via the relations between Helmholtz energy and the cavity function, Eqs. (6.259) and (6.260). For ring formation from homonuclear segments there will be m sphere–sphere contacts in the ring, rather than $m-1$ as in the chain, so that Eqs. (6.260)–(6.263) are replaced by

$$A_{\text{ring}}^{\text{hs,res}} = -kTN_{\text{ring}} \ln g^{\text{hs}}(\sigma_{12}, \sigma_{23}, \sigma_{34}, \ldots \sigma_{m1}) \tag{6.276}$$

$$y(\mathbf{r}_1, \mathbf{r}_2, \mathbf{r}_3, \ldots \mathbf{r}_m) = y(r_{12})y(r_{23})y(r_{34}) \ldots y(r_{m1}) \tag{6.277}$$

$$g^{\text{hs}}(\sigma_{12}, \sigma_{23}, \sigma_{34}, \ldots \sigma_{m1}) = g^{\text{hs}}(\sigma_{12})g^{\text{hs}}(\sigma_{23})g^{\text{hs}}(\sigma_{34}) \ldots g^{\text{hs}}(\sigma_{m1}) \tag{6.278}$$

and

$$g^{\text{hs}}(\sigma_{12}, \sigma_{23}, \sigma_{34}, \ldots \sigma_{m1}) = g^{\text{hs}}(\sigma)^{\text{m}}. \tag{6.279}$$

From (6.276) and (6.279), the contribution to the Helmholtz energy due to ring formation is

$$\left(\frac{A}{N_{\text{ring}}kT}\right)_{\text{ring}}^{\text{res}} = -m \ln g^{\text{hs}}(\sigma). \tag{6.280}$$

The corresponding contribution to the compressibility factor due to ring formation can be obtained from (6.239) as (cf. (6.238))

$$\left(\frac{p}{\rho_{\text{ring}}kT}\right)_{\text{ring}}^{\text{res}} = \left(\frac{p}{\rho_{\text{ring}}kT}\right)_{\text{ring}} - \left(\frac{p}{\rho_{\text{ring}}kT}\right)_{\text{ring}}^{\text{id}} = -m\rho \left(\frac{\partial \ln g^{\text{hs}}(\sigma)}{\partial \rho}\right), \tag{6.281}$$

where $\rho_{\text{ring}} = N_{\text{ring}}/V$. Tests of Eqs. (6.280) and (6.281) against molecular simulation results for ring molecules having 3, 4, and 5 mers show generally good agreement, with some departures at the highest densities.[261] An example of such a comparison for hard cyclic pentamers is shown in Fig. 6.33.

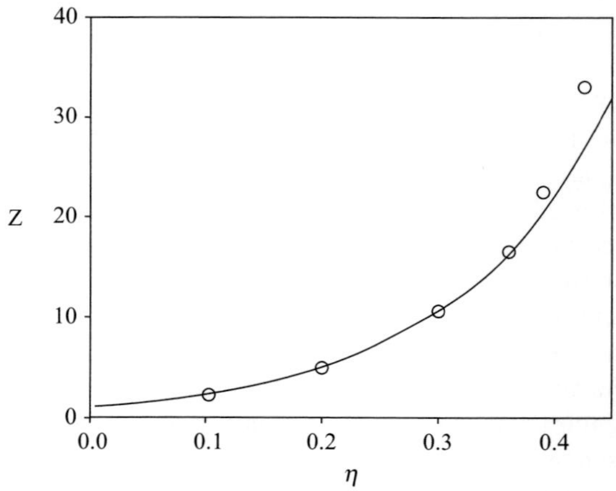

FIG. 6.33 The compressibility factor $Z = p/\rho kT$ vs reduced density, $\eta = \pi m\rho\sigma^3/6$, for a fluid composed of planar cyclic pentamers made up of tangent hard spheres of diameter σ, from the theory for rings (line) as given by (6.281), and from molecular simulation (points). (From ref. 261.)

Second-order Wertheim theory

While the first-order theory (TPT1) described above is the version most widely used, in the case where molecules possess more than one attraction site it is assumed that the different sites do not influence each other. In the case of chain molecules, for example, the thermodynamic properties will be independent of the bond angles, ω. For trimers we might expect this approximation to be good when ω is large, close to 180°, but it will clearly lead to errors when the bond angle is small. We note that very small bond angles are excluded due to steric hindrance; for hard spheres trimers can only form if $\omega > \pi/3$.

In order to account for the relative orientation of bonds in such cases it is necessary to use the second-order thermodynamic perturbation theory, TPT2, in which higher order graphs with two or more incident bonds are included.[250] In TPT2 there is an additional contribution to the Helmholtz energy that involves the 3-body correlation function for the reference system. For the case of a pure fluid composed of spherical molecules with two bonding sites, a and b, with only a–b bonding permitted (no a–a or b–b bonding), this additional contribution contains the integral[250]

$$\rho_0(\rho_0 + \rho_a)(\rho_0 + \rho_b) \int f^{\text{ab}}(12) f^{\text{ab}}(23)[g_0(123) - g_0(12)g_0(23)]\text{d}\mathbf{r}_1\text{d}\mathbf{r}_2\text{d}\mathbf{r}_3. \tag{6.282}$$

The integrand in (6.282) is sensitive to any approximation used for the 3-body correlation function. The integral vanishes in the linear approximation, (6.261), and the superposition approximation of Eq. (3.246) gives poor results.[306,307]

An alternative, but similar, expression for the second-order theory can be obtained using the approach of Zhou and Stell,[218] in which the free energy is related to the cavity function (see § 6.13.3). Consider a fluid of trimers, formed from spherical monomer molecules. For this case, (6.259) becomes

$$\frac{A^{\text{res}}_{\text{chain}}}{N_{\text{chain}}kT} = -\ln y(\mathbf{r}_1, \mathbf{r}_2, \mathbf{r}_3). \tag{6.283}$$

For tangent hard sphere trimers this becomes (see (6.260))

$$\frac{A^{\text{hs,res}}_{\text{chain}}}{N_{\text{chain}}kT} = -\ln g^{\text{hs}}(\sigma, \sigma, r_{13}). \tag{6.284}$$

The three-body correlation function can also be expressed in terms of the bond angle, ω, for the trimer, as $g_{\text{hs}}(\sigma, \sigma, 2\sigma \sin(\omega/2))$. If we now make the linear approximation, (6.261) and (6.262), these equations take the TPT1 form. However, if we avoid making this or any other approximation for the 3-body function, Eqs. (6.283) and (6.284) are second-order expressions. Although they are not identical to the TPT2 expressions of Wertheim, they are similar and are found to give similar results in applications. The chain contribution to the compressibility factor that corresponds to (6.284) is readily obtained using (6.239) as

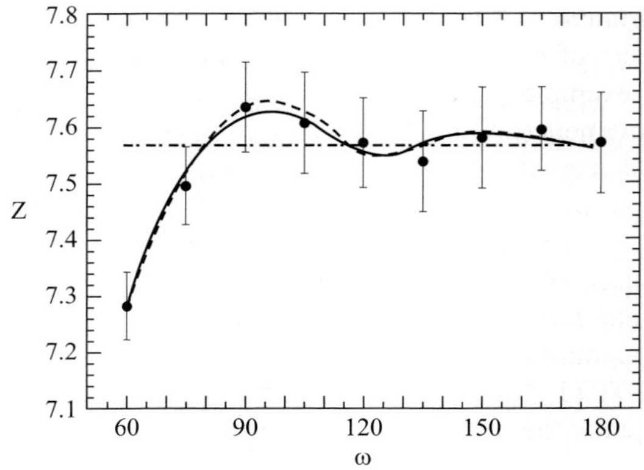

FIG. 6.34 The compressibility factor Z as a function of bond angle ω for a fluid of tangent hard sphere triatomic molecules, at reduced pressure $p^* = pv_\mathrm{m}/kT = 2.25$.
Here $v_\mathrm{m} = \pi\sigma^3/6$ is the volume of one sphere. The dash-dot (horizontal) line is Wertheim's first-order theory, TPT1, the dashed line is the second-order theory TPT2, and the solid line is the Zhou–Stell theory of Eq. (6.285). Points are molecular simulation results. Reprinted with permission from ref. 307. Copyright 1993 Taylor and Francis Ltd.

$$\left(\frac{p}{\rho_\mathrm{chain}kT}\right)^{\mathrm{hs,res}}_{\mathrm{chain}} = -\rho\frac{\partial \ln g^{\mathrm{hs}}(\sigma,\sigma,2\sigma\sin(\omega/2))}{\partial\rho}. \tag{6.285}$$

The TPT2 theory has been tested against molecular simulation data for tangent sphere hard chains,[250,307] for chains of up to 201 mers.[307] The TPT2 theory is in significantly better agreement with simulation than TPT1. The TPT2 theory, and also the Zhou–Stell theory of (6.284) and (6.285), have also been used to study the bond angle dependence of the thermodynamic properties for tangent hard sphere trimers and tetramers,[307] with 3-body hard sphere correlation functions taken from molecular simulation results.[306] A typical result for hard trimers is shown in Fig. 6.34. The two versions of the theory are in generally good agreement with each other and with the simulation results. It is seen that the contribution of the second-order term is rather small, except at small bond angles, particularly values below about 80°.

Phan *et al.*,[308] using a somewhat different version of the TPT2 theory, have studied star-like chain molecules. The TPT2 theory has also been used to study the effect of chain branching on phase equilibria and critical temperature,[309] with good agreement with experiment, and has been extended to mixtures containing linear heteronuclear chain molecules.[310]

Intramolecular association

The TPT1 theory accounts for intermolecular, but not for *intra*molecular association. Intramolecular hydrogen bonds are particularly important in many biological molecules, such as polypeptides, proteins, nucleic acids, and carbohydrates, where the configuration of the chains is often determined by such bonds. Many synthetic polymers of commercial importance also experience such intramolecular bonding; examples include carboxylic acid-based polymers, Nylon-6, and telechelic polymers (low-molecular-weight molecules containing associating groups at each end of the molecule, used in the synthesis of block and graft copolymers). In such cases, the physical properties are largely determined by the competition between intermolecular and intramolecular forces. At high density the intermolecular forces are strong and may dominate the behaviour, so that intramolecular forces have relatively less effect. However, at low density intramolecular bonding is usually dominant and has a large effect on the properties. Thus the TPT1 theory cannot account for the low density behaviour in such systems, such as the coil-up of chain molecules, p-V-T behaviour, and phase equilibria.

The theory has been extended to include intramolecular bonding by several authors,[231,299–303,311,312] using essentially equivalent approaches. For a chain molecule of tangent spheres, with association sites placed on the two spheres at the ends of the chain, intramolecular association leads to an additional term in the Helmholtz energy of the form[301,302]

$$A_{\text{assoc}} = N_{\text{chain}} kTX, \tag{6.286}$$

where

$$X = \frac{1}{1 + \frac{1}{V_{\text{mol}}} \int f_{\text{intra}}(1m) g_{\text{ref}}(1m) d(1m)}. \tag{6.287}$$

In (6.287) V_{mol} is the volume accessible to segment m of the chain with segment 1 fixed, $f_{\text{intra}} = \exp(-u_{intra}/kT) - 1$ is the Mayer function for the intramolecular association interaction, u_{intra}, between segments 1 and m, $g_{\text{ref}}(1m)$ is the intramolecular pair correlation function between segments 1 and m for the reference system of chain molecules in the absence of any association, and $d(1m)$ denotes an unweighted average over all bonding orientations and an integration over all bonding separations between segments 1 and m. A test of this extension to the theory, and a comparison of the relative contributions from inter- and intramolecular interactions, is shown in Fig. 6.35 for a fluid of flexible tangent hard sphere tetramers ($m = 4$), with association sites placed on the end segments, 1 and 4. The association sites are placed on the edge of the hard sphere segment, at right angles to the bond formed by the terminal segment and its neighbour, and can bond either to sites on other molecules or to the other site on the same molecule. The interactions are modelled as conical square well sites, Eq. (6.221). At the rather low density considered, with molecules on average rather far apart, higher values of the association energy make it easier for the molecules to form intramolecular bonds rather than intermolecular ones, so that intramolecular bonding dominates the thermodynamic behaviour. Since a fluid of ring molecules has a higher compressibility factor than a fluid of linear chains of

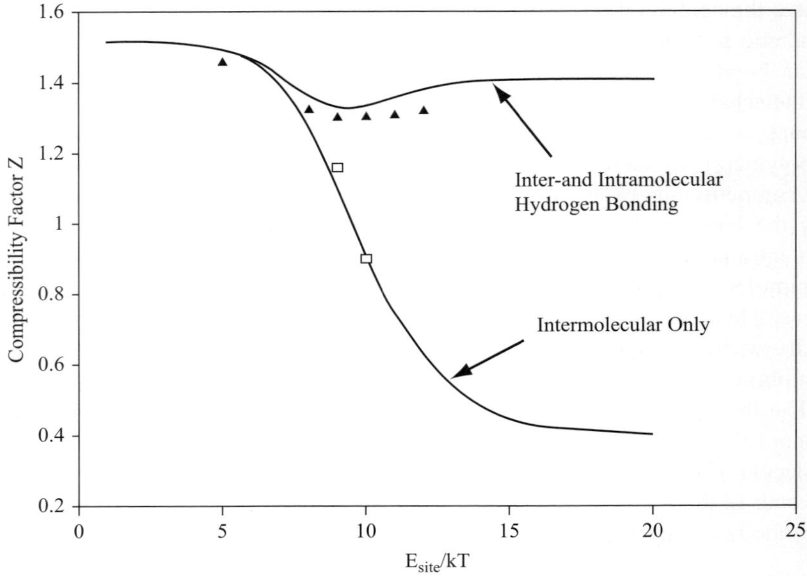

FIG. 6.35 Compressibility factor versus association energy for hard freely jointed tetramer chain molecules with association sites placed on the end segments.
Molecules can form both inter- and intramolecular association in the case of the upper set of triangle points (simulation) and line (theory), whereas for the lower line (theory) and square points (simulation) only intermolecular association can occur. The packing fraction is $\eta = 0.05$, where $\eta = \pi\rho\sigma^3/6$, where ρ is the density of segments. E_{site} is energy of intramolecular sites. Reprinted with permission from ref. 302. Copyright 1996 Elsevier.

the same length, a minimum is observed that results from a balance between inter- and intramolecular association.

Solid phases

The arguments used in deriving the Wertheim theory earlier in this section do not make any assumption as to the phase considered, and so equations for the free energy may be applied to solid, as well as fluid, phases. For flexible chains of tangent hard spheres, for example, (6.241) can be used for both fluid and solid phases to obtain fluid–solid phase equilibria, as well as the equation of state (compressibility factor versus reduced density) for the solid phase. In order to carry out such calculations one needs a hard sphere equation of state for the solid phase to calculate the residual free energy and compressibility factor, and the hard sphere pair correlation function at contact, $g^{hs}(\sigma)$, for the solid. Such calculations have been reported for tangent hard sphere chains,[313] using an equation of state for the hard sphere solid fitted to molecular dynamics data.[314] Agreement with molecular simulation results for freely jointed hard chains is generally good for both the equation of state and fluid–solid equilibria, although some deviations are noticeable at higher densities, particularly for the longer chains. Similar calculations have been reported for two-dimensional tangent hard disk chains,[315] and for chains of Lennard-Jones spheres.[316]

Liquid crystals

Prolate molecules of sufficient length are known to form liquid crystal phases.[317] However, in many cases electrostatic and dispersion interactions also play an important role in the formation of such phases. Studies have been made[318,319] of a fluid of hard spherocylindrical molecules that can dimerize through an association site placed near the end of the molecules. Such bonding is found to stabilize the nematic phase. The isotropic–nematic phase diagram exhibits a non-monotonic variation of the density of the coexisting nematic phase, which produces a novel nematic–isotropic–nematic re-entrant behaviour; such behaviour has also been observed experimentally in some liquid crystal systems.[320] The theory has also been used to study the more ordered smectic phases in mixtures of rod-like molecules, where re-entrant smectic-A behaviour is observed.[321]

6.13.5 *Application to real fluids: SAFT equation*

The success of Wertheim's theory in accounting for the contribution to thermodynamic properties of both chain (and ring) formation and association suggests that it could form the basis of a flexible equation of state that is applicable to a wide range of types of molecules. Such an approach was proposed by Chapman *et al.*,[228,256,322] and given the name *statistical associating fluid theory*. In the SAFT equation the residual Helmholtz energy is considered to be a sum of three terms,

$$A^{\text{res}} = A^{\text{res}}_{\text{seg}} + A^{\text{res}}_{\text{chain}} + A^{\text{res}}_{\text{assoc}}, \tag{6.288}$$

where the subscripts seg, chain, and assoc indicate the contributions from segment–segment interactions due to the monomers or atoms making up the molecules, chain formation (or formation of the molecular assembly from the individual segments), and molecular association. If we choose to treat the segments as hard spheres, for example, the chain term in (6.288) will be given by (6.241), and the association term by (6.199)–(6.201), with the reference system pair correlation function in (6.201) being that for the hard sphere fluid. The segment contribution is the residual Helmholtz energy for a fluid composed of the non-bonded and non-associated monomers that make up the molecules, and can be calculated from an appropriate equation of state for the monomer fluid or mixture. Among the commonly used spherical reference potentials for the segment term have been hard spheres, hard sphere plus dispersion potential treated as a perturbation, Lennard-Jones, square well, Yukawa, and a variable range attractive potential.

A widely used version of SAFT is that due to Huang and Radosz,[323] who applied the SAFT equation to over 100 real fluids, including simple inorganic substances, chain, aromatic and chlorinated hydrocarbons, ethers, alkanols, carboxylic acids, esters, ketones, amines, and polymers having molar masses up to 100,000. The segment contribution was treated as the sum of a hard sphere contribution, given by the Carnahan–Starling equation, (6.235), plus a dispersion contribution given by the Chen–Kreglewski equation of state.[324] The potential parameters (m, σ, and ε for non-associating fluids, plus two association parameters, $\varepsilon_{\text{bond}}$ and κ, the energy and volume available for bonding) were fitted to experimental vapour pressure and saturated liquid density data. Convenient empirical equations were fitted to provide estimates of these parameters for the various homologous series, so that interpolations and extrapolations could be made for fluids not included in the fitting procedure. The average deviations were on the order 0.5 to 3% for both vapour pressure and liquid density, in nearly all cases.

The many applications of the SAFT equation to real fluids have been the subject of several reviews.[193,221–223,325] We consider two examples. The first is to the phase behaviour of hydrogen fluoride (HF).[232] Because of its small molecular size, strong hydrogen bonds form in all phases of HF. It shows the strongest association known in the gas phase, with molar heat capacities exceeding 1 kJ $\text{mol}^{-1}\text{K}^{-1}$. This strong association leads to anomalously high vapour pressures at low temperature. This is believed to be due to the formation of ring aggregates in the liquid, which facilitate vapourization at lower temperatures. Using infrared absorption spectroscopy,[326] ring formation in the vapour phase has been observed, with rings having from 4 to 8 monomers. The SAFT equation has been used to study the thermodynamic properties of HF,[232] using the variable range attractive potential, $-\varepsilon\phi(r;\lambda)$, where ϕ is the square well potential and λ is the width of the well, and accounting for ring formation (see § 6.13.4). The predictions of the SAFT theory for the vapour pressure and coexisting gas and liquid densities are compared with experimental values[327,328] in Fig. 6.36. The Clausius–Clapeyron representation of the vapour pressure in Fig. 6.36(a) shows the anomalous upwards curvature of the experimental data (the opposite of that normally found for other fluids), believed due to ring aggregates. This behaviour is reproduced only when tetramer rings are included in the theory. As the ring size increases above 4 the effect on the phase behaviour becomes less significant, and little difference is seen on including 6 member rings. The inclusion of the tetramer ring formation also improves the agreement with experiment for the coexisting gas and liquid densities, as seen in Fig. 6.36(b). The theory overestimates the critical point, as can be expected from an analytical equation of state.

The second application is to the prediction of properties of linear and branched alkanes.[272,273,329–332] These calculations have included applications to heterobonded (copolymer) and branched chains,[272,273,329] to critical properties,[330,332] and to Joule-Thomson coefficients.[331] The Joule-Thomson coefficient, μ, is defined by

$$\mu = \left(\frac{\partial T}{\partial p}\right)_H, \tag{6.289}$$

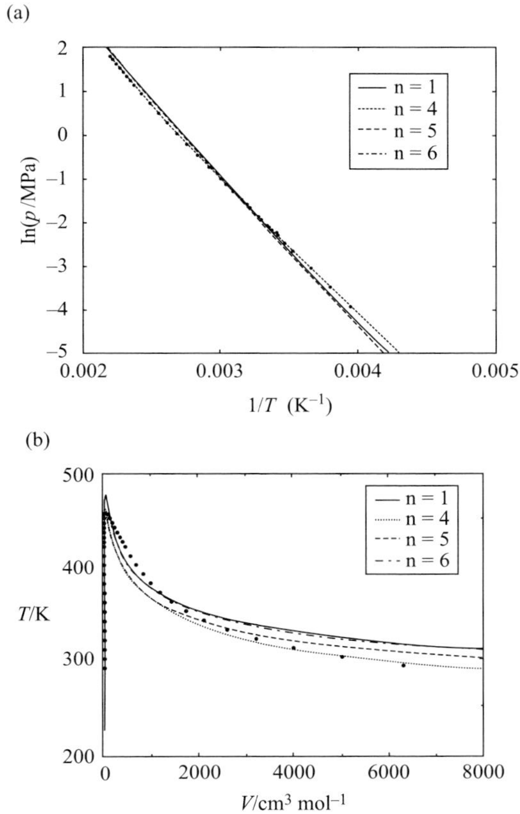

FIG. 6.36 Experimental (points) and SAFT predictions of (a) vapour pressure, and (b) coexisting gas and liquid molar volumes for hydrogen fluoride. The solid line ($n = 1$) is the prediction taking into account chain aggregates only. Other lines give the results for the theory when rings of size n monomers are accounted for. Reprinted with permission from ref. 232. Copyright 2002 Taylor and Francis Ltd.

where H is the enthalpy. The Joule-Thomson coefficient is important in expansion processes at constant enthalpy, also called throttling; such processes commonly occur in heat engines, refrigeration, and liquefaction processes. As the pressure drops the temperature can either fall or rise, depending on the initial pressure and temperature. The regions where cooling and heating take place are separated by the Joule-Thomson inversion curve (JTIC), which is the locus of states where $\mu = 0$. Direct measurement of the JTIC is difficult and often unreliable, since it requires the accurate measurement of a very small temperature difference due to relatively large pressure changes. The prediction of the JTIC is considered a very stringent test of a theoretical equation of state. In Fig. 6.37 are shown estimates of the JTIC for methane, for which a substantial amount of experimental data

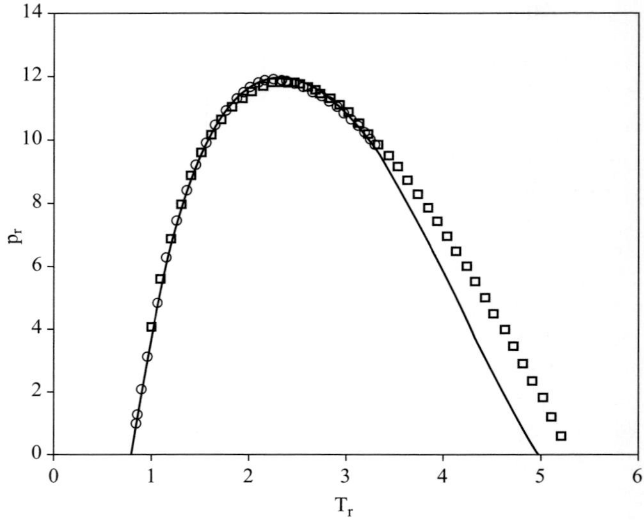

FIG. 6.37 Joule-Thomson inversion curve for methane. Points are from empirical equations of state that have been fitted to experimental data,[333,334] and the line is the SAFT predicted curve. Here $p_r = p/p_c$, $T_r = T/T_c$. (from ref. 331.)

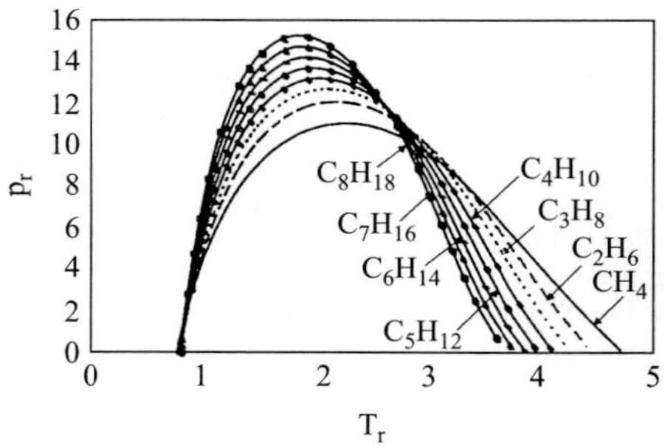

FIG. 6.38 Joule-Thomson inversion curves for the light members of the *n*-alkane series from methane to *n*-octane, predicted from the SAFT equation of state. Reprinted with permission from ref. 331. Copyright 2002 American Chemical Society.

are available. Points are from empirical equations of state for methane,[333,334] fitted to the available data, while the line is the prediction of the SAFT equation, with a Lennard-Jones treatment of the segment term. SAFT predictions of the JTIC for light members of the *n*-alkane series, up to *n*-octane, are shown in Fig. 6.38.

6.14 Nano-scale systems

In systems of nanometer dimensions, while many thermodynamic concepts and laws remain valid, others break down, so that caution is required.[335] The partition function, and hence the free energy and nearly all other total thermodynamic properties (see § 3.1.1 of Vol. 1), are always well defined however small the system, provided that the molecular number, N, and volume, V, are well defined. However, N and/or V are often not well defined for nanosystems. Consider a system consisting of a spherical nanoparticle. Where should we draw the spherical boundary that defines the volume of the nanoparticle? If we are doing a neutron scattering experiment we would probably define the volume to be the sphere defined by the outer nuclei, since the electron cloud beyond this radius is not visible to neutrons (Chapter 9). However, if the experiment involved X-ray diffraction, where scattering is from the electron cloud, we might define a larger volume to account for the electron clouds of the outer atoms. Similarly, for a nano-phase confined within a narrow pore the width H might be appropriately defined by the nuclei of atoms in the first layer of atoms in the opposing walls (the 'physical width') when doing neutron scattering, but when doing a gas adsorption experiment the pore width is usually defined as the 'effective width', $H_{\text{eff}} = H - \sigma_{ww}$, where σ_{ww} is the diameter of the wall atoms (often taken to be the Lennard-Jones σ) and is a measure of the dead space near the walls. The difference between the different definitions of pore width, and hence volume, can be large for micropores. In the case of the example just quoted, consider a slit-shaped carbon ($\sigma_{ww} = \sigma_{CC} = 0.340\,\text{nm}$) pore of physical width $H = 1.0\,\text{nm}$ (a common size for many microporous carbons) containing argon. Then $H_{\text{eff}} \approx H - 0.34 = 0.66\,\text{nm}$.. Since volume is proportional to pore width the volumes differ by $1.0/0.66 = 1.515$, i.e. by about 52%. Both definitions of H are commonly used in the literature. If instead of a slit pore (confinement in one dimension) we have a cylindrical (volume proportional to H^2) or spherical (volume proportional to H^3) one, the pore volumes corresponding to these two pore width definitions differ by $(H/H_{eff})^2$ and $(H/H_{eff})^3$, i.e. by factors of 2.29 and 3.48, respectively. Similar considerations apply to nanoparticles, nanodroplets, etc.

Nanosystems such as nanodroplets in equilibrium are usually coupled to a large reservoir with well-defined intensive properties such as temperature and, for open systems, chemical potential, so that the nanosystem values of these properties are well defined. Similarly, for a nanodroplet the normal component of the pressure tensor is well defined, but may be nonuniform. (The tangential pressure component is not uniquely defined in inhomogeneous regions, even in macrodrops—see § 8.3.) Compared to those of macrosystems, nanosystems properties such as the volume V and particle number N (and other properties that are extensive for macrosystems) can have significant relative fluctuations, so that the averages $< V >$ and $< N >$ will often be the best measures of these properties for nanosystems. Further discussion of related points is given in § 8.2.2.

Although the total thermodynamic properties can be assumed well defined by using mean values to allow for possible fluctuations, and can be defined operationally (as in the case of $< V >$ and $< N >$) in a manner that is convenient for the application at hand, the concepts of Gibbs surface thermodynamics[336] (surface tension, other excess surface properties, the Young, Laplace, and Kelvin equations, etc.) break down completely for sufficiently small systems, since it is no longer possible to divide the system into a inhomogeneous surface region and homogenous phases (see § 8.2); the whole system becomes inhomogeneous. Thus surface tension and other excess surface properties become undefined for very small systems.

The Second Law of Thermodynamics also breaks down for sufficiently small systems. The Second Law applies for large systems and sufficiently long times, and states that for an isolated system undergoing a spontaneous (irreversible) change the entropy change must be positive. However, at the nano-scale this law does not hold for short times, due to the large fluctuations in very small systems. This violation of the Second Law for small systems was well known to Maxwell and Boltzmann in the nineteenth century. Thus, in 1878 Maxwell wrote:[337]

> *Hence the Second Law of Thermodynamics is continually being violated, and that to a considerable extent in any sufficiently small group of molecules belonging to any real body.*

and Boltzmann stated:[338]

> *As soon as one looks at bodies of such small dimension that they contain only very few molecules, the validity of this theorem (the second law of thermodynamics) must cease.*

Fluctuations become increasingly important as the system is made smaller, leading to violation of the Second Law. A quantitative description of the violations of the Second Law was provided by Evans, Searles, and co-workers, who derived the fluctuation theorem, which generalizes the Second Law to include thermodynamic behaviour in very small systems and short times, and can be applied to systems far from equilibrium.[339] The fluctuation theorem is generally applicable and leads to the Second Law in the thermodynamic limit ($N \to \infty$, $\rho = N/V =$ constant). The predictions of this theorem have been tested and confirmed by both experiment[340,341] and molecular simulation.[342] This has important implications for nano-scale engines and devices; for short times such engines can operate in 'reverse'; i.e. heat energy from the surroundings can be converted into work over short time scales.

6.15 Conclusion

The thermodynamic functions can be expressed in terms of either the partition function (§ 3.1.1) or distribution functions (§§ 6.1 to 6.7). These equations serve as the starting point for predictive theories for thermodynamic properties.

Particularly powerful starting points are the test particle equations (§ 6.5) and the Kirkwood-Buff theory (§ 7.2), since these do not rely on any assumptions about the nature of the molecules, nor on the pairwise additivity approximation.

The scaled particle theory (§ 6.12) and its extensions provides a good description of the thermodynamics of fluids of hard, nonspherical bodies, particularly when the bodies are convex, and such hard body fluids serve as a reference system for perturbation theories. Perturbation theory (§ 6.11) is successful for fluids of molecules that are not very highly polar or associating. In the latter cases the pair correlation function for the fluid of interest is too different from that of simple reference fluids for rapid series convergence to be achieved. For such liquids the theory of Wertheim (§ 6.13) is the most successful at present.

The SAFT equation combines conventional treatments of the dispersion and repulsion force effects with Wertheim's theory to account for chain or ring formation and association. While this approach has been shown to be successful for a wide range of molecular types, further work is needed to develop a systematic parameter base so that this equation can be easily assimilated into engineering process simulators. Along these lines, a promising approach is a group contribution version of SAFT,[343,344] which is proving successful in process optimization and design applications. More accurate theoretical treatments are also needed to predict several more difficult thermodynamic properties, such as the solubility of solids (e.g. pharmaceuticals) and gases in liquids, and derivative properties such as the isothermal compressibility, speed of sound, interfacial tension, and Joule-Thomson coefficients.

Appendix 6A Geometry of convex bodies

6A.1 Definitions and fundamental measures

A body is termed *convex* if any line segment whose ends are inside the body lies entirely in the body. Examples of convex shapes are spheres, cubes, spherocylinders, and ellipsoids; a diatomic molecule represented by two fused hard spheres is an example of a nonconvex body. In this Appendix we summarize the most important expressions for the geometry of convex bodies. The treatment here follows that given by Kihara.[345–347] A more complete discussion is given in these refs. and in refs. 66, 348–350.

We first choose an origin O within the body and a set of body-fixed axes x, y, z. For a given direction $\omega \equiv (\theta\phi)$ there is only one plane tangential to the convex body whose normal is in the direction ω, as shown in Fig. 6A.1. This plane is called the *supporting plane* in the direction ω, and the normal distance of this plane from the origin as a function of ω is called the *supporting function*, $r_\perp(\omega)$. Let r_m be the average of $r_\perp(\omega)$ over all directions,

$$r_m \equiv \langle r_\perp \rangle_\omega \equiv \frac{1}{4\pi} \int_0^{2\pi} d\phi \int_0^{\pi} d\theta \;\; \sin\theta \; r_\perp\,(\theta\,\phi)\,. \tag{6A.1}$$

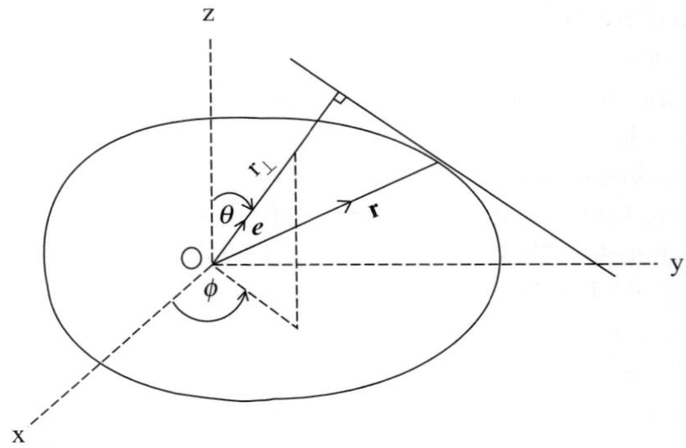

FIG. 6A.1 Supporting plane and supporting function $r_\perp(\omega)$.

For a sphere with O as the centre, we have $r_\perp = a$ and $r_m = a$, where a is the radius. In general r_m is independent of the choice of origin,[351] and is termed the *mean radius of curvature* of the convex body. Thus, for a thin rod of length ℓ, choosing O as the centre of the rod and z to lie along the rod it is easy to show that $r_\perp(\theta\phi) = (\ell/2)\,|\cos\theta|$, so that from (6A.1) the mean radius of curvature is $r_\mathrm{m} = \ell/4$.

We assume that the convex body has a smooth surface. The body is characterized by three *fundamental measures*. These are the mean radius of curvature r_m, the volume v_m, and the surface area s_m. In writing general expressions for these quantities it is convenient to introduce a unit vector $\mathbf{e}(\omega)$ in the ω direction, so that

$$r_\perp(\omega) = \mathbf{r}(\omega)\cdot\mathbf{e}(\omega) \tag{6A.2}$$

where $\mathbf{r}(\omega)$ is the position of the point where the tangent plane touches the surface (see Fig. 6A.1). It can be shown that (see § 6A.4)

$$\mathbf{e} = \frac{\partial\,\mathbf{e}}{\partial\theta}\times\frac{1}{\sin\theta}\frac{\partial\,\mathbf{e}}{\partial\phi}. \tag{6A.3}$$

Using (6A.2) and (6A.1) in (6A.3) we have

$$r_\mathrm{m} = \frac{1}{4\pi}\int \mathrm{d}\theta\, d\phi\; \mathbf{r}\cdot\frac{\partial\,\mathbf{e}}{\partial\theta}\times\frac{\partial\,\mathbf{e}}{\partial\phi}. \tag{6A.4}$$

(We remind the reader (see Appendix B) that $\mathbf{a}\cdot\mathbf{b}\times\mathbf{c}$ is unambiguously $\mathbf{a}\cdot(\mathbf{b}\times\mathbf{c})$; since $\mathbf{a}\cdot\mathbf{b}$ is a scalar $(\mathbf{a}\cdot\mathbf{b})\times\mathbf{c}$ is undefined.) The surface area and volume of the body are given by[352] (see § 6A.4 for derivation)

$$s_\mathrm{m} = \int \mathrm{d}\theta\,\mathrm{d}\phi\;\mathbf{e}\cdot\frac{\partial\,\mathbf{r}}{\partial\theta}\times\frac{\partial\,\mathbf{r}}{\partial\phi} \tag{6A.5}$$

$$v_\mathrm{m} = \frac{1}{3}\int \mathrm{d}\theta\mathrm{d}\phi\;\mathbf{r}\cdot\frac{\partial\mathbf{r}}{\partial\theta}\times\frac{\partial\mathbf{r}}{\partial\phi}. \tag{6A.6a}$$

For completeness we also have

$$1 = \frac{1}{4\pi}\int \mathrm{d}\theta\mathrm{d}\phi\ \mathbf{e}\cdot\frac{\partial\mathbf{e}}{\partial\theta}\times\frac{\partial\mathbf{e}}{\partial\phi}, \tag{6A.6b}$$

which is obvious from (6A.3).

For spheres of radius a it is easily verified from (6A.4)–(6A.6) that $r_{\mathrm{m}} = a$, $s_{\mathrm{m}} = 4\pi a^2$, and $v_{\mathrm{m}} = 4\pi a^3/3$. Expressions for the fundamental measures of some other convex bodies are given Table 6A.1.

6A.2 *Parallel body of the core*

Consider a convex body A whose surface equation or $\mathbf{r}$ function $\mathbf{r}_A(\omega)$, where, as above, ω is the direction of the supporting plane normal. We now consider a second convex body $A+\rho$, which consists of all points whose distances from the nearest point on the surface of A are less than or equal to some distance ρ (Fig. 6A.2). This larger convex body is called the *parallel body* of A with thickness ρ. For any direction ω the supporting planes of A and $A+\rho$ are parallel, and the $\mathbf{r}$ function for the parallel body is given by

$$\mathbf{r}_{A+\rho}(\omega) = \mathbf{r}_A(\omega) + \rho\mathbf{e}(\omega). \tag{6A.7}$$

Substituting this expression in (6A.4) gives the mean radius of curvature of the parallel body as

$$r_{\mathrm{m,\,A+\rho}} = \frac{1}{4\pi}\int \mathrm{d}\theta\mathrm{d}\phi\ (\mathbf{r}_A + \rho\mathbf{e})\cdot\frac{\partial\mathbf{e}}{\partial\theta}\times\frac{\partial\mathbf{e}}{\partial\phi}$$

or, using (6A.3) and (6A.6b),

$$r_{\mathrm{m},\,A+\rho} = r_{\mathrm{m},\,A} + \rho. \tag{6A.8}$$

Similarly, on substituting (6A.7) into (6A.5) and (6A.6a) we obtain the surface area and volume of the parallel body as

$$s_{\mathrm{m},\,A+\rho} = s_{\mathrm{m},\,A} + 8\pi\rho r_{\mathrm{m},\,A} + 4\pi\rho^2 \tag{6A.9}$$

$$v_{\mathrm{m},\,A+\rho} = v_{\mathrm{m},\,A} + \rho s_{\mathrm{m},\,A} + 4\pi\rho^2 r_{\mathrm{m},\,A} + \frac{4\pi}{3}\rho^3. \tag{6A.10}$$

Equations (6A.8)–(6A.10) are called Steiner's formulae.[353]

6A.3 *Excluded volume of two convex bodies*[66,347,348]

Consider two convex bodies A and B with origins O_A and O_B and fundamental measures $v_{\mathrm{m}A}$, $s_{\mathrm{m}A}$, $r_{\mathrm{m}A}$ and $v_{\mathrm{m}B}$, $s_{\mathrm{m}B}$, $r_{\mathrm{m}B}$. We suppose that the position and orientation of body A is kept fixed, and that body B is touching A (Fig. 6A.3). Keeping the orientation of B fixed, we allow B to move over the surface of A, always keeping contact with it. The locus of the centre of B forms the surface of another convex body that we call $A+B$. The volume occupied by $A+B$, $v_{\mathrm{m},A+B}$, is the region of space from which the centre of B is excluded, by virtue of the presence of A. We wish to calculate this volume. We note that the tangent plane to

Table 6A.1

Fundamental measures for convex bodies

Shape	Example	Characteristic dimensions	r_m	s_m	v_m
Sphere	Ar	Radius a	a	$4\pi a^2$	$4\pi a^3/3$
Thin rod	CO_2	Length ℓ	$\ell/4$	0	0
Circular disk	C_6H_6	Radius a	$\pi a/4$	$2\pi a^2$	0
Rectangular parallelepiped	C_2H_6	Sides $\ell_1\ell_2\ell_3$	$\frac{1}{4}(\ell_1+\ell_2+\ell_3)$	$2(\ell_1\ell_2+\ell_1\ell_3+\ell_2\ell_3)$	$\ell_1\ell_2\ell_3$
Thin hexagon	C_6H_6	Side ℓ	$\frac{3}{4}\ell$	$\sqrt{3}\,3\ell^2$	0
Regular tetrahedron	CH_4	Side ℓ	$3(\tan^{-1}\sqrt{2})\ell/2\pi$	$\sqrt{3}\ell^2$	$\sqrt{2}\ell^3/12$
Regular octahedron	SF_6	Side ℓ	$\frac{3}{\pi}\ell\cot^{-1}(\sqrt{2})$	$\sqrt{3}\,2\ell^2$	$\frac{1}{3}\sqrt{2}\ell^3$
Cylinder	C_2H_6	Length ℓ Radius a	$(\pi a+\ell)/4$	$2\pi a\,(a+\ell)$	$\pi a^2\ell$
Prolate spherocylinder*	N_2	Length ℓ Radius a	$a+\ell/4$	$2\pi a\,(2a+\ell)$	$\pi a^2(4a/3+\ell)$
Oblate spherocylinder*	C_6H_6	Length ℓ Radius a	$a+\pi\ell/8$	$4\pi a^2+\pi^2 a\ell+\pi\ell^2/2$	$\pi a(4a^2/3+\pi\ell a/2+\ell^2/2)$
Prolate axial ellipsoid	CO_2	Major semiaxis a, $\varepsilon^2=(a^2-b^2)/a^2$	$\frac{a}{2}\left(1+\frac{1-\varepsilon^2}{2\varepsilon}\ln\frac{1+\varepsilon}{1-\varepsilon}\right)$	$2\pi b^2\left(1+\frac{\sin^{-1}\varepsilon}{\varepsilon(1-\varepsilon^2)^{1/2}}\right)$	$4\pi ab^2/3$
Oblate axial ellipsoid	C_6H_6	Minor semiaxis b, $\varepsilon^2=(a^2-b^2)/a^2$	$\frac{b}{2}\left(1+\frac{\sin^{-1}\varepsilon}{\varepsilon(1-\varepsilon^2)^{1/2}}\right)$	$2\pi a^2\left(1+\frac{1-\varepsilon^2}{2\varepsilon}\ln\frac{1+\varepsilon}{1-\varepsilon}\right)$	$4\pi a^2 b/3$

* The dimensions of prolate and oblate spherocylinders are defined in Fig. 6.11(a), where $\sigma = 2a$. For a more complete listing including other shapes see the book by Hadwiger,[348] pp. 35–8. The results for nonaxial ellipsoids appear to be unknown.

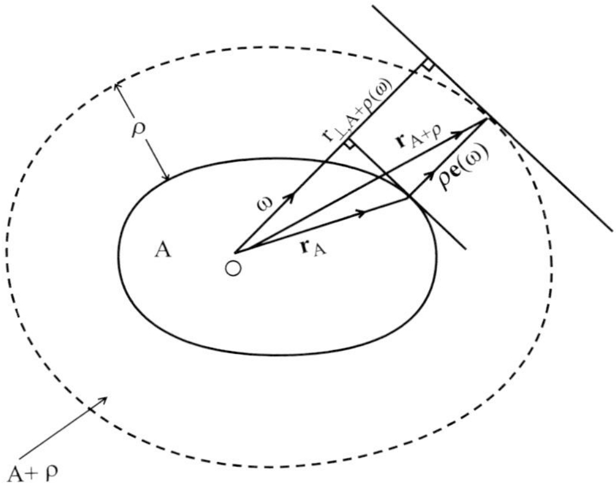

FIG. 6A.2 The parallel body of A with thickness ρ is the body A + ρ, enclosed by the dashed line; $\mathbf{e}(\omega)$ is a unit vector in the ω direction.

the body $A + B$ at O_B is parallel to the tangent plane of body A at the contact point P (see B, Fig. 6A.3). This follows because when B moves over the surface of A it can have no velocity component perpendicular to the tangent plane at contact. The bodies are rigid and the orientation ω_B of B is fixed, so that the vector from

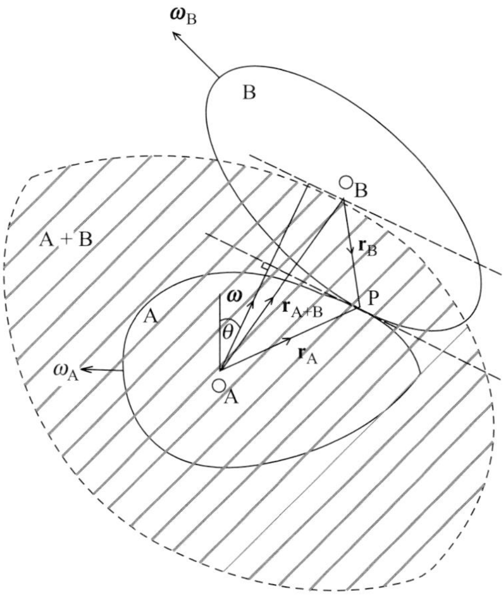

FIG. 6A.3 The excluded volume due to the presence of body A (fixed in position and orientation) is the volume $A + B$ whose surface is the locus of the centre O_B of body B as it moves with fixed orientation over the surface of A.

O_B to P is of fixed orientation; thus for an infinitesimal movement of B the points O_B and P must move along parallel paths.

The **r**-function for $A + B$ is given by

$$\mathbf{r}_{A+B}(\theta\phi) = \mathbf{r}_A(\theta\phi) + \mathbf{r}_B^*(\theta\phi), \tag{6A.11}$$

where (see Fig. 6A.3)

$$\mathbf{r}_B^*(\theta\phi) = -\mathbf{r}_B(\pi - \theta, \phi + \pi). \tag{6A.12}$$

To derive the expression for $v_{\mathrm{m},A+B}$, the excluded volume, we substitute (6A.11) into (6A.6a) to obtain

$$v_{\mathrm{m},A+B} = \frac{1}{3}\int \mathrm{d}\theta\mathrm{d}\phi\,(\mathbf{r}_A + \mathbf{r}_B^*)\cdot\left(\frac{\partial \mathbf{r}_A}{\partial\theta} + \frac{\partial \mathbf{r}_B^*}{\partial\theta}\right)\times\left(\frac{\partial \mathbf{r}_A}{\partial\phi} + \frac{\partial \mathbf{r}_B^*}{\partial\phi}\right). \tag{6A.13}$$

In order to simplify this expression we make use of the identity[354]

$$\begin{aligned}
&\frac{\partial}{\partial\theta}\left(\mathbf{r}_B^*\cdot\mathbf{r}_A\times\frac{\partial\mathbf{r}_A}{\partial\phi}\right) + \frac{\partial}{\partial\phi}\left(\mathbf{r}_B^*\cdot\frac{\partial\mathbf{r}_A}{\partial\theta}\times\mathbf{r}_A\right)\\
&\quad= \frac{\partial\mathbf{r}_B^*}{\partial\theta}\cdot\mathbf{r}_A\times\frac{\partial\mathbf{r}_A}{\partial\phi} + \mathbf{r}_B^*\cdot\frac{\partial\mathbf{r}_A}{\partial\theta}\times\frac{\partial\mathbf{r}_A}{\partial\phi} + \mathbf{r}_B^*\cdot\mathbf{r}_A\times\frac{\partial^2\mathbf{r}_A}{\partial\theta\partial\phi}\\
&\quad+ \frac{\partial\mathbf{r}_B^*}{\partial\phi}\cdot\frac{\partial\mathbf{r}_A}{\partial\theta}\times\mathbf{r}_A + \mathbf{r}_B^*\cdot\frac{\partial^2\mathbf{r}_A}{\partial\phi\partial\theta}\times\mathbf{r}_A + \mathbf{r}_B^*\cdot\frac{\partial\mathbf{r}_A}{\partial\theta}\times\frac{\partial\mathbf{r}_{\mathbf{A}}}{\partial\phi}\\
&\quad= 2\mathbf{r}_B^*\cdot\frac{\partial\mathbf{r}_A}{\partial\theta}\times\frac{\partial\mathbf{r}_A}{\partial\phi} - \mathbf{r}_A\cdot\frac{\partial\mathbf{r}_B^*}{\partial\theta}\times\frac{\partial\mathbf{r}_A}{\partial\phi} - \mathbf{r}_A\cdot\frac{\partial\mathbf{r}_A}{\partial\theta}\times\frac{\partial\mathbf{r}_B^*}{\partial\phi},
\end{aligned} \tag{6A.14}$$

where the property of the triple scalar product (B.34a), $\mathbf{a}\cdot\mathbf{b}\times\mathbf{c} = -\mathbf{b}\cdot\mathbf{a}\times\mathbf{c} = -\mathbf{c}\cdot\mathbf{b}\times\mathbf{a}$ (note that $\mathbf{a}\times\mathbf{c} = -\mathbf{c}\times\mathbf{a}$), was used in obtaining the last two terms on the right-hand side of (6A.14). Integrating both sides of (6A.14) over θ and ϕ we note that both of the terms on the left-hand side vanish. For the first term on the left-hand side we have

$$\int_0^{2\pi}\mathrm{d}\phi\int_0^{\pi}\mathrm{d}\theta\frac{\partial}{\partial\theta}\left(\mathbf{r}_B^*\cdot\mathbf{r}_A\times\frac{\partial\mathbf{r}_A}{\partial\phi}\right) = \int_0^{2\pi}\mathrm{d}\phi\left|\mathbf{r}_B^*\cdot\mathbf{r}_A\times\frac{\partial\mathbf{r}_A}{\partial\phi}\right|_{\theta=0}^{\pi}$$
$$= 0$$

since $\partial\mathbf{r}_A/\partial\phi = 0$ when $\theta = 0$ or π (Fig. 6A.4). Also we have for the second term:

$$\int_0^{\pi}\mathrm{d}\theta\int_0^{2\pi}\mathrm{d}\phi\frac{\partial}{\partial\phi}\left(\mathbf{r}_B^*\cdot\frac{\partial\mathbf{r}_A}{\partial\theta}\times\mathbf{r}_A\right) = \int_0^{\pi}\mathrm{d}\theta\left|\mathbf{r}_B^*\cdot\frac{\partial\mathbf{r}_A}{\partial\theta}\times\mathbf{r}_A\right|_{\phi=0}^{\phi=2\pi}$$
$$= 0.$$

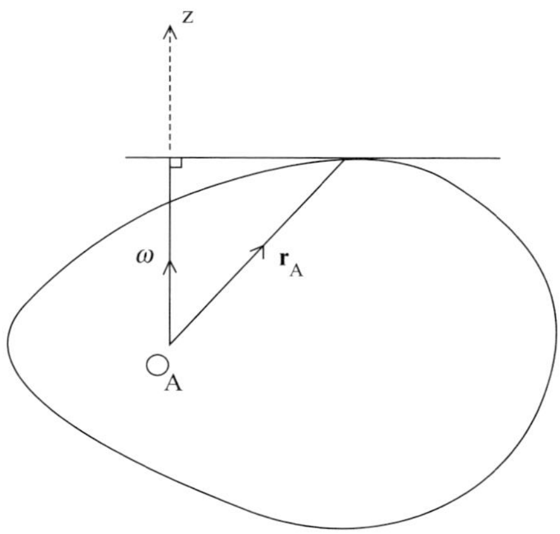

FIG. 6A.4 When ω is in the z-direction, i.e. $\theta = 0$, ϕ is undefined and the supporting plane, and hence $\mathbf{r_A}$, are independent of ϕ; similarly for $\theta = \pi$.

In this case, for any fixed θ, the function

$$\mathbf{r}_B^* \cdot \frac{\partial \mathbf{r}_A}{\partial \theta} \times \mathbf{r}_A$$

has the same value for $\phi = 0$ and $\phi = 2\pi$, causing the integral to vanish. Thus the integral of (6A.14) over θ and ϕ yields:

$$\begin{aligned}\int \mathrm{d}\theta \mathrm{d}\phi\ \mathbf{r}_A \cdot \frac{\partial \mathbf{r}_B^*}{\partial \theta} \times \frac{\partial \mathbf{r}_A}{\partial \phi} + \int \mathrm{d}\theta \mathrm{d}\phi\ \mathbf{r}_A \cdot \frac{\partial \mathbf{r}_A}{\partial \theta} \times \frac{\partial \mathbf{r}_B^*}{\partial \phi} \\ = 2 \int \mathrm{d}\theta \mathrm{d}\phi\ \mathbf{r}_B^* \cdot \frac{\partial \mathbf{r}_A}{\partial \theta} \times \frac{\partial \mathbf{r}_A}{\partial \phi}.\end{aligned} \tag{6A.15}$$

By precisely similar arguments we can show that

$$\begin{aligned}\int \mathrm{d}\theta \mathrm{d}\phi\ \mathbf{r}_B^* \cdot \frac{\partial \mathbf{r}_A}{\partial \theta} \times \frac{\partial \mathbf{r}_B^*}{\mathrm{d}\phi} + \int \mathrm{d}\theta \mathrm{d}\phi\ \mathbf{r}_B^* \cdot \frac{\partial \mathbf{r}_B^*}{\partial \theta} \times \frac{\partial \mathbf{r}_A}{\partial \phi} \\ = 2 \int \mathrm{d}\theta \mathrm{d}\phi\ \mathbf{r}_A \cdot \frac{\partial \mathbf{r}_B^*}{\partial \theta} \times \frac{\partial \mathbf{r}_B^*}{\partial \phi}.\end{aligned} \tag{6A.16}$$

Using (6A.15) and (6A.16) in (6A.13), together with (6A.6a), we have

$$\begin{aligned}v_{\mathrm{m},A+B} = v_{m,A} + \frac{1}{3} \int \mathrm{d}\theta \mathrm{d}\phi\ \mathbf{r}_B^* \cdot \frac{\partial \mathbf{r}_B^*}{\partial \theta} \times \frac{\partial \mathbf{r}_B^*}{\partial \phi} \\ + \int \mathrm{d}\theta \mathrm{d}\phi\ \mathbf{r}_B^* \cdot \frac{\partial \mathbf{r}_A}{\partial \theta} \times \frac{\partial \mathbf{r}_A}{\partial \phi} + \int \mathrm{d}\theta \mathrm{d}\phi\ \mathbf{r}_A \cdot \frac{\partial \mathbf{r}_B^*}{\partial \theta} \times \frac{\partial \mathbf{r}_B^*}{\partial \phi}.\end{aligned} \tag{6A.17}$$

The second term on the right-hand side of this equation is simply $v_{\mathrm{m},B}$. This can be seen by using (6A.12) and making the change of variables, $\theta' = \pi - \theta$, $\phi' = \phi + \pi$. Thus we have

$$\frac{1}{3}\int_0^{\pi} d\theta \int_0^{2\pi} d\phi\ \mathbf{r}_B^* \cdot \frac{\partial \mathbf{r}_B^*}{\partial \theta} \times \frac{\partial \mathbf{r}_B^*}{\partial \phi}$$

$$= \frac{1}{3}\int_0^{\pi} d\theta' \int_{\pi}^{3\pi} d\phi'\ \mathbf{r}_B(\theta'\phi') \cdot \frac{\partial \mathbf{r}_B(\theta'\phi')}{\partial \theta'} \times \frac{\partial \mathbf{r}_B(\theta'\phi')}{\partial \phi'}$$

$$= v_{\mathrm{m},B}$$

so that (6A.17) becomes

$$v_{\mathrm{m},A+B} = v_{\mathrm{m},A} + v_{\mathrm{m},B} + \int d\theta\, d\phi\ \mathbf{r}_B^* \cdot \frac{\partial \mathbf{r}_A}{\partial \theta} \times \frac{\partial \mathbf{r}_A}{\partial \phi} + \int d\theta\, d\phi\ \mathbf{r}_A \cdot \frac{\partial \mathbf{r}_B^*}{\partial \theta} \times \frac{\partial \mathbf{r}_B^*}{\partial \phi}. \tag{6A.18}$$

So far we have kept the orientation of body B fixed when calculating $v_{\mathrm{m},A+B}$. We now average both sides of (6A.18) over all possible orientations of B, to obtain the average excluded volume $\langle v_{\mathrm{m},A+B}\rangle_{\omega_B}$. The first two terms on the right-hand side of (6A.18), $v_{\mathrm{m},A}$ and $v_{\mathrm{m},B}$, are left unchanged, and the third becomes

$$\int d\theta\, d\phi\ \langle \mathbf{r}_B^* \rangle_{\omega_B} \cdot \frac{\partial \mathbf{r}_A}{\partial \theta} \times \frac{\partial \mathbf{r}_A}{\partial \phi}, \tag{6A.19}$$

where, as always, $\langle \cdots \rangle_{\omega_B} = \Omega_B^{-1} \int d\omega_B \cdots$, with $\Omega_B = 4\pi$ or $8\pi^2$ for linear or nonlinear molecules, respectively. The direction of the vector $(\partial \mathbf{r}_A/\partial\theta) \times (\partial \mathbf{r}_A/\partial\phi)$ is normal to the supporting plane at P in Fig. 6A.3 (see § 6A.4 and Fig. 6A.6); i.e. it is in the ω direction. Thus $(\partial \mathbf{r}_A/\partial\theta) \times (\partial \mathbf{r}_A/\partial\phi)$ can be rewritten as $\mathbf{ee} \cdot [(\partial \mathbf{r}_A/\partial\theta) \times (\partial \mathbf{r}_A/\partial\phi)]$, and on making this substitution (6A.19) becomes

$$\int d\theta\, d\phi\ \left\langle \mathbf{r}_B^* \cdot \mathbf{e} \right\rangle_{\omega_B} \mathbf{e} \cdot \frac{\partial \mathbf{r}_A}{\partial \theta} \times \frac{\partial \mathbf{r}_A}{\partial \phi}. \tag{6A.20}$$

We now carry out the averaging over ω_B, keeping θ and ϕ (and hence $\mathbf{e}$ and $\mathbf{r}_A$) fixed. Using (6A.12) and introducing $\mathbf{e}_B \equiv -\mathbf{e}$ as the unit $\mathbf{e}$ vector for body B, we rewrite this averaging as

$$\langle \mathbf{r}_B \cdot \mathbf{e}_B \rangle_{\omega_B} = \langle r_{\perp B} \rangle_{\omega_B}, \tag{6A.21}$$

where $r_{\perp B} = \mathbf{r}_B \cdot \mathbf{e}_B$ is the support function for body B. This average over all possible orientations of body B, keeping the contact point P in Fig. 6A.3 fixed, must be physically equivalent to averaging $r_{\perp B}$ over all possible directions $\mathbf{e}_B$ (cf. Eq. (6A.1) and Fig. 6A.1). Thus (6A.21) simply reduces to $r_{\mathrm{m},B}$, the mean radius of curvature of body B. Since $r_{\mathrm{m},B}$ is independent of θ and ϕ in (6A.20), the latter becomes, with the use of (6A.5),

$$r_{mB}\int \mathrm{d}\theta\,\mathrm{d}\phi\;\mathbf{e}\cdot\frac{\partial \mathbf{r}_A}{\partial\theta}\times\frac{\partial \mathbf{r}_A}{\partial\phi}=r_{\mathrm{m}B}\,s_{\mathrm{m}A}.$$

By similar arguments we can show that averaging the last term on the right-hand side of (6A.18) over ω_B gives $ab\sin\theta\hat{\mathbf{n}}$ (Fig. 6A.5). Thus the final expression for the average excluded volume is

$$\langle v_{\mathrm{m},A+B}\rangle_{\omega_B}=v_{\mathrm{m},A}+v_{\mathrm{m},B}+r_{\mathrm{m},B}s_{\mathrm{m},A}+r_{\mathrm{m},A}s_{\mathrm{m},B}. \tag{6A.22}$$

The mean radius of curvature and surface area for the body A+B, each averaged over ω_B, can be found by similar methods,[66,345,346] and are given by

$$\langle r_{m,A+B}\rangle_{\omega_B}=r_{m,A}+r_{m,B} \tag{6A.23}$$

$$\langle s_{m,A+B}\rangle_{\omega_B}=s_{m,A}+s_{m,B}+8\pi r_{m,A}r_{m,B} \tag{6A.24}$$

For hard spheres of equal diameter, $\sigma_A=\sigma_B=\sigma$, the expressions (6A.22)–(6A.24) reduce to $v_{\mathrm{m},A+B}=(4/3)\pi\sigma^3, r_{\mathrm{m},A+B}=\sigma$, and $s_{\mathrm{m},A+B}=4\pi\sigma^2$, as expected.

6A.4 *Proof of Eqs. (6A.3), (6A.5), and (6A.7)*

We first prove (6A.3). The unit vector $\mathbf{e}$ (Fig. 6A.1) is given by

$$\mathbf{e}=\sin\theta\cos\phi\mathbf{i}+\sin\theta\sin\phi\mathbf{j}+\cos\theta\mathbf{k} \tag{6A.25}$$

where $\mathbf{i}$,$\mathbf{j}$, and $\mathbf{k}$ are unit vectors in the x, y, and z directions, respectively. Differentiation of this expression gives

$$\frac{\partial\mathbf{e}}{\partial\theta}\times\frac{1}{\sin\theta}\frac{\partial\mathbf{e}}{\partial\phi}=(\cos\theta\cos\phi\,\mathbf{i}+\cos\theta\sin\phi\,\mathbf{j}-\sin\theta\,\mathbf{k})\times(-\sin\phi\,\mathbf{i}+\cos\phi\,\mathbf{j}). \tag{6A.26}$$

We recall that for two vectors $\mathbf{a}$ and $\mathbf{b}$ the cross product is $ab\sin\theta$; here θ is the angle between $\mathbf{a}$ and $\mathbf{b}$ and $\hat{\mathbf{n}}$ is a unit vector perpendicular to the plane defined by $\mathbf{a}$ and $\mathbf{b}$ (Fig. 6A.5). Thus we have $\mathbf{i}\times\mathbf{i}=\mathbf{j}\times\mathbf{j}=\mathbf{k}\times\mathbf{k}=\mathbf{0}$, $\mathbf{i}\times\mathbf{j}=\mathbf{k}$, $\mathbf{j}\times\mathbf{k}=\mathbf{i}$, and $\mathbf{k}\times\mathbf{i}=\mathbf{j}$. Applying these rules to (6A.26), or the general rule (see Appendix B)

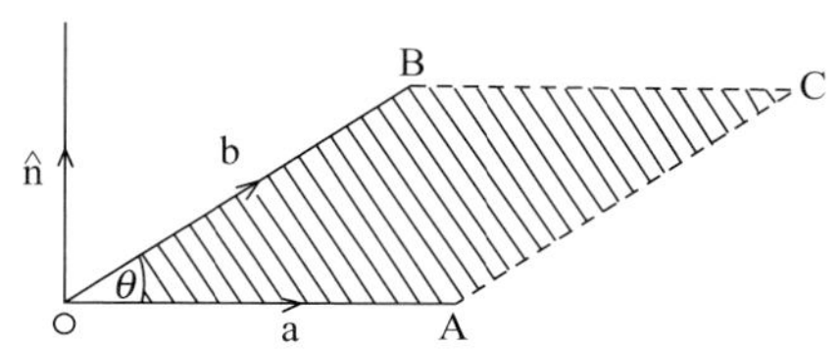

FIG. 6A.5 The cross product $\mathbf{a}\times\mathbf{b}=ab\sin\theta\hat{\mathbf{n}}$, where $ab\sin\theta$ is the area of the parallelogram OACB, and $\hat{\mathbf{n}}$ is a unit vector perpendicular to the plane of the parallelogram.

$$\mathbf{A} \times \mathbf{B} = \begin{vmatrix} \mathbf{i} & \mathbf{j} & \mathbf{k} \\ A_x & A_y & A_z \\ B_x & B_y & B_z \end{vmatrix} = (A_y B_z - A_z B_y)\,\mathbf{i} + \cdots \tag{6A.27}$$

gives

$$\frac{\partial \mathbf{e}}{\partial \theta} \times \frac{1}{\sin\theta}\frac{\partial \mathbf{e}}{\partial \phi} = \sin\theta\cos\phi\,\mathbf{i} + \sin\theta\sin\phi\,\mathbf{j} + \cos\theta\,\mathbf{k} = \mathbf{e},$$

which is (6A.3).

To derive (6A.5) and (6A.6) we first note that

$$s_m = \oint_{\mathcal{S}} \mathrm{d}\mathcal{S} = \oint_{\mathcal{S}} \mathbf{e} \cdot \mathrm{d}\mathbf{S}, \tag{6A.28}$$

where $\mathrm{d}\mathbf{S} = \mathbf{e}\,\mathrm{d}\mathcal{S}$ is a vector area element on the surface of the body located at $\mathbf{r}$, whose direction is normal to and outward from the surface and is thus the same as the direction of $\mathbf{e}$ (see Fig. 6A.1). The integral in (6A.28) extends over the surface of the body. The volume of the body is given by

$$v_m = \int_V \mathrm{d}V, \tag{6A.29}$$

where $\mathrm{d}V$ is an element of volume and the integration extends over all points within the body. This can be transformed to a surface integral by noting that

$$\int_V \nabla \cdot \mathbf{r}\,\mathrm{d}V = \oint_{\mathcal{S}} \mathbf{r} \cdot \mathrm{d}\mathbf{S} \tag{6A.30}$$

by virtue of Gauss' theorem. Since

$$\nabla \cdot \mathbf{r} \equiv \nabla_\alpha r_\alpha = \frac{\partial}{\partial x}(x) + \frac{\partial}{\partial y}(y) + \frac{\partial}{\partial z}(z) = 3,$$

we have, from (6A.29) and (6A.30),

$$v_m = \frac{1}{3}\oint_{\mathcal{S}} \mathbf{r} \cdot \mathrm{d}\mathbf{S}. \tag{6A.31}$$

To obtain the desired expression for $\mathrm{d}\mathbf{S}$ we note that $\mathbf{r}$ is a function of ω, and $\mathbf{r} = \mathbf{r}(\theta, \phi)$ is a function that defines the surface of the body.[352] If we now consider small changes $\mathrm{d}\theta$ and $\mathrm{d}\phi$ in θ and ϕ, the resultant changes in $\mathbf{r}$ will be approximately $(\partial\mathbf{r}/\partial\theta)\mathrm{d}\theta$ and $(\partial\mathbf{r}/\partial\phi)\mathrm{d}\phi$. The vector $(\partial\mathbf{r}/\partial\theta)\mathrm{d}\theta$ will be a line at constant ϕ lying in the tangent plane at $\mathbf{r}$; similarly, $(\partial\mathbf{r}/\partial\phi)\mathrm{d}\phi$ will be a line at constant θ lying in this tangent plane (Fig. 6A.6). The area of the parallelogram whose sides are formed by the vectors $(\partial\mathbf{r}/\partial\theta)\mathrm{d}\theta$ and $(\partial\mathbf{r}/\partial\phi)\mathrm{d}\phi$ can be equated to the surface element $\mathrm{d}\mathcal{S}$ (to first order in $\mathrm{d}\theta$ and $\mathrm{d}\phi$), and the vector element $\mathrm{d}\mathbf{S}$ is the cross product of these vectors,

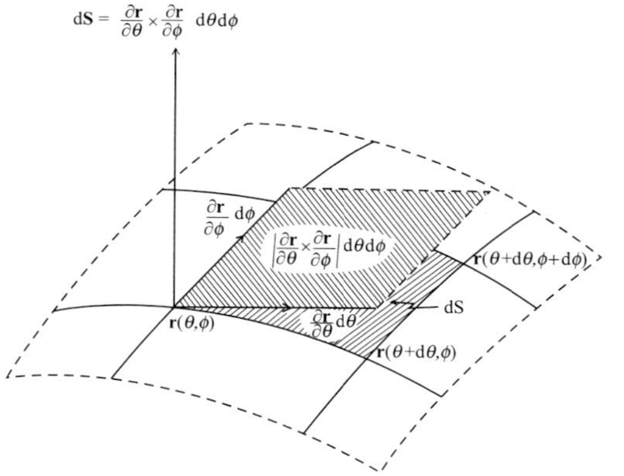

FIG. 6A.6 The area element **dS** on the surface of the convex body is bounded by the lines of constant ϕ and $\phi + d\phi$, and the lines of constant θ and $\theta + d\theta$. This can be approximated by the area of the parallelogram whose sides are $(\partial \mathbf{r}/\partial\theta)\mathrm{d}\theta$ and $(\partial \mathbf{r}/\partial\phi)\mathrm{d}\phi$ and which lies in the tangent plane at $\mathbf{r}(\theta, \phi)$. The vector area element is then given by the vector $\frac{\partial \mathbf{r}}{\partial \theta} \times \frac{\partial \mathbf{r}}{\partial \phi}\mathrm{d}\theta\mathrm{d}\phi$.

$$\mathrm{d}\mathbf{S} = \frac{\partial\, \mathbf{r}}{\partial\, \theta} \times \frac{\partial\, \mathbf{r}}{\partial\, \phi}\, \mathrm{d}\theta\, \mathrm{d}\phi. \tag{6A.32}$$

Substituting this result into (6A.28) and (6A.31) yields (6A.5) and (6A.6a).

6A.5 The second virial coefficient

As a simple illustration of the use of the expression for the excluded volume of two convex bodies (6A.22), we derive an exact expression for the second virial coefficient, due originally to Isihara.[66] For two molecules A and B the second virial coefficient is given by (3.269) as

$$B_{2AB} = \frac{1}{2} \int \mathrm{d}\mathbf{r}_{AB} \langle 1 - \exp[-\beta\, u_{AB}(12)] \rangle_{\omega_A\, \omega_B}, \tag{6A.33}$$

where $u_{AB}(12) = u_{AB}(\mathbf{r}_{AB}\omega_A\omega_B)$ is the pair potential and $\mathrm{d}\mathbf{r}_{AB}$ indicates integration over all relative positions of O_B with respect to O_A. We rewrite this as

$$B_{2AB} = \frac{1}{2} \left\langle \int \mathrm{d}\mathbf{r}_{AB} (1 - \exp[-\beta u_{AB}]) \right\rangle_{\omega_A\, \omega_B}. \tag{6A.34}$$

In evaluating the integral over $\mathbf{r}_{AB}$ in (6A.34) we note that the integrand is 1 or 0 depending on whether the centre of B enters the excluded volume $v_{m,A+B}$ (the shaded region $A + B$ of Fig. 6A.3). Thus this integral is simply $v_{\mathrm{m},A+B}$, so that (6A.34) reduces to

$$B_{2AB} = \frac{1}{2}\langle v_{\mathrm{m},A+B}\rangle_{\omega_A\omega_B} = \frac{1}{2}\langle v_{\mathrm{m},A+B}\rangle_{\omega_B}. \tag{6A.35}$$

In the last step we have used the fact that $\langle v_{\mathrm{m},A+B}\rangle_{\omega_B}$ is independent of ω_A (see Fig. 6A.3). Using (6A.22) for the average excluded volume gives

$$B_{2AB} = \frac{1}{2}(v_{\mathrm{m},A} + v_{\mathrm{m},B} + r_{\mathrm{m},B}s_{\mathrm{m},A} + r_{\mathrm{m},A}s_{\mathrm{m},B}) \tag{6A.36}$$

or, for a pure gas,

$$B_2 = 4v_{\mathrm{m}}f_B \tag{6A.37}$$

with

$$f_B = 1 + \frac{1}{4}(\gamma - 3), \tag{6A.38}$$

where $\gamma = r_{\mathrm{m}}s_{\mathrm{m}}/v_{\mathrm{m}}$ is the shape factor.

References and notes

1. Green, H.S. *The Molecular Theory of Fluids*, p. 51. North-Holland, Amsterdam (1952). This change of variables method was suggested independently by N.N. Bogoliubov in *Studies in statistical mechanics*, Vol. 1, Part A (ed. J. de Boer and G.E. Uhlenbeck). North-Holland, Amsterdam (1962).
2. See § 3.1.4 for the derivation in the canonical ensemble.
3. These arguments seem to originate with Onsager and Kirkwood; Onsager, L., *Chem. Rev.* **13**, 73 (1933); Kirkwood, J.G. *J. Chem. Phys.* **3**, 300 (1935). See also: Lebowitz, J.L., Stell, G., and Baer, S. *J. Math. Phys.* **6**, 1282 (1965).
4. See for example: Patey, G.N., and Valleau, J.P. *J. Chem. Phys.* **61**, 534 (1974); ibid., **64**, 170 (1976).
5. The approach used in this section has been referred to by some authors as the potential distribution theory or potential distribution method.
6. These expressions were derived independently by Widom, B. *J. Chem. Phys.* **39**, 2808 (1963), and Jackson, J.L., and Klein, L.S. *Phys. Fluids* **7**, 232 (1964).
7. We note that (6.31) is not restricted to uniform fluids but applies to nonuniform fluids also, provided that the potential energies $\mathcal{U}$ and $\mathcal{U}_{t\alpha}$ include interactions between the molecules and any external fields. For the nonuniform case it is possible to use this expression to prove that μ_α is constant through an interface between two phases, and to use this result to investigate the interfacial density–orientation profile. See Widom, B. *J. Stat. Phys.* **19**, 563 (1978); Rowlinson, J.S., and Widom, B. *Molecular Theory of Capillarity*, Ch. 4. Oxford University Press (1982).
8. As noted earlier, although the total partition function, Q_{cl}, of Eq. (3.77) is dimensionless, the factored parts of the partition function, such as Q_{c}, do have dimensions, and this leads to the log of a dimensional density in Eqs. (6.31), (6.33), and (6.34). The final result, (6.35), is, however, dimensionally consistent.
9. Equation (6.37) has been derived by M.J. de Oliviera (personal communication, 1981) and independently by K.S. Shing (personal communication, 1981).
10. Other non-$g(12)$ routes to the thermodynamics include the S-route [$S = -k\langle \ell nP\rangle$; see (3.13) of Vol. 1].

11. Equation (6.35) has been used in simulation studies by Adams, D.J. *Mol. Phys.* **28**, 1241 (1974); Romano, S., and Singer, K. ibid. **37**, 1765 (1979); Powles, J.G. ibid. **41**, 715 (1980). See also ref. 12.
12. The use of Eq. (6.35) in computer simulations has been extended to high densities by the use of umbrella sampling techniques: See Shing, K.S., and Gubbins, K.E. *Mol. Phys.* **43**, 717 (1981). For the use of (6.37) for simulations of pure and mixed fluids, with and without umbrella sampling, see Shing, K.S., and Gubbins, K.E. *Mol. Phys.* **46**, 1109 (1982). A somewhat different modified sampling technique suitable for high densities is described by Mezei, M. *Mol. Phys.* **40**, 901 (1980).
13. This result is easily generalized to nonadditive potentials. See Snider, N.S. *J. Chem. Phys.* **55**, 1481 (1971).
14. Eppenga, R., and Frenkel, D. *Mol. Phys.* **52**, 1303 (1984).
15. De Miguel, E., and Jackson, G. *Mol. Phys.* **104**, 3717 (2006).
16. De Miguel, E., and Jackson, G. *J. Chem. Phys.* **125**, 164109 (2006).
17. Gloor, G.J., Jackson, G., Blas, F.J., and de Miguel, E. *J. Chem. Phys.* **123**, 134703 (2005).
18. Kirkwood, J.G., and Boggs, E.M. *J. Chem. Phys.* **10**, 394 (1942).
19. The same coupling procedure is used in deriving the Kirkwood hierarchy of integral equations; see Ch. 5 of Vol. 1.
20. For examples of the use of (6.49) see Jonah, D.A., and King, M.B. *Proc. Roy. Soc.* **A323**, 361 (1971), who test various approximations for $g_{\alpha\beta}$ for atomic liquids at infinite dilution; Ben-Naim, A. *Water and Aqueous Solutions*, p. 321, Plenum, New York (1974), for an application to solution theory.
21. Steele, W.A. *J. Chem. Phys.* **39**, 3197 (1963); Streett, W.B., and Tildesley, D.J. *Proc. Roy. Soc. Lond.* **A 348**, 455 (1976); ibid. **A 355**, 239 (1977). These authors have used a harmonic coefficient $A_{\ell_1\ell_2 m}(r)$, which is related to that given here by

$$A_{\ell_1\ell_2 m}(r) = (4\pi)^{-1}A(\ell_1\ell_2 m; r).$$

In expanding $g(\mathbf{r}\omega_1\omega_2)$ we then find that $g_{ooo}(r) = g(r)$, where $g(r)$ is the centres correlation function.
22. Direct simulation of the space-fixed harmonic coefficients $g(\ell_1\ell_2\ell; r)$ is possible, but less accurate than calculating them from the simulated $g(\ell_1\ell_2 m; r)$ using (3.148); see Haile, J.M., and Gray, C.G. *Chem. Phys. Lett.* **76**, 583 (1980).
23. Tildesley, D.J., Streett, W.B., and Wilson, D.S. *Chem. Phys.* **36**, 63 (1979).
24. Monson, P.A., and Rigby, M. *Mol. Phys.* **38**, 1699 (1979).
25. Nezbeda, I. *Mol. Phys.* **33**, 1287 (1977).
26. Aviram, I., Tildesley, D.J., and Streett, W.B. *Mol. Phys.* **34**, 881 (1977).
27. Bearman, R.J. *Mol. Phys.* **34**, 1687 (1977).
28. Topol, R., and Claverie, P. *Mol. Phys.* **35**, 1753 (1978).
29. This is no longer true for nonrigid molecules; see Stillinger, F. *Adv. Chem. Phys.* **31**, 1 (1975), and also *J. Chem. Phys.* **71**, 1649 (1979).
30. Lowden, L.J., and Chandler, D. *J. Chem. Phys.* **59**, 6587 (1973).
31. Gibbs, J.W. *Elementary Principles in Statistical Mechanics*, Theorem III, p. 131, Yale University Press, New Haven (1902) (reprinted by Dover, 1960).
32. Girardeau, M.D., and Mazo, R.M., *Adv. Chem. Phys.* **24**, 187 (1973).
33. Ishihara, A. *Statistical Physics*, p. 144. Academic, New York (1971).
34. Falk, H. *Amer. J. Phys.* **38**, 858 (1970).
35. Huber, A. in *Methods and Problems of Theoretical Physics* (ed. J.E. Bowcock), North-Holland, Amsterdam (1970).
36. Münster, A. *Statistical Thermodynamics*, Vol. 2, p. 431, 744, Springer-Verlag, Heidelberg (1974).
37. Lukes, T., and Jones, R. *J. Phys. A* **1**, 29 (1968).

38. Ishihara, A. *J. Phys. A* **1**, 539 (1968).
39. Mansoori, G.A., and Canfield, F.B. *J. Chem. Phys.* **53**, 1618 (1970).
40. Rasaiah, J., and Stell, G. *Mol. Phys.* **18**, 249 (1970).
41. Henderson, R.L., and Gray, C.G. *Mol. Phys.* **30**, 881 (1975).
42. Powles, J.G., and Rickayzen, G. *Mol. Phys.* **38**, 1875 (1979). Historical references concerning the derivation of (6.114)–(6.117) are given in Ch. 1 and App. 3D of Vol. 1.
43. Twu, C.H., Gray, C.G., and Gubbins, K.E. *Mol. Phys.* **27**, 1601 (1974).
44. Gray, C.G., Wang, S.S., and Gubbins, K.E. *Chem. Phys. Lett.* **26**, 610 (1974).
45. Clancy, P., and Gubbins, K.E. *Mol. Phys.* **44**, 581 (1981).
46. Young, D.A., and Ross, M. *J. Chem. Phys.* **74**, 6950 (1981); Young, D.A. *Phys. Rev. A* **23**, 1498 (1981).
47. Estévez, G.A., Gould, H., and Cole, M.W. *Phys. Rev. A* **18**, 1222 (1978).
48. McCarty, R.D. *Hydrogen Technological Survey – Thermophysical Properties*, NASA Report SP-3089 (1975).
49. For a detailed discussion of this and other limit theorems see Münster, A. *Statistical Thermodynamics*, Vol. 1, p. 218, Springer-Verlag, Heidelberg (1969).
50. Frisch, H.L., and Helfand, E. *J. Chem. Phys.* **32**, 269 (1960).
51. Maitland, G.C., Rigby, M., Smith, E.B., and Wakeham, W.A. *Intermolecular Forces: Their Origin and Determination*, Ch. 3. Oxford University Press, Oxford (1981).
52. Jonah, D.A., and Rowlinson, J.S. *Trans. Faraday Soc.* **62**, 1067 (1966).
53. Maitland, G.C., and Smith, E.B. *Mol. Phys.* **24**, 1185 (1972).
54. Cox, H.E., Crawford, F.W., Smith, E.B., and Tindell, A.R. *Mol. Phys.* **40** 705 (1980).
55. It is possible to rearrange (3.272) to the form of (6.122) but the resulting equation is not a Laplace transform since $\bar{\phi}$ ($= \bar{u} + \bar{\varepsilon}$, where $\bar{\varepsilon}$, is the well-depth of $\bar{u}$) is temperature-dependent.
56. Macrury, T.B., Steele, W.A., and Berne, B.J. *J. Chem. Phys.* **64**, 1288 (1976).
57. Pitzer, K.S. *J. Am. Chem. Soc.* **77**, 3427 (1955).
58. Pitzer, K.S., and Curl, R.F., Jr. *J. Am. Chem. Soc.* **79**, 2369 (1957). For a review of this and similar methods see Prausnitz, J.M., Lichtenthaler, R.N., and Gomes de Azevedo, E., *Molecular Thermodynamics of Fluid-Phase Equilibria*, 3rd edn, Section 5.7. Prentice-Hall, Upper Saddle River (1999).
59. See, for example: Spurling, T.H., and Mason, E.A. *J. Chem. Phys.* **46**, 322 (1967); Spurling, T.H., De Rocco, A.G., and Storvick, T.S. ibid., **48**, 1006 (1968); Datta, K.K., and Singh, Y. ibid., **55**, 3541 (1971); Johnson, J.R., and Eubank, P.T. *Ind. Eng. Chem. Fundamentals* **12**, 156 (1973).
60. See, for example: Bose, T.K., and Cole, R.H. *J. Chem. Phys.* **52**, 140 (1970); Bose, T.K., and Cole. R.H. ibid., **54**, 3829 (1971); Bose, T.K., Sochanski, J.S., and Cole, R.H. ibid., **57**, 3592 (1972); Kirouac, S., and Bose, T.K., ibid., **59**, 3043 (1973); Hosticka, C., and Bose, T.K. ibid., **60**, 1318 (1974); Copeland, T.G., and Cole, R.H. ibid., **64**, 1741 (1976).
61. Gray, C.G. *J. Chem. Phys.* **50**, 549 (1969); Buck, U., Schleusnener, J., Malik, J., and Secrest, D. *J. Chem. Phys.* **74**, 1707 (1981).
62. Schramm, B., and Leuchs, U. *Ber. Bunsenges. Phys. Chem.* **83**, 847 (1979).
63. Murad, S., Gubbins, K.E., and Powles, J.G. *Mol. Phys.* **40**, 253 (1980).
64. Mason, E.A., and Spurling, T.H. *The Virial Equation of State*, p. 232. Pergamon, Oxford (1969); a detailed review of the earlier work is given in Hirschfelder, J.O., Curtiss, C.F., and Bird, R.B. *Molecular Theory of Gases and Liquids*, Sec. 3.8–3.10, Wiley, New York (1954).
65. Barker, J.A., and Henderson, D. *Rev. Mod. Phys.* **48**, 587 (1976).
66. Isihara, A. *J. Chem. Phys.* **18**, 1446 (1950); for a more detailed account see Isihara, A., and Hayashida, T., *J. Phys. Soc. Japan* **6**, 40, 46 (1951). Some errors in the last of these papers are corrected in Kihara, T. *J. Phys. Soc. Japan* **8**, 686 (1953).

67. Isihara, A. *J. Chem. Phys.* **19**, 397 (1951). The Isihara result can be used to determine the second virial coefficient for linear chains of hard spheres; see Williams, D.C., and Jackson, G. *Mol. Phys.* **86**, 819 (1994).
68. Wertheim, M.S. *J. Chem.. Phys.* **78**, 4625 (1983).
69. Nezbeda, I. *Czech. J. Phys. B* **335**, 752 (1985).
70. Boublik, T., and Nezbeda, I. *Coll. Czech. Chem. Communications* **51**, 2301 (1986).
71. Allen, M.P., Evans, G.T., Frenkel, D., and Mulder, B.M. *Adv. Chem. Phys.* **86**, 1 (1993).
72. Freasier, B.C. *Mol. Phys.* **39**, 1273 (1980).
73. Tildesley, D.J., and Streett, W.B. *Mol. Phys.* **41**, 85 (1980).
74. The Monte Carlo method was described by Metropolis, N.A., Rosenbluth, A.W., Rosenbluth, M.M. Teller, A.H., and Teller, E. *J. Chem. Phys.* **21**, 1087 (1953). It has been applied to virial coefficients of hard spheres by Barker, J.A., and Monaghan, J.J. *J. Chem. Phys.* **36**, 2558 (1962) and by Ree, F.H., and Hoover, W.G. *J. Chem. Phys.* **40**, 939 (1964). The application to virial coefficients of hard nonspherical molecules is described by Rigby, M. *J. Chem. Phys.* **53**, 1021 (1970).
75. Hirschfelder, J.O., Curtiss, C.F., and Bird, R.B. *Molecular Theory of Gases and Liquids*, § 6.7, Wiley, New York (1954); the calculations are due to Wang Chang, C.S., PhD. Dissertation, University of Michigan (1944).
76. Murad, S. PhD. Dissertation, Cornell University (1979).
77. De Boer, J. *Physica* **9**, 363 (1943); Margenau, H. *Phys. Rev.* **64**, 131 (1943).
78. Thakkar, A.J. *Mol. Phys.* **36**, 887 (1978). A similar calculation for the third virial coefficient of hard spheres is given by Gibson, W.G., *Mol. Phys.* **40**, 237 (1980); Boublik, T., and Vosmansky, J. *Coll. Czech. Chem. Commun.* **46**, 542 (1981).
79. See, for example: Barker, J.A. *J. Chem. Phys.* **70**, 2914 (1979), and Stratt, R.M., ibid. **70**, 3630 (1979), and references therein. Also Young, R.A. *Phys. Rev. Lett.* **45**, 638 (1980); Chandler, D., and Wolynes, P.G., *J. Chem. Phys.* **74**, 4078 (1981).
80. See, for example, ref. 79 and: Singh, Y., and Datta, K.K. *J. Chem. Phys.* **53**, 1184 (1970); Pompe, A., and Spurling, T.H. *Aust. J. Chem.* **26**, 855 (1973).
81. McCarty, M., Jr. and Babu, S.V.K. *J. Phys. Chem.* **74**, 1113 (1970); Macrury, T.B. and Steele, W.A. *J. Chem. Phys.* **61**, 3366 (1974).
82. Boublik, T. 'Perturbation Theory', in *Equations of State for Fluids and Fluid Mixtures* (ed. J.V. Sengers, R.F. Kayser, C.J. Peters, and H.J. White, Jr.), Elsevier, Amsterdam (2000).
83. In the f-expansion method, for example (§ 4.6), the reference potential u_o must be recalculated for each new temperature and each new potential by averaging $\exp[-\beta u]$ over orientations; moreover, the integrations needed to obtain A_2 in (4.67) are tedious to carry out.
84. Ananth, M.S., Gubbins, K.E., and Gray, C.G. *Mol. Phys.* **28**, 1005 (1974).
85. Flytzani-Stephanopoulos, M., Gubbins, K.E., and Gray, C.G. *Mol. Phys.* **30**, 1649 (1975).
86. Gray, C.G., Gubbins, K.E., and Twu, C.H. *J. Chem.. Phys.* **69**, 182 (1975).
87. Gubbins, K.E., and Twu, C.H. *Chem. Eng. Sci.* **33**, 863 (1978).
88. Nicolas, J.J., Gubbins, K.E., Streett, W.B., and Tildesley, D.J. *Mol. Phys.* **37**, 1429 (1979).
89. Machado, J.R.S. M.S. Thesis, Cornell University (1979).
90. Din, F. *Thermodynamic Functions of Gases*, Butterworths, London (1961).
91. Twu, C.H., and Gubbins, K.E. *Chem. Eng. Science* **33**, 879 (1978).
92. Calado, J.C.G., Gray, C.G., Gubbins, K.E., Palavra, A.M.F., Soares, V.A.M., Staveley, L.A.K., and Twu, C.H. *J. Chem. Soc., Far. Trans. I* **74**, 893 (1978).
93. Gubbins, K.E., Gray, C.G., and Machado, J.R.S. *Mol. Phys.* **42**, 817 (1981).
94. Moser, B., Lucas, K., and Gubbins, K.E. *Fluid Phase Eqba.* **7**, 153 (1981).

95. Shukla, K.P., Ram, J., and Singh, Y. *Mol. Phys.* **31**, 873 (1976); Shukla, K.P., Singh, S., and Singh, Y. *J. Chem. Phys.* **70**, 3086 (1979); Shukla, K.P., Pandey, L., and Singh, Y. *J. Phys. C* **12**, 4151 (1979).
96. Gibbs, G.M. D. Phil. Thesis, Oxford University (1979).
97. Singh, Y., and Shukla, K.P. in *Molecular-Based Study of Fluids* (ed. J.M. Haile and G.A. Mansoori), ACS Adv. Chem. Series, **204**, 365 (1983).
98. Gibbs, G.M. *Chemical Thermodynamic Data on Fluids and Fluid Mixtures*, NPL Conference, 149, IPC Scientific and Technical Press, Guildford (1979).
99. Vonau, W., Solowejtschik, W, and Winkelmann, J. *Z. Phys. Chem. Leipzig* **269**, 472 (1988).
100. Gubbins, K.E., and O'Connell, J.P. *J. Chem. Phys.* **60**, 3449 (1974).
101. See, for example, Brelvi, S.W., and O'Connell, J.P. *Amer. Inst. Chem. Eng. J.* **18**, 1239 (1972); Mathias, P.M., and O'Connell, J.P. *Adv. Chem. Ser.* **182**, 97 (1979); O'Connell, J.P. *Fluid Phase Eqba.* **6**, 21 (1981).
102. The rotationally invariant forms for $\langle Q\rangle^2$, $\langle\mu^2 Q\rangle$, and $\langle\langle Q\rangle\rangle^3$ are

$$\langle Q\rangle^2 = \frac{2}{3}\mathbf{Q}:\mathbf{Q}$$
$$= \frac{4\pi}{5}\sum_m |Q_{2m}|^2$$
$$\langle\mu^2 Q\rangle = \boldsymbol{\mu}\boldsymbol{\mu}:\mathbf{Q}$$
$$= \left(\frac{4\pi}{3}\right)(6\pi)^{1/2}\sum_{m_1m_2m}\begin{pmatrix}1 & 1 & 2\\ m_1 & m_2 & m\end{pmatrix} Q_{1m_1}Q_{1m_2}Q_{2m}$$
$$\langle\langle Q\rangle\rangle^3 = \frac{4}{3}\mathbf{Q}\cdot\mathbf{Q}:\mathbf{Q}$$
$$= -\left(\frac{4\pi}{5}\right)(14\pi)^{1/2}\sum_{m_1m_2m}\begin{pmatrix}2 & 2 & 2\\ m_1 & m_2 & m\end{pmatrix} Q_{2m_1}Q_{2m_2}Q_{2m}.$$

The equality of the spherical and Cartesian expressions for $\langle Q\rangle^2$, $\langle\mu^2 Q\rangle$, and $\langle\langle Q\rangle\rangle^3$ can be established using the general tensor transformation relations discussed in Appendix A of Vol. 1. Alternatively, we can specialize to the principal axes and use the explicit relations (2.89).
103. Nonaxial quadrupolar molecules that are *nonpolar* ($\mu = 0$) are particularly interesting, since effects due to strong dipole forces will then not mask the nonaxial Q effects. Nonaxial nonpolar molecules possess inversion symmetry, and belong to one of the four point-groups:
C_{2h}: e.g. trans-glyoxal, O=CH-HC=O
D_{2h}: e.g. ethylene, C_2H_4
D_2: e.g. gauche-diphosphine, P_2H_4
C_i: e.g. trans-dichlorodibromoethane, $C_2Cl_2Br_2H_2$
104. Buckingham, A.D. *Physical Chemistry. An Advanced Treatise*, Vol. 4 (ed. H. Eyring, D. Henderson, and W. Jost), p. 349. Academic Press, New York (1970).
105. Herzberg, G. *Molecular Spectra and Molecular Structure, II. Infrared and Raman Spectra of Polyatomic Molecules*, pp. 6, 8. Van Nostrand, New York (1945).
106. For mixtures of ethylene with other quadrupolar molecules $\langle\langle Q\rangle\rangle^3/\langle Q\rangle^3$ is not squared in the expression for A_{3A} and the nonaxial nature of C_2H_4 has a large effect on the mixture properties.
107. Zwanzig, R.W. *J. Chem. Phys.* **23**, 1915 (1955).

108. Rowlinson, J.S. *Liquids and Liquid Mixtures*, 2nd edn, p. 275. Butterworths, London (1969).
109. A misprint occurs in Eq. (4.102) in the first printing of Vol. 1. The last term on the right-hand side of this equation should include a summation over α and β.
110. McGuigan, Lupkowski, M., Paquet, D.M., and Monson, P.A. *Mol. Phys.* **67**, 33 (1989).
111. Lupkowski, M., and Monson, P.A. *Mol. Phys.* **67**, 53 (1989).
112. Fischer, J., and Lago, S. *J. Chem. Phys.* **78**, 5750 (1983).
113. Fischer, J., and Kohler, F. *Fluid Phase Eqba.* **14**, 177 (1983).
114. Fischer, J., Lustig, R., Breitenfelder-Manske, H., and Lemming, W. *Mol. Phys.* **52**, 485 (1984).
115. Bohn, M., Lustig, R., and Fischer, J. *Fluid Phase Eqba.* **25**, 251 (1986).
116. Lustig, R. *Fluid Phase Eqba.* **32**, 117 (1987).
117. Bohn, M., Lustig, R., Fischer, J., and Kohler, F. *Mol. Phys.* **64**, 595 (1988).
118. Lustig, R., Toro-Labbe, A., and Steele, W.A. *Fluid Phase Eqba.* **48**, 1 (1989).
119. Lustig, R. *Mol. Phys.* **59**, 173 (1986).
120. Lustig, R., and Steele, W.A. *Mol. Phys.* **65**, 475 (1988).
121. Sun, T.F., Schouten, J.A., Ten Seldam, C.A., and Biswas, S.N. *Mol. Phys.* **69**, 615 (1990).
122. Friedrich, A., and Lustig, R. *J. Chem. Phys.* **105**, 9597 (1996).
123. Abascal, J.L.F., and Lombardero, M. *J. Chem. Soc. Faraday Trans. 2* **78**, 965 (1982).
124. Goodwin, R.D. *J. Phys. chem. Ref. Data* **17**, 1541 (1988).
125. Kratzke, H., Niepmann, R., Spillner, E., and Kohler, F. *Fluid Phase Eqba* **16**, 287 (1984).
126. Boublik, T. *Mol. Phys.* **32**, 1737 (1976); *Fluid Phase Eqba.* **1**, 37 (1977); ibid. **3**, 85 (1979); Pavlicek, J., and Boublik, T. ibid. **7**, 1 (1981).
127. Vega, C., and Lago, S. *J. Chem. Phys.* **94**, 310 (1991).
128. Vega, C., Lago, S., and Padilla, P. *J. Phys. Chem.* **96**, 1900 (1992).
129. Vega, C., Garzón, B., Macdowell, L.G., and Lago, S. *Mol. Phys.* **85**, 679 (1995).
130. Aim, K., Pavlicek, J., and Boublik, T. *Fluid Phase Eqba.* **116**, 473 (1996).
131. Vega, C., Lago, S., and Padilla, P. *J. Phys. Chem.* **96**, 1900 (1992).
132. Vega, C., Garzón, B., Macdowell, L.G., and Lago, S. *Mol. Phys.* **85**, 679 (1995).
133. See, for example, Rigby, M. *J. Phys. Chem.* **76**, 2014 (1972); Mo, K.C., and Gubbins, K.E. *J. Chem. Phys.* **63**, 1490 (1975).
134. Allen, M.P., Evans, G.T., Frenkel, D., and Mulder, B.M. *Adv. Chem. Phys.* **86**, 1 (1993).
135. Frenkel, D., and Mulder, B.M. *Mol. Phys.* **55**, 1171 (1985).
136. Frenkel, D. *Mol. Phys.* **60**, 1 (1987); see also erratum in ibid. **65**, 493 (1988).
137. Veerman, J.A.C., and Frenkel, D. *Phys. Rev. A* **43**, 4334 (1991).
138. Frenkel, D. *Physica A* **263**, 26 (1999).
139. Perera, A., Cassou, K., Ple, F., and Dubois, S. *Mol. Phys.* **100**, 3409 (2002).
140. De Miguel, E., and Martin del Rio, E. *J. Chem.. Phys.* **118**, 1852 (2003).
141. Cao, D., Chan, K.-Y., Henderson, D., and Wang, W. *Mol. Phys.* **98**, 619 (2000).
142. Boublík, T. *Mol. Phys.* **100**, 3443 (2002).
143. Viduna, D., and Smith, W.R. *Mol. Phys.* **100**, 2815 (2002).
144. See, for example, the f-expansion (§ 4.6) and the RISM (§ 5.5) theories of Vol. 1.
145. Lippert, E., Chatzidimitriou-Dreisman, C.A., and Naumann, K.-H. *Adv. Chem. Phys.* **57**, 311 (1984).
146. Boublik, T., and Nezbeda, I. *Coll. Czech. Chem. Commun.* **51**, 2302 (1986).
147. Boublik, T. in *Equations of State for Fluids and Fluid Mixtures* (ed. J.V. Sengers, R.F. Kayser, C.J. Peters, and H.J. White), Ch. 5, p. 127, Elsevier, Amsterdam (2000).

148. An extended form of the theory that yields the pair correlation function for hard spheres has been presented: see Reiss, H., and Carsberg, R.V. *J. Chem. Phys.* **61**, 1107 (1974).
149. Reiss, H., Frisch, H.L., and Lebowitz, J.L. *J. Chem. Phys.* **31**, 369 (1959). The idea that the solubility of a solute in a given solvent could be calculated by evaluating (i) the work of forming a cavity in the solvent, and (ii) the energy change on placing the solute in the cavity, had considerably predated the work of Reiss *et al.* Kirkwood (Kirkwood, J.G. *J. Chem.. Phys.* **7**, 919 (1939)) considered the energy needed to form a cavity in a fluid. In other early studies the work of cavity formation was obtained from macroscopic arguments; Reiss *et al.* were the first to develop a systematic approach. For the earlier work see Uhlig, H.H. *J. Phys. Chem.* **41**, 1215 (1937).
150. For a review of the early work on hard spheres see Reiss, H. *Adv. Chem. Phys.* **9**, 1 (1965). See also Münster, A. *Statistical Thermodynamics*, Vol. II, p. 431. Springer-Verlag, Berlin (1974); Pierotti, R.A. *Chem. Rev.* **76**, 717 (1976); Reiss, H. in *Statistical Mechanics and Statistical Methods in Theory and Applications* (ed. V. Landman), p. 99, Plenum, New York (1977). Ben-Naim, A. *Hydrophobic Interactions*, App. A.4. Plenum, New York (1980).
151. The SPT result was obtained in 1959 (ref. 149), and preceded the solution of the PY equation by four years. The PY solution was obtained independently by Wertheim and Thiele in 1963: Wertheim, M.S. *Phys. Rev. Lett.* **10**, 321 (1963); *J. Math. Phys.* **5**, 643 (1964); Thiele, E. *J. Chem. Phys.* **39**, 474 (1963). See App. 5.A, Vol. 1, for a simplified derivation using the factorization method.
152. Gibbons, R.M. *Mol. Phys.* **17**, 81 (1969); the extension to mixtures is given in Gibbons, R.M. *Mol. Phys.* **18**, 809 (1970). The approach followed is similar to that used for hard sphere mixtures, ref. 149 and 150. An earlier paper (Ritchie, A.B. *J. Chem. Phys.* **46**, 618 (1967)) gave a somewhat different derivation that yields an expression difficult to use for numerical work. A detailed exposition of the statistical thermodynamics of convex bodies, including a derivation of SPT, is given in Boublik, T. *Mol. Phys.* **27**, 1415 (1974). In this paper Boublik also obtains expressions for the pair correlation function at contact. Yet another derivation, particularly suited to liquids that can form liquid crystals, is given by Barboy, B., and Gelbart, W.M. *J. Stat. Phys.* **22**, 685, 709 (1980). See also the papers of Cotter and Martire, ref. 168 below, who develop forms of the SPT suited to the study of liquid crystals.
153. There are at least three routes to SPT: (a) the route followed here, and the closely related test particle route (see note 156), (b) the Kirkwood route,[149,154] which uses (6.49) as a starting point (c) the cavity route,[149,150] in which the reversible work needed to form a cavity of a given size is calculated; in this last route one often introduces a correlation function $G(12)$, which is proportional to the probability that a solvent molecule is in contact with the solute for a given set of molecular orientations.
154. Ben-Naim, A. *Water and Aqueous Solutions*, p. 321, Plenum, New York (1974).
155. We must take B to be distinct here because intitially R_B is different from the characteristic radii of the other (solvent) molecules. Equations (6.26) and (6.141) still hold in this case—N_B and ρ_B are simply replaced by 1 and $1/V$, respectively.
156. The desired expression for μ_{Br}, Eq. (6.150), can also be obtained directly from the test particle equation for μ_{Br}, (6.35). We average $\exp[-\beta \mathcal{U}_{tB}]$ over all positions and orientations of the solute B molecule, holding the $(N-1)$ solvent molecules fixed. The Boltzmann factor is 0 or 1, accordingly as the molecule centre O_B lies inside the total excluded volume V_{ex} or inside the free volume V_f, respectively. The average of $\exp[-\beta \mathcal{U}_{tB}]$ is just P_1, the probability that 1 occurs. For fixed ω_B the latter is simply V_f/V, and averaging this over ω_B gives the desired result.
157. The excluded volume v_{A+B} must be positive to be physically meaningful. The work of adding body B will be zero if v_{A+B} is zero; this will occur for some minimum

(negative) R_B value that depends on the shape of the bodies and on the orientation of body B. This minimum R_B can be determined by setting the expression v_{A+B} equal to zero.

158. This follows from (a) the fact that $\mu_{Br} = \Delta A_{Br}$ = Helmholtz free energy change on transferring B from ideal gas to dense fluid—see (6.27), and (b) the thermodynamic relation $(\Delta A_{Br})_T = -p\,\Delta V = p\,v_{mB}$.
159. See Nezbeda, I. *Mol. Phys.* **33**, 1287 (1977). By including the surface terms it is possible to obtain from SPT an expression for the surface tension. This has been worked out for hard spheres.[149, 150] There is more than one definition of the surface tension for hard molecules against a wall. See Reiss *et al.*, ref. 149, and Mandell, M.J., and Reiss, H. *J. Stat. Phys.* **13**, 107, 113 (1975); Mandell, M.J. *J. Chem. Phys.* **65**, 813 (1976); Navascués, G., and Tarazona, P. *Mol. Phys.* **37**, 1077 (1979).
160. The points of nonanalyticity have geometrical significance. At $R_B = 0$ it is just possible to accommodate the centres of two molecules in the cavity of diameter σ; at

$$R_B = \sigma\left(\frac{1}{\sqrt{3}} - \frac{1}{2}\right)$$

it is just possible to accomodate the centres of three molecules in the cavity of diameter $\sigma + 2R_B = 2\sigma/\sqrt{3}$, etc.
161. An alternative route to the equation for p is to use the thermodynamic identity

$$V(\partial p/\partial V)_{TN} = \left(\partial\left(\sum_\alpha \mu_\alpha N_\alpha\right)/\partial V\right)_{TN}.$$

Using (6.160) for μ_α in the right-hand side of this expression leads to (6.167) for p. We use (6.162) in preference to this route in order to make explicit the requirement that the molecules must be of the same shape.
162. This form of the equation was first introduced by Few, G.A., and Rigby, M. *Chem. Phys. Lett.* **20**, 433 (1973).
163. Gradshteyn, I.S., and Ryzhik, I.M. *Table of Integrals, Series and Products*, 4th edn, p. 58, 59. Academic, New York (1980).
164. Monson, P.A., and Rigby, M. *Mol. Phys.* **35**, 1337 (1978).
165. Freasier, B.C., and Bearman, R.J. *Mol. Phys.* **32**, 551 (1976).
166. Nezbeda, I., Pavliček, J., and Labik, S. *Coll. Czech. Chem. Comm.* **44**, 3555 (1979).
167. Boublik, T. *Mol. Phys.* **42**, 209 (1981).
168. Cotter, M.A., and Martire, D.E. *J. Chem. Phys.* **52**, 1902, 1909 (1970); ibid. **53**, 4500 (1970). See also Cotter, M.A. *J. Chem.Phys.* **66**, 1098, 4710 (1977); ibid. 67, 4268 (1977).
169. Lasher, G. *J. Chem. Phys.* **53**, 4141 (1971)
170. Gelbart, W.M., and Baron, B.A. *J. Chem. Phys.* **66**, 207 (1977); Gelbart, W.M., and Gelbart, A. *Mol. Phys.* **33**, 1387 (1977); Baron, B.A., and Gelbart, W.M. *J. Chem. Phys.* **67**, 5795 (1977); Barboy, B., and Gelbart, W.M. *Mol. Cryst. Liq. Cryst.* **55**, 209 (1979); Gelbart, W.M., and Barboy, B. *Acc. Chem. Res.* **13**, 290 (1980); Barboy, B., and Gelbart, W.M. *J. Stat. Phys.* **22**, 685, 709 (1980).
171. Wulf, A. *J. Chem. Phys.* **67**, 2254 (1977).
172. Luckhurst, G.R., and Gray, G.W., eds, *The Molecular Physics of Liquid Crystals*, Academic, New York (1979). See the articles by M.A. Cotter.
173. Much of the early work is reviewed in papers given in ref. 150. See particularly the reviews by Reiss (1965) and Pierotti (1976). For an application to liquid water see Stillinger, F.H. *J. Solution Chem.* **2**, 141 (1973), and O'Reilly, D.E. *Phys. Rev. A* **7**, 1659 (1973).

174. Carnahan, N.F., and Starling, K.E. *J. Chem. Phys.* **51**, 635 (1969).
175. Nezbeda, I. *Chem. Phys. Lett.* **41**, 55 (1976); see also *Czech. J. Phys.* **B26**, 355 (1976).
176. The first generalized CS-type equation for hard convex bodies was that proposed by Boublik, T. *J. Chem. Phys.* **63**, 4084 (1975). We do not give this equation here since, at least for prolate spherocylinders, it is less accurate than some of the expressions proposed later (for a comparison of this equation with other equations and with simulation data see refs. 175 and 177).
177. Pavliček, J., Nezbeda, I., and Boublik, T. *Czech. J. Phys. B* **29**, 1061 (1979).
178. Nezbeda, I., Pavliček, J., and Labik, S. *Coll. Czech. Chem. Comm.* **44**, 3555 (1979).
179. Boublik, T. *Mol. Phys.* **42**, 209 (1981).
180. Barboy, B., and Gelbart, W,M. *J. Chem. Phys.* **71**, 3053 (1979). These authors expand pv_m/kT in powers of $y = \eta/(1-\eta)$, and show that the series truncated at the y^3 term gives quite good results for hard spheres, dumbells, and spherocylinders. For the last two shapes the results are, however, somewhat worse than the other expressions given here. Their equation differs from (6.184) in the omission of the final term, $C_4\eta^3/(1-\eta)^3$.
181. Boublik, T. *J. Chem.. Phys.* **53**, 471 (1970); Mansoori, G.A., Carnahan, N.F., Starling, K.E., and Leland, Jr., T.W. *J. Chem.. Phys.* **54**, 1523 (1971).
182. Viduna, D., and Smith, W.R. *Mol. Phys.* **100**, 2903 (2002).
183. Alawneh, M., and Henderson, D. *Mol. Phys.* **106**, 607 (2008).
184. Rigby, M. *Mol. Phys.* **32**, 575 (1976).
185. Boublik, T., and Nezbeda, I. *Chem. Phys. Lett.* **46**, 315 (1977). The application of this method to heteronuclear dumbells is given in Nezbeda, I., and Boublik, T. *Czech. J. Phys.* **B27**, 1071 (1977).
186. It is interesting to note that the equation proposed by Boublik in 1975 (Boublik, T. *J. Chem. Phys.* **63**, 4084), although worse than the N equation in the case of spherocyliners, is somewhat better for dumbells. Comparisons of the Boublik equation with simulation data are given in refs. 184 and 185.
187. Tildesley, D.J., and Streett, W.B. *Mol Phys.* **41**, 85 (1980).
188. Largo, J., and Solana, J.R. *Mol. Phys.* **96**, 1367 (1999).
189. Largo, J., Maeso, M.J., Solana, J.R., Vega, C., and MacDowell, L.G. *J. Chem.. Phys.* **119**, 9633 (2003).
190. Boublik, T. *J. Chem.. Phys.* **119**, 7512 (2003).
191. Nezbeda, I., and Smith, W.R. *J. Chem.. Phys.* **75**, 4060 (1981).
192. Lowden, L.J., and Chandler, D. *J. Chem. Phys.* **59**, 6587 (1973); correction, ibid. **62**, 4246 (1975); Freasier, B.C. *Chem. Phys. Lett.* **35**, 280 (1975); Jolly, D., Freasier, B.C., and Bearman, R.J. *Chem. Phys. Lett.* **46**, 75 (1977).
193. Müller, E., and Gubbins, K.E., in *Equations of State for Fluids and Mixtures* (ed. J.V. Sengers, R.F. Kayser, C.J. Peters, and H.J. White, Jr.), pp. 435–478, Elsevier, Amsterdam (2000).
194. Hobza, P., and Zahradnik, R. *Chem. Rev.* **88**, 871 (1988).
195. Thomson, Jr., C.C., and de Maine, P.A.D. *J. Phys. Chem.* **69**, 2766 (1965).
196. Alger, M.M., and Eckert, C.A. *Ind. Eng. Chem. Fund.* **22**, 249 (1983).
197. Eucken, A. *Z. Elektrochem.* **52**, 255 (1948).
198. Pimental, G.C., and McClellan, A.L. *The Hydrogen Bond*, Freeman, San Francisco (1960).
199. Vinogradov, S.N., and Linnell, R.H. *Hydrogen Bonding*, Van Nostrand Reinhold, New York (1971).
200. Franks, F., ed. *Water: A Comprehensive Treatise*, vol. 2, Plenum Press, New York (1973).
201. Schuster, P., Zundel, G., and Sandorfy, C. *The Hydrogen Bond*, vols. I, II, III, North Holland, New York (1976).

202. Prausnitz, J.M., Lichtenthaler, R.N., and Gomes de Azevedo, E. *Molecular Thermodynamics of Fluid-Phase Equilibria*, 3rd edn, Prentice-Hall, Englewood Cliffs (1999).
203. Dore, J.C., and Teixeira, J., eds. *Hydrogen Bonded Liquids*, Kluwer, Dordrecht (1991).
204. Bellisent-Funel, M.-C., and Dore, J.C., eds. *Hydrogen Bond Networks*, Kluwer, Dordrecht (1994).
205. Walsh, J.M., Greenfield, M.L., Ikonomou, G.D., and Donohue, M.D. *Chem. Eng. Communicn.* **86**, 125 (1989); Walsh, J.M., Greenfield, M.L., Ikonomou, G.D., and Donohue, M.D. *Internat. J. Thermophys.* **11**, 119 (1990).
206. Dolezalek, F. *Z. Phys. Chem.* **64**, 727 (1908).
207. Guggenheim, E.A. *Mixtures*, Oxford University Press, Oxford (1952).
208. Barker, J.A. and Fock, W. *Disc. Faraday Soc.* **15**, 188 (1953).
209. Wheeler, J.C. *J. Chem. Phys.* **62**, 433 (1975); *Ann. Rev. Phys. Chem.* **28**, 411 (1977).
210. Walker, J.S., and Vause, C.A. *Phys. Lett.* A **90**, 419 (1980).
211. Andersen, H.C. *J. Chem. Phys.* **59**, 4717 (1973); *ibid.* **61**, 4985 (1974).
212. Dahl, L.W., and Andersen, H.C. *J. Chem. Phys.* **78**, 1962, 1980 (1983).
213. Chandler, D., and Pratt, L.R. *J. Chem. Phys.* **65**, 2925 (1976); Pratt, L.R., and Chandler, D. *J. Chem. Phys.* **66**, 147 (1977).
214. Høye, J.S., and Olaussen, K. *Physica A* **104**, 435 (1980).
215. Patey, G.N., and Carnie, S.L. *J. Chem. Phys.* **78**, 5183 (1983).
216. Cummings, P.T., and Stell, G. *Mol. Phys.* **51**, 253 (1984).
217. Olaussen, K., and Stell, G. *J. Stat. Phys.* **62**, 221 (1991).
218. Stell, G., and Zhou, Y. *J. Chem. Phys.* **91**, 3618 (1989); ibid. **96**, 1504 (1992); Zhou, Y., and Stell, G. *J. Chem. Phys.* **96**, 1507 (1992).
219. Stell, G., and Zhou, Y. *Fluid Phase Eqba.* **79**, 1 (1992).
220. Wertheim, M.S. *J. Stat. Phys.* **35**, 19 (1984); ibid. **35**, 35 (1984); ibid. **42**, 459 (1986); ibid. **42**, 477 (1986).
221. Sear, R.P., and Jackson, G. in *Observation, Prediction and Simulation of Phase Transitions in Complex Fluids* (ed. M. Baus, L.F. Rull, and J.-P. Ryckaert), Kluwer Academic, Dordrecht (1995).
222. Müller, E.A., and Gubbins, K.E. *Ind. Engng. Chem. Research*, **40**, 2193 (2001).
223. Paricaud, P., Galindo, A., and Jackson, G. *Fluid Phase Eqba.*, **194–197**, 87 (2002).
224. Tan, S.P., Adidharma, H., and Radosz, M. *Ind. Eng. Chem. Research*, **47**, 8063 (2008).
225. Hansen, J.P., and McDonald, I.R. *Theory of Simple Liquids*, 3rd edn, § 4.6, Academic Press, London (2006).
226. Stell, G., in *The Equilibrium Theory of Classical Fluids* (ed. H.L. Frisch and J.L. Lebowitz), pp. 11–171, Benjamin, New York (1964).
227. Joslin, C.G., Gray, C.G., Chapman, W.G., and Gubbins, K.E. *Mol. Phys.* **62**, 843 (1987).
228. Chapman, W.G., Ph.D. Thesis, Cornell University (1988).
229. For any given state condition, the activities $\hat{z}_\alpha$ and a_α defined below Eq. (6.195) and in § 7.4 are related by a constant (of dimension inverse length):

$$\hat{z}_\alpha = a_\alpha \frac{e^{\beta\mu_\alpha^0} q_{qu,\alpha}}{\Lambda_{t,\alpha}^3 \Lambda_{r,\alpha} \Omega_\alpha}.$$

230. Although in general these three steric constraints are approximations, the first two may become exactly true for certain specific molecular models. For example, this will be the case if the molecules have rigid cores and the association points are represented by square well sites of sufficiently short range.

231. Ref. 225, § 11.10.
232. Galindo, A., Burton,S.J., Jackson, G., Visco, D.P. Jr., and Kofke, D.A. *Mol. Phys.* **100**, 2241 (2002).
233. In the usual treatment of the *f*-expansion (§ 4.6) the reference potential is taken to be $\exp[-u_{0,\alpha\beta}(r)/kT] = \langle \exp[-u_{\alpha\beta}(12)/kT] \rangle_{\omega_1\omega_2}$, so that the first-order term of (6.206) vanishes. Here we do not wish to make this specific choice of reference system.
234. Kolafa, J., and Nezbeda. I. *Mol Phys.* **61**, 161 (1987).
235. Jackson, G., Chapman, W.G., and Gubbins, K.E. *Mol. Phys.* **65**, 1, (1988).
236. Chapman, W.G., Gubbins, K.E., Joslin, C.G., and Gray, C.G. *Fluid Phase Eqba.* **29**, 337 (1986)
237. Johnson, J.K., and Gubbins, K.E. *Mol. Phys.* **77**, 1033, (1992).
238. Verlet, L., and Weis, J.-J. *Phys. Rev A.* **5**, 939 (1972).
239. Chapman, W.G. *J. Chem. Phys.* **93**, 4299 (1990).
240. Walsh, J.M., Guedes, H.J.R., and Gubbins, K.E. *J. Phys. Chem.* **96**, 10995 (1992).
241. Walsh, J.M., Koh, C.A., and Gubbins, K.E. *Fluid Phase Eqba.* **76**, 49 (1992).
242. Walsh, J.M., and Gubbins, K.E. *Mol. Phys.* **80**, 65 (1993).
243. Ghosnagi, D., and Chapman, W.G. *Mol. Phys.* **80**, 161 (1993).
244. Banaszak, M., O'Lenick, R., Chiew, Y.C., and Radosz, M. *J. Chem. Phys.* **100**, 3803 (1994).
245. Blas, F.J., and Vega, L.F. *Mol. Phys.* **92**, 135 (1997); *J. Chem. Phys.* **109**, 7405 (1998).
246. Weeks, J.D., Chandler, D., and Andersen, H.C. *J. Chem. Phys.* **54**, 5237 (1971).
247. Müller, E.A., Gubbins, K.E., Tsangaris, D.M., and de Pablo J.J. *J. Chem. Phys.* **103**, 3868 (1995).
248. Tildesley, D.J., and Streett, W.B. *Mol. Phys.* **41**, 85 (1980).
249. Johnson, J.K., Müller, E.A., and Gubbins, K.E. *J. Phys. Chem.* **98**, 6413 (1994).
250. Wertheim, M.S. *J. Chem. Phys.* **87**, 7323 (1987).
251. Dickman, R., and Hall, C.K. *J. Chem. Phys.* **89**, 3168 (1988).
252. Chapman, W.G., Jackson, G., and Gubbins, K.E. *Mol. Phys.* **65**, 1057 (1988).
253. Mansoori, G.A., Carnahan, N.F., Starling, K.E., and Leland, T.W. *J. Chem. Phys.* **54**, 1523 (1971).
254. Reed, T.M., and Gubbins, K.E. *Applied Statistical Mechanics*, Ch. 9, McGraw-Hill, New York (1973).
255. Carnahan, N.F., and Starling, K.E. *J. Chem. Phys.* **51**, 635 (1969).
256. Chapman, W.G., Gubbins, K.E., Jackson, G., and Radosz, M. *Ind. Eng. Chem. Res.* **29**, 1709 (1990).
257. Equation (6.245) is the compressibility factor of chains expressed in terms of the number of spherical segments, N, not the number of chains, N_c. The equation of state for chains is this expression multiplied by m.
258. Ben-Naim, A. *J. Chem.. Phys.* **54**, 1387 (1971); ibid. **54**, 3696 (1971).
259. Wertheim, M.S. *J. Chem.. Phys.* **85**, 2929 (1986).
260. Archer, A.L., and Jackson, G. *Mol. Phys.* **73**, 881 (1991).
261. Sear, R.P., and Jackson, G. *Mol. Phys.* **81**, 801 (1994).
262. Note that the linear approximation differs from the superposition approximation (see Eqs. (3.246) and (4.70)); the latter is symmetrical with respect to the m molecules, whereas the linear approximation is not. For the 3-body correlation function, for example the two approximations give:

$$\text{Superposition}: \quad g(123) = g(12)g(13)g(23)$$

$$\text{Linear}: \quad g(123) = g(12)g(23)$$

263. Brønsted, J.N., and Koefoed, J. *K.danske Vidensk. Selsk. (Mat. Fys. Skr.)* **22**, no. 17 (1946).

264. Rowlinson, J.S., and Swinton, F.L. *Liquids and Liquid Mixtures*, 3rd edn, Butterworths (1982).
265. Jackson, G., and Gubbins, K.E. *Pure appl. Chem*. **61**, 1021 (1989).
266. Ghonasgi, D., and Chapman, W.G. *Mol. Phys*. **83**, 145 (1994).
267. Müller, E.A., and Gubbins, K.E. *Mol. Phys*. **80**, 91 (1993).
268. Amos, M.D., and Jackson, G. *Mol. Phys*. **74**, 191 (1991); Amos, M.D., and Jackson, G. *J. Chem. Phys*. **96**,4604 (1992).
269. Sear, R.P., Amos, M.D., and Jackson, G. *Mol. Phys*. **80**, 777 (1993).
270. Shukla, K.P., and Chapman, W.G. *Mol. Phys*. **91**, 1075 (1997).
271. Banaszak, M., Chien, Y.C., and Radosz, M. *Phys. Rev. E* **48**, 3760 (1993).
272. Banaszak, M., Chen, C.K., and Radosz, M. *Macromolecules* **29**, 6481 (1996).
273. Adidharma, H., and Radosz, M. *Ind. Eng. Chem. Res*. **37**, 4453 (1998).
274. Ghonasgi, D., Llano-Restrepo, M.L., and Chapman, W.G. *J. Chem. Phys*. **98**, 5662 (1993).
275. Johnson, J.K., Müller, E.A., and Gubbins, K.E. *J. Phys. Chem*. **98**, 6413 (1994).
276. Banaszak, M., Chiew, Y.C., O'Lenick, R., and Radosz, M. *J. Chem. Phys*. **100**, 3803 (1994).
277. Ghonasgi, D., and Chapman, W.G. *AIChEJ* **40**, 878 (1994).
278. Blas, F.J., and Vega, L.F. *Mol. Phys*. **92**, 135 (1997).
279. Müller, E.A., Vega, L.F., and Gubbins, K.E. *Mol. Phys*. **83**, 1209 (1994).
280. Jog, P.K., and Chapman, W.G. *Mol. Phys*. **97**, 307 (1999).
281. Gray, C.G., and Gubbins, K.E. *Theory of Molecular Fluids*, Vol. 1, p. 316, Clarendon Press, Oxford (1984).
282. Walsh, J.M., and Gubbins, K.E. *J. Phys. Chem*. **94**, 5115 (1990).
283. Boublik, T., Vega, C., and Diaz-Pena, M. *J. Chem. Phys*. **93**, 730 (1990).
284. Boublik, T., and Nezbeda, I. *Chem. Phys. Lett*. **46**, 315 (1977).
285. Alejandre, J., and Chapela, G.A. *Mol. Phys*. **61**, 1119 (1987).
286. Ghonasgi, D., and Chapman, W.G. *J. Chem. Phys*. **100**, 6633 (1994).
287. Chang, J., and Sandler, S.I. *Chem. Eng. Sci*. **49**, 2777 (1994).
288. Sadus, R.J. *J. Phys. Chem*. **99**, 12363 (1995).
289. Sadus, R.J. *AIChE J* **45**, 2454 (1999).
290. Kushwaha, K.B., and Khanna, K.N. *Mol. Phys*. **97**, 907 (1999).
291. Blas, F.J., and Vega, L.F. *J. Chem. Phys*. **115**, 4355 (2001).
292. Gil-Villegas, A., Galindo, A., Whitehead, P.J., Mills, S.J., Jackson, G., and Burgess, A.N. *J. Chem. Phys*. **106**, 4168 (1997).
293. Galindo, A., Davies, L.A., Gil-Villegas, A., and Jackson, G. *Mol. Phys*. **93**, 241 (1998).
294. McCabe, C., Gil-Villegas, A., Jackson, G., and Del Rio, F. *Mol. Phys*. **97**, 551 (1999).
295. Walsh, J., Guedes, H.J.R., and Gubbins, K.E. *J. Phys. Chem*. **96**, 10995 (1992).
296. Müller, E.A., and Gubbins, K.E. *Ind. Eng. Chem. Res*. **34**, 3662 (1995).
297. Kraska, T., and Gubbins, K.E. *Ind. Eng. Chem. Res*. **35**, 4727 (1996).
298. Suhm, M.A. *Ber. Bunsenges. Phys. Chem*. **99**, 1159 (1995).
299. Sear, R.P., and Jackson, G. *Phys. Rev. E* **50**, 386 (1994).
300. Sear, R.P., and Jackson, G. *Mol. Phys*. **82**, 1033 (1994).
301. Ghonasgi, D., Perez, V., and Chapman, W.G. *J. Chem. Phys*. **101**, 6880 (1994).
302. García-Cuellar, A., Ghonasgi, D., and Chapman, W.G. *Fluid Phase Eqba*. **116**, 275 (1996).
303. Sear, R.P., and Jackson, G. *Mol. Phys*. **87**, 517 (1996).
304. Filipe, E.J.M., Pereira, L.A.M., Dias, L.M.B., Calado, J.C.G., Sear, R.P., and Jackson, G. *J. Phys. Chem. B* **101**, 11243 (1997).
305. García-Cuellar, A., and Chapman, W.G. *Mol. Phys*. **96**, 1063 (1999).

306. Müller, E.A., and Gubbins, K.E. *Mol. Phys*. **80**, 65 (1993).
307. Müller, E.A., and Gubbins, K.E. *Mol. Phys*. **80**, 957 (1993).
308. Phan, S., Kierlik, E., Rosinberg, M.L., Yu, H., and Stell, G. *J. Chem. Phys*. **99**, 5326 (1993).
309. Blas, F.J., and Vega, L.F. *J. Chem. Phys*. **115**, 3906 (2001).
310. Shukla, K.P., and Chapman, W.G. *Mol. Phys*. **98**, 2045 (2000).
311. Ghosnagi, D., and Chapman, W.G. *J. Chem.. Phys*. **102**, 2585 (1995).
312. Garcia-Cuéllar, A.J., and Chapman, W.G. *Mol. Phys*. **96**, 1063 (1999).
313. Vega, C., and MacDowell, L.G. *J. Chem. Phys*. **114**, 10411 (2001).
314. Hall, K.R. *J. Chem.. Phys*. **57**, 2252 (1973).
315. McBride, C., and Vega, C. *J. Chem.. Phys*. **116**, 1757 (2002).
316. Vega, C., Blas, F.J., and Galindo, A. *J. Chem. Phys*. **116**, 7645 (2002).
317. Vroege, G.J., and Lekkerkerker, H.N.W. *Rep. Prog. Phys*. **55**, 1241 (1992).
318. Sear, R.P., and Jackson, G. *Mol. Phys*. **82**, 473 (1994).
319. McGrother, S.C., Sear, R.P., and Jackson, G. *J. Chem. Phys*. **106**, 7315 (1997).
320. Cladis, P.E. *Molec. Crystals Liquid Crystals* **165**, 85 (1988).
321. Sear, R.P., and Jackson, G. *Phys. Rev. Lett*. **74**, 4261 (1995).
322. Chapman, W.G., Gubbins, K.E., Jackson, G., and Radosz, M. *Fluid Phase Eqba*. **52**, 31 (1989).
323. Huang, S.H., and Radosz, M. *Ind. Eng. Chem. Res*. **29**, 2284 (1990).
324. Chen, S.S., and Kreglewski, A. *Ber. Bunsen-Ges. Phys. Chem*. **81**, 1048 (1977).
325. Economou, I.G. *Ind. Eng. Chem. Res*. **41**, 953 (2002).
326. Suhm, M.A. *Ber. Bunsenges. Phys. Chem*. **99**, 1159 (1995).
327. Franck, E.U., and Spalthoff, W. *Z, Electrochem*. **61**, 348 (1957).
328. Vanderzee, C.E., and Rodenburg, W.W. *J. Chem. Thermodyn*. **2**, 461 (1970).
329. MacDowell, L.G., and Vega, C. *J. Chem.. Phys*. **109**, 5681 (1998).
330. Pàmies, J.C., and Vega, L.F. *Ind. Eng. Chem. Res*. **40**, 2532 (2001).
331. Colina, C.M., Turrens, L.F., Gubbins, K.E., Olivera-Fuentes, C., and Vega, L.F. *Ind. Eng. Chem. Res*. **41**, 1069 (2002).
332. Pàmies, J.C., and Vega, L.F. *Mol. Phys*. **100**, 2519 (2002).
333. Lemmon, E.W., Jacobsen, R.T., Penoncello, S.G., and Beyerlein, S.W., *Computer Programs for Calculating Thermodynamic Properties of Fluids of Engineering Interest*, Report 97-1, Center for Applied Thermodynamic Studies, University of Idaho, Moscow, ID (1997).
334. Sychev, V.V., Vasserman, A.A., Zagoruchenko, V.A., Kozlov, A.D., Spiridonov, G.A., and Tsymarny, V.A. *Thermodynamic Properties of Methane*, Hemisphere, New York (1987).
335. Lebowitz, J.L., and Percus, J.K. *Phys. Rev*. **124**, 1673 (1961); Hill, T.L. *Thermodynamics of Small Systems*, Part 1 (1963) and Part 2 (1964), Benjamin, New York; *The Chemical Physics of Atomic and Molecular Clusters* (ed. G. Scoles), North Holland, Amsterdam (1990); Binder, K. *Ann. Rev. Phys. Chem*. **43**, 33 (1992); Rowlinson, J.S. *J. Phys.: Cond. Matter* **6**, A1 (1994); Hill, T.L. *Nano Lett*. **1**, 111 (2001); Gross, D.H.R. *Microcanonical Thermodynamics; Phase Transitions in Small Systems*, World Scientific, Singapore (2001); Reiss, H., and Requera, D. *J. Phys. Chem. B* **108**, 6555 (2004); Sampayo, J.G., Malijevský, A., Müller, E.A., De Miguel, E., and Jackson, G. *J. Chem. Phys*. **132**, 141101 (2010).
336. Thompson, S.M., Gubbins, K.E., Walton, J.P.R.B., Chantry, R.A.R., and Rowlinson, J.S. *J. Chem. Phys*., **81**, 530 (1984).
337. Maxwell, J.C. *Nature* **17**, 278 (1878).
338. Broda, E. *Ludwig Boltzmann: Man, Physicist, Philosopher: L. Boltzmann, Rejoinder to the Heat Theoretical Considerations of Mr. E. Zermelo (1896)*, Ox Bow, Woodbridge (1983).

339. Evans, D.J., and Searles, D.J. *Phys. Rev.E* **50**, 1645 (1994); ibid. **52**, 5839 (1995); Evans, D.J., and Searles, D.J. *Adv. Phys.* **51**, 1529 (2002); Searles, D.J., and Evans, D.J. *J. Chem. Phys.* **113**, 3503 (2000).
340. Wang, G.M., Sevick, E.M., Mittag, E., Searles, D.J., and Evans, D.J. *Phys. Rev. Lett.* **89**, 050601 (2002).
341. Sevick, E.M., Prabhakar, R., Williams, S.R., and Searles, D.J. *Ann. Rev. phys. Chem.* **59**, 603 (2008).
342. Ayton, G., Evans, D.J., and Searles, D.J. *J. chem. Phys.* **115**, 2033 (2001); Evans, D.J., Searles, D.J., and Mittag, E. *Phys. Rev.E* **65**, 051105 (2001).
343. Lymperiadis, A., Adjiman, C. S., Galindo, A., and Jackson, G. *J. Chem. Phys.* **127**, 234903 (2007).
344. Lymperiadis, A., Adjiman, C. S., Jackson, G., and Galindo, A. *Fluid Phase Eqba.* **274**, 85 (2008).
345. Kihara, T. *Rev. Mod. Phys.* **25**, 831 (1953).
346. Kihara, T. *Physical Chemistry; An Advanced Treatise* (ed. H. Eyring, D. Henderson, and W. Jost), Vol. 5, p. 663, Academic, New York (1980).
347. Kihara, T. *Intermolecular Forces*, Ch. 7, Wiley, New York (1978).
348. Hadwiger, H. *Monatsh. Math.* **54**, 345 (1950); Hadwiger, H. *Altes und neues über konvexe korper*, Birkhäuser, Basel (1955). This latter monograph gives a clear account of convex body geometry, and includes a detailed bibliography.
349. Blaschke, W. *Vorlesungen über integralgeometrie II*, Hamburger Math. Einzelschriften, **22** (1937); reprinted by Chelsea, New York (1949).
350. Minkowski, H. *Math. Annalen* **57**, 447 (1903).
351. To see this, note that if the origin is shifted by **a** then relative to the new origin (O') $r_m{}'$ is given by (see (6A.2)) $r_m{}' = \langle \mathbf{r}' \cdot \mathbf{e} \rangle_\omega = \langle (\mathbf{r} + \mathbf{a}) \cdot \mathbf{e} \rangle_\omega = r_m + \mathbf{a} \cdot \langle \mathbf{e} \rangle_\omega = r_m$.
352. These formulae are derived from differential geometry; see, for example, Solnikoff, I.S. *Tensor Analysis*, p. 123, 151, 195, Wiley, New York (1951), and Lipschutz, M.M. *Differential Geometry*, p. 174, McGraw-Hill, New York (1969).
353. Steiner, J. *Gesammelte werke* (ed. K. Wierstrass), 2nd edn, Vol. 2, Chelsea, New York (1971); the original edition was published in 2 volumes in Berlin in 1881 and 1882.
354. A shortened version of the argument is as follows. Writing out (6A.13) will give eight terms. The terms involving

$$\mathbf{r}_A \cdot \frac{\partial \mathbf{r}_A}{\partial \theta} \times \frac{\partial \mathbf{r_A}}{\partial \phi} \quad \text{and} \quad \mathbf{r}_B^* \cdot \frac{\partial \mathbf{r}_B^*}{\partial \theta} \times \frac{\partial \mathbf{r}_B^*}{\partial \phi}$$

are just v_{mA} and v_{mB}, respectively. (To show that the first term above is just v_{mA}, use (6A.12) and change integration variables from (θ, ϕ) to $(\theta', \phi') = (\pi - \theta, \pi + \phi)$; see argument following (6A.17)). Using the permutation property (B.34a) of the triple scalar product, $\mathbf{a} \cdot \mathbf{b} \times \mathbf{c} = -\mathbf{b} \cdot \mathbf{a} \times \mathbf{c} = -\mathbf{c} \cdot \mathbf{b} \times \mathbf{a}$, we get

$$v_{m,A+B} = v_{mA} + v_{mB} + \frac{1}{3} \int \mathrm{d}\theta \mathrm{d}\phi \mathbf{r}_A \cdot \frac{\partial \mathbf{r}_B^*}{\partial \theta} \times \frac{\partial \mathbf{r}_B^*}{\partial \phi}$$
$$- \frac{1}{3} \int \mathrm{d}\theta \mathrm{d}\phi \frac{\partial \mathbf{r}_A}{\partial \theta} \cdot \mathbf{r}_B^* \times \frac{\partial \mathbf{r}_B^*}{\partial \phi} - \frac{1}{3} \int \mathrm{d}\theta \mathrm{d}\phi \frac{\partial \mathbf{r}_A}{\partial \phi} \cdot \frac{\partial \mathbf{r}_B^*}{\partial \theta} \times \mathbf{r}_B^* + (\mathbf{r}_A \leftrightarrow \mathbf{r}_B^*),$$

where $(\mathbf{r}_A \leftrightarrow \mathbf{r}_B^*)$ denotes three terms obtained from the preceding three by interchanging $\mathbf{r}_A$ and $\mathbf{r}_B^*$. We can integrate by parts the fourth and fifth terms with respect to θ and ϕ, respectively. The integrated terms vanish by virtue of $\partial \mathbf{r}_B / \partial \phi$ vanishing at $\theta = 0$ and $\theta = \pi$ (Fig. 6A.4) and the integrand value being periodic in ϕ with period 2π. The fourth and fifth terms thus each generate a term equal to the third term, plus second derivative terms that cancel by virtue of

$$\mathbf{r}_B^* \times \frac{\partial^2 \mathbf{r}_B^*}{\partial\theta\,\partial\phi} = -\frac{\partial^2 \mathbf{r}_B^*}{\partial\phi\,\partial\theta} \times \mathbf{r}_B^*$$

Thus, we find (6A.18).

7

THERMODYNAMIC PROPERTIES OF MIXTURES

Our fundamental problem. . . is to determine so far as possible how the escaping tendencies of the various components of a solution may be expected to vary, first, with the composition of the solution, and second, with the nature of the components. It is evident that this is a problem of great importance, both from the theoretical and practical standpoints,. . . The chemist is usually acquainted with but few rules for his guidance beyond the simple maxim that 'like dissolves like', which, though we make it more impressive by quoting it in Latin, is of but limited usefulness, because it leaves open the question as to what are the criteria for likeness.

Joel H. Hildebrand, Introduction to *Solubility* (1924)

The models (of mixtures) can be expected to be useful representations of only the simplest mixtures. In particular, mixtures containing electrolytes or highly polar molecules are entirely excluded from consideration.

E.A. Guggenheim, Preface to *Mixtures* (1951)

In this chapter we consider the thermodynamic properties of fluid mixtures composed of components $A, B, C, \cdots, R$. As for pure fluids, the basic equation for the configurational free energy, A_c, is given by (6.1) in the canonical ensemble, but the partition function Q_c is now given by the mixture equation (3.250) of Vol. 1. The pressure, configurational internal energy, and heat capacity at constant volume are given by (6.3)–(6.5), with the understanding that the number of molecules of each species, N_α, is held fixed. Equations for the thermodynamic functions can be derived in terms of the correlation functions (cf. §§ 6.1 to 6.7), and are in general obvious extensions of the corresponding equations for the pure fluids.

The scientific interest in mixtures springs from the extra degrees of freedom provided by the composition variables, and the consequent diversity of physical behaviour that is possible. These include liquid–liquid and fluid–fluid phase equilibria, in which two dense fluid phases coexist, solvation effects (clustering of molecules of one species about another), and new critical points (consolute points, tricritical points, etc.). Several sets of independent variables can be used when discussing mixture behaviour. The 'experimental' variables are pressure, P, temperature, T, and mole fractions, $x_A, x_B, \cdots, x_R$, where $x_\alpha \equiv N_\alpha/N$. However, for some purposes it is more convenient to use either the canonical variables $T, V, x_A, \cdots, x_R$, or the 'field' variables $T, P, \mu_A, \mu_B, \cdots, \mu_R$, where

$\mu_\alpha = (\partial A/\partial N_\alpha)_{T,V,N'}$ is the chemical potential of component α and N' denotes that all the N_β are held constant except N_α.

Mixtures are also of great practical and technological importance. Virtually all naturally occurring fluids are mixtures, and fluid processing in the chemical, oil, food, and pharmaceutical industries involve separation, purification, and reaction of mixtures. An understanding of phase equilibria is of particular importance, but the large number of possible temperatures, pressures, and compositions precludes experimental measurements for all but a few of the possible mixtures of possible interest. This is immediately apparent when one considers that the number of important chemicals produced commercially is on the order of 1000; thus the number of binary mixtures of these components is 500,000 ($^1/_2 n(n-1)$, where n is the number of pure components), the number of ternary mixtures is 166,167,000 ($10^9/3!$), and so on. Reliable prediction methods are therefore of particular importance. The attractiveness of having such methods available is further illustrated by consideration of the cost of experimental measurements. For example, the cost of a single vapour–liquid equilibrium measurement for a binary mixture (measuring the composition of the two coexisting phases at a single temperature and pressure) in an industrial laboratory has been estimated as \$2,000 and requiring 2 days[1] (1987 prices).

7.1 Qualitative behaviour

Before discussing the theory of mixtures, we first briefly review the qualitative thermodynamic behaviour of common mixtures. Since phase equilibria are of particular interest we focus to a large extent on phase diagrams; for simplicity, we restrict our discussion to binary mixtures.

7.1.1 Brief historical background

Experimental studies

The extension by van der Waals of his equation of state to mixtures in 1890[2] prompted Kamerlingh Onnes and his co-workers at Leiden to undertake an extensive series of experimental studies of binary phase diagrams. This work at Leiden, and later at Amsterdam, was carried out over the next twenty years, and laid the groundwork for much of our modern knowledge of phase behaviour in such mixtures. The phase rule, first enunciated by Gibbs in 1876,[3] was practically unnoticed at first, but its significance was later understood by van der Waals and his school, and they were able to use it to systematically interpret their experimental phase diagrams. These workers studied the phase diagrams as a function of pressure as well as temperature, and discovered a variety of types of complex behaviour. This work led to the publication of the classic books by Bakhuis Roozeboom,[4] Kuenen,[5] and van der Waals and Kohnstamm,[6] which considered binary phase diagrams in detail.

Towards the end of the nineteenth century, van der Waals[7] began applying his equation of state to investigate the A-V-x surfaces of binary fluid mixtures. He predicted various types of phase equilibria, including the existence of phase separations at temperatures *above* the critical temperatures of both components (later called 'gas–gas' or 'fluid–fluid' equilibria). The possibility of such high temperature, high pressure phase equilibria was discussed in greater detail in a paper by Kamerlingh Onnes and Keesom[8] in 1907; they called them 'limited miscibility in the gas phase'. The first experimental confirmation of van der Waals' predictions of high temperature, high pressure phase equilibria was found by Krichevskii (1940),[9] who observed that mixtures of nitrogen and ammonia separate into two fluid phases at temperatures above the critical temperatures of the two pure components, and above pressures of approximately 1000 bars. In subsequent experiments at pressures up to 15,000 bar, Krichevskii, Tsiklis, and their Russian co-workers discovered such supercritical fluid phase separations in over 20 binary mixtures. They called these phase separations 'gas–gas equilibria', and this rather misleading term gained general acceptance. Systematic studies of the effect of pressure on the liquid–liquid coexistence behaviour were first carried out by Timmermans working with Kohnstamm at Amsterdam, and later with his own group at Brussels.[10] Later, in the 1960s and 1970s, this work was extended to higher pressures and a great variety of systems by Schneider[11] at Bochum.

The history of the early Dutch work has been described in detail by Rowlinson[12] and by Levelt Sengers.[13] Reviews of both early and more modern experimental work on binary fluid phase diagrams have been given by Rowlinson,[14] Schneider,[11] de Swaan Arons,[15] Streett,[16] and Koningsveld *et al.*[17] Experimental work on excess properties of mixtures has been reviewed by Rowlinson[14] and more recently by Smith *et al.*[18]

Theoretical interpretation

Interpretations of binary phase diagrams have been of two types. The first approach makes use of empirical equations of state and is exemplified by the work of Van Konynenburg and Scott[19,20] based on van der Waals' equation of state. Van Konynenburg and Scott found that they could qualitatively predict many of the observed classes of phase behaviour for binary mixtures by using appropriate a and b parameters in the van der Waals equation; an exception was the class of systems exhibiting such phenomena as low-temperature lower critical solution temperatures (e.g., aqueous mixtures of amines or alcohols in which 'hydrogen bonding' between unlike pairs is important). Similar studies have been made by other workers for other semiempirical equations of state; for example, Dieters and Schneider[21] used the Redlich–Kwong equation to classify binary phase diagrams.

Interpretations of this first kind, based on empirical equations of state, tell little about the underlying intermolecular forces that give rise to particular types of phase behaviour. The second approach is based rigorously on the statistical thermodynamics of mixtures. Using some simplified model of the intermolecular potential, the thermodynamic functions and phase behaviour are calculated. Approximations are necessary in the theory for the dense fluid, but such approximations can be checked by comparison with molecular simulation, and are not thought to seriously affect the qualitative conclusions obtained. The object of this second approach is to determine the types of phase diagrams that may be reasonably associated with potential models of various types. Studies of this sort are described in more detail later in this chapter, but have included the following:

(a) Lebowitz and Rowlinson[22] used Percus–Yevick theory to study hard sphere mixtures. They concluded that such mixtures exhibited solid–fluid but not fluid–fluid equilibria.
(b) Gibbons[23] used scaled particle theory to study binary mixtures of hard, non-spherical, convex particles; again, no fluid–fluid equilibria were found.
(c) Mixtures of Lennard-Jones molecules have been studied by various methods, including computer simulation,[24] conformal solution theory,[24–26] and hard sphere perturbation theory.[24,25,27] Lennard-Jones mixtures show a wide range of types of fluid–fluid equilibria, but only exhibit class II (see below) liquid–liquid equilibria when the unlike pair interaction is improbably weak.
(d) Mixtures in which the potential is of the type[28–30] $u = u_0 + u_a$, where u_0 is either the hard sphere or Lennard-Jones model and u_a is some anisotropic potential (e.g. dipolar, quadrupolar, etc.). Such mixtures exhibit a much wider range of fluid phase phenomena, including liquid–liquid immiscibility.

7.1.2 *The classification of binary phase diagrams*

It is convenient to adopt the classification of binary phase diagrams shown in Fig. 7.1, which is a modification of that suggested by Van Konynenburg and Scott.[20] Their classification was based on van der Waals' equation, which is capable of describing classes I to V; in Fig. 7.1, class VI (which was not included in the original classification) has been added.[30] The classification is based on the presence or absence of three-phase lines and the way critical lines connect with these, and focuses on the fluid part of the phase diagram, neglecting solid phases; this is best seen on a *PT* projection of the usual *PTx* diagram†. In classes I, II, and VI the two components *A* and *B* have similar critical temperatures, and the gas–liquid critical line passes continuously between the pure component critical points as a function of composition; classes II and VI mixtures differ from class I in that they are more non-ideal and show liquid–liquid immiscibility. Class II behaviour is common, whereas class VI, in which closed solubility loops occur, arises less frequently (e.g. water mixed with some alcohols or amines). Mixtures of classes III, IV, and V are often composed of components with widely different critical temperatures $\left(T_B^c/T_A^c \geq 2\right)$, and the gas–liquid critical curve does not pass continuously from one pure component to the other (e.g. because the liquid–liquid immiscibility region extends to that of the gas–liquid critical curve). Included in class III are systems that exhibit 'gas–gas' immiscibility. These six classes

† In this chapter we use P, rather than p, for pressure when showing or discussing phase diagrams.

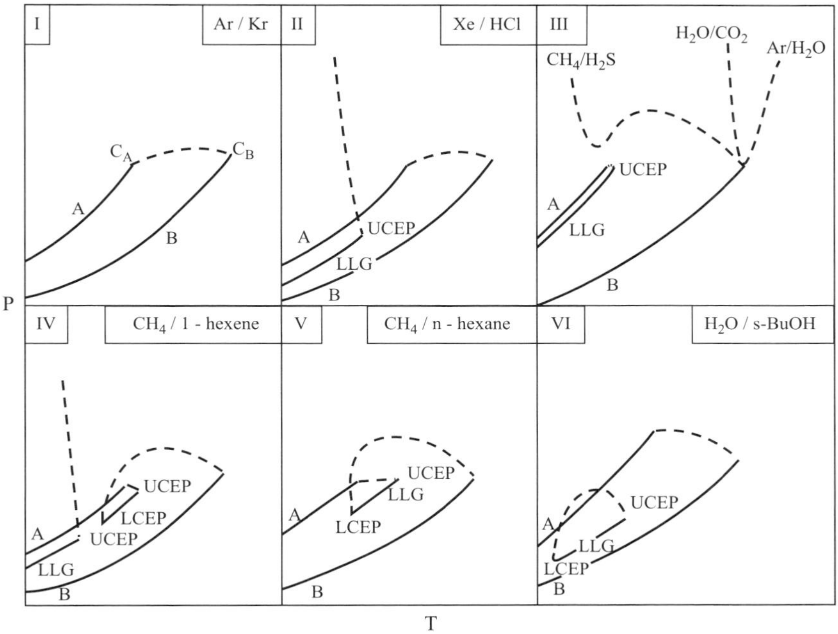

FIG. 7.1 Classification of binary phase diagrams for a mixture of components A and B. Here solid lines labeled A and B are the pure component vapour pressure curves and dashed curves show the composition dependence of critical points for the mixture. Solid lines labeled LLG are the three-phase liquid–liquid–gas coexistence lines. UCEP = upper critical end point and LCEP = lower critical end point denote the ends of critical lines. s-BuOH = secondary butanol. (From ref. 113(a).)

can be further subdivided, according to whether azeotropes are formed, etc. An azeotropic point is one where two or more phases are in thermodynamic equilibrium, and have the same composition (mole fraction).[31] The practical significance of azeotropes is that one cannot take advantage of the difference in composition of the two coexisting phases to effect separation of the components, as is normally done in distillation and liquid–liquid extraction, for example.

A more detailed discussion of these classes is given in the remainder of this section. For simplicity we omit the solid regions of the phase diagrams. The added complexities resulting from the solid phases have been well described by Rowlinson[14] and Streett.[16]

Class I

In this class there is no liquid–liquid separation, and the solution is relatively ideal ($G^E/RT < 0.5$, where $G^E \equiv G - G^{\text{id solution}}$ is the excess molar Gibbs energy relative to an ideal solution (defined in § 7.3) at the same temperature and pressue, and $R = N_A k$ is the gas constant; here N_A is Avogradro's number). The vapour–liquid critical locus connects the two pure critical points in PTx space.

In Fig. 7.2 the *PTx* diagram is shown for argon/krypton, a typical class I system in which there is no azeotrope. C_A and C_K are the critical points for the pure argon and krypton, and the critical locus for the mixture connects these points. The shaded areas at 120, 140, 160, and 180 K are isothermal cuts through the gas–liquid coexistence surfaces, the upper surface being the liquid (bubble-point) and the lower the gas (dew-point). Also shown is an isobaric cut at 20 bar. Figure 7.3 shows the *Px* and *Tx* cuts and the *PT* projections respectively for this system. In the *Px* and *Tx* cuts the tie-lines are horizontal. When these cuts intersect the critical locus, the critical point is at a maximum or minimum pressure (for the

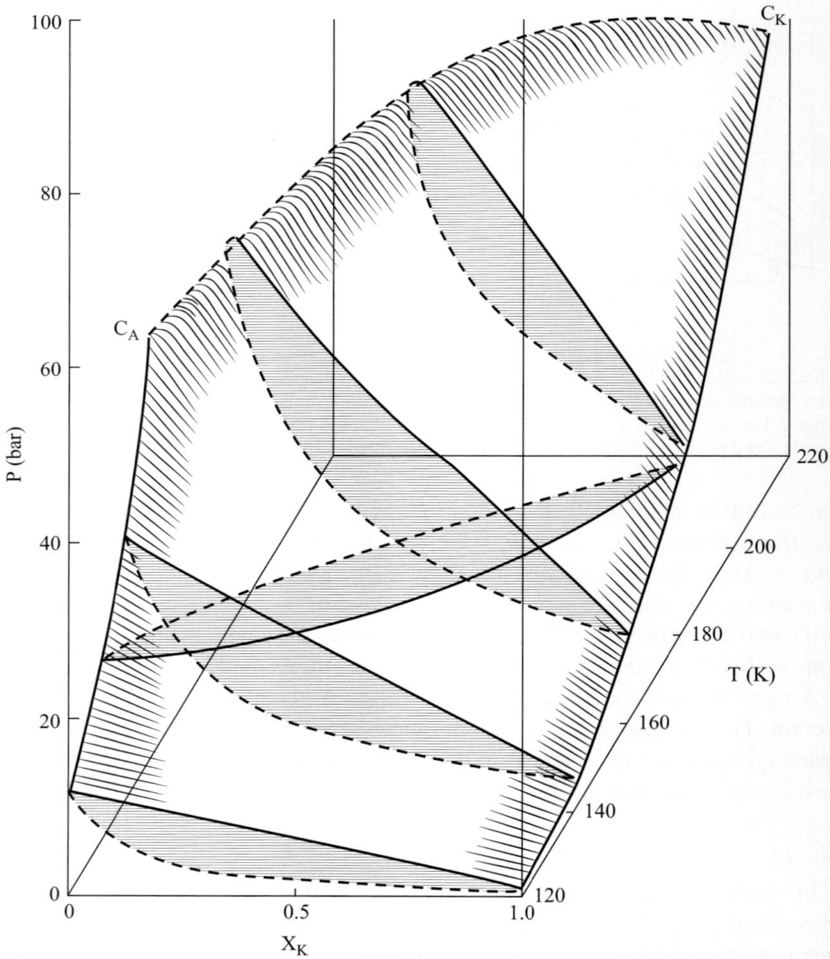

FIG. 7.2 *PTx* phase diagram for argon/krypton mixtures (class I).

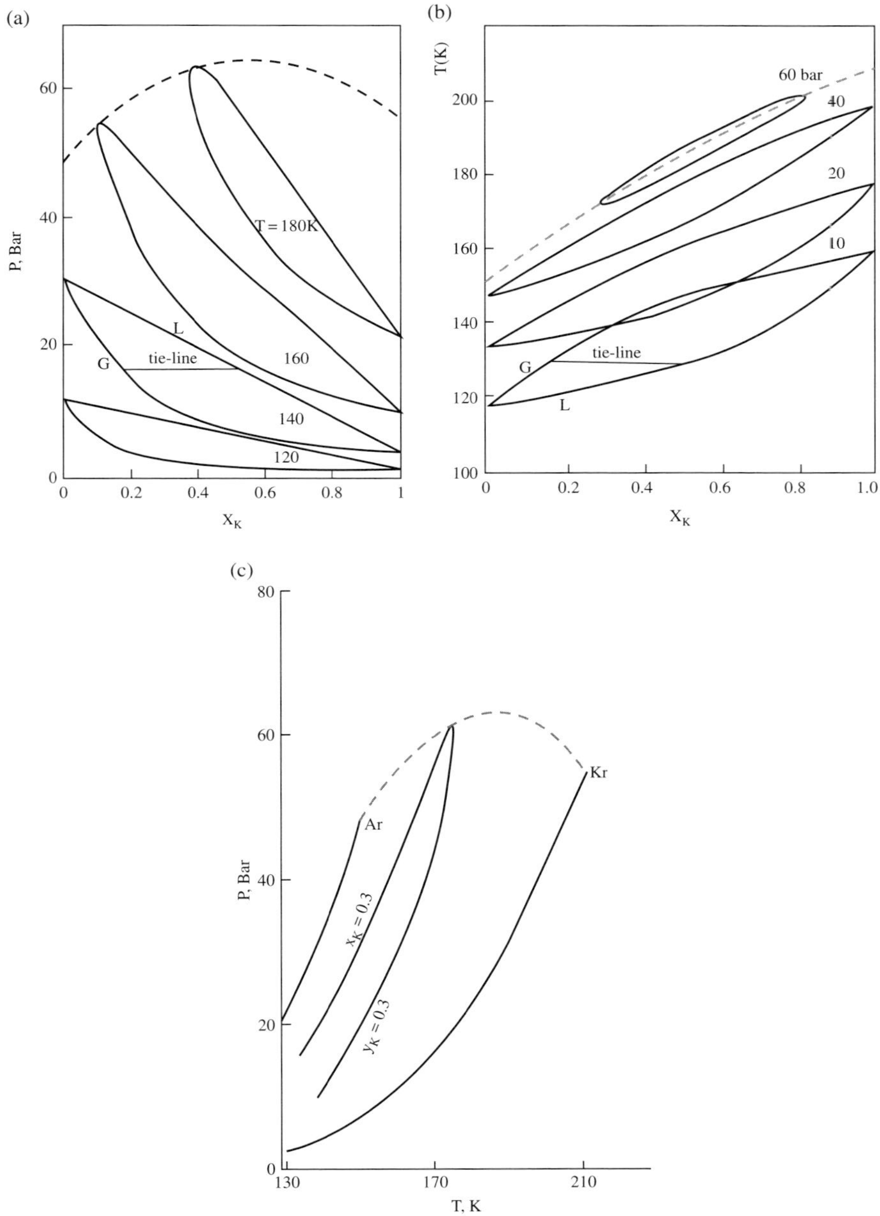

FIG. 7.3 Cuts and projections of the phase diagram for argon/krypton mixtures: (a) *Px* cuts at various temperatures; (b) *Tx* cuts at various pressures; (c) *PT* projection showing one cut at a constant composition of 30 mole % krypton. Here and hereafter G and L denote gas and liquid, and x_K and y_K are the mole fractions of krypton in the liquid and gas phases, respectively. A 'tie line' joins coexisting gas and liquid phases.

Px case) or temperature (for the *Tx* case); i.e. the critical point occurs when the tie-line just vanishes. For the *PT* projection the tie-lines are perpendicular to the *PT* plane, and the critical point is not, in general, at the maximum or minimum on a constant composition loop. Figures 7.4 and 7.5 show *PTx* plots for the systems ethane/*n*-heptane and carbon dioxide/ethane, respectively. The ethane/*n*-heptane system is similar to argon/krypton, but the greater difference in volatility leads

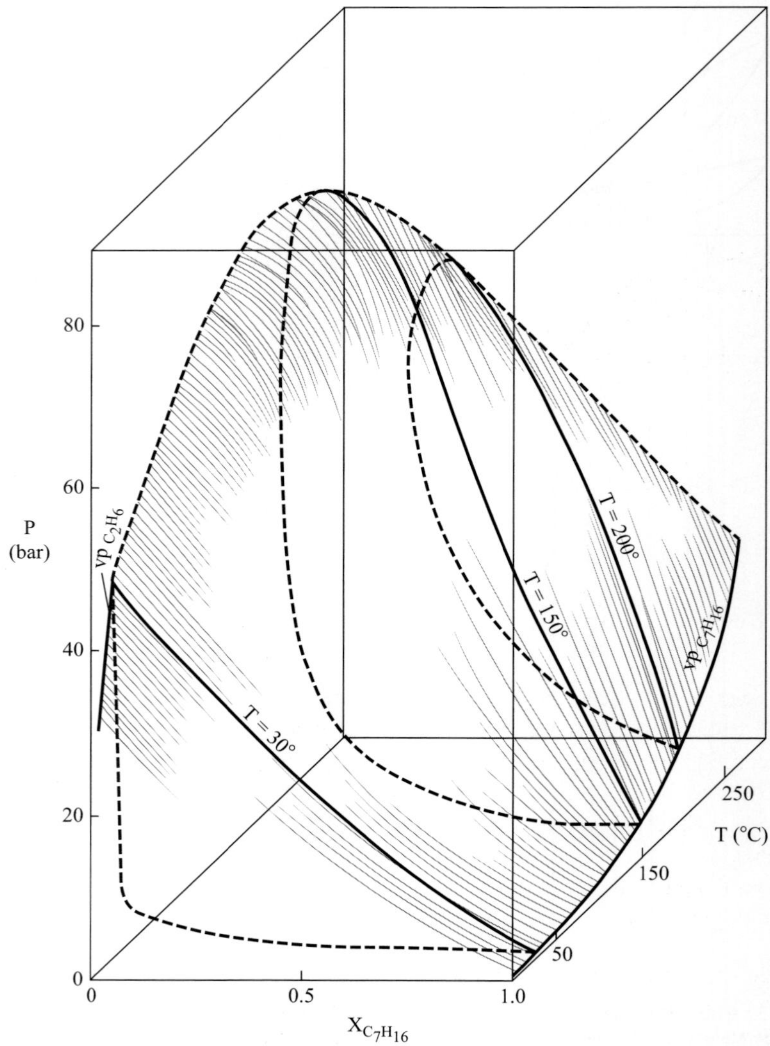

FIG. 7.4 *PTx* phase diagram for ethane/*n*-heptane mixtures. vp = vapour pressure.

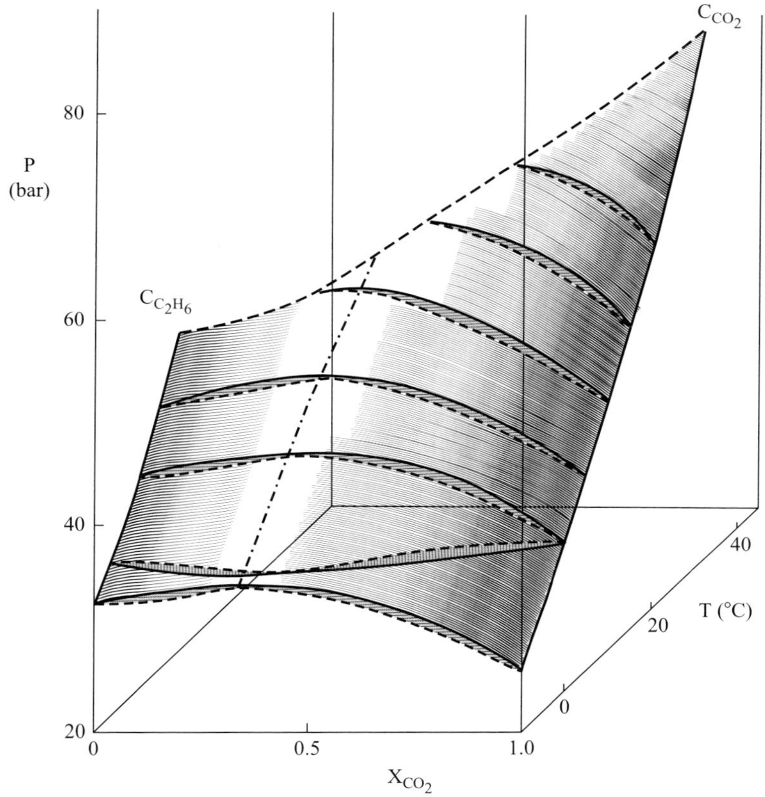

FIG. 7.5 *PTx* phase diagram for carbon dioxide/ethane mixtures, showing six isothermal cuts and one isobaric one. The dash-dot line is the locus of the azeotropic points.

to greater separation of the gas and liquid surfaces and a pronounced maximum in the critical locus. In the carbon dioxide/ethane system the gas and liquid surfaces are only slightly separated, and the system has a positive azeotrope, i.e. an azeotrope that has a positive departure ($G^E > 0$) from ideal solution behaviour. Figure 7.6 shows *PT* projections for ethane/*n*-heptane and carbon dioxide/ethane systems.

Class I can be divided into three subclasses which differ in whether azeotropes occur (Figure 7.7):

(a) No azeotrope: e.g., Ar/Kr (Fig. 7.2), C_2H_6/n-C_7H_{16} (Fig. 7.4);
(b) Positive azeotrope ($G^E > 0$): e.g. CO_2/C_2H_6 (Fig. 7.5), CH_3OH/C_6H_6, EtAc/CCl_4, CH_3OH/methyl ethyl ketone; and
(c) Negative azeotrope ($G^E < 0$): e.g. CO_2/C_2H_2, HCl/dimethyl ether, chloroform/acetone.

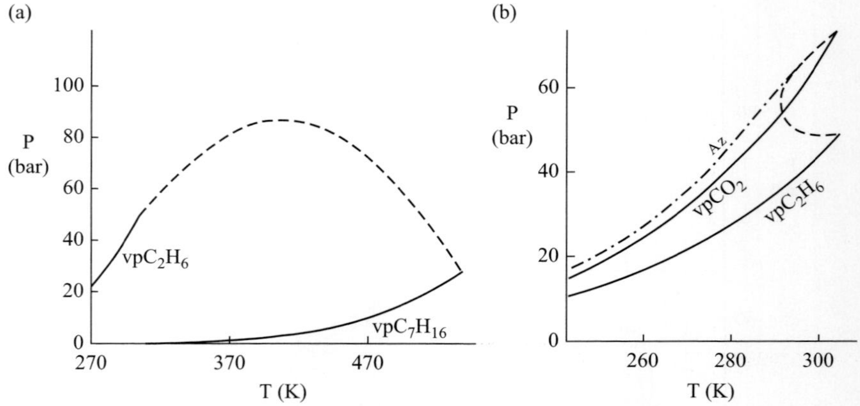

FIG. 7.6 PT projections for (a) ethane/*n*-heptane, (b) carbon dioxide/ethane systems. Here vp denotes pure component vapour pressure line and Az denotes a locus of azeotropic points.

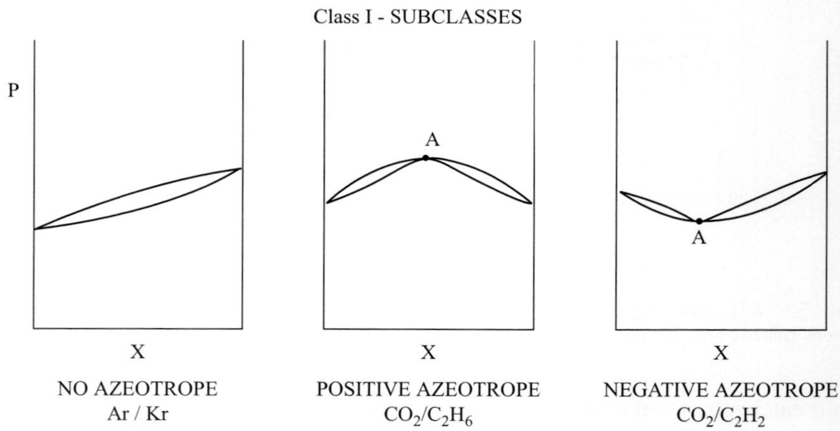

FIG. 7.7 Subclasses of class I, depending on presence or absence of azeotropes.

These azeotropes can be further subdivided as follows (see Fig. 7.8) according to whether the azeotrope is absolute or limited above, below, or both:

(i) Absolute azeotrope: e.g. CO_2/C_2H_2 (Fig. 7.8 (i));
(ii) Azeotrope limited above: e.g. chloroform/acetone, CH_3OH/C_6H_6, ethyl acetate/CCl_4, CH_3OH/methyl ethyl ketone (Fig. 7.8 (ii));
(iii) Azeotrope limited below: e.g. CO_2/ethylene, C_2H_4/C_2H_2, H_2O/C_2H_5OH, acetone/H_2O (Fig. 7.8 (iii));
(iv) Azeotrope limited above and below by different components: e.g. 2,4-dimethylpentane/2,2,3- trimethylbutane, CH_3OH/acetone (Fig. 7.8 (iv)); and

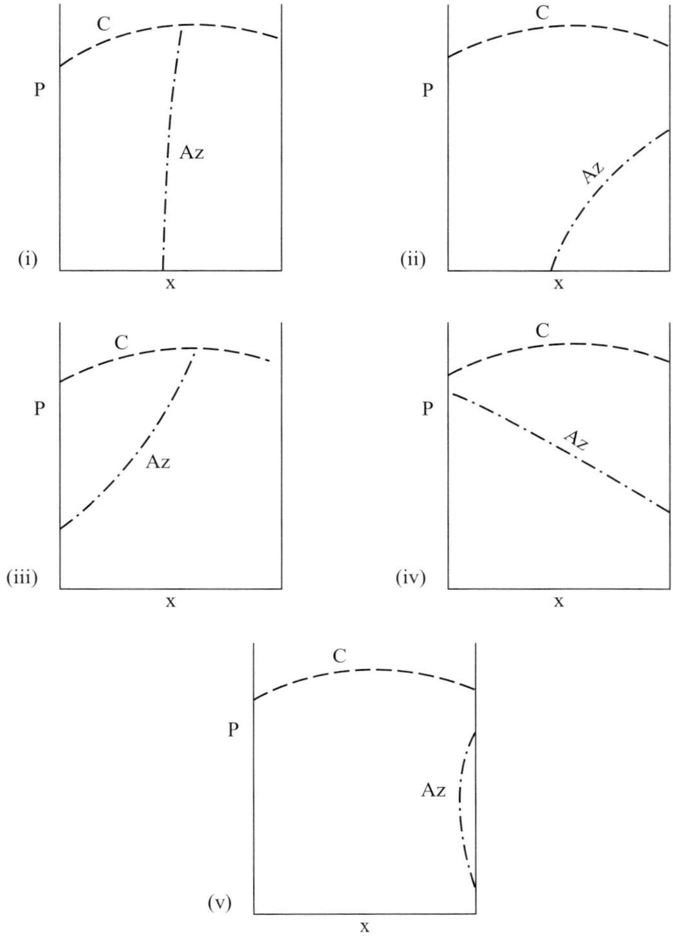

FIG. 7.8 Types of azeotropes (class I); dash-dot lines show the locus of azeotropic points. Dashed line C denotes the loci of critical points.

(v) Azeotrope limited above and below by the same component: e.g. phenol/H_2O (Fig. 7.8 (v)).

A more complete discussion of these subclasses is given by Rowlinson.[14]

Class II

In this class the solution is quite nonideal ($G^E/RT \gtrsim 0.5$) and liquid–liquid immiscibility occurs at low temperatures, the liquid–liquid region being bounded by a locus of upper critical solution points. This liquid–liquid critical line is quite distinct from the gas–liquid critical line, which connects the pure component

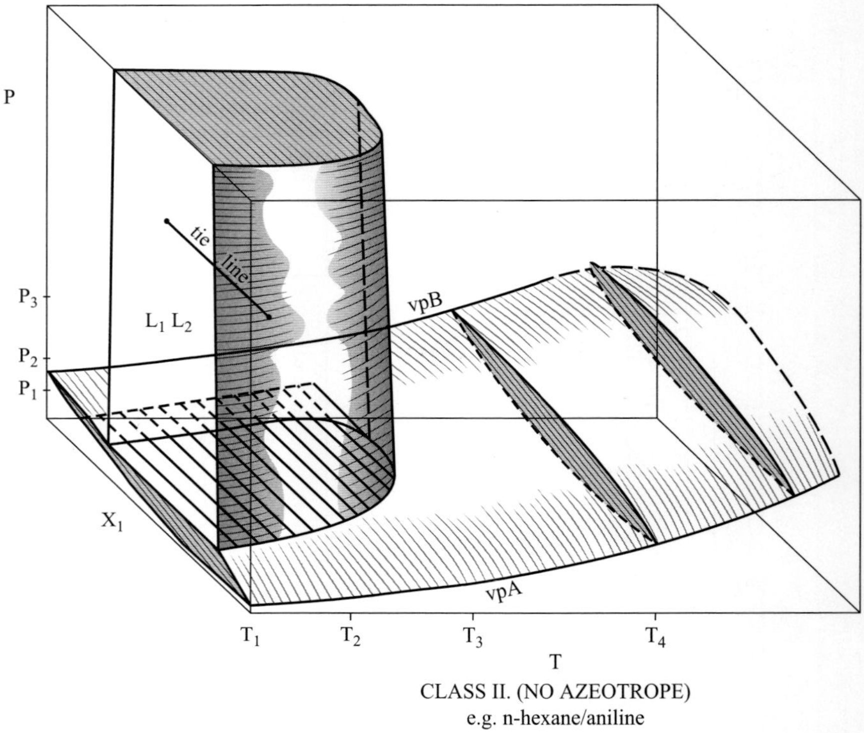

FIG. 7.9 Typical class II system (no azeotrope), e.g. n-hexane/aniline. The pressures P_1, P_2,... and temperatures T_1, T_2,... refer to isobaric and isothermal cuts through this three-dimensional diagram, and are shown in Fig. 7.10. The curved surface rising almost vertically is the liquid L_1–liquid L_2 coexistence surface (a typical tie line connecting these two phases is shown), and meets the vapour – liquid coexistence surface at the three-phase L_1L_2G surface, shown as a ruled surface. The tie-lines shown for the L_1L_2G region connect the L_1 liquid phase on the right with the L_2 liquid phase and the G phase at the left extremity of the tie-line.

critical points. A typical *PTx* diagram for a case in which there is no azeotrope is shown in Fig. 7.9. The liquid–liquid coexistence surfaces meet the gas–liquid surface in the three-phase region, where liquid phases L_1 and L_2 are in equilibrium with each other and with a gas phase (the LLG surface). The corresponding *Px*, *Tx*, and *PT* plots are shown in Fig. 7.10. The three-phase region is a curved ruled surface that passes through the tie lines connecting the three coexisting phases (liquid 1, liquid 2, and gas), and is shown enlarged in Fig. 7.11 (a).

Several subclasses are possible, depending on the presence of azeotropes and whether these are homogeneous or heterogeneous. A homogeneous azeotrope occurs when the vapour phase is in equilibrium with a single liquid phase. A heterogeneous azeotrope occurs when the vapour phase is in equilibrium with two liquid phases of different composition simultaneously, such that the overall composition of the two liquid phases equals that of the vapour phase. A heteroge-

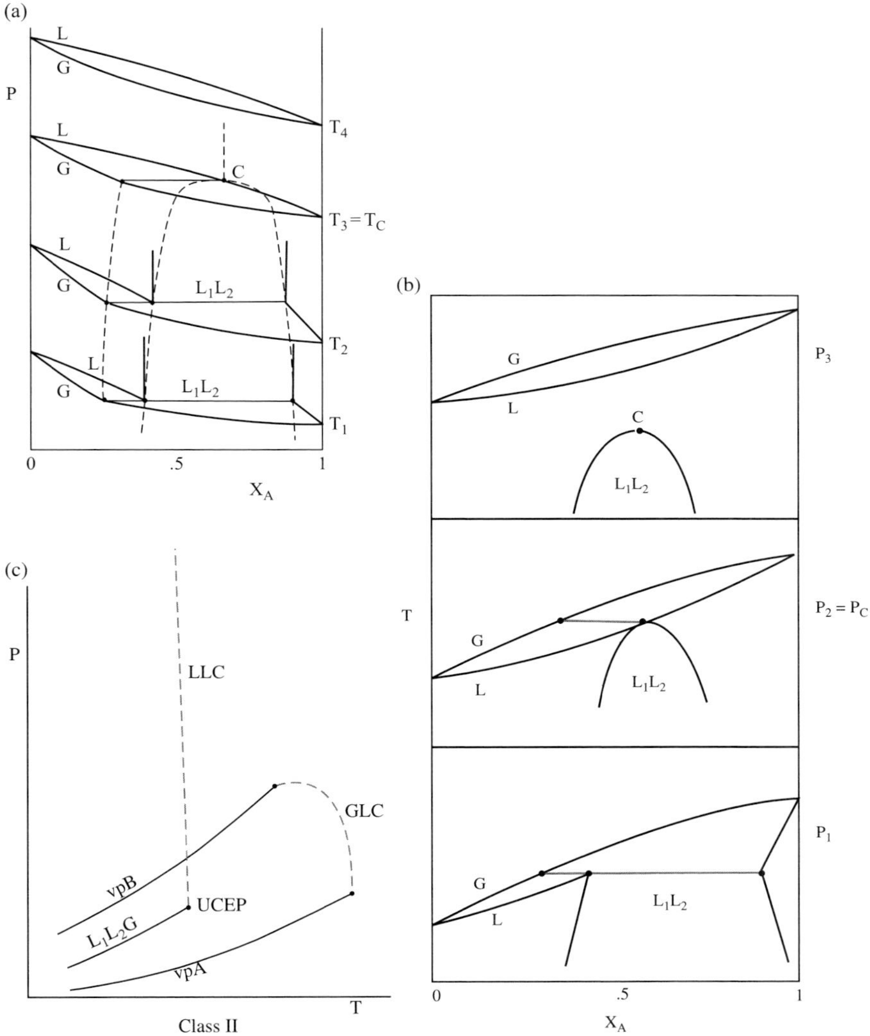

FIG. 7.10 (a) *Px*, (b) *Tx*, and (c) *PT* cuts and projections for Class II system of Fig. 7.9. GLC = gas–liquid critical line, LLC = liquid-liquid critical line, L_1L_2G = liquid-liquid-gas three phase line. The short vertical lines in (a) signify the coexisting L_1 and L_2 phases.

neous azeotrope is thus a point on the L_1L_2G surface. The *PTx* diagram for a case in which a heterogeneous azeotrope occurs is shown in Fig. 7.12. There are four possible subclasses, as shown in Fig. 7.13 (Rowlinson[14] gives a more detailed discussion):

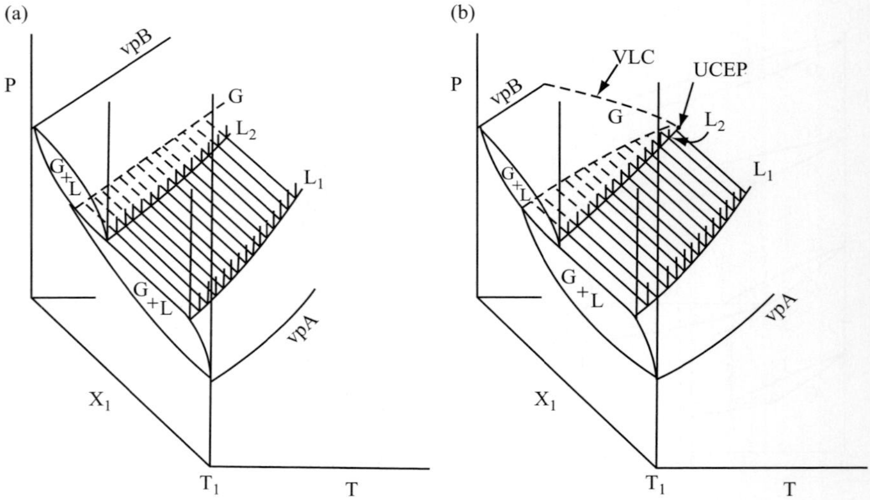

FIG. 7.11 Enlarged view of three-phase gas-liquid 1 (L_1)–liquid 2 (L_2) region for (a) class II and (b) class III system. The long and short vertical lines signify the coexisting L_1 and L_2 phases.

(a) No azeotrope: e.g. *n*-pentane/nitrobenzene, NH_3/toluene, propylene oxide/H_2O, CO_2/H_2O.
(b) Positive homogeneous azeotrope: e.g. H_2O/phenol, H_2O/nicotine (above 850 °C), H_2O/methyl ethyl ketone;
(c) Heterogeneous azeotrope: the composition of the gas lies in between that of the two liquid phases, and equals the overall composition of the two liquid phases: e.g. CH_3OH/cyclohexane, H_2O/triethylamine, H_2O/aniline, H_2O/i-C_4H_9OH, H_2O/EtAc; and
(d) Negative homogeneous azeotrope: e.g. $HC1/H_2O$, SO_2/H_2O, HBr/H_2O, acetic acid/triethylamine.

Class III

In class III the region of liquid–liquid immiscibility extends to the gas–liquid critical line. The disparity in intermolecular forces for the two constituents is particularly great for this class. Class III behaviour is very common, many of the systems met with in gas absorption being of this type.[11,14,16] The liquid ranges of the two components are often very different, but this is not always the case (e.g. many water/hydrocarbons systems are of this type). The critical line consists of two parts (see Fig. 7.1). One part starts from the critical point of pure component A and ends at a UCEP on the LLG line. The other part starts from the critical point of pure B and proceeds to higher pressures, without ever meeting the LLG

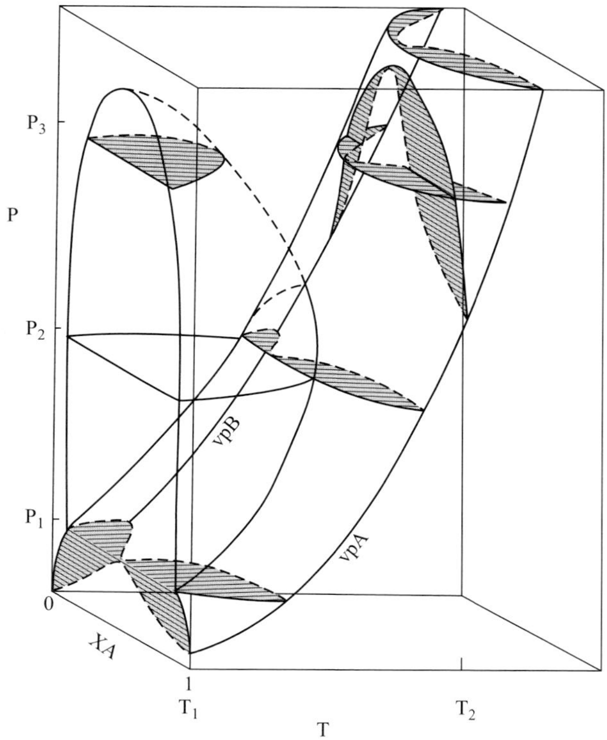

FIG. 7.12 Class II system with heterogeneous azeotrope, showing isothermal and isobaric cuts.

line or the critical point of A. In such systems there is no distinction between GL and LL immiscibility, at least at the higher pressures.

One can distinguish various subclasses of class III behaviour, although the dividing line between them is not always distinct:

(a) The critical line starting from the critical point of the less volatile component has a maximum and minimum in pressure, as shown in Fig. 7.14. In this case the disparity in intermolecular forces is usually less than in the class III systems with gas–gas immiscibility (see (c) and (d) below), and the two components have quite different relative volatilities. Examples of such systems are CH_4/H_2S, ethane/methanol, CH_4/l-heptene, CH_4/methylcyclopentane, CO_2/hexadecane. Figure 7.15 shows the *Px*, *Tx*, and *PT* plots for this system. In the systems involving CO_2 (e.g. CO_2/hexadecane) the critical line takes on a positive slope at high pressure.
(b) This subclass differs from (a) in that the components have similar volatilities, and the system has a heterogeneous azeotrope. Also, the critical line that starts

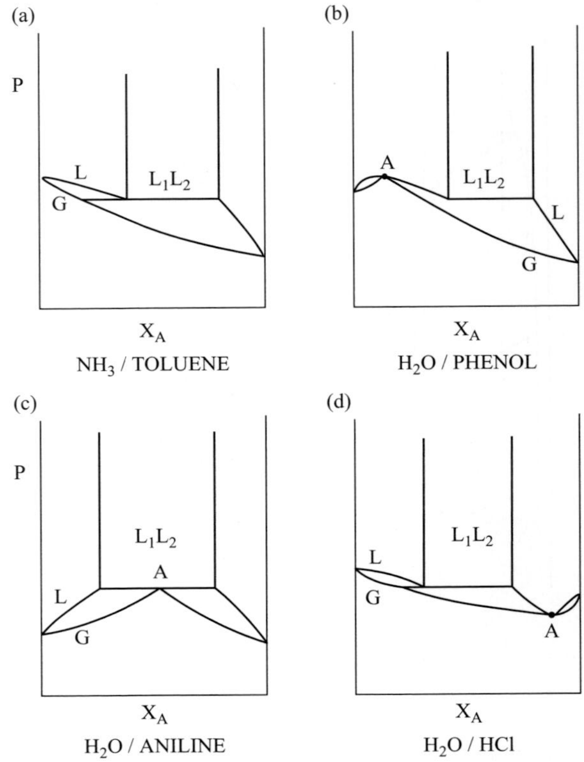

FIG. 7.13 Class II subclasses. (a) No azeotrope; (b) positive homogeneous azeotrope; (c) heterogeneous (always positive) azeotrope; (d) negative homogeneous azeotrope.

from the critical point of the less volatile component usually has a positive slope at high pressures. These systems are discussed in detail by Rowlinson[14] and Schneider;[11] see Rowlinson Fig. 6.27. Examples of this behaviour include many water/organic systems, such as H_2O/ethyl ether, H_2O/propylene, and H_2O/l-butene.

(c) As the disparity of the intermolecular forces increases, the critical line starting from the less volatile component takes on a steeper slope until phase diagrams such as that shown in Fig. 7.16 are found, where the critical line has a minimum in temperature at the same pressure, above which the critical line has a positive slope. Thus, at sufficiently high pressures the system will separate into two fluid phases even at temperatures above the critical point of either pure component. Such behaviour is called 'gas–gas immiscibility of the second kind', and was first discovered in the laboratory by Krichevskii,[9] although it had been predicted by van der Waals[7]

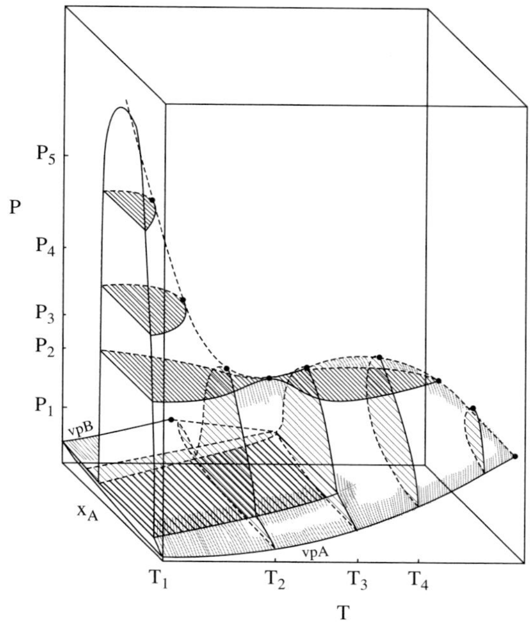

FIG. 7.14 Typical class III system, e.g. methane/1-heptene, methane/hydrogen sulphide, ethane/methanol.

much earlier. The *Px*, *Tx*, and *PT* plots for such a system are shown in Fig. 7.17. Examples of such systems are CO_2/H_2O, N_2/NH_3, C_6H_6/H_2O, Xe/H_2O, CH_4/NH_3, C_2H_6/H_2O, He/Ne, He/Ar, He/Kr/He/N_2, He/H_2, Ne/Kr, NH_3/Ar, H_2O/N_2, SO_2/N_2.

(d) For even greater disparity of the intermolecular forces the critical line starting from the less volatile component has a positive slope at all pressures, as shown, for a typical case in Fig. 7.18. This is called 'gas–gas immiscibility of the first kind'. Examples are He/CO_2, He/C_2H_4, Ar/H_2O, He/Xe, He/C_6H_6, He/cyclohexane. Most of the known systems of this type contain helium as one of the components.

Class IV

The *PTx* diagram for this class is shown in Fig. 7.19. There are two regions of immiscibility, the low-temperature region ending in a UCEP, and the high-temperature region being bounded below by a LCEP and above by the critical point of pure A (see Fig. 7.1 caption for definitions of UCEP and LCEP). This behaviour can be regarded as a variant on Class III, subclass (a). If the minimum in the critical locus shown in Fig. 7.14 occurs at sufficiently low pressures, it

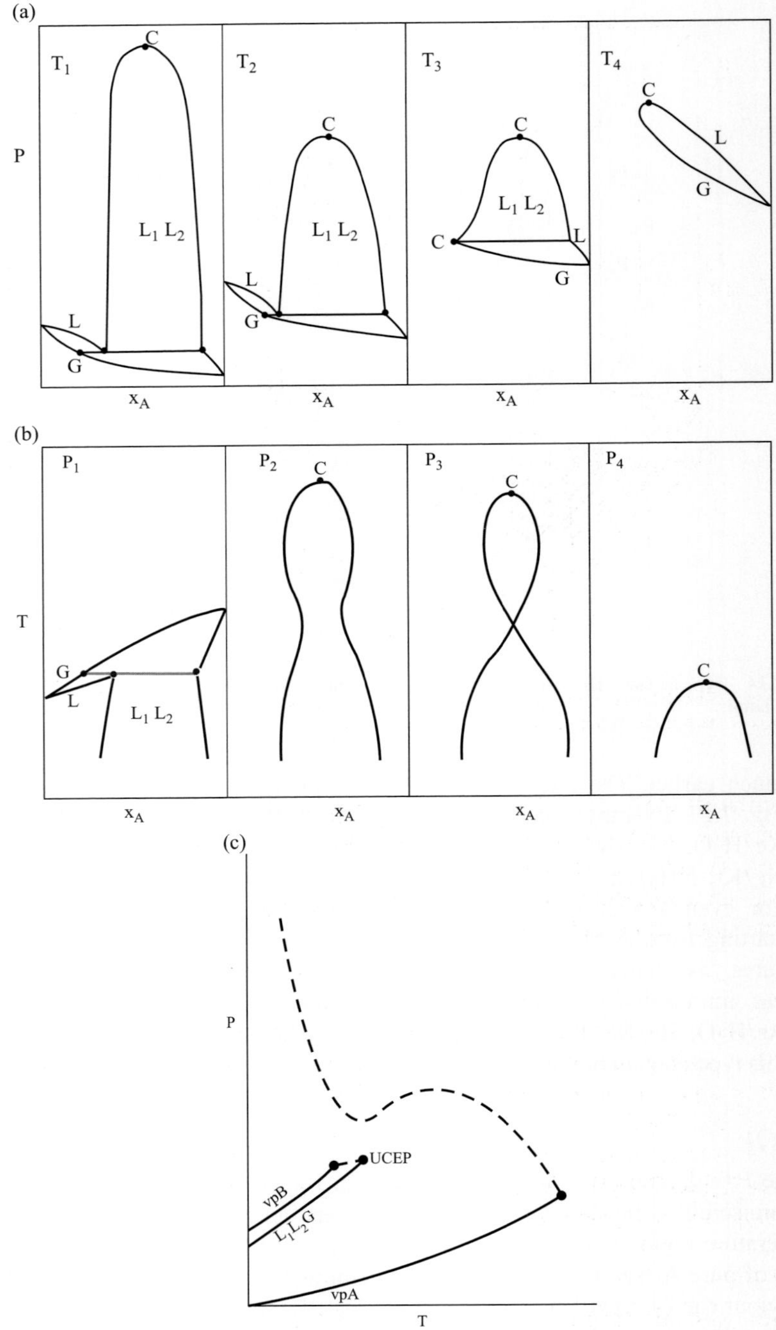

FIG. 7.15 (a) *Px*, (b) *Tx* and (c) *PT* cuts and projections for the class III system of Fig. 7.14.

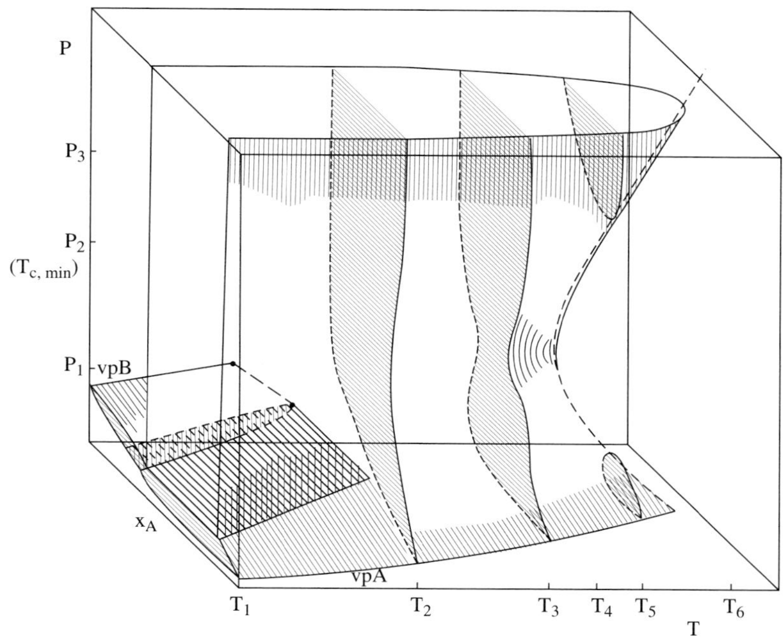

FIG. 7.16 Class III system with gas–gas immiscibility of the second kind, e.g. carbon dioxide/water. $T_{c,min}$ is the minimum temperature on the critical curve.

will intersect the LLG surface, and the portion of the critical locus at pressures below those of the three-phase line will not be observed; the system is then of class IV. class IV behaviour depends on a fine balance of intermolecular forces, and is therefore relatively uncommon. Examples are CH_4/l-hexene, benzene/polyisobutene, and cyclohexane/polystyrene.

Class V

This class, first discovered by Kuenen and Robson[32] in 1899, has liquid immiscibility at high temperatures bounded by a LCEP at lower temperatures and by a UCEP at higher temperatures (Fig. 7.20). The *Px*, *Tx*, and *PT* plots are shown in Fig. 7.21. The three-phase line is often very short, extending over only a few degrees in many cases. This behaviour is more common than Class IV. Examples are C_2H_6/C_2H_5OH, C_2H_6/n-propanol, C_2H_6/*n*-butanol, CO_2/nitrobenzene, CH_4/n-hexane, CH_4/i-octane, HC/polymers, C_3H_8/lubricating oils.

Class VI

These systems have both a LCEP and an UCEP within the liquid range, the systems being completely miscible at temperatures above the UCST and below the LCST. There are two (e.g. Fig. 7.1) or three (e.g. Fig. 7.22) distinct critical

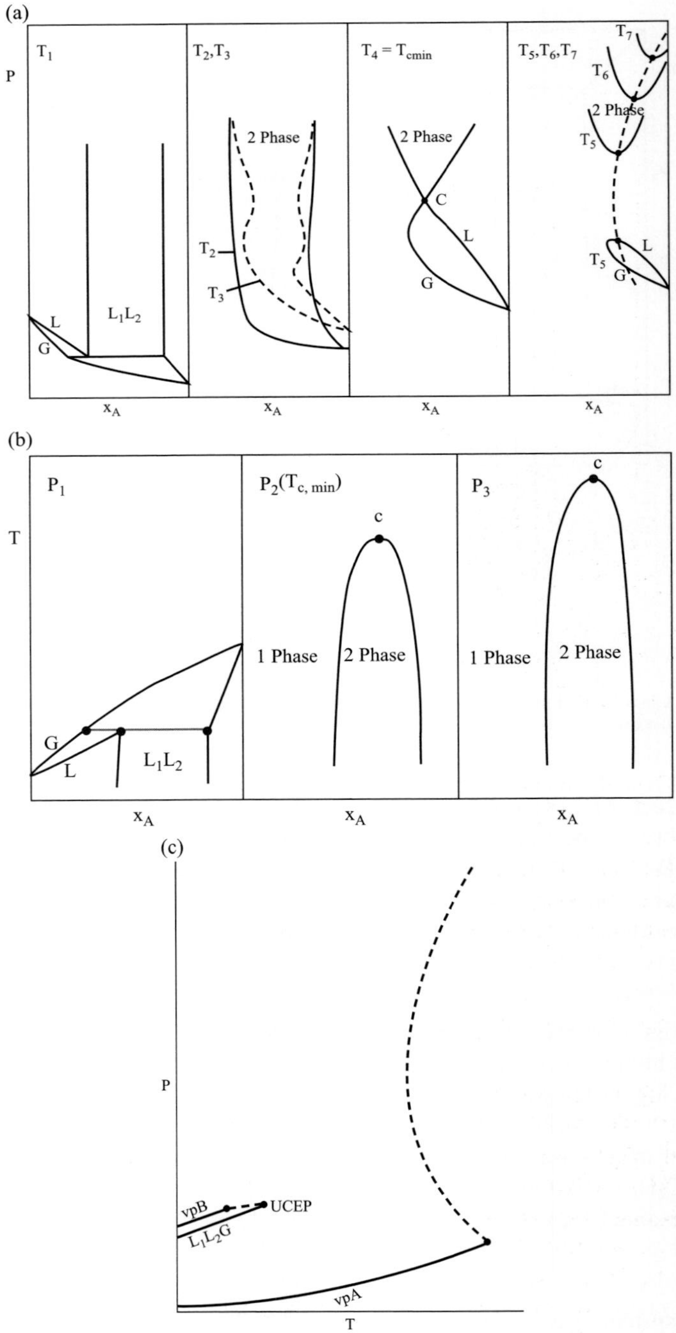

FIG. 7.17 (a) *Px*, (b) *Tx*, and (c) *PT* cuts and projections for the class III system of Fig. 7.16.

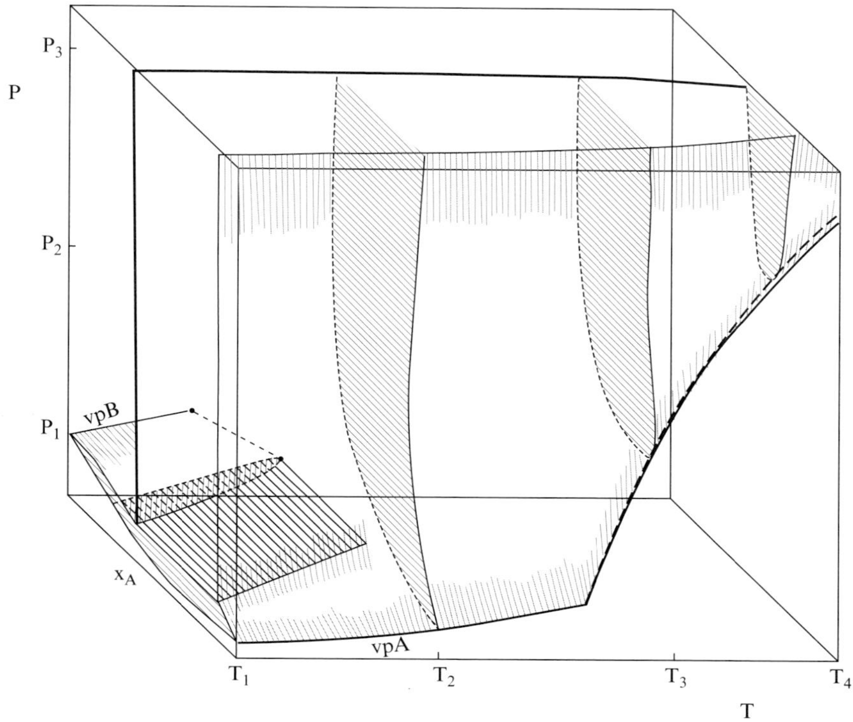

FIG. 7.18 Class III system with gas–gas immiscibility of the first kind, e.g. argon/water.

lines, depending on the subclass. One connects the gas–liquid critical lines of the two pure components, while the others are liquid–liquid critical lines for the UCEP and LCEP, respectively. These two liquid–liquid lines may meet at higher or lower pressures at a *hypercritical point* (HCP). Virtually all known examples of class VI involve water as one component. Several subclasses of this system are known, some of them discovered by Schneider:[11]

(a) The liquid–liquid immiscible region is relatively unaffected by pressure, and extends to the highest pressures attainable (Fig. 7.22). Examples are 3-methylpiperidine/H_2O and 3-methylpyridine/D_2O. The *Px*, *Tx*, and *PT* plots for such a system are shown in Fig. 7.23. For isobaric *Tx* cuts at the higher pressures the system exhibits what are called 'closed solubility loops'.
(b) The immiscibility decreases as the pressure is raised and disappears above some maximum pressure. This point in the phase diagram is the HCP. At all higher pressures the system is fully miscible. A typical *PTx* diagram is shown in Fig. 7.24. Examples are H_2O/2-butanol, methyl ethyl ketone/H_2O, and 2-butoxyethanol/H_2O.

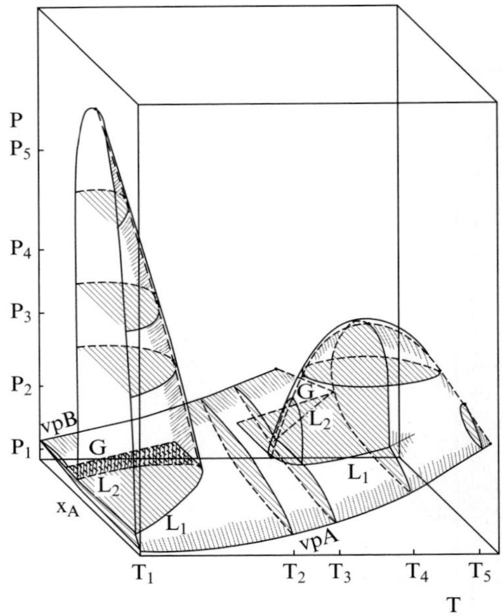

FIG. 7.19 Typical class IV system, e.g. methane/1-hexene, benzene/polyisobutene.

(c) The immiscibility decreases and disappears above some HCP, as in subclass (b). However, at higher pressures the system is immiscible again, above some second HCP. This behaviour is shown in Fig. 7.25. An example is 2-methylpyridine/D_2O.

(d) The system is completely miscible at low pressures and the vapour–liquid region is as in class I. At high pressures, however, there is a region of immiscibility, at pressures above some HCP. This is shown in Fig. 7.26. Examples are 2-methylpyridine/H_2O, 3-methylpyridine/H_2O, 4-methylpyridine/H_2O, and 4-methylpyridine/D_2O. This phenomenon has been termed 'high pressure immiscibility'.[11] Immiscible regions, bounded by an upper HCP, have been found for these systems at negative pressures[11] so that they are in fact very similar to the class (c) above (Fig. 7.25).

It is clear from these examples that these phase diagrams are extremely sensitive to small changes in the intermolecular forces. Changing H_2O to D_2O, or moving a side-group, is sufficient to change the system from one subclass to another. A detailed discussion is given by Schneider.[11]

Relation among the classes

The interrelationship among classes I to VI has been studied using empirical equations of state[19–21] and also using statistical thermodynamics.[28–30] The

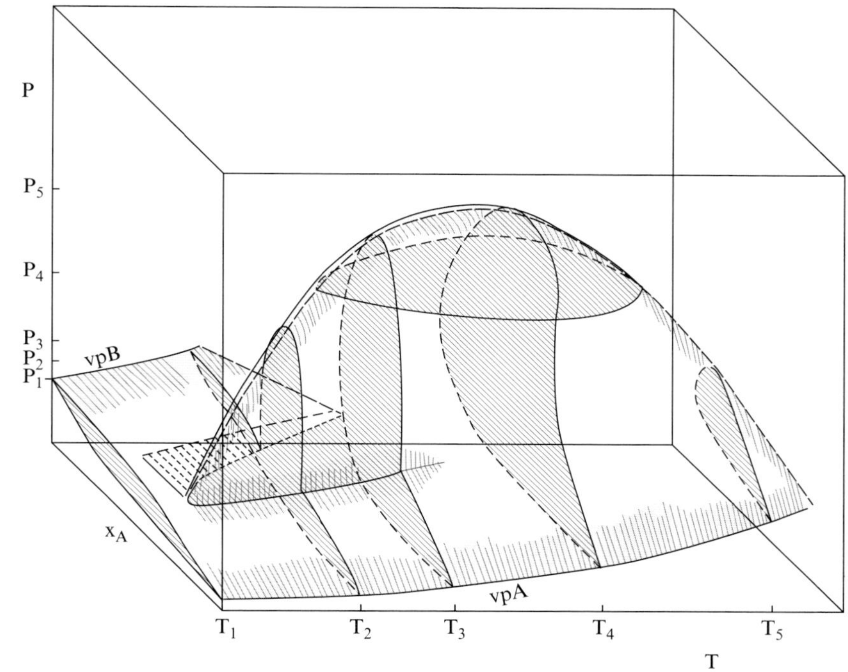

FIG. 7.20 Typical class V system, e.g. methane/*n*-hexane, methane/i-octane.

relation among classes I to V is quite well understood on theoretical grounds. All of these classes can occur in mixtures where the unlike pair intermolecular forces are relatively weak. For example, in studies of polar/nonpolar mixtures,[30] as the dipole moment of the polar molecule is increased the solution becomes increasingly nonideal, and the mixture usually passes from class I to II and then to III; classes IV and V can also occur as intermediate stages, depending on the relative sizes and interaction energies of the molecules. Thus the transition among classes can take the forms

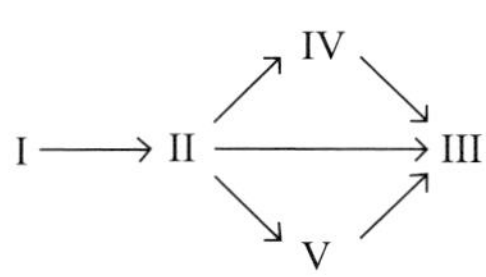

Class III is the most nonideal of these. The relation between these first five classes and class VI is less well understood. Class VI is believed to arise because of a strong unlike pair interaction, which is also strongly dependent on molecular orientation, e.g. a hydrogen bond. For such mixtures, as the nonideality increases the sequence appears to be

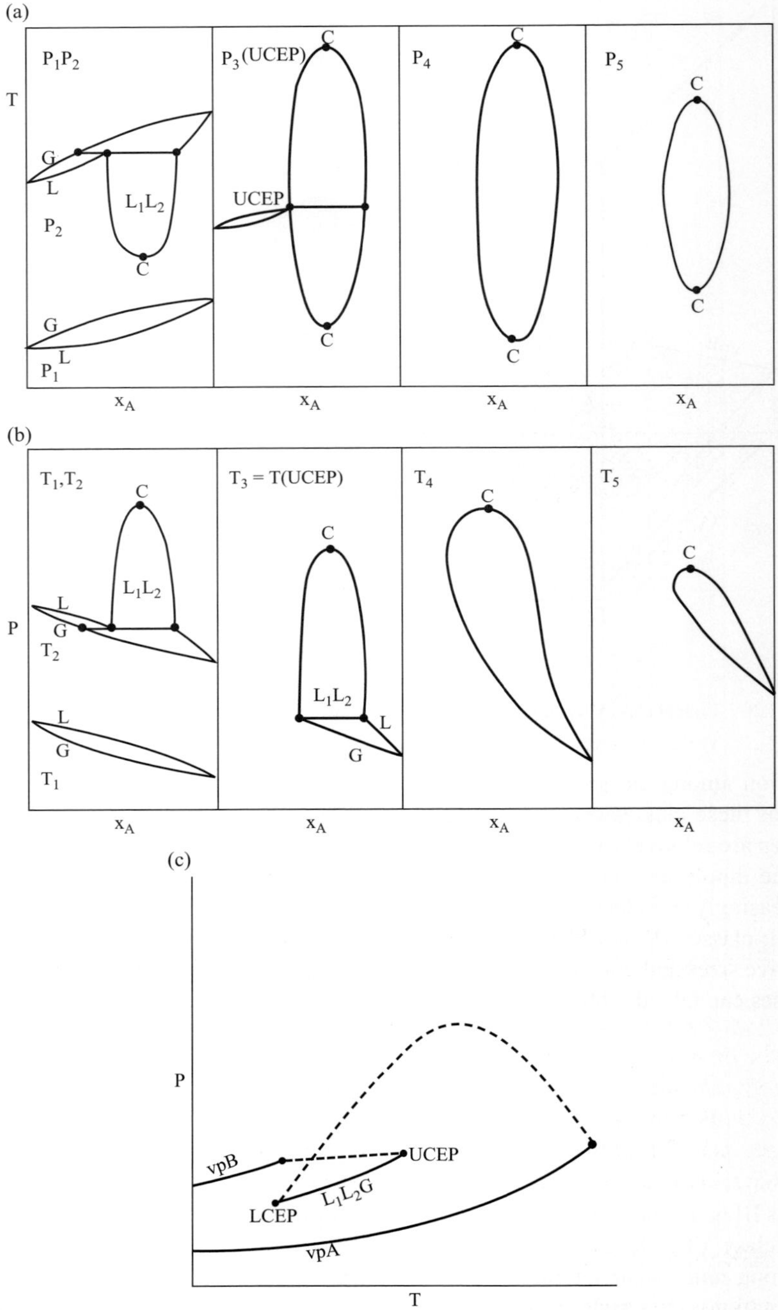

FIG. 7.21 (a) *Px*, (b) *Tx*, and (c) *PT* cuts and projections for the class V system of Fig. 7.20.

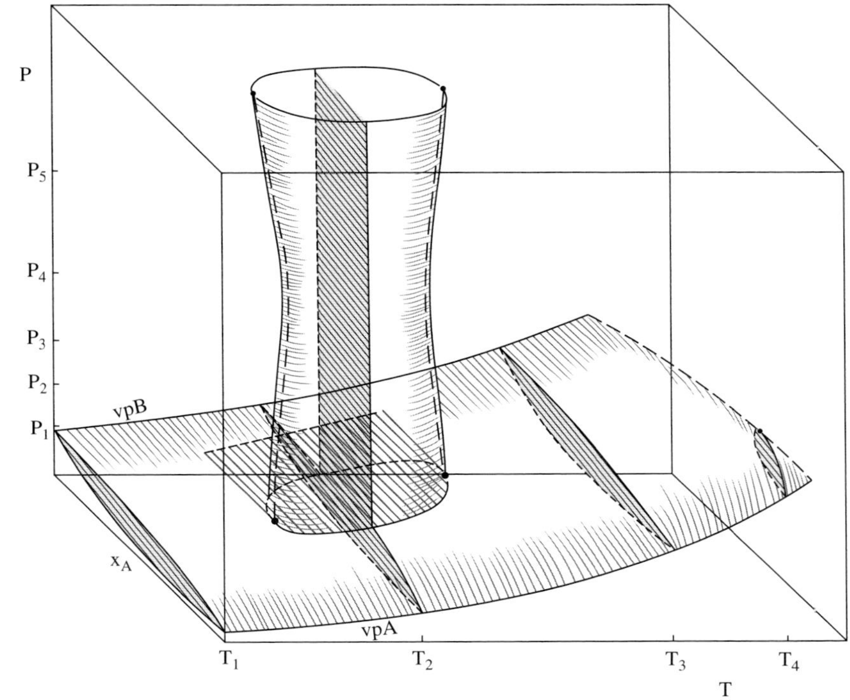

FIG. 7.22 Typical class VI system, subclass (a), e.g. H_2O/3-methylpiperidine.

I ⟶ VI

7.1.3 Conclusion

The classification described here, based on that given originally by Scott and Van Konynenburg,[19,20] contains all of the major classes of fluid phase diagrams known at present; further classes or subclasses may possibly yet remain to be discovered. For further details the reader should consult the book by Rowlinson[14] and the review papers of Schneider[11] and Streett.[16]

A significant omission in the phase diagrams given here are the melting and sublimation regions. These further complicate the diagrams, and have been omitted here for clarity. However, they can change the topology of the diagram considerably in some cases, as discussed in the reviews mentioned earlier.

7.2 Kirkwood–Buff theory (KBT)

It is possible to relate composition fluctuations in a mixture to thermodynamic properties (the partial molar volumes, isothermal compressibility, and various

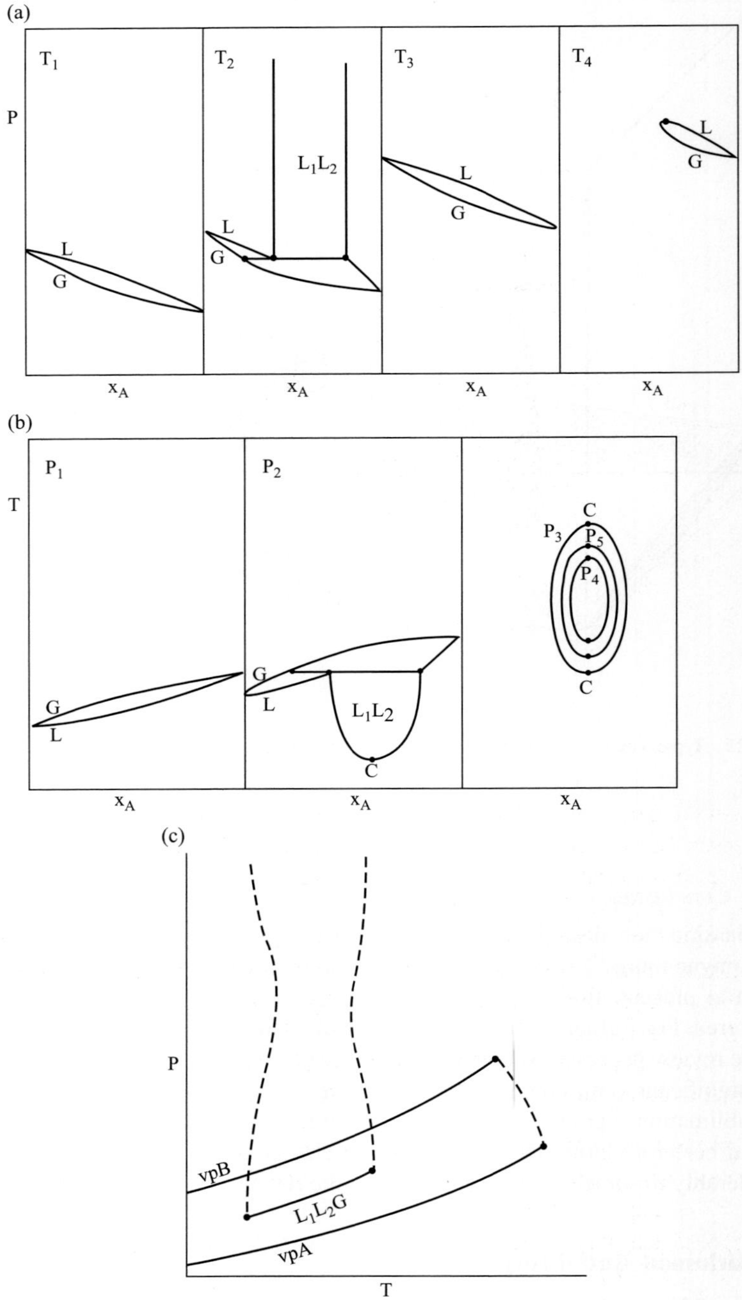

FIG. 7.23 (a) *Px*, (b) *Tx*, and (c) *PT* cuts and projections for the class VI system of Fig. 7.22.

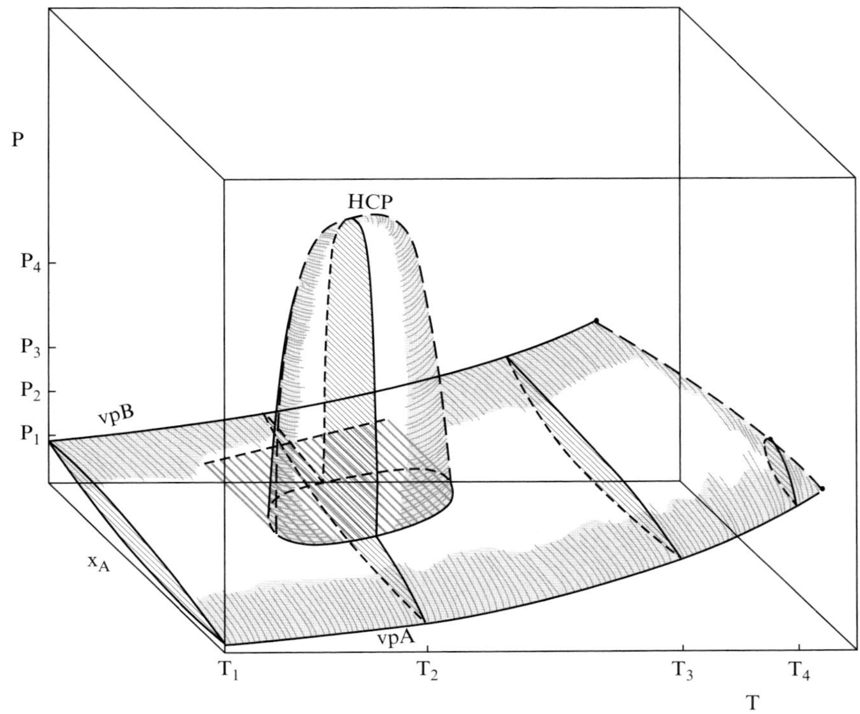

FIG. 7.24 Typical class VI system, subclass (b), e.g. H_2O/butanol, H_2O/methyl ethyl ketone.

composition derivatives of the chemical potential) in a particularly direct and simple way. Such relations were first written down by Gibbs.[33] They were put into a modern form suitable for a theory of mixtures in a classic paper by Kirkwood and Buff[34] in 1951,[35,36] and the theory is referred to as the Kirkwood–Buff theory (KBT), or sometimes as the fluctuation theory of mixtures. In this theory the thermodynamic properties are related to integrals over space of the various grand canonical centres correlation functions, $g_{\alpha\beta}(r) = \langle g_{\alpha\beta}(\mathbf{r}\omega_1\omega_2)\rangle_{\omega_1\omega_2}$, where α and β are components of the mixture (see (3.252) and (3.253) for the definition of these functions). The resulting equations provide a particularly rigorous and convenient starting point for a theory of mixtures. In particular,

- No assumption of pairwise additivity is made;
- No assumptions are made concerning the type of molecules present (spherical or non-spherical, rigid or non-rigid, etc.), so that the equations apply equally to simple or complex molecules, associating fluids, polymers, ionic mixtures, etc.; and
- The final equations only involve centres correlation functions, and not the angular ones.

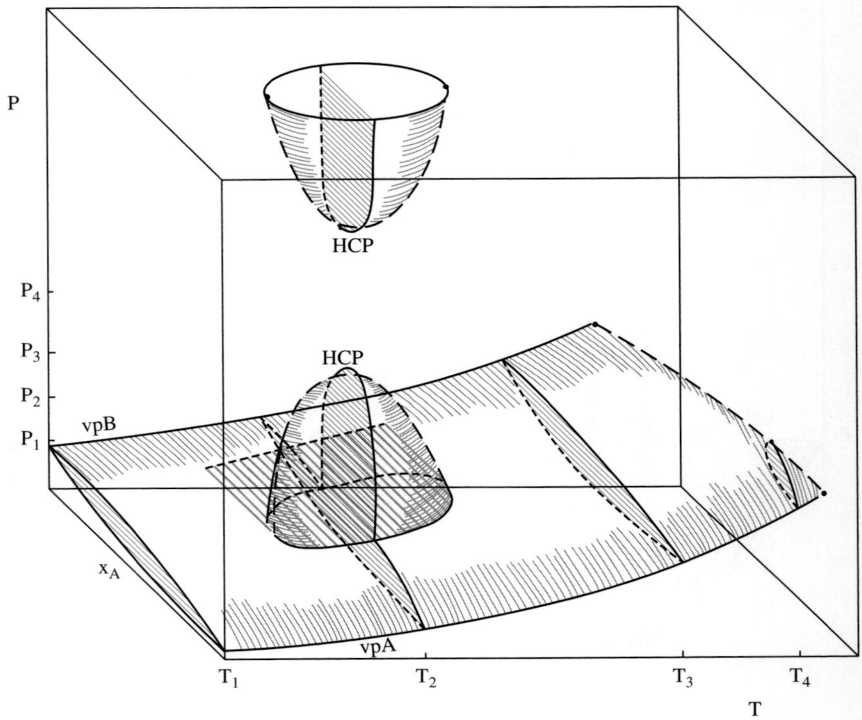

FIG. 7.25 Typical class VI system, subclass (c), e.g. D_2O/2-methylpyridine.

These advantages are the same ones that apply to the compressibility equation of state (see (3.243) and its derivation, and also § 6.3), and the KBT can be thought of as a mixture generalization of that derivation. Because of its generality the KBT forms a convenient starting point for our discussion of ideal mixtures (§ 7.3) and dilute mixtures (§ 7.5), as well as some of the approximate theories of non-ideal mixtures that we shall discuss later in this chapter.

7.2.1 The basic equations

Since we are interested in concentration fluctuations it is most convenient to adopt the grand canonical ensemble. We consider an open system of volume V, temperature T, containing components $A, B, \cdots, R$ with chemical potentials $\mu_A, \mu_B, \cdots, \mu_R$. At any instant the system contains N_A molecules of A, N_B of B, etc. The mixture generalization of (3.172) is

$$\langle N_\alpha \rangle = \Xi^{-1} \sum_{\mathbf{N}} N_\alpha \exp(\boldsymbol{\mu} \cdot \mathbf{N}/kT)\, Q(\mathbf{N}VT)$$
$$= kT \left(\frac{\partial \ln \Xi}{\partial \mu_\alpha} \right)_{TV\mu'}, \tag{7.1}$$

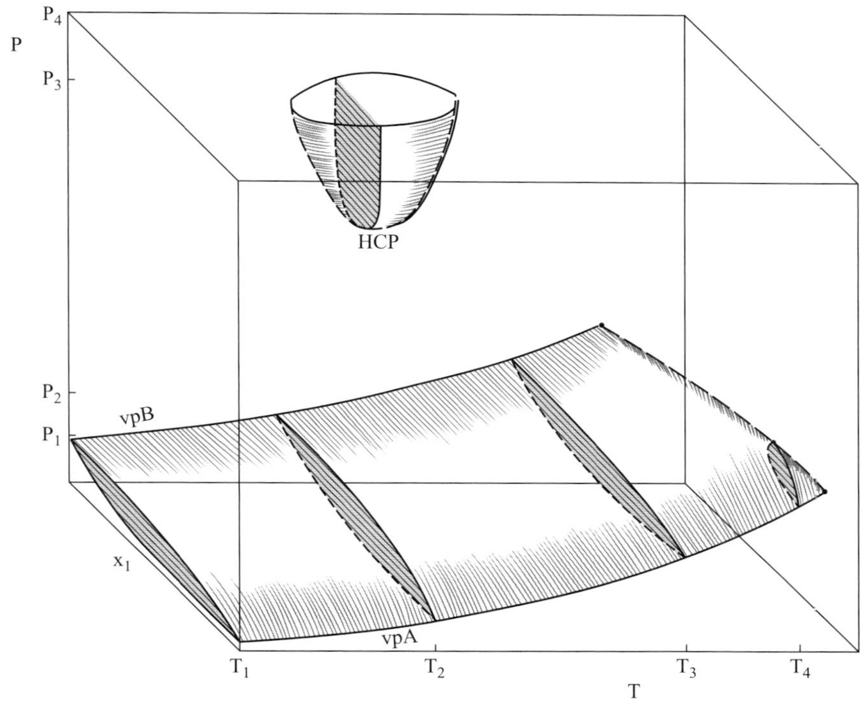

FIG. 7.26 Typical class VI system, subclass (d), e.g. H_2O/2-methylpyridine.

where $\langle N_\alpha \rangle$ is the average number of α molecules, $Q(\mathbf{N}VT)$ is the canonical partition function for the N molecule system, $\mathbf{N} \equiv (N_A, N_B, \cdots, N_R)$ and $\boldsymbol{\mu} \equiv (\mu_A, \mu_B, \cdots, \mu_R)$ represent the number of molecules and chemical potentials of all of the components, $\boldsymbol{\mu} \cdot \mathbf{N} \equiv \sum_\alpha \mu_\alpha N_\alpha$, μ' represents all μ_β except for the species with respect to which we carry out the differentiation ($\beta \neq \alpha$ here), and

$$\Xi = \sum_{\mathbf{N}} \exp(\boldsymbol{\mu} \cdot \mathbf{N}/kT)\, Q(\mathbf{N}VT) \tag{7.2}$$

is the grand partition function. From (7.1) we have

$$kT\left(\frac{\partial \langle N_\alpha \rangle}{\partial \mu_\beta}\right)_{TV\mu'} = \Xi^{-1} \sum_{\mathbf{N}} N_\alpha N_\beta \exp(\boldsymbol{\mu} \cdot \mathbf{N}/kT)\, Q(\mathbf{N}VT)$$
$$- kT \frac{\partial \ln \Xi}{\partial \mu_\beta} \Xi^{-1} \sum_{\mathbf{N}} N_\alpha \exp(\boldsymbol{\mu} \cdot \mathbf{N}/kT)\, Q(\mathbf{N}VT)$$

or

$$kT\left(\frac{\partial \langle N_\alpha \rangle}{\partial \mu_\beta}\right)_{TV\mu'} = \langle N_\alpha N_\beta \rangle - \langle N_\alpha \rangle \langle N_\beta \rangle, \tag{7.3}$$

where $\langle \cdots \rangle$ means a grand canonical average. Equation (7.3) relates the thermodynamic derivative $(\partial \langle N_\alpha \rangle / \partial \mu_\beta)_{TV\mu'}$ to the cross fluctuation in the number of molecules of species α and β, and was first derived by Gibbs.[33] That these cross fluctuations are simply related to the centres pair correlation function $g_{\alpha\beta}(r)$ can be seen as follows. The mixture generalizations of the normalization condition (3.192) for $f_\alpha(\mathbf{r}_1\omega_1)$ and $f_{\alpha\beta}(\mathbf{r}_{12}\omega_1\omega_2)$ are

$$\int \mathrm{d}\mathbf{r}_1 \mathrm{d}\omega_1 f_\alpha(\mathbf{r}_1\omega_1) = \langle N_\alpha \rangle \tag{7.4}$$

$$\int \mathrm{d}\mathbf{r}_1 \mathrm{d}\mathbf{r}_2 \mathrm{d}\omega_1 \mathrm{d}\omega_2 f_{\alpha\beta}(\mathbf{r}_{12}\omega_1\omega_2) = \langle N_\alpha N_\beta \rangle - \langle N_\alpha \rangle \delta_{\alpha\beta}, \tag{7.5}$$

where $\delta_{\alpha\beta}$ is the Kronecker delta. Thus, from (7.4) and (7.5) we have

$$\begin{aligned} \int \mathrm{d}1\mathrm{d}2[f_{\alpha\beta}(12) - f_\alpha(1)f_\beta(2)] &= \langle N_\alpha N_\beta \rangle - \langle N_\alpha \rangle \delta_{\alpha\beta} - \langle N_\alpha \rangle \langle N_\beta \rangle \\ &= \rho_\alpha \rho_\beta V H_{\alpha\beta}, \end{aligned} \tag{7.6}$$

where the last step follows from the use of (3.251) and (3.252) and the assumption of a homogeneous, isotropic fluid, $\rho_\alpha = \langle N_a \rangle / V$ is the number density of α molecules, and $H_{\alpha\beta}$ is defined by[37]

$$H_{\alpha\beta} \equiv \int \mathrm{d}\mathbf{r} h_{\alpha\beta}(r) = 4\pi \int_0^\infty \mathrm{d}r r^2 h_{\alpha\beta}(r), \tag{7.7}$$

where $h_{\alpha\beta} = g_{\alpha\beta}(r) - 1$ is the total correlation function, and $g_{\alpha\beta}(r)$ is the (grand canonical) centres pair correlation function for an $\alpha\beta$ pair,

$$g_{\alpha\beta}(r) = \langle g_{\alpha\beta}(\mathbf{r}\omega_1\omega_2) \rangle_{\omega_1\omega_2}. \tag{7.8}$$

From (7.3) and (7.6) we obtain

$$\frac{kT}{V}\left(\frac{\partial \langle N_\alpha \rangle}{\partial \mu_\beta}\right)_{TV\mu'} = kT\left(\frac{\partial \rho_\alpha}{\partial \mu_\beta}\right)_{T\mu'} = B_{\alpha\beta}, \tag{7.9}$$

where

$$B_{\alpha\beta} = B_{\beta\alpha} \equiv \rho_\alpha \rho_\beta H_{\alpha\beta} + \rho_\alpha \delta_{\alpha\beta}. \tag{7.10}$$

Equation (7.9) connects the thermodynamic properties to the centres pair correlation functions. The physical significance of $H_{\alpha\beta}$ is readily seen from (7.7) and Fig. 7.27. If an α molecule is fixed at the origin, then the average number of β molecules in a shell of thickness $\mathrm{d}r$ at distance r from the origin is $(4\pi r^2 \mathrm{d}r)\rho_\beta g_{\alpha\beta}(r)$, whereas if the origin is picked randomly, this average number is $(4\pi r^2 \mathrm{d}r)\rho_\beta$. Thus the quantity $(4\pi r^2 \mathrm{d}r)\rho_\beta h_{\alpha\beta}(r)$ gives the increase or decrease in the number of β molecules at r due to the presence of an α molecule at the origin. This quantity has been termed[38] the *affinity* of an α molecule for a β molecule, and will take both positive and negative values, depending on the value of r (see Fig. 3.7). The quantity $H_{\alpha\beta}$ gives a measure of the overall affinity of β for α.

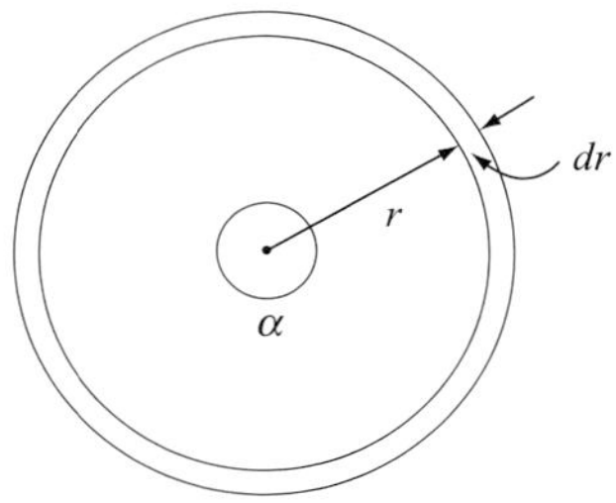

FIG. 7.27 If an α molecule is fixed at the origin, the average number of β molecules in a shell of thickness dr at r is $(4\pi r^2 \mathrm{d}r)\rho_\beta g_{\alpha\beta}(r)$.

Transformation from $(TV\boldsymbol{\mu})$ *to* $(TV\langle\mathbf{N}\rangle)$ *variables*

The derivative in (7.9) is at fixed temperature and chemical potentials, which is inconvenient except for osmotic studies. We shall usually be interested in using as independent variables $(Tp\langle\mathbf{N}\rangle)$, the temperature, pressure, and number of molecules of each component. As an intermediate step we first transform from $(TV\boldsymbol{\mu})$ to $(TV\langle\mathbf{N}\rangle)$ variables. If the chemical potentials are regarded as functions of the variables $(TV\langle\mathbf{N}\rangle)$ then

$$\left(\frac{\partial\mu_\alpha}{\partial\mu_\gamma}\right)_{TV\mu'} = \delta_{\alpha\gamma} = \sum_{\beta=A}^{R}\left(\frac{\partial\mu_\alpha}{\partial\langle N_\beta\rangle}\right)_{TV<\mathbf{N}'>}\left(\frac{\partial\langle N_\beta\rangle}{\partial\mu_\gamma}\right)_{TV\mu'}$$
$$= \sum_\beta A_{\alpha\beta}B_{\beta\gamma}, \tag{7.11}$$

where $\delta_{\alpha\gamma}$ is the Kronecker delta and $A_{\alpha\beta}$ is defined by

$$A_{\alpha\beta} \equiv \frac{V}{kT}\left(\frac{\partial\mu_\alpha}{\partial\langle N_\beta\rangle}\right)_{TV<\mathbf{N}'>} = \frac{1}{kT}\left(\frac{\partial\mu_\alpha}{\partial\rho_\beta}\right)_{T\rho'}. \tag{7.12}$$

In matrix notation (7.11) is (see Appendix B.7 of Vol. 1)

$$\mathbf{AB} = \mathbf{I}, \tag{7.13}$$

where $\mathbf{I}$ is the unit matrix. Provided that $\mathbf{B}$ is nonsingular (as it must be for the system to be stable), we can solve (7.13) for $\mathbf{A}$:

$$\mathbf{A} = \mathbf{B}^{-1} \tag{7.14}$$

or, from (B.128), (B.138), and (7.12):

$$\frac{V}{kT}\left(\frac{\partial\mu_\alpha}{\partial\langle N_\beta\rangle}\right)_{TV\langle\mathbf{N}'\rangle} = A_{\alpha\beta} = \frac{|B|_{\alpha\beta}}{\det\mathbf{B}}, \tag{7.15}$$

where det $\mathbf{B}$ is the determinant of the matrix $\mathbf{B}$ and $|B|_{\alpha\beta}$ is the cofactor of the element $B_{\alpha\beta}$ in the determinant. In (7.15) we have made use of the symmetry of

$\mathbf{B}$, $B_{\alpha\beta} = B_{\beta\alpha}$, as seen from (7.10). Equations (7.10) and (7.15) together provide a relation between $A_{\alpha\beta}$ and the $H_{\alpha\beta}$.

Transformation from $(TV\langle\mathbf{N}\rangle)$ *to* $(Tp\langle\mathbf{N}\rangle)$ *variables*

To transform the result (7.15) to $(Tp\langle\mathbf{N}\rangle)$ variables we use the thermodynamic identity,

$$\left(\frac{\partial\mu_\alpha}{\partial\langle N_\beta\rangle}\right)_{TV\langle\mathbf{N}'\rangle} = \left(\frac{\partial\mu_\alpha}{\partial\langle N_\beta\rangle}\right)_{Tp\langle\mathbf{N}'\rangle} + \left(\frac{\partial\mu_\alpha}{\partial p}\right)_{T\langle\mathbf{N}\rangle}\left(\frac{\partial p}{\partial\langle N_\beta\rangle}\right)_{TV\langle\mathbf{N}'\rangle}. \tag{7.16}$$

The derivative $(\partial\mu_\alpha/\partial p)_{T\langle\mathbf{N}\rangle}$ is given by another thermodynamic identity as

$$\left(\frac{\partial\mu_\alpha}{\partial p}\right)_{T\langle\mathbf{N}\rangle} = \bar{V}_\alpha, \tag{7.17}$$

where $\bar{V}_\alpha \equiv (\partial V/\partial\langle N\rangle_\alpha)_{Tp\langle\mathbf{N}'\rangle}$ is the partial molecular volume of component α in the mixture. Moreover, the derivative $(\partial p/\partial\langle N_\beta\rangle)_{Tp\langle\mathbf{N}'\rangle}$ that appears in (7.16) can be expressed as

$$\left(\frac{\partial p}{\partial\langle N_\beta\rangle}\right)_{TV\langle\mathbf{N}'\rangle} = -\left(\frac{\partial V}{\partial\langle N_\beta\rangle}\right)_{Tp\langle\mathbf{N}'\rangle}\left(\frac{\partial p}{\partial V}\right)_{T\langle\mathbf{N}\rangle} = \frac{\bar{V}_\beta}{V\chi}, \tag{7.18}$$

where $\chi = -V^{-1}(\partial V/\partial p)_{T\langle\mathbf{N}\rangle}$ is the isothermal compressibility. Combining (7.15)–(7.18) gives the composition derivatives of μ_α at constant $(Tp\langle\mathbf{N}'\rangle)$ as

$$\left(\frac{\partial\mu_\alpha}{\partial\langle N_\beta\rangle}\right)_{Tp\langle\mathbf{N}'\rangle} = \frac{kT}{V}\frac{|B|_{\alpha\beta}}{\det\mathbf{B}} - \frac{\bar{V}_\alpha\bar{V}_\beta}{V\chi} \qquad (\alpha,\beta = A, B, \ldots R) \tag{7.19}$$

with $B_{\alpha\beta}$ given by (7.10). Equation (7.19) is a set of R^2 equations, but involves $(R^2 + R + 1)$ thermodynamic quantities; these are the R^2 composition derivatives of the μ_α, the R partial molecular volumes $\bar{V}_\alpha$, and the isothermal compressibility χ of the mixture. However, the composition derivatives of μ_α are not independent, but are themselves interrelated by the Gibbs–Duhem equation[39]

$$\sum_\alpha \rho_\alpha\left(\frac{\partial\mu_\alpha}{\partial\langle N_\beta\rangle}\right)_{Tp\langle\mathbf{N}'\rangle} = 0 \qquad (\beta = A, B, \cdots, R). \tag{7.20}$$

Moreover, the partial molecular volumes are related by[40]

$$\sum_\alpha \rho_\alpha\bar{V}_\alpha = 1. \tag{7.21}$$

Equations (7.20) and (7.21) provide an additional $(R+1)$ relations between the thermodynamic quantities that appear in (7.19). Thus (7.19)–(7.21) comprise a set of $(R^2 + R + 1)$ equations that can be solved for the $(R^2 + R + 1)$ thermodynamic quantities $(\partial\mu_\alpha/\partial\langle N_\beta\rangle)_{Tp\langle\mathbf{N}'\rangle}$, $\bar{V}_\alpha$, and χ.

As a check on (7.19) we consider the case of a pure component. The derivative $(\partial\mu/\partial\langle N\rangle)_{Tp}$ then vanishes (μ is an intensive quantity, independent of $\langle N\rangle$ at fixed T, p) and we have $\bar{V}_\alpha = V/\langle N\rangle$, $|B|_{\alpha\beta} = 1$, $\det \mathbf{B} = \rho\left(1 + \rho \int d\mathbf{r}h\right)$, so that (7.19) reduces to the compressibility equation, (3.113),

$$\rho kT\chi = 1 + \rho \int \mathrm{d}\mathbf{r}h(r) \tag{7.22}$$

as it should.

Binary mixtures

We next consider a binary mixture, the simplest nontrivial case. Equations (7.19)–(7.21) now give seven relations among the four composition derivatives, two partial molecular volumes, and the isothermal compressibility. The determinant det **B** is given by

$$\begin{aligned}\det \mathbf{B} &= \begin{vmatrix} \rho_A^2 H_{AA} + \rho_A & \rho_A\rho_B H_{AB} \\ \rho_A\rho_B H_{AB} & \rho_B^2 H_{BB} + \rho_B \end{vmatrix} \\ &= \rho_A\rho_B\zeta,\end{aligned} \tag{7.23}$$

where

$$\zeta = 1 + \rho_A H_{AA} + \rho_B H_{BB} + \rho_A\rho_B\left(H_{AA}H_{BB} - H_{AB}^2\right). \tag{7.24}$$

The cofactors are $|B|_{AA} = \rho_B^2 H_{BB} + \rho_B$, $|B|_{BB} = \rho_A^2 H_{AA} + \rho_A$, $|B|_{AB} = |B|_{BA} = -\rho_A\rho_B H_{AB}$. Using these results in (7.19) and solving (7.19)-(7.21) gives

$$\left(\frac{\partial\mu_A}{\partial\langle N_A\rangle}\right)_{Tp\langle N_B\rangle} = kT\rho_B/V\rho_A\eta \tag{7.25}$$

$$\left(\frac{\partial\mu_A}{\partial\langle N_B\rangle}\right)_{Tp\langle N_A\rangle} = \left(\frac{\partial\mu_B}{\partial\langle N_A\rangle}\right)_{Tp\langle N_B\rangle} = -kT/V\eta \tag{7.26}$$

$$\left(\frac{\partial\mu_B}{\partial\langle N_B\rangle}\right)_{Tp\langle N_A\rangle} = kT\rho_A/V\rho_B\eta \tag{7.27}$$

$$\bar{V}_A = [1 + \rho_B(H_{BB} - H_{AB})]/\eta \tag{7.28}$$

$$\bar{V}_B = [1 + \rho_A(H_{AA} - H_{AB})]/\eta \tag{7.29}$$

$$\chi = \zeta/kT\eta, \tag{7.30}$$

where

$$\eta = \rho_A + \rho_B + \rho_A\rho_B(H_{AA} + H_{BB} - 2H_{AB}). \tag{7.31}$$

Equations (7.25)–(7.30) relate the thermodynamic properties directly to integrals over the centres pair correlation functions.

It is possible to derive alternative but equivalent forms to (7.25)–(7.27) for the composition derivatives of the chemical potential, and these will be useful in later applications. We start from the thermodynamic identity

$$\left(\frac{\partial \mu_B}{\partial \rho_B}\right)_{Tp} = \left(\frac{\partial \mu_B}{\partial \rho_B}\right)_{T\mu_A} + \left(\frac{\partial \mu_B}{\partial \mu_A}\right)_{T\rho_B} \left(\frac{\partial \mu_A}{\partial \rho_B}\right)_{Tp}. \tag{7.32}$$

The first derivative on the right side of this expression is directly related to H_{BB} by (7.9) and (7.10). We now transform the last term on the right of (7.32) to a form involving derivatives of the type $(\partial\mu_\beta/\partial\rho_\alpha)_{T\mu'}$. We can relate $(\partial\mu_B/\partial\mu_A)_{T\rho_B}$ to such derivatives by use of the thermodynamic identity

$$\left(\frac{\partial \mu_A}{\partial \rho_B}\right)_{T\mu_B} \left(\frac{\partial \rho_B}{\partial \mu_B}\right)_{T\mu_A} \left(\frac{\partial \mu_B}{\partial \mu_A}\right)_{T\rho_B} = -1. \tag{7.33}$$

Moreover, $(\partial\mu_A/\partial\rho_B)_{Tp}$ is related to $(\partial\mu_B/\partial\rho_B)_{Tp}$ by another form of the Gibbs–Duhem equation,[39]

$$\rho_A \left(\frac{\partial \mu_A}{\partial \rho_B}\right)_{Tp} + \rho_B \left(\frac{\partial \mu_B}{\partial \rho_B}\right)_{Tp} = 0. \tag{7.34}$$

Substituting for $(\partial\mu_B/\partial\mu_A)_{T\rho_B}$ and $(\partial\mu_A/\partial\rho_B)_{Tp}$ in (7.32) using (7.33) and (7.34), respectively, gives

$$\left(\frac{\partial \mu_B}{\partial \rho_B}\right)_{Tp} = \frac{\rho_A(\partial\mu_B/\partial\rho_B)_{T\mu_A}(\partial\mu_A/\partial\rho_B)_{T\mu_B}}{\rho_A(\partial\mu_A/\partial\rho_B)_{T\mu_B} - \rho_B(\partial\mu_B/\partial\rho_B)_{T\mu_A}}. \tag{7.35}$$

All of the derivatives on the right-hand side of this equation can be expressed in terms of the $H_{\alpha\beta}$ via (7.9) and (7.10). When this is done we obtain

$$\left(\frac{\partial \mu_B}{\partial \rho_B}\right)_{Tp} = \frac{kT}{\rho_B + \rho_B^2(H_{BB} - H_{AB})}. \tag{7.36}$$

Also, from (7.34) and (7.35), we have

$$\left(\frac{\partial \mu_A}{\partial \rho_B}\right)_{Tp} = -\frac{kT}{\rho_A + \rho_A\rho_B(H_{BB} - H_{AB})}. \tag{7.37}$$

The remaining two derivatives can be obtained by interchanging A and B in (7.36) and (7.37):

$$\left(\frac{\partial \mu_A}{\partial \rho_A}\right)_{Tp} = \frac{kT}{\rho_A + \rho_A^2(H_{AA} - H_{AB})} \tag{7.38}$$

$$\left(\frac{\partial \mu_B}{\partial \rho_A}\right)_{Tp} = -\frac{kT}{\rho_B + \rho_A\rho_B(H_{AA} - H_{AB})}. \tag{7.39}$$

We note from (7.37) and (7.39) that the cross derivatives $\partial\mu_A/\partial\rho_B$ and $\partial\mu_B/\partial\rho_A$ are not equal.

Another form of composition derivative that is often useful is that with respect to mole fraction, $x_\alpha = \langle N_\alpha\rangle/\langle N\rangle$, where $\langle N\rangle$ is the total number of molecules of all species. Thus the dependence of μ_A on x_A is given by

$$\left(\frac{\partial \mu_A}{\partial x_A}\right)_{Tp} = \left(\frac{\partial \mu_A}{\partial \rho_A}\right)_{Tp} \left(\frac{\partial \rho_A}{\partial x_A}\right)_{Tp} \tag{7.40}$$

and

$$\left(\frac{\partial \rho_A}{\partial x_A}\right)_{Tp} = \langle N \rangle \left(\frac{\partial \rho_A}{\partial \langle N_A \rangle}\right)_{Tp\langle N \rangle} = \rho \left[1 - \rho_A \left(\frac{\partial V}{\partial \langle N_A \rangle}\right)_{Tp\langle N \rangle}\right]. \tag{7.41}$$

The derivative $(\partial V/\partial \langle N_A \rangle)_{Tp\langle N \rangle}$ is given by the identity

$$\left(\frac{\partial V}{\partial \langle N_A \rangle}\right)_{Tp\langle N \rangle} = \bar{V}_A + \bar{V}_B \left(\frac{\partial \langle N_B \rangle}{\partial \langle N_A \rangle}\right)_{\langle N \rangle} = \bar{V}_A - \bar{V}_B, \tag{7.42}$$

where we have used $\langle N_A \rangle + \langle N_B \rangle = \langle N \rangle$. From (7.40)–(7.42), together with (7.21), we have

$$\left(\frac{\partial \mu_A}{\partial x_A}\right)_{Tp} = \rho^2 \bar{V}_B \left(\frac{\partial \mu_A}{\partial \rho_A}\right)_{Tp} \tag{7.43}$$

or, using (7.29), (7.31), and (7.38),

$$\left(\frac{\partial \mu_A}{\partial x_A}\right)_{Tp} = \frac{kT}{x_A(1 + \rho_B x_A \Delta_{AB})} = kT\left[\frac{1}{x_A} - \frac{\rho_B \Delta_{AB}}{1 + \rho_B x_A \Delta_{AB}}\right], \tag{7.44}$$

where

$$\Delta_{AB} \equiv H_{AA} + H_{BB} - 2H_{AB}. \tag{7.45}$$

Multicomponent mixtures

For a multicomponent mixture having an arbitrary number of components the solution of (7.19)–(7.21) is somewhat more tedious, but has been carried out by Kirkwood and Buff.[34] The following expressions replace the binary results given in (7.25)–(7.30):

$$\left(\frac{\partial \mu_\alpha}{\partial \langle N_\beta \rangle}\right)_{Tp\langle \mathbf{N}' \rangle} = \frac{kT}{V} \frac{|D|_{\alpha\beta}}{\rho_\alpha \rho_\beta \det \mathbf{D}} \qquad (\alpha, \beta = B, C, \ldots R) \tag{7.46}$$

$$\bar{V}_\alpha = \frac{\sum_\beta \rho_\beta |B|_{\alpha\beta}}{\sum_\beta \sum_\gamma \rho_\beta \rho_\gamma |B|_{\beta\gamma}} \tag{7.47}$$

$$\chi = \frac{\det \mathbf{B}}{kT \sum_\alpha \sum_\beta \rho_\alpha \rho_\beta |B|_{\alpha\beta}}, \tag{7.48}$$

where

$$D_{\alpha\beta} \equiv (\delta_{\alpha\beta}/\rho_\alpha) + (1/\rho_A) + H_{\alpha\beta} + H_{AA} - H_{A\alpha} - H_{A\beta}. \tag{7.49}$$

Equations (7.46)–(7.49) reduce to (7.25)–(7.30) for the binary case. In the case of the derivatives $(\partial \mu_\alpha/\partial \langle N_\beta \rangle)_{Tp\langle \mathbf{N}' \rangle}$, (7.46) gives (7.27) for the case $\alpha\beta = BB$, and, by symmetry, (7.25) when $\alpha\beta = AA$. The derivatives with $\alpha\beta = AB$ or BA are then obtained using the Gibbs–Duhem equation, (7.20).

Direct correlation function expressions[41]

For a pure fluid we have already shown that (see Appendix 3E)[37]

$$1 + \rho H = (1 - \rho C)^{-1} = \rho kT\chi, \tag{7.50}$$

which gives two useful forms of the compressibility equation. Here C is the integral over $\mathbf{r}$ of $c(r) \equiv \langle c(\mathbf{r}\omega_1\omega_2)\rangle_{\omega_1\omega_2}$, the centres direct correlation function (see § 3.1.5 and Appendix 3E of Vol. 1). The derivation of the first of equations (7.50) given in Appendix 3E carries through in a straightforward way for mixtures, for which we have (see Appendix 7A for proof)

$$H_{\alpha\beta} = C_{\alpha\beta} + \rho \sum_\gamma x_\gamma C_{\alpha\gamma} H_{\gamma\beta} \tag{7.51}$$

or, in matrix form,

$$\mathbf{H} = \mathbf{C} + \rho\mathbf{CXH}, \tag{7.52}$$

where $\mathbf{X}$ is a diagonal matrix whose nonzero elements are the mole fractions $x_A, x_B, \cdots, x_\gamma, \cdots$ and

$$C_{\alpha\beta} \equiv \int d\mathbf{r} c_{\alpha\beta}(r). \tag{7.53}$$

Here $c_{\alpha\beta}(r) \equiv \langle c_{\alpha\beta}(\mathbf{r}\omega_1\omega_2)\rangle_{\omega_1\omega_2}$ is the centres direct correlation function for an $\alpha\beta$ molecular pair, and $c_{\alpha\beta}(\mathbf{r}\omega_1\omega_2)$ is the direct correlation function given by the mixture Ornstein–Zernike equation,

$$h_{\alpha\beta}(\mathbf{r}_{12}\omega_1\omega_2) = c_{\alpha\beta}(\mathbf{r}_{12}\omega_1\omega_2) + \rho\sum_\gamma x_\gamma \int d\mathbf{r}_3 \langle c_{\alpha\gamma}(\mathbf{r}_{13}\omega_1\omega_3) h_{\gamma\beta}(\mathbf{r}_{32}\omega_3\omega_2)\rangle_{\omega_3}. \tag{7.54}$$

We note that both $\mathbf{C}$ and $\mathbf{H}$ are symmetric, i.e. $H_{\alpha\beta} = H_{\beta\alpha},\ C_{\alpha\beta} = C_{\beta\alpha}$.

Using (7.52), the KBT equations of the previous sections can be recast in terms of the integrals $C_{\alpha\beta}$, as shown by O'Connell.[41] For multicomponent mixtures this gives

$$\frac{\langle N_\alpha\rangle}{kT}\left(\frac{\partial\mu_\alpha}{\partial\langle N_\beta\rangle}\right)_{Tp\langle\mathbf{N}'\rangle} = \delta_{\alpha\beta} - \frac{x_\alpha\left[1+C_{\alpha\beta}-\sum_\gamma x_\gamma(C_{\alpha\gamma}+C_{\beta\gamma})+\sum_\gamma\sum_\eta x_\gamma x_\eta(C_{\alpha\gamma}C_{\beta\eta}-C_{\alpha\beta}C_{\gamma\eta})\right]}{1-\sum_\gamma\sum_\eta x_\gamma x_\eta C_{\gamma\eta}} \tag{7.55}$$

$$\rho\bar{V}_\alpha = \left(1 - \sum_\beta x_\beta C_{\alpha\beta}\right)\Big/\left(1 - \sum_\beta\sum_\gamma x_\beta x_\gamma C_{\beta\gamma}\right) \tag{7.56}$$

$$\frac{1}{\rho kT\chi} = 1 - \sum_\alpha\sum_\beta x_\alpha x_\beta C_{\alpha\beta}. \tag{7.57}$$

From (7.56) and (7.57) we have the useful relation[42,43]

$$\frac{\bar{V}_\alpha}{kT\chi} = \frac{1}{kT}\left(\frac{\partial p}{\partial \rho_\alpha}\right)_{T,\rho'} = \sum_\beta x_\beta(1 - C_{\alpha\beta}). \tag{7.58}$$

The activity coefficient, γ_α, defined by (see § 7.4) $\ln \gamma_\alpha = \left(\mu_\alpha - \mu_\alpha^0\right)/kT - \ell n x_\alpha$, is given by[42,43]

$$\rho\left(\frac{\partial \ln \gamma_\alpha}{\partial \rho_\beta}\right)_{T,\rho'} = 1 - C_{\alpha\beta}. \tag{7.59}$$

Integrating this expression from the pure component standard state (pure component at the same temperature and pressure as the mixture) at density ρ_β^0 to the mixture of composition x_β and density ρ_β gives:

$$\ln \gamma_\alpha = \sum_\beta \int_{\rho_\beta^0}^{\rho_\beta} \frac{(1 - C_{\alpha\beta})}{\rho} \mathrm{d}\rho_\beta. \tag{7.60}$$

7.2.2 Applications and inversion of KBT equations

The KBT equations given above enable the thermodynamic properties to be calculated provided that the Kirkwood–Buff integrals $H_{\alpha\beta}$ or the direct correlation function integrals $C_{\alpha\beta}$ are known, from theory or molecular simulation, for example. Applications have been made to the calculation of thermodynamic properties, including gas and solid solubilities in various liquid solvents[44] and supercritical solvent extraction,[44–46] where a solvent at conditions somewhat above its critical temperature is used to extract a relatively non-volatile component from a solid or liquid phase (see Appendix 7B for an example of such an application).

However, it is sometimes of interest to carry out the inverse procedure, and to calculate the integrals $H_{\alpha\beta}$ or $C_{\alpha\beta}$ from experimentally obtained thermodynamic properties. The resulting integrals are of interest, since they reflect the molecular structure of the mixture, and particularly the affinity of one species for another, in some average sense.

For pure fluids the inversion is a trivial procedure, since the only expression involved is the compressibility equation, (7.50). Thus either of the integrals H or C can be immediately obtained from experimental measurements of the isothermal compressibility. It is found that these integrals are a strong function of the density, but only a very weak function of temperature. This is not surprising, since for many non-associating liquids the structure is dominated by the repulsive forces. Moreover, the dimensionless integral ρC is found to be remarkably insensitive to the anisotropic intermolecular forces, so that ρC obeys a simple two-parameter corresponding states principle,

$$\begin{aligned} \rho C &\cong \rho C_0 \\ \rho kT\chi &= \rho kT\chi_0, \end{aligned} \tag{7.61}$$

where subscript zero indicates the values for a simple fluid of spherical molecules, such as argon. This behaviour can be understood in terms of perturbation theory.

Thus, in the *f*-expansion theory (§ 4.6) the centres pair correlation function is given by the simple result (see 4.68)

$$g(r) = g_0(r), \tag{7.62}$$

which, in the *f*-expansion for g, is valid to first order. Here g_0 is the pair correlation function for a fluid in which the molecules interact with the spherically symmetric reference potential $u_0(r)$ for the *f*-expansion, given by (see 4.57)

$$\exp[-\beta u_0(r)] = \langle \exp[-\beta u(\mathbf{r}\omega_1\omega_2)]\rangle_{\omega_1\omega_2}. \tag{7.63}$$

Equation (7.62) clearly leads to (7.61). That Eq. (7.62) is often a good approximation is suggested by simulation results for the centers pair correlation function, and is not limited to the *f*-expansion approximation, but is also suggested by the *u*-expansion[47] (see Figs. 4-12, 4-15, 4-16). Equation (7.61) has been found[48] to hold well for liquids at densities above $2\rho_c$, and also for supercritical fluids over a broad range of densities. An example of a test of (7.61) is shown in Fig. 6.14, where the dimensionless quantity $\rho kT\chi$ for carbon dioxide is compared with that for argon; results for the two fluids are in good agreement in the dense liquid region, as well as much of the supercritical region. Similar results are obtained for other molecular fluids, including fluids of molecules as anisotropic as water.

The insensitivity of the isothermal compressibility and ρC to the anisotropic intermolecular forces in liquids has been used to construct useful corresponding states correlations for these quantities.[42,49,50] Early work[49] proposed a one-parameter form of corresponding states based on a characteristic density to scale the quantity $\rho kT\chi$ for a wide range of fluids of nonspherical molecules. A more accurate and sophisticated corresponding states correlation was proposed by Huang and O'Connell,[50] who noted that when ρC was plotted against density along an isotherm, the isotherms crossed at some characteristic density, ρ^*, at which the value of ρC was $(\rho C)^*$. By introducing a characteristic temperature T^* for each fluid, they were able to develop a highly accurate three-parameter corresponding states correlation for ρC for a wide range of substances. This correlation is shown in Fig. 7.28.

For binary mixtures we have the six equations (7.25)–(7.30) for composition derivatives of the chemical potentials, the partial molecular volumes and the isothermal compressibility, and it is necessary to invert these[51] to obtain $H_{\alpha\beta}$. However, these six equations are not independent since the chemical potential derivatives and partial molecular volumes are related by the thermodynamic identities

$$\rho_B \left(\frac{\partial \mu_B}{\partial \langle N_B\rangle}\right)_{Tp\langle N_A\rangle} + \rho_A \left(\frac{\partial \mu_B}{\partial \langle N_A\rangle}\right)_{Tp\langle N_B\rangle} = 0 \tag{7.64}$$

$$\rho_B \left(\frac{\partial \mu_B}{\partial \langle N_A\rangle}\right)_{Tp\langle N_B\rangle} + \rho_A \left(\frac{\partial \mu_A}{\partial \langle N_A\rangle}\right)_{Tp\langle N_B\rangle} = 0 \tag{7.65}$$

$$\rho_A \bar{V}_A + \rho_B \bar{V}_B = 1. \tag{7.66}$$

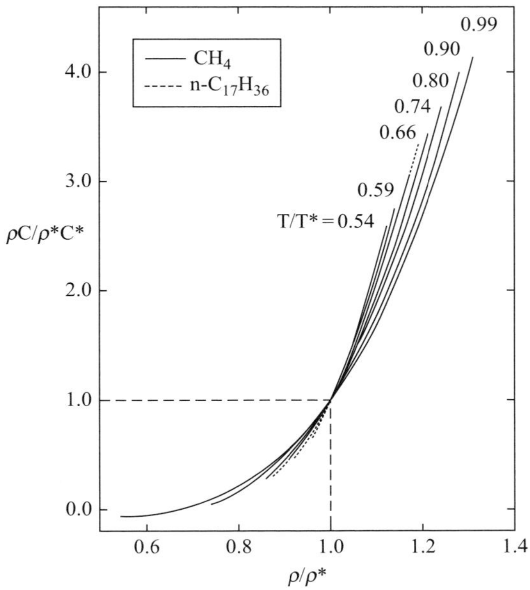

FIG. 7.28 Dimensionless plot of the direct correlation function integral, C, of Huang and O'Connell.[50] T^*, C^*, and ρ^* are parameters characteristic of the substances used to construct the corresponding states correlation. Reprinted with permission from ref. 42. Copyright 1990 Taylor and Francis Ltd.

Equations (7.64) and (7.65) are forms of the Gibbs–Duhem equation at constant temperature and pressure; (7.64) follows from the Gibbs–Duhem equation and (7.26). Equation (7.66) is readily obtained from (7.28), (7.29), and (7.31), or from the identity $V = \sum_\alpha \langle N_\alpha \rangle \bar{V}_\alpha$. Thus, on combining (7.25)–(7.30) with (7.64)–(7.66) we have three independent equations for the three integrals H_{AA}, H_{AB}, and H_{BB}.

Using the identity

$$\left(\frac{\partial \mu_B}{\partial \langle N_B \rangle}\right)_{Tp\langle N_A \rangle} = \left(\frac{\partial \mu_B}{\partial x_B}\right)_{Tp} \frac{\langle N_A \rangle}{(\langle N_A \rangle + \langle N_B \rangle)^2} \tag{7.67}$$

together with (7.27) we have

$$\left(\frac{\partial \mu_B}{\partial x_B}\right)_{Tp} = \frac{kT\rho^2}{\eta \rho_B}, \tag{7.68}$$

where we used $\rho_A = x_A \rho$. The derivative on the left side of (7.68) can be obtained experimentally by considering the vapour in equilibrium with the liquid. Then

$$\mu_B^\ell = \mu_B^{\mathrm{g}} = \mu_B^o + kT \ln(f_B/f_B{}^0), \tag{7.69}$$

where superscripts ℓ and g indicate gas and liquid phases, respectively, superscript 0 refers to some arbitrary reference state for pure component gas B at temperature T and pressure p^0, and f_B is the fugacity of the gas (see p. 806). Substituting (7.69) into (7.68) and noting that $\rho_B = x_B\rho$ gives

$$\left(\frac{\partial \ln f_B}{\partial x_B}\right)_{Tp} = \frac{\rho}{\eta x_B}. \tag{7.70}$$

Thus, by measuring the experimental partial pressure p_B versus composition, and making the pressure correction to obtain the fugacity, η can be obtained. Finally, from the Kirkwood–Buff equations for the partial molecular volumes, (7.28) and (7.29), it is easy to show that

$$\bar{V}_A\bar{V}_B = \frac{\zeta - \eta H_{AB}}{\eta^2}. \tag{7.71}$$

Thus, if experimental measurements are available for the vapour pressure versus liquid composition, the isothermal compressibility, and partial molecular volumes for the mixture, the Kirkwood–Buff integrals can be obtained by the following procedure: (a) determine η from (7.70); (b) knowing η, determine ζ from the isothermal compressibility expression, (7.30); (c) determine H_{AB} from (7.71); and (d) determine H_{AA} and H_{BB} from (7.29) and (7.28), respectively. This procedure can be extended to any number of components. Similar methods have been used to obtain the direct correlation function integrals.[42,52]

Experimental results for the $H_{\alpha\beta}$ and $C_{\alpha\beta}$ integrals have been reported by several authors, e.g. refs. 42, 43, 52–56. Examples of some of these results are shown in Figs. 7.29 and 7.30. Aqueous mixtures (with water as component A) are shown in Fig. 7.29; for mixtures with methanol, propanol, and tetrahydrofuran (THF) the water–water correlation integral H_{AA} shows increasingly positive values, and has a maximum at some composition, indicating a preference for water molecules to cluster together. Such clustering results from the hydrogen bonds between water molecules, and the hydrophobic interaction between water and the organic part of the species B molecules. The other like pair correlation integral, H_{BB}, also shows a weak maximum. The positive values in the region of the maxima for propanol and THF again indicate the strong tendency for the B molecules to cluster together, a result of the hydrophobic effect. In the case of the mixtures involving propanol and THF the cross correlation integral, H_{AB}, shows a minimum corresponding roughly to the maxima in H_{AA} and H_{BB}, as expected. In Fig. 7.30 the effect of increasing chain length of the organic cosolvent is shown. Striking differences are seen between aqueous and non-aqueous mixtures of these alcohols. In the aqueous mixtures H_{AA} and H_{BB} increase as the chain length increases, whereas in the non-aqueous mixtures (where the solvent is tetrachloromethane, TCM) the reverse is true for H_{BB}, while H_{AA} is almost unaffected. The results suggest that in aqueous mixtures the tendency to self-association may be due to hydrophobic effects, and consequent association among the hydrocarbon chains of the alcohols,

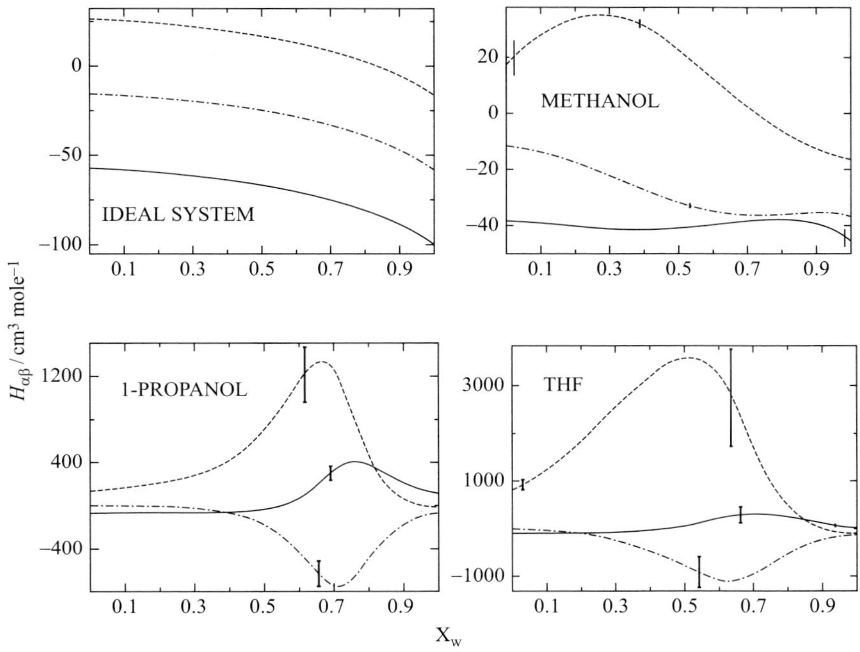

FIG. 7.29 Values of $H_{\alpha\beta}$ from experimental data plotted against mole fraction of water (component A) for three aqueous binary mixtures at 298.15 K, and also for a hypothetical ideal system: ----- H_{AA}; -·-·-·-· H_{AB}; —— H_{BB}. THF = tetrahydrofuran.Reprinted with permission from ref. 54. Copyright 1990 Taylor and Francis Ltd.

whereas in the TCM mixtures any tendency to self-association arises from H-bonding between the OH groups, which would become less effective as chain length increases.

7.3 The ideal mixture

In classical thermodynamics, an ideal mixture is usually defined as one in which the chemical potentials of all of the components obey the equation[57]

$$\mu_\alpha = \mu_\alpha^0 + kT \ln x_\alpha, \tag{7.72}$$

where μ_α^0 is the chemical potential of pure component α at the same temperature and pressure as the mixture, and x_α is the mole fraction of α in the mixture. It is assumed that (7.72) holds over all compositions, and for some finite range of temperature and pressure, so that we can differentiate with respect to these variables. This definition can be applied to gases, liquids, and solids, and it is generally required that ${\mu_\alpha}^o$ refer to pure α in the same state (gas, liquid, or solid)

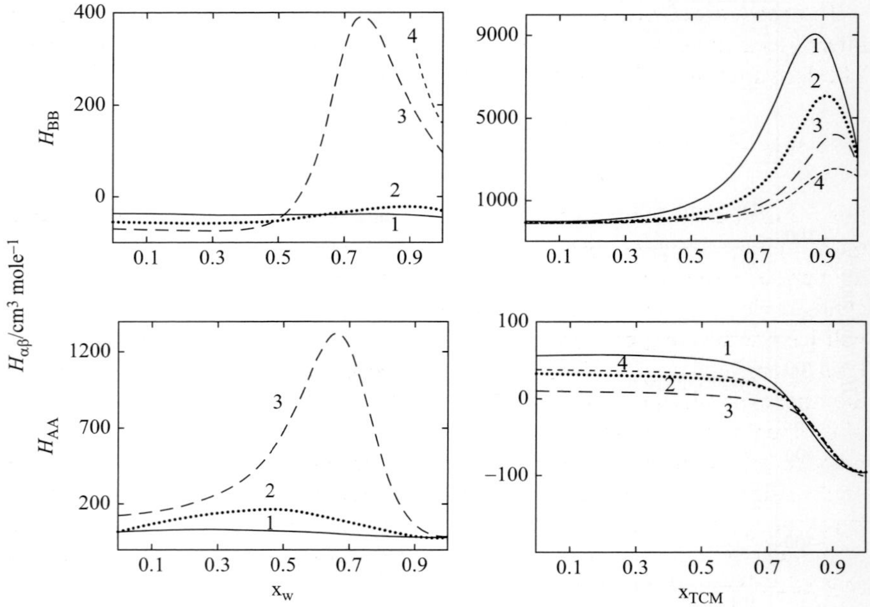

FIG. 7.30 Effect of chain length of alcohols on H_{BB} (two upper figures) and H_{AA} (two lower figures) for aqueous and non-aqueous mixtures. The left-hand figures are for water (A) mixed with alcohols (B); the right-hand figures are for mixtures of alcohols (A) with tetrachloromethane (TCM) (B). The numbers represent as follows: 1, methanol; 2, ethanol; 3, 1-propanol; 4, 1-butanol. Reprinted with permission from ref. 54. Copyright 1990 Taylor and Francis Ltd.

as the mixture. Although pure α may not exist in this state at the pressure and temperature of the mixture, the value of $\mu_\alpha{}^o$ can usually be easily calculated in the case of liquid mixtures, since it involves only a small extrapolation from experimentally accessible conditions.[57]

At the molecular level, two somewhat different definitions of an ideal solution are in common use, and we give both here. The first definition gives both a sufficient and necessary condition for ideal mixture behaviour, while the second (the 'isotopic mixture') is sufficient but not necessary.

The first and more general definition can be obtained from Eq. (7.44). The mixture will be ideal if $\Delta_{AB} = 0$, i.e. if

$$H_{AB} = \frac{1}{2}(H_{AA} + H_{BB}). \tag{7.73}$$

Integrating (7.44) from $x_A = 1$ (pure A) to x_A at constant temperature and pressure then gives (7.72). Equation (7.73) is both a sufficient and a necessary condition for the mixture to be ideal. This definition of an ideal mixture does not require that the mixture components have identical intermolecular interactions, but only that they are 'similar' in the sense that (7.73) applies.

The second definition of ideal mixtures[57] involves the assumption that the intermolecular interactions between molecules of different species are identical. The derivation starts from the mixture generalization of (6.1) and (6.2),[58]

$$A_c = -kT \ln \left\{ \frac{1}{\prod_\alpha N_\alpha ! \Omega_\alpha^{N_\alpha}} \int d\mathbf{r}^N d\omega^N \exp[-\beta \mathcal{U}(\mathbf{r}^N \omega^N)] \right\}. \tag{7.74}$$

For a mixture the intermolecular potential energy $\mathcal{U}(\mathbf{r}^N\omega^N)$ depends not only on the molecular coordinates $\mathbf{r}_1, \mathbf{r}_2, \cdots, \omega_1, \omega_2, \cdots$, but also on the way that the various molecular species are assigned to these coordinates. For example, a configuration in which an A molecule is at $(\mathbf{r}_1, \omega_1)$ and a B molecule at $(\mathbf{r}_2, \omega_2)$ will have a different potential energy from one in which A is at $(\mathbf{r}_2, \omega_2)$ while B is at $(\mathbf{r}_1, \omega_1)$ (Fig. 7.31). We now define an ideal mixture as one in which the intermolecular potential energy $\mathcal{U}(\mathbf{r}^N\omega^N)$ is independent of the assignment of different molecular species to the various locations $(\mathbf{r}_1, \omega_1)$, $(\mathbf{r}_2, \omega_2)$, etc. In practice, this implies that the like pair intermolecular interactions, as well as the unlike pair ones, for the various species are the same. This will never hold exactly in practice, but may be a good approximation for isomeric mixtures (e.g. a mixture of isomers of hexane or of xylene), and for mixtures of closely similar molecules (e.g. ethyl bromide and ethyl iodide).

With this assumption, it follows that the integral in (7.74) is independent of composition, so that the only composition dependence in this expression results

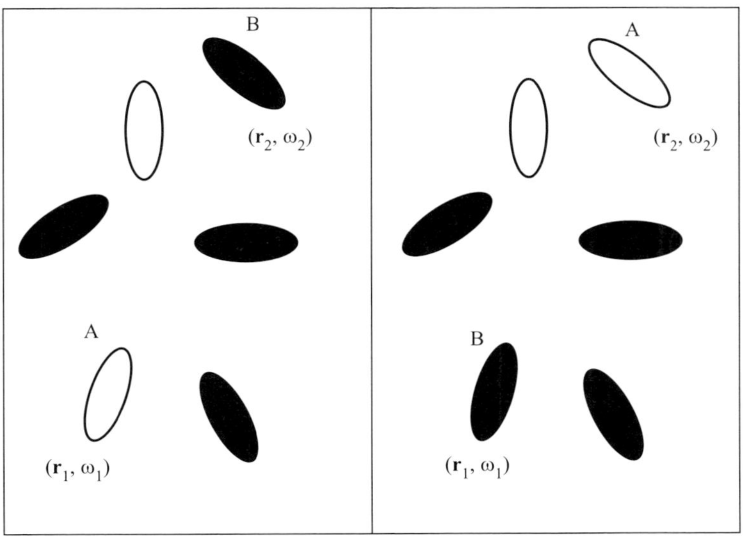

FIG. 7.31 The molecular coordinates (centres position and orientation) are the same in the two configurations shown in the left- and right-hand figures, but two of the molecular species have been switched. For a real mixture this will change the intermolecular potential energy. In an ideal 'isotopic mixture', however, the potential energy would be the same in the two cases.

from the combinatorial terms, $N_\alpha!\Omega_\alpha^{N_\alpha}$. If we call this integral $F = F(N, V, T)$, (7.74) becomes

$$A_c = -kT\ln F + kT\sum_\alpha \ln N_\alpha! + kT\sum_\alpha N_\alpha \ln \Omega_\alpha. \tag{7.75}$$

The configurational chemical potential for component β is then given by

$$\mu_{\beta c} = \left(\frac{\partial A_c}{\partial N_\beta}\right)_{TVN'} = -kT\left(\frac{\partial \ln F}{\partial N_\beta}\right)_{TVN'} + kT\sum_\alpha \frac{\partial \ln N_\alpha!}{\partial N_\beta} + kT\ln\Omega_\beta. \tag{7.76}$$

Using Stirling's approximation, (6.32), this becomes

$$\mu_{\beta c} = -kT\left(\frac{\partial \ln F}{\partial N_\beta}\right)_{TVN'} + kT\ln N_\beta + kT\ln\Omega_\beta. \tag{7.77}$$

For a system of N molecules of pure component β, at the same T and V as for the mixture, by following the derivation given above we obtain

$$A_{\beta c} = -kT\ln F + kT\ln N! + kTN\ln\Omega_\beta \tag{7.78}$$

$$\mu^0_{\beta c} = -kT\left(\frac{\partial \ln F}{\partial N_\beta}\right)_{TVN'} + kT\ln N + kT\ln\Omega_\beta. \tag{7.79}$$

We note that (a) F is the same function for the pure fluid and for the mixture, since we assumed the potential to be independent of species assignment to coordinate locations, and (b) it follows that the pure substance β at (N, V, T) will be at the same pressure as the ideal mixture at (N, V, T). Thus the chemical potential given by (7.79) is the (configurational) standard state chemical potential of Eq. (7.72). From (7.77) and (7.79) it follows that

$$\mu_\beta = \mu^o_\beta + kT\ln x_\beta, \tag{7.80}$$

where $x_\beta = N_\beta/N$ is the mole fraction of β, and we have replaced the configurational by the total chemical potential, since other contributions (kinetic energy, rotational, etc.) to μ_β and μ^0_β cancel. Equation (7.80) is the classical definition of an ideal mixture of (7.72).

This second molecular definition of an ideal mixture is more restrictive than that given by (7.73), in that it requires the intermolecular forces for different species to be the same. It is sufficient but not necessary. We shall refer to this definition as the *ideal isotopic mixture*, to distinguish it from the first definition of (7.73). Although more restrictive, it is useful in some perturbation theories (see §§ 7.7 and 7.8).

7.4 Nonideal mixtures: the activity coefficient

For a general, nonideal binary mixture of A and B, an expression for the chemical potential of component A can be obtained by integration of (7.44) over x_A from $x_A = 1$ to x_A (x_B from 0 to x_B), which gives

$$\mu_A = \mu_A^0 + kT \ln x_A + kT \ln \gamma_A, \tag{7.81}$$

where γ_A is the (dimensionless) *activity coefficient*, given by

$$\ln \gamma_A = \int_0^{x_B} \mathrm{d}x'_B \frac{\rho'_B \Delta_{AB}}{1 + \rho'_B x'_A \Delta_{AB}} = \int_0^{x_B} \mathrm{d}x'_B \frac{\rho x'_B \Delta_{AB}}{1 + \rho x'_A x'_B \Delta_{AB}}, \tag{7.82}$$

where $\rho'_B = \rho x'_B$ is the number density of B molecules and has been used to obtain the second form of (7.82), and Δ_{AB} (with dimension of volume) is defined by (7.45). A corresponding expression for the activity coefficient in terms of the direct correlation function integrals, C_{AB}, has been given by O'Connell.[59] Equation (7.82) is an exact expression for the activity coefficient, and can be used as a starting point for approximate theories. We note that the activity coefficient must obey the limits

$$\begin{aligned} \lim_{x_A \to 1} \gamma_A &= 1 \\ \lim_{\Delta_{AB} \to 0} \gamma_A &= 1. \end{aligned} \tag{7.83}$$

Equations (7.83) imply that even arbitrarily nonideal mixtures will become ideal in the sense of Raoult's Law when x_A approaches 1, i.e.

$$\lim_{x_A \to 1} \mu_A = \mu_A^o + kT \ln x_A. \tag{7.84}$$

A simpler form for the activity coefficient, suitable for some mixtures that are not highly nonideal, can be obtained as follows.[60] We first assume that the departure from ideality, (7.73), is not very large, so that

$$\rho x'_A x'_B \Delta_{AB} << 1 \tag{7.85}$$

for all compositions, and (7.82) simplifies to

$$\ln \gamma_A = \int_0^{x_B} \mathrm{d}x'_B \rho x'_B \Delta_{AB}. \tag{7.86}$$

This approximation is reasonable for a wide variety of nonionic mixtures in which the components are themselves liquids at the temperature in question. A particularly simple form results if we make the further and stronger assumption that $\rho \Delta_{AB}$ is independent of composition. The integration in (7.86) can then be carried out immediately to give

$$\begin{aligned} \ln \gamma_A &= A x_B^2 \\ \ln \gamma_B &= A x_A^2, \end{aligned} \tag{7.87}$$

where $A = (1/2)\rho \Delta_{AB}$ is treated as a constant. These equations, derived by less rigorous methods, are often referred to as the Porter, or two-suffix Margules equations. They represent a symmetric nonideal mixture (plots of the activity coefficients against mole fraction x_A are mirror images of each other), sometimes referred to as a quadratic mixture (a plot of the excess free energy against mole fraction is a quadratic curve, symmetric about $x_A = 0.5$). Equation (7.87) provides a good approximation for simple liquid mixtures, where the molecules are of similar size, shape, and chemical nature.[61] Such mixtures can have either positive or negative departures from Raoult's Law (activity coefficients greater or less than 1, respectively); A will be positive or negative in these two cases. Examples of mixtures that obey (7.87) well are argon/oxygen and benzene/cyclohexane.

In principle (7.82) can be used as a starting point for theories of the activity coefficient, by introducing approximate treatments of the correlation function integrals $H_{\alpha\beta}$. In practice, this is difficult for mixtures of liquids, because these integrals appear as a small difference of large numbers in the form of Δ_{AB} in the equations. Most applications of the Kirkwood–Buff theory have therefore been to highly nonideal mixtures, such as mixtures containing supercritical gas components or dissolved solids, associating mixtures, ionic mixtures, and reactive mixtures.[62]

The activity coefficient defined by (7.81) measures the departure of the real mixture from the ideal mixture of (7.72), in which the intermolecular forces (of arbitrary strength) are similar for the various component pairs. It is sometimes useful to measure the departure of the real mixture (which in most of the applications we will consider is a liquid) from a mixture of ideal gases, in which there are no intermolecular forces at all. This is usually done by introducing the *fugacity*, f_α, of component α, with dimensions of pressure, defined by

$$\exp\left(\frac{\mu_{\alpha r}}{kT}\right) = \frac{f_\alpha}{x_\alpha \rho kT}, \tag{7.88}$$

where $\mu_{\alpha\mathrm{r}} = \mu_\alpha - \mu_\alpha^{\mathrm{id}} = \mu_{\alpha\mathrm{c}} - \mu_{\alpha\mathrm{c}}^{\mathrm{id}}$ is the residual chemical potential, superscript 'id' indicates the ideal gas value at the same temperature, density, and composition as the real mixture, and the subscript 'c' indicates the configurational part of the chemical potential. The configurational ideal gas chemical potential is given by $\mu_{\alpha\mathrm{c}}^{\mathrm{id}} = kT \ln \rho_\alpha$ (see (6.33)[58]). For the special case of a mixture of ideal gases $\mu_{\alpha\mathrm{r}} = 0$, and (7.88) gives

$$f_\alpha^{\mathrm{id}} = x_\alpha \rho kT = x_\alpha p = p_\alpha, \tag{7.89}$$

where $p_\alpha \equiv x_\alpha p$ is the definition of the *partial pressure* of component α. Thus, the fugacity can be thought of as an 'effective partial pressure' for the real fluid, and has the units of pressure.

If (7.88) is rewritten for the pure fluid α,

$$\exp\left(\frac{\mu_{\alpha\mathrm{r}}^0}{kT}\right) = \frac{f_\alpha{}^0}{\rho^0 kT}, \tag{7.90}$$

where $\mu_{\alpha\mathrm{r}}^0$, $f_\alpha{}^0$, and ρ^0 are values for pure α at the same temperature, pressure, and state condition as the mixture, then from (7.88) and (7.90) we have

$$\exp\left(\frac{\mu_{\alpha\mathrm{r}} - \mu_{\alpha\mathrm{r}}^0}{kT}\right) = \left(\frac{f_\alpha}{x_\alpha f_\alpha{}^0}\right)\frac{\rho^o}{\rho}. \tag{7.91}$$

For an ideal solution we also have, from (7.72) and the ideal gas partition function (see § 3.1.2),

$$\exp\left(\frac{\mu_{\alpha\mathrm{r}} - \mu^0_{\alpha\mathrm{r}}}{kT}\right) = \frac{\exp\left(\frac{\mu_\alpha - \mu^0_\alpha}{kT}\right)}{\exp\left(\frac{\mu^{id}_\alpha - \mu^{0,id}_\alpha}{kT}\right)} = \frac{x_\alpha}{(x_\alpha \rho/\rho^0)} = \frac{\rho^0}{\rho}. \tag{7.92}$$

From (7.91) and (7.92) we find that

$$f_\alpha = x_\alpha f_\alpha^{\,0}, \tag{7.93}$$

which is *Raoult's Law*, and holds for an ideal mixture in the sense of (7.72). Also, from (7.88) and (7.90), together with (7.81), the fugacity and activity coefficient are related by[63]

$$f_\alpha = x_\alpha f_\alpha^{\,0} \gamma_\alpha = a_\alpha f_\alpha^{\,0}, \tag{7.94}$$

where $a_\alpha \equiv f_\alpha / f_\alpha^{\,0}$ is the *activity* of α. Activity is dimensionless. From (7.93) and (7.94) we see that $\gamma_\alpha = 1$ for an ideal mixture. For a nonideal mixture the activity coefficient provides the correction factor to Raoult's Law, (7.93).

The treatment given here provides an explicit expression for the activity coefficient, but not for the pure component fugacity $f_\alpha^{\,0}$ or chemical potential μ^0_α, which appear as unknown integration constants when (7.44) is integrated. Expressions for these quantities can be obtained from the expressions for partial molecular volume and isothermal compressibility given in § 7.2, and are derived in § 7.6.

The nonideality of mixtures can also be expressed in terms of excess properties, $M^{\mathrm{E}} \equiv M - M^{\text{id solution}}$, the excess value of the property M (where M can be Gibbs energy, entropy, enthalpy, volume, etc.) for the real solution over that of an ideal solution at the same temperature and pressure.

7.5 Dilute mixtures

Many situations of practical interest involve a dilute solute A dissolved in a solvent B, or a solvent mixture $(B, C, \cdots)$. We consider the binary case for simplicity. Starting from (7.44), in the limit as x_A tends towards 0 (keeping T and p constant), the first term on the right, $1/x_A$, will diverge, while the second term on the right will remain finite, so that in the small x_A limit we can write

$$\lim_{x_A \to 0} \left(\frac{\partial \mu_A}{\partial x_A}\right)_{T,p} = \frac{kT}{x_A}. \tag{7.95}$$

Integrating this expression at constant temperature and pressure, over a range of composition where (7.95) is expected to hold, yields

$$\mu_A = \mu_A^\infty + kT \ln x_A \qquad\qquad x_A \to 0, \tag{7.96}$$

where μ_A^∞ is an integration constant that depends on temperature and pressure. We shall refer to (7.96) as the *ideal dilute mixture law*, to be distinguished from the ideal mixture law (or Raoult's Law) form of (7.72). Superficially (7.96) resembles the ideal mixture law of (7.72) and (7.80). However, the physical meaning of μ_A^∞ is quite different from that of the constant, μ_A^0, appearing in (7.72). From (7.96) we see that $\mu_A = \mu_A^\infty$ if we put $x_A = 1$; however, (7.96) does not hold over the full concentration range usually, so that μ_A^∞ corresponds to the chemical potential for a hypothetical pure state of A that would exist *if* the ideal dilute solution law of (7.96) held over the whole composition range, $0 \leq x_A \leq 1$.

We note from (7.96) that the chemical potential of A tends to $-\infty$ as x_A tends to 0. Because of this inconvenient property of μ_A, it is often easier to work with the fugacity, defined by (7.88),

$$\lim_{x_A\to 0} \exp\left(\frac{\mu_{Ar}}{kT}\right) = \lim_{x_A\to 0}\left(\frac{f_A}{x_A\rho kT}\right). \tag{7.97}$$

For dilute solutions, we can expand f_A in powers of x_A about $x_A = 0$, at constant temperature and pressure. Since the fugacity vanishes when $x_A = 0$, this gives

$$f_A = K_A x_A + O\left(x_A^2\right), \tag{7.98}$$

where $K_A = (\partial f_A/\partial x_A)_{x_A=0}$ is the *Henry constant*, and has dimensions of pressure. We note that the solubility is inversely proportional to Henry's constant at a given state condition, so that solubility increases as K_A decreases. Thus (7.97) becomes

$$\lim_{x_A\to 0} \exp\left(\frac{\mu_{Ar}}{kT}\right) = \left(\frac{K_A}{\rho kT}\right), \tag{7.99}$$

while (7.98) becomes *Henry's Law* for sufficiently dilute mixtures,

$$\lim_{x_A\to 0} f_A = K_A x_A. \tag{7.100}$$

It follows from (7.100) that when $x_A = 1$, $K_A = f_A^\infty$, the fugacity of a hypothetical pure liquid A that would exist if Henry's Law were obeyed over the whole composition range $0 \leq x_A \leq 1$ (see Fig. 7.32). Thus (7.99) can also be written as

$$\lim_{x_A\to 0} \exp\left(\frac{\mu_{Ar}}{kT}\right) = \frac{f_A^\infty}{\rho kT}. \tag{7.101}$$

Small departures from the ideal solution law of (7.96) can be treated by expanding the final term in (7.44) in powers of x_A about $x_A = 0$. The first two terms in such an expansion are

$$\frac{\rho(1-x_A)\Delta_{AB}}{1+\rho x_A(1-x_A)\Delta_{AB}} = -B_2^\infty - 2B_3^\infty x_A - \cdots, \tag{7.102}$$

where

$$\begin{aligned} B_2^\infty &= -\rho^\infty \Delta_{AB}^\infty \\ 2B_3^\infty &= \rho^\infty\Delta_{AB}^\infty + \left(\rho^\infty\Delta_{AB}^\infty\right)^2 - \rho'^\infty\Delta_{AB}^\infty - \rho^\infty\Delta_{AB}'^\infty \end{aligned}. \tag{7.103}$$

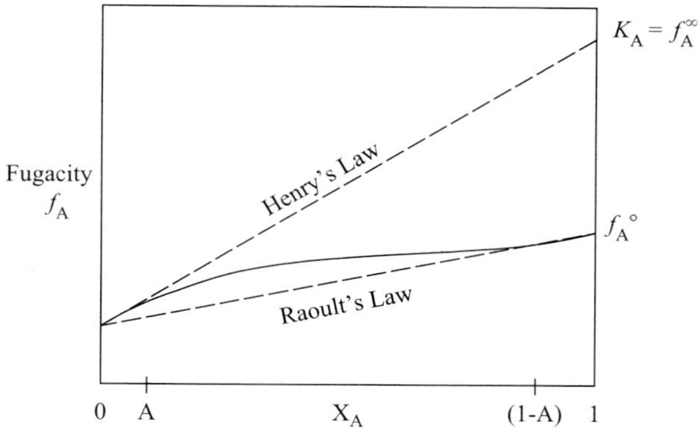

FIG. 7.32 Comparison of Raoult's Law, Eq. (7.93), and Henry's Law, Eq. (7.100), behaviour for the fugacity of a solute A in a binary mixture (schematic). The real mixture (solid line) obeys Henry's Law in the composition range $0 \leq x_A \leq A$, and obey's Raoult's Law for $(1\text{-}A) \leq x_A \leq 1$.

In (7.103) the superscript ∞ means the value at infinite dilution ($x_A \to 0$), and prime indicates derivative, $\rho' = (\partial\rho/\partial x_A)$, $\Delta'_{AB} = (\partial\Delta_{AB}/\partial x_A)$. B_2^∞ is closely related to the *osmotic second virial coefficient*, which arises in a virial (composition) expansion of the osmotic pressure. Substituting (7.102) into (7.44) and integrating over x_A for some range $x_A = x_A^{dil}$ to x_A for which (7.102) is expected to hold gives

$$\mu_A(x_A) = \mu_A\left(x_A^{\text{dil}}\right) + kT\ln\left(\frac{x_A}{x_A^{\text{dil}}}\right) + kTB_2^\infty\left(x_A - x_A^{\text{dil}}\right) + kTB_3^\infty\left(x_A^2 - x_A^{\text{dil2}}\right) + \cdots.$$

If x_A^{dil} is now chosen to be small enough that the ideal dilute solution law, (7.96), holds, and x_A^{dil} is taken to be sufficiently small compared to x_A, this becomes

$$\mu_A = \mu_A^\infty + kT\ln x_A + kT\left[B_2^\infty x_A + B_3^\infty x_A^2 + O\left(x_A^3\right)\right]. \tag{7.104}$$

The departure from the ideal dilute solution law of (7.96) can also be expressed in terms of an activity coefficient γ_A^*,

$$\mu_A = \mu_A^\infty + kT\ln x_A + kT\ln\gamma_A^*, \tag{7.105}$$

where the $*$ on γ_A^* signifies that the activity coefficient is referred to the ideal dilute solution law (as opposed to the Raoult's law ideal solution of (7.72)). From (7.104) and (7.105) we see that

$$\ln\gamma_A{*} = B_2^\infty x_A + B_3^\infty x_A^2 + O\left(x_A^3\right). \tag{7.106}$$

Truncation of this equation at the term of order x_A gives a simple equation analogous to the Porter equation, (7.87), and the coefficient B_2^∞, which is given by (7.103), only involves integrals over the centres pair correlation functions,

$g_{AB}(r)$, $g_{AA}(r)$, and $g_{BB}(r)$. The second-order term depends on B_3^∞, which involves composition derivatives of the pair correlation functions. Such derivatives involve higher order correlation functions (see § 3.4), and so are considerably more difficult to calculate.

From (7.88), (7.99), and (7.105), together with the definition of the ideal gas mixture,[63] the relation between the fugacity and the activity coefficient is

$$f_\alpha = x_\alpha K_\alpha \gamma_\alpha * = K_\alpha a_\alpha *, \tag{7.107}$$

where $a_\alpha{}^* \equiv f_\alpha / K_\alpha = f_\alpha / f_\alpha^\infty$ is the activity of α referred to the ideal dilute mixture law of (7.96). We see that $\gamma_\alpha{}^* = 1$ for an ideal dilute mixture, i.e. one that obeys Henry's Law, (7.100). For a nonideal mixture $\gamma_\alpha{}^*$ provides the correction factor to Henry's Law.

7.6 Fugacity and the Henry constant

Although the integration of (7.44) leads to the ideal mixture laws of (7.72) and (7.96), and an explicit expression for the activity coefficient, (7.82), it cannot be used to obtain an expression for the fugacity, Henry constant, or the integration constants μ_A^0 and μ_A^∞ in (7.81), (7.96), and (7.105). To obtain the fugacity, we start from the exact thermodynamic relation for the dimensionless *fugacity coefficient*,[64] $\phi_\alpha = f_\alpha / p_\alpha$, where p_α is the partial pressure of species α:

$$kT \ln \phi_\alpha = \int_V^\infty \left[\left(\frac{\partial p}{\partial N_\alpha} \right)_{T,V,N'} - \frac{kT}{V} \right] \mathrm{d}V - kT\ell\mathrm{n} \left(\frac{p}{\rho kT} \right). \tag{7.108}$$

Introducing the thermodynamic identity[65]

$$\left(\frac{\partial p}{\partial N_\alpha} \right)_{T,V,N'} = \frac{\bar{V}_\alpha}{V\chi} \tag{7.109}$$

and changing integration variables from V to ρ using $\mathrm{d}\rho / dV = -N/V^2$, (7.108) becomes

$$kT \ln \phi_\alpha = \int_0^\rho \left[\frac{\bar{V}_\alpha}{\chi} - kT \right] \frac{\mathrm{d}\rho}{\rho} - kT \ln \left(\frac{p}{\rho kT} \right). \tag{7.110}$$

The KBT expression for the fugacity coefficient can now be obtained by substituting the KBT equations for the partial molecular volume and isothermal compressibility into (7.110). For a binary mixture we would use (7.28)–(7.31). However, a more compact equation is obtained in terms of the direct correlation function integrals by using (7.58). From (7.58) and (7.110) the general expression for the fugacity coefficient is[66]

$$kT \ln \phi_\alpha = -kT \left[\sum_\beta x_\beta \int_0^\rho C_{\alpha\beta} \frac{\mathrm{d}\rho}{\rho} + \ln \left(\frac{p}{\rho kT} \right) \right]. \tag{7.111}$$

An explicit expression for the last term on the right of this equation, $\ell n(p/\rho kT)$, can be readily obtained by integration of the KBT expression for the isothermal compressibility, (7.57) or (7.48).

For a pure fluid α ($x_\alpha = 1$) at the temperature and pressure of the mixture this gives

$$kT \ln \phi_\alpha = -kT\left[\int_0^{\rho(p)} C^0_{\alpha\alpha}\frac{\mathrm{d}\rho}{\rho} + \ln\left(\frac{p}{\rho kT}\right)\right]. \tag{7.112}$$

For an infinitely dilute mixture of α ($x_\alpha \to 0$) (7.111) gives the infinite dilution limit of the fugacity coefficient, and hence the Henry constant K_α. In this limit, using (7.111), (7.97) and (7.99) gives

$$\ln\left(\frac{K_\alpha}{\rho kT}\right) = \ln \phi^\infty_\alpha = -\sum_{\beta\neq\alpha} x_\beta \int_0^\rho C^\infty_{\alpha\beta}\frac{\mathrm{d}\rho}{\rho}, \tag{7.113}$$

where superscript ∞ signifies the value at infinite dilution of α. For a binary mixture of A and B, (7.113) becomes

$$\ln\left(\frac{K_A}{\rho kT}\right) = \ln \phi^\infty_A = -\int_0^\rho C^\infty_{AB}\frac{\mathrm{d}\rho}{\rho}. \tag{7.114}$$

Integral equation theories can be used to calculate the direct correlation function integrals, $C_{\alpha\beta}$, and hence the fugacities and Henry constants from Eqs. (7.111)–(7.114). Thus, Chialvo *et al.*[67] have used GMSA theory (see § 5.4) to study the density dependence of the Henry constant and other properties at infinite dilution for a hard sphere Yukawa fluid, while Chialvo and Cummings[66] used Percus-Yevick (PY) theory to study these properties for several Lennard-Jones (LJ) mixtures. These calculations were carried out for both subcritical solvent and near-critical solvent conditions. The use of integral equation theory is an attractive alternative to full scale molecular simulations, since the total correlation functions are long range, especially for near critical conditions.

Some of the results for the residual chemical potential for LJ mixtures are shown in Fig. 7.33. Points are Monte Carlo simulation results,[68] while lines are the theoretical results from Kirkwood–Buff theory, (7.97), and (7.111), using PY theory. The theory is seen to be in excellent agreement with the simulations for dilute mixtures.

7.7 Theory of simple mixtures: spherical molecules

Before describing perturbation theories for nonspherical molecules it is helpful to first consider mixtures of simple, spherical molecules. Such mixtures are of interest in their own right, and a knowledge of successful methods of treatment forms a useful background for the study of more complex mixtures (§§ 7.8 and 7.9). In addition, perturbation treatments of complex mixtures are often based on

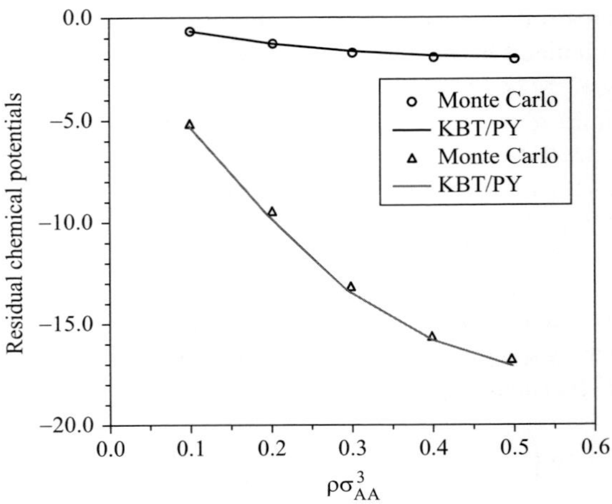

FIG. 7.33 Residual chemical potentials for solvent A (circles) and infinitely dilute solute B (triangles) for Lennard-Jones mixtures at $kT/\varepsilon_{AA} = 1.5$, $\varepsilon_{AB}/\varepsilon_{AA} = 1.5$, and $\sigma_{AB}/\sigma_{AA} = 1.518$. Points are MC simulation results,[68] lines are KBT/PY predictions. All quantities are reduced with the LJ parameters ε_{AA} and σ_{AA}. Reprinted with permission from ref. 66. Copyright 1994 Wiley-Blackwell.

a reference mixture of spherical molecules, so that an accurate theory for these is needed even to treat more complex and interesting mixtures.

There have been two successful approaches to simple mixtures. The first class of theories are the conformal solution theories, based on the idea that all intermolecular interactions in the mixture conform to the same functional form. The second class are direct extensions of the successful perturbation expansions based on a hard sphere reference system, and described in Chapter 4 for pure fluids. Both approaches are useful, and are described in this section. The conformal solution theory is very easy to use. Expansions about hard sphere fluids require more computational effort, but work better for highly nonideal mixtures, particularly those in which the component molecules show large differences in size.

7.7.1 Conformal solution theory

The central assumption on which conformal solution theory is based is that all of the interactions among the various components obey the same functional form; i.e. they are *conformal*. For simplicity, we assume that the intermolecular potentials are pairwise additive, and that the pair potential between a molecule of species α and one of species β is determined by two potential parameters, a well depth $\varepsilon_{\alpha\beta}$ and a diameter $\sigma_{\alpha\beta}$. Thus, the pair potential for an $\alpha\beta$ pair is of the form

$$u_{\alpha\beta} = \varepsilon_{\alpha\beta} f(r/\sigma_{\alpha\beta}), \tag{7.115}$$

where f is some *universal function*, the same for all pairs of components in the mixture. Examples of model mixtures of this type are a mixture of hard spheres of different sizes, or a mixture of Lennard-Jones fluids. Real mixtures that approach conformality include mixtures of inert gases (and to a good approximation, methane), or a mixture of homologous hydrocarbons of similar molecular weights.

The pure components of such conformal mixtures will obey the simple two-parameter principle of corresponding states. Corresponding states correlations for the thermodynamic properties of the mixture can also be obtained in the same way as for pure fluids, by putting the partition function in reduced form.[69] However, even for binary mixtures the number of independent variables involved is too large for such correlations to be useful. Instead, the properties of the real conformal mixture are related to those for an ideal isotopic mixture (§ 7.3), in which the species are distinguishable but all pairs have identical pair interactions, $u_x = \varepsilon_x f(r/\sigma_x)$; see Fig. 7.34. The parameters ε_x and σ_x are the same for all component pairs in this ideal reference mixture; these parameters will depend

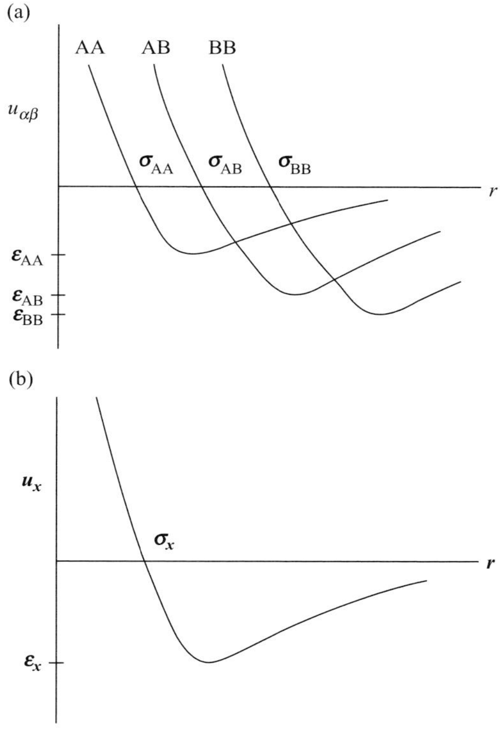

FIG. 7.34 Pair potential curves for (a) a binary conformal mixture, with potentials of the form (7.115), and (b) a conformal reference fluid. The function f in (7.115) is the same for all pair interactions in these two figures.

on composition (hence the subscript x). The principal challenge in such theories is to determine expressions for the parameters ε_x and σ_x in terms of the mixture mole fractions, the so-called *mixing rules*. Such mixing rules can be derived by expanding the properties of the mixture about those for the reference ideal isotopic mixture, as shown below. Once the mixing rules are known, it is straightforward to calculate the properties of the real mixture from those for a pure fluid with parameters ε_x and σ_x, using simple corresponding states correlations for pure fluids.

A number of such theories were proposed in the 1950s and 1960s, starting with the work of Longuet-Higgins,[70] and these have been reviewed by Rowlinson[71] and by McDonald.[72] These theories were derived before molecular simulation results became available for mixtures. They were therefore initially 'tested' by direct comparison with experimental data. However, such tests involved both choosing an intermolecular potential model and determining potential parameters by fitting to experimental thermodynamic data, a procedure that obscures any defects in the theory itself (see Fig. 1.1 and associated discussion). Once mixture simulation data became available about 1970, it was found that most of these theories gave poor results. The best of the available theories was the so-called *van der Waals-1* theory, which was in good agreement with the simulation data for mixtures that did not depart too much from the ideal isotopic mixture. We therefore focus our attention on that theory, since other theories of this kind are now mainly of historical interest. We also restrict our attention to the so-called 'one-fluid theories'. Some authors had proposed that, instead of relating the mixture properties to those of a single reference fluid, as described earlier, a reference consisting of an ideal mixture of two or three pure reference fluids be used. Such approaches were referred to as 'two-fluid' and 'three-fluid' theories.[71,72] However, comparisons with molecular simulations showed that these multi-fluid theories gave poorer results than the van der Waals one-fluid theory, and give an inadequate treatment of the critical region,[73] and we do not discuss them further.

The van der Waals-1 (vdw1) theory[74]

This form of conformal solution theory was first proposed by Leland *et al.*[75] in 1962. Since then several more rigorous approaches to deriving the theory have been put forward.[76–78] We present here the derivation due to Smith,[78] which involves a perturbation expansion of the free energy for the real mixture about that for the reference mixture in powers of the potential parameters. This method makes clear the approximations involved, and can be extended to higher orders, thus enabling improvements to be made to the vdw1 theory.

It is convenient to work in the canonical ensemble. The configurational Helmholtz energy is given by (6.1) together with (3.250), i.e.

$$A_c = -kT \ln \left(\frac{1}{\prod_\alpha N_\alpha!} Z_c \right), \tag{7.116}$$

where Z_c is the phase integral (cf. 3.53)[79] (see also endnote 58)

$$Z_c = \int d\mathbf{r}^N e^{-\beta\mathcal{U}}. \tag{7.117}$$

We now expand the configurational Helmholtz energy, A_c, about that for the ideal isotopic reference mixture, A_{cx}, in powers of the difference in the potential parameters for these two systems. A straightforward choice would be to expand in powers of $(\varepsilon_{\alpha\beta} - \varepsilon_x)$ and $(\sigma_{\alpha\beta} - \sigma_x)$, but this choice leads to poor convergence. An alternative is to expand in some combination of parameters, $\varepsilon^m_{\alpha\beta}\sigma^n_{\alpha\beta}$ and $\varepsilon^p_{\alpha\beta}\sigma^q_{\alpha\beta}$. Different choices of (m,n,p,q) lead to different theories.[72, 78] A choice that leads to good convergence is that of the vdw1 theory, $(m,n,p,q) = (1,3,0,3)$, and we shall employ this choice in the expansion here. Reasons for such a choice are discussed below. With this choice the expansion of the configurational Helmholtz energy is, to first order,

$$A_c = A_{cx} + \sum_\alpha\sum_\beta \left(\frac{\partial A_c}{\partial \sigma^3_{\alpha\beta}}\right)_x \left(\sigma^3_{\alpha\beta} - \sigma^3_x\right) + \sum_\alpha\sum_\beta \left(\frac{\partial A_c}{\partial \varepsilon_{\alpha\beta}\sigma^3_{\alpha\beta}}\right)_x \left(\varepsilon_{\alpha\beta}\sigma^3_{\alpha\beta} - \varepsilon_x\sigma^3_x\right) + \dots. \tag{7.118}$$

Assuming pairwise additivity,

$$\mathcal{U} = \frac{1}{2}\sum_\gamma\sum_\delta\sum_i\sum_{j\neq i} u_{\gamma\delta}(\mathbf{r}_{ij}\omega_i\omega_j), \tag{7.119}$$

we have, using (7.116) and (7.117), and noting that there are $N_\alpha(N_\beta - \delta_{\alpha\beta})/2 \cong N_\alpha N_\beta/2$ derivative terms involving $u_{\alpha\beta}$, each of which gives the same result on integration,

$$\frac{\partial A_c}{\partial \sigma^3_{\alpha\beta}} = \frac{1}{2}N_\alpha N_\beta \int d\mathbf{r}^N \frac{\partial u_{\alpha\beta}(r_{12})}{\partial \sigma^3_{\alpha\beta}} \frac{\exp(-\beta\mathcal{U})}{Z_c}. \tag{7.120}$$

Taking the limit as $\varepsilon_{\alpha\beta}$ goes to ε_x and $\sigma_{\alpha\beta}$ goes to σ_x for all α, β gives

$$\left(\frac{\partial A_c}{\partial \sigma^3_{\alpha\beta}}\right)_x = x_\alpha x_\beta F_x, \tag{7.121}$$

where F_x is a pure fluid integral whose value depends on the parameters ε_x and σ_x (in addition to N,T,V), and is given by

$$\begin{aligned} F_x &= \frac{1}{2}N^2 \int d\mathbf{r}^N \frac{\partial u_x(r_{12})}{\partial \sigma^3_x} \frac{\exp(-\beta\mathcal{U}_x)}{Z_{cx}} \\ &= 2\pi N\rho \int_0^\infty dr r^2 \frac{\partial u_x(r)}{\partial \sigma^3_x} g_x(r), \end{aligned} \tag{7.122}$$

where $g_x(r)$ is the pair correlation function for the reference fluid (see § 3.1.4). The final term in (7.118) is evaluated in a parallel way to give

$$\left(\frac{\partial A_c}{\partial \varepsilon_{\alpha\beta}\sigma^3_{\alpha\beta}}\right)_x = x_\alpha x_\beta G_x \tag{7.123}$$

with

$$\begin{aligned} G_x &= \frac{1}{2}N^2 \int \mathrm{d}\mathbf{r}^N \frac{\partial u_x(r_{12})}{\partial \varepsilon_x \sigma_x^3} \frac{\exp(-\beta \mathcal{U}_x)}{Z_{cx}} \\ &= 2\pi N\rho \int_0^\infty \mathrm{d}r r^2 \frac{\partial u_x(r)}{\partial \varepsilon_x \sigma_x^3} g_x(r). \end{aligned} \tag{7.124}$$

From (7.118), (7.121), and (7.123) we have

$$A_c = A_{cx} + F_x \sum_\alpha \sum_\beta x_\alpha x_\beta (\sigma^3_{\alpha\beta} - \sigma^3_x) + G_x \sum_\alpha \sum_\beta x_\alpha x_\beta (\varepsilon_{\alpha\beta}\sigma^3_{\alpha\beta} - \varepsilon_x \sigma^3_x) + \cdots \tag{7.125}$$

So far we have assumed that all pair interactions in the ideal isotopic reference mixture are the same, with parameters ε_x and σ_x, but have not specified the values of these parameters. We now choose them so as to annul the first-order terms in (7.125); i.e. we take

$$\begin{aligned} \sigma^3_x &= \sum_\alpha \sum_\beta x_\alpha x_\beta \sigma^3_{\alpha\beta} \\ \varepsilon_x \sigma^3_x &= \sum_\alpha \sum_\beta x_\alpha x_\beta \varepsilon_{\alpha\beta} \sigma^3_{\alpha\beta} \end{aligned} \tag{7.126}$$

so that (7.125) reduces to

$$A_c = A_{cx} + \cdots , \tag{7.127}$$

which holds to first order. Equations (7.127) together with (7.126) comprise the vdw1 theory, and (7.126) are known as the *van der Waals 1 mixing rules*. The name derives from the fact that the mixing rules used by van der Waals in his theory of mixtures (which relate the parameters a_x and b_x in his equation of state to quadratic sums of the pair parameters $a_{\alpha\beta}$ and $b_{\alpha\beta}$) are equivalent to (7.126).

An expression for A_{cx} can be obtained by writing it in the form

$$A_{\mathrm{c}x} = \Delta A_{\mathrm{c}x} + A^0_{\mathrm{c}x}, \tag{7.128}$$

where $\Delta A_{\mathrm{c}x}$ is the change in configurational Helmholtz energy on mixing the pure components (at constant temperature and volume[80]) to make N molecules of mixture, and $A^0_{\mathrm{c}x}$ is the configurational Helmholtz energy for N molecules of pure component (this will be the same value for each of the components, since they all have the same pair potential). The first term on the right of (7.128) is given by

$$\Delta A_{cx} = A_{cx} - \sum_{\alpha} x_\alpha A^0_{c\alpha,x}, \tag{7.129}$$

where $A^0_{c\alpha,x}$ is the configurational Helmholtz energy for N molecules of pure α after taking the $\varepsilon_{\alpha\alpha}, \sigma_{\alpha\alpha} \to \varepsilon_x, \sigma_x$ limit. Using (7.116), together with Stirling's approximation,

$$\ln N_\alpha! \cong N_\alpha \ln N_\alpha - N_\alpha \qquad \text{(large } N_\alpha) \tag{7.130}$$

(7.129) gives

$$\begin{aligned} \Delta A_{cx} &= kT \sum_{\alpha} \ln N_\alpha! - kT \sum_{\alpha} x_\alpha \ln N! \\ &= NkT \sum_{\alpha} x_\alpha \ln x_\alpha, \end{aligned} \tag{7.131}$$

where we have used the fact that Z_{cx} for the ideal mixture equals $Z^0_{c\alpha,x}$, the phase integral for pure component α. From (7.127), (7.128), and (7.131) we get

$$A_c = A^0_{cx} + NkT \sum_{\alpha} x_\alpha \ln x_\alpha. \tag{7.132}$$

We note that the expression for ΔA_{cx} is just the entropy of mixing term, $-T\Delta S$, for an ideal mixture, since the energy of mixing is zero. The term A^0_{cx} can be calculated from a suitable corresponding states correlation for pure fluids, for example a reduced equation of state for some model fluid, such as the Lennard-Jones fluid,[81] or an equation of state fitted to experimental data for some real fluid, such as argon or methane (which is approximately spherical in its interactions). The procedure is to first calculate the values of ε_x and σ_x from (7.126), and from these the reduced temperature, kT/ε_x, and either the reduced pressure, $p\sigma_x^3/\varepsilon_x$, or reduced density, $\rho\sigma_x^3$. The dimensionless thermodynamic properties of the pure reference fluid, including A^0_{cx}, can then be obtained from the equation of state. This is the procedure used when a model fluid (e.g. Lennard-Jones) is the pure reference fluid. When the properties of a real fluid are used it is usually more convenient to replace the potential parameters ε_x and σ_x by pseudo-critical constants for the mixture as reducing parameters.[82]

The derivation given above has the advantage that it can be extended to higher orders, and can be used to derive theories for other properties. The theory for the free energy has been extended to second order, and calculations have been made for mixtures of hard spheres and Lennard-Jones fluids.[78,83,84] The second-order term involves integrations over both the pair and three-body correlation functions for the pure reference fluid; the superposition approximation, (3.246), can be used for the three-body correlation functions with little error. The vdw1 conformal solution theory has been developed for the pair correlation functions for mixtures,[84] and gives good results provided the molecules do not differ too much in size. An analogous theory has also been proposed for transport properties of mixtures;[85] in this case a mixing rule for the molecular mass of the equivalent pure fluid is needed, in addition to those for the two potential parameters.

Choice of expansion parameters

The choice of $\sigma_{\alpha\beta}^3$ and $\varepsilon_{\alpha\beta}\sigma_{\alpha\beta}^3$ as expansion parameters in the above treatment seems somewhat arbitrary. Many other choices have been proposed, e.g. $\sigma_{\alpha\beta}$ and $\varepsilon_{\alpha\beta}$ (the original choice of Longuet-Higgins[70]) or $\varepsilon_{\alpha\beta}\sigma_{\alpha\beta}^{12}$ and $\varepsilon_{\alpha\beta}\sigma_{\alpha\beta}^6$ (random mixture theory for the LJ (12,6) fluid[86]), but have been found to be in poorer agreement with molecular simulation results than the vdw1 choice.

Although no formal proof exists that the vdw1 parameters are the optimum choice for such expansions, some plausibility arguments can be given, as follows.[77] For a fluid mixture in which the potentials are of the form (7.115), the pressure equation is given by the mixture generalization of (6.15),

$$\frac{p}{\rho kT} = 1 - \frac{2\pi\rho}{3kT}\sum_{\alpha}\sum_{\beta} x_\alpha x_\beta \int_0^\infty \mathrm{d}r r^3 \frac{\mathrm{d}u_{\alpha\beta}(r)}{\mathrm{d}r} g_{\alpha\beta}(r). \tag{7.133}$$

Introducing (7.115) and putting the integral in dimensionless form gives

$$\frac{p}{\rho kT} = 1 - \frac{2\pi\rho}{3kT}\sum_{\alpha}\sum_{\beta} x_\alpha x_\beta \varepsilon_{\alpha\beta}\sigma_{\alpha\beta}^3 \int_0^\infty \mathrm{d}r^* r^{*3} f'(r^*) g_{\alpha\beta}(r^*), \tag{7.134}$$

where $r^* = r/\sigma_{\alpha\beta}$ and $f' = \mathrm{d}f/\mathrm{d}r^*$. Equation (7.134) can be put in one fluid form by making the 1-fluid approximation

$$g_{\alpha\beta}(r^*) = g_x(r^*) \qquad (\text{all } \alpha,\ \beta), \tag{7.135}$$

where $g_x(r^*)$ is the pair correlation function for a pure fluid with parameters ε_x and σ_x. Equation (7.134) then takes the 1-fluid form

$$\frac{p}{\rho kT} = 1 - \frac{2\pi\rho}{3kT}\varepsilon_x\sigma_x^3 \int_0^\infty dr^* r^{*3} f'(r^*) g_x(r^*), \tag{7.136}$$

where $\varepsilon_x\sigma_x^3$ is given by

$$\varepsilon_x\sigma_x^3 = \sum_{\alpha}\sum_{\beta} x_\alpha x_\beta \varepsilon_{\alpha\beta}\sigma_{\alpha\beta}^3, \tag{7.137}$$

i.e. the second of the vdw1 mixing rules of (7.126). Thus, $\varepsilon_{\alpha\beta}\sigma_{\alpha\beta}^3$ arises as a natural expansion parameter (similar arguments can be applied using the energy or free energy equations of Chapter 6, with the same result).

In order to completely specify the equivalent pure fluid to which $g_x(r^*)$ corresponds, we need a further expression for σ_x, or some combination of σ_x and ε_x. This can be obtained by considering the pressure equation for a hard sphere mixture. This is given by the mixture generalization of (5.24),

$$\frac{p}{\rho kT} = 1 + \frac{2}{3}\pi\rho \sum_{\alpha}\sum_{\beta} x_\alpha x_\beta \sigma_{\alpha\beta}^3 g_{\alpha\beta}\left(\sigma_{\alpha\beta}^+\right), \tag{7.138}$$

where $g_{\alpha\beta}\left(\sigma_{\alpha\beta}^+\right)$ is the pair correlation function for an $\alpha\beta$ hard sphere pair at contact, as r approaches $\sigma_{\alpha\beta}$ from above. This can be put in 1-fluid form by using the approximation

$$g_{\alpha\beta}\left(\sigma_{\alpha\beta}^+\right) = g_x\left(\sigma_x^+\right) \qquad (\text{all } \alpha, \beta), \tag{7.139}$$

where g_x is the pair correlation function for a pure hard sphere fluid in which the sphere diameter is σ_x, so that (7.138) becomes

$$\frac{p}{\rho kT} = 1 + \frac{2}{3}\pi\rho\sigma_x^3 g_x\left(\sigma_x^+\right) \tag{7.140}$$

with

$$\sigma_x^3 = \sum_{\alpha}\sum_{\beta} x_\alpha x_\beta \sigma_{\alpha\beta}^3, \tag{7.141}$$

which is the first of the vdw1 mixing rules of (7.126). The above argument shows that the vdw1 choice of parameters is a reasonable one for a theory of liquid mixtures. The approximations (7.135) and (7.139) for $g_{\alpha\beta}(r^*)$ can be regarded as the zeroth-order term in a perturbation expansion of the pair correlation function about that for the equivalent pure reference fluid.

The principal advantage of the vdw1 choice of parameters and mixing rules over others that have been proposed is in its superior description of the effects on thermodynamic properties of differences in molecular size. This is most clearly seen for mixtures of hard spheres, in which size differences are the only source of nonideal mixing behaviour. For such binary mixtures, the volume decreases on mixing, i.e. $V^{\mathrm{E}} < 0$, where $V^{\mathrm{E}} \equiv V - V^{\text{id. soln.}}$ is the excess volume (see definition of excess quantities in §§ 7.1.2, 7.4. This is because small molecules can to some extent fit into spaces between large molecules, so that the mixture volume is less than the sum of the pure liquid volumes. Both the vdw1 theory and the Percus–Yevick theory predict this effect correctly.[71,76] Most other theories do not, predicting instead a positive excess volume. Getting the sign of the excess volume correct is crucial, since it ensures that the sign of the excess Gibbs energy is also correct. For hard sphere mixtures G^{E} is simply related to V^{E} by

$$G^{\mathrm{E}}(p, T) = \int_0^p V^{\mathrm{E}}(p', T)\mathrm{d}p'. \tag{7.142}$$

Thus $G^{\mathrm{E}} < 0$ for hard sphere binary mixtures.[87] While in real mixtures the picture will be complicated by the presence of attractive interactions, we know that the structure of liquids of simple spherical molecules is dominated by the repulsive forces (see § 4.1), so that these considerations for hard sphere mixtures will also be important for more realistic mixtures.

Further support for the vdw1 choice of parameters comes from the fact that (a) they give the mixture second virial coefficient correctly, and (b) as shown by Reid and Leland,[88] they can be justified by appeal to the generalized van der Waals theory of Kac, in which the attractive potential is weak compared to kT, and of infinitely long range (see endnote 223 of Chapter 4).

Comparison with simulation results

Smith[83] has compared the results of both the vdw1 theory and the second-order theory (Eq. (7.125) extended to second order) with simulation data for the equation of state, $p/\rho kT$, for hard sphere mixtures. Such comparisons make clear the ability of the theory to describe effects due solely to molecular size differences, without the added complication of differences in the attractive forces for the two species. For small size differences, e.g. $\sigma_{BB}/\sigma_{AA} = 11/10$, both vdw1 and the second-order theory are in excellent agreement with the simulation results up to the highest liquid-like densities. For much larger size ratios there are deviations between the theory and simulation data, particularly at high densities. Results for these higher size ratios are shown in Fig. 7.35; in these figures we refer to the second-order theory as 2-vdw1. For a size ratio of $\sigma_{BB}/\sigma_{AA} = 5/3$ the vdw1 theory is seen to work quite well for densities up to about $\rho^* = 0.6$–0.7, whereas for $\sigma_{BB}/\sigma_{AA} = 3$ vdw1 breaks down at lower densities, in the range $\rho^* = 0.4$–0.5; for this latter size ratio 2-vdw1 still gives good results up to high liquid-like densities. For significantly higher size ratios 2-vdw1 also breaks down for liquid densities, as seen from the result for $\sigma_{BB}/\sigma_{AA} = \infty$. In the calculations shown for 2-vdw1 in Fig. 7.35 the second-order term, which involves an integral over the three-body correlation function, was evaluated using the superposition approximation, (3.246). The results suggest that vdw1 theory can be expected to give good results for modest size differences, up to about $\sigma_{BB}/\sigma_{AA} \approx 1.5$–$2$ for liquids, the exact limit depending on the sensitivity of the property considered.

For Lennard-Jones (12,6) mixtures there have been several comparisons made of vdw1 theory with simulation results for the excess properties,[72,77,89] for parameter ratios $\varepsilon_{BB}/\varepsilon_{AA}$ and σ_{BB}/σ_{AA} that range from about 0.6 to 1.4. For these mixtures there is generally good agreement between the theory and simulation. A strong test of the theory is the calculation of chemical potentials, or equivalently the Henry constants (see (7.99)), for infinitely dilute mixtures. Such a test has been made by Shing *et al.*,[90] and some of their results are shown in Figs. 7.36 and 7.37. In these mixtures the unlike pair parameters are assumed to obey the Lorentz-Berthelot combining rules,

$$\begin{aligned}\sigma_{\alpha\beta} &= \frac{1}{2}(\sigma_{\alpha\alpha} + \sigma_{\beta\beta}) \\ \varepsilon_{\alpha\beta} &= (\varepsilon_{\alpha\alpha}\varepsilon_{\beta\beta})^{1/2}.\end{aligned} \tag{7.143}$$

The results show that the vdw1 theory gives quite good results for the parameter ranges $\varepsilon_{AB}/\varepsilon_{BB} \approx 0.25$–$2.0$ and $(\sigma_{AB}/\sigma_{BB})^3 \approx 0.5$–$2.0$. Two limiting cases are

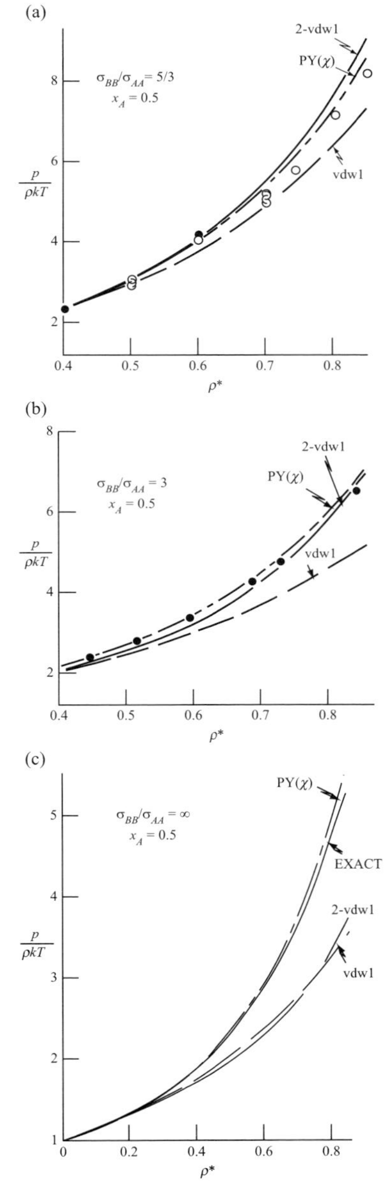

FIG. 7.35 Equation of state for an equimolar hard sphere mixture for various size ratios. Points are molecular simulation results. Curves show results for vdw1 theory, second-order vdw1 theory (2-vdw1), and Percus–Yevick equation via the compressibility route (PY(χ)). (a) $\sigma_{BB}/\sigma_{AA} = 5/3$; (b) $\sigma_{BB}/\sigma_{AA} = 3$; (c) $\sigma_{BB}/\sigma_{AA} = \infty$. In these figures the reduced density is defined as $\rho^* = \left(x_A\sigma_{AA}^3 + x_B\sigma_{BB}^3\right)\rho$, where $\rho = \rho_A + \rho_B$ is the total density. Reprinted with permission from ref. 83. Copyright 1971 Taylor and Francis Ltd.

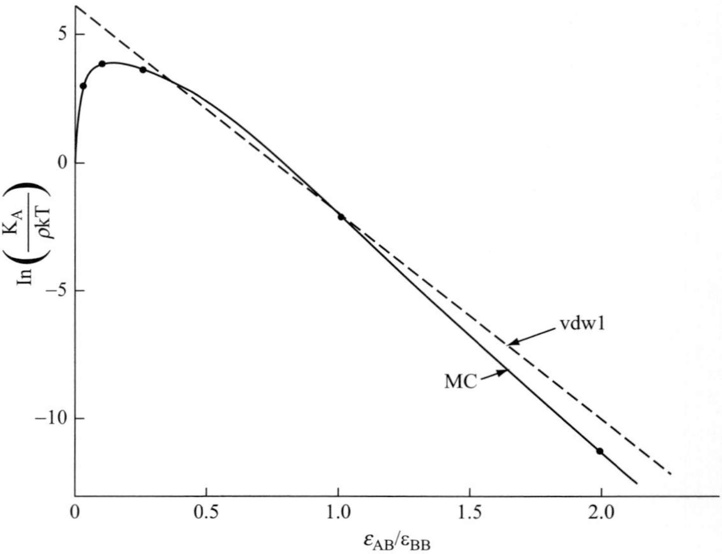

FIG. 7.36 Variation of the reduced Henry constant, $K_A/\rho kT$, with the ratio $\varepsilon_{AB}/\varepsilon_{BB}$ for LJ mixtures with $\sigma_{AB}/\sigma_{BB} = 1, kT/\varepsilon_{BB} = 1.2$, $\rho\sigma_{BB}^3 = 0.7$, $x_A = 0$, showing comparison of Monte Carlo (MC) simulation results and vdw1 theory. Reprinted with permission from ref. 90. Copyright 1982 Taylor and Francis Ltd.

included in Figs. 7.36 and 7.37: (a) when $\varepsilon_{AB}/\varepsilon_{BB} = 1$ in Fig. 7.36, and when $\sigma_{AB}/\sigma_{BB} = 1$ in Fig. 7.37, the solution is an ideal one having all pair potentials equal, so that K_A reduces to the fugacity of pure A, and (b) when $\varepsilon_{AB}/\varepsilon_{BB} = 0$ (Fig. 7.36) or $\sigma_{AB}/\sigma_{BB} = 0$ (Fig. 7.37) the A molecules behave as an ideal gas ($u_{AA} = u_{AB} = 0$) and $K_A/\rho kT = 1$. The effect of varying $\varepsilon_{AB}/\varepsilon_{AA}$ keeping $\sigma_{BB} = \sigma_{AA}$ is shown in Fig. 7.36. The maximum in the curve for the simulation results arises because of competing effects from the repulsive and attractive forces as ε_{AB} is increased. A small increase in ε_{AB} from 0 produces a very rapid increase in the short-range repulsive forces, leading to a decrease in solubility of A, and an increase in K_A. As ε_{AB} is further increased, the attractive forces lead to an increased solubility and a decrease in K_A. Although the vdw1 theory fails for very small $\varepsilon_{AB}/\varepsilon_{BB}$ ratios, it gives a reasonable approximation to the exact values over a wide range for $\varepsilon_{AB}/\varepsilon_{BB}$ values from about 0.4 up to over 2. The effect of varying the size ratio is shown in Fig. 7.37, and is a more severe test of the theory.

Panagiotopoulos and co-workers[91–93] have carried out extensive Gibbs ensemble Monte Carlo simulations for a wide range of binary and ternary mixtures showing both gas–liquid and liquid–liquid separation, and have used the results to test the vdw1 theory for such phase equilibria. They find very good agreement between the theory and simulation results for the *pxy* diagrams (here *y* is mole fraction in the vapour phase), even for highly nonideal mixtures with molecular

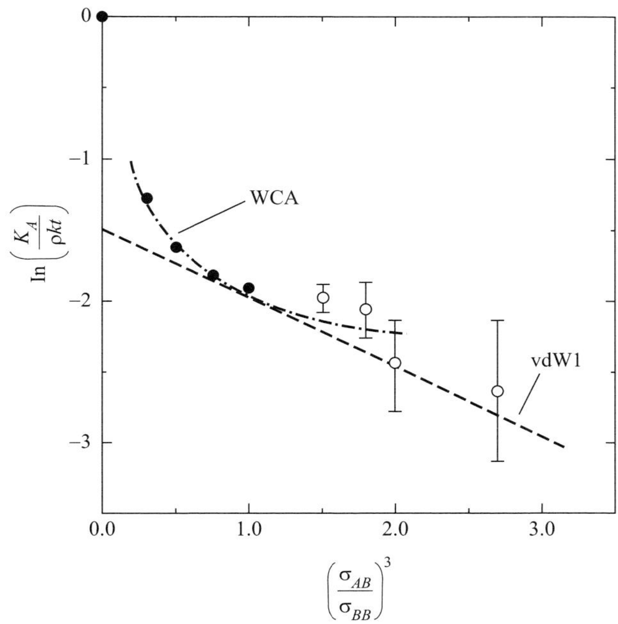

FIG. 7.37 Variation of the reduced Henry constant, $K_A/\rho kT$, with the ratio σ_{AB}/σ_{BB} for LJ mixtures with $\varepsilon_{AB}/\varepsilon_{BB} = 1$ at $kT/\varepsilon_{BB} = 1.2$, $\rho\sigma_{BB}{}^3 = 0.7$, $x_A = 0$, showing comparison of Monte Carlo (MC) simulation results (points) and vdw1 and LL (WCA for mixtures – see §7.7.2) theories. (From ref. 90.)

size ratios σ_{AA}/σ_{BB} as large as 3. Results for the pressure–density diagrams show somewhat less good agreement, although the comparisons are still quite good. Typical results are shown in Figs. 7.38 and 7.39. The unlike parameters for these mixtures obey the Lorentz–Berthelot rules of (7.143). These authors have also studied mixtures that depart from the Lorentz–Berthelot rules, with similar results. The good agreement for large size ratios is somewhat surprising in view of the results for hard sphere mixtures and for LJ mixtures at infinite dilution shown above. Apparently, calculations of gas–liquid phase diagrams are less sensitive to errors in the theory.

The extension of the vdw1 approach to the pair correlation function in mixtures has been given by Mo *et al.*.[84] The first-order expansion about the ideal isotopic reference mixture involves integrals over the three-body reference correlation function, which can be estimated to good accuracy using the superposition approximation, (3.246). The first-order term in the perturbation expansion of the pair correlation function corresponds to the second-order term in the expansion for the Helmholtz energy (see § 4.4), so that this theory yields the second-order vdw1 (2-vdw1) treatment when applied to thermodynamic properties. The theory for the pair correlation function has been tested against molecular dynamics simulation results for LJ liquid mixtures modeled on argon/krypton at $T = 116\,\mathrm{K}$ and

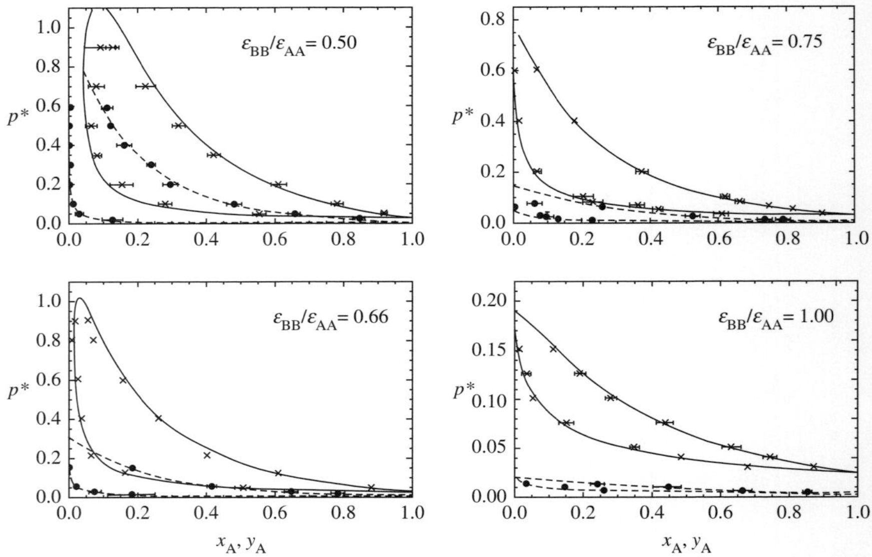

FIG. 7.38 Reduced pressure vs mole fraction in liquid (x) and vapor (y) phases for LJ mixtures with $\sigma_{BB}/\sigma_{AA} = 0.5$. Points are MC results: x, $T^* = 1.00$; •, $T^* = 0.75$. Lines are vdw1 theory: ——, $T^* = 1.00$; - - -, $T^* = 0.75$. Here $T^* = kT/\varepsilon_{AA}$ and $p^* = p\,\sigma_{AA}{}^3/\varepsilon_{AA}$. Reprinted with permission from ref. 91. Copyright 1991 Elsevier.

$V = 33.3\,\text{cm}^3\text{mole}^{-1}$. Potential parameters used were $\sigma_{AA} = 3.405$ Å, $\varepsilon_{AA}/k = 119.8$ K, and $\sigma_{KK} = 3.633$ Å and $\varepsilon_{KK}/k = 167.0$ K, and the Lorentz–Berthelot rules were used for the unlike pair parameters. These comparisons are shown in Figs. 7.40 and 7.41.

As seen from Fig. 7.40, the theory gives generally good results for equimolar mixtures, but becomes poorer for dilute solutions, as seen in Fig. 7.41. As the mole fraction of krypton increases, the peak becomes higher as expected, since N and V are fixed. In interpreting the results in Fig. 7.41 for dilute solutions, we note that in the limit $x_A \to 1$ then $u_x \to u_{AA}$ so that $g_x \to g_{AA}$. Thus, for a dilute solution of K in A the theory will work best for g_{AA} and be poorest for g_{KK}. Similarly, for a dilute solution of A in K we expect good agreement for g_{KK}, but poorer results for g_{AA}. This accounts for the essentially exact agreement for g_{AA} at $x_A = 0.9$ in Fig. 7.41; conversely, the agreement is relatively poor for g_{AA} at $x_A = 0.1$ in Fig. 7.41.

Shape factor correlations

Although the vdw1 theory is only valid for mixtures of spherical molecules, it forms the basis of semi-empirical methods for predicting the thermodynamic properties of mixtures of *nonspherical* molecules. One of the most useful methods is the shape factor correlation.[94] This is based on the use of critical constants in

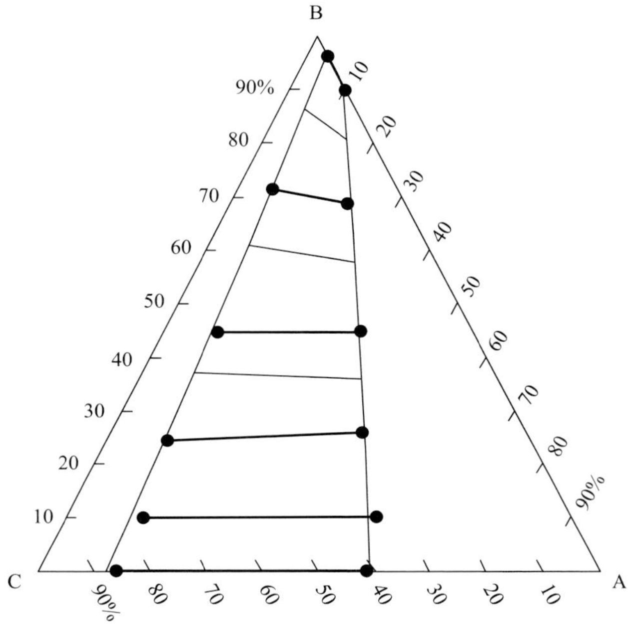

FIG. 7.39 Ternary phase diagram for a LJ mixture with parameters $\sigma_{BB}/\sigma_{AA} = 0.75$, $\sigma_{CC}/\sigma_{AA} = 0.50$, $\varepsilon_{BB}/\varepsilon_{AA} = 0.75$, and $\varepsilon_{CC}/\varepsilon_{AA} = 0.50$ at $T^* = 1.0$, $p^* = 0.20$. Points and bold tie-lines are MC simulation results, thin lines and thin tie-lines are vdw1 theory results for vapour-liquid equilibrium. Typical error bars are equal to the symbol size. Reprinted with permission from ref. 93. Copyright 1995 Elsevier.

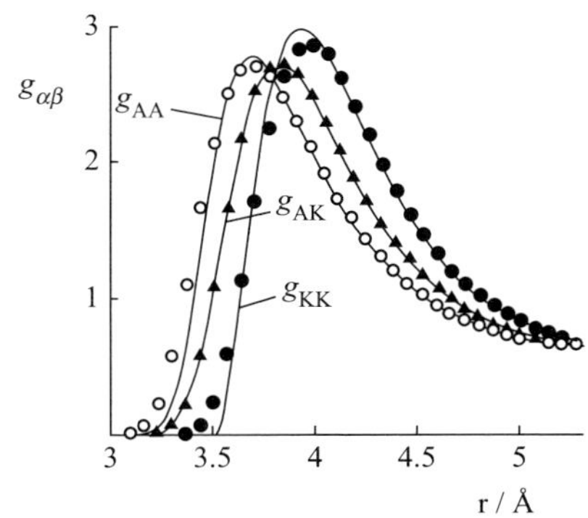

FIG. 7.40 Pair correlation functions for an equilmolar LJ liquid mixture modeled on argon and krypton, $T = 116\,\text{K}$, $V = 33.3\,\text{cm}^3\text{mole}^{-1}$. Lines are the first-order vdw1 theory (corresponding to 2-vdw1 for thermodynamic properties), points are MD data: $\circ = g_{AA}$; $\blacktriangle = g_{AK}$; $\bullet = g_{KK}$. Reprinted with permission from ref. 84. Copyright 1974 Taylor and Francis Ltd.

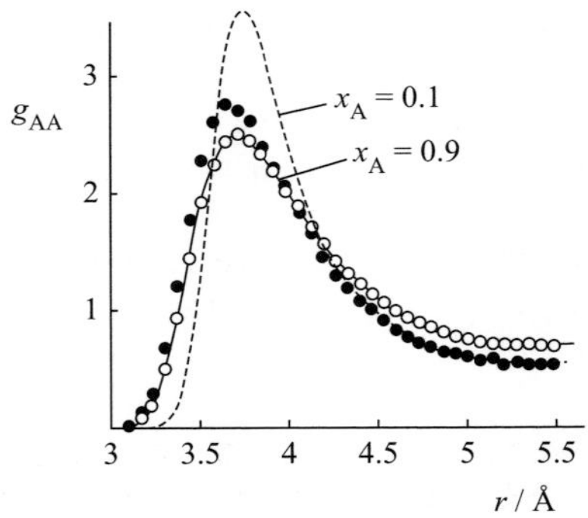

FIG. 7.41 Composition dependence of g_{AA} for argon/krypton LJ mixture at $T = 116\,\text{K}$, $V = 33.3\,\text{cm}^3\text{mole}^{-1}$. MD data are for $x_A = 0.1$ (solid circles) and $x_A = 0.9$ (open circles). Key as described in the legend to Fig. 7.40. Reprinted with permission from ref. 84. Copyright 1974 Taylor and Francis Ltd.

place of potential parameters. For a pure conformal fluid of spherical molecules corresponding states theory[94] gives

$$Z_0 = f_0\left(\frac{p}{p_c}, \frac{T}{T_c}\right), \tag{7.144}$$

where $Z_0 = p/\rho kT$ is the compressibility factor, p_c and T_c are the critical pressure and temperature, respectively, and f_0 is a universal function, obeyed by all pure fluids that obey the corresponding states law (i.e. that have two parameter potentials of the form of (7.115), with a common function f). Equation (7.144) is extended to some pure fluid α of nonspherical (non-conformal) molecules by writing

$$Z_\alpha = f_0\left(\frac{p}{p_{c\alpha}^{\text{eff}}}, \frac{T}{T_{c\alpha}^{\text{eff}}}\right), \tag{7.145}$$

where f_0 is the same function as for the simple conformal fluids that obey (7.144), and $p_{c\alpha}^{\text{eff}}$ and $T_{c\alpha}^{\text{eff}}$ are effective critical constants for fluid α that are *chosen* to make the function in (7.145) coincide with that in (7.144). These effective critical constants are written in the form

$$\begin{aligned} T_{c\alpha}^{\text{eff}} &= T_{c\alpha}\theta_\alpha(p_r, T_r) \\ p_{c\alpha}^{\text{eff}} &= p_{c\alpha}\psi_\alpha(p_r, T_r), \end{aligned} \tag{7.146}$$

where $p_r = p/p_c$, $T_r = T/T_c$, $T_{c\alpha}$ and $p_{c\alpha}$ are the true experimental critical temperature and pressure of component α, and θ_α and ψ_α are called shape factors, and depend on the reduced temperature and pressure of the fluid, as well as on the species of the component. They are obtained by fitting experimental data for the pure fluid to (7.144). Clearly this is nothing more than a fitting procedure for pure fluids, and while useful for interpolation between state points or between different members of a homologous series, it has little predictive value. Its usefulness derives primarily from its application to mixtures. The method can be extended to mixtures by assuming that the vdw1 mixing rules apply, and can then be used to determine mixture thermodynamic properties and phase diagrams based only on knowledge of the pure component parameters (critical constants and shape parameters) and the assumption of the Lorentz–Berthelot combining rules. In practice one unlike pair parameter, $\varepsilon_{\alpha\beta}$, is often fitted to mixture data to obtain better agreement with experiment. These shape factor methods have been applied successfully to a range of non-conformal mixtures that do not possess strong polarity or H-bonding. We do not dwell on these here, since they are essentially empirical in nature, but refer the reader to the literature.[94–96] Shape factor methods have also been successfully developed for transport properties,[97–100] particularly viscosities and thermal conductivities.

7.7.2 *Perturbation theory based on a hard sphere reference fluid*

Perturbation expansions about a hard sphere fluid have been described in Chapter 4 for the case of pure fluids, and have been successful for dense liquids of spherical molecules. In this section we briefly describe their extension to mixtures. The use of a hard sphere mixture as the reference fluid is valuable for two reasons: (a) at high density the structure of a simple fluid is largely determined by the repulsive forces, so that such an expansion, if carried out in a suitable way, should converge rapidly (see § 6.4 and Eq. (6.24)), and (b) the equation of state and the structure of hard sphere fluids are well understood, and we have accurate theories for their prediction. The advantages of such expansions over conformal solution theory are that (i) the components do not need to be conformal, and (ii) the treatment is not restricted to small differences in molecular size or energy parameter. In particular, they should work better when the molecular species differ greatly in size. The principal disadvantage is that these theories involve more computational effort.

Three somewhat different approaches have been proposed: the variational theory,[101] the Leonard–Henderson–Barker (LBH) theory,[102] and the Lee–Levesque (LL) theory.[103] All three theories give similar results. The variational theory is based on the Gibbs–Bogoliubov inequality of § 6.8 (see Eq. (6.112)), and involves the minimization of the right side of that equation. The variational theory is somewhat more difficult to apply than the others. The LBH and LL theories are the mixture generalizations of the Barker–Henderson (BH) and Weeks–Chandler–Andersen (WCA) theories described in § 4.1, respectively. Both involve two expansions; the first relates the fluid with the full potential u to some reference fluid with a purely repulsive potential u_0, the second an expansion of the soft repulsive force fluid about a hard sphere fluid. Since the LL theory converges somewhat more quickly for dense liquids we give a description of it below. Differences between the LL and LBH approaches are briefly discussed at the end of this section. The WCA/LL approach has also been applied to the study of the pair correlation function.[104]

The LL theory is a straightforward generalization to mixtures of the WCA theory described in Chapter 4. The pair potential between a molecule of species α and one of β is divided into a reference and a perturbation part, the reference part being the repulsive part of the full potential, shifted upwards by an amount $\varepsilon_{\alpha\beta}$ (see Fig. 4.1(c)),

$$u_{\alpha\beta}(r) = u_{\alpha\beta,0}(r) + u_{\alpha\beta,1}(r) \tag{7.147}$$

where

$$\begin{aligned} u_{\alpha\beta,0}(r) &= u_{\alpha\beta}(r) + \varepsilon_{\alpha\beta} && \text{for } r \leq r_{m,\alpha\beta} \\ &= 0 && \text{for } r > r_{m,\alpha\beta} \end{aligned} \tag{7.148}$$

$$\begin{aligned} u_{\alpha\beta,1}(r) &= -\varepsilon_{\alpha\beta} && \text{for } r \leq r_{m,\alpha\beta} \\ &= u_{\alpha\beta}(r) && \text{for } r > r_{m,\alpha\beta}. \end{aligned} \tag{7.149}$$

Expansion of the Helmholtz energy for the fluid about that of the reference fluid (see §§ 4.3 and 4.4) gives (4.16), $A = A_0 + A_1 + A_2 + \cdots$, with the first-order term given by the mixture generalization of (4.19),

$$A_1 = 2\pi N\rho \sum_\alpha \sum_\beta x_\alpha x_\beta \int_0^\infty \mathrm{d}r r^2 u_{\alpha\beta,1}(r) g_{\alpha\beta,0}(r), \tag{7.150}$$

where $g_{\alpha\beta,0}(r)$ is the pair correlation function for an $\alpha\beta$ pair in the reference fluid.

In order to complete the LL theory it is necessary to relate A_0 and $g_{\alpha\beta,0}$ to the corresponding quantities for a hard sphere mixture, in which the hard sphere diameters are $d_{\alpha\alpha}$, $d_{\alpha\beta}$, $d_{\beta\beta}, \cdots$ for the various pair interactions. The Helmholtz energy A_0 is related to that for the hard sphere mixture, A_d, using the blip function expansion[105] given in § 4.7. The parameterization (see Fig. 4.22) is the mixture generalization of (4.71), i.e.

$$\exp[-\beta u_{\alpha\beta,0\lambda}(r)] = \exp[-\beta u_{\alpha\beta,d}(r)] + \lambda\{\exp[-\beta u_{\alpha\beta,0}(r)] - \exp[-\beta u_{\alpha\beta,d}(r)]\}, \tag{7.151}$$

where $u_{\alpha\beta,d}(r)$ is the hard sphere potential for the $\alpha\beta$ pair. The first-order expansion of the Helmholtz energy for the repulsive reference system is given by (cf. (4.73))

$$A_0 = A_d - 2\pi N\rho kT \sum_\alpha \sum_\beta x_\alpha x_\beta \int_0^\infty \mathrm{d}r r^2 y_{\alpha\beta,d}(r)\{\exp[-\beta u_{\alpha\beta,0}(r)] - \exp[-\beta u_{\alpha\beta,d}(r)]\}, \tag{7.152}$$

where $y_{\alpha\beta,d}(r)$ is the 'y-function' for the hard sphere mixture (see (3.98)). So far we have not specified the hard sphere diameters in the hard sphere mixture. If we now adopt the WCA choice, i.e. we choose the hard sphere diameters so that

$$\int_0^\infty \mathrm{d}r r^2 y_{\alpha\beta,d}(r)\{\exp[-\beta u_{\alpha\beta,0}(r)] - \exp[-\beta u_{\alpha\beta,d}(r)]\} = 0,$$

$$\alpha\beta = AA, AB, BB, \cdots \tag{7.153}$$

then the first-order term in (7.152) will vanish, so that to first order

$$A_0 = A_d. \tag{7.154}$$

The reference pair correlation function appearing in (7.150) is approximated using the zeroth-order y-expansion of (4.14), so that (cf. (4.75))

$$g_{\alpha\beta,0}(r) = \exp[-\beta u_{\alpha\beta,0}(r)]y_{\alpha\beta,d}(r). \tag{7.155}$$

The procedure to calculate the free energy of the mixture is therefore to (a) first calculate the hard sphere diameters $d_{AA}, d_{AB}, d_{BB}, \cdots$ numerically from (7.153), (b) calculate the reference pair correlation function from (7.155), and (c) calculate the free energy from (7.150) and (7.154); the free energy of the hard sphere mixture can be calculated from the equation of state given by Mansoori *et al.*,[106] which is an extension to mixtures of the Carnahan–Starling equation for pure hard sphere fluids.[107] When (7.153) is used directly to obtain unlike-pair hard sphere diameters such as d_{AB} the resulting diameter will not be strictly additive; however, it is very close to $(d_{AA} + d_{BB})/2$, with a difference of about 0.1%. In actual calculations the mean value $(d_{AA} + d_{BB})/2$ is usually used for d_{AB}. The hard sphere diameters calculated from (7.153) depend on both temperature and density (see Fig. 7.42). Approximate schemes for evaluating the hard sphere diameters from (7.153), accurate enough for most purposes, and for the function $y_{\alpha\beta,d}(r)$, have been worked out.[103] The hard sphere y-function can be calculated from the Percus–Yevick theory (Ch. 5). In practice these PY results are modified empirically to match the molecular simulation results.

The LL theory has been used to calculate the thermodynamic properties of LJ mixtures at zero pressure.[103] McDonald[108] reported Monte Carlo simulations for LJ mixtures with parameters corresponding to six mixtures of simple fluids, and it is for these mixtures that results are shown in Table 7.1. The predicted molar volumes are seen to be within 1.3% of the MC values; the enthalpy values predicted by the LL theory are also within about 1.3% of the MC results. The systems studied in Table 7.1 have rather small differences in the LJ parameters for the two species. A test of the LL theory for cases where the size difference of the two species is larger is shown in Fig. 7.37. Good results are obtained even for quite large molecular size ratios; this is expected, since the correct behaviour is built into the hard sphere mixture reference system, as evidenced from comparison with simulation data for asymmetries as large as 20:1.[109]

The LBH theory[102] differs from the LL one in using a slightly different repulsive reference potential (see Fig. 4.1(b)), and a somewhat different expansion of the reference fluid properties about those of the hard sphere mixture. In this second expansion the first-order term is largely nullified by a suitable choice of

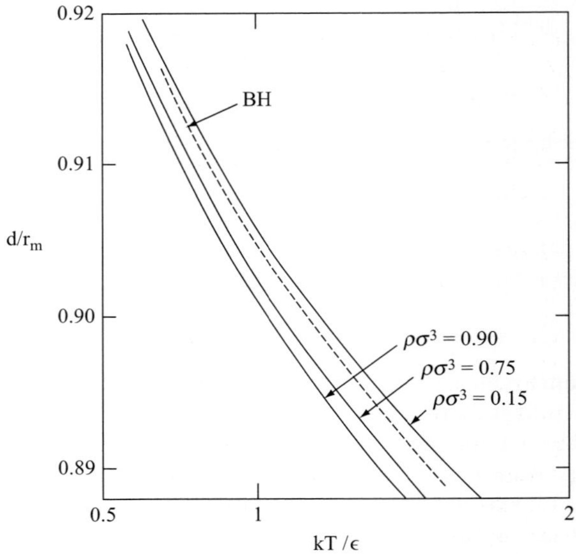

FIG. 7.42 Temperature and density dependence of the hard sphere diameter from perturbation theory. Solid lines are the WCA criterion for *d* at various temperatures and densities, from (7.153). The dashed line is the BH criterion for *d*, which depends only on temperature. Reprinted with permission from ref. 104. Copyright 1971 Taylor and Francis Ltd.

Table 7.1

Total thermodynamic properties at zero pressure for equimolar LJ mixtures obeying the Lorentz–Berthelot rules

System	T(K)	Enthalpy (J mol^{-1})		Volume (cm^3 mol^{-1})	
		MC	LL	MC	LL
Ar/Kr	115.8	-6841 ± 17	-6870	32.82 ± 0.08	32.72
Ar/CH_4	91.0	-7180 ± 12	-7047	31.81 ± 0.04	32.04
CO/CH_4	91.0	-6508 ± 16	-6423	35.47 ± 0.06	35.60
Ar/N_2	83.8	-5419 ± 9	-5369	31.66 ± 0.05	31.68
Ar/CO	83.8	-6630 ± 15	-6480	31.61 ± 0.05	31.62
O_2/N_2	83.8	-5417 ± 15	-5367	31.08 ± 0.06	31.08

† From ref. 103.

hard sphere diameter, $d_{\alpha\alpha}$, for a molecule of species α,

$$d_{\alpha\alpha} = \int_0^{\sigma_{\alpha\alpha}} \{1 - \exp[-\beta u_{\alpha\alpha,0}(r)]\}\mathrm{d}r, \tag{7.156}$$

where the reference potential is defined by

$$\begin{aligned} u_{\alpha\alpha,0}(r) &= u_{\alpha\alpha}(r) \qquad && r \le \sigma_{\alpha\alpha} \\ &= 0 && r > \sigma_{\alpha\alpha}. \end{aligned} \tag{7.157}$$

We note from (7.156) that the hard sphere diameter used in the LBH theory depends on temperature but not on density. Its temperature dependence is shown in Fig. 7.42. Although this is a simplifying feature in using the LBH theory, the expansion converges more slowly for dense liquids than that of the LL theory, and for accurate results it is desirable to include the second-order perturbation term. Rogers and Prausnitz have done this, and have successfully applied the theory to predict the gas–liquid equilibria of argon/neopentane and methane/neopentane.[110] The ratios σ_{BB}/σ_{AA} ($B =$ neopentane) for these mixtures are approximately 1.74 and 1.62, respectively.

7.8 Perturbation theory: nonspherical molecules

Perturbation theory (PT) provides a convenient and accurate approach to the prediction of the thermodynamics of mixtures, and has been widely applied to the study of phase equilibria and other properties. Theories suitable for molecular fluids have been described in some detail in Chapter 4, and typical results for pure fluids have been given in § 6.11. In this section we describe applications of PT to the thermodynamic properties of mixtures, and show comparisons with experimental data.

Two approaches have been used. The first is the PT based on the Pople reference fluid of spherical molecules, and described in § 4.5. The second is PT based on a reference fluid of nonspherical molecules (§ 4.8). Methods for treating associating liquid mixtures are discussed in § 7.9.

7.8.1 The u-expansion

The u-expansion (§ 4.5), based on the Pople reference potential of (4.23), has been widely used since the calculations are relatively straightforward. Calculations have usually been based on the Padé approximant (4.47),

$$A = A_0 + \frac{A_2}{(1 - A_3/A_2)}. \tag{7.158}$$

Expressions for A_2 and A_3 have been given for pure fluids in (4.29) and (4.30)–(4.35), respectively; the generalization for mixtures has been given elsewhere.[111–113] The second-order term is given by

$$A_2 = -\frac{1}{4}\beta\rho^2 \sum_{\alpha\beta} x_\alpha x_\beta \int d\mathbf{r}_1 d\mathbf{r}_2 g^0_{\alpha\beta}(r_{12}) \langle u^a_{\alpha\beta}(12)^2 \rangle_{\omega_1\omega_2}$$
$$-\frac{1}{2}\beta\rho^3 \sum_{\alpha\beta\gamma} x_\alpha x_\beta x_\gamma \int d\mathbf{r}_1 d\mathbf{r}_2 d\mathbf{r}_3 g^0_{\alpha\beta\gamma}(r_{12}r_{13}r_{23}) \langle u^a_{\alpha\beta}(12) u^a_{\alpha\gamma}(13) \rangle_{\omega_1\omega_2\omega_3}. \tag{7.159}$$

For potentials that are multipole-like, i.e. of the form of a sum of terms $u(l_1 l_2 l)$, where $l_1 l_2 l$ are all non-zero, A_3 for mixtures is given by

$$
\begin{aligned}
A_3 = {} & \frac{1}{12}\beta^2\rho^2 \sum_{\alpha\beta} x_\alpha x_\beta \int d\mathbf{r}_1 d\mathbf{r}_2 g^0_{\alpha\beta}(r_{12}) \langle u^a_{\alpha\beta}(12)^3 \rangle_{\omega_1\omega_2} \\
& + \frac{1}{6}\beta^2\rho^3 \sum_{\alpha\beta\gamma} x_\alpha x_\beta x_\gamma \int d\mathbf{r}_1 d\mathbf{r}_2 d\mathbf{r}_3 g^0_{\alpha\beta\gamma}(r_{12}r_{13}r_{23}) \\
& \times \langle u^a_{\alpha\beta}(12) u^a_{\alpha\gamma}(13) u^a_{\beta\gamma}(23) \rangle_{\omega_1\omega_2\omega_3}.
\end{aligned}
\tag{7.160}
$$

For more general potentials the A_3 term involves additional three- and four-body integrals[114] (see (4.30)–(4.35)).

Comparison with molecular simulation

Comparisons of the u-expansion and the Padé of (7.158) with molecular simulation results have been given in § 4.5 for pure fluids; although the series converges slowly for strongly anisotropic forces, the Padé shows good agreement even for very large multipole forces. For mixtures, there have been fewer such comparisons,[115, 116] but the general agreement between the Padé and simulation results seems similar to that for pure fluids. Mixtures studied in this way have included spherical LJ plus Stockmayer (Lennard–Jones plus a dipole–dipole term) fluids,[115] mixtures of hard spheres with dipolar hard spheres, and dipolar hard spheres with quadrupolar hard spheres.[116] In general the agreement is good for the Helmholtz energy and the configurational internal energy, and somewhat worse for derivatives of these, such as the pressure and chemical potential. Since two-phase coexistence lines are determined by equating pressures and chemical potentials for the two phases, we can expect phase equilibria to be more sensitive to errors in the Padé approximant. Results for mixtures of dipolar hard spheres (A) with hard spheres (B) at a reduced density of $\rho^* = \rho\sigma^3 = 0.800$, with the molecules of A and B of equal size σ, are shown in Figs. 7.43 and 7.44 for a range of values of the reduced dipole moment, ${\mu_A^*}^2 = \mu_A^2/kT\sigma_{AA}^3$ from 0 to 3. Generally good agreement between the Monte Carlo simulation results and the Padé of (7.158) is obtained for this range of dipole moments. Results for mixtures of dipolar hard spheres (A) with quadrupolar hard spheres (B) for values of the reduced quadrupole moment from 0 to 3 show similarly good agreement.[113]

Classification of binary phase diagrams

The Padé approximant of (7.158) has been used[112–114] to determine the relation between the intermolecular interactions for a particular model mixture and the class of binary phase diagram that results, using the classification of phase behaviour of Fig. 7.1. In these calculations, and in those described below in which comparisons are made with experimentally determined behaviour, the thermodynamic properties of the isotropic reference mixture are calculated from the vdw1 theory described in § 7.7.1. This relates the properties (A_0, U_0, etc.) of the isotropic reference mixture to those of a pure conformal fluid. The experimentally

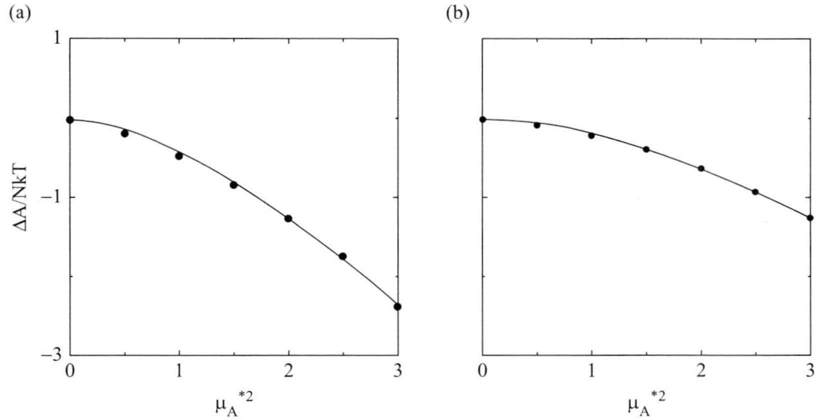

FIG. 7.43 Excess Helmholtz energy (relative to simple hard sphere mixtures), $\Delta A/NkT$ vs the square of the reduced dipole moment for mixtures of dipolar hard spheres (A) with hard spheres (B) of equal size at $\rho\sigma^3 = 0.800$, from MC simulation (points) and the Padé approximant (lines) of (7.158): (a) $x_B = 0.25$, (b) $x_B = 0.5$. (From ref. 116.)

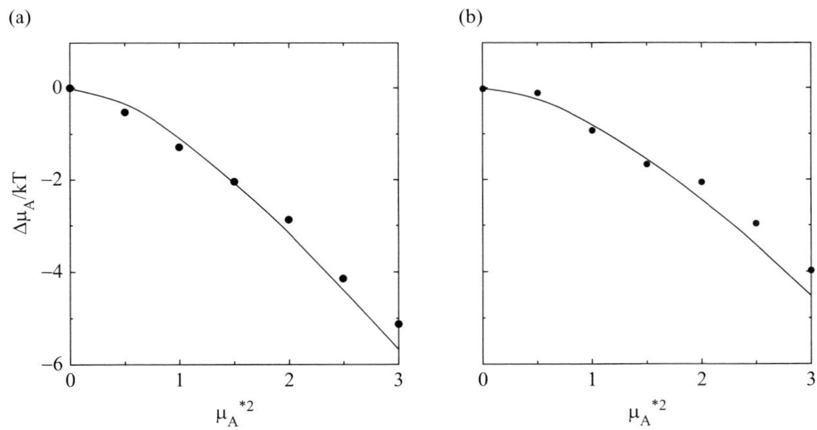

FIG. 7.44 Excess chemical potential for component A, $\Delta\mu_A/NkT$ vs the square of the reduced dipole moment for mixtures of dipolar hard spheres (A) with hard spheres (B) of equal size at $\rho\sigma^3 = 0.800$. Key as described in the legend to Fig. 7.43. (From ref. 116.)

determined[117] equation of state for argon was used to calculate these properties; the equation was put in dimensionless, corresponding states form using the usual LJ potential parameters for argon, $\varepsilon/k = 119.8\,\text{K}$, $\sigma = 0.3405\,\text{nm}$. The anisotropic potentials, u^a, appearing in the integrals in (7.159) and (7.160) can be expressed as a sum of spherical harmonic or generalized spherical harmonic terms, $u(l_1 l_2 l)$, and the averages over orientational averages easily evaluated (see

Chapter 2 and Appendix A). The integrals in (7.159) and (7.160) then reduce to standard integrals over powers of r (the exponent n depending on the potential considered) and the reference mixture correlation functions. The integrals involved in A_2 are

$$J_{\alpha\beta}^{(n)} = \int_0^\infty \mathrm{d}r^* r^{*-(n-2)} g_{\alpha\beta}^0(r^*) \tag{7.161}$$

$$L_{\alpha\beta\gamma}(nn';l) = \int_0^\infty \mathrm{d}r_{12}^* r_{12}^{*-(n-1)} \int_0^\infty \mathrm{d}r_{13}^* r_{13}^{*-(n'-1)} \int_{|r_{12}-r_{13}|}^{r_{12}+r_{13}} \mathrm{d}r_{23}^* r_{23}^* \times g_{\alpha\beta\gamma}^0\left(r_{12}^* r_{13}^* r_{23}^*\right) P_l(\cos\alpha_1). \tag{7.162}$$

In these equations $-n$ is the exponent of the separation r that occurs in the anisotropic potential u^a, $P_l(\cos\alpha_1)$ is the Legendre polynomial, and α_1 is the angle between r_{12} and r_{13}. The integrals $J_{\alpha\beta}^{(n)}$ and $L_{\alpha\beta\gamma}(nn';l)$ arise in the first and second terms in A_2, respectively. The integral $J_{\alpha\beta}^{(n)}$ occurs again in the first term in A_3; the second term in A_3 involves a new integral:

$$K_{\alpha\beta\gamma}(ll'l'';nn'n'') = \int_0^\infty \mathrm{d}r_{12}^* r_{12}^{*-(n-1)} \int_0^\infty \mathrm{d}r_{13}^* r_{13}^{*-(n'-1)} \int_{|r_{12}-r_{13}|}^{r_{12}+r_{13}} \times \mathrm{d}r_{23}^* r_{23}^{*-(n''-1)} g_{\alpha\beta\gamma}^0\left(r_{12}^* r_{13}^* r_{23}^*\right) \psi_{ll'l''}(\alpha_1\alpha_2\alpha_3), \tag{7.163}$$

where

$$\psi_{ll'l''}(\alpha_1\alpha_2\alpha_3) = \sum_{mm'm''} C(ll'l'';mm'm'') Y_{lm}(\omega_{12}) Y_{l'm'}(\omega_{13}) Y_{l''m''}(\omega_{23})^*, \tag{7.164}$$

where C is the Clebsch-Gordan coefficient (Vol. 1, Appendix A), Y_{lm} are the spherical harmonics, and α_1, α_2, and α_3, are the angles between r_{12} and r_{13}, r_{12} and r_{23}, and r_{13} and r_{23}, respectively. The limits on the third integration appearing in (7.162) and(7.163) are $r_{23}{}^* = \left|(\sigma_{\alpha\beta}/\sigma_{\beta\gamma})r_{12}{}^* - (\sigma_{\alpha\gamma}/\sigma_{\beta\gamma})r_{13}{}^*\right|$ to $(\sigma_{\alpha\beta}/\sigma_{\beta\gamma})r_{12}{}^* + (\sigma_{\alpha\gamma}/\sigma_{\beta\gamma})r_{13}{}^*$, i.e. $r_{23} = |r_{12} - r_{13}|$ to $r_{12} + r_{13}$.

The integrals of (7.161)–(7.163) depend on the composition of the mixture through the pair and triplet correlation functions, making them difficult to calculate directly. However, these can be related to corresponding pure fluid integrals to good accuracy, using a procedure suggested by the vdw1 theory.[111–113] The approximations used are

$$J_{\alpha\beta}^{(n)} = J^{(n)}\left(\rho\sigma_x^3, kT/\varepsilon_{\alpha\beta}\right) \tag{7.165}$$

$$K_{\alpha\beta\gamma}(ll'l'';nn'n'') = \left[K\left(\rho\sigma_x^3, kT/\varepsilon_{\alpha\beta}\right)K\left(\rho\sigma_x^3, kT/\varepsilon_{\alpha\gamma}\right)K\left(\rho\sigma_x^3, kT/\varepsilon_{\beta\gamma}\right)\right]^{1/3}, \tag{7.166}$$

where J and K are pure fluid integrals evaluated at the state conditions shown. Here σ_x is given by the vdw1 mixing rule of (7.126). The corresponding equation to (7.166) is used for the integral $L_{\alpha\beta\gamma}$. The pure fluid integrals have been calculated for a wide range of state conditions, using the superposition approximation for the three-body correlation function, (3.246), and molecular simulation data (and the virial series at low densities) for the pair correlation functions. The resulting values have been fitted to simple functions of reduced density and temperature,[111–113] making the calculations using the Padé straightforward.

Phase equilibria are calculated with the theory using the usual conditions for phase equilibrium for binary mixtures,

$$\begin{aligned} p'\left(T\rho' x_A' x_B' \cdots x_{R-1}'\right) &= p''\left(T\rho'' x_A'' x_B'' \cdots x_{R-1}''\right) \\ \mu_\alpha'\left(T\rho' x_A' x_B' \cdots x_{R-1}'\right) &= \mu_\alpha''\left(T\rho'' x_A'' x_B'' \cdots x_{R-1}''\right) \qquad \alpha = A, B, \cdots R, \end{aligned} \tag{7.167}$$

where $'$ and $''$ represent the two phases in equilibrium (this expression is readily generalized to three- or more phase equilibria). Of the $2R+1$ state variables involved, R must be specified; the remaining $R+1$ variables (e.g. ρ', ρ'', x_A'', $x_B'', \cdots, x_{R-1}''$ given $T, x_A', x_B', \cdots, x_{R-1}'$) can then be calculated from (7.167). The critical lines are calculated from[118]

$$\frac{1}{S}\left(\frac{\partial^2 A}{\partial x_A^2}\right)_{V,T} - \left(\frac{\partial^2 A}{\partial x_A \partial V}\right)_T = 0 \tag{7.168}$$

$$\frac{1}{S^3}\left(\frac{\partial^3 A}{\partial x_A^3}\right)_{V,T} - \frac{3}{S^2}\left(\frac{\partial^3 A}{\partial x_A^2 \partial V}\right)_T + \frac{3}{S}\left(\frac{\partial^3 A}{\partial x_A \partial V^2}\right)_T - \left(\frac{\partial^3 A}{\partial V^3}\right)_{T,x_A} = 0, \tag{7.169}$$

where

$$S = \frac{(\partial^2 A/\partial x_A \partial V)_T}{(\partial^2 A/\partial V^2)_{T,x_A}}. \tag{7.170}$$

For fixed values of x_A (7.168) and (7.169) are solved for the critical temperature and density; the critical pressure is then obtained from the Padé equation of state.

For LJ mixtures the binary phase behaviour can be predicted with the use of vdw1 theory alone.[113] When the unlike pair parameters obey the Lorentz–Berthelot combining rules of (7.143) the phase behaviour is of class I when the molecular sizes and LJ well depths for the two species are not too different (for $\varepsilon_{BB}/\varepsilon_{AA} = T_{c,B}/T_{c,A} = 1.0$–$2.0$ when $\sigma_{BB}/\sigma_{AA} \cong 1$), and of class III when the well depth ratio is above 2.0–2.2. This prediction is supported by the behaviour of the only Lennard–Jones-like liquids in nature, i.e. the inert gases. When the liquid ranges of the two constituents overlap (Ar/Kr or Kr/Xe) class I behaviour is observed with no azeotrope; when the critical temperatures are further apart (e.g. He/Ar or He/Xe) class III is found. When sets of LJ parameters are chosen so that the vapour pressures of the two components are similar, it is found that positive[119] (but not negative) azeotropes often occur. Class II mixtures can be obtained for LJ fluids by relaxing the Berthelot rule of (7.143) for ε_{AB}; if this quantity is made sufficiently weak compared to the geometric mean of the like pair parameters, liquid–liquid separation occurs at low temperatures and class II is obtained.

The Padé approximant of (7.158) has been used[112,113] to study the classification of phase behaviour for binary mixtures of the type LJ/(LJ + μμ) and LJ/(LJ + QQ), where LJ means spherical LJ, μμ means dipole–dipole, and QQ means quadrupole–quadrupole interaction. Such mixtures are found to exhibit a richer range of phase behaviour than the LJ mixtures, and behaviour of classes I, II, III, IV and V has been found for appropriate choices of parameters. Class VI has not been observed for these mixtures; this class requires a strongly orientation dependent unlike-pair interaction (see § 7.9).

Some of the results for these mixtures are shown in Table 7.2, and in Figs. 7.45–7.49. Two cases are considered in Table 7.2; in the first the reference system is an ideal mixture ($\sigma_{AA} = \sigma_{AB} = \sigma_{BB}$, $\varepsilon_{AA} = \varepsilon_{AB} = \varepsilon_{BB}$), while in the second the B molecules are twice as large as the A molecules and have twice the LJ well depth. The dipolar and quadrupolar forces considered are, on average, attractive, and so their addition to the LJ potential causes the volatility of that component to decrease, and the critical temperature and pressure to increase. Similar behaviour is observed for quadrupolar/nonpolar mixtures.[113] For the ideal solution reference case, increasing the dipole or quadrupole moment causes the system to change from class I to II and then to III. The same sequence is found for reference system (b) when the moment is added to the smaller A molecules. In this case positive azeotropes occur when the two components have similar vapour pressures. This occurs, for example, when $\mu_A{}^* \approx 2$ for the dipolar/nonpolar case with reference (b), even though the intermolecular forces for the two components are very different. This difference leads to class II behaviour, with liquid–liquid immiscibility at low temperature; the phase behaviour of such a system is shown in Fig. 7.45.

Table 7.2

Classification of binary phase diagrams for dipolar/nonpolar mixtures†

μ_A^*	μ_B^*	Class	$T_{c,B}/T_{c,A}$	Azeotrope
	(a) Reference system: $(\sigma_{BB}/\sigma_{AA})^3 = 1, (\varepsilon_{BB}/\varepsilon_{AA}) = 1$			
0–1.6	0	I	1–0.69	None
1.6–1.74	0	II	0.69–0.64	None
1.74–3.0	0	III	0.64–0.31	None
	(b) Reference system: $(\sigma_{BB}/\sigma_{AA})^3 = 2, (\varepsilon_{BB}/\varepsilon_{AA}) = 2$			
0–1.4	0	I	2.0–1.51	None
1.4–1.8	0	II	1.51–1.25	None
1.8–1.9	0	II	1.25–1.18	Positive (limited above)
1.9–2.4	0	II	1.18–0.88	Positive
2.4–3.0	0	III	0.88–0.62	Positive
0	0–1.1	I	2.0–2.32	None
0	1.1–1.2	V	2.32–2.40	None
0	1.2–3.0	III	2.40–6.45	None

† Using the Lorentz-Berthelot combining rules, (7.143), for LJ unlike pair parameters. From ref. 113.

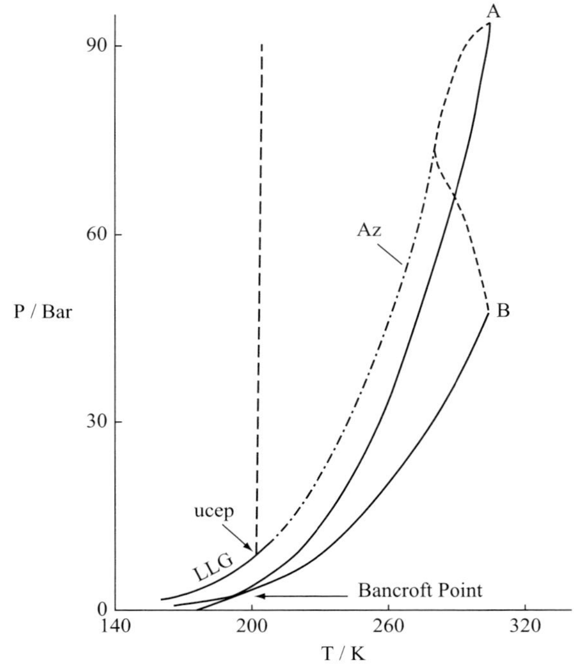

FIG. 7.45 Pressure–temperature projection for the polar/nonpolar system with $\mu_A{}^* = 2.2$, $\mu_B{}^* = 0$; reference system parameters $(\sigma_{BB}/\sigma_{AA})^3 = 2.0$, $\varepsilon_{BB}/\varepsilon_{AA} = 2.0$, $\varepsilon_{AA}/k = 119.8\,\text{K}$, $\sigma_{AA} = 3.405\,\text{Å}$. Shown are the pure component vapour pressure curves (solid lines), the liquid–liquid-gas (LLG) line, the liquid–liquid and gas–liquid critical lines (dashed), and the azeotropic line (dash-dot). A Bancroft point is one where the vapour pressures of two pure components are equal. (From ref. 113.)

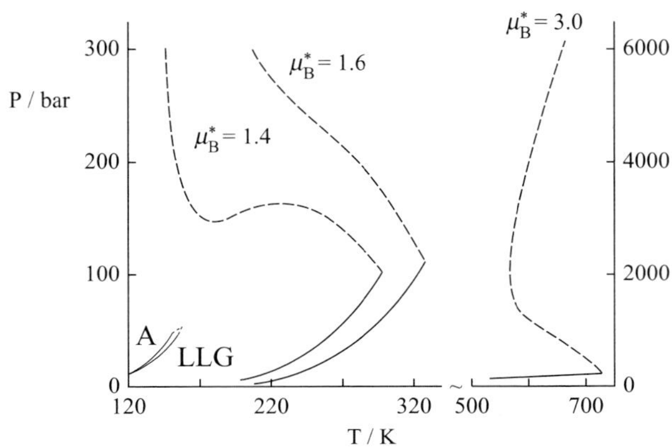

FIG. 7.46 P–T projection for the polar/nonpolar class III systems with $\mu_A{}^* = 0$, $(\sigma_{BB}/\sigma_{AA})^3 = 1.0$, $\varepsilon_{BB}/\varepsilon_{AA} = 1.5$, $\varepsilon_{AA}/k = 119.8\,\text{K}$, $\sigma_{AA} = 3.405\,\text{Å}$. The dashed lines are critical loci. The LLG line is for the case $\mu_B{}^* = 1.4$; for higher $\mu_B{}^*$ values this line lies closer to the vapor pressure curve for pure A. The pressure scale on the right is for $\mu_B^* = 3.0$. Reprinted with permission from ref. 113. Copyright 1978 Elsevier.

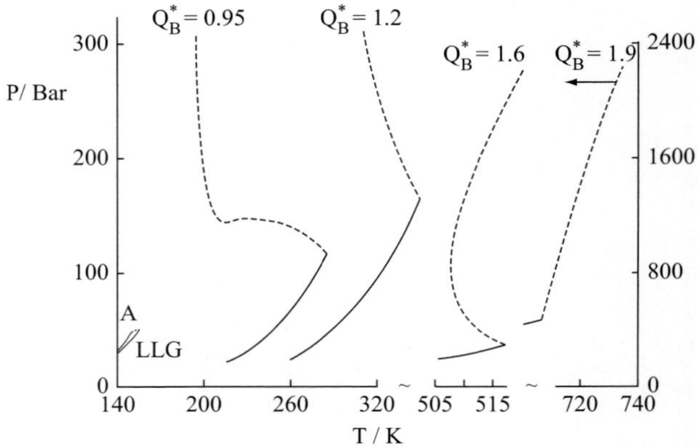

FIG. 7.47 *P–T* projection for the quadrupolar/nonplar class III systems with $Q_A{}^* = 0$, $(\sigma_{BB}/\sigma_{AA})^3 = 1.0$, $\varepsilon_{BB}/\varepsilon_{AA} = 1.5$, $\varepsilon_{AA}/k = 119.8\,K$, $\sigma_{AA} = 3.405\,Å$. The LLG curve shown is for $Q_B{}^* = 0.95$. The pressure scale to the right is for $Q_B^* = 1.9$. Reprinted with permission from ref. 113.

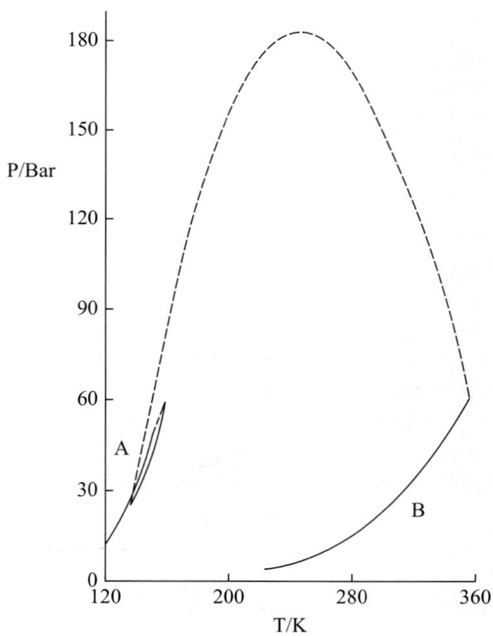

FIG. 7.48 *P–T* projection for the polar/nonpolar class V system, with $\mu_A{}^* = 0$, $\mu_B{}^* = 1.15$, $(\sigma_{BB}/\sigma_{AA})^3 = 2.0$, $\varepsilon_{BB}/\varepsilon_{AA} = 2.0$, $\varepsilon_{AA}/k = 119.8\,K$, $\sigma_{AA} = 3.405\,Å$. Reprinted with permission from ref. 113. Copyright 1978 Elsevier.

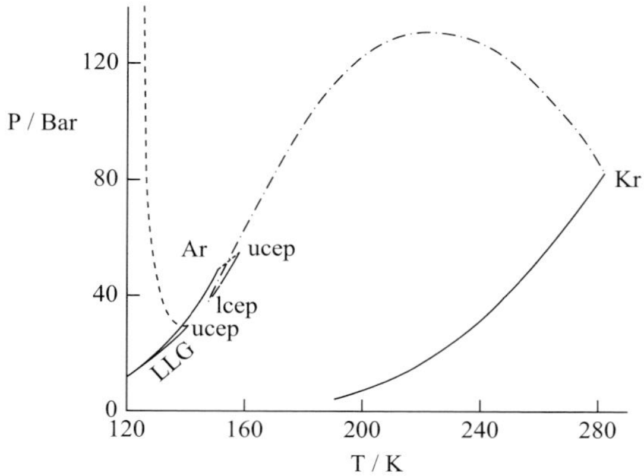

FIG. 7.49 P–T projection for the class IV polar/nonpolar system, with $\mu_A{}^* = 0$, $\mu_B{}^* = 1.44$, $(\sigma_{BB}/\sigma_{AA})^3 = 1.21$, $\varepsilon_{BB}/\varepsilon_{AA} = 1.39$, $\varepsilon_{AA}/k = 119.8\,\text{K}$, $\sigma_{AA} = 3.405\,\text{Å}$. The reference parameters used here are those for Ar and Kr. Reprinted with permission from ref. 112. Copyright 1976 American Institute of Physics.

As $\mu_A{}^*$ is increased further, component B becomes the more volatile one, and for sufficiently large $\mu_A{}^*$ the behaviour becomes class III. The mutual solubility of the two components decreases, and the L_1L_2G line approaches the vapour pressure line for component B more and more closely.

If the multipole is added to the larger B molecules for reference system (b), the system quickly passes from class I to III. For these systems the addition of the moment causes the B molecules to become even less volatile relative to the A's and $T_{c,B}/T_{c,A}$, initially 2.0, steadily increases (see Figs. 7.46 and 7.47). For the dipolar/nonpolar mixtures the system passes through class V as an intermediary stage. A typical case is shown in Fig. 7.48. For other reference systems class IV is found[112] as an intermediary stage between I and III (see Fig. 7.49). Both classes IV and V correspond to a rather fine balance of intermolecular forces, and only occur over a narrow range of potential parameters.

When the disparity of intermolecular forces for the two components is sufficiently great, class III systems exhibit 'gas–gas immiscibility' (see § 7.1). That is, compression of the gas mixture at a temperature above the critical value of either of the pure components leads to phase separation into two dense fluid phases. For the quadrupolar/nonpolar mixtures of Fig. 7.47, gas–gas equilibria is found for $Q_B{}^*$ values above 1.5. For $1.5 \leq Q_B{}^* \leq 1.8$ the gas–gas immiscibility is of the 'second kind' in the terminology of Krichevskii and co-workers; i.e. the critical line starting from the critical point of component B (the less volatile component) starts out with a negative slope, reaches a minimum temperature, and then has a positive slope, so that at sufficiently high pressures it reaches temperatures above $T_{c,B}$. For $Q_B{}^*$ values greater than 1.8 the gas–gas immiscibility is of the 'first kind'; i.e. the critical line starting from the critical point of pure B starts out with a positive slope (see curve for $Q_B{}^* = 1.9$ in Fig. 7.47). In the case of the dipolar/nonpolar mixtures shown in Fig. 7.46, gas–gas immiscibility of the second kind occurs for $\mu_B{}^* > 2.0$. Presumably the first kind is found for dipole moments larger than those shown in Fig. 7.46. Such calculations were not carried out because the pressures exceeded the range of validity of the reference equation of state. It should be noted that anisotropic intermolecular forces are not necessary for gas–gas equilibria. Thus helium–xenon and helium–argon mixtures exhibit gas–gas equilibria of the first and second types, respectively. However, most of the gas–gas equilibrium systems that have been observed[118, 120] do contain at least one component that is either strongly polar or quadrupolar; many of the systems of the second type contain either ammonia or water.[118, 120]

Calculations for other LJ reference systems have been reported by Twu *et al.*,[112] together with a detailed analysis of the resulting critical point lines and azeotropic behaviour. The effect of pressure on the upper critical solution temperature (UCST) and critical composition for liquid–liquid systems of class II are shown for typical cases in Fig. 7.50. The predicted trends agree with

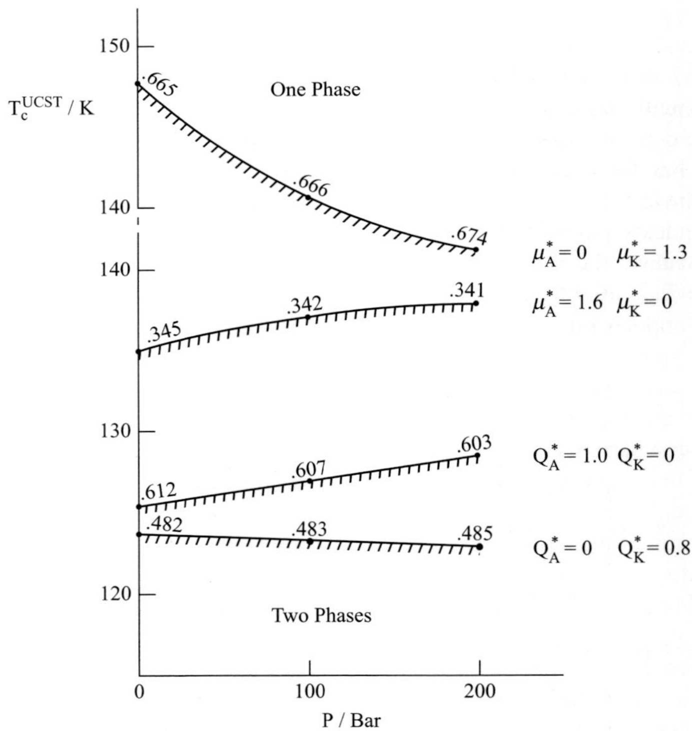

FIG. 7.50 Effect of pressure on the upper critical solution temperature for a LJ reference mixture modeled on Ar/Kr. Numbers on the curves are critical compositions x_{Ar} at the pressures indicated. Reprinted with permission from ref. 112. Copyright 1976 American Institute of Physics.

those found experimentally, and the magnitude of the slopes, dT_c/dP, are on the order $0.02\,\text{Kbar}^{-1}$, again in agreement with experiment. The UCST increases with pressure for systems in which the pure component boiling points are similar (systems with multipoles on A), and decreases with pressure when the boiling points are far apart (multipoles on B). An example of the former behaviour is methanol-cyclohexane, $(dT_c/dP) > 0$, while the latter behaviour is shown by phenol-n-hexane, $(dT_c/dP) < 0$.

In the examples shown in Table 7.2, and in Fig. 7.45–7.50 one of the components is a LJ fluid. Such mixtures are, in general, highly nonideal. The classes of phase behaviour in polar/polar, polar/quadrupolar, etc., mixtures are similar to those shown in these tables and figures. The case of quadrupole–quadrupole mixtures is of particular interest, since the two quadrupoles may be of the same or of opposite sign. The preferred orientation for an unlike molecular pair is perpendicular if the quadrupole moments are of the same sign, whereas if the moments are of opposite sign it is end to end. Rowlinson[121] has suggested that the negative azeotrope found in carbon dioxide-acetylene mixtures arises because the quadrupole moments of the two components are of opposite signs; by contrast, carbon dioxide-ethane, with quadrupole moments of the same sign, shows a positive azeotrope. Evidence for this is provided by calculations for quadrupole–quadrupole mixtures, which show that the excess Gibbs energy is lower for mixtures with quadrupoles of opposite sign than for those with quadrupoles of the same sign, and that this can lead to negative azeotropy. An example of such behaviour is shown in Fig. 7.51.

In summary, mixtures involving polar and quadrupolar molecules display five of the six classes of behaviour shown in Fig. 7.1. For polar/nonpolar or quadrupolar/nonpolar mixtures in which the reference system is a weakly nonideal LJ mixture, increasing the dipole or quadrupole moment causes a continuous transition among the classes,

$$\text{I} \rightarrow \text{II} \rightarrow (\text{V}) \rightarrow (\text{IV}) \rightarrow \text{III}. \tag{7.171}$$

Classes IV and V are observed only for certain values of the LJ parameters. A comparison of the phase diagrams for mixtures involving multipolar fluids with those for purely LJ fluids shows that the multipole forces have a major effect on the type of phase behaviour observed. By contrast, the effect of nonspherical shape forces has a more modest effect, at least for small molecules.[112]

A detailed study of the effects of various anisotropic interactions on the excess properties has been reported by Flytzani-Stephanopoulos *et al.*[111]

Comparison with experiment

There have been extensive comparisons with experimental data for mixtures composed of small molecules, such as CO, CO_2, CH_4, C_2H_6, C_2H_4, and HCl. Among the mixtures studied are HCl/Xe, HBr/Xe, and HBr/HCl;[122–128] CO_2/C_2H_6, CO_2/C_2H_4, and CO_2/C_2H_2;[129] C_2H_6/C_2H_4;[129,130] CH_4/Ar,

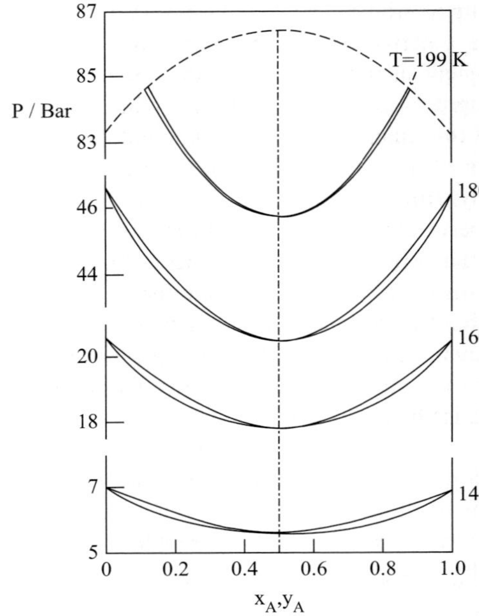

FIG. 7.51 *P–T* projection of *PTx* for a mixture of two quadrupolar molectures A and B with moments of opposite sign, $Q_A{}^* = -1.0$, $Q_B{}^* = +1.0$. The reference system is the ideal solution of case (a) in Table 7.2; $\varepsilon_{AA}/k = \varepsilon_{BB}/k = \varepsilon_{AB}/k = 119.8$ K; $\sigma_{AA} = \sigma_{BB} = \sigma_{AB} = 3.405$ Å. Reprinted with permission from ref. 112. Copyright 1976 American Institute of Physics.

CH_4/Kr; and CF_4/Kr;[128, 129] $CO_2/C_2H_6/C_2H_4$;[129] Ar/CO_2 and CO/CH_4;[123, 127, 128] Xe/C_2H_4 and Kr/C_2H_4;[126, 131] Xe/N_2O,[132] Xe/CF_4,[133] and Xe/CH_3Cl;[134] H_2/CH_4;[135–137] H_2/N_2, H_2/Ar, H_2/CO, H_2/CO_2, H_2/C_2H_4, H_2/C_2H_6;[137] N_2O/C_2H_4;[131, 138] HCl/CF_4;[139] Ar/O_2, Ar/N_2;[128, 140, 141] Ar/CO, N_2/CO, N_2/O_2;[128, 141] N_2/CH_4;[128] alcohol–alcohol (mixtures involving CH_3OH to 1-BuOH) and alcohol (CH_3OH to 1-BuOH)-alkane (methane to *n*-octane) mixtures;[142] acetone/chloroform, acetone/dimethyl formamide, chloroform/diethylether, methanol/water;[143] trichloromethane/propanone and trichloromethane/diethylether;[144] water/1-propanol, water/1-butanol, water/1,4-dioxane, methanol/1,4-dioxane, propanol/1,4-dioxane.[145]

In the calculations the intermolecular pair potential has usually been taken to be of the form of (2.3),

$$u_{\alpha\beta}(\mathbf{r}\omega_1\omega_2) = u^0_{\alpha\beta}(r) + u^{mult}_{\alpha\beta}(\mathbf{r}\omega_1\omega_2) + u^{ind}_{\alpha\beta}(\mathbf{r}\omega_1\omega_2) + u^{dis}_{\alpha\beta}(\mathbf{r}\omega_1\omega_2) + u^{ov}_{\alpha\beta}(\mathbf{r}\omega_1\omega_2) \tag{7.172}$$

with $u^0_{\alpha\beta}(r)$ taken to be the LJ potential (in some cases a $n, 6$ model has been used; see Fig. 6.13) and the multipole, induction (often omitted), dispersion, and overlap contributions represented by the first few terms in a spherical harmonic expansion. Numerous potential parameters are involved in (7.172); $\varepsilon_{\alpha\beta}$ and $\sigma_{\alpha\beta}$

(and $n_{\alpha\beta}$ if the $n, 6$ potential is used), the multipole moments (dipole, quadrupole, octopole), the polarizabilities, anisotropic dispersion parameter, and the overlap parameter (see Chapter 2). The multipole moments, polarizabilities, and anisotropic dispersion parameter are usually taken from independent experimental measurements, or from quantum mechanical calculations (Appendix D of Vol. 1). Like pair values of ε, σ, and δ (the overlap parameter) are fitted to pure component thermodynamic data, usually the saturated vapour pressure and liquid density. The unlike pair LJ parameters for the mixture can be estimated from the pure component parameters using Lorentz–Berthelot combining rules, (7.143). In this case there is no need to fit parameters to mixture data. Calculations performed in this way show that the theory with potentials of the form of (7.172) give a much better fit to experimental data than the use of isotropic potentials (LJ or $n, 6$) alone.[123, 127, 128] However, to obtain closer agreement, it is common to replace the Lorentz–Berthelot rules by

$$\begin{aligned} \sigma_{\alpha\beta} &= \frac{1}{2}\eta_{\alpha\beta}(\sigma_{\alpha\alpha} + \sigma_{\beta\beta}) \\ \varepsilon_{\alpha\beta} &= \zeta_{\alpha\beta}(\varepsilon_{\alpha\alpha}\varepsilon_{\beta\beta})^{1/2}, \end{aligned} \tag{7.173}$$

where $\eta_{\alpha\beta}$ and $\zeta_{\alpha\beta}$ are obtained from mixture data. Often $\eta_{\alpha\beta}$ is set equal to unity and it is sufficient to adjust only $\zeta_{\alpha\beta}$ in this way. It is customary to fit these mixture parameters to either vapour–liquid equilibria data, or to G^{E} (and V^{E} if both parameters are fitted) at the mid-point, $x_A = 0.5$.

We consider first some tests of the theory in which no fitting of unlike pair parameters is used. Such comparisons have been made by Lucas and colleagues.[123, 127, 128] They found that the agreement was improved by the inclusion of anisotropic intermolecular forces, the improvement being particularly noticeable for mixtures containing polar constituents. Among the mixtures studied in this way were Ar/CH_4, Kr/CH_4, Ar/N_2, Ar/O_2, N_2/CH_4, N_2/O_2, CO/CH_4, N_2/CO, Ar/CO, Xe/HBr, Xe/HCl, and HCl/HBr. Two expressions relating the unlike pair parameter $\varepsilon_{\alpha\beta}$ to the corresponding like pair parameters have been used by Lucas and co-workers, one being that of Berthelot and given by Eq. (7.143), and the other due to Kohler[146] and based on London's theory of dispersion forces:

$$\varepsilon_{\alpha\beta}\sigma_{\alpha\beta}^6 = 2\left[\varepsilon_{\alpha\alpha}\sigma_{\alpha\alpha}^6\varepsilon_{\beta\beta}\sigma_{\beta\beta}^6 \Big/ \left(\varepsilon_{\alpha\alpha}\sigma_{\alpha\alpha}^6\alpha_\beta^2 + \varepsilon_{\beta\beta}\sigma_{\beta\beta}^6\alpha_\alpha^2\right)\right]\alpha_\alpha\alpha_\beta, \tag{7.174}$$

where the α's are the molecular mean polarizabilities. They found that in most cases (7.174), together with the Lorentz rule, (7.143), for $\sigma_{\alpha\beta}$ gave better agreement with experiment. Comparisons of the theory with experimental data are shown in Figs. 7.52 and 7.53 for mixtures involving the polar fluids HCl and HBr together with xenon. In the calculations the LJ model was used for the isotropic reference fluid, and the usual anisotropic pair terms consisting of multipole (up to quadrupole–quadrupole), dispersion, overlap, and induction were included. In addition a three-body induction term was accounted for. The

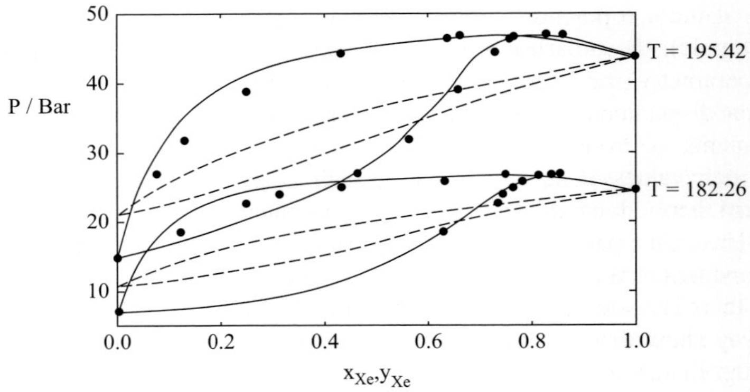

FIG. 7.52 Comparison of perturbation theory based on anisotropic intermolecular potentials (solid lines), Eq. (7.172), vdW1 theory for a fluid with isotropic potentials (dashed lines), and experimental data (points) for vapour–liquid equilibria of Xe/HCl mixtures. (From ref. 128.)

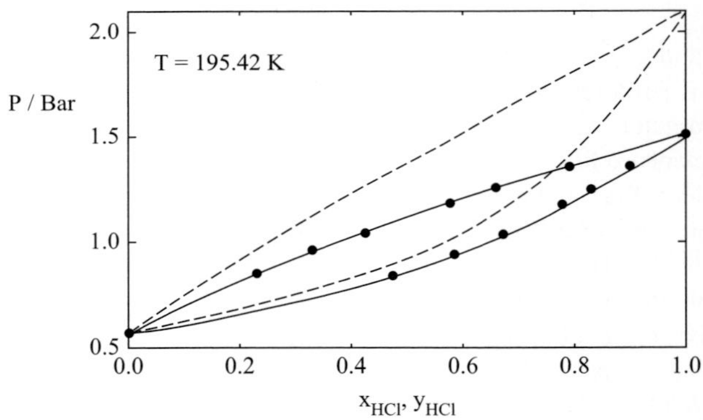

FIG. 7.53 Comparison of perturbation theory and experiment for vapour–lquid equilibria of HBr/HCl mixtures. Key as described in legend to Fig. 7.52. (From ref. 128.)

anisotropic overlap model was a spherical harmonic expansion of the two-site LJ model (§ 2.8).

The theory is seen to give quite good agreement with experiment for vapour–liquid equilibria for these two mixtures. In these calculations two potential parameters ($\varepsilon_{\alpha\alpha}$ and $\sigma_{\alpha\alpha}$) are fitted to experimental data for the pure fluids, but there is no fitting of parameters to mixture data; thus the mixture calculations are a direct test of the theory. For each of the mixtures, the Padé together with the anisotropic intermolecular potential gives a much better fit than the isotropic LJ potential alone. Results for the excess properties show a larger discrepancy between the perturbation theory and the experimental results, as for mixtures of spherical molecules; the excess properties represent a small difference between the total thermodynamic properties and those of the ideal mixture, and so their calculation is sensitive to small errors in the theory and in the intermolecular potentials.

Comparisons of the Padé calculations with the predictions of the Redlich–Kwong (RK) equation, a widely used empirical cubic equation of state, have also been made.[123] The RK predictions are very close to those of the vdw1 theory using the LJ potential shown in Figs. 7.52 and 7.53, and thus show large discrepancies with the experimental data. In engineering applications of the RK and similar equations to mixtures it is usual to fit at least one parameter to the mixture properties, thus giving a better fit. However, it is clear from these results that such fitted parameters must account for all defects in the equation of state, including neglect of anisotropic intermolecular forces. Thus such parameters have no physical significance, and are found to be state dependent.

Further improvement of the agreement between the perturbation theory of (7.158) and experiment can be obtained if one or two binary interaction parameters are fitted to mixture data, as described earlier. In Fig. 7.54 we show results for the Xe/HCl mixture, based on the modified Lorentz–Berthelot combining rules of (7.173), with $\eta_{\alpha\beta}$ set equal to unity (i.e. not fitted) and the other mixture parameter ($\zeta_{\alpha\beta}$) adjusted to fit the vapour–liquid equilibria data at the mid-point, $x_A = 0.5$. In these calculations the LJ (12,6) potential was used for Xe–Xe interactions, and for HCl/HCl and Xe/HCl interactions the models used were

$$\begin{aligned} \text{HCl/HCl}: \quad & u_{\alpha\beta}(12) = u^{0}_{\alpha\beta}(r) + u^{\mu\mu}_{\alpha\beta}(12) + u^{\mu Q}_{\alpha\beta}(12) + u^{Q\mu}_{\alpha\beta}(12) + u^{QQ}_{\alpha\beta}(12) \\ \text{Xe/HCl}: \quad & u_{\alpha\beta}(12) = u^{0}_{\alpha\beta}(r) + u^{0,\text{ind}}_{\alpha\beta}(12), \end{aligned} \tag{7.175}$$

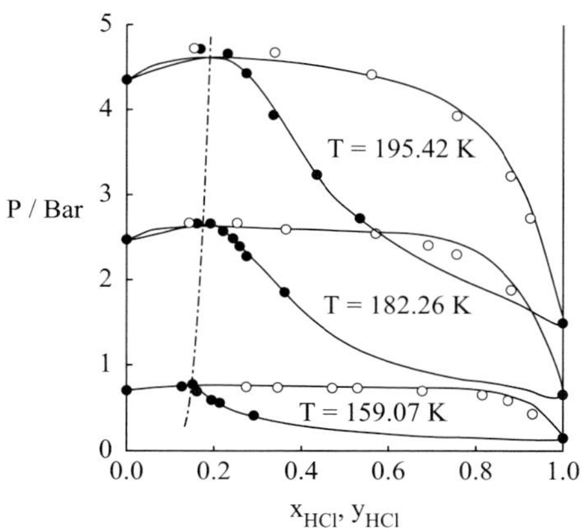

FIG. 7.54 Vapour–liquid equilibria for Xe/HCl mixtures from experiment (points) and perturbation theory (lines), Eq. (7.158). One mixture parameter is fitted to the data. The dash-dot line is the azeotropic locus. (From ref. 113.)

where u^0 is the LJ (12,6) potential, the remaining terms on the right side of the first equation are the multipole terms through quadrupole–quadrupole, and the last term in the second equation is the isotropic two-body induction potential (see § 2.5). This latter term is included in the unlike-pair interaction because the polarizability of xenon is large, $\alpha = 4.1\,\text{Å}^3$ (see Appendix D of Vol. 1). The agreement between theory and experiment is seen to be good for vapour–liquid equilibria, but is poorer for the excess volumes and enthalpies (not shown). Some improvement in the prediction of these excess properties can be achieved if many-body induction effects are included,[125] using the renormalization theory of Wertheim (see § 4.10). The poorer agreement for V^E and H^E is to be expected, since they represent derivatives of the excess Gibbs energy with respect to pressure and temperature, respectively.

Other examples of the comparison of the theory with experimental results for vapour–liquid equilibria are shown in Figs. 7.55–7.57. In these quadrupole–quadrupole mixtures both the like and unlike pair potentials included a LJ (12,6) isotropic term plus quadrupole–quadrupole, anisotropic dispersion, and anisotropic overlap terms. One unlike pair parameter, $\zeta_{\alpha\beta}$, was fitted to the experimental vapour–liquid equilibrium data for the binary mixture at a single temperature; unlike pair anisotropic overlap parameters were estimated from an arithmetic mean combining rule based on the like pair parameters, and were not fitted. The agreement with experimental data for these three systems is excellent, as it is for ethylene/ethane (not shown). No additional parameter fitting is involved in the comparison shown in Fig. 7.57 for the ternary mixture carbon dioxide/ethylene/ethane, so that this prediction is a direct test of the theory.

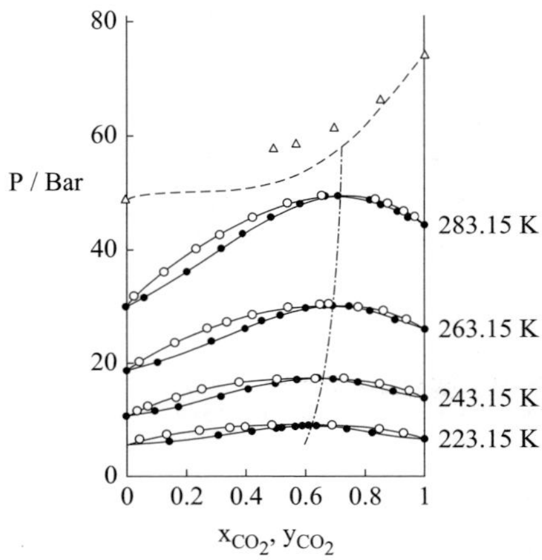

FIG. 7.55 Vapour–liquid equilibria, critical and azeotropic loci for CO_2/C_2H_6 mixtures from theory (lines) and experiment (points). Reprinted with permission from ref. 113. Copyright 1978 Elsevier.

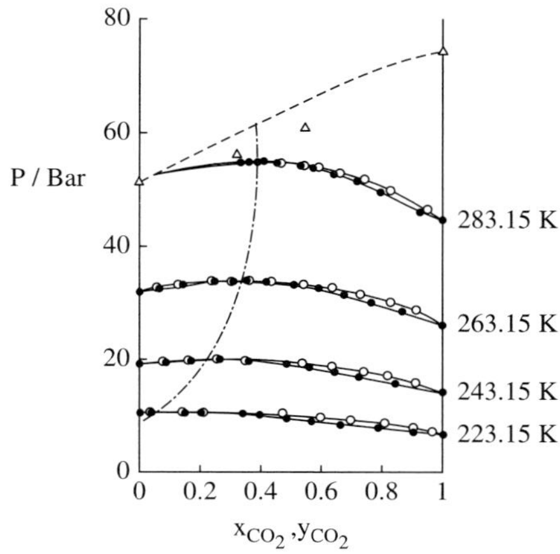

FIG. 7.56 Vapour–liquid equilibria, critical and azeotropic loci for CO_2/C_2H_4 mixtures from theory (lines) and experiment (points). Reprinted with permission from ref. 113. Copyright 1978 Elsevier.

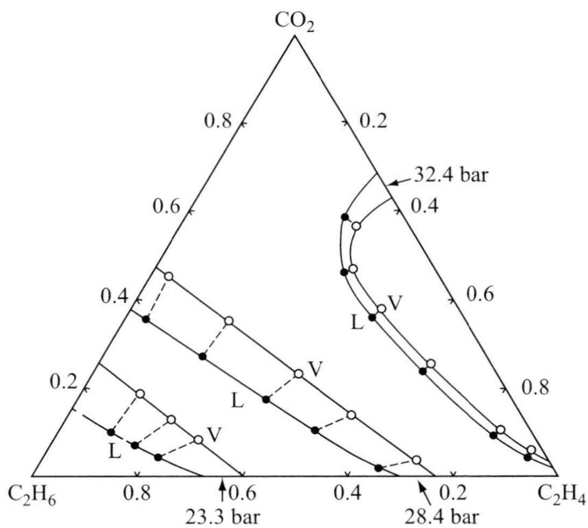

FIG. 7.57 Vapour–liquid equilibria for the ternary mixture $CO_2/C_2H_6/C_2H_4$ at 263 K from theory (lines) and experiment (points). Dashed lines are tie-lines joining coexisting gas and liquid phase. Reprinted with permission from ref. 113. Copyright 1978 Elsevier.

The calculations shown so far have been for molecules that are linear or have some high form of symmetry, so that only a single multipole moment value is involved for each ℓ value. For nonaxial molecules (see Table 6.6) the expressions for A_2 and A_3 involve combinations of the quadrupole and higher multipole moments, as shown in Eqs. (6.136)–(6.138). The effect of such nonaxial quadrupole moments has been studied for mixtures of ethylene with xenon,[131] nitrous oxide,[138] and ethane.[130] The influence of the nonaxial nature of the quadrupole moment is small for the C_2H_4/Xe and C_2H_4/C_2H_6 mixtures, but somewhat larger for the C_2H_4/N_2O mixtures.

Calculations that include the effects of quantum corrections have been made for several mixtures involving hydrogen, including H_2/CH_4,[135,136] and the mixtures H_2/Ar, H_2/N_2, H_2/CO, H_2/CO_2, H_2/C_2H_6, and H_2/C_2H_4.[137] Quantum corrections are included using the $O(\hbar^2)$ term in the expansion of the partition function (see Appendix 3D and § 6.9), and are found to be important for low temperatures (below about 200–250 K) and high pressures. The magnitude of these corrections for pure hydrogen can be seen in Figs. 6.6 and 6.7.

The influence of three-body dispersion forces on mixture properties has been studied using perturbation theory.[140,141] The magnitude of the effects is appreciable and similar to that found for pure fluids.

7.8.2 *Expansion about a fluid of nonspherical reference molecules*

The u-expansion, and the Padé approximant to it, described in the previous subsection provides a convenient way for accounting for the effect of electrostatic forces, but it is much less successful for describing the influence of nonspherical shape, for the reasons given in § 6.11.2. Nonspherical reference perturbation theories, of the type described in § 4.8, can give a good account of fluids in which the principal source of acentricity is nonspherical shape, since this is built into the reference system. They are generally less successful when both nonspherical shape and electrostatic forces are important, since the structure of the real and reference fluids can then be quite different.

Two somewhat different approaches have been used (§ 4.8). In the first, the full pair potential $u(\mathbf{r}\omega_1\omega_2)$ is divided into repulsive and attractive parts, for a fixed set of molecular orientations, as shown in Eqs. (4.89) and (4.90); in the second, which applies only to site–site interaction potentials, the site–site potential $u_{\alpha\beta}(r)$ is divided into repulsive and attractive parts, as shown in Eqs. (4.100) and (4.101).

Most applications to real mixtures have been based on theories of the first type. The free energy is given by (4.88) to first order. As for pure fluids (§ 6.11.2), the reference free energy A_0 is usually related to A_d, the free energy of a fluid of hard nonspherical molecules, through a blip function expansion, and the reference fluid pair correlation function $g_0(\mathbf{r}\omega_1\omega_2)$ is usually obtained from a zero-order expansion about a fluid of spherical molecules (e.g. the f_y expansion of § 4.6). Many mixtures composed of small inorganic and organic molecules have been studied using site–site LJ potential models, including Ar with N_2,[147–150] O_2,[147–150] CO,[149,150] CH_4,[147,149] C_2H_6,[149,150] and CF_4;[151] Kr with CH_4,[149,150] C_2H_4,[150] C_2H_6,[149,150] and CF_4;[151] Xe with C_2H_4,[150] C_2H_6,[149,150] CF_4,[149,151] and C_2F_6;[150] CH_4 with N_2,[149,150] CO,[149,150] C_2H_4,[150] C_2H_6,[149,150] and CF_4;[149,151,152] N_2 with O_2[149,150] and CO;[149,150] and CO_2/C_2H_6,[150] CS_2/CCl_4,[150,151] C_2H_4/C_2H_6,[150] C_2H_6/C_2F_6,[150]

CS_2/C_2Cl_4,[150] and CCl_4/C_5H_{12}.[149] In these applications the effects of nonspherical shape were included (via site–site LJ forces), but direct electrostatic and induction forces were neglected. Excess thermodynamic properties were calculated and compared with experiment, and for CH_4/CF_4 mixtures the vapour–liquid coexistence curve was calculated for higher temperatures. Comparisons with experimental data show the theory to be only moderately successful. Some typical results for nonpolar fluids are shown in Table 7.3. In these calculations Ar, Xe, CH_4, and CCl_4 were modeled as single-centre LJ spheres; N_2, CO, C_2H_4, CS_2, and C_2Cl_6 were modeled using two-centre LJ interactions; and CF_4 was modeled with a five-centre LJ interaction. Predicted values listed under Th were obtained with the use of the modified LB rules of (7.173), with $\eta_{\alpha\beta}$ set equal to unity. The cross energy parameter $\zeta_{\alpha\beta}$ was adjusted to fit G^E at $x_A = 0.5$. Values under Th/LB were obtained using the original combining rules of (7.143). Results based on the LB rules give results of the correct order only for the smallest and least nonspherical molecules, and they produce increasingly large discrepancies as the molecules become larger and less spherical.[153] These departures may be due in part to the neglect of anisotropic forces other than overlap, in addition to departures from the LB rules themselves. For the larger molecules in the lower part of the table, the fitted cross energy parameter $\zeta_{\alpha\beta}$ departs from unity by 10% or more in many cases; again this may reflect a compensation for the neglect of some anisotropic forces.

The calculations described above are based on the site–site LJ model. If, instead, a Kihara pair potential (§ 2.1) is used, the geometry of convex bodies (§ 6.12 and Appendix 6A) can be used to simplify the perturbation theory equations somewhat (§ 4.8). The first-order perturbation term then involves only

Table 7.3

Excess thermodynamic properties for several equimolar binary mixtures, obtained from nonspherical reference perturbation theory (Th) and experiment

Mixture	G^E			H^E			V^E		
	Exp	Th	Th/LB	Exp	Th	Th/LB	Exp	Th	Th/LB
Ar/N_2 (84)†	34	34	26	51	31	19	−0.18	−0.17	−0.19
CH_4/CO (91)	110	110	2	105	76	−84	−0.32	−0.48	−0.69
Xe/C_2H_4 (161)	145	145	−174	285	179	−299	0.35	−0.06	−0.53
CCl_4/CS_2 (298)	172	172	−935	306	298	−1250	0.32	−0.75	−1.71
C_2Cl_4/CS_2 (298)	296	296	−334	510	541	−339	0.50	0.23	−0.29
Xe/CF_4 (159)	645	645	−289	959	985	−517	1.93	0.62	−1.33

† Figures given in parentheses after mixture formulae are temperature in K. Values of G^E and H^E are given in $J mol^{-1}$; V^E values are in $cm^3\ mol^{-1}$.
Values listed under Th were obtained by fitting the parameter $\zeta_{\alpha\beta}$ to the experimental G^E at $x_A = 0.5$; Th/LB indicates predictions based on the Lorentz–Berthelot rules. (From refs. 150, 151.)

$g_{hcb}(s)$, the pair correlation function for the two hard convex bodies, where s is the shortest distance between molecular hard cores; approximate expressions have been proposed for this correlation function. Comparisons with experiment using Kihara models have been made for a range of binary mixtures, including: Ar with N_2,[154] CH_4,[155] and CO_2;[155] N_2 with O_2,[154] CH_4,[154,155] C_2H_4,[155] and C_2H_6;[155] CO_2 with C_2H_4,[155] and C_2H_6;[155] C_6H_6/C_6H_{12};[154] n-alkane mixtures covering the range C_1 to C_{16};[155–158] mixtures of lower n-alkanes with benzene[156] and ethylene;[155] and mixtures of n-alkanes with chlorinated alkanes in the range C_1 to C_4.[158] In the applications of this approach by Boublik and co-workers,[154,155,157,158] calculations have been made using a second-order form of the expansion, i.e.

$$A_c = A_0 + A_1 + A_2, \tag{7.176}$$

where A_0 is the configurational free energy of the reference fluid of molecules with interaction given by (4.89), A_1 is the mixture generalization of the first-order term appearing in (4.88), and the second-order term is calculated in the macroscopic compressibility approximation of Barker and Henderson,[159] adapted for non-spherical convex molecule fluids.[157] The reference fluid properties are related to those of the convex body fluid (corresponding to the core of the Kihara potential used) in the usual way (§ 4.8), so that the reference free energy term becomes A^{hcb}, the configurational free energy for the hard convex body fluid corresponding to the Kihara core, and A_1 and A_2 are given by[155,157]

$$A_1 = \frac{\rho}{2kT}\sum_\alpha \sum_\beta x_\alpha x_\beta \int_{D_{\alpha\beta}}^{\infty} u_1^{\alpha\beta}(s)\left\langle g_{\alpha\beta}^{\mathrm{hcb}}(s)\right\rangle S_{\alpha+s+\beta}(s)\mathrm{d}s \tag{7.177}$$

$$A_2 = \frac{\rho}{4kT}\left(\frac{\partial\rho}{\partial p}\right)^{hcb}\sum_\alpha \sum_\beta x_\alpha x_\beta \int_{D_{\alpha\beta}}^{\infty} \left[u_1^{\alpha\beta}(s)\right]^2 \left\langle g_{\alpha\beta}^{\mathrm{hcb}}(s)\right\rangle S_{\alpha+s+\beta}(s)\mathrm{d}s,$$

where s is the shortest distance between molecular cores, $D_{\alpha\beta}$ is the thickness of the representative hard convex body that is the parallel body to the cores of the Kihara molecules[155] (see Appendix 6A), $\left\langle g_{\alpha\beta}^{\mathrm{hcb}}(s)\right\rangle$ is the orientationally averaged pair correlation function for the hard convex body pair at fixed minimum distance s, $(\partial\rho/\partial p)^{\mathrm{hcb}}$ is proportional to the compressibility of the hard convex body fluid, and $S_{\alpha+s+\beta}$ is the mean surface area of the pair of molecules α and β.[155] Equation (7.176) omits a small correction term, due to nonadditive hard body diameters.[155,157]

With the fitting of one cross energy parameter, the theoretical results for the excess properties using the Kihara model[155–157] described earlier show similar agreement with experiment to that shown in Table 7.3 for the site–site LJ model; fairly good agreement with experiment is obtained for vapour–liquid equilibria.[155] However, when both the cross energy and size parameters ($\eta_{\alpha\beta}$ and $\zeta_{\alpha\beta}$ in (7.173)) were adjusted using experimental data, good agreement with experiment was found for all three excess properties,[158] G^E, H^E, and V^E, for a range of n-alkanes and chloro-alkanes and their mixtures. Moreover, the values of $\eta_{\alpha\beta}$ and $\zeta_{\alpha\beta}$ obtained were physically reasonable, showing departures from the Lorentz–Berthelot rules of a percent or less in most cases. Results for 1-chlorobutane + n-heptane mixtures are shown in Fig. 7.58. Results are shown for both the original Lorentz–Berthelot rules of (7.143), and for the modified rules of (7.173). 1-Chlorobutane possesses a significant dipole moment; the effect of dipolar forces was taken into account by adding a term of the form

$$A^{\mathrm{elec}} = A_2^{\mathrm{elec}} / \left(1 - A_3^{\mathrm{elec}} / A_2^{\mathrm{elec}}\right), \tag{7.178}$$

that is, of the Padé form of (7.158). The integrals in the perturbation terms in (7.178) were evaluated by replacing the correlation functions for the reference system by LJ $g(r)$ functions. The results in this figure show that with the modified LB rules of (7.173) the excess properties are in quite good agreement with the experimental values for H^{E} and V^{E}, although poorer for G^{E}. The values of the fitted parameters were $\eta_{\alpha\beta} = 1.00072$, $\zeta_{\alpha\beta} = 0.9992$.

Only a few calculations have been reported for real mixtures using the second non-spherical reference approach, based on division of the site–site potential functions; success with this approach has been limited. Comparisons with experimental data have been made[160] for Ar/N_2 and Ar/O_2. The energy cross interaction for the mixture was fitted to the experimental value of G^{E} for the equimolar mixture. For the first mixture the predicted excess volume was about 30% less than experiment, while the excess enthalpy was predicted to be $+8\,\mathrm{J\,mol^{-1}}$ compared to $+51\,\mathrm{J\,mol^{-1}}$ from experiment. For Ar/O_2 the theory predicted values of V^{E} of $-0.05\,\mathrm{cm^3\,mol^{-1}}$ vs $+0.14\,\mathrm{cm^3\,mol^{-1}}$ from experiment; the theory gave $+14\,\mathrm{J\,mol^{-1}}$ vs $60\,\mathrm{J\,mol^{-1}}$ from experiment for H^{E}. Tests of the theory against Monte Carlo simulations[161] for site–site models of N_2/O_2 and CO_2/C_2H_6 have shown quite good agreement for the internal energy, but poorer agreement for the pressure; excess properties were not calculated.

7.9 Associating mixtures

As for the case of pure fluids (§ 6.13), strong association between molecules (H-bonding, charge transfer, and other types of complexing) makes a large contribution to the thermodynamic properties for mixtures, and these effects are too large to be treated by the perturbation theories of the previous section. The intermolecular energies due to association are usually stronger than those due to dispersion and weak electrostatic forces by an order of magnitude or more (Table 6.10). For mixtures, such association can occur between like pairs of molecules, or between unlike pairs, or any combination of these.

As for pure fluids, association leads to unusually high melting points, and large enthalpies of vapourization, dielectric constants, and viscosities. For mixtures, association also strongly affects the excess properties and phase diagrams. Dissolving an associating liquid, such as ethanol, in a non-associating solvent, such

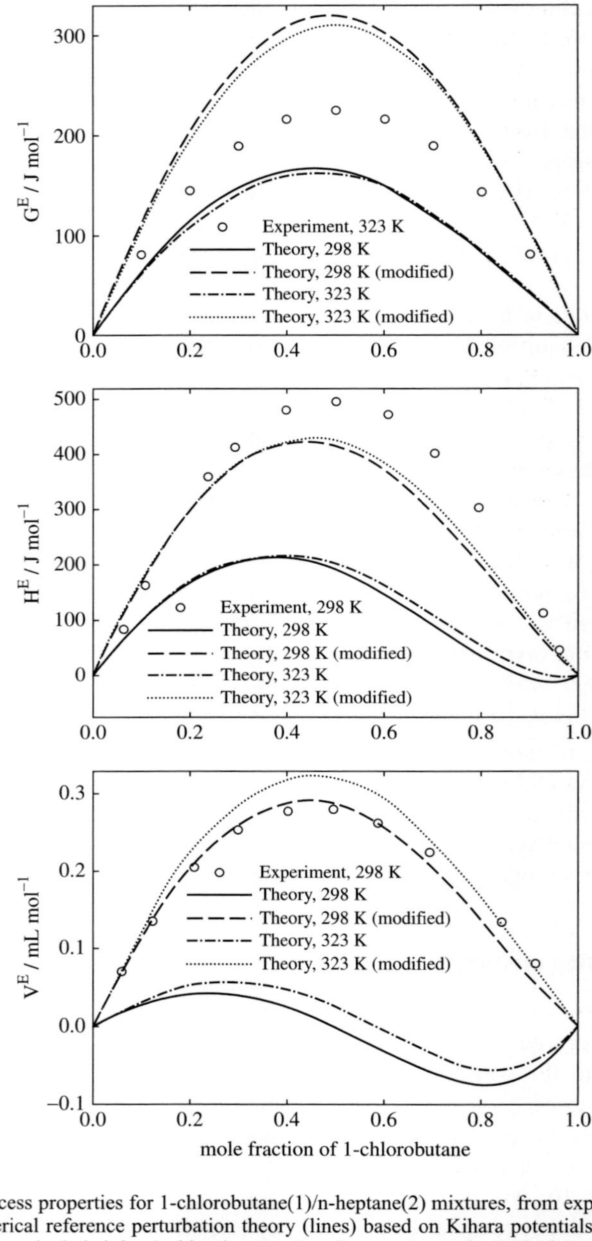

FIG. 7.58 Excess properties for 1-chlorobutane(1)/n-heptane(2) mixtures, from experiment (points) and nonspherical reference perturbation theory (lines) based on Kihara potentials. A point dipole interaction was included for 1-chlorobutane. Results are shown for both the original Lorentz–Berthelot rules of Eq. (7.143) and for the modified rules of (7.173) with $\eta_{\alpha\beta} = 1.00072$ and $\zeta_{\alpha\beta} = 0.9992$. The correction parameters are fitted to the experimental excess volume. (From ref. 158.)

as n-heptane, leads to the breaking of association bonds in the associating liquid, whose energy is released as heat, or excess enthalpy, H^E, assuming that mixing is at constant pressure and temperature. On adding a small amount of ethanol to a large amount of the solvent, n-heptane, all of the hydrogen bonds are broken, and the heat given off is large. For mixtures with large concentrations of ethanol the heat is smaller, since now the addition of n-heptane breaks only a few bonds. Thus the excess enthalpy (heat given off) is large and positive, with a maximum for a mixture that is relatively dilute in ethanol. The excess entropy, S^E, may be positive in mixtures dilute in ethanol, due to the loss of orientational order on breaking association bonds; however, it becomes negative at larger concentrations, due to the presence of bonds, and has a minimum at a concentration rich in ethanol. The excess Gibbs free energy, G^E, is large and positive, and almost symmetrical with respect to composition, due to cancellation of the asymmetries in the excess enthalpy and entropy.

In *solvating* mixtures there is an association bond between the unlike molecular pairs, even though there may be no association between either of the like pairs of molecules. The commonly quoted example of this behaviour is the system chloroform + acetone, for which an association bond forms between the electron accepting H atom of the chloroform molecule and the electron donating O atom in the carbonyl group of the acetone molecule; no hydrogen bonds form between molecular pairs of the pure components. For this system the excess enthalpy, entropy and free energy are all negative, and the excess volume, V^E, is negative except for mixtures weak in chloroform. Chloroform forms similar association interactions with the O atom in mixtures with ethers, and with the π-electrons of benzene. Since such solvation frequently leads to negative Gibbs energies, such mixtures usually exhibit negative azeotropes when the two pure components have similar vapour pressures (Fig. 7.51). For mixtures of two polar or associating liquids it is harder to generalize about the effects on excess properties or phase equilibria, since these effects depend on the relative strengths of the interactions between the various pairs of species. A detailed discussion of these effects has been given by Rowlinson and Swinton,[162] Smith *et al.*,[163] and Prausnitz *et al.*[164]

Theories of associating fluids have been discussed in § 6.13. The most successful theory at present is that due to Werheim,[165] which has been described in detail in §§ 6.13.2 and 6.13.3; equations given there are for the general mixture case. The theory takes the form of a resummed perturbation series, with a reference fluid of spherical molecules (usually taken to be either hard spheres or Lennard–Jones molecules). The most commonly used form of the theory is the first-order one, termed TPT1 (thermodynamic perturbation theory to first order). The statistical associating fluid theory (SAFT) equation for the free energy (§ 6.13.5) combines the TPT1 theory for molecular association with terms for chain formation and dispersion force effects. Both the theory and its applications have been the subject of several reviews.[166–170]

Tests of the TPT1 theory against molecular simulation results for pure fluids have been discussed in § 6.13 (see Figs. 6.23–6.26); excellent agreement was observed for the compressibility factor, internal energy, fraction of associated molecules, and vapour–liquid equilibria. Similarly good agreement with simulation data has been found for mixtures. The first such test[171] was for a solvating binary mixture of hard spheres having embedded point charges; the two components A and B had spheres of the same size, σ, and a plus and a minus charge placed 0.25σ apart. The dipoles were aligned radially, with the centre of the dipole at a distance of 0.25σ from the molecular centre. The positive charge was placed nearer the outside of the sphere for species A, while the negative charge was the outer one for B. For like pair interactions the charge–charge interaction was turned off, so that the AA and BB interactions were simple hard sphere ones, u_{hs}. However, for the unlike pair the Coulomb interactions were included, leading to a strongly attractive force when the pair were suitably aligned. Such mixtures had negative excess enthalpies, volumes, and Gibbs energies, and the theory gave an accurate description of these. Tests of the theory have also been carried out for vapour–liquid equilibria for several mixtures,[172] and some of these results are shown in Figs. 7.59 and 7.60. In these mixtures the intermolecular potential consisted of a Lennard–Jones sphere with an embedded conical square well site of the type shown in Fig. 6.22(b) and defined in Eq. (6.221). The first mixture, shown in Fig. 7.59, consists of a self-associating component in which the molecules have two association sites (component 1) with a non-associating component (2), and is a crude model of an alkanol–alkane mixture; the cross interactions for this mixture are simple Lennard–Jones forces. The mixture exhibits positive deviations from Raoult's Law (ideal solution behaviour), and a positive (minimum boiling point) azeotrope. This phase behaviour is typical of mixtures such as 1-propanol/n-heptane. The theory agrees with the simulation results within the uncertainties for both the bubble and dew point lines, and also for the fraction of monomers. The second mixture, shown in Fig. 7.60, is a solvating one, and consists of two components that do not self-associate, but for which the unlike pairs associate. The mixture exhibits negative departures from Raoult's Law and a negative (maximum boiling point) azeotrope, both of which are due to the association between the unlike pairs. Such behaviour is observed experimentally for mixtures such as acetone/chloroform and dimethyl ether/sulphur dioxide.[173] Again, the theory agrees with the simulation results within their estimated uncertainty. The theory has also been tested against simulation results for liquid–liquid equilibria with good results.[174,175]

In the remainder of this section we give several examples of applications of the theory to both hypothetical and real mixtures.

7.9.1 Classification of phase diagrams: closed solubility loops and class VI

It is possible to use the TPT1 theory to investigate the relation between observed phase behaviour and the underlying intermolecular forces, and in particular the role of association. The six main classes of phase diagram for binary fluid mixtures are shown in Fig. 7.1. The TPT1 has been used to explore the ranges

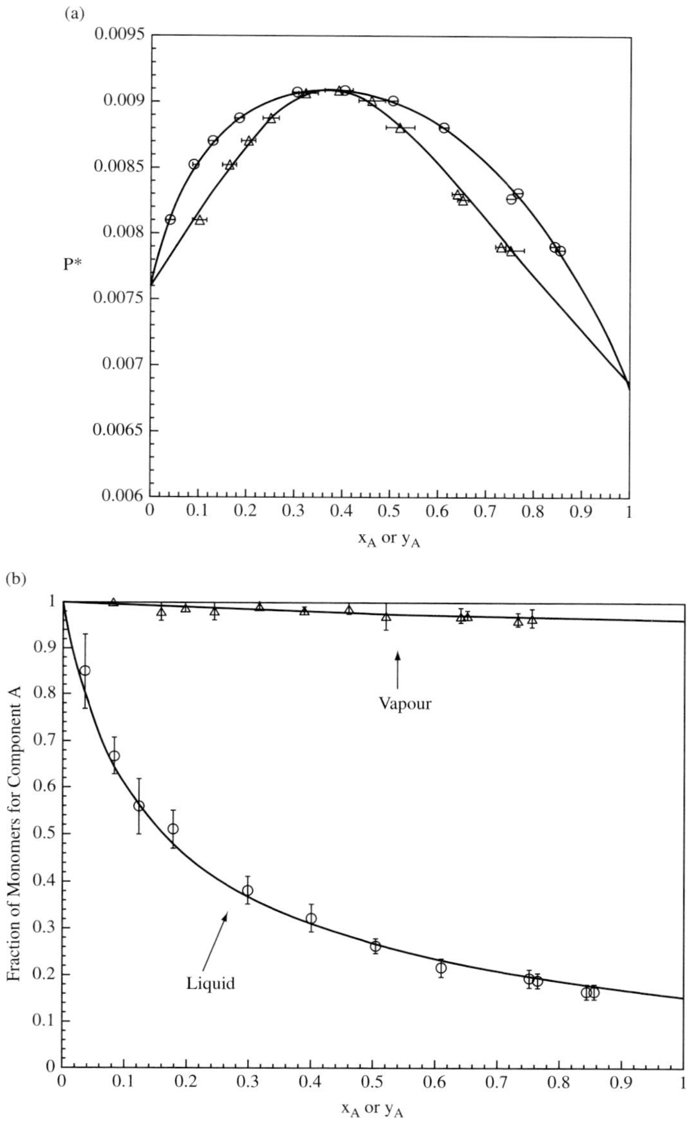

FIG. 7.59 Vapour–liquid equilibria (a) and monomer mole fraction (b) vs mole fraction of component A for a self-associating mixture in which molecules of species A self-associate, but there is no association of species B with either species B or A. Points are Gibbs ensemble Monte Carlo simulation results; lines are Wertheim theory. Molecules of species A have two conical square well bonding sites, a and b, and sites of type a on one molecule will only bond with sites of type b on another. The LJ σ parameters are the same for all pairs, and the LJ ε parameters differ for the pairs, $\varepsilon_{BB} = 1.2\varepsilon_{AA}$, and $\varepsilon_{AB} = (\varepsilon_{AA}\varepsilon_{BB})^{1/2}$; the bonding potential between a and b sites is $\varepsilon^{\text{bond}} = 8\varepsilon_{AA}$. Here $P^* = P{\sigma_{AA}}^3/\varepsilon_{AA}$. Reprinted with permission from ref. 172. Copyright 1992 Taylor and Francis Ltd.

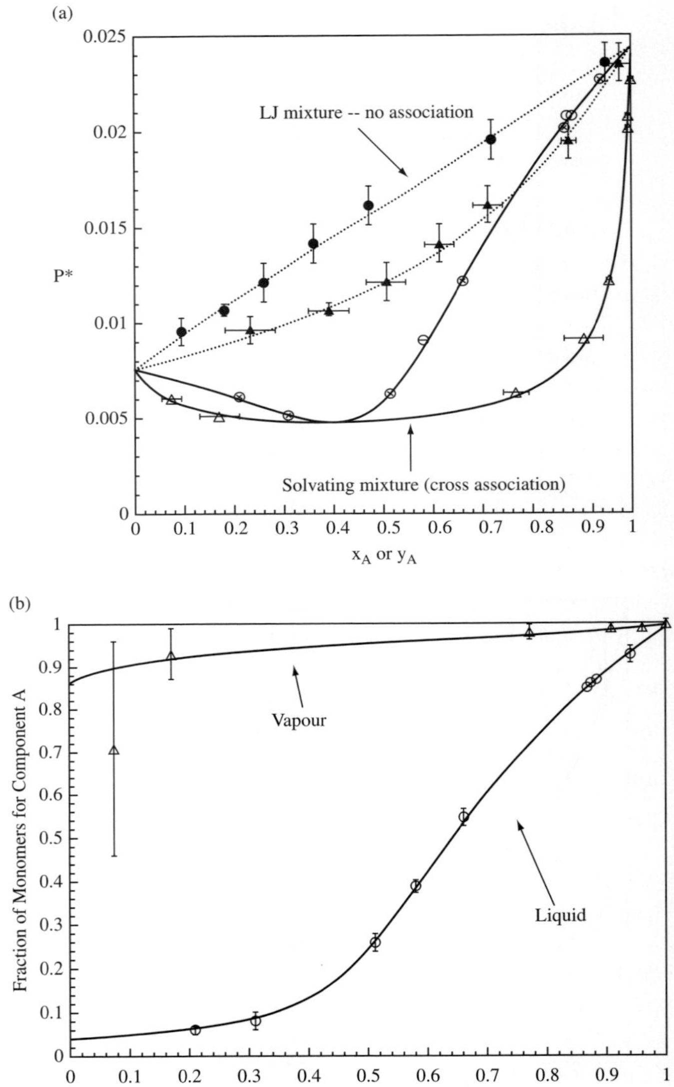

FIG. 7.60 Vapour–liquid equilibria (a) and monomer mole fraction (b) vs mole fraction of component A for a solvating mixture in which molecules of species A associate with molecules of species B, but there is no association for AA or BB pairs. Points are Gibbs ensemble Monte Carlo simulation results; lines are Wertheim theory. The like pair interactions are simple Lennard–Jones, while the unlike pair AB interaction consists of LJ plus a bonding interaction between single conical square well sites on the molecules of species A and B. The LJ σ parameters are the same for all pairs, but the LJ ε parameters differ for the pairs, $\varepsilon_{BB} = 1.2\varepsilon_{AA}$, and $\varepsilon_{AB} = (\varepsilon_{AA}\varepsilon_{BB})^{1/2}$; the bonding potential between the unlike pairs is $\varepsilon^{\text{bond}} = 8\varepsilon_{AA}$. Here $P^* = P\sigma_{AA}{}^3/\varepsilon_{AA}$. Reprinted with permission from ref. 172. Copyright 1992 Taylor and Francis Ltd.

of potential parameters for associating fluids that give rise to each of these classes.[176–181] Such studies provide a qualitative understanding of the role of different kinds of intermolecular forces in determining phase behaviour, and can be used to construct global phase diagrams. It is found that classes I to V of Fig. 7.1 can occur even in the absence of forces that are strongly orientation-dependent, such as association forces. Examples of such phase behaviour can be found among mixtures of hydrocarbons (see § 7.1.2), and for Lennard–Jones mixtures with suitably chosen intermolecular force parameters.

Binary mixtures in which closed solubility loops occur, class VI of Fig. 7.1, are of particular interest since, for the known experimental systems of this type,[182] the closed loop behaviour seems to arise from highly directional association forces between unlike pairs of molecules. This explanation of class VI phase behaviour was first proposed by Hirschfelder *et al.*[183] in 1937. The association bonds between unlike pairs depend strongly on molecular orientation, and lead to complete mixing at low temperatures, where thermal motion is reduced. As the temperature is raised thermal motion of the molecules increases, breaking these bonds and leading to demixing beginning at some lower critical solution temperature. As temperature is raised further the immiscibility of the two components at first increases, but at still higher temperatures the thermal motion leads to reduced immiscibility, which finally disappears at some upper critical solution temperature, leading to a closed solubility loop in the phase diagram (Figs. 7.22 to 7.26). The first successful theory embodying this idea was put forward by Barker and Fock,[184] who used a mean field lattice theory based on the quasi-chemical approximation, with unlike pair interactions that depended strongly on orientation. The theory produced closed solubility loops for appropriate choices of the interactions, but the predicted temperature versus composition coexistence curves were too thin (i.e. the loops gave immiscibility gaps that were too small), and the mean field approximation gave incorrect parabolic shapes near the upper and lower critical temperatures (see Fig. 7.61). Subsequently, more sophisticated lattice models, including the decorated lattice-gas model and solutions based on the renormalization group technique, as well as more phenomenological theories, were applied to these mixtures, and have been the subject of reviews.[185–187] The decorated lattice-gas and renormalization group methods have the advantage that they give the correct non-analytic behaviour near the critical points, and so reproduce the correct shape of the coexistence curves in these regions. An example of such calculations[188] based on the decorated lattice-gas model is shown in Fig. 7.61. The value $\omega = 5000$, where ω is the number of possible molecular orientations, only one of which results in bonding, gives qualitative agreement with experiment. However, it should be noted that such a high value of ω corresponds to a span in polar angle (corresponding to solid angle $4\pi/\omega$) of about $1.6°$, which is about one order of magnitude smaller than that for typical hydrogen bonds. These theoretical methods, and to some extent experimental studies on a range of class VI systems, can be used to investigate the effect of varying the strength of the unlike pair bonding on the phase behaviour. This is shown

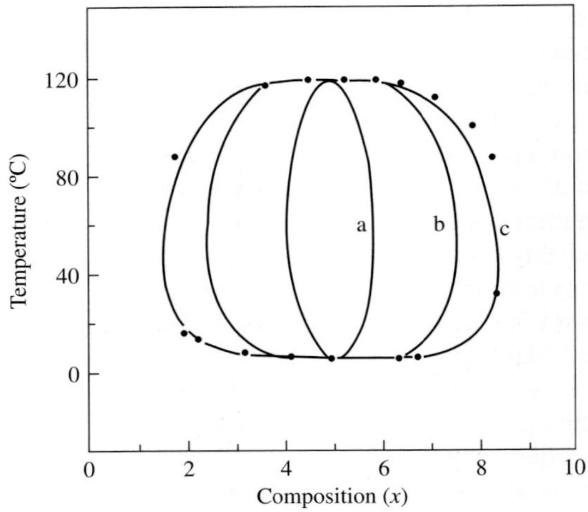

FIG. 7.61 Closed loop coexistence curve for the mixture m-toluidine-glycerine at atmospheric pressure. Points are experimental data, curve a is the Barker–Fock model solved in the quasichemical approximation, and curves b and c are decorated lattice model calculations with $\omega = 6$ (curve b) and $\omega = 5000$ (curve c). Here the molecules have ω contact points (possible orientations), only one of which can bond to a neighbouring molecule. The upper and lower critical temperatures in the model calculations were fitted to the experimental values. It should be noted that for the experimental results the composition variable is mass fraction of m-toluidine, whereas for the model calculations it is the mole fraction. Reprinted with permission from ref. 188. Copyright 1978 American Institute of Physics.

schematically[186] in Fig. 7.62(a). When the strength of the unlike pair bonding is weak (far right) the liquid–liquid coexistence region has the typical dome shape of a class II mixture (Fig. 7.9), but on increasing unlike pair bonding closed solubility loops are observed (class VI), while for sufficiently strong bonding the two critical lines join at a double critical point, beyond which the mixture is completely miscible. When bonding occurs between both like and unlike pairs, a variety of phase behaviour is possible, some of which is shown in Fig. 7.62(b). In the case shown, on increasing the unlike pair bonding the phase behaviour passes from class II to III (Figs. 7.14 and 7.15) to IV (Fig. 7.19).

An off-lattice theoretical approach, with more realistic intermolecular potential models, has been used with success by Jackson and co-workers[189–194] to study class VI mixtures. Wertheim's TPT1 theory was used to describe effects of molecular association. This off-lattice approach has several advantages over the lattice theories. In addition to providing a more realistic picture of the fluid structure, it is straightforward to study the influence of pressure on the phase diagram, and to incorporate more realistic intermolecular potentials.

In an initial study, a simple model mixture incorporating unlike pair bonding was used. Molecules of species A and B were modeled as hard spheres of equal diameter, σ, with mean field interactions between like pairs and a strongly

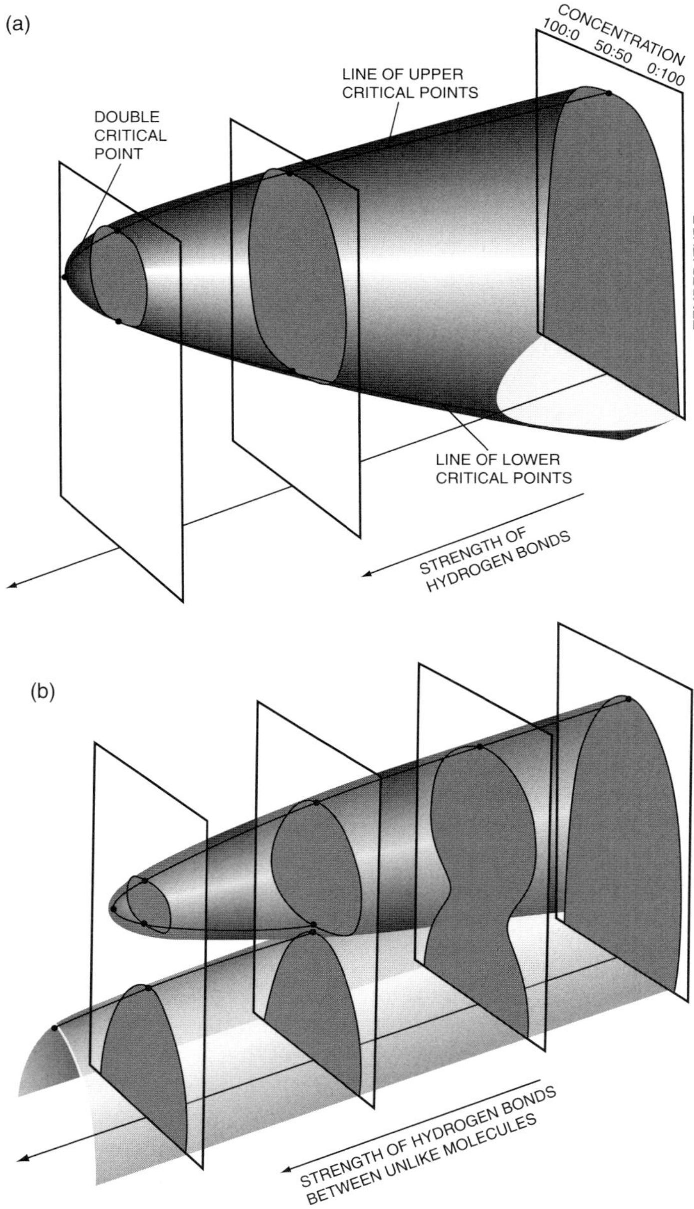

FIG. 7.62 Schematic illustration of the change in phase behaviour on increasing the strength of the unlike pair bonding (here assumed to be due to hydrogen bonding). In (a) any bonding between like pairs is weak, while in (b) there is bonding association between like pairs as well as unlike pairs; the like pair bonding is kept fixed as that between unlike pairs is varied. Reprinted with permission from ref. 186. Copyright 1987 Scientific American.

orientation-dependent bonding interaction between unlike pairs. For the bonding interaction, u_{AB}^{bond}, the off-centre square well interaction of Fig. 6.22(a) and Eq. (6.220) was used. Thus the potentials were

$$\begin{aligned} u_{AA}(r_{12}) = u_{BB}(r_{12}) &= u^{hs}(r_{12}) + u^{mf}(r_{12}) \\ u_{AB}(12) &= u^{hs}(r_{12}) + u_{AB}^{\text{bond}}(12), \end{aligned} \tag{7.179}$$

where r_{12} is the distance between the centres of the pair of molecules. AA and BB interactions have identical hard sphere and mean field interactions, $\sigma_{AA} = \sigma_{BB} = \sigma$ and $\varepsilon_{AA}^{\text{mf}} = \varepsilon_{BB}^{\text{mf}} = \varepsilon^{\text{mf}}$, while for the unlike pairs the mean field interaction is set to 0. Square well sites are placed at a distance of $\sigma/4$ from the molecular centre, with an interaction cutoff radius of 0.55σ, and sites on a molecule of species A are only allowed to bond with sites on species B. The mean field contribution to thermodynamic properties is calculated from a van der Waals term,

$$(p/\rho kT)_{\text{mf}} = -\frac{\rho}{kT} \sum_{\alpha} \sum_{\beta} x_\alpha x_\beta a_{\alpha\beta}, \tag{7.180}$$

where $a_{\alpha\beta} = -2\pi \int u_{\alpha\beta}^{\text{mf}}(r) r^2 \text{d}r \equiv \sigma^3 \varepsilon^{\text{mf}}$ is the van der Waals a term, and is a measure of the integrated strength of the mean field interaction between a pair of molecules of components α and β. With this model, for appropriate pressures, class VI behaviour was found to occur for a rather narrow range of bonding interaction strengths, as measured by $\varepsilon^* = \varepsilon^{\text{bond}}/\varepsilon^{\text{mf}}$. Some of the results for a rather high relative pressure are shown in Figure 7.63. When the strength of the bonding interaction is small (Fig 7.63(a) and (b)) the effect of the bonding on the phase diagram is small, and the phase behaviour is of class II. When the bonding strength is larger, however, the region of liquid–liquid separation decreases, particularly at lower temperatures. For $\varepsilon^* = 2.0$, Fig. 7.63(c), gas–liquid coexistence regions are seen at lower temperatures, together with gas–liquid–liquid three-phase coexistence at a temperature near $T_r = 1.2$. The effects of the bonding interaction between unlike pairs is greatest at lower temperatures, where it counteracts the unmixing tendency due to the lack of mean field attraction between the unlike pairs. For larger values of ε^* (Figs. 7.63(d) and (e)) the liquid–liquid immiscibility disappears at low temperatures, and class VI behaviour, with closed loop coexistence, is observed. As ε^* is further increased the closed loop becomes smaller, and eventually disappears (Fig. 7.63(f)), leaving only regions of gas–liquid coexistence. Calculations of the fraction of molecular pairs that are bonded show an increase with reducing temperature; the increase is particularly dramatic as the lower critical solution temperature is approached, and it is this increase that explains the lower part of the closed immiscibility loops.

This approach, in the form of the SAFT equation, has been applied to mixtures of water with alcohols,[190–192] water with n-alkyl polyoxyethylene ethers,[191] and water with polyoxyethylene polymers,[195] and direct comparisons with experiment have been made. Water is modeled as a hard sphere with four off-centre square well sites, two of which represent the H atoms (H sites) and the other two the lone

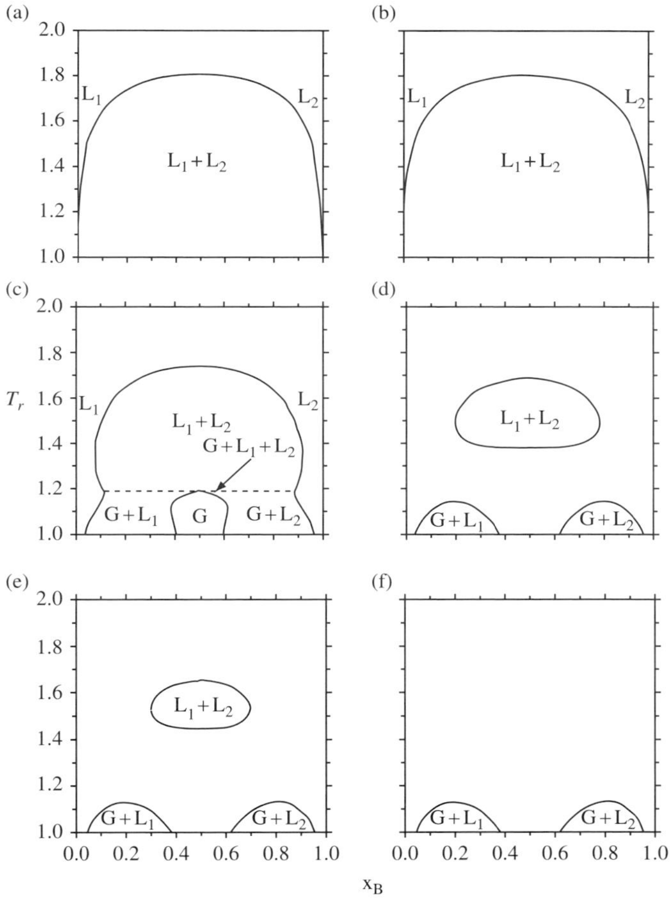

FIG. 7.63 *Txy* cuts at a constant pressure of $P_r = 10.39$ for a binary mixture of molecules represented as hard spheres of equal size, with mean field attractions between like pairs, and a bonding potential due to off-centre square well sites between unlike pairs. Here $T_r = T/T_c$ and $P_r = P/P_c$, where T_c and P_c are critical constants for the pure components. Cuts are shown for various strengths of the bonding interaction: (a) $\varepsilon^* = 0$, (b) $\varepsilon^* = 1.5$, (c) $\varepsilon^* = 2.0$, (d) $\varepsilon^* = 2.15$, (e) $\varepsilon^* = 2.195$, and (f) $\varepsilon^* = 2.25$. Reprinted with permission from ref. 189. Copyright 1991 Taylor and Francis Ltd.

pair electrons (e sites). H sites can only bond with e sites, and the bonding energy is $\varepsilon^{\text{bond}}$. The sites are arranged tetrahedrally (Fig. 7.64(a)), and a mean field van der Waals interaction of strength ε^{mf} is included to allow for the longer ranged dispersion interactions. The alcohol and alkyl polyoxyethylene ether molecules are represented as a chain of m hard sphere segments with mean field van der Waals interactions. Three square well interaction sites, one H and two e, are included on one of the terminal spheres to represent bonding by the OH group; one of the e sites is allowed only to bond to water molecules and not to like molecules.

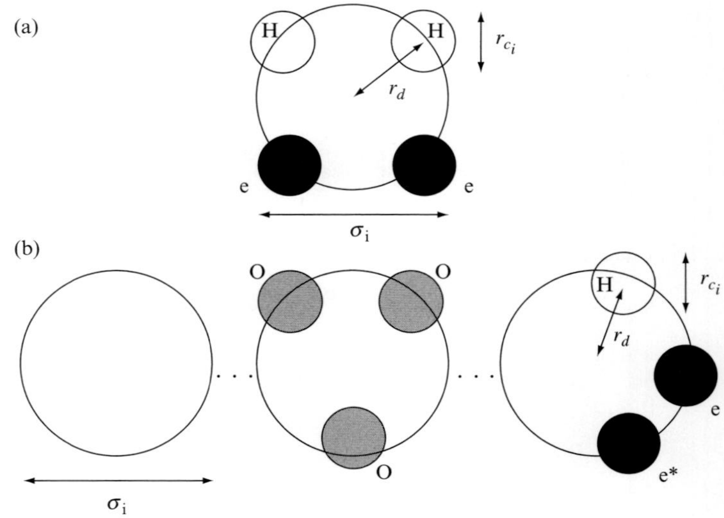

FIG. 7.64 Models for (a) water and (b) an alkyl polyoxyethylene ether, with off-centre square well bonding sites (small spheres) placed on hard sphere segments (large spheres). Bonding sites are placed at a distance r_d from the centre of the hard sphere segment, and have an interaction range of r_c. (From ref. 191.)

The ether oxygen atom is expected to form hydrogen bonds with water (but not with other ether molecules, due to steric hindrance), and three off-centre bonding sites are included on the O atom with bonding energy $\varepsilon_O^{\mathrm{hb}}$ (Fig. 7.64(b)). Results for the mixture water (1) + n-butoxyethanol (2) are shown in Fig. 7.65. The theory predicts closed solubility loops similar to those observed experimentally; the small range of mole fraction (x_2 from 0.02 to 0.19 at low pressure) over which immiscibility occurs is noteworthy, and is approximately reproduced by the theory. The pressure dependence of the closed solubility loops, including the upper hypercritical point at about 80 MPa pressure, is also quite well predicted by the theory (Fig. 7.65).

7.9.2 *Amphiphilic systems*

Amphiphilic molecules consist of two parts, one of which is attracted to the solvent while the other is not (or is considerably less attracted). In the case of aqueous solutions these parts are termed hydrophilic and hydrophobic, respectively. An example are the n-alkyl polyoxyethylene ethers, represented by the chemical structure $H(CH_2)_i(OCH_2CH_2)_jOH$, or simply C_iE_j, where i refers to the number of carbon atoms in the hydrophobic alkyl residue and j to the number of hydrophilic oxyethylene units, ($-OCH_2CH_2-$), in the molecule; the molecules also possess a terminal OH group. Block copolymers such as PVAC-PTAN act as amphiphiles in the case of supercritical carbon dioxide solutions; these

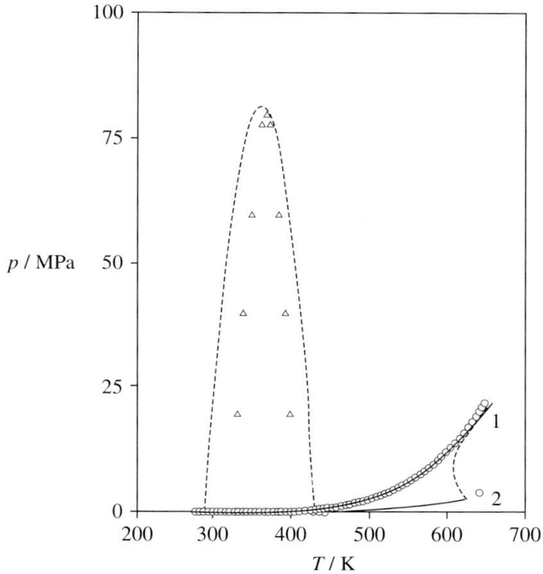

FIG. 7.65 The pT projection for the water (1) + n-butoxyethanol (2) system from experiment (points) and TPT1 theory (lines). Solid lines represent the pure component vapour pressures; dashed lines represent the vapour–liquid and liquid–liquid critical lines. Reprinted with permission from ref. 191. Copyright 1998 Taylor and Francis Ltd.

molecules have a CO_2-phobic polyvinyl acetate (PVAC) part and a CO_2-philic poly(1,1,2-tetrahydroperfluorooctyl acrylate) (PTAN) part. Such molecules tend to concentrate at interfaces. In the case of aqueous solutions, the molecules will orient themselves so that their hydrophobic tails point away from the water phase. A small concentration of such an amphiphile can result in dramatic reductions in the interfacial tension. In the case of two immiscible liquid phases, the addition of the amphiphile can cause a decrease in the immiscibility, i.e. an increase in the solubility of the dilute component in each phase. With increasing concentration of the amphiphile, the molecules dissolved in the bulk solution can aggregate to form *micelles*, self-assembled structures in which the solvent-phobic tails cluster together, with the solvent-philic head groups on the outside of the micelle, in contact with the solvent. Such micelles typically contain 20–100 amphiphile molecules, and can take a variety of shapes (spherical, cylindrical, cubic, lamellar, etc.), depending on the amphiphile architecture and concentration. The micelle formation first occurs at a *critical micelle concentration*, which depends on the temperature and density, and below which no micellization occurs. An important property of the micelle is its ability to solubilize within the micellar core solutes that are normally insoluble in the solvent. In the case of oil–water–surfactant systems it is possible, for suitable amphiphiles and concentrations, to form microemulsions containing considerable concentrations of both oil and water. These properties of amphiphilic mixtures find many applications in areas such as the manufacture of cosmetics and detergents, in drug delivery, and tertiary oil recovery, and are also of importance in biological processes such as cholesterol solubilization.

The SAFT equation (§ 6.13.5) has been used to study amphiphilic mixtures in both aqueous[166, 191, 196–200] and carbon dioxide[201] solutions. Because the structure that goes into the theory is that of the reference monomers, SAFT cannot predict the various micellar structures that occur, or the changes in structure with composition, pressure, and temperature. However, it can be used to predict phase transitions, and since it predicts aggregation through association, it can be used to estimate the critical micelle concentration.[201]

A temperature–composition plot for a hypothetical model ternary mixture of water(1)–oil(2)–amphiphile(3), calculated using the SAFT equation, is shown in Fig. 7.66. Water is modeled as shown in Fig. 7.64(a), with two off-centre H and two off-centre e sites (see above subsection); the oil is modeled as a chain of m_2 tangent hard sphere segments; and the amphiphile molecule is modeled as m_3 tangent hard sphere segments, with an e and H site on the end segment to provide the amphiphilic character. Mean field van der Waals interactions are included for all molecules. In Fig. 7.66 each of the vertices of the prism represent one of the pure

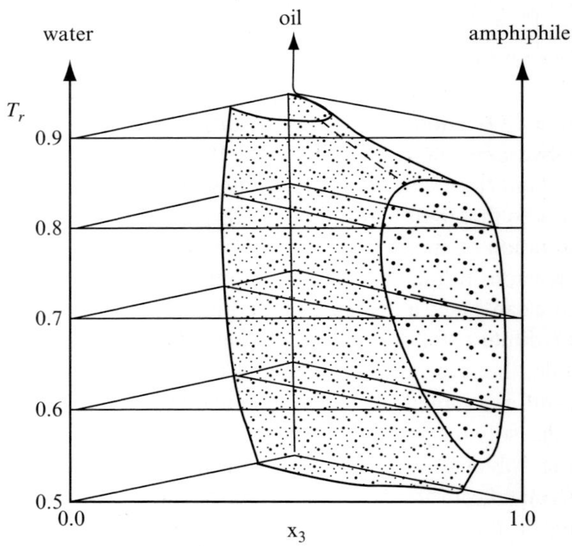

FIG. 7.66 The ternary phase diagram for a model water(1)–oil(2)–amphiphile(3) mixture for a reduced pressure $P_r = P/P_{c1} = 0.3$. Here $T_r = T/T_{c1}$ is the reduced temperature. Shaded areas represent the liquid–liquid coexistence surfaces, and the intermolecular parameters used are association energies $\varepsilon_{11} = 1.2\varepsilon_{mf}$, $\varepsilon_{33} = 0.6\varepsilon_{mf}$, where ε_{mf} is the integrated strength of the mean field interaction, ε_{13} is given by the Berthelot geometric mean rule, $m_2 = m_3 = 3$. Reprinted with permission from ref. 166. Copyright 1995 Kluwer-Academic.

components, and the opposing face represents the binary mixture devoid of that component. The water–amphiphile mixture faces the reader, the water–oil face is behind and to the left, and oil–amphiphile is behind and to the right. Shaded areas represent the liquid–liquid coexistence surfaces, and horizontal slices through the diagram represent the phase behaviour at a particular temperature. The region of liquid–liquid immiscibility disappears above an upper critical solution temperature, and there is also a lower critical solution temperature below which the mixture is miscible in all proportions, which is due to the strong unlike pair association between the water and amphiphile molecules. The immiscibility of the oil-rich and water-rich phases is seen to decrease with increasing concentration of the amphiphile, as expected.

SAFT has been used[191, 197] to study the binary mixture phase diagrams for aqueous solutions of alkyl polyoxyethylene ethers, C_iE_j, and comparisons have been made with experiment. These mixtures exhibit liquid–liquid immiscibility, with one phase being dilute in the surfactant and the other rich in surfactant, which form micelles. These systems exhibit a lower critical solution temperature, below which the mixture is completely miscible; the LCST arises due to the hydrogen bonding between water and the head groups of the surfactant molecules. In some systems a closed solubility loop occurs (class VI), with complete miscibility above some upper critical solution temperature. Some of the results for water/n-butoxyethanol (C_4E_1) are shown in Fig. 7.65. Results for other systems of this type are shown in Figs. 7.67 and 7.68. The models used for these calculations are those of Fig. 7.64, with water parameters obtained by fitting to vapour–liquid equilibrium data; hydrogen bonding parameters for the hydroxyl groups in the surfactants were determined by fitting the experimental vapour pressure of butoxyethanol, and those for the unlike pair water-hydroxyl interaction were obtained by fitting to liquid–liquid phase equilibrium data for water + 1-butanol. These parameters were taken to be transferable to other surfactant molecules. The agreement with experiment in Figs. 7.67 and 7.68 is qualitative, but agreement with experiment is good for both the LCST and UCST values, where available. For most mixtures of this type data are available only in the region of the LCST; in most cases this is due to the instability of the surfactant molecules at the higher temperatures.

7.9.3 Applications to real fluid mixtures and the SAFT equation

Most applications of the TPT1 theory to real fluid mixtures have used the SAFT equation of state, (6.288), which represents the residual Helmholtz energy as a sum of segment (monomer), chain formation, and association terms. The last two terms are obtained from the TPT1 theory (see § 6.13 for equations). One of the simplest and most widely used versions of SAFT is that of Huang and Radosz,[202] who compared the theory with experimental phase equilibrium data for 60 mixtures. For the free energy contribution from the segment term, a hard sphere repulsion term, given by the Carnahan–Starling equation, plus a dispersion

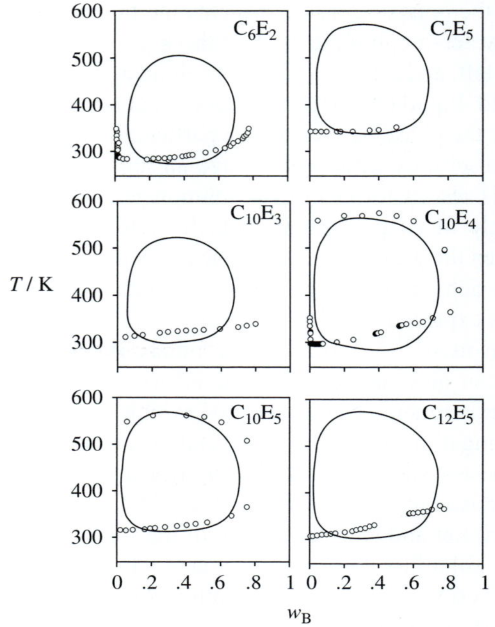

FIG. 7.67 Liquid–liquid coexistence regions for water(1)+C_iE_j(2) mixtures, plotted as temperature vs weight fraction of surfactant, w_2. Experimental results are shown as circles, and the predictions of SAFT are the lines. Reprinted with permission from ref. 197. Copyright 1998 American Chemical Society.

contribution given by the equation of state of Chen and Kreglewski[203] was used, and the chain and association terms were given by the Wertheim theory (see § 6.13). The potential parameters (m, ε, σ for non-associating molecules, and $\varepsilon^{\text{bond}}$ and κ, the bonding energy and volume available for bonding, for the association term—see § 6.13.5) were fitted to experimental vapour pressure and saturated liquid density data for the pure fluids; one unlike pair parameter, ε_{ij}, was fitted to mixture data, and the van der Waals 1 mixing rules (see § 7.7.1) were used in estimating the dispersion energy. Except in the critical region, good agreement with experiment was obtained for vapour–liquid equilibria (within a few percent for vapour pressures and liquid densities) for mixtures containing alkanols, alkanes, alkenes, acids, amines, and carbon dioxide, over the full composition range and over a range of temperatures.

There have been a large number of applications of SAFT to mixtures, including aqueous solutions, hydrocarbons, polymers, and complex petroleum fluids, and these have been the subject of several reviews.[166–170,210] Many of the applications have used the Huang–Radosz treatment and parameterization, while others have differed in their choice of reference system (e.g. Lennard–Jones in place of hard spheres) and treatment of the dispersion contributions (see § 6.13.4).

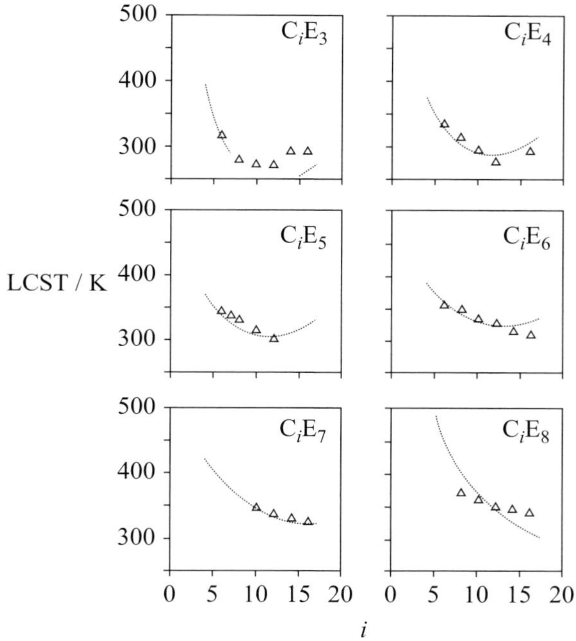

FIG. 7.68 Experimental (points) and calculated (lines) lower critical solution temperatures for water(1)+C_iE_j(2) mixtures, as a function of the number of C atoms i in the alkyl chain of the surfactant. Reprinted with permission from ref. 197. Copyright 1998 American Chemical Society.

In general the agreement with experiment is good for a wide variety of types of fluid, provided the association sites are appropriately chosen. As examples, results for three systems that provide a strong challenge to theoretical predictions are shown in Figs. 7.69–7.71. Henry constants for a range of gases dissolved in polyethylene are shown in Fig. 7.69; one temperature-independent potential parameter was fitted to the experimental data.[204] The theory is seen to give a good fit over a wide temperature range. Liquid–liquid equilibria are more difficult to predict accurately than vapour–liquid equilibria. As seen in Fig. 7.70, the theory gives a good description of the ternary data except near the critical mixing point.[205] Unified activity coefficient correlation (UNIFAC), a correlation based on the group contribution idea and widely used by chemical engineers, is seen to give significantly poorer results. An application to petroleum fluids is shown in Fig. 7.71.[206] The fitted parameters in SAFT are well-behaved, and this makes it possible to predict parameters for fluids for which no measurements are available, from fitted values for other fluids of similar structure or in the same homologous series. This provides a route to apply the theory to poorly characterized mixtures such as petroleum fluids, and has been exploited by Radosz and co-workers,[202] who correlated the potential parameters for different families, e.g.

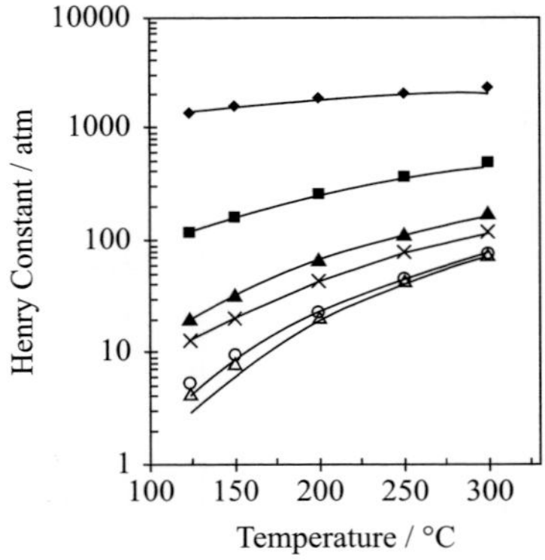

FIG. 7.69 Weight fraction Henry constants for ethylene (diamonds), n-butane (squares), n-hexane (solid triangles), n-octane (open triangles), benzene (crosses), and toluene (circles) in low-density polyethylene at 1 atm. pressure. Experimental data[204] are shown as points; lines are SAFT, using a single fitted binary interaction parameter. Reprinted with permission from ref. 170. Copyright 2002 American Chemical Society.

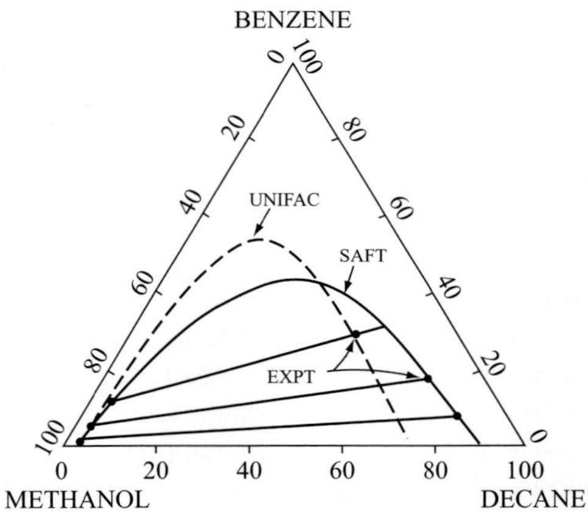

FIG. 7.70 Liquid–liquid equilibria of the decane/methanol/benzene system from experiment (points), UNIFAC correlation (dashed lines) and SAFT (solid lines). (From ref. 205.)

n-alkanes, polynuclear aromatics, etc. Successful applications have included the extraction of petroleum pitch, solubility of bitumen in carbon dioxide (Fig. 7.71), and asphaltene deposition (some of these applications are reviewed in ref. 169).

7.9.4 Electrolyte solutions

The thermodynamic properties of mixtures containing electrolytes are strongly affected by the long-range ion–ion interactions, and also by ion–solvent and ion–solute interactions, in addition to the effects due to repulsion, dispersion, chain

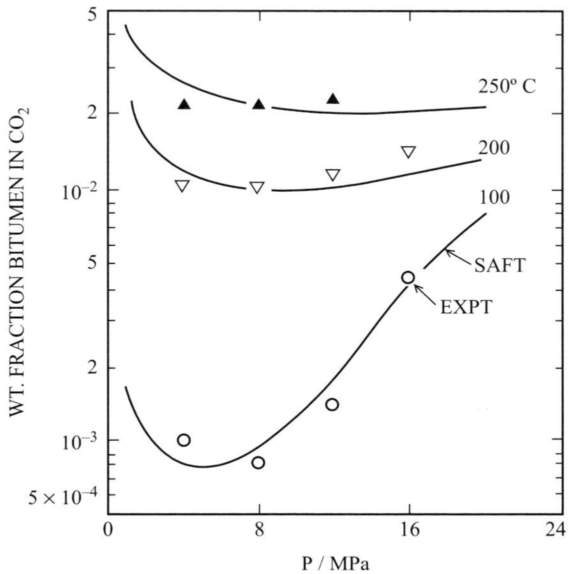

FIG. 7.71 Solubility of bitumen in compressed carbon dioxide from experiment (points) and SAFT. (From ref. 206.)

formation, and association. The SAFT equation can be extended to include such effects by including an additional free energy term for the ion–ion, ion–dipole, etc., interactions,[207] so that the Helmholtz energy can be written (cf. (6.288))

$$A^{\text{res}} = A^{\text{res}}_{\text{seg}} + A_{\text{chain}} + A_{\text{assoc}} + A_{\text{ion}}, \tag{7.181}$$

where A_{ion} is the contribution due to ion interactions. A wide range of possible theories are available for the ion–ion interaction term.[208] These include the Debye–Hückel (DH) and augmented DH theories, and integral equation methods such as the mean spherical approximation (MSA) and reference hypernetted chain (RHNC) theories (see § 5.4 of Vol. 1). The DH theory is exact in the limit of infinite dilution of the electrolyte, but fails to properly account for the effects of increased concentration of electrolyte, and treats the solvent as a continuum dielectric medium. The RHNC is usually somewhat more accurate than MSA for electrolytes, but the MSA offers the advantage of an analytic solution and consequent ease of use.

Galindo and co-workers[207,209,210] have used the MSA theory to calculate the effects of the ion–ion interactions for aqueous solutions of strong electrolytes and electrolyte mixtures. For most calculations the restricted primitive model (RPM) of electrolytes was used, in which the ions are charged hard spheres, the cations and anions having the same diameter, σ. In the RPM the ion–ion potential is given by

$$
\begin{aligned}
u_{ij}^{\text{ion}} &= +\infty \qquad \textit{if}\, r < \sigma \\
&= \frac{q_i q_j}{\varepsilon r} \qquad \textit{if}\, r > \sigma,
\end{aligned}
\tag{7.182}
$$

where ε is the dielectric constant and q_i is the charge on ion of species i, given by ez_+ for cations and ez_- for anions. The ion diameter is taken to be the average of the ion diameters, i.e.

$$\sigma = \sum_i x_i \sigma_{ii}, \tag{7.183}$$

where x_i is the fraction of ions of type i and the sum is over all ionic species. The MSA closure for solving the Ornstein–Zernike equation is (see Chapter 5)

$$
\begin{aligned}
g_{ij}(r) &= 0 \qquad \textit{if}\ r < \sigma \\
c_{ij}(r) &= -\frac{q_i q_j}{\varepsilon k T r} \qquad \textit{if}\ r > \sigma,
\end{aligned}
\tag{7.184}
$$

where $c_{ij}(r)$ is the direct correlation function. Waisman and Lebowitz solved the OZ equation for the RPM using the MSA closure to obtain an analytic solution for $g_{ij}(r)$.[211] The corresponding expression for the Helmholtz energy is[212]

$$\frac{A_{\text{ion}}}{NkT} = -\frac{3x^2 + 6x + 2 - 2(1+2x)^{3/2}}{12\pi\rho\sigma^3}, \tag{7.185}$$

where $x = \kappa\sigma$ and

$$\kappa^2 = \frac{4\pi}{\varepsilon k T} \sum_i \rho_i q_i^2 \tag{7.186}$$

is the inverse Debye length (κ here is different from that of § 7.9.3); the sum is over all ion species. In the MSA treatment of the RPM, the radial distribution function at contact is given by

$$g_{ij}(\sigma) = g_{ij}^{HS}(\sigma) - \frac{q_i q_j}{\varepsilon k T \sigma}(1 - \tau^2), \tag{7.187}$$

where

$$\tau = \frac{x^2 + x - x(1+2x)^{1/2}}{x^2}. \tag{7.188}$$

The modified SAFT theory represented by (7.181), with the ionic contribution term given by (7.185), has been applied to a variety of strong electrolytes and electrolyte mixtures.[207,209,210] Water was modeled as before by a hard sphere with four off-centre square well sites (Fig. 7.64(a)), with parameters fitted to vapour pressure and orthobaric density data. Pauling's ionic radii were used to calculate the ionic diameters, and attractive dispersion interactions between ions were neglected. Water–ion dispersion interactions were included with the variable range model (§ 6.13), consisting of a square well interaction, with fixed well width $\lambda_{ij} = 1.2$, and with the well depth fitted to electrolyte solution vapor pressure

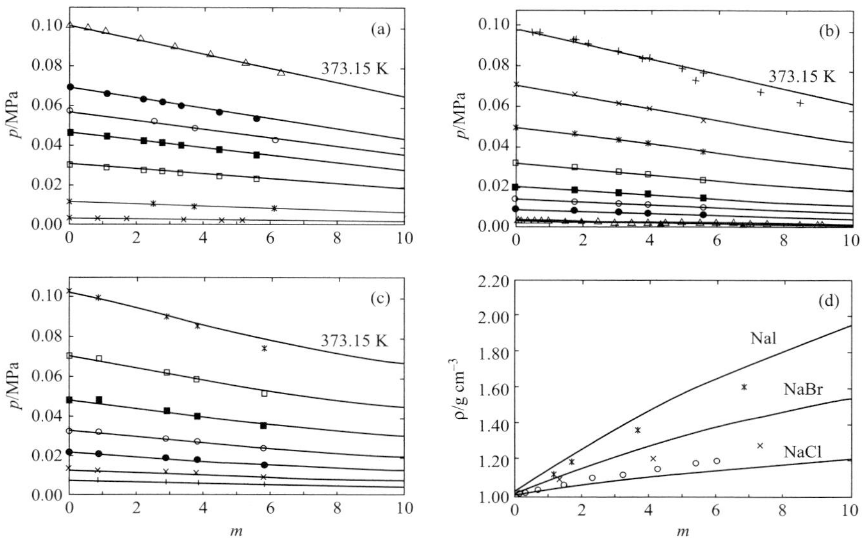

FIG. 7.72 Experimental[213] (points) and theoretical (lines) results as a function of molarity, m, for (a) vapour pressures of aqueous NaCl solutions at temperatures 373.15, 363.15, 358.01, 353.15, 343.15, 322.28, 298.74, and 273.15 K; (b) vapour pressures of aqueous NaBr solutions at 373.15, 363.91, 355.04, 344.11, 333.23, 325.68, 316.84, 298.15 and 291.15 K; (c) vapour pressures of aqueous NaI solutions at 373.57, 363.31, 353.48, 343.77, 333.97, 323.03, and 311.44 K; and (d) solution densities for NaCl at 298.15 K, NaBr at 298.15 K, and NaI at 288.15 K. Temperature sequences are read from top to bottom in (a), (b), and (c). Molarity is the electrolyte concentration in units of moles of electrolyte per litre of solution. Reprinted with permission from ref. 207. Copyright 1999. American Chemical Society.

data for the salt solution. The inclusion of the dielectric constant in the equations accounts, in a rough way, for the neglect of water–ion electrostatic and other ion interactions. Typical results for strong electrolytes are shown in Figures 7.72 and 7.73.[207,210,213,214] The theory gives an accurate account of the influence of temperature and concentration on the vapour pressure, with poorer results for the density.

7.10 Conclusions

The Kirkwood–Buff theory (§ 7.2) provides a rigorous starting point for any predictive theory of mixtures, since it does not rest on any assumption concerning the nature of the molecules, or on the pairwise additivity approximation. Moreover, it provides a clear definition of the ideal mixture (§ 7.3), and provides rigorous expressions for such properties as the activity coefficient (§ 7.4), fugacity, and Henry's constant (§ 7.6). Despite these advantages, the Kirkwood–Buff theory has seen relatively little application in prediction of mixture properties.

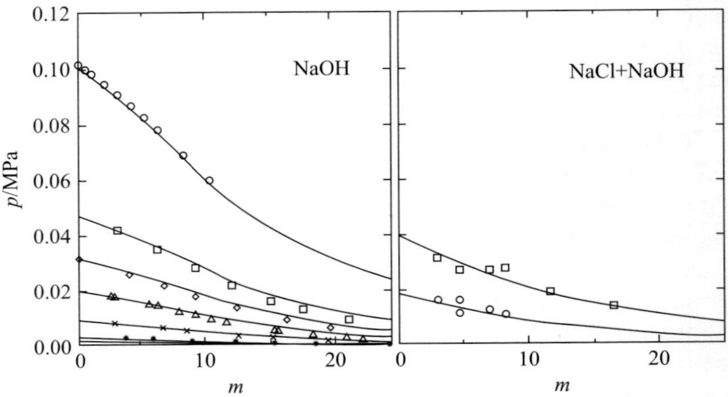

FIG. 7.73 The vapour pressure as a function of salt concentration (m = molarity) for aqueous NaOH (left) at temperatures 273 to 373 K (bottom to top), and for a mixed NaOH+NaCl aqueous solution (right) at 335 and 353 K. In the results shown in the right-hand figure, the NaCl is at a constant concentration of 4.412 molar. Symbols are experimental data[214] and the lines are the theory. (From ref. 210.)

For simple mixtures in which the molecules are near spherical and exhibit only repulsion and dispersion forces two successful theories are available (§ 7.7), conformal solution theory and hard sphere perturbation theory. The former is easier to apply in practice, but since it relies on an expansion about an ideal isotopic mixture, in which the various constituents all exhibit the same intermolecular forces, it is limited to mixtures in which the components do not differ too much in size or in their intermolecular forces. The perturbation theories based on a reference system of hard spheres are better able to account for larger differences in molecular size and intermolecular forces.

For mixtures of nonspherical molecules, having weak to moderately strong electrostatic forces, the u-expansion (§ 7.8.1) gives reasonable results, but series convergence becomes poor for strong multipolar forces, since these have a strong effect on the pair correlation function, making it significantly different from that for the reference system of spherical molecules. Nevertheless, the Padé approximant of Eq. (7.158) gives good results for thermodynamic properties even for strong electrostatic forces. A drawback of this approach is that it does not give a good description of the effects of non-spherical molecular cores. The perturbation expansion about fluids of non-spherical reference molecules is better in this respect, but does not describe electrostatic force effects well, and is more difficult to use than the u-expansion.

Wertheim's theory of associating fluids (§ 7.9) largely overcomes the shortcomings of the earlier perturbation theories, making it possible to describe chain and ring formation, and hence highly non-spherical shapes, while also accounting for strong electrostatic or other associating forces. When combined with a theory describing repulsion and dispersion interactions, as in the SAFT equation and its modifications, it is possible to describe with reasonable accuracy the thermodynamic properties of highly non-ideal liquid mixtures involving a wide range of molecular species.

Despite considerable progress over the past two decades in our ability to account for the behaviour of highly non-ideal liquid mixtures, much remains to be done to improve the predictive capabilities for phase transitions in highly non-ideal liquid mixtures, and to develop satisfactory theories for more complex fluids. Systematic methods for molecular site location and force field parameter determination are needed to make the SAFT-type methods accessible to applied scientists and engineers, and to enable them to be incorporated successfully into chemical process simulators. Improved and more sophisticated theoretical treatments are still needed for many of the more difficult phenomena, such as liquid–liquid and liquid–solid phase changes, dilute solution behaviour for solid, liquid, and gaseous solutes, electrolyte solutions, room temperature ionic liquids, and the broad group of soft matter systems—polymer and protein solutions, micellar and other self-assembly systems, liquid crystals, colloids, etc.

Appendix 7A Derivation of Eq. (7.51)

From (7.7) and (7.53) we have

$$H_{\alpha\beta} = \int d\mathbf{r} \langle h_{\alpha\beta}(\mathbf{r}\omega_1\omega_2)\rangle_{\omega_1\omega_2} = \langle \tilde{h}_{\alpha\beta}(0\omega_1\omega_2)\rangle_{\omega_1\omega_2} \tag{7A.1}$$

$$C_{\alpha\beta} = \int d\mathbf{r} \langle c_{\alpha\beta}(\mathbf{r}\omega_1\omega_2)\rangle_{\omega_1\omega_2} = \langle \tilde{c}_{\alpha\beta}(0\omega_1\omega_2)\rangle_{\omega_1\omega_2} \tag{7A.2}$$

where $\tilde{h}_{\alpha\beta}(0\omega_1\omega_2)$ and $\tilde{c}_{\alpha\beta}(0\omega_1\omega_2)$ are the Fourier transforms with $k = 0$, e.g.

$$\tilde{h}_{\alpha\beta}(\mathbf{k}\omega_1\omega_2) = \int d\mathbf{r} e^{i\mathbf{k}\cdot\mathbf{r}} h_{\alpha\beta}(\mathbf{r}\omega_1\omega_2). \tag{7A.3}$$

If we assume linear molecules we can expand $h_{\alpha\beta}(\mathbf{r}\omega_1\omega_2)$ in spherical harmonics (see (2.23) of Vol. 1),

$$h_{\alpha\beta}(\mathbf{r}\omega_1\omega_2) = \sum_{\ell_1\ell_2\ell}\sum_{m_1m_2m} h_{\alpha\beta}(\ell_1\ell_2\ell; r) C(\ell_1\ell_2\ell; m_1m_2m) Y_{\ell_1m_1}(\omega_1) Y_{\ell_2m_2}(\omega_2) Y_{\ell m}(\omega)^*. \tag{7A.4}$$

Substituting (7A.4) into (7A.1) and using (A.38) and (A.3) gives

$$H_{\alpha\beta} = (4\pi)^{-3/2} \int d\mathbf{r} h_{\alpha\beta}(000; r) = (4\pi)^{-3/2} \tilde{h}_{\alpha\beta}(000; k = 0). \tag{7A.5}$$

Similarly

$$C_{\alpha\beta} = (4\pi)^{-3/2} \int d\mathbf{r} c_{\alpha\beta}(000; r) = (4\pi)^{-3/2} \tilde{c}_{\alpha\beta}(000; k = 0). \tag{7A.6}$$

Here $\tilde{h}_{\alpha\beta}(000; k = 0)$ and $\tilde{c}_{\alpha\beta}(000; k = 0)$ are the Fourier transforms of the **r**-space harmonic coefficients for $\ell_1\ell_2\ell = 000$ in the limit $k \to 0$. These are related by the harmonic expansion of the Ornstein–Zernike equation, (3B.19). In the limit $k \to 0$ this becomes

$$\tilde{h}_{\alpha\beta}(000; k = 0) = \tilde{c}_{\alpha\beta}(000; k = 0) + (4\pi)^{-3/2} \sum_{\gamma} \rho_\gamma \tilde{c}_{\alpha\gamma}(000; k = 0) \times \tilde{h}_{\gamma\beta}(000; k = 0). \tag{7A.7}$$

Multiplying both sides of this equation by $(4\pi)^{-3/2}$ and using (7A.5) and (7A.6) gives (7.51),

$$H_{\alpha\beta} = C_{\alpha\beta} + \sum_{\gamma} \rho_\gamma C_{\alpha\gamma} H_{\gamma\beta} \tag{7A.8}$$

or, in matrix form

$$\mathbf{H} = \mathbf{C} + \rho\mathbf{CXH}, \tag{7A.9}$$

where $\mathbf{X}$ is the diagonal $R \times R$ matrix whose non-zero elements are the mole fractions. In view of the symmetry of $\mathbf{H}$ and $\mathbf{C}$, an equivalent form to (7A.9) is

$$\mathbf{H} = \mathbf{C} + \rho\mathbf{HXC}. \tag{7A.10}$$

This form is used by O'Connell.[41]

Appendix 7B Application of Kirkwood–Buff Theory: supercritical fluid extraction

Supercritical fluid extraction (SFE) is a large field[215–219] that we can only touch briefly here. Because of its benign nature there has been extensive use of supercritical CO_2 as a solvent in the food industry (e.g. to decaffeinate coffee, remove fat from snack foods, extract value-added components such as hops, spices, and essential oils, remove pesticide residues from foodstuffs), in soil remediation (e.g. to remove aromatic hydrocarbons and other pollutants), in pharmaceuticals (e.g. to isolate active agents in ointments), in the tobacco industry (to remove nicotine), etc. We discuss here how Kirkwood–Buff theory, and virial expansions, have been used to predict the solubility of a dilute component B in a supercritical solvent A (which we will take to be CO_2).

Suppose a pure solid phase of B, with low volatility, is in equilibrium with a fluid solvent A at temperature T and density ρ. Equating chemical potentials of B in the two phases gives[218] for the mole fraction x_B of B in the fluid (a measure of the solubility)

$$x_B = \left(p_B^{\text{vap}}/\rho kT\right) \mathrm{e}^{-\beta\mu_{Br}} \mathrm{e}^{\beta(p - p_B^{\text{vap}})v_B^s}, \tag{7B.1}$$

where p_B^{vap} is the vapour pressure of the solid B at the temperature T, p is the applied (supercritical) pressure, and v_B^{s} is the molecular volume of the solid phase B, which is assumed to be independent of p. The key quantity in (7B.1) is μ_{Br}, the residual chemical potential of B in the fluid solution.

Supercritical fluids are much less dense than normal liquids and a virial expansion (see § 3.6 of Vol. 1) was suggested quite early to calculate solubilities.[220] (The critical parameters for CO_2 are $v_c = 94\,\text{cm}^3/\text{mol}$, $T_c = 304.2\,\text{K}$, and $p_c = 73.8\,\text{bar}$.) Because of the difficulty of calculating virial coefficients for nonspherical molecules, this suggestion was not taken up seriously until the 1990s.[221–223] For dilute solute B in solvent A, the virial series for μ_{Br} is[222]

$$\lim_{x_B \to 0} \beta\mu_{Br} = 2B_2^{AB}(T)\rho + \frac{3}{2}B_3^{AAB}(T)\rho^2 + \frac{4}{3}B_4^{AAAB}(T)\rho^3 + \ldots, \qquad (7\text{B}.2)$$

where B_2, B_3, and B_4 are the usual second, third, and fourth pressure virial coefficients for the mixture (§§ 3.6 and 6.10). Since solubility data in supercritical solvents are often taken up to densities of about $2\rho_c$, where ρ_c is the critical density of the solvent, the questions of whether and where the series (7B.2) can be stopped loom large. Quiram *et al.*[221] stopped at the third virial coefficient (B_3), and using empirical correlations for B_2 and B_3, or direct data for B_2 and B_3, found good fits to solubility data. However, for simple monatomic LJ fluids, Joslin *et al.*[222] showed that the B_4 term was necessary to get good *ab initio* agreement at standard SFE conditions; they computed B_2, B_3, and B_4 and compared the results for μ_{Br} values obtained from simulation. Harvey[223] reached the same conclusion (stopping at B_3 is not adequate) from analysis of the virial expansions of simple model equations of state (van der Waals, Redlich-Kwong, Peng-Robinson).

It is, however, not easy to apply (7B.2) up to the B_4 term, for complex solutes B. Even for the relatively simple case of rigid linear molecules, the calculation of $B_4(T)$ would require the evaluation of 14-dimensional integrals. Using KBT Tomberli *et al.*[224] have shown that this problem can be evaded; an expression for μ_{Br} can be derived involving only B_2 and B_3 virial coefficients, which is as accurate as (7B.2) up to the B_4 term. The starting point is the expression for μ_{Br} in terms of KBT quantities $H_{\alpha\beta}$ (Eq. (7.7))

$$\lim_{x_B \to 0} \beta\mu_{Br} = -\int_0^\rho \mathrm{d}\rho' \frac{H_{AB}}{1 + \rho' H_{AA}}, \qquad (7\text{B}.3)$$

which can be derived[224] from (7.39), and where the $H_{\alpha\beta}$ depend on density. We now expand H_{AB} and H_{AA} in (7B.3) in powers of the density, using the virial expansions for the pair correlation functions $g_{\alpha\beta}(12)$ occurring in the definition (7.7) of $H_{\alpha\beta}$. In the dilute limit ($x_B \to 0$) this gives

$$H_{\alpha\beta} = -2B_2^{\alpha\beta} + \rho\left[4B_2^{\alpha\alpha}B_2^{\alpha\beta} - 3B_3^{\alpha\alpha\beta}\right] + \ldots, \quad (\alpha\beta = \text{AA or AB}) \qquad (7\text{B}.4)$$

If we substitute (7B.4) into (7B.3), we can carry out the ρ' integration analytically. If we keep only the first term in (7B.4) we find

$$\lim_{x_B \to 0} \beta \mu_{Br} = -\frac{B_2^{AB}}{B_2^{AA}} \ell n \left(1 - 2B_2^{AA} \rho\right). \tag{7B.5}$$

If we expand $\ln\left(1 - 2B_2^{AA}\rho\right)$ in (7B.5) in powers of ρ, we see that the first term agrees with the first term in (7B.2). The unexpanded form (7B.5) contains all powers of ρ and can be regarded as a partial resummation of the density series. If we keep the $O(\rho)$ term in (7B.4), the ρ' integral in (7B.3) can again be done analytically, and the result[224] involves the third virial coefficients B_3^{AAA} and B_3^{AAB}. Thus, at the (relatively modest) cost of two new lower-order virial coefficients, i.e. B_2^{AA} and B_3^{AAA}, KBT theory can give results as good as the direct virial expansion (7B.2) up to the B_4 term, for $\rho \lesssim 2\rho_c$.

For sufficiently complex molecules B in supercritical solvent $A = CO_2$, the B_2 and B_3 virial coefficients may be too difficult to calculate from an intermolecular potential u_{AB} (u_{AA} is relatively simple, and for dilute B we do not need u_{BB}). In some cases even data for the cross virial coefficients B_2^{AB} and B_3^{AAB} are unavailable. In such cases, engineering correlations[225,226] can be used to estimate the cross virial coefficients. The KBT used in this way has predicted solubilities in supercritical CO_2 in agreement with measured values for a number of species B, including common organic pollutants[224,227] (e.g. naphthalene, benzoic acid, anthracene, phenanthrene, naphthol-2, pyrene), and phenolic compounds[228] (e.g.

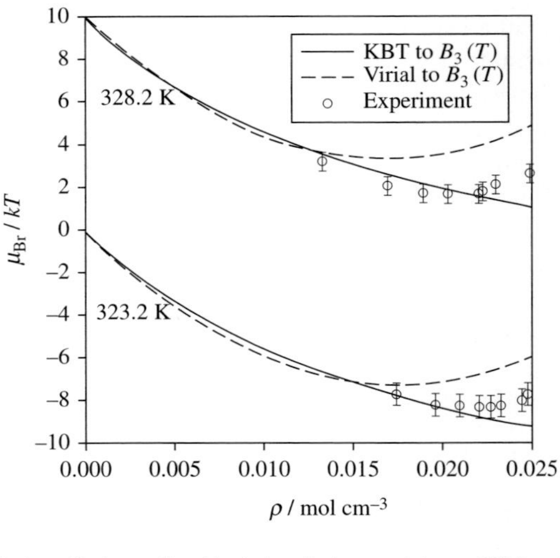

FIG. 7B.1 Theoretical predictions of residual chemical potential μ_{Br}/kT for naphthalene in supercritical CO_2 ($\rho_c = 0.01063\,\text{mol/cm}^3$) at two temperatures. Solid lines are KBT to order $B_3(T)$, dashed lines are the virial expansion to order $B_3(T)$, and the points are obtained from the solubility data of ref. 229 using Eq. (7B.1). The curves and data at $T = 328.2$ K are displaced by +10 for clarity.

caffeic acid, ferulic acid, coumaric acid). Figure 7B.1 shows the result of μ_{Br} for naphthalene using KBT (to order $B_3(T)$) compared to the corresponding virial series result, and also compared to the experiment result. We see that KBT agrees with experiment for solvent densities $\rho \lesssim 2\rho_c$ (here $\rho_c = 0.01063\ \text{mol/cm}^3$), and is an improvement over the virial series to order $B_3(T)$.

References and notes

1. Moser, B., and Kistenmacher, H. *Fluid Phase Eqba*. **34**, 189 (1987).
2. Van der Waals, J.D. *Zeit. Phys. Chem*. **5**, 133 (1890).
3. Gibbs, J.W. *Trans. Connecticut Acad*. **3**, 152 (1876).
4. Bakhuis Roozeboom, H.W. *Die Heterogenen Gleichgewichte von Standpunkte der Phasenlehre*, Vieweg, Braunschweig (1901–18).
5. Kuenen, J.P. *Theorie der Verdampfung und Verflussigung von Gemischen*, Barth, Leipzig (1906).
6. Van der Waals, J.D., and Kohnstamm, P. *Lehrbuch der Thermodynamik*, Barth, Leipzig (1923).
7. Van der Waals, J.D., and Zittinsvers, I.K. *Akad. Wet. Amst*., 133 (1894).
8. Kamerlingh-Onnes, H., and Keesom, W.H. *Commum. Phys. Lab. Univ. Leiden*, Supp. No. 15 (1907).
9. Krichevskii, I.R. *Acta Phys. Chem. URSS*, **12**, 480 (1940).
10. Timmermans, J., and Kohnstamm, P. *Proc. Sect. Sci, K. Ned. Akad. Wet. Amst*. **12**, 234 (1909–10); **15**, 1021 (1912–13); **13**, 507 (1910–11); *J. Chim. Phys*. **20**, 491(1923); Poppe, G. *Bull. Sec. Chim. Belg*. **44**, 640 (1935).
11. Schneider, G.M. *Ber. Bunsenges. Phys. Chem*. **70**, 497 (1966); *Adv. Chem. Phys*. **17**, 1 (1970); in *Water, a Comprehensive Treatise* (ed. F. Franks), Vol. 2, Chap. 6, Plenum, New York (1973); *Pure Appl. Chem*. **47**, 277 (1976); in *A Specialist Periodical Report, Chemical Thermodynamics*, Vol. 2, Chemical Society, London (1978); *J. Chem. Thermodynamics* **23**, 301 (1991).
12. Rowlinson, J.S., ed. *J.D. van der Waals: On the Continuity of the Gaseous and Liquid States*, North-Holland, Amsterdam (1988).
13. Levelt Sengers, J. *How Fluids Unmix: Discoveries by the School of Van Der Waals and Kamerlingh Onnes*, Royal Netherlands Academy of Arts and Sciences, Amsterdam (2002).
14. Rowlinson, J.S. *Liquids and Liquid Mixtures*, 2nd edn, Chap. 6, Butterworths, London (1969).
15. De Swaan Arons, J. *Fluid Phase Eqba*. **52**, 319 (1989).
16. Streett, W.B. *Canad. J. Chem. Eng*. **52**, 92 (1974); *Icarus* **29**, 173 (1976).
17. Koningsveld, R., Stockmayer, W.H., and Nies, E. *Polymer Phase Diagrams*, Oxford University Press, Oxford (2001).
18. Smith, J.M., Van Ness, H.C., and Abbott, M.M. *Introduction to Chemical Engineering Thermodynamics*, 7th edn, § 16.6, McGraw-Hill, New York (2005).
19. Van Konynenburg, P.H. Ph.D. dissertation, University of California, Los Angeles (1968).
20. Van Konynenburg, P.H., and Scott, R.L. *Phil. Trans. Roy. Soc. London A* **298**, 495 (1980).
21. Dieters, U., and Schneider, G.M. *Ber. Bunsenges. Phys. Chem*. **80**, 1316 (1976).
22. Lebowitz, J.L., and Rowlinson, J.S. *J. Chem. Phys*. **41**, 133 (1964).
23. Gibbons, R.M. *Mol. Phys*. **18**, 809 (1970).
24. McDonald, I.R. *Statistical Mechanics*, Vol. 1, Chap. 3, Chemical Society, London (1973).

25. Henderson, D., and Leonard, P.J. in *Physical Chemistry An Advanced Treatise*, Vol. VIIIB (ed. H. Eyring, D. Henderson, and W. Jost), Academic, New York (1971).
26. Hicks, C.P., and Young, C.L. *Chem. Rev.* **75**, 119 (1975); Neff, R.O., and McQuarrie, D.A. *J. Phys. Chem.* **79**, 1022 (1975).
27. Leonard, P.J., Henderson, D., and Barker, J.A. *Trans. Faraday Soc.* **66**, 2439 (1970).
28. Twu, C.H., Gubbins, K.E., and Gray, C.G. *Mol. Phys.* **29**, 713 (1975); Flytzani-Stephanopoulos, M., Gubbins, K.E., and Gray, C.G. *Mol. Phys.* **30**, 1481 (1975).
29. Melnyk, T.W., and Smith, W.R. *Chem. Phys. Lett.* **28**, 213 (1974).
30. Gubbins, K.E., and Twu, C.H. *Chem. Eng. Sci.* **33**, 863 (1978); ibid. **33**, 879 (1978); Twu, C.H., Gubbins, K.E., and Gray, C.G. *J. Chem. Phys.* **64**, 5186 (1976).
31. An azeotrope point is distinguished from a critical point by the fact that the coexisting phases are distinctly different from each other, e.g. their density and other intensive properties (other than mole fraction) are different, whereas at a critical point the coexisting phases become identical.
32. Kuenen, J.P., and Robson, W.G. *Phil. Mag.* **48**, 180 (1899).
33. Gibbs, J.W. *Elementary Principles of Statistical Mechanics*, eq. (540), Scribner, New York (1902).
34. Kirkwood, J.G., and Buff, F.P. *J. Chem. Phys.* **19**, 774 (1951); Buff, F.P., and Brout, R. *J. Chem. Phys.* **23**, 458 (1955). Rather detailed expositions of the KBT are given in Ben-Naim, A. *Molecular Theory of Solutions*, Ch. 4, Oxford University Press, Oxford (2006); Ben-Naim, A. *Statistical Thermodynamics for Chemists and Biochemists*, Ch. 6, Plenum Press, New York (1992); see also, Ben-Naim, A. *Water and Aqueous Solutions*, Ch. 4, Plenum, New York (1974); Hill, T.L. *Statistical Mechanics*, p. 113 ff, McGraw-Hill, New York (1956); and O'Connell, J.P. *Mol. Phys.* **20**, 27 (1971)..
35. Some of the relations developed in this section are also fundamental to the theory of light scattering from mixtures. For a brief discussion see Hill, T.L. *Statistical Mechanics*, pp. 115–121, McGraw-Hill, New York (1956). The first rigorous treatment of light scattering seems to have been given by Zernike in his dissertation: Zernike, F., *L'Opalescence Critique*, dissertation, Amsterdam (1915); see also *Archives Neerland (3A)* **4**, 74 (1918).
36. The McMillan–Mayer theory (McMillan, W.G. and Mayer, J.E. *J. Chem. Phys.* **13**, 276 (1945)) preceded the KBT by six years, and is formally equivalent to it. However, McMillan–Mayer theory employs an expansion in the solute fugacity that closely parallels the virial expansion for the pressure of a gas. Such an approach is convenient for dilute solutions, but is difficult to use for concentrated ones. Starting from KBT it is straightforward to derive the McMillan–Mayer expansion, as shown in § 7.5.
37. For pure fluids H defined by (7.7) is the same (within a factor ρ) as the quantity h_0 in Appendix 3E, defined by (3E.6). Similarly, C defined by (7.53) is the same as c_0/ρ of (3E.9). The notation $H_{\alpha\beta}$ and $C_{\alpha\beta}$ used here is more convenient for our present purposes.
38. Ben-Naim, A. in *Solutions and Solubilities, Part I* (ed. M.R. Dack), Ch. II, Wiley, New York (1975).
39. The Gibbs–Duhem equation is derived as follows. The Gibbs free energy G is, at constant T and p, a homogeneous function of degree 1 in the molecular numbers $\langle N_A\rangle, \ldots$, i.e. $G(k\langle N_A\rangle, k\langle N_B\rangle, \ldots k\langle N_R\rangle) = kG(\langle N_A\rangle, \langle N_B\rangle, \ldots \langle N_R\rangle)$, where k is a constant. It follows from Euler's theorem that

$$G = \sum_\alpha \langle N_\alpha\rangle \mu_\alpha, \tag{1}$$

where $\mu_\alpha = (\partial G/\partial\langle N_\alpha\rangle)_{Tp\langle\mathbf{N}'\rangle}$ is the chemical potential. Differentiation gives

$$\mathrm{d}G = \sum_\alpha \langle N_\alpha\rangle \mathrm{d}\mu_\alpha + \sum_\alpha \mu_\alpha \mathrm{d}\langle N_\alpha\rangle. \tag{2}$$

But from Gibbs' thermodynamics we also have

$$dG = -SdT + Vdp + \sum_{\alpha} \mu_{\alpha} d\langle N_{\alpha}\rangle. \quad (3)$$

At fixed T and p, Eqs. (2) and (3) give, after dividing by V,

$$\sum_{\alpha} \rho_{\alpha} d\mu_{\alpha} = 0, \quad (4)$$

which is one form of the Gibbs–Duhem equation. Equation (7.20) is obtained by dividing this expression by dN_{β}, keeping T, p, and $\mathbf{N}'$ fixed; similarly, to obtain (7.34) we divide (4) by $d\rho_{\beta}$ at constant T and p.

40. As for G (see note above), the volume is a homogeneous function of degree 1 in the $\langle N_{\alpha}\rangle$. Euler's theorem thus gives $V = \sum_{\alpha} \langle N_{\alpha}\rangle \bar{V}_{\alpha}$, where $\bar{V}_{\alpha} = (\partial V/\partial\langle N_{\alpha}\rangle)_{Tp\langle\mathbf{N}'\rangle}$.

41. O'Connell, J.P. *Mol. Phys.* **20**, 27 (1971).
42. O'Connell, J.P. in *Fluctuation Theory of Mixtures* (ed. E. Matteoli and G.A. Mansoori), p. 45, Taylor and Francis, New York (1990).
43. O'Connell, J.P. in *Supercritical Fluids* (ed. E. Kiran and J.M.H. Levelt Sengers), p. 191, Kluwer Academic, Dordrecht (1994).
44. O'Connell, J.P. in *Supercritical Fluids: Fundamentals for Application* (ed. E. Kiran and J.M.H. Levelt Sengers), Kluwer Academic, Dordrecht (1994).
45. Lee, L.L., Debenedetti, P.G., and Cochran, H.D. in *Supercritical Fluid Technology: Reviews in Modern Theory and Applications* (ed. T.J. Bruno and J.F. Ely), CRC Press, Boca Raton (1991).
46. Cochran, H.D., Lee, L.L., and Pfund, D.M. in *Fluctuation Theory of Mixtures* (ed. E. Matteoli and G.A. Mansoori), p. 69, Taylor and Francis, New York (1990).
47. Ananth, M.S., Gubbins, K.E., and Gray, C.G. *Mol. Phys.* **28**, 1005 (1974).
48. Gubbins, K.E., and O'Connell, J.P. *J. Chem. Phys.* **60**, 3449 (1974).
49. Brelvi, S.W., and O'Connell, J.P. *American Inst. Chem. Engr. J.* **18**, 1239 (1972); ibid. **21**, 171 (1975); ibid. **21**, 1024 (1975).
50. Huang, Y.-H., and O'Connell, J.P. *Fluid Phase Eqba.* **37**, 75 (1987).
51. Ben-Naim, A. *Statistical Thermodynamics for Chemists and Biochemists*, Ch. 6, Plenum Press, New York (1992); also, Ben-Naim, A. *Molecular Theory of Solutions*, § 4.4., Oxford University Press, Oxford (2006).
52. Lepori, L., Matteoli, E., Hamad, E.Z., and Mansoori, G.A. in *Fluctuation Theory of Mixtures* (ed. E. Matteoli and G.A. Mansoori), p. 175, Taylor and Francis, New York (1990).
53. Ben-Naim, A. in *Fluctuation Theory of Mixtures* (ed. E. Matteoli and G.A. Mansoori), p. 211, Taylor and Francis, New York (1990).
54. Matteoli, E., and Lepori, L. in *Fluctuation Theory of Mixtures* (ed. E. Matteoli and G.A. Mansoori), p. 259, Taylor and Francis, New York (1990).
55. Kato, T. *J. Phys. Chem.* **88**, 1248 (1984).
56. Wooley, R.J., and O'Connell, J.P. *Fluid Phase Eqba.* **66**, 233 (1991).
57. Rowlinson, J.S., and Swinton, F.L., *Liquids and Liquid Mixtures*, 3rd edn, p. 92, Butterworth, London (1982).

58. Although the total partition function, Q_{cl}, Eq. (3.77) of Vol. 1 is dimensionless, as it must be, the factored parts of the partition function, such as Q_c, do have dimensions, and this leads to the logarithm of a dimensional density in Eq. (7.74). A similar situation occurs for other thermodynamic properties, such as the chemical potential. The final result for the total Helmholtz energy and other properties is, however, dimensionally consistent.

59. O'Connell, J.P. in *Fluctutation Theory of Mixtures* (ed. E. Matteoli and G.A. Mansoori), p. 50, Taylor and Francis, New York (1990).
60. Ben-Naim, A. *Statistical Thermodynamics for Chemists and Biochemists*, p. 387, Plenum Press, New York (1992).
61. Prausnitz, J.M., Lichtenthaler, R.N., and Gomes de Azevedo, E. *Molecular Thermodynamics of Fluid-Phase Equilibria*, 3rd edn, pp. 226–7, Prentice-Hall, Englewood Cliffs (1999).
62. For example, see Matteoli, E., and Mansoori, G.A., eds, *Fluctutation Theory of Mixtures*, p. 50, Taylor and Francis, New York (1990); Perry, R.L. *Mol. Phys*. **52**, 137 (1984); Perry, R.L., Cabezas, H., Jr., and O'Connell, J.P. *Mol. Phys*. **63**, 189 (1988); Wooley, R.J., and O'Connell, J.P. *Fluid Phase Eqba*. **66**, 233 (1991); Kiran, E., and Levelt Sengers, J.M.H., eds, *Supercritical Fluids: Fundamentals for Application*, Kluwer Academic, Dordrecht (1994)—see papers by J.M.H. Levelt Sengers and J.P. O'Connell; Lee, L.L., Debenedetti, P.G., and Cochran, H.D. in *Supercritical Fluid Technology: Reviews in Modern Theory and Applications* (ed. T.J. Bruno and J.F. Ely), CRC Press, Boca Raton (1991); Chialvo, A.A., and Cummings, P.T. *Amer. Inst. Chem. Engr. J*. **40**, 1558 (1994); Chialvo, A.A., Kalyuzhnyi, Yu V., and Cummings, P.T. *Amer. Inst. Chem. Engr. J*. **42**, 571 (1996); Kwon, Y.J., and Mansoori, G.A. *J. Supercritical Fluids* **6**, 173 (1993); Hamad, E.Z., and Mansoori, G.A. *J. Phys. Chem*. **94**, 3148 (1990).
63. In deriving this we make use of the definitions of the pure ideal gas and the mixture of ideal gases (see, e.g. Denbigh, K. *The Principles of Chemical Equilibrium*, Ch. 3. Cambridge University Press, London (1961)):

$$\text{Pure gas}: \quad \mu_\alpha^{\text{id}} = \mu_\alpha^{\text{g}} + kT\ln p$$

$$\text{Mixture of gases}: \quad \mu_\alpha^{\text{id}} = \mu_\alpha^{\text{g}} + kT\ln p_\alpha,$$

where μ_α^g is the chemical potential of pure ideal gas α at some standard pressure, equal to unity in the standard units. (See ref. 58 regarding the logarithm of dimensional quantities.)
64. Equation (7.108) is derived as follows. From the thermodynamic identities $p = -(\partial A/\partial V)_{T,N}$ and $\mu_\alpha = (\partial A/\partial N_\alpha)_{T,V,N'}$ we have

$$\left(\frac{\partial \mu_\alpha}{\partial V}\right)_{T,N} = kT\left(\frac{\partial \ln f_\alpha}{\partial V}\right)_{T,N} = \frac{\partial}{\partial V}\frac{\partial A}{\partial N_\alpha} = \frac{\partial}{\partial N_\alpha}\frac{\partial A}{\partial V} = -\left(\frac{\partial p}{\partial N_\alpha}\right)_{T,V,N'}, \tag{1}$$

where the relation $\mathrm{d}\mu_\alpha = kT\mathrm{d}\ln f_\alpha$ was used. Also, since $\phi_\alpha = f_\alpha/x_\alpha p$,

$$\left(\frac{\partial \ln \phi_\alpha}{\partial V}\right)_{T,N} = \left(\frac{\partial \ln f_\alpha}{\partial V}\right)_{T,N} - \left(\frac{\partial \ln p}{\partial V}\right)_{T,N} = -\frac{1}{kT}\left(\frac{\partial p}{\partial N_\alpha}\right)_{T,V,N'} - \frac{1}{p}\left(\frac{\partial p}{\partial V}\right)_{T,N}. \tag{2}$$

Integrating (2) from $V = \infty$ to $V = V$ gives

$$kT\ln\phi_\alpha = -\int_\infty^V \left[\left(\frac{\partial p}{\partial N_\alpha}\right)_{T,V,N'} + \frac{kT}{p}\left(\frac{\partial p}{\partial V}\right)_{T,N}\right]\mathrm{d}V. \tag{3}$$

Since $p = NZkT/V = \rho kTZ$, where $Z = pV/NkT$ is the compressibility factor

$$\frac{kT}{p}\left(\frac{\partial p}{\partial V}\right) = -\frac{kT}{V} + kT\left(\frac{\partial \ln Z}{\partial V}\right). \tag{4}$$

Substituting (4) into (3) gives (7.108).

65. This formula can be obtained by expressing V as a function of T, p, N_A, N_B, ... Keeping all variables fixed except p and N_α, so that $V = f(p, N_\alpha)$ and

$$\mathrm{d}V = \left(\frac{\partial V}{\partial p}\right)_{T,N} \mathrm{d}p + \left(\frac{\partial V}{\partial N_\alpha}\right)_{T,p,N'} \mathrm{d}N_\alpha.$$

For a change at fixed V

$$0 = \left(\frac{\partial V}{\partial p}\right)_{T,N} \left(\frac{\partial p}{\partial N_\alpha}\right)_{T,V,N'} + \left(\frac{\partial V}{\partial N_\alpha}\right)_{T,p,N'}$$

Or

$$\left(\frac{\partial V}{\partial p}\right)_{T,N} \left(\frac{\partial p}{\partial N_\alpha}\right)_{T,V,N'} \left(\frac{\partial N_\alpha}{\partial V}\right)_{T,p,N'} = -1,$$

which is (7.109).
66. Chialvo, A.A., and Cummings, P.T. *Amer. Inst. Chem. Engr. J.* **40**, 1558 (1994).
67. Chialvo, A.A., Kalyuzhnyi, Yu V., and Cummings, P.T. *Amer. Inst. Chem. Engr. J.* **42**, 571 (1996).
68. Shing, K.S., Gubbins, K.E., and Lucas, K. *Mol. Phys.* **65**, 1235 (1988).
69. Reed, T.M., and Gubbins, K.E. *Applied Statistical Mechanics*, Ch. 11, McGraw-Hill, New York (1973).
70. Longuet-Higgins, H.C. *Proc. Roy. Soc.* **A205**, 247 (1951).
71. Rowlinson, J.S. *Liquids and Liquid Mixtures*, 2nd edn, Ch. 9, Butterworths, London (1969).
72. McDonald, I.R. in *Statistical Mechanics* (ed. K. Singer), Vol. 1, p. 134, Specialist Periodical Reports, Chemical Society, London (1973).
73. Rowlinson, J.S., and Swinton, F.L. *Liquids and Liquid Mixtures*, 3rd edn, p. 288, Butterworth, London (1982). See also Hicks, C.P. *J. Chem. Soc., Faraday Trans. 2* **72**, 423 (1976).
74. The somewhat confusing name *van der Waals theory* arises because the mixing rules used are the same as those used by van der Waals in his theory of mixtures; van der Waals mixing rules were actually quadratic sums over the van der Waals parameters $a_{\alpha\beta}$ and $b_{\alpha\beta}$, but these are equivalent to the vdw1 mixing rules of (7.126). It should be emphasized that this theory is not in any way tied to the van der Waals equation of state, nor is it related to 'van der Waals' theories of pure fluids, which involve this equation of state, or some variation of it.
75. Leland, T.W., Chappelear, P.S., and Gamson, B.W. *Amer. Inst. Chem. Engr. J.* **8**, 482 (1962).
76. Leland, T.W., Rowlinson, J.S., and Sather, G.A. *Trans. Faraday Soc.* **546**, 1447 (1967).
77. Henderson, D., and Leonard, P.J. *Proc. Natl. Acad. Sci. U.S.A.* **67**, 1818 (1970); ibid. **68**, 632 (1971).
78. Smith, W.R. *Canad. J. Chem. Engr.* **50**, 271 (1972).
79. The definition of the phase integral used here for spherical molecules differs from that used in previous chapters for nonspherical molecules, e.g. (3.53) of Vol. 1. Z_c used here differs from that in (3.53) by the factor $\Omega_\alpha^{N_\alpha}$. For spherical molecules this is convenient since the $\Omega_\alpha^{N_\alpha}$ term that results from integration over the orientation coordinates in the phase integral given by (3.53) cancels the $\left(\Omega_\alpha^{N_\alpha}\right)^{-1}$ factor appearing in (3.250).

80. Since the intermolecular interactions are identical for all pairs of molecules there is no volume change on mixing, and hence it makes no difference whether the mixing is done at constant volume or constant pressure; the expression for ΔA_{cx} will be the same for mixing at constant volume or constant pressure.
81. For example: Johnson, J.K., Zollweg, J.A., and Gubbins, K.E. *Mol. Phys*. **78**, 591 (1993); Kolafa, J., and Nezbeda, I. *Fluid Phase Eqba*. **100**, 1 (1994).
82. Reed, T.M., and Gubbins, K.E. *Applied Statistical Mechanics*, Ch. 11, McGraw-Hill, New York (1973); Prausnitz, J.M., Lichtenthaler, R.N., and Gomes de Azevedo, E. *Molecular Thermodynamics of Fluid-Phase Equilibria*, 3rd edn, § 5.7, Prentice-Hall, Englewood Cliffs (1999).
83. Smith, W.R. *Mol. Phys*. **22**, 105 (1971).
84. Mo, K.C., Gubbins, K.E., Jacucci, G., and McDonald, I.R. *Mol. Phys*. **27**, 1173 (1974).
85. Mo, K.C., and Gubbins, K.E. *Mol. Phys*. **31**, 825 (1976).
86. Prigogine, I., Bellemans, A. and Englert-Chwoles, A. *J. Chem. Phys*. **24**, 518 (1956).
87. For multicomponent mixtures it is possible that higher virial coefficients are negative, leading to the possibility of positive V^{E}. See Warren, P.B. *Europhys. Lett*. **46**, 295 (1999).
88. Reid, R.C., and Leland, T.W. *Amer. Inst. Chem. Engr. J*. **11**, 228 (1965).
89. Smith, W.R. in *Statistical Mechanics* (ed. K. Singer), Vol. 1, p. 108, Specialist Periodical Reports, Chemical Society, London (1973).
90. Shing, K.S., and Gubbins, K.E. *Mol. Phys*. **46**, 1109 (1982); Shing, K.S., Gubbins, K.E., and Lucas, K. *Mol. Phys*. **65**, 1235 (1988).
91. Harismiadis, V.I., Koutras, N.K., Tassios, D.P., and Panagiotopoulos, A.Z. *Fluid Phase Eqba*. **65**, 1 (1991).
92. Georgoulaki, A.M., Ntouros, I.V., Tassios, D.P., and Panagiotopoulos, A.Z. *Fluid Phase Eqba*. **100**, 153 (1994).
93. Tsang, P.C., White, O.N., Perigard, B.Y., Vega, L.F., and Panagiotopoulos, A.Z. *Fluid Phase Eqba*. **107**, 31 (1995).
94. Leland, T.W., Chappelear, P.S., and Gamson, B.W. *Amer. Inst. Chem. Engr. J*. **8**, 482 (1962); Leach, J.W., Chappelear, P.S., and Leland, T.W. *Proc. Amer. Petrol. Inst. (Div. Refining)* **46**, 223 (1966).
95. Rowlinson, J.S., and co-workers *Chem. Eng. Sci*. **24**, 1565 (1969); ibid. **24**, 1575 (1969); ibid. **28**, 521 (1973); ibid. **28**, 529 (1973); ibid. **29**, 1373 (1974).
96. Mollerup, J. *Adv. Cryogenic Engng*. **23**, 550 (1978).
97. Haile, J.M., Mo, K.C., and Gubbins, K.E. *Adv. Cryogenic Engng*. **21**, 501 (1976).
98. Murad, S., and Gubbins, K.E. *Chem. Eng. Sci*. **32**, 499 (1977); *Amer. Inst. Chem. Eng. J*. **27**, 864 (1981).
99. Hanley, H.J.M. *Amer. Chem. Soc. Symp. Series* **60**, 330 (1977).
100. Hanley, H.J.M., Gubbins, K.E., and Murad, S. *J. Phys. Chem. Ref. Data* **6**, 1167 (1977).
101. Mansoori, G.A., and Leland, T.W. *J. Chem. Phys*. **53**, 1931 (1970); Mansoori, G.A. ibid. **56**, 5335 (1972).
102. Leonard, P.J., Henderson, D. and Barker, J.A. *Trans. Faraday Soc*. **66**, 2439 (1970).
103. Lee, L.L. and Levesque, D. *Mol. Phys*. **26**, 1351 (1973).
104. Gubbins, K.E., Smith, W.R., Tham, M.K. and Tiepel, E.W. *Mol. Phys*. **22**, 1089 (1971).
105. Sung, S. and Chandler, D. *J. Chem. Phys*. **56**, 4989 (1972).
106. Mansoori, G.A., Carnahan, N.F., Starling, K.E. and Leland, T.W., Jr. *J. Chem. Phys*. **54**, 1523 (1971). See also Boublik, T. *J. Chem.. Phys*. **53**, 471.
107. Carnahan, N.F. and Starling, K.E. *J. Chem. Phys*. **51**, 635 (1969).

108. McDonald, I.R. *Chem. Phys. Lett.* **3**, 241 (1969); McDonald, I.R., and Singer, K. *Quart. Rev. Chem. Soc. London* **24**, 38 (1970); McDonald, I.R. *Mol. Phys.* **23**, 41 (1972).
109. Jackson, G., Rowlinson, J.S., and van Swol, F. *J. Phys. Chem.* **91**, 4907 (1987).
110. Rogers, B.L., and Prausnitz, J.M. *Trans. Faraday Soc.* **67**, 588 (1971).
111. Flytzani-Stephanopoulos, M., Gubbins, K.E., and Gray, C.G. *Mol. Phys.* **30**, 1649 (1975).
112. Twu, C.H., Gubbins, K.E., and Gray, C.G. *J. Chem. Phys.* **64**, 5186 (1976).
113. Gubbins, K.E., and Twu, C.H. *Chem. Eng. Sci.* **33**, 863 (1978); ibid. **33**, 879 (1978).
114. Gray, C.G., Gubbins, K.E., and Twu, C.H. *J. Chem. Phys.* **69**, 182 (1978).
115. De Leeuw, S.W., Smit, B., and Williams, C.P. *J. Chem. Phys.* **93**, 2704 (1993).
116. Jiang, S., and Pitzer, K.S. *J. Chem. Phys.* **102**, 7632 (1995).
117. Gosman, A.L., McCarty, R.D., and Hust, J.G. *Nat. Stand. Ref. Data Ser. Natl. Bur. Stand.* **27** (1969).
118. Rowlinson, J.S., and Swinton, F.L. *Liquids and Liquid Mixtures*, 3rd edn, Butterworths, London (1981).
119. A positive azeotrope is one in which the component vapor pressures (or fugacities) are greater than those predicted by Raoult's law, (7.93). (See also § 7.1.2.)
120. Schneider, G.M. *Adv. Chem. Phys.* **17**, 1 (1970); Streett, W.B. *Canad. J. Chem. Engng.* **52**, 92 (1974).
121. Rowlinson, J.S. *Liquids and Liquid Mixtures*, 2nd edn, Butterworths, London (1969).
122. Calado, J.C.G., Gray, C.G., Gubbins, K.E., Palavra, A.M.F., Soares, V.A.M., Staveley, L.A.K., and Twu, C.H. *J. Chem. Soc., Faraday Trans. I* **74**, 893 (1978).
123. Moser, B., Lucas, K., and Gubbins, K.E. *Fluid Phase Eqba.* **7**, 153 (1981).
124. Lobo, L.Q., Staveley, L.A.K., Clancy, P., and Gubbins, K.E. *J. Chem. Soc., Faraday Trans. I* **76**, 174 (1980).
125. Gray, C.G., Joslin, C.G., Venkatasubramanian, V., and Gubbins, K.E. *Mol. Phys.* **54**, 1129 (1985).
126. Clancy, P., and Gubbins, K.E. *Faraday Disc. Chem. Soc. London* **66**, 116 (1978).
127. Lucas, K. in *Proc. 8th Symposium on Thermophysical Props.*, p. 134, American Soc. Mech. Engrs., New York (1982).
128. Shukla, K.P., Lucas, K., and Moser, B. *Fluid Phase Eqba.* **17**, 19 (1984).
129. Twu, C.H., and Gubbins, K.E. *Chem. Eng. Sci.* **33**, 879 (1978).
130. Calado, J.C.G., Gomes de Azevedo, E.J.S., Clancy, P., and Gubbins, K.E. *J. Chem. Soc., Faraday Trans. I* **79**, 2657 (1983).
131. Gubbins, K.E., Gray, C.G., and Machado, J.S. *Mol. Phys.* **42**, 817 (1981).
132. Machado, J.R.S., Gubbins, K.E., Lobo, L.Q., and Staveley, L.A.K. *J. Chem. Soc., Faraday Trans. I* **76**, 2496 (1980).
133. Lobo, L.Q., McClure, D.W., Staveley, L.A.K., Clancy, P., Gubbins, K.E., and Gray, C.G. *J. Chem. Soc., Faraday Trans. 2* **77**, 425 (1981).
134. Venkatasubramanian, V., Gubbins, K.E., Gray, C.G., and Joslin, C.G. *Mol. Phys.* **52**, 1411 (1984).
135. Clancy, P., and Gubbins, K.E. *Mol. Phys.* **44**, 581 (1981).
136. Tsang, C.Y., Clancy, P., Calado, J.C.G., and Streett, W.B. *Chem. Engng. Communicn.* **6**, 365 (1980).
137. Chokappa, D., Clancy, P., Streett, W.B., Deiters, U.K., and Heintz, A. *Chem. Engng. Sci.* **40**, 1831 (1985).
138. Lobo, L.Q., Staveley, L.A.K., Clancy, P., Gubbins, K.E., and Machado, J.R.S. *J. Chem. Soc., Faraday Trans. 2* **79**, 1399 (1983).
139. Lobo, L.Q., Staveley, L.A.K., Venkatasubramanian, V., Clancy, P., Gubbins, K.E., Gray, C.G., and Joslin, C.G. *Fluid Phase Eqba.* **22**, 89 (1985).

140. Singh, Y., and Shukla, K.P. in Haile, J.M., and Mansoori, G.A., eds, *Molecular-Based Study of Fluids*, ACS Adv. Chem. Series, **204**, 365 (1983).
141. Shukla, K.P., and Y. Singh *J. Chem. Phys.* **72**, 2719 (1980).
142. Gibbs, G.M. *Chemical Thermodynamic Data on Fluids and Fluid Mixtures*, NPL Conference, 149, IPC Scientific and Technical Press, Guildford (1979); Ph.D. Thesis, Oxford (1979).
143. Winkelmann, J. *Fluid Phase Eqba.* **11**, 207 (1983).
144. Winkelmann, J. *Z. Phys. Chem. Leipzig* **266**, 257 (1985).
145. Vonau, W., Solowejtschik, W., and Winkelmann, J. *Z. Phys. Chem. Leipzig* **269**, 472 (1988).
146. Kohler, F. *Monatsh. Chem.* **88**, 857 (1957).
147. Fischer, J., and Lago, S. *J. Chem. Phys.* **78**, 5750 (1983).
148. Fischer, J., and Kohler, F. *Fluid Phase Eqba.* **14**, 177 (1983).
149. Bohn, M., Lago, S., Fischer, J., and Kohler, F. *Fluid Phase Eqba.* **23**, 137 (1985).
150. Bohn, M., Fischer, J., and Kohler, F. *Fluid Phase Eqba.* **31**, 233 (1986).
151. Bohn, M., Lustig, R., Fischer, J., and Kohler, F. *Mol. Phys.* **64**, 595 (1988).
152. Almeida, B., Bohn, M., and Fischer, J. *Fluid Phase Eqba.* **45**, 217 (1989).
153. Haslam, A.J., Galindo, A., and Jackson, G. *Fluid Phase Eqba.* **266**, 105 (2008).
154. Lago, S., and Boublik, T. *Coll. Czech. Chem. Commun.* **45**, 3051 (1980).
155. Pavlicek, J., and Boublik, T. *Fluid Phase Eqba.* **7**, 15 (1981). The surface area $S_{\alpha+s+\beta}$ in eqn. (7.177) depends on s since it is the area of the two parallel convex bodies (parallel to the cores) that represent the molecules.
156. Vega, C., Lago, S., Pospisil, R., Labik, S., and Malijevsky, A. *J. Phys. Chem.* **96**, 1895 (1992).
157. Pavlicek, J., Aim, K., and Boublik, T. *Fluid Phase Eqba.* **91**, 203 (1993).
158. Aim, K., Pavlicek, J., and Boublik, T. *Fluid Phase Eqba.* **116**, 473 (1996).
159. Barker, J.A., and Henderson, D.J. *J. Chem. Phys.* **47**, 4714 (1967).
160. Enciso, E., and Lombardero, M. *Mol. Phys.* **44**, 725 (1981).
161. Tildesley, D.J., Enciso, E., and Sevilla, P. *Chem. Phys. Lett.* **100**, 508 (1983).
162. Rowlinson, J.S., and Swinton, F.L. *Liquids and Liquid Mixtures*, 3rd edn, §§ 5.7 to 5.9, Butterworth, London (1982).
163. Smith, J.M., Van Ness, H.C., and Abbott, M.M. *Introduction to Chemical Engineering Thermodynamics*, 5th edn, § 10.11, McGraw-Hill, New York (1996).
164. Prausnitz, J.M., Lichtenthaler, R.N., and Gomes de Azevedo, E. *Molecular Thermodynamics of Fluid Phase Equilibria*, 3rd edn, ref. 61, pp. 78–93, 174–179, 354–378, 619–629.
165. Wertheim, M.S. *J. Stat. Phys.* **35**, 19 (1984); ibid. **35**, 35 (1984); ibid. **42**, 459 (1986); ibid. **42**, 477 (1986).
166. Sear, R.P., and Jackson, G. in *Observation, Prediction and Simulation of Phase Transitions in Complex Fluids* (ed. M. Baus, L.F. Rull and J.-P. Ryckaert), p. 625, Kluwer Academic, Dordrecht (1995).
167. Tan, S.P., Adidharma, H., and Radosz, M., *Ind. Eng. Chem. Research*, **47**, 8063 (2008).
168. Wei, Y.S., and Sadus, R. *Amer. Inst. Chem Eng. J.*, **46**, 169 (2000).
169. Müller, E.A., and Gubbins, K.E. *Ind. Engng. Chem. Research*, **40**, 2193 (2001).
170. Economou, I.G. *Ind. Eng. Chem. Res.***41**, 953 (2002).
171. Joslin, C.G., Gray, C.G., Chapman, W.G., and Gubbins, K.E. *Mol. Phys.* **62**, 843 (1987).
172. Johnson, J.K., and Gubbins, K.E. *Mol. Phys.* **77**, 1033 (1992).
173. Noles, J.R., and Zollweg, J.A. *Fluid Phase Eqba.* **66**, 275 (1991).
174. Davies, L.A., Jackson, G., and Rull, L.F. *Phys. Rev. Lett.* **82**, 5285 (1999).
175. Davies, L.A., Jackson, G., and Rull, L.F. *Phys. Rev. E* **61**, 2245 (2000).

176. Kraska, T. *Ber. Bunsenges. Phys. Chem.* **100**, 1318 (1996).
177. Yelash, L.V., and Kraska, T. *Phys. Chem. Chem. Phys.* **1**, 4315 (1999).
178. Nezbeda, I., Pavlíček, J., Kolafa, J., Galindo, A., and Jackson, G. *Fluid Phase Eqba.* **158–160**, 193 (1999).
179. Imre, A.R., Kraska, T., and Yelash, L.V. *Phys. Chem. Chem. Phys.* **4**, 992 (2002).
180. Tan, S.P., and Radosz, M. *Ind. Eng. Chem. Res.* **41**, 3722 (2002).
181. Tan, S.P., and Radosz, M. *Ind. Eng. Chem. Res.* **41**, 5848 (2002).
182. Although the experimentally known mixtures of this class are composed of components in which association is known to be important, several authors have shown that with appropriate intermolecular potential parameters class VI can also be predicted to occur in Lennard–Jones mixtures. (See: Boshkov, L.Z., and Mazur, V.A. *Russian J. Phys. Chem.* **60**, 16 (1986); Boshkov, L.Z. *Dokl. Akad. Nauk SSSR* **294**, 901 (1987); Lopes, J.N.C. *Molec. Phys.* **96**, 1649 (1999)). Some of these calculations are open to question,[189] because the empirical Lennard–Jones equation of state that was used was applied at temperatures below the temperature range over which the parameters were fitted; in addition, at the lower temperatures near the lower critical solution point, the predicted liquid phases are metastable with respect to the crystalline phase. (See Jackson, G. *Mol. Phys.* **72**, 1365 (1991)). Although, from a mathematical point of view, class VI mixtures appear to be possible with Lennard–Jones interactions, the parameters required to achieve this seem to be unphysical.
183. Hirschfelder, J., Stevenson, D., and Eyring, H. *J. Chem.. Phys.* **5**, 896 (1937).
184. Barker, J.A., and Fock, W. *Discuss. Faraday Soc.* **15**, 188 (1953).
185. Wheeler, J.C. *Ann. Rev. Phys. Chem.* **28**, 411 (1977).
186. Walker, J.S., and Vause, C.A. *Sci. Amer.* **256**, 98 (1987).
187. Narayann, T., and Kumar, A. *Phys. Rep.* **249**, 135 (1994).
188. Andersen, G.R., and Wheeler, J.C. *J. Chem.. Phys.* **69**, 2082 (1978).
189. Jackson, G. *Mol. Phys.* **72**, 1365 (1991).
190. Green, D.G., and Jackson, G. *J. Chem.. Phys.* **97**, 8672 (1992).
191. Nieves García-Lisbona, M., Galindo, A., Jackson, G., and Burgess, A.N. *Mol. Phys.* **93**, 57 (1998).
192. Nezbeda, I., Pavlíček, J., Kolafa, J., Galindo, A., and Jackson, G. *Fluid Phase Eqba.* **158–160**, 193 (1999).
193. Davies, L.A., Jackson, G., and Rull, L.F. *Phys. Rev. E* **61**, 2245 (2000).
194. Paricaud, P., Galindo, A., and Jackson, G. *Mol. Phys.* **101**, 2575 (2003).
195. Clark, G. N. I., Galindo, A., Jackson, G., Rogers, S., and Burgess, A. N. *Macromolecules* **41**, 6582 (2008).
196. Clements, P.J., Zafar, S., Galindo, A., Jackson, G., and McLure, I. *J. Chem. Soc. Faraday Trans.* **93**, 1331 (1997).
197. Nieves García-Lisbona, M., Galindo, A., Jackson, G., and Burgess, A.N. *J. Amer. Chem. Soc.* **120**, 4191 (1998).
198. Talanquer, V., and Oxtoby, D. *Faraday Disc.* **112**, 91 (1999).
199. Li, X.S., Lu, J.F., Li, Y.G., and Liu, J.C. *Fluid Phase Eqba.* **153**, 215 (1998).
200. Li, X.S., Lu, J.F., and Li, Y.G. *Fluid Phase Eqba.* **168**, 107 (2000).
201. Colina, C., and Gubbins, K.E. *Fluid Phase Eqba.* **194–197**, 553 (2002).
202. Huang, S.H., and Radosz, M. *Ind. Eng. Chem. Res.* **30**, 1994 (1991).
203. Chen, S.S., and Kreglewski, A. *Ber. Bunsen-Ges. Phys. Chem.* **81**, 1048 (1977).
204. Maloney, D.P., and Prausnitz, J.M. *AIChEJ* **22**, 74 (1976).
205. Yu, M.-L., and Chen, Y.-P. *Fluid Phase Eqba.* **94**, 149 (1994).
206. Huang, S.H., and Radosz, M. *Fluid Phase Eqba.* **70**, 33 (1991).
207. Galindo, A., Gil-Villegas, A., Jackson, G., and Burgess, A.N. *J. Phys. Chem. B*, **103**, 10272 (1999).

208. For reviews see Krienke, H., and Barthel, J. *Ionic Fluids*, in Sengers, J.V., Kayser, R.F., Peters, C.J., and White, Jr., H.J., eds, *Equations of State for Fluids and Fluid Mixtures*, Elsevier, Amsterdam (2000); Loehe, J.R., and Donohue, M.D. *Amer. Inst. Chem. Eng. J.* **43**, 180 (1997).
209. Gil-Villegas, A., Galindo, A., and Jackson, G. *Mol. Phys.* **99**, 531 (2001).
210. Paricaud, P., Galindo, A., and Jackson, G. *Fluid Phase Eqba.* **194–197**, 87 (2002).
211. Waisman, E., and Lebowitz, J.L. *J. Chem. Phys.* **52**, 4307 (1970).
212. Lee, L.L. *Molecular Thermodynamics of Non-ideal Fluids*, Butterworths, Boston (1988).
213. Timmermans, J. *The Physicochemical Constants for Binary Systems in Concentrated Solutions*, Interscience, New York, Vol. 3 (1960).
214. Adams, F.W., and Richards, C.W. *Ind. Eng. Chem.* **2**, 470 (1928).
215. Kiran, E., and Levelt Sengers, J.M.H. (ed.), *Supercritical Fluids*, NATO AS1 Series E, Vol. 273, Kluwer, Dordrecht (1994).
216. McHugh, M.A., and Krukonis, V.J., *Supercritical Fluid Extraction*, 2nd edn, Butterworth, New York (1994).
217. Bruno, T.J., and Ely, J.F. (ed.), *Supercritical Fluid Technology: Reviews in Modern Theory and Applications*, CRC Press, Boca Raton (1991).
218. Prausnitz, J.M., Lichtenthaler, R.N., and Gomes de Azevedo, E., *Molecular Thermodynamics of Fluid-Phase Equilibria*, 3rd edn, ref. 61, pp. 191–200.
219. Martinez, J.L. (ed.), *Supercritical Fluid Extraction of Nutraceuticals and Bioactive Compounds*, CRC Press, Boca Raton (2007).
220. Ewald, A.H., Jepson, W.B., and Rowlinson, J.S. *Faraday Soc. Disc.* **15**, 238 (1953), who employed second and third virial coefficients. Earlier workers stopped at the second virial level, e.g., Perkins, A.J. *J. Chem. Phys.* **5**, 180 (1936), and Robin, S. *J. Chim. Phys.* **48**, 501 (1951).
221. Quiram, D.J., O'Connell, J.P., and Cochran, H.D. *J. Supercrit. Fluids* **7**, 159 (1994).
222. Joslin, C.G., Gray, C.G., Goldman, S., Tomberli, B., and Li, W. *Mol. Phys.* **89**, 489 (1996).
223. Harvey, A.H. *Fluid Phase Eqba.* **130**, 87 (1997).
224. Tomberli, B., Goldman, S., and Gray, C.G. *Fluid Phase Eqba.* **187–188**, 111 (2001).
225. Tsonopoulous, C. *AIChEJ* **20**, 263 (1974); ibid. **21**, 827 (1975); ibid. **24**, 1112 (1978).
226. Orbey, H., and Vera, J.H. *AIChEJ* **29**, 107 (1983).
227. Tomberli, B., Goldman, S., Gray, C.G., Saldaña, M.D.A. and Temelli, F. *J. Supercrit. Fluids* **37**, 333 (2006).
228. Saldaña, M.D.A., Tomberli, B., Guigard, S.E., Goldman, S., Gray, C.G., and Temelli, F. *J. Supercrit. Fluids* **40**, 7 (2007).
229. Lamb, D.M., Barbara, T.M., and Jonas, J. *J. Phys. Chem.* **90**, 4210 (1986).

8
SURFACE PROPERTIES

The tension at the surface of a liquid is one of the more striking manifestations of the forces that act between molecules. . . . a direct link between capillary phenomena and the intermolecular forces had to await the development of the statistical mechanics of systems that are highly inhomogeneous on the scale of length of the range of these forces.

J.S. Rowlinson and B. Widom, *Molecular Theory of Capillarity*, Clarendon Press, Oxford (1982)

The fundamental idea of the orientation of . . . molecules in the surface and in the interior of liquids as a factor of vital importance in surface tension and related phenomena occurred to me in nearly its present form in June and July, 1916. . . . Work on oil films began in June. As a result of this extensive work, clear ideas were obtained as to the orientation of molecules in the surface of liquids.

Irving Langmuir, *Journal of the American Chemical Society*, **39**, 1848 (1917)

We have so far been concerned primarily with fluids that are uniform in their density, composition, and other point properties. However, we are surrounded by examples of nonuniform fluids—interfaces between coexisting phases (solid or liquid surfaces, drops, pores, microemulsions, films, bubbles, aerosols, etc.), fluids in the presence of electrical or other external fields, and systems not at equilibrium. For fluids in electrical fields or nonequilibrium fluids the degree of inhomogeneity is often small. In this chapter we concentrate on interfaces, where the inhomogeneity is large and all-important; in the case of liquid–vapour interfaces, for example, the density typically changes by several orders of magnitude over an interfacial layer that is only two or three molecules thick. This inhomogeneity leads to new physical effects, such as the surface tension, adsorption, preferred orientations (alignment) for molecules in the surface layer, and surface phase transitions of various types. The field of interfacial science is very broad. In this chapter we restrict our attention to effects associated with molecular orientation in the interface, and to the influence this has upon other observed properties—surface tension, phase transitions, etc.[1–9] As before, we consider small, neutral, nonspherical molecules. We omit from consideration the interfacial region for fluids of simple spherical molecules, for which a substantial literature now exists,[10–23] and also of many more complex or specialized fluids such as ionic fluids[24–28] (except for §§ 8.9.1 and 8.9.5), polymers,[29–31] liquid metals,[28,32] surfaces where chemical binding is important,[33] bilayer membranes,[34] and biological systems in general,[34] except for a brief discussion of biomembranes in § 8.9.5. We also consider surfaces involving liquid crystals[35] only briefly, in § 8.6.2.

The low-order distribution functions (see Chapter 3), $f(\mathbf{r}_1\omega_1)$, $f(\mathbf{r}_1\mathbf{r}_2\omega_1\omega_2)$, etc., provide a quantitative measure of the degree of molecular alignment in the surface region. The one-body distribution function, $f(\mathbf{r}_1\omega_1)$, which we shall call the *density-orientation profile* of the surface, is of particular importance; this function gives the average number density of molecular centres at position $\mathbf{r}_1$ with molecular orientation ω_1. For a planar surface it reduces to $f(z_1\omega_1)$ where z is the direction perpendicular to the surface. Other, cruder, measures of surface alignment are often used; for linear molecules one often uses order parameters such as the 'polarization' $F_1 \equiv \langle P_1(\cos\ \theta)\rangle$ and 'alignment' $F_2 \equiv \langle P_2(\cos\ \theta)\rangle$ (see p. 169 of Vol. 1), where $\theta \equiv \theta_1$ is the molecular polar angle relative to some space-fixed axis (the z-axis for planar surfaces), or the closely related order parameters used in work on liquid crystals. The polarization F_1 provides a crude measure of molecular alignment for polar molecules, but vanishes for nonpolar ones. For nonpolar molecules the alignment F_2 is nonzero and provides a similar measure of molecular orientation. For homogeneous, isotropic bulk fluids both F_1 and F_2, as well as other order parameters (see §§ 8.1.2, 8.1.3) are zero, while the density-orientation profile is a constant, ρ/Ω (see (3.104)).

In § 8.1 we give several integrodifferential equations for the distribution functions, and also define other measures of alignment; these can form the starting point for theories of surface alignment and structure. Rigorous equations for the surface tension and pressure tensor are derived in § 8.2 and § 8.3, respectively. In order to use these rigorous expressions it is necessary to make approximations. These approximate theories include perturbation theory, integral equation theory, gradient theory, and generalized van der Waals theory, some of which are discussed in §§ 8.5–8.8. Finally we briefly discuss experimental and simulation studies of surface orientation (§ 8.9).

8.1 Distribution functions and molecular alignment

Several measures have been used for molecular distribution and alignment in the surface, including distribution functions, order parameters, and surface potential. We define these quantities in this section, and also derive several exact integrodifferential equations for the distribution functions.

8.1.1 Distribution and correlation functions

We consider an inhomogeneous, classical fluid of rigid molecules at equilibrium. The fluid is subject to an external potential field $\mathcal{V}$, which acts on each molecule independently. As in Chapter 3, we define a configurational distribution function of order h, $f(\mathbf{r}_1\mathbf{r}_2\ldots\mathbf{r}_h;\omega_1\omega_2\ldots\omega_h) \equiv f(\mathbf{r}^h\omega^h)$. For an open system with chemical potential μ, volume V, and temperature T, this distribution function is given by (cf. (3.189))

$$f(\mathbf{r}^h\omega^h) = \frac{1}{\Xi}\sum_{N\geq h}\frac{\hat{z}^N}{(N-h)!}\int \mathrm{d}\mathbf{r}^{N-h}\mathrm{d}\omega^{N-h}\exp[-\beta(\mathcal{U}+\mathcal{V})], \tag{8.1}$$

where $\mathcal{U}(\mathbf{r}^N\omega^N)$ is the intermolecular potential energy, $\mathrm{d}\mathbf{r}^{N-h}\mathrm{d}\omega^{N-h} \equiv \mathrm{d}\mathbf{r}_{h+1}\mathrm{d}\mathbf{r}_{h+2}\ldots\mathrm{d}\mathbf{r}_N\mathrm{d}\omega_{h+1}\mathrm{d}\omega_{h+2}\ldots\mathrm{d}\omega_N$, Ξ is the grand partition function,

$$\Xi = \sum_{N\geq 0}\frac{\hat{z}^N}{N!}\int \mathrm{d}\mathbf{r}^N\mathrm{d}\omega^N\exp[-\beta(\mathcal{U}+\mathcal{V})] \tag{8.2}$$

and $\hat{z}$ is given by[36]

$$\hat{z} = \frac{e^{\beta\mu}q_{qu}}{\Lambda_t^3\Lambda_r\Omega}. \tag{8.3}$$

The external potential $\mathcal{V}$ will, in general, depend on the position and orientation coordinates of all the molecules, and can be written

$$\mathcal{V}(\mathbf{r}^N\omega^N) = \sum_{i=1}^{N} v(\mathbf{r}_i\omega_i), \tag{8.4}$$

where $v(\mathbf{r}_i\,\omega_i)$ is the external potential acting on molecule i. In many cases of interest, e.g. the vapour–liquid interface in the presence of gravity, the external potential will be independent of molecular orientations.

At an interface it is the one-body distribution function, $f(\mathbf{r}_1\omega_1) \equiv f(1)$, that is of greatest interest,

$$f(\mathbf{r}_1\omega_1) = \frac{1}{\Xi}\sum_{N\geq 1}\frac{\hat{z}^N}{(N-1)!}\int d\mathbf{r}^{N-1}d\omega^{N-1}\exp[-\beta(\mathcal{U}+\mathcal{V})]. \tag{8.5}$$

This function is proportional to the probability density for finding a molecule with centre at $\mathbf{r}_1$ and with orientation ω_1, and is usually called the density-orientation profile. It is a generic distribution function (see Vol. 1, p. 190 and p. 160), satisfying the normalization condition $\int \mathrm{d}\mathbf{r}_1\mathrm{d}\omega_1 f(\mathbf{r}_1\omega_1) = \langle N\rangle$ (or N in the canonical ensemble).

For a uniform, isotropic fluid $f(\mathbf{r}_1\omega_1)$ is independent of both $\mathbf{r}_1$ and ω_1, and we have (3.193),

$$f(\mathbf{r}_1\omega_1) = \rho/\Omega,$$

where $\rho = \langle N\rangle/V$ is the number density and Ω is 4π for linear or $8\pi^2$ for nonlinear molecules, respectively. For a nonuniform fluid such as that near a surface, however, $f(\mathbf{r}_1\omega_1)$ depends on both $\mathbf{r}_1$ and ω_1. For a planar gas–liquid surface and linear molecules, $f(\mathbf{r}_1\omega_1)$ reduces to $f(z_1\theta_1)$, which gives the probability density for finding a molecule at height z_1 in the interface with symmetry axis at an angle θ_1 relative to the z-axis. Calculated values of $f(z_1\theta_1)$ are shown in Fig. 8.1 for a diatomic Lennard-Jones (LJ) fluid, with and without the addition of a quadrupole moment.[37] In the absence of a quadrupole or other electrostatic moment (Fig. 8.1(a)) the molecules on the liquid side ($z < 0$) of the *Gibbs dividing surface* (defined formally in § 8.2.1) prefer to lie with their axes perpendicular to the surface ($\theta = 0, \pi$); on the gas side ($z > 0$) there is a weaker preference for the molecules to lie parallel to the surface. When a strong quadrupole is added to the molecules it tends to reverse these trends (Fig. 8.1(b)), so that molecules on the liquid side prefer to lie parallel to the surface, and those on the gas side prefer to lie perpendicular to it.

For bulk fluids it is sometimes useful to introduce[38] potentials of mean force (PMF) corresponding to the distribution functions and correlation functions, e.g. $g(12) \equiv \exp[-\beta w(12)]$, expressing the pair correlation function $g(\mathbf{r}_{12}\omega_1\omega_2)$ in terms of the potential of mean force, $w(\mathbf{r}_{12}\omega_1\omega_2)$, an effective (temperature- and density-dependent) pair potential for the dense fluid (for the dilute gas $w(12)$ reduces to the true pair potential $u(12)$). The name PMF stems from the theorem[38] that the mean force on molecule 1 for a fixed pair (12), i.e. $\langle \boldsymbol{f}_1 \rangle_{12}$, is given by $\langle \boldsymbol{f}_1 \rangle_{12} = -\partial w(12)/\partial \mathbf{r}_1$. PMFs can similarly be usefully introduced for inhomogeneous fluids. Thus, for example, for the density $\rho(\mathbf{r}_1) \equiv f(\mathbf{r}_1) \equiv \int d\omega_1 f(\mathbf{r}_1\omega_1)$ we write $\rho(\mathbf{r}_1) \equiv \rho g(\mathbf{r}_1) \equiv \rho \exp[-\beta w(\mathbf{r}_1)]$. For a planar interface we have $\rho = \rho(z_1)$ for the density profile and $w = w(z_1)$ for the potential of mean force. Here $-\partial w(z_1)/\partial z_1$ is the mean force $\bar{f}_1$ (in the z-direction by symmetry) on a molecule with fixed x_1, y_1, z_1, or just z_1, with all other variables averaged over. The PMF $w(z)$ is a free energy profile.

One finds $w(z)$, e.g. from a simulation, from either a simulation of $\rho(z)$ or a simulation of $\bar{f}(z)$ at a series of z values and then an integration (see ref. 39 for various specific algorithms). Various observables can be obtained directly from $w(z)$, e.g. the adsorption or binding free energy to a surface (if $w(z)$ has a minimum at the interface), and the free energy of transfer of a molecule across the surface, $\Delta\mu_T$, since $\Delta\mu_T = w(z \to -\infty) - w(z \to +\infty)$.

In Fig. 8.2 we show the PMF $w(z)$ for a CH_4 molecule interacting with the vapour–liquid surface of water, determined[40] by the mean force method in a simulation at $T = 300$ K, using a site–site potential model with charge and LJ sites for the H_2O and CH_4 molecules. There is a shallow well at the surface (depth $\sim$ ½ kcal/mol) indicating weak adsorption, and a free energy of transfer $\Delta\mu_T = 2.3 \pm 0.4$ kcal/mol, which is in good agreement with the experimental value $\Delta\mu_T^{\text{exp}} = 2.0$ kcal/mol. In Fig. 8.3 the PMF $w(z)$ for the iodide anion I^- interacting with the vapour–liquid surface of H_2O is shown, derived[41] from simulation by the same method, using potential models with and without polarizability (induction energy) included. With polarizability included in the potential model, $w(z)$ is seen to have a minimum at the interface (well depth $\sim$ 2 kcal/mol). This indicates the presence of excess iodide ion at the surface, contrary to what was once thought to be true; i.e. it was commonly believed that all ions

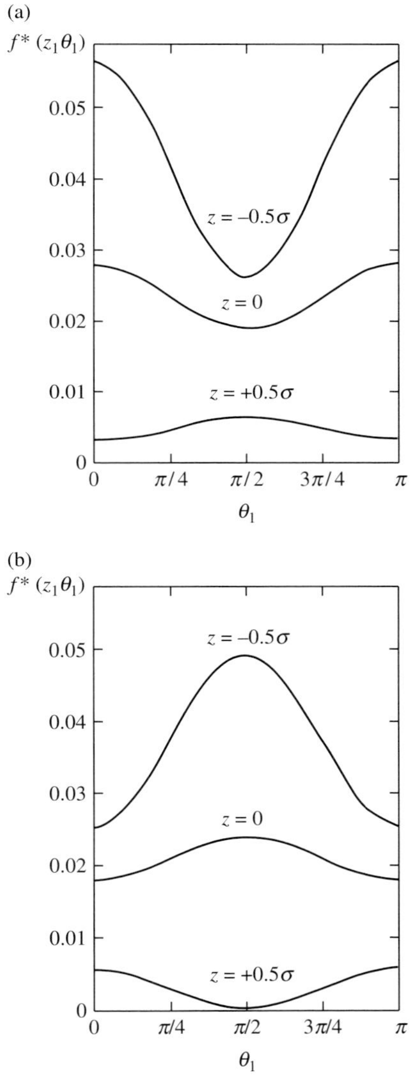

FIG. 8.1 The reduced density–orientation profile, $f^*(z\theta) = f(z\theta)\sigma^3$, for the planar gas–liquid interface for a two-centre LJ fluid with reduced bond length $l^* = l/\sigma = 0.5957$ appropriate for chlorine, at $T^* = kT/\varepsilon = 1.065$, $\rho^* = \rho\sigma^3 = 0.55$ (where ρ is the number density of the liquid phase, and ε, σ are LJ site parameters). The results are calculated from the f-expansion perturbation theory (see § 4.6). In (a) the molecules possess no quadrupole moment, whereas in (b) the molecules have a point quadrupole of strength $Q^* = Q/(\varepsilon\sigma^5)^{1/2} = 1.92$. (From ref. 37.)

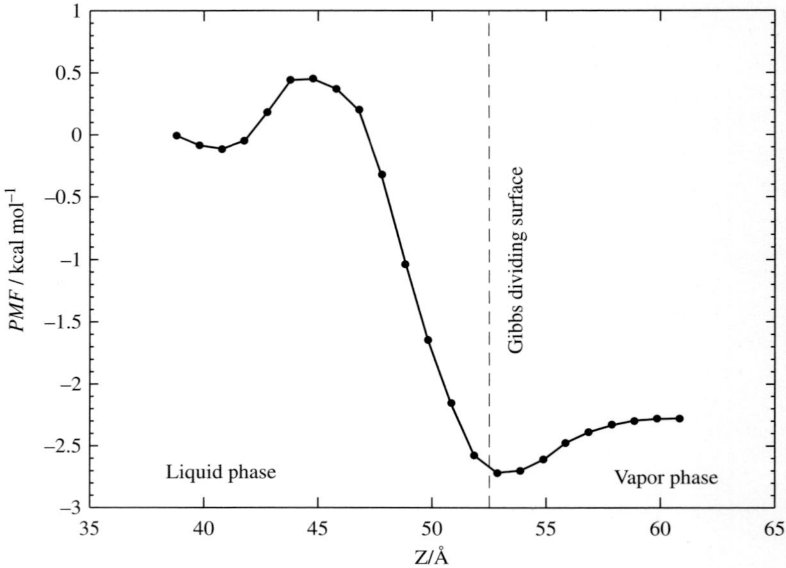

FIG. 8.2 Computed free energy profiles for a methane molecule binding to the liquid–vapour interface of water at 300 K. The Gibbs dividing surface is defined in Sec. 8.2.1. Reprinted with permission from ref. 40. Copyright 2002 American Chemical Society.

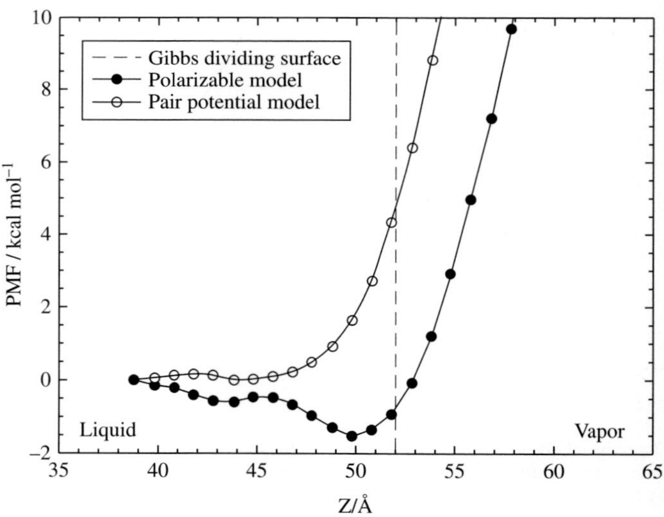

FIG. 8.3 Computed free energy profiles for iodide ion binding to the liquid–vapour interface of water at 300 K. Solid circles, polarizable model; open circles, non-polarizable model. Reprinted with permission from ref. 41. Copyright 2006 American Chemical Society.

congregated in the bulk, where they are better solvated. This surprising result is discussed further in § 8.9.1.

The angle-dependent PMF $w(\mathbf{r}\omega)$ can also be introduced via $f(\mathbf{r}\omega) = (\rho(\mathbf{r})/\Omega)\exp[-\beta w(\mathbf{r}\omega)]$. For example, for a planar water–membrane interface $w(z\omega)$ has been determined by simulation[42] for a small cationic antimicrobial peptide in a 100-mM salt solution interacting with uncharged and charged membranes (see Fig. 8.62 below).

Several rigorous integrodifferential equations for $f(1)$ can be derived.[43] They relate $f(1)$ to either the pair distribution function $f(12)$ or to the (pair) direct correlation function $c(12)$. These equations form a starting point for approximate theories of $f(1)$ in nonuniform fluids. We derive several of the more useful integrodifferential equations in the remainder of this section.

The most familiar equation is the first Yvon-Born-Green (YBG) equation, which is obtained by differentiating both sides of (8.5) with respect to $\mathbf{r}_1$. Assuming that $\mathcal{U}$ is a sum of pair potentials $u(\mathbf{r}_{ij}\omega_i\omega_j)$, Eq. (1.13), this gives

$$\frac{\partial f(1)}{\partial \mathbf{r}_1} = -\frac{\beta}{\Xi}\sum_N \frac{\hat{z}^N}{(N-1)!}\int \mathrm{d}\mathbf{r}^{N-1}\mathrm{d}\omega^{N-1}\left[\sum_{j\geq 2}\left(\frac{\partial u(\mathbf{r}_{1j}\omega_1\omega_j)}{\partial \mathbf{r}_1}\right)\right.$$
$$\left. +\frac{\partial v(1)}{\partial \mathbf{r}_1}\right]\exp(-\beta(\mathcal{U}+\mathcal{V})).$$

Noting that each of the $(N-1)$ terms in the sum over j gives the same result on integration, and the definition (8.1) of $f(12)$, we see that this becomes

$$\frac{\partial \ln f(1)}{\partial \mathbf{r}_1} = -\beta\int \mathrm{d}\mathbf{r}_2\mathrm{d}\omega_2 f(2)g(12)\frac{\partial u(12)}{\partial \mathbf{r}_1} - \beta\frac{\partial v(1)}{\partial \mathbf{r}_1}, \tag{8.6}$$

which is the first YBG equation. Here $g(12)$ is the pair correlation function in the interface defined by (see (3.95))

$$f(12) = f(1)f(2)g(12). \tag{8.7}$$

If the surface is planar and lies in the xy plane, then $\partial f(1)/\partial x = \partial f(1)/\partial y = 0$ and we need only consider the z-component of (8.6). The term $\partial u(12)/\partial z_1$ can be written

$$\frac{\partial u(12)}{\partial z_1} = \frac{\partial u(12)}{\partial r_{12}}\frac{\partial r_{12}}{\partial z_1} = -\frac{\partial u(12)}{\partial r_{12}}\frac{\partial r_{12}}{\partial z_{12}}, \tag{8.8}$$

where $z_{12} = z_2 - z_1$. Since

$$r_{12} = \left(x_{12}^2 + y_{12}^2 + z_{12}^2\right)^{\frac{1}{2}}$$

we have

$$\frac{\partial r_{12}}{\partial z_{12}} = \frac{z_{12}}{r_{12}}. \tag{8.9}$$

From (8.6)–(8.9) the first YBG equation becomes, for the planar interface,

$$\frac{\partial \ln f(1)}{\partial z_1} = \beta \int \mathbf{d r}_{12} \mathrm{d}\omega_2 f(2) g(12) \frac{z_{12}}{r_{12}} \frac{\partial u(12)}{\partial r_{12}} - \beta \frac{\partial v(1)}{\partial z_1}. \tag{8.10}$$

For molecular fluids we can obtain an angular analogue of the YBG equation by differentiating both sides of (8.5) with respect to the molecular orientation. By methods exactly analogous to the derivation of (8.6) we obtain

$$\boldsymbol{\nabla}_{\omega_1} \ln f(1) = -\beta \int \mathbf{d r}_2 \mathrm{d}\omega_2 f(2) g(12) \boldsymbol{\nabla}_{\omega_1} u(12) - \beta \boldsymbol{\nabla}_{\omega_1} v(1), \tag{8.11}$$

where $\boldsymbol{\nabla}_\omega$ is the angular gradient operator, given by (A.18b) of Vol. 1 for linear or (A.71a) for nonlinear molecules, respectively. For linear molecules $\boldsymbol{\nabla}_\omega$ can be written

$$\boldsymbol{\nabla}_\omega = \boldsymbol{\ell} \times \boldsymbol{\nabla}_{\boldsymbol{\ell}} = -\hat{\boldsymbol{\theta}} \frac{1}{\sin\theta} \frac{\partial}{\partial \phi} + \hat{\boldsymbol{\phi}} \frac{\partial}{\partial \theta}, \tag{8.12}$$

where $\boldsymbol{\ell} \equiv \ell \hat{\boldsymbol{\ell}}$ is a vector along the molecular axis (Fig. 8.4), $\boldsymbol{\nabla}_{\boldsymbol{\ell}}$ is the gradient operator with respect to $\boldsymbol{\ell}$, and $\hat{\boldsymbol{\theta}}$ and $\hat{\boldsymbol{\phi}}$ are the usual unit vectors in the directions of increasing θ and ϕ, the angles specifying the molecular (i.e. $\hat{\boldsymbol{\ell}}$) orientation. The unit vectors $\hat{\boldsymbol{\ell}}$, $\hat{\boldsymbol{\theta}}$, and $\hat{\boldsymbol{\phi}}$ form an orthogonal triplet. The last term on the right-hand side of (8.11) vanishes unless the external field depends on molecular orientation. For a homogeneous fluid (with $v = 0$) $f(1)$ is independent of both $\mathbf{r}_1$ and ω_1, and so both sides of (8.6) and (8.11) vanish. If, however, v can approach zero without $\partial f(1)/\partial \mathbf{r}$ and $\boldsymbol{\nabla}_{\omega_1} f(1)$ vanishing everywhere, as in a vapour–liquid interface in the presence of a very weak gravitational field, then (8.6) and (8.11) each provide

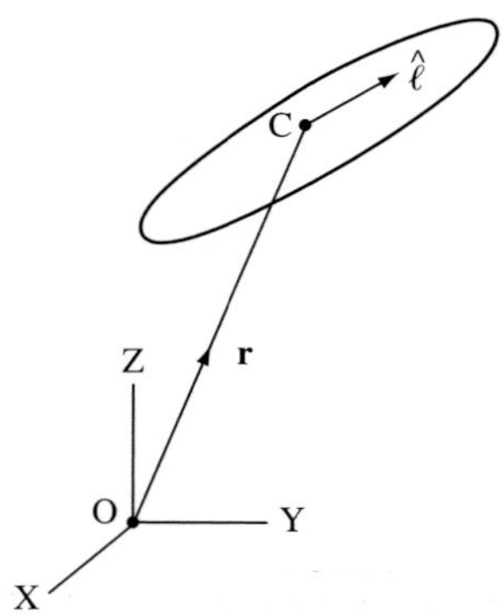

FIG. 8.4 The unit vector $\hat{\boldsymbol{\ell}}$ with orientation $\omega = \theta, \varphi$ specifies the orientation of a linear molecule. C is the molecular center.

useful routes to $f(\mathbf{r}_1\omega_1)$. While (8.6) is a straightforward generalization of the YBG equation to molecular fluids, (8.11) is specific to molecular fluids.

Further integrodifferential equations can be derived by noting that, for a given intermolecular potential energy, $f(1)$ is a unique functional of $v(1)$, and vice versa. The proof of this statement parallels that for the atomic fluid case given by Evans.[12] (The uniqueness theorem for the pair quantities $f(12)$ and $u(12)$ is discussed in § 3.1.7 of Vol. 1.) Thus we have $f(1) = f(\mathbf{r}_1\omega_1) = f(\mathbf{r}_1\omega_1; [v(1)])$; i.e. $f(1)$ is explicitly a function of $\mathbf{r}_1$ and ω_1, and implicitly a functional of $v(1)$. We consider now the behaviour of $f(1)$ under translations of coordinates (which can be interpreted either actively or passively; cf. Appendix A, p. 446 and p. 454 for the analogous discussions of rotation and inversion). First consider the simpler problem of the translational behaviour of a one-dimensional potential $v(x)$. Physically we have

$$v(x + a; [v(x)]) = v(x; [v(x + a)]), \tag{8.13}$$

i.e. the value of the potential at a shifted point $x + a$ in the original potential equals the value of the potential at the original point x in the inversely shifted potential $v(x + a)$; we recall that $v(x + a)$ is the potential shifted to the left by a (see Fig. 8.5). If some quantity f depends exclusively on v we also must have similarly

$$f(x + a; [v(x)]) = f(x; [v(x + a)]). \tag{8.14}$$

If we now take a to be a *small* translation, we can Taylor expand both sides of (8.14); to $O(a)$ we have

$$f(x) + a\frac{\partial f}{\partial x} = f(x) + \int \mathrm{d}x' \left(\frac{\delta f(x)}{\delta v(x')}\right)_{a=0} \left(a\frac{\partial v(x')}{\partial x'}\right),$$

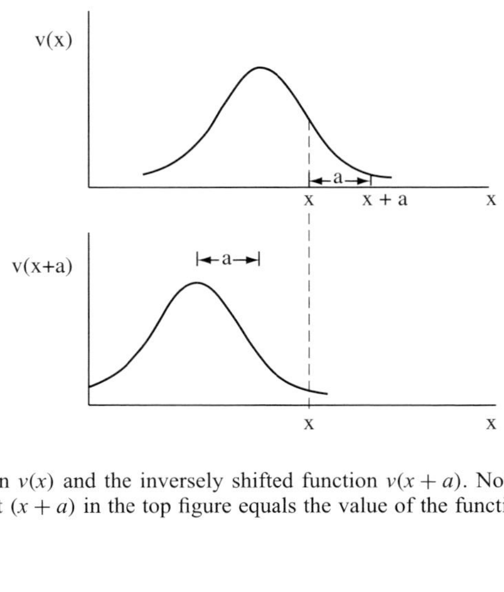

FIG. 8.5 The function $v(x)$ and the inversely shifted function $v(x + a)$. Note that the value of the function at the point $(x + a)$ in the top figure equals the value of the function at the point x in the bottom figure.

where we use a functional Taylor expansion on the right-hand side (see Appendix 8A for a review of functionals, and in particular (8A.17)). Thus we find that

$$\frac{\partial f(x)}{\partial x} = \int \mathrm{d}x' \left(\frac{\delta f(x)}{\delta v(x')}\right)_0 \frac{\partial v(x')}{\partial x'}. \tag{8.15}$$

We now derive the three-dimensional generalization of (8.15) for the real problem of interest. We carry out a translation through vector $\mathbf{s}$, at fixed orientation ω_1, so that the external field $v(\mathbf{r}_1\omega_1) \to v(\mathbf{r}_1 + \mathbf{s}, \omega_1)$. Then $f(\mathbf{r}_1\omega_1)$ must be similarly translated (cf. (8.14))

$$f(\mathbf{r}_1 + \mathbf{s}, \omega_1; [v(\mathbf{r}_1\omega_1)]) = f(\mathbf{r}_1\omega_1; [v(\mathbf{r}_1 + \mathbf{s}, \omega_1)]). \tag{8.16}$$

This follows because although $f(\mathbf{r}_1\omega_1)$ is a functional of both $\mathcal{U}$ and $v(\mathbf{r}_1\omega_1)$, the intermolecular potential energy $\mathcal{U}$ is translationally invariant, so that $f(\mathbf{r}_1\omega_1)$ is only affected by a translation through its functional dependence on $v(\mathbf{r}_1\omega_1)$. Making a functional Taylor expansion of the right-hand side of (8.16) about $\mathbf{s} = 0$ gives

$$f(\mathbf{r}_1+\mathbf{s}, \omega_1) = f(\mathbf{r}_1\omega_1) + \int \mathrm{d}\mathbf{r}_2\mathrm{d}\omega_2 \left(\frac{\delta f(1)}{\delta v(2)}\right)_{s=0} (v(\mathbf{r}_2+\mathbf{s}, \omega_2) - v(\mathbf{r}_2\omega_2)) + \dots. \tag{8.17}$$

Taking the limit as $\mathbf{s} \to 0$, so that $f(\mathbf{r}_1 + \mathbf{s}, \omega_1) - f(\mathbf{r}_1\omega_1) \to (\partial f(1)/\partial \mathbf{r}_1) \cdot \mathbf{s}$ and $v(\mathbf{r}_2 + \mathbf{s}, \omega_2) - v(\mathbf{r}_2\omega_2) \to (\partial v(2)/\partial \mathbf{r}_2) \cdot \mathbf{s}$, and noting that the direction of $\mathbf{s}$ is arbitrary, we see that (8.17) becomes

$$\frac{\partial f(1)}{\partial \mathbf{r}_1} = \int \mathrm{d}\mathbf{r}_2\mathrm{d}\,\omega_2 \left(\frac{\delta f(1)}{\delta v(2)}\right) \frac{\partial v(2)}{\partial \mathbf{r}_2}. \tag{8.18}$$

Using the first Yvon equation (see (8A.29)),

$$h(12) = -\frac{1}{\beta f(1) f(2)} \frac{\delta f(1)}{\delta v(2)} - \frac{\delta(12)}{f(1)} \tag{8.19}$$

in (8.18), with $\delta(12)$ the Dirac delta function (see (8A.25)), gives

$$\frac{\partial \ln f(1)}{\partial \mathbf{r}_1} = -\beta \int \mathrm{d}\mathbf{r}_2\mathrm{d}\,\omega_2 f(2) h(12) \frac{\partial v(2)}{\partial \mathbf{r}_2} - \beta \frac{\partial v(1)}{\partial \mathbf{r}_1}, \tag{8.20}$$

where $h(12) \equiv h(\mathbf{r}_1\mathbf{r}_2\omega_1\omega_2)$ is the total correlation function, defined by (8A.28).

Conversely, we can consider $v(1)$ to be a functional of $f(1)$, so that a displacement $\mathbf{s}$ in the argument of $f(1)$ causes $v(1)$ to be similarly displaced,

$$v(\mathbf{r}_1 + \mathbf{s}, \omega_1; [f(\mathbf{r}_1\omega_1)]) = v(\mathbf{r}_1\omega_1; [f(\mathbf{r}_1 + \mathbf{s}, \omega_1)]). \tag{8.21}$$

Making a functional expansion of the right side about $\mathbf{s} = 0$ and then taking the limit $\mathbf{s} \to 0$ as before gives

$$\frac{\partial v(1)}{\partial \mathbf{r}_1} = \int \mathrm{d}\mathbf{r}_2\mathrm{d}\,\omega_2 \left(\frac{\delta v(1)}{\delta f(2)}\right) \frac{\partial f(2)}{\partial \mathbf{r}_2}. \tag{8.22}$$

The functional derivative in (8.22) can be replaced using the second Yvon equation (see (8A.30)),

$$c(12) = \beta \frac{\delta v(1)}{\delta f(2)} + \frac{\delta(12)}{f(1)}, \tag{8.23}$$

which defines the direct correlation function $c(12) \equiv c(\mathbf{r}_1\mathbf{r}_2\omega_1\omega_2)$. From (8.23) and (8.22) we find, after rearrangement, that

$$\frac{\partial \ln f(1)}{\partial \mathbf{r}_1} = \int \mathrm{d}\mathbf{r}_2 \mathrm{d}\omega_2 c(12) \left(\frac{\partial f(2)}{\partial \mathbf{r}_2} \right) - \beta \frac{\partial v(1)}{\partial \mathbf{r}_1}. \tag{8.24}$$

Equations (8.20) and (8.24) are straightforward generalizations of the corresponding results for atomic fluids.[12,18] If the external field gradient $\partial v/\partial \mathbf{r}$ tends to 0 in (8.20) the fluid must become uniform (i.e. $\partial \ln f(1)/\partial \mathbf{r}_1 \to 0$) unless the integral $\int \mathrm{d}2 f(2)h(12)$ diverges. Wertheim[44] has shown that for a fluid with an interface in a gravitational field mgz, where m is the molecular mass, g is the gravitational acceleration, and z is measured perpendicular to the interface, this integral diverges as g^{-1} in the limit $g \to 0$. This divergence is associated with the growth of very long-ranged correlations in the xy plane. These correlations are attributed to capillary wave fluctuations at the interface.

Equations (8.6), (8.11), (8.20), and (8.24) are integrodifferential equations that can be used as starting points for approximate theories of the density–orientation function, $f(\mathbf{r}_1\omega_1)$. Each of these equations are the first members of hierarchies of coupled integrodifferential equations for the h-body distribution functions, $f(\mathbf{r}^h\omega^h)$. The higher order equations are readily written down by the methods used above; see (3.244) and (3.247) for the generalizations of (8.6) and (8.11). Equations (8.6) and (8.11) assume pairwise additivity, whereas (8.20) and (8.24) do not, and are therefore more general. While (8.6), (8.20), and (8.24) are generalizations of the corresponding atomic fluid equations, (8.11) is specific to molecular fluids. If $f(12)$ and $c(12)$ are known exactly, and if $\mathcal{U}$ is pairwise additive, all four of these equations will give identical results for $f(\mathbf{r}_1\omega_1)$.

It is possible to derive two further integrodifferential equations for $f(1)$ by considering the effect of a rotation Ω on $f(1)$ or $v(1)$ instead of a displacement $\mathbf{s}$. These equations are analogous to (8.20) and (8.24), and are (see Appendix 8C):

$$\boldsymbol{\nabla}_\Omega(1) \ln f(1) = -\beta \int \mathrm{d}\mathbf{r}_2 \mathrm{d}\omega_2 f(2) h(12) \boldsymbol{\nabla}_\Omega(2) v(2) - \beta \boldsymbol{\nabla}_\Omega(1) v(1) \tag{8.25}$$

and

$$\boldsymbol{\nabla}_\Omega(1) \ln f(1) = \int \mathrm{d}\mathbf{r}_2 \mathrm{d}\omega_2 c(12) \boldsymbol{\nabla}_\Omega(2) f(2) - \beta \boldsymbol{\nabla}_\Omega(1) v(1), \tag{8.26}$$

where $\boldsymbol{\nabla}_\Omega(1)$ is the total infinitesimal rotation operator for molecule 1,

$$\boldsymbol{\nabla}_\Omega(1) = \mathbf{r}_1 \times \boldsymbol{\nabla}_1 + \boldsymbol{\nabla}_{\omega_1}, \tag{8.27}$$

with $\mathbf{r}_1 \times \boldsymbol{\nabla}_1$ being the part due to the centre of mass and $\boldsymbol{\nabla}_{\omega_1}$ the part due to molecular orientation and given by (8.12); see Fig. 8.4.

8.1.2 *Order parameters*

Although the correlation functions $f(\mathbf{r}_1\omega_1)$ provide a complete and detailed description of the spatial and orientational structure of the fluid, it is often convenient to have available some simpler, averaged measure of molecular orientation at a point in the surface region. The orientational order parameters fulfil this need. Several definitions have been used for such order parameters, and in this section we give the most common of these, and the relations between them.

For simplicity we consider cases where the molecular orientation relative to the surface can be specified by a single angle, ϑ. Examples of such situations include linear molecules at planar or spherical surfaces. For such cases it is convenient to expand the density–orientation profile in the form

$$f(\mathbf{r}_1\omega_1) = \sum_{\ell=0}^{\infty} \hat{f}_\ell(\mathbf{r}_1) P_\ell(\cos\vartheta_1), \tag{8.28}$$

where $P_\ell(\cos\vartheta)$ is the Legendre polynomial of order ℓ (see Appendix A of Vol. 1), ϑ_1 is the polar angle of the molecule, and the coefficients $\hat{f}_\ell(\mathbf{r}_1)$ are order parameters given by

$$\hat{f}_\ell(\mathbf{r}_1) = \frac{2\ell+1}{4\pi} \int \mathrm{d}\omega_1 P_\ell(\cos\vartheta_1) f(\mathbf{r}_1\omega_1). \tag{8.29}$$

This expression is readily derived by multiplying both sides of (8.28) by $P_\ell(\cos\vartheta)$, integrating over ω_1, and using the orthogonality relation (A.9b).

The order parameters $\hat{f}_\ell(\mathbf{r}_1)$ are often calculated in computer simulation studies of surfaces, and the leading few terms in (8.28) are usually sufficient to give a good description of the molecular alignment. For symmetrical linear molecules (e.g. N_2, CO_2) the odd ℓ coefficients vanish in (8.28), so that the $\ell = 2$ term is the leading one in describing the alignment. In Fig. 8.6 are shown some results calculated from *f*-expansion perturbation theory (see Chapter 3) for the leading anisotropic coefficient $\hat{f}_2$ for a diatomic LJ fluid (bond length appropriate to Cl_2) with and without the addition of a point quadrupole Q^*. For $Q^* = 0$ there is a large positive peak in $\hat{f}_2$ at $z/\sigma \sim -0.6$, corresponding to a preference for a molecular orientation perpendicular to the surface on the liquid side; the weaker negative region for $z > 0$ indicates a preference for parallel orientations on the gas side. As Q^* is increased this orientational structure decreases, until at $Q^* = 1.92$ it is reversed; molecules on the liquid side prefer to lie parallel to the surface, and those on the gas side prefer to lie perpendicular to it (see also Fig. 8.1). The qualitative features of these results are supported by computer simulation studies.[37]

In work on the interfacial properties of nematic liquid crystals[45] or of surfactants[46] it is common to use different definitions for the order parameters. For symmetrical linear molecules, for example, one uses[47]

$$\eta(\mathbf{r}_1) = 2\pi \int_0^\pi \mathrm{d}\vartheta_1 \sin\vartheta_1 P_2(\cos\vartheta_1)[f(\mathbf{r}_1\omega_1)/f(\mathbf{r}_1)], \tag{8.30}$$

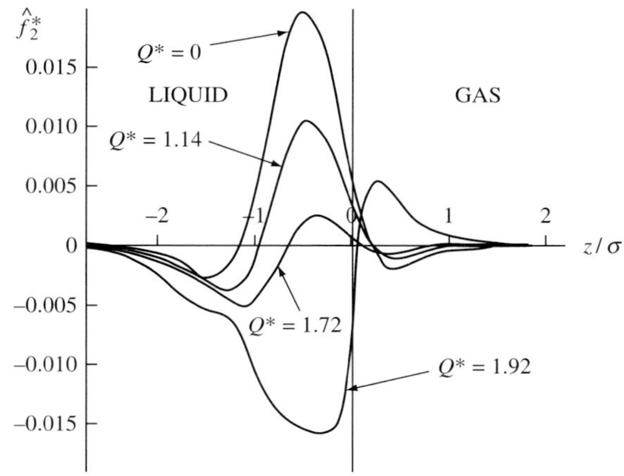

FIG. 8.6 The reduced second harmonic coefficient $\hat{f}_2^* = \hat{f}_2\sigma^3$ for the planar gas–liquid interface of a diatomic LJ fluid with added dimensionless point quadrupole moments, $Q^* = Q/(\varepsilon\sigma^5)^{1/2}$, of various strengths, at $T^* = kT/\varepsilon = 1.065$, $\rho^* = \rho\sigma^3 = 0.55$. Values are calculated from the f-expansion perturbation theory. (From ref. 37.).

which can vary from 1 (perfect alignment with $\vartheta_1 = 0$) to $-1/2$ (perfect alignment with $\vartheta_1 = \pi/2$), with $\eta = 0$ corresponding to isotropy. In this equation $f(\mathbf{r}_1) = \int d\omega_1 f(\mathbf{r}_1\omega_1)$ is the density profile, $\rho(\mathbf{r}_1)$. Later in this chapter we use (8.30) and its generalizations when discussing experimental determinations of surface potential, ellipsometry, and second harmonic generation.

For homogeneous fluids we have already used one set of order parameters, the angular order parameters F_ℓ (see p. 169 of Vol. 1). For linear molecules these are given by

$$F_\ell = \langle P_\ell(\cos\vartheta_1)\rangle = \int d\omega_1 P_\ell(\cos\vartheta_1)P(\omega_1), \tag{8.31}$$

where $P(\omega_1) = f(\mathbf{r}_1\omega_1)/\rho$, and is independent of $\mathbf{r}_1$ for homogeneous fluids. For inhomogeneous fluids the preferred molecular orientation will depend on the position chosen within the fluid, and it is convenient to define a local $F_\ell(\mathbf{r}_1)$ by

$$F_\ell(\mathbf{r}_1) = \sum_{i=1}^{n} \langle P_\ell(\cos\vartheta_i)\delta(\mathbf{r}_i - \mathbf{r}_1)\rangle = \int d\omega_1 P_\ell(\cos\vartheta_1) f(\mathbf{r}_1\omega_1), \tag{8.32}$$

where $\mathbf{r}_i$ is the position of molecule i. (In (8.32) we use the canonical ensemble definition, and can therefore speak of the specific molecule's position.) The most important of these functions are the polarization $F_1(\mathbf{r}_1)$ and alignment $F_2(\mathbf{r}_1)$,

$$F_1(\mathbf{r}_1) = \int d\omega_1 \cos\vartheta_1 f(\mathbf{r}_1\omega_1) \tag{8.33}$$

$$F_2(\mathbf{r}_1) = \frac{1}{2}\int d\omega_1 (3\cos^2\vartheta_1 - 1) f(\mathbf{r}_1\omega_1). \tag{8.34}$$

From (8.29) and (8.32) it is seen that $\hat{f}_\ell$ and F_ℓ are simply related by

$$F_\ell(\mathbf{r}_1) = \left[\frac{4\pi}{2\ell+1}\right]\hat{f}_\ell(\mathbf{r}_1) \tag{8.35}$$

and from (8.29), (8.30), and (8.32) we find (for the case when $f(\mathbf{r}, \vartheta, \phi)$ is independent of ϕ)

$$\eta(\mathbf{r}_1) = \frac{F_2(\mathbf{r}_1)}{f(\mathbf{r}_1)} = \frac{4\pi}{5}\frac{\hat{f}_2(\mathbf{r}_1)}{f(\mathbf{r}_1)}. \tag{8.36}$$

In the following sections we shall use the $\hat{f}_\ell(\mathbf{r}_1)$ coefficients in most applications, but $\eta(\mathbf{r}_1)$ will be used in applications involving surfactants or liquid crystals.

8.1.3 *Surface potential*

For fluids of molecules of any shape (including spherical) there will in general be a nonvanishing electric field $\mathbf{E}(\mathbf{r})$, and hence a change in the electrostatic potential, $\phi(\mathbf{r})$, on traversing an interface. This arises because any plane through and parallel to the interfacial region will cut the electrical charge distribution of some molecules; the molecules are distributed nonuniformly in the interface, and this leads to different average charge distributions above and below the plane. The resulting difference in potential, or *surface potential*, $\Delta\phi$, between the two coexisting phases plays an important role in many processes at aqueous surfaces, e.g. in micelles, membranes, and ionic hydration.

If the molecules are nonspherical and polar, $\Delta\phi$ will depend sensitively on the preferred alignments of the molecules in the interface. For simplicity we first discuss the simplest case of rigid point-dipolar molecules, i.e. nonpolarizable molecules that possess a dipole moment but no higher moments. In the subsection following we discuss the generalizations to include molecular polarizability and higher multipole moments.

Rigid point-dipolar molecules

For concreteness we discuss explicitly the case of a liquid–gas interface. Essentially identical formulae apply to other interfaces (e.g. liquid–solid, liquid–liquid), so that the results derived here can be applied to these other cases as needed. We also consider the simplest geometry, a planar interface between the two phases, with direction z normal to the surface and pointing from liquid to vapour. The x and y components of the average electric field will be zero, so that $\mathbf{E} = (0, 0, E_z(z))$, and (cf.(2.43) of Vol. 1)

$$E_z(z) = -\frac{\partial\phi(z)}{\partial z}, \tag{8.37}$$

where $\phi(z)$ is the macroscopic or average potential at height z in the interface. The surface potential is defined as the potential difference across the interface in moving from the bulk gas to the bulk liquid, i.e. $\Delta\phi = \phi(z \to -\infty) - \phi(z \to +\infty)$, so that from (8.37) we have

$$\Delta\phi = \int_{-\infty}^{\infty} E_z(z)\mathrm{d}z. \tag{8.38}$$

The surface potential can also be related to the *polarization profile*, $P_z(z)$ by using the macroscopic Maxwell equation, $\nabla\cdot(\mathbf{E} + 4\pi\mathbf{P}) = 0$ (see (8D.11)).[48] Integrating this equation and recalling that only the z components of $\mathbf{E}$ and $\mathbf{P}$ are nonvanishing gives $E_z = -4\pi P_z$, where we use the boundary condition that $E_z = 0$ when $P_z = 0$. Thus, from (8.38), we have

$$\Delta\phi = -4\pi \int_{-\infty}^{\infty} P_z(z)dz. \tag{8.39}$$

To obtain an explicit expression for the surface potential in terms of the density–orientation profile $f(\mathbf{r}\,\omega)$ we must derive a molecular expression for the polarization that appears in (8.39). For rigid point-dipolar molecules the macroscopic polarization is simply the average total dipole moment per unit volume (see Eq. (8D.12), with $\mathbf{\Theta}$, etc. $= 0$), so that (cf. (10A.1))

$$P_z(z) =< \sum_i \mu_{iz}\delta(\mathbf{r} - \mathbf{r}_i) >, \tag{8.40}$$

where $\mu_{iz} = \mu_i \cos\theta_i$ is the z component of the dipole moment of molecule i, θ_i is the angle between the dipole of molecule i and the normal to the interface, the sum is over all the molecules, and $< \ldots >$ indicates a canonical or grand canonical average. Assuming a pure fluid ($\mu_i = \mu$) and carrying out the ensemble average gives

$$P_z(z) = \mu \int \mathrm{d}\omega \cos\vartheta f(z\vartheta). \tag{8.41}$$

From (8.39) and (8.41) we obtain the expression for $\Delta\phi$ for rigid dipolar molecules with no higher multipoles as

$$\Delta\phi = -4\pi\mu \int_{-\infty}^{\infty} \mathrm{d}z \int \mathrm{d}\omega \cos\vartheta f(z\vartheta). \tag{8.42}$$

The surface potential is easily related to the order parameter $F_1(z)$ of (8.33) by

$$\Delta\phi = -4\pi\mu \int_{-\infty}^{\infty} \mathrm{d}z F_1(z). \tag{8.43}$$

Inclusion of higher multipoles and polarizability

The derivation of the previous subsection was for rigid point-dipolar molecules. We now relax these restrictions. We first allow for the presence of higher multipole moments, but still assume rigid (i.e. nonpolarizable) molecules. The effects of polarizability are discussed briefly at the end of this section.

Our starting point is the Maxwell equations[49] for macroscopic electrostatics. The macroscopic Maxwell equations can be written in various forms (see Appendix 8D), including a form exactly analogous to the microscopic Maxwell equations (cf. (2.39), (2.40) of Vol. 1), i.e.

$$\nabla \times \mathbf{E} = 0, \quad \nabla \cdot \mathbf{E} = 4\pi\rho_{\text{ch}}, \tag{8.44}$$

where $\mathbf{E}(\mathbf{r})$ is the macroscopic (or mean) electric field, and $\rho_{\text{ch}}(\mathbf{r})$ is the total macroscopic (or mean) charge density, at $\mathbf{r}$. In general ρ_{ch} contains contributions from both free and bound charges.[50] We assume our system has no free charges, so that ρ_{ch} arises completely from charges (nuclear and electronic) bound up in the molecules.

Because of (8.44) we can introduce the macroscopic potential $\phi(\mathbf{r})$ via $\mathbf{E} = -\nabla\phi$ (cf. microscopic (2.43)), and combining with (8.44) gives the macroscopic Poisson equation (cf. microscopic (2.46))

$$\nabla^2\phi = -4\pi\rho_{\text{ch}}. \tag{8.45}$$

For a planar interface, with z the normal direction, ϕ and ρ_{ch} depend only on z, and not on x and y, so that (8.45) reduces to

$$\frac{\mathrm{d}^2}{\mathrm{d}z^2}\phi(z) = -4\pi\rho_{\text{ch}}(z). \tag{8.46}$$

It is easy to verify that the solution to (8.46) satisfying $\phi(z \to \pm\infty) \to$ constant, corresponding to $\mathbf{E} = 0$ in the bulk fluid, is

$$\phi(z) = \phi(z_0) - 4\pi \int_{z_0}^{z} \mathrm{d}z'(z - z')\rho_{\text{ch}}(z'), \tag{8.47}$$

where z_0 is an arbitrary reference point.

As in the previous section we define the surface potential $\Delta\phi$ by $\Delta\phi = \phi(z \to -\infty) - \phi(z \to +\infty)$, or, equivalently, $\phi(z \to -\infty) - \phi(z_0 \to +\infty)$. From (8.47) and the fact that $\int_{-\infty}^{\infty} \mathrm{d}z'\rho_{\text{ch}}(z') = 0$ since the system is electrically neutral overall, we see that

$$\Delta\phi = -4\pi \int_{-\infty}^{\infty} \mathrm{d}z\, z\rho_{\text{ch}}(z), \tag{8.48}$$

where we have changed the dummy integration variable z' to z. Equation (8.48), first derived by Wilson *et al.*,[51,52] is rigorous and expresses (8.48) as the first moment of the macroscopic charge density, a kind of 'macroscopic dipole moment', sometimes referred to as a 'surface double layer'.

To relate $\Delta\phi$ to the molecular multipole moments, we express $\rho_{\text{ch}}(z)$ in terms of the macroscopic densities of dipole moment, quadrupole moment, etc. In Appendix 8D we derive the relation for uncharged molecules

$$\rho_{\mathrm{ch}}(z) = -P_z'(z) + \frac{1}{2}\Theta_{zz}''(z) + \ldots, \tag{8.49}$$

where $P_z' = \mathrm{d}P_z/\mathrm{d}z$, $\Theta_{zz}'' = \mathrm{d}^2\Theta_{zz}/\mathrm{d}z^2$, and $\mathbf{P}$, $\boldsymbol{\Theta}$, ... denote the macroscopic densities of dipole, quadrupole, ..., moments

$$\mathbf{P}(\mathbf{r}) = <\sum_i \boldsymbol{\mu}_i \delta(\mathbf{r}-\mathbf{r}_\mathrm{i})>, \tag{8.50}$$

$$\boldsymbol{\Theta}(\mathbf{r}) = <\sum_i \boldsymbol{\Theta}_i \delta(\mathbf{r}-\mathbf{r}_\mathrm{i})>, \tag{8.51}$$

and so on. In (8.50) and (8.51), the sum over i is over all the molecules in the system, $< \ldots >$ is a thermal average, and $\boldsymbol{\mu}_i$ and $\boldsymbol{\Theta}_i$ are the dipole and quadrupole moments, respectively, of molecule i. It is to be especially noted that the quadrupole moment $\boldsymbol{\Theta}$ occurring here is the one with the *nonvanishing trace* (see (2.54) of Vol. 1 for definition), and *not* the usual quadrupole moment $\mathbf{Q}$ defined by (2.58), which is traceless. The quantity $\mathbf{P}$ defined by (8.50) is usually (cf. however, Appendix 8D) called the *polarization*, and (8.51) is sometimes termed the *quadrupolarization*. $\mathbf{P}(\mathbf{r})$ is nonvanishing only in the interface region, and in that region only $P_z(z)$ is nonvanishing. $\boldsymbol{\Theta}(\mathbf{r})$ is nonvanishing everywhere. In particular in the bulk gas and bulk liquid regions, the diagonal elements $\Theta_{xx}(\mathbf{r})$, $\Theta_{yy}(\mathbf{r})$, and $\Theta_{zz}(\mathbf{r})$ are all equal to each other and independent of $\mathbf{r}$. Hence we have $\Theta_{zz} = \mathrm{Tr}(\boldsymbol{\Theta})/3 \equiv \Theta$ in the bulk. The gas and liquid values of $\Theta(\mathbf{r})$ do differ from each other, however. In fact we have simply $\Theta_\mathrm{g} = \rho_\mathrm{g}\Theta$ and $\Theta_\ell = \rho_\ell\Theta$ for gas and liquid values, respectively, where ρ_g and ρ_ℓ are the corresponding bulk number densities. Here $\Theta \equiv \mathrm{Tr}(\boldsymbol{\Theta})/3$ is the mean molecular quadrupole moment, which is defined by (cf. (2.54))

$$\Theta = \frac{1}{3}\int \mathrm{d}\mathbf{r}\rho_\mathrm{m}(\mathbf{r})r^2, \tag{8.52}$$

where $\rho_\mathrm{m}(\mathbf{r})$ is the molecular charge density; for a discrete set of charges we would write this as $\sum_i q_i r_i^2/3$ as in (2.54), but we employ (8.52) here, so as to avoid confusion with (8.50) and (8.51), where $\sum_i$ denotes a sum over molecules, and also because molecular charge distributions are at least partially continuous (see (2.54)). The electronic contribution Θ_e to $\Theta = \Theta_\mathrm{e} + \Theta_\mathrm{n}$ is proportional to the mean diamagnetic susceptibility,[53,54] and is therefore directly measurable; it has also been calculated theoretically[55] for a number of molecules. With the assumption of fixed nuclear positions, the nuclear contribution Θ_n is easily calculated[54] from the molecular equilibrium bond lengths and angles. We have assumed rigid molecules so that Θ is the same for the gas and liquid phases. The effect of a bond length shift in the liquid ($\sim 1\%$) could easily be included.

We now substitute (8.49) in (8.48). After some integrations by parts we get

$$\Delta\phi = -4\pi \int_{-\infty}^{\infty} \mathrm{d}z P_z(z) - 2\pi[\Theta_{zz}(-\infty) - \Theta_{zz}(+\infty)], \tag{8.53}$$

where $\Theta_{zz}(-\infty)$ and $\Theta_{zz}(+\infty)$ are the bulk liquid and gas values, respectively. Equation (8.53) is the required generalization[50] of (8.39) and is rigorous for rigid molecules. The higher order (octopole, etc.) multipole terms in (8.49) do not contribute to $\Delta\phi$, since they involve third- and higher order derivatives of the higher order multipole densities, which lead to z-derivatives of the bulk multipole densities after the integrations by parts. These derivatives vanish, since the bulk systems are uniform.

Equation (8.53) can be rewritten as

$$\Delta\phi = -4\pi \int_{-\infty}^{\infty} \mathrm{d}z P_z(z) - 2\pi(\rho_\ell - \rho_g)\Theta \tag{8.54}$$

when we use the results discussed above (see (8.52) and preceding discussion). A few comments on the physics of (8.54) are in order. First, only the dipole and quadrupole moments are directly relevant for the surface potential; the higher multipole moments play only an indirect role, by influencing $P_z(z)$ and the density ρ_ℓ via the intermolecular forces. Next, note that dipole alignment in the interfacial region contributes to $\Delta\phi$ (via $P_z(z)$), but it is only the *bulk mean* quadrupolar density that is relevant; hence interfacial alignment of a quadrupolar fluid like N_2 is irrelevant for $\Delta\phi$. Again, it is only the mean quadrupole moment (or trace) Θ that contributes to $\Delta\phi$; the physics behind this is given in Appendix 8E where we present a simplified derivation of the quadrupolar term $\Delta\phi_\Theta = -2\pi(\rho_\ell - \rho_g)\Theta$. Thirdly, note that $\Delta\phi$ is nonvanishing for nonpolar fluids,[56] even for atomic fluids with spherical molecules like argon, since Θ is nonvanishing for *all* real molecules (see definition (8.52)). Finally, note a paradoxical feature of (8.54). $\Delta\phi$, being a physical quantity, cannot depend on the arbitrary choice of molecular origin used to define $\boldsymbol{\mu}$ (occurring in the definition (8.50) of P_z) and $\boldsymbol{\Theta}$. The dipole moment $\boldsymbol{\mu}$ is origin-independent but $\boldsymbol{\Theta}$ is not (cf. Vol. 1, p.69). The resolution of the paradox involves writing out the averages in (8.50) and (8.51) explicitly in terms of the singlet distribution function, $f(\mathbf{r}\,\omega)$, and taking account of the change in $f(\mathbf{r}\,\omega)$ with the change of origin.

In deriving (8.54) we have assumed the molecules are uncharged (cf. (8D.8) and (8D.9)). For ionic solutions a double layer forms at the surface (an example is discussed in § 8.9.1) and in such cases we need to augment (8.54) with a term $-4\pi \int \mathrm{d}z z\rho_f(z)$, where $\rho_f(z)$ is the macroscopic density of free (i.e. unbound) charge (e.g. ionic in the example cited), which corresponds to the dipole moment of the double layer.

It is interesting to compare the magnitudes of the dipolar ($\Delta\phi_\mu$) and quadrupolar ($\Delta\phi_\Theta$) terms of (8.54) for specific systems. For H_2O, Wilson *et al.* (ref. 51 and earlier references therein) have simulated the liquid–vapour interface at $T = 325$ K using the TIP4P model[57] for the water molecule. They find $\Delta\phi_\mu =$

0.79 ± 0.04 V and $\Delta\phi_\Theta = -0.91$ V and hence a total $\Delta\phi = -0.12 \pm 0.04$ V. In this case we have $\Delta\phi_\mu > 0$ and $\Delta\phi_\Theta < 0$, with $\Delta\phi_\Theta$ slightly dominant. The net value $\Delta\phi$ is thus quite small and difficult to measure (see § 8.9). The experimental values[58] were long thought to be about $\Delta\phi \approx (0.025–0.050) \pm 0.01$ V, but a more recent re-measurement[59] gives $\Delta\phi = +0.24 \pm 0.05$ V. For methanol, Matsumoto and Kataoka[60] have simulated the liquid–vapour interface at $T = 300$ K (and other temperatures) using the TIPS model for the methanol molecules. They find $\Delta\phi_\mu = -0.155 \pm 0.039$ V. They did not calculate $\Delta\phi_\Theta$, but Yang *et al.*[61] calculated it and found $\Delta\phi_\Theta = -0.208 \pm 0.004$ V, and hence a total $\Delta\phi = -0.363 \pm 0.043$ V. In this case $\Delta\phi_\mu$ and $\Delta\phi_\Theta$ are comparable and have the same sign. The agreement with the experimental value,[62] $\Delta\phi = -0.23 \pm 0.10$ V, is satisfactory, given the uncertainties in the experimental and theoretical values (TIPS model, no polarizability included—see next paragraph).

We conclude this section with a few remarks on the further generalization needed to account for molecular polarizability. Equation (8.48) remains valid, but (8.49) will need to be amended to take account of the dipole and quadrupole moments induced in the molecules by the intermolecular forces. In principle there can be electronic induction effects via the various polarizabilities (see Appendix C of Vol. 1), and also shifts in μ and Θ due to environmental perturbations of the mean molecular bond lengths and angles, so called 'solvent shifts' (cf. ref. 63). Such induction effects are also of importance for various other phenomena (see, e.g. §§ 2.5, 4.10, 10.1.5, Chapter 11, and Appendix C). Including electronic polarizability (α) in simulations[41,64] using AMBER-type polarizable potentials gives interesting results for $\Delta\phi$ for the pure liquid water surface. The polarizable and non-polarizable models give similar results for $\Delta\phi$ (about -500 mV), but, in the polarizable case, when the contributions to $\Delta\phi$ from the fixed charges and induced dipoles are calculated, they are nearly equal. Similar results are found[64] for $\Delta\phi$ for water with dissolved ions (e.g. 1 M KCl salt solution) except that here the two contributions are of opposite sign, with the negative-induced dipole component dominant, giving a net surface potential of about -450 mV (see also § 8.9.1). It is perhaps worth noting explicitly that the higher multipole moments will play a more direct role than in the rigid molecule case (8.54), since the dipole moments induced in neighbouring molecules by the electric fields produced by the higher multipole moments of a molecule will contribute directly to the macroscopic polarization. For example, for liquid CH_4, where the permanent dipole moment vanishes and the permanent octopole moment is non-vanishing, dipole moments induced by the octopole moments of the molecules will contribute directly to the macroscopic polarization.

8.2 Surface tension

The energy needed to create a unit area of surface is called the surface (or interfacial) tension, γ. The interface will seek to arrange itself so as to minimize the surface energy (or rather the free energy—see below). This minimization

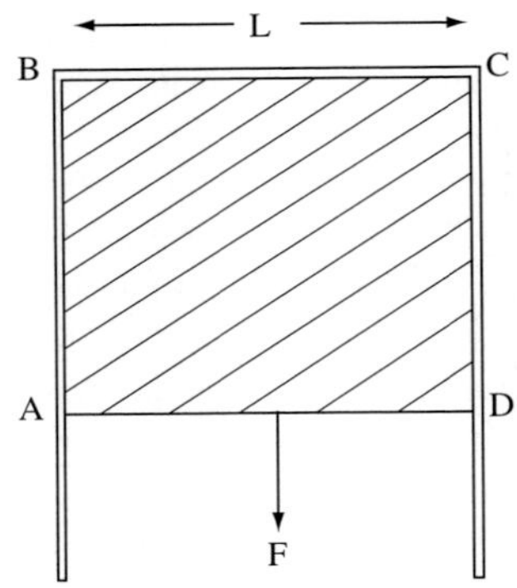

FIG. 8.7 Maxwell's diagram. A liquid film ABCD is held in a wire frame, with the lower boundary AD being movable and frictionless. A weight provides a downward force F on this movable boundary to balance the tension in the film. If the movable boundary is displaced downwards by an amount δx, the work done in increasing the surface area is $\delta w = F\delta x = \gamma \delta S = 2\gamma L \delta x$ (note that there are two gas–liquid surfaces involved). Thus the force needed to balance the tension is $F = 2\gamma L$, and γ is the tension per unit length in a single surface.

will determine the average distribution of the molecules in the interface, and for fluid–fluid interfaces the surface area will be minimized (e.g. by the formation of spherical drops or bubbles) if no external fields are present. An alternative interpretation of surface tension is as the tractional force acting across a unit length of line in the surface. The relation between the two interpretations is easily seen from Fig. 8.7, which shows an experiment where a film of liquid is held in a wire frame, one side of which is movable. At a molecular level, surface tension occurs because of the fluid inhomogeneity. A molecule located in the surface layer (Fig. 8.8) with a range of intermolecular interaction ℓ_{int} will, on average, experience stronger attractions due to molecules on the side of the denser (e.g. liquid) phase than on the side of the less dense (e.g. gas) phase. The surface free energy or surface tension is a measure of the work needed to move molecules from the interior of the bulk fluid to the surface, against this net inward attractive force. For a planar liquid–vapour interface, for example, this net inward attraction causes the transverse pressure p_{T} to vary with the height z in the interface, p_{T} usually being less (and sometimes even negative, i.e. a true tension) than the normal pressure p_{N} in the inhomogeneous zone. To visualize this,[65] note that molecules in the surface region are on average being pulled into the bulk, to be replaced by other molecules just below. Because molecules tend to leave the surface region the mean spacing between molecules in the transverse direction tends to be a little larger than in the bulk, corresponding to the attractive region of the potential; as a consequence the surface molecules are transversely attracted to each other more strongly to try to overcome this 'too large' separation. Hence a transverse tension tends to exist in the surface region. It is easily shown (see § 8.2.5) that the surface tension is related to the difference in the two pressures, $p_{\text{N}} - p_{\text{T}}$.

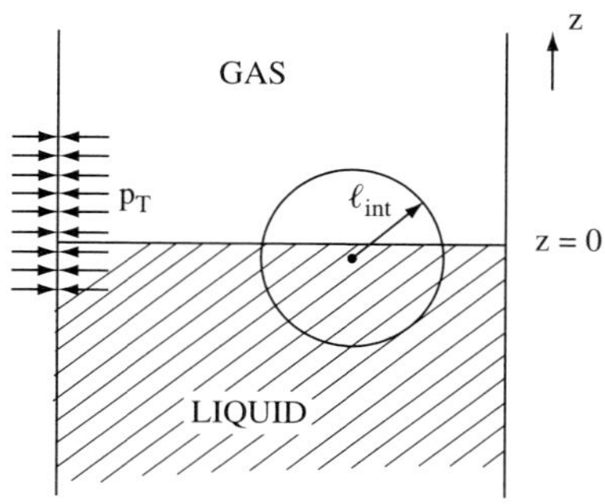

FIG. 8.8 A planar gas–liquid surface lying in the xy plane. The plane $z = 0$ is drawn at some arbitrary location in the surface region, and ℓ_{int} is the range of the intermolecular forces for a molecule in the region of the interface. The transverse pressure, $p_{\text{T}} = p_{xx} = p_{yy}$, will vary with height z in the interfacial region. In the bulk phase (homogeneous) regions $p_{\text{T}} = p_{\text{N}}$, where $p_{\text{N}} = p_{zz}$ is the normal pressure component.

The surface tension leads to many striking surface effects, such as the rise (or fall, e.g. mercury) of liquids in capillaries, the contact angle of a liquid on a solid, wetting, bubble and drop formation, and the formation of monolayers on liquid surfaces. In this section we first summarize some thermodynamic relations for surfaces, and then derive formulae that relate the surface tension to the intermolecular forces and distribution functions. These equations are exact (except for the Fowler model, discussed in § 8.2.2). Approximate equations for γ are given in later sections.

Several expressions are available in statistical mechanics for the gas–liquid interfacial tension,[66] and we consider three of these. The first, or *virial route* value, derived for atomic fluids by Kirkwood and Buff[67] (see also Buff[68]) relates γ to the pair distribution function $f(\mathbf{r}_1\mathbf{r}_2)$ in the interface. The second is the *direct correlation function route* value, derived for atomic fluids by Triezenberg and Zwanzig;[69] it relates γ to the direct correlation function $c(\mathbf{r}_1\mathbf{r}_2)$ in the interface. The third is the *mechanical route* value, which for a flat interface relates γ to the difference $[p_N - p_T(z)]$ between the normal and tangential components of the pressure tensor. We give below the generalization of the virial and direct correlation function equations to molecular fluids. We also give the equation resulting from the mechanical route; the general form of this equation is independent of the nature of the molecules, although the expression for the pressure tensor itself will depend on the type of intermolecular forces involved.

8.2.1 *Thermodynamics of surfaces*

Before proceeding to derive the statistical mechanical expressions for the surface tension, we first summarize some of the most important thermodynamic relations for surface properties, using the methods of Gibbs.[70] We restrict our attention to the planar interface. More comprehensive treatments, including the extension to curved surfaces, are given elsewhere.[13, 18, 70–73]

We consider an equilibrium fluid mixture at temperature T and total volume V, containing N_A molecules of component A, N_B of $B, \ldots, N_R$ of R. The system contains two phases, I and II, separated by a plane surface of area $\mathcal{S}$. For these canonical variables the appropriate thermodynamic potential is the Helmholtz free energy A, which must be a minimum at equilibrium. The change dA that results from small changes in the temperature, volume, surface area, and composition is given by

$$\mathrm{d}A = S\mathrm{d}T - p\mathrm{d}V + \gamma\mathrm{d}\mathcal{S} + \sum_{\alpha} \mu_\alpha \mathrm{d}N_\alpha, \tag{8.55}$$

where $\gamma \mathrm{d}\mathcal{S}$ is the surface contribution, i.e. the work that must be done to increase the surface area by d$\mathcal{S}$, keeping constant T, V, and the N_α; i.e. since V is fixed, γ represents the work done to change the *shape* of the system. If no external fields are present, the minimum free energy will correspond to the minimum surface area; for a given volume of each phase this minimum will be for a sphere. Thus a planar fluid–fluid surface will only occur if there is a gravitational field present. In much of what follows we implicitly assume that such a field is present, and is strong enough to maintain a planar fluid surface, but is also weak enough that the density and orientation profiles in the surface region are unaffected by it.

For an open system it is convenient to use the grand canonical variables T, V, $\mu_A, \ldots \mu_N$ (see § 3.3 of Vol. 1), and the appropriate thermodynamic potential is then the grand potential Ω, defined by

$$\Omega = A - \sum_{\alpha} \mu_\alpha N_\alpha. \tag{8.56}$$

Differentiating this expression and using (8.55) gives

$$\mathrm{d}\Omega = -S\mathrm{d}T - p\mathrm{d}V + \gamma\mathrm{d}\mathcal{S} - \sum_{\alpha} N_\alpha \mathrm{d}\mu_\alpha. \tag{8.57}$$

From (8.55) and (8.57) we have the following relations for the surface tension in terms of the potentials,

$$\gamma = \left(\frac{\partial A}{\partial \mathcal{S}}\right)_{TV\mathbf{N}} = \left(\frac{\partial \Omega}{\partial \mathcal{S}}\right)_{TV\boldsymbol{\mu}}, \tag{8.58}$$

where $\mathbf{N} = N_A, \ldots, N_R$ and $\boldsymbol{\mu} = \mu_A, \ldots, \mu_R$. The free energy A is a homogeneous function of first order in V, $\mathcal{S}$, and $\mathbf{N}$ (i.e. it is extensive with respect to these variables). Thus, we can apply Euler's theorem to integrate (8.55), which gives

$$A = -pV + \gamma\mathcal{S} + \sum_{\alpha} \mu_\alpha N_\alpha. \tag{8.59}$$

Similarly, Ω is homogeneous and first order in V and $\mathcal{S}$, so that

$$\Omega = -pV + \gamma\mathcal{S}, \tag{8.60}$$

which also follows from (8.56) and (8.59). We note that for a homogeneous, bulk phase $\Omega = -pV$.

The definition of the surface tension in (8.58) is independent of the location of any mathematical dividing surface between the two phases. In discussing other properties of the surface layer it is convenient to follow Gibbs in introducing the concept of such a surface.[70] The choice of location of this surface is somewhat arbitrary since the inhomogeneous region is of finite thickness. We shall choose it to lie somewhere in the transition zone, at right angles to the direction of the density gradient. Having chosen the dividing surface, the volume of the system V can be written as the sum of two parts, $V = V^I + V^{II}$, where V^I and V^{II} are the volumes of the phases I and II. Here V^I contains the bulk phase I together with a part of the surface zone; similarly V^{II} contains bulk phase II and the rest of the transition zone. Surface properties are now defined by comparison of the real two-phase system with a hypothetical system in which the two phases are strictly homogeneous right up to the dividing surface (see Fig. 8.9). Thus the surface excess number of molecules of α, N^s_α, is defined by

$$N_\alpha = N^I_\alpha + N^{II}_\alpha + N^s_\alpha, \tag{8.61}$$

where $N^I_\alpha = \rho^I_\alpha V^I$ is the number of molecules of component α in phase I in the hypothetical system, ρ^I_α is the number density of α molecules in the bulk I phase, $N^{II}_\alpha = \rho^{II}_\alpha V^{II}$ is the number of α molecules in the hypothetical II phase, and N_α is the total number of α molecules in the whole system. The *adsorption* of α, Γ_α, is the surface excess number divided by the surface area,

$$\Gamma_\alpha = N^s_\alpha/\mathcal{S} = \int_{-\infty}^{0} [\rho_\alpha(z) - \rho^I_\alpha]\mathrm{d}z + \int_0^{\infty} [\rho_\alpha(z) - \rho^{II}_\alpha]\mathrm{d}z, \tag{8.62}$$

where $\rho_\alpha(z)$ is the density profile in the real system. The Γ_α will depend strongly on the choice of dividing surface.[74] However, the dividing surface can be chosen to make one of the Γ_α, say Γ_A, vanish,

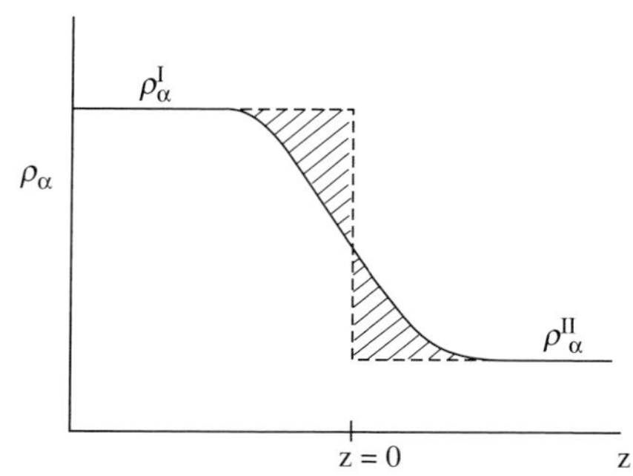

FIG. 8.9 Choice of the Gibbs dividing surface, $z = 0$, between phases I and II. The solid curve is the density profile for component α in the real system. The dashed lines show the density profile for the hypothetical system, in which the phase remains homogeneous up to the dividing surface. If the choice $\Gamma_A = 0$ is made (see Eq. (8.63)), the two shaded areas are equal.

$$\Gamma_A = N_A^s/\mathcal{S} = \int_{-\infty}^{0} [\rho_A(z) - \rho_A^I]\mathrm{d}z + \int_{0}^{\infty} [\rho_A(z) - \rho_A^{II}]\mathrm{d}z = 0, \tag{8.63}$$

which defines $z = 0$. With this choice Γ_B, Γ_C, etc., are non-zero in general. For a pure fluid there is only one Γ, and this is zero if the dividing surface is chosen according to (8.63). This choice is usually used for pure fluids, and is called the *equimolar dividing surface* (or *Gibbs dividing surface*). For mixtures, another choice often made is to choose $z = 0$ so that

$$\sum_\alpha \mu_\alpha N_\alpha^s = 0 \tag{8.64}$$

since this yields a simple relation between the surface free energy and the surface tension (see below). The two definitions coincide for a pure fluid.

The introduction of the Gibbs dividing surface enables clearly defined values to be assigned to other thermodynamic functions for the interfacial zone. For example, we define

$$A^s = A - A^I - A^{II} \tag{8.65}$$

$$U^s = U - U^I - U^{II} \tag{8.66}$$

$$S^s = S - S^I - S^{II} \tag{8.67}$$

$$\Omega^s = \Omega - \Omega^I - \Omega^{II}, \tag{8.68}$$

where A^s, U^s, S^s, and Ω^s are the surface excess Helmholtz energy, internal energy, entropy, and grand potential, respectively. From (8.59) and (8.65), together with the bulk phase expressions

$$A^I = -pV^I + \sum_\alpha \mu_\alpha N_\alpha^I \text{ and } A^{II} = -p\, V^{II} + \sum_\alpha \mu_\alpha N_\alpha^{II},$$

we obtain

$$A^s = \gamma\mathcal{S} + \sum_\alpha \mu_\alpha N_\alpha^s. \tag{8.69}$$

If the dividing surface defined by (8.64) is used this simplifies to

$$\gamma = A^s/\mathcal{S}. \tag{8.70}$$

Similarly, from (8.60) and (8.68) we have

$$\gamma = \Omega^s/S, \tag{8.71}$$

which holds for any choice of the dividing surface. From these last two equations we see again (cf. (8.58)) that γ has the interpretation of density of surface excess free energy, as well as the interpretation of tractional force per unit length across a line in the surface.

The surface excess entropy and internal energy can be related to the surface tension (=surface free energy density) and adsorption through use of the usual thermodynamic identities,

$$S^s = -\frac{\partial A^s}{\partial T} \tag{8.72}$$

$$U^s = A^s + TS^s. \tag{8.73}$$

We give the results only for pure fluids; the extension to the mixture case is straightforward.[75] We first introduce the surface excess densities, i.e. the surface excess properties per unit area,

$$a^s = A^s/\mathcal{S}, \quad s^s = S^s/\mathcal{S}, \quad u^s = U^s/\mathcal{S}. \tag{8.74}$$

For a pure fluid and the equimolar dividing surface, we find from (8.70), (8.72), and (8.73) that

$$a^s = \gamma \tag{8.75}$$

$$s^s = -\frac{\mathrm{d}\gamma}{\mathrm{d}T} \tag{8.76}$$

$$u^s = \gamma - T\frac{\mathrm{d}\gamma}{\mathrm{d}T}. \tag{8.77}$$

Surface tension usually falls with increasing temperature, so that the surface excess entropy is positive, implying that the surface region is less ordered than the bulk, and u^s will be greater than $a^s = \gamma$. Some systems show negative values of s^s, so that the surface tension rises with increasing temperature for a restricted temperature region, although γ always falls with rising temperature as the critical point is approached. Examples of such behaviour can be found in the vapour–liquid interface for metals,[14] the liquid–vapour interface for nematic liquid crystals,[76] and possibly at the vapour–liquid interface for mixtures in which strong adsorption of a surface-active solute occurs.[77] In these cases the negative s^s implies greater spatial (e.g. liquid metals) or orientational (liquid crystals, some liquid mixtures) order in the surface than in the bulk. Examples both of the usual, positive surface entropy behavior[78,79] and of a system showing negative surface entropy[80] are shown in Fig. 8.10. The behaviour for carbon dioxide, in which the surface tension and surface entropy both decrease monotonically with increasing temperature, is typical of nonpolar liquids. As we discuss later in this chapter, molecular fluids such as CO_2 are more orientationally ordered at the surface than in the bulk. This means that the greater surface positional disorder more than compensates the orientational ordering. Water is unusual in having both a high surface tension and in having a maximum in the surface entropy at about 200 °C; below this temperature s^s decreases as the temperature falls. This latter behaviour is presumably related to the break-up of orientational ordering in the surface as the temperature is raised.

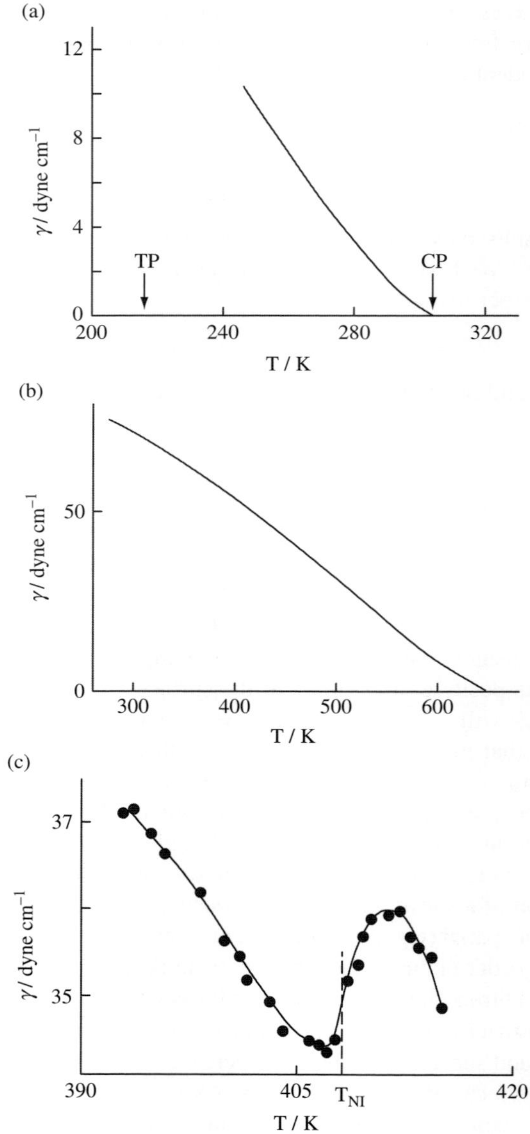

FIG. 8.10 Vapour–liquid surface tension for (a) carbon dioxide[78] (CP and TP denote the critical and triple points), (b) water, and (c) p-azoxyanisole (PAA),[80] a liquid crystal. Carbon dioxide exhibits the usual behaviour for molecular liquids, with both γ and s^s falling monotonically as the temperature rises. The surface entropy is positive, indicating a bulk liquid more ordered than the surface. For PAA, the bulk orientational order decreases very rapidly as the nematic-isotropic liquid transition (T_{NI}) is approached, so that for temperatures near T_{NI} the surface has more orientational order than the bulk; consequently, the surface entropy is negative and γ increases with temperature[77] in this region.

The surface tension and the adsorption are related through an equation known as the *Gibbs adsorption equation*, which we now derive. From (8.55) and the corresponding expressions for changes in A in the bulk phases, dA^I and dA^{II}, we find that

$$dA^s = -S^s dT + \gamma d\mathcal{S} + \sum_\alpha \mu_\alpha dN_\alpha^s. \tag{8.78}$$

Another expression for dA^s can be obtained by differentiation of (8.69):

$$dA^s = \gamma d\mathcal{S} + \mathcal{S} d\gamma + \sum_\alpha \mu_\alpha dN_\alpha^s + \sum_\alpha N_\alpha^s d\mu_\alpha. \tag{8.79}$$

From (8.78) and (8.79) it follows that

$$S^s dT + \mathcal{S} d\gamma + \sum_\alpha N_\alpha^s d\mu_\alpha = 0 \tag{8.80}$$

or, on dividing throughout by the surface area,

$$s^s dT + d\gamma + \sum_\alpha \Gamma_\alpha d\mu_\alpha = 0, \tag{8.81}$$

which is the Gibbs–Duhem equation for the surface, or Gibbs adsorption equation. It can be used to relate the adsorption to changes in the interfacial tension, temperature, and composition of one of the phases. As an example of the use of this equation, consider a binary mixture of components A and B, with the dividing surface chosen so that $\Gamma_A = 0$ (see Eq. (8.63)). At constant temperature (8.81) can then be rewritten as

$$\Gamma_B = -\left(\frac{\partial \gamma}{\partial x_B^I}\right)_T \left(\frac{\partial \mu_B}{\partial x_B^I}\right)_T^{-1}, \tag{8.82}$$

which provides the connection between Γ_B and experimentally measurable quantities. For the case of a liquid phase I in equilibrium with an ideal gas phase II, for which $\mu_B = \mu_B^0(T) + kT\ln(p_B/p^0)$, where p_B is the vapour pressure and $\mu_B{}^0$ and p^0 are independent of x_B^I (see Chapter 7), this equation becomes

$$\Gamma_B = -\frac{1}{kT}\left(\frac{\partial \gamma}{\partial x_B^I}\right)_T \left(\frac{\partial \ln p_B}{\partial x_B^I}\right)_T^{-1}. \tag{8.83a}$$

Thus, a solute B that lowers the interfacial tension will be concentrated at the interface ($\Gamma_B > 0$), and vice versa.

The case of a dilute solute B is of interest. Here, we know from both thermodynamics and statistical mechanics that μ_B can be written as[81] $\mu_B = kT\ln(C_B\rho_B)$, where C_B is a solute-solvent property with the dimensions of inverse density, which depends on T and p but is independent of ρ_B. Thus for fixed T and p, we have $\mathrm{d}\mu_B = kT\mathrm{d}\rho_B/\rho_B$ and hence from (8.81) $\mathrm{d}\gamma = -\Gamma_B kT\mathrm{d}\rho_B/\rho_B$, which we rewrite as

$$\rho_B \frac{\partial \gamma}{\partial \rho_B} = -\Gamma_B kT. \tag{8.83b}$$

Since B is assumed dilute, we can expand the decrease $\gamma_0 - \gamma$ in pure liquid surface tension γ_0 due to the solute as $\gamma_0 - \gamma \equiv \pi_B = \hat{A}\,\rho_B + O\left(\rho_B^2\right)$, where π_B is the spreading pressure due to B and $\hat{A}$ is a constant that describes the effect of the dilute solute (no solute–solute interactions) on surface tension, so that equation (8.83b) gives[82]

$$\Gamma_B = \pi_B/kT = \hat{A}\,\rho_B/kT. \tag{8.83c}$$

We see again from (8.83c) that $\Gamma_B > 0$ corresponds to a decrease in surface tension due to the solute B. A molecular picture for this behaviour can be given along the lines of that for surface tension of a pure fluid given in the introduction to § 8.2. Suppose $\Gamma_B > 0$ so that B prefers the surface to the bulk. Because they prefer the surface, the B molecules will not leave the surface as often as the A molecules, so that the 'too large' molecular separations at the surface are less frequent than with pure A. The surface tension is thereby reduced.

8.2.2 *Thermodynamics of nano-scale systems*

Implicit in the thermodynamic analysis of surfaces of Gibbs, described in the previous section, is the idea of a well-defined interface between two bulk phases. In very small systems, such as small drops, small molecular clusters, or fluids or solids confined within narrow pores (as in porous carbons, zeolites, etc.), the whole system may be highly inhomogeneous, so that there is no well-defined surface separating two phases. In such nano-scale systems Gibbs' thermodynamics of surfaces no longer applies, and surface tension and other surface excess properties are undefined. This can readily be seen from Eq. (8.58), $\gamma = (\partial A/\partial \mathcal{S}) = (\partial \Omega/\partial \mathcal{S})$, for example. The free energy is always well defined, since it is directly related to the partition function ($A = -kT\ln Q,\ \Omega = -kT\ln \Xi$, etc.), which is well defined for systems of arbitrary size, even as small as a single molecule (see § 6.14 for further discussion). However, surface tension (and other surface excess properties) are only valid concepts when a well-defined interface is present, so that $\mathcal{S}$ is defined.

In molecular simulations of nano-scale systems, this breakdown of the Gibbs' treatment manifests itself in thermodynamic inconsistencies, and for very small systems in unphysical behaviour of the surface tension and other excess surface properties. For example, in a molecular dynamics study[83] of small liquid drops of Lennard-Jones molecules (in equilibrium with vapour) ranging in size from about 40 to 2000 molecules (reduced drop radius, $R_\mathrm{e}^* = R_\mathrm{e}/\sigma \approx 2.4$ to 8.4, where R_e is the equimolar radius of the drop and σ is the Lennard-Jones

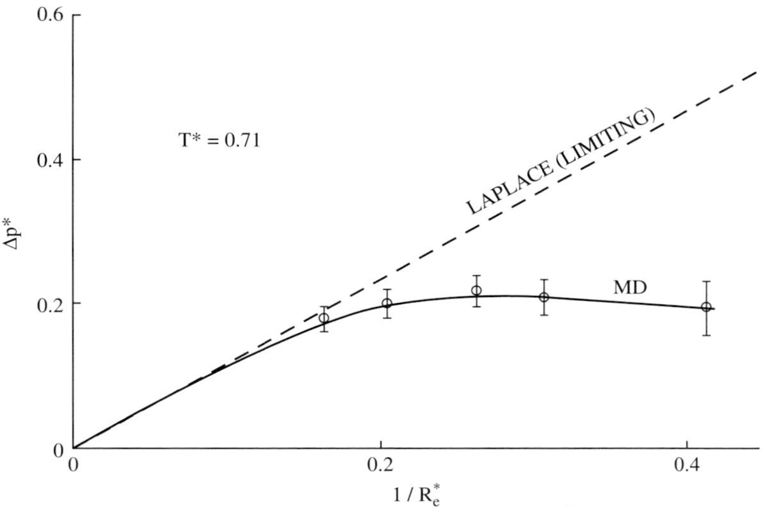

FIG. 8.11 The reduced pressure drop, $\Delta p^* = p_\ell^* - p_g^*$, where $p^* = p\sigma^3/\varepsilon$, versus inverse reduced drop radius at $T^* = 0.71$ from the Laplace equation (with $\gamma_\infty^* = 0.581 \pm 0.029$), given by the dashed line, and from MD simulation results given by the points and solid line. Reprinted with permission from ref. 83. Copyright 1984 American Institute of Physics.

diameter), it was found that for reduced drop radii R_e^* below about 6 the surface tension calculated via the virial route and the mechanical route (see §§ 8.2.3 and 8.2.5 below) gave different results, indicating thermodynamic inconsistency (these two routes should give the same result). Moreover, the equations of Laplace for the difference in pressure between the liquid and gas phases, due to curvature of the drop surface (see (8.175)), and the Kelvin equation for the vapour pressure of the drop, both derived using the Gibbs' analysis, break down for R_e^* values below about 8 (see Figure 8.11). The Laplace and Kelvin equations are

$$p_\ell - p_g = \frac{2\gamma_s}{R_s} \qquad \text{(Laplace)}$$

$$\rho_\ell kT \ln\left(\frac{p_g(R)}{p_g(\infty)}\right) = \frac{2\gamma_s}{R_s} \qquad \text{(Kelvin)},$$

where $p_g(R)$ and $p_g(\infty)$ are the vapour pressures over the liquid drop of radius R and over the planar surface, respectively, ρ_ℓ is the density of the bulk liquid, and γ_s is the surface tension referred to the surface of tension, R_s. The Gibbs–Thomson equation, the analogue of the Kelvin equation for the effect of surface curvature on the freezing temperature for small systems, likewise breaks down for very small systems.[84]

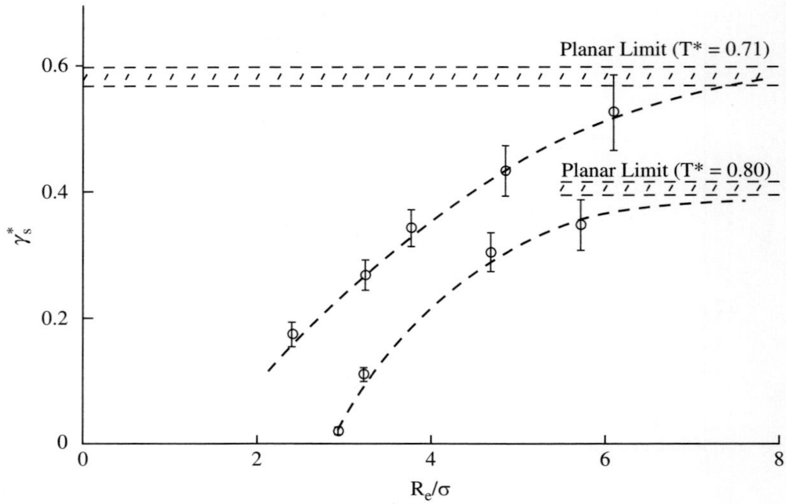

FIG. 8.12 Variation of surface tension $\gamma_s^* = \gamma_s\sigma^2/\varepsilon$ with drop radius at $T^* = 0.71$ and 0.80. The values plotted are obtained by the virial route. Reprinted with permission from ref. 83. Copyright 1984 American Institute of Physics.

For very small drops ($R_e^* \lesssim 1.5$–6, the value depending on the temperature, with the breakdown more severe for higher temperatures), the surface tension calculated from the virial route becomes negative, an unphysical result (see Fig. 8.12).

8.2.3 *The virial route*

We consider a one-component two-phase fluid system of N molecules in thermal equilibrium at temperature T in a total volume V. The dividing surface $\mathcal{S}$ between the phases lies in the xy plane at $z = 0$. The thermodynamic definition of the surface tension γ is (8.58),

$$\gamma = \left(\frac{\partial A}{\partial \mathcal{S}}\right)_{TVN} = -\frac{kT}{Z}\left(\frac{\partial Z}{\partial \mathcal{S}}\right)_{TVN}, \tag{8.84}$$

where Z is the configurational partition function (cf.(3.53))

$$Z = \int_0^V \mathrm{d}\mathbf{r}^N \int \mathrm{d}\omega^N \exp[-\beta(\mathcal{U} + \mathcal{V})]. \tag{8.85}$$

The integrals over $\mathbf{r}_1 \ldots \mathbf{r}_N$ in (8.85) extend over the dimensions of the vessel and therefore depend on the surface area $\mathcal{S}$. To calculate $\partial Z/\partial \mathcal{S}$ from (8.85) we use the method due to Bogoliubov[85] and Green[86] for scaling the centre of mass coordinates; i.e. we introduce dimensionless $\mathbf{r}_i'$, where

$$x_i = Lx_i', \quad y_i = Ly_i', \quad z_i = L_1 z_i', \tag{8.86}$$

where $L = \mathcal{S}^{1/2}$ and $L_1 = V/\mathcal{S}$. From (8.85) and (8.86) we have (note V is held fixed),

$$\begin{aligned}\left(\frac{\partial Z}{\partial \mathcal{S}}\right)_{TVN} &= \frac{\partial}{\partial \mathcal{S}}\left(V^N \int_0^1 \mathrm{d}\mathbf{r}'^{\mathrm{N}} \int \mathrm{d}\omega^N \exp\left[-\beta(\mathcal{U}+\mathcal{V})\right]\right) \\ &= -\beta V^N \int_0^1 \mathrm{d}\mathbf{r}'^{\mathrm{N}} \int \mathrm{d}\omega^N \left(\frac{\partial \mathcal{U}}{\partial \mathcal{S}} + \frac{\partial \mathcal{V}}{\partial \mathcal{S}}\right) \exp\left[-\beta(\mathcal{U}+\mathcal{V})\right].\end{aligned} \tag{8.87}$$

From (8.84), (8.85), and (8.87) we have, on reverting to the $\mathbf{r}_i$ variables,

$$\gamma = \int_0^V \mathrm{d}\mathbf{r}^{\mathrm{N}} \int \mathrm{d}\omega^N \left(\frac{\partial \mathcal{U}}{\partial \mathcal{S}} + \frac{\partial \mathcal{V}}{\partial \mathcal{S}}\right) \frac{1}{Z} \exp\left[-\beta(\mathcal{U}+\mathcal{V})\right]$$

or

$$\gamma = \left\langle \frac{\partial \mathcal{U}}{\partial \mathcal{S}} \right\rangle + \left\langle \frac{\partial \mathcal{V}}{\partial \mathcal{S}} \right\rangle. \tag{8.88}$$

For a free fluid surface (e.g. vapour/liquid) under the influence of gravity the external field term in (8.88) is usually neglected. In the case of a solid–fluid interface the fluid–solid potential is often treated as an external field, so that the $\partial \mathcal{V} / \partial \mathcal{S}$ term is then important. We therefore retain this term for generality, and shall later drop it where appropriate.

We now assume pairwise additivity for the intermolecular potential,

$$\mathcal{U} = \sum_{i<j} u(\mathbf{r}_{ij}\omega_i\, \omega_j) \tag{8.89}$$

and use (8.4) for $\mathcal{V}$, so that (8.88) becomes

$$\gamma = \frac{1}{2}N(N-1)\left\langle \frac{\partial u(12)}{\partial \mathcal{S}} \right\rangle + N\left\langle \frac{\partial v(1)}{\partial \mathcal{S}} \right\rangle. \tag{8.90}$$

Writing out the averages $\langle \partial u(12)/\partial \mathcal{S}\rangle$ and $\langle \partial v(1)/\partial \mathcal{S}\rangle$ in terms of the distribution functions $f(12) = f(\mathbf{r}_1\mathbf{r}_2\omega_1\omega_2)$ and $f(1) = f(\mathbf{r}_1\omega_1)$, respectively, gives

$$\gamma = \frac{1}{2}\Omega^2 \int \mathrm{d}\mathbf{r}_1 \mathrm{d}\mathbf{r}_2 \left\langle f(12)\frac{\partial u(12)}{\partial \mathcal{S}} \right\rangle_{\omega_1\omega_2} + \Omega \int \mathrm{d}\mathbf{r}_1 \left\langle f(1)\frac{\partial v(1)}{\partial \mathcal{S}} \right\rangle_{\omega_1}. \tag{8.91}$$

Since we have assumed the interface to be planar, the fluid is homogeneous in the x and y directions and (8.91) can be written

$$\gamma = \frac{1}{2}\Omega^2 \mathcal{S} \int_{-\infty}^{\infty} \mathrm{d}z_1 \int \mathrm{d}\mathbf{r}_{12} \left\langle f(12)\frac{\partial u(12)}{\partial \mathcal{S}} \right\rangle_{\omega_1\omega_2} + \Omega \mathcal{S} \int_{-\infty}^{\infty} \mathrm{d}z_1 \left\langle f(1)\frac{\partial v(1)}{\partial \mathcal{S}} \right\rangle_{\omega_1}. \tag{8.92}$$

The derivatives with respect to surface area in (8.92) can be evaluated (see Appendix 8B) with the aid of (8.86). They are

$$\begin{aligned} \mathcal{S}\frac{\partial u(12)}{\partial \mathcal{S}} &= \frac{1}{2}\left[r_{12}\frac{\partial u(12)}{\partial r_{12}} - 3z_{12}\frac{\partial u(12)}{\partial z_{12}}\right] \\ &= -P_2(\cos\theta_{12})r_{12}\frac{\partial u(12)}{\partial r_{12}} + \frac{3}{2}\sin\theta_{12}\cos\theta_{12}\frac{\partial u(12)}{\partial \theta_{12}} \end{aligned} \tag{8.93}$$

and

$$\mathcal{S}\frac{\partial v(1)}{\partial \mathcal{S}} = \frac{1}{2}\left[x_1\frac{\partial v(1)}{\partial x_1} + y_1\frac{\partial v(1)}{\partial y_1} - 2z_1\frac{\partial v(1)}{\partial z_1}\right], \tag{8.94}$$

where $P_2(\cos\theta_{12}) = \frac{1}{2}(3\cos^2\theta_{12} - 1)$. In these expressions $\partial u(12)/\partial r_{12}$ is taken at fixed $(\theta_{12}, \phi_{12}, \omega_1\omega_2)$, $\partial u(12)/\partial z_{12}$ is at fixed $(x_{12},\ y_{12},\ \omega_1\omega_2)$, $\partial u(12)/\partial\theta_{12}$ is at fixed $(r_{12}, \phi_{12}, \omega_1\omega_2)$, $\partial v(1)/\partial x_1$ is at fixed $(y_1,\ z_1,\ \omega_1)$, etc. The second form of (8.93) is more convenient in general. From (8.92)–(8.94) we have

$$\begin{aligned} \gamma = &-\frac{1}{2}\Omega^2\int_{-\infty}^{\infty} dz_1 \int d\mathbf{r}_{12} P_2(\cos\theta_{12}) r_{12}\left\langle f(12)\frac{\partial u(12)}{\partial r_{12}}\right\rangle_{\omega_1\omega_2} \\ &+\frac{3}{4}\Omega^2\int_{-\infty}^{\infty} dz_1 \int d\mathbf{r}_{12}\sin\theta_{12}\cos\theta_{12}\left\langle f(12)\frac{\partial u(12)}{\partial \theta_{12}}\right\rangle_{\omega_1\omega_2} \\ &+\frac{1}{2}\Omega\int_{-\infty}^{\infty} dz_1 \left\langle f(1)\left[x_1\frac{\partial v(1)}{\partial x_1} + y_1\frac{\partial v(1)}{\partial y_1} - 2z_1\frac{\partial v(1)}{\partial z_1}\right]\right\rangle_{\omega_1}. \end{aligned} \tag{8.95}$$

For a free fluid surface under gravity the final external field term is usually neglected and this equation becomes

$$\begin{aligned} \gamma = &-\frac{1}{2}\Omega^2\int_{-\infty}^{\infty} dz_1 \int d\mathbf{r}_{12} P_2(\cos\theta_{12}) r_{12}\left\langle f(12)\frac{\partial u(12)}{\partial r_{12}}\right\rangle_{\omega_1\omega_2} \\ &+\frac{3}{4}\Omega^2\int_{-\infty}^{\infty} dz_1 \int d\mathbf{r}_{12}\sin\theta_{12}\cos\theta_{12}\left\langle f(12)\frac{\partial u(12)}{\partial \theta_{12}}\right\rangle_{\omega_1\omega_2}, \end{aligned} \tag{8.96}$$

which is the generalization of the Kirkwood–Buff equation to molecular fluids. This equation was first derived independently by Gray and Gubbins[87] and by Davis.[88] A derivation from density functional theory has also been given.[89] For a fluid of nonspherical molecules the second term on the right of (8.96) involving $\partial u(12)/\partial\theta_{12}$ does not vanish in general, because $u(12)$ is a function of θ_{12}. (This term does vanish in the Fowler model of the surface, as shown below.) It is often convenient to replace $f(12)$ in (8.96) by the pair correlation function $g(12)$ using (8.7). The surface tension will therefore also depend on $f(1) = f(z_1\omega_1)$, and hence on whether there are preferred molecular orientations in the interface. The fact that γ depends on the surface molecular orientations can also be seen intuitively[90] as follows. The surface tension depends on the surface energy and free energy through (8.77) and (8.75), and since molecules at the interface will tend to align to minimize these surface energies, γ will be thereby affected.

For a fluid of spherical molecules with pair potential $u(r)$ the term $\partial u/\partial\theta$ vanishes and (8.96) reduces to

$$\gamma = \frac{1}{2}\int_{-\infty}^{\infty} \mathrm{d}z_1 \int \mathrm{d}\mathbf{r}_{12} \frac{(x_{12}^2 - z_{12}^2)}{r_{12}} f(z_1\mathbf{r}_{12}) \frac{\mathrm{d}u(r_{12})}{\mathrm{d}r_{12}}, \tag{8.97}$$

where we have made use of the fact that the integrals over x_{12} and y_{12} are equal, from the symmetry of the surface. Equation (8.97) is the familiar Kirkwood–Buff expression for the free surface of a fluid of spherical molecules. Croxton[14] has extended (8.97) to the case in which quantum corrections are included. Grant and Desai[91,92] have generalized both (8.96) and (8.97) to the case of many-body interactions.

The Fowler model[67,87,93]

In this simple model of the vapour–liquid interface it is assumed that (a) the transition layer has zero thickness, and (b) the vapour density is negligible. These assumptions imply a pair distribution function of the form

$$\begin{aligned} f(z_1\mathbf{r}_{12}\omega_1\omega_2) &= 0, && z_1 \text{ or } z_2 > 0 \\ &= \Omega^{-2}\rho_\ell g(\mathbf{r}_{12}\omega_1\omega_2), && z_1 \text{ and } z_2 < 0, \end{aligned} \tag{8.98}$$

where ρ_ℓ is liquid density and $g(\mathbf{r}_{12}\omega_1\omega_2)$ is the pair correlation function for the bulk liquid. To obtain the Fowler equation we substitute (8.98) into the general equation (8.96) for γ. We first show that the second term on the right-hand side of (8.96) vanishes in this model. In the Fowler model the integrand in this term vanishes if either z_1 or z_2 is greater than zero, while for the non-vanishing regions (i.e. z_1 and $z_2 < 0$) the integrand contains the quantity

$$Q(r_{12}) \equiv \left\langle g(\mathbf{r}_{12}\omega_1\omega_2)\frac{\partial u(\mathbf{r}_{12}\omega_1\omega_2)}{\partial\theta_{12}}\right\rangle_{\omega_1\omega_2}, \tag{8.99}$$

which is independent of both z_1 and ω_{12} in the Fowler model. A simple rotational and inversion invariance argument can be used to show that Q must vanish.[87] Consider the transformation (see Fig. 8.13)

$$\omega_1 \to \omega_2, \quad \omega_2 \to \omega_1, \quad \omega_{12} \to -\omega_{12}. \tag{8.100}$$

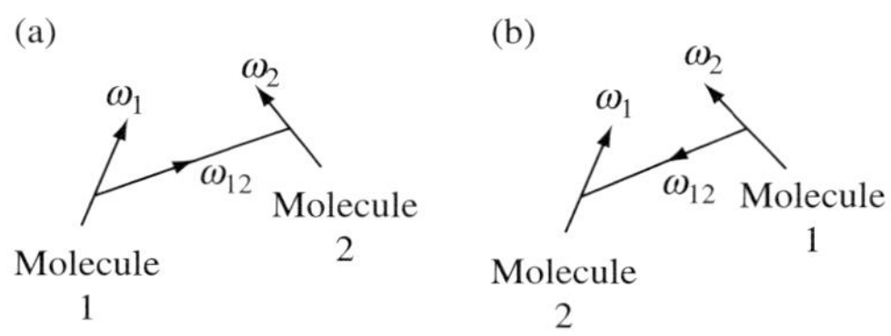

FIG. 8.13 The transformation (a) → (b) of (8.100).

In the Fowler model (but not in the more general case) this transformation should leave Q unchanged. It also leaves $g(12)$ and $u(12)$ unchanged, so that the right side of (8.99) becomes

$$\begin{aligned} Q(r_{12}) &= \langle g(\mathbf{r}_{12}\omega_1\omega_2)\partial u(\mathbf{r}_{12}\omega_1\omega_2)/\partial(\pi - \theta_{12})\rangle_{\omega_1\omega_2} \\ &= -\langle g(\mathbf{r}_{12}\omega_1\omega_2)\partial u(\mathbf{r}_{12}\omega_1\omega_2)/\partial\theta_{12}\rangle_{\omega_1\omega_2} \\ &= -Q(r_{12}). \end{aligned} \tag{8.101}$$

Thus Q must vanish, and hence the second term on the right of (8.96) vanishes.

From (8.96), (8.98), and (8.101) we have

$$\gamma = -\frac{1}{2}\rho_\ell^2 \int_0^\infty \mathrm{d}r_{12} r_{12}^3 \left\langle g(\mathbf{r}_{12}\omega_1\omega_2)\frac{\partial u(\mathbf{r}_{12}\omega_1\omega_2)}{\partial r_{12}}\right\rangle_{\omega_1\omega_2} \int_{-r_{12}}^0 \mathrm{d}z_1 \int' \mathrm{d}\omega_{12} P_2(\cos\theta_{12}), \tag{8.102}$$

where we have used the fact that $\langle g(12)\partial u(12)/\partial r_{12}\rangle_{\omega_1\omega_2}$ is independent of z_1 and ω_{12} provided that z_1 and z_2 are less than zero. The prime on the integral over ω_{12} indicates that the integration is limited to regions where z_1 and z_2 are less than zero. For $|z_1| < r_{12}$ the integration over θ_{12} must be limited to angles greater than or equal to the value θ_m shown in Fig. 8.14, since otherwise the integrand in the first term of (8.96) must vanish. The lower limit on the z_1 integral is $-r_{12}$ since the unrestricted integral $\int \mathrm{d}\omega_{12} P_2(\cos\theta_{12})$ vanishes. Carrying out the integral over ω_{12} in (8.102) gives

$$\begin{aligned} \int' \mathrm{d}\omega_{12} P_2(\cos\theta_{12}) &= \frac{1}{2}\int_0^{2\pi} \mathrm{d}\phi_{12} \int_{\theta_m}^{\pi} \mathrm{d}\theta_{12} \sin\theta_{12}(3\cos^2\theta_{12} - 1) \\ &= -\pi \int_{\cos\theta_m}^{\cos\pi} \mathrm{d}\cos\theta_{12}(3\cos^2\theta_{12} - 1) \\ &= -\pi(\cos^3\theta_{12} - \cos\theta_{12})\Big|_{\cos\theta_m}^{\cos\pi} \\ &= \pi(\cos^3\theta_m - \cos\theta_m) \\ &= -\pi\left[\left(\frac{z_1}{r_{12}}\right)^3 - \frac{z_1}{r_{12}}\right], \end{aligned}$$

where we have used $\cos\theta_m = -z_1/r_{12}$ in the last step (see Fig. 8.14). Carrying out the integration over z_1 we have

$$\begin{aligned} \int_{-r_{12}}^0 \mathrm{d}z_1 \int' \mathrm{d}\omega_{12} P_2(\cos\theta_{12}) &= -\pi \int_{-r_{12}}^0 \mathrm{d}z_1 \left[\left(\frac{z_1}{r_{12}}\right)^3 - \frac{z_1}{r_{12}}\right] \\ &= \pi\left(\frac{1}{4}r_{12} - \frac{1}{2}r_{12}\right) \\ &= -\frac{1}{4}\pi r_{12}. \end{aligned} \tag{8.103}$$

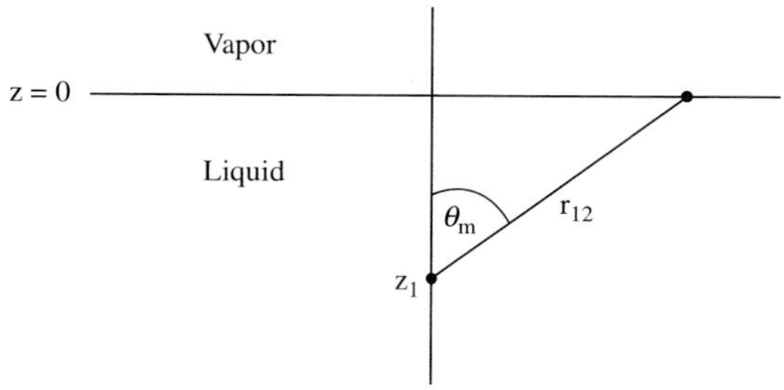

FIG. 8.14 The Fowler model of the vapour–liquid interface, with bulk liquid below $z = 0$ and an ideal vapour above. If molecule 1 is at z_1, the minimum value of θ_{12} to be used in the integral over ω_{12} in (8.102) is the angle θ_m shown.

From (8.102) and (8.103) we obtain the Fowler approximation for the surface tension of the vapour–liquid interface as

$$\gamma = \frac{\pi}{8}\rho_\ell^2 \int_0^\infty \mathrm{d}r_{12} r_{12}^4 \langle g(\mathbf{r}_{12}\omega_1\omega_2) u'(\mathbf{r}_{12}\omega_1\omega_2)\rangle_{\omega_1\omega_2}, \tag{8.104}$$

where $u' = \partial u/\partial r_{12}$. For atomic fluids this reduces to

$$\gamma = \frac{\pi}{8}\rho_\ell^2 \int_0^\infty \mathrm{d}r r^4 g(r) u'(r). \tag{8.105}$$

Freeman and McDonald[94] have evaluated the integral in (8.105) by Monte Carlo computer simulation, and have thus obtained γ for Lennard-Jones fluids in the Fowler approximation. They have compared their calculated values with experimental results for γ for a variety of simple liquids, using LJ parameters fitted from bulk liquid thermodynamic properties (energy and pressure). These comparisons are shown in Table 8.1 for both the surface tension and the surface excess internal energy u^s, where u^s is given by (8.77); in the Fowler model this expression can be written as

$$u^s = -\frac{\pi}{2}\rho_\ell^2 \int_0^\infty \mathrm{d}r r^3 u(r) g(r) \tag{8.106}$$

for atomic fluids. It can be seen from Table 8.1 that the Fowler model is in quite good agreement with experiment[95] for γ near the triple point. The results for u^s, and for γ at higher temperatures, are much poorer, however. At higher temperatures the interfacial thickness increases, so that the Fowler approximation is expected to become poorer. Moreover, u^s and γ are rigorously related through the thermodynamic identity (8.77), and since $\mathrm{d}\gamma/\mathrm{d}T$ will be poorly predicted in the Fowler model we expect the results for u^s to be poor also. More importantly, the values of u^s obtained from (8.77) disagree with those from (8.106), so that the Fowler model is thermodynamically inconsistent.

Monte Carlo simulation results[96–98] for a free Lennard-Jones surface suggest that the good agreement between the Fowler model and experiment (Table 8.1)

Table 8.1

Surface tension and surface excess internal energy for simple liquids from experiment and from the Fowler model, using the Lennard-Jones potential

Fluid	$T(K)$	γ (dyne cm^{-1})		u^s (erg cm^{-2})	
		Expt.*	Fowler	Expt.*	Fowler
Ar	83.8	13.4	13.7	35.0	27.6
	90.0	11.9	13.2	34.4	26.0
	100.0	9.4	12.2	33.3	23.3
	110.0	7.1	11.1	31.7	20.9
	120.0	5.0	10.3	29.8	18.2
	130.0	3.0	9.0	27.1	15.3
Kr	115.8	16.4	16.7	—	33.8
Xe	161.3	19.2	18.7	50.1	42.2
Ne	24.6	5.7	6.8	—	10.6
CH_4	91.0	17.0	15.4	—	32.2
N_2	63.2	12.2	10.3	27.6	23.7
O_2	54.4	—	14.9	—	37.6
CO	68.1	12.5	11.3	—	22.9

*From ref. 95. Values of u^s are referred to the Gibbs dividing surface.
From ref. 94.

for γ at the triple point is fortuitous, and probably arises from a cancellation of errors due to the inaccuracies of the potential and of the Fowler model. In Table 8.2 several simulation results for the free surface are compared with the Fowler model result, experiment, and perturbation theory.[99] We see that both the surface tension and excess surface energy for the free surface are larger than for the Fowler model surface in the case of the Lennard-Jones fluid. When the more accurate Barker–Fisher–Watts pair potential[100] (Fig. 2.4 of Vol. 1) is used, together with the three-body dispersion term (Axilrod–Teller–Muto, see (2.300) of Vol. 1), the free surface results are in quite good agreement with experiment.

Comparisons of the Fowler model with experiment for molecular fluids, based on (8.104), have been made using perturbation theory.[87] Several authors (see, e.g. Berry *et al.*[101] and Freeman and McDonald[94]) have attempted to improve the Fowler model by using more realistic approximations for $\rho(z)$. These have included linear, cubic, and exponential profiles. Such models introduce the surface width as an adjustable, temperature-dependent parameter, and so are of limited use for predictive purposes.

8.2.4 *The direct correlation function route*

It is possible to express the surface tension as an integral over the direct correlation function. For *atomic* fluids this equation was first given by Yvon[102] based on a leading virial coefficient type of argument, and was later presented without proof by Buff and Lovett,[95] but the first detailed derivation was given by Triezenberg and Zwanzig[69] based on a calculation of the change in the grand potential due to a density fluctuation in the surface. Lovett *et al.*[103] have presented a different

Table 8.2

*Comparison of results for surface properties of liquid argon at the triple point, 83.8 K**

	γ (dyne cm^{-1})	u^s (erg cm^{-2})	s^s (erg cm^{-2}K^{-1})
MC, LJ†	18.3 ± 0.3	38.9 ± 0.8	0.244
MC, LJ‡	16.5 ± 2.6	—	—
Perturbation theory	19.7	—	—
BFW + ATM	14.1	34.7	0.244
Fowler Model	13.7	27.6	—
Experiment	13.4	35.0	0.255

*Here $s^s = -\mathrm{d}\gamma/\mathrm{d}T$ is the surface excess entropy. MC, LJ is a Monte Carlo calculation for the free surface, as is the entry perturbation theory, which is calculated using the theory of Toxvaerd.[99] The Fowler model result is obtained form the Monte Carlo calculation of Freeman and McDonald.[94] The BFW + ATM result is for a free surface of a fluid in which the molecules interact with the Barker–Fisher–Watts pair potential for argon[100] with the addition of the Axilrod–Teller–Muto three-body dispersion term. This potential is known to give an accurate account of the bulk properties of argon. These values were calculated from the simulation results for the Lennard-Jones fluid using perturbation theory. Experimental values are from Buff and Lovett.[95]

† From Miyazaki *et al.*[98] who calculate surface tension from the free energy to form the surface.

‡ From Lee *et al.*[96] who calculate surface tension from the surface stress.

From ref. 98.

derivation in which they calculate the pressure change across a curved surface. The extension of the direct correlation function route to the case of *molecular* fluids has been carried out by several authors using different methods. Grant and Desai[92] start from the generalized Kirkwood–Buff equation derived in the previous section and apply dynamical sum rules. Gubbins[104] (see also Gubbins and Thompson[105]) used a generalization of the method of Lovett *et al.*,[103] and Sluckin[89] based his derivation on density functional theory.

In this section we derive the direct correlation function expression for γ for molecular fluids, using a generalization of the derivation of Lovett *et al.*[103] We consider a system with a gas–liquid interface lying in the xy plane; the $z = 0$ plane is located at some convenient point in the interfacial region (see Fig. 8.15). In this system the density–orientation profile $f_0(z\omega)$ is proportional to the probability density of finding a molecule at height z with orientation ω. We now introduce an external potential $v(\mathbf{r}\omega)$ that acts on the transition region in such a way as to give the surface (at $z = 0$) a spherical curvature of radius R (see Fig. 8.15(b)). We further suppose that $v(\mathbf{r}\omega)$ is such that the preferred molecular orientation is unchanged in this deformation of the surface; i.e. there is no local rotation of axes.[106] We shall only be interested in the limiting behaviour as $R \to \infty$, so v can be arbitrarily weak. In this limit we shall assume that the density–orientation profile $f(\mathbf{r}\omega)$ in the curved interface equals the planar density–orientation profile $f_0(z_0\omega)$ shifted by an appropriate amount. Thus for a point (x, y, z) in the curved interface region the density-orientation profile is given by

$$f(xyz\omega) = f_0(z_0\omega), \tag{8.107}$$

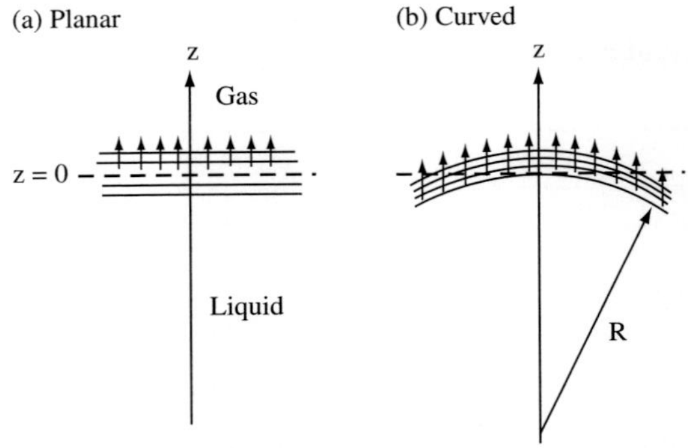

FIG. 8.15 The gas–liquid interface (a) before and (b) after applying the external potential $v(\mathbf{r}\omega)$. Arrows represent, schematically, molecular orientations in the interface, which are assumed to be unaffected by curvature of the surface for large R.

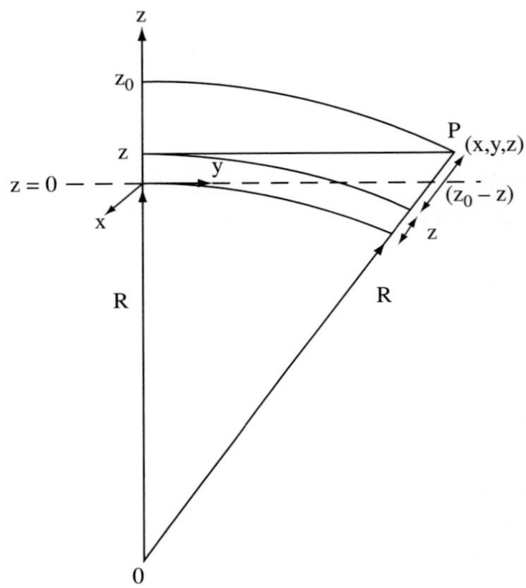

FIG. 8.16 Geometry of the curved surface.

where $f_0(z_0\omega)$ is the density–orientation profile in the planar interface at the height z_0 shown in Fig. 8.16.

To obtain the expression for the surface tension we first relate $v(\mathbf{r}\omega)$ to the density–orientation profile in the planar ($v = 0$) system, and then use this to evaluate the pressure difference across the curved surface. This pressure difference, $\Delta p = p_\ell - p_g$, where p_ℓ and p_g are the pressures in the liquid and gas phases, respectively, is related to the surface tension through the Laplace equation[18]

$$p_\ell - p_g = \frac{2\gamma}{R}. \tag{8.108}$$

Strictly R here should be the radius of the surface of tension R_s and γ the surface tension γ_s corresponding to this specific choice of radius, but since we shall only be concerned with the limit $R \to \infty$ this distinction is immaterial.

The external field will be a functional of the change in the density–orientation profile $\Delta f(\mathbf{r}\omega) = f(xyz\omega) - f_0(z\omega)$. Expanding $v(\mathbf{r}\omega)$ in a functional Taylor series in Δf gives (see (8A.17)), to first order:

$$v(\mathbf{r}_1\omega_1) = \int \mathrm{d}\mathbf{r}_2 \mathrm{d}\omega_2 \left(\frac{\delta v(\mathbf{r}_1\omega_1)}{\delta f(\mathbf{r}_2\omega_2)} \right)_{R=\infty} \Delta f(\mathbf{r}_2\omega_2). \tag{8.109}$$

Using the second Yvon equation, (8A.30), for the derivative $\delta v(1)/\delta f(2)$ we have

$$v(\mathbf{r}_1\omega_1) = kT \int \mathrm{d}\mathbf{r}_2 \mathrm{d}\omega_2 \left[c_0(12) - \frac{\delta(12)}{f_0(1)} \right] \Delta f(\mathbf{r}_2\omega_2). \tag{8.110}$$

The change $\Delta f(\mathbf{r}\omega)$ in $f(\mathbf{r}\omega)$ at $\mathbf{r} = (xyz)$ on curving the interface is given to first order by

$$\begin{aligned} \Delta f(\mathbf{r}\omega) &= f(xyz\omega) - f_o(z\omega) \\ &= f_o(z_o\omega) - f_o(z\omega) \\ &= (z_o - z)\frac{\partial f_o(z\omega)}{\partial z}. \end{aligned} \tag{8.111}$$

From the geometry of Fig. 8.16 we have, on letting $\tilde{R} = R + z$,

$$\tilde{R}^2 + x^2 + y^2 = [\tilde{R} + (z_o - z)]^2$$

or

$$\begin{aligned} (z_0 - z) &= \tilde{R}\left[1 + \frac{x^2 + y^2}{\tilde{R}^2}\right]^{1/2} - \tilde{R} \\ &= \frac{1}{2}\frac{x^2 + y^2}{\tilde{R}} + O(R^{-3}) \xrightarrow{\tilde{R}\to\infty} \frac{1}{2}\frac{x^2 + y^2}{R}. \end{aligned} \tag{8.112}$$

From (8.110), (8.111), and (8.112) we have, for $R \to \infty$,

$$v(\mathbf{r}_1\omega_1) = \frac{1}{2} kT \int \mathrm{d}\mathbf{r}_2 \mathrm{d}\omega_2 \left[c_o(12) - \frac{\delta(12)}{f_o(1)} \right] \frac{x_2^2 + y_2^2}{R} \frac{\partial f_o(z_2\omega_2)}{\partial z_2}. \tag{8.113}$$

The external field $v(\mathbf{r}\omega)$ gives rise to a force $-\partial v(1)/\partial \mathbf{r}_1$ acting on the molecules at $(\mathbf{r}_1\omega_1)$. From (8.113) this force vanishes when $\partial f_o/\partial z$ is zero, i.e. outside the interfacial region. The force varies on moving from the liquid to the gas side of the interface, and leads to a pressure difference $\Delta p = p_\ell - p_g$ between the two phases (here $p = p_N$, the normal component of the pressure; see below for definition of p_N). This external force difference exactly balances the pressure difference existing between the two phases, as determined by the well-known hydrostatic equilibrium condition (see (8.136) and (8.154) below). For points on the z-axis of Fig. 8.16 the latter can be written $\partial p/\partial z_1 = F_z(\mathbf{r}_1)$, where

$\mathbf{F}(\mathbf{r}_1) \equiv \int \mathrm{d}\omega_1 f(1)(-\partial v(1)/\partial \mathbf{r}_1)$ is the external force per unit volume at $\mathbf{r}_1$. The pressure difference $\Delta p = p_\ell - p_g$ between the two phases can thus be obtained by integrating $F_z(\mathbf{r}_1)$ along the z_1-axis ($x_1 = y_1 = 0$) from $z_1 = -\infty$ to $z_1 = \infty$:

$$\Delta p \equiv p(z_1 = -\infty) - p(z_1 = \infty) = \int_{-\infty}^{\infty} \mathrm{d}z_1 F_z(\mathbf{r}_1),$$

or, using the explicit value of $F_z(\mathbf{r}_1)$,

$$\Delta p = -\int_{-\infty}^{\infty} \mathrm{d}z_1 \mathrm{d}\omega_1 f_o(z_1\omega_1) \frac{\partial}{\partial z_1} v(\mathbf{r}_1\omega_1), \tag{8.114}$$

where we have used $f(1) = f_0(z_1\omega_1)$, which is valid for points $\mathbf{r}_1$ on the z-axis (see Fig. 8.16). We now substitute (8.113) for $v(\mathbf{r}\omega)$ in (8.114), and note that the term $\delta(12)\left(x_2^2 + y_2^2\right)$ does not contribute, since it vanishes for points $\mathbf{r}_1$ on the z-axis. Thus we get

$$\Delta p = -\frac{1}{2R} kT \int_{-\infty}^{\infty} \mathrm{d}z_1 \mathrm{d}\omega_1 f_o(1) \frac{\partial}{\partial z_1} \int \mathrm{d}\mathbf{r}_2 \mathrm{d}\omega_2 c_o(12) \left(x_{12}^2 + y_{12}^2\right) f_0'(2), \tag{8.115}$$

where $f_o(i) \equiv f_o(z_i\omega_i)$ and $f_0'(i) = \partial f_o(z_i\omega_i)/\partial z_i$, and we have replaced x_2 and y_2 by x_{12} and y_{12} (since $x_1 = y_1 = 0$). Provided that the direct correlation function $c_0(12)$ vanishes at the limits of integration on z_1, as it should, (8.115) can be integrated by parts to give

$$\Delta p = \frac{1}{2R} kT \int_{-\infty}^{\infty} \mathrm{d}z_1 \int \mathrm{d}\mathbf{r}_2 \mathrm{d}\omega_1 \mathrm{d}\omega_2 f_0'(1) f_0'(2) c_o(12) \left(x_{12}^2 + y_{12}^2\right). \tag{8.116}$$

From this expression and the Laplace equation, (8.108), we finally obtain the direct correlation function expression for the surface tension

$$\gamma = \frac{1}{4} kT \int_{-\infty}^{\infty} \mathrm{d}z_1 \int \mathrm{d}\mathbf{r}_2 \mathrm{d}\omega_1 \mathrm{d}\omega_2 f'(1) f'(2) c(12) \left(x_{12}^2 + y_{12}^2\right), \tag{8.117}$$

where we have dropped the subscript 0 on f and c for convenience. We note that the integral over $\mathrm{d}\mathbf{r}_2$ can be written as one over $\mathrm{d}z_2 \mathrm{d}x_{12} \mathrm{d}y_{12}$. The integral (8.117) is a planar geometry one, and the result is therefore independent of x_1 and x_2.

Equation (8.117) is the generalization to molecular fluids of the Triezenberg–Zwanzig equation for atomic fluids. It is more general than the generalized Kirkwood–Buff expression, (8.96), since no assumption of pairwise additivity or of the type of molecules is made. For atomic fluids, Schofield[107] showed that (for pairwise additive potentials) the Triezenberg–Zwanzig and Kirkwood–Buff equations are equivalent. Schofield's proof made use of dynamical sum rules.

The use of these time-dependent properties in the proof was unsatisfying, since surface tension is a static property, and an alternative proof of the equivalence that relied only on equilibrium properties has been given by Waldor and Wolf.[108] The Schofield proof has been generalized to the case of molecular fluids by Grant and Desai.[92]

8.2.5 *The mechanical route*

The surface tension can be related to the pressure tensor in the interface by calculating the reversible work required to deform the surface. The derivation is based on purely macroscopic arguments, and we refer to this as the mechanical route to γ. We consider two fluid phases $I \equiv \alpha$ and $II \equiv \beta$ separated by an interface lying in the xy plane. Although the pressure is a scalar quantity p in either of the bulk phases, it will be a tensor $\mathbf{p}$ in the interfacial region because of its inhomogeneity; i.e. the pressure on an element of area will depend on the element's orientation, and, for an element arbitrarily oriented, there will be shear (or tangential) pressure. Because of the symmetry of the system and its rotational invariance about the z-axis, the xyz-axes are the principal axes so that shear pressure vanishes for these axes; the xx and yy components of $\mathbf{p}$ are equal, i.e. $p_{xx}(z) = p_{yy}(z) = p_{\mathrm{T}}(z)$, the tangential pressure; also $p_{zz} = p_{\mathrm{N}}$, the normal pressure. From the condition of hydrostatic equilibrium, or mechanical stability ($\boldsymbol{\nabla} \cdot \mathbf{p} = 0$), it follows that $p_{\mathrm{N}} = p_\alpha = p_\beta$ will be a constant, independent of z, provided that external forces can be neglected. (If they cannot, we have $\boldsymbol{\nabla} \cdot \mathbf{p} = \mathbf{F}$, as used in § 8.2.4.)

The derivation of the expression for γ is as follows (see, for example, Rowlinson and Widom[18]). We consider the deformation of the system of height ℓ and surface area $\mathcal{S}$ illustrated in Fig. 8.17, which is carried out at constant temperature and total volume. Therefore, the work done on the system δW in this process equals the increase in the Helmholtz energy dA. Once an expression is available for δW we can obtain the equation for γ from (8.84). We first displace the xz face of the box by dℓ, which results in an increase in the surface area of the interface from $\mathcal{S}$ to $\mathcal{S} + \mathrm{d}\mathcal{S}$. The work done on the system in this process is

$$\delta W_1 = -\mathrm{d}\mathcal{S} \int_{-\ell/2}^{\ell/2} p_{_T}(z)\,\mathrm{d}z. \tag{8.118}$$

The system is next compressed in the z direction by an amount $\mathrm{d}\mathcal{S}/\ell$, so that the final volume remains unchanged. The work done on the system in this second process is

$$\delta W_2 = p_N \mathcal{S} \mathrm{d}\mathcal{S}/\ell = p_N \ell \mathrm{d}\mathcal{S} \tag{8.119}$$

since $\mathcal{S} = \ell^2$. The total work done is $\delta W_1 + \delta W_2$, i.e.

$$\delta W = \mathrm{d}A = \mathrm{d}\mathcal{S} \int_{-\ell/2}^{\ell/2} [p_N - p_{_T}(z)]\mathrm{d}z \tag{8.120}$$

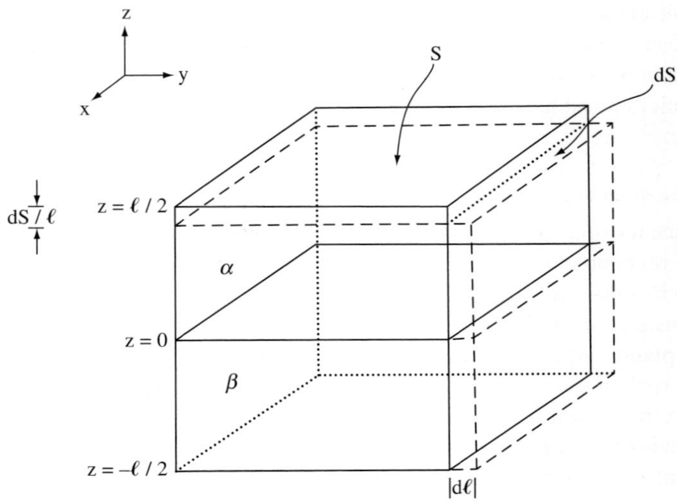

FIG. 8.17 Deformation of a two-phase system. The original system is the cubic box shown as solid lines, of side ℓ. The system is first expanded by $d\ell$ in the y direction, and then compressed by $d\mathcal{S}/\ell$ in the z direction, where $d\mathcal{S}$ is the increase in surface area in the first expansion. The volume (and temperature) are left unchanged in this process. The interface is centred around $z = 0$.

so that from (8.84) and (8.120) we have

$$\gamma = \int_{-\infty}^{\infty} [p - p_T(z)]\,dz, \tag{8.121}$$

where $p = p_N$ is just the bulk-phase pressure. Since $p_N = p_T$ in the homogeneous bulk phase regions, the integration limits $-\ell/2$ and $\ell/2$ have been replaced by $-\infty$ and ∞, respectively. Equation (8.121) provides the mechanical route to γ; it remains valid if an external field is present in the z direction, due for example to a solid wall or electric field. However, p_N in (8.120) and in the interfacial region then becomes z-dependent. Expressions analogous to (8.121) have been derived for curved surfaces[18, 83] (see (8.174) for the case of a spherical surface).

The derivation of (8.121) rests on the assumption that we can define local values of the pressure tensor, even though the fluid density in the interface is changing rapidly over distances comparable with the range of the intermolecular forces. In addition, we must have a molecular expression for the pressure tensor for (8.121) to be useful. This expression is obtained in the next section (see also Appendix E of Vol. 1 for a brief discussion).

8.3 Pressure tensor

The pressure is the force per unit area acting across a surface element within the fluid, and is the sum of a kinetic contribution due to motion of molecules across the surface (i.e. convective momentum transport) and a configurational contribution due to intermolecular forces. For a homogeneous fluid the pressure is the same in all directions, and it is therefore the scalar quantity p. In an inhomogeneous fluid it is a second rank tensor $\mathbf{p}$, since the pressure depends on both the orientation of the surface and the direction relative to the surface. Thus $p_{\alpha\beta}$ gives the force per unit area in the β-direction on a surface pointing in the α-direction. The kinetic contribution to $\mathbf{p}$ at a point $\mathbf{r}$ in the fluid, assumed to be in equilibrium but not necessarily homogeneous, is $\mathbf{p}_K(\mathbf{r}) = \rho(\mathbf{r})kT\mathbf{1}$, the same as for an ideal gas; it is well defined, and arises from the rate of transport of momentum of molecules that cross the surface element. The configurational contribution, $\mathbf{p}_C$, is not well defined, however, because there is no unambiguous way to decide which intermolecular forces contribute to the stress across a particular element of surface in the fluid. This arbitrariness in the definition of $\mathbf{p}$ was clearly recognized by Kirkwood and co-workers[109, 110] in their studies of transport processes, but its implications for surface properties have only been explored more recently.[10, 18, 111–113] Of course, physically observable properties such as the surface tension must have a unique value, and so must be invariant with respect to the choice of definition of $\mathbf{p}$. This has important consequences for the validity of such concepts as surface tension and surface of tension. For a planar interface, for example, the surface tension given by the mechanical route, (8.121), is found to be invariant to the choice of pressure tensor, but the surface of tension is not.

In this section we first derive a general expression for the microscopic stress tensor and pressure tensor in an inhomogeneous system (§ 8.3.1). We then consider some common choices of the definition of $\mathbf{p}$ and the special case of a planar interface (§ 8.3.2). The derivation follows that given by Schofield and Henderson[113] for atomic fluids, but extends it to the molecular case.[114] The principal difference in the molecular case is that the pair intermolecular force no longer lies along the line joining the molecular centres, and this gives rise to additional terms in the expression for $\mathbf{p}$.

8.3.1 General definition of the stress and pressure tensors

We first derive[113] the microscopic law of momentum conservation at some point $\mathbf{r}$ in the fluid at time t, and from this obtain an expression for the microscopic stress tensor $\boldsymbol{\sigma}(\mathbf{r}, t)$. The pressure tensor $\mathbf{p}(\mathbf{r})$ at point $\mathbf{r}$ is then given by the ensemble average of minus the stress tensor, $\mathbf{p}(\mathbf{r}) = -\langle\boldsymbol{\sigma}(\mathbf{r}, t)\rangle$.

The force on a body is given by Newton's law as the rate of change of its linear momentum. We therefore start from the local density of linear momentum at point $\mathbf{r}$ at time t, $\mathbf{J}(\mathbf{r}, t)$, which is given by

$$\mathbf{J}(\mathbf{r}, t) = \sum_i \mathbf{p}_i(t)\delta[\mathbf{r} - \mathbf{r}_i(t)], \tag{8.122}$$

where $\mathbf{r}_i(t)$ and $\mathbf{p}_i(t)$ are the location of the molecular center (chosen to be the centre of mass in this section) and the momentum of molecule i, respectively, at the instant t. The rate of change of the α component of $\mathbf{J}$ is

$$\begin{aligned}\dot{J}^{\alpha}(\mathbf{r},t) &= -\frac{\partial}{\partial r^{\beta}}\sum_i p_i^{\alpha}\dot{r}_i^{\beta}\delta(\mathbf{r}-\mathbf{r}_i)+\sum_i \dot{p}_i^{\alpha}\delta(\mathbf{r}-\mathbf{r}_i)\\ &= -\frac{\partial}{\partial r^{\beta}}\sum_i \frac{p_i^{\alpha}p_i^{\beta}}{m_i}\delta(\mathbf{r}-\mathbf{r}_i)+\sum_i \dot{p}_i^{\alpha}\delta(\mathbf{r}-\mathbf{r}_i),\end{aligned} \tag{8.123}$$

where we have used the usual summation convention for repeated indices and the property

$$\frac{\partial}{\partial \mathbf{r}_i}f(\mathbf{r}-\mathbf{r}_i) = -\frac{\partial}{\partial \mathbf{r}}f(\mathbf{r}-\mathbf{r}_i) \tag{8.124}$$

(Cartesian tensor notation is summarized in Appendix B of Vol. 1). Equation (8.123) is the microscopic conservation law for linear momentum. The first and second terms on the right of (8.123) give the kinetic and configurational contributions to $\dot{J}^{\alpha}$, respectively. We now derive a more convenient and explicit form for the second term on the right in (8.123). Since this term gives the force due to potential fields it can be written as

$$\sum_i \dot{p}_i^{\alpha}\delta(\mathbf{r}-\mathbf{r}_i) = -\sum_i\left[\frac{\partial}{\partial r_i^{\alpha}}\mathcal{U}(\mathbf{r}^N\omega^N)+\frac{\partial}{\partial r_i^{\alpha}}\mathcal{V}(\mathbf{r}^N)\right]\delta(\mathbf{r}-\mathbf{r}_i), \tag{8.125}$$

where $\mathcal{V}$ is an external field assumed to depend only on the positions of the molecular centres of mass. Note that the sum of the intermolecular forces, $\sum_i -\nabla_i\mathcal{U}$, vanishes. Physically, this is true since otherwise the system's centre of mass, in the absence of external forces, would spontaneously accelerate, thereby violating conservation of total momentum. (Alternatively,[115] for the case of pairwise additive intermolecular potentials, the intermolecular forces cancel in pairs due to Newton's third law.) Since $\sum_i -\nabla_i\mathcal{U}$ vanishes, we can add any multiple of it to the right side of (8.125), e.g. $\delta(\mathbf{r}-\mathbf{r}_0)\sum_i -\nabla_i\mathcal{U}$, where $\mathbf{r}_0$ is an arbitrary point in the fluid. Thus we get

$$\sum_i\left[\frac{\partial}{\partial r_i^{\alpha}}\mathcal{U}(\mathbf{r}^N\omega^N)\right]\delta(\mathbf{r}-\mathbf{r}_i) = \sum_i\left[\frac{\partial}{\partial r_i^{\alpha}}\mathcal{U}(\mathbf{r}^N\omega^N)\right][\delta(\mathbf{r}-\mathbf{r}_i)-\delta(\mathbf{r}-\mathbf{r}_0)]. \tag{8.126}$$

The last expression can be written as a line integral

$$\begin{aligned}\delta(\mathbf{r}-\mathbf{r}_i)-\delta(\mathbf{r}-\mathbf{r}_0) &= \int_{C_{0i}} \mathrm{d}\tilde{\ell}^{\beta}\frac{\partial}{\partial\tilde{\ell}^{\beta}}\delta(\mathbf{r}-\tilde{\boldsymbol{\ell}})\\ &= -\frac{\partial}{\partial r^{\beta}}\int_{C_{0i}}\mathrm{d}\tilde{\ell}^{\beta}\delta(\mathbf{r}-\tilde{\boldsymbol{\ell}}),\end{aligned} \tag{8.127}$$

where the integral is over any contour C_{0i} from $\mathbf{r}_0$ to $\mathbf{r}_i$. If we now substitute (8.125), (8.126), and (8.127) into (8.123), the conservation law becomes

$$j^{\alpha}(\mathbf{r},t) = \frac{\partial}{\partial r^{\beta}}\sigma^{\alpha\beta}(\mathbf{r},t) - \rho(\mathbf{r},t)\frac{\partial}{\partial r^{\alpha}}v(\mathbf{r}), \tag{8.128}$$

where

$$\sigma^{\alpha\beta}(\mathbf{r},t) = \sigma_K^{\alpha\beta}(\mathbf{r},t) + \sigma_C^{\alpha\beta}(\mathbf{r},t) \tag{8.129}$$

is the microscopic stress tensor with the kinetic and configurational parts given by

$$\sigma_{\mathrm{K}}^{\alpha\beta}(\mathbf{r},t) = -\sum_i \frac{p_i^{\alpha}p_i^{\beta}}{m_i}\delta(\mathbf{r}-\mathbf{r}_i) \tag{8.130}$$

$$\sigma_{\mathrm{C}}^{\alpha\beta}(\mathbf{r},t) = \sum_i \left[\frac{\partial}{\partial r_i^{\alpha}}\mathcal{U}(\mathbf{r}^N\omega^N)\right]\int_{C_{0i}} d\tilde{\ell}^{\beta}\,\delta(\mathbf{r}-\tilde{\ell}) \tag{8.131}$$

and $\rho(\mathbf{r},t)$ is the number density at point $\mathbf{r}$ at time t, given by

$$\rho(\mathbf{r},t) = \sum_i \delta[\mathbf{r}-\mathbf{r}_i(t)]. \tag{8.132}$$

The form of the last term in (8.128) follows when we use (8.4) for $\mathcal{V}(\mathbf{r}^N)$ with $v(\mathbf{r}_{\mathrm{i}})$ independent of ω_{i}. We note that the arbitrariness in the choice of contour in (8.131) arises because the conservation law (8.128) involves the divergence of the stress tensor, $\nabla\cdot\boldsymbol{\sigma}$, and not $\boldsymbol{\sigma}$ itself. Thus we can define any new stress tensor $\boldsymbol{\sigma}'$, provided $\nabla\cdot(\boldsymbol{\sigma}'-\boldsymbol{\sigma}) = 0$.

The macroscopic pressure tensor $\mathbf{p}$ is defined as minus the ensemble average of the microscopic stress tensor,

$$\mathbf{p}(\mathbf{r}) = -\langle\boldsymbol{\sigma}(\mathbf{r},t)\rangle, \tag{8.133}$$

where the average on the right-hand side is independent of t since the system is in equilibrium. Thus we can take $t = 0$. We have, from (8.130), the fact that the $\mathbf{p}_i$ are uncorrelated with the $\mathbf{r}_i$, and using (3D.30)

$$\begin{aligned}\langle\boldsymbol{\sigma}_K(\mathbf{r},t)\rangle &= -\left\langle\sum_i \frac{\mathbf{p}_i\mathbf{p}_i}{m_i}\delta(\mathbf{r}-\mathbf{r}_i)\right\rangle = -kT\mathbf{1}\left\langle\sum_i \delta(\mathbf{r}-\mathbf{r}_i)\right\rangle \\ &= -kT\rho(\mathbf{r})\mathbf{1} \equiv -\mathbf{p}_K(\mathbf{r}),\end{aligned} \tag{8.134}$$

where $\rho(\mathbf{r})$ is the average number density at $\mathbf{r}$ (cf. (8.132)) and $\mathbf{1}$ is the unit tensor. Thus the general expression for the pressure tensor is

$$\mathbf{p}(\mathbf{r}) = \rho(\mathbf{r})kT\mathbf{1} - \left\langle\sum_i\left[\frac{\partial}{\partial\mathbf{r}_i}\mathcal{U}(\mathbf{r}^N\omega^N)\right]\int_{C_{0i}} \mathrm{d}\tilde{\ell}\,\delta(\mathbf{r}-\tilde{\ell})\right\rangle. \tag{8.135}$$

If we take the ensemble average of (8.128) and note that $\langle \dot{\mathbf{J}}_i \rangle = 0$ (see (E.23)), and use the fact that $\mathbf{p}$ is a symmetric tensor,[116] we obtain the condition of hydrostatic equilibrium,

$$\nabla \cdot \mathbf{p} = -\rho(\mathbf{r})\nabla v(\mathbf{r}), \tag{8.136}$$

where $(\nabla \cdot \mathbf{p})_\alpha = \nabla_\beta p_{\beta\alpha}$ (see (B.31)). In the absence of external fields this simplifies to

$$\nabla \cdot \mathbf{p}(\mathbf{r}) = 0. \tag{8.137}$$

Substituting (8.134) for $\mathbf{p}_\mathrm{K}(\mathbf{r})$ and using (8.124) to obtain $\nabla \cdot \mathbf{p}_\mathrm{C}$ in (8.136) or (8.137) gives the first member of the BBGKY hierarchy of equations (see § 3.4.3).

Equations (8.129), (8.131), and (8.135) hold for a general N-body intermolecular potential. For pairwise additive intermolecular forces, Eq. (1.13), the intermolecular potential contribution to the configurational term in the momentum conservation equation (cf. (8.126)) becomes

$$\begin{aligned}\sum_i \left[\frac{\partial}{\partial \mathbf{r}_i}\mathcal{U}(\mathbf{r}^N\omega^N)\right]\delta(\mathbf{r}-\mathbf{r}_i) &= \frac{1}{2}\sum_i \left[\frac{\partial}{\partial \mathbf{r}_i}\sum_{j\neq k} u(jk)\right]\delta(\mathbf{r}-\mathbf{r}_i)\\ &= \frac{1}{2}\sum_{i\neq j}\frac{\partial u(ij)}{\partial \mathbf{r}_i}\delta(\mathbf{r}-\mathbf{r}_i) + \frac{1}{2}\sum_{i\neq j}\frac{\partial u(ji)}{\partial \mathbf{r}_i}\delta(\mathbf{r}-\mathbf{r}_i)\\ &= \frac{1}{2}\sum_{i\neq j}\frac{\partial u(ij)}{\partial \mathbf{r}_i}[\delta(\mathbf{r}-\mathbf{r}_i)-\delta(\mathbf{r}-\mathbf{r}_j)],\end{aligned} \tag{8.138}$$

where $u(ij) \equiv u(\mathbf{r}_{ij}\omega_i\omega_j)$, and in the last step we interchanged the dummy indices i and j in the second term on the right and made use of

$$\frac{\partial u(ij)}{\partial \mathbf{r}_i} = -\frac{\partial u(ij)}{\partial \mathbf{r}_j}. \tag{8.139}$$

As before (cf. (8.127)) we can write the difference in delta functions as a line integral,

$$\sum_i \left[\frac{\partial}{\partial \mathbf{r}_i}\mathcal{U}(\mathbf{r}^N\omega^N)\right]\delta(\mathbf{r}-\mathbf{r}_i) = -\frac{\partial}{\partial \mathbf{r}}\cdot\frac{1}{2}\sum_{i\neq j}\frac{\partial u(ij)}{\partial \mathbf{r}_{ij}}\int_{C_{ij}} \mathrm{d}\tilde{\boldsymbol{\ell}}\,\delta(\mathbf{r}-\tilde{\boldsymbol{\ell}}), \tag{8.140}$$

where C_{ij} is a contour from $\mathbf{r}_\mathrm{i}$ to $\mathbf{r}_\mathrm{j}$, and we have used $\nabla_{ij}u(ij) = -\nabla_i u(ij)$ (recall $\mathbf{r}_{ij} = \mathbf{r}_j - \mathbf{r}_i$). The configurational stress tensor (8.131) becomes

$$\boldsymbol{\sigma}_\mathrm{C}(\mathbf{r},t) = \frac{1}{2}\sum_{i\neq j}\frac{\partial u(ij)}{\partial \mathbf{r}_{ij}}\int_{C_{ij}} \mathrm{d}\tilde{\boldsymbol{\ell}}\,\delta(\mathbf{r}-\tilde{\boldsymbol{\ell}}) \tag{8.141}$$

and the pressure tensor for the pairwise additive case is now (cf. (8.135))

$$\mathbf{p}(\mathbf{r}) = \rho(\mathbf{r})kT\mathbf{1} - \frac{1}{2}\left\langle \sum_{i\neq j} \frac{\partial u(ij)}{\partial \mathbf{r}_{ij}} \int_{C_{ij}} \mathrm{d}\tilde{\boldsymbol{\ell}}\delta(\mathbf{r}-\tilde{\boldsymbol{\ell}}) \right\rangle. \tag{8.142}$$

The principal difference between this expression for molecular fluids and the corresponding one for the atomic fluid case[113] is that for molecules the intermolecular pair force is no longer directed along the line of centres, and this gives rise to additional terms in $\mathbf{p}$ due to the θ and ϕ components of $\nabla_{ij}u(ij)$. In spherical coordinates the gradient operator is (see (A.18a))

$$\frac{\partial}{\partial \mathbf{r}} = \hat{\mathbf{r}}\frac{\partial}{\partial r} + \hat{\boldsymbol{\theta}}\frac{1}{r}\frac{\partial}{\partial \theta} + \hat{\boldsymbol{\phi}}\frac{1}{r\sin\theta}\frac{\partial}{\partial \varphi}, \tag{8.143}$$

where $\hat{\mathbf{r}}, \hat{\boldsymbol{\theta}}$, and $\hat{\boldsymbol{\phi}}$ are the unit vectors.

The pressure tensor of (8.142) can be conveniently written in terms of the pair distribution function as follows. We first introduce a distance vector $\boldsymbol{\ell}$ defined relative to the location of the centre of molecule i,

$$\boldsymbol{\ell}(\mathbf{r}_{ij}) = \tilde{\boldsymbol{\ell}}(\mathbf{r}_i, \mathbf{r}_j) - \mathbf{r}_i, \qquad 0 \leq \boldsymbol{\ell} \leq \mathbf{r}_{ij} \tag{8.144}$$

so that (8.142) can be rewritten as

$$\mathbf{p}(\mathbf{r}) = \rho(r)kT\mathbf{1} - \frac{1}{2}\left\langle \sum_{i\neq j} \frac{\partial u(ij)}{\partial \mathbf{r}_{ij}} \int_{C_{ij}} \mathrm{d}\boldsymbol{\ell}\delta(\mathbf{r}-\boldsymbol{\ell}-\mathbf{r}_i) \right\rangle.$$

Using the relations $f(x) = \int \mathrm{d}x'\delta(x-x')f(x')$ and $f(x)\delta(x-a) = f(a)\delta(x-a)$ we rewrite the above relation as

$$\begin{aligned}
\mathbf{p}(\mathbf{r}) &= \rho(\mathbf{r})kT\mathbf{1} - \frac{1}{2}\int \mathrm{d}\mathbf{r}'_{12}\mathrm{d}\omega'_1\mathrm{d}\omega'_2 \frac{\partial u(1'2')}{\partial \mathbf{r}'_{12}} \int_{C_{1'2'}} \mathrm{d}\boldsymbol{\ell} \\
&\quad \times \sum_{i\neq j} \left\langle \delta(\mathbf{r}-\boldsymbol{\ell}-\mathbf{r}_i)\delta\left(\mathbf{r}'_{12}-\mathbf{r}_{ij}\right)\delta\left(\omega'_1-\omega_i\right)\delta\left(\omega'_2-\omega_j\right)\right\rangle \\
&= \rho(\mathbf{r})kT\mathbf{1} - \frac{1}{2}\int \mathrm{d}\mathbf{r}'_{12}\int \mathrm{d}\omega'_1\int \mathrm{d}\omega'_2 \frac{\partial u(1'2')}{\partial \mathbf{r}'_{12}} \int_{C_{1'2'}} \mathrm{d}\boldsymbol{\ell} \\
&\quad \times \sum_{i\neq j} \left\langle \delta(\mathbf{r}-\boldsymbol{\ell}-\mathbf{r}_i)\delta\left(\mathbf{r}-\boldsymbol{\ell}+\mathbf{r}'_{12}-\mathbf{r}_j\right)\delta\left(\omega'_1-\omega_i\right)\delta\left(\omega'_2-\omega_j\right)\right\rangle.
\end{aligned}$$

From (3.105) and (3.108b) the ensemble average can be written in terms of the pair distribution function $f(\mathbf{r}_1\mathbf{r}_2\omega_1\omega_2)$,

$$\mathbf{p}(\mathbf{r}) = \rho(\mathbf{r})kT\mathbf{1} - \frac{1}{2}\int \mathrm{d}\mathbf{r}_{12}\mathrm{d}\omega_1\mathrm{d}\omega_2 \frac{\partial u(12)}{\partial \mathbf{r}_{12}} \int_{C_{12}} \mathrm{d}\boldsymbol{\ell} f(\mathbf{r}-\boldsymbol{\ell}, \mathbf{r}-\boldsymbol{\ell}+\mathbf{r}_{12}, \omega_1, \omega_2), \tag{8.145}$$

which is the general expression for $\mathbf{p}$ for a fluid of nonspherical molecules that obey pairwise additivity.[116–119] The integration path C_{12} in the integral over $\boldsymbol{\ell}$ must go from position $\mathbf{r}_1(\boldsymbol{\ell}=0)$ to $\mathbf{r}_2(\boldsymbol{\ell}=\mathbf{r}_{12})$, but is otherwise arbitrary.

We now consider several special or limiting cases, for an interface of general shape. One obvious choice of integration path is a straight line joining $\mathbf{r}_1$ to $\mathbf{r}_2$, so that

$$\boldsymbol{\ell} = \lambda(\mathbf{r}_2 - \mathbf{r}_1) = \lambda\mathbf{r}_{12}, \quad 0 \leq \lambda \leq 1 \tag{8.146}$$

and (8.145) becomes

$$\mathbf{p}_{IK}(\mathbf{r}) = \rho(\mathbf{r})kT\mathbf{1} - \frac{1}{2}\int \mathrm{d}\mathbf{r}_{12}\mathbf{r}_{12}\mathrm{d}\omega_1\mathrm{d}\omega_2\frac{\partial u(12)}{\partial \mathbf{r}_{12}}\int_0^1 \mathrm{d}\lambda f(\mathbf{r}-\lambda\mathbf{r}_{12},$$
$$\mathbf{r}+(1-\lambda)\mathbf{r}_{12},\omega_1,\omega_2). \tag{8.147}$$

We refer to this definition as the Irving–Kirkwood pressure tensor, since the straight line path was first proposed by Irving and Kirkwood.[110] With this choice, the force between a pair of molecules 'acts across' an element of area $\mathrm{d}S$ only if the line between their centres passes through $\mathrm{d}S$. Other choices are possible; one such choice, due to Harasima,[10] is considered in the subsection below.

For the case of atomic fluids, where u depends only on $\mathbf{r}_{12}$, (8.145) simplifies to the result of Schofield and Henderson[113]

$$\mathbf{p}(\mathbf{r}) = \rho(\mathbf{r})kT\mathbf{1} - \frac{1}{2}\int \mathrm{d}\mathbf{r}_{12}\hat{\mathbf{r}}_{12}u'(r_{12})\int_{C_{12}} \mathrm{d}\boldsymbol{\ell}f(\mathbf{r}-\boldsymbol{\ell},\mathbf{r}-\boldsymbol{\ell}+\mathbf{r}_{12}). \tag{8.148}$$

Finally, if $\mathbf{r}$ is in a region where the fluid is homogeneous the distribution function is independent of $\mathbf{r}$ and the integral over $\boldsymbol{\ell}$ is independent of the path,

$$\int_{C_{12}} \mathrm{d}\boldsymbol{\ell}f(\mathbf{r}-\boldsymbol{\ell},\mathbf{r}-\boldsymbol{\ell}+\mathbf{r}_{12},\omega_1,\omega_2) = \mathbf{r}_{12}f(\mathbf{r}_{12},\omega_1,\omega_2) = \mathbf{r}_{12}\rho^2\Omega^{-2}g(\mathbf{r}_{12},\omega_1,\omega_2) \tag{8.149}$$

so that (8.145) becomes (cf. (E.30))

$$\mathbf{p} = \rho kT\mathbf{1} - \frac{1}{2}\rho^2\int \mathrm{d}\mathbf{r}_{12}\mathbf{r}_{12}\left\langle\frac{\partial u(12)}{\partial \mathbf{r}_{12}}g(12)\right\rangle_{\omega_1\omega_2}. \tag{8.150}$$

The scalar pressure, $\mathrm{p} = (1/3)\mathrm{Tr}(\mathbf{p})$, is

$$p = \rho kT - \frac{1}{6}\rho^2\int \mathrm{d}\mathbf{r}_{12}\left\langle r_{12}u'(\mathbf{r}_{12}\omega_1\omega_2)g(\mathbf{r}_{12}\omega_1\omega_2)\right\rangle_{\omega_1\omega_2}, \tag{8.151}$$

which is the pressure equation, (3.242).

8.3.2 The planar surface

We now consider the particular case of a flat surface lying in the xy plane, in the presence of a weak external field (e.g. gravity) that acts in the z direction (see Fig. 8.8). From the axial symmetry of the system about the z-axis and the translational symmetry in the xy plane it follows that $p_{xx} = p_{yy} = p_{\mathrm{T}}(z)$ and $p_{zz} = p_{\mathrm{N}}(z)$, where p_{T} and p_{N} are the tangential and normal pressures, respectively, while the off-diagonal components $p_{\alpha\beta} = 0$ $(\alpha \neq \beta)$. Thus we have

$$\mathbf{p}(z) = p_{\mathrm{T}}(z)[\mathbf{ii} + \mathbf{jj}] + p_{\mathrm{N}}(z)\mathbf{kk}, \tag{8.152}$$

where $\mathbf{i}$, $\mathbf{j}$, $\mathbf{k}$, are unit vectors in the x, y, and z directions, respectively. Also, from the condition of hydrostatic equilibrium (8.136) we have

$$\frac{\partial p_{xx}}{\partial x} = \frac{\partial p_{yy}}{\partial y} = 0, \tag{8.153}$$

consistent with the translational symmetry mentioned above, and

$$\frac{\partial p_{zz}}{\partial z} = -\rho(z)\frac{\partial v(z)}{\partial z}. \tag{8.154}$$

If the external field is very weak the term on the right side of (8.154) is often negligible (as for many systems subjected to a gravitational field), and the z-dependence of p_N can be neglected. In such cases p_{N} is just the hydrostatic pressure p for the bulk phase.

For such an interface the general expression for $\mathbf{p}$, (8.145), becomes

$$\mathbf{p}(z) = \rho(z)kT\mathbf{1} - \frac{1}{2}\int \mathrm{d}\mathbf{r}_{12}\mathrm{d}\omega_1\mathrm{d}\omega_2\frac{\partial u(12)}{\partial \mathbf{r}_{12}}\int_{C_{12}} \mathrm{d}\boldsymbol{\ell} f(z - \ell_z, \mathbf{r}_{12}, \omega_1, \omega_2), \tag{8.155}$$

where we have made use of the fact that $f(12)$ can be written $f(z_1\mathbf{r}_{12}\omega_1\omega_2)$ for a planar interface. The Irving–Kirkwood pressure tensor, (8.147), becomes

$$\mathbf{p}_{\mathrm{IK}}(z) = \rho(z)kT\mathbf{1} - \frac{1}{2}\int \mathrm{d}\mathbf{r}_{12}\mathrm{d}\omega_1\mathrm{d}\omega_2\mathbf{r}_{12}\frac{\partial u(12)}{\partial \mathbf{r}_{12}}\int_0^1 \mathrm{d}\lambda f(z - \lambda z_{12}, \mathbf{r}_{12}, \omega_1, \omega_2) \tag{8.156}$$

with components

$$\begin{aligned} p_{\mathrm{N,IK}} &= \rho(z)kT - \frac{1}{2}\int \mathrm{d}\mathbf{r}_{12}\mathrm{d}\omega_1\mathrm{d}\omega_2 z_{12}\frac{\partial u(12)}{\partial z_{12}} \\ &\times \int_0^1 \mathrm{d}\lambda f(z - \lambda z_{12}, \mathbf{r}_{12}, \omega_1, \omega_2) \end{aligned} \tag{8.157}$$

$$\begin{aligned} p_{\mathrm{T,IK}}(z) = \rho(z)kT - \frac{1}{4}\int \mathrm{d}\mathbf{r}_{12}\mathrm{d}\omega_1\mathrm{d}\omega_2 \left[x_{12}\frac{\partial u(12)}{\partial x_{12}} + y_{12}\frac{\partial u(12)}{\partial y_{12}}\right] \\ \times \int_0^1 \mathrm{d}\lambda f(z - \lambda z_{12}, \mathbf{r}_{12}, \omega_1, \omega_2). \end{aligned} \tag{8.158}$$

Note that the kinetic and configurational terms in (8.157) both vary with z but the sum does not.

An alternative integration path for the integral over $\boldsymbol{\ell}$ in (8.155) has been proposed by Kirkwood and Buff[67] and by Harasima,[10] and this yields what is usually called the Harasima pressure tensor $\mathbf{p}_{\mathrm{H}}$. In this case two molecules contribute to the stress across an element of area d**S** only if one of them lies in a cylinder normal to d**S**, while the other lies on the other side of the infinite plane which contains d**S**. In this case the line joining the molecule centres will not, in general, pass through d**S**. In terms of the choice of contour C_{12} from molecule 1 to molecule 2, this corresponds to first going from $(x_1y_1z_1)$ parallel to the interface to the point $(x_2y_2z_1)$, then perpendicular to the interface from z_1 to z_2, keeping x_2 and y_2 fixed (see Fig. 8.18). This choice of C_{12} can be defined by $\int_{C_{12}} \mathrm{d}\boldsymbol{\ell} = \boldsymbol{\ell}$, with

$$\boldsymbol{\ell} = \boldsymbol{\ell}_1 + \boldsymbol{\ell}_2, \tag{8.159}$$

where

$$\boldsymbol{\ell}_1 = \alpha \mathbf{s}_{12} \qquad 0 \le \alpha \le 1; \ell_{1z} = 0 \tag{8.160}$$

$$\boldsymbol{\ell}_2 = \beta z_{12}\mathbf{k} \qquad 0 \le \beta \le 1; \ell_{2z} = \beta z_{12}, \tag{8.161}$$

where $\mathbf{s}_{12}$ is the projection of $\mathbf{r}_{12}$ onto the xy plane that passes through $\mathbf{r}_1$. Substituting (8.159)–(8.161) into (8.155) gives the Harasima pressure tensor,

$$\mathbf{p}_H(z) = \rho(z)kT\mathbf{1} - \frac{1}{2}\int \mathrm{d}\mathbf{r}_{12}\mathrm{d}\omega_1\mathrm{d}\omega_2 \frac{\partial u(12)}{\partial \mathbf{r}_{12}}[\mathbf{s}_{12} f(z, \mathbf{r}_{12}, \omega_1, \omega_2). + z_{12}\mathbf{k}\int_0^1 \mathrm{d}\beta f(z - \beta z_{12}, \mathbf{r}_{12}, \omega_1, \omega_2)] \tag{8.162}$$

with the components

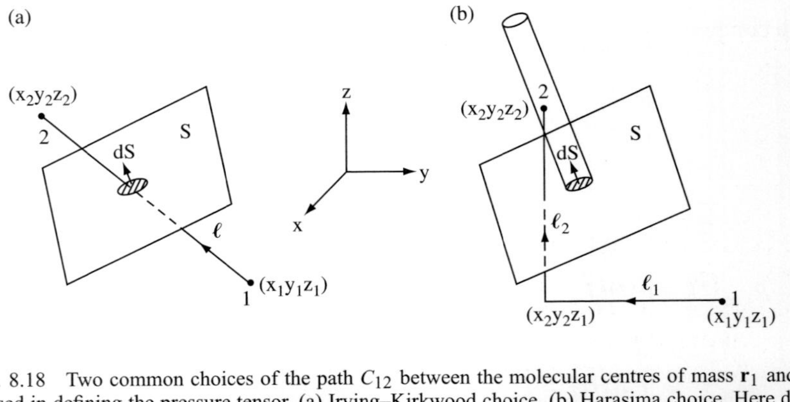

FIG. 8.18 Two common choices of the path C_{12} between the molecular centres of mass $\mathbf{r}_1$ and $\mathbf{r}_2$ used in defining the pressure tensor. (a) Irving–Kirkwood choice, (b) Harasima choice. Here d**S** is an element of area in some infinite plane **S**; the interface (assumed planar) lies in the xy plane.

$$p_{\mathrm{N,H}} = \rho(z)kT - \frac{1}{2}\int \mathrm{d}\mathbf{r}_{12}\mathrm{d}\omega_1\mathrm{d}\omega_2 z_{12}\frac{\partial u(12)}{\partial z_{12}}\int_0^1 \mathrm{d}\beta f(z-\beta z_{12},\mathbf{r}_{12},\omega_1,\omega_1) \tag{8.163}$$

$$p_{\mathrm{T,H}}(z) = \rho(z)kT - \frac{1}{4}\int \mathrm{d}\mathbf{r}_{12}\mathrm{d}\omega_1\mathrm{d}\omega_2\left[x_{12}\frac{\partial u(12)}{\partial x_{12}} + y_{12}\frac{\partial u(12)}{\partial y_{12}}\right]f(z,\mathbf{r}_{12},\omega_1,\omega_2). \tag{8.164}$$

From (8.157) and (8.163) we see that the normal component of the pressure tensor is the same for the two definitions of $\mathbf{p}$. That p_{N} must be invariant to the choice of path follows from (8.154) together with the fact that p_{N} is uniquely defined in the bulk phase regions. However, (8.158) and (8.164) will give different results for the tangential pressure. Although p_{T} has no physical significance and is arbitrary, integrals over p_{T}, such as surface tension (see below), may be observables and therefore unambiguous. The difference between (8.158) and (8.164) is given by

$$\Delta p_T(z) \equiv p_{\mathrm{T,H}}(z) - p_{\mathrm{T,IK}}(z) = \frac{1}{2}\int \mathrm{d}\mathbf{r}_{12}\mathrm{d}\omega_1\mathrm{d}\omega_2 x_{12}\frac{\partial u(12)}{\partial x_{12}} \times \int_0^1 \mathrm{d}\lambda[f(z-\lambda z_{12},\mathbf{r}_{12},\omega_1,\omega_2) - f(z,\mathbf{r}_{12},\omega_1,\omega_2)]. \tag{8.165}$$

Expanding $f(z-\lambda z_{12},\mathbf{r}_{12},\omega_1\omega_2)$ about $f(z,\mathbf{r}_{12},\omega_1\omega_2)$ in a Taylor series in powers of λz_{12}, and carrying out the integration over λ gives

$$\Delta p_T(z) = \frac{1}{2}\int \mathrm{d}\mathbf{r}_{12}\mathrm{d}\omega_1\mathrm{d}\omega_2 x_{12}\frac{\partial u(12)}{\partial x_{12}}\sum_{n=1}^{\infty}\frac{(-z_{12})^n}{(n+1)!}\frac{\partial^n}{\partial z^n}f(z,\mathbf{r}_{12},\omega_1,\omega_2). \tag{8.166}$$

From (8.166) and the homogeneity of the phases on both sides of the surface layer it follows that

$$\int_{-\infty}^{\infty} \mathrm{d}z\Delta p_T(z) = 0 \tag{8.167}$$

and we see from (8.121) that the surface tension is invariant to the choice of contour used for the pressure tensor, as it should be. We also see from (8.121) that, since $\gamma > 0$ and for the common case $p_{\mathrm{N}} \sim 0$, we have $\int \mathrm{d}z p_T(z) < 0$, i.e. p_{T} corresponds to a true tension, irrespective of the arbitrariness in the definition. If we substitute the Irving–Kirkwood pressure tensor expressions, (8.157) and (8.158), into the mechanical expression (8.121) for the surface tension, we find that

$$\gamma = \frac{1}{4}\int_{-\infty}^{\infty} \mathrm{d}z \int \mathrm{d}\mathbf{r}_{12}\mathrm{d}\omega_1\mathrm{d}\omega_2 \left[x_{12}\frac{\partial u(12)}{\partial x_{12}} + y_{12}\frac{\partial u(12)}{\partial y_{12}} - 2z_{12}\frac{\partial u(12)}{\partial z_{12}}\right] \times \int_0^1 \mathrm{d}\lambda f(z - \lambda z_{12}, \mathbf{r}_{12}, \omega_1, \omega_2).$$

This can be rewritten in the simpler form

$$\gamma = \frac{1}{4}\int_{-\infty}^{\infty} \mathrm{d}z \int \mathrm{d}\mathbf{r}_{12}\mathrm{d}\omega_1\mathrm{d}\omega_2 \left[x_{12}\frac{\partial u(12)}{\partial x_{12}} + y_{12}\frac{\partial u(12)}{\partial y_{12}} - 2z_{12}\frac{\partial u(12)}{\partial z_{12}}\right] \times f(z, \mathbf{r}_{12}, \omega_1, \omega_2), \tag{8.168}$$

where we make use of the relation

$$\int_{-\infty}^{\infty} \mathrm{d}z \int_0^1 \mathrm{d}\lambda f(z - \lambda z_{12}, \mathbf{r}_{12}, \omega_1, \omega_2) = \int_{-\infty}^{\infty} \mathrm{d}z f(z, \mathbf{r}_{12}, \omega_1, \omega_2), \tag{8.169}$$

which follows by interchanging the order of integration over z and λ. Equation (8.168) is identical to the equation (8.96) obtained for the surface tension using the virial route (see § 8.2.2), as can be seen by transforming the variables in the integrand from Cartesian (x_{12}, y_{12}, z_{12}) to spherical polar $(r_{12}, \theta_{12}, \phi_{12})$ coordinates (see (8.93)). Thus the virial and mechanical routes to γ are equivalent. The same expression for γ, (8.168), is of course obtained if we use the Harasima form of the pressure tensor in (8.121) in place of the Irving–Kirkwood one.

It also follows from (8.166) that integrals of the type

$$\int_{-\infty}^{\infty} \mathrm{d}z f(z) \Delta p_{\mathrm{T}}(z)$$

are not zero for arbitrary $f(z)$. In particular, the so-called *surface of tension*, z_{s}, for a planar surface is defined by[18, 120]

$$z_{\mathrm{s}} = \frac{1}{\gamma}\int_{-\infty}^{\infty} [p_{\mathrm{N}} - p_{\mathrm{T}}(z)]\, z\mathrm{d}z \tag{8.170}$$

and will depend on the choice of contour used in calculating the pressure tensor. Thus, in contrast to the situation for the surface tension, difference values for z_s will be obtained using the Irving–Kirkwood and Harasima definitions of the pressure tensor.[121–123] In fact, z_{s} can be placed anywhere in the surface region by choosing different contours in the definition of $\mathbf{p}$.

Computer simulation and theoretical calculations of the pressure tensor, surface tension, and surface of tension have confirmed these predictions. These studies

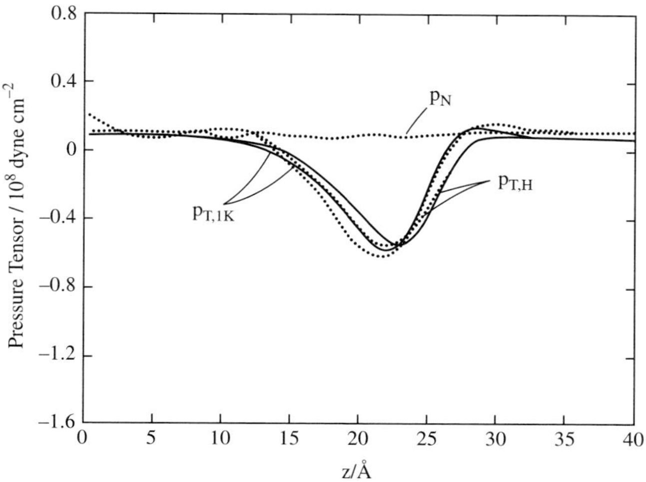

FIG. 8.19 Pressure tensor components for a LJ mixture modelled on argon-krypton at $T =$ 115.77K, $N_{\mathrm{Ar}}/N = 0.5$. The dotted lines are smoothed molecular dynamics simulation results and the full lines are mean field theory. All results are for a truncated, shifted LJ model with cutoff distance $r_c = 2.5\sigma_{\alpha\beta}$. (From ref. 124.)

have included the planar gas–liquid interface for pure[123] and mixed[124] Lennard-Jones fluids and for a pure Stockmayer (Lennard-Jones plus dipole) fluid,[125] and the spherical gas–liquid interface for Lennard-Jones[83] and Stockmayer[126] drops. The results of these various studies are qualitatively similar, and we show in Fig. 8.19 results for both the Irving–Kirkwood and Harasima pressure tensor components for a planar interface for a Lennard-Jones fluid. The normal pressure is seen to be constant through the interfacial region, within the statistical errors of the simulation. The tangential component, on the other hand, shows a pronounced minimum of about -50 bar, corresponding to a strong tension in the interfacial region. The Harasima and Irving–Kirkwood p_{T} values differ as expected, the Harasima curve being shifted more towards the vapour side of the surface (higher z) relative to the Irving–Kirkwood curve. There is also a small positive peak in $p_{\mathrm{T,IK}}$ on the vapour side, as has also been found for pure Lennard-Jones fluids.[123] The mean field theory (see § 8.6) predicts these features well. The results for the pure Lennard-Jones fluid at other state conditions[123] and for the Stockmayer fluid[125] are qualitatively similar to those shown in Fig. 8.19.

The values of the surface of tension, and hence $\delta = z_{\mathrm{e}} - z_{\mathrm{s}}$, the distance separating the equimolar surface and the surface of tension, depends on the choice of pressure tensor, and some simulation values for a Lennard-Jones fluid are shown in Table 8.3. The surface of tension lies on the liquid side of the interface. The Harasima $p_{\mathrm{T,H}}$ is found to yield smaller values of δ than the Irving–Kirkwood $p_{\mathrm{T,IK}}$, so that the IK surface of tension lies somewhat deeper in the liquid than the H one. Similar results are found for Lennard-Jones mixtures.[124]

Table 8.3

Surface tensions (γ) and surfaces of tension (z_s) for a pure truncated and shifted Lennard-Jones ($r_c = 3\sigma$) fluid from molecular dynamics simulation†

T^*	γ^*	δ^*_{IK}	δ^*_H
0.723 ± 0.001	0.680	1.02	0.71
1.013 ± 0.005	0.237	2.65	2.37

† $T^* = kT/\varepsilon$, $\gamma^* = \gamma\sigma^2/\varepsilon$, $\delta^* = (z_e - z_s)/\sigma$. Here γ is the same whether calculated from the mechanical or virial route, and is invariant to the choice of pressure tensor, but δ is not.
From ref. 123.

8.3.3 *The spherical surface*

The pressure tensor for a spherical interface has been considered in detail by Schofield and Henderson[113,127] and by Henderson[21] for atomic fluids, and by Walton and Gubbins[114] and Tjatjopoulos and Mann[128] for molecular fluids, and we consider it only very briefly here. From symmetry considerations there are only two independent principal axes components of the pressure tensor, the normal one, $p_N(r) = p_{rr}(r)$, and the tangential one, $p_T(r) = p_{\theta\theta}(r) = p_{\phi\phi}(r)$; here (r, θ, ϕ) are the usual spherical polar coordinates. Again, by symmetry, $\mathbf{p}(\mathbf{r})$ depends only on $r = |\mathbf{r}|$. Thus the pressure tensor is of the form

$$\mathbf{p}(r) = p_N(r)\hat{\mathbf{r}}\hat{\mathbf{r}} + p_T(r)[\hat{\boldsymbol{\theta}}\hat{\boldsymbol{\theta}} + \hat{\boldsymbol{\phi}}\hat{\boldsymbol{\phi}}] \tag{8.171}$$

and on using this in (8.137), the condition of hydrostatic equilibrium (here we neglect the external field), we get a relation between $p_N(r)$ and $p_T(r)$,

$$p_T(r) = p_N(r) + \frac{r}{2}\frac{\mathrm{d}p_N(r)}{\mathrm{d}r} \tag{8.172}$$

or, on integration

$$p_N(r) = \frac{2}{r^2}\int_0^r p_T(r')r'\mathrm{d}r'. \tag{8.173}$$

Expressions for $p_N(r)$ and $p_T(r)$ for spherical surfaces have been given for molecular fluids by Walton and Gubbins[114] for both the Irving–Kirkwood and Harasima choices of contour. For a spherical surface the mechanical route (cf. § 8.2.4) gives for the surface tension[18]

$$\gamma = \frac{1}{R_s^2}\int_0^\infty [p_N(r) - p_T(r)]r^2\mathrm{d}r, \tag{8.174}$$

where R_s is the surface of tension radius, which is also related to γ by Laplace's equation,[18]

$$p^{\alpha} - p^{\beta} = \frac{2\gamma}{R_s}, \tag{8.175}$$

where p^{α} is the pressure at the centre of the drop (of phase α) and p^{β} is the (bulk) pressure in the surrounding phase. Examination of Eqs. (8.174) and (8.175) shows that *both* γ and R_s depend on the choice of contour used to define the pressure tensor, so that neither is well defined when the mechanical route is used. This is in contrast to the planar case, where γ was well defined, but R_s was not. These points are discussed in detail in the literature.[18,83,113,126]

Computer simulation studies[83,126,129] of the pressure tensor, surface tension, and surface of tension of liquid drops surrounded by vapour have confirmed these general conclusions. In Fig. 8.20 are shown typical results for the Irving–Kirkwood pressure tensor for Lennard-Jones (cutoff and shifted, $r_c = 2.5\sigma$) drops of several sizes. In such drops the pressures within the drop and outside differ even in the bulk phases because of the tension of the curved surface, the liquid pressure being higher than the gas (cf. Laplace's equation, (8.108) and (8.175)). For the larger drops $p_T < p_N$ at all r, so that the surface is under tension. For the $N = 54$ drop there is a region near $r/\sigma = 2$ where $p_T > p_N$, i.e. the interface is under compression, a property also predicated by theories.[130] The Harasima pressure tensor results[83] for such drops are of similar form to the Irving–Kirkwood ones. There are significant uncertainties in the computer simulation results for the

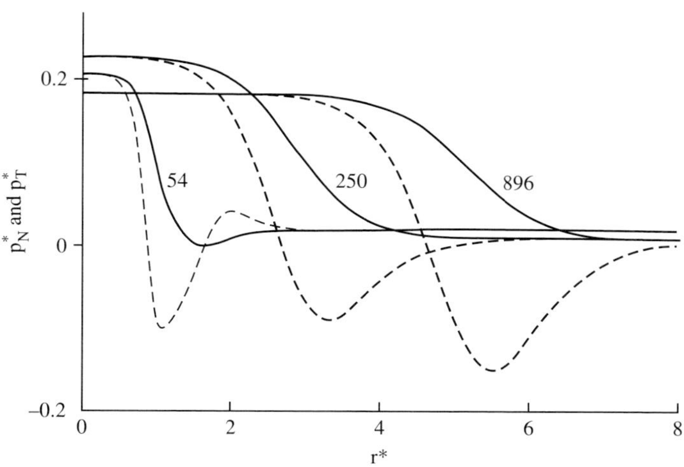

FIG. 8.20 The normal (———) and transverse (– – –) components of the Irving–Kirkwood pressure tensor for Lennard-Jones drops of several sizes at $T^* = 0.71$, from molecular dynamics calculations. Here $r^* = r/\sigma$ and $p^* = p\sigma^3/\varepsilon$. The numbers on the curves give N, the total number (liquid + vapour) of molecules in the system. (From ref. 83.)

surface tension and surface of tension, because of statistical errors. However, the surface tension appears to fall as the drop curvature increases (i.e. as radius decreases).[83, 126, 129–131]

The pressure tensor and surface tension for an interface of cylindrical symmetry have also been studied.[132]

8.4 Density functional theory

So far we have confined our attention to the statistical mechanics of systems at equilibrium. In this section we introduce two quantities, $\Omega[f']$ and $\mathcal{A}[f']$, which are functionals of an arbitrary distribution function f' that does not necessarily correspond to an equilibrium state. These functionals are defined so that they possess two important properties: (a) for a given intermolecular potential and external potential, both $\Omega[f']$ and $\mathcal{A}[f']$ take their minimum values when $f' = f$, the distribution function for the equilibrium state of the system at the given state condition, and (b) these minimum values, $\Omega[f]$ and $\mathcal{A}[f]$, are the thermodynamic grand potential Ω and intrinsic Helmholtz energy A_i (i.e. the total Helmholtz energy minus a contribution from the external field; A_i is defined more precisely below), respectively, for the equilibrium state. From these properties a variational principle is derived for $\Omega[f']$ that can be used to determine the density–orientation profile and other properties of the inhomogeneous system. This approach, called *density functional theory*, has been a particularly fruitful one, and will form the starting point for several of the more approximate theories in the sections that follow. The *classical* density functional theory described here, first introduced by Ebner, Saam, and Stroud in 1976,[133] has its roots in quantum mechanical density functional theory, in which the energy of an electronic system containing n electrons is expressed as a functional of the total electron density, $\rho(\mathbf{r})$, at a point $\mathbf{r}$. Thus density functional theory provides a unified framework for treating both quantal and classical systems. The treatment here follows that of Mermin[134] for the electron gas, and of Ebner *et al.*,[133] Evans,[12] and Rowlinson and Widom[18] for classical atomic fluids.[135] The extension to molecular fluids is straightforward, and has been given by Chacón *et al.*[136] and Telo da Gama.[45]

We consider an open, single-component system of volume V, temperature T, and chemical potential μ, and define an N-body equilibrium configurational distribution function $f(\mathbf{r}^N\omega^N)$ for the equilibrium state by

$$f(\mathbf{r}^N\omega^N) = \Xi^{-1}\exp[-\beta(\mathcal{U} + \mathcal{V} - \mu N)], \tag{8.176}$$

where Ξ is the grand partition function,

$$\Xi = \mathrm{Tr}_{\mathrm{cl}}\exp[-\beta(\mathcal{U} + \mathcal{V} - \mu N)] \tag{8.177}$$

and $\mathrm{Tr}_{\mathrm{cl}}$ is the classical trace, given by (cf. (3.190))[36]

$$\mathrm{Tr}_{\mathrm{cl}}(\cdots) = \sum_N \left[\frac{q_{qu}}{\Lambda_t^3\Lambda_r\Omega}\right]^N \frac{1}{N!}\int(\cdots)\mathrm{d}\mathbf{r}^N\mathrm{d}\omega^N. \tag{8.178}$$

We note that for any configurational variable, O_N, which may depend on N, we have $\langle O_N\rangle = \mathrm{Tr}_{\mathrm{cl}}(f(\mathbf{r}^N\omega^N)O_N)$ (analogous to the quantum expression (B.59) of Vol. 1), so that with $O_N = 1$ we see that $f(\mathbf{r}^N\omega^N)$ obeys the normalization condition

$$\mathrm{Tr}_{\mathrm{cl}}f(\mathbf{r}^N\omega^N) = 1 \tag{8.179}$$

and also that the reduced configurational distribution function of order h (cf. (3.189)) is related to $f(\mathbf{r}^N\omega^N)$ by[36]

$$\begin{aligned} f(\mathbf{r}^h\omega^h) &= \sum_N \left[\frac{q_{\mathrm{qu}}}{\Lambda_t^3\Lambda_r\Omega}\right]^N \frac{1}{(N-h)!}\int \mathrm{d}\mathbf{r}^{N-h}\mathrm{d}\omega^{N-h} f(\mathbf{r}^N\omega^N) \\ &= \frac{1}{\Xi}\sum_N \left[\frac{\exp[\beta\mu]q_{\mathrm{qu}}}{\Lambda_t^3\Lambda_r\Omega}\right]^N \frac{1}{(N-h)!}\int \mathrm{d}\mathbf{r}^{N-h}\mathrm{d}\omega^{N-h}\exp[-\beta(\mathcal{U}+\mathcal{V})]. \end{aligned} \tag{8.180}$$

We now define the *grand potential functional* $\Omega[f'(\mathbf{r}^N\omega^N)] \equiv \Omega[f']$ by[134]

$$\Omega[f'] = \mathrm{Tr}_{\mathrm{cl}}f'(\mathcal{U}+\mathcal{V}-\mu N + kT\ell\mathrm{n}f'), \tag{8.181}$$

where $f' \equiv f'(\mathbf{r}^N\omega^N)$ is an N-body distribution function for the system at (V, T, μ), but not necessarily for the equilibrium state under these conditions. The function f' obeys the normalization condition (8.179),

$$\mathrm{Tr}_{\mathrm{cl}}f'(\mathbf{r}^N\omega^N) = 1. \tag{8.182}$$

When the system *is* at thermodynamic equilibrium we have $f' = f$, the equilibrium distribution function, and from (8.176) and (8.181) we find that[137]

$$\Omega[f] = -kT\ln\Xi = \Omega, \tag{8.183}$$

where $\Omega = A - \mu\langle N\rangle = -pV + \gamma\mathcal{S}$ is the grand potential (see (8.56) and (8.60)). Moreover, we can show that $\Omega[f]$ is the minimum of $\Omega[f']$, i.e.

$$\Omega[f] < \Omega[f'] \quad \text{for } f \neq f'. \tag{8.184}$$

The proof is based on the Gibbs inequality (see § 6.8), and is as follows. From (8.181) and (8.183) together with (8.182), we have

$$\Omega[f'] - \Omega[f] = \mathrm{Tr}_{\mathrm{cl}}f'(\mathcal{U}+\mathcal{V}-\mu N + kT\ln f' + kT\ln\Xi). \tag{8.185}$$

Replacing Ξ using (8.176) gives

$$\Omega[f'] - \Omega[f] = kT\,\mathrm{Tr}_{\mathrm{cl}}(f'\ln f' - f'\ln f) \tag{8.186}$$

$$= kT\,\mathrm{Tr}_{\mathrm{cl}}f\left[\frac{f'}{f}\ln\frac{f'}{f} - \frac{f'}{f} + 1\right], \tag{8.187}$$

where we note that the last two terms in the bracket cancel from the normalization condition (8.182). The right-hand side of (8.187) must be positive if $f' \neq f$ since

$x \ln x > (x-1)$ for $x \neq 1$. When $x = 1$, i.e. $f' = f$, the right side of (8.187) vanishes. Thus (8.184) follows.

It will be convenient in what follows to write the definition of $\Omega[f']$ in a somewhat different form. We first define a second functional, the *intrinsic free energy* functional $\mathcal{A}[f']$ by

$$\mathcal{A}[f'] \equiv \mathrm{Tr}_{\mathrm{cl}} f'(\mathcal{U} + kT \ln f'). \tag{8.188}$$

If the the system is at thermodynamic equilibrium, this becomes, on using (8.176) for f and (8.4) for $\mathcal{V}$,

$$\begin{aligned}
\mathcal{A}[f] &= \mathrm{Tr}_{\mathrm{cl}} f(-kT \ln \Xi + \mu N - \mathcal{V}) \\
&= -kT \ell\mathrm{n} \Xi + \mu \langle N \rangle - \mathrm{Tr}_{\mathrm{cl}} f \mathcal{V} \\
&= \Omega + \mu \langle N \rangle - \mathrm{Tr}_{\mathrm{cl}} f \textstyle\sum_i v(\mathbf{r}_i \omega_i) \\
&= A - \int \mathbf{dr} \mathrm{d}\omega f(\mathbf{r}\,\omega) v(\mathbf{r}\,\omega) \\
&= A_{\mathrm{i}}
\end{aligned} \tag{8.189}$$

where $A = \Omega + \mu \langle N \rangle$ is the total Helmholtz energy, $f(\mathbf{r}\,\omega)$ is the equilibrium singlet distribution function, or density–orientation profile, and $A_{\mathrm{i}} \equiv A - \int \mathbf{dr} \mathrm{d}\omega f(\mathbf{r}\,\omega) v(\mathbf{r}\,\omega)$ is the *intrinsic* Helmholtz energy. The intrinsic Herlmholtz free energy is the total minus the *explicit* external potential term; it still depends implicitly on the external potential. By an argument similar to that used above for $\Omega[f']$, it is easy to show that $\mathcal{A}[f]$ is the minimum of $\mathcal{A}[f']$, i.e.

$$\mathcal{A}[f] < \mathcal{A}[f'] \quad \text{for } f \neq f'. \tag{8.190}$$

Thus we have shown that the functionals $\Omega[f']$ and $\mathcal{A}[f']$ take their minimum values at equilibrium, and that these minimum values are the familiar thermodynamic potentials Ω and A_{i}. From (8.181) and (8.188) we can rewrite $\Omega[f']$ as

$$\begin{aligned}
\Omega[f'] &= \mathcal{A}[f'] + \mathrm{Tr}_{\mathrm{cl}} f' \mathcal{V} - \mu \mathrm{Tr}_{\mathrm{cl}} f' N \\
&= \mathcal{A}[f'] + \int \mathbf{dr} \mathrm{d}\omega f'(\mathbf{r}\,\omega) v(\mathbf{r}\,\omega) - \mu \langle N \rangle' \\
&= \mathcal{A}[f'] + \int \mathbf{dr} \mathrm{d}\omega f'(\mathbf{r}\,\omega) v(\mathbf{r}\,\omega) - \mu \int \mathbf{dr} \mathrm{d}\omega f'(\mathbf{r}\,\omega),
\end{aligned} \tag{8.191}$$

where we have used (8.4) for $\mathcal{V}$, and $f'(\mathbf{r}\,\omega)$ is a one-body density-orientational profile corresponding to $f'(\mathbf{r}^N \omega^N)$, given by

$$f'(\mathbf{r}\,\omega) = \sum_N \left[\frac{q_{\mathrm{qu}}}{\Lambda_t^3 \Lambda_r \Omega} \right]^N \frac{1}{(N-1)!} \int \mathbf{dr}^{N-1} \mathrm{d}\omega^{N-1} f'(\mathbf{r}^N \omega^N). \tag{8.192}$$

In the above discussion Ω and $\mathcal{A}$ are regarded as functionals of the full N-body distribution function $f'(\mathbf{r}^N \omega^N)$. It is possible to show (cf. §3.1.7, where we show the one-to-one correspondence between the pair potential $u(12)$ and the pair distribution function $f(12)$) that only one external potential $v(\mathbf{r}\,\omega)$ can give rise to a given equilibrium one body density–orientation profile $f(\mathbf{r}\,\omega)$ and to the corresponding N-body function, $f(\mathbf{r}^N\omega^N)$. The proof is given by Evans.[12] It follows that $f(\mathbf{r}^N\omega^N)$ is a functional of $f(\mathbf{r}\,\omega)$, and we can therefore equally well take $\Omega[f']$ and $\mathcal{A}[f']$ to be functionals of the singlet density–orientation profile $f'(\mathbf{r}\,\omega)$ in what follows.

Since Ω is a minimum at equilibrium, we can write

$$\left.\frac{\delta\Omega[f']}{\delta f'(\mathbf{r}\,\omega)}\right|_{f'=f(\mathbf{r}\,\omega)} = 0, \tag{8.193}$$

where the differentiation is at fixed $v(\mathbf{r}\,\omega)$ and (μ, V, T), and since this minimum value is just the grand potential we have

$$\Omega[f] = \Omega. \tag{8.194}$$

From (8.191) and (8.193) we find

$$\mu_{\rm i}(\mathbf{r}\,\omega) + v(\mathbf{r}\,\omega) = \mu, \tag{8.195}$$

where $\mu_{\rm i}(\mathbf{r}\,\omega)$ is the equilibrium intrinsic chemical potential, defined by

$$\mu_{\rm i}(\mathbf{r}\,\omega) \equiv \left.\frac{\delta\mathcal{A}}{\delta f'(\mathbf{r}\,\omega)}\right|_{f'=f(\mathbf{r}\,\omega)}. \tag{8.196}$$

Equations (8.195) and (8.196) are the fundamental ones for the density functional approach to inhomogeneous fluids. Equation (8.195) states that the chemical potential μ is made up of an intrinsic contribution $\mu_{\rm i}$ and an external field contribution v. Given an expression for $\mathcal{A}$, the equilibrium density–orientation profile $f(\mathbf{r}\,\omega)$ can be determined by minimization of $\mathcal{A}$; this is usually done by iterative methods. The intrinsic chemical potential and grand potential can then be obtained from (8.196) and (8.191), respectively. In the sections that follow we shall consider several approximate forms for $\mathcal{A}$.

We note that, despite appearances, $\mu_{\rm i}$ in (8.196) has the correct intensive and dimensional properties, as can be seen from the definition of a functional derivative, Eq. (8A.6), which for the case at hand becomes

$$\delta\mathcal{A} = \int_0^V \mathrm{d}\mathbf{r} \int \mathrm{d}\omega \left(\frac{\delta\mathcal{A}[f']}{\delta f'(\mathbf{r}\,\omega)}\right) \delta f'(\mathbf{r}\,\omega).$$

For the special case of an atomic fluid in the limit when the external field vanishes, so that the fluid is uniform, we have $\mu_{\rm i} = \mu$, and restricting the variations of f' to variations of the equilibrium f by varying N, we find the above equation gives $(\delta\mathcal{A}/\delta f')|_{f'=f} = (1/V)(\partial A/\partial\rho)_{V,T} = (\partial A/\partial N)_{V,T} = \mu$, the standard equilibrium relation for uniform fluids.

8.5 Perturbation theory

In perturbation theory (see Chapter 4) the properties of the system of interest, in which the molecules interact with intermolecular potential energy $\mathcal{U}(\mathbf{r}^N\omega^N)$, are related through an expansion to those of a reference system in which the potential is $\mathcal{U}_0(\mathbf{r}^N\omega^N)$. As for the homogeneous fluids considered in Chapter 4, we will want to choose a reference fluid that is well understood (i.e. one whose thermodynamic properties and low-order distribution functions are known, from molecular simulation, theory or experiment), but that also has a structure similar to that of the fluid of interest (see §6.4 and endnote 96 of Chapter 4). For the cases considered here there is the added consideration that the system of interest contains an interface, and is therefore highly nonuniform. The reference system must exhibit a similar degree of nonuniformity if the perturbation expansion is to be successful. Two somewhat different approaches have been used in dealing with this nonuniformity, and are described in the subsection below.

For nonuniform fluids composed of spherical molecules, the first use of perturbation theory was that of Toxvaerd[99,138] and this was followed by the work of Lee *et al.*[96] and Abraham.[139] The first applications to nonuniform molecular fluids were those by Gray and Gubbins[87] (for the surface tension) and by Haile *et al.*[140] (for the density–orientation profile).

8.5.1 *General expansion*

In this section we describe two somewhat different approaches that have been used to apply perturbation theory to nonuniform fluids. In the first, the potentials $\mathcal{U}$ and $\mathcal{U}_0$ are assumed to be sufficiently similar that both the real and reference systems exhibit phase separation, and similar degrees of nonuniformity. The Taylor expansion methods of Chapter 4 can then be used. In the second approach one assumes a reference system with potential $\mathcal{U}_0$ in which the density–orientation profile $f(\mathbf{r}\,\omega)$ is the same as that present in the system of interest. Such a reference fluid will not be an equilibrium one, but we assume (see §8.4) that it is nevertheless possible to define grand potential and free energy functionals for such a fluid. A functional perturbation expansion is used to relate the properties of the real and reference fluids. The Taylor expansion method cannot be used for reference molecules that interact with a purely repulsive potential, since there is no fluid–fluid phase transition;[141] the second approach must then be used.

Taylor expansion

We consider a system of two fluid phases having N molecules in volume V at temperature T. We suppose that N, V, and T are such that the reference system is in a corresponding two-phase region, although the relative amounts of the two phases will differ for the two systems. Thus for vapour–liquid equilibrium the temperature must lie between the triple and critical points for both the real and reference systems; this implies that the perturbation potential must not be so large that there is not a substantial overlap of the coexistence regions for the real and the reference systems.

We assume that pairwise additivity of the intermolecular potentials holds, and that external fields can be neglected. The perturbation expansion for the surface tension is obtained by expanding the Helmholtz free energy A of the two-phase system and then using (8.58) to obtain the series for the surface tension.[142] We consider a fluid with pair potential $u(\mathbf{r}_{12}\omega_1\omega_2;\lambda)$, where λ is some parameter such that (cf. (4.10), (4.11))

$$\begin{aligned} u(\mathbf{r}_{12}\omega_1\omega_2;\lambda=0) &= u_o(\mathbf{r}_{12}\omega_1\omega_2),\\ u(\mathbf{r}_{12}\omega_1\omega_2;\lambda=1) &= u(\mathbf{r}_{12}\omega_1\omega_2). \end{aligned} \tag{8.197}$$

Expanding the Helmholtz free energy for this system in powers of λ, and subsequently putting $\lambda=1$, yields (4.16), $A_c = A_0 + A_1 + A_2 + \ldots$. For a planar interface in the xy plane, we have

$$A_1 = \frac{1}{2}\int d\mathbf{r}_1 d\mathbf{r}_2 \left\langle \left[\frac{\partial u(12;\lambda)}{\partial\lambda}\right]_{\lambda=0} f_0(z_1\mathbf{r}_{12}\omega_1\omega_2)\right\rangle_{\omega_1\omega_2} \tag{8.198}$$

$$\begin{aligned} A_2 = {}& \frac{1}{4}\int d\mathbf{r}_1 d\mathbf{r}_2 \left\langle \left[\frac{\partial^2 u(12;\lambda)}{\partial\lambda^2}\right]_{\lambda=0} f_o(z_1\mathbf{r}_{12}\omega_1\omega_2)\right\rangle_{\omega_1\omega_2}\\ &+\frac{1}{4}\int d\mathbf{r}_1 d\mathbf{r}_2 \left\langle \left[\frac{\partial u(12;\lambda)}{\partial\lambda}\right]_{\lambda=0} f_1(z_1\mathbf{r}_{12}\omega_1\omega_2)\right\rangle_{\omega_1\omega_2}, \end{aligned} \tag{8.199}$$

where $u(12;\lambda) \equiv u(\mathbf{r}_{12}\omega_1\omega_2;\lambda)$. Equations (8.198) and (8.199) are a formally straightforward extension of the usual expressions for uniform systems, (4.19) and (4.20), to the inhomogeneous case; $f_0\,(z_1\mathbf{r}_{12}\omega_1\omega_2)$ is the inhomogeneous angular pair distribution function for the reference system, and $f_1\,(z_1\mathbf{r}_{12}\omega_1\omega_2)$ is the first-order perturbation term in the expansion of $f(z_1\mathbf{r}_{12}\omega_1\omega_2)$.

The expansion for the surface tension is

$$\gamma = \gamma_o + \gamma_1 + \gamma_2 + \cdots. \tag{8.200}$$

By applying (8.58) to (8.198) and using Green's method (see § 8.2.2) we obtain[142]

$$\gamma_1 = -\frac{1}{2}\int_0^\infty \mathrm{d}r_{12} r_{12}^2 \int \mathrm{d}\omega_{12} \int \mathrm{d}z_1 z_1 \left\langle \left[\frac{\partial u(12;\lambda)}{\partial\lambda}\right]_{\lambda=0} \left[\frac{\partial f_o(z_1\mathbf{r}_{12}\omega_1\omega_2)}{\partial z_1}\right]_{\mathbf{r}_{12}}\right\rangle_{\omega_1\omega_2}. \tag{8.201}$$

The general expression for γ_2 can be derived by similar methods, and involves reference distribution functions up to fourth order.

The distribution functions can be expanded in a similar fashion. For the density–orientation profile we have

$$f(\mathbf{r}_1\omega_1) = f_o(\mathbf{r}_1\omega_1) + f_1(\mathbf{r}_1\omega_1) + f_2(\mathbf{r}_1\omega_1) + \cdots, \tag{8.202}$$

where f_0 is the profile for the reference system and $f_k(\mathbf{r}_1\omega_1) = (k!)^{-1}(\partial^k f(\mathbf{r}_1\omega_1)/\partial\lambda^k)_{\lambda=0}$. Differentiation of (8.5), using the methods of Chapter 4, gives the first-order term as[143]

$$f_1(\mathbf{r}_1\omega_1) = -\beta\Omega \int \mathrm{d}\mathbf{r}_2 \left\langle \left[\frac{\partial u(12;\lambda)}{\partial\lambda}\right]_{\lambda=0} f_o(12) \right\rangle_{\omega_2}$$
$$-\frac{1}{2}\beta\Omega^2 \int \mathrm{d}\mathbf{r}_2\mathrm{d}\mathbf{r}_3 \left\langle \left[\frac{\partial u(23;\lambda)}{\partial\lambda}\right]_{\lambda=0} f_o(123) \right\rangle_{\omega_2\omega_3} \tag{8.203}$$
$$+\frac{1}{2}\beta\Omega^2 f_o(\mathbf{r}_1\omega_1) \int \mathrm{d}\mathbf{r}_1\mathrm{d}\mathbf{r}_2 \left\langle \left[\frac{\partial u(12;\lambda)}{\partial\lambda}\right]_{\lambda=0} f_o(12) \right\rangle_{\omega_1\omega_2},$$

where $\Omega = 4\pi$ or $8\pi^2$ for linear or nonlinear molecules, respectively.

Functional perturbation theory

It is also possible to develop perturbation theory starting from the density functional formalism of §8.4.[12,18,144] We begin by deriving a relation between the grand potential Ω and the two-body distribution function. The grand partition function Ξ is a functional of the intermolecular potential energy $\mathcal{U}$. If we assume pairwise additivity, (1.13), then Ξ is a functional of the pair potential $u(12)$. To maintain generality in what follows we shall regard Ξ as a functional of $s(u)$, where $s(u)$ is some arbitrary function of $u(12)$. Examples of $s(u)$ in which we shall be interested later include $s(u) = u$, appropriate for the u-expansion, and $s(u) = \exp(-\beta u)$, appropriate for the f-expansion. We now consider the change $\delta\Xi$ in Ξ due to some small change δs in s, keeping $\mathcal{V}$ fixed. From (8.2) we have

$$\delta\Xi = -\beta \sum_N \frac{\hat{z}^N}{N!} \int \mathrm{d}\mathbf{r}^N \mathrm{d}\omega^N \exp[-\beta(\mathcal{U}+\mathcal{V})] \left[\sum_{i<j} u_s(ij)\delta s(ij)\right], \tag{8.204}$$

where

$$u_s(ij) = \frac{\partial u(ij)}{\partial s(u(ij))} \tag{8.205}$$

and $s(ij) = s(u(ij))$. All of the terms in the sum over i and j give the same result after integration, so that, using the definition of the two-body distribution function, (8.1), we find that

$$\delta\Xi = -\frac{1}{2}\beta\Xi \int \mathrm{d}\mathbf{r}_1\mathrm{d}\mathbf{r}_2\mathrm{d}\omega_1\mathrm{d}\omega_2 u_s(12) f(12)\delta s(12), \tag{8.206}$$

so that from (8A.6) we have

$$\frac{\delta \ln \Xi}{\delta s(12)} = -\frac{1}{2}\beta u_s(12) f(12) \tag{8.207}$$

or, since $\Omega = -kT \ln \Xi$,

$$\frac{\delta\Omega}{\delta s(12)} = \frac{\delta\Omega[f]}{\delta s(12)} = \frac{1}{2}u_s(12)f(12), \tag{8.208}$$

where $\Omega[f]$ is the grand potential functional of §8.4. Thus, for pairwise additive potentials, the grand potential is a convenient generating functional for the equilibrium pair distribution function. Moreover, $\Omega[f]$ is a functional of the equilibrium density–orientation profile $f(\mathbf{r}\,\omega)$ and is given by (8.191). From (8.191) and (8.208) we have

$$\frac{\delta\mathcal{A}[f]}{\partial s(12)} = \frac{1}{2}u_s(12)f(12), \tag{8.209}$$

where the functional derivative gives the change in $\mathcal{A}$ due to a small change in $s(12)$, keeping $f(1)$ and $v(1)$ fixed.[145]

We now introduce a perturbation parameter λ such that $u(12;\lambda)$ is $u(12)$ when $\lambda = 1$ and $u_0(12)$ when $\lambda = 0$, where u and u_0 are the pair potentials for the full and reference systems, respectively (cf. (8.197) and § 4.3). We assume $s(12;\lambda) = s(u(12;\lambda))$ can be taken[146] to depend linearly on λ,

$$s(12;\lambda) = s_0(12) + \lambda s_1(12), \tag{8.210}$$

where $s_1(12) = s(12) - s_0(12)$ is the perturbation in $s(u(12))$. On functionally integrating (8.209) and using (8.210) we have

$$\begin{aligned}\delta\mathcal{A}(\lambda) &= \frac{1}{2}\int \mathrm{d}\mathbf{r}_1\mathrm{d}\mathbf{r}_2\mathrm{d}\omega_1\mathrm{d}\omega_2 u_s(12;\lambda)f(12;\lambda)\delta s(12;\lambda)\\ &= \frac{1}{2}\mathrm{d}\lambda\int \mathrm{d}\mathbf{r}_1\mathrm{d}\mathbf{r}_2\mathrm{d}\omega_1\mathrm{d}\omega_2 u_s(12;\lambda)f(12;\lambda)s_1(12).\end{aligned} \tag{8.211}$$

Integrating over λ from $\lambda = 0$ to $\lambda = 1$ we obtain

$$\mathcal{A}[f] = \mathcal{A}_0[f] + \frac{1}{2}\int_0^1 \mathrm{d}\lambda \int \mathrm{d}\mathbf{r}_1\mathrm{d}\mathbf{r}_2\mathrm{d}\omega_1\mathrm{d}\omega_2 u_s(12;\lambda)f(12;\lambda)s_1(12). \tag{8.212}$$

Here $\mathcal{A}[f] = A_i$ is the intrinsic Helmholtz energy for the system of interest with potential $u(12)$ and density–orientation profile $f(\mathbf{r}\,\omega)$, while $\mathcal{A}_0[f]$ is the corresponding quantity for the reference fluid with potential $u_0(12)$, but with *the same density–orientation profile* $f(\mathbf{r}\,\omega)$ as the system of interest. The two-body distribution function $f(12;\lambda)$ is for a system with pair potential $u(12;\lambda)$ and the density–orientation profile $f(\mathbf{r}\,\omega)$ is for the system of interest. Expanding $u_s(12;\lambda)f(12;\lambda)$ in powers of λ gives,

$$A_i = \mathcal{A}[f] = \mathcal{A}_0[f] + \frac{1}{2}\int \mathrm{d}\mathbf{r}_1\mathrm{d}\mathbf{r}_2\mathrm{d}\omega_1\mathrm{d}\omega_2 u_{s,0}(12)f_0(12)s_1(12) + O\left(s_1^2\right), \tag{8.213}$$

where $u_{s,0}(12) = u_s(12;\lambda = 0)$ and $f_0(12)$ is the pair distribution function for the reference system with profile $f(\mathbf{r}\,\omega)$. We note that both $\mathcal{A}_0$ and $f_0(12)$ are for the reference fluid with profile $f(\mathbf{r}\,\omega)$. We can imagine this to be accomplished physically by some external field that induces the profile $f(\mathbf{r}\,\omega)$. Alternatively, we can suppose the system to be a nonequilibrium one. The latter approach implies that quantities such as $\mathcal{A}_0$ and $f_0(12)$ can be defined for systems that are not equilibrium ones. The same assumption underlies the density functional theory of §8.4.

The quantities $\mathcal{A}_0$ and $f_0(12)$ are usually calculated from a mean density approximation. In the simplest version, they are given by[12, 144]

$$\mathcal{A}_o[f] = \int \mathrm{d}\mathbf{r} a_o(f(\mathbf{r}\,\omega)) \tag{8.214}$$

$$f_o(12) = f(\mathbf{r}_1\omega_1)f(\mathbf{r}_2\omega_2)g_o(\mathbf{r}_{12}\omega_1\omega_2;\bar{\rho}), \tag{8.215}$$

where $a_0[f(\mathbf{r}\,\omega)]$ is the Helmholtz free energy density of a spatially homogeneous fluid of molecules with pair potential $u_0(12)$ with a constant density $\rho(\mathbf{r}) = f(\mathbf{r}) = \int \mathrm{d}\omega f(\mathbf{r}\,\omega)$ equal to the number density of the real fluid at $\mathbf{r}$, and with the same distribution of molecular orientations, $f(\mathbf{r}\,\omega)$, as the real fluid at point $\mathbf{r}$. In (8.215) $f(\mathbf{r}\,\omega)$ is the density–orientation profile for the real fluid, while $g_0(\mathbf{r}_{12}\omega_1\omega_2;\bar{\rho})$ is the pair correlation function for the reference fluid at some mean density $\bar{\rho}$, e.g. $\bar{\rho} = 1/2[\rho(\mathbf{r}_1) + \rho(\mathbf{r}_2)]$. From (8.213)–(8.215) we can obtain an expression for the grand potential Ω by using (8.191). The density–orientation profile $f(\mathbf{r}\,\omega)$ is then found by minimizing Ω as in (8.193). Once $f(\mathbf{r}\,\omega)$ is known the free energy and grand potential can be determined from (8.213)–(8.215) and (8.193). The surface tension can be obtained from (8.60),

$$\gamma \mathcal{S} = \Omega + pV. \tag{8.216}$$

We shall often choose $s(u)$ to be simply the pair potential u itself, so that $u_s = 1$ and $s_1 = u_1$, the perturbation potential, so that (8.212) simplifies to

$$A_i = \mathcal{A}_0[f] + \frac{1}{2}\int_0^1 \mathrm{d}\lambda \int \mathrm{d}\mathbf{r}_1\mathrm{d}\mathbf{r}_2\mathrm{d}\omega_1\mathrm{d}\omega_2\, f(12;\lambda)u_1(12). \tag{8.217}$$

For a uniform fluid this expression reduces to (6.24), the well-known free energy equation.

8.5.2 *The u- and f-expansions*

In these expansions (see §§4.5 and 4.6) the reference fluid is composed of spherical molecules that have both attractive and repulsive intermolecular forces, and we can use either of the expansion methods described in the previous section. We shall choose the Taylor expansion procedure, so that the reference fluid is a two-phase system. In the u-expansion[87, 140, 142, 143] the reference pair potential $u_0(r)$ is defined as a simple unweighted average of $u(12)$ over molecular orientations, (4.23), and the property of interest is expanded in powers of the anisotropic part of the potential, $u_a(12) = u(12) - u_0(12)$. This choice of reference is particularly convenient when the full potential has a central part that can be identified with $u_0(\mathrm{r})$. Often the Lennard-Jones model is used for $u_0(r)$; one then needs to know the surface tension and low-order correlation functions for this fluid in order to use perturbation theory. In the *f*-expansion[37,105] the reference potential is obtained from an unweighted average of the Boltzmann factor, $\exp(-\beta u)$, over molecular orientations, and is given by (4.57). This reference potential varies with the temperature, and so is less convenient to use than the *u*-expansion reference. However, the *f*-expansion offers advantages when electrostatic forces are present, and gives the low-density limit of the pair correlation function and thermodynamic properties correctly.

Surface tension

We illustrate the use of perturbation theory for the surface tension, using the *u*-expansion.[87,142,147,148] With this reference potential $u(12;\lambda) = u_0(r_{12}) + \lambda u_a(12)$, so that $\partial u(12;\lambda)/\partial\lambda = u_a(12)$. From (4.23) we see that $\langle u_a(12)\rangle_{\omega_1\omega_2} = 0$, so that the first-order term γ_1 given by Eq. (8.201) vanishes and

$$\gamma = \gamma_o + \gamma_2 + \gamma_3 + \cdots . \tag{8.218}$$

The general expressions for γ_2 and γ_3 in the *u*-expansion have been derived[142] but are intractable as they stand because the pair and triplet reference correlation functions in the surface region are not usually known. The simplest way around this difficulty is to use the Fowler approximation, Eq. (8.98), i.e. to assume a step-function change in density at the vapour–liquid interface for the reference system. In this approximation γ_2 and γ_3 are given by[142]

$$\begin{aligned}\gamma_2^{(F)} &= \frac{\pi}{4}\beta\rho^2 \int_0^\infty \mathrm{d}r_{12} r_{12}^3 g_o(r_{12}) \left\langle u_a(12)^2 \right\rangle_{\omega_1\omega_2} \\ &+ \frac{\pi^2}{2}\beta\rho^3 \int_0^\infty \mathrm{d}r_{12} r_{12} \int_0^\infty \mathrm{d}r_{13} r_{13} \int_{|r_{12}-r_{13}|}^{(r_{12}+r_{13})} \mathrm{d}r_{23} r_{23} \\ &\times [r_{12}+r_{13}+r_{23}] g_o(r_{12}r_{13}r_{23}) \langle u_a(12)u_a(13)\rangle_{\omega_1\omega_2\omega_3}\end{aligned} \tag{8.219}$$

$$\begin{aligned}\gamma_3^{(F)} &= -\frac{\pi}{12}\beta^2\rho^2 \int_0^\infty \mathrm{d}r_{12} r_{12}^3 g_o(r_{12}) \left\langle u_a(12)^3 \right\rangle_{\omega_1\omega_2} \\ &- \frac{\pi^2}{6}\beta^2\rho^3 \int_0^\infty \mathrm{d}r_{12} r_{12} \int_0^\infty \mathrm{d}r_{13} r_{13} \int_{|r_{12}-r_{13}|}^{(r_{12}+r_{13})} \mathrm{d}r_{23}\, r_{23} \\ &\times [r_{12}+r_{13}+r_{23}] g_o(r_{12}r_{13}r_{23}) \langle u_a(12)u_a(13)u_a(23)\rangle_{\omega_1\omega_2\omega_3}.\end{aligned} \tag{8.220}$$

Equation (8.219) is valid for an arbitrary potential, but (8.220) is restricted to 'multipole-like' $u_a(12)$, i.e. anisotropic potentials that contain only $\ell \neq 0$ spherical harmonics in both ω_1 and ω_2. The more general expression for γ_3 contains additional four-body terms. For 'multipole-like' $u_a(12)$ we have $\langle u_a(ij)\rangle_{\omega_i} = \langle u_a(ij)\rangle_{\omega_j} = 0$, and the second term in the expression for $\gamma_2^{(F)}$ vanishes. In (8.219) and (8.220) $g_0(r_{12})$ and $g_0(r_{12}r_{13}r_{23})$ are the pair and triplet correlation functions for the bulk reference liquid.

When $u_a(12)$ contains electrostatic force terms the convergence of (8.218) is usually slow when these forces are strong. Typical results are shown in Fig. 8.21 for the surface tension of a Stockmayer fluid ($u = u_0 + u_a$, where u_0 is the Lennard-Jones potential and u_a is the dipole–dipole potential). The second-order theory gives γ values that are too high, while the values from the third-order theory are too low. Haile *et al.*[142] proposed a Padé approximant to the γ series,

$$\gamma = \gamma_o + \gamma_2/(1 - \gamma_3/\gamma_2), \tag{8.221}$$

which is analogous to the corresponding expression proposed by Stell *et al.*[149] for the free energy of a bulk phase. For multipolar u_a the γ_2 and γ_3 terms are of opposite sign. The second-order theory is seen to fail for $\mu^* \equiv \mu/(\varepsilon\sigma^3)^{1/2} >\sim 0.6$, while the third-order series fails for $\mu^* >\sim 0.9$. The Padé approximant, (8.221),

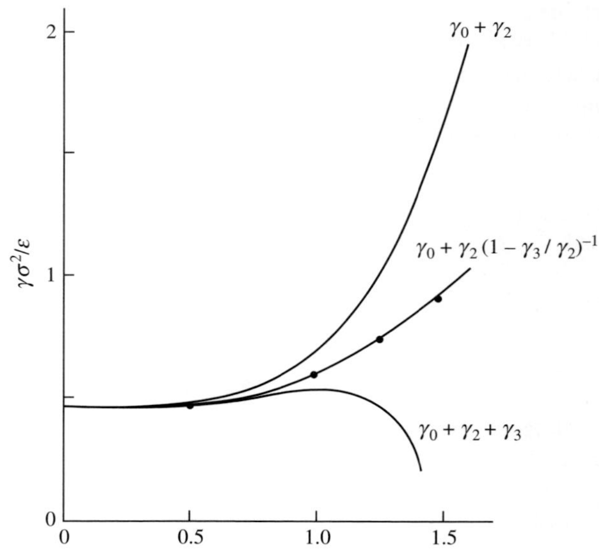

FIG. 8.21 The effect of dipolar forces on the surface tension (in the Fowler approximation) for a Stockmayer fluid at $kT/\varepsilon = 1.273$, $\rho\sigma^3 = 0.85$ from the u-expansion to second and third order, Eq. (8.218), and from the Padé approximant, (8.221). The points are Monte Carlo values. Here $\mu^* \equiv \mu/(\varepsilon\sigma^3)^{1/2}$. (From refs. 142, 148.)

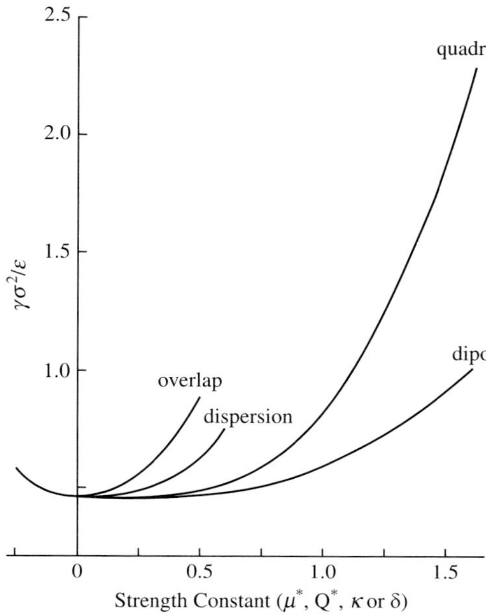

FIG. 8.22 The effect of various types of anisotropic intermolecular interactions on the surface tension at $kT/\varepsilon = 1.273$, $\rho\sigma^3 = 0.85$, as calculated from the u-expansion. Results for dipole and quadrupole interactions are from the Padé approximant, (8.221); those for overlap and dispersion are from the second-order theory. κ and δ are defined in the text. (From ref. 142.)

gives good results up to $\mu^* = 1.5$. For quadrupolar forces the curves are similar, the convergence being poor for $Q^* \equiv Q/(\varepsilon\sigma^5)^{1/2} > 1$.

The effects of various types of anisotropic potentials on γ are illustrated in Fig. 8.22. The surface tension is increased in all cases. The results for dipole–dipole and quadrupole–quadrupole forces are calculated from the Padé approximant (8.221). The third-order term γ_3 is more difficult to calculate in the case of anisotropic overlap and dispersion forces, since these potentials contain $\ell = 0$ spherical harmonics. The results shown for the overlap and dispersion cases are therefore from the second-order theory. The anisotropic overlap and dispersion potentials consist of the first few spherical harmonic terms $u_a(\ell_1\ell_2\ell)$ and are of the form (see §§2.6, 2.7)

$$u_{\mathrm{ov}} = u_{\mathrm{ov}}(202) + u_{\mathrm{ov}}(022) \tag{8.222}$$

$$u_{\mathrm{dis}} = u_{\mathrm{dis}}(202) + u_{\mathrm{dis}}(022) + u_{\mathrm{dis}}(224). \tag{8.223}$$

The anisotropic overlap potential contains a strength parameter δ, while the dispersion potential has a strength parameter κ. The overlap parameter must lie in the range $-0.25 \leq \delta \leq 0.5$ in order that the total overlap potential remain positive, while κ usually lies in the range 0 to 0.5 for small molecules.

Comparisons of (8.221) with experimental data have been made for several liquids,[87, 142, 147] including HBr, HCl, and CO_2, using intermolecular potentials fitted to bulk liquid properties. The theory gives good agreement with the experimental surface tension (within a few percent) at low temperatures, near the triple and boiling points, but becomes less accurate as the temperature is increased, presumably due to the failure of the Fowler approximation as the transition layer becomes more diffuse.

Density–orientation profile

Both the u- and f-expansions have been used to study the density–orientation profile. In both cases the reference system potential is independent of molecular orientations, so that the reference term in (8.202) for the expansion of $f(\mathbf{r}_1\omega_1)$ reduces to

$$f_0(\mathbf{r}_1\omega_1) = \frac{1}{\Omega}\rho_0(\mathbf{r}_1) \tag{8.224}$$

as can be seen from (8.5). Here $\rho_0(\mathbf{r}_1)$ is given by

$$\rho_0(\mathbf{r}_1) = \int \mathrm{d}\omega_1 f_0(\mathbf{r}_1\omega_1) \tag{8.225}$$

and is the density profile for the reference system.

In the u-expansion we have $u(12;\lambda) = u_0(r_{12}) + \lambda u_a(12)$, with u_0 given by (4.23), and the first-order term of (8.203) reduces to[140, 143]

$$f_1(\mathbf{r}_1\omega_1) = -\frac{\beta}{\Omega}\rho_0(\mathbf{r}_1)\int \mathrm{d}\mathbf{r}_{12}\rho_0(\mathbf{r}_2)g_0(\mathbf{r}_1\mathbf{r}_2)\langle u_a(12)\rangle_{\omega_2}. \tag{8.226}$$

We immediately note that this first-order term will vanish for multipolar and other multipole-like potentials, where $\langle \mathrm{u_a}(12)\rangle_{\omega_2} = 0$. Thus the first-order u-expansion predicts that multipole forces have no effect on orientational structure at the surface, in contrast to the simulation results discussed later. However, anisotropic dispersion and overlap forces will influence orientational structure according to (8.226).

In the f-expansion $u_0(r)$ is given by (4.57) and $u(12;\lambda)$ is given by

$$\exp[-\beta u(12;\lambda)] = \exp[-\beta u_0(r_{12})] + \lambda\{\exp[-\beta u(12)] - \exp[-\beta u_0(r_{12})]\} \tag{8.227}$$

and the first-order term of (8.204) becomes[37]

$$f_1(\mathbf{r}_1\omega_1) = \frac{1}{\Omega}\rho_0(\mathbf{r}_1)\int \mathrm{d}\mathbf{r}_{12}\rho_0(\mathbf{r}_2)g_0(\mathbf{r}_1\mathbf{r}_2)\langle f_a(12)\rangle_{\omega_2}, \tag{8.228}$$

where

$$f_a(12) = \exp\{-\beta[u(12) - u_0(12)]\} - 1. \tag{8.229}$$

We note from (8.228) that the first-order f-expansion predicts that multipolar forces *will* affect orientational structure at the surface, in contrast to the u-expansion. For both the u- and f-expansions the density profile $\rho(\mathbf{r}_1)$, given by

$$\rho(\mathbf{r}_1) = \int \mathrm{d}\omega_1 f(\mathbf{r}_1\omega_1), \tag{8.230}$$

is the same for the real and the reference systems to first order, so that the Gibbs dividing surface is unchanged.

Calculations based on these expansions have been carried out for a planar surface[37, 105, 140, 143, 150] for the following intermolecular pair potential models: (a) AALJ, (b) AALJ + Q, (c) ABLJ, (d) ABLJ + μ, (e) SLJ + μ, (f) SLJ + anisotropic dispersion, and (g) SLJ + anisotropic overlap. Here AALJ is a homonuclear diatomic (AA) molecule consisting of two LJ sites, ABLJ is a heteronuclear (AB) molecule with two LJ sites, SLJ is a spherical LJ molecule, and μ and Q represent added point dipole and point quadrupole terms, respectively. In carrying out these calculations it is convenient to expand $f(z\omega) = f(z\theta)$ in terms of harmonic coefficients $\hat{f}_\ell(z)$, using (8.28). Both computer simulation studies and theoretical calculations show that this expansion converges rapidly for the models studied; the first $\ell \neq 0$ harmonic coefficient usually gives the major contribution to the anisotropic part of $f(z\theta)$, and is sufficient to give the preferred molecular orientation in the surface. For the symmetrical linear molecules (AALJ, AALJ + Q, etc.) the odd harmonic terms $\hat{f}_1$, $\hat{f}_3$, etc. in (8.28) vanish, and the leading anisotropic coefficient is $\hat{f}_2(z)$.

For the AALJ model both the u- and f-expansions are in qualitative agreement with each other and with the computer simulation results, as seen in Fig. 8.23. The results show that for the AALJ potential the molecules prefer to align perpendicular to the surface ($\theta = 0$) on the liquid side ($z < 0$) of the Gibbs dividing surface; on the gas side ($z > 0$) there is a weaker preference for a parallel orientation. Qualitatively similar results have been found for the ABLJ fluid, with parameters modeled on HCl.[105] When a dipole or quadrupole is added to the AALJ model, the u-expansion to first order (see (8.226)) predicts that it has no influence on the density–orientation profile, in contradiction to the simulation results; thus, at first order the u-expansion gives qualitatively incorrect results. The f-expansion, to first non-vanishing order, does predict an effect due to the multipole (see (8.228)), and the predicted trend is usually in qualitative agreement with simulation results.[37] The addition of the quadrupole causes the molecules on the liquid side to tend to lie parallel to the surface. This latter effect competes with the preference for the perpendicular orientation for the AALJ model, and dominates the structure at large Q* values. On the gas side of the surface there is a weaker tendency for the perpendicular orientation.

Perturbation expansions in which the the reference fluid is composed of spherical molecules that interact with a purely repulsive pair potential $u_0(r)$ have also been used.[144, 151–154] Usually u_0 is taken to be the repulsive part of the u-expansion reference potential of Eq. (4.23), i.e.

$$u_o(r) \equiv \langle u(12)\rangle^{\mathrm{rep}}_{\omega_1\omega_2}. \tag{8.231}$$

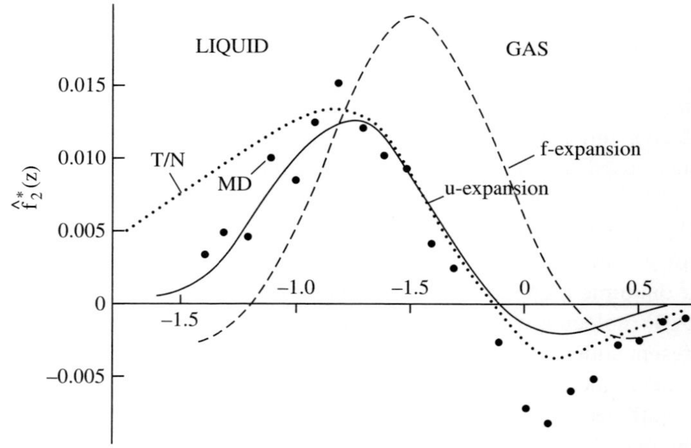

FIG. 8.23 Reduced second harmonic coefficient, $\hat{f}_2^*(z) = \hat{f}_2(z)\sigma^3$, for the AALJ model for Cl_2 from the u-expansion (solid line), f-expansion (dashed line), Tarazona–Navascués (T/N) perturbation theory (dotted line), and molecular dynamics (MD) simulation (points) at $T^* = kT/\varepsilon = 1.065$, $\rho^* = \rho\sigma^3 = 0.55$, $\ell^* = \ell/\sigma = 0.596$. Here ε, σ are site LJ parameters and ℓ is the bond length. (From refs. 37 and 154.)

Results from a theory of this type, the Tarazona–Navascués theory,[154] are also included in Fig. 8.23. The theory is again in qualitative agreemend with the other expansions and with the MD results.

8.6 Local mean field density functional theory

In this theory (also called local mean field theory or generalized van der Waals theory)[155] the effects of the strong, short-range (usually repulsive) intermolecular forces and the longer range (usually attractive) forces are treated separately. The simplest form of the theory is *local* mean field theory, and is the one we treat in detail here; we describe non-local treatments in the next section. In the local theory the free energy density $a_0(\mathbf{r})$ of the reference fluid of molecules with short-range forces is treated in a local approximation; i.e. $a_0(\mathbf{r})$ is taken to be the free energy density of a *uniform* fluid of the same density as the local density $\rho(\mathbf{r})$ at point $\mathbf{r}$ in the *non-uniform* fluid in which the molecules interact with the strong, short-range forces. The long-range forces are treated in a mean field approximation; i.e. the inhomogeneous pair correlation function $g(\mathbf{r}_1\omega_1\mathbf{r}_2\omega_2)$ is set equal to unity in the perturbation term containing the long-range attractive potential. The theory has been developed by Sullivan[156,157] for atomic fluids in the study of adsorption of gases on solids. Telo da Gama[45] has generalized the theory to molecular fluids, and has used it to study the interfacial properties of nematic liquid crystals. The main advantages of mean field theory are its simplicity and its ability to include the effects of both non-spherical cores (through

a_o)[158] and of orientation-dependent attractive forces of arbitrary strength. Thus, in contrast to the perturbation theories of the previous section, which usually rely on some expansion in the orientational order parameter, mean field theory can be used to study highly ordered non-uniform fluids such as surfactants at interfaces or liquid crystals. The neglect of intermolecular correlations arising from the long-range forces leads to some error in the predicted properties, parti cularly the density of the coexisting phases; in vapour–liquid equilibrium calculations the predicted liquid density is generally too low, often by 10–20%. Nevertheless, the mean field approximation is thermodynamically self-consistent, and hence it does permit the possibility of fluid phase transitions, critical points, and wetting layers. A more serious defect is that the theory fails to predict the short-range fluid density oscillations near the wall in fluid–solid systems. This error arises from the local approximation used in calculating $a_o(\mathbf{r})$, and can be corrected by using non-local extensions of the theory.[159–164] In common with many other approximate theories considered in this chapter (see, however, § 8.7), local mean field theory fails to predict the solid–fluid transition. Despite these drawbacks, mean field theory has proved useful in providing qualitatively correct descriptions of a variety of surface phenomena for fluid systems. For a homogeneous fluid it reduces to the well-known van der Waals model, which has played a similarly useful role in investigating fluid phase equilibria.

To account for interface orientational ordering of molecules with multipole-like potentials, it is necessary to go beyond the strict mean field theory (which is linear in the anisotropic intermolecular potential $u_a(12)$). Extensions to include intermolecular pair correlations have been developed by Teixeira and Telo da Gama[165] and by Yang *et al.*[166–168] References 165, 166, and 167 are based on u-expansions to second order in $u_a(12)$, and ref. 168 is based on an f-expansion. Applications of the latter to properties of the liquid–vapour interface of water are described in §§ 8.6.4 and 8.9.

8.6.1 *Local mean field theory*

The theoretical equations are most easily derived[45, 46, 169, 170] within the framework of the density functional formalism of §§8.4 and 8.5.1. We first write an expression for the grand potential functional by combining (8.191) for Ω and (8.217) for $A_i = \mathcal{A}[f]$; when (8.214) is used for $\mathcal{A}_0$ this gives

$$\Omega[f'] = \int \mathrm{d}\mathbf{r} a_o(f'(\mathbf{r}\omega)) + \frac{1}{2}\int_0^1 \mathrm{d}\lambda \int \mathrm{d}\mathbf{r}_1 \mathrm{d}\mathbf{r}_2 \mathrm{d}\omega_1 \mathrm{d}\omega_2 f'(12;\lambda) u_1(12) + \int \mathrm{d}\mathbf{r}\mathrm{d}\omega f'(\mathbf{r}\omega) v(\mathbf{r}\omega) - \mu \int \mathrm{d}\mathbf{r}\mathrm{d}\omega f'(\mathbf{r}\omega). \tag{8.232}$$

Equation (8.232) is based on the assumption of pairwise additivity, but is otherwise general. We now introduce the two approximations of the local mean field theory: (i) the free energy density at a point $\mathbf{r}$ in the fluid of reference molecules, $a_0(\mathbf{r})$, is set equal to that of a *uniform* fluid of molecules with pair potential $u_0(12)$

and of a constant density ρ equal to the number density of the real fluid at **r**, and with the same density–orientation profile $f'(\mathbf{r}\,\omega)$ as that of the real fluid (with pair potential $u(12)$) at **r**, and (ii) the pair distribution function $f'(12;\lambda)$, appearing in the second term on the right of (8.232), is approximated by its large r limit, i.e. by (8.215) with $g(12) = 1$. Pair correlations arising from attractive forces are thus neglected. Thus (8.232) becomes

$$\Omega[f'] = \int d\mathbf{r}a_o(f'(\mathbf{r}\omega)) + \frac{1}{2}\int d\mathbf{r}_1 d\mathbf{r}_2 d\omega_1 d\omega_2 f'(\mathbf{r}_1\omega_1)f'(\mathbf{r}_2\omega_2)u_1(12) \\ + \int d\mathbf{r}d\omega f'(\mathbf{r}\omega)v(\mathbf{r}\omega) - \mu\int d\mathbf{r}d\omega f'(\mathbf{r}\omega), \tag{8.233}$$

which is the grand potential in the local mean field approximation; in this equation a_o is understood to be the free energy density in the uniform fluid approximation. The density–orientation profile $f(\mathbf{r}\omega)$ for the equilibrium system is calculated by minimizing $\Omega\,[f']$, i.e. by solving (8.195), which becomes in this approximation (see (8.196) and (8.233)),

$$\mu = v(\mathbf{r}_2\omega_2) + \mu_o(f(\mathbf{r}_2\omega_2)) + \int d\mathbf{r}_1 d\omega_1 f(\mathbf{r}_1\omega_1)u_1(\mathbf{r}_{12}\omega_1\omega_2), \tag{8.234}$$

where μ_0 is the chemical potential[45] of the reference fluid of molecules with pair potential $u_0(12)$ and density–orientation profile $f(\mathbf{r}\,\omega)$. In most applications the reference fluid potential is the short-ranged repulsive interaction of the real fluid, while $u_1(12)$ is the long-ranged (usually attractive) potential of the real molecules. Often $u_0(12)$ is approximated by a hard core potential, e.g. hard spheres or hard spherocylinders. It should be noted that in using (8.233) it is implicitly assumed that the reference fluid does not itself undergo a phase transition.[45,156] For a reference system of hard spheres this will always be the case, except for very high densities where a freezing transition occurs. Sufficiently elongated hard core molecules, however, can exhibit a nematic-isotropic phase transition[171] and this theory cannot be used in regions of the phase diagram where such a reference fluid transition occurs.

As examples of the use of this theory for fluids in which the molecules are highly anisotropic in their interactions, we now consider applications to a model fluid exhibiting a liquid crystal phase, to a liquid–liquid interface in the presence of an amphiphilic component (a component concentrated at the interface), and to the liquid–vapour interface of water.

8.6.2 Interfacial properties of a model liquid crystal

The simplest ordered liquid crystal phase, the nematic, is characterized by positional disorder but long-range orientational order; the molecules tend to align themselves along some preferred direction **n**, called the nematic director. In the bulk fluid, with no external fields or surfaces, the free energy is independent of the direction **n**, but when a surface is present this is no longer the case; the molecules then adopt orientations that minimize the free energy. Experimental measurements of molecular orientation in the vapour–liquid and solid–liquid interfaces of nematics have revealed a variety of surface-induced orientational transitions close to the nematic–isotropic phase transition temperature (for a discussion of the experimental results see refs. 35, 169).

The local mean field theory has been used to study a simple model nematic in which the pair potential consists of a central hard sphere term, $u_0(r)$, as the short-range potential, and a longer-ranged attractive dispersion term u_1 that depends on molecular orientations (cf. (2.223)):[45,46,169,172]

$$\begin{aligned} u_1(\mathbf{r}_{12}\omega_1\omega_2) &= 0, \qquad\qquad\qquad\qquad\qquad\qquad\qquad\qquad r < \sigma \\ &= -A\left[\frac{\sigma}{r_{12}}\right]^6 - B\left[\frac{\sigma}{r_{12}}\right]^6 P_2\{\cos(\omega_1,\omega_2)\} \\ &\quad + C\left[\frac{\sigma}{r_{12}}\right]^6 [P_2\{\cos(\omega_1,\omega_2)\} + P_2\{\cos(\omega_2,\omega_{12})\}], \quad r > \sigma, \end{aligned} \tag{8.235}$$

where σ is the hard-core diameter, ω_{12} is the orientation of $\mathbf{r}_{12}$, P_2 is the second-order Legendre polynomial (see (A.8)), and (ω_a, ω_b) is the angle between ω_a and ω_b. A and B are positive constants proportional to the square of the mean polarizability $\alpha = (\alpha_{||} + 2\alpha_{\perp})/3$ and the polarizability anisotropy $(\alpha_{||} - \alpha_{\perp})$, respectively, cf. (2.223), (2.226), and (2.230)–(2.232). In the LJ model (see (2.235)) we take $A = 4\varepsilon$, where ε is the usual LJ well depth. The B term causes the molecules to tend to align with their symmetry axes parallel. C is proportional to $\alpha(\alpha_{||} - \alpha_{\perp})$ and can therefore be positive (prolate molecules) or negative (oblate molecules). When $C = 0$ in this potential model the intermolecular potential reduces to that of Maier and Saupe.[173] Such a potential does not depend on the orientation of the molecules relative to the intermolecular vector $\mathbf{r}_{12}$, a defect that causes the model to fail to predict any preferred molecular orientation at the vapour–isotropic liquid interface; for the nematic–isotropic and nematic–vapour interfaces the model predicts the same qualitative molecular orientation as the underlying nematic phase, although order parameter η now depends on z. The inclusion of the last C term in (8.235) has no effect on the thermodynamic properties of the system; the phase diagram is thus independent of the value of C. These terms do, however, provide coupling between rotational and translational degrees of freedom, and strongly affect the surface properties, leading to a preferred molecular orientation at the interface.

The potential model of (8.235) accounts only for anisotropic dispersion forces, and neglects anisotropic repulsion and other interactions. However, the anisotropic dispersion forces are believed to mimic the dominant anisotropic contributions to the effective mean field potential for a liquid crystal, so that this model should provide a useful insight into the qualitative behaviour of orientational ordering at interfaces.[172]

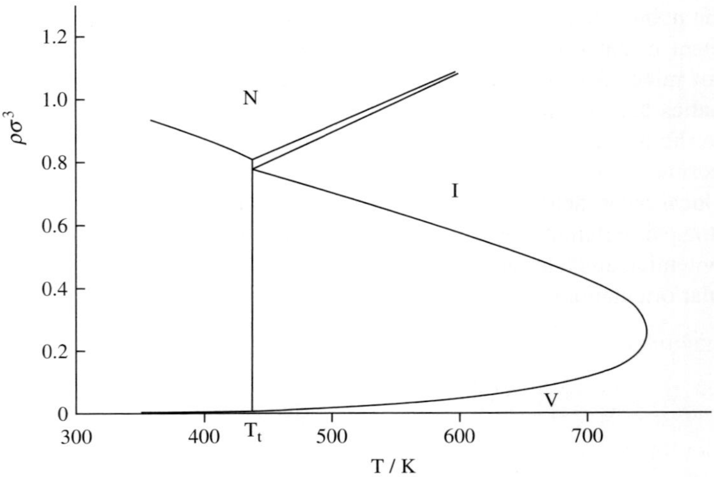

FIG. 8.24 The density–temperature fluid phase diagram of a model nematic liquid crystal, based on the pair potential (8.235) and local mean field theory, showing the nematic (N), isotropic liquid (I) and vapour (V) phases. These phases coexist in pairs along the lines, and all three coexist at the triple point $T_t = 436.3\,\text{K}$. (There is a solid phase to the left of the vertical line at T_t.) Here $A/k = 2000\,\text{K}$ and $B/A = 0.3$. These values are fairly typical of real liquid crystals. (From ref. 45.)

The phase diagram of this model fluid exhibits three fluid phases—a nematic, an isotropic liquid, and a vapour phase.[45,169] These phases can coexist in pairs, or all three can coexist at a triple point, T_t (see Fig. 8.24). Below T_t the vapour coexists with the nematic phase, while above T_t vapour and isotropic liquid coexist. The nematic–isotropic liquid transition is first order, the density difference being less than 5% at the triple point and decreasing rapidly as the temperature rises. At the lower temperatures this predicted density difference is about one order of magnitude larger than that observed in real liquid crystals. The theory thus predicts a transition that is too strongly first order, a defect that arises from the failure of the mean field approximation to predict the correct fluid structure.

For C values in the range $-0.1175 \le C/A \le 0.0552$ the nematic–vapour interface is completely wet by the isotropic liquid. Under these conditions a macroscopically thick film of isotropic liquid intrudes between the vapour and nematic liquid, and the surface tensions obey the relation[21,45,174]

$$\gamma_{\text{NV}} = \gamma_{\text{NI}} + \gamma_{\text{IV}}, \tag{8.236}$$

where the subscripts indicate the two-phase interface involved.[175,176] This phenomenon is termed triple-point wetting. When $C/A > 0.0552$ or $C/A < -0.1175$ the above equality does not usually hold, but instead we have the inequality

$$\gamma_{\text{NV}} < \gamma_{\text{NI}} + \gamma_{\text{IV}}, \tag{8.237}$$

so that only partial wetting occurs. The wetting transitions at the limiting values $C/A = -0.1175$ and $C/A = 0.0552$ are weakly first order.[172] For sufficiently large magnitudes of the parameter C the isotropic liquid–vapour interface is completely wet by the nematic phase. This occurs for $C/A > 0.8477$ and $C/A < -2.1249$ (with **n** perpendicular and parallel to the interface, respectively), and is indicated by the equality

$$\gamma_{\mathrm{IV}} = \gamma_{\mathrm{NI}} + \gamma_{\mathrm{NV}}. \tag{8.238}$$

Such complete or partial wetting of the IV interface by the nematic phase has been observed experimentally (e.g. for PAA, 5CB, and 8CB). The wetting transition at these limiting C values is again first order, and more strongly so than the wetting of the nematic–vapour phase by isotropic liquid. The first-order nature and strength of the transition can be judged by calculating the variation of surface tension with wetting-layer thickness, when minima are observed for both a thin film and an infinitely thick film, the minima being separated by a metastable surface tension barrier; the height of this barrier is an indicator of how strongly first order the transition is.

Preferred molecular orientation at the various interfaces between the I, N, and V phases are usually discussed in terms of orientational order parameters, such as η of Eq. (8.30). However, since more than one direction is involved in the system, more than one order parameter is needed. When one of the phases is the nematic, it is found[45,46] that if the xz plane is chosen to contain the nematic director **n** then the density–orientation profile and free energy depend in general on three orientational order parameters $\eta(z)$, $\nu(z)$, and $\sigma(z)$, with $\eta(z)$ given by (8.30) and $\nu(z)$ and $\sigma(z)$ defined by

$$\nu(z_1) = \int_0^{2\pi} \mathrm{d}\phi_1 \int_0^{\pi} \mathrm{d}\theta_1 \sin\theta_1 \sin 2\theta_1 \cos\phi_1 (f(z_1\omega_1)/f(z_1)) \tag{8.239}$$

$$\sigma(z_1) = \int_0^{2\pi} \mathrm{d}\phi_1 \int_0^{\pi} \mathrm{d}\theta_1 \sin^3\theta_1 \cos 2\phi_1 (f(z_1\omega_1)/f(z_1)). \tag{8.240}$$

These order parameters indicate the fraction of molecules aligned in a particular direction. Thus $\eta = 1$ when all of the molecules are aligned along the z-axis, $\sigma = 1$ when all molecules are aligned along the x-axis, and $\nu = 1$ when all molecules are parallel to the bisector of the xz-plane. When the nematic director **n** in the bulk nematic phase is in the z direction, i.e. perpendicular to the interface, the only non-zero order parameter is η, since the molecules in the vapour or isotropic phase will have an orientational distribution independent of the angle ϕ. Similarly, if the nematic director for the bulk nematic phase is parallel to the interface, ν vanishes while η and σ are non-zero. When all of the molecules are oriented along the x-axis we have $\eta = -\frac{1}{2}$ and $\sigma = 1$. For other orientations of the nematic director all three order parameters will be non-zero. For an equilibrium nematic–vapour or nematic–isotropic liquid interface both the nematic director **n** in the nematic phase and the distribution of molecular orientations in the interface will be that which minimizes the free energy, as determined by the mean field theory.

Examples of the interfacial density and order parameter profiles observed at the transition point are shown in Fig. 8.25 for $C/A = -2.2149$. For this C/A value thin and thick nematic films coexist.[177] Since C is negative, **n** is parallel to the interface and the orientational order in the film is characterized by two parameters, $\eta(z)$ and $\sigma(z)$. The thin nematic film shown in Fig. 8.25(a) has a uniaxial structure and is characterized by the single-order parameter $\eta(z)$; the molecular orientations are found to be completely disordered with respect to directions parallel to the

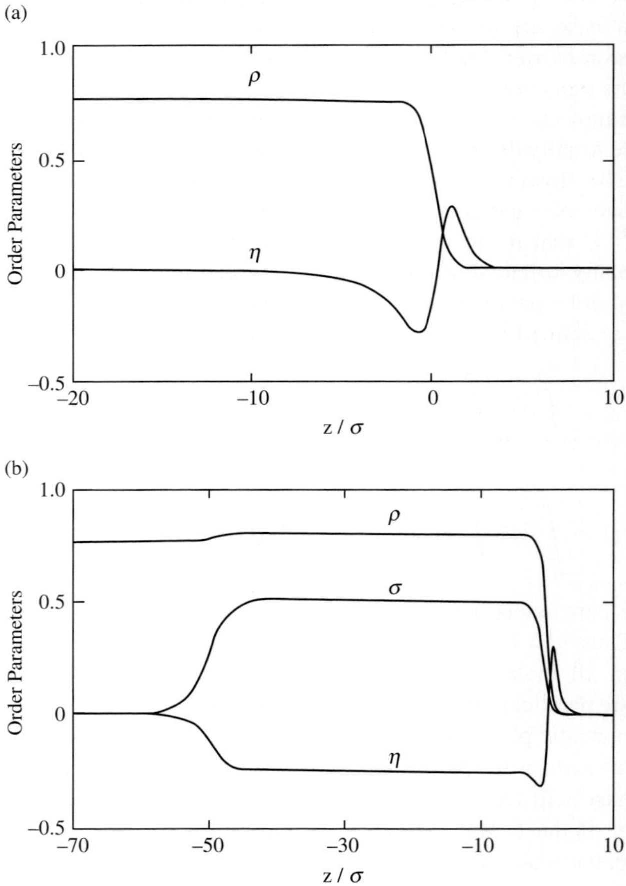

FIG. 8.25 Density and orientation parameter profiles at the $I-V$ interface for $T = T_t$ and $C/A = -2.2149$. (a) Thin nematic film; (b) thick nematic wetting film that coexists with the structure in (a). (From ref. 172.)

surface. By contrast, the thick nematic film of Fig. 8.25(b) has a biaxial structure. The wetting transition is thus accompanied by a uniaxial to biaxial symmetry-breaking transition.[172]

Although the surface tension usually decreases with rising temperature, there is a discontinuous jump, $\Delta\gamma = \gamma_{\mathrm{IV}} - \gamma_{\mathrm{NV}}$, on raising the temperature through T_t, corresponding to the first-order nematic to isotropic transition in the bulk liquid. The sign and magnitude of $\Delta\gamma$ depends on the wetting properties of the interface at T_t.[45] The discontinuity $\Delta\gamma$ has its largest magnitude for complete wetting of the interface. For complete wetting of the nematic–vapour interface by the isotropic liquid $\Delta\gamma = -\gamma_{\mathrm{NI}}$ (see (8.236)), while for complete wetting of the isotropic liquid–vapour interface by the nematic liquid $\Delta\gamma = \gamma_{\mathrm{NI}}$ (see (8.238)). An example of the behaviour of the surface tension for the latter case is shown in Fig. 8.26, where $C/A = 1.2$, well above the value needed for complete wetting.[178] For $T < T_t$ there is a monotonic decrease in γ_{NV}, followed by the positive jump $\Delta\gamma = \gamma_{\mathrm{NI}}$ at T_t. For $T > T_t$ the surface tension first *increases* with temperature, then goes through a maximum and decreases in the usual way. This non-monotonic variation of γ_{IV} can be shown to be a consequence of the approach to complete wetting, and has been observed experimentally[80,179] (see Fig. 8.10(c)). However, the model of (8.235) shows no evidence of an increase in γ_{NV} with temperature near T_t,[172] as has been sometimes observed experimentally[80,179] (Fig. 8.10(c)); this appears to be a deficiency of the model.

The model of (8.235) has also been used as a basis to study interfacial smectic ordering.[180]

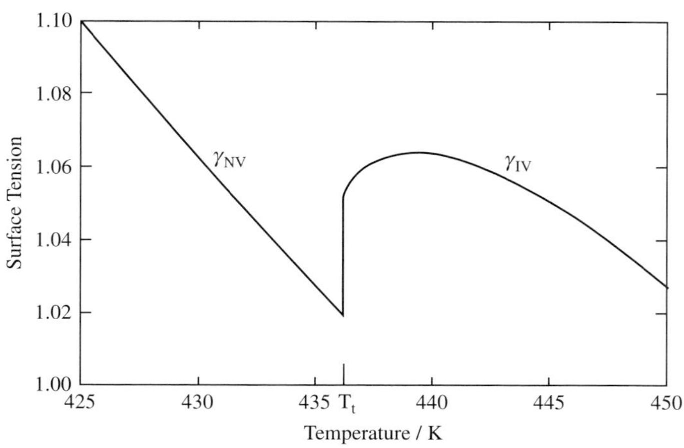

FIG. 8.26 Temperature dependence of the vapour–liquid reduced surface tension $\gamma\sigma^2/kT$ for the model of (8.235) with $C/A = 1.2$. For $T < T_t$ the surface tension is that of the interface between vapour and nematic liquid, while for $T > T_t$ it is for the vapour–isotropic liquid interface. (From ref. 172.)

8.6.3 Model surfactant at a liquid–liquid interface

The surface tension of an oil–water interface can be greatly reduced by the addition of a small amount of a suitable third component (usually called a *surfactant* or *amphiphile*), which is concentrated in the surface layer between the two phases. The surfactant molecules usually possess a polar head group, which is attracted to the water phase, and a hydrocarbon part that is attracted to the oil phase (Fig. 8.27). If the surfactant is present only as a two-dimensional monolayer between the phases, and is sufficiently dilute that surfactant–surfactant interactions can be neglected, the reduction in surface tension is given by the surface pressure of the two-dimensional film[4, 181] (see (8.83c))

$$\gamma_0 - \gamma = \Gamma kT, \tag{8.241}$$

where γ_0 and γ are the surface tensions without and with surfactant, respectively, and Γ is the adsorption (see (8.62)), i.e. the number of molecules per unit area (which is equal to the excess number per unit area for an insoluble monolayer). This equation, which was originally derived by Langmuir,[4] is simply a force balance relation, and does not provide any insight into the relation between the observed γ and Γ values and the molecular nature of the surfactant molecules. Also, (8.241) will break down at higher concentrations, where surfactant–surfactant interactions must be considered.

The local mean field theory has been used to study a very simple model of such surfactant behaviour, in which components A and B are the oil and water constituents, and C is the surfactant.[169, 170] The intermolecular potential model must be such that the binary mixture of A with B splits into two liquid phases, with little or no mutual solubility of A and B. The AC and BC interactions must be such that the C molecules concentrate at the liquid–liquid interface; this can be achieved by providing the C molecules with a hydrophilic (water-liking) end and a hydrophobic (water-disliking) end. The simplest model that satisfies these requirements is one in which the A, B, and C molecules all possess hard spherical cores of diameter σ, and attractive interactions given by

$$u_{AA}(r) = u_{BB}(r) = -\varepsilon \left[\frac{\sigma}{r}\right]^6 \tag{8.242}$$

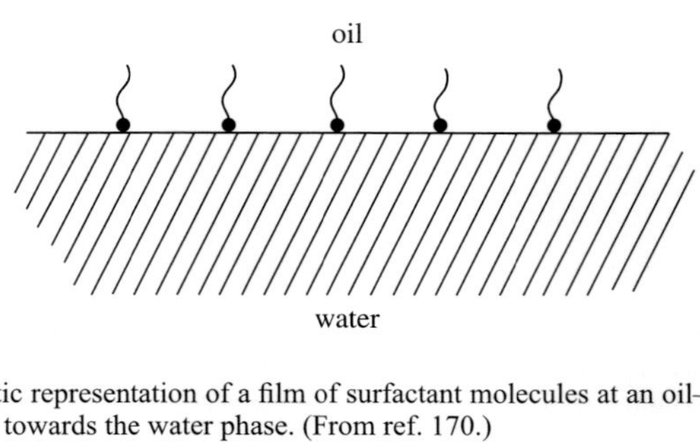

FIG. 8.27 Schematic representation of a film of surfactant molecules at an oil–water interface, their polar heads being towards the water phase. (From ref. 170.)

$$u_{\mathrm{AB}}(r) = 0 \tag{8.243}$$

$$u_{\mathrm{AC}}(\mathbf{r}\omega_{\mathrm{C}}) = -\beta\varepsilon\left[\frac{\sigma}{r}\right]^6 \cos(\mathbf{r}, \omega_{\mathrm{C}}) \tag{8.244}$$

$$u_{\mathrm{BC}}(\mathbf{r}\omega_{\mathrm{C}}) = \beta\varepsilon\left[\frac{\sigma}{r}\right]^6 \cos(\mathbf{r}, \omega_{\mathrm{C}}) \tag{8.245}$$

$$u_{\mathrm{CC}}(r) = 0. \tag{8.246}$$

In the absence of C molecules the above potential ensures that the AB mixture exhibits liquid–liquid immiscibility below an upper critical end point; it is a symmetric mixture of class III (see Chapter 7). The AC and BC interactions are strongly dependent on the orientation ω_C of the C molecules, their strength being determined by the parameter β (not to be confused with $(kT)^{-1}$). In the case of the AC interactions, the orientation dependence derives from the cosine of the angle between the intermolecular vector $\mathbf{r} = \mathbf{r}_{\mathrm{AC}}$ and ω_C; u_{AC} is attractive when ω_C is parallel to $\mathbf{r}_{\mathrm{AC}}$ and repulsive when ω_C is anti-parallel to $\mathbf{r}_{\mathrm{AC}}$. Similarly, for the BC interaction, u_{BC} is repulsive when ω_C is parallel to $\mathbf{r} = \mathbf{r}_{\mathrm{BC}}$ and attractive when ω_C and $\mathbf{r}_{\mathrm{BC}}$ are anti-parallel. The C molecules thus possess a hydrophilic and a hydrophobic end (see Fig. 8.28). In this simplest model there are no attractions between the C molecules. In the local mean field theory, using this model it is found that the free energy (and hence the bulk AB phase diagram) is unaffected by the value of β. The surface properties depend strongly on β, however.

Typical results for the density profiles and orientational order parameter are shown in Fig. 8.29. Because of the symmetry of the model, the density of A in one phase is equal to the density of B in the other, while the density of C is the same in both phases. The large values of η near $z = 0$ show that the C molecules are almost completely aligned in the surface layer. The η curve is quite broad, indicating that the orientational correlations are fairly long-ranged. As x_{C} increases the interface broadens and η decreases somewhat, while ρ_{C} increases. The addition of the surfactant has a striking effect on the surface tension, as shown in Fig. 8.30.

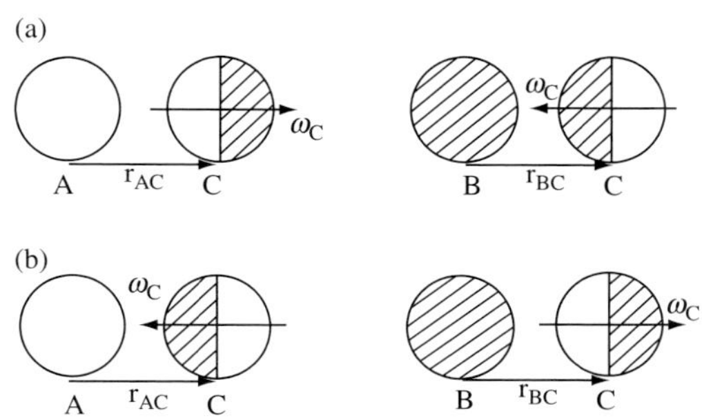

FIG. 8.28 Model of a symmetric ternary mixture of two immiscible liquids A ('oil') and B ('water') and an amphiphilic component C. The interactions of C with A and B are strongly anisotropic: (a) attractive configurations, (b) repulsive configurations. (From ref. 170.)

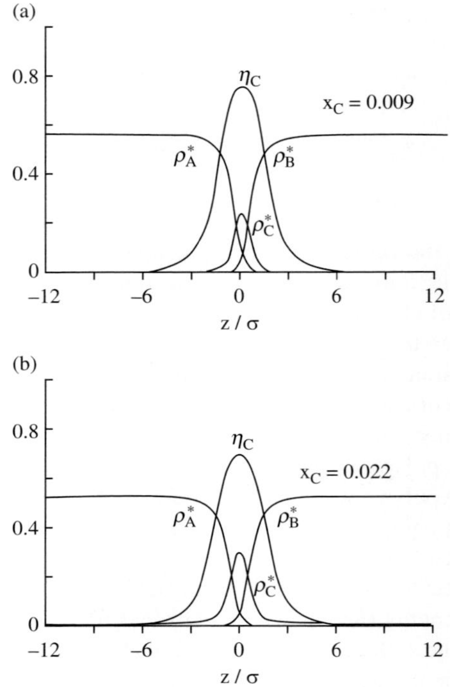

FIG. 8.29 Density ($\rho^* = \rho\sigma^3$) and orientational order parameter η_C profiles at the (A-rich–B-rich) liquid–liquid interface of a model of a symmetric ternary mixture at $kT/\varepsilon = 0.4$ and $\rho\sigma^3 = 0.563$ for two concentrations of the amphiphilic component C. (a) $x_C = 0.009$, (b) $x_C = 0.022$. The concentration of C at the interface increases with x_C and the orientational order decreases. Here $\beta = 2$. (From ref. 170.)

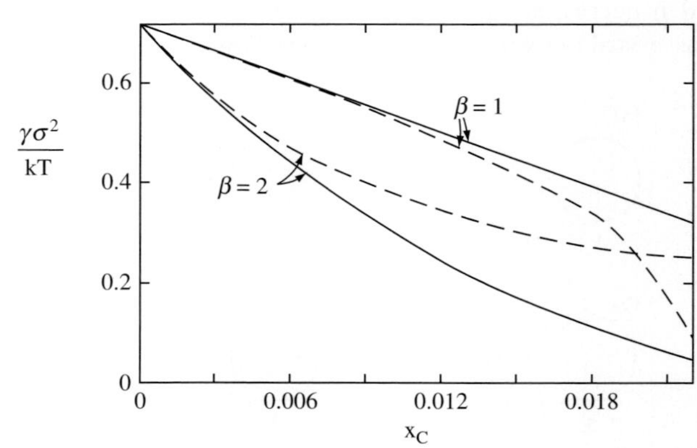

FIG. 8.30 The surface tension γ for the liquid–liquid interface of the model ternary mixture described by (8.242)–(8.246) as a function of the bulk composition of the surfactant (C) component, at $kT/\varepsilon = 0.4$, $\rho\sigma^3 = 0.563$. Solid lines are the results of mean field theory, dashed lines are the Langmuir equation, (8.241). (From ref. 170.)

Also shown in this figure are the results of the Langmuir equation, (8.241), which is seen to break down for $x_C \gtrsim 0.010$ when $\beta = 1$ and for $x_C \gtrsim 0.004$ when $\beta = 2$. When $\beta = 2$ and $x_C = 0.022$ the mean field theory predicts that γ has fallen by 94% of its value at $x_C = 0$. For β values much above 2, the theory predicts that γ is negative. This is probably an indication that micelles should form for $\beta > 2$ in this model—a possibility not accounted for in the theory.

This model of the water–surfactant–oil system has been extended by Guerra *et al.*[182] to asymmetric water–oil systems, enabling them to study the elastic properties of the amphiphilic film and the stability of lamellar phases and droplet microemulsions.

8.6.4 *Liquid–vapour interface of water*

For H_2O–H_2O interactions, a number of multipoles are required to represent accurately the strong anisotropic electrostatic forces. For concreteness, we start with the TIP4P model,[57] which has been used in simulations and in density functional theory (DFT)[166–168] for the H_2O liquid–vapour interface. There is a central LJ core centred on the O atom and point charges on the O and H sites. For the purposes of DFT we write the H_2O–H_2O pair potential $u(12)$ as

$$u(12) = u_{\text{ref}}(r_{12}) + u_{\text{pert}}(12), \tag{8.247}$$

where $u_{\text{ref}}(r_{12})$ is the repulsive part of the LJ core potential† (the part for $r_{12} < \sigma$), and

$$u_{\text{pert}}(12) = u_{\text{att}}(r_{12}) + u_a(12), \tag{8.248}$$

where $u_{\text{att}}(r_{12})$ is the attractive part of the LJ core potential (the part for $r_{12} > \sigma$) and $u_a(12)$ is the anisotropic electrostatic part of the potential represented as the multipole series

$$u_a(12) = u_{11}(12) + u_{12}(12) + u_{21}(12) + u_{22}(12) + \dots \tag{8.249}$$

with $u_{\ell_1\ell_2}(12)$ given by (2.171) of Vol. 1 in terms of spherical harmonics. In the u-expansion version[166,167] of DFT the terms in (8.249) up to $u_{22}(12)$ (i.e. quadrupole–quadrupole terms) are kept, and in the f-expansion version[168] terms up to $u_{55}(12)$ are kept.

As mentioned in the introduction to § 8.6, for pure multipole interactions it is necessary to go beyond the strict mean field theory version of DFT by including pair correlations to low order, in order to obtain orientational ordering of the interface molecules. As in earlier sections, we minimize the grand potential Ω with respect to the singlet distribution function $f(1) \equiv f(\mathbf{r}_1\omega_1)$, with pair correlation terms now included in Ω. The final results are simply expressed by introducing the conditional orientational single molecule distribution function

† In practice[166–168] a hard sphere potential is often used for $u_{\text{ref}}(r_{12})$.

$P(\omega_1|\mathbf{r}_1) \equiv f(\mathbf{r}_1\omega_1)/\rho(\mathbf{r}_1)$, which is normalized to unity (i.e. $\int d\omega P(\omega|\mathbf{r}) = 1$), where $\rho(\mathbf{r}_1) = \int d\omega f(\mathbf{r}\,\omega)$ is the local number density and $\int d\omega = \Omega = 8\pi^2$). Physically, $P(\omega|\mathbf{r})$ is the probability density to find a molecule with orientation ω, given that its position is $\mathbf{r}$. The external potential $\mathcal{V}$ is assumed to vanish in this section.

DFT leads to a pair of coupled equations for the single-molecule distribution function $P(\omega|\mathbf{r})$ and the single-molecule effective (or mean) potential $v_{\text{eff}}(\mathbf{r}\,\omega)$. The first equation is

$$P(\omega|\mathbf{r}) = \frac{e^{-\beta v_{\text{eff}}(\mathbf{r}\omega)}}{\int d\omega e^{-\beta v_{\text{eff}}(\mathbf{r}\omega)}} \tag{8.250}$$

and applies to both the *f*- and *u*-expansion versions of the DFT. In the *f*-expansion version, $v_{\text{eff}}(\mathbf{r}\omega)$ is expressed back in terms of $P(\omega|\mathbf{r})$ by

$$v_{\text{eff}}(\mathbf{r}_1\omega_1) = -kT \int d\mathbf{r}_2 d\omega_2 \rho(\mathbf{r}_2) P(\omega_2|\mathbf{r}_2) e^{-\beta[u_{\text{ref}}(r_{12}) - \bar{u}_a(12)]}[e^{-\beta u_a(12)} - 1], \tag{8.251}$$

where $\bar{u}_a(12) \equiv \bar{u}_a(\mathbf{r}_1\mathbf{r}_2)$ is defined by

$$e^{-\beta\bar{u}_a(12)} = \int d\omega_1 d\omega_2 P(\omega_1|\mathbf{r}_1) P(\omega_2|\mathbf{r}_2) e^{-\beta u_a(12)} \tag{8.252a}$$

with $u_a(12) \equiv u_a(\mathbf{r}_{12}\omega_1\omega_2)$ defined in (8.249). The pair of coupled nonlinear equations (8.250) and (8.251) are characteristic of a mean field type theory. The local density $\rho(\mathbf{r})$ in (8.251) is found from a third equation that couples ρ to P, i.e.

$$\mu = \mu_{\text{ref}}(\rho(\mathbf{r}_1)) + kT \int d\omega_1 P(\omega_1|\mathbf{r}_1) \ln[8\pi^2 P(\omega_1|\mathbf{r}_1)]$$
$$+ \int d\mathbf{r}_2 \rho(\mathbf{r}_2) e^{-\beta u_{\text{ref}}(r_{12})}[u_{\text{att}}(r_{12}) + \bar{u}_a(12)], \tag{8.252b}$$

where $\mu_{\text{ref}}(\rho)$ is the chemical potential of the repulsive reference fluid. The chemical potential μ is determined by satisfying the conditions for liquid–vapour coexistence. In practice[168] these equations are solved by expanding all orientationally dependent quantities in spherical harmonic series (such as (8.249)) and, for H_2O, truncating the series at $\ell_{\max} = 5$. The resulting coupled algebraic equations for the expansion coefficients are solved by iteration. Similar techniques are described for bulk fluid correlation functions in Vol. 1, Chapter 5.

In the *u*-expansion version of DFT, (8.250) is closed with (instead of (8.251))

$$v_{\text{eff}}(\mathbf{r}_1\omega_1) = \int d\mathbf{r}_2 d\omega_2 \left[u_{\text{pert}}(12) - \frac{1}{2}\beta u_a(12)^2\right]$$
$$+ \beta \int d\mathbf{r}_2 d\omega_2 \frac{d\omega_1'}{8\pi^2} \rho(\mathbf{r}_2) P\left(\omega_1'|\mathbf{r}_1\right) u_a(12) u_a(1'2), \tag{8.253a}$$

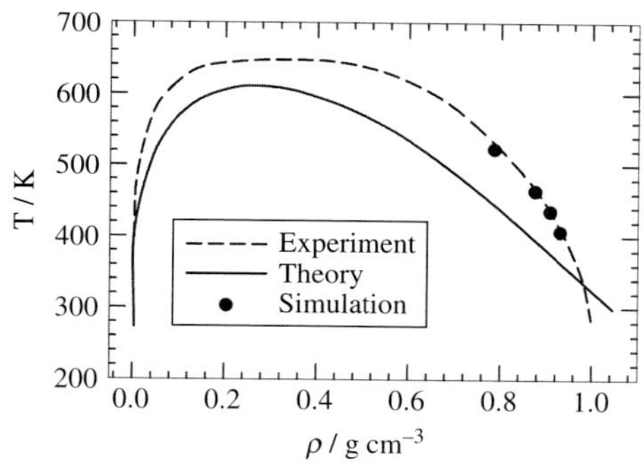

FIG. 8.31 The bulk vapour–liquid coexistence curve for water, comparing theory[168] with experiment.[183] The points are computer simulation results for the TIP4P model.[184] (From ref. 168.)

where $u_{\text{pert}}(12)$ is defined by (8.248) and $u_a(1'2) \equiv u_a\left(\mathbf{r}_{12}\omega_1'\omega_2\right)$. Here the local density $\rho(\mathbf{r})$ is determined from a third equation that couples ρ to v_{eff}, i.e.

$$\begin{aligned}\mu = {} & \mu_{\text{ref}}(\rho(\mathbf{r}_1)) - kT\ln\langle e^{-\beta v_{\text{eff}}(\mathbf{r}_1\omega_1)}\rangle_{\omega_1} \\ & + \frac{1}{2}\beta\Omega^2\int d\mathbf{r}_2\rho(\mathbf{r}_2)\left\langle\langle u_a(12)P(\omega_2|\mathbf{r}_2)\rangle_{\omega_2}^2\right\rangle_{\omega_1} \\ & - \frac{1}{2}\beta\Omega^2\int d\mathbf{r}_2\rho(\mathbf{r}_2)\left\langle\langle u_a(12)P(\omega_1|\mathbf{r}_1)\rangle_{\omega_1}^2\right\rangle_{\omega_2},\end{aligned} \tag{8.253b}$$

where $\Omega = 8\pi^2$ and $\langle\ldots\rangle_\omega \equiv (1/\Omega)\int d\omega(\ldots)$. The simpler u-expansion is essentially an expansion in u_a/kT and thus is expected to be valid at high temperatures. For H_2O the u-expansion DFT gives useful results[166, 167] for $T \gtrsim 400$ K. For lower temperatures, including room temperature, the f-expansion is required.[168]

An example of the results from DFT/f-expansion is shown in Fig. 8.31, where the bulk vapour–liquid coexistence curve predicted by DFT is presented, and compared to experiment and available computer simulation results for the TIP4P model. The theoretical critical temperature T_c and density ρ_c are found to be 609 K and 0.277 gcm^{-3}, to be compared with the experimental values of 647 K and 0.3227 gcm^{-3}, respectively. Overall, the theory agrees reasonably well with experiment and simulation, the major discrepancy being in the shape of the liquidus curve at high temperatures. In Fig. 8.32 is shown the variation of vapour–liquid surface tension γ with temperature. The DFT results are considerably higher than experiment, but are in quite reasonable agreement with the single simulation result available for the TIP4P model at $T = 325$ K. At this temperature we have $\gamma(\text{DFT}) = 170$ dyncm^{-1}, $\gamma(\text{sim}) = 149 \pm 18$ dyncm^{-1}, and $\gamma(\text{expt}) = 68$ dyncm^{-1}. Other DFT results are available for the density and order parameter profiles.[166–168] In § 8.9 we give some DFT results for observables more sensitive to molecular preferred orientations at the liquid–vapour interface.

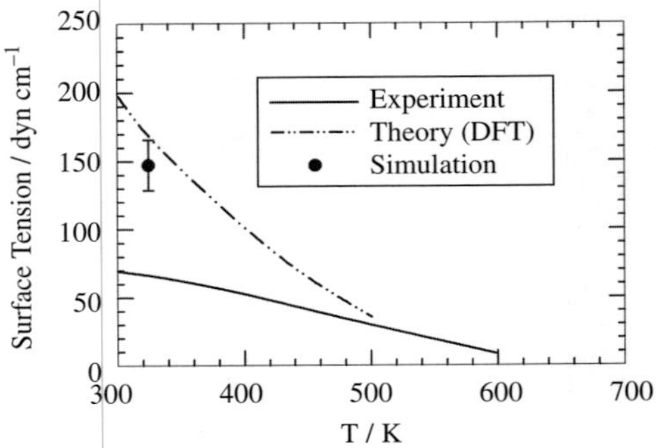

FIG. 8.32 The variation of vapour–liquid surface tension γ of water with temperature. Experimental results from ref. 185 and DFT from ref. 168. The single point with its associated error bar is the molecular dynamics result.[184] (From ref. 168.)

8.7 Non-local density functional theory

The local density approximation embodied in (8.214) is expected to give good results when the variation in density is small over the range of the intermolecular forces. This is the case for a fluid in a weak external field such as gravity, for example. For fluid–fluid interfaces, where the density variations are greater, it can still give reasonable results, as seen in the previous section. In the case of crystalline solids or of a fluid close to a solid wall, however, the density profile shows very strong short-range structure, and the local density approximation breaks down. As an example, results for a Lennard-Jones fluid between two parallel walls[186] are shown in Fig. 8.33. The simulation results show a strongly oscillating density profile with very high peaks near the walls. The local mean field theory misses this oscillating structure. In such a case it is clear that the free energy density $a_0(\mathbf{r})$ will depend not only on the local density $\rho(\mathbf{r})$ at that point, but also on the quite different densities in neighbouring regions. Moreover, there will be points in the inhomogeneous region where the local density far exceeds any value typical of a homogeneous liquid (e.g. $\rho\sigma^3 = 4.3$ at the peak near the walls in Fig. 8.33), so that it is not possible to estimate the value of the free energy for such a homogeneous liquid.

A variety of procedures have been proposed to introduce non-local effects and to thus formulate an approximate Helmholtz free energy functional, and these have been the subject of several reviews.[187–198] One approach is to expand the free energy functional in the gradient and higher derivatives of the local number density. However, this fails for highly inhomogeneous systems, such as a crystal or a fluid near a wall. An alternative, and more useful, approach is to replace the local density, $\rho(\mathbf{r})$, by one or more *weighted (or 'smoothed') densities* that incorporate density variations near the point $\mathbf{r}$. It is then assumed that the number density and free energy density at $\mathbf{r}$ can be obtained by using these weighted densities in a suitable free energy expression for the *uniform* fluid or material. We

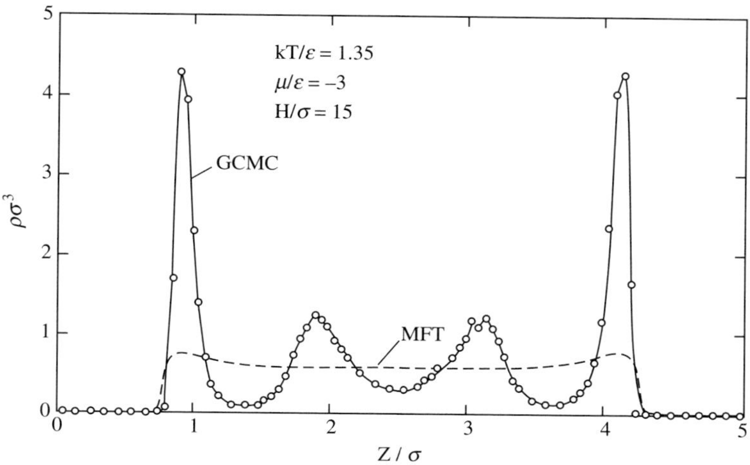

FIG. 8.33 The fluid density profile for a Lennard-Jones fluid (with the potential cutoff and shifted at $r_c = 2.5\sigma$) confined between two parallel walls separated by a distance $H = 5\sigma$, at a reduced temperature $kT/\varepsilon = 1.35$ and chemical potential $\mu/\varepsilon = -3$ (here σ and ε are the fluid–fluid interaction parameters). The fluid–wall potential is averaged over the surface, and the fluid–fluid and fluid–wall parameters approximate those for ethylene in a graphite slit. Points and solid lines are grand canonical Monte Carlo (GCMC) results; the dashed line is local mean field theory (MFT), with the hard sphere diameter chosen to match the liquid–vapour coexistence curves for the bulk homogenous fluid predicted by the MFT to the simulation values. (From ref. 186.)

shall assume that the reference system is a mixture of species, A, B, ..., i, ... R. We first write the Helmholtz energy functional as (cf. (8.214))

$$\mathcal{A}_0[f] = \int d\mathbf{r} a_0^{\mathrm{id}}(\mathbf{r}) + \int d\mathbf{r} a_0^{\mathrm{r}}[\{\bar{f}_\alpha(\mathbf{r}\,\omega)\}] \tag{8.254}$$

or, for spherical molecules,

$$\mathcal{A}_0[\rho] = \int d\mathbf{r} a_0^{\mathrm{id}}(\mathbf{r}) + \int d\mathbf{r} a_0^{\mathrm{r}}[\{\bar{\rho}_\alpha(\mathbf{r})\}], \tag{8.255}$$

where a_0^{id} is the ideal gas part of the free energy density, for which the local treatment is exact,

$$a_0^{\mathrm{id}}(\mathbf{r}) = kT \sum_i \rho_i(\mathbf{r}) \left[\ln\left(\rho_i(\mathbf{r})\Lambda_{t,i}^3\right) - 1\right] \tag{8.256}$$

and $\rho_i(\mathbf{r})$ is the number density of species i at point $\mathbf{r}$, $\Lambda_{t,i}$ is the thermal de Broglie wavelength (Eq. (3.79) of Vol. 1) for species i, $a_0^{\mathrm{r}} = a_0 - a_0^{\mathrm{id}}$ is the residual (non-ideal gas) part of the free energy per molecule, for which a non-local treatment is required, and $\{\bar{f}_\alpha\} = \bar{f}_1, \bar{f}_2, \bar{f}_3, \ldots$ and $\{\bar{\rho}_\alpha\} = \bar{\rho}_1, \bar{\rho}_2, \bar{\rho}_3, \ldots$ represent one or more smoothed or 'coarse grained' one-body distribution functions or densities. The smoothed densities can be defined in terms of the number densities and weight functions (we consider spherical molecules for simplicity of notation), which can be either scalars, w_i^{α}, or vectors, $\mathbf{w}_i^{\alpha}$,

$$\bar{\rho}_\alpha(\mathbf{r}) = \sum_i \int \mathrm{d}\mathbf{r}' \rho_i(\mathbf{r}') w_i^{\alpha}(\mathbf{r} - \mathbf{r}') \tag{8.257a}$$

$$\bar{\boldsymbol{\rho}}_\alpha(\mathbf{r}) = \sum_i \int \mathrm{d}\mathbf{r}' \rho_i(\mathbf{r}') \mathbf{w}_i^{\alpha}(\mathbf{r} - \mathbf{r}'). \tag{8.257b}$$

The vector nature of the weight functions and corresponding smoothed densities in (8.257b) occurs because in some versions of DFT account is taken of the fact that the density changes in the vicinity of the point of interest depend on the direction considered. Such vector weight functions and densities have the usual three components, e.g. (ρ_x, ρ_y, ρ_z).

The problem is thus to specify an appropriate form for the weighting functions, $w_i^{\alpha}(\mathbf{r})$. Once these are known the weighted densities $\bar{\rho}_\alpha(\mathbf{r})$ corresponding to a given set of local density profiles $\{\rho_i(\mathbf{r})\}$ can be calculated from (8.257). The residual free energy is then calculated from a suitable equation of state for the bulk (uniform) reference fluid, using this set of weighted densities. Finally, Ω is minimized to find the equilibrium density profile using (8.193). In the case of pure fluids, if the reference fluid is taken to be one of hard spheres the Percus–Yevick (PY) compressibility equation, (5.22) (equivalent to the earlier scaled particle theory equation—see § 6.12), can be used, or for more accurate work (see Fig. 5.2) one can use the Carnahan–Starling formula,[199] (5.27). For hard sphere mixtures, either the mixture generalization of the $\mathrm{PY_c}$ (equivalent to the scaled particle theory equation, (6.167)) or the more accurate mixture generalization of the Carnahan–Starling equation[200,201] can be used.

Since little is known of the properties of the inhomogeneous reference system, the $w_i^{\alpha}(\mathbf{r})$ are generally chosen to give good results for the homogeneous reference fluid, whose properties are well known. To successfully describe the short-range fluid structure, the definition of the $w_i^{\alpha}(\mathbf{r})$, and hence $\bar{\rho}_\alpha(\mathbf{r})$, must incorporate in some way the effects of the short-range structure in the reference system with pair potential $u_0(r)$. A commonly used approach is to choose the $w_i^{\alpha}(\mathbf{r})$ in such a way as to ensure that the nonlocal theory gives a good description of the direct correlation function, $c_0(\mathbf{r}_1\mathbf{r}_2)$, for the *uniform, homogeneous* reference fluid; it is then implicitly assumed that this will lead to good results for the inhomogeneous fluid. In density functional theory the direct correlation function is given by a second functional derivative of the residual Helmholtz free energy functional (we continue to assume spherical molecules for the moment), (8A.46),

$$c_0(\mathbf{r}_1\mathbf{r}_2) = -\frac{1}{kT}\frac{\delta^2 \mathcal{A}_0^r}{\delta\rho(\mathbf{r}_1)\delta\rho(\mathbf{r}_2)}, \tag{8.258}$$

where $\mathcal{A}_0{}^r$ is the last term on the right-hand side of (8.255). The weighting function needs to be chosen so that the resulting expression for $\mathcal{A}_0^r$ yields, via (8.258), a direct correlation function that is as close as possible to the correct one for the homogeneous reference fluid over a range of state conditions. For a reference fluid of hard spheres, the PY result,[202] (5.16), is quite accurate, and has frequently been chosen as the criterion for determining the weight functions.

Early versions (e.g. refs. 161–164) of DFT introduced a single weighted density, and chose the weighting function in (8.257) to best approximate the PY result for the direct correlation function. These theories were very successful for some applications, including fluids against walls and in pores, but were difficult to extend to mixtures or to non-spherical molecules. Subsequently, in an influential paper, Rosenfeld,[203–205] proposed a more powerful version of DFT for hard spheres inspired by the link between scaled particle theory (§ 6.12) and the PY theory for the direct correlation function, $c(r)$, given by (5.16) of Vol. 1; in particular, PY theory, when used in the compressibility equation, gives an equation of state identical to that from scaled particle theory. Rosenfeld's approach does not rely on any bulk homogenous fluid properties as input, but is based on a geometric analysis, and starts from the observation that $c(r)$ can be expressed in terms of the geometrical properties (radius, surface area, volume) of the spheres. This version of DFT is now referred to as *fundamental measure theory* (FMT) as a result (see Appendix 6A, and Table 6A.1). Rosenfeld showed that consideration of the (exact) low-density (second virial) limit of the Helmholtz energy functional, and in particular of the Mayer f function, for a hard sphere mixture with radii R_i, leads naturally to the definition of six weight functions for each of the species, of which four are scalars and two are vectors,

$$\begin{aligned}
w_i^3(r) &= \Theta(R_i - r) \\
w_i^2(r) &= \delta(R_i - r) \\
w_i^1(r) &= \frac{w_i^2(r)}{4\pi R_i} \\
w_i^0(r) &= \frac{w_i^2(r)}{4\pi R_i^2} \\
\mathbf{w}_i^2(\mathbf{r}) &= \frac{\mathbf{r}}{r}\delta(R_i - r) \\
\mathbf{w}_i^1(\mathbf{r}) &= \frac{\mathbf{w}_i^2(\mathbf{r})}{4\pi R_i},
\end{aligned} \tag{8.259}$$

where $\Theta(r)$ and $\delta(r)$ are the Heaviside and Dirac delta functions, respectively (see Appendix B.3 of Vol. 1). For the uniform bulk fluid the last two (vector) densities vanish. The smoothed densities corresponding to these weight functions are then obtained from Eqs. (8.257). We note that not all of these weight functions have the units of inverse volume; thus the corresponding smoothed 'densities' of (8.257) will not have the usual density units.[206]

Rosenfeld obtains an expression for the residual free energy density functional, a_0^r, by requiring that it obey the low-density limit and also obeys the large sphere thermodynamic limit (as in scaled particle theory, see (6.144)),

$$\mu_{0,i}^r = p_0 v_{mi} \qquad R_i \to \infty, \tag{8.260}$$

where $\mu_{0,i}^r$ is the residual chemical potential for component i in the reference fluid. This results in the expression,[207,208]

$$\beta a_0^r = -\bar{\rho}_0 \ln(1-\bar{\rho}_3) + \frac{\bar{\rho}_1\bar{\rho}_2 - \bar{\boldsymbol{\rho}}_1 \cdot \bar{\boldsymbol{\rho}}_2}{1-\bar{\rho}_3} + \frac{\bar{\rho}_2^3 - 3\bar{\rho}_2\bar{\boldsymbol{\rho}}_2 \cdot \bar{\boldsymbol{\rho}}_2}{24\pi(1-\bar{\rho}_3)^2}, \tag{8.261}$$

where $\bar{\rho}_i$ and $\bar{\boldsymbol{\rho}}_i$ are the weighted densities of (8.257), with the latter corresponding to the use of the vector weight functions of Eq. (8.259).

For the bulk homogenous fluid, Rosenfeld's DFT produces the scaled particle theory result (§ 6.12) and the PY (compressibility) result for the direct correlation function as output. A related and somewhat simpler form of DFT has been proposed by Kierlik and Rosinberg,[209] who tailor their functional to the PY result for $c_{ij}(r)$, the two-body direct correlation function for the uniform hard sphere mixture, as input, and show that the minimum number of weighting functions needed to achieve this is four, and that they are related to the fundamental measures of the spheres (see Table 6A.1 and (6.154)–(6.156)). The resulting expressions for the weighting functions are

$$\begin{aligned}
w_i^3(r) &= \Theta(R_i - r) \\
w_i^2(r) &= \delta(R_i - r) \\
w_i^1(r) &= \left(\tfrac{1}{8\pi}\right)\delta'(R_i - r) \\
w_i^0(r) &= -\left(\tfrac{1}{8\pi}\right)\delta''(R_i - r) + \left(\tfrac{1}{2\pi r}\right)\delta'(R_i - r),
\end{aligned} \tag{8.262}$$

where $\delta'(R_i - r)$ and $\delta''(R_i - r)$ are the first and second derivatives of the Dirac delta function $\delta(R_i - r)$, respectively. We note that the weight functions w_i^3 and w_i^2 are identical in the Rosenfeld FMT and Kierlik–Rosinberg DFT, but that the other weight functions differ in the two theories. Both theories yield the PY (pair) direct correlation correlation function for uniform hard sphere mixtures, and it has been shown that both yield the same uniform fluid Helmholtz energy density functional.[210] Since the hard sphere potential is additive, the uniqueness theorem (§ 3.1.7, p. 178, Vol. 1) implies that both theories yield the same structure (as given by the two-body correlation functions) and thermodynamics within the PY approximation for the uniform hard sphere mixture. However, for *non-uniform* fluids the two theories will, in general, give different results. Although the FMT and Kierlik–Rosinberg theories give similar results when applied to surface properties and confinement effects for simple atomic fluids, the FMT is of more general scope, since it can be extended to other intermolecular potentials, including convex nonspherical bodies.[211]

A limitation of both the FMT and Kierlik–Rosinberg DFT is that the underlying direct correlation function, $c(r)$, and the equation of state for the uniform hard sphere fluid is the Percus–Yevick result. This overestimates the pressure for higher densities (see Fig. 5.2). It is possible[205,208] to preserve the structure of the FMT but to replace the PY compressibility equation of state by the empirical but accurate Mansoori–Carnahan–Starling–Leland (MCSL) equation of state[212] for hard sphere mixtures, thus improving the accuracy of the theory. The MCSL equation reduces to the Carnahan–Starling equation for pure hard spheres (see (5.27) and Fig. 5.2 of Vol. 1). This results in a modified expression for the residual free energy density[205,208]

$$\beta a_0^r = -\bar{\rho}_0 \ln(1-\bar{\rho}_3) + \frac{\bar{\rho}_1\bar{\rho}_2 - \bar{\boldsymbol{\rho}}_1 \cdot \bar{\boldsymbol{\rho}}_2}{1-\bar{\rho}_3}$$
$$+ \left(\bar{\rho}_2^3 - 3\bar{\rho}_2\bar{\boldsymbol{\rho}}_2 \cdot \bar{\boldsymbol{\rho}}_2\right) \frac{\bar{\rho}_3 + (1-\bar{\rho}_3)^2\ln(1-\bar{\rho}_3)}{36\pi\bar{\rho}_3^2(1-\bar{\rho}_3)^2}, \qquad (8.263)$$

which replaces the original FMT expression, (8.261).

Nonlocal DFT has been widely and successfully used to study adsorption at solid walls and in narrow pores, and confinement effects on phase transitions, selective adsorption from mixtures, heats of adsorption, and solvation pressures. In Fig. 8.34 we show results for both the original FMT and the FMT-MCSL modification of it. While the FMT-MCSL version gives quantitative agreement with the MC results for $c(r)$ for the uniform hard sphere fluid and for the density profile for hard spheres near a hard wall, the original FMT predicts values of $c(r)$ that are too small, and a contact density at the wall that is too high.

In most applications of FMT to adsorption the system is treated as a single-component one, the wall interactions being incorporated through an external field term (the fluid–wall interactions). Such treatments work well for very simple pore geometries (e.g. slit and cylindrical pores) but become unwieldy when trying to account for wall roughness, pore tortuosity, etc. A simpler approach to including wall roughness is to treat the system as a multi-component one, in which the wall atoms are treated as 'quenched' components, with given density distributions, rather than with an external potential (such 'quenched-annealed' models have also been widely used in integral equation theories of fluid–wall systems, see § 8.8). The free energy functional for such 'quenched-annealed' mixtures (the fluid being the 'annealed' component) takes a simple form.[214] Application of this quenched solid form of FMT to adsorption in nano-porous materials, incorporating wall roughness, has provided very good agreement with experimental results for adsorption isotherms and capillary condensation in pores.[215,216] An example is shown in Fig. 8.35, for krypton adsorption on a mesoporous silica material, SBA-15, with roughly cylindrical pores of 9 nm diameter. In this calculation the Kr–Kr Lennard-Jones parameters were fitted to properties of bulk krypton, and the density profile of the solid silica (which was treated as a single component) was approximated as a linear ramp function,[217] with the width of the ramp at the surface, which represents the surface roughness, fitted to X-ray diffraction data on this material.[218]

In addition to predicting adsorption behaviour for pores of a given geometry and width, non-local DFT has been widely used to carry out the inverse

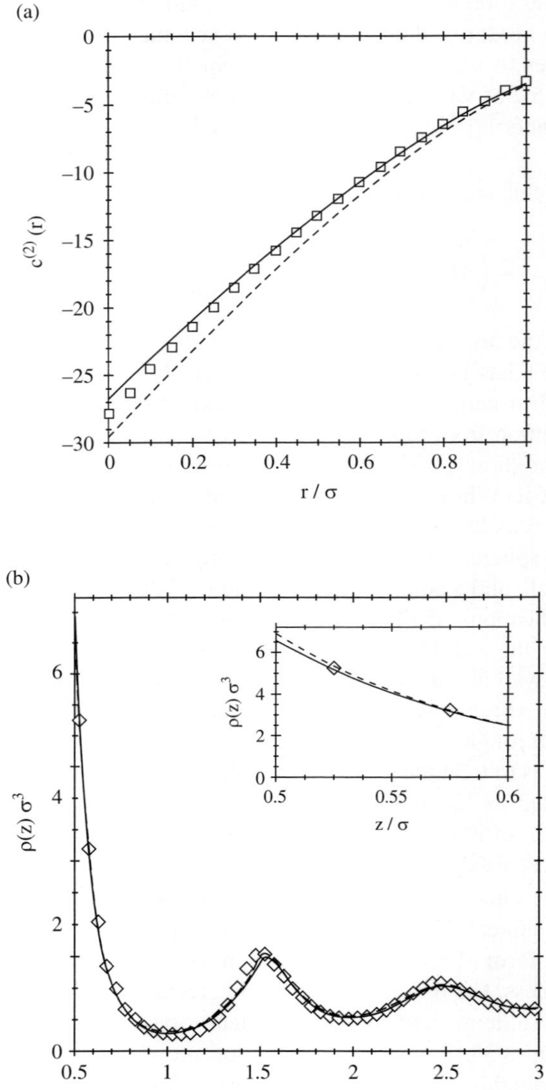

FIG. 8.34 (a) The two-body direct correlation function for the uniform hard sphere fluid for a packing fraction $\eta = 0.4189$; (b) density profile of a pure hard sphere fluid against a planar hard wall for $\eta = 0.4257$. In both figures points are from Monte Carlo simulation,[213] dashed lines are from the original FMT, and solid lines are the FMT-MCSL version of the theory. Reprinted with permission from ref. 205. Copyright 2002 Institute of Physics Publishing.

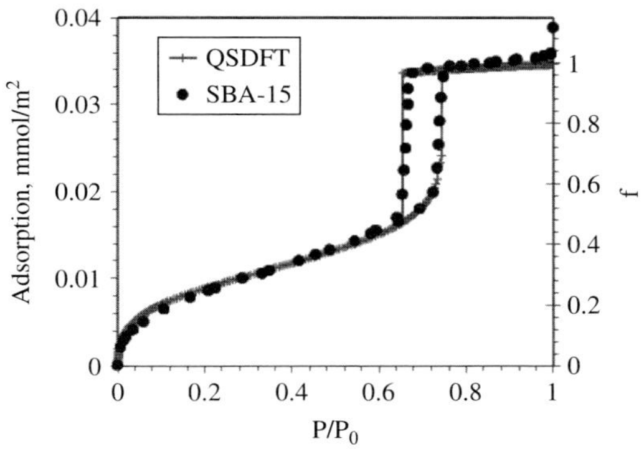

FIG. 8.35 Prediction of krypton adsorption on a mesoporous silica material, SBA-15, with 9 nm diameter pores at 119 K. Lines are the quenched-annealed version of FMT; points are experimental data.[218] The sharply rising region to the right shows capillary condensation from a gas-like to a liquid-like phase in the pore on increasing the pressure (adsorption curve); the nearly vertical region shown at lower pressure represents capillary evaporation on decreasing the pressure from that for the filled pore (desorption curve). Here P is the pressure of the bulk gas phase in contact with the SBA-15 material, and P_0 is the vapor pressure of bulk krypton at 119 K. Reprinted with permission from ref. 215. Copyright 2006 American Chemical Society.

calculation, using experimental adsorption isotherms to estimate the pore size distribution of a nano-porous material.[219–221] Such DFT-based methods have been routinely used in commercial adsorption instruments to determine pore size distributions.

Confinement in very narrow pores results in reduced dimensionality. Reducing the pore width causes the system to change from a 3-dimensional one to 2-dimensional for narrow slit shaped pores, to 1-dimensional for cylindrical pores of width just sufficient to contain a line of molecules and to 0-dimensional for a spherical cavity sufficiently small as to accommodate just one molecule. Thus, it is desirable that any theory used to describe adsorption be able to accurately describe this dimensional crossover to 2, 1, or 0 dimensions. Unfortunately, the FMT does not meet this criterion. If the last term on the right-hand side of the FMT free energy functional (8.261) is omitted, the resulting modified FMT gives correctly the exact results for systems of 0 and 1 dimension, but this results in considerably poorer results for the 3-dimensional system. There have been several proposals to resolve this problem by modifying the free energy functional (8.261).[222–227] A relatively simple modification of the free energy functional that accounts for dimensional crossover quite well is[222]

$$\beta a_0^r = -\bar{\rho}_0 \ln(1-\bar{\rho}_3) + \frac{\bar{\rho}_1\bar{\rho}_2 - \bar{\boldsymbol{\rho}}_1\cdot\bar{\boldsymbol{\rho}}_2}{1-\bar{\rho}_3} + \frac{\bar{\rho}_2^3}{24\pi(1-\bar{\rho}_3)^2}\left(1-\left|\frac{\bar{\boldsymbol{\rho}}_2}{\bar{\rho}_2}\right|^2\right)^3, \tag{8.264}$$

which differs from (8.261) only in the last term on the right. This modified term is zero for systems of 0 dimensions, and is small for 1-dimensional systems, and thus serves as an interpolation term accounting in a rough way for crossover. Such crossover effects become significant for adsorption in small spherical and cylindrical cavities.[227]

Nonlocal density functional theory has also been applied to freezing of fluids,[195] nucleation,[189] polymers,[192–194, 228–230] and other forms of soft matter,[193] and to dynamical properties.[194, 231] Accounting for dimensional crossover is essential in freezing studies, since the atoms are tightly confined by their neighbours, experiencing a quasi-zero dimensional situation. The original FMT of (8.261) is unable to predict freezing due to its failure to even qualitatively describe the 0-dimensional limit; the final term on the right-hand side of (8.261) tends to minus infinity in the 0-dimensional limit, so that the solid is always the stable phase. The solid–fluid transition, can, however, be described using the modified functionals described above.

The FMT can be extended to non-spherical convex bodies,[232, 233] and there have been several applications to mixtures involving rods, platelets, spheres, and polymer molecules.[196, 234, 235]

8.8 Integral equation theory

In using integral equation theory for the study of inhomogeneous fluids one can either start from one of the integrodifferential equations of §8.1.1, or from the Ornstein–Zernike equation for inhomogeneous fluids (see (3.117) and footnote in Vol. 1). Usually the main objective is to calculate the one-body distribution function or density–orientation profile, $f(\mathbf{r}_1\omega_1)$. Whichever of the two approaches is used, it is necessary to have a closure relation, i.e. an approximation for either $g(12)$ or $c(12)$ in the interfacial region. In the case of the integrodifferential equations of §8.1.1 it is common to relate the pair correlation (or direct correlation) function to the corresponding function for the uniform fluid, usually at some mean density intermediate between the densities of the two phases. If instead the starting point is the generalized Ornstein–Zernike equation, one of the closures described in Chapter 5 of Vol. 1 can be used, e.g. PY, HNC, LHNC, RHNC, MSA. Once $f(1)$ is known, the surface tension can be calculated from either the virial route, (8.95), or the direct correlation function route, (8.117), by using the solution for $f(1)$ together with the approximate closure relation for $g(12)$ or $c(12)$. Such theories often give good results for the fluid structure for a fluid–fluid interface, and also for a fluid–solid interface provided the bulk fluid is not near a phase transition. With such simple closures these theories cannot describe wetting or drying (wetting of a surface by the gas phase) phenomenon, in contrast to the density functional theories described in §§8.6 and 8.7.

In studies of fluid–fluid interfaces, it is common to neglect the external potential term in the integral equations. Much work has been based on the first YBG equation, (8.6), although for molecular fluids its angular analogue, (8.11), may be more convenient. In studies of fluid–solid interfaces, the external potential (due to the fluid–wall forces) must be included, and approximations based on the Ornstein–Zernike equation are often used. A common starting point is the so-called 'wall-particle' Ornstein–Zernike equation, discussed below. In view of the somewhat different approaches used, we discuss fluid–fluid and fluid–solid interfaces separately in what follows. The application of integral equation theories to atomic fluids has been extensively discussed by other authors,[14, 17, 18, 138, 188, 195] so that we concentrate on molecular fluid interfaces in what follows.

8.8.1 *Fluid–fluid interfaces*

Both the first YBG equation, (8.6), and its angular analogue, (8.11), have been used to study the planar vapour–liquid interface for a Stockmayer fluid.[150] In order to solve these equations for the density–orientation profile, the functions $f(1)$, $g(12)$, and $u(12)$ are expanded in spherical harmonics, leading to a set of coupled integral equations. The calculations are based on Toxvaerd's approximation[236] for the pair correlation function in the interface,

$$g(z_1 z_{12} r_{12} \omega_1 \omega_2) = A(z_1 z_{12}) g_\ell(r_{12} \omega_1 \omega_2) + B(z_1 z_{12}) g_g(r_{12} \omega_1 \omega_2), \qquad (8.265)$$

where

$$A(z_1 z_{12}) = [\alpha \rho(z_1) + (1 - \alpha)\rho(z_1 + z_{12}) - \rho_g]/(\rho_\ell - \rho_g) \qquad (8.266)$$

$$B(z_1 z_{12}) = -[\alpha \rho(z_1) + (1 - \alpha)\rho(z_1 + z_{12}) - \rho_\ell]/(\rho_\ell - \rho_g). \qquad (8.267)$$

In these equations g_ℓ and g_g are the angular pair correlation functions for the bulk liquid and bulk gas. The liquid phase $g_\ell(12)$ was calculated from the LHNC theory and the vapour phase $g_g(12)$ was set equal to the dilute gas value. The angular analogue of the YBG equation, (8.11), was found to be more tractable than the usual YBG equation, (8.6), when calculating the density–orientation profile. The theory gives good agreement with MD results for the density–orientation profile for the Stockmayer fluid (Lennard-Jones plus dipole–dipole term),[150] correctly predicting the preferred molecular orientation, parallel to the liquid surface, on the liquid side of the Gibbs dividing surface. Some corresponding results for the surface tension are shown in Fig. 8.36. The theory seems to be in good agreement with the simulation results.

8.8.2 *Solid–fluid interfaces*

The interfacial properties of a fluid near a solid wall depend strongly on both the fluid–fluid and fluid–wall intermolecular forces, and it is the relative strengths and nature of these two interactions that determine much of the most interesting phenomena. We consider systems in which the fluid–wall interactions are of the usual physical kind (dispersion, repulsion, electrostatic, and induction) and

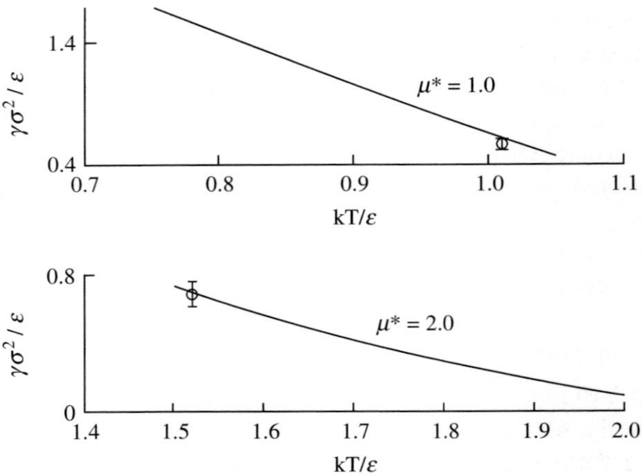

FIG. 8.36 Surface tension for the Stockmayer fluid for two different dipole strengths from the YBG integral equation theory (lines) and from molecular dynamics simulation (points). (From ref. 150.)

exclude from consideration strong bonding fluid–wall forces of the chemisorption type,[237,238] where significant electron rearrangement and transfer occurs.

In realistic treatments of real fluid–wall systems the inhomogeneous nature of the molecular arrangement in the solid should be accounted for. However, it is sometimes reasonable to approximate the wall as a structureless continuum, by integrating the fluid–wall pair potential $u_{fw}(\mathbf{r}_1\omega_1\mathbf{r}_2\omega_2)$, where 1 and 2 are molecules in the fluid and solid phases, respectively, over the position $\mathbf{r}_2$ and orientation ω_2 of the molecule in the solid phase, assuming a continuous distribution of wall molecules of uniform number density ρ_{w}. For simple wall geometries this integration often yields closed analytic forms for the fluid–wall potential; such potentials for structureless walls have been considered in detail by Steele.[239–241] This approximation should be a good one far from the wall, where the wall structure is not 'noticed'; this approximiation is also justified close to the wall if the fluid molecular size is large relative to the interatomic spacing in the wall, provided the fluid–wall interaction is not too large.

The wall–fluid OZ equation

A fluid in contact with a solid wall can be viewed as the limiting case of a homogeneous fluid mixture in which the concentration of one component (the 'wall' molecules) goes to 0, while the size or diameter of its molecules goes to infinity.[242] Consider a uniform fluid mixture of two components A and W having a total density ρ. The Ornstein–Zernike (OZ) equation for a uniform mixture is (cf. (3.117) and (5.278))

$$h_{\alpha\beta}(12) = c_{\alpha\beta}(12) + \rho \sum_{\gamma} x_{\gamma} \int \langle c_{\alpha\gamma}(13)h_{\gamma\beta}(32)\rangle_{\omega_3} \mathbf{dr}_3, \tag{8.268}$$

where the sum is over the components A and W, and $x_\gamma = N_\gamma/N$ is the mole fraction of γ. Since $h_{\alpha\beta}(12) = h_{\beta\alpha}(21)$ and $c_{\alpha\beta}(12) = c_{\beta\alpha}(21)$ we can interchange the $\alpha 1$ and $\beta 2$ molecules in (8.268) to obtain an alternative form for the mixture OZ equation,

$$h_{\alpha\beta}(12) = c_{\alpha\beta}(12) + \rho \sum_\gamma x_\gamma \int \langle h_{\alpha\gamma}(13) c_{\gamma\beta}(32)\rangle_{\omega_3} \mathbf{dr}_3, \tag{8.269}$$

which is more useful for our present purposes. We shall suppose the W molecules to be spherical in their interactions, whereas the A molecules are in general nonspherical. We now consider the limiting case (in this order) where (1) the mole fraction of the W molecules goes to 0, and (2) their size tends to infinity. In this limit W represents a structureless planar surface (the wall), and A the molecular fluid in contact with it. The mixture OZ equations decouple when $x_A \to 1$ and $x_W \to 0$, since the sum on the right of (8.269) contains only the term for A, and the equations for the fluid–wall correlations ($\alpha\beta$ = AW and WA) are

$$h_{AW}(12) = c_{AW}(12) + \rho \int \langle h_{AA}(13) c_{AW}(32)\rangle_{\omega_3} \mathbf{dr}_3 \tag{8.270}$$

$$h_{WA}(12) = c_{WA}(12) + \rho \int \langle h_{WA}(13) c_{AA}(32)\rangle_{\omega_3} \mathbf{dr}_3. \tag{8.271}$$

These two equations are equivalent, although one or the other may be more convenient in a particular application. In these equations $h_{WA}(12)$, in the limit of $R_W \to \infty$ (where R_W is the radius of the W molecule), has the physical significance of a singlet correlation function such that $h_A(z_2\omega_2) + 1$ is proportional to the probability density that a fluid molecule A is at distance z_2 from the surface of the wall with orientation ω_2 (note that we have translated the origin of the z coordinate from the centre of the W molecule to its surface). Thus we have

$$h_{WA}(12) \equiv h_A(z_2\omega_2) = g_{WA}(z_2\omega_2) - 1 = \frac{f_A(z_2\omega_2)}{\rho/\Omega_A} - 1, \tag{8.272}$$

where $f_A(z_2\omega_2)$ is the one-body distribution function giving the density of molecules at distance z_2 from the wall with orientation ω_2, and $f_A(z_2 = \infty, \omega_2) = \rho/\Omega_A$ is the value of this function at infinite distance from the wall, i.e. in the homogeneous, bulk fluid. Similarly, we have $h_{AW}(12) = h_A(z_1\omega_1)$, with a similar interpretation, and $c_{AW}(12) = c_A(z_1\omega_1)$ and $c_{WA}(12) = c_A(z_2\omega_2)$ depend only on the coordinates of the fluid molecule relative to the wall. The functions $h_{AA}(13)$ and $c_{AA}(32)$ appearing under the integrals in (8.270) and (8.271) are the pair correlation functions for the pure, homogeneous bulk fluid far from the wall, because of the order of the limits $x_W \to 0$ and $R_W \to \infty$ taken above. Also in these equations we have $\rho = \rho_b$, the bulk fluid density of A molecules. Thus

(8.270) and (8.271) become (we now drop the A subscript)

$$h(z_1\omega_1) = c(z_1\omega_1) + \rho \int \langle h(12)c(z_2\omega_2)\rangle_{\omega_2} \mathbf{dr}_2 \tag{8.273}$$

$$h(z_1\omega_1) = c(z_1\omega_1) + \rho \int \langle h(z_2\omega_2)c(21)\rangle_{\omega_2} \mathbf{dr}_2. \tag{8.274}$$

Although these two equations are equivalent, (8.274) is more useful in practice; it is usually referred to as the *wall–fluid Ornstein–Zernike equation*. It should be noted that the proper interpretation of $c(z_2\omega_2)$ is $c_{WA}(12)$ in the limit $x_W \to 0$, $R_W \to \infty$; it should not be confused with the one-body direct correlation function defined in (8A.38).

In order to use (8.274) one must introduce a closure relation between the fluid–wall functions $h(z_1\omega_1)$ and $c(z_1\omega_1)$, and must in addition have a means to calculate the direct correlation function for the bulk fluid, $c(12)$. For fluids of spherical molecules a variety of such closures have been studied,[243] including PY, HNC, RHNC (renormalized hypernetted chain theory—see §5.4.9), and GMSA (see §5.4.4). For a hard sphere fluid against a hard featureless wall the PY theory, Eq. (5.4), gives good results (see Fig. 8.37), similar to those obtained for the pair correlation function for the bulk hard sphere fluid (Fig. 5.1). As for the bulk fluid, the value of $\rho(z)$ from PY theory lies below the simulation value at $z = 0$ (contact with the wall).

When attractive forces are present in the fluid–fluid or fluid–wall interactions conventional integral equation closures often give poorer results.[243] In contrast

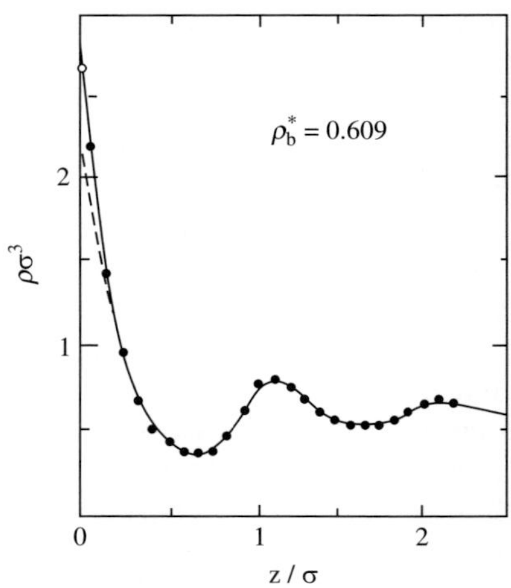

FIG. 8.37 The density profile for hard spheres near a hard, featureless wall from (8.274) together with PY closure (dashed line) and Monte Carlo simulations (solid line and points) for a bulk fluid density of $\rho_b\sigma^3 = 0.609$. Here the local density $\rho(z)$ is given by $\rho(z) = \rho_b g(z)$, where $g(z) = \langle g(z,\omega)\rangle_\omega$ and $g(z,\omega) = 1 + h(z,\omega)$ is calculated from ((8.274). (From ref. 244.)

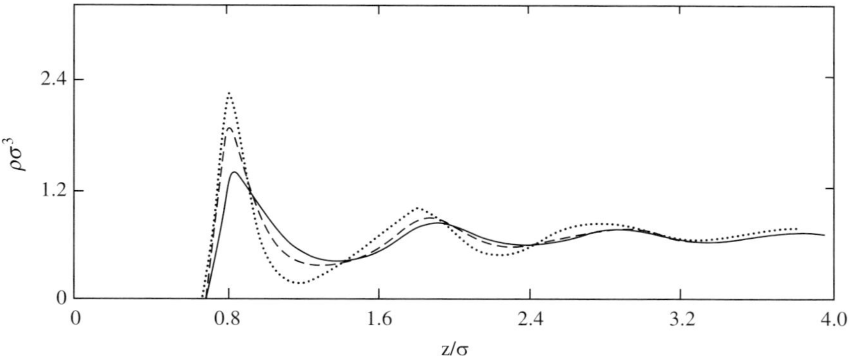

FIG. 8.38 The density profile for a LJ (12, 6) fluid near a LJ (9, 3) wall at $T^* = 1.0$, $\rho_b^* = 0.65$, with potential parameters chosen to approximate argon on graphite: solid line, MC simulation results; dashed line, PY and wall–fluid OZ; dotted line, HNC and wall–fluid OZ equation. The local density $\rho(z)$ is obtained as explained in the caption of Fig. 8.37. (From ref. 250.)

to the situation for monatomic bulk liquids (see §4.1), in fluid–wall systems attractive forces often have a profound effect on the structure of the fluid in the neighbourhood of the wall. That this will be so can be seen from the exact sum rule for a fluid at a hard wall[245]

$$p = \rho(0^+)kT, \tag{8.275}$$

where p is the pressure in the bulk fluid, and $\rho(0^+)$ is the fluid density at contact with the wall. Since attractive forces will have a large effect on the pressure, they will also have a large effect on the density profile near the wall. A typical result using PY and HNC closures is shown in Fig. 8.38 for a Lennard-Jones (12,6) fluid modeled on argon near a structureless LJ (9,3) wall modeled on graphite; the theory is seen to overestimate the structure. In contrast to the density functional theories, simple closures of the wall–fluid OZ equation, such as the PY, HNC, and MSA, cannot accommodate two coexisting phases near the wall, and so cannot describe wetting and contact angle phenomena.[246,247] That such a failure must occur can be seen most clearly by examining the grand potential functionals $\Omega[f]$ corresponding to these simple closures. It is found that the functionals corresponding to such simple closures for the bulk fluid can have only a single minimum when examined as a function of density, corresponding to a one-phase system.[248] A proper description of wetting and contact angle phenomena will require closures that more correctly describe the pair correlations near the wall. The MHNC (modified hypernetted chain) theory of Nieminen and Ashcroft,[249] which involves the use of the YBG equation combined with the OZ equation for the nonuniform fluid, does not suffer from the above defects, but is computationally more involved than the theories considered in this subsection.

The wall–fluid OZ equation has been successfully applied to *molecular* fluids against both neutral and charged walls, using a variety of more sophisticated closures. As an example, we consider the case of dipolar hard spheres against a neutral wall, which may be a hard wall or an attractive one. Monte Carlo simulations[251,252] show that for the hard wall case the molecules near the wall prefer to lie with their dipoles parallel to the wall, but for walls with attractive forces this orientational preference changes and depends on the strength and range of the attractive forces. Although the MSA closure fails to predict any orientational preference for this case (the structure is the same as that for hard spheres), a closure based on a modification[253,254] of the optimized random phase approximation[255] (ORPA, see § 5.4.2) gives good results, as shown in Fig. 8.39; the theory correctly predicts the molecular orientation near the wall.

Dipolar hard spheres against a wall with attractive and orientation dependent forces have been studied[256] using a fluid–wall potential of the form

$$u_{\mathrm{fw}}(z_1\omega_1) = u_{\mathrm{hs}}(z_1) - AkT\exp(-\lambda z_1)P_1(\cos\theta_1), \tag{8.276}$$

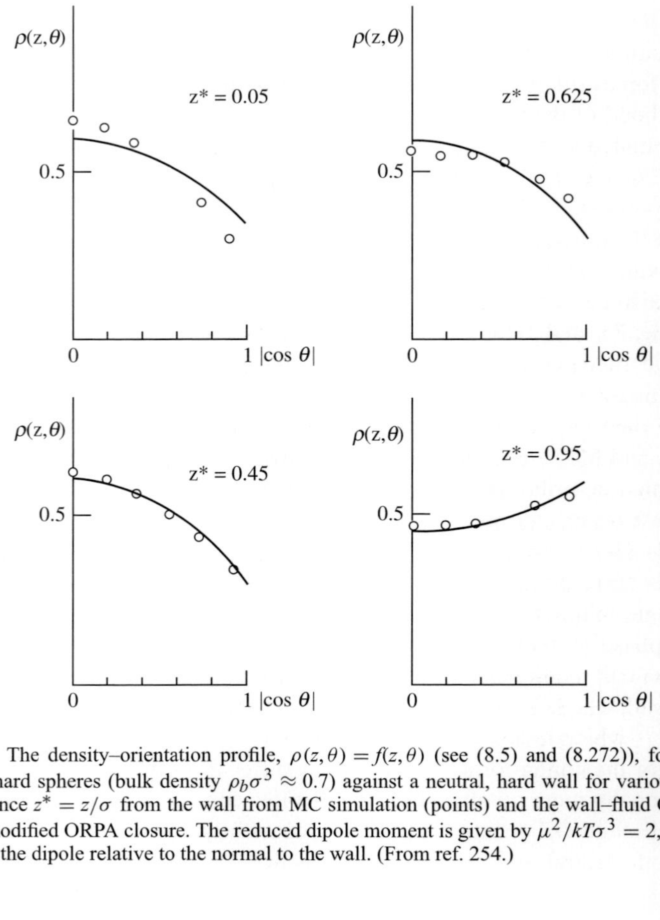

FIG. 8.39 The density–orientation profile, $\rho(z,\theta) = f(z,\theta)$ (see (8.5) and (8.272)), for a fluid of dipolar hard spheres (bulk density $\rho_b\sigma^3 \approx 0.7$) against a neutral, hard wall for various values of the distance $z^* = z/\sigma$ from the wall from MC simulation (points) and the wall–fluid OZ equation with a modified ORPA closure. The reduced dipole moment is given by $\mu^2/kT\sigma^3 = 2$, and θ is the angle of the dipole relative to the normal to the wall. (From ref. 254.)

where A and λ determine the strength and range of the potential, respectively. The wall–fluid total correlation function that appears on the left side of (8.274) is written

$$h(1) = h_0(1) + \delta h(1), \tag{8.277}$$

where $h_0(1)$ is the function for the reference system, in which the attractive fluid–wall potential is turned off (i.e. $A = 0$), and is calculated from the extended ORPA theory.[254] Since from (8.274) $h_0(1)$ obeys

$$h_0(1) = c_0(1) + \rho \int \langle h_0(2)c(21)\rangle_{\omega_2} \mathrm{d}\mathbf{r}_2, \tag{8.278}$$

it follows that $\delta h(1)$ obeys

$$\delta h(1) = \delta c(1) + \rho \int \langle \delta h(2)c(21)\rangle_{\omega_2} \mathrm{d}\mathbf{r}_2, \tag{8.279}$$

where $c(12)$ is the direct correlation function for the bulk fluid of dipolar hard spheres and $\delta c(1) = c(1) - c_0(1)$. Since $h_0(1) = -1$ for $z_1 < 0$, then $\delta h(1)$ must satisfy

$$\delta h(1) = 0, \quad z_1 < 0. \tag{8.280}$$

In order to solve (8.279) a closure between $\delta h(1)$ and $\delta c(1)$ is needed, together with a theory for $c(12)$. Russier *et al.*[252] have considered several possible closures, including the MSA,

$$\delta c(1) = -\beta u_{fw}(1), \quad z_1 > 0 \tag{8.281}$$

together with the MSA for $c(12)$, which they call the MSA/MSA theory, and another modified ORPA theory which they call DIP (dipole). Both the theory and simulations show that when the attractive fluid–wall force in (8.276) is relatively weak, the orientational structure near the wall is only slightly changed from the case of the hard wall, with the most likely orientation near the wall being $\theta = 80°$ when $A = 1$, only a little different from the hard wall case ($\theta = 90°$). As A is increased there is competition between the orientating effects due to the hard sphere and attractive terms in (8.276), and for $A = 5$ the most probable dipole orientation near the wall is 35°, while the second layer at $z = 1$ is also clearly oriented. The likelihood of dipoles pointing towards the wall (e.g. $\theta = 155°$) is very small because of the strong orientating effect of the attractive fluid–wall term. The DIP approximation gives good qualitative results in general. The MSA/MSA theory gives much poorer results. The polarization profile $P(z) \equiv P_z(z)$ and electric potential profile $\phi(z)$ are shown in Figs. 8.40 and 8.41, respectively, for this same system. Here $P(z)$ is given by (8.41) and $\phi(z)$ by (cf. (8.39))

$$\phi(z) = 4\pi \int_0^z P(z')\mathrm{d}z'. \tag{8.282}$$

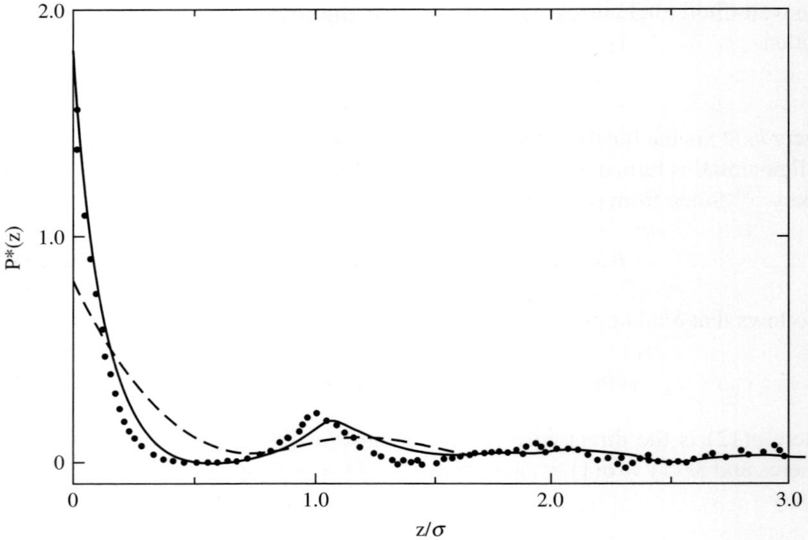

FIG. 8.40 The reduced polarization profile, $P^* = (\sigma^3/\mu)P$, for a dipolar hard sphere fluid ($\mu^2/kT\sigma^3 = 2$, $\rho_b\sigma^3 = 0.7$) against a wall with the fluid–wall potential (8.276) with $A = 5$, $\lambda = 1/\sigma$:—DIP theory, - - MSA/MSA, • MC. (From ref. 252.)

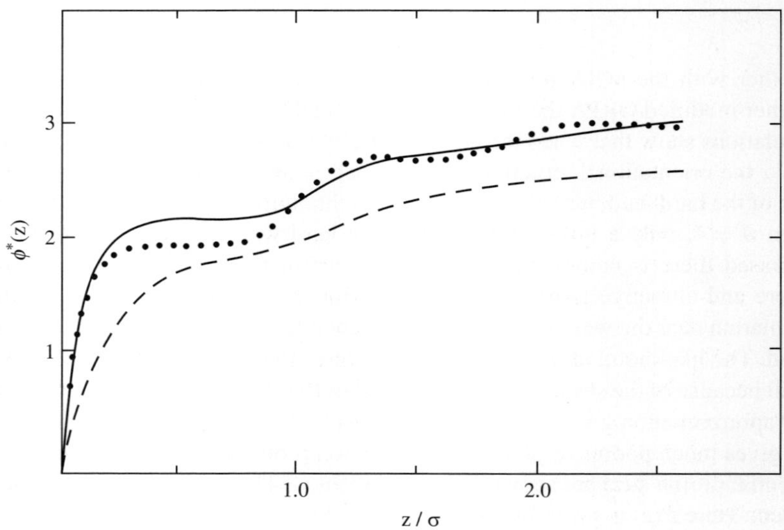

FIG. 8.41 The reduced electrical potential, $\phi^* = (\sigma^3/4\pi\mu)\phi$, for the same system as shown in Fig. 8.40. Key as described in the legend to Fig. 8.40. (From ref. 252.)

From Fig. 8.40 it can be seen that molecular orientation in the first three molecular layers has an appreciable effect on the polarization; this leads to the steep rises in $\phi(z)$ in Fig. 8.41 near z/σ values of 0, 1, and 2. Again, the DIP theory gives considerably better results than the MSA/MSA. From the results of Fig. 8.41 one can estimate that the overall potential drop would be about 1.2 V for molecules of diameter $\sigma = 4$Å at room temperature.

By considering the special case of a mixture of two dipolar species, it is possible to derive a wall–fluid OZ equation for the case where there is an electric field due to the wall.[257] One first allows the density of the W molecules to go to zero, then lets their radius go to infinity, as before, but in this case the polarity on the W molecules is kept; specifically, μ_W/σ_{DW}^3 and the direction of μ_W are kept constant while taking the limit, where $\sigma_{DW} = (\sigma_D + \sigma_W)/2$, and σ_D and σ_W are the diameters of the fluid (dipole) and 'wall' molecules, respectively. It is then easy to show that the electric field $\mathbf{E}_W$ due to the wall is given by[257,258]

$$\mathbf{E}_W = (3\hat{\mathbf{r}}\hat{\mathbf{r}} - \mathbf{1}) \cdot \boldsymbol{\mu}_W/{\sigma_{DW}}^3, \tag{8.283}$$

where $\hat{\mathbf{r}}$ is a unit vector along a line joining the D and W molecules and $\mathbf{1}$ is the unit tensor. Thus the electric field $\mathbf{E}_W$ is independent of the distance z from the wall, and its magnitude is determined by the strength and direction of the wall dipole, which is located at $z = -\infty$. The resulting wall–fluid OZ equation provides a convenient starting point for studying the effect of the electric field on the behaviour of a dipolar fluid near the wall, and it has been solved for dipolar hard spheres near a hard wall, using various closures including QHNC (see §5.4.9) and higher corrections to it.[259]

For a fluid of interacting site model (ISM) molecules (see §2.1.2) near a wall[260–264] it is convenient to consider the site density profiles $\rho_i(z)$, with $\rho_i(z)$ proportional to the probability density of finding a site i at a normal distance z from the wall, averaged over all orientations of the molecule while keeping the position of site i fixed, and averaged over all positions and orientations of the other molecules.[261] The site density profiles contain less information than the full density–orientation profile $f(z\omega)$, but are easier to compute. The relation between the site and full density–orientation profile is analogous to that between the site–site pair correlation function and the full two-body correlation function (§3.1.6). For such an ISM fluid one can derive a wall analogue of the 'OZ-like' equation for such fluids in the bulk, and then apply a wall analogue of the RISM equation[261] (see §5.5). As for the bulk fluid case, it is easiest to derive the OZ-like equation for the initial binary mixture of spherical ('wall') and polyatomic molecules by the method described in §5.5.3, where the polyatomic component of the mixture is in turn regarded as a mixture of its constituent particles (the sites). The OZ-like equation is then derived by distinguishing between *inter*molecular and *intra*molecular correlations. One then takes the infinite dilution followed by the infinite size limit for the spherical wall molecules to obtain the wall OZ-like equation,[261]

$$\mathbf{h}_W = \boldsymbol{\omega} \cdot \mathbf{c}_W + \rho\boldsymbol{\omega} \cdot \mathbf{c}_B \cdot \mathbf{h}_W, \tag{8.284}$$

where the symbol $\cdot$ denotes matrix multiplication and convolution, $\mathbf{c}_\mathrm{B}$ denotes the $m \times m$ matrix whose elements are the bulk fluid site–site direct correlation functions $c_{\mathrm{B},ij}$, and $\boldsymbol{\omega}$ is the matrix of intramolecular site–site correlation functions, with ij elements $\omega_{ij} = \omega_{ji}$ given by (cf. (5.240)–(5.242))

$$\omega_{ij}(r) = \delta_{ij}\delta(\mathbf{r}) + (1 - \delta_{ij})s_{ij}(r) \tag{8.285}$$

$$s_{ij}(r) = \frac{\delta(r - \ell_{ij})}{4\pi\ell_{ij}^2}, \tag{8.286}$$

where ℓ_{ij} is the fixed bond distance between sites i and j in a molecule. The functions $\mathbf{h}_\mathrm{W}$ and $\mathbf{c}_\mathrm{W}$ denote column matrices whose elements are $h_{\mathrm{W},i}(r)$ and $c_{\mathrm{W},i}(r)$, respectively, where i is a site in the molecule. The functions $h_{\mathrm{W},i}(r)$ and $c_{\mathrm{W},i}(r)$ are wall–site total and direct correlation functions; the latter can be regarded as being defined by (8.284). If the radius of the wall particles is now allowed to become very large the variable $r(= R_\mathrm{W} + z)$ can be replaced by z, the distance of the site from the planar wall, so that we can write $h_{\mathrm{W},i}(z)$ and $c_{\mathrm{W},i}(z)$. From the probabilistic interpretation of the pair correlation function, it follows that

$$\rho(h_{\mathrm{W},i}(z) + 1) \equiv \rho_i(z) \tag{8.287}$$

is the density of sites of type i at the normal distance z from the surface; i.e. it is the site density profile for sites of type i. For molecules in which all the sites are equivalent, e.g. a homonuclear diatomic, we have $h_{\mathrm{W},i} = h_\mathrm{W}$, $c_{\mathrm{W},i} = c_\mathrm{W}$, $c_{\mathrm{B},ij} = c_\mathrm{B}$, and the wall–fluid OZ-like equation (8.284) reduces to the scalar equation

$$h_\mathrm{W} = \omega_\mathrm{T} \cdot c_\mathrm{W} + m\rho\omega_\mathrm{T} \cdot c_\mathrm{B} \cdot h_\mathrm{W}, \tag{8.288}$$

where $\omega_\mathrm{T}(\mathbf{r}) \equiv \sum_{j=1}^{m} \omega_{ij}(\mathbf{r})$ and is independent of i, m is the number of equivalent sites per molecule, and $\cdot$ now simply denotes a convolution.

The wall–fluid OZ-like equation, (8.284) or (8.288), must be closed by introducing some relation between $h_{\mathrm{W},i}(z)$ and $c_{\mathrm{W},i}(z)$. For fused hard sphere molecules against a hard wall it is reasonable to adopt the analogue of the RISM closure (§5.5),

$$c_{\mathrm{W},i}(z) = 0, \qquad z > \sigma_{\mathrm{W},i}/2, \tag{8.289}$$

where $\sigma_{\mathrm{W},i}/2$ is the closest that the site i can approach the hard wall, together with the exact condition on $g_{\mathrm{W},i}$ imposed by the hard wall,

$$g_{\mathrm{W},i}(z) \equiv h_{\mathrm{W},i}(z) + 1 = 0, \qquad z < \sigma_{\mathrm{W},i}/2. \tag{8.290}$$

Equations (8.289) and (8.290) together constitute the 'wall-RISM' approximation.

Tests of the wall-RISM approximation against Monte Carlo simulation results for hard homonuclear diatomics (dumbbells) near a hard wall are shown in Fig. 8.42. While the theory is in qualitative agreement with the simulations, there are significant quantitative differences, particularly near the first peak and the

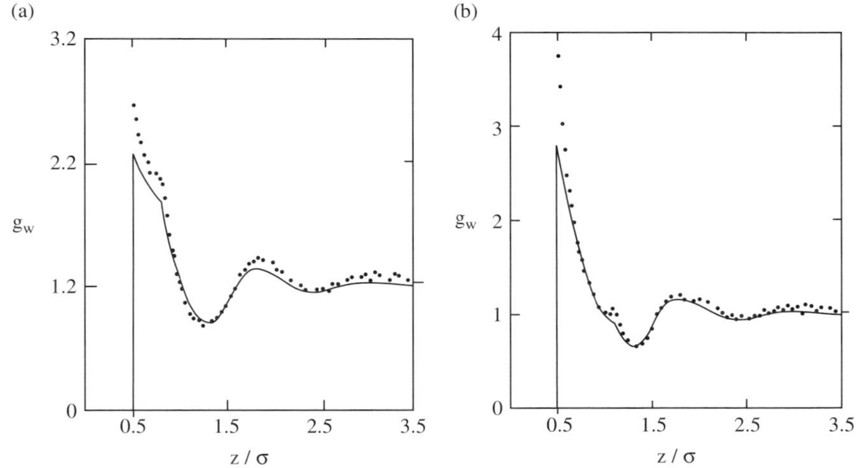

FIG. 8.42 Comparison of Monte Carlo (points) and wall-RISM (solid curve) results for the wall–site correlation function of hard homonuclear diatomics near a hard wall. Here the bulk density is $\rho_b\sigma^3 = 0.40$ and the elongations are: (a) $\ell/\sigma = 0.3$, (b) $\ell/\sigma = 0.6$. (From ref. 261.)

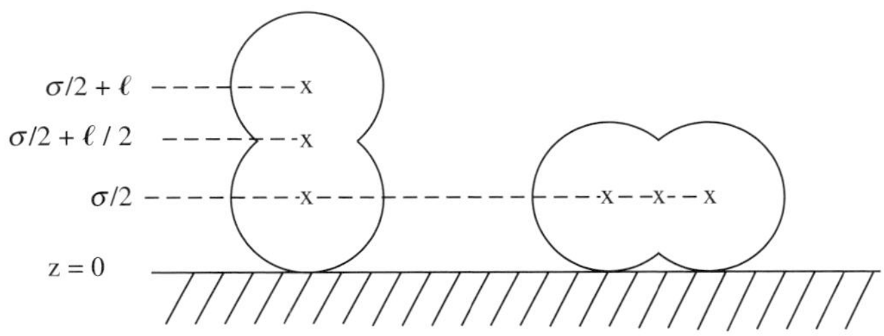

FIG. 8.43 Two possible orientations of a hard homonuclear diatomic molecule in contact with a hard planar wall: (a) perpendicular, (b) parallel. (From ref. 261.)

cusp[265] at $z = \sigma/2 + \ell$. These differences are similar to, but more pronounced than, those found using RISM for the bulk fluid. The small magnitude of g_W at the cusp, and the fact that both theory and simulation predict that as ℓ/σ or the density increases the cusp decreases in magnitude relative to the peak at $\ell/\sigma = 1$, strongly suggests that the molecules close to the wall prefer an orientation parallel to it (see Fig. 8.43), and this is confirmed by the simulation results. The wall-RISM approximation can be extended to homonuclear Lennard-Jones diatomics by using the PY-like closure (cf. (5.6))

$$c_W(z) = g_W(z)[1 - \exp(\beta u_{fw}(z))] \tag{8.291}$$

with (8.288), in place of the closure (8.289). Equation (8.291) reduces to the closure used previously, (8.289), in the case of hard dumbbells. However, the resulting wall–site correlation function for a homonuclear Lennard-Jones diatomic fluid against a wall of the same molecules is in rather poor agreement with molecular simulation results.[260]

8.9 Experimental and simulation studies of surface orientation

For atomic fluids, a longstanding and difficult experimental problem[266] has been the determination of the density profile $\rho(z)$ at liquid–gas and liquid–solid interfaces. For molecular fluids one is interested in the more detailed quantity $f(z\omega)$, which gives the distribution of density and orientation through the interface. At the present time methods for the direct experimental determination of $f(z\omega)$ do not exist (excluding computer experiments), so that attention has focused on reduced quantities such as $\rho(z)$ and the order parameters F_ℓ (see § 8.1.2).

In the next subsection we discuss the principal methods used to study molecular orientation at vapour–liquid interfaces. Largely the same methods have been applied to study the liquid–solid interface and in § 8.9.2 we give some representative references. There is a vast literature concerning molecular orientation in adsorbed monolayers (§ 8.9.3) but we had to confine ourselves to outlining the methods used and again giving a few representative references. In §§ 8.9.4 and 8.9.5 we briefly discuss work on liquid–liquid and water–membrane interfaces, respectively. Simulations of the various interfaces are also numerous and we give some corresponding references in the various subsections.

8.9.1 Orientation at vapour–liquid interfaces

We first discuss four of the principal techniques used to gain experimental information on molecular orientation at vapour–liquid interfaces. These are surface thermodynamics, surface potential, ellipsometry, and second harmonic generation (and the closely related sum frequency generation). The first method is the least direct, and the last method the most direct, in the sense of the data being directly interpretable in terms of surface molecular orientations. Studies have been done on pure vapour–liquid interfaces, and on interfaces with adsorbates, e.g. surfactants. To keep the discussion within reasonable bounds, we use as the example in each of the experiments discussed mainly the pure water vapour–liquid interface. At the end of this subsection we discuss X-ray and neutron reflectivity, which, for vapour–liquid interfaces, have been mainly used to study large adsorbate molecules.

Surface thermodynamics

Earlier in this chapter (see Eq. (8.76) and following discussion) we discussed how the surface entropy s^s, derived experimentally from the temperature derivative of the surface tension, can be interpreted in terms of the surface order or disorder, in part due to the preferred molecular orientations at the surface.

The case of water was discussed in § 8.6.4 in connection with density functional theory. Good[267] has determined the surface entropy s^s for 240 liquids, from the surface tension. For most of the hydrogen-bonded liquids, the surface entropy per molar area is significantly lower than for the simpler liquids—see Table 8.4. Low values of s^s suggest preferred orientations of the molecules in the surface region, and simplified models have been suggested. Good[267] proposed a layer model, with the molecules totally aligned in each layer, and Gannon and Faber[268] proposed a non-local mean field model involving the order parameters $F_\ell(z) = \langle P_\ell(\cos\theta)\rangle$, and their gradients $F'_\ell(z)$, where θ is the polar angle of the molecule (assumed axial) relative to the surface normal, taken as the z direction. F_ℓ is determined by the singlet distribution function $f(z\theta)$, see (8.31). These models neglect[269] the contribution to the entropy due to the angular correlations between the molecules. This can be seen[270] from the relation $s^s = -\mathrm{d}\gamma/\mathrm{d}T$ and

Table 8.4

Molar surface entropy, $S^s_m = s^s A_m$ for various liquids.

Liquid	$S^s_m/\mathrm{J\,mol^{-1}}K^{-1}$
1. Water	9.8
2. Formic acid	11.3
3. Methanol	10.0
4. Formamide	9.1
5. Methyl amine	10.8
6. Hydrogen cyanide	13.5
7. Argon	21.5
8. Nitrogen	22.3
9. Methane	20.2
10. Carbon tetrachloride	22.2
11. Benzene	25.9
12. Cyclohexane	26.6
13. n-decane	26.7
14. Nitrobenzene	24.3
15. Iodobenzene	26.2
16. Acetone	23.7
17. Carbon disulphide	22.7
18. Carbon dioxide	22.6
19. Benzonitrile	24
20. Ethyl acetate	23.4
21. Diethyl ether	23
22. Methyl iodide	21.9

Here A_m is the area of N_A surface molecules, taken by Good[267] to be given approximately by $A_\mathrm{m} = 1.1 V_\mathrm{m}^{2/3} N_\mathrm{A}^{1/3}$, where V_m = molar volume, N_A = Avogadro's number, and the factor 1.1 is a 'packing factor'. Liquids 1-6 are hydrogen-bonded, with S^s_m values taken from Good[267]. Good states (without examples) that the nonhydrogen-bonded liquid values are usually in the range $S^s_\mathrm{m} = 24 \pm 4\mathrm{J\,mol^{-1}\,K^{-1}}$. Liquids 7-22 confirm Good's rule. [Surface tension derivative data from refs. 273–277 and molar volume data from refs. 274, 276–278.]

(8.96), which involves the pair correlation function $g(\mathbf{r}_1\omega_1\mathbf{r}_2\omega_2)$ in the interface, as well as the singlet distribution, because of the relation $f(12) = f(1)f(2)g(12)$. Perturbation theory methods for including angular correlation effects in γ, and hence in s^s, are described in § 8.5.2. Simulation results for γ for water are given in ref. 271.

For most materials surface tension decreases with increasing temperature, so that $s^s > 0$. A few, reviewed by Schlossman[272] (see also discussion below (8.77)), have $s^s < 0$, indicating strong surface ordering, and include: (i) some liquid metals, where layering occurs at the vapour–liquid interface; (ii) liquid crystals, where smectic layering occurs at the vapour–liquid interface, and nematic ordering at the isotropic liquid–vapour interface (see § 8.6.2); and (iii) the n-alkanes C_{14}-C_{50}, where C_n refers to $CH_3(CH_2)_{n-2}CH_3$, in the small temperature range 2°–3° above the bulk freezing temperature. In these cases *surface freezing* occurs, where the top, single layer has orientational and positional order. By contrast, many materials exhibit *surface melting* a few degrees below the bulk freezing temperature.

Surface entropy is at best a qualitative indicator of surface orientation, useful for testing theories to be sure, and we turn now to more direct indicators.

Surface potential

In § 8.1.3 we derived various expressions for the surface potential $\Delta\phi$, which is the potential difference across the interface in moving from the gas to the liquid. From (8.54) and (8.43) we have

$$\Delta\phi = -4\pi\,\mu \int_{-\infty}^{\infty} dz F_1(z) - 2\pi(\rho_\ell - \rho_g)\,\Theta, \tag{8.292}$$

where μ is the magnitude of the permanent dipole moment, Θ is the mean quadrupole moment of a single molecule, and $F_1(z)$ is the local order parameter (8.32) for $l = 1$:

$$F_1(z) = \int d\omega\, f(z\theta)\cos\theta, \tag{8.293}$$

where $d\omega = \sin\theta d\theta d\phi$, with θ, ϕ the polar angles of the dipole moment $\boldsymbol{\mu}$ with respect to the surface normal z, taken as pointing from the liquid to the gas. For axial molecules, $\boldsymbol{\mu}$ points along the symmetry axis of the molecule. In deriving (8.292) we assumed the molecules to be rigid, so that the contribution to the surface potential due to interaction-induced dipole and quadrupole moments has been neglected. We note that the term $\Delta\phi_\Theta = -2\pi(\rho_\ell - \rho_g)\Theta$ should be subtracted from the experimental $\Delta\phi$ value before an interpretation in terms of F_1 and preferred orientations is carried out.

Bearing in mind the above restrictions, (8.292) can be used to give some indication of surface orientation for polar fluids. The surface potential of water has long been studied experimentally, and the results have been controversial. The early workers[58,279] could not agree even on the sign, but more recently[59] the value $\Delta\phi \sim +240 \pm 50\,\text{mV}$ has been obtained. As discussed in § 8.1.3, the measurements of $\Delta\phi$ are indirect.[280] The reason for the necessity of indirect measurements is not hard to find; $\phi(z)$ is essentially the mean electric potential field at points in space, through which the molecules can freely pass. Hence $\Delta\phi$ will depend on the detailed charge distribution of the molecules, in particular the interior part reflected by Θ. Since Θ produces no electric field external to charge distribution (see Vol. 1, p. 50, and Vol. 2, Appendix 8E), it is not 'seen' by real test charges that probe only the exterior regions of the molecular charge distribution. Typically, in electrochemistry experiments used to probe $\phi(z)$, low-energy ions are used. Such an ion is not an ideal test charge since the fluid molecules will interact with it, and tend to push it away, so that the internal charge distribution of the molecules is not probed very effectively.

If we take the water result $\Delta\phi > 0$ seriously, and note that the $\Delta\phi_\Theta$ contribution is negative, at least according to various models[281] (see below for TIP4 and SPC/E values), this means $\Delta\phi_\mu$ is large and positive, which in turn naively suggests preferred orientations of the surface H_2O molecules with the protons on average inward, i.e. towards the liquid side of the interface. However, because of (a) all the restrictions and difficulties discussed above, (b) the fact that H_2O is a nonlinear molecule, and (c) the fact that preferred orientations in general depend on position z (cf. Figs. 8.23, 8.25, 8.29), more complicated structures are quite possible.[51,269]

The difficulties in obtaining reliable experimental results for $\Delta\phi$ for H_2O have been mirrored in the results obtained from theoretical calculations[167,282] and from computer simulations.[51,269,281] In both instances results with either sign of $\Delta\phi(H_2O)$ have been obtained.

In § 8.1.3 we briefly discussed the simulation results of Wilson *et al.*[51] for TIP4P water at roughly room temperature (i.e. $\Delta\phi_\mu = 0.79 \pm 0.04\,\text{V}$, $\Delta\phi_\Theta = -0.91\,\text{V}$, $\Delta\phi = \Delta\phi_\mu + \Delta\phi_\Theta = -0.12 \pm 0.04\,\text{V}$), where the simulated value has the opposite sign from the experimental value. Sokhan and Tildesley[281] find a similar result from a simulation of SPC/E water (i.e. $\Delta\phi_\mu = 0.21\,\text{V}$, $\Delta\phi_\Theta = -0.76\,\text{V}$, $\Delta\phi = -0.55\,\text{V}$), and give a table listing results of other simulations, almost all of which give a sign of $\Delta\phi$ opposite from the experimental value. In Fig. 8.44 are shown the temperature dependences of $\Delta\phi_\mu$, $\Delta\phi_\Theta$, and $\Delta\phi$ from the SPC/E simulation.[281] As noted already, the sign of $\Delta\phi$ disagrees with experiment, but, interestingly, the temperature derivative of $\Delta\phi$ agrees with experiment[281] near room temperature. A density functional calculation[167] of $\Delta\phi$ also gives the wrong sign.

For the case of methanol, simulation and experiment are in reasonable agreement, for both magnitude and sign (see § 8.1.3).

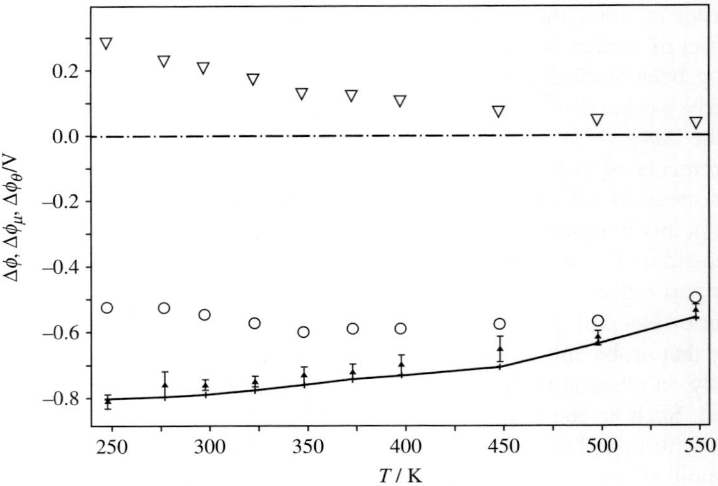

FIG. 8.44 The temperature dependence of the surface potential. Circles, full potential ($\Delta\phi$); triangles down, dipole contribution ($\Delta\phi_\mu$); solid line and triangles up, quadrupole contribution ($\Delta\phi_\theta$) estimated directly and as a difference between the total potential and the dipole contribution. Reprinted with permission from ref. 281. Copyright 1997 Taylor and Francis Ltd.

For mixtures of methanol + water, the excess $\Delta\Delta\phi \equiv \Delta\phi(\text{mixture}) - \Delta\phi(\text{pure water})$ has been measured (Case and Parsons[62]) and simulated[59] for mole fraction x_{CH_3OH} ranging from 0 (pure water) to unity (pure methanol). Qualitative agreement is found:[59] $\Delta\Delta\phi$ is negative for all x_{CH_3OH}, with a rapid decrease near $x_{CH_3OH} = 0$ and a weak minimum near $x_{CH_3OH} = 0.5$.

Ellipsometry

In ellipsometry[283–285] one studies the change in the polarization state of a reflected beam of light, as compared to the polarization state of the incident beam—see Fig. 8.45. For an idealized, flat, perfectly sharp interface between nonadsorbing media, a linearly polarized incident beam gives rise to a linearly polarized reflected beam. For a real, nonsharp, interface, the P and S incident polarizations suffer different phase shifts, so that in general a linearly polarized incident beam now gives rise to an elliptically polarized reflected beam.

Consider a beam incident at the Brewster angle $\theta_i = \theta_B$, where $\tan\theta_B = n_\ell/n_g$ with n_ℓ and n_g the liquid and gas phase refractive indices, respectively. By definition of θ_B there is no P-polarization (Fig. 8.45) in the reflected beam for a perfectly sharp and flat interface. In reality, a small amount of P-polarization arises due to the finite width of the interface (either intrinsic or due to surface roughness). One defines a coefficient of ellipticity $\bar{\rho}$ proportional to the amplitude of the P-polarized reflected light, and shows that to $O(\sigma/\lambda)$, $\bar{\rho}$ is given by[283]

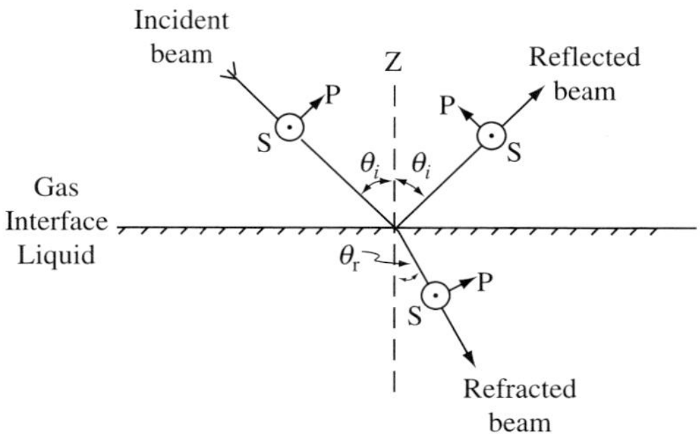

FIG. 8.45 Geometry of a reflection experiment from a liquid surface. P and S denote polarization states parallel and perpendicular, respectively, to the plane of incidence (defined by the incident and reflected beams). θ_i and θ_r are the angles of incidence and refraction, respectively.

$$\bar{\rho} = \frac{\pi}{\lambda} \frac{(\varepsilon_g + \varepsilon_\ell)^{1/2}}{\varepsilon_g - \varepsilon_\ell} \eta, \tag{8.294}$$

where λ is the wavelength, σ the interface thickness, and $\eta = \eta_o + \Delta\eta$, where

$$\eta_o = \int_{-\infty}^{\infty} dz(\varepsilon(z) - \varepsilon_g)(\varepsilon(z) - \varepsilon_\ell)/\varepsilon(z) \tag{8.295}$$

and, to first order in the anisotropy $\Delta\varepsilon$,

$$\Delta\eta = -\frac{1}{3} \int_{-\infty}^{\infty} dz \left(1 + 2\frac{\varepsilon_g \varepsilon_\ell}{\varepsilon(z)^2}\right) \Delta\varepsilon(z) \tag{8.296}$$

with $\varepsilon(z) = (\varepsilon_{zz}(z) + 2\varepsilon_{xx}(z))/3$ the mean and $\Delta\varepsilon(z) \equiv \varepsilon_{zz}(z) - \varepsilon_{xx}(z)$ the anisotropy[286] of the high-frequency dielectric tensor $\boldsymbol{\varepsilon}(z)$ in the interface region, and ε_g, ε_ℓ the bulk gas and liquid dielectric constants, respectively. The first term (8.295) was derived nearly a century ago by Lorenz, Rayleigh, Drude, and others,[283,287] and is present even for spherical molecules. The second term,[287] $\Delta\eta$, of more recent vintage, arises from the anisotropy of the interface, due mainly[288] to the alignment of the molecules in the interface. The expression (8.294) is valid[283] only for thin interfaces, i.e. where $\sigma/\lambda \ll 1$. The quantities ε_ℓ, ε_g, ε_{xx} and ε_{yy} are all 'high-frequency' values, corresponding to wavelength λ, and are related to the appropriate refractive index n by $\varepsilon = n^2$ (see § 10.4). In writing (8.296) we have also assumed weak anisotropy ($\Delta\varepsilon/\varepsilon \ll 1$) and kept only the $O(\Delta\varepsilon)$ term.

To relate $\Delta\eta$ to surface orientations we must make a theoretical model for $\Delta\varepsilon(z)$, the high-frequency dielectric constant anisotropy in the highly nonuniform interface region. This is a complicated matter if one aims for complete rigour. For uniform systems approximate theories for high-frequency dielectric constants are discussed in Chapter 10, see § 10.4. To keep the discussion reasonably short and simple we calculate $\Delta\varepsilon(z)$ for one of the simplest models,[167,269] namely the Clausius–Mossotti (or Lorentz–Lorenz) model (see (10.53) of Chapter 10) giving the relation between the principal axes components of the dielectric tensor $\varepsilon_{\alpha\alpha}(z)$ (where $\alpha\alpha = xx$ (or yy by symmetry) and zz) and the corresponding components of the mean polarizability tensor $\langle\alpha_{\alpha\alpha}\rangle$, i.e.

$$\varepsilon_{\alpha\alpha}(z) = \left(1 + \frac{8\pi}{3}\rho(z)\langle\alpha_{\alpha\alpha}\rangle\right) \Big/ \left(1 - \frac{4\pi}{3}\rho(z)\langle\alpha_{\alpha\alpha}\rangle\right). \tag{8.297}$$

Here $\langle\alpha_{\alpha\alpha}\rangle$ denotes an angle average, weighted by the conditional distribution function $P(\omega|\mathbf{r}) \equiv f(\mathbf{r}\omega)/\rho(\mathbf{r})$, of the space-fixed axes $\alpha\alpha$ component of the single-molecule polarizability tensor where $f(\mathbf{r}\omega)$ $(= f(z\omega)$ here) is the singlet distribution function, $\rho(\mathbf{r})$ $(= \rho(z))$ is the local density, the x, y plane is parallel to the surface, and the z axis is perpendicular to the surface. Note that $\langle\alpha_{\alpha\alpha}\rangle$ depends on z. Equation (8.297) has been derived for uniform fluids, and is assumed to hold locally for nonuniform fluids. Introducing the mean $\alpha = (\langle\alpha_{zz}\rangle + 2\langle\alpha_{xx}\rangle)/3$, which is independent of z, and the anisotropy $\overline{\Delta\alpha}(z) = \langle\alpha_{zz}\rangle - \langle\alpha_{xx}\rangle$ in the standard way,[289] we can write

$$\langle\alpha_{zz}\rangle = \alpha + \frac{2}{3}\overline{\Delta\alpha}(z) \quad , \quad \langle\alpha_{xx}\rangle = \alpha - \frac{1}{3}\overline{\Delta\alpha}(z). \tag{8.298}$$

We now use (8.297) and (8.298) to derive an expression for $\Delta\varepsilon(z) = \varepsilon_{zz}(z) - \varepsilon_{xx}(z)$ to $O(\overline{\Delta\alpha})$; as earlier, we assume weak anisotropy and keep only the leading term, which is

$$\Delta\varepsilon(z) = \frac{4\pi\ \rho(z)\overline{\Delta\alpha}(z)}{\left(1 - \frac{4\pi}{3}\rho(z)\alpha\right)^2}. \tag{8.299}$$

If we substitute (8.299) into (8.296) and also use again the Clausius–Mossotti relation in the form (see also (10.53))

$$\frac{\varepsilon(z) - 1}{\varepsilon(z) + 2} = \frac{4\pi}{3}\rho(z)\alpha, \tag{8.300}$$

we find for $\Delta\eta$

$$\Delta\eta = -\frac{1}{9}\int_{-\infty}^{\infty} \mathrm{d}z(\varepsilon(z) - 1)(\varepsilon(z) + 2)\left(1 + 2\frac{\varepsilon_\ell \varepsilon_g}{\varepsilon(z)^2}\right)\frac{\overline{\Delta\alpha}(z)}{\alpha}. \tag{8.301}$$

To express $\overline{\Delta\alpha}(z) \equiv \langle\alpha_{zz} - \alpha_{xx}\rangle$ in terms of orientational order parameters we use the relation between the space-fixed components $\alpha_{\alpha\beta}$ and body-fixed components $\alpha_{\alpha'\beta'}$ of α, where $x'y'z'$ are taken to be the principal axes of $\boldsymbol{\alpha}$. This relation is (see Appendix A of Vol. 1, Eq. (A.213b))

$$\alpha_{\alpha\beta} = D_{\alpha\alpha'}(\omega)D_{\beta\beta'}(\omega)\alpha_{\alpha'\beta'}, \tag{8.302}$$

where $D_{\alpha\alpha'}(\omega) \equiv \cos\theta_{\alpha\alpha'}$ denotes the direction cosine between space-fixed axis α and body-fixed axis α', and the summation convention is used. We thus need to evaluate $\langle D_{\alpha\alpha'}(\omega)D_{\beta\beta'}(\omega)\rangle$, where the average is over orientations ω, at fixed position z, using the distribution function $P(\omega|z) = f(z\omega)/\rho(z)$. The result depends on the molecular geometry (e.g. axial, C_{2V} symmetry, etc.), so we now specialize to C_{2V} symmetry in order to derive an expression valid for water, which we discuss below.

The principal axes $x'y'z'$ for H_2O are depicted in Vol. 1 (see Appendix D, p. 582), with origin there chosen to be the centre of mass. Here we choose the origin on the O-atom (the charge density is nearly spherical with respect to the O-atom) with z' the two-fold axis and $z'x'$ the plane of the molecule. Note that in the orientational averaging $\langle D_{\alpha\alpha'}(\omega)D_{\beta\beta'}(\omega)\rangle$, $P(\omega|z) \equiv P(\phi\theta\chi|z)$ is independent of azimuthal angle ϕ here, due to the axial symmetry around the space-fixed z-axis; P will depend only on θ and χ for H_2O (and only on θ for an axial molecule)—see Fig. 8.46.

We will find that for ellipsometry, and for other experiments such as surface potential (which depends on $\langle D_{\alpha\alpha'}(\omega)\rangle$) and second harmonic generation (which depends on $\langle D_{\alpha\alpha'}(\omega)\, D_{\beta\beta'}(\omega)\, D_{\gamma\gamma'}(\beta)\rangle$), the relevant orientational order parameters are $\eta_{ln}(z)$,

$$\begin{aligned} \eta_{\ell 0}(z) &= \left\langle D^{\ell}_{00}(\omega)\right\rangle \equiv \langle P_\ell(\cos\theta)\rangle \\ \eta_{\ell n}(z) &= \left\langle D^{\ell}_{0n}(\omega)^* + D^{\ell}_{0\underline{n}}(\omega)^*\right\rangle, \qquad n \neq 0, \end{aligned} \tag{8.303}$$

where $\underline{n} = -n$, and $\ell = 2$ is relevant for ellipsometry, $\ell = 1$ for surface potential, and $\ell = 3, 2, 1$ for second harmonic generation. Here $D^{\ell}_{mn}(\omega)^*$ is a generalized

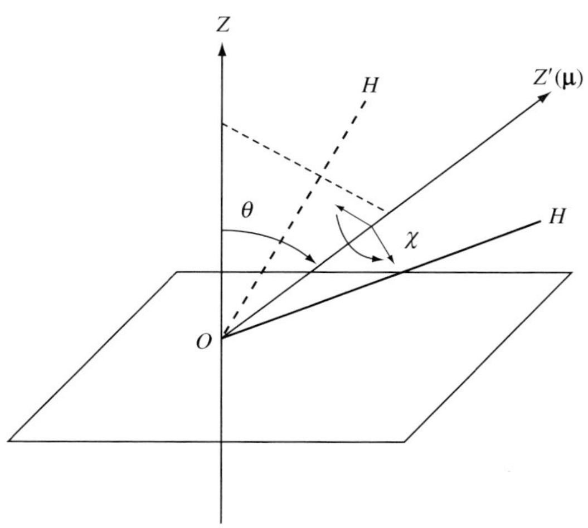

FIG. 8.46 Illustration of Euler angles θ and χ for a water molecule. χ is the dihedral angle between the zOz' plane and the HOH plane. (From ref. 167.)

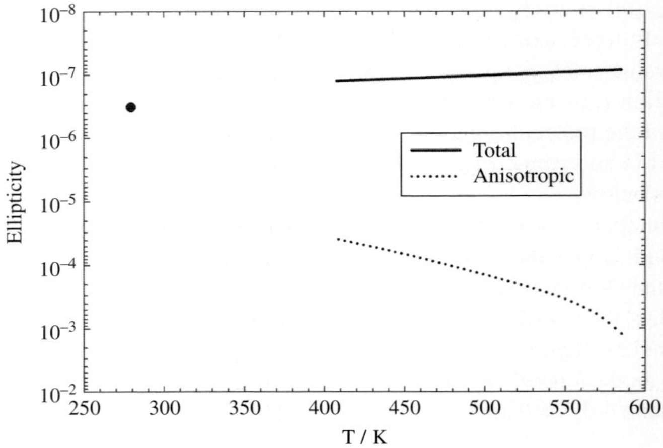

FIG. 8.47 The total ellipticity coefficient $\bar{\rho}$ and the contribution to $\bar{\rho}$ from $\Delta\eta$ as a function of temperature. The point gives the experimental result from ref. 290. (From ref. 167.)

spherical harmonic (Vol. 1, Appendix A, p. 462, and Eq. (A. 106) for an expression for $D^{\ell}_{0n}(\varphi\theta\chi)^*$ in terms of $Y_{\ell n}(\theta\chi)$, where $Y_{\ell n}$ is an ordinary spherical harmonic).

In terms of the $\eta_{\ell n}(z)$, for H_2O we find[166]

$$\overline{\Delta\alpha}(z) = \Delta\alpha_{\|}\eta_{20}(z) + \frac{1}{2}\sqrt{\frac{2}{3}}\Delta\alpha_{\perp}\,\eta_{22}(z) \tag{8.304}$$

where $\eta_{20}(z) = \langle P_2(\cos\theta)\rangle$, $\eta_{22}(z) = (3/2)^{1/2}\langle\sin^2\theta\,\cos 2\chi\rangle$, and $\Delta\alpha_{\|} \equiv \alpha_{z'z'} - \alpha$ and $\Delta\alpha_{\perp} \equiv \alpha_{x'x'} - \alpha_{y'y'}$ are polarizability anisotropies. The values of $\alpha_{x'x'}$, $\alpha_{y'y'}$, and $\alpha_{z'z'}$ for H_2O are listed in Vol. 1, Appendix D, p. 582 and yield $\alpha = 1.470 \times 10^{-24}\,\text{cm}^3$, $\Delta\alpha_{\|} = -0.00233 \times 10^{-24}\,\text{cm}^3$, and $\Delta\alpha_{\perp} = 0.133 \times 10^{-24}\,\text{cm}^3$.

The order parameters $\eta_{\ell n}(z)$ have been calculated[167, 168] by DFT for H_2O, using the u-expansion version (Sec. 8.6.4). Unfortunately, the theory is only valid for $T > 400\,\text{K}$, and Fig. 8.47 shows the results for the ellipticity (8.294), obtained from (8.295) and (8.301) with $\overline{\Delta\alpha}(z)$ given by (8.304) and $\varepsilon(z)$ from $\rho(z)$ using (8.300). In the calculations it is found that both η_o and $\Delta\eta$ are negative, with $\Delta\eta$ about three orders of magnitude smaller than η_o, which is reflected in Fig. 8.47. Even though the theory is valid only for $T > 400\,\text{K}$, a smooth extrapolation of the calculated value in Fig. 8.47 to $T \sim 300\,\text{K}$ indicates reasonable agreement with experiment,[290] and also with the simulated result of Matsumoto and Kataoka.[269]

The distribution function $f(z, \theta, \chi)$ and order parameter profiles $\eta_{\ell n}(z)$ are analysed in refs. 166,167 (and in ref. 168 at lower temperatures using the f-expansion version of DFT) and lead to the conclusions:

(i) The most probable dipole alignment is parallel to the interface on the liquid side, and perpendicular on the vapour side;
(ii) The dipole-down (i.e. protons towards the liquid side) orientation is slightly preferred over the dipole-up orientation; and
(iii) The molecular plane tends to align perpendicular to the interface on the liquid side, and parallel on the vapour side.

Because $\Delta\eta$ is so much smaller than η_o for water, it is not really possible to test these predictions using ellipsometry data.

Experimental ellipsometry results have been obtained by Beaglehole and others[283] for various other liquids, e.g. CS_2 and the liquid crystal 5CB[291] (pentylcyanobiphenyl). For CS_2, again $\Delta\eta$ is very small, making it difficult to learn about the preferred orientations from the experiment. For 5CB in the isotropic phase, the measurements[292] unambiguously indicate surface orientational ordering, which increases as the temperature approaches the isotropic–nematic transition temperature. The sign of $\Delta\eta$ is found to be negative, indicating preferred molecular orientations normal to the surface, on average through the interface. Earlier studies[268] of the temperature dependence of the surface tension had suggested surface ordering, in the manner discussed in § 8.9.1. Surface ordering has also been found[293] in mixtures of polar and nonpolar molecules (e.g. acetonitrile + carbon disulphide) near the critical point.

Second harmonic generation (SHG) and sum frequency generation (SFG).

We discuss SHG (for reviews see refs. 294–298) in some detail, and the closely related SFG (for reviews see refs. 297–307) briefly at the end of this section. In the previous section it was assumed that the liquid was a linear medium, so that the outgoing reflected and refracted beams of Fig. 8.45 vary harmonically at the frequency ω of the incident beam. In reality all media are nonlinear,[308] at least to some extent, and using a high-power laser as the source one can observe the harmonics, at frequencies $2\omega, 3\omega$, etc., in the outgoing beams.

To see the molecular origin of harmonic generation we recall the relation (Eq. (C.6) of Vol. 1) for the dipole moment $\boldsymbol{\mu}$ induced in an isolated molecule in an externally applied electrostatic field $\mathbf{E}$:

$$\boldsymbol{\mu} = \boldsymbol{\alpha} \cdot \mathbf{E} + \frac{1}{2}\boldsymbol{\beta} : \mathbf{EE} + \frac{1}{6}\boldsymbol{\gamma} \vdots \mathbf{EEE} + \dots . \tag{8.305}$$

Here $\alpha_{\alpha\beta}$ is the usual second-rank polarizability, $\beta_{\alpha\beta\gamma}$ the third-rank hyperpolarizability, $\gamma_{\alpha\beta\gamma\delta}$ the fourth-rank hyperpolarizability, etc. The tensor contractions are defined by $(\boldsymbol{\alpha} \cdot \mathbf{E})_\alpha = \alpha_{\alpha\beta}E_\beta$ (summation convention used), etc.; see Appendix B of Vol. 1. There are additional terms in (8.305) involving $\nabla\mathbf{E}$ (see (C.6)), which we neglect for now. We now take the applied field $\mathbf{E}$ to be time-dependent, with frequency ω : $\mathbf{E} = \mathbf{E}_\omega \cos \omega t$, with $\mathbf{E}_\omega$ being the amplitude. To avoid technicalities, we simplify the discussion by neglecting (a) dispersion,[308,309] i.e. the dependence on frequency of the various molecular polarizabilities; implicit in the neglect of dispersion is the neglect of absorption (due to electronic resonances) so that $\boldsymbol{\beta}$ is real; and (b) the rotational and vibrational motions of the molecules. The latter would render the various polarizabilities time-dependent, at the rotational and vibrational frequencies (low compared to ω); such time dependences are in fact observable in Raman and hyper-Raman light scattering[309,310] (see also Chapter 11). Using the identity $\cos^2 \omega t = (1 + \cos 2\omega t)/2$ we see that the $\boldsymbol{\beta}$ term in (8.305) will generate a contribution $\boldsymbol{\mu}_{2\omega} \cos 2\omega t$ to $\boldsymbol{\mu}$, with frequency 2ω and amplitude $\boldsymbol{\mu}_{2\omega}$ given by

$$\boldsymbol{\mu}_{2\omega} = \frac{1}{4}\boldsymbol{\beta} : \mathbf{E}_\omega \, \mathbf{E}_\omega. \tag{8.306}$$

This term is the molecular origin of second harmonic generation (SHG). Similarly γ gives rise to third harmonic generation.[308,309] When considering a macroscopic medium, and in complete analogy with the usual linear term $\boldsymbol{\mu}_\omega = \boldsymbol{\alpha} \cdot \mathbf{E}_\omega$, (8.306) gives rise to incoherent nonlinear optics effects (e.g. hyper-Raman scattering[309,310]) as mentioned above, in addition to the coherent effects (reflection and refraction) in which we are interested.

To discuss reflection and refraction, we consider the macroscopic analogue of (8.305). We first consider a medium in a static field $\mathbf{E}$. The macroscopic polarization $\mathbf{P}$, which is the mean dipole moment per unit volume (see § 10.1) is given by

$$\mathbf{P} = \boldsymbol{\chi}^{(1)} \cdot \mathbf{E} + \frac{1}{2}\boldsymbol{\chi}^{(2)} : \mathbf{EE} + \frac{1}{6}\boldsymbol{\chi}^{(3)} \vdots \mathbf{EEE} + \ldots, \tag{8.307}$$

which generalizes (10.1) in allowing for molecules and anisotropic effects, so that the macroscopic susceptibilities $\boldsymbol{\chi}^{(i)}$ are tensors (cf. (10A.35)). Strictly speaking $\mathbf{E}$ in (8.307) is the macroscopic Maxwell field, which differs in general from the applied field $\mathbf{E}_0$ (see § 10.2), but for simplicity we shall drop the distinction for the time being. Later we shall briefly mention local field corrections, which are connected with this distinction.

If the field $\mathbf{E}$ is now taken to be a harmonic wave of frequency ω, $\mathbf{E} = \mathbf{E}_\omega \cos(\omega t - \mathbf{k} \cdot \mathbf{r})$, then the $\boldsymbol{\chi}^{(2)}$ term in (8.307) gives rise to SHG in the same way as (8.306) arises from (8.305). Again neglecting dispersion and absorption, so that $\boldsymbol{\chi}^{(2)}$ is constant and real, we find

$$\mathbf{P}_{2\omega} = \frac{1}{4}\boldsymbol{\chi}^{(2)} : \mathbf{E}_\omega \mathbf{E}_\omega. \tag{8.308}$$

The amplitudes of the reflected and refracted SHG waves are proportional to $\mathbf{P}_{2\omega}$ (see ref. 308), and the SHG beam intensity is proportional to $|\mathbf{P}_{2\omega}|^2$.

The macroscopic $\boldsymbol{\chi}^{(2)}$ is related to the microscopic $\boldsymbol{\beta}$ in the same way that $\boldsymbol{\chi}^{(1)}$ (denoted simply $\boldsymbol{\chi} \equiv (\boldsymbol{\varepsilon} - \mathbf{1})/4\pi$ in Chapters 10 and 11) is related to $\boldsymbol{\alpha}$ (see (10.43) and retain only the 'high-frequency' or α term, or see (10.53) and assume $\rho\alpha_1$ is small). The simplest relation occurs at low densities and/or for small values of the molecular polarizabilities, where $\mathbf{P}_{2\omega}$ is the sum of single-molecule contributions, see (8.306). Hence we get

$$\boldsymbol{\chi}^{(2)} = \rho\langle\boldsymbol{\beta}\rangle \tag{8.309}$$

for a uniform medium of number density ρ. Here $\langle\cdots\rangle$ is a thermal average over molecular orientations. An important selection rule is evident from (8.309); SHG is nonzero only if[311] (a) the molecules lack a centre of inversion, so that $\boldsymbol{\beta} \neq 0$ (see discussion below Eq. (C.6)); and (b) the medium lacks a centre of inversion, so that $\langle\boldsymbol{\beta}\rangle \neq 0$. Hence (8.309) vanishes for *bulk* gases and liquids, but not for some bulk crystals,[308] such as quartz. At gas and liquid *interfaces*, inversion symmetry is broken so that $\langle\boldsymbol{\beta}\rangle \neq 0$ for polar and other non-centrosymmetric molecules, due to preferred molecular orientations.

Because the gas–liquid interface region is non-uniform, $\boldsymbol{\chi}^{(2)}$ depends on z in this region, and the relevant quantity is[308] the 'surface susceptibility' $\boldsymbol{\chi}^S \equiv \int dz\boldsymbol{\chi}^{(2)}$, i.e. the bulk susceptibility integrated through the interface. Thus we have

$$\boldsymbol{\chi}^S = \int_{-\infty}^{\infty} \mathrm{d}z\rho(z)\langle\boldsymbol{\beta}\rangle, \tag{8.310}$$

where

$$\langle\boldsymbol{\beta}\rangle = \int \mathrm{d}\omega\frac{f(z\omega)}{\rho(z)}\boldsymbol{\beta}(\omega) \tag{8.311}$$

with ω here the molecular orientation (not frequency). These relations are analogous to those found in the preceding section for linear reflection. The difference between (8.310) and (8.301) is that the latter includes local field correction terms (via $\varepsilon(z)$ originating from the Clausius–Mossotti formula), whereas for simplicity we have omitted such factors in (8.310); ref. 281 discusses them. In general the space-fixed axes components $\beta_{\alpha\beta\gamma}$ of $\boldsymbol{\beta}$ are related to the body-fixed axes components $\beta_{\alpha'\beta'\gamma'}$ by (cf. (A.216), and (8.302))

$$\beta_{\alpha\beta\gamma} = D_{\alpha\alpha'}(\omega)D_{\beta\beta'}(\omega)D_{\gamma\gamma'}(\omega)\beta_{\alpha'\beta'\gamma'}, \tag{8.312}$$

where $D_{\alpha\alpha'}$ is the direction cosine matrix (see Fig. A23 and Eq. (A.123) of Vol. 1), specifying the orientation ω of $x'y'z'$ with respect to XYZ and the summation convention is used. Thus in general we will get a complicated orientational order parameter involving $\langle D_{\alpha\alpha'}(\omega)D_{\beta\beta'}(\omega)D_{\gamma\gamma'}(\omega)\rangle\beta_{\alpha'\beta'\gamma'}$. This can be simplified considerably for molecules with symmetry.

Again we restrict ourselves to molecules with C_{2V} symmetry so that we can compare theory, simulation, and experiment for H_2O below. For C_{2V} symmetry, the nonvanishing body-fixed (principal axes) components of $\beta_{\alpha'\beta'\gamma'}$ are

$$\begin{aligned} &\beta_{z'z'z'} \equiv \beta_0 \\ &\beta_{z'x'x'} = \beta_{x'z'x'} = \beta_{x'x'z'} \equiv \beta_1 \\ &\beta_{z'y'y'} = \beta_{y'z'y'} = \beta_{y'y'z'} \equiv \beta_2, \end{aligned} \tag{8.313}$$

where we have assumed the static limit, i.e. we neglect dispersion effects. In the static case the Kleinman symmetry[312] holds, i.e. $\beta_{z'x'x'} = \beta_{x'z'x'}$ (and similarly for terms with $x' \to y'$); for dynamic components this symmetry is not exact but holds approximately away from electronic resonances (see below). When the Kleinman symmetry does not apply there are five independent $\beta_{\alpha'\beta'\gamma'}$ components, instead of three as assumed in (8.313).

Using the components (8.313) and evaluating the nonzero space-fixed components $\langle\boldsymbol{\beta}\rangle$, taking account of the axial symmetry around z for the interface, gives[167]

$$\begin{aligned}\langle\beta_{zxx}\rangle &= \langle\beta_{zyy}\rangle \qquad \text{(and permutations)}\\ &= \frac{1}{5}\beta_0(\eta_{10}-\eta_{30}) + \frac{1}{20}\beta_1\left(4\,\eta_{10}+6\,\eta_{30}-\sqrt{30}\,\eta_{32}\right)\\ &\quad + \frac{1}{20}\beta_2\left(4\,\eta_{10}+6\,\eta_{30}+\sqrt{30}\,\eta_{32}\right)\end{aligned} \tag{8.314}$$

$$\begin{aligned}\langle\beta_{zzz}\rangle &= \frac{1}{5}(3\,\eta_{10}+2\,\eta_{30}) + \frac{3}{10}\beta_1\left(2\,\eta_{10}-2\,\eta_{30}+\sqrt{\frac{10}{3}}\eta_{32}\right)\\ &\quad + \frac{3}{10}\beta_2\left(2\,\eta_{10}-2\,\eta_{30}-\sqrt{\frac{10}{3}}\eta_{32}\right),\end{aligned} \tag{8.315}$$

where $\eta_{ln}(z)$ are the order parameters (8.303). Explicitly we have $\eta_{10} = \langle\cos\theta\rangle$, $\eta_{30} = \langle P_3\cos\theta\rangle$, and $\eta_{32} = (15/2)^{1/2}\langle\sin^2\theta\cos\theta\cos 2\chi\rangle$. The two independent surface susceptibility components χ^s_{zzz} and χ^s_{zxx} are obtained from (8.314) and (8.315) using (8.310). If the Kleinman symmetry is not assumed, there is a third independent component[281] χ^s_{xzx}.

The DFT (u-expansion version) of § 8.6.4 has been used[167] to calculate the order parameters $\eta_{ln}(z)$ and susceptibility components χ^s_{zzz} and χ^s_{zxz}. Kleinman symmetry[314] is assumed and the theoretically calculated values[314] of the static hyperpolarizability components are used: $\beta_0 = -14.44$ au, $\beta_1 = -9.95$ au, and $\beta_2 = -5.61$ au, where au designates atomic units (see Appendix D, § D.3, of Vol. 1 for a discussion of au; here 1 au $= 3.206 \times 10^{-53}\,\mathrm{C^3\,m^3\,J^{-2}}$). This DFT is restricted to $T > 400$ K and the results for χ^s_{zzz} and χ^s_{zxz} are shown in Fig. 8.48 (see upper insert) as a function of temperature in comparison with simulated results.[281] The agreement is good; the DFT uses TIP4P water (multipole expanded) and the simulation uses SPC/E water,[315] with no local field correction factors applied to either DFT or simulation.

Also shown in Fig. 8.48 is a comparison of simulation and experiment[313] for the susceptibility components, this time with local field corrections applied to the simulated results (they give about a factor of 2 enhancement). Again, agreement is good, with one exception: the simulated values of χ^s_{xzx} and χ^s_{zxx} are nearly equal (i.e. Kleinman symmetry is nearly valid), whereas the experimental values differ significantly. The difference has been attributed[313,316] to a bulk quadrupolar

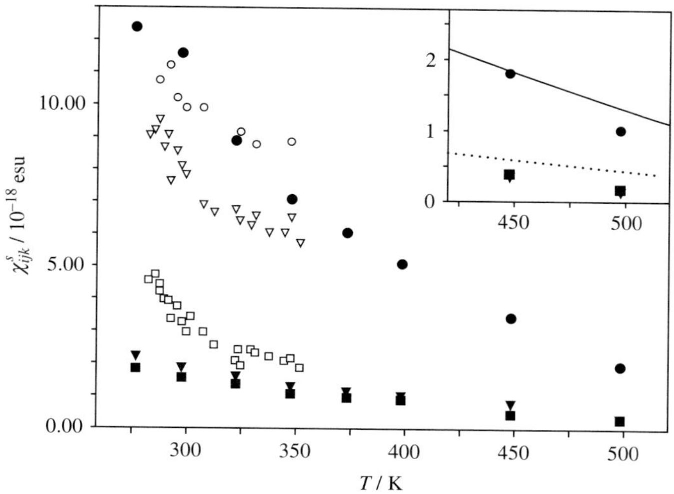

FIG. 8.48 The temperature dependence of the nonlinear susceptibility components. Circles (χ^s_{zzz}), triangles (χ^s_{xzx}), squares (χ^s_{zxx}). The solid symbols are the results of the MD simulations (ref. 281), and the open symbols the results of SHG experiments (ref. 313). The lines on the inset are the results of DFT theory (ref. 167). Reprinted with permission from ref. 281. Copyright 1997 Taylor & Francis Ltd.

nonlinear susceptibility contribution to the SHG signal (see (8D.12), and (C.19a) with the $O(E^2)$ term in $\mathbf{Q}(\mathbf{E})$ (included in the dots in (C.19a)) obtained from the **B** term in (C.7)).

We have focused on the liquid–vapour interface of water in this section in order to have a uniform theme throughout § 8.9. Very many other liquid–vapour and liquid–liquid systems have been studied, including pure liquids, adsorbates, small and large molecules (including surfactants), etc., and we refer to the various reviews[294–298] for details. SHG, together with its close cousin SFG (see next paragraph), is emerging as a method of choice for studying the structure and dynamics of liquid interfaces. The major limitation currently is to noncentrosymmetric molecules (so that $\boldsymbol{\beta} \neq 0$; in principle interaction-induced hyperpolarizability can occur for centrosymmetric molecules in the inhomogeneous surface region, but the effect is too small to be seen).

We end this section with a brief discussion of sum frequency generation (SFG) applied to interfaces.[297–305] If two beams of light, one with frequency ω_1 and the other with ω_2, impinge on a noncentrosymmetric molecule, we see from (8.305) that the induced dipole due to $\boldsymbol{\beta}$ has terms varying at frequencies $\omega_1 + \omega_2$ and $\omega_1 - \omega_2$. Focusing on the sum frequency term, we see that the dipole amplitude $\boldsymbol{\mu}_{\omega_1+\omega_2}$ is given by $(1/4)\boldsymbol{\beta} : \mathbf{E}_{\omega_1}\mathbf{E}_{\omega_2}$, analogous to (8.306). A medium containing such molecules then has an induced polarization with sum frequency component $\mathbf{P}_{\omega_1+\omega_2} = (1/4)\boldsymbol{\chi}^{(2)} : \mathbf{E}_{\omega_1}\mathbf{E}_{\omega_2}$, analogous to (8.308). The rest of the formal development, applied to interfaces, also parallels that for SHG. The main advantage is

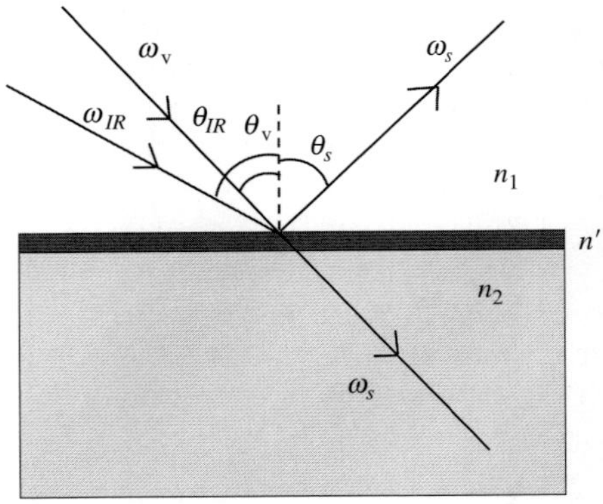

FIG. 8.49 Geometry for SFG from an interface in the reflected direction. n_1 and n_2 denote the refractive indices of the two bulk fluids, and n' denotes schematically the mean refractive index in the surface zone. The subscripts IR, v, and s denote infrared, visible, and scattered light, respectively. (From ref. 300.)

that $\chi^{(2)}(\omega_1, \omega_2, \omega_1 + \omega_2)$ depends on two input frequencies, one of which can be tuned. (In general $\chi^{(2)}(\omega_1, \omega_2, \omega_3)$ can depend on three frequencies.) In the usual setup one chooses $\omega_1 = \omega_{IR}$ (an infrared frequency) and $\omega_2 = \omega_{VIS}$ (a visible frequency), so that the sum frequency is $\omega_{SF} = \omega_{IR} + \omega_{VIS}$, and one varies ω_{IR}, keeping ω_{VIS} fixed. This enables one to enhance the outgoing signal if one tunes ω_{IR} through a molecular vibrational resonance, and also aids in identifying the scattering molecular species since such resonances characterize the molecules. The scattering geometry is indicated schematically in Fig. 8.49.

Among the many liquid–vapour and liquid–liquid interfaces studied[297–305] we mention only the liquid–vapour interface for pure water.[317] By applying SFG to the water surface and tuning ω_{IR} through the water vibrational resonances, Du *et al.*[317] were able to identify red-shifted OH bond frequencies, due to bulk water-like surroundings of some bonds, and unshifted OH bond frequencies, due to OH bonds apparently free—see the schematic Fig. 8.50. The preferred H_2O orientation at the surface inferred from analysis of the SFG experiment agrees with that inferred from the SHG experiment,[313] and also agrees with that deduced from DFT theory (see earlier discussion of ellipsometry experiments).

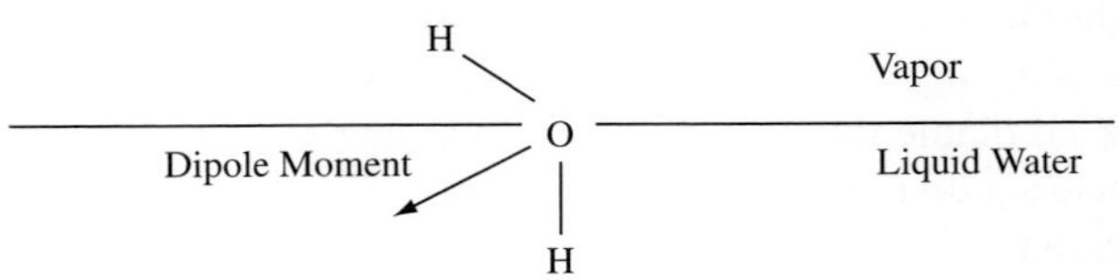

FIG. 8.50 Schematic of average orientation of an H_2O molecule at the vapor–liquid interface of water. (From ref. 298.)

We mention a few other studies of small molecule molecular orientations at air–liquid surfaces using SFG and SHG. For pure liquids, methanol (CH_3OH),[298] dichloroethane (CH_2Cl_2),[296] hexane (C_6H_{12}),[296] and hydrogen chloride (HCl)[318] have been studied. In the last case the authors did not explicitly discuss the preferred HCl orientations at the liquid surface, but from the weak vibrational frequency perturbation found (compared to the gas) one might expect the orientational distribution to be nearly isotropic. (The case of HCl dissolved in H_2O, where ionization occurs, is discussed below.) Solute molecule orientation studies at the liquid water surface include[301] acetonitrile (CH_3CN) and ammonia (NH_3), and phenol (C_6H_5OH).[298]

We conclude this subsection with a few remarks on the surface-ions paradox mentioned in § 8.1.1, which SHG and SFG experiments have helped to resolve. We follow mainly the excellent reviews 41, 319–321. Refer again to Fig. 8.3, which shows, from simulation, a stable minimum in the potential of mean force $w(z)$ for an iodide ion I^- at an air–water surface. Traditionally,[322] one thinks of ions as preferentially absorbed in the bulk and avoiding the surface, due to a more favourable solvation energy in the bulk where an ion has more neighbouring water molecules. The traditional view has been challenged by several simulation results[41,319,320] such as that of Fig. 8.3, as well as by some experimental facts[41,319,320] of which we highlight three.

(1) *SHG[323] and SFG[324] experiments*. These experiments on various saltwater solutions, e.g. NaF, NaCl, NaBr, and NaI, show that the large anions Br^- and I^- are present in excess in the outer layer of an air–water surface. Since the surface tension is also found experimentally to increase[319,320] for all salt solutions (approximately linearly with salt concentration above about 0.01 M), there might appear to be a contradiction with the rigorous Gibbs adsorption equation (8.83), which links positive surface excess to a decrease in surface tension. We explain below (see paragraph (1′)) why there is, in fact, no contradiction.

(2) *Ion-specific effects*. The traditional view[322] takes no account of differences in $\Delta\gamma$ values for differing ions (e.g. I^- versus Br^-, or I^- versus Na^+). Here we write $\Delta\gamma \equiv \gamma(\text{solution}) - \gamma(\text{pure water})$. In fact, for fixed cation (e.g. Na^+) and varying anion, for a given bulk concentration of salt, the experimental $\Delta\gamma$ values increase in the order NaI < NaBr < NaCl < NaF (which is part of a Hofmeister[325] series). We note that this ordering is opposite to that of the anion size (R) and polarizability ($\alpha \sim R^3$; Eq. (C.5) of Vol. 1), which is $F^- < Cl^- < Br^- < I^-$. The experimental ordering for surface potential[319,320] increments $\Delta(\Delta\phi)$ follows that of anion size, e.g. KF < KCl < KBr < KI. For alkali-halide salt solutions, usually $\Delta(\Delta\phi)$ is positive (KF is exceptional.[319])

(3) *Acids*. When certain acids (e.g. HCl) are added to water, it has long been known[306,319] that $\Delta\gamma < 0$. This is direct evidence (from the Gibbs relation) that ions (here H^+, or Cl^-, or both) accumulate in excess at the surface. This

is in direct conflict with the traditional view[322] that ions are repelled from the surface.

These three facts have now been rationalized[319–321] as follows:

(1′) The large anions I^- and Br^- do indeed accumulate in the outer layer of the liquid–air interface (see argument below). Due to electroneutrality, the salt cation (e.g. Na^+) accumulates in the next (liquid side) layer, thus depleting the anion in this region of the interface. Thus a double layer forms. The result is that the total anion amount, integrated over all surface layers, is in fact depleted compared to the bulk. This is consistent with the observed $\Delta\gamma > 0$ and the Gibbs relation. The (usual) observed sign ($\Delta(\Delta\phi) > 0$) of the surface potential increment is easy to understand since there are three positive contributions. For the sake of argument assume that $\Delta\phi(\text{pure } H_2O) < 0$, as observed in most simulations. The double layer, with anion in the outer layer, gives a positive contribution to $\Delta\phi$. Another positive contribution is due to the relatively large dielectric constant of the outer layer containing anions (recall α for anions is relatively large), which screens (reduces) the inward-pointing surface electric field generated by pure H_2O. The third contribution to $\Delta\phi$ is also positive, and is due to (see (8.54)) the more negative quadrupole moment Θ of anions compared to that for H_2O. These three positive contributions will increase $\Delta\phi$ compared to pure water. (We have considered the three principal effects of the ions on $\Delta\phi$ and have neglected higher order effects such as ion-induced dipoles and ion-field effects on the water molecules' orientations.)
To understand why large anions prefer the outer surface region over the bulk (contrary to intuition), note again that the polarizability α of the larger anions is large, and can therefore interact strongly with the surface electric field E (corresponding to the surface potential), giving an induction energy $-(1/2)\alpha E^2$ (see Eq. (C.2) of Vol. 1). This attractive surface energy can compete with the bulk solvation energy if α is large enough. Thus it is plausible that the larger anions I^- and Br^- can populate the outer surface region, whereas smaller anions (e.g. F^-) and most cations (which have relatively small α) cannot.

(2′) To rationalize ion specificity on the surface tension, consider the example $\Delta\gamma(\text{NaI}) < \Delta\gamma(\text{NaBr})$ from paragraph (2) above. Because $\alpha(I^-) > \alpha(Br^-)$, the I^- ion in the outer surface layer will have a greater induction energy with the surface electric field than will the ion Br^-. The outer surface layer for I^- will thus be more populated than for Br^-. The overall surface depletion for I^- will therefore be less than for Br^-. From the Gibbs relation the surface tension increase for I^- will thus be less than for Br^- as observed. To rationalize the relative surface potential increments, consider the example $\Delta\Delta\phi(\text{KBr}) < \Delta\Delta\phi(\text{KI})$ from paragraph (2) above. The three positive contributions to $\Delta\Delta\phi$ are all larger for I^- than for Br^-. Thus, since

$\alpha(I^-) > \alpha(Br^-)$, a stronger double layer arises for $K^+ I^-$ than for $K^+ Br^-$ (the surface induction energy for I^- is larger than that for Br^-), and also a stronger screening of the (negative) pure water surface potential occurs for I^- than for Br^-. Further, the Θ values are negative for both ions and since $|\Theta(I^-)| > |\Theta(Br^-)|$, the Θ contribution to $\Delta\phi$ will be more positive for I^- than for Br^- (see (8.54)). Thus the order $\Delta\Delta\phi(\text{KBr}) < \Delta\Delta\phi(\text{KI})$ is reasonable physically.

(3′) When the strong acid HI is added to water, dissociation produces H^+ and I^- ions. The H^+ then protonates an H_2O molecule producing a hydronium ion H_3O^+, as well as other species.[326] The anion I^- populates the outer surface layer as described earlier. In contrast to most cations, H_3O^+ can also populate the outer surface layer for the following reason.[320,327] Unlike H_2O molecules that can form four hydrogen bonds with neighbours, H_3O^+ can form only three. The H_3O^+ molecules in the bulk liquid form defects in the hydrogen bond network formed by the H_2O's. Like many defects in solids, H_3O^+ migrates to the surface where it is more easily accommodated. Both I^- and H_3O^+ populations are now enhanced in the outer surface layer, leading to an overall enhancement of ions in the total surface region (all layers), compared to the bulk. This surface excess leads, via the Gibbs relation, to a decrease in the surface tension.

X-ray and neutron reflectivity

Because of the long wavelength ($\lambda \sim 5000$ Å) ordinary light reflectivity studies (as opposed to ellipsometry studies—see Ellipsometry section above) are not widely used to explore surfaces except[328–333] for liquids near a critical point or for thick adsorbed layers (see, however, ref. 331). On the other hand reflectivity with X-rays[272,334–343] and neutrons[334,335,337,339,342–350] is widely used for Angstrom-scale liquid surface studies. The main advantage of X-rays (over neutrons) is the tremendous flux available from modern synchrotron sources. The main advantage of neutrons is the option to use isotope substitution, e.g. H → D, because the different isotopes have different scattering lengths b (e.g., $b_H = -3.74 \times 10^{-5}$ Å and $b_D = +6.67 \times 10^{-5}$Å). The value for oxygen is $b_O = +5.80 \times 10^{-5}$Å. Thus by mixing light and heavy water with 8.1% mole fraction of D_2O, the mean scattering length of the sample is zero, so that no specular reflection occurs. If there is a surface species adsorbed to this null reflecting water (nrw), the scattering is entirely due to the surface species. In addition, isotope substitution within the adsorbed molecules can highlight specific subregions of the interfacial region.

The reflection geometry for specular reflection ($\alpha_{\text{refl}} = \alpha$) is shown schematically in Fig. 8.51 for a vapour–liquid interface. Because of the very rapid drop in scattering intensity with increasing angle of incidence α, the experiments are done at near grazing incidence angles. (Note that in this field, in contrast to

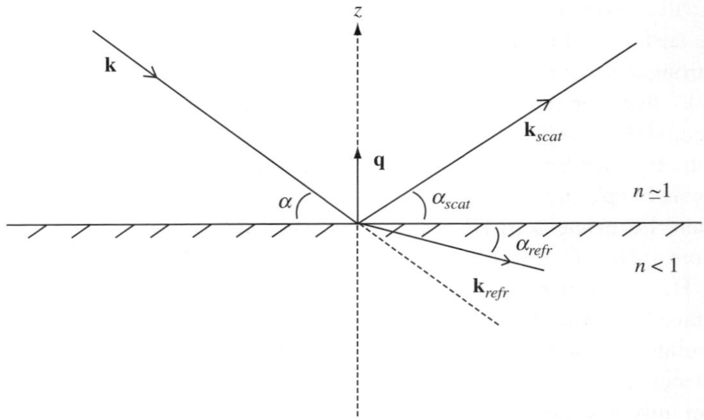

FIG. 8.51 In grazing incidence reflectivity the angle of incidence α is small, as are the angles of scattering (α_{scat}) and refraction (α_{refr}). For a vapour–liquid interface we assume indices of refraction $n \simeq 1$ (vapour) and $n < 1$ (liquid) for X-rays and neutrons. For specular reflection ($\alpha_{\text{scat}} = \alpha$) the transfer wave vector $\mathbf{q} \equiv \mathbf{k}_{\text{scat}} - \mathbf{k}$ is parallel to z, which is perpendicular to the flat surface in the x, y plane: here $\mathbf{k}$ and $\mathbf{k}_{\text{scat}}$ are the wave vectors of the incident and scattered radiation, respectively. Since k_z is reversed in specular reflection we have $q = 2k \sin\alpha \equiv (4\pi/\lambda)\sin\alpha \doteq (4\pi/\lambda)\alpha$, where λ is the wave length of the incident radiation. For n < 1, total external reflection ($\alpha_{\text{refr}} = 0$) can occur when $\cos\alpha = n$ (Snell's law), or $\alpha_c = \cos^{-1}(n) \doteq (1 - n^2)^{1/2}$, and $q_c \doteq 2k\alpha_c$. For off-specular scattering the detector registers scattered radiation at angles $\alpha_{\text{scat}} \neq \alpha$; in this case $\mathbf{q}$ has in-plane components q_x and q_y, as well as a z-component q_z.

classical optics, the angle of incidence is defined with respect to the surface, not the normal (cf. Fig. 8.45)). For X-rays and neutrons (except for hydrogen for neutrons, which has a negative scattering length), the scattering medium has an index of refraction less than unity ($n < 1$ indicated in Fig. 8.51). Thus total external reflection occurs for $\alpha < \alpha_c$, where α_c is the critical angle (defined in Fig. 8.51's legend). For both X-rays and neutrons α_c is typically on the order of 0.1–0.3°.

In the kinematic (or first Born) approximation the reflectivity $R(q)$ for the ideal case of pure specular scattering is given by[335,340]

$$R(q)/R_F(q) = \frac{1}{(\Delta\rho_s)^2}\left|\int_{-\infty}^{\infty} \mathrm{d}z e^{iqz}\rho_s'(z)\right|^2, \tag{8.316}$$

where $q \equiv |\mathbf{q}|$ is the magnitude of the wave vector transfer (Fig. 8.51), related to the angle of incidence α by $q = 2k\sin\alpha \simeq 2k\alpha$ for small α, where $k \equiv |\mathbf{k}|$ is the magnitude of the incident wave vector. We have assumed the surface is uniform in the x and y directions in Fig. 8.51; otherwise, off-specular scattering can also occur, in which case q should be replaced by q_z—see discussion in Fig. 8.51's legend. Off-specular scattering (also called diffuse scattering) is important in its own right and is also widely used[337,343] for surface studies. Surface roughness due to any cause (e.g. capillary waves) gives rise to off-specular scattering, and in-plane structure (e.g. from a Langmuir monolayer) does as well. In (8.316) $R_F(q)$ is the reflectivity for Fresnel interfaces (ideally sharp and flat). General expressions are available[334,351] for $R_F(q)$ for X-rays and neutrons, but the simple expression $R_F(q) \simeq (q_c/2q)^4$ is valid for $q \gg q_c$, where q_c is the critical value corresponding to α_c, i.e. $q_c \doteq 2k\alpha_c$. The quantity $\rho_s' \equiv d\rho_s(z)/dz$ is the derivative of the scattering length density $\rho_s(z)$, i.e. electron scattering lengths for X-rays and nuclear scattering lengths for neutrons, and can be simply related to the density $\rho_\alpha(z)$ of atomic sites α constituting the scattering molecules by $\rho_s(z) = \sum_\alpha b_\alpha \rho_\alpha(z)$, where b_α is the scattering length for site α . For neutrons b_α is the nuclear scattering length, and for X-rays b_α is the electron scattering length, which is approximately equal to $Z_\alpha r_e$, where Z_α is the atomic number and $r_e \equiv e^2/m_e c^2 (= 2.82 \times 10^{-13}\text{cm})$ is the classical (Thomson) electron radius, with e and m_e the charge and mass of an electron, respectively, and c the speed of light. An example with neutrons is discussed in § 8.9.3. Note that $\rho_s'(z)$ is nonzero only in the surface region, so that only this region contributes to $R(q)$. If $\rho_s'(z)$ is assumed to the have a delta function shape $\delta(z)$ (corresponding to a sharp interface with step-function shape for $\rho_s(z)$), $R(q)/R_F(q)$ is unity, as expected. If $\rho_s'(z)$ is assumed to have a Gaussian (or sum of Gaussians) shape, $R(q)/R_F(q)$ is also Gaussian (or a sum of Gaussians). The quantity $\Delta\rho_s \equiv \rho_s^{(\ell)} - \rho_s^{(g)} \simeq \rho_s^{(\ell)}$ is the difference in bulk scattering densities on the two sides of the interface, e.g. liquid and gas. The expression (8.316) is strictly valid only for $q \gg q_c$, but holds down to $q \simeq q_c$ in favourable cases;[334] below q_c we have $R_F(q) = 1$ and $R(q) = 1$. In cases where the first Born approximation is not valid, more exact methods can be used.[334–336,352]

By measuring $R(q)$ over as large a range of q as possible, one extracts information about $\rho_s(z)$. A model is needed to relate $\rho_s(z)$ to the molecular site densities $\rho_\alpha(z)$. For vapour–liquid interfaces most studies have been carried out for large adsorbates (e.g. surfactants, see § 8.9.3), but there have been a few for small molecules. For example, with X-rays, the density profiles have been examined for the atomic liquids Hg and other liquid metals (which show an oscillating profile, or surface layering[353]) and He,[354] and for the molecular liquids ethanol,[355] methanol,[356] carbon tetrachloride,[356] glycerol,[357] and glycerol + water mixtures.[358] Fewer studies have been done with neutrons, but in Fig. 8.52 we show the surface excess Γ (see § 8.2.1) for phenol at an air–water interface, determined by neutron reflectivity,[359] as a function of bulk phenol concentration. Also shown for comparison are the values obtained from surface tension[360] and SHG[361] measurements. Simulation results[362] are also available for this interface.

The pure liquid water surface has been studied by both X-ray[334] and neutron[344] reflectivity methods. Interface thicknesses σ have been extracted from the data in both cases, each giving about $\sigma = (2.7–2.8)$ Å. The X-ray experimentalists have attempted to decompose σ into (intrinsic) diffuseness (σ_D) and roughness (σ_R) (due to capillary waves[363]) contributions, according to $\sigma^2 = \sigma_D^2 + \sigma_R^2$. The

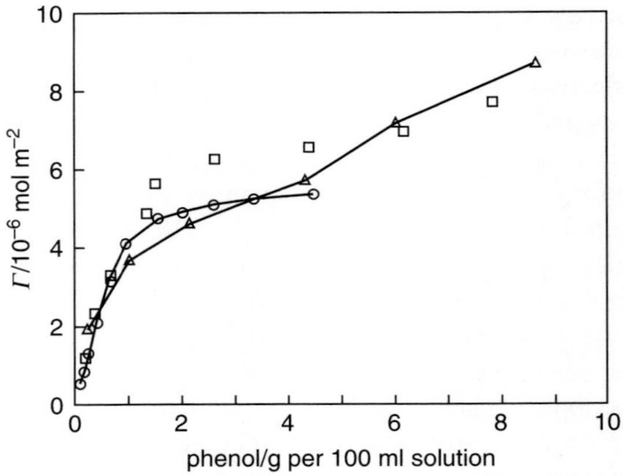

FIG. 8.52 Comparison of surface excesses of phenol at the air–water interface determined by surface tension (○), neutrons (△), SHG (□). Reprinted with permission from ref. 345. Copyright 1998 Royal Society of Chemistry.

decomposition is found to be not unique (it depends on experimental resolution). It has been argued theoretically[364] that such a decomposition of σ is inherently non-unique.

As just discussed, some authors have estimated the broadening of the intrinsic molecular position distribution function, or local density, due to roughness caused by thermal capillary waves. The same roughness will also cause a broadening of the intrinsic local distribution of molecular orientations (denoted $P(\omega|z)$ earlier). We refer to ref. 365 for a review of calculations of this effect.

Information about molecular orientations is in principle contained in the scattering site density $\rho_s(z)$, and one needs a model to extract it. This has not yet been possible for small molecules, but has been done for large molecules (e.g. surfactants at an air–water interface)—see § 8.9.3.

Other X-ray methods

We mentioned X-ray and neutron off-specular (or diffuse) scattering in the last section, and additional X-ray probes of the surface include[334] surface diffraction (or grazing-incidence diffraction, G1D), X-ray fluorescence, X-ray standing waves[365] (XSW), and X-ray absorption fine structure spectroscopy (EXAFS) and near-edge X-ray absorption fine structure (NEXAFS).[366]

Other simulations

In addition to the water[367] and methane surface simulations already mentioned, we note a few other representative works: methanol using polarizable potentials,[368] acetone (CH_3COCH_3),[369] benzyl alcohol ($C_6H_5CH_2OH$),[370] and alcohol + water mixtures.[371]

8.9.2 *Liquid–solid interfaces*

The methods used for liquid–vapour interfaces, as well as some of the surface science techniques described in Table 8.5, have also been used to study molecular orientation at liquid–solid interfaces. Some representative papers are thermodynamics,[330,372–378] ellipsometry,[330,372,373,379] SHG,[295,380] SFG,[300,302,332,373,380,381] and X-ray,[336,337] and neutron[345,346,350,372,382–384] reflectivity. Additional methods include IR[372,373,375,376,380,385] (infrared) absorption, ordinary,[373,386] resonant,[386] and surface enhanced[373,376,386] Raman scattering (SERS), total internal reflection fluorescence (TIRF),[387] surface X-ray (or grazing incidence) diffraction (GID),[376] surface force apparatus (SFA),[388–391] surface plasmon resonance (SPR),[391–394] scanning tunneling microscopy (STM),[332,376] atomic force microscopy (AFM),[332,395–397] and dual polarization interferometry (DPI).[398]

As an example, in Fig. 8.53 is shown the preferred orientations of the protein lysozyme at the silica–water interface for two pH values, as inferred from neutron reflectivity[399] and dual polarization interferometry[398] experiments.

Computer simulations, sometimes in conjunction with specific experiments, are numerous for liquid–solid interfaces.[375,385,400–402]

8.9.3 *Adsorbed monolayers and multilayers*

In this section we discuss the most important techniques used in studies of molecular orientation for adsorbed molecules. The systems studied are small and large molecules at vacuum–solid, air–solid, and air–liquid interfaces. Adsorption at liquid–solid interfaces (e.g. by proteins and other polymers) was dealt with in the preceding section, and adsorption at liquid–liquid interfaces (e.g. oil–water) is discussed in the next. Adsorption at water–membrane interfaces is discussed separately in § 8.9.5.

All the techniques mentioned previously for liquid–vapour and liquid–solid interfaces have been applied to the adsorption problem: thermodynamics,[328,404–411] surface potential,[407,409,412] ellipsometry,[405,406,409,413] Brewster angle microscopy (BAM),[414] SHG,[295,296,298,303,413] SFG,[413] neutron reflectivity,[345,347,349,409] and X-ray reflectivity.[337,338] In addition to these methods, many of the pioneering surface science techniques[403,415,416] (see Table 8.5) were introduced to study adsorption at vacuum–solid interfaces, and many of them have been used to study molecular orientations at interfaces. As an example, in Fig. 8.54 is shown an STM image of benzene molecules lying flat on a graphite surface.

In most of these techniques one does not measure directly the orientational distribution function or the orientational order parameters. Rather, one infers the preferred molecular orientation(s) by comparing the results of the measurements

Table 8.5

Selection of most frequently used surface science techniques

Acronym	Name	Description	Primary surface information
—	Adsorption or selective chemisorption	Atoms or molecules are physisorbed; the amount of gas adsorbed is a measure of total surface area. Chemisorption of atoms or molecules yields surface concentration of selected elements or adsorption sites.	Surface area, adsorption site concentration
AD	Atom or helium diffraction	Monoenergetic beams of atoms are scattered from ordered surfaces and detected as a function of scattering angle, giving structural information on the outermost layer of the surface. Method is extremely sensitive to surface ordering and defects.	Surface structure
AES	Auger electron spectroscopy	Core-hole excitations are created, usually by 1–10 keV incident electrons, and Auger electrons of characteristic energies are emitted through a 2-electron process as excited atoms decay to their ground state. AES gives information on the near-surface chemical composition.	Chemical composition
AFM	Atomic force microscopy	Similar to STM. An extremely delicate mechanical probe is used to scan the topography of a surface by measuring forces exerted by surface atoms. Light interference is used to measure the deflection of the mechanical surface probe. This is designed to provide STM-type images of insulating surfaces or to detect mechanical properties at the molecular level	Surface structure
ELS or EELS	Electron energy loss spectroscopy	Monoenergetic electrons $\sim 5-50$ eV are scattered off a surface and the energy losses are measured. This gives information on the electronic excitation of the surface and the adsorbed molecules (see HREELS)	Electronic structure, surface structure
ESCA	Electron spectroscopy for chemical analysis	Now generally called XPS (see XPS).	Composition, oxidation state
EXAFS	Extended X-ray absorption fine structure	Monoenergetic photons excite a core hole. Modulation of the absorption cross-section with energy at 100–500 eV above the excitation threshold yields information on radial distances to neighbouring atoms. The cross-section can be monitored by fluorescence as core holes decay or by the attenuation of the transmitted photon beam. EXAFS is one of many, fine-structure, techniques	Local surface structure and coordination numbers
FEM	Field emission microscopy	A strong electric field (on the order of V/Å) is applied to the tip of a sharp, single-crystal wire. Electrons tunnel into the vacuum and are accelerated along radial trajectories by Coulomb repulsion. When the electrons impinge on a fluorescent screen, variations of the electric field strength across the surface of the tip are displayed.	Surface structure

FIM	Field ionization microscopy	A strong electric field (on the order of V/Å) is created at the tip of a sharp, single-crystal wire. Gas atoms, usually He, are polarized and attracted to the tip by the strong electric field, then ionized by electrons tunneling from the gas atoms into the tip. These ions, accelerated along radial trajectories by Coulomb repulsion, map variations in the electric field strength across the surface, showing the surface topography with atomic resolution	Surface structure and surface diffusion
FTIR	Fourier transform infrared spectroscopy	Broad-band IRAS experiments are performed, and the IR absorption spectrum is deconvoluted by using a Doppler-shifted source and Fourier analysis of the data. This technique is not restricted to surfaces.	Bonding geometry and strength
HREELS	High-resoluation electron energy loss spectroscopy	A monoenergetic electron beam, usually ~ 2–10 eV, is scattered off a surface and the energy losses below ~ 0.5 eV to bulk and surface phonons and vibrational excitations of adsorbates are measured as a function of angle and energy (also called EELS).	Bonding geometry, surface atom vibrations
IRAS	Infrared reflection absorption spectroscopy	IR photons are reflected off a surface and the attenuation of IR intensity is measured as a function of frequency yielding a spectrum of the vibrational excitations of adsorbed molecules. Recent improvements in sensitivity allow IRAS measurements to be made on single-crystal surfaces.	Molecular structure
ISS	Ion scattering spectroscopy	Ions are inelastically scattered from a surface, and the chemical composition of the surface is determined from the momentum transfer to surface atoms. The energy range is ~ 1 keV to 10 MeV, and the lower energies are more surface sensitive. At higher energies this technique is also known as Rutherford backscattering (RBS).	Surface structure, composition
LEED	Low-energy electron diffraction	Monoenergetic electrons below ~500 eV are elastically back-scattered from a surface and detected as a function of energy and angle. This gives information on the structure of the near surface region.	Atomic and molecular surface structure
—	Neutron diffraction	Neutron diffraction is not an explicitly surface-sensitive technique, but experiments on large surface-area samples have provided important structural information on adsorbed molecules and surface phase transition. See text for neutron reflectivity.	Molecular structure
NMR	Nuclear magnetic resonance	NMR is not an explicitly surface-sensitive technique, but NMR data on large surface-area samples ($\geq 1\ m^2$) have provided useful information on molecular adsorption geometries. The nucleus magnetic moment interacts with an externally applied magnetic field and provides spectra highly dependent on the nuclear environment of the sample. The signal intensity is directly proportional to the concentration of the active species. This method is limited to the analysis of magnetically active nuclei.	Chemical state
RHEED	Reflection high-energy electron diffraction	Monoenergetic electrons of ~ 1–20 keV are elastically scattered from a surface at glancing incidence and detected as a function of angle and energy for small forward-scattering angles. Backscattering is less important at high energies and glancing incidence is used to enhance surface sensitivity.	Surface structure, structure of thin films

(cont.)

Table 8.5 *(Continued)*

Acronym	Name	Description	Primary surface information
SFG	Sum frequency generation (refs. 297–307)	Similar to SHG, but the laser output is split into a visible laser beam and a tunable IR beam. The two beams meet at the surface. The sum frequency signal is monitored as a function of IR frequency. In this way the vibrational spectrum of adsorbed molecules is obtained	Molecular structure
SHG	Second harmonic generation (refs. 294–298)	A surface is illuminated with a high-intensity laser, and photons are generated at the second harmonic frequency through nonlinear optical processes. For many materials only the surface region has the appropriate symmetry to produce an SHG signal. The nonlinear polarizability tensor depends on the nature and geometry of adsorbed atoms and molecules	Electronic structure, molecular orientation
SIMS	Secondary ion mass spectrometry	Ions and ionized clusters ejected from a surface during ion bombardment are detected with a mass spectrometer. Surface chemical composition and some information on bonding can be extracted from SIMS ion fragment distributions.	Surface composition
STM	Scanning tunneling microscopy	The topography of a surface is measured by mechanically scanning a probe over a surface. The distance from the probe to the surface is measured by the probe-surface tunneling current. Angstrom resolution of surface features is routinely obtained.	Atomic surface structure
TEM	Transmission electron microscopy	TEM can provide surface information for carefully prepared and oriented buk samples. Real images have been formed of the edges of crystals where surface planes and surface diffusions have been observed. Diffraction patterns of reconstructed surfaces, superimposed in the bulk diffraction pattern, have also provided surface structural information.	Surface structure
TDS	Thermal desorption spectroscopy	An adsorbate-covered surface is heated, usually at a linear rate, and the desorbing atoms or molecules are detected with a mass spectrometer. This gives information on the nature of adsorbate species and some information on adsorption energies and the surface structure.	Composition, heat of adsorption, surface structure
TPD	Temeprature programmed desorption	Similar to TDS, except the surface may be heated at a non-uniform rate to obtain more selective information on adsorption energies	Composition, heat of adsorption, surface structure
UPS	Ultraviolet photoemission spectroscopy	Electrons photoemitted from valence and conduction bands are detected as a function of energy to measure the electronic density of states near the surface. This gives information on the bonding of adsorbates to the surface.	Valence band structure

XPS	X-ray photoemission spectroscopy	Electrons photoemitted from atomic core levels are detected as a function of energy. Shifts of core-level energies give information on the chemical environment of the atoms.	Composition, oxidation state
XRD	X-ray diffraction	X-ray diffraction has been carried out at extreme glancing angles of incidence where total reflection ensures surface sensitivity. This provides structural information that can be interpreted by well-known methods. An extremely high X-ray flux is required to obtain useful data from single-crystal surfaces. Bulk X-ray diffraction is used to determine the structure of organometallic clusters, which provide comparisons to molecules adsorbed on surfaces. X-ray diffraction has also given structural information on large surface-area samples. See text for X-ray reflectivity.	Surface structure

From ref. 403, where references for these techniques are given.

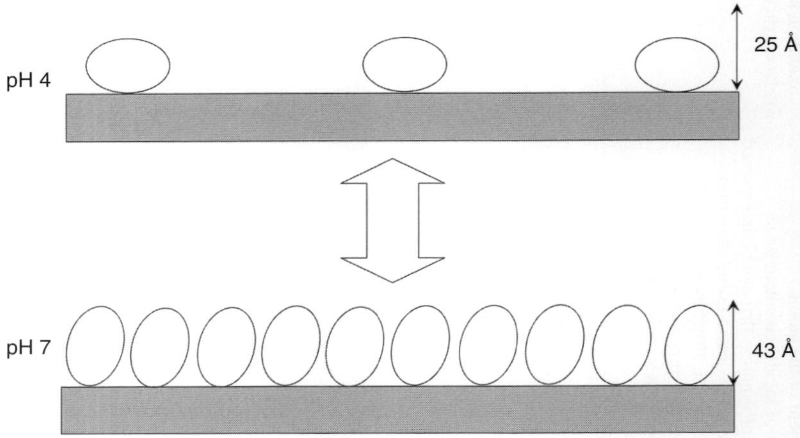

FIG. 8.53 Preferred orientations of the protein lysosyme at the water–silica interface for two pH values. (From ref. 398.)

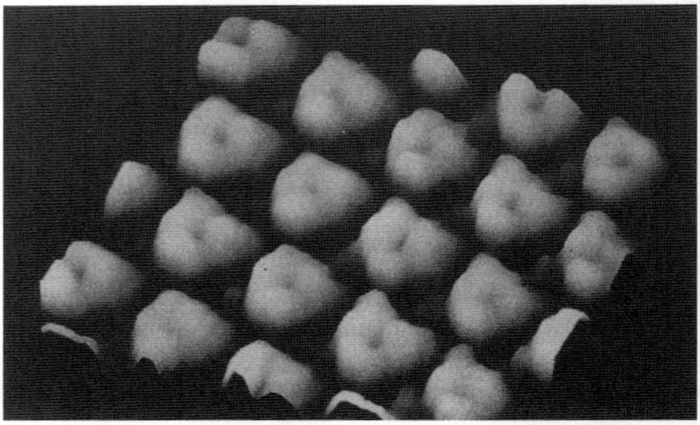

FIG. 8.54 STM image of benzene molecules lying flat on a graphite surface. Reprinted with permission from ref. 417. Copyright 1997 Princeton University Press.

with those predicted by a molecular model. We illustrate with an example from neutron reflectivity. Lu and co-workers[347] have carried out reflectivity experiments for many surfactants at an air–water interface. The measured reflectivity R as a function of angle of incidence, or equivalently q, the magnitude of the wave vector transfer due to the scattering, is related to the scattering length density $\rho_s(z)$ by (8.316). We write $\rho_s(z)$ in terms of the atomic site number densities $\rho_\alpha(z)$ as

$$\rho_{\mathrm{s}}(z) = \sum_{\alpha} b_\alpha \rho_\alpha(z), \tag{8.317}$$

where b_α is the scattering length for nuclear site α. Consider the specific surfactant $C_{14}TAB$, which denotes $C_{14}H_{29}N(CH_3)_3Br$ (or tetradecyl trimethyl ammonium bromide). Instead of an all-atom model of $C_{14}TAB$, course-grained models have been used. In one,[347] the scattering by $C_{14}TAB$ at the water interface is represented by a three-group model, i.e.

$$\rho_s(z) = b_w\rho_w(z) + b_h\rho_h(z) + b_t\rho_t(z), \tag{8.318}$$

where b_w, b_h and b_t are average scattering lengths for the water, head group, and tail group of the surfactant, respectively, and $\rho_w(z)$, $\rho_h(z)$, and $\rho_t(z)$ are the corresponding number densities. Alternatively,[418] in a less coarse grained model one can take

$$\rho_s(z) = b_w\rho_w(z) + b_h\rho_h(z) + \sum_{j=1}^{14} b_j\rho_j(z), \tag{8.319}$$

where j labels the different methylene (CH_2) and terminal methyl (CH_3) groups of the chain. If one also assumes the chains are fully stretched and rigid, as in Fig. 8.55, the Fourier transform of $\rho_s(z)$, required for $R(q)$, can be carried out analytically if simple forms are assumed for $\rho_w(z)$, $\rho_h(z)$, and $\rho_j(z)$, e.g. Gaussians for $\rho_h(z)$ and $\rho_j(z)$, and a hyperbolic tangent profile for the water density $\rho_w(z)$. The main fitting parameter in the model is the tilt angle (θ_o in Fig. 8.55). A best-fit to the data yields $\theta_o \sim 45°$, i.e. a significant tilt. The rigid model of Fig. 8.55 is crude, but gives results qualitatively in agreement with refined models.[328,418] We have also greatly oversimplified the description of the experimental analysis (which involves H $\rightarrow$ D substitution for the water and the $C_{14}TAB$) in order to sketch the principle of how orientational information is extracted from the experiment.

A class of monolayers much studied is termed self-assembled monolayers (SAMs). In contrast to Langmuir monolayers, which occur on liquid surfaces and are relatively weakly bound (physisorbed), SAMs occur on solid surfaces and are strongly bound (chemisorbed). The systems most often studied are thiols on gold and silanes on silicon oxide.[412,419] Thiols contain a thiol group (SH) and many have the structure $R(CH_2)_nSH$, where R is an end group (e.g. CH_3) and n denotes the number of methylene (CH_2) groups. In Fig. 8.56 is shown a schematic thiol-on-gold SAM. An AFM image of an actual system is shown on the cover of ref. 412. Computer simulations of these systems have also been carried out.[412]

An unusual case[420] of (relatively) small molecule adsorption on a water surface is C_{60}. The molecule is hydrophobic but with the aid of benzene as a solvent to induce spreading, monolayer and multilayer $\pi - A$ curves (surface pressure versus area) were observed for C_{60} and other fullerenes. Mono- and multilayer adsorption of C_{60} at a solid–liquid interface (i.e. mica/toluene $+C_{60}$ solution) has also been observed[421] using the SFA, as has[420] C_{60} adsorbed on a vapour–liquid (silver-coated glass) interface using SERS.

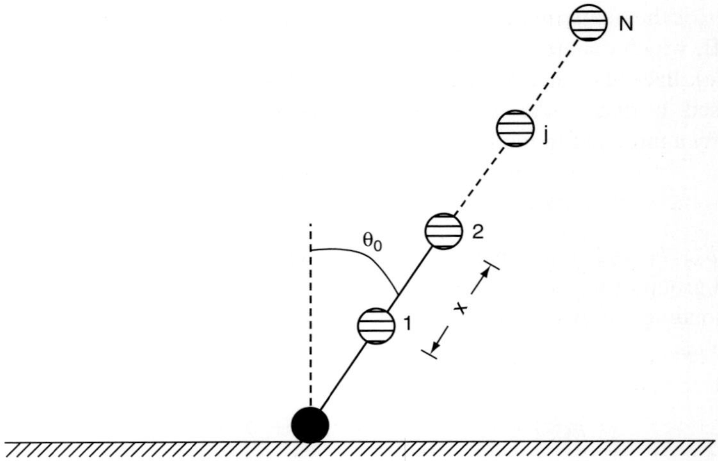

FIG. 8.55 Simplified structural model for C_NTAB (where $N = 14$ in the case discussed in the text) at air–water interface. Each bond length is fixed at x, and each bond tilts at angle θ_o. Reprinted with permission from ref. 418. Copyright 1994 Elsevier.

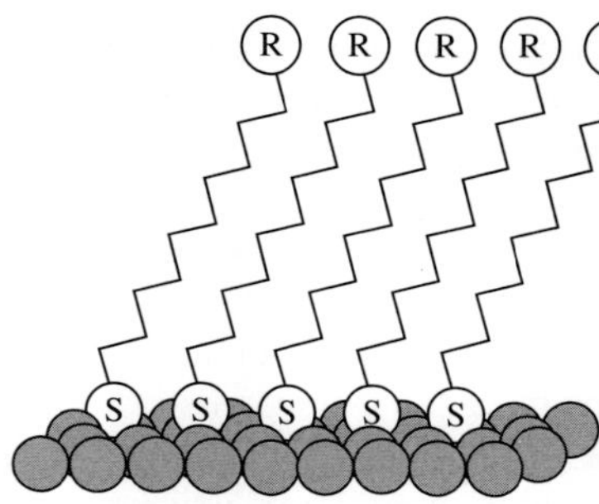

FIG. 8.56 Schematic of thiols containing alkyl chains on gold (111). Only a row of molecules is shown. In reality a densely packed monolayer is formed. Reprinted with permission from ref. 412. Copyright 2006 Wiley–VCH.

We end this section by noting a few entry points to the extensive simulation literature in this field; see, e.g. refs. 347, 422–426.

8.9.4 *Liquid–Liquid Interfaces*

The interfaces between two immiscible liquids are studied for their intrinsic interest, as well as for various applications.[427–430] Such an interface (e.g. water–oil) can serve as a water–membrane mimetic where the interfacial potential and interfacial tension can be controlled by the use of electrolytes and surfactants. In electrochemistry, the interface of two immiscible electrolyte solutions (ITIES) are widely studied, e.g. for use in electrochemical extraction.[431] ITIES commonly studied include electrolyte solutions of water–nitrobenzene ($C_6H_5NO_2$) and water–1,2 dichloroethane ($C_2H_4Cl_2$). Like the liquid–solid interface, the liquid–liquid interface is a 'buried' one, and hence difficult to access experimentally. Nevertheless, in recent years great progress has been made with a variety of techniques including thermodynamics,[432–434] surface potential, SHG,[295–298] SFG,[299,300,303,305] neutron[342,344,345,434] and X-ray[336–338,342,435] reflectivity, and various others of the techniques of Table 8.5.

Computer simulations are also being used increasingly to test theories and to augment experiments, showing molecular details not yet accessible experimentally. Some general reviews are refs. 428 and 436–439. Examples of some specific liquid–liquid interfaces simulated are water–benzene (C_6H_6),[440] water–carbon tetrachloride (CCl_4),[441] water–dodecane ($C_{12}H_{26}$),[437] water–chloroform (CCl_3H),[442] and the ITIES system water–2-heptanone ($C_7H_{14}O$).[443]

Liquid–liquid interfaces containing surfactants have also been simulated, e.g. water–SDS–trichloroethylene (C_2Cl_3H),[444] where sodium dodecylsulphate (SDS: $CH_3(CH_2)_{10}CH_2SO_4Na$) is a standard anionic surfactant, water–methanol–alkanes[445] where methanol (CH_3OH) behaves as a surfactant in these systems, and water–PC–carbon tetrachloride[446] where the phospholipid molecule PC (phosphatidylcholine) is either DLPC (di-laureoyl PC) or DPPC (di-palmitoyl PC).

The concentration of various small molecules, including anesthetics (e.g. nitrous oxide, ethanol, xenon, trifluoroethane), at the water–hexane interface, and at a water–membrane interface, have also been simulated.[447]

8.9.5 Water–membrane interfaces

By membranes we mean biomembranes or one of their models or mimetics (micelles,[448] bicelles (bilayer disks), liposomes, vesicles, planar lipid bilayers, lipid monolayers,[448] supported membranes). We do not discuss synthetic membranes used in industry[449] (e.g. for dialysis, or ion exchange). Biomembranes surround cells (the plasma membranes), and internal organelles (e.g. the mitochondrion) of animals and plants. Because of their biological importance, the literature on water–membrane interfaces is almost overwhelming. Fortunately there are good reviews—for general reviews, see refs. 450–453 (experimental), and refs. 437, 454, 455 (theoretical). More specialized reviews are cited below.

We first note some of the problems of interest, together with techniques used to investigate them. For simplicity we start with a simple planar monolipid bilayer, and ignore the complications occurring in real biomembranes such as (for the plasma membrane) peripheral and integral membrane proteins, thickness variations near proteins, asymmetry between the inner and outer leaflets, embedded cholesterol, cytoskeleton attachments to the inner leaflet, mixtures of different lipids, lipid domain formation (so-called rafts), and carbohydrate attachments to the proteins and lipids of the outer leaflet (glycocalyx). In Fig. 8.57 we show the

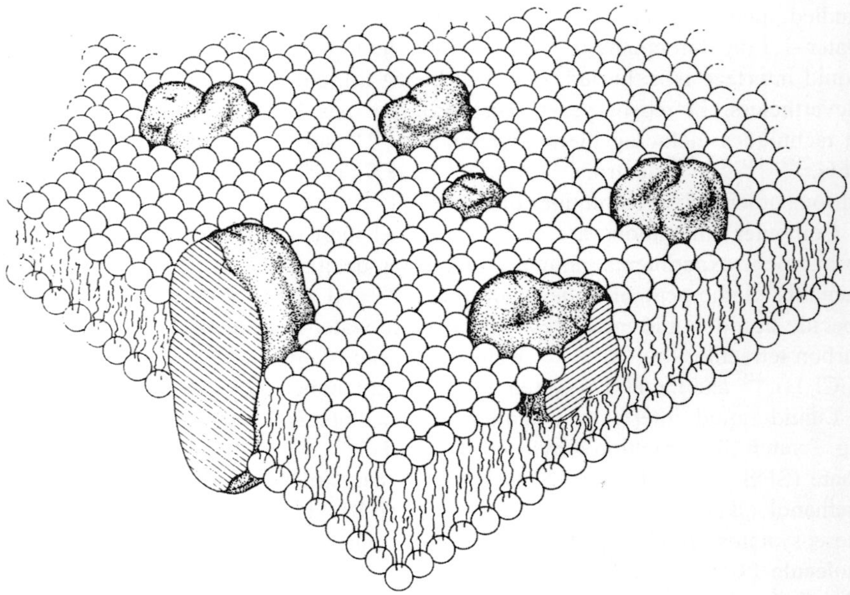

FIG. 8.57 The Singer–Nicolson fluid mosaic model of a biomembrane. A few proteins are shown, both integral (membrane spanning) and peripheral. The membrane thickness (typically 40–60 Å) and area per lipid (typically 50–60 Å^2) are experimental observables. Reprinted with permission from ref. 458. Copyright 1984 Springer.

standard simplified Singer–Nicolson[456] fluid mosaic model. There are numerous refinements[457] that incorporate the points above, and others such as shape undulations.

The aqueous solutions on both sides of a cell membrane contain various solutes, including salt ions such as K^+, Na^+, and Cl^-, with Na^+ and Cl^- more concentrated on the outside and K^+ more concentrated on the inside. A transmembrane electrical potential difference exists, of order 100 mV with the inside negative with respect to the outside. The potential is maintained by a membrane protein Na^+/K^+ pump that pumps out Na^+ ions and pumps in K^+ ions. In nerve and muscle cells, the pores of specific proteins (ion channels[459]) open and close during excitation, which generates a propagating electrical signal. In Fig. 8.57 the lipid head groups are typically about 5–10 Å wide, and the head-to-head tail group (total hydrocarbon region) is typically about 30 Å wide. Some lipid heads are negatively charged (e.g. PG or phosphatidylglycerol) and others (e.g. PC or phosphatidylcholine) are neutral but zwitterionic (large dipole moment). Many important biological processes such as respiratory electron transfer and proton transfer reactions in mitochondria, signaling, and ion and nutrient transport occur at least partly at the water–membrane interface and hence depend on the properties

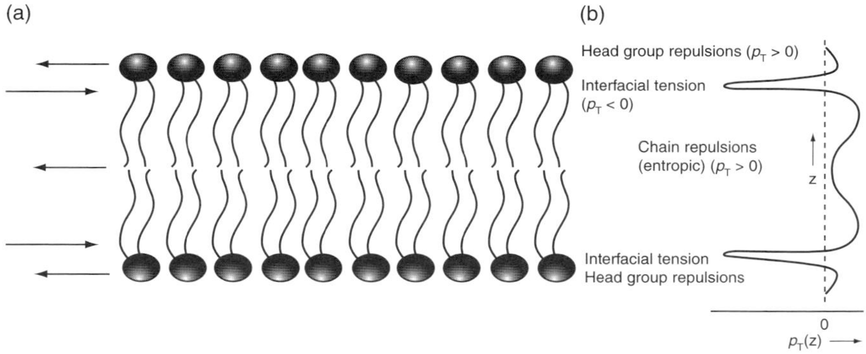

FIG. 8.58 Lateral pressure profile of a lipid bilayer. (a) Schematic illustration of a cross-section through a symmetric lipid bilayer with an indication of the forces that act within the layer. (b) The resulting lateral pressure or stress profile is $p_T(z)$ (see § 8.3.2). Each interfacial tension region for surface tension is typically about 50 dyn/cm; assuming a head group is about 5 Å wide, this corresponds to an average pressure $\bar{p}_T \sim -1000$ atm in each head group region. The attractive forces are balanced by the repulsive forces since the net tension in a free bilayer is widely believed to vanish at equilibrium.[475] Assuming a head-to-head width of about 30 Å and neglecting the smaller head group repulsion regions, we find a mean repulsive pressure $\bar{p}_T \sim 330$ atm in the chain region. Adapted, with permission, from ref. 457. Copyright 2005 Springer.

of the interface. We briefly discuss the states of the interfacial water, ions, and other solutes (e.g. glucose, peptides, drugs).

Water is necessary for the very existence and stability of the membrane planar bilayer structure.[460] The hydrophobic effect (the hydrocarbon tails structure themselves to avoid the water as far as possible) squeezes the head groups together[461] with a tension $\gamma \sim 50\,\text{dyn/cm}$ (see Fig. 8.58). Because the heads are charged or polar, the head group region is hydrophilic, and the water molecules can penetrate the interface to the level of the beginning of the tail region. This is illustrated in Fig. 8.59 where we compare the water density profile (from simulation) at water–oil and water–membrane interfaces. The water hydrates significantly the head group region (typically 15–25 water molecules per lipid head group). To maintain osmotic balance in cells, water can enter/leave by various means: simple diffusion across the membrane,[457,462] diffusion through ion channel and other transporter pores,[463] and via specialized water channels (aquaporin proteins[463]). Water is thus bound, and is also oriented[464] at the interface. The orientation contributes to the interface potential.[465–467] The bound water molecules are responsible for the hydration force;[468–471] i.e. at short distances of separation between two bilayers, there is a strong repulsive force between the bilayers due to the free energy cost to dehydrate the bilayers. The principal techniques[450–452] used to determine the water–membrane interface structural properties are X-ray and neutron scattering, NMR and IR spectroscopy, and osmotic stress. Sometimes[472,473] specific simulations are done in conjunction with the experiments to assist the interpretation. Dynamical properties are measured using an array of other techniques.[474]

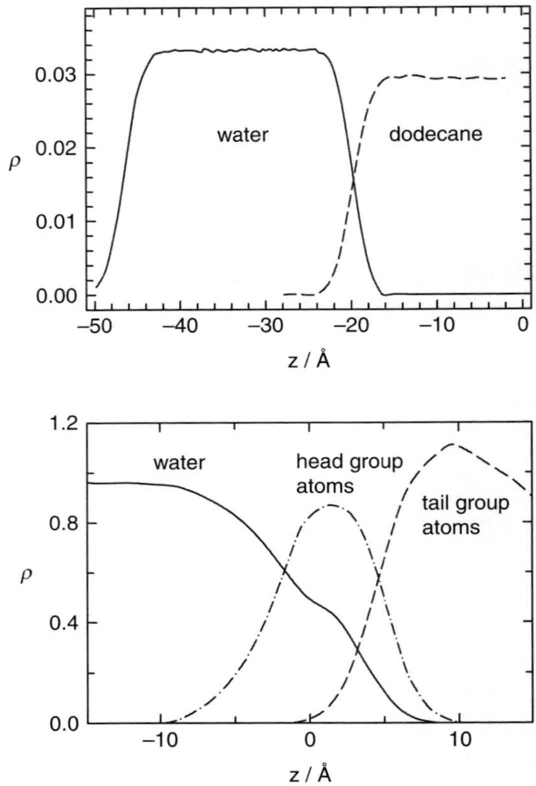

FIG. 8.59 Density profiles of water oxygen atoms and methyl–methylene groups of dodecane at the water–dodecane interface (upper) or heavy atoms in headgroups and tails of POPC membrane (lower) (see n. 559 for the abbreviation POPC). $Z = 0$ is at the equimolar surface of water for the lower panel. The profile for tail atoms of POPC extends to the centre of the bilayer, approximately 18 Å from the interface. Note that water penetrates the headgroup region of the membrane strongly but its hydrocarbon core only weakly. The units for density are molecules/$Å^3$ in the upper panel and arbitrary units in the lower panel. (From ref. 437.)

Ions bind at the interface, not only to the negatively charged head groups, but also to the zwitterionic ones. This binding affects the surface charge and potential,[465–467] and also the membrane structure, hydration, membrane fluidity, membrane phase transitions, conductance, membrane adhesion, fusion, and other properties.[476,477] In Fig. 8.60 we show schematically the electrical potential $\phi(z)$ in the interface region, assuming negatively charged head groups. The counter ions (e.g. Na^+) and co-ions (e.g. Cl^-) form a double layer outside the membrane, with $\phi(z)$ changing rapidly (exponentially) over the Debye screening length (~ 10 Å for 100 mM salt solution) near the surface (note $\Delta\phi_{12}$ on the figure). The rapid increase $\Delta\phi_{32}$ on the inside is due to the net dipole potential of the hydrated head groups. With few exceptions,[478] ions cannot penetrate significantly into the

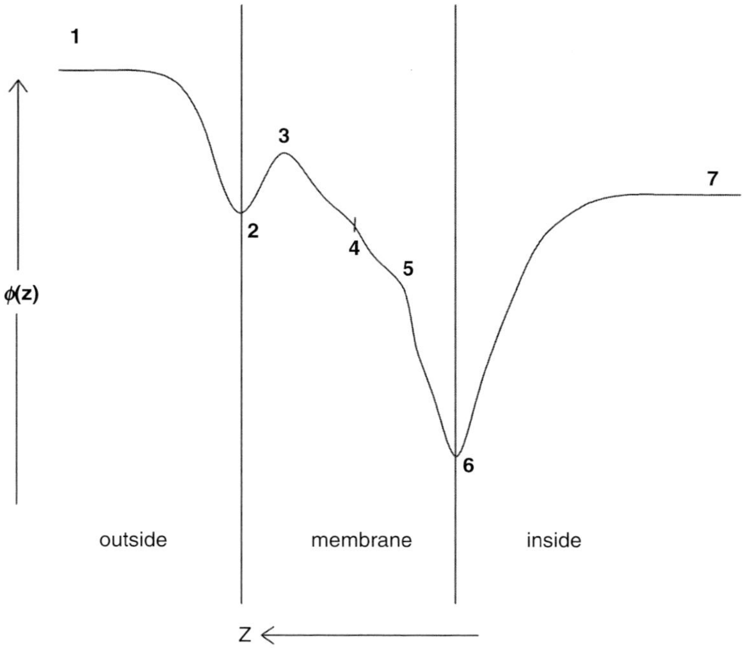

FIG. 8.60 Schematic of electrical potential $\phi(z)$ across a membrane. Points 1 and 7 are in the bulk solutions, and point 4 is the mid-point of the membrane. Using the notation $\Delta\phi_{21} = \phi(z_2) - \phi(z_1)$, etc., we designate $\Delta\phi_{41}$ as the interface (or surface) potential and $\Delta\phi_{71}$ as the transmembrane potential. Other terminologies include those of (a) membranologists (ref. 467) where $\Delta\phi_{21} =$ surface potential and $\Delta\phi_{32} =$ dipole (or internal) potential, and (b) electrochemists (ref. 483) where $\Delta\phi_{21} =$ outer potential, $\Delta\phi_{42} =$ surface potential, and $\Delta\phi_{41} =$ inner potential. (Adapted from ref. 467.)

hydrocarbon (central) part of the membrane without assistance from ion channels[459] or pumps.[479] A simple order of magnitude calculation[480,481] shows why. We use the macroscopic Born model[482] of ionic solvation, in which the ion is represented by a sphere of charge q and radius R inside an infinite dielectric medium of dielectric constant ε. Ions clearly prefer to be solvated in water ($\varepsilon_W \sim 80$) compared to a membrane ($\varepsilon_M \sim 2$). In the Born model the free energy difference (transfer free energy) between the two media is $\Delta\mu = (q^2/2R)(1/\varepsilon_M - 1/\varepsilon_W)$. For K^+, we take $q = 1e$ and $R = 1.33$ Å and find $\Delta\mu \sim 70\,kT$ at physiological temperatures. Thus the ions find a nearly impenetrable barrier to cross the water–membrane interface. The model neglects molecular details and the fact that the membrane is of finite thickness, but gives the correct order of magnitude of the effect. Nutrients such as glucose and amino acids also require specialized transporter proteins to cross the membrane.[463]

In contrast to ions and nutrients, water and small solutes such as oxygen, carbon dioxide, nitrous oxide, methanol, urea, toxins, and many drugs can cross the membrane by simple diffusion. An empirical rule found by Overton[484] is that the membrane permeability of a species is proportional to its solubility in oil (similar dielectrically to a membrane interior). Another empirical rule found by Overton[485] and Meyer[486] independently is that the potency of an anesthetic is correlated with its solubility in oil. The molecular basis of these rules, and their exceptions, is an active area of research.[487] Whether the seat of anesthetic and alcohol[488] action is the cell membrane or a protein embedded in it has long been debated.[489,490] A current view is that Na^+ ion channels are bound and blocked by an anesthetic. However, it is possible an anesthetic may act indirectly[491] on an ion channel by binding to the membrane and thereby disturbing the interface potential and/or the lateral pressure profile acting on the ion channel protein. For example, NMR experiments have shown that halothane (CF_3CHBr)[492] and ethanol (CH_3CH_2OH)[493] partition at the water–membrane interface. Computer simulations are being used increasingly to study water–membrane partitioning of anesthetics and alcohols. In ref. 487 the authors review simulation studies of the partitioning of nearly thirty molecules, including ethanol, nitrous oxide, Xe, fluoroform (CHF_3), and butane. Recent simulation partitioning studies include the local anesthetic lidocaine,[494] the inhaled anesthetic halothane,[495] and the alcohols ethanol,[496,497] propanol,[497] and butanol.[497]

Another important class of water–membrane interface active solutes is peptides. In particular, certain small peptides (usually 10–50 amino acid residues), which are polycationic and amphiphilic, have antimicrobial activity.[498–502] Hundreds of such natural peptides[502] have been harvested from bacteria, plants, insects, amphibians, and mammals (including humans), and thousands more have been synthesized. Examples of natural ones are magainin from frog skin, melittin from bee venom, lactoferricin B from cow's milk, protamine from fish sperm cells, nisin from a bacterium, alamethicin from a fungus, protegrin from pig white blood cells, defensin from human white blood cells, indolicidin from cow white blood cells, cardiotoxin 1 from cobra venom, gramicidin A from a bacterium, and cecropin from pig intestine and moths. The mechanisms of antibacterial action are not known with certainty; it has been hypothesized[500] that at least some of the peptides act by binding to a bacterial membrane and disrupting it, either by aggregating and forming pores (see Fig. 8.61) or by carpeting the membrane and then acting like a detergent to dissolve the membrane. Other peptides may cross the membrane and attack the internal cell machinery. Similarly, the mechanism by which some of these peptides select bacterial membranes for action, and leave intact mammalian membranes (e.g. magainin is selective, melittin is not) is not understood; it is hypothesized that the cationic peptides bind easily to the negatively charged bacterial membranes (the outer leaflet of a mammalian membrane is uncharged), or perhaps somehow exploit the stronger transmembrane electric field of a bacterial membrane, or its lack of cholesterol. Antimicrobial peptides are under study as potential replacements or supplements to traditional antibiotics, necessitated by the development of bacterial resistance to traditional antibiotics. Rational design of such peptides requires a molecular understanding of their mechanism(s) of action and selectivity. In addition to antimicrobial peptides,

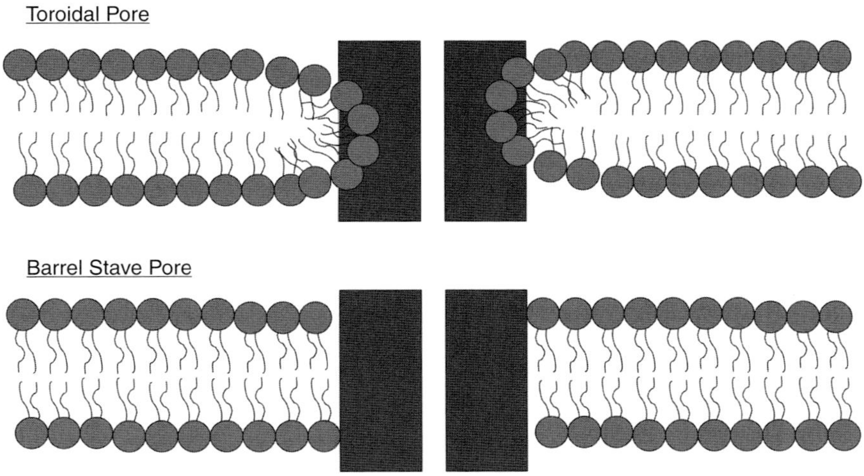

FIG. 8.61 Schematic of two proposed models (toroidal and barrel stave) for pores induced by peptide aggregation in a membrane. (From ref. 522.)

many other peptide–membrane interactions are being studied, for example toxins,[503] fusion peptides,[504, 505] hormones, and signal peptides.[506] Techniques being used to study the various peptide–membrane interactions include thermodynamics (including differential scanning calorimetry (DSC) and isothermal titration calorimetry (ITC)),[504, 507–512] surface potential,[512] NMR,[507, 513–516] X-ray scattering,[508, 510, 517] neutron scattering,[501, 508, 510, 517] circular dichroism (CD),[501, 507] fluorescence,[518] infrared (including FTIR and reflection methods),[519, 520] surface plasmon resonance (SPR),[519] electron spin resonance (ESR),[521] reversed phase high performance liquid chromatography (RP-HPLC),[519] and electron microscopy.[507]

Simulations are also being used to study peptide interactions at the water–membrane interface. For reviews, see refs. 523, 524 and for representative recent work see refs. 525, 526. As an example, in Fig. 8.62 we show a snapshot from a simulation of lactoferricin B adsorbing at the water–POPC interface, where POPC stands for the lipid palmitoyl-oleoyl PC, with PC designating the head group phosphatidylcholine.

Some miscellaneous studies of solutes at the water–membrane interface include drugs, by X-ray,[527] NMR,[528] thernmodynamics (calorimetry),[529] and molecular dynamics (MD);[530] toxins, by neutron and X-ray scattering,[531] MD,[532] the combination Monte Carlo simulation–neutron reflectivity,[533] and the combination FTIR-MD,[534] and electrochemical methods (transmembrane potential, and current-voltage measurements);[535, 536] organic (toxic) molecules by MD,[537] NMR,[538] and SHG;[539] disaccharides of glucose by MD; hormones by MD;[540] tryptophan analogues by MD;[541] synthetic molecules by MD;[542] various small molecules by Monte Carlo simulation (H_2O, O_2, CO, CO_2, NO, $CHCl_3$).[543]

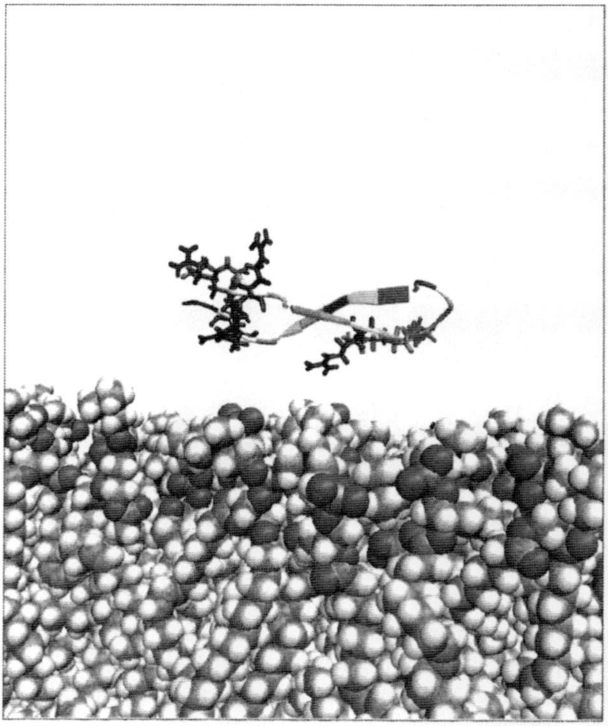

FIG. 8.62 Representation of a snapshot from a simulation of interaction of lactoferricin B (bovine lactoferricin) with POPC membrane in 100 mM salt solution at 37 °C. Lactoferricin B is a cationic (charge +8e) antimicrobial peptide with 25 amino acid residues and POPC is a neutral membrane. For clarity the water molecules and salt ions in the solution above the membrane are not shown. The simulation yields the potential of mean force $w(z, \omega)$ as a function of the peptide distance z from the membrane surface and peptide orientation ω. From $w(z, \omega)$, observables such as the free energy of binding (or adsorption) can be calculated.[526] Reprinted with permission from ref. 526 (original in colour). Copyright 2008 American Physical Society.

Proteins interacting with membranes is a subject in itself—for representative reviews, see ref. 544.

We end this section with additional references to techniques being used to study the water–membrane interface, in addition to those listed already in connection with specific problems discussed: thermodynamics,[471,545] X-ray and neutron scattering, X-ray standing waves (XSW),[546] SFG,[547] fluorescence, osmotic stress,[548,549] and SFA,[549,550] atomic force microscopy (AFM),[551] optical tweezers,[552] IR,[553] NMR,[554] quartz crystal microbalance (QCM), ellipsometry, electron crystallography,[555] and Raman spectroscopy.

In Fig. 8.63 we show the density profiles across the membrane of the various groups for a partially hydrated POPC membrane as determined by combined X-ray and neutron diffraction.[557]

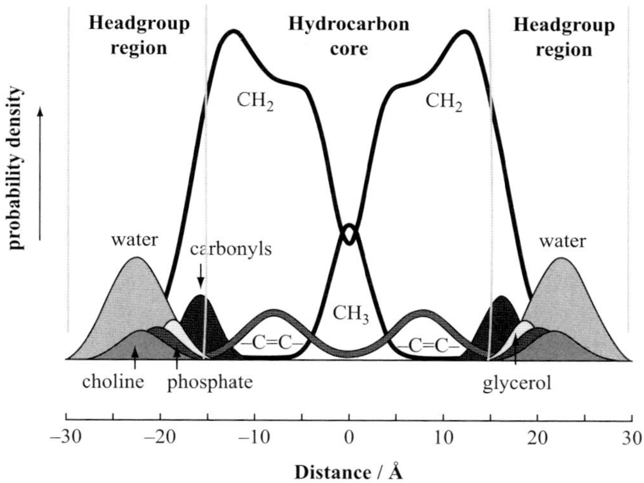

FIG. 8.63 The structure of a bilayer of unsaturated DOPC (one double C = C bond in each lipid tail) at 23 °C at low hydration (5.4 waters per headgroup). The figure shows density profiles of the principal structural groups. Fragments shown are the methyls (CH_3), methylenes (CH_2), double bonds (C = C), carbonyls, the glycerol backbone, the phosphate and choline groups of the headgroup, and water. The hydrocarbon thickness (32 Å) is about 5 Å greater than for a fully hydrated bilayer (ref. 451). Reprinted with permission from ref. 556. Copyright 2006 Springer. Based on work of M.C. Weiner and S.H. White (ref. 557).

General reviews of water–membrane simulations are available.[558] Simulations have been carried out for lipids with both charged (e.g. PG and PS) and uncharged (e.g. PC and PE) head groups; the abbreviations PG, POPG, etc., are spelled out in endnote 559. Some specific lipid membranes simulated include POPG, DPPS, POPS, DMPC, DPPC, DOPC, POPC, and DLPE. Membranes with two- and three-component lipid mixtures,[560] including cholesterol,[561] have also been simulated. Several joint simulation and experimental studies of membranes have been carried out to interpret, for example, NMR,[562] IR,[563] and X-ray[564] diffraction experimental results. Various transporter proteins embedded in membranes have been simulated,[565] e.g. ion channels (reviewed in refs. 565 and 566). Specific ion channels simulated include KcsA (a bacterial K^+ channel, the first ion selective channel for which the structure was determined by X-ray diffraction), Kv1.2 (a eukaryotic voltage-gated K^+ channel), KirBac1.1 (an inward rectifier bacteria K^+ channel), ClC (a voltage-gated Cl^- channel found in vertebrates and bacteria), and MscS (the mechanosensitive channel of small conductance, weakly anion selective, and found in bacteria and plants). Other membrane protein transporters[565] simulated include members of the aquaporin family (e.g. AQP1[567] (a water transporter in humans), GlpF[567,568] (a glycerol transporter in *E. coli* bacteria), AqpZ[568] (a water transporter in humans), OmpF[569] (a nonspecific porin from the *E. coli* outer membrane), LacY[570] (a lactose/H^+ cotransporter from *E. coli*), AmtB[571] (an NH_3 transporter from *E. coli*), and SecYEβ[572] (a transporter of proteins from an archaebacterium). The membrane-bound enzyme OmpT[573] (a protease from the *E. coli* outer membrane) has also been simulated.

8.10 Conclusions

The formal statistical mechanics needed to treat surface properties is now largely in place (§§8.1–8.4), but the development of satisfactory approximations to make these expressions numerically tractable is considerably more difficult than for homogeneous fluids. The use of non-local approximations in the density functional formalism (§8.7) is particularly promising, and is being used for a variety of applications, including adsorption of fluids on walls and in pores, for solid properties, and freezing and melting. The approximate theories described in §§8.5–8.8 are readily applied to simple interfacial systems—simple molecules and simple interface geometries, but may be considerably more difficult to apply to complex geometries and intermolecular potentials. In such cases molecular simulation (Monte Carlo and molecular dynamics) is the preferred approach. With simulations care is needed to ensure that the system is large enough and the real time long enough to capture the phenomena of interest.

Several approximate theoretical techniques have been omitted from this chapter, either because they do not appear to have sufficiently rigorous foundations or because they are only appropriate for certain limited types of interfacial systems. Two of the more rigorous approaches that should be mentioned are lattice fluid theory and quenched-annealed models. Lattice fluid theories[574] have proven very useful in studies of adsorption and phase changes in porous media.[575–577] Quenched-annealed models[578] provide a way of studying fluids confined within random matrices, such as many porous materials or gels. The confining disordered medium is treated as a model fluid that has been quenched, so that the molecules composing it are frozen in space, while the confined fluid phase is 'annealed', i.e. is under the influence of the forces due to the quenched matrix. The advantage of such a model is that the fluid used for the quenched matrix can be chosen so that its properties are well understood, providing a direct link to the equilibrium statistical mechanics of fluids. The resulting equations can be solved using either integral equation or density functional theory. Such models have proven successful for treating adsorption and phase changes in random porous media,[578,579] and fluid mixtures in gels.[580]

Appendix 8A Functionals

8A.1 Functionals[581–584]

A functional is a generalized function of a function. Just as a function $f(x)$ has a numerical value for each numerical input x, so a functional $F[f]$ has a numerical value for each *function* input f. Equivalently, a functional can be defined as a function of an infinite number of variables. Examples of functionals are[585]

$$F[f] = \int_a^b \mathrm{d}x f(x) \tag{8A.1}$$

$$F[f] = \int_a^b \mathrm{d}x e^{f(x)} \tag{8A.2}$$

$$F[f] = \int_a^b \mathrm{d}x A(x) f(x). \tag{8A.3}$$

In the first and third examples, F depends linearly on $f(x)$, and is said to be a *linear functional*. $F[f]$ is a *continuous* functional if small variations $\delta f(x)$ in $f(x)$ lead to small variations δF in F.

The above definitions generalize immediately to three or more variables; e.g., if $f = f(x, y, z) \equiv f(\mathbf{r})$, we have

$$F[f] = \int_V \mathrm{d}\mathbf{r} A(\mathbf{r}) f(\mathbf{r}) \tag{8A.4}$$

as the generalization of (8A.3). Here V denotes the region of space over which the integration over $\mathbf{r}$ is carried out. We shall usually omit the V in what follows.

8A.2 Functional differentiation

Functional derivative techniques have been used extensively[586–592] in statistical mechanics (see, e.g. refs. 588–592 for reviews). In this section we give a brief outline of the method. In §8A.4 we give some important examples of their use.

Functional derivatives can be defined in two ways, by analogy with either the ordinary differential or the derivative of a function, respectively. In the first method, analogous to the differential $\mathrm{d}f$ of a function $f = f(x_1 \ldots x_n)$, where

$$\mathrm{d}f = \sum_{i=1}^{n} \left(\frac{\partial f}{\partial x_i} \right) \mathrm{d}x_i, \tag{8A.5}$$

we define the functional derivative $\delta F / \delta f(x)$ of a functional $F[f]$ by

$$\delta F = \int_a^b \mathrm{d}x \left(\frac{\delta F}{\delta f(x)} \right) \delta f(x), \tag{8A.6}$$

i.e. as the coefficient of the linear term in the variation $\delta F = F[f + \delta f] - F[f]$ of F, given a small variation δf of $f(x)$. Alternatively,[593] by analogy with the definition of an ordinary derivative, we can define $\delta F / \delta f(x')$ by

$$\frac{\delta F}{\delta f(x')} = \lim_{\varepsilon \to 0} \frac{1}{\varepsilon} (F[f(x) + \varepsilon \delta(x - x')] - F[f(x)]). \tag{8A.7}$$

As a first example we consider the case (8A.3) of linear functionals. Either definition gives

$$\frac{\delta F}{\delta f(x)} = A(x). \tag{8A.8}$$

For the special case of (8A.1), where $A(x) = 1$, we have

$$\frac{\delta F}{\delta f} = 1. \tag{8A.9}$$

Another special case is $A(x) = \delta(x' - x)$, so that[594]

$$\frac{\delta f(x')}{\delta f(x)} = \delta(x' - x). \tag{8A.10}$$

More generally, if $F[f]$ is defined by

$$F[f] = \int_a^b \mathrm{d}x G(f(x)), \tag{8A.11}$$

where G is an arbitrary function of f (e.g. as in (8A.2)), we find that[595]

$$\frac{\delta F}{\delta f} = \frac{\mathrm{d}G}{\mathrm{d}f}. \tag{8A.12}$$

Thus for $F[f] = \int \mathrm{d}x f(x)^2$, for example, we get $\delta F/\delta f = 2f$.

If $\phi = \phi(F)$ is a function and $F = F[f]$ a functional, then a chain rule holds,

$$\frac{\delta \phi}{\delta f} = \frac{\mathrm{d}\phi}{\mathrm{d}F}\frac{\delta F}{\delta f}. \tag{8A.13}$$

Hence we find, for example,

$$\frac{\delta}{\delta f(x)} e^{-\int \mathrm{d}x' f(x')^2} = -2f(x) e^{-\int \mathrm{d}x' f(x')^2}. \tag{8A.14}$$

Similarly, if $F[H]$ and $H[f]$ are functionals, then the chain rule is

$$\frac{\delta F}{\delta f(x)} = \int \mathrm{d}x' \frac{\delta F}{\delta H(x')}\frac{\delta H(x')}{\delta f(x)}. \tag{8A.15a}$$

A third form of chain rule is given by (8.15).

Putting $F = f(x'')$ in (8A.15a) and using (8A.10) gives

$$\int \mathrm{d}x' \frac{\delta f(x'')}{\delta H(x')}\frac{\delta H(x')}{\delta f(x)} = \delta(x'' - x), \tag{8A.15b}$$

thereby defining the functional derivative inverse of $\delta f/\delta H$ as $\delta H/\delta f$. Here f is regarded as a functional of H in $\delta f/\delta H$, and conversely for $\delta H/\delta f$. Equation (8A.15b) is analogous to matrix and operator equations of the type

$\sum_k A_{ik}(A^{-1})_{kj} = \delta_{ij}$, or $AA^{-1} = 1$ (unit operator), respectively. In all such cases the inverse is assumed to exist.

Second and higher functional derivatives $\delta^2 F/\delta f(x)\delta f(x')$, etc., are defined by iteration and are symmetric, e.g.

$$\frac{\delta}{\delta f(x')}\frac{\delta}{\delta f(x)}F = \frac{\delta}{\delta f(x)}\frac{\delta}{\delta f(x')}F. \tag{8A.16}$$

Functional Taylor series are defined by

$$\begin{aligned} F[f] = F[0] &+ \int \mathrm{d}x \left(\frac{\delta F}{\delta f(x)}\right)_0 f(x) \\ &+ \frac{1}{2!}\iint \mathrm{d}x\mathrm{d}x' \left(\frac{\delta^2 F}{\delta f(x)\delta f(x')}\right)_0 f(x)f(x') \\ &+ \ldots, \end{aligned} \tag{8A.17}$$

where $F[0] = F[f=0]$, $(\delta F/\delta f)_0 = (\delta F/\delta f)_{f=0}$, etc.

The above equations generalize immediately to the case of multi-variables. For three variables $(x, y, z) \equiv \mathbf{r}$, for example, (8A.10), (8.15a), and (8.15b) become

$$\frac{\delta f(\mathbf{r}')}{\delta f(\mathbf{r})} = \delta(\mathbf{r}' - \mathbf{r}) \tag{8A.18}$$

$$\frac{\delta F}{\delta f(\mathbf{r})} = \int \mathrm{d}\mathbf{r}' \frac{\delta F}{\delta H(\mathbf{r}')}\frac{\delta H(\mathbf{r}')}{\delta f(\mathbf{r})} \tag{8A.19a}$$

$$\int \mathrm{d}\mathbf{r}' \frac{\delta f(\mathbf{r}'')}{\delta H(\mathbf{r}')}\frac{\delta H(\mathbf{r}')}{\delta f(\mathbf{r})} = \delta(\mathbf{r}'' - \mathbf{r}). \tag{8A.19b}$$

8A.3 Functional integration

Functional integrals of the type

$$\int \mathrm{d}[f]F[f] \tag{8A.20}$$

are becoming of increasing importance in statistical mechanics. Here $F[f]$ is a functional of $f(x)$, and $\int \mathrm{d}[f]\ldots$ indicates schematically an integration over all functions f. Precise definitions of the integral are given in various reviews.[596–603] As a simple example, the mean position of a Brownian particle at time t, starting from the origin at time $t = 0$, is

$$\left\langle \int_0^t \mathrm{d}t'\mathbf{v}(t') \right\rangle = \int \mathrm{d}[\mathbf{v}]P[\mathbf{v}] \int_0^t \mathrm{d}t'\mathbf{v}(t'),$$

where $\mathbf{v}(t)$ is the velocity. The averaging (or functional integration) is over all functions $\mathbf{v}(t)$. Here $P[\mathbf{v}]$ is the probability distribution functional for $\mathbf{v}(t')$.

We shall not need functional integrals in this book.

8A.4 Examples

We illustrate the use of functionals and their derivatives with examples involving the various equilibrium correlation functions.

The first and second Yvon equations

The total and direct pair correlation functions,[604] $h(\mathbf{r}_1\mathbf{r}_2\omega_1\omega_2) \equiv h(12)$ and $c(\mathbf{r}_1\mathbf{r}_2\omega_1\omega_2) \equiv c(12)$, can be expressed in terms of functional derivatives of the singlet distribution function $f(\mathbf{r}_1\omega_1) \equiv f(1)$. These equations were first obtained for atomic fluids by Yvon.[18,605] We give here the generalization to the molecular case, which is straightforward.

To obtain the expression for $h(12)$, the first Yvon equation, we note that the singlet distribution function $f(1)$ is a functional of the external potential $\mathcal{V}$. For a small change in $\mathcal{V}$,

$$\delta\mathcal{V} = \sum_{i=1}^{N} \delta v(i) \tag{8A.21}$$

the corresponding change in $f(1)$ is, from (8.4),

$$\begin{aligned}\delta f(\mathbf{r}_1\omega_1) = &-\frac{1}{\Xi^2}\delta\Xi\sum_N \frac{\hat{z}^N}{(N-1)!}\int \mathrm{d}\mathbf{r}^{N-1}\mathrm{d}\omega^{N-1}\exp[-\beta(\mathcal{U}+\mathcal{V})]\\ &-\frac{\beta}{\Xi}\sum_N \frac{\hat{z}^N}{(N-1)!}\int \mathrm{d}\mathbf{r}^{N-1}\mathrm{d}\omega^{N-1}\exp[-\beta(\mathcal{U}+\mathcal{V})]\\ &\times[\delta v(\mathbf{r}_1\omega_1) + (N-1)\delta v(\mathbf{r}_2\omega_2)].\end{aligned} \tag{8A.22}$$

To evaluate the first term on the right side of (8A.22) we note from (8.2) that

$$\begin{aligned}\frac{\delta\Xi}{\Xi} &= -\frac{\beta}{\Xi}\sum_N \frac{\hat{z}^N}{(N-1)!}\int \mathrm{d}\mathbf{r}^N\mathrm{d}\omega^N\exp[-\beta(\mathcal{U}+\mathcal{V})]\delta v(\mathbf{r}_1\omega_1)\\ &= -\beta\int \mathrm{d}\mathbf{r}_1\mathrm{d}\omega_1\delta v(\mathbf{r}_1\omega_1)f(\mathbf{r}_1\omega_1).\end{aligned} \tag{8A.23}$$

From (8A.22), (8A.23), and (8.4) we find that

$$\begin{aligned}\delta f(\mathbf{r}_1\omega_1) &= \beta f(\mathbf{r}_1\omega_1)\int \mathrm{d}\mathbf{r}_2\mathrm{d}\omega_2\delta v(\mathbf{r}_2\omega_2)f(\mathbf{r}_2\omega_2)\\ &-\beta\delta v(\mathbf{r}_1\omega_1)f(\mathbf{r}_1\omega_1) - \beta\int \mathrm{d}\mathbf{r}_2 d\omega_2\delta v(\mathbf{r}_2\omega_2)f(\mathbf{r}_1\mathbf{r}_2\omega_1\omega_2)\\ &= -\beta\int \mathrm{d}\mathbf{r}_2\mathrm{d}\omega_2\delta v(\mathbf{r}_2\omega_2)[f(\mathbf{r}_1\mathbf{r}_2\omega_1\omega_2) - f(\mathbf{r}_1\omega_1)f(\mathbf{r}_2\omega_2) + f(\mathbf{r}_2\omega_2)\delta(12)],\end{aligned} \tag{8A.24}$$

where δ (12) is the Dirac delta function,

$$\delta(12) = \delta(\mathbf{r}_1 - \mathbf{r}_2)\delta(\omega_1 - \omega_2) \tag{8A.25}$$

and (see (A.32) and (A.85))

$$\begin{aligned} \delta(\omega) &= \delta(\cos\theta)\delta(\phi) && (\textit{linear molecule}) \\ &= \delta(\phi)\delta(\cos\theta)\delta(\chi) && (\textit{nonlinear molecule}) \end{aligned} \tag{8A.26}$$

is the angular Dirac delta function. From (8A.24) and (8A.6) we have

$$\frac{\delta f(\mathbf{r}_1\omega_1)}{\delta v(\mathbf{r}_2\omega_2)} = -\beta\,[f(\mathbf{r}_1\mathbf{r}_2\omega_1\omega_2) - f(\mathbf{r}_1\omega_1)f(\mathbf{r}_2\omega_2) + f(\mathbf{r}_2\omega_2)\delta(12)]. \tag{8A.27}$$

If we now introduce the total pair correlation function $h(12) \equiv g(12) - 1$ by

$$f(\mathbf{r}_1\omega_1)f(\mathbf{r}_2\omega_2)h(\mathbf{r}_1\mathbf{r}_2\omega_1\omega_2) = f(\mathbf{r}_1\mathbf{r}_2\omega_1\omega_2) - f(\mathbf{r}_1\omega_1)f(\mathbf{r}_2\omega_2), \tag{8A.28}$$

then (8A.27) becomes, after rearrangement,

$$h(12) = -\frac{1}{\beta f(1)f(2)}\frac{\delta f(1)}{\delta v(2)} - \frac{\delta(12)}{f(1)}. \tag{8A.29}$$

Equation (8A.29) is the first Yvon equation. It relates the total correlation function to the functional derivative of $f(1)$ with respect to $v(2)$.

The second Yvon equation is the inverse of (8A.29), and defines the pair function $c(\mathbf{r}_1\mathbf{r}_2\omega_1\omega_2) = c(12)$ in terms of the functional derivative $\delta v(1)/\delta f(2)$. It is

$$c(12) = \beta\frac{\delta v(1)}{\delta f(2)} + \frac{\delta(12)}{f(1)}. \tag{8A.30}$$

That $c(12)$ is in fact the usual direct correlation function can be shown by comparison of (8A.30) with the Ornstein–Zernike (OZ) equation. For inhomogeneous fluids the OZ equation is (see (3.117) of Vol. 1 and footnotes there)

$$h(12) = c(12) + \int c(13)f(3)h(32)\mathrm{d}3, \tag{8A.31}$$

where $c(ij)$ is the pair direct correlation function. The OZ equation (8A.31) can be rewritten as

$$\int \mathrm{d}3[\delta(13) + f(3)h(13)][\delta(32) - f(2)c(32)] = \delta(12), \tag{8A.32}$$

which is easily verified by multiplying out the terms on the left-hand side and integrating over 3. (The **k**-space version of (8A.32) was used in Vol. 1, Eq. (5.13).) Comparison of (8A.32) and (8A.19b) shows that the two bracketed terms [...] in (8A.32) are functional inverses. From the first Yvon equation (8A.29) the first bracket is given by $\delta(13) + f(3)h(13) = -(1/\beta f(1))\delta f(1)/\delta v(3)$. The second bracket [...] is therefore given by $\delta(32) - f(2)c(32) = -\beta f(2)\delta v(3)/\delta f(2)$, which is just a rearrangement of the second Yvon equation, (8A.30). Thus the function $c(12)$ defined by the second Yvon equation (8A.30) is just the direct correlation function defined by the OZ equation (8A.31).

The hierarchy of equations for the direct correlation functions

We have already defined the two-body direct correlation function $c(12)$ in terms of the functional derivative $\delta v(1)/\delta f(2)$ through the second Yvon equation, (8A.30). We can further define one-, three-, etc., body direct correlation functions,[18,606,607] and these are related through a hierarchy of functional derivative relations.

We consider first the one-body function $c(\mathbf{r}_1\omega_1) \equiv c(1)$. This is defined as follows. For an ideal gas ($\mathcal{U} = 0$) we have[608] for $f(1)$

$$f(1) = \hat{z}\exp[-\beta v(1)] \tag{8A.33}$$

with $\hat{z}$ given by (8.3). Alternatively $\hat{z}$ can be determined by the normalization condition $\int \mathrm{d}1 f(1) = \langle N \rangle$. For a real fluid $f(1)$ must depend on the intermolecular forces in addition to the external field. We define $c(1)$ by writing[18,607]

$$f(1) = \hat{z}\exp[-\beta v(1) + c(1)] \tag{8A.34}$$

or, upon rearranging,

$$c(1) = \beta v(1) + \ln[f(1)/\hat{z}]. \tag{8A.35}$$

We can think of $-c(1)/\beta$ as the additional effective one-body potential that determines the density–orientation profile, $f(\mathbf{r}_1\omega_1)$. The total effective potential, $v(1) - kTc(1)$, is the potential of mean force $w(1)$ (§ 8.1.1). To demonstrate the relation between $c(1)$ and the two-body function $c(12)$ we differentiate (8A.35) with respect to $f(2)$:

$$\frac{\delta c(1)}{\delta f(2)} = \beta\frac{\delta v(1)}{\delta f(2)} + \frac{\delta(12)}{f(1)}$$

or, using the second Yvon equation, (8A.30),

$$c(12) = \frac{\delta c(1)}{\delta f(2)}. \tag{8A.36}$$

The higher order direct correlation functions are obtained by further differentiation,[607] e.g.

$$c(123) = \frac{\delta^2 c(1)}{\delta f(3)\delta f(2)} = \frac{\delta c(12)}{\delta f(3)}. \tag{8A.37}$$

We can also express the direct correlation functions in terms of the free energy functional $\mathcal{A}[f']$ introduced in §8.4. With the aid of (8.3) we can rewrite (8A.35) as

$$c(1) = \beta[v(1) - \mu] + \ln f(1) + \ln(\Lambda_t^3 \Lambda_r \Omega / q_{\mathrm{qu}})$$

or, with the aid of (8.195) and (8.196),

$$c(1) = -\beta \left. \frac{\delta \mathcal{A}}{\delta f'(1)} \right|_{f'=f} + \ln f(1) + \ln \left(\Lambda_t^3 \Lambda_r \Omega / q_{\text{qu}} \right), \tag{8A.38}$$

which expresses the one-body direct correlation function, $c(1)$, in terms of a functional derivative of $\mathcal{A}$. We can obtain the corresponding expression for c(12) by applying (8A.36) to (8A.38),

$$c(12) = -\beta \left. \frac{\delta^2 \mathcal{A}}{\delta f'(1) \delta f'(2)} \right|_{f'=f} + \frac{1}{f(1)} \delta(12), \tag{8A.39}$$

where the generalization of (8A.10) has been used to evaluate the last term on the right-hand side of (8A.39). In density functional theory, where an approximate expression is available for the free energy functional $\mathcal{A}$, (8A.38) and (8A.39) provide a route to calculating the direct correlation functions (see, for example, §8.7).

Equations (8A.38) and (8A.39) can be written more compactly by expressing $\mathcal{A}$ as the sum of an ideal gas ($\mathcal{U} = 0$) part, $\mathcal{A}^{\text{id}}$, and a residual part, $\mathcal{A}^{\text{r}}$, that contains the effect of the intermolecular forces. From (8.189) with $\mathcal{U} = 0$ we find that[591,597]

$$\mathcal{A}^{\text{id}}[f'] = kT \int d\mathbf{r} d\omega f'(\mathbf{r}\omega) \left[\ln \left(\frac{\Lambda_t^3 \Lambda_r \Omega}{q_{qu}} f'(\mathbf{r}\omega) \right) - 1 \right]. \tag{8A.40}$$

Using this expression in (8A.38) and (8A.39) gives

$$c(1) = -\beta \left. \frac{\delta \mathcal{A}^r}{\delta f'(1)} \right|_{f'=f} \tag{8A.41}$$

and

$$c(12) = -\beta \left. \frac{\delta^2 \mathcal{A}^r}{\delta f'(1) \delta f'(2)} \right|_{f'=f}. \tag{8A.42}$$

Using (8A.41), (8A.42), etc., we can derive from (8A.19a) a functional perturbation expression for $\mathcal{A}^r[f]$ in terms of $\mathcal{A}^r[f_0]$, where we assume $f_0(1) \equiv \rho / \Omega$ corresponds to a uniform fluid:

$$\mathcal{A}^r[f] = \mathcal{A}^r[f_0] - kT \int d1 c_0(1)(f(1) - f_0) - \frac{1}{2} kT \int d1 d2 c_0(12)(f(1) - f_0)(f(2) - f_0) + \ldots . \tag{8A.43}$$

Such expansions have been used, for example, in theories of freezing, and require as input the direct correlation functions $c_0(1) =$ constant, $c_0(12)$, $c_0(123)$, etc., for the uniform fluid.

Expressions for the distribution functions in terms of the partition function

The distribution functions $f(1)$, $f(12)$, etc., can be written as functional derivatives of the grand partition function Ξ. The expression for $f(1)$ can be obtained directly from (8A.23). From (8A.23) and (8A.6) we have

$$f(1) = -\frac{1}{\beta}\frac{\delta \ln \Xi}{\delta v(1)}. \tag{8A.44}$$

Differentiating again gives

$$\begin{aligned} -\frac{1}{\beta^2}\frac{\delta^2 \ln \Xi}{\delta v(2)\delta v(1)} &= \frac{1}{\beta}\frac{\delta f(1)}{\delta v(2)} \\ &= -[f(12) - f(1)f(2) + f(2)\delta(12)] \end{aligned} \tag{8A.45}$$

from (8A.24).

Integral equation theory

The PY and HNC (and also other) integral equations for the pair correlation function can be derived using a functional Taylor expansion. The method is due to Percus.[587,606] We consider an initially uniform, isotropic fluid, and take the external field $\mathcal{V}$ to be due to a molecule fixed at $(\mathbf{r}_0\omega_0)$. The other molecules then move in the force field due to this fixed molecule 0. Thus, the external field experienced by molecule i at $(\mathbf{r}_i\omega_i)$ is just the pair potential,

$$v(i) = v(\mathbf{r}_i\omega_i) = u(\mathbf{r}_0\mathbf{r}_i\omega_0\omega_i). \tag{8A.46}$$

Similarly, the one-body distribution function in the presence of the external field, $f(1|v)$, is related to the number density of molecules at $(\mathbf{r}_1\omega_1)$ when a molecule is known to be fixed at $(\mathbf{r}_0\omega_0)$ and so is essentially the unperturbed fluid two-body distribution function,

$$f(1|v) = f(01)/f(0) = \frac{\rho}{\Omega}g(\mathbf{r}_{01}\omega_0\omega_1), \tag{8A.47}$$

where ρ is the density of the fluid. Physically, since the fluid distribution functions are independent of mass, the pair distribution function in the unperturbed fluid is the same as the distribution function for molecules around a fixed (or infinitely heavy) molecule introduced into the fluid. In the absence of the field the one-body distribution function takes its usual form, (3.104),

$$f(1|v=0) = f(1) = \frac{\rho}{\Omega}. \tag{8A.48}$$

We now consider some functional[609] $\mathrm{F}(1|v)$ of the one-body distribution function f, which is in turn a functional of the external field v. Expanding $\mathrm{F}(1|v)$ in terms of f about $f(v=0)$ gives, to first order:

$$F(1|v) = F(1|0) + \int \left(\frac{\delta F(1|v)}{\delta f(2|v)}\right)_{v=0} [f(2|v) - f(2|v=0)]\mathrm{d}2. \tag{8A.49}$$

The functional derivative in (8A.49) can be written, using (8A.15a), as

$$\frac{\delta F(1|v)}{\delta f(2|v)} = \int \frac{\delta F(1|v)}{\delta v(3)} \frac{\delta v(3)}{\delta f(2|v)} \mathrm{d}3, \tag{8A.50}$$

where $\delta v(3)/\delta f(2|v)$ is given by the second Yvon equation, (8A.30), as

$$\frac{\delta v(3)}{\delta f(2|v)} = \beta^{-1}\left[c(23) - \frac{\Omega}{\rho}\delta(23)\right]. \tag{8A.51}$$

So far the functional $F(1|v)$ is arbitrary and no approximation has been made beyond (8A.49). To proceed further we must make a specific choice for F, and for some choices the approximate relation (8A.49) will give us a useful approximate relation. We discuss the two classic choices.

Percus–Yevick equation To derive the PY equation we take F to be

$$F(1|v) = f(1|v)\exp[\beta v(1)] \tag{8A.52}$$

so that

$$\frac{\delta F(1|v)}{\delta v(3)} = \frac{\delta f(1|v)}{\delta v(3)}\exp[\beta v(1)] + \beta f(1|v)\frac{\delta v(1)}{\delta v(3)}\exp[\beta v(1)].$$

Using (8A.27) for the first term on the right, and taking the limit as $v \to 0$ gives

$$\left(\frac{\delta F(1|v)}{\delta v(3)}\right)_{v=0} = -\frac{\beta\rho^2}{\Omega^2}h(13) - \beta f(3)\delta(13) + \beta f(1)\delta(13). \tag{8A.53}$$

From (8A.50), (8A.51), and (8A.53) we have

$$\left(\frac{\delta F(1|v)}{\delta f(2|v)}\right)_{v=0} = \frac{\rho}{\Omega}\left[h(12) - \rho\int \mathrm{d}\mathbf{r}_3\langle h(13)c(23)\rangle_{\omega_3}\right] = \frac{\rho}{\Omega}c(12), \tag{8A.54}$$

where the last step follows from the OZ equation (3.117). We now substitute (8A.54) into (8A.49) and use (8A.46)–(8A.48) together with (8A.52) to get

$$g(10)\exp[\beta u(10)] = 1 + \rho\int \mathrm{d}\mathbf{r}_2\langle c(12)h(20)\rangle_{\omega_2}.$$

The left-hand side is just $y(10)$, while the right-hand side is $1 + h(10) - c(10)$ from the OZ equation. Thus, this equation reduces to

$$y(12) = 1 + h(12) - c(12)$$

or

$$c(12) = g(12)[1 - \exp(\beta u(12))], \tag{8A.55}$$

which is the PY approximation (see (5.4) and (5.6)).

Hypernetted chain equation To derive the HNC equation we choose F to be

$$F(1|v) = \ln[f(1|v)/\exp(-\beta v(1))]. \tag{8A.56}$$

By a procedure similar to that used above for the PY approximation we find that

$$\left(\frac{\delta F(1|v)}{\delta v(3)}\right)_{v=0} = -\beta\frac{\rho}{\Omega}h(13). \tag{8A.57}$$

Using this result in (8A.50) gives, on using (8A.51),

$$\left(\frac{\delta F(1|v)}{\delta f(2|v)}\right)_{v=0} = c(12) \tag{8A.58}$$

and from (8A.49), (8A.52), and (8A.58) we have

$$\ln g(12) = -\beta u(12) + h(12) - c(12)$$

or

$$c(12) = h(12) - \beta u(12) - \ln g(12), \tag{8A.59}$$

which is the HNC approximation (see (5.7)). The PY (8A.55) and HNC (8A.59) approximations are to be regarded as approximate closures for the OZ integral equation (8A.31). Solutions for uniform fluids are discussed in Vol. 1, Chapter 5.

Both the PY and HNC theories have been extended[610,611] by carrying out the functional Taylor expansion of (8A.49) to second order. These theories are referred to as PY2 and HNC2.

Appendix 8B Derivatives with respect to surface area

In this appendix we derive Eqs. (8.93) and (8.94) for the derivatives of $u(12)$ and $v(1)$ with respect to surface area.

8B.1 The term $\partial u/\partial \mathcal{S}$

The surface derivative of $u(12)$ can be written

$$\begin{aligned}\mathcal{S}\frac{\partial u(\mathbf{r}_{12}\omega_1\omega_2)}{\partial \mathcal{S}} &= \mathcal{S}\frac{\partial u(\mathbf{r}_{12}\omega_1\omega_2)}{\partial \mathbf{r}_{12}}\cdot\frac{\partial \mathbf{r}_{12}}{\partial \mathcal{S}} \\ &= \mathcal{S}\frac{\partial u(12)}{\partial \mathbf{r}_{12}}\cdot\left[\frac{\partial x_{12}}{\partial \mathcal{S}}\mathbf{i} + \frac{\partial y_{12}}{\partial \mathcal{S}}\mathbf{j} + \frac{\partial z_{12}}{\partial \mathcal{S}}\mathbf{k}\right].\end{aligned} \tag{8B.1}$$

From (8.86) we have

$$x_{12}^2 = (x_2 - x_1)^2 = \mathcal{S}\left(x_2' - x_1'\right)^2 \tag{8B.2}$$

so that

$$2x_{12}\frac{\partial x_{12}}{\partial \mathcal{S}} = \left(x_2' - x_1'\right)^2 = \frac{1}{\mathcal{S}}x_{12}^2$$

$$\mathcal{S}\frac{\partial x_{12}}{\partial \mathcal{S}} = \frac{1}{2}x_{12}.$$

Similarly we find

$$\mathcal{S}\frac{\partial y_{12}}{\partial \mathcal{S}} = \frac{1}{2}y_{12} \tag{8B.3}$$

$$\mathcal{S}\frac{\partial z_{12}}{\partial \mathcal{S}} = -z_{12}. \tag{8B.4}$$

From (8B.1)–(8B.4) we have

$$\begin{aligned}\mathcal{S}\frac{\partial u(12)}{\partial \mathcal{S}} &= \frac{\partial u(12)}{\partial \mathbf{r}_{12}} \cdot \frac{1}{2}(x_{12}\mathbf{i} + y_{12}\mathbf{j} - 2z_{12}\mathbf{k}) \\ &= \frac{1}{2}\frac{\partial u(12)}{\partial \mathbf{r}_{12}} \cdot (\mathbf{r}_{12} - 3z_{12}\mathbf{k}) \\ &= \frac{1}{2}\left[r_{12}\left(\frac{\partial u(12)}{\partial r_{12}}\right)_{\theta_{12}\phi_{12}\omega_1\omega_2} - 3z_{12}\left(\frac{\partial u(12)}{\partial z_{12}}\right)_{x_{12}y_{12}\omega_1\omega_2}\right],\end{aligned} \tag{8B.5}$$

which is Eq. (8.93). Since $u(12)$ is usually expressed in terms of the variables $(\mathbf{r}_{12}\omega_1\omega_2)$, it is more convenient to replace $\partial u(12)/\partial z_{12}$ in (8B.5) by derivatives with respect to spherical coordinates. Thus we have

$$\left(\frac{\partial u}{\partial z}\right)_{xy} = \left(\frac{\partial u}{\partial r}\right)_{\theta\phi}\left(\frac{\partial r}{\partial z}\right)_{xy} + \left(\frac{\partial u}{\partial \theta}\right)_{r\phi}\left(\frac{\partial \theta}{\partial z}\right)_{xy} + \left(\frac{\partial u}{\partial \phi}\right)_{r\theta}\left(\frac{\partial \phi}{\partial z}\right)_{xy}. \tag{8B.6}$$

Using the usual coordinate relations

$$x = r\sin\theta\cos\phi, \quad y = r\sin\theta\sin\phi, \quad z = r\cos\theta \tag{8B.7}$$

and

$$r = (x^2 + y^2 + z^2)^{1/2}, \tag{8B.8}$$

we find that

$$\begin{aligned}\tfrac{\partial r}{\partial z} &= \cos\theta \\ \tfrac{\partial \theta}{\partial z} &= -\tfrac{1}{r}\sin\theta \\ \tfrac{\partial \phi}{\partial z} &= 0,\end{aligned} \tag{8B.9}$$

where the first equation follows from (8B.8), the last follows becuase ϕ depends only on x and y, and the middle equation of (8B.9) follows from noting that

$\theta = \theta(z, r)$ so that

$$\left(\frac{\partial\theta}{\partial z}\right)_{xy} = \left(\frac{\partial\theta}{\partial z}\right)_{r} + \left(\frac{\partial\theta}{\partial r}\right)_{z}\left(\frac{\partial r}{\partial z}\right)_{xy}$$
$$= -\frac{1}{r\sin\theta} + \left(\frac{z}{r^2\sin\theta}\right)\cos\theta$$
$$= -\frac{1}{r\sin\theta}(1-\cos^2\theta)$$
$$= -\frac{1}{r}\sin\theta.$$

From (8B.6) and (8B.9) we have

$$\left(\frac{\partial u}{\partial z}\right)_{xy} = \cos\theta\left(\frac{\partial u}{\partial r}\right)_{\theta\phi} - \frac{1}{r}\sin\theta\left(\frac{\partial u}{\partial\theta}\right)_{r\phi}. \qquad (8B.10)$$

From (8B.5), (8B.7), and (8B.10) we have

$$\mathcal{S}\frac{\partial u(12)}{\partial\mathcal{S}} = \frac{1}{2}\left[r_{12}\frac{\partial u(12)}{\partial r_{12}} - 3z_{12}\cos\theta_{12}\frac{\partial u(12)}{\partial r_{12}} + 3\frac{z_{12}}{r_{12}}\sin\theta_{12}\frac{\partial u(12)}{\partial\theta_{12}}\right]$$
$$= -P_2(\cos\theta_{12})r_{12}\frac{\partial u(12)}{\partial r_{12}} + \frac{3}{2}\sin\theta_{12}\cos\theta_{12}\frac{\partial u(12)}{\partial\theta_{12}}, \qquad (8B.11)$$

where $P_2(\cos\theta_{12}) = \frac{1}{2}(3\cos^2\theta_{12} - 1)$ is the second-order Legendre polynomial.

8B.2 The term $\partial v/\partial\mathcal{S}$

The surface derivative of $v(1)$ is given by

$$\mathcal{S}\frac{\partial v(1)}{\partial\mathcal{S}} = \mathcal{S}\frac{\partial v(1)}{\partial\mathbf{r}_1}\cdot\frac{\partial\mathbf{r}_1}{\partial\mathcal{S}}. \qquad (8B.12)$$

From (8.86) we find that

$$\mathcal{S}\frac{\partial\mathbf{r}_1}{\partial\mathcal{S}} = \mathcal{S}\left(\frac{\partial x_1}{\partial\mathcal{S}}\mathbf{i} + \frac{\partial y_1}{\partial\mathcal{S}}\mathbf{j} + \frac{\partial z_1}{\partial\mathcal{S}}\mathbf{k}\right)$$
$$= \mathcal{S}\left(\frac{1}{2}\mathcal{S}^{-1/2}x_1'\mathbf{i} + \frac{1}{2}\mathcal{S}^{-1/2}y_1'\mathbf{j} - V\mathcal{S}^{-2}z_1'\mathbf{k}\right) \qquad (8B.13)$$
$$= \frac{1}{2}(x_1\mathbf{i} + y_1\mathbf{j} - 2z_1\mathbf{k}) = \frac{1}{2}(\mathbf{r}_1 - 3z_1\mathbf{k}).$$

From (8B.12) and (8B.13) we have

$$\mathcal{S}\frac{\partial v(1)}{\partial\mathcal{S}} = \frac{1}{2}\left[x_1\frac{\partial v(1)}{\partial x_1} + y_1\frac{\partial v(1)}{\partial y_1} - 2z_1\frac{\partial v(1)}{\partial z_1}\right], \qquad (8B.14)$$

which is Eq. (8.94).

Appendix 8C Derivation of the angular gradient equations (8.25) and (8.26)

To derive (8.25) and (8.26), which are the 'rotational' analogues of the 'translational' equations (8.20) and (8.24), we begin with the rotational analogue of (8.16). We apply a rotation Ω (which can be interpreted actively or passively—see Appendix A of Vol. 1, p. 446) to all coordinates (i.e. to all the $\mathbf{r}_i$ and all the ω_i). For the same reason that (8.16) is valid we must have

$$f(\mathbf{r}_1 + \Delta\mathbf{r}_{1\Omega}, \omega_1 + \Delta\omega_{1\Omega}; [v(\mathbf{r}_1\omega_1)]) = f(\mathbf{r}_1\omega_1; [v(\mathbf{r}_1 + \Delta\mathbf{r}_{1\Omega}, \omega_1 + \Delta\omega_{1\Omega})]), \tag{8C.1}$$

where $\Delta\mathbf{r}_{1\Omega}$ and $\Delta\omega_{1\Omega}$ are the changes undergone by $\mathbf{r}_1$ and ω_1, respectively, due to the rotation. Equation (8C.1) expresses the physical fact that f for the rotated configuration in the original potential has the same value as f for the original configuration but in the inversely rotated potential; see the analogous interpretation for translations following (8.13).

We now take Ω to be a small rotation. We recall Eq. (A.16a) (and (A.71b) for general shape molecules), which tells us how to perform a small rotation of angle α about axis $\mathbf{n}$; i.e. the small change in a general function $F(\omega)$ of some orientation variable ω due to this small rotation is $-\alpha\mathbf{n}\cdot\nabla_\omega F$, where ∇_ω is the angular gradient operator.

In our case, the function F depends on both the orientation of $\mathbf{r}_1$ (a point), for which the relevant rotation operator is $\mathbf{r}_1 \times \nabla_1$ (see (A.18b)), and on ω_1 (a spin or molecular orientation variable), for which the relevant operator is ∇_{ω_1} (see (A.18c) for the case of linear molecules). The total infinitesimal rotation operator, for rotating both variables $\mathbf{r}_1$ and ω_1, is

$$\nabla_\Omega(1) = \mathbf{r}_1 \times \nabla_1 + \nabla_{\omega 1}, \tag{8C.2}$$

i.e. the sum of an orbital and a spin part.[612] To see why the separate infinitesimal rotation operators in (8C.2) add, note that to $O(\alpha)$, applying the product operator $(1 - \alpha\mathbf{n}\cdot(\mathbf{r}_1 \times \boldsymbol{\nabla}_1)(1 - \alpha\mathbf{n}\cdot\boldsymbol{\nabla}_{\omega_1})$ to a function $F(\mathbf{r}_1\omega_1)$ of both variables $\mathbf{r}_1$ and ω_1 (which we do to change the function by a small amount in both variables) is equal to the result of applying $(1 - \alpha\mathbf{n}\cdot((\mathbf{r}_1 \times \boldsymbol{\nabla}_1) + \boldsymbol{\nabla}_{\omega_1})$. Alternatively, simply regard the term $(-\alpha)\mathbf{n}\cdot(\mathbf{r}_1 \times \boldsymbol{\nabla}_1)F + (-\alpha)\mathbf{n}\cdot\boldsymbol{\nabla}_{\omega_1}F$ as the first correction term in a Taylor expansion in both variables that change by small amounts $(-\alpha)$.

When we take the rotation Ω in (8C.1) to be the small rotation of angle α about direction $\mathbf{n}$, we see that to $O(\alpha)$ (8C.1) becomes (cf. how (8.14) becomes (8.15))

$$f(1) - \alpha\mathbf{n}\cdot\nabla_\Omega(1)f(1) = f(1) + \int \mathrm{d}2 \left(\frac{\delta f(1)}{\delta v(2)}\right)_{\alpha=0} (-\alpha\mathbf{n}\cdot\nabla_\Omega(2))v(2),$$

where $\mathrm{d}2 = \mathrm{d}\mathbf{r}_2\mathrm{d}\omega_2$, or, since $\mathbf{n}$ is arbitrary,

$$\nabla_\Omega(1)f(1) = \int \mathrm{d}2 \left(\frac{\delta f(1)}{\delta v(2)}\right)_{\alpha=0} \nabla_\Omega(2)v(2). \tag{8C.3}$$

Using the first Yvon equation (8.19) in (8C.3) gives

$$\nabla_\Omega(1)\ln f(1) = -\beta \int \mathrm{d}2 f(2)h(12)\nabla_\Omega(2)v(2) - \beta\nabla_\Omega(1)v(1), \tag{8C.4}$$

which is (8.25).

To derive the inverse relation, (8.26), we regard $v(1)$ as a unique functional of $f(1)$. In place of (8C.1) we have an equation with f and v interchanged on both sides. Taking Ω to be a small rotation thus leads to the inverse of (8C.3),

$$\nabla_\Omega(1)v(1) = \int \mathrm{d}2 \left(\frac{\delta v(1)}{\delta f(2)}\right)_{\alpha=0} \nabla_\Omega(2)f(2). \tag{8C.5}$$

Substituting $\delta v(1)/\delta f(2)$ from the second Yvon equation (8.23) gives

$$\nabla_\Omega(1)\ln f(1) = \int \mathrm{d}2 c(12)\nabla_\Omega(2)f(2) - \beta\nabla_\Omega(1)v(1), \tag{8C.6}$$

which is (8.26).

Appendix 8D The Maxwell equations for macroscopic electrostatics

In this appendix we derive (8.49), which is based on the derivation of the macroscopic Maxwell equations from the microscopic ones. The derivation of the complete set of macroscopic Maxwell equations, including magnetic and electrodynamic effects, is somewhat complicated;[613] here we are interested in deriving just the electrostatic equations (8.44). Our treatment is based on that of Jackson,[614] but with the difference that our averaging, denoted as always $< \ldots >$, denotes the usual statistical mechanical averaging, whereas the averaging of Jackson denotes a coarse-grained spacial averaging, i.e. over all points in a small neighbourhood of a point **r** of interest.

We begin with the Maxwell equations for microscopic electrostatics, (2.39) and (2.40), which, in order to distinguish them from the macroscopic equations (8.44), we write in the new notation

$$\nabla \times \mathbf{E}_{\mathrm{mic}} = 0, \qquad \nabla \cdot \mathbf{E}_{\mathrm{mic}} = 4\pi\rho_{\mathrm{mic}}, \tag{8D.1}$$

where $\mathbf{E}_{\mathrm{mic}}(\mathbf{r})$ is the electric field at **r** due to some fixed configuration of the molecules of the system. This fixed, or microscopic, configuration of the molecules gives rise to a charge density $\rho_{\mathrm{mic}}(\mathbf{r})$, which can be decomposed into separate contributions from each of the molecules $i = 1, 2, \ldots$, N; i.e.

$$\rho_{\mathrm{mic}}(\mathbf{r}) = \sum_i \sum_{c_i} q_c \delta(\mathbf{r} - \mathbf{r}_c), \tag{8D.2}$$

where Σ_i denotes a sum over the molecules, and $\sum_{c_i}$ denotes a sum over charges c belonging to molecule i; q_c is the charge on charge c. For simplicity of notation we have assumed in (8D.2) that the charge density of the molecules is totally discrete. As discussed in Chapter 2 of Vol. 1 (see p. 56) the molecular charge density in reality is partly discrete (that due to the nuclei) and partly continuous (that due to the electrons).

The macroscopic field $\mathbf{E}(\mathbf{r})$ is defined by

$$\mathbf{E}(\mathbf{r}) = \langle \mathbf{E}_{\text{mic}}(\mathbf{r}) \rangle , \tag{8D.3}$$

i.e. as the average electric field at point $\mathbf{r}$, where the average is the usual thermal average over the configurations (positions and orientations) of the molecules. Similarly the macroscopic charge density $\rho_{\text{ch}}(\mathbf{r})$ is defined by

$$\rho_{\text{ch}}(\mathbf{r}) = \langle \rho_{\text{mic}}(\mathbf{r}) \rangle. \tag{8D.4}$$

We note for future use the obvious relation

$$\boldsymbol{\nabla} \langle f \rangle = \langle \boldsymbol{\nabla} f \rangle, \tag{8D.5}$$

where f is any quantity that depends on the fixed point $\mathbf{r}$ in space, and on the molecular (or microscopic) variables; here $\nabla \equiv \partial/\partial\mathbf{r}$.

If we now simply average the microscopic equations (8D.1), and use (8D.5), we get the macroscopic equations (8.44). From (8D.4) and (8D.2) we find an explicit expression for the macroscopic charge density

$$\rho_{\text{ch}}(\mathbf{r}) = \left\langle \sum_{i} \sum_{c_i} q_c \delta(\mathbf{r} - \mathbf{r}_c) \right\rangle. \tag{8D.6}$$

We assume $\mathbf{r}_c$ and $\mathbf{r}_i$ (the position of molecule i) are not too different (see Fig. 8D.1), so that we can Taylor expand $\delta(\mathbf{r} - \mathbf{r}_c) \equiv \delta(\mathbf{r}_c - \mathbf{r})$ around $\mathbf{r}_c = \mathbf{r}_i$:

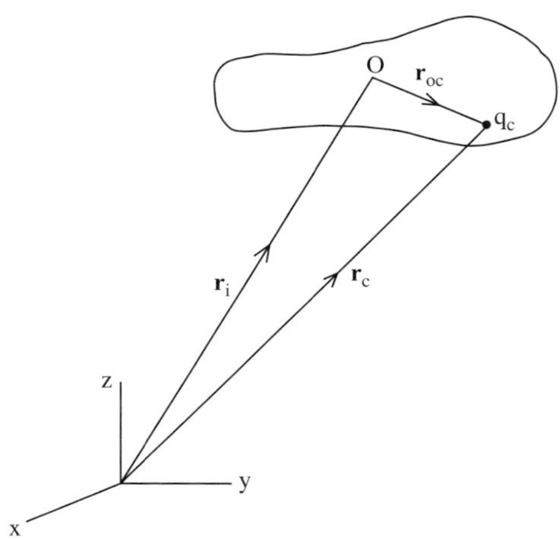

FIG. 8D.1 Geometry for Taylor expansion of a function of the variable $\mathbf{r}_c$. Here $\mathbf{r}_i$ is the position of the origin O chosen for molecule i, and q_c is the charge at point $\mathbf{r}_c$ in the molecule. The vectors $\mathbf{r}_i$ and $\mathbf{r}_c$ refer to some arbitrary set of space-fixed axes xyz.

$$\delta(\mathbf{r}_c - \mathbf{r}) = \delta(\mathbf{r}_i - \mathbf{r}) + \mathbf{r}_{0c} \cdot \nabla_i \delta(\mathbf{r}_i - \mathbf{r}) + \frac{1}{2}\mathbf{r}_{oc}\mathbf{r}_{oc} : \nabla_i \nabla_i \delta(\mathbf{r}_i - \mathbf{r}) + \ldots, \tag{8D.7}$$

where $\nabla_i \equiv \partial/\partial \mathbf{r}_{\mathrm{i}}$ and $\mathbf{r}_{oc} \equiv \mathbf{r}_c - \mathbf{r}_i$. Substituting (8D.7) into (8D.6), and using the fact that $\nabla_{\mathrm{i}}\delta(\mathbf{r}_{\mathrm{i}} - \mathbf{r}) = -\nabla\delta(\mathbf{r}_{\mathrm{i}} - \mathbf{r})$ and the relation (8D.5), we get

$$\rho_{\mathrm{ch}}(\mathbf{r}) = \left\langle \sum_i \delta(\mathbf{r}_i - \mathbf{r})) \sum_{c_i} q_c \right\rangle - \nabla \cdot \left\langle \sum_i \delta(\mathbf{r}_i - \mathbf{r}) \sum_{c_i} q_c \mathbf{r}_{oc} \right\rangle$$

$$+ \frac{1}{2}\nabla\nabla : \left\langle \sum_i \delta(\mathbf{r}_i - \mathbf{r}) \sum_{c_i} q_c \mathbf{r}_{oc}\mathbf{r}_{oc} \right\rangle + \ldots. \tag{8D.8}$$

We recognize the quantities $\sum_c q_c$, $\sum_c q_c \mathbf{r}_{0c}$, and $\sum_c q_c \mathbf{r}_{0c}\mathbf{r}_{0c}$ as the net charge q_i, dipole moment $\boldsymbol{\mu}_i$, and quadrupole moment $\boldsymbol{\Theta}_i$ of molecule i, respectively (see (2.54)). We assume the molecules are uncharged ($q_i = 0$). By introducing the macroscopic densities of dipole, quadrupole, etc., moment, denoted $\mathbf{P}(\mathbf{r})$, $\boldsymbol{\Theta}(\mathbf{r})$, etc., respectively (see definitions (8.50) and (8.51)) we rewrite (8D.8) as

$$\rho_{\mathrm{ch}}(\mathbf{r}) = -\nabla \cdot \mathbf{P}(\mathbf{r}) + \frac{1}{2}\nabla\nabla : \boldsymbol{\Theta}(\mathbf{r}) + \ldots. \tag{8D.9}$$

For a planar interface region, where $\rho_{\mathrm{ch}}(\mathbf{r})$, $\mathbf{P}(\mathbf{r})$, $\boldsymbol{\Theta}(\mathbf{r})$, etc., depend only on z, (8D.9) reduces to

$$\rho_{\mathrm{ch}}(z) = -P_z'(z) + \frac{1}{2}\Theta_{zz}''(z) + \ldots, \tag{8D.10}$$

where $P_z' \equiv \mathrm{d}P_z/\mathrm{d}z$ and $\Theta_{zz}'' \equiv \mathrm{d}^2\Theta_{zz}/\mathrm{d}z^2$. Equation (8D.10) is the one needed in the text (see (8.49)).

Using (8D.9) we can rewrite the macroscopic Maxwell equation $\nabla\cdot\mathbf{E} = 4\pi\rho_{\mathrm{ch}}$ as

$$\nabla\cdot(\mathbf{E} + 4\pi\mathbf{P}_{\mathrm{eff}}) = 0, \tag{8D.11}$$

where $\mathbf{P}_{\mathrm{eff}}$ is an effective polarization

$$\mathbf{P}_{\mathrm{eff}} = \mathbf{P} - \frac{1}{2}\nabla \cdot \boldsymbol{\Theta} + \ldots. \tag{8D.12}$$

The form (8D.11) is often used (see, e.g., text below (8.38)).

Appendix 8E Simplified derivation of quadrupolar surface potential term

Here we derive the quadrupolar term $\Delta\phi_\Theta = -2\pi(\rho_\ell - \rho_g)\Theta$ in the surface potential (8.54) by a simple method[615] that illuminates why only the mean or trace part (Θ) of the quadrupole moment $\boldsymbol{\Theta} = \sum_i q_i \mathbf{r}_i \mathbf{r}_i$ (see (2.54) of Vol.1) contributes to $\Delta\phi$. We show that $\Delta\phi_\Theta$ originates from the contact term of the molecular quadrupolar field.

Contact terms for dipolar fields are well known[615,616]—see Appendix 10E—but have been discussed for quadrupolar fields only more recently.[615,617] Consider the potential $\phi(\mathbf{r})$ due to a pure quadrupole at the origin, given by Eq. (2.52) of Vol. 1 as

$$\phi(\mathbf{r}) = \frac{1}{2}\mathbf{\Theta} : \nabla\nabla\frac{1}{r}, \tag{8E.1}$$

where $\mathbf{\Theta}$ is the complete (i.e. 'traceful') quadrupole moment given by (2.54). Previously we were interested only in the long-range part of $\phi(\mathbf{r})$, where $\nabla\nabla(1/r) = (3\hat{\mathbf{r}}\hat{\mathbf{r}} - \mathbf{1})/r^3$, $r \neq 0$, see (2.56), where $\hat{\mathbf{r}} = \mathbf{r}/r$ is a unit vector along $\mathbf{r}$. The complete expression for $\nabla\nabla(1/r)$, valid for all $\mathbf{r}$ including $\mathbf{r} = 0$, is

$$\nabla\nabla\frac{1}{r} = -\frac{4\pi}{3}\delta(\mathbf{r})\mathbf{1} + (3\hat{\mathbf{r}}\,\hat{\mathbf{r}} - \mathbf{1})/r^3 \tag{8E.2a}$$

$$\equiv \nabla\nabla^{(0)}\frac{1}{r} + \nabla\nabla^{(2)}\frac{1}{r}, \tag{8E.2b}$$

where $\delta(\mathbf{r})$ is the Dirac delta function. The term $\nabla\nabla^{(0)}(1/r) = -(4\pi/3)\delta(\mathbf{r})\mathbf{1}$ is the contact term, and contributes only for $\mathbf{r} = 0$. To derive (8E.2a) heuristically,[618] note that any symmetric second-rank tensor $\mathbf{A}$ can be written as (see (A.217) or (C.21) of Vol. 1)

$$\mathbf{A} = \mathbf{A}^{(0)} + \mathbf{A}^{(2)}, \tag{8E.3}$$

where the $\ell = 0$ part $\mathbf{A}^{(0)}$ is the isotropic or trace part of $\mathbf{A}$, i.e.

$$\mathbf{A}^{(0)} \equiv \left(\frac{\mathrm{Tr}(\mathbf{A})}{3}\right)\mathbf{1}, \tag{8E.4}$$

where $\mathrm{Tr}(\mathbf{A})$ denotes the trace $\sum_\alpha A_{\alpha\alpha}$ and $\mathbf{1}$ is the unit tensor. The $\ell = 2$ part $\mathbf{A}^{(2)}$ is the anisotropic or traceless part,

$$\mathbf{A}^{(2)} \equiv \mathbf{A} - \mathbf{A}^{(0)}. \tag{8E.5}$$

The original second-rank symmetric tensor $\mathbf{A}$, with six independent components $A_{\alpha\beta}$, has been decomposed into an $\mathbf{A}^{(0)}$ part with one independent component, and an $\mathbf{A}^{(2)}$ part with five. The notation $\ell = 0, 2$ reminds us that $\mathbf{A}^{(0)}$ and $\mathbf{A}^{(2)}$ are irreducible tensors, the Cartesian analogues of the spherical tensors Y_0 and Y_2 (see § A.4.3 of Vol. 1). We have previously seen that for $r \neq 0$, $\nabla\nabla(1/r) = (3\hat{\mathbf{r}}\,\hat{\mathbf{r}} - \mathbf{1})/r^3 \equiv \nabla\nabla^{(2)}(1/r)$ since $(3\hat{\mathbf{r}}\,\hat{\mathbf{r}} - \mathbf{1})$ is an $\ell = 2$ (symmetric traceless) tensor. To obtain the contact term $\nabla\nabla^{(0)}(1/r)$, note that $\mathrm{Tr}[\nabla\nabla(1/r)] \equiv \nabla^2(1/r)$, which is equal to $-4\pi\delta(\mathbf{r})$, see Eq. (B.81) of Vol. 1.

We write out the double contraction in (8E.1) using the relation for symmetric tensors (which have no $\ell = 1$ part)

$$\mathbf{A} : \mathbf{B} = \mathbf{A}^{(0)} : \mathbf{B}^{(0)} + \mathbf{A}^{(2)} : \mathbf{B}^{(2)}. \tag{8E.6}$$

The cross terms $\mathbf{A}^{(0)} : \mathbf{B}^{(2)}$ and $\mathbf{A}^{(2)} : \mathbf{B}^{(0)}$ are easily shown to vanish. This gives

$$\phi(\mathbf{r}) = \phi_0(\mathbf{r}) + \phi_2(\mathbf{r}), \tag{8E.7}$$

where

$$\phi_0(\mathbf{r}) = \frac{1}{2}\boldsymbol{\Theta}^{(0)} : \nabla\nabla^{(0)}\frac{1}{r}, \tag{8E.8}$$

$$\phi_2(\mathbf{r}) = \frac{1}{2}\boldsymbol{\Theta}^{(2)} : \nabla\nabla^{(2)}\frac{1}{r}. \tag{8E.9}$$

The traceless quadrupole moment $\boldsymbol{\Theta}^{(2)}$ is related to the standard traceless quadrupole moment $\mathbf{Q}$ (see Eq. (2.58) of Vol. 1) by $\boldsymbol{\Theta}^{(2)} = (2/3)\mathbf{Q}$. As we shall see presently, it is $\boldsymbol{\Theta}^{(0)}$ that is relevant for the surface potential problem.

Using the definitions of $\boldsymbol{\Theta}^{(0)}$ and $\nabla\nabla^{(0)}(1/r)$ we write (8E.8) explicitly as

$$\phi_0(\mathbf{r}) = \frac{1}{2}(\Theta\mathbf{1}) : \left(-\frac{4\pi}{3}\delta(\mathbf{r})\mathbf{1}\right),$$

or

$$\phi_o(\mathbf{r}) = -2\pi\Theta\delta(\mathbf{r}), \tag{8E.10}$$

using $\mathbf{1} : \mathbf{1} = 3$, with the mean quadrupole moment Θ defined by $\Theta = \mathrm{Tr}(\boldsymbol{\Theta})/3 = (1/3)\sum_i q_i r_i^2$. We shall see that (8E.10) is the central result.

We now consider the field due to N identical quadrupolar molecules, with orientations $\omega_1, \omega_2, \ldots, \omega_N$ and positions $\mathbf{r}_1, \mathbf{r}_2, \ldots, \mathbf{r}_N$, in a volume V. The macroscopic or mean potential $\bar{\phi}(\mathbf{r})$ at $\mathbf{r}$ in a uniform bulk fluid due to these molecules is

$$\bar{\phi}(\mathbf{r}) = \left\langle \sum_i \phi(\mathbf{r} - \mathbf{r}_i) \right\rangle, \tag{8E.11}$$

where $\phi(\mathbf{r} - \mathbf{r}_i)$ is given by (8E.7) and $\langle \ldots \rangle$ is a configurational average over the positions and orientations of the molecules. The contribution $\langle \phi_2(\mathbf{r} - \mathbf{r}_i) \rangle$ is easily seen to vanish. Since $\bar{\phi}(\mathbf{r})$ is independent of $\mathbf{r}$ for a uniform bulk fluid, consider $\bar{\phi}(0)$. The average $\langle \phi_2(-\mathbf{r}_i) \rangle$ involves the unweighted average over positional orientations $\hat{\mathbf{r}}_i$ of $(3\hat{\mathbf{r}}_i\hat{\mathbf{r}}_i - 1)$, which is zero. This is easily seen from a calculation, giving $\langle \hat{\mathbf{r}}_i\hat{\mathbf{r}}_i \rangle_{\mathbf{r}_i} = (1/3)\mathbf{1}$, or by recalling that the corresponding spherical harmonic $Y_2(\hat{\mathbf{r}}_i)$ has zero orientational average. The contact term, on the other hand, gives a nonvanishing result, since from (8E.10) the unweighted positional average over $\mathbf{r}_i$ is

$$\langle \phi_0(\mathbf{r} - \mathbf{r}_i) \rangle_{\mathbf{r}_i} = \int \frac{d\mathbf{r}_i}{V}(-2\pi\Theta\delta(\mathbf{r} - \mathbf{r}_i)) = -\frac{2\pi\Theta}{V}. \tag{8E.12}$$

The result (8E.12) is independent of orientation ω_i, so that the average over ω_i can be omitted. Since each of the N identical molecules gives the contribution (8E.12) to $\bar{\phi}(\mathbf{r})$, we have

$$\bar{\phi}(\mathbf{r}) = -2\pi\rho\Theta, \tag{8E.13}$$

where $\rho = N/V$ is the number density. Note that $\bar{\phi}(\mathbf{r})$ is independent of $\mathbf{r}$, as expected.

For a two-phase system with a vapour–liquid interface, application of (8E.13) separately to the bulk gas and liquid phases gives for the interface potential $\Delta\bar{\phi} \equiv \bar{\phi}_\ell - \bar{\phi}_g$

$$\Delta\bar{\phi} = -2\pi(\rho_\ell - \rho_g)\Theta, \tag{8E.14}$$

as in the text, Eq. (8.54), where there the simpler notation $\Delta\phi$ was used for $\Delta\bar{\phi}$.

Thus we see that the surface potential arises entirely from the contact term of the quadrupolar potential; $\mathbf{\Theta}^{(2)}$ (or $\mathbf{Q}$) does not contribute. We have assumed here purely quadrupolar molecules for simplicity. If the molecular octopole and higher moments are taken into account, they are found to give zero contribution. The octopolar field is (see (2.52)) $\phi(\mathbf{r}) = -(1/6)\mathbf{O}\vdots\nabla\nabla\nabla(1/r)$, where $\mathbf{O}$ is the complete octopole moment given in (2.54). The third-rank symmetric tensors $\mathbf{O}$ and $\nabla\nabla\nabla(1/r)$ have ten independent components, and each can be decomposed into $\ell = 3$ and $\ell = 1$ parts, with seven and three independent components, respectively. Again the full contraction $\mathbf{O}\vdots\nabla\nabla\nabla(1/r)$ splits into a sum of contributions from irreducible parts, here $\ell = 3$ and $\ell = 1$. The $\ell = 3$ noncontact term arising from $\nabla\nabla\nabla(1/r)$ gives no contribution as before, since it involves an angular average of $Y_3(\hat{\mathbf{r}})$. The $\ell = 1$ contact term, here involving $\nabla\delta(\mathbf{r})$, also does not contribute, since $\nabla\delta(\mathbf{r})$ vanishes when integrated. The higher multipoles are found similarly not to contribute. If the molecules have dipole moments or net charges, their contributions to $\Delta\bar{\phi}$ are more easily found by the method given in the text as there are no contact terms. Again we see that only the monopole, dipole, and quadrupole moments of the molecules contribute to the surface potential, and that only the trace part of the quadrupole moment contributes. The quadrupolar contribution is remarkable in its being independent of the nature of the intermolecular forces, and independent of the structure of the gas–liquid interface. In contrast the monopole and dipole terms do depend on these details.

References and notes

1. Hardy[2] speculated on the existence of surface alignment in 1912. Indirect experimental evidence for its existence, based on liquid–liquid interfacial tensions and on the properties of thin oil films on water, was obtained by Harkins *et al.*[3] and Langmuir.[4] Electrochemical evidence for oriented organic molecules, adsorbed at a mercury–water interface, was obtained by Gouy.[5] Early theoretical work was based on simplified models[6–8] or thermodynamic[9] arguments. More recent evidence for molecular alignment is based on (a) statistical mechanical theories (see §§ 8.5 to 8.8), (b) computer 'experiments' (see §§ 8.5 and 8.9), and (c) direct experiments (see § 8.9).
2. Hardy, W.B. *Proc. Roy. Soc.* **A86**, 634 (1912); **A88**, 303 (1913).
3. Harkins, W.D., Brown, F.E., and Davies, E.C.H. *J. Am. Chem. Soc.* **39**, 354 (1917); Harkins, W.D., Davies, E.C.H., and Clark, G.L. ibid. **39**, 541 (1917). See also Harkins, W.D. *The Physical Chemistry of Surface Films*, Reinhold, New York (1952), p. 18. Other early work is reviewed by de Boer, J.H. *The Dynamical Character of Adsorption*, Oxford University Press, Oxford, (1953), pp. 150, 168, 217.
4. Langmuir, I. *Met. Chem. Eng.* **15**, 468 (1916); *J. Am. Chem. Soc.* **39**, 1848 (1917).
5. Gouy, G. *Ann. Chim. Phys.* **8**, 294 (1908).

6. Frenkel, J. *Kinetic Theory of liquids*, Oxford University Press, Oxford (1946), p. 353; Dover reprint (1955).
7. Fletcher, N.H. *Phil. Mag.*, Series 8, **18**, 1287 (1968); *Physics and Chemistry of Ice*, ed. E. Whalley, S.J. Jones, and J.W. Gold, Roy. Soc. Can., Ottawa (1973), p. 132.
8. Sparnaay, M.J. *The Electrical Double Layer*, Pergamon Press, Oxford (1972), p. 120.
9. Sanfeld, A., *Introduction to the Thermodynamics of Charged and Polarized Layers*, Wiley, New York (1968), p. 151.
10. Harasima, A. *Adv. Chem. Phys.* **1**, 203 (1958).
11. Abraham, F.F. *Phys. Rep.* **53**, 95, (1979).
12. Evans, R. *Adv. Phys.* **28**, 143 (1979).
13. Navascués, G. *Rep. Progr. Phys.* **42**, 1131 (1979).
14. Croxton, C.A. *Statistical Mechanics of the Liquid Surface*, Wiley, Chichester (1980).
15. Dickenson, E., and Lal, M. *Adv. Mol. Relaxation Processes* **17**, **1** (1980).
16. Jhon, N.S., Dahler, J.S., and Desai, R.C. *Adv. Chem. Phys.* **46**, 279 (1981).
17. Nicholson, D., and Parsonage, N.G. *Computer Simulation and the Statistical Mechanics of Adsorption*, Academic Press, London (1982).
18. Rowlinson, J.S., and Widom, B. *Molecular Theory of Capillarity*, Chapter 4, Clarendon Press, Oxford (1982).
19. Davis, H.T., and Scriven, L.E. *Adv. Chem. Phys.* **49**, 357 (1982).
20. Bedeaux, D. *Adv. Chem. Phys.* **64**, 47 (1986).
21. Croxton, C.A., ed., *Fluid Interfacial Phenomena*, Wiley, Chichester (1986). See the reviews on the intrinsic interface (J.K. Percus and G.O. Williams), wetting (D.E. Sullivan and M.M. Telo da Gama), capillary waves (R.C. Desai and M. Grant), nematic liquid crystals (T.J. Sluckin and A. Poniewierski), statistical mechanics of drops (J.R. Henderson), and experimental studies of liquid interfaces (D. Beaglehole).
22. Davis, H.T. *Statistical Mechanics of Phases, Interfaces and Thin Films*, VCH, Weinheim (1996).
23. Hansen, J.-P. and McDonald, I.R. *Theory of Simple Liquids*, 3rd edn, Chapter 6, Academic Press, London (2006).
24. Blum, L., ref. 21.
25. See, e.g. ref. 9.
26. Henderson, D. *Prog. in Surf. Sci.* **13**, 197 (1973).
27. Carnie, S.L., and Torrie, G.M. *Adv. Chem. Phys.* **56**, 141 (1984).
28. March, N.H., and Tosi, M.P. *Coulomb Liquids*, Chapter 8, Academic Press, London (1984).
29. Croxton, C.A., ref. 21.
30. Lipotov, Y., and Feinerman, A. *Adv. Colloid Int. Sci.* **11**, 195 (1979).
31. Fredrickson, G.H. *The Equilibrium Theory of Inhomogeneous Polymers*, Clarendon Press, Oxford (2006).
32. Rice, S.A., Gryko, J. and Mohanty, U. ref. 21.
33. Weinberg, W.H. *Ann. Rev. Phys. Chem.* **34**, 217 (1983).
34. Brown, G.H., and Walker, J.J. *Liquid Crystals and Biological Structures*, Academic Press, New York (1979); Israelachvili, J. *Intermolecular and Surface Forces*, 2nd edn, Academic Press, New York (1991); Merz, K.M., and Roux, eds, *Biological Membranes*, Birkhauser, Boston (1996).
35. De Gennes, P.G., and Prost, J. *The Physics of Liquid Crystals*, 2nd edn, Clarendon Press, Oxford (1993). See also Sluckin, T., and Poniewierski, A., ref. 21.
36. A factor Ω^{-1} was inadvertently omitted on the right-hand side of (3.191) in the first printing of Vol. 1.
37. Thompson, S.M., Gubbins, K.E., and Haile, J.M. *J. Chem. Phys.*, **75**, 1325 (1981).
38. Hill, T.L. *Statistical Mechanics*, McGraw Hill, New York (1956), p.193.
39. Chipot, C., and Pohorille, A., eds, *Free Energy Calculations*, Springer, Berlin (2007).

40. Dang, L.X. *J. Phys. Chem. B* **106**, 10388 (2002). The potentials used were AMBER type with 4 H_2O sites; Case, D.A. *et al.*, AMBER6, University of California, San Francisco (1999).
41. Chang, T-M., and Dang, L.X. *Chem. Rev.* **106**, 1305 (2006).
42. Vivcharuk, V., Tomberli, B., Tolokh, I.S., and Gray, C.G., ref. 526.
43. Gubbins, K.E. *Chem. Phys. Lett.* **76**, 329 (1980). Equations (15) and (16) of this paper contain an error. Corrected equations are given by Tarazona, P., and Evans, R. *Chem. Phys. Lett.* **97**, 279 (1983). See also Appendix 8C at the end of this chapter.
44. Wertheim, M.S. *J. Chem. Phys.* **65**, 2377 (1976).
45. See, for example: Telo da Gama, M.M. *Mol. Phys.* **52**, 585 (1984); ibid. **52**, 611.
46. Thurtell, J.H., Telo da Gama, M.M., and Gubbins, K.E. *Mol. Phys.* **54**, 321 (1985). The results for the nematic liquid–vapour surface tension in this paper are incorrect. Correct values are given in ref. 169.
47. Even for symmetrical linear molecules and a planar interface, other order parameters are needed for interfaces involving liquid crystals as one of the phases,[31] because the molecules have preferred orientations in the bulk liquid crystal phase. In the absence of external fields three independent orientational order parameters are needed, η, ν, and σ, with η given by (8.30) and ν and σ given by:

$$\nu(\mathbf{r}_1) = \int d\omega_1 \sin 2\theta_1 \cos\varphi_1 [f(\mathbf{r}_1\omega_1)/f(\mathbf{r}_1)]$$

$$\sigma(\mathbf{r}_1) = \int d\omega_1 \sin^2\theta_1 \cos 2\varphi_1 [f(\mathbf{r}_1\omega_1)/f(\mathbf{r}_1)].$$

When the molecules in the bulk nematic are aligned in the z direction (perpendicular to the interface), only η is nonzero. If the bulk alignment is parallel to the surface, ν is zero while η and σ can be nonzero. All three order parameters are required to describe other orientations for the bulk molecules. For *uns*ymmetrical linear molecules, different definitions of η, ν, and σ must be used (see Telo da Gama, M.M., and Gubbins, K.E., *Proc. 9th Thermophysical Props. Conf.*, Boulder, Co, June 1985).
48. The polarization profile P_z (z) used here should be distinguished from the 'polarization' F_1 defined in (8.33). The two quantities are related by (8.43) for rigid dipolar fluids.
49. Jackson, J.D., *Classical Electrodynamics*, 2nd edn, Wiley, New York (1975), p. 226.
50. Ref. 49, p. 230
51. Wilson, M.A., Pohorille, A., and Pratt, L.R. *J. Chem. Phys.* **90**, 5211 (1989).
52. Our (8.48) corresponds to Eq. (1) of Wilson *et al.*;[51] however, they use the opposite convention to us for the direction of positive z.
53. Flygare, W.H. *Molecular Structure and Dynamics*, Prentice-Hall, Englewood Cliffs, NJ (1978), p. 384.
54. Flygare, W.H., and Benson, R.C. *Mol. Phys.* **20**, 225 (1971), and references therein.
55. See for example, Snyder, L.C., and Basch, H. *Molecular Wave Functions and Properties*, Wiley, New York (1972).
56. We estimate $\Delta\phi \approx -2\pi\Theta\rho_\ell$ for liquid argon at the triple point ($T_t = 84\,\mathrm{K}$) as follows. The liquid density is $\rho_\ell = 2.14 \times 10^{28}$ molecules/m^3 (see Hunter, P.J., and Rowlinson, J.S., in *Simple Dense Fluids*, ed. H.L. Frisch and Z.W. Salsburg, Academic Press, New York (1968), p. 1). The quadrupole moment Θ is proportional to the diamagnetic susceptibility (per atom) χ_d since $\chi_d = -(r_e/6)\left\langle \sum_i r_i^2 \right\rangle$, where $r_e \equiv e^2/m_ec^2 = 2.82 \times 10^{-13}$ cm is the classical electron radius, e is(negative) electron charge, $\sum_i$ denotes a sum over the electrons of the atom, with radial coordinates r_i, and $< \ldots >$ is an expectation value over the ground electronic

state of the atom. From the experimental value $\chi_d = -3 \times 10^{-29}$ cm^3/atom (Van Vleck, J.H. *The Theory of Electric and Magnetic Susceptibilities*, Oxford University Press, London (1932), p. 215), we thus find that $\left\langle \sum_i r_i^2 \right\rangle = 6.38 \times 10^{-16}$ cm^2 and hence $\Theta = -3.4 \times 10^{-39}$ Cm^2. To obtain $\Delta\phi$ in volts (SI) we use the esu formula $\Delta\phi \approx -2\pi\theta\rho_\ell$ divided by $4\pi\varepsilon_0 = 1.11 \times 10^{-10}$ C^{2}N^{-1}m^{-2}, with Θ and ρ_ℓ in SI as given above. We find that $\Delta\phi = 4.57$V, a surprisingly large value. We are not aware of any measurements.

A similar calculation (data again from Frisch-Salsburg and Van Vleck) for liquid He at $T = 5$ K gives the more modest value $\Delta\phi = 0.32\,V$. This is about $10 \times 2 = 20$ times smaller than that for Ar, with the 10 due to the much smaller quadrupole moment and the 2 due to the smaller density.

For liquid N_2 at $T = 77$ K we find that $\Delta\phi = 4.01\,V$. Here $\Theta \equiv \Theta_n + \Theta_e$ has both nuclear (Θ_n) and electronic (Θ_e) contributions. From the equilibrium bond length (= 1.098 Å) we calculate $\Theta_n = +22.4 \times 10^{-40}$ Cm^2, and a quantum calculation (ref. 55, or Amos, R.D. *Mol. Phys.* **39**, 1 (1980)) gives the estimate $\Theta_e = -59.2 \times 10^{-40}$ Cm^2, so that $\Theta = -36.8 \times 10^{-40}$ Cm^2. The liquid density is $\rho_\ell = 1.74 \times 10^{28}$ molecules/m^3.

57. The TIP4P water model has a central LJ core centred on the O atom, and point charges on the O and H sites. See Jorgensen, W.L., Chandrasekhar, J., Madura, J.D., Impey, R.W., and Klein, M.L. *J. Chem. Phys.* **79**, 926 (1983). (TIP4P stands for transferable intermolecular potential—4 points (sites)).
58. Gomer, R., and Tryson, G. *J. Chem. Phys.* **66**, 4413 (1977); Madden, W.G., Gomer, R., and Mandell, M.J. *J. Phys. Chem.* **81**, 2652 (1977); Farrell, J.R., and McTigue, P. *J. Electroanal. Chem.* **139**, 37 (1982); Borazio, A., Farrell, J.R., and McTigue, P. ibid. **193**, 103 (1985); Farrell, J.R., and McTigue, P. ibid. **163**, 129 (1984).
59. Barraclough, C.G., McTigue, P.T., and Ng, Y.L. *J. Electroanal. Chem.* **329**, 9 (1992).
60. Matsumoto, M., and Kataoka, Y. *J. Chem. Phys.* **90**, 2398 (1989); ibid. **95**, 7782 (1991).
61. Yang, B., Sullivan, D.E., and Gray, C.G. *J. Chem. Phys.* **95**, 7777 (1991).
62. Randles, J.E.B. *Phys. Chem. Liquids* **7**, 107 (1977); Case, B., and Parsons, R. *Trans. Farday Soc.* **63**, 1224 (1967).
63. See, e.g. Alessi, N., Tolokh, I.S., Goldman, S., and Gray, C.G. *Mol. Phys.* **103**, 2381 (2005) for a calculation of solvent shift of the mean vibrational frequency of liquid N_2.
64. Wick, C.D., Dang, L.X., and Jungwirth, P. *J. Chem. Phys.* **125**, 024706 (2006).
65. Adam, N.K. *The Physics and Chemisty of Surfaces*, 3rd edn, Oxford University Press, Oxford (1941). See preface to the Dover reprint (1968), and references there to work of R.C. Brown (1947) and C. Gurney (1947). See also Berry, M.V. *Phys. Ed.* **6**, 79 (1971); Walton, A.J. *Phys. Ed.* **8**, 491 (1972), and *Three Phases of Matter*, 2nd edn, Oxford University Press, Oxford (1983), p. 448.
66. A good discussion of various routes to the surface tension, with emphasis on the most promising routes for molecular simulations, is given in Gloor, G.J., Jackson, G., Blas, F.J. and de Miguel, E. *J. Chem. Phys.* **123**, 134703 (2005).
67. Kirkwood, J.G., and Buff, F.P. *J. Chem. Phys.* **17**, 338 (1949).
68. Buff, F.P. *Z. Electochim.* **56**, 311 (1952).
69. Triezenburg, D.G., and Zwanzig, R. *Phys. Rev. Lett.* **28**, 1183 (1972).
70. Gibbs, J.W. *The Collected works of J. Willard Gibbs*, Longmans, Green, New York (1928); Dover reprint (1961).
71. An alternative approach is to define the surface region as a third phase, by introducing two dividing surfaces that separate the surface phase from the bulk phases I and II, respectively. This approach has been used by Bakker[72] and Guggenheim,[73] among others. The surface phase cannot be defined unambiguously, so the method offers no

distinct advantage over the more widely used Gibbs approach.
72. Bakker, G. *Kapillarität und Oberflächenspannung, Vol. 6 of Handbuch der Experimentalphysik*, ed. W. Wien, F. Harms, and H. Lenz, Chapter 10, Akad. Verlags., Leipzig (1928).
73. Guggenheim, E.A. *Trans. Faraday Soc.* **36**, 397 (1940); Thermodynamics, 5th edn, North-Holland, Amsterdam (1967), p. 45.
74. As an example, E.A. Guggenheim and N.K. Adam (*Proc. Roy. Soc.* A **139**, 218 (1933) have shown by calculation that for an equimolar mixture of water and alcohol the adsorption of alcohol varies from 9.5×10^{-11} to 42.5×10^{-11} mol cm^{-2} as the dividing surface is moved by about 2.5 Å. On shifting the dividing surface by 10 Å the adsorption becomes negative, -130×10^{-11} mol cm^{-2} (R. Defay and I. Prigogine, *Surface Tension and Adsorption*, p. 25, Longman, Green & Co., London (1966)).
75. Gubbins, K.E., and Haile, J.M. *Improved Oil Recovery by Surfactant and Polymer Flooding*, ed. D.O. Shah and R.S. Schechter, Academic Press, New York (1977).
76. Chandrasekhar, S., *Liquid Crystals*, Cambridge University Press, Cambridge (1977), pp. 90–97.
77. Lee, D.J., Telo de Gama, M.M., and Gubbins, K.E. *J. Phys. Chem.* **89**, 1514 (1985).
78. Jasper, J.J., *J. Phys. Chem. Ref. Data*, **1**, 841 (1972).
79. Volyak, L.D. *Dokl. Akud. Nauk. SSSR* **74**, 307 (1950); Grigull, K., and Bach, J., *Brensst.-Wärme-Kraft* **18**, 73 (1966). See also p. 11 of Rowlinson and Widom.[18]
80. Krishnaswamy, S., and Shashidhar, R. *Proceedings of the International Liquid Crystals Conference*, Bangalore, December 1973, Pramana supplement I, p. 247; *Molec. Cryst. Liq. Cryst.* **35**, 253 (1976); ibid. 38, 353 (1977).
81. Widom, B. *Statistical Mechanics*, Cambridge University Press, Cambridge (2002), p.170.
82. Langmuir, I. *Proc. Roy. Soc.* **A 170**, 1 (1934); Adamson, A.W. *Physical Chemistry of Surfaces*, 5th edn, Wiley, New York (1990), p. 88.
83. Thompson, S.M., Gubbins, K.E., Walton, J.P.R.B., Chantry, R.A.R., and Rowlinson, J.S. *J. Chem. Phys.* **81**, 530 (1984).
84. Alba-Simionesco, C., Coasne, B., Dosseh, G., Dudziak, G., Gubbins, K.E., Radhakrishnan, R., and Śliwinska-Bartkowiak, M. *J. Phys.: Cond. Matter* **18**, R15–R68 (2006).
85. Bogoliubov, N.N. *J. Phys. USSR* **10**, 256 (1946); see also Bogoliubov, N.N. in *Studies in Statistical Mechanics*. Vol. 1, Part A, ed. J. de Boer and G.E. Uhlenbeck, North-Holland, Amsterdam (1962).
86. Green, H.S., *Proc. Roy. Soc. London* **A159**, 229 (1947).
87. Gray, C.G., and Gubbins, K.E. *Mol. Phys.* **30**, 179 (1975).
88. Davis, H.T. *J. Chem. Phys.* **62**, 3412 (1975).
89. Sluckin, T.J. *Mol. Phys.* **43**, 817 (1981).
90. Weisskopf, V. *Am. J. Phys.* **53**, 19, 1131 (1985); Turner, G.J. *Am. J. Phys.* **53**, 1130 (1985).
91. Grant, M., and Desai, R.C. *J. Chem. Phys.* **72**, 1482 (1980).
92. Grant, M., and Desai, R.C. *Mol. Phys.* **43**, 1031 (1981).
93. Fowler, R.H. *Proc. R. Soc. London* **A159**, 229 (1937).
94. Freeman, K.S.C., and McDonald, I.R. *Mol. Phys.* **26**, 529 (1973).
95. Buff, F.P., and Lovett, R.A., in *Simple Dense Fluids*, ed. H.L. Frisch and Z.W. Salsburg, Academic, New York (1968), Chapter 2.
96. Lee, J.K., Barker, J.A., and Pound, G.M., *J. Chem. Phys.* **60**, 1976 (1974).
97. Chapela, G.A., Saville, G., and Rowlinson, J.S. *Faraday Disc. Chem. Soc.* **59**, 22 (1975).
98. Miyazaki, J., Barker, J.A., and Pound, G.M. *J. Chem. Phys.* **64**, 3364 (1976).
99. Toxvaerd, S. *J. Chem. Phys.* **55**, 3116 (1971).
100. Barker, J.A., Fisher, R.A., and Watts, R.O. *Mol. Phys.* **21**, 657 (1971).

101. Berry, M.V., Durrans, R.F., and Evans, R. *J. Phys.* **A5**, 166 (1975).
102. Yvon, *J. Proc. IUPAP Symp. on Thermodynamics*, Brussels (1948), p. 9.
103. Lovett, R., De Haven, P.W., Vieceli, J.J., and Buff, F.P. *J. Chem. Phys.* **58**, 1880 (1973).
104. Gubbins, K.E. *Revista de Fisica, ser.* **A3**, 129 (1981).
105. Gubbins, K.E., and Thompson, S.M. *Faraday Symp. Chem. Soc.* **16**, 60 (1981).
106. It is also possible to carry out the derivation using an external potential that causes the preferred molecular orientation to change with the surface curvature; see Sluckin[89] and Tarazona and Evans (Tarazona, P., and Evans, R. *Chem. Phys. Lett.* **97**, 279 (1983)). This rotation of axes leads to no extra energy proportional to area, so that the formula for surface tension is unchanged.
107. Schofield, P. *Chem. Phys. Lett.* **62**, 413 (1979).
108. Waldor, M.H., and Wolf, D.E. *J. Chem. Phys.* **85**, 6082 (1986).
109. Kirkwood, J.G., and Buff, F.P. *J. Chem. Phys.* **17**, 338 (1945).
110. Irving, J.H., and Kirkwood, J.G. *J. Chem. Phys.* **18**, 817 (1950).
111. Berry, M.V. *Surface Science*, **1**, 291 (1975), I.A.E.A., Vienna.
112. Hemingway, S.J., Henderson, J.R., and Rowlinson, J.S. *Faraday Symp. Chem. Soc.* **16**, 33 (1981).
113. Schofield, P., and Henderson, J.R. *Proc. Roy. Soc. Lond.* **A379**, 231 (1982).
114. Walton, J.P.R.B., and Gubbins, K.E. *Mol. Phys.* **55**, 679 (1985); see also ibid. **58**, 1013 (1986).
115. Here is a more formal argument. Since $\mathcal{U}(\mathbf{r}^N\omega^N)$ is invariant under arbitrary translations $\mathbf{r}_i \to \mathbf{r}_i + \mathbf{a}$, (all i, arbitrary $\mathbf{a}$), and since the generator of translations in the direction $\hat{\mathbf{a}}$ is $\sum_i \hat{\mathbf{a}} \cdot \nabla_\mathrm{i}$ we see that $\sum_i \hat{\mathbf{a}} \cdot \nabla_i \mathcal{U} = \hat{\mathbf{a}} \cdot \sum_i \nabla_i \mathcal{U} = 0$. Since $\hat{\mathbf{a}}$ is arbitrary, we get $\sum_i \nabla_i \mathcal{U} = 0$.
116. In general the pressure tensor can be non-symmetric, although McLennan[117] has shown that a symmetric tensor can always be construced by including in the expression for linear momentum a contribution from the internal angular momentum. (Recall that $\nabla \cdot \mathbf{p} = \mathbf{F}$ (Eq. (8.136)) does not uniquely define $\mathbf{p}$.) The non-symmetry of the pressure tensor has also been discussed by Schofield and Henderson,[113] Evans,[118] and Evans and Streett.[119]
117. McLennan, J.A. *Physica* **32**, 689 (1966).
118. Evans, D.J. *Mol. Phys.* **32**, 1171 (1976); see also *J. Stat. Phys.* **20**, 547 (1979).
119. Evans, D.J., and Streett, W.B. *Mol. Phys.* **36**, 161 (1978).
120. Tolman, R.C. *J. Chem. Phys.* **16**, 758 (1948).
121. An alternative definition of the surface of tension z_s is given by Tolman (Tolman, R.C. *J. Chem. Phys.* **17**, 118, 333 (1949)).
122. Rowlinson, J.S. *J. Chem. Soc. Faraday Trans.II* **79**, 77 (1983).
123. Walton, J.P.R.B., Tildesley, D.J., Rowlinson, J.S., and Henderson, J.R. *Mol. Phys.* **48**, 1357 (1983).
124. Lee, D.J., Telo da Gama, M.M., and Gubbins, K.E. *Mol. Phys.* **53**, 1113 (1984).
125. Eggebrecht, J., Thompson, S.M., and Gubbins, K.E. *J. Chem. Phys.*, **86**, 2299 (1987).
126. Shreve, A.P., Walton, J.P.R.B., and Gubbins, K.E. *J. Chem. Phys.* **85**, 2178 (1986).
127. Henderson, J.R., and Schofield, P. *Proc. Roy. Soc. London* **A380**, 211 (1982).
128. Tjatjopoulos, G.J., and Mann, J.A. *Mol. Phys.* **60**, 1425 (1987).
129. Rusanov, A.I., and Brodskaya, E.N. *J. Coll. Interf. Sci.* **62**, 542 (1977).
130. Falls, A.H., Scriven, L.E., and Davis, H.T., *J. Chem. Phys.* **75**, 3986 (1981); ibid. **78**, 7300 (1983).
131. Lee, D.J., Telo da Gama, M.M., and Gubbins, K.E. *J. Chem. Phys.*, **85**, 490 (1986).
132. Henderson, J.R., and Rowlinson, J.S. *J. Phys. Chem.* **88**, 6484 (1984).
133. Ebner, C., Saam, W.F., and Stroud, D. *Phys. Rev. A* **14**, 2264 (1976).
134. Mermin, N.D. *Phys. Rev.* **A137**, 1441 (1965).

135. The history of the early use of functionals of this type is discussed on p. 94 of Rowlinson and Widom.[18]
136. Chacón, E., Tarazona, P., and Navascués, G. *J. Chem. Phys*. **79**, 4426 (1983); *Mol. Phys*. **51**, 1475 (1984).
137. In §3.3.1 we neglected surface effects and thus derived $\Omega = -pV = -kT \ln \Xi$ (see 3.183)). When surface effects are important, an additional $\gamma \mathrm{d}S$ term must be included in (3.174), and this leads to the expressions $\Omega = -pV + \gamma S = -kT \ln \Xi$ for inhomogeneous systems.
138. Toxvaerd, S. *Statistical Mechanics*, ed. K. Singer, Vol. 2, Ch. 4, Specialist Periodical Reports, Chemical Society, London (1975).
139. Abraham, F.F. *J. Chem. Phys*. **63**, 157 (1975).
140. Haile, J.M. Gubbins, K.E., and Gray, C.G. *J. Chem. Phys*. **64**, 1852 (1976).
141. We omit from consideration very elongated molecules that can develop a nematic–isotropic interface.
142. Haile, J.M., Gray, C.G., and Gubbins, K.E. *J. Chem. Phys*. **64**, 2569 (1976).
143. Thompson, S.M., and Gubbins, K.E. *J. Chem. Phys*. **70**, 4947 (1979). The vertical scales used in Figs. 3, 4, and 5 of this paper are incorrect. They should be reduced by a factor of 2.
144. Chacón, E., Tarazona, P., and Navascués, G. *J. Chem. Phys*. **79**, 4426 (1983).
145. Keeping $f(1)$ fixed means that $f(1)$ is not the equilibrium profile for the potential $u(12) + \delta u(12)$.
146. Smith, W.R. *Can. J. Phys*. **52**, 2022 (1974).
147. Haile, J.M. Ph.D. thesis, University of Florida (1976).
148. Gubbins, K.E., Haile, J.M., and McDonald, I.R. *J. Chem. Phys*. **66**, 364 (1977).
149. Stell, G., Rasaiah, J.C., and Narang, H. *Mol. Phys*. **27**, 1393 (1974).
150. Eggebrecht, J.E., Gubbins, K.E., and Thompson, S.M. *J. Chem. Phys*., **86**, 2286 (1987); Eggebrecht, J.E., Thompson, S.M., and Gubbins, K.E. ibid. **86**, 2299 (1987).
151. Chacón, E., Tarazona, P., and Navascués, G. *Mol. Phys*. **54**, 1073 (1985).
152. Tarazona, P., and Navascués, G. *Mol. Phys*. **47**, 1021 (1982)
153. Chacón, E., Tarazona, P., and Navascués, G. *Mol. Phys*. **51**, 1475 (1984).
154. Tarazona, P., and Navascués, G. *Mol. Phys*. **47**, 145 (1982).
155. Since the terms 'van der Waals theory' and 'generalized van der Waals theory' have been used to describe different theories (including forms of density gradient theory) by various authors we avoid their use here.
156. Sullivan, D.E. *Phys. Rev. B* **20**, 3991 (1979).
157. Sullivan, D.E. *Faraday Symp. Chem. Soc*. **16**, 191 (1981).
158. For an example of the use of nonspherical cores with this approach see Poniewerski, A., and Holyst, R. *Phys. Rev. Lett*. **61**, 2461 (1988).
159. Nordholm, S., Johnston, M., and Freasier, B.C. *Aust. J. Chem*. **33**, 2139 (1980).
160. Johnson, M., and Nordholm, S. *J. Chem. Phys*. **75**, 1953 (1981).
161. Tarazona, P. *Phys. Rev.A* **31**, 2672 (1985); see also erratum, ibid. **32**, 3148 (1985). It should be noted that the expressions for the weighting functions w_1 and w_2 are incorrect in *both* the paper *and* the erratum. Correct expressions are given in the appendix of ref. 162.
162. Tarazona, P., Marini Bettolo Marconi, U., and Evans, R. *Mol. Phys*. **60**, 573 (1987).
163. Meister, T.F., and Kroll, D.M. *Phys. Rev. A* **31**, 4055 (1985). This theory has been extended to inhomogeneous mixtures of hard spheres: see Groot, R.D. *Mol. Phys*. **60**, 45 (1987).
164. Curtin, W.A., and Ashcroft, N.W. *Phys. Rev. A* **32**, 2909 (1985); Curtin, W.A., and Ashcroft, N.W. *Phys. Rev. Lett*. **56**, 2775 (1986); Curtin, W.A. *Phys. Rev. Lett*. **59**, 1228 (1987); Curtin, W.A. *J. Chem. Phys*. **88**, 7050 (1988).

165. Teixeira, P.I., and Telo da Gama, M.M. *J. Phys.: Cond. Matter* **3**, 111 (1991); ibid. **14**, 12159 (2002). See also Frodl, P., and Dietrich, S. *Phys. Rev. A* **45**, 7330 (1992); *Phys. Rev. E* **48**, 3741 (1993).
166. Yang, B., Sullivan, D.E., Tjipto-Margo, B., and Gray, C.G. *J. Phys.: Cond. Matter* **3**, F109 (1991).
167. Yang, B., Sullivan, D.E., Tjipto-Margo, B., and Gray, C.G. *Mol. Phys*. **76**, 709 (1992).
168. Yang, B., Sullivan, D.E., and Gray, C.G. *J. Phys.:Cond. Matter* **6**, 4823 (1994).
169. Telo da Gama, M.M., and Thurtell, J.H. *Faraday Symp. Chem. Soc.* **20**, *J. Chem. Soc. Faraday Trans. II* **82**, 1721 (1986).
170. Telo da Gama, M.M., and Gubbins, K.E. *Mol. Phys*. **59**, 227 (1986).
171. Cotter, M.A. in *The Molecular Physics of Liquid Crystals*, ed. G.R. Luckhurst and G.W. Gray, Academic Press, London (1979), p. 169.
172. Tjipto-Margo, B., Sen, A.K., Mederos, L., and Sullivan, D.E. *Mol. Phys*. **67**, 601 (1989).
173. Maier, W., and Saupe, A., *Z. Naturforsch*. **13a**, 564 (1958); see also ibid. **14a**, 882 (1959); ibid. **15a**, 287.
174. See ref. 18, Ch.8.
175. A surface phase transition occurs when, for a given set of values of the thermodynamic variables, two interfaces of different structures but of equal surface tension can exist (see ref. 176). A wetting phase transition (sometimes called a Cahn transition since Cahn[176] was the first to predict it) is the particular case in which a transition occurs between the wetting and nonwetting of an $\alpha\beta$ surface (here a NV surface) by a third phase γ (here the I phase). Such a transition is between two alternative structures of the $\alpha\beta$ interface—one in which it contains a layer of phase γ, the other in which it does not. In normal wetting the three bulk phases are in thermodynamic equilibrium, and the γ layer at the interface is of macroscopic thickness. In *prewetting* the temperature is somewhat above that at which wetting occurs, so that the γ phase is no longer in equilibrium with the α and β phases, but a microscopically thin layer of incipient γ phase forms at the $\alpha\beta$ interface. For a detailed discussion of wetting and prewetting see ref. 174 and the review by Sullivan and Telo da Gama in ref. 21.
176. Cahn, J.W. *J. Chem. Phys*. **66**, 3667 (1977).
177. Strictly, the thick film shown in Fig. 8.25(b) is metastable with respect to an infinitely thick wetting film, which coexists with the thin film shown in Fig. 8.25(a). However, in practice the free energy difference between a thick film such as that shown and one of infinite thickness is negligible (see ref. 172).
178. A first-order prewetting (see endnote 175) transition occurs at $T > T_t$ over a finite range of *C/A* values beyond 0.8477. The extent of the prewetting line is very short.[172]
179. Gannon, M.G.J., and Faber, T.E. *Phil. Mag. A* **37**, 117 (1978).
180. Mederos, L., and Sullivan, D.E. *Phys. Rev. A* **39**, 854 (1989).
181. Gaines, G.L., *Insoluble Monolayers at Liquid–Gas Interfaces*, Wiley, New York (1966).
182. Guerra, C., Somoza, A.M., and Telo da Gama, M.M. *J. Chem. Phys*. **109**, 1152 (1998).
183. Haar, L., Gallagher, J.J., and Kell, G.S. *NBS/NRC Steam Tables: Thermodynamic and Transport Properties and Computer Programs for Vapour and Liquid States of Water in SI Units*, McGraw Hill, New York (1984)
184. Wilson, M.A., Pohorille, A., and Pratt, L.R. *J. Phys. Chem*. **91**, 4873 (1987).
185. Floriano, M.A., and Angell, C.A. *J. Phys. Chem*. **94**, 4199 (1990), and references therein; Pellicer, J., Garcia-Morales, J., Guanter, V., and Hernández, L. *Amer. J. Phys*. **70**, 705 (2002), and references therein.
186. Walton, J.P.R.B., and Quirke, N. *Chem. Phys. Lett*., **129**, 382 (1986).
187. Evans, R., in *Fundamentals of Inhomogeneous Fluids*, ed. D. Henderson, Dekker, New York (1992), p. 85.

188. Davis, H.T. *Statistical Mechanics of Phases, Interfaces and Thin Films*, VCH, New York (1996).
189. Oxtoby, D.W. *Ann. Rev. Materials Res.* **32**, 39 (2002).
190. Löwen, H. *J. Phys.:Cond. Matter* **15**, V1 (2003).
191. Frink, L.J.D., Salinger, A.D., Sears, M.P., Weinhold, J.D., Frischknecht. *J. Phys.:Cond. Matter* **14**, 12167 (2002).
192. A good overview of the development of density functional theory up to 2002 can be obtained from a Special Issue of *Journal of Physics: Condensed Matter*, **14**, Issue 46 (2002).
193. Wu, J. *Amer. Inst. Chem. Eng. J.* **52**, 1169 (2006).
194. Wu, J., and Li, Z. *Ann. Rev. phys. Chem.* **58**, 85 (2007).
195. Hansen, J-P., and McDonald, I.R. *Theory of Simple Liquids*, 3rd edn, Chapter 6, Academic Press, London (2006).
196. Schmidt, M. *J. Phys.: Cond. Matter* **15**, S101 (2003).
197. Tarazona, P., Cuesta, J.A., and Martínez-Ratón. *Lect. Notes Phys.* **753**, 247 (2008).
198. Roth, R. *J. Phys.:Cond. Matter*, **22**, 063102 (2010).
199. Carnahan, N.F., and Starling, K.E. *J. Chem. Phys.* **51**, 365 (1969); ibid. **53**, 600 (1970).
200. Mansoori, G.A., Carnahan, N.F., Starling, K.E., and Leland, T.W. *J. Chem. Phys.* **54**, 1523 (1971).
201. Reed, T.M., and Gubbins, K.E. *Applied Statistical Mechanics*, McGraw-Hill, New York (1973), pp. 255–8.
202. Wertheim, M.S. *Phys. Rev. Lett.* **10**, 321 (1963); Thiele, E. *J. Chem. Phys.* **39**, 474 (1963). See also Rosenfeld, Y. *J. Chem. Phys.* **89**, 4272 (1988).
203. Rosenfeld, Y. *Phys. Rev. Lett.* **63**, 980 (1989).
204. Rosenfeld, Y., Levesque, D., and Weis, J-J. *J. Chem. Phys.* **92**, 6818 (1990).
205. A useful summary of Rosenfeld's theory is given by Roth, R., Evans, R., Lang, A., and Kahl, G. *J. Phys.: Cond. Matter* **14**, 12063 (2002). See also Hansen and McDonald.[195]
206. Although the weight functions do not all have the same units, this is taken care of in the construction of the ansatz for the free energy density, (8.261), so that all is consistent finally.
207. We note that Rosenfeld's original papers[203,204] contains two sign errors in the equation for the residual Helmholtz energy density. The corrected expression is given in Roth *et al.*,[205] and independently by Yu and Wu.[208]
208. Yu, Y-X., and Wu, J. *J. Chem. Phys.* **117**, 10156 (2002).
209. Kierlik, E., and Rosinberg, M.L. *Phys. Rev. A* **42**, 3382 (1990).
210. Phan, S., Kierlik, E., Rosinberg, M-L., Bildstein, B., and Kahl, G. *Phys. Rev. E* **48**, 618 (1993).
211. Rosenfeld, Y. *Phys. Rev. E* **50**, R3318 (1994); *Mol. Phys.* **86**, 637 (1995).
212. Mansoori, G.A., Carnahan, N.F., Starling, K.E., and Leland, T.W., Jr. *J. Chem. Phys.* **54**, 1523 (1971).
213. Groot, R.D., van der Eerden, J.P., and Faber, N.M. *J. Chem. Phys.* **87**, 2263 (1987); Groot, R.D., Faber, N.M., and van der Eerden, J.P. *Mol. Phys.* **62**, 861 (1987).
214. Schmidt, M. *Phys. Rev. E* **66**, 041108 (2002); ibid. **68**, 021106 (2003); *J. Phys.: Cond. Matter* **17**, S3481 (2005).
215. Ravikovitch, P.I., and Neimark, A.V. *Langmuir* **22**, 11171 (2006).
216. Ravikovitch, P.I., and Neimark, A.V. *Studies Surf. Sci. Catal.* **160**, 9 (2006).
217. MacElroy, J.M.D., and Raghaven, K. *J. Chem. Phys.* **93**, 2068 (1990).
218. Hofmann, T., Wallacher, D., Huber, P., Birringer, R., Knorr, K., Schreiber, A., and Findenegg, G.H. *Phys. Rev. B* **72** (2005).
219. Lastoskie, C., Gubbins, K.E., and Quirke, N. *J. Phys. Chem.* **97**, 4786 (1993); *Langmuir* **9**, 2693 (1993).

220. Olivier, J.P. *J. Porous Matls*. **2**, 217 (1995).
221. Ravikovitch, P.I., and Neimark, A.V. *J. Phys. Chem. B* **105**, 6817 (2001); *Langmuir* **18**, 1550 (2002).
222. Rosenfeld, Y., Schmidt, M., Löwen, H., and Tarazona, P. *Phys. Rev. E* **55**, 4245 (1997).
223. Tarazona, P., and Rosenfeld, Y. *Phys. Rev. E* **55**, R4873 (1997).
224. Gonzalez, A. White, J.A., and Evans, R. *J. Phys.: Cond. Matter* **9**, 2375 (1997).
225. Cuesta, J.A., Martínez-Ratón, Y., and Tarazona, P. *J. Phys.: Cond. Matter* **14**, 11965 (2002).
226. Tarazona, P. *Physica A* **306**, 243 (2002).
227. González, A., White, J.A., Román, F.L., and Velasco, S. *J. Chem. Phys*. **125**, 064703 (2006).
228. Ye, Z., Chen, H., Cai, J., Liu, H., and Hu, Y. *J. Chem. Phys*. **125**, 124705 (2006).
229. Chen, H., Ye, Z., Peng, C., Liu, H., and Hu, Y. *J. Chem. Phys*. **125**, 204708 (2006).
230. Löwen, H. *J. Phys.: Cond. Matter* **14**, 11897 (2002).
231. Löwen, H. *J. Phys.: Cond. Matter* **15**, V1 (2003).
232. Rosenfeld, Y. *Phys. Rev. E* **50**, R3318 (1994).
233. Rosenfeld, Y. *Mol. Phys*. **86**, 637 (1995).
234. Esztermann, A., Reich, H., and Schmidt, M. *Phys. Rev. E* **73**, 011409 (2006).
235. Esztermann, A., and Schmidt, M. *Phys. Rev. E* **70**, 022501 (2004).
236. Toxvaerd, S. *Mol. Phys*. **26**, 91 (1973).
237. Somorjai, G., *Introduction to Surface Chemistry and Catalysis*, Ch. 6, Wiley, New York (1994).
238. See, for example: van Santen, R.A., and Niemantsverdriet, J.W. *Chemical Kinetics and Catalysis*, Ch. 6, Plenum Press, New York (1995).
239. Steele, W.A. *The Interaction of Gases with Solid Surfaces*, Ch. 2, Pergamon, Oxford (1974).
240. Steele, W.A. *Surf. Sci*. **36**, 317 (1973).
241. Steele, W.A. *J. de Phys*. **38**, C4-61 (1977).
242. This idea seems to originate in the work on scaled particle theory by Helfand *et al*. (Helfand, E., Frisch, H.L. and Lebowitz, J.L. *J. Chem. Phys*. **34**, 1037 (1961)), later extended to fluid mixtures by Lebowitz *et al*. (Lebowitz, J.L., Helfand, E., and Praestgaard, E. *J. Chem. Phys*. **43**, 774 (1965)). Its use in connection with the OZ equation dates from the work of Perram and White (Perram, J.W., and White, L.R. *Faraday Disc. Chem. Soc. London* **59**, 29 (1975)). A more explicit derivation and use of the wall–fluid OZ equation was given a little later by Henderson *et al*. (Henderson, D., Abraham, F.F., and Barker, J.A. *Mol. Phys*. **31**, 1291 (1976)). The derivation given here follows those of Perram and White and of Henderson, Abraham, and Barker, based on the idea of 'blowing up a molecule', but alternative derivations are possible, notably those of Percus (Percus, J.K. *J. Stat. Phys*. **15**, 423 (1976) and of Grimson *et al*. (Grimson, M.L., Rickayzen, G., and Richmond, P. *Mol. Phys*. **39**, 61 (1980)). These latter authors show how to derive the wall–fluid OZ equation by starting from density functional theory (§ 8.4) and assuming the effect of the wall on the nearby fluid structure to be small. For a review of applications of the OZ equation (both the fluid–wall OZ and the full OZ equation for the pair correlation function) see Henderson, D., and Plischke, M. *J. Phys. Chem*. **92**, 7177 (1988).
243. Reviews of the use of integral equation theory to study adsorption of spherical molecules on walls, with and without attractive forces present, have been given in Chapters 6 and 7 of Nicholson and Parsonage (ref. 17) and in Chapter 15 of Lee (Lee, L.L. *Molecular Thermodynamics of Nonideal Fluids*, Butterworths, Boston, 1988).
244. Waisman, E., Henderson, D., and Lebowitz, J.L. *Mol. Phys*. **32**, 1373 (1976).

245. Fisher, I.Z. *Statistical Theory of Liquids*, University of Chicago Press, Chicago (1964), p. 109.
246. Evans, R., Tarazona, P., and Marini Bettolo Marconi, U. *Mol. Phys.* **50**, 993 (1983).
247. Henderson, J.R. *Mol. Phys.* **52**, 1467 (1984).
248. Grimson, M.J., and Rickayzen, G. *Mol. Phys.* **42**, 767 (1981).
249. Nieminen, R.M., and Ashcroft, N.W. *Phys. Rev. A* **24**, 560 (1981).
250. Smith, L.S., and Lee, L.L. *J. Chem. Phys.* **71**, 4085 (1979).
251. Levesque, D., and Weis, J.J. *J. Stat. Phys.* **40**, 29 (1985).
252. Russier, V., Rosinberg, M.L., Badiali, J.P., Levesque, D., and Weis, J.J. *J. Chem. Phys.* **87**, 5012 (1987).
253. Badiali, J.P., and Boudh-Hir, M.E. *Mol. Phys.* **53**, 1399 (1984).
254. Badiali, J.P. *Mol. Phys.* **55**, 939 (1985). See also Badiali, J.P. *J. Chem. Phys.* **90**, 4401 (1989).
255. Andersen, H.C., and Chandler, D. *J. Chem. Phys.* **57**, 1918 (1972).
256. Russier, V., Rosinberg, M.L., Badiali, J.P., Levesque, D., and Weis, J.J. *J. Chem. Phys.* **87**, 5012 (1987).
257. Isbister, D.J., and Freasier, B.C. *J. Stat. Phys.* **20**, 331 (1979).
258. Eggebrecht, J.M., Isbister, D.J., and Rasaiah, J.C. *J. Chem. Phys.* **73**, 3980 (1980).
259. e.g. Rasaiah, J.C., Isbister, D.J., Stell, G. *J. Chem. Phys.* **75**, 4707 (1981).
260. Thompson, S.M., Gubbins, K.E., Sullivan, D.E., and Gray, C.G. *Mol. Phys.* **51**, 21 (1984).
261. Sullivan, D.E., Barker, R., Gray, C.G., Streett, W.B., and Gubbins, K.E. *Mol. Phys.* **44**, 597 (1981).
262. Sokolowski, S., and Patrykiejew, A. *Mol. Phys.* **50**, 1311 (1983).
263. Borštnik, B., and Janežič, D. *Chem. Phys.* **130**, 195 (1989).
264. Borštnik, B., and Janežič, D. *J. Mol. Liquids* **48**, 293 (1991).
265. This cusp occurs because when a site at position z is held fixed, the range of permissible orientations is limited by the presence of the wall when $z < \sigma/2 + \ell$, whereas no limitation on the orientation occurs if z is greater than this value.
266. Ref. 21. See the reviews by D. Beaglehole (p. 523) and S.A. Rice, J. Gryko, and U. Mohanty (p. 255).
267. Good, R.J. *J. Phys. Chem.* **61**, 810 (1957); see also *J. Coll. Int. Sci.* **59**, 398 (1977).
268. Gannon, M.G., and Faber, T.E., ref. 179. For extensions, see the work of Sluckin, T.J.[89]
269. Matsumoto, M., and Kataoka, Y. *J. Chem. Phys.* **88**, 3233 (1988).
270. Alternatively, see refs. 37 of Ch. 3, and Nettleton, R.E., and Green, M.S. *J. Chem. Phys.* **29**, 1365 (1958), for expressions for the entropy involving $f(1)$, $f(12)$, $f(123)$, etc.
271. Alejandre, J., Tildesley, D.J., and Chapela, G.A. *J. Chem. Phys.* **102**, 4574 (1995); Taylor, R., Dang, L.X., and Garrett, B.C. *J. Phys. Chem.* **100**, 11720 (1996), Rivera, J.L., Starr, F., Paricaud, P. and Cummings, P.T. *J. Chem. Phys.* **125**, 094712 (2006); Lu, Y.J., and Wei, B. *Appl. Phys. Lett.* **89**, 164106 (2006).
272. Schlossman, M.L. in *Encyclopedia of Applied Physics*, Vol. 20, ed. G.L. Trigg, VCH, New York (1997), p. 311.
273. Timmermans, J. *Physico-chemical Constants of Pure Organic Compounds*, Elsevier, New York (1950), Vol. 1, and (1965), Vol. 2.
274. Partington, J. R. *An Advanced Treatise on Physical Chemistry*, Vol. 2, Longmans, London (1951), p. 194.
275. Buff, F. P., and Lovett, R. A. in *Simple Dense Fluids*, ed. H.L. Frisch and Z.W. Salsburg, Academic, New York (1968), p. 17.
276. Reid, R. C., Prauznitz, J. M., and Roling, B. E. *The Properties of Gases and Liquids*, 4th edn, McGraw-Hill, New York (1987), p. 634, and p. 54.

277. Kőrösi, G., and Kováts, E. S. *J. Chem. Eng*. Data **26**, 323 (1981).
278. Rowlinson, J. S., and Swinton, F. L. *Liquids and Liquid Mixtures*, 3rd edn, Butterworth, London (1982).
279. For a review of the earlier work see Randles, J.E.B., ref. 62.
280. The question of whether $\Delta\phi$ can be observed directly has been discussed for over a century. See Gibbs, J.W. *Scientific Papers*, Vol. 1, p. 429, letter to W.D. Bancroft (1899), Longmans, London (1906), reprinted by Dover, New York (1961); Guggenheim, E.A. *J. Phys. Chem.* **33**, 842 (1929), see also *Thermodynamics*, 5th edn, North Holland, Amsterdam (1967), p. 300; Oppenheim, I. *J. Phys. Chem.* **68**, 2959 (1964); Zhou, Y., Stell, G., and Friedman, H.L. *J. Chem. Phys.* **89**, 3836 (1988); Pratt, L.R. *J. Phys. Chem.* **96**, 25 (1992).
281. Sokhan, V.P., and Tildesley, D.J. *Mol. Phys.* **92**, 625 (1997).
282. Stillinger, F.H., and Ben-Naim, A. *J. Chem. Phys.* **47**, 4431 (1967); Croxton, C.A. *Physica* A**106**, 239 (1981).
283. Beaglehole, D., ref. 21, p. 523.
284. Azzam, R. M. A., and Bashara, N. M. *Ellipsometry and Polarized Light*, North Holland, Amsterdam (1977); Lekner, J. *Theory of Reflection*, Martinus Nijhoff, Dordrecht (1987).
285. In reflectometry one focuses more on the intensity of the reflected beam: see, e.g. Sluckin, T.J. *Mol. Phys.* **47**, 267 (1982).
286. Since $\varepsilon_{\alpha\alpha} = n_\alpha^2$, $\Delta\varepsilon$ is closely related to the birefringence, $\Delta n = n_z - n_x$ (cf. Eq. (10.158)).
287. Lekner, J. *Mol. Phys.* **49**, 1385 (1983), and references therein to earlier work of F. Buff and F. Abèles. See also Lekner, ref. 284 above, p. 144.
288. There is a small contribution, present even for spherical molecules, due to the density inhomogeneity in the interface. See Castle, P., and Lekner, J. *Physica A* **101**, 99 (1980).
289. This definition of $\overline{\Delta\alpha}$ differs from that $\overline{\Delta\alpha}^{MK}$ of refs. 61 and 269 by $\overline{\Delta\alpha} = (3/2)\overline{\Delta\alpha}^{MK}$.
290. Beaglehole, D. *J. Physique Coll. C*10 **44**, 147 (1983).
291. The general member of the cyanobiphenyl family is denoted nCB, which has a core of two phenyl groups surrounded by a cyano group on one side and an alkyl chain with *n* carbon atoms on the other.
292. Beaglehole, D. *Mol. Cryst. Liq. Cryst.* **89**, 319 (1982).
293. Mukhopadhyay, A., and Law, B.M. *Phys. Rev. E* **63**, 011507 (2000).
294. Shen, Y. R. *Ann. Rev. Phys. Chem.* **40**, 327 (1989).
295. Corn, R. M., and Higgins, D. A. *Chem. Rev.* **94**, 107 (1994).
296. Brevet, P. F., and Girault, H. H. *Prog. Coll. Polym. Sci.* **103**, 1 (1997). Brevet, P. F. *Surface Second Harmonic Generation*, Presses Polytechniques Universitaires Romandes Lausanne (1997).
297. Eisenthal, K. B. *Ann. Rev. Phys. Chem.* **43**, 627 (1992).
298. Eisenthal, K. B. *Chem. Rev.* **96**, 1343 (1996); *Chem. Rev.* **106**, 1462 (2006).
299. Conboy, J. C., Messmer, M. C., Walker, R. A., and Richmond, G. L. *Prog. Coll. Polym. Sci.* **103**, 10 (1997).
300. Miranda, P. B., and Shen, Y. R. *J. Phys. Chem.* B **103**, 3292 (1999).
301. Schultz, M. J., Schnitzer, C., Simonelli, D., and Baldelli, S. *Int. Rev. Phys. Chem.* **19**, 123 (2000).
302. Shen, Y. R. in *Water in Biomaterials Surface Science*, ed. Morra, M., Wiley, New York (2001), p. 215; Shen, Y. R., and Ostroverkhov, V. *Chem. Rev.* **106**, 1140 (2006).
303. Richmond, G. L. *Ann. Rev. Phys. Chem.* **52**, 357 (2001).
304. Shultz, M. J., Baldelli, S., Schnitzer, C., and Simonelli, D. *J. Phys. Chem. B* **106**, 5313 (2002).

305. Richmond, G. L., and Leich, M. A. *Faraday Discuss*. **129**, 1(2005).
306. Gopalakrishnan, S., Liu, D., Allen, H. C., Kuo, M., and Schultz, M. J. *Chem. Rev.* **106**, 1155 (2006).
307. Perry, H., Neipert, C., Space, B., and Moore, P. B. *Chem. Rev*. **106**, 1234 (2006).
308. Shen, Y.R. *The Principles of Nonlinear Optics*, Wiley, New York (1984). Second harmonic generation from surfaces is discussed in Ch. 25.
309. Bogaard, M.P., and Orr, B.J. MTP International review of Science, Physical chemistry, Series two, Vol. 2, *Molecular Structure and Properties*, ed. A.D. Buckingham, Butterworth, London (1975), Chapter 5.
310. Long, D. A. *Raman Spectroscopy*, McGraw Hill, New York (1977), Chapter 8.
311. Actually, usually small bulk SHG can always arise (see ref. 308 and Richmond, G.L., Robinson, J.M., and Shannon, V.L. *Prog. Surf. Sci*. **28**, 1 (1988).), from the following sources: (a) the field gradient term (1/3)**B**:**ΔEE** in (C.6) of Vol. 1, (b) induced quadrupole and magnetic dipole moments (see (C.19a)), (c) collision-induced polarizability effects: Ye, P., and Shen, Y.R. *Phys. Rev. B* **28**, 4288 (1983). Other work (see Richmond, G.L., Robinson, J.M., and Shannon, V.L. *Prog. Surf. Sci*. **28**, 1 (1988). and Sipe, J.E., Mizraki, V., and Stegman, G.I. *Phys. Rev. B*, **35**, 9091 (1987); Guyot-Sionnest, P., and Shen, Y.R. *Phys. Rev. B*, **38**, 7985 (1988).) has shown that (i) additional terms, arising from the rapid variation of $\chi^{(2)}(z)$ through the interface, occur, so that (8.310) is only approximate, and (ii) it can be complicated to separate the bulk and surface contributions to SHG.
312. Kleinman, D. A. *Phys. Rev*. **126**, 1977 (1962).
313. Goh, M. C., Hicks, J. M., Kemnitz, K., Pinto, G. R., Bhattacharyya, K., Eisenthal, K. B., and Heinz, T. F. *J. Phys. Chem*. **92**, 5074 (1988); Goh, M. C., and Eisenthal, K. B. *Chem. Phys. Lett*. **157**, 101 (1989).
314. Maroulis, G. *J. Chem. Phys*. **94**, 1182 (1991). The effect of dispersion on $\boldsymbol{\beta}$ is discussed theoretically by Luo, Y., Agren, H., Vahtras, O., Jorgensen, P., Spirko, V., and Hettema, H. *J. Chem. Phys*. **98**, 7159 (1993). Most components for $\lambda = 640$ nm, typical of the experiments, are not significantly different from the static values. An exception is the pair $\beta_{x'x'z'} = \beta_{x'z'x'}$, which differ by about 25% for wavelength $\lambda = 640$ nm (see Table 1 in ref. 281).
315. Berendsen, H.J.C., Grigera, J.R., and Straatsma, T.P. *J. Phys. Chem*. **91**, 6269 (1987). The SPC/E (simple point charge/extended) model for H_2O has a LJ core centred on the O site and point charges on the O and H sites. In comparison, the TIP4P model is a 4-site one, with the negative O-charge moved off the LJ O-site, along the dipole towards the H sites.
316. Tamburello-Luca, A. A., Hébert, P., Brevet, P. F., and Girault, H. H. *J. Chem. Soc. Faraday Trans*. **91**, 1763 (1995). See, however, Fordyce, A. J., *et al. Mol. Phys*. **99**, 677 (2001), who attribute the breakdown of Kleinman symmetry to dispersion.
317. Du, Q., Superfine, R., Freysz, E., and Shen, Y.R. *Phys. Rev. Lett*. **70**, 2313 (1993); Du, Q., Freysz, E., and Shen, Y.R. *Science* **264**, 826 (1994).
318. Baldelli, S., Schnitzer, C., and Shultz, M.J. *J. Chem. Phys*. **108**, 9817 (1998).
319. Jungwirth, P., and Tobias, D.J. *Chem. Rev*. **106**, 1259 (2006).
320. Petersen, P.B., and Saykally, R.J. *Ann. Rev. Phys. Chem*. **57**, 333 (2006).
321. Jungwirth, P., and Winter, B. *Ann. Rev. Phys. Chem*. **59**, 343 (2008).
322. Onsager, L., and Samaras, N.N.T. *J. Chem. Phys*. **2**, 528 (1934). These authors use continuum dielectric theory and explain ion repulsion from surfaces in terms of the repulsive force on an ion near the surface on the liquid side, due to repulsion from the ion's image charge.
323. Petersen, P.B., and Saykally, R.J. *Chem. Phys. Lett*. **397**, 51 (2004).
324. Liu, D., Ma, G., Levering, L.M. and Allen, H.C. *J. Phys. Chem. B***108**, 2252 (2004); Raymond, E.A., and Richmond, G.L. *J. Phys. Chem. B* **108**, 5051 (2004); Brown,

E.C., Mucha, M., Jungwirth, P., and Tobias, D.J. *J. Phys. Chem. B* **109**, 7934 (2005).
325. The papers of F. Hofmeister and collaborators were published in *Arch. für exper. Path. und Pharm.* in the period 1887–98. Two of the most important are translated into English in Kunz, W., Henle, J., and Ninham, B.W. *Curr. Opin. Coll. Int. Sci.* **9**, 19 (2004). For reviews of the various Hofmeister series, see Kunz, W., Nostro, P.L., and Ninham, B.W. *Curr. Opin. Coll. Int. Sci.* **9**, 1 (2004); Cacace, M.G., Landau, E.M., and Ramsden, J.J. *Quart. Rev. Biophys.* **30**, 241 (1997); Collins, K.D., and Washbaugh, M.W. *Quart. Rev. Biophys.* **18**, 323 (1985).
326. Beck, T.L., Paulaitis, M.E., and Pratt, L.R. *The potential distribution theorem and models of molecular solutions*, Cambridge University Press, Cambridge (2006), p. 208. These authors discuss the Zundel $\left(H_5O_2^+\right)$ and Eigen $\left(H_9O_4^+\right)$ species.
327. Petersen, M.K., Iyengar, S.S., Day, T.J.F., and Voth, G.A. *J. Phys. Chem. B* **108**, 14804 (2004).
328. Langevin, D., ed., *Light Scattering by Liquid Surfaces and Complementary Techniques*, Dekker, New York (1992).
329. Meunier, J. ref. 328, p. 333.
330. Baszkin, A., and Norde, W. (eds), *Physical Chemistry of Biological Interfaces*, Dekker, New York (2000).
331. Lheveder, C., Hénon, S., and Meunier, J. ref. 330, p. 559. The technique Brewster angle microscopy (BAM), designed to overcome the low reflectivity problem, is described in this reference and ref. 333.
332. Binks, B. P., ed., *Modern Characterization Methods of Surfactant Systems*, Dekker, New York (1999).
333. Hénon, S., and Meunier, J., ref. 332, p. 109.
334. Schlossman, M.L., and Pershan, P.S., ref. 328, p. 365.
335. Als-Nielson, J., Jacquemain, D., Kjaer, K., Leveiller, F., Lahav, M., and Leiserowitz, L. *Phys. Rep.* **246**, 251 (1994).
336. Dietrich, S., and Haase, A. *Phys. Rep.* **260**, 1 (1995).
337. Daillant, J., and Gibaud, A. (eds), *X-Ray and Neutron Reflectivity*, Springer, Berlin (1999).
338. Daillant, J., and Alba, M. *Rep. Prog. Phys.* **63**,1725 (2000).
339. Lu, J. R., and Thomas, R. K. ref. 330, p. 609.
340. Als-Nielson, J., and McMorrow, D. *Elements of Modern X-Ray Physics*, Wiley, New York (2001), p. 61.
341. Penfold, J. *Rep. Prog. Phys.* **64**, 777 (2001).
342. Schlossman, M. L. *Curr. Opin. Coll. Int. Sci.*, **7**, 235 (2002).
343. Dalgliesh, R. *Curr. Opin. Coll. Int. Sci.* **7**, 244 (2002).
344. Penfold, J., and Thomas, R. K. *J. Phys. Cond. Matter* **2**, 1369 (1990).
345. Lu, J. R., and Thomas, R. K. *J. Chem. Soc., Faraday Trans.* **94**, 995 (1998).
346. Thomas, R. K. ref. 332, p. 417.
347. Lu, J. R., Thomas, R. K., and Penfold, J. *Adv. Coll. Int. Sci.* **84**, 143 (2000).
348. Penfold, J. *Curr. Opin. Coll. Int. Sci.* **7**, 139 (2002).
349. Penfold, J., Thomas, R. K., and Taylor, D. J. F. *Curr. Opin. Coll. Int. Sci.* **11**, 337 (2006).
350. Thomas, R. K. *Ann. Rev. Phys. Chem.* **55**, 391 (2004).
351. For grazing incidence X-rays, the value of $R(q)$ is independent of polarization direction (ref. 334). For neutrons, we neglect magnetic scattering due to interactions with the neutron spin so that polarization is irrelevant.
352. Asmussen, A., and Riegler, H. *J. Chem. Phys.* **104**, 8159 (1996).
353. DiMasi, E., Tostmann, H., Ocko, B.M., Pershan, P.S., and Deutsch, M. *Phys. Rev. B* **58**, R13419 (1998); DiMasi, E., Tostmann, H., Shpyrko, O.G., Deutsch, M., Pershan, P.S., and Ocko, B.M. *J. Phys.: Cond. Matter* **12**, A209 (2000).

354. Lurio, L.B., Rabedeau, T.A., Pershan, P.S., Silvera, I.F., Deutsch, M., Kosowsky, S.D., and Ocko, B.M. *Phys. Rev. Lett.* **68**, 2628 (1992).
355. Sanyal, M.K., Sinha, S.K., Huang, K.G., and Ocko, B.M. *Phys. Rev. Lett.* **66**, 628 (1991).
356. Braslau, A., Pershan, P.S., Swislow, G., Ocko, B.M., and Als-Nielson, J., *Phys. Rev. A* **38**, 2457 (1988).
357. Seydel, T., Tolan, M., Ocko, B.M., Seeck, O.H. Weber, R., DiMasi, E., and Press, W. *Phys. Rev. B* **65**, 184207 (2002).
358. Streit, S., Sprung, C., Gutt, C., and Tolan, M. *Physica B* **357**, 110 (2005).
359. Li, Z. X., Thomas, R. K., Rennie, A. R., and Penfold, J. *J. Phys. Chem. B* **102**, 185 (1998).
360. Fu, Y., and Bartell, F.E. *J. Phys. Chem.* **54**, 537 (1950).
361. Hicks, J.M., Kemnitz, K., Eisenthal, K.B., and Heinz, T. F. *J. Phys. Chem.* **90**, 560 (1986).
362. Sokhan, V. P., and Tildesley, D. J. *Faraday Discuss.* **104**, 193 (1996).
363. See, for example, ref. 18, p. 115.
364. Gelfand, M.P., and Fisher, M.E. *Physica A* **166**, 1 (1990).
365. Caffrey, M., and Wang, J. *Ann. Rev. Biophys. Biomol. Struct.* **24**, 351 (1995).
366. Wilson, K.R., Rude, B.S., Smith, J., Cappa, C., Co, D.T., Schaller, R.D., Larsson, M., Catalano, T., and Saykally, R.J. *Rev. Sci. Inst.* **75**, 725 (2004).
367. Liquid–vapour interface results for water from simulations using various empirical force fields (nonpolarizable and polarizable) and *ab initio* methods (based on electron density functional theory) are compared in Euo, I-F.W., Munday, C.J., Eggimann, B.L., McGrath, M.J., Siepmann, J.I., Chen, B., Vieceli, J. and Tobias, D.J. *J. Phys. Chem. B* **110**, 3738 (2006).
368. Dang, L.X., and Chang, T-M. *J. Chem. Phys.* **119**, 9851 (2003).
369. Yeh, Y.L., Zhang, C., Held, H., Mebel, A.M., Wei, Z., Lin, S.H., and Shen, Y.R. *J. Chem. Phys.* **114**, 1837 (2001).
370. Dietter, J., and Morgner, H. *Chem. Phys.* **241**, 55 (1999).
371. Chen, B., Siepmann, J.I., and Klein, M.L. *J. Amer. Chem. Soc.* **124**, 12232 (2002), and references therein.
372. Malmsten, M. (ed.) *Biopolymers at Interfaces*, Dekker, New York (2003). Techniques reviewed include spectroscopies (IR, CD, SERS, fluorescence, TIRF) by V. Hlady and J. Buijs (p. 453), thermodynamics (differential scanning calorimetry (DSC), isothermal titration calorimetry (ITC)) by K. Fromell, K. Caldwell and S-C Huang (p. 517), ellipsometry and reflectivity by M. Malmsten (p. 539), surface plasmon resonance (SPR) by E-K. Sinner, K. Kobayashi, T. Lehmann, T. Newmann, F. Yu, W. Knoll, B. Prein, and J. Rühe (p. 583) and neutron reflectivity by J.R. Lu (p. 609).
373. Lipkowski, J., and Ross, P. N. (eds), *Adsorption of Molecules at Metal Electrodes*, VCH, New York (1992). Techniques reviewed include thermodynamics by J. Lipkowski and L. Stolberg (p. 171), reflectance spectroscopy and ellipsometry by W. Plieth, W. Kozlowski, and T. Twomey (p. 239), Raman spectroscopy by B. Pettinger (p. 285), IR spectroscopy by R.J. Nicols (p. 347), and SHG by R.M. Corn (p. 391).
374. Parfitt, G. D., and Rochester, C. H. (eds), *Adsorption from Solution at the Solid/Liquid Interface*, Academic Press, London (1983).
375. Schmickler, W., *Interfacial Electrochemistry*, Oxford University Press, New York (1996).
376. Weaver, M. J., and Gao, X. *Ann. Rev. Phys. Chem.* **44**, 459 (1993).
377. Jones, R. A. L., and Richards, R. W. *Polymers at Surfaces and Interfaces*, Cambridge University Press, Cambridge (1999), p. 187.
378. Damadaran, S. *Curr. Opin. Coll. Int. Sci.* **9**, 328 (2004).
379. Keddie, J. L. *Curr. Opin. Coll. Int. Sci.* **6**, 102 (2001).
380. Clark, R. J. H., and Hester, R. E. *Spectroscopy for Surface Science*, **26** of *Advances in Spectroscopy*, Wiley, Chichester (1998).

381. Chen, Z., Shen, Y. R., and Somorjai, G.A. *Ann. Rev. Phys. Chem.* **53**, 437 (2002).
382. Penfold, J. *Curr. Opin. Coll. Int. Sci.* **7**, 139 (2000).
383. Lu, J. R., Zhao, X., and Yaseen, M. *Curr. Opin. Coll. Int. Sci.* **12**, 9 (2007).
384. Steitz, R., Gutberlet, T., Hauss, T., Klösgen, B., Krastev, R., Schemmel, S., Simonsen, A. C., and Findenegg, G. H. *Langmuir* **19**, 2409 (2003).
385. Lipkowski, J., and Ross, P. N., eds, *Structure of Electrified Interfaces*, VCH, New York (1993).
386. Melendres, C. A. in *Equilibrium Structure and Properties of Surfaces and Interfaces*, ed. A. Gonis and G.M. Stocks, Plenum, New York (1992), p. 319.
387. Rondelez, F., Ausserré, D., and Hervet, H. *Ann. Rev. Phys. Chem.* **38**, 317 (1987).
388. Israelachvili, J. *Intermolecular and Surface Forces*, 2nd edn, Academic, London (1992), p. 169.
389. Kurihara, K. in *Nano-surface Chemistry*, ed. M. Rosoff, Dekker, New York (2002), p. 1.
390. Kékicheff, P. in *Electrostatic Effects in Soft Matter and Biophysics*, ed. C. Holm, P. Kékicheff, and R. Podgornik, Kluwer, Dordrecht (2001), p. 205.
391. Claesson, P. M., Ederth, T., Bergeron, V., and Rutland, M.W. *Adv. Coll. Int. Sci.* **67**, 119 (1996).
392. Knoll, W. *Ann. Rev. Phys. Chem.* **49**, 569 (1998).
393. Schuck, P. *Ann. Rev. Biophys. Biomol. Struct.* **26**, 541 (1997).
394. Sundberg, E. J., Andersen, P. S., Gorshkova, I. I., and Schuck, P. in *Protein Interactions*, ed. P. Schuck, Springer, New York (2007), p. 97.
395. Manne, S., and Gaab, H. E. *Curr. Opin. Coll. Int. Sci.* **2**, 145 (1997).
396. Warr, G. G. *Curr. Opin. Coll. Int. Sci.* **5**, 88 (2000).
397. Mackie, A. R. *Curr. Opin. Coll. Int. Sci.* **9**, 357 (2004); Fukuma, T., Ueda, Y., Yoshioka, S., and Asakawa, H. *Phys. Rev. Lett.* **104**, 016101 (2010).
398. Freeman, N., in *Proteins at Solid–Liquid Interfaces*, ed. P. Déjardin, Springer, Berlin (2006), p. 75.
399. Lu, J.R., Zhao, X., and Yaseen, M. *Curr. Opin. Coll. Inter. Sci.* **12**, 9 (2007).
400. Benjamin, I. in *Modern Methods for Multidimensional Dynamics Computations in Chemistry*, ed. D. L. Thompson, World Scientific, Singapore (1998), p. 101.
401. Mundy, C. J., Balsubramanian, S., Bagchi, K., Siepmann, J. I., and Klein, M. L. *Faraday Discuss.* **104**, 17 (1996).
402. Wang, J. C., and Fichthorn, K. A. *Colloids Surf.* A**206**, 267 (2002).
403. Somorjai, G.A., and Rupprechter, G. *J. Chem. Educ.* **75**, 161 (1998).
404. McConnell, H. M. *Ann. Rev. Phys. Chem.* **42**, 171 (1991).
405. Knobler, C. M. *Adv. Chem. Phys.* **77**, 397 (1990).
406. Swalen, J. D. *Ann. Rev. Mater. Sci.* **21**, 373 (1991).
407. Petty, M. C. *Langmuir-Blodgett Films*, Cambridge University Press, Cambridge (1996).
408. Bruch, L.W., Diehl, R. D., and Venables, J. A. *Rev. Mod. Phys.* **79**, 1381 (2007).
409. Möhwald, H. *Ann. Rev. Phys. Chem.* **41**, 441 (1990); *Rep. Prog. Phys.* **56**, 653 (1993); *Structure and Dynamics of Membranes*, Vol. **1A**, ed. R. Lipkowski and E. Saskmann, North Holland, Amsterdam, (1995), p. 161.
410. Andelman, D., Brochard, F., Knobler, C., and Rondelez, F. in *Micelles, Membranes, Microemulsions and Monolayers*, ed. W.M. Gelbart, A. Ben-Shaul, and D. Roux, Springer, New York (1994), p. 559.
411. Birdi, K. S. *Lipid and Biopolymer Monolayers at Liquid Interfaces*, Plenum, New York (1989).
412. Butt, H.-J., Graf, K., and Kappl, M. *Physics and Chemistry of Interfaces*, Wiley–VCH, Weinheim (2006), p. 222.
413. Bain, C. *Curr. Opin. Coll. Int. Sci.* **3**, 287 (1998).

414. Hénon, S., and Meunier, J., *Rev. Sci. Inst.* **62**, 936 (1991).
415. Somorjai, G. A. *Chem. Rev.* **96**, 1223 (1996).
416. Woodruff, D. P., and Delchar, T. A. *Modern Techniques of Surface Science*, 2nd edn, Cambridge University Press, Cambridge (1994).
417. Ball, P. *Made to Measure*, Princeton University Press, Princeton (1997), p. 394.
418. Denton, A.R., Gray, C.G., and Sullivan, D.E. *Chem. Phys. Lett.* **219**, 310 (1994).
419. Ulman, A. *Chem. Rev.* **96**, 1533 (1996); Schreiber, F. *Prog. Surf. Sci.* **65**, 151 (2000); Schwartz, D.K. *Ann. Rev. Phys. Chem.* **52**, 107 (2001); Whitesides, G.M., Kriebel, J.K., and Love, J.C. *Science Progress* **88**, 17 (2005).
420. Stine, K.J., and Moore, B.G. in *Nano-surface Chemistry*, ed. M. Rosoff, Dekker, New York (2002), p. 59.
421. Campbell, S.E., Luengo, G., Srdanov, V.I., Wudl, F., and Israelachvili, J.N. *Nature* **382**, 520 (1996).
422. Puibasset, J., and Pellenq, J. M. *Phys. Chem. Chem. Phys.* **6**, 1933 (2004).
423. Shelley, J. C., and Bérard, D. R. *Rev. Comput. Chem.* **12**, 137 (1998).
424. Early Langmuir monolayer simulations are reviewed in Kaganer, V. M., Möhwald, H., and Dutta, P. *Rev. Mod. Phys.* **71**, 779 (1999). For representative later work see refs. 425 and 426.
425. Opps, S. B., Yang, B., Gray, C. G., and Sullivan, D. E. *Phys. Rev.* E**63**, 041602 (2001).
426. Rodriguez, J., and Laria, D. *J. Phys. Chem. C* **111**, 908 (2007).
427. Volkov, A. G., and Deamer, D. W. (eds), *Liquid–Liquid Interfaces*, CRC Press, Boca Raton (1996).
428. Benjamin, I. *Ann. Rev. Phys. Chem.* **48**, 407 (1997).
429. *Dynamics and Structure of the Liquid–Liquid Interface*, *Faraday Discussions* **129** (2005).
430. Girault, H., Kornyshev, A. A., Monroe, C. W., and Urbakh, M. (eds), *Functionalized Liquid–Liquid Interfaces*, *J. Phys. Cond. Matter* **19**, 370301, Special Issue, (2007).
431. Koryta, J. *Ions, Electrodes and Membranes*, 2nd edn, Wiley, Chichester (1991), p. 149.
432. MacRitchie, F. *Chemistry at Interfaces*, Academic Press, San Diego (1990), p. 140.
433. Volkov, A. G. *Liquid Interfaces in Chemical, Biological and Pharmaceutical Applications*, Dekker, New York (2001). Various aspects of thermodynamics are discussed by Kakiuchi, T. (p.105), Samec, Z. (p. 415), and Murtomäki, L., Manzanares, J. A., Mafé, S., and Koutturi, K. (p. 533).
434. Floor, G. J., Cohen Stuart, M.A., Scheutjens, J.M.H.M., Cosgrove, T., and Vincent, B. *Polymers at Interfaces*, Chapman and Hall, London (1993), p. 401.
435. Schlossman, M. L., and Tikhonov, A. M. *Ann. Rev. Phys. Chem.* **59**, 153 (2008).
436. Pohorille, A., and Wilson, M. A. *J. Molec. Struc.* **284**, 271 (1993).
437. Pratt, L. R., and Pohorille, A. *Chem. Rev.* **102**, 2671 (2002).
438. Benjamin, I. in *Modern Methods for Multidimensional Dynamics Computations in Chemistry*, ed. D. L. Thompson, World Scientific, Singapore (1998), p. 101.
439. Benjamin, I. *Chem. Rev.* **106**, 1212 (2006).
440. Jedlovszky, P., Keresztúri, A., and Horvai, G. *Faraday Discuss.* **129**, 35 (2005).
441. Senapati, S., and Berkowitz, M.L. *Phys. Rev. Lett.* **87**, 176101 (2001); Jedlovszky, P., Vincze, A., and Horvai, G. *Phys. Chem. Chem. Phys.* **6**, 1874 (2004); Moreira, N.H., and Skaf, M.S. *Prog. Colloid Polym. Sci.* **128**, 81 (2004).
442. Schnell, B., Schurhammer, R., and Wipft, G. *J. Phys. Chem.* **108**, 2285 (2004).
443. Fernandes, P.A., Cordeiro, M.N.D.S., and Gomes, J.A.N.F. *J. Phys. Chem. B* **105**, 981 (2001).
444. Luo, M., and Dai, L.L. *J. Phys.: Cond. Matter* **19**, 375109 (2007).
445. Rivera, J.L., McCabe, C., and Cummings, P.T. *Phys. Rev. E* **67**, 011603 (2003).

446. Dominguez, H., Smondyrev, A.M., and Berkowitz, M.L. *J. Phys. Chem. B* **103**, 9582 (1999).
447. Pohorille, A., Nem, M.H., Schweighofer, K., and Wilson, M.A. in *Current Topics in Membranes*, Vol. 48, ed. D.W. Deamer, A. Kleinzeller, and D.M. Fambrough, Academic Press, San Diego (1999), p. 49.
448. Jones, M.N., and Chapman, D. *Micelles, Monolayers and Biomembranes*, Wiley–Liss, New York (1995).
449. Helfferich, F. *Ion Exchange*, McGraw Hill, New York (1962), Dover reprint (1995).
450. Milhaud, J. *Biochim. Biophys. Acta* **1663**, 19 (2004).
451. Nagle, J. F., and Tristram-Nagle, S. *Biochim. Biophys. Acta* **1469**, 159 (2000).
452. Tristram-Nagle, S., and Nagle, J. F. *Chem. Phys. Lipids* **127**, 3 (2004).
453. Disalvo, E. A., Martini, F., Almaleck, H., Diaz, S., and Gordillo, G. J. *Argentine Chem. Soc.* **92**, 1 (2004).
454. Berkowitz, M. L., Bostick, D. L., and Pandit, S. *Chem. Rev.* **106**, 1527 (2006).
455. Marrink, S. J., Tielman, D. P., van Buuren, A. R., and Berendsen, H. J. C., *Faraday Disc.* **103**, 191 (1996).
456. Singer, S.J., and Nicolson, G.L. *Science* **175**, 720 (1972).
457. Mouritsen, O. G. *Life–as a Matter of Fat*, Springer, Berlin (2005), p. 84.
458. Gennis, R. B. *Biomembranes*, Springer, New York (1989), p. 8.
459. Hille, B. *Ion Channels of Excitable Membranes*, 3rd edn, Sinauer, Sunderland MA (2001).
460. Fukada, K., and Gershfeld, N.L. *J. Phys. Chem. B* **41**, 8225 (1997).
461. Marsh, D. *Biochim. Biophys. Acta* **1286**, 183 (1996); Feller, S.E., and Pastor, R.W. *Biophys. J.* **71**, 1350 (1996).
462. Marrink, S. J. and Berendsen, H.J.C. *J. Phys. Chem.* **98**, 4155 (1994).
463. Zenthen, T., and MacAulay, N. *Int. Rev. Cytol.* **215**, issue on *Molecular Mechanisms of Water Transport across Biological Membranes*, ed. T. Zenthen and W.D. Stein, Academic Press, San Diego (2002), p. 203.
464. Feller, S.E., Zhang, Y., and Pastor, R.W. *J. Chem. Phys.* **103**, 10267 (1995); Hyvönen, M.T., Rantala, T.T., and Ala-Korpela, M. *Biophys. J.* **73**, 2907 (1997); Jedlovszky, P., and Mezei, M. *J. Phys. Chem. B* **105**, 3614 (2001).
465. Andelman, D. in *Handbook of Biological Physics*, Vol. 1, ed. R. Lipkowski and E. Sackmann, Elsevier, Amsterdam (1995), p. 603.
466. McLaughlin, S. *Ann. Rev. Biophys. Biophys. Chem.* **18**, 113 (1989).
467. Honig, B. H., Hubbell, W. L., and Flewelling, R. F. *Ann. Rev. Biophys. Biophys. Chem.* **15**, 163 (1986).
468. Rand, R. P. ref. 463, p. 33.
469. Leckband, D., and Israelachvili, J. *Quart. Rev. Biophys.* **34**, 105 (2001).
470. Perera, L., Essmann, U., and Berkowitz, M. *Langmuir* **12**, 2625 (1996).
471. Manciu, M., and Ruckenstein, E. *Adv. Coll. Int. Sci.* **112**, 109 (2004).
472. Benz, R.W., Castro-Román, F., Tobias, D.J., and White, S.H., *Biophys. J.* **88**, 805 (2005).
473. Klauda, J.B., Fučerka, N., Brooks, B.R., Pastor, R.W., and Nagle, J.F. *Biophys. J.* **90**, 2796 (2006).
474. Vereb, G., Szöllösi, J., Matkó, J., Nagy, P., Farkas, T., Vigh, L., Mátyus, L., Waldmann, T.A., and Damjanovich, S. *PNAS* **100**, 8053 (2003).
475. Patra, M. *Eur. Biophys.J.* **35**, 79 (2005); Feller, S.E., and Pastor, R.W. *J. Chem. Phys.* **111**, 1281 (1999); Goetz, R., and Lipowsky, R. *J. Chem. Phys.* **108**, 7397 (1998).
476. Tatulian, S.A. in *Surface Chemistry and Electrochemistry of Membranes*, ed. T.S. Sørensen, Dekker, New York (1999), p. 871.
477. Langner, M., and Kubica, K. *Chem. Phys. Lipids* **101**, 3 (1999).

478. One exception is large hydrophobic anions (ref. 467). Protons also have anomalously large permeability through membranes ($\sim 10^5$ larger than Na^+), which is not understood (see Deamer, D.W., and Nichols, J.W. *J. Membrane Biol.* **107**, 91 (1989)).
479. Stein, W.D. *Channels, Carriers and Pumps*, Academic Press, San Diego (1990).
480. Parsegian, V.A. *Nature* **221**, 844 (1969).
481. Nossal, R.J., and Lecar, H. *Molecular and Cell Biophysics*, Addison-Wesley, Redwood City, CA (1991), p. 196.
482. Born, M. *Z. Phys.* **1**, 45 (1920).
483. Schmickler, W. ref. 375, p. 11.
484. Overton, E. *Vierteljahrsschr. Naturforsch. Ges. Zürich* **44**, 88 (1899).
485. Overton, E. *Studien Über die Narkose Zugleich ein Betrag zur Allemeinen Pharmakologie*, Gustav Fischer, Jena (1901).
486. Meyer, H.H. *Arch. Exp. Pathol. Pharmakol.* **42**, 109 (1899).
487. Pohorille, A., New, M.H., Schweighofer, K., and Wilson, M.A. *Current Topics in Membranes* **48**, 49 (1999).
488. The influence of various short-chain alcohols on membrane thermal and mechanical properties has been studied recently by Ly, H.V., and Longo, M.L. *Biophys. J.* **87**, 1013 (2004).
489. Sandberg, W. S., Dubois, B. W., and Miller, K. W. *Current Topics in Membranes* **48**, 311 (1999).
490. Eckenhoff, R. G., and Johansson, J. S. *Pharmacol. Rev.* **49**, 343 (1997); Vemparala, S., Domene, C., and Klein, M.L. *Acc. Chem. Res.* **43**, 103 (2010).
491. Cantor, R. *Biophys. J.* **36**, 2339 (1997).
492. Baber, J., Ellena, J.F., and Cafiso, D.S. *Biochemistry* **34**, 6533 (1995); Eckenhoff, R.G. *PNAS* **93**, 2807 (1996).
493. Feller, S.E., Brown, C.A., Nizza, D.T., and Gawrisch, K. *Biophys. J.* **82**, 1396 (2002).
494. Högberg, C.-J., and Lyubartsev, A.P. *Biophys. J.* **94**, 525 (2008).
495. Chau, P-L., Hoang, P.M.N., Picand, S., and Jedlovszky, P. *Chem. Phys. Lett.* **438**, 294 (2007); Pickholz, M., Saiz, L., and Klein, M.L. *Biophys. J.* **88**, 1524 (2005); Tang, P., and Xu, Y. *PNAS* **99**, 16035 (2002); Koubi, L., Tarek, M., Bandyopadhyay, S., Klein, M.L., and Scharf, D. *Biophys. J.* **81**, 3339 (2001); Venparala, S., Saiz, L., Eckenhoff, R.G., and Klein, M.L. *Biophys. J.* **91**, 2815 (2006).
496. Patra, M., Salonen, E., Terama, E., Vattulainen, I., Fuller, R., Holopainen, J., and Karttunen, M. *Biophys. J.* **90**, 1121 (2006); Chanda, J., Chakraborty, S., and Bandyopadhyay, S. *J. Phys. Chem. B* **110**, 3791 (2006); Kranenburg, M., and Smit, B. *FEBS Lett.* **568**, 15 (2004); Wilson, M.A., and Pohorille, A. *J. Phys. Chem. B* **101**, 3130 (1997).
497. Dickey, A.N., and Faller, R. *Biophys. J.* **92**, 2366 (2007).
498. Devine, D. A., and Hancock, R. E.W. (eds), *Mammalian Host Defense Peptides*, Cambridge University Press, Cambridge (2004).
499. Zasloff, M. *Nature* **415**, 389 (2002).
500. Shai, Y. *Biochim. Biophys. Acta* **1462**, 55 (1999); in *Protein–Lipid Interactions*, ed. C.R. Mateo, J. Gómez, J. Villalaín, and J.M. González-Ros, Springer, Berlin (2006), p. 177.
501. White, S. H., Winley, W. C., and Selsted, M.E. *Curr. Opin. Struct. Biol.* **5**, 521 (1995); White, S. H., and Winley, W. C. *Curr. Opin. Struc. Biol.* **4**, 79 (1994).
502. Tossi, A., Antimicrobial peptide sequences database, available at http://www.bbcm.univ.trieste.it/~tossi/pag1.htm
503. Huang, W-N., Sue, S-C., Wang, D-S., Wu, P-L., and Wu, W-G. *Biochem.* **42**, 7457 (2003).
504. Tamm, L.K. *Current Topics in Membranes* **52**, 191 (2002), (review).

505. Znelev, D.V., and Needham, D. *Current Topics in Membranes* **52**, 437 (2002), (review).
506. Cafiso, D.S. *Current Topics in Membranes* **48**, 197 (1999).
507. Segrest, J.P., Jones, M.K., Mishra, V.K., and Anantharamaiah, G.M. *Current Topics in Membranes* **52**, 397 (2002), and references therein (review).
508. White, S.H., and Wimley, W.C. *Biochim. Biophys. Acta* **1376**, 339 (1998); *Curr. Opin. Struc. Biol.* **4**, 79 (1994), (reviews).
509. Wieprecht, T., and Seelig, J., *Current Topics in Membranes* **52**, 31 (2002), (review).
510. Laggner, P., and Lohner, K. in *Lipid bilayers*, ed. J. Katsaras and T. Gutberlet, Springer, Berlin (2001), p. 233 (review).
511. Kessel, A., And Be-Tal, N. *Current Topics in Membranes* **52**, 205 (2002) (review).
512. McIntosh, T.J., Vidal, A., and Simon, S.H. *Current Topics in Membranes* **52**, 309 (2002) (review).
513. Bechinger, B. *J. Membr. Biol.* **156**, 197 (1997); *Curr. Opin. Chem. Biol.* **4**, 639 (2000).
514. Dave, P.C., Billington, E., Pan, Y.L., and Strauss, S.K. *Biophys. J.* **89**, 2434 (2005).
515. Gawrisch, K., and Koenig, B.W. *Current Topics in Membranes* **52**, 163 (2002).
516. Davis, J.H., and Auger, M. *Prog. Nucl. Mag. Reson. Spectros.* **35**, 1 (1999).
517. Lösche, M. *Current Topics in Membranes* **52**, 117 (2002).
518. London, E., and Ladokhin, A.S. *Current Topics in Membranes* **52**, 89 (2002).
519. Blondelle, S.E., Lohner, K., and Aiguilar, M-I *Biochim. Biophys. Acta* **1462**, 89 (1999).
520. Jackson, M., Moore, D.J., Mantsch, H.N., and Mendelsohn, R. in *Handbook of Vibrational Spectroscopy*, ed. J.M. Chalmers and P.R. Griffiths, Wiley, Chichester (2002).
521. Cafiso, D.S. *Current Topics in Membranes* **52**, 3 (2002).
522. Dempsey, C.E. *Biochim. Biophys. Acta* **1031**, 143 (1990).
523. Tielman, D.P., and Sansom, M.S.P. *Int. J. Quantum Chem.* **83**, 166 (2001).
524. Ash, W.L., Zlomislic, M.R., Oloo, E.O., and Tielman, D.P. *Biochim. Biophys. Acta* **1666**, 158 (2004).
525. Jang, H., Ma, B., Woolf, T.B., and Nussinov, R. *Biophys. J.* **91**, 2848 (2006).
526. Vivcharuk, V., Tomberli, B., Tolokh, I.S., and Gray, C.G. *Phys Rev E* **77**, 031913 (2008). This paper simulates peptide–membrane interactions for uncharged (mammalian-like) membranes. A later paper deals with charged (bacterial-like) membranes, *Phys. Rev. E* **80**, 031911 (2009).
527. Cavalcanti, L.P., Konovalov, O., and Haas, H. *Chem. Phys. Lipids* **150**, 58 (2007).
528. Xu, Y., Yushnanov, V.E., and Tang, P. *Biosci. Rep.* **22**, 175 (2002); Gaede, H.C., and Gawrisch, K. *Biophys. J.* **85**, 1734 (2003), and Meier *et al.*, ref. 529.
529. Matos, C., Lina, J.L.C., Lopes, A., and Bastos, M. *Biophys. J.* **86**, 946 (2004); Meier, M., Blatter, X.L., Seelig, A., and Seelig, J. *Biophys. J.* **91**, 2943 (2006).
530. Song, C., and Nerdal, W. *Chem. Phys. Lipids* **149**S, S23 (2007); Lee, B.W., Fuller, R., Sum, A.K., Vattulainen, I., Patra, M., and Karttunen, M. *Fluid Phase Eqba.* **225**, 63 (2004).
531. Miller, C.E., Majewski, J., Kjaer, K. Weygand, M., Faller, R., Satija, S., and Kuhl, T.L. *Colloids Surf.* B**40**, 159 (2005).
532. Yu, K., Fu, W., Liu, H., Luo, X., Chen, K.X., Ding, J., Shen, J., and Jiang, H. *Biophys. J.* **86**, 3542 (2004); Gao, M., and Schulten, K. *Biophys. J.* **90**, 3267 (2006).
533. Miller, C.E., Majewski, J., Faller, R., Satija, S., and Kuhl, T.L. *Biophys. J.* **86**, 3700 (2004).
534. Huang, W.N., Sue, S.C., Wang, D.S., Wu, P.L., and Wu, W.G. *Biochem.* **42**, 7457 (2003).
535. *Biological Physics, Physics in Canada* **60** (2), 2004 special issue, ed. J. Katsaras and J.H. Davis.

536. Vachon, V., Schwartz, J-L., Masson, L., Brousseau, R., and Laprude, R., ref. 535, p. 101.
537. Mukhopadhyay, P., Vogel, H.J., and Tielman, D.P. *Biophys. J.* **86**, 337 (2004).
538. Hoff, B., Strandberg, E., Ulrich, A.S., Tielman, D.P., and Posten, C. *Biophys. J.* **88**, 1818 (2005).
539. Liu, Y., Yan, E.C.Y., and Eisenthal, K.B. *Biophys. J.* **80**, 1004 (2001).
540. Pascutti, P.G., El-Jaik, L.J., Bisch, P.M., Mundim, K.C., and Ito, A.S. *Eur. Biophys. J.* **28**, 499 (1999).
541. Grossfield, A. and Woolf, T.B. *Langmuir* **18**, 198 (2002).
542. Srinivas, G., and Klein, M.L. *Nanotechnol.* **15**, 1289 (2004).
543. Jedlovsz, P., and Mezei, M. *J. Am. Chem. Soc.* **122**, 5125 (2000).
544. White, S.H., and Wimley, W.C. *Ann. Rev. Biophys. Biomolec. Struc.* **28**, 319 (1999); Jensen, M.O., and Mouritsen, O.G. *Biochim. Biophys. Acta* **1666**, 205 (2004); Tamm, L.K. (ed.), *Protein–Lipid interactions*, Wiley–VCH, Weinheim (2005); Mateo, C.R., Gómez, J., Villalaín, J., and González-Ros, J.M. (eds) *Protein–Lipid interactions*, Springer, Berlin (2006); Cafiso, D.S. *Curr. Opin. Struc. Biol.* **1**, 185 (1991).
545. Brezesinski, G., and Möhwald, H. *Adv. Colloid Inter. Sci.*, **100–102**, 563 (2003); Marsh, D. *Biochim. Biophys, Acta* **1286**, 183 (1996).
546. Caffrey, M., and Wang, M. *Ann. Rev. Biophys. Biomolec. Struc.* **24**, 351 (1995); Caffrey, M. in *Electrostatic Effects in Soft Matter and Biophysics*, (ed. C. Holm, P. Kékicheff and R. Podgornik), Kluwer, Dordrecht (2001), p. 1.
547. Liu, J., and Conboy, J.C. *Langmuir* **21**, 9091 (2005).
548. Rand, R.P. *Ann. Rev. Biophys. Bioeng.* **10**, 277 (1981).
549. McIntosh, T.J., and Simon, S.A. *Ann. Rev. Biophys. Biomolec. Struc.* **23**, 27 (1994).
550. Israelachvili, J. *Acc. Chem. Res.* **20**, 415 (1987); Benz, M., Gutsmann, T., Chen, N., Tadmor, R., and Israelachvili, J. *Biophys. J.* **86**, 870 (2004).
551. Yuan, C., and Johnston, L.J. *J. Microsc.* **205**, 136 (2002); Benz *et al.*, ref. 550; Diociaiuti, M., Ruspantini, I., Giordani, C., Bordi, F., and Christolini, P. *Biophys. J.* **86**, 321 (2004).
552. Sheetz, M.P. *Laser Tweezers in Cell Biology*, Academic Press, San Diego (1998). See in particular the articles by J. Dai and M.P. Sheetz (p. 157) and A. Kusumi, Y. Sako, T. Fujiwara, and M. Tomishige (p. 173).
553. Arsov, Z., and Quaroni, L. *Chem. Phys. Lipids* **150**, 35 (2007); Snyder, R.G., Tu, K., Klein, M.L., Mendelssohn, R., Strauss, H.L., and Sun, W. *J. Phys. Chem. B* **106**, 6273 (2002); Lewis, R.N.A.H., and McElhaney, R.N. *Chem. Phys. Lipids* **96**, 9 (1998). (See also other papers in this issue of *Chem. Phys. Lipids* devoted to IR of membranes and lipids.)
554. Lindblom, G., and Gröbner, G. *Curr. Opin. Coll. Int. Sci.* **11**, 24 (2006).
555. Hite, R.K., Raunser, S., and Walz, T. *Curr. Opin. Struc. Biol.* **17**, 289 (2007).
556. Lee, A.G. in *Protein–Lipid Interactions*, ed. C.R. Mateo, J. Gómez, J. Vallalaín, and J.M. González-Ros Springer, Berlin (2006), p. 141.
557. Wiener, M.C., and White, S.H. *Biophys. J.* **61**, 434 (1992).
558. Feller, S.E., ed., *Computational Modeling of Membrane Bilayers, Current Topics in Membranes* **60** (2008); Ciccotti, G., and Binder, K., eds, *Computer Simulations in Condensed Matter: From Materials to Chemical Biology, Vol. 2*, Springer, Berlin (2006). See especially the articles by F. Schmid (p. 211) and B. Smit, M. Kranenburg, M.M. Sperotto, and M. Venturoli (p. 259); Nielson, S.G., and Klein, M.L. in *Bridging Time Scales: Molecular Simulations for the Next Decade*, ed. P. Nielaba, M. Mareschal, and G. Cicotti, Springer, Berlin (2002), p. 27; Scott, H.L. *Current Opin Struc. Biol.* **12**, 495 (2002); Saiz, L., and Klein, M.L. *Acc. Chem. Res.* **35**, 482 (2002); Saiz, L., Bandyopadhy, S., and Klein, M.L. *Biosci. Rep.* **22**, 151 (2002).

559. Uncharged (zwitterionic) lipid head groups include PC (phosphatidylcholine (lecithin)) and PE (phosphatidylethanolamine). Some corresponding complete lipid molecules are POPC (palmitoyl-oleoyl PC), DPPC (di-palmitoyl PC), DMPC (di-myristoyl PC), DOPC (di-oleoyl PC), and DLPE (di-laureoyl PE). Charged (anionic) lipid head groups include PG (phosphatidylglycerol) and PS (phosphatidylserine). Some corresponding complete lipid molecules are POPG (palmitoyl-oleoyl PG), POPS (palmitoyl-oleoyl PS); and DPPS (di-palmitoyl PS).
560. Gurtovenko, A.A., Miettinen, M., Karttunen, M., and Vattulainen, I. *J. Phys. Chem. B* **109**, 21126 (2005); Jang, H., Ma, B., Woolf, T.B., and Nussinov, R. *Biophys. J.* **91**, 2848 (2006); Bennum, S.V., Longo, M., and Faller, R. *J. Phys. Chem. B* **111**, 9504 (2007).
561. Falck, E., Patra, M., Karttunen, M., Hyvönen, M.T., and Vattulainen, I. *Biophys. J.* **87**, 1076 (2004); Bhide, S., Zhang, Z., and Berkowitz, M. *Biophys. J.* **92**, 1284 (2007); Aittoniemi, J., Niemelä, P.S., Hyvönen, M.T., Karttunen, M., And Vattulainen, I. *Biophys. J.* **92**, 1125 (2007).
562. Zaraiskaya, T., and Jeffrey, K.R. *Biophys. J.*, **88**, 4017 (2005).
563. Snyder *et al.*, ref. 553.
564. Sachs, J.N., Petrache, H.I., and Woolf, T.B. *Chem. Phys. Lipids* **126**, 211 (2004); Benz, R.W., Castro-Román, F., Tobia, D.J., and White, S.H. *Biophys. J.* **88**, 805 (2005); Klauda, J.B., Kučerka, N., Brooks, B.R., Pastor, R.W., and Nagle, J.F. *Biophys. J.* **90**, 2796 (2006).
565. Roux, B., and Schulten, K. *Structure* **12**, 1343 (2004); Gumbart, J., Wang, Y., Aksimentiev, A., Tajkhorshid, E., and Schulten, K. *Curr. Opin. Struc. Biol.* **15**, 423 (2005).
566. Roux, B. *Ann. Rev. Biophys. Biomolec. Struc.* **34**, 153 (2005).
567. de Groot, B.L., and Grubmüller, H. *Curr. Opin. Struc. Biol.* **15**, 176 (2005).
568. Wang, Y., Schulten, K., and Tajkhorshid, E. *Structure* **13**, 1107 (2005).
569. Tielman, D.P., and Berendsen, H.J.C. *Biophys. J.* **74**, 2786 (1998); Im, W., and Roux, B. *J. Mol. Biol.* **319**, 1177 (2002).
570. Jensen, M.O., Yin, Y., Tajkhorshid, E., and Schulten, K. *Biophys. J.* **93**, 92 (2007).
571. Lamoureux, G., Klein, M.L., and Bernèche, S. *Biophys. J.: Biophys. Lett.* L **82** (2007).
572. Gumbart, J., and Schulten, K. *Biophys. J.* **90**, 2356 (2006).
573. Baaden, M., and Sansom, M.S.P. *Biophys. J.* **87**, 2942 (2004).
574. Klapp, S.H.L., and Schoen, M. *Rev. Comput. Chem.*, Vol. 24, Ch. 4–6, Wiley, New York (2007).
575. Kierlik, E., Monson, P.A., Rosinberg, M.L., Sarkisov, L., and Tarjus, G. *Phys. Rev. Lett.* **87**, 055701 (2001).
576. Detcheverry, F., Kierlik, E., Rosinberg, M.L., and Tarjus, G. *Phys. Rev. E* **68**, 061504 (2003).
577. Woo, H.J., and Monson, P.A. *Phys. Rev. E* **67**, 041207 (2003)
578. Klapp, S.H.L., and Schoen, M. *Rev. Comp. Chem.*, Vol. 24, Ch. 7, Wiley, New York (2007).
579. Schmidt, M. *J. Phys.: Cond. Matter* **17**, S3481 (2005).
580. Archer, A.J., Schmidt, M., and Evans, R. *Phys. Rev. E* **73**, 011506 (2006).
581. Volterra, V. *Theory of Functionals*, Blackie and Sons, London (1930); Dover, New York (1959).
582. Killingbeck, J., and Cole, G.H.A. *Mathematical Techniques and Physical Applications*, Academic Press, New York (1971), pp. 290–7.
583. Elsgolc, L.E. *Calculus of Variations*, Pergamon, London (1961).
584. Parr, R.G., and Yang, W. *Density-Functional Theory of Atoms and Molecules*, Oxford University Press, Oxford (1989), pp. 246–54.

585. The notation $F[f(x)]$ will sometimes be used in place of $F[f]$; however, in addition to brevity, the latter has the advantage of stressing that in general F depends on f over its whole range, and not just on f at a particular point. (For an exceptional case, see note 594 below.)
586. Functional techniques were extensively used in quantum field theory prior to their use in statistical mechanics. See, for example: Wentzel, G., *Introduction to the Quantum Theory of Fields*, Interscience, New York (1949); Friedrichs, K.O. *Mathematical Aspects of the Quantum Theory of Fields*, Interscience, New York (1953); Primas, H. *Int. J. Quantum Chem.* **1**, 493 (1967).
587. The first use of functional derivative methods in statistical mechanics seems to have been by Bogoliubov and Percus. See Bogoliubov, N.N. *Studies in Statistical Mechanics*, ed. J. deBoer and G.E. Uhlenbeck, Vol. I, North-Holland, Amsterdam (1962), and Percus, J.K. *Phys. Rev. Lett.*, **8**, 462 (1962). Other reviews are given by Lebowitz, J.L., and Percus, J. K. *J. Math. Phys.* **4**, 116 (1963); Verlet, L. *Physica*, **32**, 304 (1966).
588. Henderson, D., and Davison, S.G. *Physical Chemistry, an Advanced Treatise, Vol. II Statistical Mechanics*, ed. H. Eyring, Academic Press, New York (1967), pp. 374–8.
589. Münster, A. *Statistical Thermodynamics, Vol. I*, Springer-Verlag, Berlin (1969), pp. 358–64.
590. Ben-Naim, A. *Water and Aqueous Solutions*, Plenum, New York (1974), pp. 443–53.
591. Hansen, J.-P., and McDonald, I.R. *Theory of Simple Liquids*, 3rd edn, Academic Press, London (2006), pp. 50–3.
592. Balescu, R. *Equilibrium and Nonequilibrium Statistical Mechanics*, Wiley, New York (1975), p. 250.
593. Roman, P. *Introduction to Quantum Field Theory*, Wiley, New York (1969), p. 220. For more refined definitions, which treat the δ-function in (8A.7) more rigorously, see Schweber, S.S. *An Introduction to Relativistic Quantum Field Theory*, Harper and Row Evanston, (1961), p. 193, and Hanggi, P. in *Stochastic Processes Applied to Physics*, ed., L. Presquera and M.A. Rodriguez, World Scientific, Singapore (1985).
594. Recall that $f(x') = \int \mathrm{d}x\delta(x'-x)f(x)$, so that $\delta f(x') = \int \mathrm{d}x\delta(x'-x)\delta f(x)$. Here we regard $F[f] = f(x')$ as that functional that gives the value of f at point x'. Note also that (8A.10) is intuitively reasonable; i.e., the variation in f at x' due to a variation in f at x is zero unless $x' = x$. Furthermore we have $\delta f(x')/\delta f(x) = \delta f(x)/\delta f(x')$.
595. If $G = G(f, f')$, i.e. depends on the derivative f', as well as on f, (8A.12) generalizes to[582]

$$\frac{\delta F}{\delta f} = \frac{\partial G}{\partial f} - \frac{\mathrm{d}}{\mathrm{d}x}\frac{\partial G}{\partial f'}$$

i.e. the Euler-Lagrange variational derivative, familiar from classical mechanics.
596. Schulman, L.S. *Techniques and Applications of Path Integration*, Wiley, New York (1981).
597. Feynman, R.P., and Hibbs, A.R. *Quantum Mechanics and Path Integrals*, McGraw Hill, New York (1965).
598. Wiegel, F.W. *Phys. Reports* **16C**, 57 (1975); *Introduction to Path-Integral Methods in Physics and Polymer Science*, World Scientific, Singapore (1986).
599. Revzen, M. *Am. J. Phys.* **38**, 611 (1970).
600. Gel'fand, I.M., and Yaglom, A.M. *J. Math. Phys.* **1**, 48 (1960).
601. Brush, S.G. *Rev. Mod. Phys.* **33**, 79 (1961).
602. Montroll, E.W. *Commun. Pure Appl. Math.* **V**, 415 (1952).
603. Rowlinson, J.S. and Widom, B., ref. 18, p. 298.
604. The correlation functions $h(12)$ and $c(12)$ are defined in §3.1.5, while $f(1)$ is defined in §3.1.4.

605. Yvon, J. *Nuovo Cimento (suppl.)* **9**, 144 (1958).
606. Percus, J.K. *The Equilibrium Theory of Classical Fluids*, ed. H.L. Frisch and J.L. Lebowitz, Benjamin, New York (1964), p. II-33.
607. Evans, R.A. *Adv. Phys.*, **28**, 143 (1979).
608. To derive (8A.33) we use (8.5) together with (8.4). Noting that $\mathcal{U} = 0$ for the ideal gas and letting $N' = N - 1$ we have

$$f(1) = \frac{\hat{z}}{\Xi} \exp[-\beta v(1)] \sum_{N'} \frac{\hat{z}^{N'}}{N'!} [\int \mathrm{d}\mathbf{r} \mathrm{d}\omega \exp(-\beta v)]^{N'} = \hat{z} \exp[-\beta v(1)],$$

since the sum over N' is just the grand partition function.
609. We could expand $f(1|v)$ in terms of v about $v = 0$,

$$f(1|v) = f(1|0) + \int \left(\frac{\delta f(1|v)}{\delta v(2)} \right)_{v=0} v(2)\mathrm{d}2 + \ldots.$$

However, this expansion is not useful for potentials having a repulsive core because v is large and approaches infinity at short range. Thus the series will not converge at short range.
610. Verlet, L. *Physica* **30**, 95 (1964).
611. Verlet, L. *Physica* **31**, 959 (1965).
612. In quantum mechanics the angular momentum operator is proportional to the operator that generates rotations, so that (8C.2) can be interpreted as $\mathbf{J} = \mathbf{L} + \mathbf{S}$; i.e. the total angular momentum operator is the sum of an orbital part and a spin part.
613. The contributions of the Dutch school, beginning with Lorentz a century ago, and continuing to this day, have been especially noteworthy; see, e.g. refs. 2, 13, and 16 of Chapter 10, and also de Groot, S.R. *The Maxwell Equations*, North Holland, Amsterdam (1969). The latter is volume 4 in the series *Studies in Statistical Mechanics*, ed. J. de Boer and G.E. Uhlenbeck, North Holland.
614. Jackson, J.D. *Classical Electrodynamics*, 2nd edn, Wiley, New York (1975), p. 226. Jackson follows the development of Russakoff, G. *Am. J. Phys.* **38**, 1188 (1970).
615. Gray, C.G., Karl, G., and Novikov, V.A. *Am. J. Phys.* **77**, 807 (2009).
616. The contact term in the magnetic dipole field was first discussed by Fermi, E. *Z. Phys.* **60**, 320 (1930), in connection with hyperfine interactions in atoms. See also Casimir, H.B.G. *On the Interaction between Atomic Nuclei and Electrons, Teyler's tweede Genootschap*, E.F. Bohn, Haarlem, Netherlands (1936), esp. 2nd edn, Freeman (1963), p. ix. For modern discussions, see Jackson, J.D. *Classical Electrodynamics*, 3rd edn, Wiley, New York (1999), pp 188–91, and Gray, C.G., Karl, G., and Novikov, V.A. *Am. J. Phys.* **78**, 936 (2010).
617. Karl, G., and Novikov, V.A. *Phys. Rev. C* **74**, 024001 (2006).
618. More rigorous arguments can be given: see, e.g. Frahm, C.P. *Am. J. Phys.* **51**, 826 (1983).

9
STRUCTURE FACTOR

> The men of experiment are like the ant, they only collect and use; the reasoners resemble spiders, who make cobwebs out of their own substance. But the bee takes the middle course: it gathers its material from the flowers of the garden and field, but transforms and digests it by a power of its own. Not unlike this is the true business of philosophy (science); for it neither relies solely or chiefly on the powers of the mind, nor does it take the matter which it gathers from natural history and mechanical experiments and lay up in the memory whole, as it finds it, but lays it up in the understanding altered and digested. Therefore, from a closer and purer league between these two faculties, the experimental and the rational (such as has never been made), much may be hoped.
>
> Francis Bacon, *Novum organum* (1620)

> X-rays will prove to be a hoax.
>
> William Thomson, Lord Kelvin (1899)

The microscopic structure of simple atomic fluids is now quite well understood. Neutron scattering studies, in particular, have yielded much important information on the rare gas liquids[1] neon,[2,3] argon,[4,5] krypton,[6–14] and xenon.[15] Using highly accurate *ab initio* pair potentials, and incorporating quantum diffraction effects (in the case of Ne) and three-body forces (in the case of Kr and Xe), researchers have been able to demonstrate essentially exact agreement between theory (whether based on integral equations or computer simulation) and experiment.

For molecular liquids, in contrast, the picture is quite different. Our understanding is not nearly so complete, and the investigation and analysis of the factors affecting the structure of these fluids remains an active field of research.

The structure factors of molecular fluids show important qualitative differences from those of atomic fluids. The most noticeable changes are as follows:[16]

(a) The height of the first peak in $S(k)$ is smaller than for monatomic fluids at the same reduced density and temperature, and subsequent maxima and minima tend to be less pronounced;
(b) $S(k)$ usually oscillates slowly at large k, whereas for monatomic fluids the oscillations quickly die out; and
(c) One often observes new maxima and minima in $S(k)$ that are not present for monatomic fluids.

Observations (a) and (b) are associated largely with intramolecular effects, and are present even if the molecules rotate freely. Behaviour of type (c), however, often arises because of correlations in the orientations of neighbouring molecules. Examples of behaviour of type (c) are the subsidiary maximum on the high-k side of the main neutron diffraction peak for bromine,[17] the oscillations in $S(k)$ for $k = 2$–$4\,\text{Å}^{-1}$ for carbon tetrachloride,[18] and the bump on the high-k side of the main peak for water.[19] These phenomena cannot be explained on the basis of *any* isotropic intermolecular potential. In this chapter we discuss the effects of angular correlations on the structure factor for molecular fluids.

Structure factors[20] may be measured using the techniques of neutron,[21] X-ray,[22,23] γ-ray,[24,25] or electron diffraction.[26] In § 9.1 we consider the general theory of scattering, and derive expressions for the scattering differential cross-section for neutrons, X-rays, γ-rays, and electrons. In § 9.2 we show how these quantities are related to $S(k)$. In subsequent sections we consider various theoretical treatments that relate $S(k)$ to the intermolecular pair potential $u(\mathbf{r}\omega_1\omega_2)$. These include perturbation theory (§ 9.3), integral equations (§ 9.4), and computer simulations (§ 9.5).

9.1 Scattering theory

In this section we consider the processes of neutron, X-ray, γ-ray, and electron scattering by molecules. In each case, the principal result is an expression for the scattered intensity, expressed in terms of the relevant structure factor (neutrons, Eq. (9.28); X-rays, Eq. (9.44); electrons, Eq. (9.48)). These are all of the form

$$S(k) \sim \left\langle \sum_{ij} \mathcal{A}_i \mathcal{A}_j \exp(i\mathbf{k}\cdot\mathbf{r}_{ij}) \right\rangle, \tag{9.1}$$

where $\mathbf{r}_{ij}$ is the vector joining two scattering centres i and j, $\mathcal{A}_i$ is some scattering amplitude, and $\mathbf{k} = \mathbf{k}_0 - \mathbf{k}_S$ is the scattering vector, with $\mathbf{k}_0$ and $\mathbf{k}_S$ the wavevectors of the incident and scattered waves.

To understand this result qualitatively, we note that (9.1) can be written as a superposition of scattered intensities

$$S(k) \sim \left\langle \left| \sum_{i} \mathcal{A}_i \exp(i\mathbf{k}\cdot\mathbf{r}_{i}) \right|^2 \right\rangle, \tag{9.2}$$

that is, as the sum of terms with amplitudes $\mathcal{A}_i \exp(i\varphi_i)$, where $\varphi_i = \mathbf{k}\cdot\mathbf{r}_i$ is a site-dependent phase shift. Referring to Fig. 9.1, we see that, for a given scattering geometry, this shift represents the change in the path length relative to some origin of coordinates. It therefore simply reflects the spatial modulation of incident and scattered waves over the dimensions of the sample.

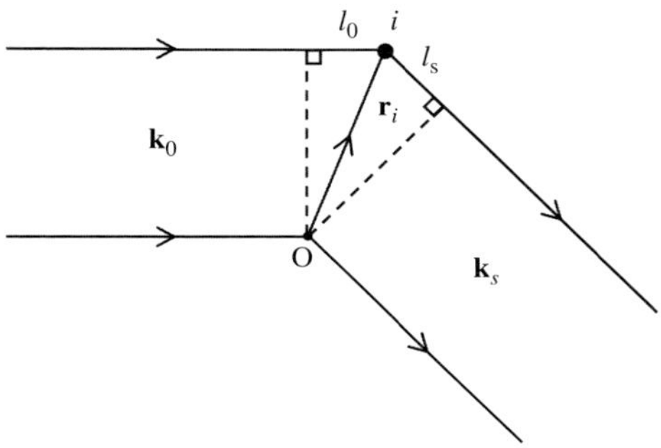

FIG. 9.1 Phase lag for scattering at a site i. The lengths $l_0 = \hat{\mathbf{k}}_0 \cdot \mathbf{r}_i$ and $l_S = -\hat{\mathbf{k}}_S \cdot \mathbf{r}_i$, where $\hat{\mathbf{k}}_0$ and $\hat{\mathbf{k}}_S$ are unit vectors along the incident and scattered waves, respectively. The resultant phase shift is thus $\varphi_i = k_0 l_0 + k_S l_S = \mathbf{k} \cdot \mathbf{r}_i$, where $\mathbf{k}$ is the scattering wavevector.

9.1.1 *Neutron scattering*[27]

We begin our discussion of neutron scattering by discussing the processes involved in scattering from a single nucleus, and then consider how the scattering is modified when several scattering centres are present—specifically, the nuclei of molecules in a fluid.

Scattering by a single nucleus

To understand the process by which a nucleus scatters incident neutrons, it is necessary to have some understanding of the strong nuclear force.[28] The fundamental strong interaction is that between quarks, and arises from the exchange of spin 1 particles called gluons. Nucleons are made up of three quarks: a neutron of two down quarks each of charge $-\frac{1}{3}$ and an up quark of charge $\frac{2}{3}$; a proton of one down and two up quarks. The gluon exchange force increases as quark separation increases, so that quarks are confined. The nuclear force between nucleons arises from the exchange of mesons, which are made up of a quark–antiquark pair: it thus bears essentially the same relationship to the strong interaction as the van der Waals force between molecules does to the electrostatic interaction.

Let us first consider the simplest possible systems consisting of two nucleons: *pp*, *nn*, and *np*. The interaction in the *np* system is directly relevant to neutron scattering by hydrogen and hydrogen-containing molecules. There is evidence (e.g. the deuteron quadrupole moment) to show that the strong nuclear force $V(\mathbf{r})$ is not purely central. Anisotropy arises because the mesons exchanged have negative parity, being formed from fermion–antifermion pairs. Conservation of parity and angular momentum in the nucleon–nucleon interaction therefore gives rise to a spin-dependent tensor force proportional[28] to $Y_{2m}(\theta, \phi)$. The long-range part of the strong nuclear force is the pseudoscalar one-pion exchange potential[28,29]

$$V_{ps}(\mathbf{r}) = \left(\frac{1}{3}\boldsymbol{\sigma}_1 \cdot \boldsymbol{\sigma}_2 + S_{12}\left[\frac{1}{3} + (\mu r)^{-1} + (\mu r)^{-2}\right]\right) \boldsymbol{\tau}_1 \cdot \boldsymbol{\tau}_2 \frac{\alpha \hbar c \exp(-\mu r)}{r}. \tag{9.3}$$

Here $\boldsymbol{\sigma}$ and $\boldsymbol{\tau}$ denote Pauli spin and isospin[30] operators, respectively. $S_{12} = 3(\boldsymbol{\sigma}_1 \cdot \hat{\mathbf{r}})(\boldsymbol{\sigma}_2 \cdot \hat{\mathbf{r}}) - \boldsymbol{\sigma}_1 \cdot \boldsymbol{\sigma}_2$. $\mu^{-1} = 1.4$ fm is the range of the potential and is given by $\hbar/m_\pi c$, where m_π is the pion mass. $\alpha \approx 0.075$ is the pion–nucleon coupling constant. Scalar and vector meson exchange are also possible.[28,29] Since these processes involve heavy mesons such as ω and ρ, the range of such forces is less than that of the pseudoscalar interaction.

In the non-relativistic limit the two-nucleon wavefunction corresponding to the potential $V(\mathbf{r})$ satifies the Schrödinger equation

$$-\frac{\hbar^2}{2m}\nabla^2\psi + V(\mathbf{r})\psi = E\psi, \tag{9.4}$$

where E is the energy and m is the reduced mass. We are interested in positive energy scattering solutions in which an incident neutron of momentum $\hbar\mathbf{k}_0$ is scattered elastically by a proton at the origin. The solution of (9.4) is then[31]

$$\psi(\mathbf{r}) = \psi_0(\mathbf{r}) + \psi_S(\mathbf{r}), \tag{9.5}$$

where the incident wave is given by

$$\psi_0(\mathbf{r}) = \exp(i\mathbf{k_0} \cdot \mathbf{r}) \tag{9.6}$$

and the scattered wave satisfies

$$\psi_S(\mathbf{r}) = -\frac{2m}{\hbar^2}\int \mathrm{d}\mathbf{r}' \,\frac{\exp(ik_0|\mathbf{r}-\mathbf{r}'|)}{4\pi|\mathbf{r}-\mathbf{r}'|}V(\mathbf{r}')\psi(\mathbf{r}'). \tag{9.7}$$

At large distances from the scattering centre $|\mathbf{r}-\mathbf{r}'| \simeq r - r'(\hat{\mathbf{r}} \cdot \hat{\mathbf{r}}')$, and therefore $k_0|\mathbf{r}-\mathbf{r}'| \simeq k_0 r - \mathbf{k}_S \cdot \mathbf{r}'$, where $\mathbf{k}_S = k_0\hat{\mathbf{r}}$ is a vector of magnitude k_0 in the scattering direction $\mathbf{r}$. Therefore in the limit $r \to \infty$

$$\psi_S(\mathbf{r}) = f(\theta,\phi)\frac{\exp(ik_0 r)}{r}, \tag{9.8}$$

where the scattering amplitude

$$f(\theta,\phi) = -\frac{2m}{\hbar^2}\frac{1}{4\pi}\int \mathrm{d}\mathbf{r}' \,\exp(-i\mathbf{k}_S \cdot \mathbf{r}')V(\mathbf{r}')\psi(\mathbf{r}'). \tag{9.9}$$

In the first Born approximation (weak scattering), we can replace $\psi(\mathbf{r}')$ in (9.9) by $\psi_0(\mathbf{r}')$. This yields

$$f(\theta,\phi) = -\frac{2m}{\hbar^2}\frac{1}{4\pi}\int \mathrm{d}\mathbf{r}' \,\exp(i\mathbf{k} \cdot \mathbf{r}')V(\mathbf{r}'), \tag{9.10}$$

where $\mathbf{k} = \mathbf{k}_0 - \mathbf{k}_S$ is the momentum (in units of $\hbar$) transferred to the nucleus; note that $k = 2k_0 \sin\frac{1}{2}\theta$ depends on the scattering angle θ. Thus the scattering amplitude is related to the Fourier transform of the potential.

If we write the nucleon–nucleon interaction in the form

$$V(\mathbf{r}) = \sum_{lm} V_{lm}(r)C_{lm}(\theta,\phi), \tag{9.11}$$

where $C_{lm}(\theta, \phi)$ is a reduced spherical harmonic (Appendix A), then from the Rayleigh partial wave expansion (B.89),

$$\exp(i\mathbf{k} \cdot \mathbf{r}') = \sum_{lm} (2l+1) i^l j_l(kr) C_{lm}(\theta_k, \phi_k) C_{lm}(\theta', \phi')^*, \tag{9.12}$$

where $j_l(x)$ denotes a spherical Bessel function[32] of order l, we find

$$f(\theta, \phi) = -\frac{2m}{\hbar^2} \sum_{lm} i^l C_{lm}(\theta_k, \phi_k) \int_0^\infty \mathrm{d}r' \, r'^2 j_l(kr') V_{lm}(r'). \tag{9.13}$$

Now $V_{lm}(r')$ differs from 0 only for $r' \lesssim \mu^{-1} \approx 10^{-15}$ m, while in typical neutron scattering experiments on molecular fluids $k \sim 10^{11}$ m^{-1}. The relative magnitude of the terms of order l in (9.13) is therefore of order $j_l(k/\mu) \sim (k/\mu)^l \sim 10^{-4}$. Thus the anisotropic components of the *np* potential do not contribute to the scattering amplitude. This is also true for scattering from more complex nuclei, and may be explained physically as follows. An element of the scattering potential $V(\mathbf{r})$ that transforms as a spherical harmonic $Y_{lm}(\theta, \phi)$ scatters incoming neutrons such that a change of angular momentum of order $[l(l+1)]^{1/2}\hbar$ occurs. For a potential of range μ^{-1}, this means that the incident and scattered momenta must differ by $\hbar k \gtrsim [l(l+1)]^{1/2}\hbar\mu$, else the necessary torque cannot be exerted. Thus if $k \ll \mu$, the situation that applies in slow neutron scattering studies of molecular fluids, only *s*-wave scattering by the isotropic ($l = 0$) component of the potential is possible.

The most general form for the isotropic component of the two-nucleon potential is a combination of spin and isospin exchange forces:[28]

$$V_{00}(r) = \sum_{i,j=0}^{1} V^{ij} (\boldsymbol{\sigma}_1 \cdot \boldsymbol{\sigma}_2)^i (\boldsymbol{\tau}_1 \cdot \boldsymbol{\tau}_2)^j. \tag{9.14}$$

In this equation each coefficient $V^{ij}(r) = \hbar c \sum_\alpha \lambda_\alpha^{(ij)} \exp(-\mu_\alpha r)/r$ is a sum of terms representing exchange of mesons of mass m_α; $\mu_\alpha = m_\alpha c/\hbar$. $\lambda_\alpha^{(ij)}$ is a dimensionless coupling constant. Values for the m_α and $\lambda_\alpha^{(ij)}$ are given in ref. 29. We therefore obtain for the *np* *s*-wave scattering amplitude

$$f(\theta, \phi) = -a, \tag{9.15}$$

where

$$a = \frac{2mc}{\hbar} \sum_\alpha \sum_{ij} \frac{\lambda_\alpha^{(ij)}}{\mu_\alpha^2} \langle S | (\boldsymbol{\sigma}_1 \cdot \boldsymbol{\sigma}_2)^i | S \rangle \, \langle T | (\boldsymbol{\tau}_1 \cdot \boldsymbol{\tau}_2)^j | T \rangle \tag{9.16}$$

is called the scattering length. The scattering is isotropic for small k/μ. Here S and T denote the spin and isospin states of the system. For the $l = 0$ state of $^1(np)$, $\boldsymbol{\sigma}_1 \cdot \boldsymbol{\sigma}_2 = -3$ and $\boldsymbol{\tau}_1 \cdot \boldsymbol{\tau}_2 = +1$, while for $^3(np)$, $\boldsymbol{\sigma}_1 \cdot \boldsymbol{\sigma}_2 = +1$ and $\boldsymbol{\tau}_1 \cdot \boldsymbol{\tau}_2 = -3$ (the Pauli exclusion principle requires that for the exchange of two fermions $l + S + T = \text{odd}$). Thus the scattering length is different for the singlet and triplet states. Inserting the appropriate meson masses and meson–nucleon coupling constants into (9.16), we obtain[28,33]

$$\begin{aligned} {}^1a &= -23.71 \ \text{fm}, \\ {}^3a &= \ \ 5.43 \ \ \text{fm}. \end{aligned} \tag{9.17}$$

The negative scattering length of ${}^1(np)$ is related to the fact that the singlet state of the deuteron is unbound.[28] These results show that the scattering is very different for singlet and triplet states, and as a consequence hydrogen has a very large incoherent neutron cross-section (*vide infra*).

The scattering length can be interpreted[34] as the radius of a hard sphere that would produce the same scattering at infinity as the actual scattering potential. In theory, a could be calculated for *any* nucleus if a good model for the interaction potential were available. In practice, however, a is regarded as an experimentally determined parameter. A compilation of scattering lengths for the isotopes is given by Lovesey.[21] For the heavier nuclei negative scattering lengths are very rare. This can be explained on the basis of a simple model in which $V(\mathbf{r})$ is approximated by a square well of depth V_0 and width R. For $kR \ll 1$ the scattering length is readily shown to be[34]

$$a = R\left(1 - \frac{\tan KR}{KR}\right), \tag{9.18}$$

where $K = (2mV_0)^{1/2}/\hbar$. Except for $KR < \pi/2$, the regions where $a < 0$ are very small (Fig. 9.2). For all nuclei except hydrogen it turns out that $KR > 3$, and as

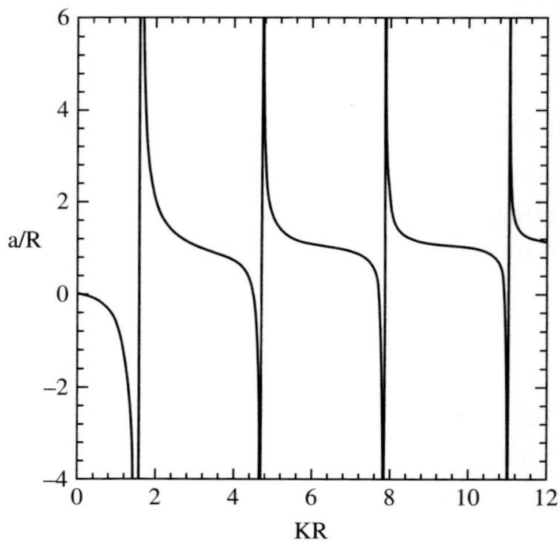

FIG. 9.2 Neutron scattering length a for a square well potential as a function of $KR = (2mV_0)^{1/2}R/\hbar$, as given by (9.18).

a result there are only three or four known cases of negative a. If we assume that the values of KR are distributed randomly over the different isotopes, we can predict a distribution of values of a in good agreement with what is observed.[35] a may also contain an imaginary component, which corresponds to absorption of the incident neutron. This can arise when the neutron couples to internal modes of motion of the nucleus, or, in the case of np, through radiative capture. For most nuclei the probability of absorption at typical incident energies is small. An observable consequence, however, is neutron inelastic scattering,[21] in which the absorbed neutron is reemitted at a slightly different energy. The theoretical description of inelastic scattering by fluids leads naturally to the introduction of the dynamic structure factor[36] $S(k, \omega)$, which is the Fourier transform of the Van Hove correlation function[37] $G(\mathbf{r}, t)$, and which contains dynamical information concerning the molecules of the fluid. $S(k, \omega)$ is discussed further in Chapter 11.

The scattered neutron flux (i.e. the current density, or number of neutrons per unit area per unit time) associated with the scattered wave is

$$\mathbf{j}_S = -\frac{i\hbar}{2m}\left(\psi_S^*\nabla\psi_S - \psi_S\nabla\psi_S^*\right). \tag{9.19}$$

The radial (indeed, the only) component of this flux is

$$\begin{aligned}\mathbf{j}_S \cdot \hat{\mathbf{r}} &= -\frac{i\hbar}{2m}\left(\psi_S^*\frac{\partial}{\partial r}\psi_S - \psi_S\frac{\partial}{\partial r}\psi_S^*\right) \\ &= \frac{\hbar k_0 a^2}{mr^2}.\end{aligned} \tag{9.20}$$

The differential scattering cross-section $d\sigma$ is defined as the number of neutrons scattered into an element of area $d\mathbf{S} = dS\,\hat{\mathbf{r}} = r^2 d\Omega\,\hat{\mathbf{r}}$ per unit time, divided by the incident flux $j_0 = \hbar k_0/m$, i.e.

$$\mathrm{d}\sigma = \frac{\mathbf{j}_S \cdot \mathrm{d}\mathbf{S}}{j_0} = a^2 d\Omega. \tag{9.21}$$

The differential scattering cross-section for elastic scattering by a single nucleus is therefore given by

$$\frac{\mathrm{d}\sigma}{\mathrm{d}\Omega} = a^2. \tag{9.22}$$

Scattering by several nuclei

Let us now consider the scattering from an assembly of nuclei. When the scattering is weak, so that each nucleus is subject to the same incident flux, and multiple scattering can be neglected, the total differential cross-section is easily calculated.[38] Consider a system of nuclei at points $\mathbf{r}_\alpha$, $\alpha = 1, 2, \ldots, N_{\text{nuc}}$. Since the nuclei are bound, the appropriate scattering length is not a_α but $b_\alpha = (m_\text{N}/m_\alpha)a_\alpha$, where m_N is the neutron mass. The scattered wave is the superposition of waves scattered from each nucleus, i.e.

$$\psi_S(\mathbf{r}) = -\sum_\alpha b_\alpha \exp(i\mathbf{k}_0 \cdot \mathbf{r}_\alpha) \frac{\exp(ik_0|\mathbf{r}-\mathbf{r}_\alpha|)}{|\mathbf{r}-\mathbf{r}_\alpha|}$$
$$\approx -\frac{\exp(ik_0 r)}{r} \sum_\alpha b_\alpha \exp(i\mathbf{k} \cdot \mathbf{r}_\alpha). \tag{9.23}$$

The calculation of the scattering cross-section $d\sigma/d\Omega$ now proceeds as for a single scattering centre. Since $\mathbf{k}$ is angle dependent, the scattered flux also contains an angular component, which does not, however, contribute to $\mathbf{j}_S \cdot d\mathbf{S}$ or to the cross-section. The radial component of the flux is again computed using (9.20). We obtain a result similar to (9.22), but with a^2 replaced by $|\sum_\alpha b_\alpha \exp(i\mathbf{k}\cdot\mathbf{r}_\alpha)|^2 = \sum_\alpha b_\alpha b_\beta \exp(i\mathbf{k}\cdot\mathbf{r}_{\alpha\beta})$, where $\mathbf{r}_{\alpha\beta} = \mathbf{r}_\alpha - \mathbf{r}_\beta$ is the vector joining nuclei α and β. The cross-section so derived must be averaged over all possible nuclear spin states and all nuclear configurations. For the fluids with which we shall be concerned these averages can be performed independently, and the final result is

$$\frac{d\sigma}{d\Omega} = \sum_{\alpha\beta} \overline{b_\alpha b_\beta} \left\langle \exp(i\mathbf{k}\cdot\mathbf{r}_{\alpha\beta})\right\rangle. \tag{9.24}$$

Angular brackets denote a (classical or quantum, as appropriate) average over nuclear positions, while an overline denotes an average over all possible nuclear spin states.

For our purposes it is natural to assign nuclei to the particular molecule λ to which they belong. We therefore rewrite Eq. (9.24) as

$$\frac{d\sigma}{d\Omega} = \sum_{\lambda\lambda'}\sum_{\alpha\beta} \overline{b_\alpha b_\beta} \left\langle \exp(i\mathbf{k}\cdot\mathbf{r}_{\lambda\alpha\lambda'\beta})\right\rangle, \tag{9.25}$$

where the summations over λ and λ' are over all N molecules in the system, summations over α and β are over all nuclei in a single molecule, and $\mathbf{r}_{\lambda\alpha\lambda'\beta} = \mathbf{r}_{\lambda\alpha} - \mathbf{r}_{\lambda'\beta}$ is the vector joining nucleus α in molecule λ to nucleus β in molecule λ'. The angular brackets now denote an average in the canonical ensemble. This cross-section can be split into coherent and incoherent terms,[39] given by

$$\left(\frac{d\sigma}{d\Omega}\right)_{\text{inc}} = \sum_{\lambda\lambda'}\sum_{\alpha\beta} \left(\overline{b_\alpha b_\beta} - \bar{b}_\alpha \bar{b}_\beta\right) \left\langle \exp(i\mathbf{k}\cdot\mathbf{r}_{\lambda\alpha\lambda'\beta})\right\rangle$$
$$= N\sum_\alpha \left(\overline{b_\alpha^2} - \bar{b}_\alpha^2\right) \equiv N\sum_\alpha b_\alpha^{\text{inc}\,2}, \tag{9.26}$$

and

$$\left(\frac{d\sigma}{d\Omega}\right)_{\text{coh}} = N\left(\sum_\alpha \bar{b}_\alpha\right)^2 S(k), \tag{9.27}$$

where $S(k)$ is the static neutron structure factor

$$S(k) = N^{-1}\left(\sum_{\alpha}\bar{b}_{\alpha}\right)^{-2}\sum_{\lambda\lambda'}\sum_{\alpha\beta}\bar{b}_{\alpha}\bar{b}_{\beta}\langle\exp(i\mathbf{k}\cdot\mathbf{r}_{\lambda\alpha\lambda'\beta})\rangle. \tag{9.28}$$

Some authors[40] prefer a different definition of $S(k)$, in which $\left(\sum_{\alpha}\bar{b}_{\alpha}\right)^2$ in Eqs. (9.27) and (9.28) is replaced by $\sum_{\alpha}\bar{b}_{\alpha}^2$.

In writing (9.26) we have assumed that the spin states of different nuclei are uncorrelated. $\bar{b}_{\alpha}$ and b_{α}^{inc} are the mean coherent and incoherent bound scattering lengths, respectively, for nucleus α. Typical values of $\bar{b}_{\alpha}$ lie in the range[21] 5–10 fm. For some elements such as carbon, which is composed almost exclusively of a single isotope ^{12}C with zero nuclear spin, the scattering is almost entirely coherent, while for others such as hydrogen the scattering is almost totally incoherent:

$$\bar{b}_H^2 = \left(\frac{1}{4}{}^1b + \frac{3}{4}{}^3b\right)^2 = 14\,\text{fm}^2,$$

$$b_H^{\text{inc}\,2} = \left(\frac{1}{4}{}^1b^2 + \frac{3}{4}{}^3b^2\right)^2 - \bar{b}_H^2 = 637\,\text{fm}^2. \tag{9.29}$$

The incoherent scattering is independent of k and simply adds an isotropic background to the coherent scattering. Because hydrogen has a very large incoherent scattering cross-section, neutron scattering experiments on hydrogenic molecules often substitute[21] D for H.

9.1.2 X-ray and γ-ray scattering

For an incident X-ray photon with $k = 10\,\text{Å}^{-1}$, $\hbar\omega_0 = \hbar ck_0 \approx 20\,\text{keV}$. This energy is large compared to the binding energy $-Z^2/2n^2$ of all but the innermost K-shell electrons for $Z \leq 50$, i.e. for the first five rows of the periodic table. Thus for molecules containing elements lighter than Xe, the scattering may be computed as if the electrons were unbound.

To a good approximation low-energy X-ray scattering cross-sections can be calculated using classical mechanics.[41] The electric field radiated by an array of charges q_i at $\mathbf{r}_i$ is given, in SI units, by

$$\mathbf{E}_S(\mathbf{r},t) = \frac{1}{4\pi\epsilon_0 c^2}\sum_i q_i\left[\frac{\hat{\mathbf{r}}\times(\hat{\mathbf{r}}\times\mathbf{a}_i)}{R_i}\right]_{\text{ret}}. \tag{9.30}$$

In this equation $\mathbf{R}_i = \mathbf{r} - \mathbf{r}_i$, and the suffix 'ret' indicates that the retarded time $t - R_i/c$ is to be used to compute the right-hand side. $\mathbf{a}_i$ is the acceleration experienced by charge q_i. This is given by $\mathbf{a}_i = (q_i/m_i)\mathbf{E}_0(\mathbf{r}_i,t)$, where $\mathbf{E}_0(\mathbf{r},t) = \mathbf{n}_0E_0\exp(i\mathbf{k}_0\cdot\mathbf{r} - i\omega_0 t)$ is the incident electric field of amplitude E_0 and polarization $\mathbf{n}_0$.

At large distances from the scattering array, $R_i \approx r - (\hat{\mathbf{r}} \cdot \hat{\mathbf{r}}_i) r_i$, and thus

$$\mathbf{E}_S(\mathbf{r}, t) \to \frac{E_0}{4\pi\epsilon_0 c^2} \hat{\mathbf{r}} \times (\hat{\mathbf{r}} \times \mathbf{n}_0) \frac{\exp(ik_0 r - i\omega_0 t)}{r} \sum_i \frac{q_i^2}{m_i} \exp(i\mathbf{k} \cdot \mathbf{r}_i). \quad (9.31)$$

We may regard the massive nucleus as essentially fixed. The sum in (9.31) is therefore over the electrons only. Thus we can write

$$\mathbf{E}_S(\mathbf{r}, t) = E_0 \, \mathbf{f}(\theta, \phi) \, \frac{\exp(ik_0 r - i\omega_0 t)}{r}, \quad (9.32)$$

where the vector scattering amplitude

$$\mathbf{f}(\theta, \phi) = \hat{\mathbf{r}} \times (\hat{\mathbf{r}} \times \mathbf{n}_0) \, r_e \sum_i \exp(i\mathbf{k} \cdot \mathbf{r}_i), \quad (9.33)$$

and

$$r_e = \frac{e^2}{4\pi\epsilon_0 mc^2} = 2.82 \text{ fm} \quad (9.34)$$

is the classical electron radius.[41]

The differential scattering cross-section is given by

$$d\sigma = \frac{|E_S(\mathbf{r}, t)|^2 \hat{\mathbf{r}} \cdot \mathrm{d}\mathbf{S}}{|E_0|^2} \quad (9.35)$$

and therefore

$$\begin{aligned} \frac{\mathrm{d}\sigma}{\mathrm{d}\Omega} &= |\mathbf{f}(\theta, \phi)|^2 \\ &= \left(1 - (\mathbf{n_0} \cdot \hat{\mathbf{r}})^2\right) r_e^2 \left\langle \left| \sum_i \exp(i\mathbf{k} \cdot \mathbf{r}_i) \right|^2 \right\rangle, \end{aligned} \quad (9.36)$$

where the angular brackets denote averages over the electronic coordinates. The scattering is therefore not isotropic, but depends on the angle between the direction of observation and the polarization of the incident beam. For unpolarized incident radiation, we can show that

$$\frac{\mathrm{d}\sigma}{\mathrm{d}\Omega} = \frac{1}{2}(1 + \cos^2\theta) \, r_e^2 \left\langle \left| \sum_i \exp(i\mathbf{k} \cdot \mathbf{r}_i) \right|^2 \right\rangle, \quad (9.37)$$

and the scattering is then a maximum in the forward ($\theta = 0$) and reverse ($\theta = \pi$) directions. Equation (9.37) is the classical Rayleigh-Thomson formula for the scattering of radiation by free charges.

We note that a classical approach is valid only at low frequencies where the momentum of the incident photon can be ignored. When $\hbar k_0 \sim mc$, quantum dynamical corrections must be applied. The most important of these is associated with Compton inelastic scattering, which reduces the wavevector of the scattered photon according to[41]

$$\frac{k_S}{k_0} = \left[1 + \frac{\hbar k_0}{mc}(1 - \cos\theta)\right]^{-1}, \tag{9.38}$$

where $\hbar/mc = 3.86 \times 10^{-13}$ m is the reduced Compton wavelength. The effect is to reduce the scattered photon count by a factor k_S/k_0. For X-rays with $k_0 = 10^{11}$ m, the correction is about 4% and can safely be ignored.

In certain cases, particularly near resonance, it is important to consider the discrete spectrum of energy levels available to molecular electrons. A more sophisticated treatment of the scattering, including the so-called 'anomalous dispersion' terms associated with transitions involving excited molecular states, is then required.[24,25]

We will assume that the description of the scattering given by (9.37) is adequate for our purposes. If we now assign electrons to particular molecules, we can replace the average $\left\langle \left|\sum_i \exp(i\mathbf{k}\cdot\mathbf{r}_i)\right|^2 \right\rangle$ by

$$\left\langle \left|\sum_\lambda F_\lambda(\mathbf{k}) \exp(i\mathbf{k}\cdot\mathbf{r}_\lambda)\right|^2 \right\rangle = \sum_{\lambda\lambda'} F_\lambda(\mathbf{k}) F_{\lambda'}(\mathbf{k})^* \langle \exp(i\mathbf{k}\cdot\mathbf{r}_{\lambda\lambda'})\rangle, \tag{9.39}$$

where $\mathbf{r}_{\lambda\lambda'} = \mathbf{r}_\lambda - \mathbf{r}_{\lambda'}$ is the vector joining the centres[42] of molecules λ and λ', and the brackets now denote averages over molecular positions and orientations. We have defined the molecular form factor

$$F_\lambda(\mathbf{k}) = \int d\mathbf{r}\, n_\lambda(\mathbf{r}) \exp(i\mathbf{k}\cdot\mathbf{r}), \tag{9.40}$$

which is the Fourier transform of the electron density $n_\lambda(\mathbf{r})$ for molecule λ. We have used the Born-Oppenheimer approximation to separate electronic and nuclear motion, and we have also assumed that the electronic structure of a molecule is not perturbed by the presence of neighbouring molecules. Equation (9.39) applies only to molecules in states of zero orbital angular momentum.[43]

In the analysis of X-ray scattering data it is often convenient to further assign electrons to particular *atoms* (or nuclei) α within a molecule λ. This scheme is particularly appropriate for molecules containing the heavier elements, since core electrons greatly outnumber valence electrons, and can be unambiguously assigned. We can then write

$$\left\langle \left|\sum_i \exp(i\mathbf{k}\cdot\mathbf{r}_i)\right|^2 \right\rangle \approx \sum_{\lambda\lambda'} \sum_{\alpha\beta} F_\alpha(k) F_\beta(k) \langle \exp(i\mathbf{k}\cdot\mathbf{r}_{\lambda\alpha\lambda'\beta})\rangle, \tag{9.41}$$

where $\mathbf{r}_{\lambda\alpha\lambda'\beta} = \mathbf{r}_{\lambda\alpha} - \mathbf{r}_{\lambda'\beta}$ is the vector joining nucleus α of molecule λ to nucleus β of molecule λ'. Here

$$F_\alpha(k) = 4\pi \int_0^\infty \mathrm{d}r\, r^2 n_\alpha(r) \frac{\sin kr}{kr} \tag{9.42}$$

is the (real) form factor for nucleus α, i.e. the Hankel transform of the atomic electron density $n_\alpha(r)$. A compilation of atomic form factors for different values of k has been given by Klug and Alexander.[44] At $k = 0$, $F_\alpha(k) = Z_\alpha$, the atomic number. For elements heavier than carbon, at $k = 10\,\text{Å}^{-1}$ $F_\alpha(k)$ is an appreciable fraction of Z_α.

For molecules formed from the lighter elements, particularly those which engage in multiple bonding, this prescription must be modified. Consider for example nitrogen, N_2: of the fourteen electrons, six are engaged in bonding between the two atoms. In this case Egelstaff[45] has shown that the X-ray scattering can be approximately described if a third scattering 'centre' representing these bonding electrons is placed at the mid-point of the molecule.[46]

From (9.37) and (9.41), we see that the X-ray scattering elastic cross-section is purely coherent, and can be written as

$$\frac{\mathrm{d}\sigma}{\mathrm{d}\Omega} = \frac{1}{2}(1 + \cos^2\theta) N \left(\sum_\alpha \beta_\alpha(k) \right)^2 S_X(k), \tag{9.43}$$

where $S_X(k)$ is the X-ray scattering structure factor

$$S_X(k) = N^{-1} \left(\sum_\alpha b_{X\alpha}(k) \right)^{-2} \sum_{\lambda\lambda'} \sum_{\alpha\beta} b_{X\alpha}(k) b_{X\beta}(k) \langle \exp(i\mathbf{k} \cdot \mathbf{r}_{\lambda\alpha\lambda'\beta}) \rangle, \tag{9.44}$$

and $b_{X\alpha}(k) = r_e F_\alpha(k)$ is the X-ray coherent scattering length for nucleus (atom) α. For elements between carbon and xenon, at $k = 10\,\text{Å}^{-1}$ $b_{X\alpha}$ lies in the range 4–60 fm. For large k, Thomas–Fermi theory predicts[47] $b_{X\alpha}(k) \sim k^{-3/2}$. Thus for most elements X-ray scattering cross-sections are of magnitude similar to, or greater than, those expected for neutron scattering. For the lightest elements, however, $b_{X\alpha}(k)$ is rather small. Thus for H at $k = 10\,\text{Å}^{-1}$, $b_{XH} = 0.04$ fm, making hydrogen atoms invisible to X-rays.

As a probe of the structure of molecular fluids X-rays offer both advantages and disadvantages over neutrons.[48] A particular advantage is the relative abundance and strength of X-ray sources, including synchrotrons[49] and, potentially, X-ray lasers.[23] Additionally, precise measurements of elastic X-ray scattering may yield information on the distribution of valence electrons in molecules, especially when combined with neutron data.[45] Disadvantages include high attenuation associated with a large absorption cross-section, and insensitivity to molecular shape for scattering from molecules with approximately spherical electron clouds such as the hydrogen halides, water, and ammonia.

Small-angle γ-ray scattering,[24,25] especially using the isotope ^{241}Am ($\hbar\omega_0 = 59.5$ keV, $T_{1/2} = 432$ years), is also useful as a probe of fluid structure. For soft γ-rays the scattering cross-section is identical to that for X-rays.

9.1.3 *Electron scattering*

Molecular structure may also be probed using electron diffraction.[26] In this technique a low-energy beam of electrons of well-defined wavelength is scattered by the Coulomb potential of the molecular nuclei and electrons.

In the first Born approximation the scattering amplitude for electron diffraction by a molecule is given by (9.10), where now m is the electron mass, and $V(\mathbf{r})$ is the electron–molecule interaction. This approximation is valid provided the velocity of the incident electron is much greater than a typical molecular electron velocity.[50] Since most molecular electrons are closely bound to individual atoms, this translates into $k_0 a_0 \gg Z_\alpha/n^2$, where n is the principal quantum number for electrons bound to nucleus α of atomic number Z_α, and $a_0 = 4\pi\epsilon_0\hbar^2/me^2 = 0.53$ Å is the Bohr radius. Assuming small-angle scattering with $\theta = 0.1$ rad and $k_0 = 100\,\text{Å}^{-1}$ (so that $k = 10\,\text{Å}^{-1}$), this condition is satisfied for $Z_\alpha \ll 50n^2$, i.e. for the first five rows of the periodic table. In these circumstances the electron does not appreciably perturb the molecular electronic structure. Exchange effects can be neglected.[50] The appropriate interaction potential energy is then $V(\mathbf{r}) = -e\phi(\mathbf{r})$, where $\phi(\mathbf{r})$ is the molecular electrostatic potential. This is related to the molecular charge density $\rho(\mathbf{r})$ by Poisson's equation (2.46). Since $\rho(\mathbf{r}) = \sum_\alpha eZ_\alpha\delta(\mathbf{r} - \mathbf{r}_\alpha) - en(\mathbf{r})$, where $n(\mathbf{r})$ is the molecular electronic probability density, we have

$$\begin{aligned} f(\theta, \phi) &= \frac{2me}{4\pi\hbar^2}\int d\mathbf{r}'\, \exp(i\mathbf{k}\cdot\mathbf{r}')\phi(\mathbf{r}') \\ &= -\frac{2me}{4\pi\epsilon_0\hbar^2k^2}\int d\mathbf{r}'\, \exp(i\mathbf{k}\cdot\mathbf{r}')\rho(\mathbf{r}') \\ &= -\frac{2}{a_0k^2}\left(\sum_\alpha Z_\alpha - F(\mathbf{k})\right), \end{aligned} \tag{9.45}$$

where $F(\mathbf{k})$ is the molecular form factor defined in (9.40), and we have used SI units here and above. If one assumes, as in X-ray scattering, that electrons may be assigned to particular atoms within the molecule, then the scattering amplitude becomes

$$f(\theta) = -\frac{2}{a_0k^2}\sum_\alpha [Z_\alpha - F_\alpha(k)], \tag{9.46}$$

where $F_\alpha(k)$ is the atomic form factor defined by (9.42).

Following the discussion in §§ 9.1.1 and 9.1.2, and noting that $k^2 = k_0^2(1 - \cos\theta)$, we can now express the elastic scattering cross-section of an assembly of molecules as

$$\frac{d\sigma}{d\Omega} = (1 - \cos\theta)^{-2}N\left(\sum_\alpha b_{e\alpha}(k)\right)^2 S_e(k), \tag{9.47}$$

where

$$S_e(k) = N^{-1}\left(\sum_{\alpha} b_{e\alpha}(k)\right)^{-2} \sum_{\lambda\lambda'}\sum_{\alpha\beta} b_{e\alpha}(k) b_{e\beta}(k) \langle \exp(i\mathbf{k}\cdot\mathbf{r}_{\lambda\alpha\lambda'\beta})\rangle \tag{9.48}$$

is the electron diffraction structure factor, and

$$b_{e\alpha}(k) = 2\left(a_0 k_0^2\right)^{-1}(Z_\alpha - F_\alpha(k)) \tag{9.49}$$

is the scattering length. Clearly the scattering is very anisotropic, peaking sharply in the forward direction. It is entirely coherent. For a 4-keV electron with $k_0 = 10^{12}\,\mathrm{m}^{-1}$, $b_{e\alpha}(k) \approx 20Z_\alpha$ fm at typical values of k. This is substantially larger than neutron and X-ray scattering lengths. Because the scattering is very intense, electron diffraction may be used to study the intermolecular structure in dense gases, providing information on the second structural virial coefficient (9.69). The study of *liquid* structure requires high-energy electrons with $E > 10\,\mathrm{keV}$; low-energy electrons do not penetrate beyond the surface layers.[26]

For more sophisticated treatments of electron–molecule scattering, and a discussion of its applications, see Lane[51] and Christophorou.[52]

9.2 Structure factor for a molecular fluid

In the preceding section we have seen that measurements of the neutron, X-ray, or electron differential scattering cross-sections yield information concerning the static structure factor $S(k)$. In this section we show how $S(k)$ is related to the intermolecular pair correlation function $g(\mathbf{r}\omega_1\omega_2)$. For definiteness we consider formulae appropriate to neutron scattering; however, with the appropriate modifications, the results also apply to structure factors derived from X-ray and electron diffraction measurements.

Our starting point is Eq. (9.28). We may split $S(k)$ up into intramolecular ($\lambda = \lambda'$) and intermolecular ($\lambda \neq \lambda'$) terms, viz.

$$S(k) = S_{\text{intra}}(k) + S_{\text{inter}}(k), \tag{9.50}$$

where

$$S_{\text{intra}}(k) = \left(\sum_{\alpha} \bar{b}_\alpha\right)^{-2} \sum_{\alpha\beta} \bar{b}_\alpha \bar{b}_\beta \left\langle \exp(i\mathbf{k}\cdot\mathbf{d}_{\alpha\beta})\right\rangle \tag{9.51}$$

and

$$S_{\text{inter}}(k) = N^{-1}\left(\sum_{\alpha} \bar{b}_\alpha\right)^{-2} \sum_{\lambda\neq\lambda'}\sum_{\alpha\beta} \bar{b}_\alpha \bar{b}_\beta \left\langle \exp(i\mathbf{k}\cdot\mathbf{r}_{\lambda\alpha\lambda'\beta})\right\rangle. \tag{9.52}$$

In (9.51) $\mathbf{d}_{\alpha\beta}$ is the vector joining nuclear sites α and β within a particular molecule.

$S_{\text{intra}}(k)$ is often denoted $f_1(k)$ in the literature. In liquids it can usually be calculated with sufficient accuracy using bond lengths and angles appropriate to the ideal gas,[53] and neglecting intermolecular vibrational coupling. In the rigid molecule approximation, using the Rayleigh expansion (B.89), (9.51) reduces to

$$S_{\text{intra}}(k) = \left(\sum_{\alpha} \bar{b}_{\alpha}\right)^{-2} \sum_{\alpha\beta} \bar{b}_{\alpha}\bar{b}_{\beta}\frac{\sin kd_{\alpha\beta}}{kd_{\alpha\beta}}, \tag{9.53}$$

where $d_{\alpha\beta}$ is the distance between nuclei α and β. If we wish to take intramolecular vibrations into account in computing $S_{\text{intra}}(k)$, we must average (9.53) over the vibrational motion of the molecule. An approximate treatment yields

$$S_{\text{intra}}(k) = \left(\sum_{\alpha} \bar{b}_{\alpha}\right)^{-2} \sum_{\alpha\beta} \bar{b}_{\alpha}\bar{b}_{\beta}\frac{\sin k\bar{d}_{\alpha\beta}}{k\bar{d}_{\alpha\beta}} \exp\left(-\frac{1}{2}\Delta_{\alpha\beta}^2 k^2\right), \tag{9.54}$$

where $\bar{d}_{\alpha\beta}$ and $\Delta_{\alpha\beta}$ are the mean and root mean square fluctuations in $d_{\alpha\beta}$, respectively. We have neglected corrections of order[54] $O(\Delta^2/\bar{d}^2)$ and anharmonic[55] effects. The exponential correction is termed the Debye–Waller factor, and is important at large k. As an example, for CCl_4 at 300 K, $\Delta \approx 0.06$ Å; thus molecular non-rigidity affects $S_{\text{intra}}(k)$ for $k \gtrsim 10$ Å^{-1}.

$S_{\text{inter}}(k)$, often denoted $D_M(k)$ by experimentalists, is associated with correlations between the relative positions and orientations of different molecules of the fluid. For the purposes of calculating this quantity we may usually ignore internal rotation and vibration, and treat the molecules as if they were rigid. Vibrational excitation and intermolecular vibrational coupling are expected to make some contribution to the structure factor at large k; however, $S_{\text{inter}}(k)$ usually decays rapidly to 0 as $k \to \infty$, and consequently vibrational contributions to the intermolecular structure factor are generally small.[56]

Equation (9.52) can be written as

$$S_{\text{inter}}(k) = \left(\sum_{\alpha} \bar{b}_{\alpha}\right)^{-2} \sum_{\alpha\beta} \bar{b}_{\alpha}\bar{b}_{\beta}S_{\alpha\beta}(k), \tag{9.55}$$

where the *partial structure factor* is given by

$$S_{\alpha\beta}(k) = (N-1)\langle \exp(i\mathbf{k}\cdot\mathbf{r}_{1\alpha 2\beta})\rangle = \rho g_{\alpha\beta}(k). \tag{9.56}$$

Here $g_{\alpha\beta}(k)$ is the Fourier transform of the partial, or site–site, correlation function $g_{\alpha\beta}(r)$, i.e.

$$g_{\alpha\beta}(k) = \int d\mathbf{r}\, \exp(i\mathbf{k}\cdot\mathbf{r}) g_{\alpha\beta}(r) = 4\pi \int_0^{\infty} dr\, r^2 \frac{\sin kr}{kr} g_{\alpha\beta}(r), \tag{9.57}$$

and $\rho = N/V$ is the molecular number density. Partial correlation functions are discussed in detail in § 3.1.6. Here we simply note that $g_{\alpha\beta}(r) \equiv g_{\alpha\beta}(r_{\alpha\beta})$ is defined as the average of the pair correlation function $g(\mathbf{r}\omega_1\omega_2)$ over molecular orientations keeping the distance between sites α and β fixed, i.e.

$$g_{\alpha\beta}(r_{\alpha\beta}) = \langle g(\mathbf{r}\omega_1\omega_2)\rangle_{\omega_1\omega_2}, \qquad r_{\alpha\beta} \text{ constant.} \tag{9.58}$$

The concept of partial structure factors has proven to be a very useful tool for the analysis of scattering data.[58] In favourable cases, the individual $S_{\alpha\beta}(k)$ may be determined from neutron measurements using the technique of isotopic substitution. One assumes that the $S_{\alpha\beta}(k)$ themselves are isotope-independent; this is normally an excellent approximation,[59] although recently differences have been measured.[62] Suppose each molecule contains s atoms. Since $S_{\alpha\beta}(k) = S_{\beta\alpha}(k)$, measurement of $S_{\text{inter}}(k)$ at $\frac{1}{2}s(s+1)$ different isotopic mole fractions then suffices to determine the $S_{\alpha\beta}(k)$ uniquely. For X-rays the analogous technique of anomalous scattering[64] may be used to accomplish the same result.

Fourier transformation of the $S_{\alpha\beta}(k)$ yields the individual partial correlation functions $g_{\alpha\beta}(r)$. However, as discussed in §§ 3.1.6 and 3.2.3, $g(\mathbf{r}\omega_1\omega_2)$ contains *more* information than is contained in the set of $g_{\alpha\beta}(r)$. Consequently, it is impossible to determine the full angular dependence of $g(\mathbf{r}\omega_1\omega_2)$ from scattering experiments. For molecules that contain a scattering site at the molecular centre of mass, we can obviously determine $S_S(k) = 1 + \rho \int d\mathbf{r} \exp(i\mathbf{k}\cdot\mathbf{r}) g_S(r)$, where $g_S(r) = \langle g(\mathbf{r}\omega_1\omega_2)\rangle_{\omega_1\omega_2} = (4\pi)^{-1/2} g(000;00;r)$ is the centres correlation function. Soper[65–67] has discussed approximate methods for reconstructing additional spherical harmonic components of $g(\mathbf{r}\omega_1\omega_2)$ from experiment.

It is sometimes useful, e.g. for the perturbational treatment discussed in § 9.3, to express $S_{\text{inter}}(k)$ directly in terms of $g(\mathbf{r}\omega_1\omega_2)$. For this purpose we return to (9.52). We write

$$\mathbf{r}_{\lambda\alpha\lambda'\beta} = \mathbf{r}_{\lambda\lambda'} + \mathbf{d}_{\lambda\alpha} - \mathbf{d}_{\lambda'\beta}, \tag{9.59}$$

where $\mathbf{r}_{\lambda\lambda'}$ is the vector between molecular centres (defined arbitrarily) and $\mathbf{d}_{\lambda\alpha}$ is the vector from the centre of molecule λ to nucleus α of λ. Then

$$\begin{aligned} S_{\text{inter}}(k) &= (N-1)\left(\sum_\alpha \bar{b}_\alpha\right)^{-2} \sum_{\alpha\beta} \bar{b}_\alpha \bar{b}_\beta \langle \exp[i\mathbf{k}\cdot(\mathbf{r} + \mathbf{d}_{1\alpha} - \mathbf{d}_{2\beta})]\rangle \\ &= \rho \int d\mathbf{r} \exp(i\mathbf{k}\cdot\mathbf{r}) \langle g(\mathbf{r}\omega_1\omega_2) F(\mathbf{k}\omega_1) F(\mathbf{k}\omega_2)^* \rangle_{\omega_1\omega_2}, \end{aligned} \tag{9.60}$$

where the form factor

$$F(\mathbf{k}\omega_\lambda) = \left(\sum_\alpha \bar{b}_\alpha\right)^{-1} \sum_\alpha \bar{b}_\alpha \exp(i\mathbf{k}\cdot\mathbf{d}_{\lambda\alpha}) \tag{9.61}$$

is proportional to the scattering amplitude for molecule λ in orientation ω_λ. Equation (9.60) can also be written in terms of the pair correlation function $h(\mathbf{r}\omega_1\omega_2) = g(\mathbf{r}\omega_1\omega_2) - 1$ as

$$S_{\text{inter}}(k) = \rho \int d\mathbf{r} \exp(i\mathbf{k}\cdot\mathbf{r}) \langle h(\mathbf{r}\omega_1\omega_2) F(\mathbf{k}\omega_1) F(\mathbf{k}\omega_2)^* \rangle_{\omega_1\omega_2}. \tag{9.62}$$

Forward scattering terms ($\mathbf{k} = 0$) have been dropped, since scattering is never observed in the direction of the incident beam.

If we define $h(r) = \langle h(\mathbf{r}\omega_1\omega_2)F(\mathbf{k}\omega_1)F(\mathbf{k}\omega_2)^*\rangle_{\omega_1\omega_2}$, which is a function only of the molecular centres separation r, then (9.62) can be written concisely as

$$S_{\text{inter}}(k) = 4\pi\rho \int_0^\infty \mathrm{d}r\, r^2 \frac{\sin kr}{kr} h(r). \tag{9.63}$$

The inverse of this relation is

$$\rho h(r) = (2\pi^2)^{-1} \int_0^\infty \mathrm{d}k\, k^2 \frac{\sin kr}{kr} S_{\text{inter}}(k). \tag{9.64}$$

$g(r) = h(r) + 1$ is often spoken of somewhat loosely as the 'pair correlation function' of the liquid, although in reality it is obviously only a single component of the full pair correlation function $g(\mathbf{r}\omega_1\omega_2)$. Note that $g(r)$ is clearly *not* the same as the centres correlation function $g_S(r)$, except trivially in the case of a monatomic fluid.

Measurements of the isothermal pressure derivative of $S_{\text{inter}}(k)$ provide information on three-body correlations in molecular fluids.[16] We derive an explicit formula for $(\partial S(k)/\partial p)_T$ in terms of the triplet correlation function in Appendix 9A.

Special cases

We close this section with a brief discussion of some special cases in which the expression for the structure factor simplifies.

Small k For $k \to 0$ we have $S_{\text{intra}}(0) = 1$ and $F(0\omega_i) = 1$, so that

$$S(0) = 1 + \rho \int \mathrm{d}\mathbf{r}\, h_S(r) = \rho kT\chi, \tag{9.65}$$

where χ is the isothermal compressibility (3.113).

Large k In the other extreme limit $k \to \infty$, $S_{\text{inter}}(k) \to 0$, and thus

$$S(\infty) = S_{\text{intra}}(\infty) = \frac{\sum_\alpha \bar{b}_\alpha^2}{\left(\sum_\alpha \bar{b}_\alpha\right)^2}. \tag{9.66}$$

Free rotation If the potential energy is independent of molecular orientation, or nearly so, the molecules will rotate freely. Equations (9.50) and (9.62) then give

$$S^{\text{free}}(k) = S_{\text{intra}}(k)\left(1 + \rho \int \mathrm{d}\mathbf{r}\, \exp(i\mathbf{k}\cdot\mathbf{r}) h_S(r)\right). \tag{9.67}$$

Dilute gases For gases, we can insert the virial expansion for $h(\mathbf{r}\omega_1\omega_2)$ (§ 3.6) into (9.62) to obtain the virial expansion of the structure factor,

$$S(k) = \sum_{n=1}^{\infty} S_n(k)\rho^{n-1}, \tag{9.68}$$

where $S_1(k) \equiv S_{\text{intra}}(k)$,

$$S_2(k) = \int d\mathbf{r} \exp(i\mathbf{k}\cdot\mathbf{r})\langle(\exp[-\beta u(\mathbf{r}\omega_1\omega_2)] - 1)\, F(\mathbf{k}\omega_1)F(\mathbf{k}\omega_2)^*\rangle_{\omega_1\omega_2}, \quad (9.69)$$

etc. are called the *structural virial coefficients*. Note that $S_2(0) = -2B_2$, the second pressure virial coefficient (3.269). Thus studies of the wavelength and temperature dependence of neutron, X-ray and electron scattering from compressed gases can in principle yield information on the intermolecular pair potential. In practice, however, $S_2(k)$ is normally relatively featureless and insensitive to the form of $u(\mathbf{r}\omega_1\omega_2)$.[68,69]

9.3 Perturbation theory

Often the anisotropic component of the intermolecular potential is comparatively weak. In these circumstances, a perturbational analysis can yield useful predictions concerning the behaviour of $S(k)$.

For the general case in which angular correlations between molecules occur, we can separate (9.62) into an isotropic part and a part associated with these correlations. To this end, we first write the pair potential as the sum of isotropic and anisotropic parts, as in (4.22):

$$u(\mathbf{r}\omega_1\omega_2) = u_0(r) + u_a(\mathbf{r}\omega_1\omega_2). \quad (9.70)$$

Triplet and higher order potentials, if present, can be similarly divided. Next, we let $g_0(r) = h_0(r) + 1$ be the pair correlation function for a system with potential $u_0(r)$, and define $g_a(\mathbf{r}\omega_1\omega_2)$ by

$$g(\mathbf{r}\omega_1\omega_2) = g_0(r) + g_a(\mathbf{r}\omega_1\omega_2). \quad (9.71)$$

Note that, in general, $g_0(r) \neq g_S(r)$. Substituting (9.71) into (9.62) now gives

$$S_{\text{inter}}(k) = S_0(k) + S_a(k), \quad (9.72)$$

where

$$S_0(k) = S_{\text{intra}}(k) S_{\text{inter}}^{\text{mono}}(k), \quad (9.73)$$

with

$$S_{\text{inter}}^{\text{mono}}(k) = \rho \int d\mathbf{r} \exp(i\mathbf{k}\cdot\mathbf{r}) h_0(r), \quad (9.74)$$

and

$$S_a(k) = \rho \int d\mathbf{r} \exp(i\mathbf{k}\cdot\mathbf{r})\langle g_a(\mathbf{r}\omega_1\omega_2) F(\mathbf{k}\omega_1)F(\mathbf{k}\omega_2)^*\rangle_{\omega_1\omega_2}. \quad (9.75)$$

We can relate $g_a(\mathbf{r}\omega_1\omega_2)$ to the pair correlation function $g_0(r)$ for the system with isotropic interactions by treating $u_a(\mathbf{r}\omega_1\omega_2)$ as a perturbation. To first order, from (4.25),

$$g_a(\mathbf{r}_{12}\omega_1\omega_2) = -\beta u_a(\mathbf{r}_{12}\omega_1\omega_2)g_0(r) - \beta\rho\int d\mathbf{r}_3\, g_0(r_{12}r_{13}r_{23})\langle u_a(\mathbf{r}_{13}\omega_1\omega_3) + u_a(\mathbf{r}_{23}\omega_2\omega_3)\rangle_{\omega_3}. \quad (9.76)$$

For spherical harmonic contributions to u_a of order $l \neq 0$ in the orientation ω_3, the integral over the isotropic system triplet correlation function vanishes; this term will therefore be zero for neutral molecules interacting with multipole forces only, and is expected to be small provided that anisotropic overlap and dispersion forces are not large. Accordingly, we approximate (9.76) by

$$g_a(\mathbf{r}\omega_1\omega_2) = -\beta u_a(\mathbf{r}\omega_1\omega_2)g_0(r). \quad (9.77)$$

Substituting (9.77) into (9.75) yields

$$S_a(k) = -\beta\rho\int d\mathbf{r}\,\exp(i\mathbf{k}\cdot\mathbf{r})g_0(r)\langle u_a(\mathbf{r}\omega_1\omega_2)F(\mathbf{k}\omega_1)F(\mathbf{k}\omega_2)^*\rangle_{\omega_1\omega_2}. \quad (9.78)$$

We now expand the angle-dependent quantities in (9.78) in terms of spherical harmonics. For the potential u_a we have (for molecules of a general shape; see (2.20))

$$u_a(\mathbf{r}\omega_1\omega_2) = \sum_{l_1l_2l} u(l_1l_2l), \quad (9.79)$$

where

$$u(l_1l_2l) = \sum_{m_1m_2m}\sum_{n_1n_2} u(l_1l_2l;n_1n_2;r) \times C(l_1l_2l;m_1m_2m)D^{l_1}_{m_1n_1}(\omega_1)^*D^{l_2}_{m_2n_2}(\omega_2)^*Y_{lm}(\omega)^*. \quad (9.80)$$

From (B.89),

$$\exp(i\mathbf{k}\cdot\mathbf{r}) = 4\pi\sum_{lm} i^l j_l(kr)Y_{lm}(\omega)Y_{lm}(\omega_k)^*, \quad (9.81)$$

while for $F(\mathbf{k}\omega_i)$, using (9.61)

$$F(\mathbf{k}\omega_i) = 4\pi\left(\sum_\alpha \bar{b}_\alpha\right)^{-1}\sum_\alpha \bar{b}_\alpha\sum_{lm} i^l j_l(kd_\alpha)Y_{lm}(\omega_{i\alpha})^*Y_{lm}(\omega_k). \quad (9.82)$$

In (9.82), $\omega_{i\alpha}$ specifies the orientation of nucleus α of molecule i relative to a set of space-fixed axes. We can express the spherical harmonic $Y_{lm}(\omega_{i\alpha})$ in terms of the orientation ω_i of the molecule and the orientation ω'_α of the nucleus in body-fixed axes using (A.43),

$$Y_{lm}(\omega_{i\alpha}) = \sum_n D^l_{mn}(\omega_i)^*Y_{ln}\left(\omega'_\alpha\right). \quad (9.83)$$

Using (9.79)–(9.83) in (9.78), together with the orthogonality properties of the spherical harmonics (A.39) and rotation matrices (A.93), and noting the sum rule (A.90) for products of three spherical harmonics, we find straightforwardly

$$S_a(k) = -(4\pi)^{1/2}\beta\rho \sum_{l_1 l_2 l}\sum_{n_1 n_2} \frac{i^l(-)^{l_2}(2l+1)^{1/2}}{(2l_1+1)(2l_2+1)} C(l_1 l_2 l; 000)$$

$$\times I_{l_1 l_2 l; n_1 n_2}(k) F_{l_1 n_1}(k)^* F_{l_2 n_2}(k)^*. \tag{9.84}$$

Here we have defined the Hankel transform

$$I_{l_1 l_2 l; n_1 n_2}(k) = \int_0^\infty \mathrm{d}r\, r^2 j_l(kr) g_0(r) u(l_1 l_2 l; n_1 n_2; r) \tag{9.85}$$

and

$$F_{ln}(k) = i^l[4\pi(2l+1)]^{1/2} \frac{\sum_\alpha \bar{b}_\alpha j_l(kd_\alpha) Y_{ln}(\omega'_\alpha)}{\sum_\alpha \bar{b}_\alpha}. \tag{9.86}$$

We note that (9.84) predicts that $S_a(0) = 0$, as there is no term in $l = 0$. This follows directly from (9.65), since to first order in u_a, $g_S(r) = g_0(r)$.

Equation (9.84) is valid for molecules of arbitrary shape. For particular shapes and intermolecular interactions we simply substitute the appropriate values for $u(l_1 l_2 l; n_1 n_2; r)$ (given in §§ 2.4–2.7) and for $F_{ln}(k)$. Explicit expressions for the latter quantity, obtained from (9.86), are given in Appendix 9B for some common molecular geometries.

Combining Eqs. (9.50), (9.67), and (9.72)–(9.74), we find that to first order in perturbation theory, the structure factor of a molecular fluid is

$$S(k) = S^{\text{free}}(k) + S_a(k), \tag{9.87}$$

where the free-rotation structure factor $S^{\text{free}}(k)$ is given by

$$S^{\text{free}}(k) = S_{\text{intra}}(k)\left[1 + S^{\text{mono}}_{\text{inter}}(k)\right], \tag{9.88}$$

and the anisotropic correction term is given by (9.84).

9.3.1 Calculations on model fluids and comparison with experiment

In this section we examine the predictions of Eq. (9.87) for some model fluids, and compare with experimental data from neutron scattering studies on real liquids. Our calculations assume a Lennard-Jones potential $u_0(r)$; the pair correlation function $g_0(r)$ was obtained from Verlet's molecular dynamics calculations.[70]

Figures 9.3 and 9.4 show typical results[16] for a homonuclear diatomic (AA) fluid at the state condition $\rho^* \equiv \rho\sigma^3 = 0.850$, $T^* \equiv kT/\epsilon = 0.786$, corresponding to a dense liquid near room temperature. We have included contributions from quadrupole–quadrupole, overlap, and dispersion interactions. The quadrupolar interaction is of the form $u_{\text{mult}}(224)$ with $Q^* \equiv Q/(\epsilon\sigma^5)^{1/2} = 1$ (§ 2.4), the overlap interaction contains $u_{\text{ov}}(202)$ and $u_{\text{ov}}(022)$ terms with $\delta_2 = 0.2$ (§ 2.7), and the dispersion interaction contains $u_{\text{dis}}(202)$, $u_{\text{dis}}(022)$, and $u_{\text{dis}}(224)$ terms appropriate to a polarizability anisotropy $\kappa \equiv \gamma/3\alpha = 0.267$ (§ 2.6). The reduced bond lengths are $2r_A* \equiv 2r_A/\sigma = 0.3$ (Fig. 9.3, appropriate for N_2) and 0.6

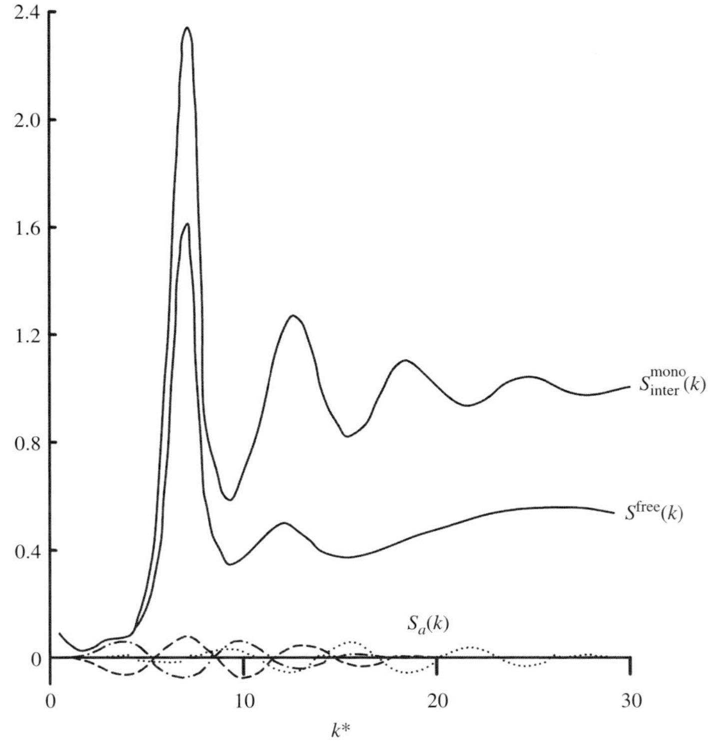

FIG. 9.3 Contributions to $S(k)$ for homonuclear AA diatomics. The reduced bond length is $2r_A^* = 0.30$; state conditions and other parameters are as specified in the text. Contributions to $S_a(k)$ shown are from quadrupole–quadrupole (dotted line), dispersion (dash-dot line), and overlap (dashed line) interactions. (From ref. 16.)

(Fig. 9.4, appropriate for Br_2). We have plotted $S^{\text{free}}(k)$ and $S_a(k)$, the two contributions to $S(k)$ in Eq. (9.87), as well as $1 + S_{\text{inter}}^{\text{mono}}(k)$ (the value $S(k)$ would assume in the limit $r_A \to 0$) as a function of $k^* \equiv k\sigma$.

If anisotropic components of the potential are neglected, $S_a(k) = 0$ and $S(k) = S^{\text{free}}(k)$. We note that even in the absence of orientational correlations this differs markedly from $1 + S_{\text{inter}}^{\text{mono}}(k)$; the effect of the factor $S_{\text{intra}}(k)$ in (9.88) is to damp the peaks in this function, causing, in particular, a very substantial reduction in the height of the first peak.

From Figs. 9.3 and 9.4 it is seen that overlap and dispersion contributions to $S_a(k)$ have similar k-dependence, but are of opposite sign, leading to some cancellation. The quadrupole–quadrupole contribution persists to larger k values than the overlap and dispersion contributions. Increasing the bond length magnifies the various contributions to $S_a(k)$ at small k, but reduces their effect at larger k.

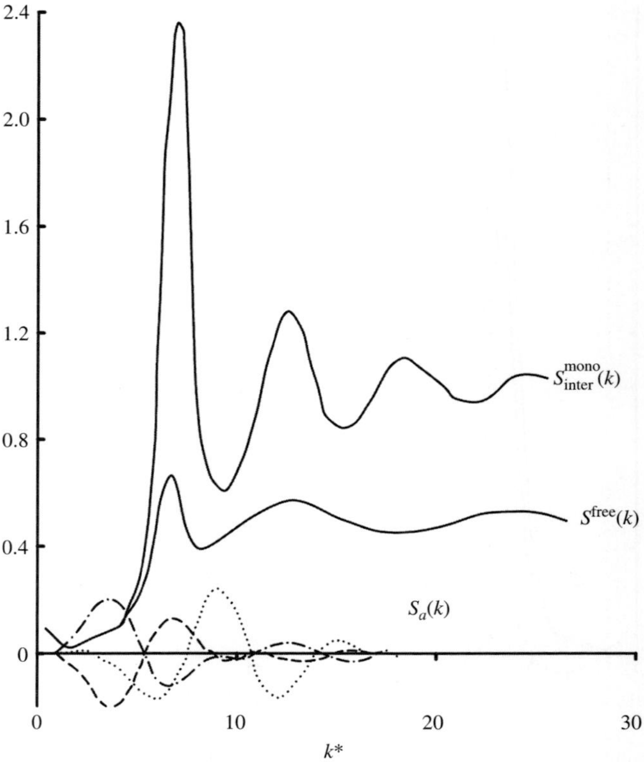

FIG. 9.4 Contributions to $S(k)$ for homonuclear AA diatomics. The reduced bond length is $2r_A^* = 0.60$; state conditions and other parameters are as specified in the text. Contributions to $S_a(k)$ shown are from quadrupole–quadrupole (dotted line), dispersion (dash-dot line), and overlap (dashed line) interactions. (From ref. 16.)

Calculations on heteronuclear AB diatomics[16] show that dipole–dipole interactions contribute to $S_a(k)$ only for relatively small k. From (9B.2) it is apparent that this contribution to $S_a(k)$ will vanish if $\bar{b}_A = \bar{b}_B$, and will only be appreciable if the scattering lengths for the A and B nuclei are significantly different.

Calculations have also been performed[16] on tetrahedral AB_4 molecules. Figures 9.5 and 9.6 show results for $S^{\text{free}}(k)$ and $S_a(k)$ for a Lennard-Jones fluid with octopolar ($u_{\text{mult}}(336)$; $\Omega^* \equiv \Omega/(\epsilon\sigma^7)^{1/2} = 0.5$ (§ 2.4)) plus overlap ($u_{\text{ov}}(303) + u_{\text{ov}}(033)$; $\delta_3 = 0.4$ (§ 2.7)) interactions. The state conditions are $\rho^* = 0.850$ and $T^* = 0.786$, and we have assumed $\bar{b}_A = \bar{b}_B$. The reduced bond lengths are $r_B^* \equiv r_B/\sigma = 0.3$ (Fig. 9.5, appropriate for CH_4) and 0.4 (Fig. 9.6, appropriate for $GeBr_4$). We note the increasing suppression of the first peak in $S^{\text{free}}(k)$ as r_B^* increases. For small k the dominant contribution to $S_a(k)$ comes

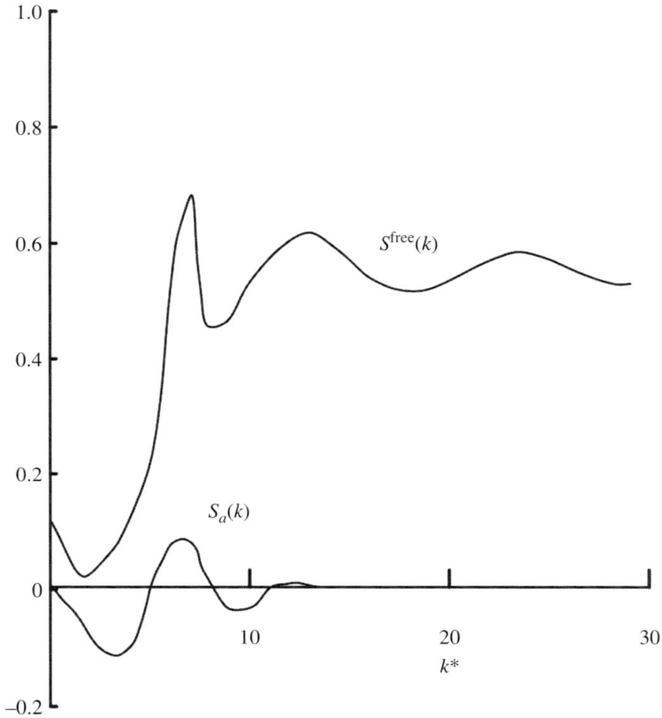

FIG. 9.5 Contributions to $S(k)$ for tetrahedral AB_4 molecules. The reduced bond length is $r_B^* = 0.30$; state conditions and other parameters are as specified in the text. Contributions to $S_a(k)$ shown are from octopole–octopole (dotted line) and overlap (dashed line) interactions. (From ref. 16.)

from the overlap interaction, while at large k the octopole interaction dominates. As r_B^* is increased, $S_a(k)$ is increased at small k but reduced at large k. Changing the ratio $\bar{b}_A/\bar{b}_B$ has relatively little effect on $S_a(k)$, causing only a slight reduction in $S_a(k)$ at small k. An increase in $\bar{b}_A/\bar{b}_B$ increases the height of the first peak in $S^{free}(k)$, and decreases that of the second. These predicted trends are supported by experimental data.[71]

Figures 9.7 and 9.8 illustrate the use of the theory[16] for liquids that have been studied experimentally. The most noteworthy features of the experimental $S(k)$ for liquid Br_2 (Fig. 9.7) are the maximum at $k = 2.2\,\text{Å}^{-1}$ and the minimum at $k = 3.0\,\text{Å}^{-1}$; these features cannot be explained on the basis of any purely isotropic potential model, i.e. in terms of $S^{free}(k)$ alone. The calculated $S_a(k)$ includes only the quadrupole–quadrupole contribution to the potential energy, since it was found that the dispersion and overlap terms could not explain the observed dependence on k (see, however, § 9.5). The calculations assume $Q = 8.2 \times 10^{-26}$ esu, while

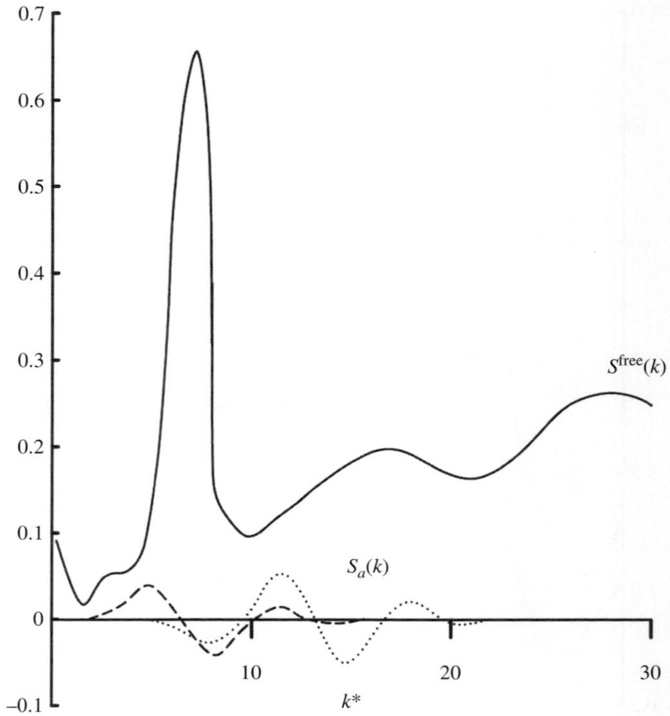

FIG. 9.6 Contributions to $S(k)$ for tetrahedral AB_4 molecules. The reduced bond length is $r_B^* = 0.40$; state conditions and other parameters are as specified in the text. Contributions to $S_a(k)$ shown are from octopole–octopole (dotted line) and overlap (dashed line) interactions. (From ref. 16.)

the Lennard-Jones parameters ϵ and σ were fit to the isothermal compressibility and the height of the first peak in $S(k)$, respectively. The gas-phase bond length r_B was used. $S_a(k)$ is seen to have a pronounced maximum at $k = 2.2$ Å^{-1} and a minimum at $k = 2.9$ Å^{-1}, resulting in excellent agreement between the computed and experimental $S(k)$.

For CCl_4 (Fig. 9.8) the most interesting feature of the data is the behaviour of $S(k)$ in the range $k = 2$–4 Å^{-1}. The maxima observed at $k = 2.4$ and 3.2 Å^{-1} and the minimum at $k = 2.8$ Å^{-1} cannot be explained on the basis of free rotation. Gubbins *et al.*[16] found that anisotropic overlap could not account for the observed behaviour, although this result is at odds with later MD studies by McDonald *et al.*[72] (§ 9.5). If an octopole–octopole contribution to $S_a(k)$ is added to $S^{\text{free}}(k)$, however, the computed $S(k)$ can be fit accurately to experiment if we assume $\Omega = 26 \times 10^{-34}$ esu; unfortunately, this is substantially larger than the value derived from collision-induced absorption measurements,[73] $\Omega = 16 \pm 3 \times 10^{-34}$ esu.

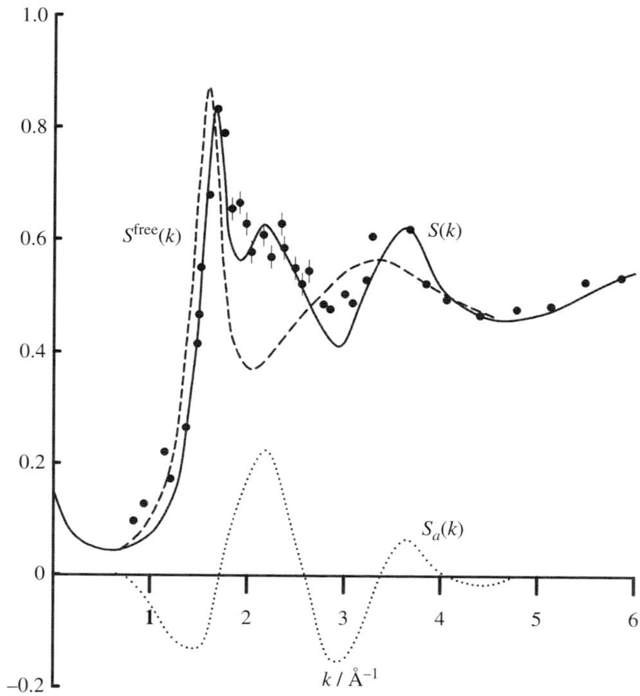

FIG. 9.7 Structure factor for liquid bromine at 300 K. Full line, predicted $S(k)$; dashed line, predicted $S^{\text{free}}(k)$; dotted line, predicted $S_a(k)$ for quadrupole–quadrupole interaction; points, experimental data. Parameters used are $\epsilon/k = 336\,\text{K}$, $\sigma = 4.17\,\text{Å}$, $2r_A = 2.28\,\text{Å}$, and $Q = 8.2 \times 10^{-26}$ esu. (From ref. 16.)

9.4 Integral equations

Integral equation approaches to the pair correlation function $g(\mathbf{r}\omega_1\omega_2)$ are discussed in Chapter 5. For molecular fluids one has two choices: (i) simple closures (PY, MSA, LHNC, etc.) may be used to augment the molecular Ornstein–Zernike (OZ) equation (3.117), which is then solved for the spherical harmonic expansion coefficients $g(l_1 l_2 l; n_1 n_2; r)$ of (3.142) (§ 5.4); (ii) the reference interaction site model, (RISM), or its extensions, may be used. In this case the site–site Ornstein–Zernike (SSOZ) equation (5.238) (or later variants) is combined with site closure relations for the partial correlation functions $h_{\alpha\beta}(r)$ and $c_{\alpha\beta}(r)$ (§ 5.5). Once either $g(\mathbf{r}\omega_1\omega_2)$ or the partial correlation function $g_{\alpha\beta}(r)$ is known, it is a simple matter to calculate $S_{\text{inter}}(k)$ or $S_{\alpha\beta}(k)$ by Fourier transformation.

The first approach has only recently been used to calculate $S(k)$ for real fluids. Martín and co-workers solved the OZ equation using a reference hypernetted chain (RHNC) closure,[74] in which the bridge function is approximated by that

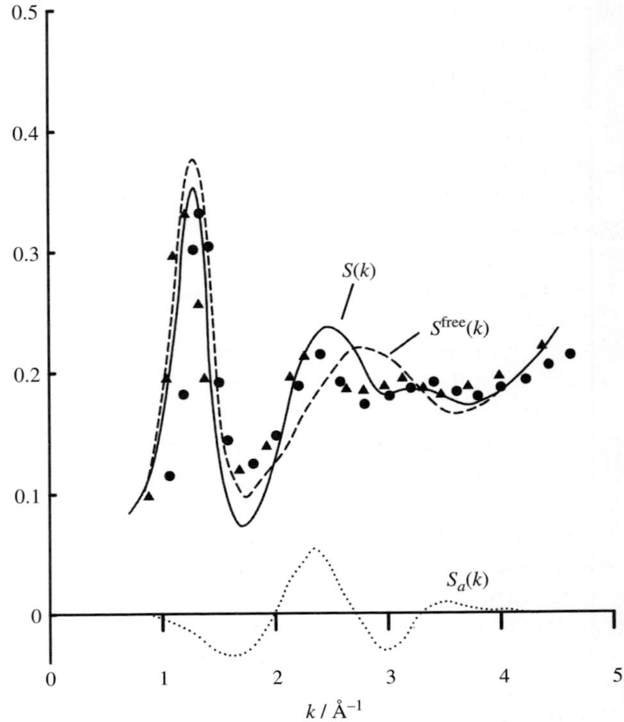

FIG. 9.8 Structure factor for liquid carbon tetrachloride at 295 K. Full line, predicted $S(k)$; dashed line, predicted $S^{\text{free}}(k)$; dotted line, predicted $S_a(k)$ for octopole–octopole interaction; points, experimental data. Parameters used are $\epsilon/k = 226\,\text{K}$, $\sigma = 4.93\,\text{Å}$, $r_B = 1.74\,\text{Å}$, $\Omega = 26 \times 10^{-34}$ esu, $\bar{b}_C = 6.63$ fm, $\bar{b}_{Cl} = 9.58$ fm. (From ref. 16.)

of a parametrized hard-dumbbell fluid. This approach yielded very good agreement with MC simulations of partial correlation functions and structure factors for several systems, including the homonuclear diatomics[75] N_2, Cl_2, and Br_2 (although the error was found to increase with the bond length), and the heteronuclear species HCl[76,77] and HI.[77] However, the procedure is computationally expensive.[75]

Patey and co-workers have computed the centre's structure factor $S_S(k)$ for model fluids of hard spherocylinders[78] and ellipsoids[79] in the Percus-Yevick (PY) and HNC approximations. The two closures yield qualitatively similar $S_S(k)$ values, although the differences become more pronounced at higher densities. Interestingly, for the spherocylinders $S_S(k)$ shows oscillatory structure not present for the ellipsoids, suggesting correlations are sensitive to the precise details of the anisotropy in short-range forces.

The first calculations of structure factors using RISM were reported by Chandler and co-workers[80–82] more than thirty years ago, and provided a qualitative explanation for the rather imprecise $S(k)$ data then available for molecular fluids. This early work employed very crude molecular models, representing carbon tetrachloride,[80] carbon disulphide,[80] carbon diselenide,[80] benzene,[80] acetonitrile,[81] and chloroform[82] as fused hard spheres. PY-type approximations were invoked as closures to the SSOZ equation. By treating the sphere diameters as adjustable parameters, generally good agreement with neutron or X-ray data could be obtained. This was taken as an indication that liquid structure is dominated by short-range repulsive intermolecular forces, and that the effects of longer range electrostatic or dispersion interactions could largely be ignored. This is now known to be incorrect: there is a great deal of evidence to suggest that in fact dipolar,[83,84] quadrupolar,[57] and dispersion[85] forces often have a significant impact on $g(\mathbf{r}\omega_1\omega_2)$ (*vide infra*, and cf. also § 9.5 and Chapters 4, 5, and 10).

Montague *et al.*[86] compiled an extensive tabulation of X-ray and neutron data on tetrahedral liquids, and compared these with RISM/PY calculations in which the molecules were represented as a collection of fused hard spheres situated at the nuclear sites. In the case of phosphorus, P_4, and arsenic, As_4, they also considered the effect of adding an auxiliary site at the molecular centre of mass. When sphere diameters were varied so as to obtain the best overall agreement with experimental structure factors, they found that the RISM formalism gave a satisfactory description of the experimental data in most cases, but that there were significant discrepancies in others. For example, Figure 9.9 shows $S_{\text{inter}}(k)$ for liquid phosphorus at 323 K, (a) without and (b) with an extra site. Agreement with experiment is poor, due partly no doubt to the unrealistic nature of the molecular model.[86] Furthermore, even though the fifth site is completely embedded

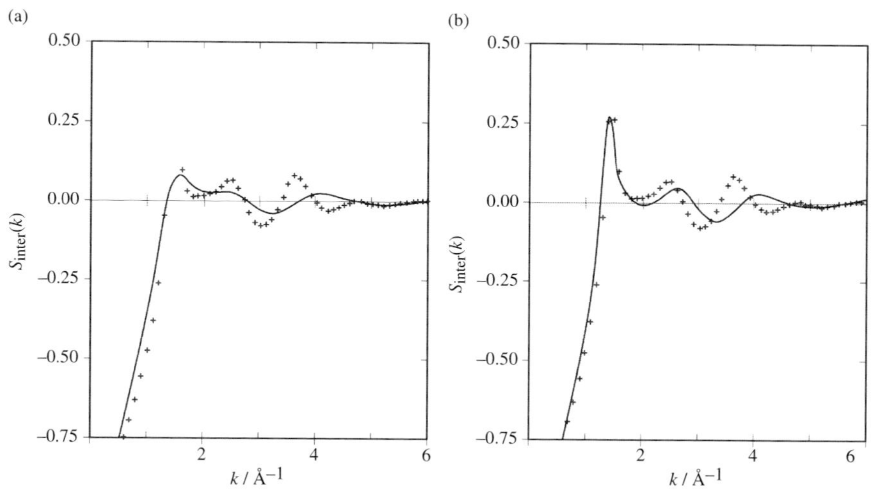

FIG. 9.9 Intermolecular structure factor for liquid phosphorus, P_4, at 323 K. Comparison of RISM calculations (full line) with experiment (crosses). (a) Four-site model; (b) five-site model with auxiliary site at molecular centre of mass. (From ref. 86.)

within the contours of the molecule, and can therefore have no direct effect on the correlation functions, it *does* lead to a change in the solution of the RISM equation, and produces a profound change in the computed structure factor. This shows that the well-known deficiencies of RISM in the long-wavelength limit (Appendix 10D.2) extend also to the regime of nonzero k.

9.4.1 Improvements to RISM

In principle, RISM estimates of $S(k)$ can be improved by using a more realistic site–site potential, incorporating softness into the repulsive forces and including also attractive long-range dispersion and electrostatic components. Since such calculations do not employ a 'reference' hard-core fluid to provide a zeroth-order estimate of the liquid structure, they should strictly speaking be called ISM rather than RISM. In practice, the latter term is widely used, with some authors[87–89] preferring XRISM, for 'extended RISM', especially when electrostatic forces are included.

Narten and co-workers[85,90,91] have calculated structure factors for several molecular liquids using exp-6 site–site potentials. The RISM SSOZ equations were solved using PY closures. A comparison with X-ray data for liquid nitrogen,[85] N_2, and methane,[90] CH_4, showed good agreement except at small k, suggesting the theory fails at large intermolecular separations. The calculations also demonstrated that the attractive dispersion forces exert a significant influence on the liquid structure. For ethylene,[90] C_2H_4, a six-site exp-6 potential gave a very poor description of the intermolecular carbon–carbon correlation function; this probably suggests that electrostatic interactions should also be incorporated into $u_{\alpha\beta}(r)$, since the molecule has a fairly large quadrupole moment.

One can also try to obtain a better $S(k)$ by changing the closure. As discussed in Chapter 5, the HNC approximation is generally superior to PY when attractive forces are present; for polar fluids it often yields accurate estimates of orientational correlations and such related quantities as the dielectric constant (Chapter 10). But it must be stressed that a solution of the SSOZ equations cannot always be found,[88,89,92] especially when the interaction sites carry partial charges; this must be considered a serious shortcoming of the RISM method.

The structure of supercritical sulphur hexafluoride, SF_6, has been studied by Strauss *et al.*[93] using neutron diffraction. RISM/HNC calculations using a site–site Lennard-Jones/electrostatic potential show fair agreement with experiment over a range of densities at 398 K. The coordination number of the first peak of $g_{SS}(r)$ decreases from 8 at $\rho = 1.85\,\mathrm{g\,cm^{-3}}$ to 4 at $0.85\,\mathrm{g\,cm^{-3}}$, suggesting a reduction in short-range order as the density is lowered. Reverse Monte Carlo calculations[94] of the S–S–S angular distribution were also performed, and show that at the highest density the fluid structure resembles that of a bcc lattice.

Strauss *et al.*[95] used RISM/HNC calculations to refine a site–site 12-6-1 potential for deuteromethane, CD_4, which accurately reproduces neutron scattering data at the supercritical temperature of 370 K over a range of densities. Subsequently,

Bertagnolli *et al.*[96] studied the structure of supercritical deuteroethane, C_2D_6, using neutron diffraction, and compared this with RISM/HNC calculations using a two-site Lennard-Jones potential; although these could satisfactorily reproduce the structure of the fluid at low densities, they were rather less accurate at higher densities.

Bermejo *et al.*[97] have performed RISM/HNC (and also RISM/PY) calculations on carbon tetrachloride, CCl_4, using a potential containing site–site Lennard-Jones and, optionally, electrostatic terms. The atomic charges were chosen so as to reproduce the molecular octopole moment. From the insensitivity of the partial correlation functions to these terms, it was concluded that octopolar interactions have only a small effect on orientational correlations, in agreement with the results of MD simulations.[72] Alvarez *et al.*[98] found similar results for the quadrupolar molecule tetrachloroethylene, C_2Cl_4. However, the calculated $S_{\text{inter}}(k)$ did not agree well with experiment; it is not clear whether this represents a failure of the integral equation per se, the closure, or the assumed intermolecular potential.

Bausenwein *et al.*[89] have studied neutron scattering from deuterated ammonia, ND_3, in the supercritical regime. They also performed RISM/HNC calculations using a variety of one-, four- and five-site interaction potentials. No potential could be found that gave an adequate description of the neutron data at all state conditions. Since the more recent potential models[99,100] are known to be quite accurate, the results imply a failure of RISM. Even more seriously, Bertagnolli and Tödheide[101] find that iterative solution of RISM/HNC fails to converge at densities near the region of the liquid-gas coexistence curve.

Fraser *et al.*[84] have applied RISM/HNC to a six-site 12-6-1 model of the strongly polar molecule acetonitrile, CH_3CN. Figure 9.10 shows a comparison of RISM partial correlation functions with exact results from MD calculations. Agreement is on the whole good. However, some peak heights and trough depths are too small: thus it appears the integral equation somewhat underestimates pair correlations. Examination of features in the RISM $g_{\alpha\beta}(r)$ reveals a preference for $\rightleftarrows$ and $\rightarrow\rightarrow$ configurations, both of which are favoured by dipolar forces. The figure also shows the effect of switching off the electrostatic interactions. Several of the $g_{\alpha\beta}(r)$ change significantly, proving that long-range forces strongly affect the liquid structure. However, $g_{C_2C_2}(r)$ and other correlation functions involving the cyano carbon C_2 hardly change. The molecular centre of mass lies close to C_2; thus the harmonics $g(l0l; n_1n_2; r)$ (Eq. (3.142)), controlling correlations between the intermolecular vector and the molecular axes, are insensitive to multipolar forces. To see the reason for this we note that $g(l0l; n_1n_2; r) \approx -\beta u(l0l; n_1n_2; r)$, and that such harmonics are absent from the electrostatic interaction of two neutral molecules (cf. § 2.4).

Figure 9.11 shows the neutron coherent scattering cross-section for deuteroacetonitrile, CD_3CN, which from (9.27) is proportional to $S(k)$. We note that $S(k)$ is quite insensitive to the electrostatic forces betwen the molecules: apart from small differences at low k, the two RISM curves, calculated with and

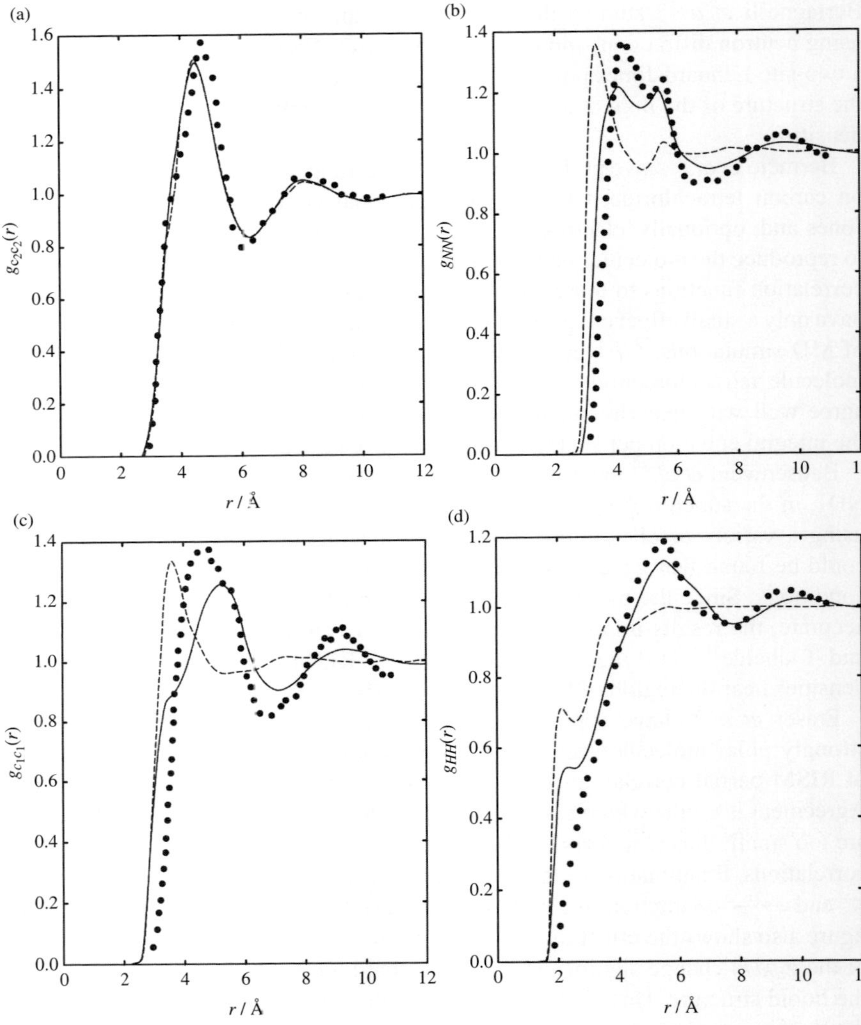

FIG. 9.10 Partial correlation functions for liquid acetonitrile at 291 K. Full line, RISM/HNC with six-site Lennard-Jones/electrostatic potential; dashed line, ditto excluding electrostatic forces; dots, molecular dynamics. C_1 and C_2 are methyl and cyano carbons, respectively. (From ref. 84.)

without partial atomic charges, essentially agree (although neither is in particularly good accord with experiment). This is despite the sensitivity of the $g_{\alpha\beta}(r)$ to electrostatic forces noted above. An explanation has been given by Fraser *et al.*[102] These authors also point out that in the mean spherical approximation (MSA), in cases where the scattering lengths are site-independent, Coulombic interactions have absolutely *no* effect on RISM structure factors. This obvious shortcoming is closely related to the trivial prediction for the dielectric constant

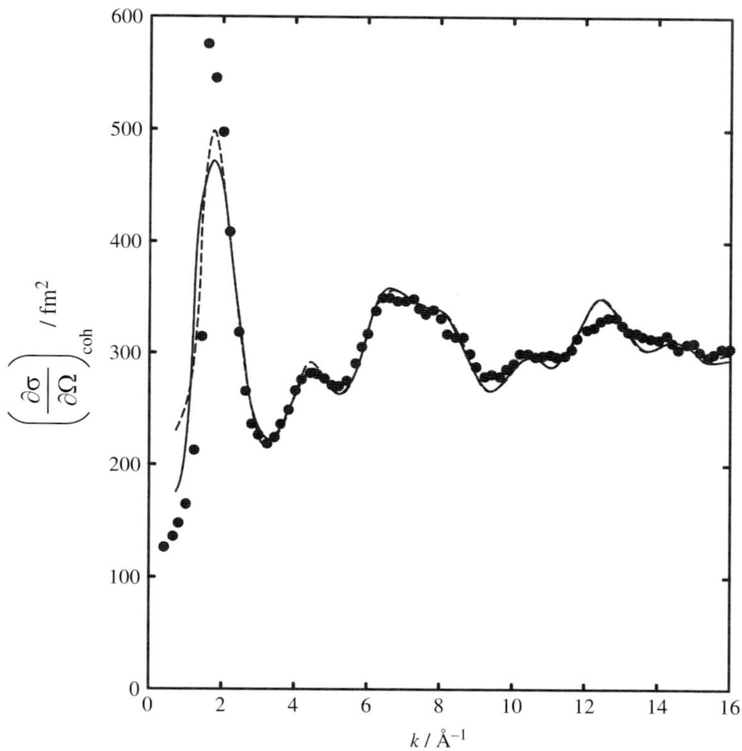

FIG. 9.11 Neutron coherent scattering cross-section for liquid CD_3CN at 293 K. Full line, RISM/HNC with six-site Lennard-Jones/electrostatic potential; dashed line, ditto excluding electrostatic forces; dots, experiment. (From ref. 84.)

given by RISM/MSA (Appendix 10D.2). It demonstrates that RISM analysis of structure factors *cannot* be used to decide unequivocally between different potential models.

The inadequacies of RISM/PY and RISM/HNC estimates of structure factors for N_2, Cl_2, and Br_2 have been demonstrated by Martín and co-workers; discrepancies are particularly marked at low k.[75]

Bausenwein *et al.*[88] and Bertagnolli and Tödheide[101] have performed RISM/HNC and RISM/PY calculations on supercritical CO_2 in an attempt to determine the form of the intermolecular potential. These efforts have so far proven unsuccessful; and in light of the above remarks, further attempts seem destined to fail also.

There have been a few[103–105] calculations using proper site–site integral equations (see Appendix 10D). Lue and Blankschtein have calculated partial structure factors for diatomic fluids[104] and water[105] using HNC and modified HNC closures. They find poor agreement with simulated $g_{\alpha\beta}(r)$ values, and note that the structure factors may even become negative for separations on the order of the molecular diameter.[104]

9.5 Computer simulations and comparison with experiment[20]

Computer simulations of $S(k)$ possess of course the advantage that no approximations need be introduced once the form of the intermolecular potential and molecular parameters such as bond lengths and angles have been selected. For this reason most analyses of experimental $S(k)$ data in the past decade have employed Monte Carlo (MC) or molecular dynamics (MD) simulations to extract information concerning intermolecular correlations and, ultimately, information on the intermolecular potential.

Technical difficulties arise in the comparison of simulation and experiment.[40] The quantity obtained directly from simulations is not the structure factor itself, but the pair correlation function $g(\mathbf{r}\omega_1\omega_2)$ or the partial correlation functions $g_{\alpha\beta}(r)$. These must be Fourier transformed to obtain $S(k)$. Since one can only simulate finite systems, one normally imposes periodic boundary conditions. This places a lower limit on the value of k for which $S(k)$ can be calculated accurately: if the simulation cell is a cube of side L, this limit is about $4\pi/L$, typically 0.5 Å^{-1}. A method proposed by Salacuse *et al.*[106] is largely successful in eliminating size effects for simple fluids, but has not yet been tested for more complex molecular liquids. The inverse problem, that of computing an experimental pair correlation function from (often noisy) $S(k)$ data of finite range, has been studied by Root *et al.*[107] and Soper[108] using maximum entropy methods.

In this section we review simulations of the static structure factor for several molecules: halogens, deuterium, hydrogen halides, carbon monoxide, carbon dioxide, hydrogen sulphide, sulphur dioxide, carbon tetrachloride, dichlorodifluoromethane, boron trifluoride, ammonia, acetonitrile, dimethylsulphoxide, methanol, water and water mixtures, and hydrocarbons.

9.5.1 Halogens

The structure factor for the halogens displays a uniform progression along the sequence F_2, Cl_2, Br_2, I_2, in which a shoulder on the large k side of the first peak of $S(k)$ becomes progressively more prominent.[109] As discussed in the introduction to this chapter and § 9.3, this feature is attributable to anisotropy in the intermolecular potential. For F_2, which has the smallest bond length ($d =$ 1.43 Å) and is almost spherical, this feature is absent;[110] for Cl_2 ($d = 1.99$ Å), it is weakly apparent,[111–116] for Br_2 ($d = 2.28$ Å) it is much stronger,[17,117–119] while for I_2 ($d = 2.66$ Å) it appears as a separate peak of magnitude comparable to the first maximum.[118,120] Although the early perturbational analysis of Gubbins *et al.*[16] identified quadrupolar interactions as the dominant mechanism behind this effect (cf. § 9.3), more recent simulations suggest that anisotropic repulsive (overlap) terms may be more important. MD calculations by Rodger *et al.* on

Cl_2[115,116] and Br_2 and I_2[118] have employed sophisticated intermolecular potentials derived in part from *ab initio* calculations on isolated halogen molecules, and in part by fitting to the low-temperature crystal structure. These potentials include atomic overlap, dispersion, and electrostatic (distributed multipole[121]) terms, and represent considerable improvements over earlier, cruder potentials of the Lennard-Jones plus point quadrupole[16] or two-centre Lennard-Jones plus point quadrupole[117,122] type, which do not correctly predict the structure of the solid phase.[111] These calculations strongly suggest that it is the hard-core repulsive interactions that dominate, and that the effect of electrostatic interactions is negligible. Subsequent studies on Cl_2[112–114] and Br_2,[17] comparing MD simulations with newer and more accurate neutron data, have tended to confirm this conclusion, while suggesting that the anisotropic terms in the potentials should be lowered.[17] Figure 9.12 shows a comparison between simulation and experiment for Br_2. These calculations also suggest[112] that, except at very small or large k, nearest neighbour effects are primarily responsible for the observed appearance

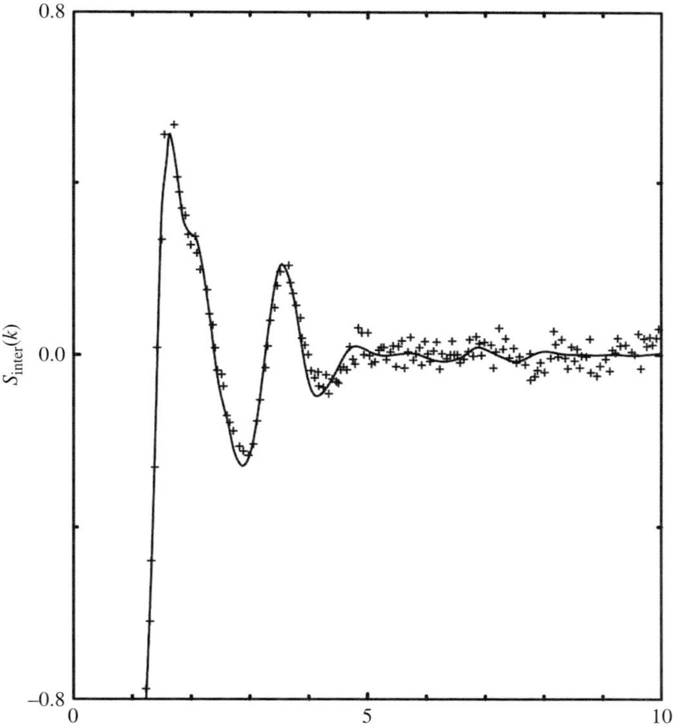

FIG. 9.12 Structure factor for liquid bromine at 296 K. Full line, MD $S(k)$ calculated using potential of ref. 118; crosses, experiment. (From ref. 17.)

of $S(k)$ (at small k, the structure of the fluid at larger r is probed, so that $S(k)$ is sensitive to correlations with next nearest neighbours, etc., while at large k intramolecular effects dominate).

There is still considerable disagreement as to which is the most probable relative molecular orientation in the liquid phase. For Cl_2, calculations by Ricci *et al.*[111] employing two-centre Lennard-Jones plus point quadrupole potentials suggest T-shaped configurations dominate, while more sophisticated atom–atom potentials predict cross-configurations; however, de Santis *et al.*[112] found that two-centre Lennard-Jones (with or without a central point quadrupole) and atom–atom potentials from refs. 115 and 116 all predict deformed T-shapes. Misawa[123,124] has suggested that the preferred orientation is a staggered parallel configuration. These studies show that it is very difficult, if not impossible, to explain the features of $S(k)$ in terms of particular configurations occurring in the liquid, and it has been suggested[111] that collision-induced absorption and light scattering (Chapter 11) are probably better tools than neutron or X-ray scattering for probing intermolecular structure, since these yield more information on the intermolecular harmonic coefficients $g(ll'm; r)$ (§ 3.2).

Neutron scattering studies of liquid I_2 at various temperatures along the coexistence curve show that strong orientational correlations between neighbouring molecules persist up to the vicinity of the critical point.[125]

9.5.2 *Deuterium*

Zoppi and co-workers have studied neutron scattering by deuterium, D_2, under various state conditions near the critical[126,127] and triple[128] points, and have performed both classical[126,127] and quantum path integral[127] MC simulations of $S(k)$ and its thermodynamic derivatives $(\partial S/\partial T)_\rho$ and $(\partial S/\partial \rho)_T$. D_2 is nearly spherical and the pair interactions can be described by a central potential. Classical simulations qualitatively reproduce the observed trends of $S(k)$ with density and temperature, but to obtain quantitative agreement a treatment of quantum diffraction effects is required. Exchange effects, as usual, can be neglected.[129]

Near the critical point ($T_c = 38\,\text{K}$) excellent agreement with experiment is obtained by treating D_2 as a classical rigid rotor. This is at first sight rather surprising, since the rotational temperature $\theta_r = 43\,\text{K}$. The explanation is that the incident neutrons have sufficiently high energies to excite many rotational transitions.[127] Some vibrational excitation also occurs, and the improved statistics of the data collected at the triple point require a more detailed analysis of molecular motion including the effects of nonrigidity.[130]

As discussed in Appendix 9A, $(\partial S/\partial \rho)_T$ contains information on three-body correlations. Zoppi *et al.* have used their results to evaluate various theories for the triplet correlation function. None of these could quantitatively reproduce the experimental data over the whole range of k. However, the comparison was done at a very low temperature, and the disagreement may be due in part to the classical nature of the approximations.[128]

9.5.3 *Hydrogen halides*

Studies of neutron and X-ray scattering by the hydrogen halides (HF, HCl, HBr, and HI) and their deuterated analogues have yielded considerable information concerning the structure of these liquids, and in particular the degree and nature of hydrogen bonding that they display.

Deuterium fluoride, DF, has been studied experimentally by Deraman *et al.*,[131] Pfleiderer *et al.*,[132] and McLain *et al.*[133] under a variety of state conditions in both the liquid and supercritical regions.

As in the solid, the liquid is extensively associated, forming branched, long-chain, zig-zag polymers. Hydrogen bonding within these chains is very strong; Fourier transformation of the neutron $S(k)$ reveals clear evidence for a D–F peak at 1.7 Å, and shows further structure up to 6 Å that characterizes the distribution of adjacent molecules. Structure at larger values of r suggests the formation of cyclic species containing between four and eight monomers, with the most probable ring size strongly dependent on the state conditions.

The results have been compared with MC and MD simulations using various models for the intermolecular potential of different degrees of sophistication. These include both two- and three-centre effective potentials,[134–139] some incorporating polarizability, as well as *ab initio* potentials of the Car-Parrinello type.[140,141] The results establish unequivocally that hydrogen bonding is an electrostatic effect. Three-centre models, which correctly reproduce both the dipole and quadrupole moment of DF, are more successful than two-centre models, which can duplicate only the dipole. The polarizable models,[138,139] which add a scalar polarizability centred on the F atom, are marked by an increased tendency for head-to-tail alignment of neighbouring molecules, resulting from a larger effective molecular dipole moment, and by an increase in the ring size of cyclic clusters (although these are less likely to form).[139] They also provide a more accurate description of the first (H-bonding) peak in the pair correlation function, and in general yield a clearly superior description of the structure and its dependence on state conditions, explaining, for example, the observed elongation of the H-bond length at higher temperatures and lower densities.[138]

Significant changes in the pair correlation functions are apparent as the state conditions change.[132,133,140] Most notably, as the temperature is raised, there is a reduction in the average number of H bonds, coordination number, and chain length.[140] These structural changes appear to be largely due to a rearrangement of the deuterium atom positions within the fluid,[133] with H bonds becoming on the average less linear; the observed F–F distances hardly change at all.

The *ab initio* simulations show excellent agreement with experiment.[140] They demonstrate that, at most, each molecule forms two hydrogen bonds, with F much more likely to doubly H-bond than D; as a result, branching of the chains occurs mostly at the F atoms. The angle between two DF molecules is found to be quite variable, typically falling in the range 75°–150°, with a maximum at the tetrahedral angle.[140] Notably, they suggest a shortening of the hydrogen bond in passing from the gaseous into the liquid phase,[141] a feature not predicted by any of the empirical models.

The neutron scattering spectrum of hydrogen chloride, HCl, has been measured by Soper and Egelstaff.[142] Using the technique of isotopic substitution, these authors were able to measure the partial structure factors $S_{\mathrm{HH}}(k)$, $S_{\mathrm{HCl}}(k)$, and $S_{\mathrm{ClCl}}(k)$. The structure displayed by $S_{\mathrm{HCl}}(k)$ is particularly interesting: this function is relatively weak, and exhibits a double peak at 1.8 and 2.8 Å. MD simulations[143,144] suggest this can only be explained by invoking intermolecular H–Cl interactions that are partly covalent in nature. However, this conclusion cannot be regarded as definite, since the electrostatic contribution to the pair potentials used in these calculations was not altogether satisfactory, the assumed distribution of charges greatly overestimating the molecular octopole moment.[144]

Early neutron studies[145] of deuterium bromide, DBr, suggested weak hydrogen bonding. More recently, however, careful measurements by Andreani *et al.*[146] of partial structure factors for liquid HBr and DBr along the coexistence curve suggest that although there is a significant degree of orientational correlation in the liquid, the tendency is for neighbouring molecules to align with their axes at an angle of 45°; some pairs even align antiparallel. These results cannot be explained on the basis of a simple electrostatic model, and argue against a significant degree of H-bonding. The relative lack of structure in $g_{\mathrm{HBr}}(r)$—only a weak shoulder is visible at 2.6 Å—tends to confirm this. The lack of a reliable intermolecular potential has restricted further study of this system by computer simulation.[146]

Andreani *et al.*[147] have measured the neutron scattering from liquid hydrogen and deuterium iodide and HI/DI mixtures at 253 K and the orthobaric density. The partial structure factors thus obtained were compared with MD simulations using a central-force Lennard-Jones potential; the results of this comparison are are shown in Fig. 9.13. In HI the centre of mass nearly coincides with the position of the iodine nucleus, so that $S_{\mathrm{II}}(k)$ is essentially the Fourier transform of the centres' correlation function. This quantity is significantly influenced by anisotropy in the intermolecular pair potential only when there is a contribution from overlap forces (cf. § 4.5, and especially Fig. 4.12); the excellent agreement between experiment and simulation therefore suggests that the short-range interaction of two HI molecules is almost isotropic. $S_{\mathrm{HI}}(k)$ also shows little evidence of sensitivity to orientational correlations, suggesting that no significant H-bonding occurs. This conclusion is reinforced by the absence of a peak in $g_{\mathrm{HI}}(r)$ in the vicinity of $r = 3.9$ Å as would be expected if a hydrogen bond were formed.[147] In contrast, the hydrogen–hydrogen structure factor $S_{\mathrm{HH}}(k)$ is poorly reproduced, indicating a degree of correlation between the molecular axis vectors of neighbours consistent with the observed Kirkwood g-factor.[147] Soper *et al.*[65] have attempted to reconstruct $g(\mathbf{r}\omega_1\omega_2)$ from the measured partial structure factors. Their analysis shows that of the coefficients that contribute to $g(\mathbf{r}\omega_1\omega_2)$ via (3.143), the isotropic component $g(000;r)$ is the largest, and $g(110;r)$ and $g(112;r)$ are also appreciable, while $g(101;r)$ and $g(202;r)$ are much smaller (Fig. 9.14). These results offer further support for the idea that HI is a nonassociated liquid, since it is coefficients

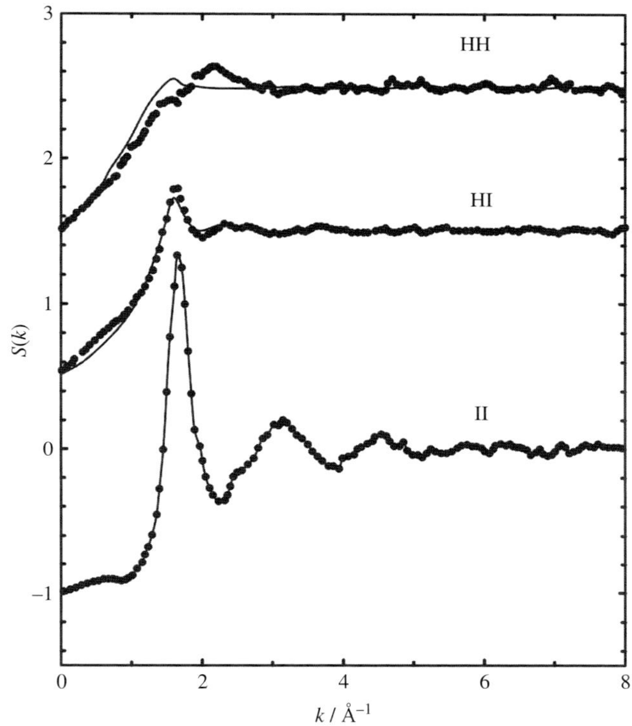

FIG. 9.13 Partial structure factors for liquid hydrogen iodide at 253 K. Points, neutron data; full lines, MD simulations using Lennard-Jones potential. (From ref. 65.)

of the form $g(l0l; r)$ that determine correlations between the orientations of the intermolecular vector and the HI axis, and therefore the extent of H-bonding to be expected. At the same time, the relatively large values of $g(110; r) \sim h_\Delta(r)$ and $g(112; r) \sim h_D(r)$ are consistent with a significant correlation in the relative orientations of the molecular axes of two neighbouring molecules, and account for the relatively high dielectric constant of the liquid.

9.5.4 *Carbon monoxide*

Fracassi *et al.*[148] have simulated the partial correlation functions for CO using constant pressure MD and employing a three-site electrostatic, overlap, and dispersion model of the intermolecular potential. Their results predict a considerable degree of structure in the $g_{\alpha\beta}(r)$, but disagree sharply with neutron $S(k)$ data obtained by Bellissent-Funel *et al.*[149] The experimental data can be reproduced quite accurately assuming a central Lennard-Jones potential. Howe *et al.*[150] have performed MC simulations of $S(k)$ using a simple hard-sphere model and also find good agreement with experiment. These results suggest that

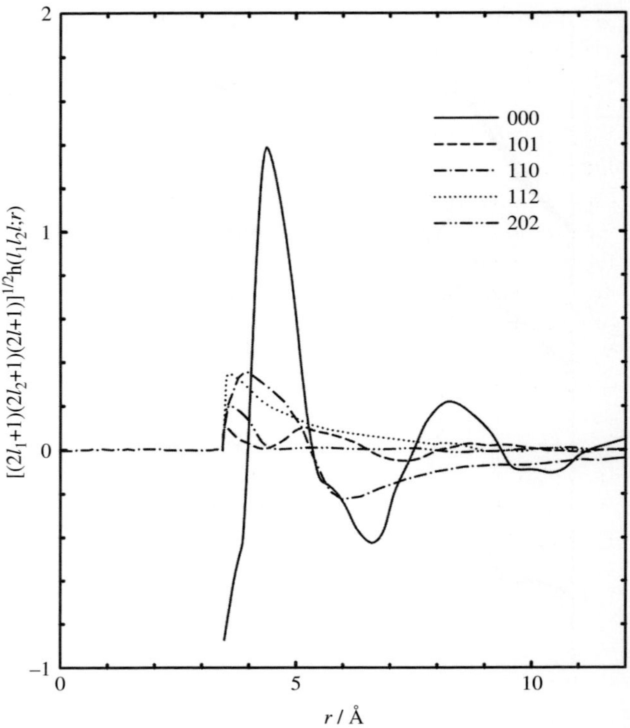

FIG. 9.14 Space-fixed spherical harmonic expansion coefficients $h(l_1 l_2 l; r)$ of the pair correlation function $h(\mathbf{r}\omega_1\omega_2) = g(\mathbf{r}\omega_1\omega_2) - 1$ for liquid HI at 253 K. (From ref. 65.)

orientational correlations in CO are either absent or very weak. Support for this conclusion comes from a reexamination of early neutron scattering studies[151] on the isoelectronic species N_2; careful analysis reveals that orientational structure originally thought to be present is in fact absent.[149]

9.5.5 *Carbon dioxide*

MD simulations of the intermolecular structure factor for liquid carbon dioxide, CO_2, have been reported by Böhm *et al.*[57] Their potential contains atom–atom exchange, electrostatic, and dipersion interactions; the exchange and electrostatic terms were calculated by *ab initio* methods, while the dispersion term was obtained by fitting to the measured second virial coefficient of the gas.[152] Figure 9.15 shows a comparison of the calculated $S_{\text{inter}}(k)$ with neutron data obtained by van Tricht *et al.*[153] near the triple point. There is good agreement over the whole range of k values studied. The simulations reproduce the shoulder on the high-k side of the first peak, a feature not predicted by RISM.[153] Böhm *et al.* point out that a short-range site–site potential cannot provide a realistic description of the static structure of liquid CO_2: the observed orientational correlations are attributable to the electrostatic (essentially quadrupolar) interaction of neighbouring molecules.[57] More recently, Adya and Wormald[154] have studied neutron scattering by CO_2 as a function of temperature. As the temperature is raised, a peak in the correlation function $g(r)$ (defined through Eq. (9.64)) at 3.3 Å associated with nearest-neighbour orientational correlations becomes weaker and eventually disappears near the critical temperature, presumably when quadrupolar ordering can no longer compete with randomizing thermal fluctuations. The absence of structure in $g(r)$ at larger r at *any* temperature shows that next-nearest-neighbour correlations, if present, are very weak.

The structure of supercritical CO_2 has been studied over a wide range of k by Ishii *et al.*[155] using neutron diffraction, and by Nishikawa and Takematsu[156] using X-rays. The radial distribution function shows remarkable differences from that of the subcritical liquid. The supercritical fluid appears to consist of a mixture of liquid-like high-density and gas-like low-density regions. A few degrees above T_c, and at densities slightly above ρ_c, $g(r)$ remains positive up to $r = 14$ Å, suggesting that the liquid-like regions are composed of clusters of 80–90 molecules. The peaks in $g(r)$ are somewhat broadened; this is probably attributable to the

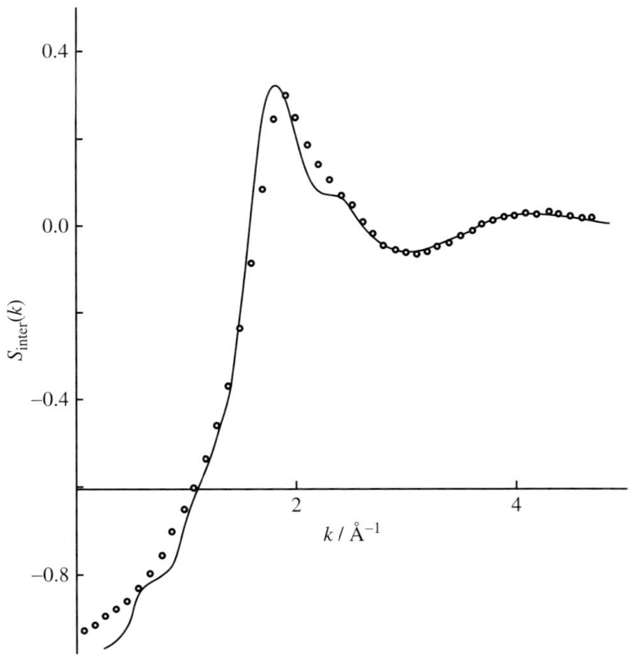

FIG. 9.15 Intermolecular structure factor for liquid CO_2 near the triple point. Full line, MD simulation using site–site potential including overlap, electrostatic, and dispersion interactions; points, neutron data. (From ref. 57.)

finite sizes of the molecular clusters, which results in a wider range of molecular environments than would be expected for the bulk liquid.[156] MD simulations using a three-site Lennard-Jones plus point quadrupole potential yield good agreement with experiment, and show that T-shaped orientations predominate. Interestingly, however, the shoulder on the high-k side of the main peak in $S(k)$ is absent, suggesting that these correlations are weaker in the high-temperature fluid than in the liquid.[155]

At high densities, the structure of the supercritical fluid more closely resembles that of the liquid.[155]

Monte Carlo simulations of the structure of supercritical CO_2 using a simple modified central LJ plus quadrupole–quadrupole interaction model have been surprisingly successful.[157]

Reverse Monte Carlo modeling of the structure factors has been performed for the gaseous, liquid and, supercritical phases.[158]

9.5.6 *Hydrogen sulphide*

Andreani *et al.*[159] have performed MD simulations on liquid hydrogen sulphide (H_2S) at 208 K using an intermolecular potential proposed by Forester *et al.*[160] consisting of a Lennard-Jones term (centred on S) plus a four-site electrostatic interaction. The computed partial structure factors and correlation functions agree well with data obtained from neutron studies on H_2S, D_2S and H_2S/D_2S mixtures. Particularly noteworthy is the lack of any structure in $g_{HS}(r)$ and $g_{HH}(r)$ attributable to hydrogen bonding, or to the tetrahedral short-range order characteristic of H_2O under similar conditions. This observation is supported by measurements of $g_{SS}(r)$ by Santoli *et al.*,[161] which shows no evidence of a second peak at the expected distance. These authors did, however, find evidence for weak orientational correlation in the liquid, as reflected in a doubly structured first peak in $g_{HS}(r)$; this appears to result from dipolar ordering.

9.5.7 *Sulphur dioxide*

Neutron scattering studies on SO_2 have been performed[162] at various temperatures. The experimental $g(r)$ obtained by transformation of the intermolecular structure factor (Eq. (9.64)) shows sharp peaks at 3.5, 4.5, and 5.6 Å, as well as oscillations extending to 18 Å, which are attributable to strong short- and long-range S–O and O–O orientational correlations. These correlations, which are supported by results from vibrational Raman spectroscopy,[163] are difficult to explain in terms of simple pair potentials. MD simulations using site–site Lennard-Jones and electrostatic (point charge or central dipole plus quadrupole) potentials fail to predict the observed degree of structure in $g(r)$, displaying only humps where clear peaks are visible in the experimental data.[164] The position of the second maximum is badly misplaced. This suggests that some anisotropic component in $u(\mathbf{r}\omega_1\omega_2)$ has been omitted. Although good agreement can be obtained[162] using a potential[165] derived from studies of the lattice dynamics of solid SO_2, which incorporates a distributed molecular dipole moment, this cannot accurately predict the thermodynamics of the liquid. It would seem that, at present, there is no satisfactory model for the SO_2–SO_2 intermolecular potential.

9.5.8 *Carbon tetrachloride*

There have been many experimental studies of the structure of liquid CCl_4: see, e.g., refs. 18, 166–168 and references therein. Unfortunately, there is often significant disagreement between the different sets of data, and this has led to some difficulties when comparisons are made with theoretical models. Nevertheless, there is clear evidence for orientational correlations in the structure even at temperatures as high as 430 K.[18]

Both the nature of these correlations, and the form of the interactions that cause them, are still in question. While perturbational calculations[16] suggest electrostatic (octopolar) interactions as the primary source (cf. § 9.3), MD simulations appear to identify short-range anisotropic overlap as a more likely candidate.[72] Figure 9.16 shows a comparison of $S_{\mathrm{inter}}(k)$ at 285 K simulated using a five-site Lennard-Jones potential with two sets of neutron data. Except for the intensity of the first peak at 1.3 Å^{-1}, the calculations reproduce the measured data within experimental error; the error in peak intensity is probably attributable to truncation of the MD correlation function rather than to inadequacies in the intermolecular potential. McDonald *et al.*have investigated the effect of adding an electrostatic site–site potential consistent with the molecular octopole moment, and find this results in a negligible change in $S_{\mathrm{inter}}(k)$.[72]

Misawa has measured the temperature dependence of $S_{\mathrm{inter}}(k)$ along the liquid–vapour coexistence curve over the range 250–430 K.[18] He finds two principal effects as the temperature increases: (a) the first peak broadens and shifts to lower k, resulting from a decrease in the bulk density; and (b) the second small peak at 2.2 Å^{-1} is broadened, attributable to a decrease in the number, and a broadening in the distribution, of orientationally correlated molecules. Chang *et al.*have simulated the temperature dependence of the partial correlation functions for C–C, C–Cl, and Cl–Cl interactions, and their results are consistent with this explanation.[171]

RMC simulations by Pusztai and McGreevy suggest the average coordination number for Cl–Cl interactions is less than four;[168] this would appear to suggest that the preferred intermolecular orientation is corner to corner, rather than corner to face, as might perhaps have been expected on the basis of packing considerations.[71]

9.5.9 *Dichlorodifluoromethane*

Hall *et al.*have investigated neutron diffraction from liquid dichlorodifluoromethane (CCl_2F_2) at 153 K, and have performed MD simulations using a pair potential consisting of site–site Lennard-Jones plus, optionally, electrostatic terms.[172] The simulated correlation function, which reflects mainly Cl-F and

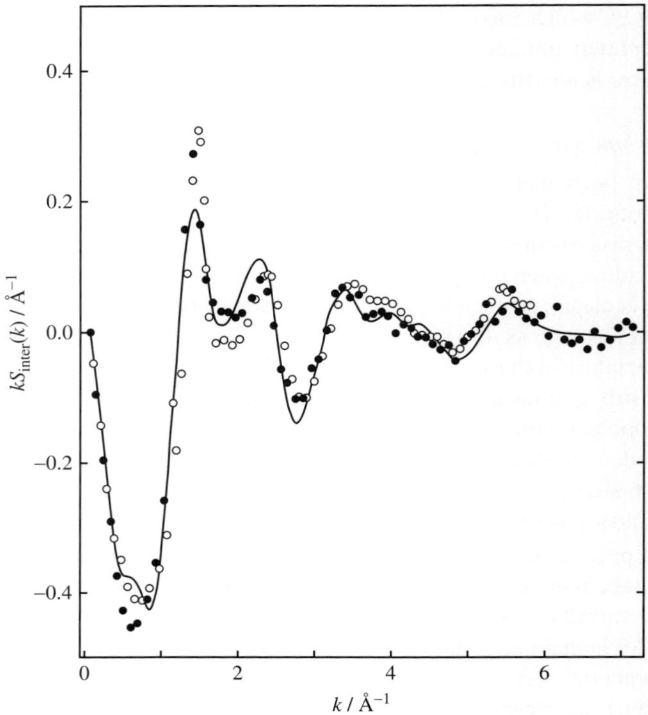

FIG. 9.16 Intermolecular structure factor for liquid CCl_4 at 285 K. Full line, MD simulation using site–site Lennard-Jones potential; filled and open circles, two different sets of neutron data. (From ref. 72.)

Cl-Cl correlations, yields a good description of the experimental data. Simulated partial correlation functions remain essentially unchanged when the atomic charges are turned off, which is to be expected since the negatively charged fluorine and chlorine atoms of one molecule cannot approach closely the positively charged carbon atom of a second. The local structure of the liquid is similar to that encountered in CCl_4. Further ordering is found to exist, however, due to the different polarizabilities (reflected in the different Lennard-Jones ϵ values) of Cl and F: thus Cl-Cl close contacts are favored over Cl-F, and these in turn are favored over F-F.

9.5.10 *Boron trifluoride*

The structure of boron trifluoride, BF_3, has been investigated by McLain *et al.*using neutron diffraction, and the results have been interpreted using RMC.[173] There is evidence for partial association through an intermolecular Lewis acid–base interaction. In the liquid phase order exists only in the first coordination sphere; the dominant structural motif of the solid phase, a pseudo trigonal bipyramidal [3 + 2] coordination at B, is partially retained. Some vestiges of this motif persist even in the supercritical fluid, although there is a much wider distribution in the position and population of the structural sites.

9.5.11 *Ammonia*

Narten has measured the structure factor of liquid ammonia (NH_3) using X-ray diffraction.[174] Since X-ray scattering from hydrogen is extremely weak, this experiment yields only the N–N (centres) correlation function. Neutron scattering from deuterated ammonia has been measured by Chieux and Bertagnolli in the subcritical liquid at 208 and 295 K,[175] and by Bausenwein *et al.*in the supercritical fluid.[89] More recently, Ricci *et al.*have performed isotopic substitution experiments at 213 and 273 K, and have extracted atom–atom correlation functions.[176] The data on $g_{NH}(r)$ and $g_{HH}(r)$ are particularly valuable for the light they shed on the nature of orientational correlations and the extent of hydrogen bonding.

Several computer simulations, using intermolecular potentials of varying degrees of sophistication, have been performed.[99,100,134,177,178] These include site–site electrostatic models supplemented by a central Lennard-Jones interaction[177] or atom–atom dispersion and overlap terms,[99,100,134] and a single-centre model including Lennard-Jones, dipole, quadrupole, and polarizability terms.[178] The effect of molecular nonrigidity associated with the NH_3 inversion mode has also been considered.[99] All simulations predict a $g_{NN}(r)$ in good agreement with experiment. The calculated $g_{HH}(r)$ and $g_{NH}(r)$ are relatively uniform, and the latter does not exhibit the structure indicative of H-bonding shown, for example, by $g_{OH}(r)$ in water (*vide infra*). Experimentally,[176] a weak feature in $g_{NH}(r)$ at $r = 2.25$ Å, just before the first interference maximum, is evidence for the formation of less than two H-bonds per molecule (which should be contrasted with three in the solid state). Another weak feature in $g_{HH}(r)$ at $r = 2.7$ Å supports this assignment. These results suggest a rather weak degree of molecular association and orientational correlation.[99,176] The calculated coordination number of 12, and the experimental value[176] of 14, are also consistent with a close-packed liquid structure, rather than an open hydrogen-bonded network.[99]

Bausenwein *et al.*have performed RMC simulations on supercritical NH_3 at 449 K at various densities.[89] As the density increases, configurations in which the symmetry axes of neighbouring molecules are parallel become less likely; this can be understood in terms of the increased importance of packing considerations when less free space is available.

9.5.12 *Acetonitrile*

Several groups[83,57,179,180] have reported MD simulations of acetonitrile, or methyl cyanide, CH_3CN. Figure 9.17 shows a comparison of the X-ray intermolecular structure factor $S_{X\text{inter}}(k)$ calculated by Böhm *et al.*[57] at 288 K with experimental data obtained by Bertagnolli and Zeidler.[181] The simulations employ a

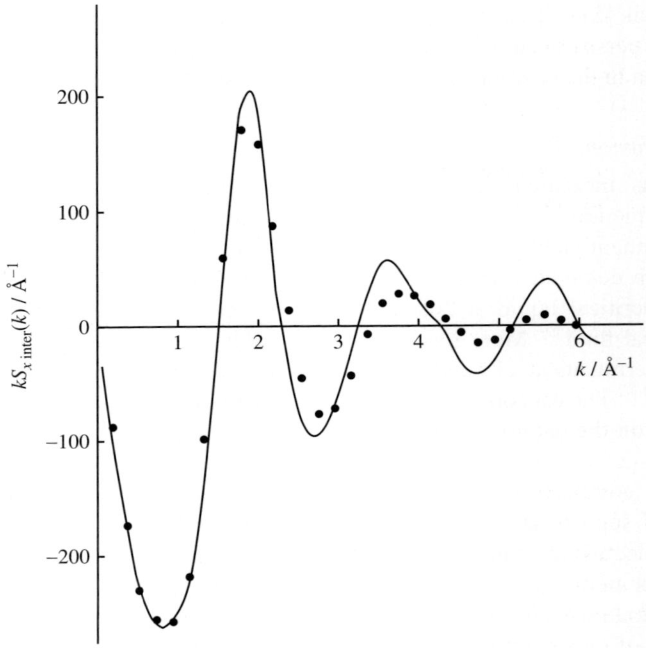

FIG. 9.17 X-ray intermolecular structure factor for liquid CH_3CN at 288 K. Full line, MD simulation using six-site potential including electrostatic terms; circles, experiment. (From ref. 57.)

six-site atom–atom potential containing overlap, dispersion, and electrostatic terms. Agreement is good for $k < 2\text{Å}^{-1}$, but for larger k the simulated function exhibits more pronounced extrema than the experimental curve. This may be attributable in part to the neglect of molecular nonrigidity. CH_3CN is rather flexible, especially because of low-frequency C-C-N bending motion;[57] molecular vibration would tend to broaden the maxima and minima in the partial correlation functions $g_{\alpha\beta}(r)$ and in the experimental structure factor. Similar results are found[180] for the neutron structure factor $S_{\text{inter}}(k)$. However, the intermolecular contribution is only about 5% of the total neutron scattered intensity for $k > 2\text{Å}^{-1}$, and since this is comparable to the experimental error, it is difficult to draw definite conclusions.

Ohba and Ikawa have performed MD simulations using a six-site potential incorporating Lennard-Jones and, optionally, electrostatic interactions.[83] They find that the computed $S(k)$ displays little evidence of electrostatic correlations despite the large molecular dipole moment $\mu = 3.91$ D. However, these are much in evidence in other quantities, e.g. intermolecular correlation functions describing reorientational motion. Böhm *et al.*[180] and Jorgensen and Briggs[179] have simulated the atom–atom partial correlation functions $g_{\alpha\beta}(r)$ using site–site potentials (in the latter case the methyl group is represented by a single site).

Both groups find that electrostatic interactions cause a considerable degree of orientational correlation. The first peak in $g_{\mathrm{NMe}}(r)$ (Me = methyl), in particular, sharpens and moves to smaller r when these are included in the pair potential. The results suggest a strong head-to-tail parallel alignment of neighbouring molecules. However, the computed $S(k)$ is not very sensitive to the details of the atom–atom correlations, in agreement with the results of Ohba and Ikawa. These simulations also reveal shortcomings in the RISM calculations of Hsu and Chandler,[81] which suggest that orientational correlations in liquid acetonitrile are entirely due to short-range interactions and are insensitive to dipolar coupling.

9.5.13 *Dimethyl sulphoxide*

The most reliable structural information on liquid dimethylsulphoxide, $(CH_3)_2SO$, comes from the neutron scattering studies of Luzar *et al.*[182] These authors achieved a partial separation of the atomic correlations using isotopic substitution of the methyl hydrogen atoms. Both $g_{X\mathrm{H}}(r)$, where $X =$ C, S, or O, and $g_{\mathrm{HH}}(r)$ are relatively structureless. The absence of distinct small-r peaks or shoulders implies a rather low degree of directionality between molecules at room temperature, despite the large molecular dipole moment of 3.9 D.

MD simulations using various site–site Lennard-Jones/electrostatic potentials have also been performed.[182] These potentials successfully duplicate most features of the $g_{\alpha\beta}(r)$, although no model could be found that gave an entirely satisfactory account of the short-range interactions. It is particularly noteworthy that a sharp peak in $g_{\mathrm{OC}}(r)$, which has been attributed by some[183] to hydrogen bonding between the methyl hydrogens and the oxygen atom of the polar SO group, is reproduced also by a model in which the methyl group carries no charge. This peak apparently represents packing caused by the Lennard-Jones cores. Furthermore, the estimated nearest-neighbour coordination number is 12. These results suggest that $(CH_3)_2SO$ is a tightly packed, nonassociated liquid.

9.5.14 *Methanol*

MC[184–186] and MD[187–189] simulations of the structure of liquid methanol, CH_3OH, have been performed at various temperatures. The intermolecular potential is modelled as a site–site Lennard-Jones/electrostatic interaction with sites on C, O and the hydroxyl H. The partial charges imply a dipole moment significantly larger than the gas-phase value, a device for taking approximate account of induction forces.

Sharp first peaks in the O-O, O-H, and H-H partial correlation functions indicate the presence of hydrogen bonding. Visual inspection of configurations shows that the H-bonding structure is dominated by winding chains. Early studies suggested an average chain length of about eight monomers at room temperature,[187] with some much longer chains also present;[188] the chain size was found to be strongly sensitive to temperature, rising as T drops.[188] However, more recent investigations by Yamaguchi *et al.*, which employ a potential refinement technique in which the intermolecular potential is adjusted to give better agreement with neutron partial structure factors, show that the chains are actually rather shorter: on average they contain about six methanol units.[185] It would therefore appear that the unrefined potentials somewhat overemphasize the degree of hydrogen bonding.

Each hydroxyl group has slightly less than two nearest neighbours, the average O-O and O-H distances being 2.8 and 1.8 Å, respectively. On average, each molecule forms about two hydrogen bonds.[185]

Sarkar and Joarder have argued that the experimental data suggest the presence of hexamer molecular clusters rather than chains.[190] This model of the liquid structure would explain certain other properties, such as the low viscosity, which are hard to reconcile with the presence of long molecular chains. However, Svishchev and Kusalik[189] have performed a very careful MD study of orientational correlations, and their results, as well as the studies of Yamaguchi *et al.*,[185] appear to confirm that the chain picture is correct. The chains show a preference for parallel packing, with the dominant interaction between chains associated with oxygen-methyl contacts.

In the moderately dense supercritical fluid, analysis of the neutron scattering spectrum through potential refinement[186] shows that the average chain length drops to about three, while the average number of hydrogen bonds is only slightly reduced; these are mostly associated with clusters of three to five molecules. As the density is lowered, a greater tendency to form ring structures is observed. Very similar results are found in the subcritical fluid at elevated temperatures. Perhaps not surprisingly, the raw (unrefined) potentials, which are designed to reproduce the structure of the liquid under ambient conditions, provide a very poor description of the experimental results in this regime.[186,191]

Quantum effects on the structure of liquid methanol have been discussed by Egelstaff and co-workers, who find that substitution of the hydroxyl hydrogen by deuterium leads to small but measurable changes at low temperatures.[63]

Neutron diffraction studies and simulations have also been performed on higher alcohols, including ethanol, CH_3CH_2OH;[191,192] the nature of the hydrogen-bonded structure is similar to that of methanol.

9.5.15 Water

The intermolecular structure of water has been extensively studied. Neutron,[19,193–204] γ-ray,[60] X-ray,[19,198,205–210] and electron[211] diffraction experiments have been performed under a variety of state conditions. Since X-rays do not scatter appreciably from hydrogen atoms, X-ray diffraction yields only the O-O, or centres, correlation function.[212] However, in combination with isotopic substitution analysis, the neutron data have supplied information on the O-H and H-H distributions, and therefore on the orientational correlations and extent of molecular association. There have been studies of the dense gas,[198] the subcritical liquid at temperatures both above[197,202] and below the normal boiling point, supercooled states,[199,201] and the supercritical fluid.[194,196] The effects of both pressure[207–209] and temperature[193,205,206] have been measured. Differences between the structure of H_2O and D_2O have been recorded.[60] Analysis of scattering data has yielded information on anharmonicity in intermolecular OH-O vibrations,[213] intermolecular effects on the intramolecular structure,[200,203,214] nonadditive terms in the intermolecular potential energy,[215] and three-body correlations.[66]

Several simulations of the structure of water have been performed using a variety of interaction potentials. Corongiu and Clementi[61] reported MD simulations on both light (H_2O) and heavy (D_2O) liquid water with the Nieser–Corongiu–Clementi (NCC) flexible *ab initio* potential.[216] This is acknowledged to be the most sophisticated water potential currently available, yielding excellent fits to many equilibrium and nonequilibrium properties: it supplements the successful Matsuoka–Clementi–Yoshimine (MCY) potential[217] with many-body induction interactions, and additionally allows for anharmonic O-H stretching and H-O-H bending. The study covered a temperature range 238–368 K, spanning supercooled, ambient, and high-temperature states. The density was held constant at $0.997\,\mathrm{g\,cm^{-3}}$. Water geometry in the liquid, pair correlation functions, coordination numbers, and X-ray and neutron structure factors were compared with experiment.

Water geometry

The simulations predict an increase in the O-H bond length of about 0.02 Å and a decrease in the H-O-H bond angle of about 3° in the dense liquid (presumably associated with hydrogen bonding). Neutron scattering experiments by Thiessen and Narten,[203] Soper and Phillips,[200] and Ichikawa *et al.*[214] all suggest bond length increases, but only Thiessen and Narten predict bond angle narrowing.

Pair correlation functions

Figure 9.18 shows a comparison of the computed site–site partial correlation functions for H_2O and D_2O at 305 K with experimental X-ray and neutron data obtained by Narten[19] and Soper and Phillips,[200] respectively. There are notable differences between the two experimental sets of data; the simulations appear to validate Soper and Phillips' results.

As the temperature is lowered, the first peak in $g_{OO}(r)$ sharpens and moves to smaller r as the liquid structure becomes more rigid. The first minimum also deepens, and eventually a well-structured second peak appears, indicating the emergence of a well-defined next-nearest-neighbour shell. The position of this peak at 4.5 Å is usually assumed to be indicative of local tetrahedral ordering. In the supercooled state a third peak is also evident. The NCC potential yields a more structured $g_{OO}(r)$ than is obtained using MCY.[218] The principal effect of

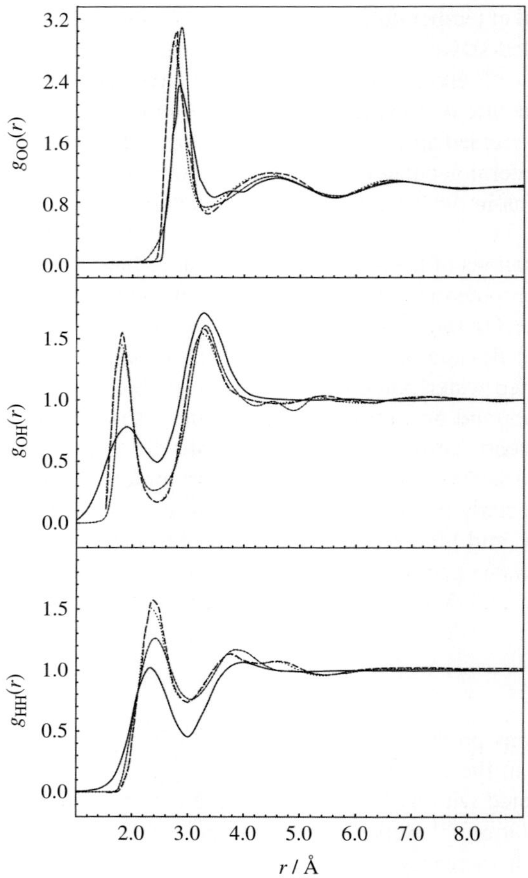

FIG. 9.18 Partial correlation functions for light and heavy water. Simulated results at 305 K (dots, H_2O; dashes, D_2O) are compared with X-ray (full line) and neutron (dotted full line) data at 298 K. (From ref. 61.)

allowing the molecules to vibrate is to further localize the first shell of water molecules. This happens because the molecular nonrigidity allows the formation of stronger hydrogen bonds, reducing the amplitude of intermolecular O–H vibrations. However, the changes in the pair correlation functions are not very pronounced.[216,219]

The peak in $g_{OH}(r)$ at 1.9 Å and the peak in $g_{HH}(r)$ at 2.3 Å reveal the existence of relatively strong and directional correlations between nearest-neighbour water molecules, a structural manifestation of hydrogen bonding. As the temperature increases, these peaks weaken and move to higher r. Tromp *et al.*have measured neutron scattering from water at elevated temperatures.[194] At 673 K, in the supercritical regime, the peaks have completely disappeared, suggesting the complete absence of H-bonding. More recent and very careful measurements by Soper *et al.*confirm this result.[195] However, MD simulations[220] using the SPC potential[221] suggest that a tetrahedrally coordinated H-bonded structure should persist even under supercritical conditions. It seems likely that the SPC model overemphasizes angle-dependent two-body electrostatic forces in an attempt to allow for effective many-body terms in the potential;[222] however, simulations using a polarizable water potential predict essentially identical partial correlation functions.[223] The question of the extent of H-bonding in the supercritical fluid must therefore currently be considered open. On the one hand, Soper has pointed out that the observed partial structure factors can be simulated accurately using a short-ranged potential;[196] on the other, Cummings and Chialvo have argued that $g_{OH}(r)$ may be a poor indicator of the degree of H-bonding because of its orientationally averaged nature.[224]

Soper has shown how the measured $g_{\alpha\beta}(r)$ can be used to estimate $g(\mathbf{r}\omega_1\omega_2)$.[66] His analysis reveals that, while hydrogen bonding produces a pronounced *directionality* in the organization of neighbouring molecules around any given molecule, favouring especially tetrahedral coordination, the lobes of this distribution are rather broad. Trigonal and perhaps pentagonal clusters are also expected. This suggests that local 'ice-like' clusters do not occur to any great extent in the liquid. However, these results may be an artefact of the calculational procedure, which by its construction minimizes structure in the molecular correlations.

There are only minor differences[60,61] between the correlation functions for light and heavy water, with the latter appearing slightly more structured. These differences have also been investigated theoretically by Kuharski and Rossky,[225] who performed quantum Monte Carlo path integral calculations on ST2 water.[226] Since they were concerned exclusively with quantum dynamical effects in the molecular rotation, they did not consider the effects of intramolecular vibrations. Consequently, their predictions are in only qualitative accord with experiment.[60,225]

The effect of pressure on the liquid structure has been studied, both theoretically[227] and experimentally.[193,197,205–209] Rather significant structural changes are seen, especially at high density. Figure 9.19 demonstrates the behaviour observed[207] in $g(r)$ (defined by Eq. (9.64)) as a function of pressure p at a constant temperature of 298 K. As p increases, the first peak in $g(r)$ first moves to smaller r; however, for $p > 2$ kbar this trend is reversed. No high-pressure simulations have yet been attempted with the NCC potential; the experimental estimate of about -0.03 Å kbar^{-1} at $p = 0$ is *not* well reproduced by MCY, which considerably underestimates the effect.[227] The second peak weakens and eventually disappears, suggesting a reduction in tetrahedral ordering. At high pressures the coordination number for the first shell rises significantly;[227] however, the simultaneous appearance of a shoulder on the high side of the first peak rules out any possibility that the structure is approaching that of a close-packed simple liquid. A possible explanation of the observed pressure effects has been suggested

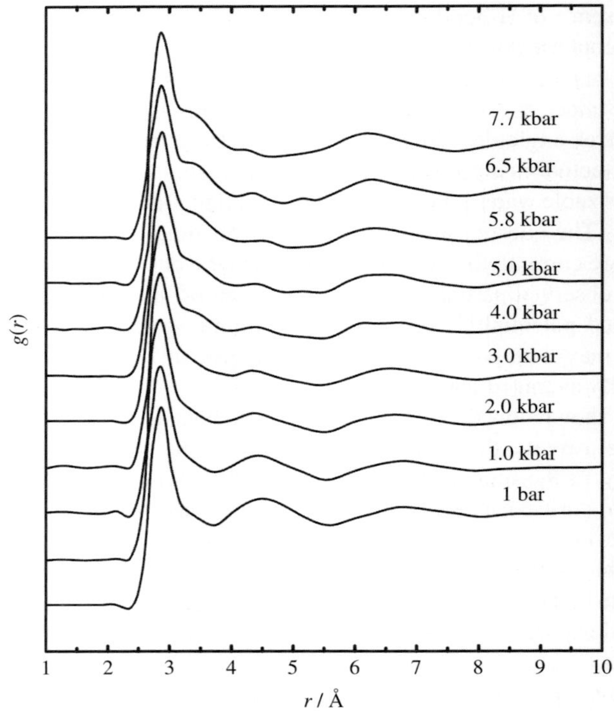

FIG. 9.19 Effect of pressure on the water pair correlation function, as determined by X-ray diffraction. (From ref. 207.)

by Okhulkov *et al.*[207] At normal pressures, the water structure resembles that of ice I_h. It is proposed that additional ice-like structures (ice III, ice V, and ice VI), which are stabilized at higher p, also form and exist in equilibrium with the ice I_h-like structure. A multiphase model of this sort can be accurately fit to experiment.[207]

Bellissent-Funel and Bosio[193] find that at high pressures the temperature dependence of $g(r)$ is reduced, suggesting that the relatively mobile, and therefore temperature-sensitive, water network found at ambient pressures is replaced by more rigid ice-like structures. Conversely, Postorino *et al.*[197] have found evidence that at high temperatures the effects of pressure on the water structure are magnified. At 573 K, hydrogen bonding is almost entirely destroyed in the vicinity of the coexistence curve. Increasing the pressure causes the reappearance of a peak in $g_{\mathrm{OH}}(r)$ at 1.9 Å, suggesting that H-bonds have reformed. However, examination of $g_{\mathrm{OO}}(r)$ implies that the local molecular arrangement is far from tetrahedral; the emergence of a shoulder at 3.5 Å may instead indicate the presence of interstitial, non-H-bonded molecules.

The structure of D_2O steam at 580 K has been investigated by Buontempo *et al.*[198] using neutron diffraction. $g(r)$ is quite structureless, consisting of a single peak at 3.16 Å ($\rho = 1.55 \times 10^{-3}$ Å^{-3}) or 3.09 Å ($\rho = 0.93 \times 10^{-3}$ Å^{-3}), with a shoulder at ≈ 2 Å. It was suggested that the shift of the peak from its position in the dense liquid,[194] 3.5 Å, reflects an apparent breakdown of structure at low ρ, allowing a closer molecular approach. Simulations have recently been performed by Martí *et al.*[228] using an SPC-like model (but with a reduced dipole moment). The computed $g(r)$ was consistent with the neutron data. The 2 Å shoulder was shown to correspond to the decay of $g_{\mathrm{OD}}(r)$ after its first maximum. The simulations suggest the observed shift in the maximum of $g(r)$ may not necessarily be related to a change in intermolecular distances, but simply result from a superposition of the three partial structure factors $g_{\mathrm{OO}}(r)$, $g_{\mathrm{OD}}(r)$, and $g_{\mathrm{DD}}(r)$. The absence of a peak at 4.5 Å in $g_{\mathrm{OO}}(r)$ appears to rule out a significant tetrahedral ordering of oxygen atoms, but the structure of $g_{\mathrm{OD}}(r)$ does suggest there are some residual H-bonds. At least 95% of the H-bonding molecules are dimers; larger molecular groupings are very rare. Simulations have also been performed by Fontana *et al.*using a polarizable potential.[229]

De Santis and Rocca have used the experimental data to infer the existence of nonadditive, many-body terms in the water potential.[215] A detailed understanding of the structure of steam awaits further experimental investigation.

Coordination numbers

The coordination number

$$N_{\alpha\beta}(r) = 4\pi\rho \int_0^r \mathrm{d}r'\, r'^2 g_{\alpha\beta}(r') \tag{9.89}$$

yields the number of atoms of type β within a distance r of atoms of type α. For $r = r_{\min}^{(1)}$, the position of the first minimum in $g_{\alpha\beta}(r)$, this is the number of atoms lying in the first coordination shell. A study of the temperature dependence of $N_{\mathrm{OO}}\left(r_{\min}^{(1)}\right)$ shows a sharp increase at a well-defined transition temperature of about 305 K. Below this temperature $N_{\mathrm{OO}}\left(r_{\min}^{(1)}\right) \approx 4$, as expected for a hydrogen-bonded tetrahedral network. Above it $N_{\mathrm{OO}}\left(r_{\min}^{(1)}\right)$ increases sharply, pointing to a major structural reorganization of the liquid. The calculated temperature dependence is consistent with measurements of $\partial S_{X\mathrm{inter}}(k)/\partial T$ by Bosio *et al.*[206] Conversely, $N_{\mathrm{OH}}\left(r_{\min}^{(1)}\right)$ *decreases* slightly above 305 K (see erratum to ref. 61). These results are consistent with a decrease in hydrogen bonding and molecular association above the transition temperature: if the O-H interactions become relatively less important, $N_{\mathrm{OH}}\left(r_{\min}^{(1)}\right)$ will drop, while the structure will begin to resemble that of a close-packed liquid, with a concomitantly larger $N_{\mathrm{OO}}\left(r_{\min}^{(1)}\right)$.

X-ray and neutron structure factors

Figure 9.20(a) shows the simulated light water X-ray intermolecular structure factor at different temperatures. Above the supercooled region, both the intensity and position of the first peak increase almost linearly with temperature, whereas there is a decrease for the second peak; eventually, at temperatures near the boiling point, the two peaks merge. This is precisely the behaviour observed experimentally,[206] and is indicative of a gradual loss of orientational correlation. We note that the two-body MCY and flexible MCY[219] potentials do *not* correctly reproduce the intensities of these peaks.[216]

Figure 9.20(b) compares the simulated $S_{X\text{inter}}(k)$ for H_2O and D_2O with experimental data[19] for H_2O. Although the oscillations are somewhat out of phase beyond the second minimum, the agreement is generally satisfactory, and is superior to that achieved by any other interaction potential.[61]

A comparison of simulated and experimental[203] neutron $S_{\text{inter}}(k)$ appears in Fig. 9.21. Agreement is good except for the case of pure H_2O; discrepancies here probably reflect inaccuracies in the experimental data caused by the large incoherent scattering cross-section of H. Simulations of $S_{\text{inter}}(k)$ for D_2O have also been performed at the much higher temperature of 523 K by Buontempo *et al.*,[202] using the *ab initio* NEMO potential of Wallqvist *et al.*[230] At this temperature $S_{\text{inter}}(k)$ resembles that of a simple fluid such as argon, with no remaining evidence of orientational correlations.

In summary, we see that the NCC flexible potential is clearly capable of an accurate description of the structure of liquid water over an extended temperature range. Its performance exceeds that of all other models proposed to date.[230]

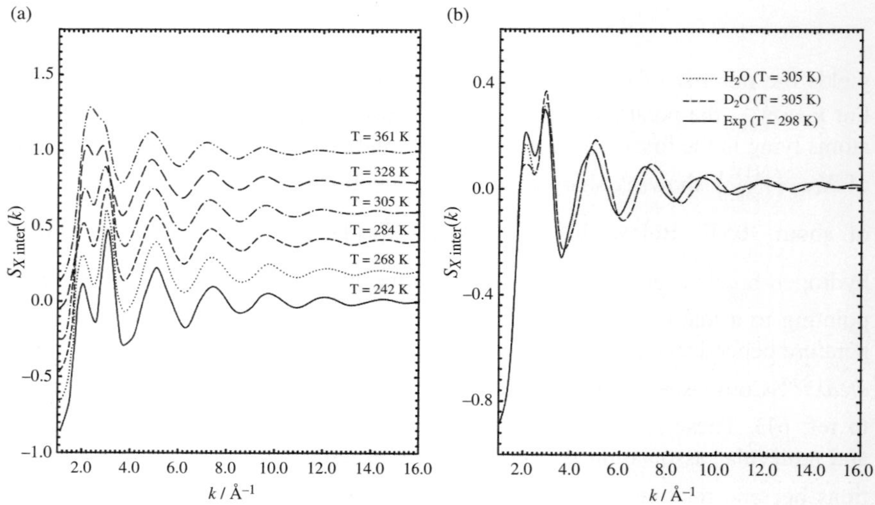

FIG. 9.20 (a) Temperature dependence of computed intermolecular X-ray structure factor for water. (b) Comparison of simulated (dots, H_2O; dashes, D_2O) X-ray structure factor at 305 K with experiment (full line) at 298 K. (From ref. 61.)

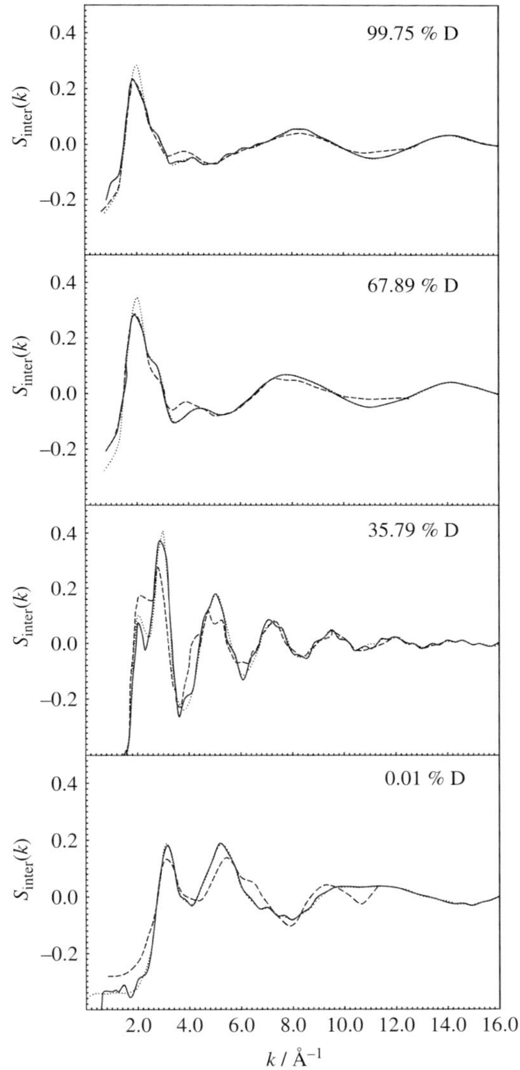

FIG. 9.21 Neutron intermolecular structure factor for water samples with different D_2O content. Dotted line, simulations based on H_2O potential; full line, simulations based on D_2O potential; dashed line, experiment. (From ref. 61.)

9.5.16 *Aqueous solutions*

The structure of supercritical aqueous solutions of the rare gases He, Ar, and Kr has been examined by Botti *et al.*[231] They performed neutron diffraction measurements using H/D isotopic substitution, and compared these with simulations employing a potential-refinement algorithm.[232] This yielded estimates of the partial correlation functions representing H–H, H–O, and O–O interactions, as well as those governing water–solute interactions. There is clear evidence for a water deficit around each solute atom. Comparison with data for pure supercritical water shows that the solvation of the rare gas atoms influences the microscopic structure even at low solute concentrations; in particular, the spatial distribution of water molecules is less isotropic than in the pure fluid. Each solute produces a compression of the water–water spatial density function, which is visible as a shift towards smaller r of the nearest-neighbour peaks of $g_{OO}(r)$ and $g_{OH}(r)$. This is most apparent for He, which is the only gas that penetrates the first water–water coordination shell.

Soper and co-workers[67,233,234] have performed neutron scattering studies of dilute alcohol/water mixtures. Using isotopic substitution analysis they have isolated partial correlation functions describing the correlation of water hydrogens with each other ($g_{HH}(r)$) and with the apolar alkyl groups. For methanol, CH_3OH, Soper and Finney[67] have attempted to reconstruct the orientational pair correlation function $g(\mathbf{r}\omega_1\omega_2)$ describing CH_3OH-H_2O correlations. Examination of the spherical harmonic components of this function demonstrates the existence of a definite hydration shell of water molecules at a distance of 3.7 Å from the methanol C atom; the plane of H_2O prefers to lie tangential to the CH_3OH-H_2O axis, and there is a weak alignment of the water and methanol dipole moments.

At a mole fraction of 10% methanol, $g_{HH}(r)$ is remarkably similar to that for pure water. Thus the data do *not* support the widely held view[235] that apolar head groups such as CH_3 enhance the water structure. The hydrophobic solute is apparently localized in a cavity in the H-bonded water network, and the roughly tetrahedral local coordination found in pure water is preserved. Similar results are also found[233] for dilute aqueous mixtures of ethanol and 2-methyl-2-propanol. For the latter species the evidence suggests that water–alcohol hydrogen bonds form preferentially over alcohol–alcohol hydrogen bonds; alcohol–alcohol contacts are limited to the nonpolar head groups.[234]

Simulations have been performed with a variety of potentials.[236–239] The results depend strongly on the parameters adopted. Potentials that emphasize the acceptor character[237,239] of CH_3OH (i.e. its ability to hydrogen bond to water H atoms) predict a significant degree of structural reorganization in dilute solutions, while those that emphasize its donor properties[238] (i.e. its ability to hydrogen bond to water O atoms) do not. The *ab initio* potential of Hernandéz-Cobos and Ortega-Blake[236] yields generally good agreement with the experimental results, and in particular reproduces the measured hydrogen–hydrogen and carbon–(water) oxygen partial correlation functions, although the number of water molecules in the cage surrounding the methanol CH_3 group is somewhat larger than that found experimentally. This simulation does, however, suggest that methanol-induced changes in the water structure are indeed present, but that shifts in water–water H–H correlations are largely masked by compensating methanol–water H–H correlations.

Luzar and Chandler have reported MD simulations on concentrated dimethylsulphoxide (DMSO)/water mixtures.[240] Water–water interactions were modelled by the SPC[221] potential, while water–DMSO and DMSO–DMSO interactions were represented by site–site Lennard-Jones plus electrostatic terms. The results have been compared with neutron diffraction data obtained by Soper and Luzar.[241] Both simulations and experiment suggest that the tetrahedral water structure is not strongly affected by the presence of DMSO, even at mole fractions of 35%. As the DMSO mole fraction increases, the correlation of nearest neighbour waters becomes more pronounced, counterbalancing the reduced number of water molecules present in the bulk solution. As a result, the water–water coordination number for the first shell remains constant, at four. In the second and successive coordination shells, however, the average number of water–water H-bonds is markedly reduced: their place is taken by water–DMSO H-bonds. The simulations suggest each DMSO molecule can form a maximum of two such bonds, but that the most likely number of bonds is less than this in concentrated solutions.

Soper and Luzar have examined the structure of aqueous solutions of DMSO using hydrogen/deuterium isotopic substitution.[242] The partial correlation functions thus obtained were used in an RMC-like simulation to refine the DMSO–DMSO and DMSO–H_2O intermolecular potentials. The picture of the DMSO hydration structure that emerges is very similar to that predicted by Luzar and Chandler, and appears to preclude a significant degree of hydrophobic association.

Aqueous ionic solutions are somewhat outside the scope of this work; for recent theoretical and experimental treatments see, e.g., the studies by DeMille and Molinero[243] and Mancinelli *et al.*,[244] respectively.

9.5.17 *Hydrocarbons*

There have been several experimental studies of the structure of hydrocarbons, using X-ray,[90,245–249] neutron,[95,96,249–252] and electron[253] diffraction, and in many cases MC and MD simulations have been reported and compared with experiment.

Methane

MD simulations of the structure of liquid CH_4 have been performed by Murad *et al.*using a site–site exp-6 potential.[254] Although these results have not been compared directly with experiment, there is indirect evidence[90] that the potential accurately reproduces the centres correlation function $g_{CC}(r)$. A comparison of the simulated neutron intermolecular structure factor $S_{inter}(k)$ for CD_4 with the free-rotation prediction $S^{free}(k)$ (§ 9.2) suggests that orientational correlations are rather weak,[254] as is to be expected given the weak H-H and (to a lesser extent) H-C components of the intermolecular potential. This is supported by an RMC analysis of neutron scattering measurements on CD_4 at 370 K, which indicate no detectable orientational correlations in the supercritical fluid, the structure of which is almost independent of density.[95]

Guarini *et al.*[255] have shown that density analysis of the structure factor of the dilute gas can, in principle, yield information on three-body contributions to the potential, as has been done for the rare gases.[12,15]

Neopentane

MC simulations of liquid neopentane, $(CH_3)_4C$, have been compared with X-ray diffraction data at 298 K for three different site–site potential models (Lennard-Jones, repulsive Lennard-Jones, and exp-6; each methyl group is treated as a single site). There is a substantial difference between the computed partial correlation functions for the three models, especially $g_{\mathrm{MeMe}}(r)$. Comparison with experiment suggests the exp-6 model is most satisfactory, and that attractive forces have an important effect on the liquid structure, particularly on the degree of orientational correlation between neighbours. Overall the $g_{\alpha\beta}(r)$ tend to resemble those for CCl_4 (*vide supra*). Differences between $g_{\mathrm{MeMe}}(r)$ and $g_{\mathrm{HH}}(r)$ for methane[254] reflect significantly different degrees of orientational correlation in the two species.

Ethane

Sandler *et al.*[256] have measured the X-ray structure factor for liquid ethane, CH_3CH_3, at 105 and 181 K. The carbon–carbon (or methyl–methyl) intermolecular pair correlation function was obtained by Fourier inversion, and compared with the results of MC simulation. No two-site exp-6 interaction model, with the interaction sites situated at the carbon atoms, could be found that reproduced the experimental data. The results could, however, be fit to a single-site pair potential containing anisotropic dispersion and overlap interactions of the form discussed in §§ 2.6 and 2.7. This potential was more strongly anisotropic than the two-site model, with strongly localized attractive regions encouraging T and — — orientations. Of course, X-ray measurements are insensitive to interactions involving hydrogen atoms; more detailed information on the intermolecular potential, in the form of a refined six-site model, results from a combined MC/RMC analysis[257] of neutron data.[96]

Propane

Habenschuss and Narten[245] measured X-ray scattering from liquid propane, $CH_3CH_2CH_3$, at temperatures near the triple and boiling points. Near the triple point, the structure is similar to that of ethane. Whereas this structure is retained by ethane over the whole liquid range,[256] however, propane shows a qualitative structural change exemplified by an almost featureless $g(r)$ near the boiling point. (Similar structural changes have been observed for the higher *n*-alkanes.[246]) This behaviour has been explained by MD simulations[258] using a three-site Lennard-Jones potential. The sites are located not at the carbon atoms but at the geometric centres of the CH_x subunits, making the molecule somewhat more elongated and increasing orientational correlations. Figure 9.22 shows a

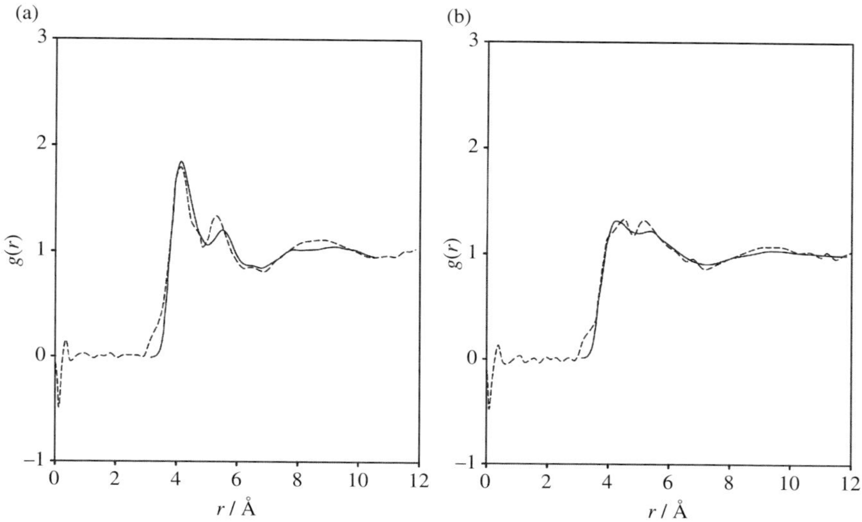

FIG. 9.22 Pair correlation function (transform of $S_{X\,\mathrm{inter}}(k)$) for liquid propane at (a) 95 K (near the triple point) and (b) 230 K (near the boiling point). Solid lines, MD simulation using three-site Lennard-Jones potential; dashed lines, experiment. (From ref. 258.)

comparison of experiment and simulation at two temperatures. The split in the first peak of $g(r)$ near the triple point indicates the presence of T orientations;[245,258] these are sensitive to temperature and have almost disappeared at the boiling point. Simulations have also been performed using a purely repulsive potential, and indicate that the effect of attractive forces on the liquid structure is small.[258]

Neutron scattering from supercritical deuteropropane has been measured by Pfleiderer *et al.*[252] Comparison with MC simulations employing a variety of united-atom potentials showed that an optimized, shifted, three-site Lennard-Jones model proposed by Ungerer *et al.*[259] gave the best description of the fluid structure and its variation with density.

Butane

Butane, $CH_3CH_2CH_2CH_3$, is the simplest hydrocarbon capable of exhibiting *trans/gauche* rotational isomerism, and as such it has been the subject of both theoretical[260,261] and experimental[246] studies. These have attempted to determine, in particular, the degree to which the relative amounts of the two conformers present depend on the thermodynamic state. The consensus seems to be that the equilibrium shifts slightly towards the *gauche* state in going from the ideal gas to the dense liquid.[261] However, Bañón *et al.*[260] point out that this conclusion is sensitive to small changes in the intermolecular potential, and the matter cannot be regarded as completely settled.

Hexane

Partial structure factors for deuterated room-temperature liquid *n*-hexane have been obtained by neutron diffraction experiments, and simulated successfully by MD.[262]

Higher alkanes

The intra- and intermolecular structure of the chain-like higher normal alkanes $CH_3(CH_2)_nCH_3$ has been investigated using X-ray and neutron diffraction. Simulations have also been performed. Here technical difficulties arise due to the flexible nature of the molecules; for a discussion of the strategies that have been used to counter these, see Almarza *et al.*[263]

X-ray studies have been performed[247] on heptane ($n = 5$), decane ($n = 8$), pentadecane ($n = 13$), and eicosane ($n = 18$). From a study of peak heights in the pair correlation function one can rule out the possibility of a large fraction of extended chains in the liquid, as proposed in 'bundle' models and as found in the crystal. Analysis of the data has also yielded information on the average torsional angles for *trans* and *gauche* rotational isomer segments, and on *trans/gauche* populations. The average *gauche* torsion angle barely differs from the nominal 60° of a simple threefold rotational potential. The average *trans* torsion angle of 170° implies that the *trans* segments are twisted out of plane by 10°, probably as a result of interactions with neighbouring *gauche* bonds. These values are not significantly different from those found in the gas, suggesting that intermolecular interactions have a very small effect on conformational geometries. The fraction of *gauche* segments in the liquid phase is $45 \pm 8\%$ (with only a weak dependence on n). This is slightly higher than in the gas phase, but overall it appears that the conformational equilibrium of alkanes is only minimally perturbed by intermolecular forces. Support for this conclusion comes from MD[264,265] and MC[266] simulations.

Neutron studies on decane and eicosane have been reported by Londono *et al.*[250] These authors used isotopic substitution to obtain a clean separation of intramolecular and intermolecular correlations. The observed $g_{HH}(r)$ shows broad features at 2.5, 5, and 7 Å that can be interpreted as the first three shells of neighbours of the intermolecular H-H structure. For $r < 10$ Å (decane) and $r < 15$ Å (eicosane), $g_{HH}(r) < 1$. Thus there is a screening of intermolecular correlations over a distance that is roughly the end-to-end separation of a single chain. This is also apparent in the C-C correlations,[248] and is the phenomenon of the de Gennes 'correlation hole'.[267] All of these features are accurately reproduced by MD simulations using a pair potential containing C-C, C-H, and H-H exp-6 interactions.[268]

Simulations[268] have demonstrated that 'united atom' models, which treat CH_x units as single sites, can be made to yield an excellent description of the liquid C-C structure if the site–site interaction parameters are carefully chosen.

Benzene

The structure of liquid benzene, C_6H_6, has been studied using X-ray,[249] neutron,[249,251,269,270] and electron[253] diffraction. The intermolecular structure factor shows a feature at 1.9 Å^{-1} that is associated with orientational correlations between nearest neighbours. From an analysis of the neutron data, Misawa and Fukunaga have determined the most probable relative configuration to be that of Fig. 9.23.[251] This is essentially a perpendicular orientation in which two H atoms of one molecule point towards the centre of the ring of a second, where the π-electron density is high. The centre-to-centre separation is 5.1 Å. Calculations of the electrostatic potential energy surface for the benzene dimer support this conclusion, but also reveal the existence of other local minima lying within 1 kJ mol^{-1};[271] it thus seems likely that several orientations contribute to the observed $S_{\text{inter}}(k)$.

Anderson *et al.* have reported MD simulations using a flexible model of benzene with a twelve-site exp-6 pair potential.[272] Their results suggest that the atom–atom distribution functions are rather structureless; this is not unexpected in view of the large number of distances contributing to $g_{\text{CC}}(r)$ (36), $g_{\text{HH}}(r)$ (36), and $g_{\text{CH}}(r)$ (72). The correlation functions are relatively insensitive to molecular vibration, reflecting the rigidity of the benzene structure.

MC simulations using various different site–site potentials have been performed by Bartell *et al.*[273] and compared with electron diffraction data[253] on supercooled droplets at 160 K. Neither six- nor twelve-site Lennard-Jones models yield a satisfactory description of $S_{e\,\text{inter}}(k)$. By including electrostatic interactions one can reproduce the observed peak at 1.9 Å^{-1}, but not the structure visible at higher k. Price and Stone[271] have shown that the precise representation of the molecular charge distribution has a major effect on the calculated electrostatic

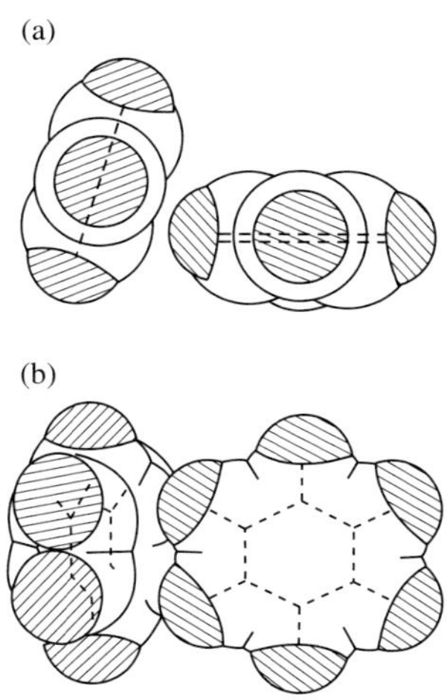

FIG. 9.23 Preferred orientation of two benzene molecules: (a) top view, (b) side view. (Reprinted with permission from ref. 251. Copyright 1990, American Institute of Physics.)

potential, and it would appear that at present there is no entirely satisfactory model for C_6H_6–C_6H_6 interactions.

At elevated temperatures, the translational and orientational order observed under ambient conditions is decreased. In the supercritcal fluid, translational order extends only to the nearest-neighbour shell, and orientational ordering is completely isotropic.[270]

Appendix 9A Pressure derivative of $S(k)$

In this Appendix we show how the isothermal pressure derivative of the intermolecular structure factor is related to three-body correlations.

Our starting point is Eq. (9.62). Differentiating with respect to pressure at constant temperature we find

$$(\partial[\rho^{-1}S_{\text{inter}}(k)]/\partial p)_T = \int \mathbf{dr}\, \exp(i\mathbf{k}\cdot\mathbf{r})\langle(\partial h(\mathbf{r}\omega_1\omega_2)/\partial p)_T F(\mathbf{k}\omega_1)F(\mathbf{k}\omega_2)^*\rangle_{\omega_1\omega_2}. \quad (9A.1)$$

From (3.239) we have

$$\beta^{-1}\chi(\partial[\rho^2 g(12)]/\partial\rho)_T = 2g(12) + \rho\int \mathbf{dr}_3\, \langle g(123) - g(12)\rangle_{\omega_3}, \quad (9A.2)$$

where $g(12) \equiv g(\mathbf{r}\omega_1\omega_2)$ and $g(123) \equiv g(\mathbf{r}_{12}\mathbf{r}_{13}\mathbf{r}_{23}\omega_1\omega_2\omega_3)$ are the pair and triplet correlation functions, respectively, and

$$\chi = \rho^{-1}(\partial\rho/\partial p)_T = \beta\rho^{-1}\left(1 + \rho\int \mathbf{dr}\, \langle h(\mathbf{r}\omega_1\omega_2)\rangle_{\omega_1\omega_2}\right) \quad (9A.3)$$

is the isothermal compressibility. Using (9A.2) and (9A.3) it is easily shown that

$$(\partial h(\mathbf{r}\omega_1\omega_2)/\partial p)_T = \beta\int \mathbf{dr}_3\, \langle g(123) - g(12)[1 + h(13) + h(23)]\rangle_{\omega_3}. \quad (9A.4)$$

Let us now write

$$g(123) = g(12)g(13)g(23)\exp\tau(123), \quad (9A.5)$$

where the function $\tau(123)$ is defined as a sum of many-body cluster diagrams.[274] Combining (9A.1), (9A.4), and (9A.5) we obtain

$$(\partial[\rho^{-1}S_{\text{inter}}(k)]/\partial p)_T = \beta\int \mathbf{dr}\, \exp(i\mathbf{k}\cdot\mathbf{r})\langle g(\mathbf{r}\omega_1\omega_2)\Gamma(\mathbf{r}\omega_1\omega_2)F(\mathbf{k}\omega_1)F(\mathbf{k}\omega_2)^*\rangle_{\omega_1\omega_2}, \quad (9A.6)$$

where $\Gamma(\mathbf{r}\omega_1\omega_2) \equiv \Gamma(12)$ is given by

$$\Gamma(12) = \int \mathbf{dr}_3\, \langle g(13)g(23)[\exp\tau(123) - 1] + h(13)h(23)\rangle_{\omega_3}. \quad (9A.7)$$

Equations (9A.6) and (9A.7) constitute the desired result.

Many-body forces affect $S(k)$ particularly for k values below the first minimum, and make a substantial contribution to the structure factor and its derivative in this range. A comparison of the predictions of (9A.6) with experiment[6] has revealed the inadequacy of simple theories for $\tau(123)$, while recent studies of the pressure (or, equivalently, density) dependence of $S(k)$ have provided clear evidence in some cases for the presence of terms in the many-body interaction beyond the simple Axilrod-Teller triple-dipole potential.[14]

Appendix 9B Body-fixed spherical harmonic coefficients of molecular form factors

In this Appendix we list the nonvanishing coefficients of the form factors $F_{ln}(k)$, as defined through (9.86), for some common molecular geometries. These quantities are used in perturbational calculations of $S(k)$, as described in § 9.3.

Homonuclear diatomics (AA)

$$\begin{aligned} F_{l0}(k) &= i^l(2l+1)j_l(kd_A), \qquad && l \text{ even},\\ &= 0, && l \text{ odd}. \end{aligned} \tag{9B.1}$$

Heteronuclear diatomics (AB)

$$F_{l0}(k) = i^l(2l+1)\frac{\bar{b}_A j_l(kd_A) + (-)^l \bar{b}_B j_l(kd_B)}{\bar{b}_A + \bar{b}_B}. \tag{9B.2}$$

Linear triatomics (BAB)

$$\begin{aligned} F_{l0}(k) &= i^l(2l+1)\frac{\bar{b}_A\delta_{l0} + 2\bar{b}_B j_l(kd_B)}{\bar{b}_A + 2\bar{b}_B}, \qquad && l \text{ even},\\ &= 0, && l \text{ odd}. \end{aligned} \tag{9B.3}$$

Tetrahedral molecules (AB_4)

$$\begin{aligned} F_{00}(k) &= \frac{\bar{b}_A + 4\bar{b}_B j_0(kd_B)}{\bar{b}_A + 4\bar{b}_B},\\ F_{32}(k) &= \frac{14}{3}(10)^{1/2}\frac{\bar{b}_B}{\bar{b}_A + 4\bar{b}_B} j_3(kd_B). \end{aligned} \tag{9B.4}$$

Octahedral molecules (AB_6)

$$F_{00}(k) = \frac{\bar{b}_A + 6\bar{b}_B j_0(kd_B)}{\bar{b}_A + 6\bar{b}_B},$$

$$F_{40}(k) = \frac{63}{2}\frac{\bar{b}_B}{\bar{b}_A + 6\bar{b}_B} j_4(kd_B),$$

$$F_{44}(k) = \frac{9}{2}\left(\frac{35}{2}\right)^{1/2}\frac{\bar{b}_B}{\bar{b}_A + 6\bar{b}_B} j_4(kd_B). \tag{9B.5}$$

The other non-vanishing coefficients can be obtained using the symmetry property $F_{l\underline{n}}(k) = (-)^{l+n} F_{ln}(k)^*$. The above equations for the F_{ln} are for the following choices of body-frame polar axis: for linear molecules, the axis coincides with the internuclear axis, pointing from B to A in the case of AB diatomics; for tetrahedral molecules, it coincides with a C_3 axis; for octahedral molecules, the C_4 axis. In (9B.1)–(9B.5) d_A and d_B are the distances of the corresponding nuclei from the molecular centre, which is taken as the centre of mass. It is important to be consistent in the choice of molecular centre, since this also affects the parameters of the isotropic reference potential $u_0(r)$ and the values of the spherical harmonic expansion coefficients $u(l_1 l_2 l; n_1 n_2; r)$.

References and notes

1. Barocchi, F., Bafile, U., and Magli, R., *J. Phys. Cond. Matt.* **8**, 9111 (1996).
2. Sesé, L. M., *Mol. Phys.* **89**, 1783 (1996).
3. Neumann, M., and Zoppi, M., *Phys. Rev.* **E65**, 031203 (2002).
4. Pfleiderer, T., Waldner, I., Bertagnolli, H., Tödheide, K., Kirchner, B., Huber, H., and Fischer, H. E., *J. Chem. Phys.* **111**, 2641 (1999).
5. Bomont, J.-M., Bretonnet, J.-L., Pfleiderer, T., and Bertagnolli, H., *J. Chem. Phys.* **113**, 6815 (2000).
6. Fontana, R., and Magli, R., *J. Mol. Struct.* **296**, 215 (1993).
7. Barocchi, F., Chieux, P., Fontana, R., and Magli, R., *Phys. Scripta* **T57**, 27 (1995).
8. Formisano, F., Benmore, C. J., Bafile, U., Barocchi, F., Egelstaff, P. A., Magli, R., and Verkerk, P., *Phys. Rev. Lett.* **79**, 221 (1997).
9. Barocchi, F., Chieux, P., Fontana, R., Magli, R., Meroni, A., Parola, A., Reatto, L., and Tau, M., *J. Phys. Cond. Matt.* **9**, 8849 (1997).
10. Bomont, J.-M., Jakse, N., and Bretonnet, J.-L., *Phys. Rev.* **B57**, 10217 (1998).
11. Benmore, C. J., Formisano, F., Magli, R., Bafile, U., Verkerk, P., Egelstaff, P. A., and Barocchi, F., *J. Phys. Cond. Matt.* **11**, 3091 (1999).
12. Guarini, E., Casanova, G., Bafile, U., and Barocchi, F., *Phys. Rev.* **E60** 6682 (1999).
13. Jakse, N., Bomont, J.-M., Charpentier, I., and Bretonnet, J.-L., *Phys. Rev.* **E62**, 3671 (2000).
14. Guarini, E., Magli, R., Tau, M., Barocchi, F., Casanova, G., and Reatto, L., *Phys. Rev.* **E63**, 052201 (2001).
15. Formisano, F., Barocchi, F., and Magli, R., *Phys. Rev.* **E58**, 2648 (1998).
16. Gubbins, K. E., Gray, C. G., Egelstaff, P. A., and Ananth, M. S., *Mol. Phys.* **25**, 1353 (1973).

17. Filabozzi, A., Buontempo, U., Delaplane, R., and Ricci, F. P., *Mol. Phys.* **82**, 427 (1994).
18. Misawa, M., *J. Chem. Phys.* **91**, 5648 (1989).
19. Narten, A. H., and Levy, H. A., *J. Chem. Phys.* **55**, 2263 (1971); Narten, A. H., *J. Chem. Phys.* **56**, 5681 (1972).
20. A review [Andreani, C., Dore, J. C., and Ricci, F. P., *Rep. Prog. Phys.* **54**, 731 (1991)] discusses the measurement of structure factors using the techniques of neutron and X-ray diffraction and their interpretation by computer simulation. The emphasis is on diatomic molecules.
21. Lovesey, S. W., *Theory of Neutron Scattering from Condensed Matter*, Vol. 1, Oxford (1984).
22. Azàroff, L. V., Kaplow, R., Kato, N., Weiss, R. J., Wilson, A. J. C., and Young, R. A., *X-Ray Diffraction*, McGraw-Hill (1974).
23. Agarwal, B. K., *X-Ray Spectroscopy*, 2nd. edn., Springer-Verlag (1991).
24. Bui, C., and Milazzo, M., *Nuovo Cimento* **D11**, 655 (1989).
25. Kane, P. P., Kissel, L., Pratt, R. H., and Roy, S. C., *Phys. Rep.* **140**, 75 (1986).
26. Rymer, T. B., *Electron Diffraction*, Methuen (1970).
27. We will neglect magnetic scattering, caused by the interaction of the magnetic moment of the neutron with nuclear, atomic, or molecular spin and orbital angular momenta [Balcar, E. and Lovesey, S. W., *Theory of Magnetic Neutron and Photon Scattering*, Oxford (1989)]. This is normally much weaker than scattering caused by nuclear forces, but makes a measurable contribution to the cross-section in some cases: see e.g. Deraman, M., Dore, J. C., Powles, J. G., and Schweizer, *J., Mol. Phys.* **53**, 557 (1984).
28. Jelley, N. A., *Fundamentals of Nuclear Physics*, Cambridge (1990).
29. Machleidt, R., Holinde, K. and Elster, Ch., *Phys. Rep.* **149**, 1 (1987).
30. Isospin (or isotopic, isobaric or *i*-spin) is an additional internal degree of freedom of baryons that takes two values, and distinguishes protons from neutrons, which are eigenstates of τ_z with quantum numbers +1 and -1, respectively. Raising and lowering operators, constructed from the x and y components of $\boldsymbol{\tau}$, transform a neutron into a proton, and vice versa.
31. Morse, P. M., and Feshbach, H., *Methods of Theoretical Physics*, Part 2, p. 1068.
32. Ref. 31, pp. 1465, 1573.
33. An accurate calculation of the scattering lengths should also include two-meson exchange contributions; see ref. 29. In addition, higher Born corrections should be included (ref. 31, p. 1076).
34. Golub, R., Richardson, D. J., and Lamoreaux, S. K., *Ultra-cold Neutrons*, Adam Hilger, Bristol (1991).
35. Peshkin, M., and Ringo, G. R., *Am. J. Phys.* **39**, 324 (1971).
36. Egelstaff, P. A., Gray, C. G., Gubbins, K. E., and Mo, K. C., *J. Stat. Phys.* **13**, 315 (1975).
37. Van Hove, L., *Phys. Rev.* **95**, 249 (1954).
38. For a more general treatment including the effects of attenuation and multiple inelastic scattering, see V. F. Sears, *Adv. Phys.* **24**, 1 (1975).
39. This division is to some extent arbitrary, and in some circumstances is not meaningful (as when quantum-mechanical exchange forces operate between the nuclei). The terms 'coherent' and 'incoherent' mean somehat different things to workers in different fields. As used in this chapter, 'coherent' scattering refers to the anisotropic (angle-dependent) scattering resulting from constructive interference of outgoing waves centred on different nuclei occupying the same spin state (relative to the neutron); while 'incoherent' scattering is the isotropic scattering resulting from random deviations in the spin states of nuclei throughout the fluid. See Lovesey, ref. 21, and Golub *et al.*, ref. 34.

40. Egelstaff, P. A., in *Methods of Experimental Physics* **23B**, (ed. D. L. Price and K. Sköld), Academic Press (1987), p. 405.
41. Jackson, J. D., *Classical Electrodynamics*, 2nd. edn, Wiley (1975).
42. While the choice of what constitutes the molecular 'centre' is obviously arbitrary, it makes sense to locate this as close as possible to the centre of electron density. For water, for example, we would naturally choose the oxygen atom as centre. This removes most of the dependence on the direction of the scattering vector **k** from both the molecular form factor and the structure factor.
43. The great majority of molecular electronic ground states have zero orbital angular momentum. For these molecules elastic X-ray scattering is purely coherent. There are a few exceptions, such as NO ($^2\Pi$); for such species the elastic scattering also contains an incoherent piece. For a discussion of coherent and incoherent scattering by ^2P H, see ref. 22.
44. Klug, H. P., and Alexander, L. E., *X-Ray Diffraction Procedures for Polycrystalline and Amorphous Materials*, 2nd. edn, Wiley (1974).
45. Egelstaff, P. A., *Phys. Chem. Liq.* **11**, 353 (1982).
46. In a similar manner, apparent discrepancies between the neutron and X-ray scattering spectra of chlorine can be resolved by postulating a third scattering site at the molecular centre. See Andreani, C., Colognesi, D., Filabozzi, A., and Menzinger, F., *Mol. Phys.* **82** (1994) 1181.
47. Dyson, N. A., *X-Rays in Atomic and Nuclear Physics*, 2nd. edn, Cambridge (1990).
48. Egelstaff, P. A., in *Molecular Liquids: New Perspectives in Physics and Chemistry* (ed. J. J. C. Teixeira-Dias), Kluwer Academic, Dordrecht (1992), pp. 1, 29.
49. Hellemans, A., *Science* **266**, 26 (1994).
50. Landau, L. D., and Lifshitz, E. M., *Quantum Mechanics: Non-relativistic Theory*, 2nd. edn, Pergamon (1965), p. 531.
51. Lane, N. F., *Rev. Mod. Phys.* **52**, 29 (1980).
52. Christophorou, L. G. (ed.), *Electron-Molecule Interactions and Their Applications*, Academic Press (1984), Vols. 1 & 2.
53. Typically bond lengths change by about 10^{-2} Å in passing to the liquid phase. These changes may be measured using such techniques as EXAFS (extended X-ray absorption fine structure), which is capable of an accuracy on of the order of 10^{-3} Å. For a recent application to iodine, see Buontempo, U., Filipponi, A., Postorino, P., and Zaccari, R., *J. Chem. Phys.* **108**, 4131 (1998).
54. Zemach, A. C., and Glauber, R. J., *Phys. Rev.* **101**, 118 (1956).
55. Kuchitsu, K., and Bartell, L. S., *J. Chem. Phys.* **35**, 1945 (1961), and refs. therein.
56. There are some exceptions, e.g. acetonitrile,[57] where it is apparently necessary to include the effects of molecular nonrigidity to obtain a proper description of $S_{\text{inter}}(k)$.
57. Böhm, H. J., Meissner, C., and Ahlrichs, R., *Mol. Phys.* **53**, 651 (1984).
58. *Methods in the Determination of Partial Structure Factors* (ed. J.-B. Suck, P. Chieux, D. Raux, and C. Riekel), World Scientific (1993).
59. For example, there are only minor differences between the structure of light and heavy water.[60,61] Differences for heavier elements, e.g. ^{35}Cl and ^{37}Cl, are expected to be completely negligible under normal conditions.
60. Root, J. H., Egelstaff, P. A., and Hime, A., *Chem. Phys.* **109**, 437 (1986).
61. Corongiu, G., and Clementi, E., *J. Chem. Phys.* **97**, 2030, 8818 (1992).
62. Egelstaff and co-workers have shown that quantum effects do lead to slightly different structures for deuterated fluids, which may have a measurable effect on the scattering cross-section at low temperatures. For example, in comparing the scattering of neutrons from CD_3OH and CD_3OD at -30° C, they amount to about 15% of the total change in $S(k)$. These effects may be largely eliminated by comparing spectra of deuterated and undeuterated samples at slightly different temperatures.[63]

63. Benmore, C. J., Tomberli, B., Neuefeind, J., and Egelstaff, P. A., *Appl. Phys.* **A74**, S1670 (2002); Tomberli, B., Egelstaff, P. A., Benmore, C. J., and Neuefeind, J., *J. Phys. Cond. Matter* **13**, 11405, 11421 (2001).
64. The technique of anomalous X-ray scattering [Fuoss, P. H., Eisenberger, P., Warburton, W. K., and Bienenstock, A., *Phys. Rev. Lett.* **46**, 1537 (1981)] utilizes incident radiation, which is tuned near an atomic absorption edge so that it interacts resonantly with the electrons of that particular atomic species. The scattering power of each chemical component can therefore be varied individually and in principle the $S_{\alpha\beta}(k)$ and $g_{\alpha\beta}(r)$ can be determined. The method has been applied successfully to $GeBr_4$ [Ludwig, K. F., Warburton, W. K., Wilson, L., and Bienenstock, A. I., *J. Chem. Phys.* **87**, 604 (1987)].
65. Soper, A. K., Andreani, C., and Nardone, M., *Phys. Rev.* **E47**, 2598 (1993).
66. Soper, A. K., *J. Chem. Phys.* **101**, 6888 (1994).
67. Soper, A. K., and Finney, J. L., *Phys. Rev. Lett.* **71**, 4346 (1993).
68. Egelstaff, P. A., Hawkins, R. K., Litchinsky, D., de Longgi, P. A., and Suck, J.-B., *Mol. Phys.* **53**, 389 (1984); Deraman, M., Dore, J. C., and Powles, J. G., *Mol. Phys.* **52**, 173 (1984); Powles, J. G., Dore, J. C., Deraman, M. B., and Osae, E. K., *Mol. Phys.* **50**, 1089 (1983).
69. Murad, S., Gubbins, K. E. and Powles, J. G., *Mol. Phys.* **40**, 253 (1980).
70. Verlet, L., *Phys. Rev.* **165**, 201 (1968).
71. Egelstaff, P. A., Page, D. I., and Powles, J. G., *Mol. Phys.* **20**, 881 (1971).
72. McDonald, I. R., Bounds, D. G., and Klein, M. L., *Mol. Phys.* **45**, 521 (1982).
73. Dagg, I. R., Missio, M., Anderson, A., Smith, W., and Read, L. A. A., *Can. J. Phys.* **67**, 507 (1989).
74. For a description of the RHNC equation see, for example, Lado, F., Foiles, S. M., and Ashcroft, N. W., *Phys. Rev.* **A28**, 2374 (1983).
75. Martín, C., Lombardero, M., Alvarez, M., and Lomba, E., *J. Chem. Phys.* **102**, 2092 (1995).
76. Lombardero, M., Martín, C., Lomba, E., and Lado, F., *J. Chem. Phys.* **104**, 6710 (1996).
77. Alvarez, M., Lomba, E., Martín, C., and Lombardero, M., *J. Chem. Phys.* **103**, 3680 (1995).
78. Perera, A., and Patey, G. N., *J. Chem. Phys.* **89**, 5861 (1988).
79. Perera, A., Kusalik, P. G., and Patey, G. N., *J. Chem. Phys.* **87**, 1295 (1987).
80. Lowden, L. J., and Chandler, D., *J. Chem. Phys.* **61**, 5228 (1974).
81. Hsu, C. S., and Chandler, D., *Mol. Phys.* **36**, 215 (1978).
82. Hsu, C. S., and Chandler, D., *Mol. Phys.* **37**, 299 (1979).
83. Ohba, T., and Ikawa, S., *Mol. Phys.* **73**, 999 (1991).
84. Fraser, K. J., Dunn, L. A., and Morriss, G. P., *Mol. Phys.* **61**, 775 (1987).
85. Narten, A. H., Johnson, E. and Habenschuss, A., *J. Chem. Phys.* **73**, 1248 (1980).
86. Montague, D. G., Chowdhury, M. R., Dore, J. C., and Reed, J., *Mol. Phys.* **50**, 1 (1983).
87. Hirata, F., and Rossky, P. J., *Chem. Phys. Lett.* **83**, 329 (1981); Hirata, F., Pettitt, B. M., and Rossky, P. J., *J. Chem. Phys.* **77**, 509 (1982); Pettitt, B. M., and Rossky, P. J., *J. Chem. Phys.* **77**, 1451 (1982); Pettitt, B. M., and Rossky, P. J., *J. Chem. Phys.* **78**, 7296 (1983).
88. Bausenwein, T., Bertagnolli, H., Gutwerk, D., Tödheide, K., and Chieux, P., *Mol. Phys.* **76**, 127 (1992).
89. Bausenwein, T., Bertagnolli, H., David, A., Goller, K., Zweier, H., Tödheide, K., and Chieux, P., *J. Chem. Phys.* **101**, 672 (1994).
90. Habenschuss, A., Johnson, E., and Narten, A. H., *J. Chem. Phys.* **74**, 5234 (1981).
91. Narten, A. H., and Habenschuss, A., *J. Chem. Phys.* **75**, 3073 (1981).

92. Enciso, E., *Mol. Phys.* **56**, 129 (1985).
93. Strauss, G., Zweier, H., Bertagnolli, H., Bausenwein, T., Tödheide, K., and Chieux, P., *J. Chem. Phys.* **101**, 662 (1994).
94. Reverse Monte Carlo (RMC) calculations are simulations performed without a potential [Pusztai, L., and Tóth, G., *J. Chem. Phys.* **94**, 3042 (1991); McGreevy, R. L., and Pusztai, L., *Mol. Simul.* **1**, 359 (1988)]. An initial guess is made for the positions and orientations of an ensemble of molecules. The pair correlation function $g(r)$ is computed, Fourier-transformed, and compared with the experimental structure factor. Molecular moves are then generated randomly and $S(k)$ is recomputed. If this results in worse agreement with experiment the moves are rejected, while if the agreement improves they are accepted. If desired, the acceptance algorithm may be modified to allow for experimental error. Iteration of this procedure eventually leads to duplication of $S(k)$ over the whole range of k for which data are available, and it is assumed that at this point relative molecular configurations are reproduced with the correct probabilities. Examination of the liquid structure thus generated reveals information on preferred relative orientations, mean molecular separations, and many-body correlations. A readable, and informative (though quite uncritical) review of RMC has been given by McGreevy [McGreevy, R. L., *J. Phys. Cond. Matt.* **13**, R877 (2001)]. For *simple* liquids, it can be shown that, provided there are no many-body terms in the potential, the procedure is unique [Henderson, R. L., *Phys. Lett.* **49A**, 197 (1974); Evans, R., *Mol. Simul.* **4**, 409 (1990); and § 3.1.7 of Vol. 1], and can indeed yield information on the liquid structure *beyond* that contained in $g(r)$ (for example, triplet correlations). However, its application to *molecular* liquids has been criticized [Head-Gordon, T., and Stillinger, F. H., *J. Chem. Phys.* **98**, 3313 (1993)] on the grounds that reproduction of $g(r)$ alone, rather than the full pair correlation function $g(\mathbf{r}\omega_1\omega_2)$, will not necessarily generate many-body correlations correctly. Head-Gordon and Stillinger cite as an example a potential that accurately describes the X-ray $g(r)$ for water, yet generates a highly defective model of the liquid structure. Further criticisms have been voiced by De Santis and Rocca [De Santis, A., and Rocca, D., *Mol. Simul.* **17**, 143 (1996)], Colognesi *et al.* [Colognesi, D., De Santis, A., and Rocca, D., *Mol. Phys.* **88**, 465 (1996)] and Jedlovszky *et al.* [Jedlovszky, P., Bakó, I., Pálinkás, G., Radnai, T., and Soper, A. K., *J. Chem. Phys.* **105**, 245 (1996)]. Attempts to improve the method, e.g. by reducing its bias towards sampling the relatively more disordered states of the system [Tóth, G., and Baranyai, A., *J. Chem. Phys.* **107**, 7402 (1997)] appear to be more or less ad hoc, and the technique, and its method of application, continue to generate controversy [see, e.g., da Silva, F. L. B., Olivares-Rivas, W., Degrève, L., and Åkesson, T., *J. Chem. Phys.* **114**, 907 (2001), and refs. therein]. RMC may, however, prove to be a valuable tool for revealing inaccuracies in experimental data [Jedlovszky, P., and Vallauri, R., *J. Chem. Phys.* **105**, 2391 (1996)]. A molecular dynamics analogue, RMD, has been proposed [Tóth, G., and Baranyai, A., *J. Chem. Phys.* **114**, 2027 (2001)]. Simulations of liquid argon and water show that the method is equivalent to, and competitive with, RMC; it may even offer some advantages for complex molecular systems. In recent developments, a hybrid Monte Carlo procedure, which incorporates information about the intermolecular potential as well as experimental data, has been described [Rutledge, G. C., *Phys. Rev.* **E63**, 021111 (2001)]. Such a method can obviously yield a better description of the fluid thermodynamics than conventional RMC, and presumably therefore of the structure also; as yet, however, there appear to have been very few practical applications.
95. Strauss, G., Bassen, A., Zweier, H., Bertagnolli, H., Tödheide, K., Soper, A. K., and Turner, J., *Phys. Rev.* **E53**, 3505 (1996).
96. Bertagnolli, H., Waldner, I., Tödheide, K., and Fischer, H., *Mol. Phys.* **94**, 325 (1998).

97. Bermejo, F. J., Enciso, E., Alonso, J., Garcia, N., and Howells, W. S., *Mol. Phys.* **64**, 1169 (1988).
98. Alvarez, M., Bermejo, F. J., Chieux, P., Enciso, E., Alonso, J., and Garcia, N., *J. Phys. Cond. Matter* **1**, 8595 (1989).
99. Hannongbua, S. V., Ishida, T., Spohr, E., and Heinzinger, K., Z. *Naturforsch.* **A43**, 572 (1988).
100. Sagarik, K. P., Ahlrichs, R., and Brode, S., *Mol. Phys.* **57**, 1247 (1986).
101. Bertagnolli, H., and Tödheide, K., *J. Phys. Cond. Matter* **8**, 9293 (1996).
102. Fraser, K. J., Morriss, G. P., and Dunn, L. A., *Mol. Phys.* **57**, 1233 (1986).
103. Rossky, P. J., and Chiles, R. A., *Mol. Phys.* **51**, 661 (1984).
104. Lue, L., and Blankschtein, D., *J. Chem. Phys.* **102**, 4203 (1995).
105. Lue, L., and Blankschtein, D., *J. Chem. Phys.* **102**, 5427 (1995).
106. Salacuse, J. J., Denton, A. R., and Egelstaff, P. A., *Phys. Rev.* **E53**, 2382 (1996); Salacuse, J. J., Denton, A. R., Egelstaff, P. A., Tau, M., and Reatto, L., *Phys. Rev.* **E53**, 2390 (1996); Denton, A. R., and Egelstaff, P. A., *Zeit. Physik* **B103**, 343 (1997). The authors distinguish two different kinds of size effects: (a) explicit, or ensemble, effects, caused by the suppression of density fluctuations upon fixing the number of particles in the canonical and microcanonical ensembles; and (b) implicit size effects, resulting from the breaking of orientational symmetry due to imposed (usually periodic) boundary conditions. They argue that the former are more important; the latter are *not* corrected for. It seems likely, however, that they may be significant for highly anisotropic molecules such as CS_2.
107. Root, J. H., Egelstaff, P. A., and Nickel, B. G., in *Maximum Entropy and Bayesian Methods in Science and Engineering* (ed. G. J. Erickson and C. R. Smith), Kluwer Academic, Dordrecht (1988), p. 395.
108. Soper, A. K., *Chem. Phys.* **107**, 61 (1986).
109. Howe, M. A., *Mol. Phys.* **69**, 161 (1990).
110. Andreani, C., Cilloco, F., and Osae, E. K., *Mol. Phys.* **57**, 931 (1986).
111. Ricci, F. P., Rocca, D., and Vallauri, R., *Mol. Phys.* **60**, 1245 (1987).
112. De Santis, A., Gregori, A., and Rocca, D., *Mol. Phys.* **79**, 645 (1993).
113. Bellissent-Funel, M.-C., Buontempo, U., Petrillo, C., and Ricci, F. P., *Mol. Phys.* **71**, 253 (1990).
114. Bellissent-Funel, M.-C., Buontempo, U., Petrillo, C., and Ricci, F. P., *Mol. Phys.* **74**, 1209 (1991).
115. Rodger, P. M., Stone, A. J., and Tildesley, D. J., *Mol. Phys.* **63**, 173 (1988).
116. Rodger, P. M., Stone, A. J., and Tildesley, D. J., *J. Chem. Soc. Faraday II* **83**, 1689 (1987).
117. Andreani, C., Cilloco, F., Nencini, L., Rocca, D., and Sinclair, R. N., *Mol. Phys.* **55**, 887 (1985).
118. Rodger, P. M., Stone, A. J., and Tildesley, D. J., *Chem. Phys. Lett.* **145**, 365 (1988).
119. Szornel, C., Egelstaff, P. A., Whalley, E., and McLaurin, G., *Can. J. Phys.* **71**, 507 (1993).
120. Van der Marel, C., Bras, W., and van der Lugt, W., *Mol. Phys.* **64**, 445 (1988); ibid., **66**, 863 (1989).
121. Stone, A. J., and Alderton, M., *Mol. Phys.* **56**, 1047 (1985); Stone, A. J., *Chem. Phys. Lett.* **83**, 233 (1981).
122. Murthy, C. S., Singer, K., and Vallauri, R., *Mol. Phys.* **49**, 803 (1983).
123. Misawa, M., *J. Chem. Phys.* **90**, 6563 (1989).
124. Misawa, M., *J. Chem. Phys.* **91**, 2575 (1989).
125. Bellissent-Funel, M. C., Buontempo, U., Nardone, M., Ricci, F. P., and Ricci, M. A., *Phys. Rev.* **B50**, 6047 (1994).
126. Zoppi, M., Magli, R., Howells, W. S., and Soper, A. K., *Phys. Rev.* **A39**, 4684 (1989).

127. Neumann, M., and Zoppi, M., *Phys. Rev.* **A44**, 2474 (1991).
128. Zoppi, M., Bafile, U., Magli, R., and Soper, A. K., *Phys. Rev.* **E48**, 1000 (1993); *J. Mol. Struct.* **296**, 243 (1993).
129. See § 1.2.2 for a discussion of quantum effects.
130. Zoppi, M., *Physica* **B183**, 235 (1993).
131. Deraman, M., Dore, J. C., Powles, J. G., Holloway, J. H., and Chieux, P., *Mol. Phys.* **55**, 1351 (1985).
132. Pfleiderer, T., Waldner, I., Bertagnolli, H., Tödheide, K., and Fischer, H. E., *J. Chem. Phys.* **113**, 3690 (2000).
133. McLain, S. E., Benmore, C. J., Siewenie, J. E., Molaison, J. J., and Turner, J. F. C., *J. Chem. Phys.* **121**, 6448 (2004).
134. Klein, M. L., and McDonald, I. R., *J. Chem. Phys.* **71**, 298 (1979); Hinchliffe, A., Bounds, D. G., Klein, M. L., McDonald, I. R., and Righini, R., *J. Chem. Phys.* **74**, 1211 (1981); Klein, M. L., and McDonald, I. R., *J. Chem. Phys.* **74**, 4214 (1981).
135. Cournoyer, M. E., and Jorgensen, W. L., *Mol. Phys.* **51**, 119 (1984).
136. Jedlovsky, P., and Vallauri, R., *Mol. Phys.* **93**, 15 (1998).
137. Jedlovsky, P., and Vallauri R., *Mol. Phys.* **92**, 331 (1997).
138. Jedlovsky, P., Mezei, M., and Vallauri, R., *J. Chem. Phys.* **115**, 9883 (2001).
139. Jedlovsky, P. and Vallauri, R., *J. Chem. Phys.* **107**, 10166 (1997).
140. Kreitmeir, M., Bertagnolli, H., Mortensen, J. J., and Parrinello, M., *J. Chem. Phys.* **118**, 3639 (2003).
141. Röthlisberger, U., and Parrinello, M., *J. Chem. Phys.* **106**, 4658 (1997).
142. Soper, A. K., and Egelstaff, P. A., *Mol. Phys.* **42**, 399 (1981).
143. Klein, M. L., and McDonald, I. R., *Mol. Phys.* **42**, 243 (1981).
144. McDonald, I. R., O'Shea, S. F., Bounds, D. G., and Klein, M. L., *J. Chem. Phys.* **72**, 5710 (1980).
145. Powles, J. G., Dore, J. C., Osae, E. K., Clarke, J. H., Chieux, P., and Cummings, S., *Mol. Phys.* **44**, 1131 (1981).
146. Andreani, C., Menzinger, F., Ricci, M. A., Soper, A. K., and Dreyer, J., *Phys. Rev.* **B49**, 3811 (1994).
147. Andreani, C., Nardone, M., Ricci, F. P., and Soper, A. K., *Phys. Rev.* **A46**, 4709 (1992).
148. Fracassi, P. F., Cardini, G., O' Shea, S. F., Impey, R. W., and Klein, M. L., *Phys. Rev.* **B33**, 3441 (1986).
149. Bellissent-Funel, M.-C., Buontempo, U., Petrillo, C., and Ricci, F. P., *Mol. Phys.* **71**, 239 (1990).
150. Howe, M. A., Wormald, C. J., and Neilson, G. W., *Mol. Phys.* **66**, 847 (1989).
151. Dore, J. C., Walford, G., and Page, D. I., *Mol. Phys.* **29**, 565 (1975).
152. Böhm, H. J., Ahlrichs, R., Scharf, P., and Schiffer, H., *J. Chem. Phys.* **81**, 1389 (1984).
153. Van Tricht, J. B., Fredrikze, H., and van der Laan, J., *Mol. Phys.* **52**, 115 (1984).
154. Adya, A. K., and Wormald, C. J., *Mol. Phys.* **74**, 735 (1991).
155. Ishii, R., Okazaki, S., Okada, I., Furusaka, M., Watanabe, N., Misawa, M., and Fukunaga, T., *J. Chem. Phys.* **105**, 7011 (1996).
156. Nishikawa, K., and Takematsu, M., *Chem. Phys. Lett.* **226**, 359 (1994).
157. Mognetti, B. M., Oettel, M., Virnau, P., Yelash, L., and Binder, K., *Mol. Phys.* **107**, 331 (2009).
158. Temleitner, L., and Pusztai, L., *J. Phys. Cond. Matt.* **19**, 335203 (2007).
159. Andreani, C., Merlo, V., Ricci, M. A., and Soper, A. K., *Mol. Phys.* **73**, 407 (1991).
160. Forester, T. R., McDonald, I. R., and Klein, M. L., *Chem. Phys.* **129**, 225 (1989).
161. Santoli, G., Bruni, F., Ricci, F. P., Ricci, M. A., and Soper, A. K., *Mol. Phys.* **97**, 777 (1999).

162. Alvarez, M., Bermejo, F. J., Chieux, P., Enciso, E., Garcia-Hernandez, M., Garcia, N., and Alonso, J., *Mol. Phys.* **66**, 397 (1989).
163. Kamova, M., *J. Raman Spec.* **8**, 225 (1979).
164. Sokolić, F., Guissani, Y., and Guillot, B., *Mol. Phys.* **56**, 239 (1985).
165. Rastogi, A., Anderson, A., and Leech, J. W., *Can. J. Phys.* **57**, 2120 (1979).
166. Hyodo, S., Tatsuta, N., and Okamoto, T., *J. Chem. Phys.* **92**, 882 (1990).
167. Bermejo, F. J., Dore, J. C., Howells, W. S., Chieux, P., and Enciso, E., *Physica* **B156&157**, 154 (1989).
168. Pusztai, L., and McGreevy, R. L., *Mol. Phys.* **90**, 533 (1997). Rather different conclusions were reached by Jedlovsky,[169] also on the basis of RMC studies; however, these results have been criticized on a number of grounds.[170]
169. Jedlovsky, P., *J. Chem. Phys.* **107**, 7433 (1997).
170. Jóvári, P., Mészáros, G., Pusztai, L., and Sváb, E., *J. Chem. Phys.* **114**, 8082 (2001).
171. Chang, T.-M., Peterson, K. A. and Dang, L. X., *J. Chem. Phys.* **103**, 7502 (1995).
172. Hall, C. D., Johnson, K. A., Burgess, A. N., Winterton, N., and Howells, W. S., *Mol. Phys.* **76**, 1061 (1992); ibid., **74**, 27 (1991).
173. McLain, S. E., Siewenie, J. E., Benmore, C. J., and Turner, J. F. C., *J. Chem. Phys.* **119**, 6671 (2003).
174. Narten, A. H., *J. Chem. Phys.* **66**, 3117 (1977).
175. Chieux, P., and Bertagnolli, H., *J. Phys. Chem.* **88**, 3726 (1984).
176. Ricci, M. A., Nardone, M., Ricci, F. P., Andreani, C., and Soper, A. K., *J. Chem. Phys.* **102**, 7650 (1995).
177. Impey, R. W., and Klein, M. L., *Chem. Phys. Lett.* **104**, 579 (1984).
178. Caillol, J. M., Levesque, D., Weis, J. J., Perkyns, J. S., and Patey, G. N., *Mol. Phys.* **62**, 1225 (1987).
179. Jorgensen, W. L., and Briggs, J. M., *Mol. Phys.* **63**, 547 (1988).
180. Böhm, H. J., McDonald, I. R., and Madden, P. A., *Mol. Phys.* **49**, 347 (1983).
181. Bertagnolli, H., and Zeidler, M. D., *Mol. Phys.* **35**, 177 (1978); Bertagnolli, H., Chieux, P., and Zeidler, M. D., *Mol. Phys.* **32**, 759, 1731 (1976).
182. Luzar, A., Soper, A. K., and Chandler, D., *J. Chem. Phys.* **99**, 6836 (1993).
183. Rao, B. G., and Singh, U. C., *J. Am. Chem. Soc.* **112**, 3803 (1990).
184. Jorgensen, W. L., *J. Phys. Chem.* **90**, 1276 (1986).
185. Yamaguchi, T., Hidaka, K., and Soper, A. K., *Mol. Phys.* **96**, 1159 (1998); *ibid.*, **97**, 603 (1999).
186. Yamaguchi, T., Benmore, C. J., and Soper, A. K., *J. Chem. Phys.* **112**, 8976 (2000).
187. Haughney, M., Ferrario, M., and McDonald, I. R., *J. Phys. Chem.* **91**, 4934 (1987).
188. Matsumoto, M., and Gubbins, K. E., *J. Chem. Phys.* **93**, 1981 (1990).
189. Svishchev, I. M., and Kusalik, P. G., *J. Chem. Phys.* **100**, 5165 (1994).
190. Sarkar, S., and Joarder, R. N., *J. Chem. Phys.* **99**, 2032 (1993).
191. For simulations using a *polarizable* model of an alcohol (ethanol), see González, M. A., Enciso, E., Bermejo, F. J., and Bée, M., *J. Chem. Phys.* **110**, 8045 (1999). In principle, such a model can better account for the liquid structure over a range of thermodynamic states.
192. Benmore, C. J., and Loh, Y. L., *J. Chem. Phys.* **112**, 5877 (2000).
193. Bellissent-Funel, M.-C., and Bosio, L., *J. Chem. Phys.* **102**, 3727 (1995).
194. Tromp, R. H., Postorino, P., Neilson, G. W., Ricci, M. A., and Soper, A. K., *J. Chem. Phys.* **101**, 6210 (1994); Postorino, P., Tromp, R. H., Ricci, M. A., Soper, A. K, and Neilson, G. W., *Nature* **366**, 668 (1993).
195. Soper, A. K., Bruni, F., and Ricci, M. A., *J. Chem. Phys.* **106**, 247 (1997).
196. Soper, A. K., *J. Phys. Cond. Matter* **8**, 9263 (1996).
197. Postorino, P., Ricci, M. A., and Soper, A. K., *J. Chem. Phys.* **101**, 4123 (1994).

198. Buontempo, U., Postorino, P., Ricci, M. A., and Soper, A. K., *Mol. Phys.* **81**, 217 (1994).
199. Bosio, L., Teixeira, J., Dore, J. C., Steytler, D. C., and Chieux, P., *Mol. Phys.* **50**, 733 (1983).
200. Soper, A. K., and Phillips, M. G., *Chem. Phys.* **107**, 47 (1986).
201. Bellissent-Funel, M.-C., Teixeira, J., Bosio, L., Dore, J. C., and Chieux, P., *Europhys. Lett.* **2**, 241 (1986).
202. Buontempo, U., Postorino, P., Ricci, M. A., and Soper, A. K., *Europhys. Lett.* **19**, 385 (1992).
203. Thiessen, W. E., and Narten, A. H., *J. Chem. Phys.* **77**, 2656 (1982).
204. Soper, A. K., *Physica* **136B&C**, 322 (1986).
205. Radnai, T., and Ohtaki, H., *Mol. Phys.* **87**, 103 (1996).
206. Bosio, L., Chen, S.-H., and Teixeira, J., *Phys. Rev.* **A27**, 1468 (1983).
207. Okhulkov, A. V., Demianets, Yu. N., and Gorbaty, Yu. E., *J. Chem. Phys.* **100**, 1578 (1994).
208. Gaballa, G. A., and Neilson, G. W., *Mol. Phys.* **50**, 97 (1983).
209. Gorbaty, Yu. E., and Demianets, Yu. N., *Mol. Phys.* **55**, 571 (1985).
210. Yamanaka, K., Yamaguchi, T., and Wakita, H., *J. Chem. Phys.* **101**, 9830 (1994).
211. Kálmán, E., Pálinkás, G., and Kovács, P., *Mol. Phys.* **34**, 505 (1977).
212. However, X-ray studies yield more accurate information on this quantity than neutron experiments, since only 10% of the neutron scattering intensity is associated with $g_{OO}(r)$.[210]
213. The mean OH-O distance and rms amplitude have been estimated at 2.91 ± 0.02 and 0.17 ± 0.01 Å, respectively, at 298 K. See Iijima, T., *Chem. Phys. Lett.* **217**, 503 (1994).
214. Ichikawa, K., Kameda, Y., Yamaguchi, T., Wakita, H., and Misawa, M., *Mol. Phys.* **73**, 79 (1991).
215. De Santis, A., and Rocca, D., *J. Chem. Phys.* **105**, 7227 (1996).
216. Corongiu, G., *Int. J. Quant. Chem.* **42**, 1209 (1992).
217. The MCY pair potential is a site–site model including Coulombic and exponential repulsive (O-O, H-H) and attractive (O-H) terms. The parameters are obtained by fitting to *ab initio* calculations on the water dimer. See Matsuoka, O., Clementi, E., and Yoshimine, M., *J. Chem. Phys.* **64**, 1351 (1976), and Lie, G. C., Clementi, E. and Yoshimine, M., *J. Chem. Phys.* **64**, 2314 (1976).
218. Detrich, J., Corongiu, C., and Clementi, E., *Chem. Phys. Lett.* **112**, 426 (1984).
219. Lie, G. C., and Clementi, E., *Phys. Rev.* **A33**, 2679 (1986).
220. Cummings, P. T., Cochran, H. D., Simonson, J. M., Mesmer, R. E., and Karaborni, S., *J. Chem. Phys.* **94**, 5606 (1991).
221. Berendsen, H. J. C., Postma, J. P. M., van Gunsteren, W. F., and Hermans, J., in *Intermolecular Forces*, (ed. B. Pullman), Reidel (1981), p. 331.
222. Chialvo, A. A., and Cummings, P. T., *J. Chem. Phys.* **101**, 4466 (1994).
223. Mountain, R. D., *J. Chem. Phys.* **103**, 3084 (1995).
224. Cummings, P. T., and Chialvo, A. A., J. Phys. Cond. Matter **8**, 9281 (1996).
225. Kuharski, R. A., and Rossky, P. J., *J. Chem. Phys.* **82**, 5164 (1985).
226. Stillinger, F. H., and Rahman, A., *J. Chem. Phys.* **60**, 1545 (1974).
227. Impey, R. W., Klein, M. L., and McDonald, I. R., *J. Chem. Phys.* **74**, 647 (1981).
228. Martí, J., Padró, J. A., and Guàrdia, E., *Mol. Phys.* **86**, 263 (1995).
229. Fontana, A., Postorino, P., Ricci, M. A., and Soper, A. K., in *Hydrogen Bond Networks*, (ed. M. C. Bellissent-Funel and J. C. Dore), Kluwer Academic, Dordrecht (1994), p. 69.
230. See, for example, Sprik, M., and Klein, M. L., *J. Chem. Phys.* **89**, 7556 (1988); Ahlström, P., Wallqvist, A., Engström, S., and Jönsson, B., *Mol. Phys.* **68**, 563

(1989); and Wallqvist, A., Ahlström, P., and Karlström, G., *J. Phys. Chem.* **94**, 1649 (1990).
231. Botti, A., Bruni, F., Isopo, A., Modesti, G., Oliva, C., Ricci, M. A., Senesi, R., and Soper, A. K., *J. Chem. Phys.* **118**, 235 (2003).
232. Soper, A. K., *Chem. Phys.* **202**, 295 (1996); *Mol. Phys.* **99**, 1503 (2001). The method, empirical potential structure refinement (EPSR) is related to RMC; see the review by McGreevy, ref. 94.
233. Turner, J., and Soper, A. K., *J. Chem. Phys.* **101**, 6116 (1994).
234. Finney, J. L., Bowron, D. T., and Soper, A.K., *J. Phys. Cond. Matter* **12**, A123 (2000).
235. This picture was proposed to explain such phenomena as the observed heat of mixing and negative excess entropy of aqueous solutions of organic compounds. See, for example, Franks, F., and Reid, D. S., in *Water: A Comprehensive Treatise*, Vol. 2 (ed. F. Franks), Plenum (1973), Chapter 5.
236. Hernández-Cobos, J., and Ortega-Blake, I., *J. Chem. Phys.* **103**, 9261 (1995).
237. Ferrario, M., Haughney, M., McDonald, I. R., and Klein, M. L., *J. Chem. Phys.* **93**, 5156 (1990).
238. Okazaki, S., Touhara, H., and Nakanishi, K., *J. Chem. Phys.* **81**, 890 (1984).
239. Jorgensen, W. L. and Madura, J. D., *J. Am. Chem. Soc.* **105**, 1407 (1983).
240. Luzar, A., and Chandler, D., *J. Chem. Phys.* **98**, 8160 (1993).
241. Soper, A. K., and Luzar, A., *J. Chem. Phys.* **97**, 1320 (1992).
242. Soper, A. K., and Luzar, A., *J. Phys. Chem.* **100**, 1357 (1996).
243. DeMille, R. C., and Molinero, V., *J. Chem. Phys.* **131**, 034107 (2009).
244. Mancinelli, R., Botti, A., Bruni, F., Ricci, M. A., and Soper, A. K., *J. Phys. Chem.* **B111**, 13570 (2007).
245. Habenschuss, A., and Narten, A. H., *J. Chem. Phys.* **85**, 6022 (1986).
246. Habenschuss, A., and Narten, A. H., *J. Chem. Phys.* **91**, 4299 (1989).
247. Habenschuss, A., and Narten, A. H., *J. Chem. Phys.* **92**, 5692 (1990).
248. Narten, A. H., Habenschuss, A., Honnell, K. G., McCoy, J. D., Curro, J. G., and Schweizer, K. S., *J. Chem. Soc. Faraday Trans.* **88**, 1791 (1992).
249. Bartsch, E., Bertagnolli, H., Schulz, G., and Chieux, P., *Ber. Bunsenges. Phys. Chem.* **89**, 147 (1985).
250. Londono, J. D., Annis, B. K., Turner, J. Z., and Soper, A. K., *J. Chem. Phys.* **101**, 7868 (1994).
251. Misawa, M., and Fukunaga, T., *J. Chem. Phys.* **93**, 3495 (1990).
252. Pfleiderer, T., Bertagnolli, H., Tödheide, K. and Soper, A. K., *J. Chem. Phys.* **115**, 331 (2001).
253. Valente, E. J., and Bartell, L. S., *J. Chem. Phys.* **80**, 1451 (1984).
254. Murad, S., Evans, D. J., Gubbins, K. E., Streett, W. B., and Tildesley, D. J., *Mol. Phys.* **37**, 725 (1979).
255. Guarini, E., Bafile, U., Barocchi, F., Cilloco, F., and Magli, R., *Mol. Phys.* **94**, 289 (1998).
256. Sandler, S. I., Lombardo, M. G., Wong, D. S.-H., Habenschuss, A., and Narten, A. H., *J. Chem. Phys.* **77**, 2144 (1982).
257. Bassen, A., Waldner, I., and Bertagnolli, H., *Phys. Chem. Chem. Phys.* **1**, 737 (1999).
258. Gupta, S., Yang, J., and Kestner, N. R., *J. Chem. Phys.* **89**, 3733 (1988).
259. Ungerer, P., Beauvais, C., Delhommelle, J., Boutin, A., Rousseau, B., and Fuchs, A. H., *J. Chem. Phys.* **112**, 5499 (2000).
260. Bañón, A., Adán, F. S., and Santamaría, J., *J. Chem. Phys.* **83**, 297 (1985).
261. Edberg, R., Evans, D. J., and Morriss, G. P., *J. Chem. Phys.* **84**, 6933 (1986).
262. Venturi, G., Formisano, F., Cuello, G. J., Johnson, M. R., Pellegrini, E., Bafile, U., and Guarini, E., *J. Chem. Phys.* **131**, 034508 (2009).
263. Almarza, N. G., Enciso, E., and Bermejo, F. J., *J. Chem. Phys.* **96**, 4625 (1992).

264. Toxvaerd, S., *J. Chem. Phys.* **87**, 6140 (1987).
265. Rigby, D. H., and Roe, R. J., *Macromol.* **22**, 2259 (1989).
266. Boyd, R. H., *Macromol.* **22**, 2477 (1989).
267. De Gennes, P.-G., *J. Phys. (Paris)* **31**, 235 (1970).
268. Smith, G. D., and Yoon, D. Y., *J. Chem. Phys.* **100**, 649 (1994).
269. Felici, R., Cilloco, F., and Bosi, P., *Mol. Phys.* **70**, 455 (1990).
270. Tassaing, T., Cabaço, M. I., Danten, Y., and Besnard, M., *J. Chem. Phys.* 113, 3757 (2000).
271. Price, S. L., and Stone, A. J., *J. Chem. Phys.* **86**, 2859 (1987).
272. Anderson, J., Ullo, J. J., and Yip, S., *J. Chem. Phys.* **86**, 4078 (1987).
273. Bartell, L. S., Sharkey, L. R., and Shi, X., *J. Am. Chem. Soc.* **110**, 7006 (1988).
274. Haymet, A. D. J., Rice, S. A., and Madden, W. G., *J. Chem. Phys.* **74**, 3033 (1981); Abe, R., *Prog. Theor. Phys.* **21**, 421 (1959).

10

DIELECTRIC PROPERTIES

> From a long view of the history of mankind—seen from, say, ten thousand years from now—there can be little doubt that the most significant event of the nineteenth century will be judged as Maxwell's discovery of the laws of electrodynamics.
>
> Richard P. Feynman, *The Feynman Lectures on Physics* (1964)

In this chapter we review the theory of the dielectric constant and some closely related phenomena, the Kerr effect and field gradient-induced birefringence, all of which are concerned with the response of a system to an electric field.[1] Our focus is on those aspects that shed light on the equilibrium structure (particularly the orientational structure) of the fluid.

10.1 Introduction

When an external electric field is applied to a nonconducting molecular fluid, a polarization, or separation of electric charges, results. This may arise either from the orientation of permanent molecular dipole moments or from the induction of moments in polarizable molecules. Provided the applied field is spatially uniform on a molecular scale, and also sufficiently weak, the response is both local (as we demonstrate in Appendix 10A) and linear.[3] That is, the polarization, or mean dipole moment per unit volume, is simply proportional to the mean electric field at that position. For an isotropic system, the constitutive relation can be written

$$\mathbf{P}(\mathbf{r}) = \chi \mathbf{E}(\mathbf{r}) = \frac{\epsilon - 1}{4\pi} \mathbf{E}(\mathbf{r}), \tag{10.1}$$

where χ is the susceptibility, and ϵ is the dielectric constant, or permittivity, of the fluid.[8] These are intensive quantities, which depend only on the density and temperature. Note that $\mathbf{E}(\mathbf{r})$ (sometimes called the Maxwell field) differs from the applied field $\mathbf{E}_0(\mathbf{r})$, because the polarization is itself a field source.

In general, the dielectric constant is a function of the frequency of the applied field. Our emphasis in this chapter is on the static dielectric constant, which describes the response of a system to a slowly varying field, and which is related to the equilibrium structure of the fluid; the frequency dependence of ϵ contains additional information on the molecular dynamics.[9]

The theories described in this chapter are for the most part classical, although we briefly discuss quantum corrections in low-density gases in § 10.6. The quantum theory of the dielectric constant for dense fluids is in its infancy; we give a short introduction in Appendix 11A.

In anisotropic systems, the relation between the polarization and the electric field is tensorial. Although we shall mostly be concerned with isotropic fluids, our discussions of the Kerr effect and electric field gradient induced-birefringence require that we also consider the dielectric response of systems exhibiting uniaxial and biaxial dielectric anisotropies (§ 10.8).

10.2 Microscopic and macroscopic theories

In this section we give a brief overview of the preliminary steps involved in the calculation of the dielectric constant. Some of the issues that arise are considered in greater detail in Appendix 10A.

First, one uses linear response theory to calculate $\mathbf{P}$ as a functional of the external field $\mathbf{E}_0$. This yields

$$\mathbf{P}(\mathbf{r}) = \int_V d\mathbf{r}' \, \boldsymbol{\chi}_0(\mathbf{r}, \mathbf{r}') \cdot \mathbf{E}_0(\mathbf{r}'), \tag{10.2}$$

where $\boldsymbol{\chi}_0(\mathbf{r}, \mathbf{r}')$ is the linear response coefficient, a function of microscopic variables. This function depends upon the shape of the system;[10] in general, it is both nonlocal and tensorial even for an isotropic fluid.

Second, one relates $\mathbf{E}$ to $\mathbf{E}_0$. This relation may be written

$$\mathbf{E} = \mathbf{E}[\mathbf{E}_0; \epsilon], \tag{10.3}$$

where $\mathbf{E}[\mathbf{E}_0; \epsilon]$ denotes a functional of $\mathbf{E}_0$, which also depends parametrically on ϵ. This too depends on the system shape.[10] Formally, this can also be accomplished microscopically, either directly,[15] or by deriving the macroscopic Maxwell equations from the underlying microscopic relations.[2,16]

If we eliminate $\mathbf{E}_0$ between these two equations, we obtain $\mathbf{P}$ as a function of $\mathbf{E}$. Provided certain conditions are satisfied (the applied field must not vary appreciably over molecular dimensions, and the system must be large enough that surface effects may be neglected), the resulting expression is local. Comparison with (10.1) yields a shape-independent, microscopic expression for ϵ. Fully microscopic treatments of this kind have been given both for rigid polar and polarizable fluids[13,17–19] and for solids.[2]

Here we will follow a much simpler semimacroscopic approach, in which macroscopic electrostatics is used for the second step. One starts from the assumption that a shape-independent dielectric constant exists, and computes it using simple sample geometries. Although this is less fundamental, it quickly generates the desired result, providing rigorous molecular expressions for ϵ.

We will therefore begin with the macroscopic relation[20]

$$\mathbf{E}(\mathbf{r}) = \mathbf{E}_0(\mathbf{r}) + \int_V d\mathbf{r}' \, \mathbf{T}(\mathbf{r} - \mathbf{r}') \cdot \mathbf{P}(\mathbf{r}'), \tag{10.4}$$

where the second-rank tensor $\mathbf{T}(\mathbf{r}) = \nabla\nabla(1/r)$. The physical significance of this result is clear. $\mathbf{P}(\mathbf{r}')\mathrm{d}\mathbf{r}'$ is the dipole moment of the volume element $\mathrm{d}\mathbf{r}'$, so that $\mathbf{T}(\mathbf{r}-\mathbf{r}')\cdot\mathbf{P}(\mathbf{r}')\mathrm{d}\mathbf{r}'$ is the contribution of the polarized element $\mathrm{d}\mathbf{r}'$ to the macroscopic field at $\mathbf{r}$. The integration extends over the sample volume; together with the external field $\mathbf{E}_0$, this gives the total field at $\mathbf{r}$.

From (A.218), the symmetric tensor $\mathbf{T}(\mathbf{r})$ contains both isotropic and traceless anisotropic parts, $\mathbf{T} = \mathbf{T}^{(0)} + \mathbf{T}^{(2)}$. Using (8E.2), we see that

$$\mathbf{T}^{(0)}(\mathbf{r}) = -(4\pi/3)\delta(\mathbf{r})\mathbf{1}, \tag{10.5}$$

while from (2.56)

$$\mathbf{T}^{(2)}(\mathbf{r}) = (3\hat{\mathbf{r}}\hat{\mathbf{r}} - \mathbf{1})r^{-3}. \tag{10.6}$$

Equation (10.4) can now be written rather more usefully as

$$\mathbf{E}(\mathbf{r}) = \mathbf{E}_0(\mathbf{r}) - \frac{4\pi}{3}\mathbf{P}(\mathbf{r}) + \int_V \mathrm{d}\mathbf{r}'\,\mathbf{T}^{(2)}(\mathbf{r}-\mathbf{r}')\cdot\mathbf{P}(\mathbf{r}'). \tag{10.7}$$

10.3 Rigid polar fluids

With the preliminaries completed, we now turn to the calculation of ϵ for a polar fluid. We begin our discussion with nonpolarizable molecules, since the theory for this case is much simpler.[4,11–13,17,18,21] We will consider only pure fluids, although the extension to mixtures is trivial.

10.3.1 Expressions involving the total correlation function

As discussed earlier, the susceptibility χ and dielectric constant ϵ, defined through (10.1), are bulk properties, independent of the position within the sample (even when the applied field is moderately nonuniform), and also independent of its shape. We therefore simplify the calculation by considering (a) a spherical system, and (b) a uniform applied field (for a discussion of the response to a nonuniform field, see Appendix 10A).

In the presence of a uniform applied field $\mathbf{E}_0$, the system potential energy for the N molecules, $\mathcal{U} = \mathcal{U}(\mathbf{r}^N\omega^N)$, is

$$\mathcal{U} = \mathcal{U}_0 - \boldsymbol{\mu}\cdot\mathbf{E}_0. \tag{10.8}$$

Here $\mathcal{U}_0 = \mathcal{U}_0(\mathbf{r}^N\omega^N)$ is the energy in the absence of $\mathbf{E}_0$, and $\boldsymbol{\mu} = \boldsymbol{\mu}(\omega^N)$ is the total dipole moment of the system. Since the molecules are assumed to be rigid, we have

$$\boldsymbol{\mu} = \sum_i \boldsymbol{\mu}_i, \tag{10.9}$$

where $\boldsymbol{\mu}_i$ is the permanent dipole moment of molecule i.

Since $\mathbf{E}_0$ is uniform and the sample is spherical, the polarization $\mathbf{P}$ and the macroscopic field $\mathbf{E}$ will also be uniform.[22] $\mathbf{P}$ is the average dipole moment per unit volume, so that

$$\mathbf{P} = \frac{1}{V}\langle \boldsymbol{\mu} \rangle_{E_0}, \tag{10.10}$$

where $\langle \cdots \rangle_{E_0}$ indicates a thermal average (canonical or grand canonical) in the presence of $\mathbf{E}_0$ (i.e. over an ensemble with potential (10.8)). For sufficiently small field strengths,

$$\mathbf{P} = (\partial \mathbf{P}/\partial \mathbf{E}_0)_0 \cdot \mathbf{E}_0 \equiv \boldsymbol{\chi}_0 \cdot \mathbf{E}_0, \tag{10.11}$$

where the subscript on the partial derivative indicates that this should be evaluated at zero applied field.

To calculate $\boldsymbol{\chi}_0 = (\partial \mathbf{P}/\partial \mathbf{E}_0)_0$ we apply the general fluctuation–response relation (3.249) for parameter differentiation of an average $\langle B \rangle_\lambda$,

$$\frac{\partial}{\partial \lambda}\langle B \rangle_\lambda = \left\langle \frac{\partial B}{\partial \lambda} \right\rangle_\lambda - \beta \left(\left\langle B \frac{\partial \mathcal{U}}{\partial \lambda} \right\rangle_\lambda - \langle B \rangle_\lambda \left\langle \frac{\partial \mathcal{U}}{\partial \lambda} \right\rangle_\lambda \right). \tag{10.12}$$

In our case $B = \boldsymbol{\mu}$ and $\lambda = \mathbf{E}_0$. The first term on the right vanishes here, since for rigid molecules $\boldsymbol{\mu}$ is independent of $\mathbf{E}_0$. In taking the limit $\lambda \to 0$, the third term will also vanish, since $\langle \boldsymbol{\mu} \rangle_0$ vanishes for an isotropic system. The derivative $\partial \mathcal{U}/\partial \mathbf{E}_0$ is given from (10.8) as

$$\partial \mathcal{U}/\partial \mathbf{E}_0 = -\boldsymbol{\mu}. \tag{10.13}$$

Therefore

$$\boldsymbol{\chi}_0 = (\beta/V)\langle \boldsymbol{\mu}\boldsymbol{\mu} \rangle, \tag{10.14}$$

where $\langle\ \rangle \equiv \langle\ \rangle_0$. For an isotropic system, $\boldsymbol{\chi}_0$ must be proportional to the unit tensor, $\boldsymbol{\chi}_0 = \chi_0 \mathbf{1}$, where $\chi_0 = \frac{1}{3}\mathrm{Tr}\,\boldsymbol{\chi}_0$. Thus we can rewrite (10.14) as

$$\chi_0 = (\beta/3V)\langle \mu^2 \rangle, \tag{10.15}$$

where $\mu^2 = \boldsymbol{\mu} \cdot \boldsymbol{\mu}$.

From (10.9) we have

$$\mu^2 = \sum_i \mu_i^2 + \sum_{i \neq j} \boldsymbol{\mu}_i \cdot \boldsymbol{\mu}_j. \tag{10.16}$$

For rigid molecules each $\boldsymbol{\mu}_i$ is fixed in magnitude, and because the molecules are identical,

$$\langle \mu^2 \rangle = N\mu_1^2 + N(N-1)\langle \boldsymbol{\mu}_1 \cdot \boldsymbol{\mu}_2 \rangle. \tag{10.17}$$

Thus from (10.15) and (10.17) we get

$$\chi_0 = \left(\beta\rho\mu_1^2/3\right)(1 + (N-1)\langle \cos\gamma_{12} \rangle), \tag{10.18}$$

where γ_{12} is the angle between the dipole moments of molecules 1 and 2. We can write this result in terms of the pair correlation function $g(\mathbf{r}\omega_1\omega_2)$. We have

$$\chi_0 = \left(\beta\rho\mu_1^2/3\right) g_1, \tag{10.19}$$

where $g_1 = 1 + G_1$ is the Kirkwood g-factor,[23] and

$$G_1 = \rho \int d\mathbf{r} \langle g(\mathbf{r}\omega_1\omega_2) \cos \gamma_{12} \rangle_{\omega_1\omega_2}, \tag{10.20}$$

where $\langle\ \rangle_{\omega_1\omega_2}$ denotes an orientational average. Since $g(\mathbf{r}\omega_1\omega_2)$ contains shape-dependent terms of order V^{-1},[10] G_1 and g_1 depend on the geometry and surroundings of the sample.

The angular correlation parameter G_1 can also be written in terms of the total correlation function $h(\mathbf{r}\omega_1\omega_2) = g(\mathbf{r}\omega_1\omega_2) - 1$ as

$$G_1 = \rho \int d\mathbf{r} \langle h(\mathbf{r}\omega_1\omega_2) \cos \gamma_{12} \rangle_{\omega_1\omega_2}, \tag{10.21}$$

since $\langle \cos \gamma_{12} \rangle_{\omega_1\omega_2} = 0$.

The susceptibility χ is defined from (10.1) as $\chi = P/E$, or

$$\chi = (P/E_0)(E_0/E) = \chi_0(E_0/E), \tag{10.22}$$

so that we now need the relation between the applied field E_0 and the macroscopic field E. This relation is shape dependent, as the following examples demonstrate.

Sphere in vacuo

Here (10.7) yields

$$\mathbf{E} = \mathbf{E}_0 - (4\pi/3)\mathbf{P}, \tag{10.23}$$

since the integral vanishes for uniform $\mathbf{P}$ and spherical V. Using (10.1), we then find

$$\mathbf{E} = \frac{3}{\epsilon + 2}\mathbf{E}_0. \tag{10.24}$$

From (10.19), (10.22), and (10.24) we therefore have

$$\chi = (\beta\rho\mu_1^2/9)(\epsilon + 2)g_1. \tag{10.25}$$

Since $(\epsilon - 1)/4\pi = \chi$, we can rewrite (10.25) as

$$\frac{\epsilon - 1}{\epsilon + 2} = yg_1, \tag{10.26}$$

where $y = 4\pi\beta\rho\mu_1^2/9$ is dimensionless.

Infinite sample

The analogue of (10.26) for an infinite sample is

$$\frac{(\epsilon - 1)(2\epsilon + 1)}{9\epsilon} = yg_1. \tag{10.27}$$

A direct derivation of this result is given in Appendix 10A. Here we give a simpler derivation that employs Kirkwood's artifice of treating the sample as a sphere embedded in a continuous medium of the same dielectric constant.[23]

Our expression for χ_0 is formally unchanged from that of the previous section, Eq. (10.19). In place of (10.24), however, we find that

$$\mathbf{E} = \frac{2\epsilon + 1}{3\epsilon}\mathbf{E}_0, \tag{10.28}$$

which immediately leads to the desired result (10.27). To derive (10.28), suppose that the spherical sample is first empty of fluid. If the external medium is uniformly polarized with macroscopic field $\mathbf{E}$ at large distances from the cavity, then the field $\mathbf{E}_c$ inside the cavity is given from macroscopic electrostatics by[4]

$$\mathbf{E}_c = \frac{3\epsilon}{2\epsilon + 1}\mathbf{E}. \tag{10.29}$$

When the fluid is returned to the cavity, it is $\mathbf{E}_c$ which plays the role of an externally applied field $\mathbf{E}_0$; and this completes the derivation.

In Appendix 10A we derive two further expressions for the dielectric constant of an infinite system, Eqs. (10A.63) and (10A.64); in practice, these results are rarely used in calculations of ϵ.

Sphere in a dielectric medium

More generally, when a sphere of volume V is surrounded by a medium of dielectric constant ϵ', one finds by a simple extension of the above argument

$$\frac{(\epsilon - 1)(2\epsilon' + 1)}{3(\epsilon + 2\epsilon')} = y g_1. \tag{10.30}$$

Equations (10.26) and (10.27) are special cases of (10.30), derived by setting $\epsilon' = 1$ and ϵ, respectively.

Other cases have also been discussed in the literature, including slab[13] and ellipsoidal[24] geometries; in practice, these are less useful for computations.

Asymptotic behaviour of the total correlation function

The above formulae relate ϵ to integrals over the pair correlation function. A result due to Høye and Stell[18] instead relates ϵ to the asymptotic behaviour of this function. In an infinite system, at large r, $h(\mathbf{r}\omega_1\omega_2)$ is proportional to the dipole–dipole potential $u(\mathbf{r}\omega_1\omega_2)$,

$$h(\mathbf{r}\omega_1\omega_2) \sim -\frac{1}{\epsilon}\left(\frac{\epsilon - 1}{3y}\right)^2 \beta u(\mathbf{r}\omega_1\omega_2). \tag{10.31}$$

A derivation of this result is given in Appendix 10A. It is not very useful for the purposes of computation, since it is not clear in precisely what range it applies.

For other geometries, at large r $h(\mathbf{r}\omega_1\omega_2)$ also contains a constant term of order V^{-1}, as given by (10.126).

10.3.2 *Shape-independent expressions for* ϵ

It is possible to derive expressions for ϵ that involve only the short-range correlations in the fluid, thereby making the shape independence manifest.[12,13,17] In Appendix 10A, we show that

$$\frac{\epsilon - 1}{\epsilon + 2} = y\left(1 - \rho \int d\mathbf{r} \langle \Delta c(\mathbf{r}\omega_1\omega_2) \cos \gamma_{12} \rangle_{\omega_1\omega_2}\right)^{-1}, \tag{10.32}$$

where (cf. (3.119))

$$\Delta c(\mathbf{r}\omega_1\omega_2) = c(\mathbf{r}\omega_1\omega_2) + \beta u(\mathbf{r}\omega_1\omega_2) \tag{10.33}$$

is the short-range part of the direct correlation function. This result was first derived by Ramshaw.[17] Unlike the results presented in the last section, it holds for *any* sample geometry, since the integrand is short-ranged.

Suppose we now decompose the total correlation function into long- and short-range parts according to

$$h(\mathbf{r}\omega_1\omega_2) = \frac{\mathcal{H}(\mathbf{r}\omega_1\omega_2)}{r^3} + \Delta h(\mathbf{r}\omega_1\omega_2), \tag{10.34}$$

where $\Delta h(\mathbf{r}\omega_1\omega_2)$ is defined[25] by an Ornstein-Zernike relation,

$$\Delta h(\mathbf{r}_{12}\omega_1\omega_2) = \Delta c(\mathbf{r}_{12}\omega_1\omega_2) + \rho \int d\mathbf{r}_3 \langle \Delta c(\mathbf{r}_{13}\omega_1\omega_3)\Delta h(\mathbf{r}_{32}\omega_3\omega_2)\rangle_{\omega_3}, \tag{10.35}$$

and is $\sim r^{-6}$ for large r. From (10.32) and (10.35) it follows that[11]

$$\frac{\epsilon - 1}{\epsilon + 2} = y\left(1 + \rho \int d\mathbf{r} \langle \Delta h(\mathbf{r}\omega_1\omega_2) \cos \gamma_{12} \rangle_{\omega_1\omega_2}\right). \tag{10.36}$$

This result closely resembles (10.26). Note, however, that whereas the latter applies only to a spherical sample *in vacuo*, (10.36) holds for any shape.

These results have played a useful role in establishing the shape independence of ϵ.[11,17] However, since $\Delta c(\mathbf{r}\omega_1\omega_2)$ and $\Delta h(\mathbf{r}\omega_1\omega_2)$ are not directly available from simulations, they are not, in practice, especially useful for computational purposes.

Site–site formalism

The dielectric constant can also be expressed in terms of the site–site correlation functions $h_{\alpha\beta}(r)$ of § 3.1.6.

From (10A.64), for an infinite system

$$\frac{\epsilon - 1}{3\epsilon} = y\left(1 + \frac{1}{3}\rho \tilde{h}_\Delta(0) + \frac{2}{3}\rho \lim_{k\to 0} \tilde{h}_D(k)\right). \tag{10.37}$$

Now, we can readily establish[26] that the coefficient of k^2 in the small-k expansion of

$$\Delta\tilde{h}(k) = \sum_{\alpha\beta} q_\alpha \tilde{h}_{\alpha\beta}(k) q_\beta, \tag{10.38}$$

where $\tilde{h}_{\alpha\beta}(k) = \int d\mathbf{r} \exp(i\mathbf{k}\cdot\mathbf{r}) h_{\alpha\beta}(r)$ is the Fourier transform of $h_{\alpha\beta}(r)$, and q_α is the charge assigned to site α, satisfies

$$\Delta\tilde{h}^{(2)} = \frac{1}{9}\mu_1^2 \left(\tilde{h}_\Delta(0) + 2 \lim_{k\to 0} \tilde{h}_D(k) \right). \tag{10.39}$$

It follows from (10.37) and (10.39) that

$$\frac{\epsilon - 1}{3\epsilon} = y\left(1 + \left(3\rho/\mu_1^2\right) \Delta\tilde{h}^{(2)}\right). \tag{10.40}$$

For dipolar fluids the functions $h_{\alpha\beta}(r)$ are short-ranged,[26,27] $h_{\alpha\beta}(r) \sim r^{-6}$; therefore (10.40) applies for any system geometry.

10.4 Polarizable fluids

The rigid polar model described in § 10.3 ignores the contributions to the polarization that arise from electronic degrees of freedom. In practice, all real molecules are polarizable, and their electron clouds will distort in the presence of an applied electric field, generating a field-dependent dipole moment.[28] This affects the polarization directly, and is present even for atomic fluids. For molecular fluids, there is a second, indirect, effect resulting from the modification of the dipole moment by intermolecular interactions; the mean dipole moment of a molecule in a dense fluid is often quite different from that in the gas phase. Accurate calculation of the dielectric constant of real fluids generally requires that the effects of polarizability be included in some manner.[4,29]

We begin our analysis by computing the linear response function $\boldsymbol{\chi}_0$. Without loss of generality, we may suppose that the applied field $\mathbf{E}_0$ is uniform, and that the sample is spherical. The analysis then parallels that of the rigid polar case exactly, except that in place of (10.14) we now find

$$\boldsymbol{\chi}_0 = (\beta/V)\langle \boldsymbol{\mu}\boldsymbol{\mu} \rangle + (1/V)\langle \boldsymbol{\alpha} \rangle, \tag{10.41}$$

where $\boldsymbol{\alpha}$ is total polarizability of the sample, given by

$$\boldsymbol{\alpha} = \partial\boldsymbol{\mu}/\partial\mathbf{E}_0. \tag{10.42}$$

For an isotropic fluid we therefore have $\boldsymbol{\chi}_0 = \chi_0 \mathbf{1}$, where

$$\chi_0 = (\beta/3V)\langle \mu^2 \rangle + (1/V)\langle \alpha \rangle, \tag{10.43}$$

and $\alpha = \frac{1}{3}\mathrm{Tr}\,\boldsymbol{\alpha}$.

The expressions that relate the macroscopic and applied fields are formally unchanged from the rigid polar case. It then follows that, for a sphere *in vacuo*,

$$\frac{\epsilon - 1}{\epsilon + 2} = (4\pi\beta/9V)\langle\mu^2\rangle + (4\pi/3V)\langle\alpha\rangle. \tag{10.44}$$

Similarly, for an infinite system (or, equivalently, for a sphere inside a continuous medium of the same ϵ),

$$\frac{(\epsilon - 1)(2\epsilon + 1)}{9\epsilon} = (4\pi\beta/9V)\langle\mu^2\rangle + (4\pi/3V)\langle\alpha\rangle. \tag{10.45}$$

In each case the appropriate ensemble averages must be taken. The left-hand sides of (10.44) and (10.45) are sometimes referred to as the Clausius-Mossotti and the Onsager or Kirkwood functions, respectively.

Equations (10.44) and (10.45) generalize (10.26) and (10.27) for a nonpolarizable system. We note two differences. First, a second term, involving the system polarizability, has been added to the term involving dipole moment fluctuations. Second, many-body terms are present in both $\boldsymbol{\mu}$ and $\boldsymbol{\alpha}$. In the dipole–induced-dipole (DID) model,[30] the system dipole moment and polarizability are given by

$$\boldsymbol{\mu} = \sum_i \boldsymbol{\mu}_i + {\sum_{ij}}' \boldsymbol{\alpha}_i \cdot \mathbf{T}_{ij} \cdot \boldsymbol{\mu}_j + {\sum_{ijk}}' \boldsymbol{\alpha}_i \cdot \mathbf{T}_{ij} \cdot \boldsymbol{\alpha}_j \cdot \mathbf{T}_{jk} \cdot \boldsymbol{\mu}_k + \ldots \tag{10.46}$$

and

$$\boldsymbol{\alpha} = \sum_i \boldsymbol{\alpha}_i + {\sum_{ij}}' \boldsymbol{\alpha}_i \cdot \mathbf{T}_{ij} \cdot \boldsymbol{\alpha}_j + {\sum_{ijk}}' \boldsymbol{\alpha}_i \cdot \mathbf{T}_{ij} \cdot \boldsymbol{\alpha}_j \cdot \mathbf{T}_{jk} \cdot \boldsymbol{\alpha}_k + \ldots, \tag{10.47}$$

where $\boldsymbol{\mu}_i$ and $\boldsymbol{\alpha}_i$ denote the dipole and polarizability of an iolated molecule, and $\mathbf{T}_{ij}$ is the dipolar tensor representing the interactions of atoms i and j. The primes on the sums indicate that the terms $i = j, j = k$ are omitted. More generally, additional terms resulting from higher order multipolar induction and dispersion and overlap mechanisms must be included; these terms are considered further in connection with collision-induced absorption and light scattering spectra in Chapter 11. It follows that (10.44) and (10.45) can no longer be written in terms of the pair correlation function alone; written out explicitly, the averages $\langle\mu^2\rangle$ and $\langle\alpha\rangle$ would involve h-body correlation functions $g(1\ldots h)$ of all orders h (cf. § 3.1.4).

The Clausius–Mossotti equation

An approximate relation for the dielectric constant follows if we ignore orientational and positional correlations in the ensemble averages in Eq. (10.44), and many-body terms in the expansions (10.46) and (10.47). The result is the well-known Clausius–Mossotti equation,

$$\frac{\epsilon - 1}{\epsilon + 2} = \frac{4\pi}{3}\rho\left(\beta\mu_1^2/3 + \alpha_1\right). \tag{10.48}$$

It is possible to give a simplified derivation of this formula using the concept of the Lorentz local field $\mathbf{E}_\mathrm{L}$, which is the mean field experienced by a molecule in the fluid. This differs from the Maxwell field $\mathbf{E}$, which is the mean field at a fixed point, because of intermolecular correlations; and most particularly, because molecules possess impenetrable cores, and thus cannot overlap completely. As we show in Appendix 10B, the relation between the two fields is approximately

$$\mathbf{E}_\mathrm{L} = \mathbf{E} + \frac{4\pi}{3}\mathbf{P}. \tag{10.49}$$

Combining this result with Eq. (10.1), we find that

$$\mathbf{P} = \frac{3}{4\pi}\frac{\epsilon - 1}{\epsilon + 2}\mathbf{E}_\mathrm{L}. \tag{10.50}$$

Within the approximations discussed above, we expect

$$\mathbf{P} = \rho\left(\beta\mu_1^2/3 + \alpha_1\right)\mathbf{E}_\mathrm{L}, \tag{10.51}$$

and combining the last two equations yields (10.48) directly.

The Clausius–Mossotti relation is strictly valid only in the absence of all intermolecular interactions. For polar liquids, especially, Eq. (10.48) is quite inaccurate. However, for nonpolar liquids, and especially atomic fluids such as Ar, where $\mu_1 = 0$, the approximation

$$\frac{\epsilon - 1}{\epsilon + 2} = \frac{4\pi}{3}\rho\alpha_1 \tag{10.52}$$

does surprisingly well. Similarly, as we discuss in the next section, the approximation

$$\frac{\epsilon_\infty - 1}{\epsilon_\infty + 2} = \frac{4\pi}{3}\rho\alpha_{1\infty} \tag{10.53}$$

for the high-frequency limit of ϵ of a *polar* fluid is also a reasonable approximation. Here $\alpha_{1\infty}$ is the high-frequency single-molecule mean polarizability. We use a version of (10.53) in § 8.9.1 (c).

On a historical note, the original Clausius–Mossotti relation[31,32] was Eq. (10.52) for nonpolar liquids. It was Debye[33] who derived the extension for polar liquids.

Frequency-dependent dielectric response

Although the focus of this chapter is on the static dielectric constant, it is instructive at this point to consider also the response to a frequency-dependent field $\mathbf{E}_0(\omega)$. In general, $\epsilon(\omega) = \epsilon'(\omega) + i\epsilon''(\omega)$ is complex: the real part is approximately given by $\epsilon'(\omega) = n'(\omega)^2$, where $n'(\omega)$ is the refractive index, while the imaginary part, $\epsilon''(\omega)$, is closely related to the absorption coefficient $A(\omega)$. $\epsilon'(\omega)$ and $\epsilon''(\omega)$ are themselves related via the Kramers–Kronig relations. These relationships are discussed more fully in Chapter 11.

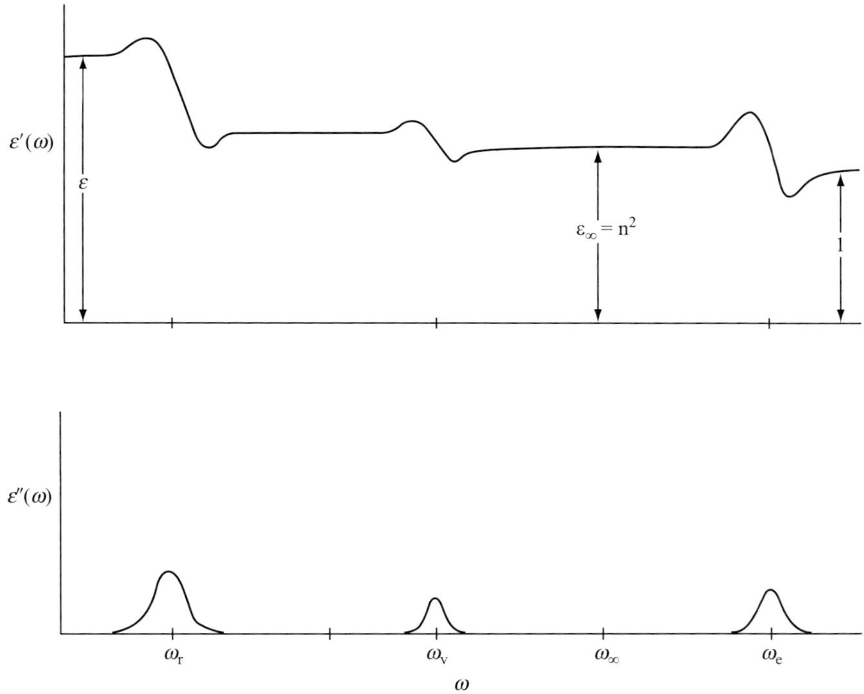

FIG. 10.1 Schematic illustration of the behaviour of the real and imaginary parts of the dynamic dielectric constant $\epsilon(\omega) = \epsilon'(\omega) + i\epsilon''(\omega)$. The rotational, vibrational, and electronic absorption bands are shown as broad bands without the line structure that is present, for example, in low-density gases. For small molecules ω_r is a far-infrared ($\sim 10\,\mathrm{cm}^{-1}$), ω_v an infrared ($\sim 10^3\,\mathrm{cm}^{-1}$), and ω_e an ultraviolet ($\sim 10^5\,\mathrm{cm}^{-1}$) frequency.

Here we wish to point out that $\epsilon'(\omega)$ shows a dispersion that is linked to the absorption bands of the sample. This is illustrated schematically in Fig. 10.1. At frequencies above the far-infrared rototranslational absorption bands, $\epsilon'(\omega)$ is reduced from its static value. At such frequencies, the molecular dipole moments cannot respond to changes in the applied field, and no longer contribute to the dielectric response of the system. Only the polarizability term will then contribute to (10.44) and (10.45). Provided the frequency is still below the vibrational[28] and optical bands, this will be essentially unchanged from its static value. The high-frequency dielectric constant ϵ_∞ is therefore given by

$$\frac{\epsilon_\infty - 1}{\epsilon_\infty + 2} = (4\pi/3V)\langle\alpha\rangle \tag{10.54}$$

for a sphere *in vacuo*, and

$$\frac{(\epsilon_\infty - 1)(2\epsilon_\infty + 1)}{9\epsilon_\infty} = (4\pi/3V)\langle\alpha\rangle \tag{10.55}$$

for an infinite system.

We can therefore rewrite (10.44) and (10.45) as

$$\frac{\epsilon - 1}{\epsilon + 2} - \frac{\epsilon_\infty - 1}{\epsilon_\infty + 2} = (4\pi\beta/9V)\langle\mu^2\rangle \tag{10.56}$$

and

$$\frac{(\epsilon - 1)(2\epsilon + 1)}{9\epsilon} - \frac{(\epsilon_\infty - 1)(2\epsilon_\infty + 1)}{9\epsilon_\infty} = (4\pi\beta/9V)\langle\mu^2\rangle, \tag{10.57}$$

respectively. For many polar liquids, ϵ lies in the range 5–20, while ϵ_∞ is typically about 2.[29]

10.5 Perturbation theory

Chapter 4 introduced the ideas and techniques of perturbation theory, and showed how one could relate the properties of a fluid to those of a simpler, reference fluid, by expanding in powers of a perturbational parameter λ, which is assumed to be small. The emphasis in that chapter was on the prediction of thermodynamic properties, particularly the free energy, from which other quantities such as the internal energy, entropy, and pressure can be obtained using standard thermodynamic relations. In this section we discuss the perturbational calculation of the dielectric constant. Suitable choices for λ then include $\beta\mu_1^2/\sigma^3$ and α_1/σ^3, where μ_1 and α_1 are the molecular dipole moment and polarizability, and σ is a typical molecular dimension.

We begin with the rigid polar case. Our starting point is Kirkwood's equation (10.27) for ϵ, expressed in terms of the dimensionless parameter $y = 4\pi\beta\rho\mu_1^2/9$ and $g_1 = 1 + G_1$, with G_1 defined by (10.20). As shown in Appendix 10C, we may develop a perturbational expansion for G_1 in powers of $\lambda = \beta u_a$, where u_a is the anisotropic component of the intermolecular pair potential $u(\mathbf{r}\omega_1\omega_2)$. This yields

$$G_1 = G_1^{(1)} + G_1^{(2)} + G_1^{(3)} + \ldots, \tag{10.58}$$

where $G_1^{(n)} \propto \lambda^n$. When u_a is multipole-like,[34] $G_1^{(1)} = 0$, and the leading term in G_1 is of order λ^2, which is proportional to y^2 for a dipolar fluid. Approximately, therefore,

$$\frac{(\epsilon - 1)(2\epsilon + 1)}{9\epsilon} = y(1 + \phi y^2), \tag{10.59}$$

where $\phi = G_1^{(2)}/y^2$. In Appendix 10C we derive an explicit expression for ϕ for a fluid whose molecules possess dipoles and linear quadrupoles. Results for more general $u(\mathbf{r}\omega_1\omega_2)$ have also been calculated.[35–37]

For a fluid of dipolar hard spheres, Tani *et al.*[38] showed that (10.59) did not yield accurate results, overestimating ϵ for all but the smallest values of μ_1^2 and y. However, they pointed out that for small y it could be cast in the form

$$\epsilon = 1 + 3y + 3y^2 + 3y^3(\phi - 1) + \dots, \tag{10.60}$$

and that this equation, truncated at order y^3, gave surprisingly accurate estimates of ϵ even for $y = 3$, well outside its radius of convergence. They noted that, over a wide range of state conditions, $\phi \approx 1$,[39] and they suggested that (10.60) converges rapidly. Later, Matyushov and Ladanyi[41] showed that the range of applicability of perturbation theory could be increased still further up to $y = 4$ by replacing (10.60) by

$$\epsilon = 1 + 3y + 3y^2 + \frac{2}{(\phi - 1)^2}(\exp[3(\phi - 1)^3 y^3/2] - 1), \tag{10.61}$$

which is again equivalent to order y^3.

Goldman[40] and Goldman and Joslin[35,36,42,43] have investigated the application of (10.60) to a variety of polar fluids and fluid mixtures. In all cases they find excellent agreement with computer simulation and/or experiment. Figure 10.2 shows a plot[35] of ϵ versus y for TIP4P water at a density of about 1 g cm^{-3}. Theory and simulation agree well up to $y = 2.5$, at which point ϵ is overestimated. The theory has also been successfully applied to water at high temperatures and pressures[43] and to supercritical mixtures of water and benzene.[36]

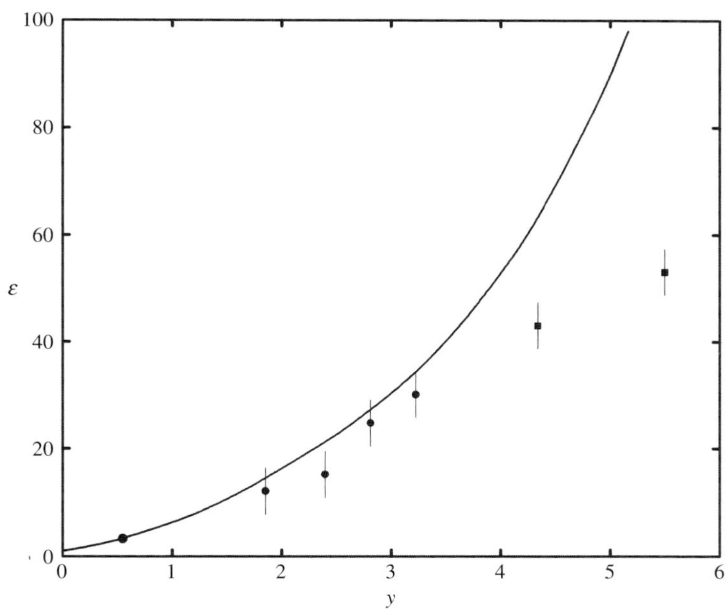

FIG. 10.2 The dielectric constant of TIP4P water vs y. Perturbation theory (solid line) is compared with MC and MD simulation data (points) at $\rho\sigma^3 = 1.055$, corresponding to a density $\approx$ 1 g cm^{-3}. (From ref. 35.)

Despite these undoubted successes, we should probably be cautious of predictions based on (10.60) (or variants thereof). If we include terms of order y^4, y^5, and y^6 in the expansion of ϵ, we find that

$$\epsilon = 1 + 3y + 3y^2 + 3y^3(\phi - 1) + 3y^4(2\phi - 1) + 3y^5(-3\phi + 5) \\ +3y^6(\phi^2 - 4\phi - 3) + O(y^7). \tag{10.62}$$

If we take $y = 2$, we are well within the claimed limits of validity of the theory. Setting $\phi = 1$, as discussed earlier, we obtain

$$\epsilon = 1 + 6 + 12 + 0 + 48 + 192 - 1152 + \ldots, \tag{10.63}$$

from which it is apparent that we do *not* obtain sensible results if we continue past the term of order y^3. This is because we are well outside the radius of convergence of (10.62), which is about $y = 0.23$. The approximate vanishing of the term of order y^3 is seen to be essentially accidental, with no particular implications for terms of higher order in y. The apparent success of the perturbational approach may be in large measure fortuitous, resulting from a chance cancellation of terms.

An extension of (10.60) to polarizable fluids that uses the ideas contained in Wertheim's renormalization theory has been proposed. Starting from (10.45), Valiskó *et al.*[44] find that

$$\epsilon = 1 + 3\bar{y} + 3\bar{y}^2 + 3\bar{y}^3(\bar{\phi} - 1) + \ldots, \tag{10.64}$$

where

$$\begin{aligned} \bar{y} &= y(1 + a), \qquad & y &= 4\pi\beta\rho\mu^2/9, \\ \bar{\phi} &= \phi(1 - (1 + a^{-1})^{-3}), & a &= \alpha_1/(\beta\mu^2/3). \end{aligned} \tag{10.65}$$

Here a is a numerically small, dimensionless parameter that reflects the relative importance of polarizability and dipolar orientation as they affect single-molecule contributions to the polarization. μ is the mean, or renormalized dipole moment; this typically differs significantly from the bare dipole moment μ_1 because of the many-body nature of induction. It may be calculated self-consistently from a perturbational expansion of the Helmholtz free energy as discussed in § 4.10. Comparison with Monte Carlo simulations of the polarizable Stockmayer (dipolar Lennard-Jones) fluid shows that (10.64) yields accurate estimates of ϵ for low and moderate values of μ_1 and α_1; the range of validity appears to be about the same as for the nonpolarizable case, if y is replaced by $\bar{y}$.

Perturbation theory can also be used to calculate the high-frequency dielectric constant ϵ_∞ defined by (10.54). In the DID approximation, Eq. (10.47), and neglecting anisotropy in the polarizability, we find

$$\frac{\epsilon_\infty - 1}{\epsilon_\infty + 2} = (4\pi/3)\rho\alpha_1(1 + S), \tag{10.66}$$

where

$$S = \sum_{n=2}^{\infty} s_n \alpha_1^n \tag{10.67}$$

is a power series in the molecular polarizability α_1, with coefficients

$$s_n = \frac{1}{3N} \sum_{\alpha=1}^{3} \langle \boldsymbol{\lambda}_\alpha^T \mathcal{T}^n \boldsymbol{\lambda}_\alpha \rangle. \tag{10.68}$$

Here $\mathcal{T}$ is the $3N \times 3N$ matrix with elements $\mathcal{T}_{i\alpha,j\beta} = T_{\alpha\beta}(\mathbf{r}_{ij})$, and $\boldsymbol{\lambda}_\alpha^T = (\mathbf{e}_\alpha^T\ \mathbf{e}_\alpha^T \ldots \mathbf{e}_\alpha^T)$ (N terms), where $\mathbf{e}_1^T = (1\ 0\ 0)$, $\mathbf{e}_2^T = (0\ 1\ 0)$, $\mathbf{e}_3^T = (0\ 0\ 1)$; $\boldsymbol{\lambda}_\alpha$ is the corresponding column vector. In low-density gases, S is positive (cf. § 10.6), but for liquids it can be negative. It is typically rather small, usually 0.01–0.1.

Cichocki and Felderhof have computed the first forty terms of the series (10.67) for a fluid of polarizable hard spheres by Monte Carlo simulation.[45] They find that terms beyond the first are normally quite small, so that to a good approximation $S = s_2\alpha_1^2$. Figure 10.3 shows s_2 as a function of the volume fraction $\eta = (\pi/6)\rho\sigma^3$.

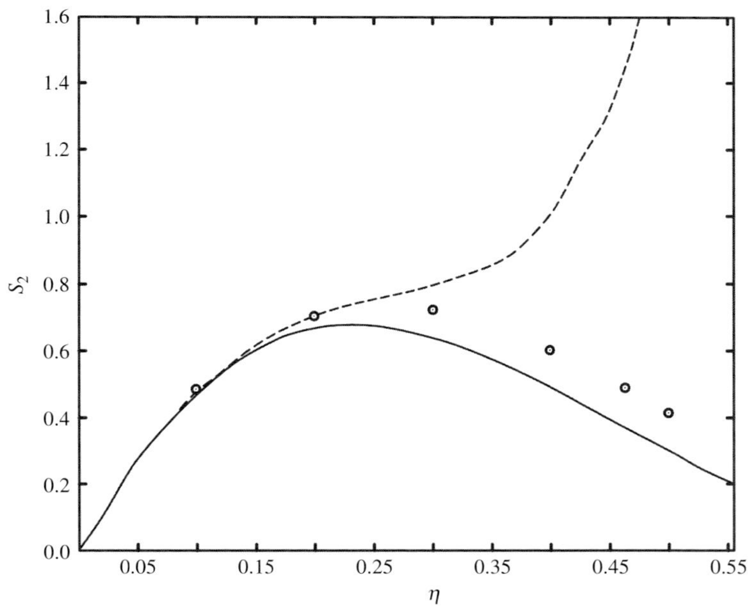

FIG. 10.3 The coefficient s_2 in the α_1-expansion of S for a fluid of polarizable hard spheres as a function of the volume fraction η. The solid line is based on the ladder approximation for $g_0(r_{12}r_{13}r_{23})$, the dashed line on the Kirkwood superposition approximation, and the circles are Monte Carlo results. (From ref. 46.)

We may express s_2 as a sum of two- and three-body integrals over the reference pair and triplet correlation functions of the fluid. From (10.68), we have

$$s_2 = 2\rho \int d\mathbf{r}_{12}\, g_0(r_{12}) r_{12}^{-6} \tag{10.69}$$

$$+2\rho^2 \int d\mathbf{r}_{12} d\mathbf{r}_{23}\, [g_0(r_{12}r_{13}r_{23}) - g_0(r_{12})g_0(r_{23})] r_{12}^{-3} r_{23}^{-3} P_2(\cos\theta_2),$$

where θ_2 is the angle at molecule 2 in the triangle formed by molecules 1, 2, and 3. The two terms in (10.69) are of opposite sign, and cancel substantially at liquid densities; this explains the maximum in s_2 near $\eta = 0.25$. Figure 10.3 also shows results computed using two approximations for the triplet correlation function, namely the Kirkwood superposition approximation (3.246), which is very poor, and the ladder approximation of Bławzdziewicz *et al.*,[46] essentially an HNC closure of the generalized Ornstein–Zernike equation,[47] which is much more accurate.

Hinsen and Felderhof[48] have considered the effect of adding quadrupole (C_1) polarizability.[49] If one takes $C_1/\alpha_1\sigma^2 = 0.06$, which is appropriate for the rare gases, S increases by approximately 10–20%.

10.6 Dielectric virial expansions

In a dilute fluid, we can expand the Clausius–Mossotti function of (10.44) in powers of the density,

$$\frac{\epsilon - 1}{\epsilon + 2} = A_\epsilon \rho + B_\epsilon \rho^2 + C_\epsilon \rho^3 + \ldots. \tag{10.70}$$

$A_\epsilon, B_\epsilon, C_\epsilon$, etc., are called the first, second, third, etc., dielectric virial coefficients, and reflect one-, two-, three-body, etc., contributions to the dielectric response of the system. Expressions for these coefficients have been derived by various means, both classical[17,50–52] and quantum.[53–56] Here we will follow a classical procedure described by Joslin,[52] postponing a discussion of quantum corrections until the end of this section.

Suppose a dielectric sphere *in vacuo* is subject to a uniform applied field $\mathbf{E}_0$. The sample contains N molecules in a volume V and is maintained at a temperature T. The polarization induced by the field may be written in terms of the configurational partition function $Q_N(V, T, \mathbf{E}_0)$,

$$\mathbf{P} = (kT/V)\partial \ln Q_N(V, T, \mathbf{E}_0)/\partial \mathbf{E}_0, \tag{10.71}$$

where the derivative is to be taken at constant N, V, and T. We can expand Q_N in powers of the activity z according to[57]

$$\ln Q_N(V, T, \mathbf{E}_0) = \frac{V}{\lambda^3} \sum_{l=1}^{N} b_l z^l - N \ln z, \tag{10.72}$$

where z is related to the number density $\rho = N/V$ by the implicit equation

$$\sum_{l=1}^{N} l b_l z^l = \rho\lambda^3. \tag{10.73}$$

Here λ is the de Broglie wavelength (§ 1.2.2). b_l is a cluster integral over the coordinates and orientations of l molecules, and is given by[58]

$$b_l = \frac{1}{l!\lambda^{3l-3}V}\left\langle \int \mathbf{d}\mathbf{r}_1 \ldots \mathbf{d}\mathbf{r}_l \, U_l(1\ldots l)\right\rangle_{\omega_1\ldots\,\omega_l}, \tag{10.74}$$

where $\langle\ldots\rangle_{\omega_1\ldots\omega_l}$ denotes an average over internal degrees of freedom, and the U_l are Ursell functions (see below).

Using (10.71)–(10.73) we find that

$$\mathbf{P} = (kT/\lambda^3)\sum_{l=1}^{N}(\partial b_l/\partial \mathbf{E}_0)z^l, \tag{10.75}$$

and eliminating z in favour of ρ, we have in the limit $\mathbf{E}_0 \to 0$ (where $b_1 \to 1$)

$$\begin{aligned}\mathbf{P} = (kT/\lambda^3)\Big(&(\partial b_1/\partial \mathbf{E}_0)\,\rho\lambda^3\\ &+[(\partial b_2/\partial \mathbf{E}_0) - 2b_2(\partial b_1/\partial \mathbf{E}_0)]\,(\rho\lambda^3)^2\\ &+\Big[(\partial b_3/\partial \mathbf{E}_0) - 4b_2(\partial b_2/\partial \mathbf{E}_0) + \left(8b_2^2 - 3b_3\right)(\partial b_1/\partial \mathbf{E}_0)\Big]\,(\rho\lambda^3)^3\\ &+\ldots\,\Big).\end{aligned} \tag{10.76}$$

The U_l required for the evaluation of the derivatives in (10.76) can be written as

$$\begin{aligned}U_1(1) &= \exp[-\beta\mathcal{U}(1)],\\ U_2(12) &= \exp[-\beta\mathcal{U}(12)] - \exp[-\beta\mathcal{U}(1,2)],\\ U_3(123) &= \exp[-\beta\mathcal{U}(123)]\\ &\quad - \exp[-\beta\mathcal{U}(12,3)] - \exp[-\beta\mathcal{U}(23,1)] - \exp[-\beta\mathcal{U}(31,2)]\\ &\quad +2\exp[-\beta\mathcal{U}(1,2,3)],\end{aligned} \tag{10.77}$$

where we have adopted the notation

$$\mathcal{U}(ijk, lm, n, \ldots) = \mathcal{U}(ijk) + \mathcal{U}(lm) + \mathcal{U}(n) + \ldots, \tag{10.78}$$

in which $\mathcal{U}(i\ldots j)$ is the potential energy of a system of interacting particles $i\ldots j$ in the presence of the field. This can be written as

$$\mathcal{U}(i\ldots j) = \mathcal{U}_0(i\ldots j) - \boldsymbol{\mu}(i\ldots j)\cdot\mathbf{E}_0 - \frac{1}{2}\boldsymbol{\alpha}(i\ldots j) : \mathbf{E}_0\mathbf{E}_0, \tag{10.79}$$

where $\mathcal{U}_0(i\ldots j)$ is the corresponding energy in the absence of the field, and $\boldsymbol{\mu}(i\ldots j)$ and $\boldsymbol{\alpha}(i\ldots j)$ denote the dipole moment and polarizability, respectively. From (10.79), the partial derivative of the energy with respect to the field is

$$\partial\mathcal{U}(i\ldots j)/\partial\mathbf{E}_0 = -(\boldsymbol{\mu}(i\ldots j) + \boldsymbol{\alpha}(i\ldots j)\cdot\mathbf{E}_0). \tag{10.80}$$

Therefore, in the limit $\mathbf{E}_0 \to 0$, for an isotropic system, the derivatives of the cluster integrals are given by

$$\partial b_l/\partial\mathbf{E}_0 = \frac{\beta}{l!\lambda^{3l-3}V}\left\langle\int d\mathbf{r}_1\ldots d\mathbf{r}_l\,\mathcal{A}_l(1\ldots l)\right\rangle_{\omega_1\ldots\omega_l}\mathbf{E}_0, \tag{10.81}$$

where

$$\begin{aligned}
\mathcal{A}_1(1) &= \phi(1),\\
\mathcal{A}_2(12) &= \phi(12)\exp[-\beta\mathcal{U}_0(12)] - \phi(1) - \phi(2),\\
\mathcal{A}_3(123) &= \phi(123)\exp[-\beta\mathcal{U}_0(123)]\\
&\quad -(\phi(12)+\phi(3))\exp[-\beta\mathcal{U}_0(12)] - (\phi(23)+\phi(1))\exp[-\beta\mathcal{U}_0(23)]\\
&\quad -(\phi(31)+\phi(2))\exp[-\beta\mathcal{U}_0(31)] + 2(\phi(1)+\phi(2)+\phi(3)),
\end{aligned} \tag{10.82}$$

and we have defined

$$\phi(i\ldots l) = (\beta/3)\mu(1\ldots l)^2 + \alpha(1\ldots l). \tag{10.83}$$

From (10.1) and (10.24), the relation between $\mathbf{P}$ and ϵ for the chosen geometry is

$$\mathbf{P} = \frac{3}{4\pi}\frac{\epsilon-1}{\epsilon+2}\mathbf{E}_0. \tag{10.84}$$

Combining Eqs. (10.76), (10.81), and (10.84) now yields the virial expansion up to third order in the density, and generates explicit statistical-mechanical expressions for the coefficients A_ϵ, B_ϵ, and C_ϵ, which we discuss below.

10.6.1 *First dielectric virial coefficient, A_ϵ*

The first virial coefficient, which describes ideal gas behaviour, is given by

$$A_\epsilon = A_\epsilon^\mu + A_\epsilon^\alpha, \tag{10.85}$$

where

$$A_\epsilon^\mu = 4\pi\beta\mu_1^2/9, \qquad A_\epsilon^\alpha = 4\pi\alpha_1/3. \tag{10.86}$$

Measurement of A_ϵ as a function of temperature affords a standard method of measuring μ_1 and α_1, as discussed in Appendix D.

10.6.2 *Second dielectric virial coefficient, B_ϵ*

For the second virial coefficient we find

$$B_\epsilon = B_\epsilon^\mu + B_\epsilon^\alpha, \tag{10.87}$$

where

$$B_\epsilon^\mu = (2\pi\beta/9)\left\langle \int \mathbf{r}_{12} \exp[-\beta\mathcal{U}(12)]\Delta\mu(12)^2 \right\rangle_{\omega_1\omega_2},$$

$$B_\epsilon^\alpha = (2\pi/3)\left\langle \int \mathbf{dr}_{12} \exp[-\beta\mathcal{U}(12)]\Delta\alpha(12) \right\rangle_{\omega_1\omega_2}. \quad (10.88)$$

These expressions were first derived by Buckingham and Pople[51] using a different method. We have defined the pair functions

$$\Delta\mu(12)^2 = \mu(12)^2 - \mu(1)^2 - \mu(2)^2, \qquad \Delta\alpha(12) = \alpha(12) - \alpha(1) - \alpha(2), \quad (10.89)$$

where $\mu(1) = \mu(2) = \mu_1$ and $\alpha(1) = \alpha(2) = \alpha_1$, which are *irreducible* in the sense that $\lim_{r_{12}\to\infty} \Delta\mu(12) = \lim_{r_{12}\to\infty} \Delta\alpha(12) = 0$.

B_ϵ describes the leading departures from ideality in a dilute gas. It is much more sensitive to anisotropic forces than is the pressure virial coefficient B_2, discussed in §§ 3.6 and 6.10. In principle, measurements of B_ϵ can yield information on molecular multipole moments. For example, Hourri *et al.*[59] have determined the quadrupole moment of carbon dioxide from studies on CO_2–SF_6 mixtures, and their estimate is in good agreement with the accepted value. However, the results are very sensitive to the assumed form of the intermolecular potential and to other molecular parameters, and in general measurements of electric field gradient-induced birefringence (discussed in § 10.8) provide a much more direct and reliable route to such quantities.

10.6.3 *Third dielectric virial coefficient, C_ϵ*

The third virial coefficient, C_ϵ, is given by

$$C_\epsilon = C_\epsilon^\mu + C_\epsilon^\alpha, \quad (10.90)$$

where

$$\begin{aligned} C_\epsilon^\mu = (2\pi\beta/27)\Big\langle \int & \mathbf{dr}_{12}\mathbf{dr}_{13}\, \Delta\mu(123)^2 \exp[-\beta\mathcal{U}(123)] \\ & +(\Delta\mu(12)^2 + \Delta\mu(23)^2 + \Delta\mu(31)^2) \\ & \times \exp[-\beta(\mathcal{U}(12) + \mathcal{U}(23) + \mathcal{U}(31))](\exp[-\beta\Delta\mathcal{U}(123)] - 1) \\ & +\Delta\mu(12)^2 \exp[-\beta\mathcal{U}(12)](\exp[-\beta\mathcal{U}(23)] - 1)(\exp[-\beta\mathcal{U}(31)] - 1) \\ & +\Delta\mu(23)^2 \exp[-\beta\mathcal{U}(23)](\exp[-\beta\mathcal{U}(31)] - 1)(\exp[-\beta\mathcal{U}(12)] - 1) \\ & +\Delta\mu(31)^2 \exp[-\beta\mathcal{U}(31)](\exp[-\beta\mathcal{U}(12)] - 1)(\exp[-\beta\mathcal{U}(23)] - 1)\Big\rangle_{\omega_1\omega_2\omega_3}, \end{aligned} \quad (10.91)$$

and

$$C_\epsilon^\alpha = (2\pi/9)\Big\langle \int \mathbf{dr}_{12}\mathbf{dr}_{13}\, \Delta\alpha(123)\exp[-\beta\mathcal{U}(123)]$$
$$+(\Delta\alpha(12) + \Delta\alpha(23) + \Delta\alpha(31))$$
$$\times \exp[-\beta(\mathcal{U}(12) + \mathcal{U}(23) + \mathcal{U}(31))](\exp[-\beta\Delta\mathcal{U}(123)] - 1)$$
$$+\Delta\alpha(12)\exp[-\beta\mathcal{U}(12)](\exp[-\beta\mathcal{U}(23)] - 1)(\exp[-\beta\mathcal{U}(31)] - 1)$$
$$+\Delta\alpha(23)\exp[-\beta\mathcal{U}(23)](\exp[-\beta\mathcal{U}(31)] - 1)(\exp[-\beta\mathcal{U}(12)] - 1)$$
$$+\Delta\alpha(31)\exp[-\beta\mathcal{U}(31)](\exp[-\beta\mathcal{U}(12)] - 1)(\exp[-\beta\mathcal{U}(23)] - 1)\Big\rangle_{\omega_1\omega_2\omega_3}. \quad (10.92)$$

Here $\mathcal{U}(123)$ is the total three-body interaction energy, and

$$\Delta\mathcal{U}(123) = \mathcal{U}(123) - \mathcal{U}(12) - \mathcal{U}(23) - \mathcal{U}(31) \quad (10.93)$$

is its irreducible triplet part. The irreducible triplet squared dipole moment and polarizability are

$$\Delta\mu(123)^2 = \mu(123)^2 - (\mu(12)^2 + \mu(3)^2)$$
$$-(\mu(23)^2 + \mu(1)^2) - (\mu(31)^2 + \mu(2)^2)$$
$$+2(\mu(1)^2 + \mu(2)^2 + \mu(3)^2) \quad (10.94)$$

and

$$\Delta\alpha(123) = \alpha(123) - (\alpha(12) + \alpha(3))$$
$$-(\alpha(23) + \alpha(1)) - (\alpha(31) + \alpha(2))$$
$$+2(\alpha(1) + \alpha(2) + \alpha(3)), \quad (10.95)$$

and vanish when any one of the three interacting molecules is removed to infinity.

10.6.4 *Virial expansion of the high-frequency dielectric constant*

Each virial coefficient is the sum of two contributions, one associated with fluctuations in the dipole moment, the other with averages of the polarizability. We recall from § 10.4 that the high-frequency dielectric constant of (10.54) contains only polarizability terms. Thus at high frequencies, the Clausius–Mossotti function has the virial expansion

$$\frac{\epsilon_\infty - 1}{\epsilon_\infty + 2} = A_\epsilon^\alpha\rho + B_\epsilon^\alpha\rho^2 + C_\epsilon^\alpha\rho^3 + \ldots. \quad (10.96)$$

Because $\epsilon_\infty = n_\infty^2$, where n_∞ is the high-frequency refractive index, the virial coefficients $A_\epsilon^\alpha, B_\epsilon^\alpha, C_\epsilon^\alpha$, etc. are often called the first, second, third, etc. refractivity virial coefficients, and given the symbols A_r, B_r, C_r, etc.

Strictly speaking, the series (10.96) describes the dielectric response of a gas only when the applied field is spatially uniform on the scale of a typical intermolecular separation.[60] That is, the density of the gas and wavelength of the field must satisfy

$$\rho\lambda^3 \gg 1. \tag{10.97}$$

At an infrared wavelength of 1 μ and a pressure of 1 atm, the left-hand side of (10.97) exceeds 10^7. It seems that, in practice, nonlocal effects would be very hard to measure.

10.6.5 The effect of intermolecular forces

In the hypothetical case in which there are no molecular interactions, $\mathcal{A}_l(1\ldots l)=0$ for all $l>1$. All virial coefficients beyond the first, A_ϵ, then vanish, and the Clausius–Mossotti equation,

$$\frac{\epsilon-1}{\epsilon+2} = 4\pi\beta\rho\mu_1^2/9 + 4\pi\rho\alpha_1/3, \tag{10.98}$$

follows as a rigorous statistical-mechanical result.

In practice, of course, intermolecular forces are present in all real fluids. The Clausius–Mossotti equation then applies only at very low densities, and experiments at higher densities reveal the effects of higher terms in the virial series.

With the exception of A_ϵ, all of the dielectric virial coefficients are sensitive to the details of the intermolecular potential. In this section we discuss theoretical and experimental results for various models of atomic and molecular fluids. Our emphasis is primarily on B_ϵ, since there have been relatively few measurements[61–65] of C_ϵ, and almost nothing is known about the higher coefficients.[66]

Atomic gases

For atomic gases

$$B_\epsilon = B_r = (2\pi/3)\int d\mathbf{r}\,\exp[-\beta u(r)]\Delta\alpha(r), \tag{10.99}$$

where $u(r)$ is the pair potential. In the DID model for $\Delta\alpha(r)$, we have from (10.47)

$$\Delta\alpha^{\text{DID}}(r) = \frac{4\alpha_1^3 r^{-6}}{(1+\alpha_1 r^{-3})(1-2\alpha_1 r^{-3})}, \tag{10.100}$$

from which it is apparent that $B_\epsilon > 0$. Expanding the denominator in powers of $\alpha_1 r^{-3}$, and integrating term by term, we obtain[67]

$$B_\epsilon = (32\pi^2/9)\sum_{n\geq 0}\alpha_1^{n+3}(2^{n+1}+(-)^n)J_{3n+6}, \tag{10.101}$$

where $J_n = \int_0^\infty \mathrm{d}r\, r^{2-n} \exp[-\beta u(r)]$. This is an asymptotic series, valid for sufficiently small α_1; it is usually quite well approximated by the first term.

For a fluid of polarizable hard spheres of diameter σ, we can sum (10.101) to obtain a simple closed-form expression,[67]

$$B_\epsilon/\sigma^6 = (32\pi^2/27)\,\alpha^{*2} \ln\left(\frac{1+\alpha^*}{1-2\alpha^*}\right), \tag{10.102}$$

where $\alpha^* = \alpha_1/\sigma^3$. Equation (10.102) is valid for $\alpha^* < 0.5$; for larger values of α^*, the DID approximation yields a divergent result for $\Delta\alpha(r)$ for $r > \sigma$, and therefore for B_ϵ. Typical values of α^* for the rare gases are in the range 0.01–0.1.

The effects of quadrupole polarizability (C_1) have also been investigated.[68] For a fluid of hard spheres, we find to lowest order

$$B_\epsilon/\sigma^6 \approx (32\pi^2/9)\,\alpha^{*3}(1 + (3C^*/\alpha^*)), \tag{10.103}$$

where $C^* = C_1/\sigma^5$. For the rare gases $C^*/\alpha^* \approx 0.06$, suggesting a small correction of order 15–20%.

For the rare gases, the DID approximation gives values for B_ϵ that are larger than observed. For He and Ne, even the sign is wrong, since B_ϵ is actually negative.[62,64,69] To obtain agreement with experiment, we must include the dispersion and overlap contributions to $\Delta\alpha(r)$. The leading dispersion term[70–72] is comparable in magnitude to $\Delta\alpha^{\mathrm{DID}}(r)$. It is a function of the frequency-dependent polarizability $\alpha_1(i\omega)$ and tensor hyperpolarizability $\gamma_{1\,\alpha\beta\gamma\delta}(i\omega, 0, 0)$. It may be written in terms of static quantities as

$$\Delta\alpha^{\mathrm{dis}}(r) = \sigma_\alpha \frac{5\gamma_1 C_6}{9\alpha_1 r^6}, \tag{10.104}$$

where γ_1 is the second atomic hyperpolarizability defined in (C.6), and C_6 is the coefficient of the leading term in the dispersion energy. σ_α is a dimensionless number close to unity, and whose precise value depends on the relative degree of dispersion in the two polarizabilities. For He, $\sigma_\alpha = 1.19$.[71] Overlap contributions are also significant; these are negative, and operate only at short range, decaying roughly exponentially with r. Meinander[73] has incorporated overlap effects into a semiempirical model for $\Delta\alpha(r)$ that yields reasonable agreement with experiment for Ne, Ar, Kr, and Xe.

For He and Ar, very accurate *ab initio* calculations of $\Delta\alpha(r)$ that lead to B_ϵ of the correct sign and magnitude have recently appeared.[53–55,74]

The third dielectric virial coefficient has been calculated for a fluid of polarizable hard spheres. The result is[75]

$$C_\epsilon/\sigma^9 = -\frac{8\pi^3}{27}(7 - 3\ln 2)\alpha^{*3} - \frac{\pi^3}{135}(509 - 120\ln 2)\alpha^{*4} + (540.87 \pm 0.02)\alpha^{*5} + (420.77 \pm 0.03)\alpha^{*6} + \ldots. \tag{10.105}$$

The term of order α^{*3} is dominant. We can show that for a fluid of hard spheres carrying both dipole and quadrupole polarizability, there is no term in C_ϵ of order $\alpha_1^2 C_1$.[68] Any surviving corrections must therefore be of order $\alpha_1^2 C_1^2$, $\alpha_1^3 C_1$, or higher, and are expected to be negligible in practice.

Nonpolar molecular gases

Here B_ϵ^μ is nonzero due to collisional induction of a pair dipole moment; it is normally the dominant contribution to B_ϵ.[51] If one considers only quadrupolar induction, then to leading order in $\boldsymbol{\alpha}_1$ and $\mathbf{Q}_1$, and neglecting anisotropy in the intermolecular potential,[76]

$$B_\epsilon^\mu = (16\pi^2\beta/3)\langle Q^2\alpha^2\rangle J_8, \tag{10.106}$$

where $\langle Q^2\alpha^2\rangle = (2/9)[(\mathbf{Q_1} : \mathbf{Q_1})(\boldsymbol{\alpha}_1 : \boldsymbol{\alpha}_1) - (3/5)(\mathbf{Q_1} : \boldsymbol{\alpha}_1)^2]$; for linear or symmetric top molecules, this simplifies to $Q_1^2\left(\alpha_1^2 + (4/45)\gamma_1^2\right)$ (here γ_1 is the polarizability anisotropy, cf. Appendix C, Vol. 1). Higher order terms have also been computed, but are usually small.[77] The effect of molecular shape has also been considered.[77]

To calculate $B_\epsilon^\alpha = B_r$, one requires $\Delta\alpha(12)$. In view of the difficulty in calculating this quantity for atomic gases, it is not surprising that it has been computed *ab initio* only for H_2 and N_2.[78] The results have not yet been applied to B_r. Neglecting anisotropy in the potential, the DID approximation yields to lowest order[79]

$$B_\epsilon^\alpha = (32\pi^2\alpha_1/3)\left(\alpha_1^2 + 2/9\gamma_1^2\right)J_6. \tag{10.107}$$

For quasispherical molecules there is good agreement with experiment. Thus for C_2H_4 and SF_6, the theoretical result is about 10% too large, while for C_2H_6 it is about 15% too small.[64] For strongly prolate molecules such as CO_2, (10.107) greatly overestimates the experimental B_ϵ^α.[64]

Polar gases

Whereas for nonpolar gases B_ϵ is usually positive (with He and Ne the only exceptions), for polar gases B_ϵ can be of either sign. It is usually one or two orders of magnitude larger than for nonpolar gases; e.g. at 323 K we have[80] (in units of $\text{cm}^6\,\text{mol}^{-2}$) $B_\epsilon(CHF_3) = 1125$ and $B_\epsilon(CH_3F) = -1307$, whereas[61] $B_\epsilon(CO_2) = 50$ and $B_\epsilon(CH_4) = 7$. B_ϵ^μ is normally much larger than B_ϵ^α.[61]

For rigid polar fluids, $B_\epsilon^\alpha = 0$, and

$$B_\epsilon^\mu = (4\pi\beta/9)\int d\mathbf{r}\,\langle \exp[-\beta\mathcal{U}(12)]\,\boldsymbol{\mu}_1\cdot\boldsymbol{\mu}_2\rangle_{\omega_1\omega_2}. \tag{10.108}$$

If we assume that the only anisotropic component of $\mathcal{U}(12)$ is the dipole–dipole potential $-\boldsymbol{\mu}_1\cdot\mathbf{T}(\mathbf{r}_{12})\cdot\boldsymbol{\mu}_2$, then[51]

$$B_\epsilon = 72\left(4\pi\beta\mu_1^2/9\right)^2\sum_{n=1}^{\infty}\left(\beta\mu_1^2\right)^{2n}\left(\frac{2^n(n+1)!}{(2n+3)!}\right)^2 J_{6n+3}\sum_{r=0}^{n}(3r-n)\binom{2r}{r}. \tag{10.109}$$

All terms in this equation are positive, so that $B_\epsilon > 0$. For a fluid of dipolar hard spheres, (10.109) reduces to[81]

$$B_\epsilon/\sigma^6 = 12(4\pi\mu^{*2}/9)^2 \sum_{n=1}^{\infty} \mu^{*4n} \frac{1}{n} \left(\frac{2^n(n+1)!}{(2n+3)!}\right)^2 \sum_{r=0}^{n} (3r-n)\binom{2r}{r}, \quad (10.110)$$

where $\mu^{*2} = \beta\mu_1^2/\sigma^3$. This is a strongly increasing function of μ^*. B_ϵ has also been calculated for a dipolar sticky hard sphere fluid, and is found to increase with both μ^* and the stickiness of the fluid.[82]

The effect of higher multipoles on B_ϵ has also been considered.[83] For linear molecules one finds the following perturbational (high-temperature) expansion:

$$B_\epsilon = \beta^3 B_\epsilon^{(3)} + \beta^4 B_\epsilon^{(4)} + \ldots, \quad (10.111)$$

where

$$B_\epsilon^{(3)} = -(4\pi/9)^2\mu_1^2 \left(\frac{18}{5}\mu_1^2 Q_1^2 J_8 + \frac{72}{5}\mu_1 Q_1^2 \Omega_1 J_{10} + \frac{162}{7} Q_1^2 \Omega_1^2 J_{12} + \frac{72}{3}\mu_1 Q_1 \Omega_1 \Phi_1 J_{12} + \frac{792}{7} Q_1 \Omega_1^2 \Phi_1 J_{14} + \frac{572}{3} \Omega_1^2 \Phi_1^2 J_{16} + \ldots\right) \quad (10.112)$$

and

$$B_\epsilon^{(4)} = (4\pi/9)^2\mu_1^2 \left(\frac{6}{25}\mu_1^6 J_9 + \frac{72}{35}\mu_1^4 Q_1^2 J_{11} + \frac{234}{35}\mu_1^2 Q_1^4 J_{13} + \ldots\right), \quad (10.113)$$

where μ_1, Q_1, Ω_1, and Φ_1 are the molecular dipole, quadrupole, octopole, and hexadecapole moments, respectively. However, for many molecules at room temperature, equation (10.111), truncated at the term of order β^4, is a rather poor approximation to the true B_ϵ.[83]

B_ϵ has also been computed numerically for multipolar hard spheres.[83] These calculations show that the effect of higher moments on B_ϵ is significant: in particular, quadrupolar interactions can lead to large negative values of B_ϵ, especially for highly polar molecules (see Fig. 10.4). They also demonstrate that the computed B_ϵ will, typically, fail to converge as successive multipoles are added. This argues strongly against the validity of a single-centre multipole expansion of the molecular interaction energy in calculations of dielectric properties.

The dielectric response of an interacting pair of molecules is largely determined by the forces operating at close distances, typically a few molecular diameters. A one-centre multipole expansion of the potential, however, may converge only at much longer range, in regions lying outside a sphere containing all the molecular charge, as discussed in § 2.4. In the presence of significant molecular anisotropy, it may fail to yield an adequate representation of the pair interaction for some molecular orientations.

This suggests we would do better to seek an alternative, more rapidly convergent, mode of expansion. A procedure suggested by Stone may be more appropriate for this purpose. Stone's distributed multipole analysis (DMA)[84] is

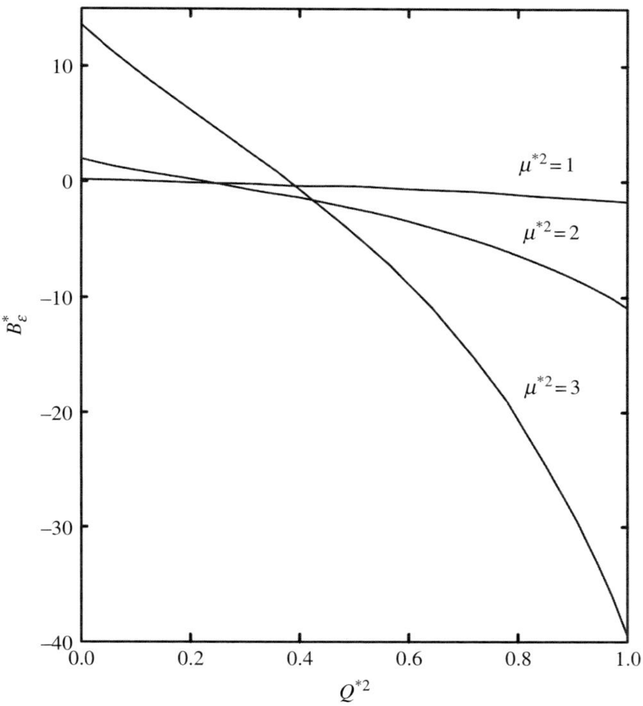

FIG. 10.4 The reduced second dielectric virial coefficient $B^*_\epsilon = B_\epsilon/\sigma^6$ for a fluid of dipolar-quadrupolar hard spheres as a function of the quadrupole strength $Q^{*2} = \beta Q_1^2/\sigma^5$, for different values of the dipole strength $\mu^{*2} = \beta\mu_1^2/\sigma^3$. The values of μ^{*2} and Q^{*2} shown are typical of many small polar molecules at room temperature. (From ref. 83.)

a generalization of Mulliken population analysis. Multipole moments Q_l, from the charge upwards, are first calculated from the overlap integrals of the atomic basis of the molecular wavefunction. These are then relocated to a limited number of sites within the molecule, usually the nuclei and bond centres. Such a procedure requires very little computational effort if the one-electron density matrix is available from an *ab initio* calculation, especially if a basis of Gaussian orbitals is employed. Most importantly, it yields an intermolecular potential that converges rapidly to the correct value for *all* accessible regions of space. Calculations for molecules such as HCl, HF, CO, HCN, H_2O, and CH_3F show that in DMA the effects of moments beyond Q_2 are often quite small. Converged estimates of B_ϵ have been obtained, which (with the inclusion of induction effects through an analogous model of distributed polarizabilities[85]) are in reasonable agreement with experiment.[86]

DMA makes the interesting prediction that for HCN, B_ϵ should change sign at 120 K because of the presence of two closely separated minima, one with a parallel, and one an antiparallel, alignment of dipoles.[86] As yet, however, this result has not been verified experimentally.

The effects of molecular shape have been considered by several authors. Perturbational expansions,[51,87] which yield B_ϵ as an infinite series in parameters associated with nonspherical components of the intermolecular potential, are normally too slowly convergent to prove useful in practice. Numerical studies have been performed for the LJ + δ_2 model[80] (cf. § 2.7), for the dipolar Gay-Berne model,[88] and for fluids of dipolar hard ellipsoids[89] and polar hard dumbbells.[90] Figure 10.5 shows $B_\epsilon^* = B_\epsilon/(8ab^2)^2$ for a fluid of dipolar hard ellipsoids as a function of dipole strength $\mu^{*2} = \beta\mu_1^2/8ab^2$ for various values of the aspect ratio b/a, where $2a$ is the length of each ellipsoid along its axis of revolution, and $2b$ is its transverse size. Oblate molecules, with $a < b$, have positive B_ϵ, while prolate molecules, with $a > b$, have negative B_ϵ (except when $a \approx b$ and μ^{*2} is large, i.e. when shape effects are relatively unimportant, and dipolar ordering takes over). These signs, which agree with what is observed experimentally,[80]

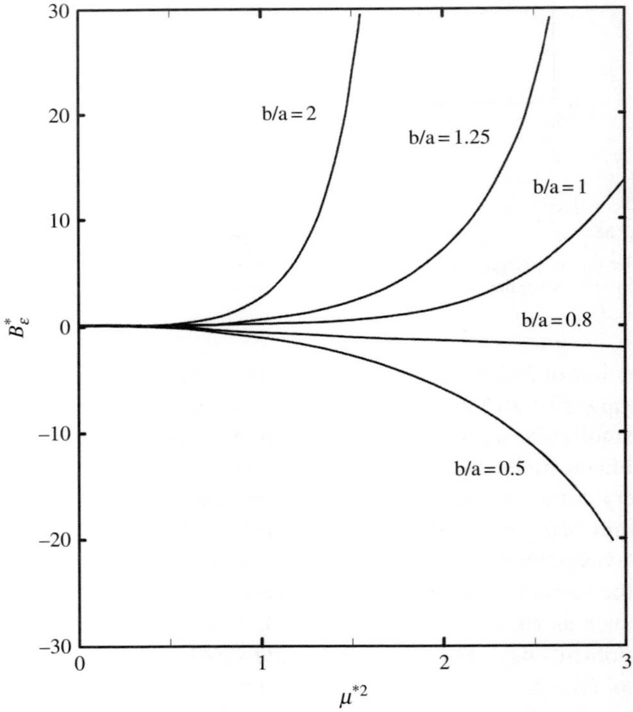

FIG. 10.5 The reduced second dielectric virial coefficient $B_\epsilon^* = B_\epsilon/(8ab^2)^2$ for a fluid of dipolar hard ellipsoids as a function of dipole strength $\mu^{*2} = \beta\mu_1^2/8ab^2$, for different aspect ratios b/a. (From ref. 89.)

can be rationalized by considering the sign of $\boldsymbol{\mu}_1 \cdot \boldsymbol{\mu}_2$ at the distance of closest molecular approach.

The influence of polarizability on the coefficients B_ϵ and B_r has also been studied.[24,51,67,79] For an isotropically polarizable fluid of dipolar molecules, neglecting higher multipoles and ignoring shape effects, a perturbational expansion in powers of α_1 yields[51]

$$B_\epsilon = B_\epsilon^{(0)} + \alpha_1 B_\epsilon^{(1)} + \alpha_1^2 B_\epsilon^{(2)} + \ldots, \tag{10.114}$$

where $B_\epsilon^{(0)}$ is given by Eq. (10.109), and

$$\begin{aligned} B_\epsilon^{(1)} &= (64\pi^2/27)\left(\beta^2\mu_1^4 J_6 + \frac{2}{5}\beta^3\mu_1^6 J_9 + \ldots\right), \\ B_\epsilon^{(2)} &= (32\pi^2/3)\,\beta\mu_1^2 J_6 + \ldots, \end{aligned} \tag{10.115}$$

and

$$\begin{aligned} B_r = (32\pi^2/3)\,\alpha_1^3 \Big(& J_6 + \frac{1}{3}\beta^2\mu_1^4 J_{12} + \frac{1}{25}\beta^4\mu_1^8 J_{18} \\ & + \frac{29}{11025}\beta^6\mu_1^{12} J_{24} + \ldots\Big) + \ldots. \end{aligned} \tag{10.116}$$

These expressions are reasonably accurate for small values of μ_1 and α_1, but break down[67] for larger, more realistic, values of these parameters. B_ϵ and B_r have been computed numerically[67] for a fluid of polarizable dipolar hard spheres. Figures 10.6 and 10.7 show the variation of $B_\epsilon^* = B_\epsilon/\sigma^6$ and $B_r^* = B_r/\sigma^6$ with $\alpha^* = \alpha_1/\sigma^3$ for various values of $\mu^{*2} = \beta\mu_1^2/\sigma^3$. It is apparent that both B_ϵ and B_r are strongly increasing functions of α_1, and that $B_\epsilon \approx B_\epsilon^\mu \gg B_\epsilon^\alpha \equiv B_r$.

The influence of anisotropy in the polarizability can be large, especially for positive γ_1. Compared to the isotropic case, at fixed μ_1 and α_1, B_ϵ is reduced when $\gamma_1 < 0$ and increased when $\gamma_1 > 0$.[67] These facts may be understood in terms of an increased tendency for the dipoles of an interacting pair of molecules to align parallel to one another when $\alpha_\parallel$ is large. For strongly polar molecules, for large positive γ_1 B_r may be increased over its isotropic value by an order of magnitude, while for large negative γ_1 B_r may actually become negative.[67]

Accurate experimental data on B_ϵ are available only for a comparatively small number of compounds.[61] Table 10.1 shows a comparison of the numerical calculations of Joslin and Goldman[67] with the data of Sutter and Cole[65] for CHF_3. The calculations assume the molecules are anisotropically polarizable dipolar hard spheres with $\mu_1 = 1.65$ D, $\alpha_1 = 3.6 \times 10^{-24}$ cm^3, $\gamma_1 = -0.22 \times 10^{-24}$ cm^3, and $\sigma = 4.0$ Å. Agreement is very satisfactory, perhaps surprisingly so in view of the relative crudeness of the molecular model, which ignores quadrupolar and shape effects. The table also shows B_ϵ calculated ignoring polarizability; this is clearly a very poor approximation.

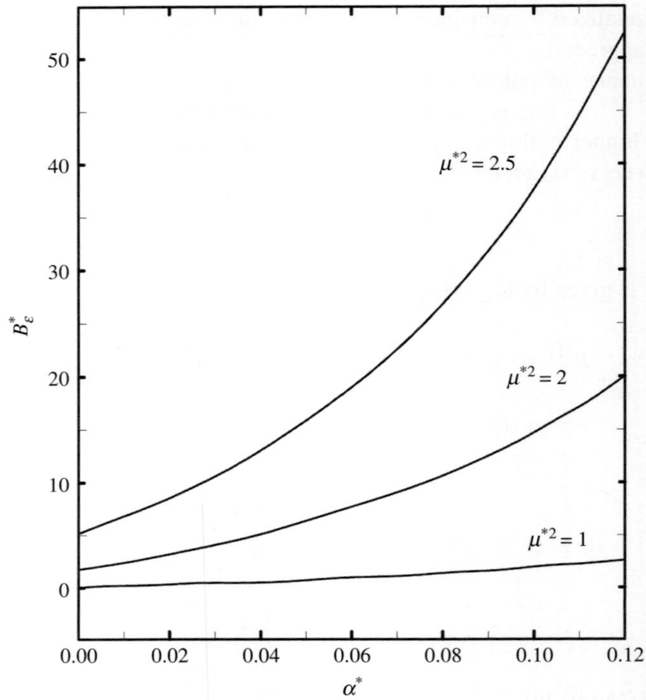

FIG. 10.6 The reduced second dielectric virial coefficient $B^*_\epsilon = B_\epsilon/\sigma^6$ for a fluid of isotropically polarizable dipolar hard spheres as a function of $\alpha^* = \alpha_1/\sigma^3$ for different values of the dipole strength $\mu^{*2} = \beta\mu_1^2/\sigma^3$. The values of μ^{*2} and α^* shown are typical of many small polar molecules at room temperature. (From ref. 67.)

In cases for which the effects of both molecular shape and polarizability are important, the distributed polarizability analysis (DPA) of Stone and co-workers[85] can be employed. For example, DPA yields a good description of the temperature dependence of B_ϵ for HCl.[86]

The third virial coefficient C_ϵ has been calculated for a fluid of dipolar hard spheres, both analytically[52,91,92] and numerically.[92] A perturbational expansion in powers of $\mu^{*2} = \beta\mu_1^2/\sigma^3$ yields[92]

$$C_\epsilon/\sigma^9 = (4\pi/9)^3 \Big(-0.9375\mu^{*6} - 0.096389\mu^{*8} + 0.00017068\mu^{*10}$$
$$-0.016263\mu^{*12} - 0.00100\mu^{*14} - 0.00059\mu^{*16}$$
$$-0.00025\mu^{*18} + 0.00010\mu^{*20} + \dots\Big), \tag{10.117}$$

which is accurate for $\mu^{*2} < 3$. For $\mu^{*2} < 3.7$, the numerical calculations show that $C_\epsilon < 0$, while for larger μ^{*2}, it is positive.[92] Sutter and Cole[65] have measured

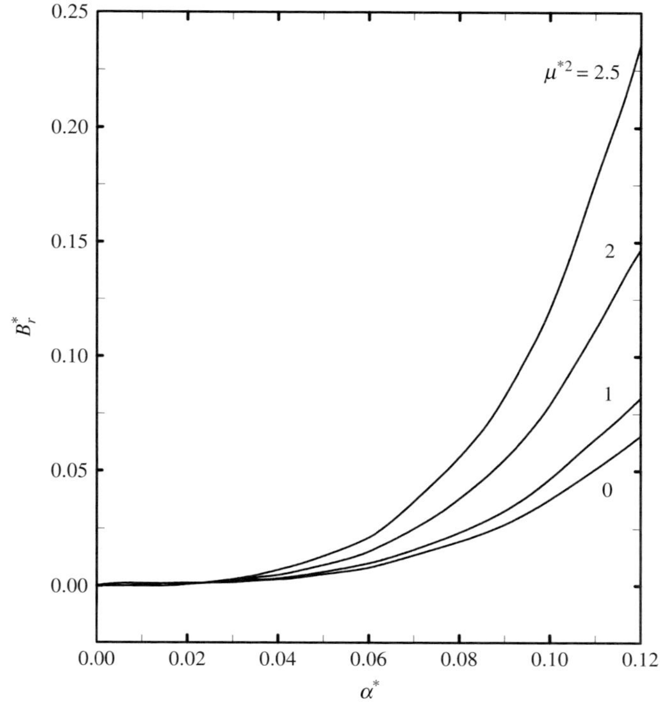

FIG. 10.7 The reduced second refractivity virial coefficient $B_r^* = B_r/\sigma^6$ for a fluid of isotropically polarizable dipolar hard spheres as a function of $\alpha^* = \alpha_1/\sigma^3$ for different values of the dipole strength $\mu^{*2} = \beta\mu_1^2/\sigma^3$. (From ref. 67.)

Table 10.1

Comparison of calculated B_ϵ (units of $cm^6\ mol^{-2}$) with experiment for CHF_3 at various temperatures

$T(K)$	$B_\epsilon^{calc}(\alpha \to 0)$	B_ϵ^{calc}	B_ϵ^{exp}
303	135	1330	1315 ± 100
323	105	1140	1125 ± 50
370	60	840	905 ± 20
417	35	645	705 ± 10

The theoretical model assumes the molecule is an anisotropically polarizable dipolar hard sphere. The table also shows the value of B_ϵ computed when the effects of polarizability are ignored. From ref. 67.

C_ϵ for CHF_3 at various temperatures. Compared to their experimental values, results calculated from (10.117) are about a factor of 2 too small. The calculations can be brought into agreement with experiment if the molecular dipole moment is increased by 10–15%, reflecting, presumably, the influence of polarizability.[92]

No studies have yet appeared dealing with the effects on C_ϵ of higher multipole moments, shape, or polarizability.

10.6.6 Quantum corrections to the virial coefficients

Quantum corrections to the dielectric constant are of two types. Symmetry, or exchange, effects arise from the bosonic or fermionic character of the particles of the fluid. However, these only come into play when the interparticle separation approaches the de Broglie wavelength, so that in real fluids they are almost entirely suppressed by the repulsive molecular cores. Much larger are the dynamic, or diffraction, effects, which are associated with the quantization of molecular translational and rotational motion. Provided the temperature is not too low, these can be calculated using the Wigner–Kirkwood expansion of the configurational distribution function[93–95] for a fluid subjected to an external electric field. Assuming the intermolecular potential is nowhere infinitely steep, the virial coefficients can then be expanded as

$$A_\epsilon = A_\epsilon^{(0)} + \hbar^2 A_\epsilon^{(2)} + \hbar^4 A_\epsilon^{(4)} + \ldots,$$

$$B_\epsilon = B_\epsilon^{(0)} + \hbar^2 B_\epsilon^{(2)} + \hbar^4 B_\epsilon^{(4)} + \ldots,$$

$$C_\epsilon = C_\epsilon^{(0)} + \hbar^2 C_\epsilon^{(2)} + \hbar^4 C_\epsilon^{(4)} + \ldots, \tag{10.118}$$

etc. The leading terms on the right of (10.118) are the classical expressions discussed in the previous section, while the terms of order $\hbar^2$, $\hbar^4$, etc., represent quantum corrections of successively higher order. For potentials with hard molecular cores, or other discontinuities, there will also be terms in odd powers of $\hbar$.[96]

Atomic gases

For atomic gases, for which only translational degrees of freedom exist, $A_\epsilon = A_\epsilon^{(0)}$, because the translation of a free particle is not quantized (more precisely, in a system of volume V one would expect corrections of order λ^3/V, where λ is the de Broglie wavelength).

Quantum corrections enter at the level of the second virial coefficient, because the repulsive overlap of two interacting atoms serves to confine the relative translational motion. Moszynski *et al.*[53] have calculated the corrections of order $\hbar^2$ and $\hbar^4$. These are given by

$$\hbar^2 B_\epsilon^{(2)} = -(\lambda^2/36) \int d\mathbf{r} \exp(-v) \left(v'^2 \Delta\alpha - 2v' \Delta\alpha'\right),$$

$$\hbar^4 B_\epsilon^{(4)} = -(\lambda^4/720\pi) \int d\mathbf{r} \exp(-v) \left(a_0 \Delta\alpha + a_1 \Delta\alpha' + a_2 \Delta\alpha''\right), \tag{10.119}$$

where $\lambda = (2\pi\beta\hbar^2/m)^{1/2}$ (with m the atomic mass), $v = \beta u(r)$, $\Delta\alpha$ is the pair polarizability, primes denote differentiation with respect to r, and

$$\begin{aligned} a_0 &= v''^2 + v'^2/r^2 + 10v'^3/9r - 5v'^4/36, \\ a_1 &= -4v'/r^2 - 10v'^2/3r + 5v'^3/9, \\ a_2 &= -2v''. \end{aligned} \tag{10.120}$$

Such corrections are potentially important at low temperatures. Even for He, however, they are negligible above 75 K. For He the semiclassical expansion begins to diverge below 40 K, although it can be made to agree with exact quantum calculations quite closely down to 10 K by forming a Padé approximant.[54] For Ar, quantum effects are unimportant in the temperature range for which experimental data are available.[55]

Linear molecules

The leading corrections to the first and second virial coefficients of polar linear molecules have been calculated by MacRury and Steele.[56] For A_ϵ one finds

$$\hbar^2 A_\epsilon^{(2)}/A_\epsilon^{\mu(0)} = -\beta\hbar^2/6I = -\theta_r/3T, \tag{10.121}$$

where θ_r is the rotational temperature, which is listed for several molecules in Table 1.1. This correction, which arises from quantization of the molecular rotation, is usually quite small, amounting to less than 2% for HCl at room temperature.

For B_ϵ, there are both translational and rotational quantum corrections. For molecules with small moments of inertia, the latter invariably dominate, often comprising 90% of the total. MacRury and Steele estimate that for such molecules the corrections may reach 10–20% at room temperature, about twice what is expected for the second pressure virial coefficient.

Nonlinear molecules

For nonlinear molecules, the leading correction to the first virial coefficient is given by

$$\hbar^2 A_\epsilon/A_\epsilon^{\mu(0)} = -\frac{\beta\hbar^2}{12\mu_1^2}\left(\frac{\mu_y^2+\mu_z^2}{I_{xx}} + \frac{\mu_z^2+\mu_x^2}{I_{yy}} + \frac{\mu_x^2+\mu_y^2}{I_{zz}}\right), \tag{10.122}$$

where I_{xx}, I_{yy}, and I_{zz} are the principal components of the inertia tensor $\mathbf{I}$. A simple proof of this relation, based on the relationship that exists between the dielectric constant and infrared absorption, is given in Appendix 11A.

10.7 Computer simulation

Over the course of the past thirty years considerable progress has been made in computer simulation of the dielectric constant.[97] Classical simulations of ϵ

have been reported for rigid polar,[98–134] polarizable atomic,[45,135] and polar-polarizable[136–151] fluids. Quantum simulations have also been performed for water.[152,153]

There are two basic approaches to the simulation of ϵ. In one method, we apply an external electric field to the sample, and measure the resulting polarization. The dielectric constant is then computed from the linear response coefficient.[5,119,125,127] In the other method, ϵ is calculated from equilibrium fluctuations in the sample dipole moment. The relative merits of the two methods have been debated.[5,119,127] In either case, the precise relationships employed depend on the geometry of the sample, as discussed in §§ 10.3 and 10.4.

Computer simulations are performed on a system containing a finite number, N, of molecules. Usually N lies in the range 100–1000, although continuing increases in computer power have recently allowed the study of much larger systems, with N up to 100,000. Many of the molecules lie close to the boundaries of the simulation cell, making surface effects very important. To eliminate these effects, we may surround the cell with an infinitely replicating lattice of identical cells; in this way every molecule can be made to behave as if it were in the bulk of the fluid. However, the additional intermolecular interactions thus introduced must also be accounted for. This can be accomplished by introducing a cutoff distance r_c beyond which the pair potential vanishes. This method works very well for systems with short-range forces, but presents some problems when the potential is long ranged, as in a dipolar fluid. These problems become particularly acute when one tries to calculate ϵ, because truncation of the potential suppresses the dipolar fluctuations. Irrespective of the shape of the sample, for a rigid polar fluid the fluctuation formula for ϵ becomes[101]

$$\frac{\epsilon - 1}{\epsilon + 2} = (4\pi\beta/9V)\langle\mu^2\rangle_{\text{cell}}, \tag{10.123}$$

which is reminiscent of the result for a dielectric sphere *in vacuo*. Accurate determination of ϵ is then no longer possible, except at low dipole strengths, since the right-hand side of (10.123) saturates, attaining the limiting value of unity.

10.7.1 Reaction fields

As a way around this difficulty, Barker and Watts[154] suggested that instead of entirely ignoring interactions beyond r_c, one could include them approximately by treating this portion of the fluid as a dielectric continuum of dielectric constant ϵ. This is the basis of the *reaction field* (RF) method.[118,145] In its original form, the method is somewhat inconvenient because it requires some prior knowledge of ϵ, necessitating an iterative solution.[155] A later refinement by Neumann[101] avoids this complication. Neumann considered the effect of ascribing some *arbitrary* dielectric constant to the continuum, and showed that this effectively incorporates an additional term into the pair potential given by

$$u(r\omega_1\omega_2)^{RF} = -\frac{2(\epsilon' - 1)}{2\epsilon' + 1}\frac{\boldsymbol{\mu}_1 \cdot \boldsymbol{\mu}_2}{r_c^3}, \qquad r < r_c, \tag{10.124}$$

and that in the limit $r_c \to \infty$ the appropriate fluctuation formula is then

$$\frac{(\epsilon - 1)(2\epsilon' + 1)}{3(\epsilon + 2\epsilon')} = (4\pi\beta/9V)\langle\mu^2\rangle_{\epsilon'}, \tag{10.125}$$

where $\langle\mu^2\rangle_{\epsilon'}$ is the mean square dipole moment of the central simulation cell. This result is reminiscent of Eq. (10.30). For $\epsilon' > \epsilon$ the right-hand side of (10.125) is $O(\epsilon)$, allowing an accurate estimate of the dielectric constant; in practice, one often chooses the so-called 'tin-foil' boundary condition $\epsilon' = \infty$.

A proof of (10.125) can be given using perturbation theory. In the limit $r_c \to \infty$, we may regard the potential of (10.124) as a small perturbation of order $O\left(V_c^{-1}\right)$ on the true pair potential $u(\mathbf{r}\omega_1\omega_2)$, where $V_c = 4\pi r_c^3/3$ is the volume of the truncation sphere. If we calculate the resulting change in the Δ component of the pair correlation function $h(\mathbf{r}\omega_1\omega_2)$, we find to order $O\left(V_c^{-1}\right)$[14] that

$$\Delta h_\Delta(r) = h_\Delta(r, \epsilon') - h_\Delta(r, 1) = \frac{2}{y\rho V_c}\frac{(\epsilon - 1)^2(\epsilon' - 1)}{(\epsilon + 2)(\epsilon + 2\epsilon')}g_S(r), \tag{10.126}$$

where $g_S(r)$ is the isotropic component of the pair correlation function of the unperturbed system. Then

$$\begin{aligned}
(4\pi\beta/9V)\langle\mu^2\rangle_{\epsilon'} &= y\left(1 + \frac{1}{3}\rho\int d\mathbf{r}\, h_\Delta(r, \epsilon')\right) \\
&= y\left(1 + \frac{1}{3}\rho\int d\mathbf{r}\, h_\Delta(r, 1)\right) + \frac{1}{3}y\rho\int d\mathbf{r}\, \Delta h_\Delta(r) \\
&= \frac{\epsilon - 1}{\epsilon + 2} + \frac{2(\epsilon - 1)^2(\epsilon' - 1)}{3(\epsilon + 2)(\epsilon + 2\epsilon')} \\
&= \frac{(\epsilon - 1)(2\epsilon' + 1)}{3(\epsilon + 2\epsilon')},
\end{aligned} \tag{10.127}$$

using (10.123) and the fact that $\lim_{r_c\to\infty} V_c^{-1}\int_{r<r_c} d\mathbf{r}\, g_S(r) = 1$.

The value chosen for r_c is clearly arbitrary, provided only that second-order corrections to $\Delta h_\Delta(r)$ are negligible. This is generally satisfied if we choose $r_c = L/2$, where L is the size of the central simulation cell. Each particle then interacts only with the nearest image of every other particle in the periodic array, the so-called 'minimum-image' (MI) convention.

There are subtleties in applying RF corrections in systems where the molecules have separated charges, such as the SPC and TIP4P models of water; these are discussed by Hünenberger and van Gunsteren.[156]

10.7.2 Ewald sums

An alternative solution sets $r_c = R$, where R is the radius of a pseudospherical lattice[157] of cells. For $R \to \infty$ the resulting dipolar energy can be summed using a variant of the Ewald sum (EW) procedure used to compute the lattice energy of an ionic crystal.[98,158–160] Suppose the central simulation cell is a cube of side L, containing dipoles $\boldsymbol{\mu}_i$ at points $\mathbf{r}_i$, $1 \le i \le N$, and define $\mathbf{r}_{ij} = \mathbf{r}_i - \mathbf{r}_j$. Let $\mathbf{n}$ be a lattice vector specifying the position of an image cell relative to the central cell; then $\mathbf{r}_{ij}^{(n)} = \mathbf{r}_{ij} + \mathbf{n}L$ measures the distance between dipoles in different cells. In Appendix 10D, we show that the dipole–dipole contribution to the energy of the system is

$$\mathcal{U}_{DD} = \mathcal{U}_A + \mathcal{U}_B + \sum_{i<j} \frac{4\pi \boldsymbol{\mu}_i \cdot \boldsymbol{\mu}_j}{3L^3}, \tag{10.128}$$

where $\mathcal{U}_A$ and $\mathcal{U}_B$ are given by

$$\begin{aligned}\mathcal{U}_A = \sum_{\mathbf{n}} \sum_{i<j} \Bigg\{ & \left(\frac{\boldsymbol{\mu}_i \cdot \boldsymbol{\mu}_j}{r_{ij}^{(n)3}} - \frac{3\boldsymbol{\mu}_i \cdot \mathbf{r}_{ij}^{(n)} \mathbf{r}_{ij}^{(n)} \cdot \boldsymbol{\mu}_j}{r_{ij}^{(n)5}} \right) \\ & \times \left(\operatorname{erfc}\left[\alpha r_{ij}^{(n)}/L\right] + \frac{2\alpha r_{ij}^{(n)}}{\sqrt{\pi} L} \exp\left[-\alpha^2 r_{ij}^{(n)2}/L^2\right] \right) \\ & - \boldsymbol{\mu}_i \cdot \mathbf{r}_{ij}^{(n)} \mathbf{r}_{ij}^{(n)} \cdot \boldsymbol{\mu}_j \frac{4\alpha^3}{\sqrt{\pi} r_{ij}^{(n)2} L^3} \exp\left[-\alpha^2 r_{ij}^{(n)2}/L^2\right] \Bigg\}\end{aligned} \tag{10.129}$$

and

$$\begin{aligned}\mathcal{U}_B = \frac{2\pi}{L^3} \sum_{\mathbf{n} \ne 0} & \left\{ \left| \sum_i \frac{\boldsymbol{\mu}_i \cdot \mathbf{n}}{n} \exp(2\pi i \mathbf{n} \cdot \mathbf{r}_i / L) \right|^2 - \sum_i \frac{(\boldsymbol{\mu}_i \cdot \mathbf{n})^2}{n^2} \right\} \\ & \times \exp\left[-\pi^2 n^2/\alpha^2\right].\end{aligned} \tag{10.130}$$

Terms with $i = j$ have been omitted from $\mathcal{U}_{DD}$, since the self-energy of a dipole with its periodic array of images vanishes.[161] α is an arbitrary dimensionless parameter that does not affect the value of $\mathcal{U}_{DD}$, and is chosen to speed convergence. $\operatorname{erfc}(x) = 2\pi^{-1/2} \int_x^\infty dt \exp(-t^2)$ is the complementary error function. Since both (10.129) and (10.130) are absolutely convergent, the lattice vector sums can be taken in any order.

The Ewald potential $\mathcal{U}_{DD}$ is anisotropic, and there is therefore a tendency for the dipoles to align preferentially along one of the axes of the simulation cell. In an extreme case, this could result in a ferroelectric configuration of the system. Smith and Pettitt[162] have investigated the degree to which such anisotropy can be expected to perturb the fluid structure; they show that such effects are negligible for simulations of hundreds of particles at ambient temperatures and liquid densities, provided $\epsilon > 10$.

Traditionally, the sum in $\mathcal{U}_A$ has been restricted to the central simulation cell $\mathbf{n} = 0$. If the resulting error in evaluating $\mathcal{U}_A$ is to be $O(\delta)$, we must choose[163] $\alpha \approx 2(-\ln \delta)^{1/2}$. To limit the error[164] in the sum $\mathcal{U}_B$ to $O(\delta)$ we must then sum to $n_{\max} \approx -(2/\pi)\ln \delta$. Typically $\delta \approx 10^{-4}$, implying $\alpha \approx 6$ and $n_{\max} = 6$. This results in a total of $O(N^2)$ and $O(N[-\ln \delta]^{3/2})$ operations to calculate $\mathcal{U}_A$ and $\mathcal{U}_B$, respectively. It is evident this procedure is not optimal, since for large N a disproportionate amount of time is spent evaluating $\mathcal{U}_A$. An improved algorithm in which equal amounts of time are spent evaluating both sums has been suggested by Perram *et al.*[163] This procedure requires $O(N^{3/2})$ operations, giving some increase in computational efficiency. Duan and Krasny[165] have shown that it is in fact possible to achieve $O(N \ln N)$ performance, although the benefits are only realized for rather large N.

More recently even faster algorithms have been suggested.[166] These include the particle mesh Ewald method (PME),[167] which relies on computationally efficient fast Fourier transforms, and requires only $O(N \ln N)$ operations, and the fast multipole method (FM),[168] which requires only $O(N)$. Petersen[169] has compared the relative efficiency of Ewald, PME, and FM approaches, and has concluded that the PME method is significantly faster than either of the others in the important decade $10^4 < N < 10^5$. Below $N = 10^4$, the Ewald method is preferred.

If Eqs. (10.128)–(10.130) are used directly, then ϵ is again given by (10.123). In practice, a reaction-field term is always added to the dipolar energy $\mathcal{U}_{DD}$. This is given from (10.124) as

$$\begin{aligned}\mathcal{U}_{EW}^{RF} &= \sum_{\mathbf{n}} \sum_{i<j} u(r_{ij}\omega_1\omega_j)^{RF}, \qquad r_{ij} < R \\ &= -N_{cells} \sum_{i<j} \frac{2(\epsilon' - 1)}{2\epsilon' + 1} \frac{\boldsymbol{\mu}_i \cdot \boldsymbol{\mu}_j}{R^3} \\ &= -\sum_{i<j} \frac{8\pi(\epsilon' - 1)}{3(2\epsilon' + 1)} \frac{\boldsymbol{\mu}_i \cdot \boldsymbol{\mu}_j}{L^3},\end{aligned} \tag{10.131}$$

where $4\pi R^3/3 = N_{\text{cells}} L^3$ is the volume of the entire system of N_{cells} copies of the central simulation cell. As discussed earlier, it then follows that the correct fluctuation formula is (10.125). Most applications of the EW/RF geometry have taken $\epsilon' = \infty$; note that $\mathcal{U}_{EW}^{RF}$ then exactly cancels the last term in (10.128).

10.7.3 *Comparison of minimum image and Ewald sum simulations*

It was once believed that simulations using the minimum image and Ewald sum geometries gave different values for the dielectric constant. It is now recognized that these discrepancies are not real, and result from incorrect applications of the theory. For example, some early MI studies[170] substituted the mean square moment of the RF truncation sphere in place of that of the entire simulation cell, as required in (10.125). Early EW calculations used the wrong fluctuation formula,[171,172] or inappropriate values[173,174] for α. Disagreement between MI simulations by Patey *et al.*[105] and EW studies of the same fluid by de Leeuw *et al.*[98,106] was shown to result from an error in the latter work.[175]

Although many authors appear to regard the two methods as essentially distinct,[176] both are, in fact, special cases of a more general simulation algorithm in which the cutoff radius r_c assumes an essentially arbitrary value, intermediate between $L/2$ (its value in MI/RF geometry) and $R \to \infty$ (its value in EW/RF). Clearly, the EW method sums more terms in the dipolar energy; but as shown by Gray *et al.*,[104] MI and EW simulations yield essentially identical results for systems of reasonable size, and it is now accepted that they predict the same value of ϵ in the thermodynamic limit.[103,158] In practice, both methods are widely used, and which one is preferred is largely a matter of personal taste.

10.7.4 Size dependence of ϵ

Gray *et al.*[104] have studied the N-dependence of ϵ for a Stockmayer (dipolar Lennard-Jones) fluid in both MI and EW geometries. They found that for $\mu^{*2} = \mu_1^2/\epsilon\sigma^3 = 1$, $T^* = kT/\epsilon = 1$, and $\rho^* = \sigma^3 = 0.8$, and using tin-foil boundary conditions, both sets of results have converged to the infinite-system limit, $\epsilon = 8.6 \pm 0.1$, for $N = 108$. More strongly polar systems require a larger N,[107] but even when $\mu^{*2} = 3$, $T^* = 1.15$, and $\rho^* = 0.822$, ϵ has attained its limiting value of 67 for $N = 256$.[103] There is some evidence[107] that the size dependence of ϵ may be reduced by taking $\epsilon' \approx \epsilon$.

For an infinite system, the probability distribution function for the dimensionless quantity $\mathcal{G} = \boldsymbol{\mu}^2/N\mu_1^2$, where $\boldsymbol{\mu}$ is the system dipole moment, should obey

$$P(\mathcal{G}) = \frac{2\kappa^{3/2}}{\pi^{1/2}}\mathcal{G}^{1/2}\exp(-\kappa\mathcal{G}), \qquad (10.132)$$

where $\kappa^{-1} = \frac{3}{2}g_1$, and $g_1 = \langle\mathcal{G}\rangle$ is the Kirkwood g-factor. Kusalik *et al.*[131] have examined fluctuations in various simulation geometries, and suggest that deviations from (10.132) may be used to determine the approach to the infinite-system limit. They have also argued that improved estimates of g_1, and therefore of ϵ, may be achieved by fitting simulation data to an equation of this form; however, recent studies by Heinz *et al.*[133] do not support this conclusion.

10.7.5 Simulations on a hypersphere

Alternative methods of eliminating surface effects have been devised. Simulations on the surface of a four-dimensional hypersphere[114–117] appear to be especially useful for inhomogeneous systems, where application of the MI or EW methods is made more difficult by the low symmetry. Such methods appear to be competitive with more traditional approaches.[116]

10.7.6 Results and comparisons with integral equations

Computer simulations, performed properly, generate essentially exact results for a given potential model. They therefore provide a benchmark against which approximate theories can be judged, and show under what conditions, if any, these apply. Section 10.5 contains a comparison of perturbational estimates of ϵ for polar and polarizable fluids with simulation results. Here we discuss the predictions of various integral equation theories for ϵ, including the mean spherical approximation (MSA), and the hypernetted chain (HNC) and its derivatives, the reference (RHNC), linearized (LHNC), and quadratic (QHNC) hypernetted chain aproximations. The results are compared with Monte Carlo (MC) and molecular dynamics (MD) simulations over a range of state conditions, and for several different potential models, incorporating dipoles and higher multipoles, shape effects, and polarizability. Because of its importance, and the attention it has received, we also consider results for one real fluid, water.

Chapter 5 discusses integral equation methods in general. These are reviewed, especially as they apply to calculations of the dielectric constant, in Appendix 10E. This appendix also discusses integral equation theories for interaction-site models of polar fluids, in both their original and diagrammatically proper[177] versions, and explains why these have not, as yet, yielded useful results.

Dipolar hard spheres

For this fluid the MSA has been solved analytically[178] (see § 5.4.2, Vol. 1). The dielectric constant is given by (10E.3) and (10E.4), and is a function of $y = 4\pi\rho^*\mu^{*2}/9$ only, where $\rho^* = \rho\sigma^3$ is the reduced density, and $\mu^{*2} = \beta\mu_1^2/\sigma^3$ is the reduced dipolar strength. For the LHNC,[179] QHNC,[179] RHNC,[180] and HNC[181] approximations, numerical results are available. In these theories, ϵ depends separately on ρ^* and μ^{*2}; indeed it has been shown[179,182] that the MSA prediction for ϵ is the limit of the LHNC result as $\rho \to 0$ for fixed y.

Figure 10.8 shows the predictions of the MSA,[178] LHNC,[179] QHNC,[179] and RHNC[180] equations for $\rho^* = 0.8$ and for various dipole strengths. These are compared with the simulation results of Caillol and Levesque.[116] The MSA is reasonably accurate for small μ^{*2}, but underestimates ϵ for $\mu^{*2} > 1$. LHNC and QHNC both badly overestimate ϵ, with QHNC, if anything, performing worse (a possible explanation is given in Appendix 10E). Overall the RHNC performs best, and appears to give a good description of ϵ for $\mu^{*2} < 2$. For larger μ^{*2}, RHNC overestimates ϵ, and is in error by about 50% at $\mu^{*2} = 2.75$.

Lado *et al.*[183] have investigated an optimized RHNC closure, in which the reference fluid is composed of hard spheres whose diameter is chosen so as to minimize a free-energy functional. While thermodynamic quantities such as the pressure and compressibility were significantly improved, the dielectric constant was found to be hardly affected.

Soft central potentials

Figure 10.9 compares the dielectric constant for dipolar hard spheres,[180] dipolar soft spheres,[180] and the dipolar Lennard-Jones (Stockmayer) fluid,[184] calculated in the RHNC approximation at $\rho^* = 0.8$ and $T^* = kT/\epsilon = 1.35$, for various

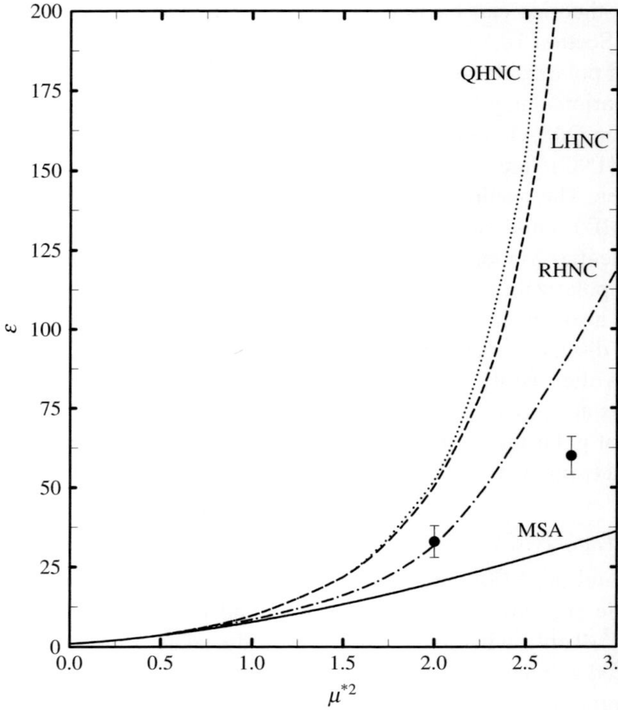

FIG. 10.8 Comparison of theory (lines) and simulation (points) for the dielectric constant ϵ of a fluid of dipolar hard spheres for $\rho^* = 0.8$ at various values of μ^{*2}.

values of $\mu^{*2} = \beta\mu_1^2/\sigma^3$. The soft-sphere potential is the repulsive part of the LJ interaction. For $\mu^{*2} < 2$, all three fluids are predicted to have very similar dielectric constants; this is confirmed by computer simulation.[108,116] Thus, for weakly or moderately polar fluids, the form of the central potential has little effect on ϵ. Similarly, Henderson *et al.*[134] have shown that the addition of a Yukawa tail $\propto \exp(-\lambda r)/r$ to a dipolar hard sphere fluid does not change the MSA prediction for ϵ, even though thermodynamic properties are strongly affected. This result is also confirmed by simulation.[134]

For the Stockmayer fluid, simulations show that the RHNC approximation yields a good description of both ϵ and the harmonics $h_S(r)$, $h_\Delta(r)$, and $h_D(r)$, while the LHNC and QHNC approximations are significantly less accurate.[184]

Quadrupolar forces[185]

Quadrupole–dipole and quadrupole–quadrupole interactions favour T-shaped orientations of molecular pairs; this leads to a reduction in $G_1 \sim \langle \cos\gamma_{12} \rangle$, and therefore in ϵ. Figure 10.10 shows the effect of adding an axial quadrupole moment to a fluid of dipolar hard spheres, as calculated in the MSA and LHNC.[186]

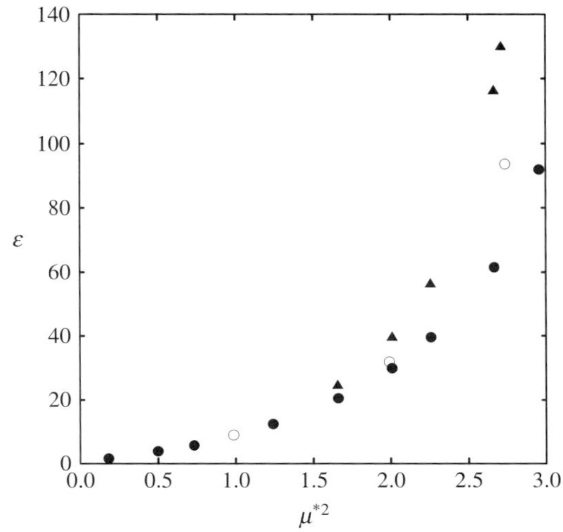

FIG. 10.9 RHNC approximation for ϵ for various polar fluids at $\rho^* = 0.8$ and $T^* = 1.35$. Open circles—dipolar hard spheres; triangles—dipolar soft spheres; filled circles—Stockmayer fluid.

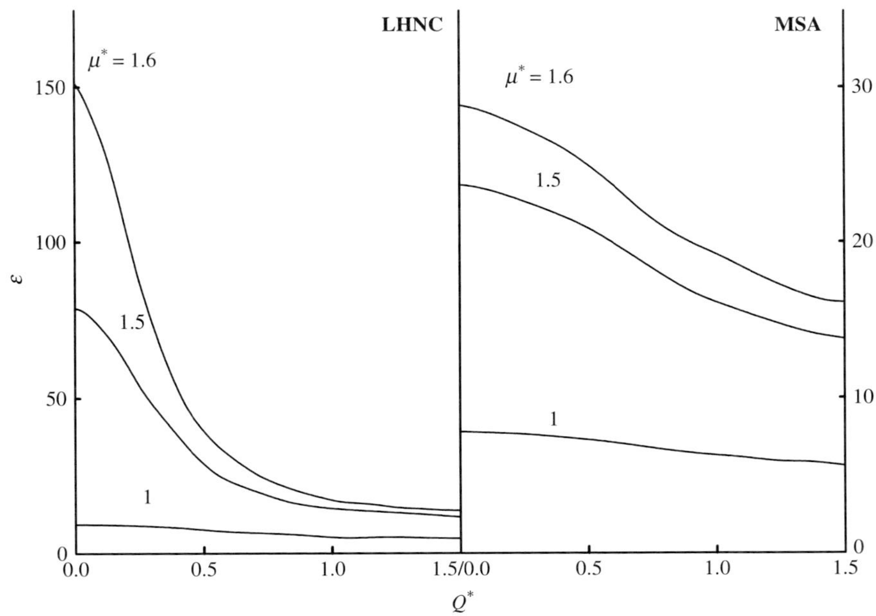

FIG. 10.10 Dipolar–quadrupolar hard sphere fluid. The reduction of ϵ with increasing quadrupole moment according to the MSA and LHNC at $\rho^* = 0.8$ for several values of μ^*. Reduced moments are defined as $\mu^* = (\beta\mu_1^2/\sigma^3)^{1/2}$, $Q^* = (\beta Q_1^2/\sigma^5)^{1/2}$. (From ref. 186.)

Comparison with simulation[186,187] shows that for dipolar–quadrupolar hard spheres, the MSA, LHNC, QHNC, and RHNC equations appear to give reasonable values for ϵ, and its trend with $Q^* = \left(\beta Q_1^2/\sigma^5\right)^{1/2}$ (although for some combinations of μ^* and Q^*, the QHNC approximation has no solutions). The RHNC equation is found to give the best approximation[187] for the components of the pair correlation function $h(12)$.

The dependence of ϵ on μ^{*2} in the LHNC approximation is shown in Fig. 10.11 for various values of Q^*. We also show the Onsager approximation (10.45), which neglects dipolar correlations, setting $G_1 = 0$. For large Q^*, where these correlations are effectively destroyed by quadrupolar forces, the two theories approximately agree; and the success of the Onsager approximation for many real fluids may largely reflect this effect.

Molecular shape

It is widely recognized that short-ranged repulsive forces play a dominant role in determining the structure of simple liquids.[188] The shape of the molecular core

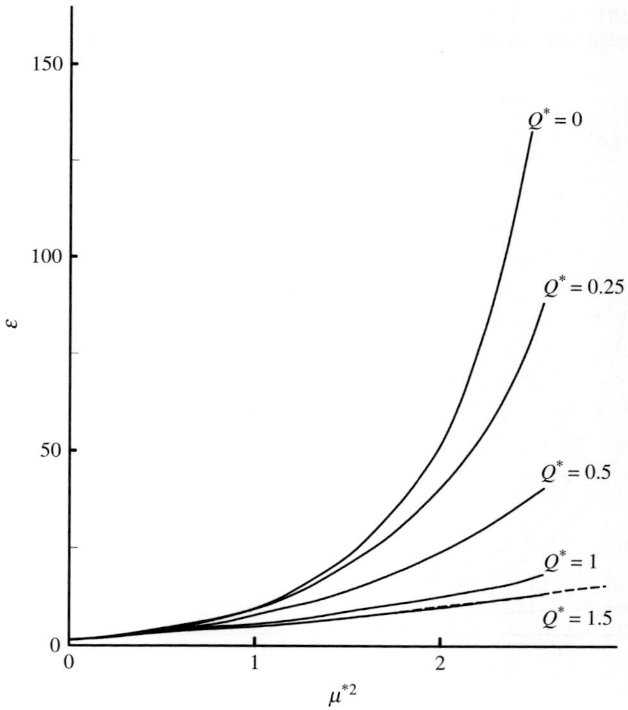

FIG. 10.11 Dipolar-quadrupolar hard sphere fluid. Variation of the LHNC ϵ with μ^{*2} at $\rho^* = 0.8$ for several values of the quadrupole moment. The dashed line is the Onsager approximation. (From ref. 186.)

affects G_1, and hence ϵ, by influencing the molecular packing. Prolate molecules will pack so that $G_1 < 0$, oblate molecules so that $G_1 > 0$;[51] only for rather strongly polar molecules do electrostatic considerations take precedence.[89]

If we altogether neglect dipolar influences on the structure, the Kirkwood expression (10.27) becomes

$$\frac{(\epsilon - 1)(2\epsilon + 1)}{9\epsilon} = y\left(1 + \frac{1}{3}\rho\tilde{h}^0_\Delta(0)\right), \tag{10.133}$$

where $\tilde{h}^0_\Delta(0) = \int d\mathbf{r}\, h^0_\Delta(\mathbf{r}\omega_1\omega_2)$ is the zero-wavevector transform of the reference fluid pair correlation function. For a fluid of dipolar asymmetric fused hard spheres, (10.133) was shown[189] to be a useful zeroth-order approximation for ϵ.

The interplay of dipolar and shape effects on fluid structure has been studied using perturbation theory,[37] the interaction-site model (discussed in Appendix 10E), the HNC approximation,[190] and, of course, computer simulation.[109,111,112,126,130,189]

MC simulations of ϵ have been reported for various model diatomic fluids, including fused hard spheres with central point dipoles[109] or with point charges at the sphere centres,[112] and two-centre LJ molecules with central point dipoles.[111] They confirm, as expected, that ϵ decreases as the molecular elongation increases.

Dipolar hard ellipsoids have been studied, both by simulation[126] and in the HNC approximation;[190] the two sets of results agree reasonably well. The HNC calculations show that, over a range of densities, for prolate ellipsoids with $a/b = 3$, $G_1 < 0$, and is a monotonically decreasing function of μ_1; while for oblate ellipsoids with $b/a = 3$, $G_1 > 0$, and is a monotonically increasing function of μ_1. In either case, ϵ itself is a monotonically increasing function of μ_1.

MD simulations[130] of acetonitrile, CH_3CN, using a three-site LJ plus point-charge model, show the importance of molecular shape effects on ϵ for real fluids. At 310 K, $\epsilon \approx 30$; a Stockmayer fluid with the same y would have $\epsilon \approx 200$. The molecular quadrupole moment is very small; short-range anisotropy in the potential causes this reduction in ϵ.

Polarizable fluids

Adding polarizability to a dipolar fluid changes its dielectric response in two ways: the molecules are polarized directly by the applied field, and indirectly by the fields of their neighbours. The indirect mechanism effectively leads to a renormalization of the dipole moment, whereby the mean body-fixed molecular dipole μ_1' exceeds the bare, gas-phase moment μ_1. Monte Carlo simulations[142] of a fluid of polar-polarizable hard spheres show that, for $\mu^* = 1$ and $\alpha^* = 0.03 - 0.06$ (typical for many fluids), the increase at liquid densities ranges from 10 to 30%. Anisotropy in the polarizability strongly influences the magnitude of the effect, with α_{zz} having much the greatest influence, assuming the dipole points along z.

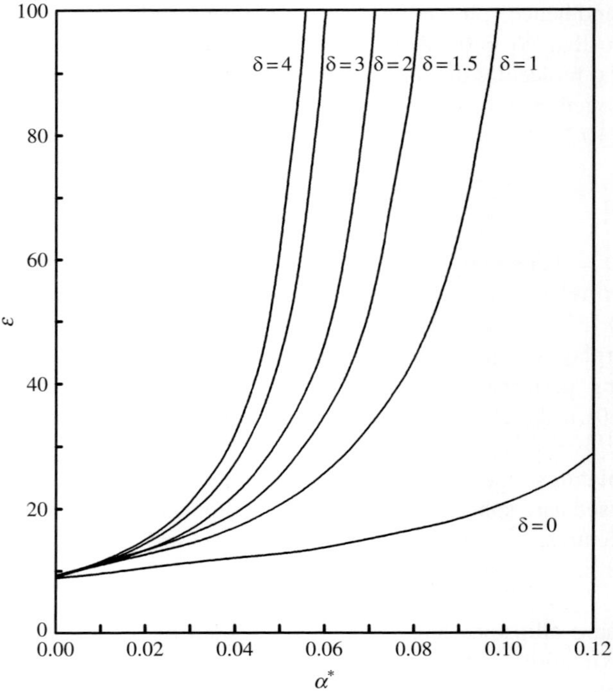

FIG. 10.12 Anisotropically polarizable dipolar hard spheres in the SSC approximation. ϵ is shown as a function of α^* at $\rho^* = 0.8$ and $\mu^* = 1$, for various values of the anisotropy ratio $\delta = \alpha_{zz}/\alpha_{xx}$. (From ref. 21.)

Integral equation approaches for polarizable fluids are based on this idea of renormalization, as discussed in Appendix 10E. Figure 10.12 shows the dielectric constant for a fluid of anisotropically polarizable dipolar hard spheres, as calculated in the single superchain (SSC) approximation, essentially a renormalized reference LHNC equation[11,19,21] (see § 5.4.7, Vol. 1). Comparison of SSC results for the polarizable Stockmayer fluid with MD simulations shows that the integral equation performs well at small μ^*, but overestimates ϵ at larger dipole strengths.[141]

The HNC,[138] RHNC,[138] LHNC,[140] and QHNC[140] approximations have been solved for a polarizable, dipolar–quadrupolar Lennard-Jones fluid. Table 10.2 shows the dielectric constants and mean dipole moments obtained from the HNC and RHNC using parameters appropriate for liquid ammonia. The results agree quite closely, and compare favourably with those from simulation.[138]

MD simulations of polarizable systems typically run 2–3 times more slowly than those of their nonpolarizable counterparts, depending on the precise method used to update the induced molecular dipoles at each time step. In contrast, MC

Table 10.2

Dielectric constant and mean dipole moment for liquid ammonia in the HNC and RHNC approximations

	HNC	RHNC	MD
ϵ	15.0	15.0	19 ± 2
μ_1'/μ_1	1.27	1.26	1.32

Calculations performed at $\rho = 0.632\,\text{g cm}^{-3}$ and $T = 277.5\,\text{K}$, assuming $\mu_1 = 1.47\,\text{D}$, $Q_1 = -2.3 \times 10^{-26}$ esu, $\alpha_1 = 2.22\,\text{Å}^3$, and $\gamma_1 = 0.288\,\text{Å}^3$. MD results are shown for comparison.
From ref. 138

simulations may be slowed by a much larger factor of $O(N)$. Various methods of speeding up the calculations have been proposed, but some of these are known to bias the sampling inappropriately, leading to skewed estimates of ϵ.[191]

Water

The dielectric constant of water is larger than that of most other fluids; at room temperature, $\epsilon \approx 80$. This is caused[192] by hydrogen bonding, an essentially electrostatic interaction between the rather substantial positive and negative charges located, respectively, near the H atoms of one molecule and in the O lone pairs of another. Because these charges reside near the surface of the molecule, and because of their essentially tetrahedral distribution, efficient fourfold coordination of other molecules, which can themselves hydrogen-bond in a similar fashion, becomes possible. This leads to long-range ordering in the liquid, which in turn produces large values of g_1 and ϵ.

Simulations employing nonpolarizable point-charge potential models, including SPC,[193] SPCE,[194] TIP3P,[195] and TIP4P,[196] can reproduce ϵ within 10–15% over a wide temperature range.[99,100,119,120,124,197] ϵ is found to be quite sensitive to the potential parameters, especially the HOH bond angle, small departures from the tetrahedral value causing a sharp increase. Thus ϵ for TIP3P water, with $\angle\text{HOH} = 104.5^\circ$, is almost 50% larger than ϵ for SPCE water, which has $\angle\text{HOH} = 109.5^\circ$; this reflects a uniform increase in the orientational order parameter $h_\Delta(r)$ for all r,[197] and explains the observation by Smith and Haymet[132] that flexibility has a greater effect on ϵ than on other physical properties of the fluid.

The success of such models is at least partly due to their implicit inclusion of the effects of polarizability, derived as they are by fitting to thermodynamic and structural properties of the dense fluid. Nevertheless, careful studies at room temperature have revealed their limitations,[197] and much effort is currently devoted to the explicit inclusion of polarizability. In principle, this may be accomplished within the framework of *ab initio* treatments,[198] renormalization theory,[136] or the

Drude[137,199] or other fluctuating-charge models.[150] For example, Zhu *et al.*[148] have reported MD simulations of ϵ at $1\,\mathrm{g\,cm^{-3}}$ and 298 K using their MST-FP (modified ST2 flexible/polarizable) model. Induction increases the mean molecular dipole moment from a gas-phase value of 1.83 D to 2.45 D, which is close to the generally accepted liquid state value of 2.4 D. However, the dielectric constant obtained from the simulations was $\epsilon = 59 \pm 1$, which is about 25% too low. Using the very sophisticated polarizable NCC potential,[198] Soetens *et al.*[149] obtained $\epsilon = 100 \pm 8$ under the same conditions, which is 25% too high. Rick *et al.*[150] have performed simulations using a fluctuating-charge model that is supposed to mimic the effects of polarizability. Although they found very good agreement with the experimental ϵ, the molecular polarizability tensor is incorrect (there are too few charge sites, none of which lie outside the molecular plane),[200] and the fluid structure is poorly reproduced. Mountain[151] simulated a site-polarizable ST2-like model of water over a range of state conditions, and found that ϵ was typically overestimated by 30%. Bearing in mind that many models incorporate physically unreasonable polarizability anisotropies,[201] it can reasonably be argued[151] that the explicit introduction of polarizability has not yet led to any improvement in the simulated dielectric constant.

Quantum simulations of flexible SPC models of water have been reported.[152] These simulations were performed using the path-integral molecular dynamics method, which represents quantum particles as classical quasiparticles or 'beads' on a cyclic chain or 'necklace',[202] and is a discretized, approximate, version of Feynman's path-integral formulation of quantum mechanics.[203] The quantum fluid was found to be less structured and hydrogen-bonded than its classical analogue, and this was reflected in a significant reduction in ϵ.

Sharma *et al.*[153] performed quantum simulations of water using Car-Parrinello *ab initio* MD, and found $\epsilon = 67 \pm 6$ at $T = 330\,\mathrm{K}$, in agreement with the experimental value of $\epsilon = 68$ for this temperature.

Various integral equation theories of ϵ have appeared. Most use very simple models for the water molecule, and although good agreement with simulation or experiment is often claimed, this must be regarded as largely fortuitous. For example, Carnie and Patey[204] solved the MSA, SSC $\equiv$ LHNC, and QHNC for a polarizable, dipolar–quadrupolar hard sphere fluid using parameters appropriate for H_2O. At 25° C the LHNC estimate was $\epsilon = 80$, with the QHNC and MSA values about 10% higher and lower, respectively. Subsequently Kusalik and Patey[205] solved the RHNC approximation for a polarizable, dipolar–quadrupolar–octopolar hard sphere fluid. At 25°C they found $\epsilon = 81.8$, which is an improvement over the value obtained when the octopole is neglected (98.1).

Lombardero *et al.*[206] have studied SPCE water in the RHNC approximation, using a hard sphere reference fluid. Although the predicted dielectric constant agrees well with simulation results near the critical point, it is much too low at room temperature; it appears that the choice of an isotropic reference system inhibits the development of the tetrahedral order that characterizes the real fluid.

10.8 Anisotropic fluids

In §§ 10.3 and 10.4 we derived fluctuation formulae for the dielectric constant ϵ of an isotropic fluid. Here, we present analogous results for an anisotropic fluid, where the dielectric response is described by a second-rank tensor $\boldsymbol{\epsilon}$. Such anisotropy can be either uniaxial (when two of the principal components of $\boldsymbol{\epsilon}$ are equal, but differ from the third), or biaxial (when all three are different). As examples of each we consider the anisotropy induced by a uniform electric field (the Kerr effect[207]) and by an electric field gradient (the Buckingham effect[208]). Both of these phenomena are manifest as an optical birefringence, measurement of which yields information on molecular parameters such as the polarizability anisotropy[216] and quadrupole moment.[215]

10.8.1 Uniaxial anisotropy–the Kerr effect

We will first consider the case where there is uniaxial symmetry, so that there is a single preferred direction specified by the vector $\mathbf{n}$; all second-rank tensors then diagonalize in the same principal axes. This is sufficient for a discussion of the Kerr effect. Our results may also be applied to the nematic phases of liquid crystals.[217]

The application of an external vector field $\mathbf{F} = F\mathbf{n}$ imposes uniaxial symmetry on a fluid, such that any symmetric second-rank tensor $\mathbf{T}$ can be written as the sum of isotropic and traceless anisotropic parts, i.e.

$$\mathbf{T} = T^{(0)}\mathbf{1} + \frac{1}{3}T^{(2)}(3\mathbf{nn} - \mathbf{1}), \tag{10.134}$$

where $T^{(0)} = (T_\parallel + 2T_\perp)/3$ and $T^{(2)} = T_\parallel - T_\perp$, with $T_\parallel$ and $T_\perp$ measuring the response parallel and perpendicular to $\mathbf{n}$, respectively. In particular, this is true of the susceptibilities $\boldsymbol{\chi}_0$ and $\boldsymbol{\chi}$, and the dielectric tensor $\boldsymbol{\epsilon}$.

For simplicity we will consider the case of a dielectric sphere *in vacuo* subject to a uniform applied field $\mathbf{E}_0$. Shape independence (§ 10.2) then guarantees that our results will be valid in any geometry. Derivation of the fluctuation formulae proceeds as in the isotropic case, but with two differences: first, the relation between the mean and applied fields must take into account the anisotropy of the dielectric tensor; second, in our calculation of the linear response function $\boldsymbol{\chi}_0$, the final average over directions of the polarizing field must be performed with proper regard for the anisotropy of the Hamiltonian.

The relation between $\mathbf{E}$ and $\mathbf{E}_0$ is calculated using (10.7). The assumption of uniformity, coupled with the homogeneity of the sample, implies that the integral on the right of this equation vanishes, just as in the isotropic case. Therefore

$$\mathbf{E} = \mathbf{E}_0 - \frac{4\pi}{3}\mathbf{P}, \tag{10.135}$$

and using the constitutive relation

$$\mathbf{P} = \frac{1}{4\pi}(\boldsymbol{\epsilon} - \mathbf{1}) \cdot \mathbf{E}, \tag{10.136}$$

we find that

$$\frac{1}{3}(\boldsymbol{\epsilon} + 2\mathbf{1}) \cdot \mathbf{E} = \mathbf{E}_0. \tag{10.137}$$

Now, from (10.11) we recall that

$$\mathbf{P} = \boldsymbol{\chi}_0 \cdot \mathbf{E}_0. \tag{10.138}$$

Combining (10.137) and (10.138), we have

$$\mathbf{P} = \frac{1}{3}\boldsymbol{\chi}_0 \cdot (\boldsymbol{\epsilon} + 2\mathbf{1}) \cdot \mathbf{E}, \tag{10.139}$$

and comparison with (10.136) yields the desired result

$$(\boldsymbol{\epsilon} - \mathbf{1}) \cdot (\boldsymbol{\epsilon} + 2\mathbf{1})^{-1} = \frac{4\pi}{3}\boldsymbol{\chi}_0. \tag{10.140}$$

In the principal axis system $(\mathbf{l}, \mathbf{m}, \mathbf{n})$, we can write

$$\frac{\epsilon_\parallel - 1}{\epsilon_\parallel + 2} = \frac{4\pi}{3}\chi_{0\parallel}$$

$$\frac{\epsilon_\perp - 1}{\epsilon_\perp + 2} = \frac{4\pi}{3}\chi_{0\perp}, \tag{10.141}$$

where $\chi_{0\parallel} = \mathbf{n} \cdot \boldsymbol{\chi}_0 \cdot \mathbf{n}$ and $\chi_{0\perp} = \mathbf{l} \cdot \boldsymbol{\chi}_0 \cdot \mathbf{l} = \mathbf{m} \cdot \boldsymbol{\chi}_0 \cdot \mathbf{m}$. Using (10.134), we can also write our results in the form

$$\frac{1}{3}\left(\frac{\epsilon_\parallel - 1}{\epsilon_\parallel + 2} + 2\frac{\epsilon_\perp - 1}{\epsilon_\perp + 2}\right) = \frac{4\pi}{3}\chi_0^{(0)} \tag{10.142}$$

and

$$\frac{\epsilon_\parallel - 1}{\epsilon_\parallel + 2} - \frac{\epsilon_\perp - 1}{\epsilon_\perp + 2} = \frac{4\pi}{3}\chi_0^{(2)}. \tag{10.143}$$

A microscopic expression for $\boldsymbol{\chi}_0 = (\partial \mathbf{P}/\partial \mathbf{E}_0)_0$ can be obtained using statistical mechanics. Applying the parameter differentiation expression (10.12), we find that the dielectric response is described by the tensor

$$\boldsymbol{\chi}_0 = (1/V)\langle \boldsymbol{\pi} + \beta \Delta\boldsymbol{\xi}\Delta\boldsymbol{\xi}\rangle_F, \tag{10.144}$$

where the subscript indicates that the ensemble average must be evaluated in the presence of the external field, $\boldsymbol{\pi}$ is the system polarizability, and $\Delta\boldsymbol{\xi} = \boldsymbol{\xi} - \langle\boldsymbol{\xi}\rangle$ represents the instantaneous departure of the system dipole moment from its average value. (We reserve the notation $\boldsymbol{\alpha}$, $\boldsymbol{\mu}$ for the values of these quantities in the limit $F \to 0$.) The irreducible components of $\boldsymbol{\chi}_0$ are then

$$\chi_0^{(0)} = (1/3V)\langle \mathbf{1} : (\boldsymbol{\pi} + \beta \Delta\boldsymbol{\xi}\Delta\boldsymbol{\xi})\rangle_F,$$

$$\chi_0^{(2)} = (1/2V)\langle (3\mathbf{nn} - \mathbf{1}) : (\boldsymbol{\pi} + \beta \Delta\boldsymbol{\xi}\Delta\boldsymbol{\xi})\rangle_F. \tag{10.145}$$

To proceed further, we will assume that $\mathbf{F}$ is weak. We may then expand the field-dependent term in the Hamiltonian as

$$\mathcal{U}_1 = -\mathbf{m} \cdot \mathbf{F} - \frac{1}{2}\mathbf{a} : \mathbf{FF} + \ldots, \tag{10.146}$$

where $\mathbf{m}$ and $\mathbf{a}$ are system properties that play the role of a generalized dipole moment and polarizability, respectively; indeed, if $\mathbf{F}$ is an electric field, $\mathbf{m} = \boldsymbol{\mu}$ and $\mathbf{a} = \boldsymbol{\alpha}$. It then follows that

$$\begin{aligned}\chi_0^{(0)} &= \chi_0^{(0)}[0] + \frac{1}{2}\chi_0^{(0)}[2]F^2 + \ldots,\\ \chi_0^{(2)} &= \frac{1}{2}\chi_0^{(2)}[2]F^2 + \ldots,\end{aligned} \tag{10.147}$$

where

$$\chi_0^{(0)}[0] = \lim_{F\to 0} \chi_0^{(0)} = (1/V)\langle \alpha + (1/3)\beta\boldsymbol{\mu}\cdot\boldsymbol{\mu}\rangle, \tag{10.148}$$

exactly as obtained in § 10.4 under the assumption of isotropy, and

$$\begin{aligned}\chi_0^{(0)}[2] &= (1/3V)\langle \mathbf{1} : \mathbf{C} : \mathbf{nn}\rangle,\\ \chi_0^{(2)}[2] &= (1/2V)\langle (3\mathbf{nn} - \mathbf{1}) : \mathbf{C} : \mathbf{nn}\rangle.\end{aligned} \tag{10.149}$$

We have introduced the tensor

$$C_{\alpha\beta,\gamma\delta} = C^{\alpha}_{\alpha\beta,\gamma\delta} + \beta C^{\mu}_{\alpha\beta,\gamma\delta}, \tag{10.150}$$

where

$$\begin{aligned}C^{\alpha}_{\alpha\beta,\gamma\delta} &= \lim_{F\to 0} \partial^2\langle \pi_{\alpha\beta}\rangle/\partial F_\gamma \partial F_\delta,\\ C^{\mu}_{\alpha\beta,\gamma\delta} &= \lim_{F\to 0} \partial^2\langle \Delta\xi_\alpha \Delta\xi_\beta\rangle/\partial F_\gamma \partial F_\delta.\end{aligned} \tag{10.151}$$

We note that $C_{\alpha\beta,\gamma\delta}$ is symmetric under interchange of α and β, and γ and δ, but not under other permutations of its indices.

To derive explicit expressions for the components of $\mathbf{C}$, we will write

$$\begin{aligned}\xi_\alpha &= \mu_\alpha + a'_{\alpha,\beta}F_\beta + \frac{1}{2}b'_{\alpha,\beta\gamma}F_\beta F_\gamma + \ldots,\\ \pi_{\alpha\beta} &= \alpha_{\alpha\beta} + b_{\alpha\beta,\gamma}F_\gamma + \frac{1}{2}c_{\alpha\beta,\gamma\delta}F_\gamma F_\delta + \ldots,\end{aligned} \tag{10.152}$$

where $\mathbf{a}'$, $\mathbf{b}'$, $\mathbf{b}$, and $\mathbf{c}$ are generalized (hyper)polarizabilities. Using (10.12) twice, we then find

$$\begin{aligned}C^{\alpha}_{\alpha\beta,\gamma\delta} &= \langle c_{\alpha\beta,\gamma\delta}\rangle + \beta\left(\langle b_{\alpha\beta,\gamma}m_\delta\rangle + \langle b_{\alpha\beta,\delta}m_\gamma\rangle\right)\\ &+ \beta\left(\langle \alpha_{\alpha\beta}a_{\gamma\delta}\rangle - \langle\alpha_{\alpha\beta}\rangle\langle a_{\gamma\delta}\rangle\right) + \beta^2\left(\langle \alpha_{\alpha\beta}m_\gamma m_\delta\rangle - \langle\alpha_{\alpha\beta}\rangle\langle m_\gamma m_\delta\rangle\right)\end{aligned} \tag{10.153}$$

and

$$
\begin{aligned}
C^{\mu}_{\alpha\beta,\gamma\delta} = {} & \left\langle a'_{\alpha,\gamma} a'_{\beta,\delta} \right\rangle - \left\langle a'_{\alpha,\gamma} \right\rangle \left\langle a'_{\beta,\delta} \right\rangle + \left\langle a'_{\alpha,\delta} a'_{\beta,\gamma} \right\rangle - \langle a'_{\alpha,\delta} \rangle \left\langle a'_{\beta,\gamma} \right\rangle \\
& + \left\langle b'_{\alpha,\gamma\delta} \mu_\beta \right\rangle + \left\langle b'_{\beta,\gamma\delta} \mu_\alpha \right\rangle \\
& + \beta \left(\left\langle a'_{\alpha,\gamma} \mu_\beta m_\delta \right\rangle - \left\langle a'_{\alpha,\gamma} \right\rangle \langle \mu_\beta m_\delta \rangle + \left\langle a'_{\beta,\gamma} \mu_\alpha m_\delta \right\rangle - \left\langle a'_{\beta,\gamma} \right\rangle \langle \mu_\alpha m_\delta \rangle \right. \\
& \left. + \langle a'_{\alpha,\delta} \mu_\beta m_\gamma \rangle - \langle a'_{\alpha,\delta} \rangle \langle \mu_\beta m_\gamma \rangle + \left\langle a'_{\beta,\delta} \mu_\alpha m_\gamma \right\rangle - \left\langle a'_{\beta,\delta} \right\rangle \langle \mu_\alpha m_\gamma \rangle \right) \\
& + \beta^2 \left(\left\langle \mu_\alpha \mu_\beta a'_{\gamma\delta} \right\rangle - \langle \mu_\alpha \mu_\beta \rangle \left\langle a'_{\gamma\delta} \right\rangle + \langle \mu_\alpha \mu_\beta m_\gamma m_\delta \rangle \right. \\
& \left. - \langle \mu_\alpha \mu_\beta \rangle \langle m_\gamma m_\delta \rangle - \langle \mu_\alpha m_\gamma \rangle \langle \mu_\beta m_\delta \rangle - \langle \mu_\alpha m_\delta \rangle \langle \mu_\beta m_\gamma \rangle \right). \qquad (10.154)
\end{aligned}
$$

Since the quantities appearing on the left and right of (10.149) are scalars, we can average over all orientations of **n**. This yields expressions for the field-dependent corrections to the susceptibility in terms of invariant combinations of **C**. We find that

$$
\begin{aligned}
\chi_0^{(0)}[2] &= (1/9V) C_{\alpha\alpha,\beta\beta}, \\
\chi_0^{(2)}[2] &= (1/15V)(3C_{\alpha\beta,\alpha\beta} - C_{\alpha\alpha,\beta\beta}). \qquad (10.155)
\end{aligned}
$$

Equation (10.155) describes the response to a static, or slowly varying, electric field. To calculate the dielectric response at high frequencies, we simply make the replacement $\mathbf{C} \to \mathbf{C}^{\alpha}$ (cf. § 10.4).

Kerr effect

Perhaps the most important practical application of these formulae is associated with the Kerr effect, the development of optical anisotropy in the presence of an electric field, first observed in 1875.[207] From (10.143), (10.147), and (10.155), the difference in refractive indices for light of frequency ω traversing a sample with polarizations parallel and perpendicular to the field can be written as

$$
\lim_{F\to 0} (n_{\parallel} - n_{\perp})/\bar{F}^2 = \frac{3}{2}\frac{1}{n}\left(\frac{n^2+2}{3}\right)^2 \left(\frac{\epsilon+2}{3}\right)^2 K(\omega), \qquad (10.156)
$$

where $\bar{F} = 3/(\epsilon+2)F$ is the magnitude of the static field within the sample, $n = n(\omega) = \epsilon^{1/2}(\omega)$ is the refractive index, and

$$
K(\omega) = \frac{2\pi}{45V}\left(3C^{\alpha}_{\alpha\beta,\alpha\beta} - C^{\alpha}_{\alpha\alpha,\beta\beta}\right) \qquad (10.157)
$$

is the Kerr constant.[218] This can be written explicitly as

$$
\begin{aligned}
K(\omega) = \frac{2\pi}{45V} & \left(\langle 3\gamma_{\alpha\beta,\alpha\beta}(\omega) - \gamma_{\alpha\alpha,\beta\beta}(\omega) \rangle + 2\beta \left\langle \left(3\beta_{\alpha\beta,\alpha}(\omega) - \beta_{\alpha\alpha,\beta}(\omega)\right) \mu_\beta \right\rangle \right. \\
& \left. + \; 3\beta \left\langle \alpha^{(2)}_{\alpha\beta}(\omega) \left(\alpha^{(2)}_{\alpha\beta}(0) + \beta\, \mu_\alpha \mu_\beta \right) \right\rangle \right). \qquad (10.158)
\end{aligned}
$$

Here $\boldsymbol{\alpha}(0)$ is the static polarizability, while $\boldsymbol{\alpha}(\omega)$, $\boldsymbol{\beta}(\omega)$, and $\boldsymbol{\gamma}(\omega)$ denote the polarizability and first and second hyperpolarizabilities of the system at the frequency of the light.[220,221]

Kerr virial series For low-density gases, we may derive a virial expansion of $K(\omega)$ using cluster expansions for the system dipole moment and polarizabilities, as was done for the dielectric constant in § 10.6. This yields

$$K(\omega) = A_K\rho + B_K\rho^2 + C_K\rho^3 + \dots, \tag{10.159}$$

where $A_K = A_K(\omega, T)$ is the first virial coefficient, or isolated molecule contribution, $B_K = B_K(\omega, T)$ is the second virial coefficient, or two-body contribution, and so on.

The one-body term A_K is obtained by neglecting all intermolecular correlations in (10.158). The result can be written in the form

$$A_K = \frac{2\pi}{45}\Big(A_0 + \beta A_1 + \beta^2 A_2\Big), \tag{10.160}$$

where

$$\begin{aligned} A_0 &= 3\gamma_{1\,\alpha\beta,\alpha\beta}(\omega) - \gamma_{1\,\alpha\alpha,\beta\beta}(\omega), \\ A_1 &= 2\big(3\beta_{1\,\alpha\beta,\alpha}(\omega) - \beta_{1\,\alpha\alpha,\beta}(\omega)\big)\mu_{1\,\beta} + 3\alpha^{(2)}_{1\,\alpha\beta}(\omega)\alpha^{(2)}_{1\,\alpha\beta}(0), \\ A_2 &= 3\,\alpha^{(2)}_{1\,\alpha\beta}(\omega)\mu_{1\,\alpha}\mu_{1\,\beta}. \end{aligned} \tag{10.161}$$

For strongly polar molecules, the term of order β^2 is expected to dominate at normal temperatures.[222] For spherical tops, and for atoms, on the other hand, only the temperature-independent term contributes, and A_K is determined entirely by the second hyperpolarizability.[223]

Temperature studies using (10.160) to determine molecular polarizabilities and hyperpolarizabilities have been critically reviewed.[216]

Quantum corrections to A_K have been calculated by Buckingham and Orr.[224] They are particularly important for hydrides, and at room temperature amount to 25, 8, and 4% for H_2, HF, and HCl, respectively.

An overview of theoretical and experimental studies of the second Kerr virial coefficient B_K has been given by Couling and Graham.[225] In the most general case, the calculation of B_K is rather complex. Here we will consider only systems exhibiting axial symmetry; we will further neglect the effects of hyperpolarizability, which typically contribute only about 1% of the total.[225,226]

For nonpolar molecules it is generally possible to neglect collision-induced effects, in which case[219]

$$\begin{aligned} B_K &= (2\pi\beta/15V)\Big\langle \boldsymbol{\alpha}_1^{(2)}(\omega) : \boldsymbol{\alpha}_2^{(2)}(0)\Big\rangle \\ &= (4\pi\beta\gamma_1(\omega)\gamma_1(0)/45)\int d\mathbf{r}\,\langle \exp[-\beta u(12)]P_2(\cos\gamma_{12})\rangle_{\omega_1\omega_2}. \end{aligned} \tag{10.162}$$

Here $\gamma_1(\omega) = \alpha_{\parallel 1}(\omega) - \alpha_{\perp 1}(\omega)$ denotes, not the hyperpolarizability, but rather the anisotropy in the polarizability, and γ_{12} is the angle between the axes of molecules 1 and 2. B_K therefore depends on P_2-type orientational correlations between the molecules.

For polar molecules, we find similarly[89,219]

$$B_K = B_K^{\gamma} + B_K^{\mu}, \tag{10.163}$$

where B_K^{γ} is given by an expression identical to (10.162), and

$$B_K^{\mu} = \left(4\pi\beta^2\gamma_1(\omega)\mu_1^2/45\right) \times \int d\mathbf{r}\, \langle \exp[-\beta u(12)]\,[P_2(\cos\gamma_{12}) + 2P_1(\cos\gamma_{12})]\rangle_{\omega_1\omega_2}. \tag{10.164}$$

Normally $\gamma_1 \ll \beta\mu_1^2$, so that $B_K \approx B_K^{\mu}$. Since the term proportional to $P_1(\cos\gamma_{12})$ dominates except for very small μ_1, B_K might be expected to show a very similar dependence on shape and dipole strength as the second dielectric virial coefficient B_{ϵ}. In particular, it should be positive for oblate molecules, and negative for prolate molecules. However, both CH_3F (which is prolate) and CHF_3 (which is oblate) have $B_K > 0$, and the coefficient is an order of magnitude larger than predicted by (10.164). Buckingham *et al.*[227] have shown that this is because of significant positive contributions from collision-induced terms. By including such contributions within the DID model, and by incorporating shape effects in a realistic manner, Couling and Graham[225] find that the theoretical and experimental predictions for B_K can be brought into agreement.

Expressions for the third virial coefficient, C_K, have been derived only for atomic gases. The DID model does not yield an accurate description of these three-body effects,[226,228] presumably because of its neglect of short-range overlap contributions to the induced polarizability.

Dense fluids For liquids, further reduction of (10.158) is difficult, especially when collision-induced effects are included. Two-body contributions involve the orientational correlation parameters G_1 (polar molecules only) and G_2, as well as mixed translational-orientational terms such as $G_2' = \left\langle r_{12}^{-3}\, P_2\left(\cos\theta_1'\right)\right\rangle$ (where θ_1' is the angle between the symmetry axis of molecule 1 and the intermolecular (12) axis), which derives from the collisional term $\left\langle \Delta\boldsymbol{\alpha}^{(2)}(12) : \boldsymbol{\alpha}_1^{(2)}\right\rangle$. Three-body contributions include terms such as $\langle 3\cos\gamma_{12}\cos\gamma_{13} - \cos\gamma_{23}\rangle$, deriving from the average $\left\langle \boldsymbol{\alpha}_1^{(2)} : \boldsymbol{\mu}_2\boldsymbol{\mu}_3\right\rangle$.

For nonpolar fluids, an approximate treatment developed by Ladanyi and Keyes[229] yields results in reasonable agreement with experiment. Hyperpolarizability is neglected. The analysis incorporates the effects of both orientational and translational-orientational correlations of the type discussed above, with the latter manifest as a renormalization of the molecular polarizability anisotropy. The result, expressed relative to the Kerr constant for a hypothetical perfect gas of the same density as the liquid, can be written as

$$K/K_0 = \left(\frac{\gamma_{\text{eff}}(\omega)}{\gamma(\omega)}\right)\left(\frac{\gamma_{\text{eff}}}{\gamma}\right) g_2. \qquad (10.165)$$

Here $g_2 = 1 + G_2$ is the angular correlation parameter, which is available from simulation,[230,231] perturbation theory (Appendix 10C), or integral equation theories (though not the interaction-site model; see § 10E.2). γ_{eff} is the effective polarizability in the liquid state; this is expressed in terms of first-order DID correlation terms such as G_2'. The physical interpretation of this quantity is that the shape of a central molecule excludes other molecules from a nonspherical region, thereby changing its effective polarizability anisotropy.[232,233] Because of the r^{-3} weighting, the anisotropy is in fact *reduced* for prolate fluids.

For liquid CS_2, MC simulations with a site–site hard sphere potential yield $\gamma_{\text{eff}}/\gamma \approx 0.6$ and $g_2 \approx 1.1 \pm 0.1$; the small value of G_2 does not mean that orientational correlations are small, but rather that there is extensive cancellation between nearest-neighbour, next-nearest neighbour, etc., alignments.[230]

10.8.2 Biaxial anisotropy—the Buckingham effect[208]

In 1921, Debye[234] pointed out that a nonuniform electric field imposes biaxial anisotropy on an anisotropically polarizable fluid. While the effect is actually present for all molecules (through the effect of higher polarizabilities),[214,235] it is particularly important for quadrupolar species, because the system potential energy then contains a term proportional to the field gradient $\nabla\mathcal{E}$. From (C.8),

$$\mathcal{U}_1 = -\frac{1}{3}\mathbf{Q} : \nabla\mathcal{E}, \qquad (10.166)$$

where $\mathbf{Q}$ is the quadrupole moment of the system; we use the symbol $\mathcal{E}$ to denote the perturbing field, to distinguish it from the mean field $\mathbf{E}$ within the fluid. There is therefore a tendency for molecules to align preferentially with respect to the field gradient. Since quadrupolar molecules are in general anisotropically polarizable, there is a resulting anisotropy in the polarizability of the system as a whole; and this is manifest as an optical birefringence. Although the effect is rather weak, it can be measured by careful experimental design; such measurements provide one of the very few ways of directly determining molecular quadrupole moments[210] (see also Appendix D of Vol. 1).

Classical theory

To calculate the magnitude of this effect, we consider the dielectric response of an isolated sphere of fluid lying midway between two long, straight wires, charged to the same voltage, to a uniform applied field $\mathbf{E}_0(\omega)$ (typically supplied by a laser). We suppose the sphere sits at the origin of a coordinate system (x, y, z). If the wires lie in the xz-plane and run parallel to z, then for small x and y the electrostatic potential they generate can be written as

$$\phi(x,y) = -\frac{1}{2}\mathcal{E}'(x^2 - y^2). \tag{10.167}$$

If x, y, and z measure displacements along directions defined by the unit vectors $\mathbf{e}_1$, $\mathbf{e}_2$, and $\mathbf{e}_3$, the associated electric field is then

$$\boldsymbol{\mathcal{E}} = \mathcal{E}'(x\mathbf{e}_1 - y\mathbf{e}_2). \tag{10.168}$$

This is nonuniform. The field gradient tensor is

$$\nabla\boldsymbol{\mathcal{E}} = \mathcal{E}'(\mathbf{e}_1\mathbf{e}_1 - \mathbf{e}_2\mathbf{e}_2), \tag{10.169}$$

so that $\mathcal{E}' = \nabla_x\mathcal{E}_x = -\nabla_y\mathcal{E}_y$.

In the presence of this nonuniformity, the dielectric response is, in general, nonlocal, and the relation between the polarization and the Maxwell field is modified. If we are only interested in the long-wavelength response, it is sufficient to take[236]

$$\frac{1}{4\pi}(\boldsymbol{\epsilon} - \mathbf{1}) \cdot \mathbf{E} = \mathbf{P}^{(0)} - \nabla \cdot \mathbf{P}^{(1)}, \tag{10.170}$$

where

$$\mathbf{P}^{(n)} = V^{-1}\left\langle \int \mathrm{d}\mathbf{r}\, \mathbf{r}^n \boldsymbol{\xi}(\mathbf{r}) \right\rangle, \qquad n \geq 0, \tag{10.171}$$

and $\boldsymbol{\xi}$ is the microscopic dipole density within the sample, a function of the field $\boldsymbol{\mathcal{E}}(\mathbf{r})$. (As in our treatment of the Kerr effect, we reserve the symbol $\boldsymbol{\mu}$ for the dipole moment in the limit $\boldsymbol{\mathcal{E}} \to 0$.) The effect of the field gradient is therefore to augment (10.1) with a term proportional to the first moment of $\boldsymbol{\xi}$, although more generally, in cases where the gradient is nonuniform, one expects additional contributions from higher moments also.[237]

Using (10.137) and (10.170), we can now write

$$(\boldsymbol{\epsilon} - \mathbf{1}) \cdot (\boldsymbol{\epsilon} + 2\mathbf{1})^{-1} = \frac{4\pi}{3}\boldsymbol{\chi}_0, \tag{10.172}$$

where

$$\boldsymbol{\chi}_0 = \frac{\partial}{\partial \mathbf{E}_0}\left(\mathbf{P}^{(0)} - \nabla \cdot \mathbf{P}^{(1)}\right) \tag{10.173}$$

is the modified electric susceptibility of the system. At optical frequencies, $\boldsymbol{\chi}_0$ contains contributions only from the electronic motions. In terms of the electronic polarizability density $\pi_{\alpha\beta} = \partial\xi_\alpha/\partial E_{0\beta}$ and its first moment $\pi^\gamma_{\alpha\beta} = V^{-1}\int d\mathbf{r}\, r_\gamma \pi_{\alpha\beta}$, we can write

$$\chi_{0\,\alpha\beta} = \langle\pi_{\alpha\beta}\rangle_{\mathcal{E}} - \nabla_\gamma\left\langle\pi^\gamma_{\alpha\beta}\right\rangle_{\mathcal{E}}, \tag{10.174}$$

where the ensemble averages must be taken in the presence of the field.

When the field gradient is small, we may expand $\pi_{\alpha\beta}$ and $\pi^\gamma_{\alpha\beta}$ in powers of $\boldsymbol{\mathcal{E}}$. We will consider only the dominant contributions of order β, as the temperature-independent term in the birefringence is usually rather small.[238] Using (10.12), and the multipole expansion (C.8), we find to first order

$$\langle \pi_{\alpha\beta} \rangle_{\mathcal{E}} = \alpha \delta_{\alpha\beta} + \frac{1}{3} \beta \langle \alpha_{\alpha\beta} Q_{\gamma\delta} \rangle \mathcal{E}_{\gamma\delta} \tag{10.175}$$

and

$$\left\langle \pi^{\gamma}_{\alpha\beta} \right\rangle_{\mathcal{E}} = \beta \left\langle \alpha^{\gamma}_{\alpha\beta} \mu_\delta \mathcal{E}_\delta \right\rangle, \tag{10.176}$$

where $\alpha^{\gamma}_{\alpha\beta} = V^{-1} \int \mathrm{d}\mathbf{r}\, r_\gamma \alpha_{\alpha\beta}$. The susceptibility is therefore

$$\chi_{0\,\alpha\beta} = \alpha \delta_{\alpha\beta} + \Delta \chi_{0\,\alpha\beta,\gamma\delta} \mathcal{E}_{\gamma\delta}, \tag{10.177}$$

where $\alpha = \frac{1}{3}\alpha_{\alpha\alpha}$ is the polarizability density of the unperturbed system, and $\Delta \boldsymbol{\chi}_0$ is a fourth-rank tensor that describes the anisotropy induced by the nonuniform field, with elements

$$\Delta \chi_{0\,\alpha\beta,\gamma\delta} = \frac{1}{3} \beta \langle \alpha_{\alpha\beta} Q_{\gamma\delta} \rangle - \beta \left\langle \alpha^{\gamma}_{\alpha\beta} \mu_\delta \right\rangle. \tag{10.178}$$

The dielectric tensor will diagonalize in the axes $(\mathbf{e}_1, \mathbf{e}_2, \mathbf{e}_3)$, so that

$$\boldsymbol{\epsilon} = \epsilon_1 \mathbf{e}_1 \mathbf{e}_1 + \epsilon_2 \mathbf{e}_2 \mathbf{e}_2 + \epsilon_3 \mathbf{e}_3 \mathbf{e}_3, \tag{10.179}$$

where ϵ_1, ϵ_2, and ϵ_3 denote principal components. Using (10.169), (10.172), and (10.177)–(10.179), we find that

$$\frac{\epsilon_3 - 1}{\epsilon_3 + 2} = \frac{4\pi}{3} \alpha(\omega), \tag{10.180}$$

so that ϵ_3 is unaffected by the field gradient. This is also true of the average $\frac{1}{2}[(\epsilon_1 - 1)/(\epsilon_1 + 2) + (\epsilon_2 - 1)/(\epsilon_2 + 2)]$. However, a difference first order in the field gradient develops between ϵ_1 and ϵ_2, such that

$$\frac{\epsilon_1 - 1}{\epsilon_1 + 2} - \frac{\epsilon_2 - 1}{\epsilon_2 + 2} = \frac{4\pi}{3} \Delta \chi_0\, \mathcal{E}', \tag{10.181}$$

where

$$\begin{aligned} \Delta \chi_0 &= (\mathbf{e}_1 \mathbf{e}_1 - \mathbf{e}_2 \mathbf{e}_2) : \Delta \boldsymbol{\chi}_0 : (\mathbf{e}_1 \mathbf{e}_1 - \mathbf{e}_2 \mathbf{e}_2) \\ &= \frac{2}{15} (3 \Delta \chi_{0\,\alpha\beta,\alpha\beta} - \Delta \chi_{0\,\alpha\alpha,\beta\beta}). \end{aligned} \tag{10.182}$$

The second form of the result follows from averaging over all possible orientations of $\mathbf{e}_1$ and $\mathbf{e}_2$, and noting that

$$\langle (e_{1\alpha} e_{1\beta} - e_{2\alpha} e_{2\beta})(e_{1\gamma} e_{1\delta} - e_{2\gamma} e_{2\delta}) \rangle = \frac{1}{15} (3 \delta_{\alpha\gamma} \delta_{\beta\delta} + 3 \delta_{\alpha\delta} \delta_{\beta\gamma} - 2 \delta_{\alpha\beta} \delta_{\gamma\delta}). \tag{10.183}$$

To calculate the induced birefringence we note that $\epsilon_m = n_m^2$, $m = 1, 2, 3$. The difference in refractive indices for light traversing the sample with polarizations along $\mathbf{e}_1$ and $\mathbf{e}_2$ is then given given immediately as

$$\lim_{\mathcal{E}'\to 0}(n_1 - n_2)/\bar{\mathcal{E}}' = \frac{1}{n}\left(\frac{n^2+2}{3}\right)^2\left(\frac{2\epsilon+3}{5}\right)\mathcal{Q}(\omega), \quad (10.184)$$

where ϵ and n are the dielectric constant and refractivity of the system when $\mathcal{E}' = 0$, $\bar{\mathcal{E}}' = 5/(2\epsilon+3)\mathcal{E}'$ is the magnitude of the field gradient within the sample,[209] and

$$\mathcal{Q}(\omega) = 2\pi\,\Delta\chi_0 = \frac{4\pi}{15}\beta\left(\alpha_{\alpha\beta}(\omega)Q_{\alpha\beta} - \left[3\alpha^{\alpha}_{\alpha\beta}(\omega) - \alpha^{\beta}_{\alpha\alpha}(\omega)\right]\mu_\beta\right), \quad (10.185)$$

or equivalently

$$\mathcal{Q}(\omega) = \frac{4\pi}{15}\beta\left(\alpha^{(2)}_{\alpha\beta}(\omega)Q_{\alpha\beta} - 3\alpha^{(2)\,\alpha}_{\alpha\beta}(\omega)\mu_\beta\right), \quad (10.186)$$

where $\boldsymbol{\alpha}^{(2)}(\omega)$ is the anisotropic part of the polarizability density at the frequency of the light.

We note that, under a change of origin $O \to O'$, $O' = (R_x, R_y, R_z)$, $Q_{\alpha\beta}$ and $\alpha^{\alpha}_{\alpha\beta}$ transform according to

$$\begin{aligned} Q'_{\alpha\beta} &= Q_{\alpha\beta} - \frac{3}{2}\mu_\alpha R_\beta - \frac{3}{2}\mu_\beta R_\alpha + \mu_\gamma R_\gamma \delta_{\alpha\beta}, \\ \alpha^{(2)\,\alpha\prime}_{\alpha\beta} &= \alpha^{(2)\,\alpha}_{\alpha\beta} - \alpha^{(2)}_{\alpha\beta}R_\alpha. \end{aligned} \quad (10.187)$$

It follows that our expression for $\mathcal{Q}(\omega)$ is independent of the choice of origin even for polar fluids (where its constituent parts are not).

Dilute gases

In a dilute gas, $\mathcal{Q}(\omega)$ can be expanded as a virial series in the number density. For sufficiently small ρ, only the single-body terms contribute, and then

$$\mathcal{Q}(\omega) = \frac{4\pi}{15}\beta\rho\left(\alpha^{(2)}_{1\,\alpha\beta}(\omega)Q_{1\,\alpha\beta} - 3\alpha^{(2)\,\alpha}_{1\,\alpha\beta}(\omega)\mu_{1\,\beta}\right). \quad (10.188)$$

We may define an ‘effective quadrupole centre’, or EQC, within each molecule, where the first moment of the molecular polarizability vanishes. We then obtain the particularly simple result

$$\mathcal{Q}(\omega) = \frac{4\pi}{15}\beta\rho\,\boldsymbol{\alpha}^{(2)}_1(\omega) : \mathbf{Q}^{\mathrm{EQC}}_1, \quad (10.189)$$

where $\mathbf{Q}^{\mathrm{EQC}}_1$ is now referred to this centre. For nonpolar species, the EQC coincides with the centre of mass.

Although the birefringence induced by a field gradient is typically some two or three orders of magnitude smaller than in the Kerr effect,[215,241] with modern apparatus it is relatively easily measured. If the anisotropy in the molecular polarizability is available from other sources (such as studies of the depolarization of scattered light; see Chapter 11), observations of $\mathcal{Q}(\omega)$ allow us to directly determine both the sign and magnitude of $\mathbf{Q}^{\mathrm{EQC}}_1$. Ideally, measurements should be performed over a range of temperatures, to enable separation of the small temperature-independent term.[242] Quadrupoles have been measured in this way for a variety of molecules,[215] including N_2, Cl_2, CO, CO_2, and BF_3,[243] CS_2,[244] C_2H_6, C_3H_6, N_2O, and CH_3F,[245] and C_6H_6 and C_6F_6[246] (see also Appendix D, Vol. 1).

For polar molecules, a correct (origin-independent) description of the birefringence was first obtained by Buckingham and Longuet-Higgins in 1968.[212] These authors used time-dependent scattering theory to compute the scattering of polarized light by a thin lamina of dielectric material; this produces a phase shift proportional to the field gradient and that may be interpreted as a change in the refractive index. Their results are equivalent to ours, although the derivation is considerably more complex. Their analysis shows, additionally, that the position of the EQC can be related to the molecular hyperpolarizabilities $\mathbf{A}$ and $\mathbf{G}'$,[212] which describe, respectively, the dipole induced by a field gradient and a time-varying magnetic field. For linear molecules, the EQC occurs at the point at which[213] $A_{\parallel} + 2A_{\perp} + 10c/\omega\, G' = 0$; in some instances this may actually lie outside the molecule.[245,247]

Imrie, Raab, and de Lange[248–250] have presented an alternative treatment of the effect using an eigenvalue theory of wave propagation based on Maxwell's equations. In its original formulation,[248] this was incorrect. The errors have subsequently been traced to the implicit assumption that light propagates as a plane wave—which in fact it cannot in the presence of a nonuniform field. Once this approximation is removed, the correct result is obtained.[250]

Quantum corrections

Quantum dynamical corrections to the birefringence, which account for the quantized nature of the molecular rotation, have been calculated by Buckingham and Pariseau.[211] Their magnitude depends on the ratio θ_r/T, where θ_r is the rotational temperature defined in § 1.2.2. For most species the effect is unimportant, although it is larger for hydrides; for H_2 at room temperature, the correction is -25%.

Dense fluids

In gases, the effects of pressure appear to be very small.[213] For He, Ne, and Ar, virial series corrections to $\mathcal{Q}(\omega)$ have been calculated;[251] they are never more than a few parts per thousand under typical experimental conditions, and lie far outside the limits of what can currently be measured.

In quadrupolar liquids, $\mathcal{Q}(\omega)$ contains contributions involving the average $\left\langle \boldsymbol{\alpha}_1^{(2)} : \mathbf{Q}_2 \right\rangle$; for linear molecules this reduces to $\gamma_1 Q_1 \langle P_2(\cos\gamma_{12})\rangle$, which is proportional to the orientational correlation parameter G_2. Some quadrupole moments have been extracted from measurements performed on liquids and solutions,[252] but in view of the rather ad hoc treatment of many-body interactions employed in the analysis, these must be considered of rather uncertain reliability.

Appendix 10A Dielectric response in nonuniform fields

In this appendix we consider the dielectric response of an infinite system of rigid polar molecules to a *nonuniform* electric field $\mathbf{E}_0(\mathbf{r})$. This helps to establish under what conditions the dielectric response is local,[253] as expressed by (10.1). It also yields an alternative derivation of Kirkwood's formula for the dielectric constant, (10.27), and establishes additional results for ϵ in the form of the Høye and Stell formula, (10.31), and Ramshaw's expression, (10.32).

10A.1 Microscopic theory

Using linear response theory we may derive a microscopic relation expressing the polarization $\mathbf{P}(\mathbf{r})$ as a linear functional of $\mathbf{E}_0(\mathbf{r})$. In terms of the microscopic polarization density

$$\mathcal{P}(\mathbf{r};\mathbf{r}^N\boldsymbol{\omega}^N) = \sum_i \boldsymbol{\mu}_i\delta(\mathbf{r}-\mathbf{r}_i), \tag{10A.1}$$

we can write

$$\mathbf{P}(\mathbf{r}) = \langle\mathcal{P}(\mathbf{r};\mathbf{r}^N\boldsymbol{\omega}^N)\rangle_{E_0} = \int d\mathbf{r}'\,\boldsymbol{\chi}_0(\mathbf{r},\mathbf{r}')\cdot\mathbf{E}_0(\mathbf{r}'), \tag{10A.2}$$

where the kernel is the functional derivative

$$\boldsymbol{\chi}_0(\mathbf{r},\mathbf{r}') = \delta\langle\mathcal{P}(\mathbf{r})\rangle_{E_0}/\delta\mathbf{E}_0(\mathbf{r}'). \tag{10A.3}$$

Using (10.12), and writing the field-dependent term in the system potential energy as the integral

$$\mathcal{U}_1(\mathbf{r}^N\boldsymbol{\omega}^N) = -\int d\mathbf{r}\,\mathcal{P}(\mathbf{r};\mathbf{r}^N\boldsymbol{\omega}^N)\cdot\mathbf{E}_0(\mathbf{r}), \tag{10A.4}$$

we find

$$\boldsymbol{\chi}_0(\mathbf{r},\mathbf{r}') = \beta\langle\mathcal{P}(\mathbf{r};\mathbf{r}^N\boldsymbol{\omega}^N)\mathcal{P}(\mathbf{r}';\mathbf{r}^N\boldsymbol{\omega}^N)\rangle. \tag{10A.5}$$

Since an infinite system is translationally invariant,[254] we may write

$$\boldsymbol{\chi}_0(\mathbf{r},\mathbf{r}') = \boldsymbol{\chi}_0(\mathbf{r}-\mathbf{r}'), \tag{10A.6}$$

where

$$\boldsymbol{\chi}_0(\mathbf{r}) = \beta\langle\mathcal{P}(\mathbf{0};\mathbf{r}^N\boldsymbol{\omega}^N)\mathcal{P}(\mathbf{r};\mathbf{r}^N\boldsymbol{\omega}^N)\rangle. \tag{10A.7}$$

Substituting (10A.1) into (10A.7), and separating the self ($i=j$) and distinct ($i\neq j$) terms, we now find

$$\boldsymbol{\chi}_0(\mathbf{r}) = \beta\Big\langle\sum_i \boldsymbol{\mu}_i\boldsymbol{\mu}_i\delta(\mathbf{r}_i)\delta(\mathbf{r}-\mathbf{r}_i)\Big\rangle + \beta\Big\langle\sum_{i\neq j}\boldsymbol{\mu}_i\boldsymbol{\mu}_j\delta(\mathbf{r}_i)\delta(\mathbf{r}-\mathbf{r}_j)\Big\rangle, \tag{10A.8}$$

or equivalently

$$\boldsymbol{\chi}_0(\mathbf{r}) = \rho\beta\mu_1^2\left(\langle\mathbf{n}_1\mathbf{n}_1\rangle_{\omega_1}\delta(\mathbf{r}) + \rho\,\langle g(\mathbf{r}\omega_1\omega_2)\mathbf{n}_1\mathbf{n}_2\rangle_{\omega_1\omega_2}\right), \tag{10A.9}$$

where $\mathbf{n}_1$ and $\mathbf{n}_2$ are unit vectors along the directions of $\boldsymbol{\mu}_1$ and $\boldsymbol{\mu}_2$, and $g(\mathbf{r}\omega_1\omega_2)$ is the pair correlation function.

To reduce (10A.9) further, we expand $g(\mathbf{r}\omega_1\omega_2) = 1 + h(\mathbf{r}\omega_1\omega_2)$ in Cartesian tensors according to (5.60), and average over all orientations of $\mathbf{n}_1$ and $\mathbf{n}_2$ using the definitions (5.61), (5.62). This yields

$$\boldsymbol{\chi}_0(\mathbf{r}) = \frac{3}{4\pi}y\left[\left(\delta(\mathbf{r}) + \frac{1}{3}\rho h_\Delta(r)\right)\mathbf{1} + \frac{1}{3}\rho h_D(r)\,\boldsymbol{\mathcal{D}}(\hat{\mathbf{r}})\right], \tag{10A.10}$$

where $\boldsymbol{\mathcal{D}}(\hat{\mathbf{r}}) = 3\hat{\mathbf{r}}\hat{\mathbf{r}} - \mathbf{1}$, and, as usual, the dimensionless constant $y = 4\pi\rho\beta\mu_1^2/9$.

Since the relation between $\mathbf{P}$ and $\mathbf{E}_0$ is a convolution, it is natural to express our results in Fourier space. Defining

$$\mathcal{F}(\mathbf{k}) = \int d\mathbf{r}\, \exp(i\mathbf{k}\cdot\mathbf{r})\mathcal{F}(\mathbf{r}) \tag{10A.11}$$

for $\mathcal{F} = \mathbf{P}, \mathbf{E}_0, \boldsymbol{\chi}_0$, we find

$$\mathbf{P}(\mathbf{k}) = \boldsymbol{\chi}_0(\mathbf{k})\cdot\mathbf{E}_0(\mathbf{k}). \tag{10A.12}$$

To compute the transform of (10A.10), we use the Rayleigh expansion of § B.5 in the form

$$\exp(i\mathbf{k}\cdot\mathbf{r}) = \sum_{l=0}^{\infty}(2l+1)i^l j_l(kr)P_l(\cos\theta), \tag{10A.13}$$

where $\cos\theta = \hat{\mathbf{k}}\cdot\hat{\mathbf{r}}$, and $P_l(\cos\theta)$ denotes a Legendre polynomial,[255] and $j_l(kr)$ a spherical Bessel function,[255] of order l. Noting that

$$P_2(\cos\theta) = \frac{1}{2}(3\cos^2\theta - 1) = \frac{1}{6}\left(3\hat{k}_\alpha\hat{k}_\beta - \delta_{\alpha\beta}\right)\left(3\hat{r}_\alpha\hat{r}_\beta - \delta_{\alpha\beta}\right) \tag{10A.14}$$

and

$$\left\langle\left(3\hat{r}_\alpha\hat{r}_\beta - \delta_{\alpha\beta}\right)\left(3\hat{r}_\gamma\hat{r}_\delta - \delta_{\gamma\delta}\right)\right\rangle = \frac{1}{5}(3\delta_{\alpha\gamma}\delta_{\beta\delta} + 3\delta_{\alpha\delta}\delta_{\beta\gamma} - 2\delta_{\alpha\beta}\delta_{\gamma\delta}), \tag{10A.15}$$

we then obtain

$$\boldsymbol{\chi}_0(\mathbf{k}) = \frac{3}{4\pi}y\left[\left(1 + \frac{1}{3}\rho\tilde{h}_\Delta(k)\right)\mathbf{1} + \frac{1}{3}\rho\tilde{h}_D(k)\,\boldsymbol{\mathcal{D}}(\hat{\mathbf{k}})\right], \tag{10A.16}$$

where $\tilde{h}_\Delta(k)$ and $\tilde{h}_D(k)$ are the Hankel transforms (cf. § 3B.5)

$$\tilde{h}_\Delta(k) = 4\pi\int_0^\infty \mathrm{d}r\, r^2 j_0(kr)h_\Delta(r),$$

$$\tilde{h}_D(k) = -4\pi\int_0^\infty \mathrm{d}r\, r^2 j_2(kr)h_D(r). \tag{10A.17}$$

The inverse of (10A.16) is

$$\boldsymbol{\chi}_0^{-1}(\mathbf{k}) = \frac{4\pi}{3} y^{-1} \left[\left(1 - \frac{1}{3}\rho\tilde{c}_\Delta(k)\right) \mathbf{1} - \frac{1}{3}\rho\tilde{c}_D(k)\, \boldsymbol{\mathcal{D}}(\hat{\mathbf{k}}) \right], \tag{10A.18}$$

where $\tilde{c}_\Delta(k)$ and $\tilde{c}_D(k)$ are the Hankel transforms of the corresponding components of the direct correlation function $c(\mathbf{r}\omega_1\omega_2)$. To derive this result, we proceed as follows. In Fourier space, the Ornstein–Zernike equation (10E.1) is

$$\tilde{h}(\mathbf{k}\omega_1\omega_2) = \tilde{c}(\mathbf{k}\omega_1\omega_2) + \rho \left\langle \tilde{c}(\mathbf{k}\omega_1\omega_3)\tilde{h}(\mathbf{k}\omega_3\omega_2) \right\rangle_{\omega_3} . \tag{10A.19}$$

Expanding the transforms according to

$$\begin{aligned} \tilde{h}(\mathbf{k}\omega_1\omega_2) &= \tilde{h}_S(k) + \tilde{h}_\Delta(k)\Delta(\omega_1\omega_2) + \tilde{h}_D D(\omega_1\omega_2\omega_k) + \ldots, \\ \tilde{c}(\mathbf{k}\omega_1\omega_2) &= \tilde{c}_S(k) + \tilde{c}_\Delta(k)\Delta(\omega_1\omega_2) + \tilde{c}_D D(\omega_1\omega_2\omega_k) + \ldots, \end{aligned} \tag{10A.20}$$

where

$$\Delta(\omega_1\omega_2) = \mathbf{n}_1 \cdot \mathbf{n}_2, \qquad D(\omega_1\omega_2\omega_k) = \mathbf{n}_1 \cdot \boldsymbol{\mathcal{D}}(\hat{\mathbf{k}}) \cdot \mathbf{n}_2, \tag{10A.21}$$

and using the results

$$\begin{aligned} \langle \Delta(\omega_1\omega_2)\Delta(\omega_3\omega_2)\rangle_{\omega_3} &= \frac{1}{3}\Delta(\omega_1\omega_2), \\ \langle \Delta(\omega_1\omega_2)D(\omega_3\omega_2\omega_k)\rangle_{\omega_3} &= \frac{1}{3}D(\omega_1\omega_2\omega_k), \\ \langle D(\omega_1\omega_2\omega_k)D(\omega_3\omega_2\omega_k)\rangle_{\omega_3} &= \frac{1}{3}\left(D(\omega_1\omega_2\omega_k) + 2\Delta(\omega_1\omega_2)\right), \end{aligned} \tag{10A.22}$$

we find that

$$\begin{aligned} \tilde{h}_\Delta(k) &= \tilde{c}_\Delta(k) + \frac{1}{3}\rho\, \tilde{c}_\Delta(k)\tilde{h}_\Delta(k) + \frac{2}{3}\rho\, \tilde{c}_D(k)\tilde{h}_D(k), \\ \tilde{h}_D(k) &= \tilde{c}_D(k) + \frac{1}{3}\rho\, \tilde{c}_\Delta(k)\tilde{h}_D(k) + \frac{1}{3}\rho\, \tilde{c}_D(k)\tilde{h}_\Delta(k) \\ &\quad + \frac{1}{3}\rho\, \tilde{c}_D(k)\tilde{h}_D(k). \end{aligned} \tag{10A.23}$$

Now

$$\boldsymbol{\mathcal{D}}(\hat{\mathbf{k}}) \cdot \boldsymbol{\mathcal{D}}(\hat{\mathbf{k}}) = \boldsymbol{\mathcal{D}}(\hat{\mathbf{k}}) + 2\,\mathbf{1}. \tag{10A.24}$$

Equation (10A.18) then follows immediately using (10A.16), (10A.23), and (10A.24).

For our purposes it is convenient to split the correlation functions into long- and short-range parts. To this end, we write

$$h(\mathbf{r}\omega_1\omega_2) = \lambda\frac{\beta\mu_1^2}{r^3}D(\omega_1\omega_2\omega) + \Delta h(\mathbf{r}\omega_1\omega_2),$$

$$c(\mathbf{r}\omega_1\omega_2) = \frac{\beta\mu_1^2}{r^3}D(\omega_1\omega_2\omega) + \Delta c(\mathbf{r}\omega_1\omega_2), \qquad (10A.25)$$

where the terms $\Delta h(\mathbf{r}\omega_1\omega_2)$ and $\Delta c(\mathbf{r}\omega_1\omega_2)$ vary as r^{-6} for large r, and λ is a dimensionless number that is a function of the state conditions.[256] Using (10A.17) together with the result[255]

$$\int_0^\infty \mathrm{d}x\, x^{-1} j_2(x) = \frac{1}{3}, \qquad (10A.26)$$

we find that

$$\tilde{h}_D(k) = -\lambda\frac{4\pi\beta\mu_1^2}{3} + \Delta\tilde{h}_D(k),$$

$$\tilde{c}_D(k) = -\frac{4\pi\beta\mu_1^2}{3} + \Delta\tilde{c}_D(k), \qquad (10A.27)$$

where $\Delta\tilde{h}_D(k)$ and $\Delta\tilde{c}_D(k)$ are the Hankel transforms of $\Delta h_D(r)$ and $\Delta c_D(r)$. Substituting (10A.27) into (10A.16) and (10A.18), we now find that

$$\boldsymbol{\chi}_0(\mathbf{k}) = \frac{3}{4\pi}y\left[\left(1+\frac{1}{3}\rho\tilde{h}_\Delta(k)\right)\mathbf{1} - \left(\lambda y - \frac{1}{3}\rho\Delta\tilde{h}_D(k)\right)\boldsymbol{\mathcal{D}}(\hat{\mathbf{k}})\right],$$

$$\boldsymbol{\chi}_0^{-1}(\mathbf{k}) = \frac{4\pi}{3}y^{-1}\left[\left(1-\frac{1}{3}\rho\tilde{c}_\Delta(k)\right)\mathbf{1} + \left(y - \frac{1}{3}\rho\Delta\tilde{c}_D(k)\right)\boldsymbol{\mathcal{D}}(\hat{\mathbf{k}})\right]. \qquad (10A.28)$$

Small-k limit

As the Hankel transforms of short-range functions, $\Delta\tilde{h}_D(k)$ and $\Delta\tilde{c}_D(k)$ must vanish in the limit $k \to 0$. Indeed, for any reasonable potential model it is easy to show from (10A.17) that

$$\Delta\tilde{h}_D(k) \sim \Delta\tilde{c}_D(k) \sim \beta\varepsilon\sigma^3(k\sigma)^2, \qquad k\sigma \ll 1, \qquad (10A.29)$$

where ε and σ are energy and range parameters that characterize the short-range intermolecular interactions. At small wavevectors the susceptibility and its inverse therefore take on the limiting values

$$\boldsymbol{\chi}_0 = \frac{3}{4\pi}y\left[\left(1+\frac{1}{3}\rho\tilde{h}_\Delta(0)\right)\mathbf{1} - \lambda y\boldsymbol{\mathcal{D}}(\hat{\mathbf{k}})\right],$$

$$\boldsymbol{\chi}_0^{-1} = \frac{4\pi}{3}y^{-1}\left[\left(1-\frac{1}{3}\rho\tilde{c}_\Delta(0)\right)\mathbf{1} + y\boldsymbol{\mathcal{D}}(\hat{\mathbf{k}})\right]. \qquad (10A.30)$$

Note that a dependence on the *direction* of $\mathbf{k}$ remains even as the magnitude of this quantity shrinks to 0; this is associated with the long-range character of the dipolar field.

10A.2 Macroscopic theory

Polarization of the system by the applied field produces a mean, or macroscopic, field given by (10.7). In Fourier space

$$\mathbf{E}(\mathbf{k}) = \mathbf{E}_0(\mathbf{k}) - \frac{4\pi}{3}\mathbf{P}(\mathbf{k}) + \mathbf{T}(\mathbf{k}) \cdot \mathbf{P}(\mathbf{k}), \tag{10A.31}$$

where the transform of the dipolar tensor (10.6) is

$$\mathbf{T}(\mathbf{k}) = -\frac{4\pi}{3}\boldsymbol{\mathcal{D}}(\hat{\mathbf{k}}), \tag{10A.32}$$

which follows from (10A.13)–(10A.15) and (10A.26). The form of (10A.31) implies that we can write

$$\mathbf{E}(\mathbf{k}) = \boldsymbol{\chi}^{-1}(\mathbf{k}) \cdot \mathbf{P}(\mathbf{k}), \tag{10A.33}$$

where $\boldsymbol{\chi}^{-1}(\mathbf{k})$ satisfies

$$\boldsymbol{\chi}^{-1}(\mathbf{k}) = \boldsymbol{\chi}_0^{-1}(\mathbf{k}) - \frac{4\pi}{3}\left(\mathbf{1} + \boldsymbol{\mathcal{D}}(\hat{\mathbf{k}})\right). \tag{10A.34}$$

Inverting (10A.33), we find that

$$\mathbf{P}(\mathbf{k}) = \boldsymbol{\chi}(\mathbf{k}) \cdot \mathbf{E}(\mathbf{k}), \tag{10A.35}$$

or alternatively

$$\mathbf{P}(\mathbf{k}) = \frac{1}{4\pi}(\boldsymbol{\epsilon}(\mathbf{k}) - \mathbf{1}) \cdot \mathbf{E}(\mathbf{k}), \tag{10A.36}$$

where

$$\boldsymbol{\epsilon}(\mathbf{k}) = \mathbf{1} + 4\pi\boldsymbol{\chi}(\mathbf{k}) \tag{10A.37}$$

defines a generalized, wavevector-dependent tensor permittivity.[257] Written in terms of this quantity, our result for $\boldsymbol{\chi}_0^{-1}(\mathbf{k})$ becomes

$$\boldsymbol{\chi}_0^{-1}(\mathbf{k}) = \frac{4\pi}{3}\left[(\boldsymbol{\epsilon}(\mathbf{k}) - \mathbf{1})^{-1} \cdot (\boldsymbol{\epsilon}(\mathbf{k}) + 2\,\mathbf{1}) + \boldsymbol{\mathcal{D}}(\hat{\mathbf{k}})\right]. \tag{10A.38}$$

Denoting the components of $\boldsymbol{\epsilon}(\mathbf{k})$ parallel and perpendicular to $\mathbf{k}$ as $\epsilon_{\parallel}(k)$ and $\epsilon_{\perp}(k)$, respectively, we have

$$\boldsymbol{\epsilon}(\mathbf{k}) = \frac{1}{3}\left(\epsilon_{\parallel}(k) + 2\epsilon_{\perp}(k)\right)\mathbf{1} + \frac{1}{3}\left(\epsilon_{\parallel}(k) - \epsilon_{\perp}(k)\right)\boldsymbol{\mathcal{D}}(\hat{\mathbf{k}}). \tag{10A.39}$$

If we define the anisotropy parameter

$$z(k) = \frac{\epsilon_{\parallel}(k) - \epsilon_{\perp}(k)}{(\epsilon_{\parallel}(k) - 1)(\epsilon_{\perp}(k) - 1)} \tag{10A.40}$$

we now find

$$\boldsymbol{\chi}_0^{-1}(\mathbf{k}) = \frac{4\pi}{3}\left[\left(\frac{\epsilon_\perp(k)+2}{\epsilon_\perp(k)-1} - z(k)\right)\mathbf{1} + (1-z(k))\,\boldsymbol{\mathcal{D}}(\hat{\mathbf{k}})\right]. \tag{10A.41}$$

Inversion of this result yields finally

$$\begin{aligned}\boldsymbol{\chi}_0(\mathbf{k}) = \frac{3}{4\pi}\Bigg[&\frac{(\epsilon_\perp(k)-1)(2\epsilon_\parallel(k)+1)+\epsilon_\parallel(k)-\epsilon_\perp(k)}{9\epsilon_\parallel(k)}\mathbf{1}\\ &-\frac{(\epsilon_\parallel(k)-1)(\epsilon_\perp(k)-1)}{9\epsilon_\parallel(k)}(1-z(k))\,\boldsymbol{\mathcal{D}}(\hat{\mathbf{k}})\Bigg].\end{aligned} \tag{10A.42}$$

Small-k limit

Comparison of the coefficients of the tensor $\boldsymbol{\mathcal{D}}(\hat{\mathbf{k}})$ in (10A.28) and (10A.41) shows that

$$yz(k) = \frac{1}{3}\rho\Delta\tilde{c}_D(k), \tag{10A.43}$$

so that in general $\boldsymbol{\epsilon}(\mathbf{k})$ is anisotropic. However, for small k, (10A.29), (10A.40), and (10A.43) imply that

$$\epsilon_\parallel(k) - \epsilon_\perp(k) \sim y\rho\Delta\tilde{c}_D(k) \sim \frac{\beta\mu_1^2}{\sigma^3}\,\beta\varepsilon\,(\rho\sigma^3)^2\,(k\sigma)^2, \tag{10A.44}$$

so that this anisotropy disappears as $k \to 0$. In this limit, the susceptibility and its inverse can be written as

$$\begin{aligned}\boldsymbol{\chi}_0 &= \frac{3}{4\pi}\left[\frac{(\epsilon-1)(2\epsilon+1)}{9\epsilon}\mathbf{1} - \frac{(\epsilon-1)^2}{9\epsilon}\boldsymbol{\mathcal{D}}(\hat{\mathbf{k}})\right],\\ \boldsymbol{\chi}_0^{-1} &= \frac{4\pi}{3}\left[\frac{\epsilon+2}{\epsilon-1}\mathbf{1} + \boldsymbol{\mathcal{D}}(\hat{\mathbf{k}})\right],\end{aligned} \tag{10A.45}$$

where $\epsilon = \epsilon_\parallel(0) = \epsilon_\perp(0)$.

It may also be noted that, since the anisotropy in the permittivity arises from molecular interactions, it also vanishes for an ideal gas even when k is nonzero.

10A.3 Locality of the dielectric response

If we transform (10A.35) to **r**-space, we obtain

$$\mathbf{P}(\mathbf{r}) = \int d\mathbf{r}'\,\boldsymbol{\chi}(\mathbf{r}-\mathbf{r}')\cdot\mathbf{E}(\mathbf{r}'), \tag{10A.46}$$

so that in general the relation between the polarization and the Maxwell field is not local. In this section, we investigate the conditions under which our results are consistent with (10.1); loosely speaking, the dielectric constant as normally defined may be said to ‘exist’ in this limit.

We begin by considering the case in which the applied field is harmonic,

$$\mathbf{E}_0(\mathbf{r}) = \mathbf{E}_0 \cos(\mathbf{k}_0 \cdot \mathbf{r} + \varphi_0). \tag{10A.47}$$

Fourier transformation of (10A.47) yields

$$\mathbf{E}_0(\mathbf{k}) = \mathbf{E}_0 \frac{1}{2}(2\pi)^3 \left[\delta(\mathbf{k} - \mathbf{k}_0) \exp(-i\varphi_0) + \delta(\mathbf{k} + \mathbf{k}_0) \exp(i\varphi_0)\right], \tag{10A.48}$$

and on substituting (10A.48) into (10A.12), and transforming back to **r**-space, we find that

$$\mathbf{P}(\mathbf{r}) = \boldsymbol{\chi}_0(\mathbf{k}_0) \cdot \mathbf{E}_0(\mathbf{r}). \tag{10A.49}$$

Similarly, the Maxwell field is found by transforming (10A.31); this yields

$$\begin{aligned}\mathbf{E}(\mathbf{r}) &= \mathbf{E}_0(\mathbf{r}) - \frac{4\pi}{3}\left(\mathbf{1} + \boldsymbol{\mathcal{D}}(\hat{\mathbf{k}}_0)\right) \cdot \mathbf{P}(\mathbf{r}) \\ &= \left[\mathbf{1} - \frac{4\pi}{3}\left(\mathbf{1} + \boldsymbol{\mathcal{D}}(\hat{\mathbf{k}}_0)\right) \cdot \boldsymbol{\chi}_0(\mathbf{k}_0)\right] \cdot \mathbf{E}_0(\mathbf{r}). \end{aligned} \tag{10A.50}$$

It follows that

$$\mathbf{P}(\mathbf{r}) = \boldsymbol{\chi}(\mathbf{k}_0) \cdot \mathbf{E}(\mathbf{r}), \tag{10A.51}$$

where

$$\boldsymbol{\chi}(\mathbf{k}_0) = \left[\boldsymbol{\chi}_0^{-1}(\mathbf{k}_0) - \frac{4\pi}{3}\left(\mathbf{1} + \boldsymbol{\mathcal{D}}(\hat{\mathbf{k}}_0)\right)\right]^{-1}. \tag{10A.52}$$

The dielectric response is therefore local, since the polarization is strictly proportional to the applied field.

For small k_0, we have at once from (10A.28), (10A.29), and (10A.52) that

$$\boldsymbol{\chi}(\mathbf{k}_0) = \left(\chi + O(k_0\sigma)^2\right) \mathbf{1} + O(k_0\sigma)^2 \boldsymbol{\mathcal{D}}(\hat{\mathbf{k}}_0), \tag{10A.53}$$

where $\chi = \chi(0)$ is the zero-wavector limit of the isotropic part of the susceptibility. With negligible error even at optical frequencies,[258] we therefore have

$$\mathbf{P}(\mathbf{r}) = \chi \mathbf{E}(\mathbf{r}), \tag{10A.54}$$

so that the response is both independent of k_0 and isotropic.

More generally, the field may be represented as a sum of harmonics,

$$\mathbf{E}_0(\mathbf{r}) = \sum_{n<N} \mathbf{E}_0^{(n)}(\mathbf{r}) = \sum_{n<N} \mathbf{E}_0^{(n)} \cos(\mathbf{k}_n \cdot \mathbf{r} + \varphi_n), \tag{10A.55}$$

with the upper limit N reflecting the length scale on which $\mathbf{E}_0(\mathbf{r})$ changes (approximately k_N^{-1}). The response to such a field is just the sum of the responses to the individual Fourier components, so that now

$$\mathbf{P}(\mathbf{r}) = \sum_{n<N} \boldsymbol{\chi}_0(\mathbf{k}_n) \cdot \mathbf{E}_0^{(n)}(\mathbf{r}). \tag{10A.56}$$

Similarly, the Maxwell field is given by

$$\mathbf{E}(\mathbf{r}) = \sum_{n<N} \left[\mathbf{1} - \frac{4\pi}{3} \left(\mathbf{1} + \mathcal{D}(\hat{\mathbf{k}}_n) \right) \cdot \boldsymbol{\chi}_0(\mathbf{k}_n) \right] \cdot \mathbf{E}_0^{(n)}(\mathbf{r}). \tag{10A.57}$$

In this case, we see that $\mathbf{P}(\mathbf{r})$ will be a simple multiple of $\mathbf{E}(\mathbf{r})$ only if $\boldsymbol{\chi}_0(\mathbf{k}_n)$ is (essentially) independent of k_n for all $n < N$. This will be true if $k_N\sigma \ll 1$—that is, if the applied field varies slowly on a molecular scale. We then find

$$\mathbf{P}(\mathbf{r}) = \chi\mathbf{E}(\mathbf{r}) = \frac{\epsilon - 1}{4\pi}\mathbf{E}(\mathbf{r}), \tag{10A.58}$$

where $\epsilon = 1 + 4\pi\chi$ is the small-k limit of the isotropic part of $\boldsymbol{\epsilon}(\mathbf{k})$; this is (10.1) of the text, and the desired result of this section.

10A.4 Comparison of macroscopic and microscopic results

The macroscopic relations for $\boldsymbol{\chi}_0(\mathbf{k})$ and its inverse obtained in § 10A.2 can now be compared with the microscopic results of § 10A.1. In this way, we generate statistical-mechanical expressions for the components of the dielectric tensor $\boldsymbol{\epsilon}(\mathbf{k})$.

As we have seen from §10A.3, the limit $k \to 0$ is of particular importance, since the dielectric constant ϵ is just the zero-wavevector limit of $\epsilon_{\|}(k)$ and $\epsilon_{\perp}(k)$. Comparison of the isotropic parts of $\boldsymbol{\chi}_0 = \lim_{k\to 0} \boldsymbol{\chi}_0(\mathbf{k})$ as given by (10A.30) and (10A.45) yields

$$\begin{aligned}\frac{(\epsilon-1)(2\epsilon+1)}{9\epsilon} &= y\left(1 + \frac{1}{3}\rho\tilde{h}_{\Delta}(0)\right) \\ &= y\left(1 + \rho\int d\mathbf{r}\ \langle h(\mathbf{r}\omega_1\omega_2)\Delta(\omega_1\omega_2)\rangle_{\omega_1\omega_2}\right).\end{aligned} \tag{10A.59}$$

This is Kirkwood's formula for the dielectric constant of an infinite system, Eq. (10.27) of the text.

Comparison of the isotropic parts of $\boldsymbol{\chi}_0^{-1}$ in the same pair of equations yields a complementary relation,

$$\begin{aligned}\frac{\epsilon-1}{\epsilon+2} &= y\left(1 - \frac{1}{3}\rho\tilde{c}_{\Delta}(0)\right)^{-1} \\ &= y\left(1 - \rho\int d\mathbf{r}\ \langle c(\mathbf{r}\omega_1\omega_2)\Delta(\omega_1\omega_2)\rangle_{\omega_1\omega_2}\right)^{-1}.\end{aligned} \tag{10A.60}$$

This result, Eq. (10.32) of the text, was first derived by Ramshaw.[17]

Comparison of the anisotropic parts of $\boldsymbol{\chi}_0$ yields a relationship between ϵ and the parameter λ that governs the long-range behaviour of $h(\mathbf{r}\omega_1\omega_2)$. We find that

$$\frac{1}{\epsilon}\left(\frac{\epsilon-1}{3y}\right)^2 = \lambda, \tag{10A.61}$$

which is Eq. (10.31) of the text. This formula has previously[259] been given by various authors, including Nienhuis and Deutch[13] and Høye and Stell.[18]

This result can also be written as

$$\frac{(\epsilon - 1)^2}{3\epsilon y} = -\rho \lim_{k \to 0} \tilde{h}_D(k), \tag{10A.62}$$

as we see if we compare the anisotropic parts of (10A.16) and (10A.45) directly in the limit $k \to 0$.

If we now combine (10A.59) with (10A.62), we obtain two further results,

$$\frac{\epsilon - 1}{3} = y\left(1 + \frac{1}{3}\rho\tilde{h}_\Delta(0) - \frac{1}{3}\rho \lim_{k \to 0} \tilde{h}_D(k)\right), \tag{10A.63}$$

and

$$\frac{\epsilon - 1}{3\epsilon} = y\left(1 + \frac{1}{3}\rho\tilde{h}_\Delta(0) + \frac{2}{3}\rho \lim_{k \to 0} \tilde{h}_D(k)\right). \tag{10A.64}$$

Neither of these formulae is especially useful in and of itself; however, (10A.64) can be used[26] to express ϵ in terms of the site–site correlation functions $h_{\alpha\beta}(r)$ (§ 10.3.2).

Appendix 10B The Clausius–Mossotti equation

Here we present a simple derivation of the Clausius–Mossotti formula, Eq. (10.48), for the dielectric constant of nonpolar fluids, viz.

$$\frac{\epsilon - 1}{\epsilon + 2} = \frac{4\pi}{3}\rho\alpha_1, \tag{10B.1}$$

where ρ is the fluid molecular number density and α_1 is the mean (isotropic part of the) polarizability of a single molecule. For simplicity we assume spherical molecules, as in liquid Ar, so that α_1 is just the polarizability α.

Equation (10B.1) is derived from the Lorentz local field $\mathbf{E}_\mathrm{L}$, where

$$\mathbf{E}_\mathrm{L} = \mathbf{E} + \frac{4\pi}{3}\mathbf{P}, \tag{10B.2}$$

where $\mathbf{E}$ is the Maxwell field, and $\mathbf{P}$ is the macroscopic polarization of the medium. It is important to note the distinction between the three macroscopic fields $\mathbf{E}_0$, $\mathbf{E}$, and $\mathbf{E}_\mathrm{L}$ that arise in dielectric theory. $\mathbf{E}_0$ is the applied or external field, here taken to be uniform for simplicity ($\mathbf{E}$ and $\mathbf{E}_\mathrm{L}$ will also then be uniform). The Maxwell field $\mathbf{E}(\mathbf{r})$ is the mean field at a point $\mathbf{r}$ in space, which can be chosen arbitrarily in this case; $\mathbf{E}$ consists of $\mathbf{E}_0$ plus the average of the fields at $\mathbf{r}$ due to the polarized molecules. The local field $\mathbf{E}_\mathrm{L}$ is the mean field *seen by a particular molecule*. This differs from $\mathbf{E}$ since the molecules can pass freely through points $\mathbf{r}$ in space, but cannot penetrate each other; as a result, the average fields at $\mathbf{r}$ and at a particular molecular centre will differ.

We first show how one obtains (10B.1) from (10B.2), and then give a simple derivation of (10B.2). Since $\mathbf{E}_\mathrm{L}$ is the mean field seen by a single molecule, the mean polarization of that molecule will be $\alpha\mathbf{E}_\mathrm{L}$. The mean dipole moment per unit volume (polarization) $\mathbf{P}$ due to the N identical molecules is thus $N\alpha\mathbf{E}_\mathrm{L}/V$, i.e.

$$\mathbf{P} = \rho\alpha\mathbf{E}_\mathrm{L}. \tag{10B.3}$$

Eliminating $\mathbf{E}_\mathrm{L}$ between (10B.2) and (10B.3) and rearranging, we find that

$$\mathbf{P} = \frac{\rho\alpha}{1 - (4\pi/3)\rho\alpha}\mathbf{E}. \tag{10B.4}$$

Comparing (10B.4) with the definition (10.1) of the dielectric constant ϵ, i.e.

$$\mathbf{P} = \frac{\epsilon - 1}{4\pi}\mathbf{E}, \tag{10B.5}$$

gives a molecular expression for ϵ,

$$\epsilon - 1 = \frac{4\pi\rho\alpha}{1 - (4\pi/3)\rho\alpha}. \tag{10B.6}$$

When this result is rearranged and solved for α, we obtain the Clausius–Mossotti formula (10B.1).

We conclude by giving a simple derivation of the Lorentz local field, Eq. (10B.2). Usually[4] this field is derived by surrounding a particular molecule by a small imaginary cavity, and calculating the mean field produced by those molecules outside the cavity, but neglecting the contribution from those inside. The following simpler derivation[260] deserves to be better known. As mentioned earlier, the difference between the two fields $\mathbf{E}$ and $\mathbf{E}_\mathrm{L}$ is that, because the molecules avoid each other, $\mathbf{E}_\mathrm{L}$ (the mean field sensed by a given molecule) does not contain a contribution from the fields internal to other molecules. When the fluid is polarized, each of the molecules has some average dipole moment $\boldsymbol{\mu}$. We make the approximation (mean field approximation) that, in calculating the mean fields $\mathbf{E}_L$ and $\mathbf{E}$, as the molecules move about, each retains a fixed moment $\boldsymbol{\mu}$.

Recall that the field $\mathbf{E}_\mu(\mathbf{r})$ due to a dipole $\boldsymbol{\mu}$ at the origin has the form

$$\mathbf{E}_\mu(\mathbf{r}) = \boldsymbol{\mu}\cdot(3\hat{\mathbf{r}}\hat{\mathbf{r}} - \mathbf{1})r^{-3} - \frac{4\pi}{3}\boldsymbol{\mu}\delta(\mathbf{r}), \tag{10B.7}$$

where $\hat{\mathbf{r}} = \mathbf{r}/r$, $\mathbf{1}$ is the unit tensor, and $\delta(\mathbf{r})$ is the Dirac delta function (see Appendix 8E for a derivation). The first term is the familiar long-range r^{-3} term and the second is the so-called contact term. Figure 10B.1 illustrates the two terms. We have included the field lines produced by a point dipole, (a), and also a dipole of finite size, (b), as the contact term appears only at the origin. Note that in (b) the direct internal field line points in the direction opposite to that of $\boldsymbol{\mu}$; this explains the sign of the contact term.

As the molecules move around in the fluid, the difference between $\mathbf{E}$ and $\mathbf{E}_\mathrm{L}$ is due to the contact term $\mathbf{E}_\mathrm{C} = -(4\pi/3)\boldsymbol{\mu}\delta(\mathbf{r})$ in (10B.7); the former senses the complete field, while the latter senses all but the contact field. Therefore

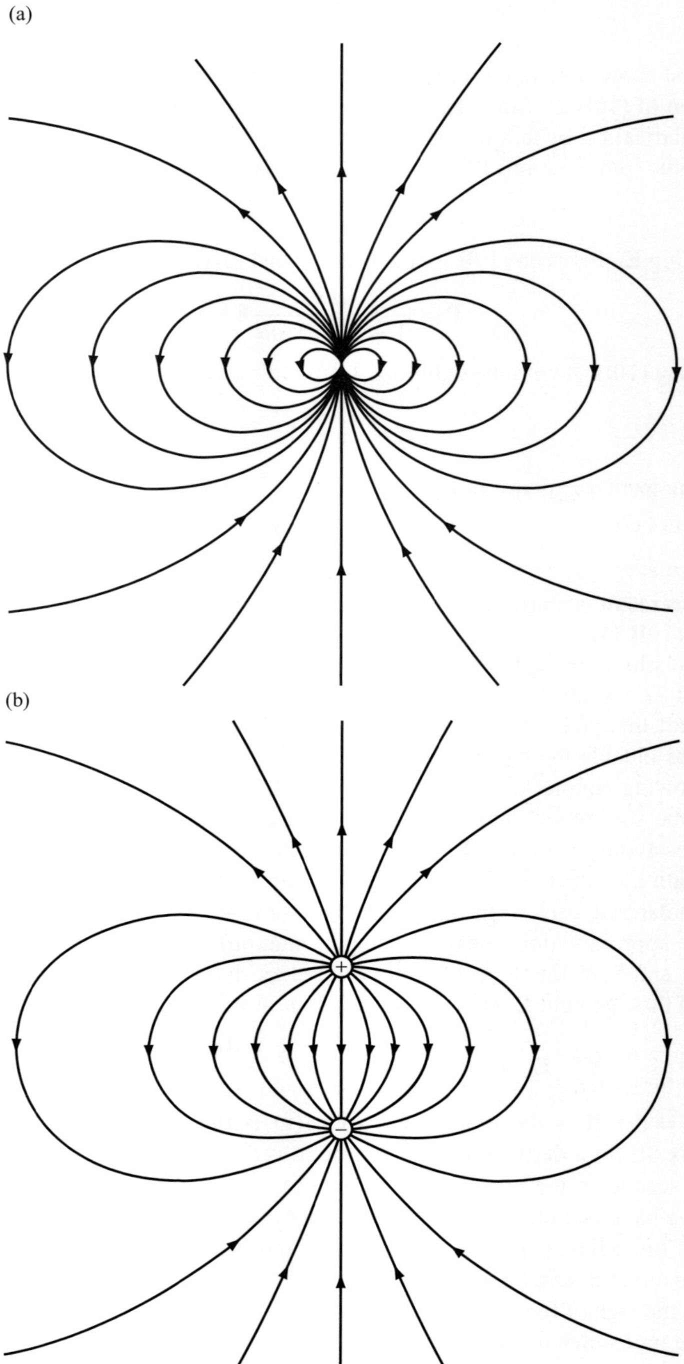

FIG. 10B.1 Electric field lines of a point (a) and finite (b) dipole. (From ref. 261.)

$$\mathbf{E}_{\rm L} = \mathbf{E} - N\langle \mathbf{E}_{\rm C} \rangle, \tag{10B.8}$$

where the factor N comes from the fact that a given molecule sees $N-1 \approx N$ others, and $\langle \mathbf{E}_{\rm C} \rangle$ is an average of $\mathbf{E}_{\rm C}$:

$$\langle \mathbf{E}_{\rm C} \rangle = \frac{1}{V} \int d\mathbf{r} \left(-\frac{4\pi}{3} \boldsymbol{\mu} \delta(\mathbf{r}) \right) = -\frac{4\pi \boldsymbol{\mu}}{3V}, \tag{10B.9}$$

where V is the system volume. Hence

$$\mathbf{E}_{\rm L} = \mathbf{E} + \frac{4\pi}{3} \rho \boldsymbol{\mu} = \mathbf{E} + \frac{4\pi}{3} \mathbf{P}, \tag{10B.10}$$

as in (10B.2); we note that this result applies for polar as well as nonpolar molecules.

Appendix 10C Perturbation theory for the angular correlation parameters G_l

In this appendix we show how the angular correlation parameters for a fluid of linear molecules, viz.

$$G_l = \rho \int d\mathbf{r} \, \langle g(\mathbf{r}\omega_1\omega_2) P_l(\cos\gamma_{12}) \rangle_{\omega_1\omega_2}, \qquad l = 1, 2 \ldots, \tag{10C.1}$$

may be evaluated using perturbation theory. In (10C.1), $g(\mathbf{r}\omega_1\omega_2)$ is the pair correlation function, while γ_{12} is the angle between the symmetry axes of molecules 1 and 2. G_1 enters into the theory of the dielectric constant (§§ 10.3, 10.5), and G_2 appears in treatments of both the Kerr effect (§ 10.8) and depolarized light scattering (§ 11.2).

We employ the u-expansion of $g(\mathbf{r}\omega_1\omega_2)$ (cf. § 4.5) in powers of the anisotropic part of the potential $u_a(\mathbf{r}\omega_1\omega_2) = u(\mathbf{r}\omega_1\omega_2) - u_0(r)$, where $u_0(r) = \langle u(\mathbf{r}\omega_1\omega_2) \rangle_{\omega_1\omega_2\omega}$ defines the reference fluid, for which the distribution functions are presumed known. By definition, $\langle u_a(\mathbf{r}\omega_1\omega_2) \rangle_{\omega_1\omega_2\omega} = 0$. For simplicity, we will consider only multipole-like potentials,[34] for which the stronger conditions $\langle u_a(\mathbf{r}\omega_1\omega_2) \rangle_{\Omega} = 0$ also hold, where Ω is *any one* of ω_1, ω_2, and ω. We then find that there is no first-order correction to any of the G_l, and in second order, only two- and three-body terms contribute. We obtain[37]

$$G_l = G_{lA} + G_{lB}, \tag{10C.2}$$

where

$$G_{lA} = \frac{1}{2} \beta^2 \rho \int d\mathbf{r}_{12} \, g_0(r_{12}) \langle u_a(\mathbf{r}_{12}\omega_1\omega_2)^2 P_l(\cos\gamma_{12}) \rangle_{\omega_1\omega_2} \tag{10C.3}$$

and

$$G_{lB} = \beta^2 \rho^2 \int d\mathbf{r}_{12} d\mathbf{r}_{13} \, g_0(r_{12}r_{13}r_{23}) \times \langle u_a(\mathbf{r}_{13}\omega_1\omega_3) u_a(\mathbf{r}_{23}\omega_2\omega_3) P_l(\cos\gamma_{12}) \rangle_{\omega_1\omega_2\omega_3}, \tag{10C.4}$$

where $g_0(r_{12})$ and $g_0(r_{12}r_{13}r_{23})$ are the pair and triplet correlation functions for the reference system, respectively. G_{lA} measures the direct angular correlation between molecules 1 and 2, while G_{lB} reflects indirect correlations caused by their interaction with molecule 3.

To evaluate the angular averages in (10C.3) and (10C.4) we expand the potential $u_a(\mathbf{r}\omega_1\omega_2)$ in spherical harmonics (cf. (2.23)),

$$u_a(\mathbf{r}\omega_1\omega_2)$$
$$= \sum_{l_1 l_2 l} \sum_{m_1 m_2 m} u(l_1 l_2 l; r) C(l_1 l_2 l; m_1 m_2 m) Y_{l_1 m_1}(\omega_1) Y_{l_2 m_2}(\omega_2) Y_{lm}(\omega)^*. \qquad (10C.5)$$

Explicit expressions for $u(l_1 l_2 l; r)$ for multipolar interactions are given in § 2.4. We can now show that

$$\langle u_a(\mathbf{r}_{12}\omega_1\omega_2)^2 P_L(\cos\gamma_{12})\rangle_{\omega_1\omega_2}$$
$$= (4\pi)^{-3} \sum_{\substack{l_1 l_2 l \\ l_1' l_2'}} (-)^{l_2+l_2'} \left[\left(2l_1'+1\right)\left(2l_2'+1\right)\right]^{1/2} (2l+1)$$
$$\times C\left(Ll_1'l_1;000\right) C\left(Ll_2'l_2;000\right) \begin{Bmatrix} l_2' & l_1' & l \\ l_1 & l_2 & L \end{Bmatrix} u\left(l_1 l_2 l; r_{12}\right) u\left(l_1' l_2' l; r_{12}\right), \qquad (10C.6)$$

where $\{\cdots\}$ denotes a $6j$ symbol (§ A.5). In the special case when u_a consists of only one spherical harmonic term $(l_1 l_2 l)$, then (10C.6) vanishes for odd values of L. We also find that

$$\langle u_a(\mathbf{r}_{13}\omega_1\omega_3) u_a(\mathbf{r}_{23}\omega_2\omega_3) P_L(\cos\gamma_{12})\rangle_{\omega_1\omega_2\omega_3}$$
$$= (4\pi)^{-3} \sum_{l_3 l} (2l+1)(2L+1)^{-1} u(Ll_3 l; r_{13}) u(Ll_3 l; r_{23}) P_l(\cos\hat{\alpha}_3), \qquad (10C.7)$$

where $\hat{\alpha}_3$ is the angle at molecule 3 of the triangle formed by molecules 1, 2, and 3.

Since for multipolar interactions $u(l_1 l_2 l; r) = u(l_1 l_2 l)/r^{l+1}$ (see § 2.4), it now follows that, to second order in the perturbation,

$$G_{LA} = \frac{1}{2}(4\pi)^{-2}\beta^2\rho \sum_{\substack{l_1 l_2 l \\ l_1' l_2'}} (-)^{l_2+l_2'} \left[\left(2l_1'+1\right)\left(2l_2'+1\right)\right]^{1/2} (2l+1)$$
$$\times C\left(Ll_1'l_1;000\right) C\left(Ll_2'l_2;000\right) \begin{Bmatrix} l_2' & l_1' & l \\ l_1 & l_2 & L \end{Bmatrix} u(l_1 l_2 l) u\left(l_1' l_2' l\right) J_{2l+2}, \qquad (10C.8)$$

and

$$G_{LB} = \frac{1}{2}(4\pi)^{-1}\beta^2\rho^2 \sum_{l_3 l} (2l+1)(2L+1)^{-1} u(Ll_3 l)^2 I_{l+1\, l+1\, l}. \qquad (10C.9)$$

Here J_n and $I_{nn'l}$ are integrals over the reference fluid correlation functions defined by

$$J_n = \int_0^\infty \mathrm{d}r_{12}\, g_0(r_{12}) r_{12}^{-(n-2)}, \tag{10C.10}$$

$$I_{nn'l} = \int_0^\infty \mathrm{d}r_{12}\, r_{12} \int_0^\infty \mathrm{d}r_{13}\, r_{13}^{-(n-1)} \times \int_{|r_{12}-r_{13}|}^{r_{12}+r_{13}} \mathrm{d}r_{23}\, r_{23}^{-(n'-1)} g_0(r_{12}r_{13}r_{23}) P_l(\cos\hat{\alpha}_3). \tag{10C.11}$$

In the special case when u_a arises from dipolar and quadrupolar forces, (10C.8) and (10C.9) reduce to

$$G_1 = \frac{16\pi^2}{9}\beta^2\mu^4\rho^2 I_{332} - \frac{8\pi}{5}\beta^2\mu^2 Q^2\rho J_8 + \frac{8\pi^2}{3}\beta^2\mu^2 Q^2\rho^2 I_{443} \tag{10C.12}$$

and

$$G_2 = \frac{4\pi}{75}\beta^2\mu^4\rho J_6 + \frac{8\pi}{35}\beta^2\mu^2 Q^2\rho J_8 + \frac{8\pi^2}{5}\beta^2\mu^2 Q^2\rho^2 I_{443} + \frac{16\pi}{35}\beta^2 Q^4\rho J_{10} + \frac{112\pi^2}{25}\beta^2 Q^4\rho^2 I_{554}. \tag{10C.13}$$

Calculation of the integrals J_n requires the pair correlation function for the isotropic reference fluid. This may be obtained either from simulation or, in the special case when $u_0(r)$ is a Lennard-Jones potential, from an equation due to Goldman.[262] In calculating the integrals $I_{nn'l}$ the superposition approximation, or a variant thereof,[263] is normally employed. Note that the dipolar three-body integral I_{332} is conditionally convergent, and care must be exercised in its evaluation.[264]

Equation (10C.12) for G_1 has been generalized in various ways. Joslin[192] showed that for nonlinear molecules it still applies provided we make the replacements $Q^2 \to (\boldsymbol{\mu}\cdot\mathbf{Q}\cdot\boldsymbol{\mu})^2/\mu^4$ and $\frac{2}{3}\mathbf{Q}:\mathbf{Q}$, respectively, in the two- and three-body terms. Goldman and Joslin extended the theory to nonlinear multipoles of all orders,[35] and later generalized this result to a multicomponent mixture.[36] Finally, Gray and Gubbins[37] have considered the perturbational effects of anisotropic overlap and dispersion forces.

Appendix 10D Ewald sums for ionic and dipolar systems

In this appendix we derive expressions for the lattice energies of ionic and dipolar systems. The results are used in computer simulations of the dielectric constant, as discussed in § 10.7. For simplicity, we treat a simple cubic lattice consisting of a replicated array of the central simulation cell. Our analysis is similar in some respects to that of de Leeuw *et al.*,[97,98] but differs in that the lattice, though large, is *not* infinite in extent. This is because a reaction field must, as a rule, be applied if ϵ is to be simulated accurately—meaning that the structure must be surrounded by a dielectric continuum.

We consider the Coulombic lattice sum

$$\mathcal{U}_{CC} = \sum_{\mathbf{n}} \sum_{i<j} \frac{q_i q_j}{|\mathbf{r}_{ij} + \mathbf{n}L|} \tag{10D.1}$$

and its dipolar counterpart

$$\mathcal{U}_{DD} = -\sum_{\mathbf{n}} \sum_{i<j} \boldsymbol{\mu}_i \boldsymbol{\mu}_j : \nabla\nabla \frac{1}{|\mathbf{r}_{ij} + \mathbf{n}L|}. \tag{10D.2}$$

Here L is the lattice parameter, and $\mathbf{n}$ is a vector, with integer components, specifying the coordinates of the centre of each image cell relative to the central cell. q_i, q_j, $\boldsymbol{\mu}_i$, and $\boldsymbol{\mu}_j$ are the charges and dipole moments at positions $\mathbf{r}_i$ and $\mathbf{r}_j$, and $\mathbf{r}_{ij} = \mathbf{r}_i - \mathbf{r}_j$. We assume the charged system is electrically neutral, i.e. $\sum_i q_i = 0$. Both $\mathcal{U}_{CC}$ and $\mathcal{U}_{DD}$ then contain contributions of order $O(n^{-3})$. This slow convergence makes term-by-term evaluation of (10D.1) and (10D.2) impractical, and we therefore seek an alternative procedure. We will concentrate first on evaluating $\mathcal{U}_{CC}$, since $\mathcal{U}_{DD}$ can be derived subsequently by differentiation.

Since we generally wish to use these expressions in connection with fluctuation formulae appropriate to spherical geometry (cf. § 10.7), we must perform the lattice sums over successive shells of a pseudospherical lattice[157] of large but finite radius $R = n_{\max}L$, i.e.

$$\mathcal{U}_{CC}(R) = \sum_{\mathbf{n}:0 \le n \le n_{\max}} \sum_{i<j} \frac{q_i q_j}{|\mathbf{r}_{ij} + \mathbf{n}L|}. \tag{10D.3}$$

We can write this as

$$\mathcal{U}_{CC}(R) = \lim_{s \to 0} \mathcal{U}_{CC}(R, s), \tag{10D.4}$$

where $s > 0$ and

$$\mathcal{U}_{CC}(R, s) = \sum_{\mathbf{n}:0 \le n \le n_{\max}} \sum_{i<j} \frac{q_i q_j}{|\mathbf{r}_{ij} + \mathbf{n}L|} \exp(-sn^2). \tag{10D.5}$$

The limit of this representation $\mathcal{U}_{CC}(s) \equiv \lim_{R \to \infty} \mathcal{U}_{CC}(R, s)$ may now be written as

$$\mathcal{U}_{CC}(s) = \sum_{\mathbf{n}} \sum_{i<j} \frac{q_i q_j}{|\mathbf{r}_{ij} + \mathbf{n}L|} \exp(-sn^2). \tag{10D.6}$$

This sum is absolutely convergent, and the sum over lattice vectors may be taken to infinity in any order we choose. As we shall see, this allows us to make use of a transformation due to Ewald[159,160] that permits rapid evaluation of $\mathcal{U}_{CC}(s)$. We then simply use

$$\mathcal{U}_{CC} = \lim_{s\to 0} \mathcal{U}_{CC}(s) \tag{10D.7}$$

to compute the desired lattice sum.

We therefore wish to calculate $\lim_{s\to 0} \sum_{i<j} q_i q_j \psi(\mathbf{r}_{ij}, s)$, where

$$\psi(\mathbf{r}, s) = \sum_{\mathbf{n}} \frac{\exp(-sn^2)}{|\mathbf{r} + \mathbf{n}L|} \tag{10D.8}$$

may be thought of as the 'screened' Coulombic potential experienced by a charge. To evaluate this potential we make use of the identity

$$x^{-2p} = \frac{1}{\Gamma(p)} \int_0^\infty \mathrm{d}t\, t^{p-1} \exp(-tx^2), \qquad p > 0, \tag{10D.9}$$

which is just the definition of the Γ-function.[255] Setting $x = |\mathbf{r} + \mathbf{n}L|/L$ and $p = 1/2$, we find that

$$\psi(\mathbf{r}, s) = \frac{1}{\pi^{1/2} L} \int_0^\infty \mathrm{d}t\, t^{-1/2} \sum_{\mathbf{n}} \exp[-sn^2 - t(\mathbf{r}/L + \mathbf{n})^2]. \tag{10D.10}$$

We now split the integral over t into two parts: one for which $\alpha^2 \le t < \infty$, and the other for which $0 \le t \le \alpha^2$. Here α is an arbitrary, positive, dimensionless parameter. This yields

$$\psi(\mathbf{r}, s) = \psi_A(\mathbf{r}, s) + \psi_B(\mathbf{r}, s), \tag{10D.11}$$

where

$$\psi_A(\mathbf{r}, s) = \frac{1}{\pi^{1/2} L} \int_{\alpha^2}^\infty \mathrm{d}t\, t^{-1/2} \sum_{\mathbf{n}} \exp[-sn^2 - t(\mathbf{r}/L + \mathbf{n})^2] \tag{10D.12}$$

and

$$\psi_B(\mathbf{r}) = \frac{1}{\pi^{1/2} L} \int_0^{\alpha^2} \mathrm{d}t\, t^{-1/2} \sum_{\mathbf{n}} \exp[-sn^2 - t(\mathbf{r}/L + \mathbf{n})^2]. \tag{10D.13}$$

We first consider $\psi_A(\mathbf{r}, s)$. Making the substitution $x = t^{1/2}|\mathbf{r}/L + \mathbf{n}|$ in (10D.12) yields

$$\psi_A(\mathbf{r}, s) = \sum_{\mathbf{n}} \exp(-sn^2) \frac{\operatorname{erfc}(\alpha|\mathbf{r} + \mathbf{n}L|/L)}{|\mathbf{r} + \mathbf{n}L|}. \tag{10D.14}$$

Here $\operatorname{erfc} x = 2\pi^{-1/2} \int_x^\infty \mathrm{d}t \exp(-t^2)$ is the complementary error function.[255] For large x, $\operatorname{erfc} x \approx \exp(-x^2)/\pi^{1/2} x$. Therefore, when $n \gg \alpha^{-1}$, the lattice sum in (10D.14) contains contributions of order $O(\exp[-(\alpha^2 + s)n^2]/(\alpha n^2))$; it is therefore absolutely (and rapidly) convergent for $\alpha > 0$ for all $s \ge 0$. We can therefore take the limit $s \to 0$ for this part of the sum without difficulty.

We next consider $\psi_B(\mathbf{r}, s)$. This function is in fact singular, $O(s^{-1})$, as we show below, and further mathematical manipulations are needed to extract its behaviour in the limit $s \to 0$. We first write

$$\psi_B(\mathbf{r}, s) = \frac{1}{\pi^{1/2}L} \int_0^{\alpha^2} \mathrm{d}t\, t^{-1/2} \exp\left[-\frac{str^2}{(s+t)L^2}\right] \times \sum_{\mathbf{n}} \exp\left[-(s+t)\left(\mathbf{n} + \frac{t\mathbf{r}}{(s+t)L}\right)^2\right]. \quad (10D.15)$$

To evaluate the lattice sum in (10D.15) we make use of the identity

$$\sum_{\mathbf{n}} \exp[-\zeta(\mathbf{n}+\mathbf{m})^2] = (\pi/\zeta)^{3/2} \sum_{\mathbf{n}} \exp[-\pi^2 n^2/\zeta + 2\pi i \mathbf{n}\cdot\mathbf{m}], \quad (10D.16)$$

a form of the Jacobi imaginary transformation for theta-functions.[160] Here $\mathbf{m}$ and $\zeta > 0$ are arbitrary vector and scalar parameters, respectively. This transformation was first used in the context of lattice sums by Ewald.[159] Setting $\mathbf{m} = t\mathbf{r}/(s+t)L$ and $\zeta = s + t$, we then find that

$$\psi_B(\mathbf{r}, s) = \psi_B^s(\mathbf{r}, s) + \psi_B^{ns}(\mathbf{r}, s), \quad (10D.17)$$

where

$$\psi_B^s(\mathbf{r}, s) = \frac{\pi}{L} \int_0^{\alpha^2} \mathrm{d}t\, t^{-1/2} (s+t)^{-3/2} \exp\left[-\frac{str^2}{(s+t)L^2}\right] \quad (10D.18)$$

and

$$\psi_B^{ns}(\mathbf{r}, s) = \frac{\pi}{L} \int_0^{\alpha^2} dt\, t^{-1/2} (s+t)^{-3/2} \exp\left[-\frac{str^2}{(s+t)L^2}\right] \times \sum_{\mathbf{n}\neq 0} \exp\left[-\frac{\pi^2 n^2}{s+t} + \frac{2\pi i t \mathbf{n}\cdot\mathbf{r}}{(s+t)L}\right]. \quad (10D.19)$$

Let us consider first $\psi_B^s(\mathbf{r}, s)$. If we substitute $x^2 = str^2/[(s+t)L^2]$, we find that

$$\psi_B^s(\mathbf{r}, s) = \frac{1}{r}\left(\frac{\pi}{s}\right)^{3/2} \operatorname{erf}\left[\frac{\alpha r}{L}\left(\frac{s}{s+\alpha^2}\right)^{1/2}\right], \quad (10D.20)$$

where $\operatorname{erf} x = 2\pi^{-1/2} \int_0^x \mathrm{d}t \exp(-t^2)$ is the error function.[255] For small x, $\operatorname{erf} x = 2\pi^{-1/2}(x - x^3/3 + O(x^5))$, and therefore for small s

$$\psi_B^s(\mathbf{r}, s) = \frac{2\pi}{L}\left[s^{-1} - \frac{1}{2\alpha^2} - \frac{r^2}{3L^2} + O(s)\right]. \quad (10D.21)$$

We note that $\psi_B^s(\mathbf{r}, s)$ is singular at $s = 0$. However, the singular piece does not depend on r, and hence does not contribute to $\mathcal{U}_{CC}(s)$ by virtue of the charge neutrality condition $\sum_i q_i = 0$. We therefore experience no difficulty in taking the limit $s \to 0$ in (10D.7).

Next we consider $\psi_B^{ns}(\mathbf{r}, s)$. Substituting $x^2 = (\pi\mathbf{n} + is\mathbf{r}/L)^2[t(s+t)]^{-1}$, we find that

$$\psi_B^{ns}(\mathbf{r}, s) = \sum_{\mathbf{n}\neq 0} \frac{\exp[(\omega^2 - \pi^2 n^2)/s]}{|\pi\mathbf{n}\mathbf{L} + is\mathbf{r}|} 2\pi s^{-1/2} D(\omega/s^{1/2}), \tag{10D.22}$$

where $\omega^2 = (\pi\mathbf{n} + is\mathbf{r}/L)^2\alpha^2(s+\alpha^2)^{-1}$ and

$$D(x) = \exp(-x^2)\int_0^x \mathrm{d}t \exp(t^2) \tag{10D.23}$$

is Dawson's integral.[255] For large x, this has the asymptotic expansion $D(x) = (2x)^{-1}(1 + O(x^{-1}))$. Thus for small s,

$$\begin{aligned}\psi_B^{ns}(\mathbf{r}, s) &= \sum_{\mathbf{n}\neq 0} \frac{\pi \exp[(\omega^2 - \pi^2 n^2)/s]}{\omega|\pi\mathbf{n}L + is\mathbf{r}|}[1 + O(s)] \\ &= \sum_{\mathbf{n}\neq 0} \frac{\exp[-\pi^2 n^2/\alpha^2 + 2\pi i\mathbf{n}\cdot\mathbf{r}/L]}{\pi n^2 L}[1 + O(s)].\end{aligned} \tag{10D.24}$$

Thus the beauty of Ewald's transformation (10D.16) is that it permits *both* parts of the lattice sum (10D.11) to be written as rapidly and absolutely convergent series. The value of α may be chosen so that approximately equal amounts[163] of computer time are spent evaluating (10D.14) and (10D.24), or as discussed in § 10.7.

Finally, collecting together our results yields the Coulombic lattice energy as

$$\mathcal{U}_{CC} = \sum_{i<j} q_i q_j \bar{\psi}(\mathbf{r}_{ij}), \tag{10D.25}$$

where $\bar{\psi}(\mathbf{r})$ is that part of $\psi(\mathbf{r}, s)$ that is both $O(s^0)$ and r-dependent, i.e.

$$\begin{aligned}\bar{\psi}(\mathbf{r}) &= \sum_{\mathbf{n}} \frac{\operatorname{erfc}(\alpha|\mathbf{r} + \mathbf{n}L|/L)}{|\mathbf{r} + \mathbf{n}L|} \\ &\quad + \sum_{\mathbf{n}\neq 0} \frac{\exp(-\pi^2 n^2/\alpha^2 + 2\pi i\mathbf{n}\cdot\mathbf{r}/L)}{\pi n^2 L} - \frac{2\pi r^2}{3L^3}.\end{aligned} \tag{10D.26}$$

Equation (10D.26) has been applied to simulations of point-charge models of dielectric fluids such as TIP4P and SPC water.[124]

The dipolar analogue of (10D.25) is

$$\mathcal{U}_{DD} = -\sum_{i<j} \boldsymbol{\mu}_i \cdot \bar{\mathbf{T}}(\mathbf{r}_{ij}) \cdot \boldsymbol{\mu}_j, \tag{10D.27}$$

where the effective **T**-tensor for simulations in Ewald sum geometry is given by $\bar{\mathbf{T}}(\mathbf{r}) = \nabla\nabla\bar{\psi}(\mathbf{r})$, or

$$\begin{aligned}
\bar{\mathbf{T}}(\mathbf{r}) \\
&= \sum_{\mathbf{n}} \left\{ \left(\frac{3\mathbf{r}^{(n)}\mathbf{r}^{(n)}}{r^{(n)5}} - \frac{\mathbf{1}}{r^{(n)3}} \right) \left(\operatorname{erfc}[\alpha r^{(n)}/L] + \frac{2\alpha r^{(n)}}{\sqrt{\pi}L} \exp[-\alpha^2 r^{(n)2}/L^2] \right) \right. \\
&\quad \left. - \mathbf{r}^{(n)}\mathbf{r}^{(n)} \frac{4\alpha^3}{\sqrt{\pi} r^{(n)2} L^3} \exp[-\alpha^2 r^{(n)2}/L^2] \right\} \\
&\quad - \frac{4\pi}{L^3} \sum_{\mathbf{n}\neq 0} \frac{\mathbf{nn}}{n^2} \exp(-\pi^2 n^2/\alpha^2 + 2\pi i \mathbf{n}\cdot\mathbf{r}/L) - \frac{4\pi}{3L^3}\mathbf{1}, \qquad \text{(10D.28)}
\end{aligned}$$

where $\mathbf{r}^{(n)} = \mathbf{r} + \mathbf{n}L$. These formulae are equivalent to (10.128)–(10.130) of § 10.7.

Appendix 10E Integral equations and the dielectric constant

Chapter 5 discusses integral equation theories for the angular pair correlation function $g(\mathbf{r}\omega_1\omega_2)$ and the site–site pair correlation functions $g_{\alpha\beta}(r)$ of molecular fluids. In this appendix, we review the principal features of several approximate closures for the molecular Ornstein–Zernike equation, whose predictions for ϵ are compared with simulation in § 10.7. We also investigate the shortcomings of the site–site Ornstein–Zernike equation, or interaction site model, in both its original and diagrammatically proper versions, and we explain why it is so difficult to obtain nontrivial predictions for ϵ using this method.

10E.1 The molecular Ornstein–Zernike equation

The molecular Ornstein-Zernike (OZ) integral equation,

$$h(12) = c(12) + \rho \int d\mathbf{r}_3 \, \langle c(13)h(32)\rangle_{\omega_3}, \qquad \text{(10E.1)}$$

formally relates the total and direct correlation functions, $h(12) \equiv h(\mathbf{r}\omega_1\omega_2)$ and $c(12) \equiv c(\mathbf{r}\omega_1\omega_2)$. Solution of (10E.1) requires that we supply a closure relation, in the form of an approximate relation for $c(12)$ in terms of $h(12)$. Various closures have been studied, including the mean spherical approximation (MSA), and the hypernetted chain (HNC) approximation and its derivatives.

Mean spherical approximation

The MSA is usually applied to fluids of molecules with hard, spherical cores. It consists of the closures

$$\begin{aligned}
h(12) &= -1, \qquad & r_{12} < \sigma, \\
c(12) &= -\beta u(12), \qquad & r_{12} > \sigma, \qquad \text{(10E.2)}
\end{aligned}$$

where $u(12)$ is the potential outside the core. The first of these is exact, while the second is an approximation that only becomes exact in the limit $r_{12} \to \infty$.

The MSA has been solved exactly for a few simple cases, including the dipolar hard sphere fluid, for which $u(12) = -\mu_1^2 D(12)/r_{12}^3$. For ϵ one finds that[178]

$$\epsilon = \frac{(1+4\xi)^2(1+\xi)^4}{(1-2\xi)^6}, \tag{10E.3}$$

where ξ is given implicitly by

$$\frac{(1+4\xi)^2}{(1-2\xi)^4} - \frac{(1-2\xi)^2}{(1+\xi)^4} = 3y. \tag{10E.4}$$

Details of the solution can be found in §§ 5.4.2–5.4.5.

MSA solutions have also been obtained for dipolar–quadrupolar hard spheres,[186] polarizable dipolar–quadrupolar hard spheres,[204] and the dipolar Yukawa fluid.[134]

Hypernetted chain approximation

The HNC approximation is the closure

$$h(12) = c(12) + \beta u(12) + \ln[1 + h(12)]. \tag{10E.5}$$

HNC solutions must be obtained numerically, generally by expanding in a suitable set of spherical harmonic basis functions, transforming between **r**- and **k**-space, and iterating until convergence is achieved. Efficient algorithms for solving integral equations in nonspherical systems are discussed by Anta *et al.*,[265] Lado *et al.*,[266] and Kinoshita and Harada.[267]

HNC calculations of ϵ have been reported for dipolar hard spheres,[181] dipolar hard ellipsoids,[190] the polarizable dipolar–quadrupolar LJ fluid,[138] a polarizable dipolar–quadrupolar two-centre LJ fluid,[147] and a fluid with site–site LJ plus Coulombic interactions and central polarizability.[268]

Quite generally, we can write $u(12) = u_0(r_{12}) + u_a(12)$, where $u_0(r_{12}) = \langle u(12)\rangle_{\omega_1\omega_2\omega}$ is the average potential at fixed separation r_{12}. This potential may be considered to define a 'reference' fluid, and if the HNC approximation is also applied directly to this, we find that

$$h_0 = c_0 + \beta u_0 + \ln g_0. \tag{10E.6}$$

If we now write $h(12) = h_0(r_{12}) + h_a(12)$ and $c(12) = c_0(r_{12}) + c_a(12)$, it then follows that

$$h_a = c_a + \beta u_a + \ln(1 + h_a/g_0). \tag{10E.7}$$

Equations (10E.6) and (10E.7) consitute the reference hypernetted chain (RHNC) approximation. Since the reference fluid structure is generally rather poorly predicted by (10E.6), in practice g_0 is usually taken directly from simulation.

RHNC estimates of ϵ are available for dipolar hard spheres,[180,183] dipolar soft spheres,[108,180,184] dipolar LJ fluids,[184] dipolar–quadrupolar hard spheres,[187] polarizable dipolar–quadrupolar–octopolar hard spheres,[205] polarizable dipolar–quadrupolar LJ fluids,[138] polarizable two-centre dipolar-quadrupolar LJ fluids,[147] and SPCE water.[206]

The linearized hypernetted chain (LHNC) closure is derived from the HNC approximation by expanding the logarithm and retaining only the term linear in h_a. This yields the decoupled pair of equations

$$\begin{aligned} h_0 &= c_0 + \beta u_0 + \ln g_0, \\ c_a + \beta u_a &= h_a h_0 / g_0, \end{aligned} \tag{10E.8}$$

from which it is apparent that the anisotropic part of the potential does not affect the radial structure of the fluid. The linearity of the closure implies that, when u_a can be represented as a finite sum of spherical harmonic basis functions, so too can h_a and c_a. In particular, when there is only dipolar anisotropy in the potential, one finds that

$$\begin{aligned} h_a(12) &= h_\Delta(r_{12})\Delta(12) + h_D(r_{12})D(12), \\ c_a(12) &= c_\Delta(r_{12})\Delta(12) + c_D(r_{12})D(12). \end{aligned} \tag{10E.9}$$

The quadratic hypernetted chain (QHNC) closure results when we also keep the quadratic term in the ln expansion. Then

$$h(12) = c(12) + \beta u(12) + \ln g_0 + h_a/g_0 - \frac{1}{2}(h_a/g_0)^2. \tag{10E.10}$$

Because h_a^2 contains a spherically symmetric component, no decoupling of isotropic and anisotropic components occurs. If, however, the reference fluid is also treated in the HNC approximation, we obtain the decoupled pair of equations

$$\begin{aligned} h_0 &= c_0 + \beta u_0 + \ln g_0, \\ c_a + \beta u_a &= h_a h_0 / g_0 + \frac{1}{2}(h_a/g_0)^2, \end{aligned} \tag{10E.11}$$

which consitute the reference QHNC (RQHNC) approximation, which is often used in place of (10E.10).

In the QHNC and RQHNC, h_a and c_a can no longer be represented as finite sums of harmonics. In particular, (10E.9) is no longer exact for a dipolar fluid; its use[179] introduces further errors into the theory, and leads to poor results for ϵ,[269] even precluding solutions in certain instances.[186]

LHNC and QHNC solutions for ϵ are available for several potential models: dipolar hard spheres,[179] dipolar soft spheres,[184] dipolar–quadrupolar hard spheres,[186] polarizable dipolar hard spheres,[21] polarizable dipolar–quadrupolar hard spheres,[204] and polarizable dipolar–quadrupolar LJ fluids.[140]

10E.2 Site–site Ornstein–Zernike equations[270]

As we saw in § 5.5.1, the site–site Ornstein–Zernike equation, or interaction site model (ISM) can be written in **k**-space as

$$\tilde{\boldsymbol{\omega}} + \rho\tilde{\mathbf{h}} = (\mathbf{1} - \rho\tilde{\boldsymbol{\omega}}\tilde{\mathbf{c}})^{-1}\tilde{\boldsymbol{\omega}}. \tag{10E.12}$$

Here $\tilde{\mathbf{h}}$ and $\tilde{\mathbf{c}}$ are matrices whose elements are Fourier transforms of the total and direct site–site correlation functions $h_{\alpha\beta}(r)$ and $c_{\alpha\beta}(r)$, and $\tilde{\boldsymbol{\omega}}$ is the intramolecular correlation function, with elements $\omega_{\alpha\beta}(k) = \sin kl_{\alpha\beta}/kl_{\alpha\beta}$, where $l_{\alpha\beta} = |\mathbf{l}_\alpha - \mathbf{l}_\beta|$ is the distance between the sites α and β in a single molecule.

By analogy with (3.119), one might reasonably suppose that the site–site direct correlation functions behave asymptotically as

$$c_{\alpha\beta}(r) \sim -\beta u_{\alpha\beta}(r), \qquad r \to \infty \tag{10E.13}$$

where $u_{\alpha\beta}(r)$ is the site–site potential. This asymptotic decay will always result when standard closures such as the MSA or HNC are employed. For charged sites, the long-range part of the site–site potential is the Coulomb term $q_\alpha q_\beta/r$. In **k**-space, this results in a small-k divergence

$$c_{\alpha\beta}(k) \sim -4\pi\beta q_\alpha q_\beta/k^2, \qquad k \to 0. \tag{10E.14}$$

One can show that,[271,272] for *any* closure of (10E.12) satisfying (10E.14),

$$\lim_{k\to 0} k^{-2}\sum_{\alpha\beta} q_\alpha \tilde{h}_{\alpha\beta}(k) q_\beta = -y\mu_1^2/[\rho(1+3y)], \tag{10E.15}$$

where $\mu_1^2 = \sum_{\alpha\beta} q_\alpha \mathbf{l}_\alpha \cdot \mathbf{l}_\beta q_\beta$, and $y = 4\pi\rho\beta\mu_1^2/9$. This result holds for molecules of any symmetry. Substituting (10E.15) into (10.40) now yields

$$\epsilon = 1 + 3y \tag{10E.16}$$

for the dielectric constant of rigid polar fluids in the ISM. This result is both trivial and incorrect. We have seen in Chapters 5 and 8 that the ISM often provides a reasonable qualitative description of site–site correlations at short range; its inability to correctly describe the dielectric response is associated with the fact that ϵ is a long-wavelength property, linked, as we have seen in (10.31), to the asymptotic decay of the correlation function.

To obtain nontrivial predictions for the dielectric constant, one is faced with two options: either modify the closure in some way to obtain a more correct theory, or modify the ISM OZ equation itself. The first route has been propounded by Stell and co-workers:[273–275] the ISM OZ equation is left intact, and either (a) the closure for $c_{\alpha\beta}(r)$ is chosen so as to reproduce the observed dielectric constant (but there is then no independent means for calculating $c_{\alpha\beta}(r)$, and the method is not predictive of ϵ); or (b) a very complex, and nonituitive, *nonlocal* closure is invoked,[275] coupling all of the molecular interaction sites—once again of very limited predictive ability.

The second, and potentially more fruitful, avenue has been explored by Chandler and co-workers.[177,276] The basic problem with the standard ISM is that the interaction graphs contributing to the site–site direct correlation function $c_{\alpha\beta}(r)$ cannot be obtained from those in the site–site total correlation function $h_{\alpha\beta}(r)$ by a process of topological reduction, as is required in a consistent theory. A proper topological reduction of the interaction site cluster series leads not to the standard ISM equation (10E.12), but rather to a generalization that can be written as

$$\tilde{\boldsymbol{\omega}} + \rho\tilde{\mathbf{h}} = (\mathbf{1} - \rho\tilde{\boldsymbol{\Omega}}\tilde{\mathbf{c}}_0)^{-1}\tilde{\boldsymbol{\Omega}}. \tag{10E.17}$$

Here $\mathbf{c}_0$ is a matrix with elements given by a set of direct correlation functions, and $\boldsymbol{\Omega} = \boldsymbol{\omega} + \Delta\boldsymbol{\Omega}$, where $\Delta\boldsymbol{\Omega}$ is a functional of another set of topologically distinct direct correlation functions that form the elements of matrices $\mathbf{c}_r$, $\mathbf{c}_l$, and $\mathbf{c}_b$. The standard ISM integral equation (10E.12) can be obtained by neglecting $\Delta\boldsymbol{\Omega}$ and associating the ISM $\mathbf{c}$ with $\mathbf{c}_0$. This procedure is equivalent to taking the low-density limit $\rho \to 0$, which explains why the result $\epsilon = 1 + 3y$ is then obtained. If $\Delta\boldsymbol{\Omega}$ is *not* neglected, the modified ISM equation yields nontrivial results for ϵ when appropriate closures are employed for $\mathbf{c}_0$, $\mathbf{c}_r$, $\mathbf{c}_l$, and $\mathbf{c}_b$.[276]

Unfortunately, the added complexity of the modified equation has so far precluded its successful application in real cases of interest. Recently, too, Lue and Blankschtein[277] have drawn attention to a feature of the modified ISM equation that, in practice, severely restricts its ability to make sensible predictions for ϵ. The constraints of connectivity between different sites on the same molecule impose restrictions[278] on the site–site correlation functions $h_{\alpha\beta}(r)$ and $c_{\alpha\beta}(r)$. However, the most general closures do not satisfy these constraints; indeed it is very difficult to formulate closures that *do*. When they are not met, the resulting correlation functions do not satisfy local electroneutrality constraints, an essential requirement for any theory that aims to predict the dielectric constant.[276] Instead, they satisfy the Stillinger–Lovett moment conditions, constraints applicable to *ionic* fluids. The predicted ϵ is therefore infinite.

The modified ISM equation has been put into a form in which it more closely resembles the standard equation.[279] Kalyuzhnyi and Cummings[280] have shown that it is in fact a special case—the limit of complete association—of a more general two-density integral equation theory of associating fluids developed by Wertheim.[281]

References and notes

1. The topics discussed here and in Chapter 11 (which deals with the spectroscopic properties of fluids) can be regarded as parts of the general field of molecular optics, which also encompasses microscopic theories of diffraction, nonlinear optics, magneto-optics, optical activity, and derivation of the macroscopic Maxwell equations and constitutive relations. Ref. 2 contains a critical and historical general discussion of this subject, with particular emphasis on the early work of Lorentz, Hoek, and others.
2. Van Kranendonk, J., and Sipe, J. E., *Prog. Optics* **15**, 247 (1977).

3. Nonlinear terms are typically very small at normal field strengths. See refs. 4–7, and also our discussion of the Kerr effect and field gradient-induced birefringence in § 10.8.
4. Böttcher, F. J. F., *Theory of Electric Polarization*, 2nd. edn., Elsevier, Amsterdam, Vols. 1 (1973) and 2 (1978).
5. Petersen, H. G., de Leeuw, S. W., and Perram, J. W., *Mol. Phys.* **66**, 637 (1989).
6. Hinsen, K., Bratz, A., and Felderhof, B. U., *J. Chem. Phys.* **97**, 9299 (1992).
7. R. L. Fulton, *J. Chem. Phys.* **130**, 204503 (2009).
8. Our equations are written in the electrostatic system of units. To convert esu to SI see, e.g. Buckingham A. D., *MTP International Review of Science*, Phys. Chem. Series One, **2**, 241 (1972).
9. Madden, P., and Kivelson, D., *Adv. Chem. Phys.* **56**, 467 (1984).
10. The shape dependence of (10.2) and (10.3) arises from a long-range shape-dependent feature in the pair correlation function, $g(12)$, of order $O(V^{-1})$. See refs. 11–14, and also § 10.7.
11. Wertheim, M. S., *Ann. Rev. Phys. Chem.* **30**, 471 (1979).
12. Adelman, S. A., and Deutch, J. M., *Adv. Chem. Phys.* **31**, 103 (1975).
13. Nienhuis, G., and Deutch, J. M., *J. Chem. Phys.* **55**, 4213 (1971); ibid. **56**, 5511 (1972).
14. De Leeuw, S. W., Perram, J. W., and Smith, E. R., *Proc. Roy. Soc.* **A373**, 57 (1980).
15. Mandel, M., and Mazur, P., *Physica* **24**, 116 (1958).
16. Mazur, P., *Adv. Chem. Phys.* **1**, 309 (1958).
17. Ramshaw, J. D., *J. Chem. Phys.* **57**, 2684 (1972); ibid. **66**, 3134 (1977); **68**, 5199 (1978).
18. Høye, J. S., and Stell, G., *J. Chem. Phys.* **61**, 562 (1974).
19. Wertheim, M. S., *Mol. Phys.* **25**, 211 (1973); ibid. **26**, 1425 (1973); **33**, 95 (1977); **34**, 1109 (1977); **36**, 1217 (1978); **37**, 83 (1979).
20. The result (10.4) follows from the physical argument given in the text. It also follows by solving the macroscopic Maxwell equations $\nabla \cdot (\mathbf{E} + 4\pi\mathbf{P}) = 0$, $\nabla \times \mathbf{E} = 0$ together with the boundary condition that $\mathbf{E} = \mathbf{E}_0$ when $\mathbf{P} = 0$. The two Maxwell equations are solved either by introducing the potential ϕ ($\mathbf{E} = -\nabla\phi$), or by applying $\nabla\times$ to the second equation and using the identity $\nabla \times \nabla\times = \nabla\nabla\cdot - \nabla^2$. This gives $\nabla^2\mathbf{E} = -4\pi\nabla\nabla \cdot \mathbf{P}$, with solution (cf. Poisson's equation (2.46) and solution (2.49)) $\mathbf{E} = \mathbf{E}_0(\mathbf{r}) + \int d\mathbf{r}'[\nabla'\nabla' \cdot \mathbf{P}(\mathbf{r}')]|\mathbf{r}' - \mathbf{r}|^{-1}$, where $\mathbf{E}_0$ satisfies the homogeneous equation $\nabla^2\mathbf{E}_0 = 0$, which is equal to (10.4) when integration by parts is carried out.
21. Stell, G., Patey, G. N., and Høye, J. S., *Adv. Chem. Phys.* **48**, 183 (1981).
22. This is a well-known result from macroscopic electrostatics; see e.g. ref. 4.
23. Kirkwood, J., *J. Chem. Phys.* **7**, 911 (1939).
24. Schröer, W., *Adv. Chem. Phys.* **63**, 719 (1985).
25. This should not be confused with the correlation function $\Delta h(\mathbf{r}\omega_1\omega_2)$ defined through (10A.25); the two functions are not the same.
26. Høye, J. S., and Stell, G., *J. Chem. Phys.* **65**, 18 (1976).
27. Chandler, D., *J. Chem. Phys.* **67**, 1113 (1977).
28. In general the molecular polarizability contains both electronic and nuclear contributions. The latter are associated with molecular nonrigidity, and are usually quite small (cf. § D.2).
29. Buckingham, A. D., *Disc. Faraday Soc.* **43**, 205 (1967); *Proc. Roy. Soc.* **A238**, 235 (1956).
30. Silberstein, L., *Phil. Mag.* **33**, 92, 521 (1917).
31. Mossotti, O. F., *Mem. Mat. Fis. Modena* **24** (Pt. 2), 49 (1850).

32. Clausius, R., *Die Mechanische Wärmetheorie*, Band 2, F. Vieweg und Sohn, Braunschweig (1879).
33. Debye, P., *Phys. Zeit.* **13**, 97 (1912); *Polar Molecules*, Chemical Catalog Co., New York (1929).
34. A multipole-like potential is one consisting of a sum of spherical harmonic terms $u(l_1 l_2 l)$ in which l_1, l_2, and l are all nonzero. All multipolar potentials are of this type, but dispersion, overlap, and induction potentials usually are not. See also § 4.5.
35. Goldman, S., and Joslin, C. G., *J. Chem. Phys.* **99**, 3021 (1993).
36. Goldman, S., and Joslin, C. G., *Ber. Bunsenges. Phys. Chem.* **99**, 204 (1995).
37. Gray, C. G., and Gubbins, K. E., *Mol. Phys.* **30**, 1481 (1975).
38. Tani, A., Henderson, D., Barker, J. A., and Hecht, C. E., *Mol. Phys.* **48**, 863 (1983).
39. For a dipolar hard sphere fluid at low densities $\phi = 17/16$, as can be seen from Eq. (10.117) for C_ϵ (we recall that ϕ arises in the expansion of the Kirkwood, rather than the Clausius–Mossotti, function, so that $\lim_{\rho\to 0}\phi = C_\epsilon \rho^3/y^3 + 2 = -15/16 + 2 = 17/16$). Calculations show that ϕ is only weakly dependent on density,[38,40] so that $\phi \approx 1$ for all state conditions.
40. Goldman, S., *Mol. Phys.* **71**, 491 (1990).
41. Matyushov, D. V., and Ladanyi, B. M., *J. Chem. Phys.* **110**, 994 (1999).
42. Goldman, S., and Joslin, C. G., *J. Phys. Chem.* **97**, 12349 (1993).
43. Goldman, S., Joslin, C. G., and Wasserman, E. A., *J. Phys. Chem.* **98**, 6231 (1994).
44. Valiskó, M., Boda, D., Liszi, J., and Szalai, I., *Mol. Phys.* **100**, 3239 (2002); ibid. **101**, 2309 (2003).
45. Cichocki, B., and Felderhof, B. U., *J. Chem. Phys.* **90**, 4960 (1989).
46. Bławzdziewicz, J., Cichocki, B., and Szamel, G., *J. Chem. Phys.* **91**, 7467 (1989).
47. Bławzdziewicz, J., Cichocki, B., and Holyst, R., *Physica* **157A**, 857 (1989).
48. Hinsen, K., and Felderhof, B. U., *J. Chem. Phys.* **94**, 5655 (1991); ibid. **96**, 5554 (1992).
49. The quadrupole–quadrupole polarizability measures the quadrupole moment induced by a field gradient, $Q_{\alpha\beta} = C_{\alpha\beta,\gamma\delta}\nabla_\gamma E_\delta$, and is given by $C_{\alpha\beta,\gamma\delta} = (2/3)\sum_{n>0}\langle 0|Q_{\alpha\beta}|n\rangle\langle n|Q_{\gamma\delta}|0\rangle/(E_n - E_0)$. For an atom, $C_1 = (1/5)C_{\alpha\beta,\alpha\beta}$. See Buckingham, A. D., *Adv. Chem. Phys.* **12**, 107 (1967).
50. Kirkwood, J. G., *J. Chem. Phys.* **4**, 592 (1936).
51. Buckingham, A. D., and Pople, J. A., *Trans. Faraday Soc.* **51**, 1029, 1179 (1955).
52. Joslin, C. G., *Mol. Phys.* **42**, 1507 (1981).
53. Moszynski, R., Heijmen, T. G. A., Wormer, P. E. S., and van der Avoird, A., *J. Chem. Phys.* **104**, 6997 (1996).
54. Moszynski, R., Heijmen, T. G. A., and van der Avoird, A., *Chem. Phys. Lett.* **247**, 440 (1995).
55. Rizzo, A., Hättig, C., Fernández, B., and Koch, H., *J. Chem. Phys.* **117**, 2609 (2002).
56. MacRury, T. B., and Steele, W. A., *J. Chem. Phys.* **61**, 3352, 3366 (1974).
57. Rushbrooke, G. S., *Introduction to Statistical Mechanics*, Clarendon Press, Oxford (1962), p. 264 *et seq.*
58. Huang, K., *Statistical Mechanics*, Wiley, New York (1963), Ch. 14.
59. Hourri, A., St-Arnaud, J. M., and Bose, T. K., *J. Chem. Phys.* **106**, 1780 (1997).
60. Bullough, R. K., *J. Phys.* **A3**, 708, 726, 751 (1970).
61. Sutter, H., in *Specialist Periodical Reports, Dielectric and Related Molecular Processes*, Vol. 1 (ed. M. Davies), Chem. Soc., London (1972), p. 65.
62. Vidal, D., and Lallemand, M., *J. Chem. Phys.* **64**, 4293 (1976); ibid. **66**, 4776 (1977).
63. Weber, L. A., *J. Chem. Phys.* **65**, 446 (1976).
64. Achtermann, H. J., Magnus, G., and Bose, T. K., *J. Chem. Phys.* **94**, 5669 (1991).
65. Sutter, H., and Cole, R. H., *J. Chem. Phys.* **52**, 132 (1970).

66. The fourth refractivity virial coefficient D_r, i.e. the high-frequency contribution to the fourth dielectric virial coefficient D_ϵ, has been measured for the gases C_2H_4, C_2H_6, and SF_6.[64] There have as yet been no theoretical studies of this quantity.
67. Joslin, C. G., and Goldman, S., *Mol. Phys.* **79**, 1351 (1993).
68. Logan, D. E., and Madden, P. A., *J. Chem. Phys.* **96**, 5552, 4812 (1992).
69. Huot, J., and Bose, T. K., *J. Chem. Phys.* **95**, 2683 (1991).
70. Buckingham, A. D., and Clarke, K. L., *Chem. Phys. Lett.* **57**, 321 (1978).
71. Champagne, M. H., Li, X., and Hunt, K. L. C., *J. Chem. Phys.* **112**, 1893 (2000).
72. Hunt, K. L. C., and Li, X., in *Collision- and Interaction-induced Spectroscopy* (ed. G. C. Tabisz and M. N. Neuman), NATO ASI Series C **452**, Kluwer Academic, Dordrecht (1995), p. 61.
73. Meinander, N., *Chem. Phys. Lett.* **228**, 295 (1994).
74. Fernández, B., Hättig, C., Koch, H., and Rizzo, A., *J. Chem. Phys.* **110**, 2872 (1999); Hättig, C., Larsen, H., Olsen, J., Jørgensen, P., Koch, H., Fernández, B., and Rizzo, A., *J. Chem. Phys.* **111**, 10099 (1999); Koch, H., Hättig, C., Larsen, H., Olsen, J., Jørgensen, P., Fernández, B., and Rizzo, A., *J. Chem. Phys.* **111**, 10108 (1999).
75. Joslin, C. G., unpublished.
76. Gray, C. G., and Gubbins, K. E., *Mol. Phys.* **42**, 843 (1981).
77. Huot, J., and Bose, T. K., *J. Chem. Phys.* **94**, 3849 (1991).
78. Bounds, D. G., Hinchliffe, A., and Spicer, C. J., *Mol. Phys.* **42** 73 (1981); Bounds, D. G., *Mol. Phys.* **38**, 2099 (1979).
79. Buckingham, A. D., *Trans. Faraday Soc.* **52**, 747 (1956).
80. Copeland, T. G., and Cole, R. H., *J. Chem. Phys.* **64**, 1741, 1747 (1976).
81. Buckingham, A. D., and Joslin, C. G., *Mol. Phys.* **40**, 1513 (1980).
82. Joslin, C. G., and Gray, C. G., *Mol. Phys.* **57**, 1197 (1986).
83. Joslin, C. G., *Mol. Phys.* **47**, 693 (1982).
84. Stone, A. J., and Alderton, M., *Mol. Phys.* **56**, 1047 (1985); Stone, A. J., *Chem. Phys. Lett.* **83**, 233 (1981). One drawback of the approach is that the distributed moments are not available from experiment, as in the one-centre case (Appendix D).
85. Stone, A. J., *Mol. Phys.* **56**, 1065 (1985); LeSueur, C. R., and Stone, A. J., *Mol. Phys.* **78**, 1267 (1993); ibid. **83**, 293 (1994); Williams, G. J., and Stone, A. J., *J. Chem. Phys.* **119** 4620 (2003).
86. Stone, A. J., Tantirungrotechai, Y., and Buckingham, A. D., *Phys. Chem. Chem. Phys.* **2**, 429 (2000).
87. Muckenfuss, C., Curtiss, C. F., and Bird, R. B., *J. Chem. Phys.* **23**, 1542 (1955).
88. Buckingham, A. D., and Tantirungrotechai, Y., *Can. J. Chem.* **77**, 1946 (1999).
89. Joslin, C. G., *Mol. Phys.* **47**, 771 (1982).
90. Ohba, M. and Nomura, H., *Mol. Phys.* **70**, 729 (1990); ibid. **65**, 1431 (1988).
91. Rushbrooke, G. S., and Shrubsall, R. G., *Mol. Phys.* **55**, 599 (1985).
92. Joslin, C. G., and Goldman, S., *Mol. Phys.* **79**, 499 (1993).
93. Wigner, E. P., *Phys. Rev.* **40**, 749 (1932).
94. Kirkwood, J. G., *Phys. Rev.* **44**, 31 (1933).
95. Fujiwara, Y., Osborn, T. A., and Wilk, S. F. J., *Phys. Rev.* **A25**, 14 (1982).
96. See, e.g., Sinha, S. K., and Singh, Y., *J. Math. Phys.* **18**, 367 (1977), and refs. therein.
97. For a review, see de Leeuw, S. W., Perram, J. W., and Smith, E. R., *Ann. Rev. Chem. Phys.* **37**, 245 (1986).
98. De Leeuw, S. W., Perram, J. W., and Smith, E. R., *Proc. Roy. Soc.* **A373**, 27 (1980).
99. Neumann, M., *J. Chem. Phys.* **82**, 5663 (1985).
100. Neumann, M., *J. Chem. Phys.* **85**, 1567 (1986).
101. Neumann, M., *Mol. Phys.* **50**, 841 (1983).
102. Neumann, M., and Steinhauser, O., *Mol. Phys.* **39**, 437 (1980).
103. Neumann, M., Steinhauser, O., and Pawley, G. S., *Mol. Phys.* **52**, 97 (1984).

104. Gray, C. G., Sainger, Y. S., Joslin, C. G., Cummings, P. T., and Goldman, S., *J. Chem. Phys.* **85**, 1502 (1986).
105. Patey, G. N., Levesque, D., and Weis, J. J., *Mol. Phys.* **45**, 733 (1982).
106. De Leeuw, S. W., Perram, J. W., Quirke, N., and Smith, E. R., *Mol. Phys.* **42**, 1197 (1981).
107. Kusalik, P. G., *J. Chem. Phys.* **93**, 3520 (1990).
108. Kusalik, P. G., *Mol. Phys.* **67**, 67 (1989).
109. Lomba, E., Lombardero, M., and Abascal, J. L. F., *Mol. Phys.* **68**, 1067 (1989).
110. Smith, E. R., and Wielopolski, P. A., *Mol. Phys.* **61**, 1063 (1987).
111. De Leeuw, S. W., and Quirke, N., *J. Chem. Phys.* **81**, 880 (1984).
112. Morriss, G. P., *Mol. Phys.* **47**, 833 (1982).
113. Hesse-Bezot, C., Bossis, G., and Brot, C., *J. Chem. Phys.* **80**, 3399 (1984).
114. Caillol, J. M., and Levesque, D., *J. Chem. Phys.* **94**, 597 (1991).
115. Caillol, J. M., *J. Chem. Phys.* **96**, 1455 (1992).
116. Caillol, J. M., and Levesque, D., *J. Chem. Phys.* **96**, 1477 (1992).
117. Caillol, J. M., *J. Chem. Phys.* **96**, 7039 (1992).
118. A different reaction field model has recently been described by Alper, H., and Levy, R. M., *J. Chem. Phys.* **99**, 9847 (1993), and has been applied successfully to dielectric simulations of liquid water.
119. Alper, H. E., and Levy, R. M., *J. Chem. Phys.* **91**, 1242 (1989).
120. Anderson, J., Ullo, J. J., and Yip, S., *J. Chem. Phys.* **87**, 1726 (1987).
121. Cummings, P. T., and Blum, L., *J. Chem. Phys.* **85**, 6658 (1986).
122. Lee, P. H., and Ladanyi, B. M., *J. Chem. Phys.* **87**, 4093 (1987).
123. Lee, P. H., and Ladanyi, B. M., *J. Chem. Phys.* **91**, 7063 (1989).
124. Gordon, H., and Goldman, S., *Mol. Simul.* **2**, 177 (1989).
125. Pollock, E. L., and Alder, B. J., *Physica* **102A**, 1 (1980).
126. Zarragoicochea, G. J., Levesque, D., and Weis, J. J., *Mol. Phys.* **74**, 629 (1991).
127. Kusalik, P. G., *Mol. Phys.* **81**, 199 (1994).
128. Kusalik, P. G., *Mol. Phys.* **73**, 1349 (1991).
129. De Leeuw, S. W., Smit, B., and Williams, C. P., *J. Chem. Phys.* **93**, 2704 (1990).
130. Edwards, D. M. F., Madden, P. A., and McDonald, I. R., *Mol. Phys.* **51**, 1141 (1984).
131. Kusalik, P. G., Mandy, M. E., and Svishchev, I. M., *J. Chem. Phys.* **100**, 7654 (1994); Kusalik, P. G., *Mol. Phys.* **80**, 225 (1993).
132. Smith, D. E., and Haymet, A. D. J., *J. Chem. Phys.* **96**, 8450 (1992).
133. Heinz, T. N., van Gunsteren, W. F., and Hünenberger, P. H., *J. Chem. Phys.* **115**, 1125 (2001). These authors review various methods that have been proposed to improve the convergence of simulations of ϵ; none is found to be particularly successful.
134. Henderson, D., Boda, D., Szalai, I., and Chan, K.-Y., *J. Chem. Phys.* **110**, 7348 (1999).
135. Cichocki, B., and Felderhof, B. U., *J. Chem. Phys.* **94**, 5655 (1991).
136. Watanabe, K., and Klein, M. L., *Chem. Phys.* **131**, 157 (1989).
137. Sprik, M., *J. Chem. Phys.* **95**, 6762 (1991).
138. Caillol, J. M., Levesque, D., Weis, J. J., Perkyns, J. S., and Patey, G. N., *Mol. Phys.* **62**, 1225 (1987).
139. Patey, G. N., Levesque, D., and Weis, J. J., *Mol. Phys.* **57**, 337 (1986).
140. Caillol, J. M., Levesque, D., Weis, J. J., Kusalik, P. G., and Patey, G. N., *Mol. Phys.* **55**, 65 (1985).
141. Pollock, E. L., Alder, B. J., and Patey, G. N., *Physica* **108A**, 14 (1981).
142. Patey, S. N., Torrie, G. M., and Valleau, J. P., *J. Chem. Phys.* **71**, 96 (1979).
143. Patey, G. N., Klein, M. L., and McDonald, I. R., *Chem. Phys. Lett.* **73**, 375 (1980).
144. Mooij, G. C. A. M., de Leeuw, S. W., Smit, B., and Williams, C. P., *J. Chem. Phys.* **97**, 5113 (1992).

145. Ruocco, G., and Sampoli, M., *Mol. Phys.* **82**, 875 (1994). This paper generalizes the RF method to polarizable systems.
146. Edwards, D. M. F., and Madden, P. A., *Mol. Phys.* **51**, 1163 (1984). This is a generalization of the acetonitrile simulation of ref. 130 that includes polarizability; it yields an ϵ in good agreement with experiment.
147. Martín, C., Lomba, E., Lombardero, M., Lado, F., and Høye, J. S., *J. Chem. Phys.* **100**, 1599 (1994).
148. Zhu, S.-B., Singh, S., and Robinson, G. W., *J. Chem. Phys.* **95**, 2791 (1991). The potential is a five-site exp-6-1 with Morse-like hydrogen-bonding interactions between H atoms and lone pairs, and an anharmonic Urey–Bradley force field governing intramolecular degrees of freedom. The partial charges vary according to the strength of the electric field experienced by the molecule, thereby effectively rendering it polarizable.
149. Soetens, J.-C., Martins Costa, M. T. C., and Millot, C., *Mol. Phys.* **94**, 577 (1998).
150. Rick, S. W., Stuart, S. J., and Berne, B. J., *J. Chem. Phys.* **101**, 6141 (1994). See also Liu, Y.-P., Kim, K., Berne, B. J., Friesner, R. A., and Rick, S. W., *J. Chem. Phys.* **108**, 4739 (1998).
151. Mountain, R. D., *J. Chem. Phys.* **105**, 10496 (1996).
152. Lobaugh, J., and Voth, G. A., *J. Chem. Phys.* **106**, 2400 (1997).
153. Sharma, M., Resta, R., and Car, R., *Phys. Rev. Lett.* **98**, 247401 (2007).
154. Barker, J. A., and Watts, R. O., *Mol. Phys.* **26**, 789 (1973); Watts, R. O., *Mol. Phys.* **28**, 1069 (1974).
155. Watts, R. O., *Mol. Phys.* **28**, 1069 (1974). A more recent and efficient procedure is described by Smith, P. E., and van Gunsteren, W. F., *J. Chem. Phys.* **100**, 3169 (1994): this also yields the fluctuations in $\boldsymbol{\mu}$ as well as ϵ itself.
156. Hünenberger, P. H., and van Gunsteren, W. F., *J. Chem. Phys.* **108**, 6117 (1998).
157. By ‘pseudospherical’ we mean that the lattice sum is over all lattice vectors $\mathbf{n} = (n_1, n_2, n_3)$ such that $n^2 = n_1^2 + n_2^2 + n_3^2 \leq n_{max}^2$.
158. Neumann, M., and Steinhauser, O., *Chem. Phys. Lett.* **95** , 417 (1983).
159. Ewald, P. P., *Ann. Phys.* **64**, 263 (1921); Kornfeld, H., *Z. Phys.* **22**, 27 (1924).
160. Born, M., and Huang, K., *Dynamical Theory of Crystal Lattices*, Clarendon Press, Oxford (1954).
161. This is easily seen by summing over shells with the same value of $n = \left(n_1^2+n_2^2+n_3^3\right)^{1/2}$. For such a shell the self-energy of dipole i is $\mathcal{U}_{\text{self}} = S_1 - 3S_2$, where $S_1 = \sum \mu_i^2/n^3$ and $S_2 = \sum(\boldsymbol{\mu}_i \cdot \mathbf{n})^2/n^5 = \left(\mu_{ix}^2 \sum n_1^2 + \mu_{iy}^2 \sum n_2^2 + \mu_{iz}^2 \sum n_3^2\right)/n^5$, since $\sum n_1 n_2 = 0$, etc. By symmetry $\sum n_1^2 = \sum n_2^2 = \sum n_3^2 = \frac{1}{3}\sum n^2$, and hence $S_2 = \frac{1}{3}\sum \mu_i^2/n^3 = \frac{1}{3}S_1$. Thus $\mathcal{U}_{\text{self}} = 0$.
162. Smith, P. E., and Pettitt, B. M., *J. Chem. Phys.* **105**, 4289 (1996). Ewald artifacts are also discussed by Figueirido, F., Del Buono, G. S., and Levy, R. M., *J. Chem. Phys.* **103**, 6133 (1995).
163. Perram, J. W., Petersen, H. G., and de Leeuw, S. W., *Mol. Phys.* **65**, 875 (1988).
164. For a more detailed analysis of errors in dipolar Ewald sums see Wang, Z., and Holm, C., *J. Chem. Phys.* **115**, 6351 (2001).
165. Duan, Z.-H., and Krasny, R., *J. Chem. Phys.* **113**, 3492 (2000).
166. Greengard, L. F., *The Rapid Evaluation of Potential Fields in Particle Systems*, MIT Press (Cambridge, 1988); Science **265**, 909 (1994).
167. Deserno, M., and Holm, C., *J. Chem. Phys.* **109**, 7678, 7694 (1998); Luty, B. A., Tironi, I. G., and van Gunsteren, W. F., *J. Chem. Phys.* **103**, 3014 (1995); York, D., and Yang, W., *J. Chem. Phys.* **101**, 3298 (1994); Darden, T., York, D., and Pedersen, L., *J. Chem. Phys.* **98**, 10089 (1993).

168. Challacombe, M., White, C., and Head-Gordon, M., *J. Chem. Phys.* **107**, 10131 (1997); Greengard, L., and Rokhlin, V., *J. Comput. Phys.* **73**, 325 (1987).
169. Petersen, H. G., *J. Chem. Phys.* **103**, 3668 (1995).
170. Adams, D. J., Adams, E. M., and Hills, G. J., *Mol. Phys.* **38**, 387 (1979).
171. Jansoone, V. M., *Chem. Phys.* **3**, 78 (1974).
172. Adams, D. J., and McDonald, I. R., *Mol. Phys.* **32**, 931 (1976).
173. Adams, D. J., *Mol. Phys.* **40**, 1261 (1980).
174. Adams, D. J., and Adams, E. M., *Mol. Phys.* **42**, 907 (1981).
175. De Leeuw, S. W., Perram, J. W., and Smith, E. R., *Proc. Roy. Soc.* **A388**, 177 (1983).
176. In ref. 97, de Leeuw *et al.* argue that because $\mathcal{U}_{EW}^{RF}$ is not truncated in EW/RF, it is of a different nature than in MI/RF. However, $\mathcal{U}_{EW}^{RF}$ *is* formally truncated, at $r_c = R$, the radius of the pseudospherical lattice. As shown in (10.131), this sum can for computational purposes be reduced to an untruncated sum over the central simulation cell.
177. Chandler, D., Silbey, R., and Ladanyi, B. M., *Mol. Phys.* **46**, 1335 (1982).
178. Wertheim, M. S., *J. Chem. Phys.* **55**, 4291 (1971).
179. Patey, G. N., Levesque, D., and Weis, J. J., *Mol. Phys.* **38**, 219 (1979).
180. Fries, P. H., and Patey, G. N., *J. Chem. Phys.* **82**, 429 (1985).
181. Lomba, E., Martín, C., and Lombardero, M., *Mol. Phys.* **77**, 1005 (1992).
182. Rushbrooke, G. S., *Mol. Phys.* **43**, 975 (1981).
183. Lado, F., Lombardero, M., Enciso, E., Lago, S., and Abascal, J. L. F., *J. Chem. Phys.* **85**, 2916 (1986).
184. Lee, L. Y., Fries, P. H., and Patey, G. N., *Mol. Phys.* **55**, 751 (1985).
185. The effect of quadrupolar forces on ϵ has also been studied using perturbation theory; see § 10.5.
186. Patey, G. N., Levesque, D., and Weis, J. J., *Mol. Phys.* **38**, 1635 (1979).
187. Perkyns, J. S., Fries, P. H., and Patey, G. N., *Mol. Phys.* **57**, 529 (1986).
188. Andersen, H. C., *Ann. Rev. Phys. Chem.* **26**, 145 (1975).
189. Martina, E., Deutch, J. M., and Streett, W. B., *J. Chem. Phys.* **67**, 3157 (1977).
190. Perera, A., and Patey, G. N., *J. Chem. Phys.* **91**, 3045 (1989).
191. Předota, M., Cummings, P. T., and Chialvo, A. A., *Mol. Phys.* **99**, 349 (2001); ibid. **100**, 2703 (2002).
192. Joslin, C. G., *Chem. Phys. Lett.* **91**, 452 (1982).
193. Berendsen, H. J. C., Postma, J. P. M., van Gunsteren, W. F., and Hermans, J., in *Intermolecular Forces*, (ed. B. Pullman), Springer, Reidel (1981), p. 331.
194. Berendsen, H. J. C., Grigera, J. R., and Straatsma, T. P., *J. Phys. Chem.* **91**, 6269 (1987).
195. Neria, E., Fischer, S., and Karplus, M., *J. Chem. Phys.* **105**, 1902 (1996) describe a slightly modified version of the TIP3P potential; see also references to earlier work in this paper.
196. Jorgensen, W. L., Chandrasekhar, J., Madura, J. D., Impey, R. W., and Klein, M. L., *J. Chem. Phys.* **79**, 926 (1983).
197. Höchtl, P., Boresch, S., Bitomsky, W., and Steinhauser, O., *J. Chem. Phys.* **109**, 4927 (1998).
198. Corongiu, G., *Int. J. Quant. Chem.* **42**, 1209 (1992).
199. Sprik, M., and Klein, M. L., *J. Chem. Phys.* **89**, 7556 (1988).
200. By letting partial charges reside, e.g., in the oxygen lone pairs, the situation might be improved, at the cost of added complexity in the intermolecular potential.
201. Lamoureux, G., MacKerell, A. D., and Roux, B., *J. Chem. Phys.* **119**, 5185 (2003).
202. Parrinello, M., and Rahman, A., *J. Chem. Phys.* **80**, 860 (1984).
203. Feynman, R. P., and Hibbs, A. R., *Quantum Mechanics and Path Integrals*, McGraw-Hill, New York (1965).

204. Carnie, S. L., and Patey, G. N., *Mol. Phys.* **47**, 1129 (1982).
205. Kusalik, P. G., and Patey, G. N., *Mol. Phys.* **65**, 1105 (1988).
206. Lombardero, M., Martín, C., Jorge, S., Lado, F., and Lomba, E., *J. Chem. Phys.* **110**, 1148 (1999). See also Richardi, J., Millot, C., and Fries, P. H., ibid. 1138, who reach essentially the same conclusion for several other water models, and also for the H-bonded solvent methanol.
207. Kerr, J., *Phil. Mag.* **50**, 337, 446 (1875).
208. The modern statistical-mechanical theory of field gradient-induced birefringence was developed largely by Buckingham and co-workers. For early references, see e.g. refs. 209 and 210. Subsequent developments in the theory are described in refs. 211 (quantum effects), 212, and 213 (the nontrivial extension to polar molecules), and 214 (the much smaller effect exhibited by spherical molecules). Buckingham also designed an experiment to measure the magnitude of the effect;[210] essentially the same experimental arrangement is used today.[215]
209. Buckingham, A. D., *J. Chem. Phys.* **30**, 1580 (1959).
210. The first quadrupole moment determined from birefringence studies was that of carbon dioxide; see Buckingham, A. D., and Disch, R. L., *Proc. Roy. Soc.* **A273**, 275 (1963).
211. Buckingham, A. D., and Pariseau, M., *Trans. Faraday Soc.* **62**, 1 (1966).
212. Buckingham, A. D., and Longuet-Higgins, H. C., *Mol. Phys.* **14**, 63 (1968).
213. Buckingham, A. D., Disch, R. L., and Dunmur, D. A., *J. Am. Chem. Soc.* **90**, 3104 (1968).
214. Buckingham, A. D., and Jamieson, M. J., *Mol. Phys.* **22**, 117 (1971).
215. Ritchie, G. L. D., in *Optical, Electric and Magnetic Properties of Molecules* (ed. D. C. Clary and B. J. Orr), Elsevier, Amsterdam (1997), p. 67.
216. Orr, B. J., in *Nonlinear Behaviour of Molecules, Atoms and Ions in Electric, Magnetic or Electromagnetic Fields* (ed. L. Neel), Elsevier, Amsterdam (1979), p. 227.
217. Chaikin, P. M., and Lubensky, T. C., *Principles of Condensed Matter Physics*, Cambridge University Press, Cambridge (1995).
218. Various definitions of the Kerr constant are used. Our K is related to the ${}_mK$ defined by Buckingham[219] by ${}_mK = (V_m/9)K$, where V_m is the molar volume.
219. Buckingham, A. D., *Proc. Phys. Soc.* **68**, 910 (1955).
220. In a standard notation, $\boldsymbol{\alpha}(\omega) = \boldsymbol{\alpha}(-\omega;\omega)$, $\boldsymbol{\beta}(\omega) = \boldsymbol{\beta}(-\omega;\omega,0)$, and $\boldsymbol{\gamma}(\omega) = \boldsymbol{\gamma}(-\omega;\omega,0,0)$. See Willetts, A., Rice, J. E., Burland, D. M., and Shelton, D. P., *J. Chem. Phys.* **97**, 7590 (1992).
221. Recent coupled-cluster calculations of the tensors $\boldsymbol{\beta}(\omega)$ and $\boldsymbol{\gamma}(\omega)$ are described in Rizzo, A., Coriani, S., Fernández, B., and Christiansen, O., *Phys. Chem. Chem. Phys.* **4**, 2884 (2002).
222. Buckingham, A. D., and Pople, J. A., *Proc. Phys. Soc.* **68**, 905 (1955).
223. In molecules there are contributions to $\boldsymbol{\gamma}$ from both the nuclear and electronic motions; for CH_4, CF_4 and SF_6, the former range from 6 to 35% of the total. See Shelton, D. P., and Palubinskas, J. J., *J. Chem. Phys.* **104**, 2482 (1996).
224. Buckingham, A. D., and Orr, B. J., *Proc. Roy. Soc.* **A305**, 259 (1968).
225. Couling, V. W., and Graham, C., *Mol. Phys.* **93**, 31 (1998).
226. Dunmur, D. A., and Jessup, N. E., *Mol. Phys.* **37**, 697 (1979); Dunmur, D. A., Hunt, D. C., and Jessup, N. E., *Mol. Phys.* **37**, 713 (1979).
227. Buckingham, A. D., Galwas, P. A., and Fan-Chen, L., *J. Mol. Struct.* **100**, 3 (1983).
228. Gray, C. G., and Ralph, H., *Phys. Lett.* **A33**, 165 (1970).
229. Ladanyi, B. M., and Keyes, T., *Mol. Phys.* **34**, 1643 (1977); Can. J. Phys. **59**, 1421 (1981); Keyes, T., and Ladanyi, B. M., *Mol. Phys.* **37**, 1643 (1979).
230. Ladanyi, B. M., Keyes, T., Tildesley, D. J., and Streett, W. B., *Mol. Phys.* **39**, 645 (1980).

231. Talbot, J., Kivelson, D., Allen, M. P., Evans, G. T., and Frenkel, D., *J. Chem. Phys.* **92**, 3048 (1990).
232. Buckingham, A. D., Stiles, P. J., and Ritchie, G. L. D., *Trans. Faraday Soc.* **67**, 577 (1971).
233. Freasier, B., Hamer, N., and Isbister, D., *Mol. Phys.* **38**, 1661 (1979).
234. Debye, P., *Phys. Zeit.* **22**, 302 (1921).
235. Coriani, S., Hättig, C., and Rizzo, A., *J. Chem. Phys.* **111**, 7828 (1999).
236. For the modification of the constitutive relationship in the presence of a nonuniform field, see, e.g. Robinson, F. N. H., *Macroscopic Electromagnetism*, Pergamon, 1973; Munn, R. W., *J. Chem. Phys.* **100**, 6203 (1994); Barron, L. D., *Molecular Light Scattering and Optical Activity*, 2nd. edn., Cambridge University Press, Cambridge (2004); and ref. 237.
237. Graham, E. B., Pierrus, J., and Raab, R. E., *J. Phys.* **B25**, 4673 (1992); Raab, R. E., and de Lange, O. L., *Multipole Theory in Electromagnetism*, Oxford University Press, Oxford (2005).
238. For most species it is typically 1–10% of the total, although for some molecules it is larger: 35–40% for C_2H_6,[215] 30–40% for H_2.[239] It derives principally from an interaction with the magnetic field associated with the light wave,[212] and is determined by the same tensor that governs the magnitude of the Faraday effect (see also refs. 239, 240). It is the only contribution for atoms, or for tetrahedral or octahedral species such as CH_4 and SF_6, which do not possess a quadrupole moment.[214]
239. Coriani, S., Hättig, C., Jørgensen, P., Rizzo, A., and Ruud, K., *J. Chem. Phys.* **109**, 7176 (1998).
240. Amos, R. D., *Chem. Phys. Lett.* **85**, 123 (1982).
241. See, e.g., Rizzo, A., Cappelli, C., Jansík, B., Jonsson, D., Sałek, P., Coriani, S., and Ågren, H., *J. Chem. Phys.* **121**, 8814 (2004).
242. Watson, J. N., Craven, I. E., and Ritchie, G. L. D., *Chem. Phys. Lett.* **274**, 1 (1997).
243. Graham, C., Imrie, D. A., and Raab, R. E., *Mol. Phys.* **93**, 49 (1998).
244. Battaglia, M. R., Buckingham, A. D., Neumark, D., Pierens, R. K., and Williams, J. H., *Mol. Phys.* **43**, 1015 (1981).
245. Buckingham, A. D., Graham, C., and Williams, J. H., *Mol. Phys.* **49**, 703 (1983).
246. Battaglia, M. R., Buckingham, A. D., and Williams, J. H., *Chem. Phys. Lett.* **78**, 421 (1981).
247. Coriani, S., Halkier, A., Jonsson, D., Gauss, J., Rizzo, A., and Christiansen, O., *J. Chem. Phys.* **118**, 7329 (2003); Rizzo, A., Coriani, S., Halkier, A., and Hättig, C., ibid. **113**, 3077 (2000).
248. Imrie, D. A., and Raab, R. E., *Mol. Phys.* **74**, 833 (1991).
249. Raab, R. E., and de Lange, O. L., *Mol. Phys.* **101**, 3467 (2003).
250. De Lange, O. L., and Raab, R. E., *Mol. Phys.* **102**, 125 (2004).
251. Marchesan, D., Coriani, S., and Rizzo, A., *Mol. Phys.* **101**, 1851 (2003).
252. Vrbancich, J., and Ritchie, G. L. D., *J. Chem. Soc. Faraday* II **76**, 648 (1980).
253. The most general proof for the existence of ϵ requires that one consider the dielectric response of a finite sample of arbitrary shape; for this see refs. 5, 6, 13, 17–19.
254. For a *finite* system, the response function $\boldsymbol{\chi}_0(\mathbf{r}, \mathbf{r}')$ can be written as the sum of the translationally invariant infinite-system value plus a shape-dependent correction term of order $O(V^{-1})$; see e.g. ref. 11.
255. Abramowitz, M., and Stegun, I. A., *Handbook of Mathematical Functions*, NBS, Washington, DC (1964).
256. The graphical expansion of the direct correlation function $c(12)$ contains a single Mayer graph, $f(12)$, which is not at least doubly connected; and for large r, $f(12) \to -\beta u(12)$. In contrast, the total correlation function $h(12)$ contains additional

singly connected terms such as $\rho \int d\mathbf{r}_3 \langle f(13)f(32)\rangle_{\omega_3}$, and therefore exhibits somewhat more complex behavior at large r.

257. The behaviour of $\boldsymbol{\epsilon}(\mathbf{k})$ can be quite exotic for some systems. For example, its longitudinal component may actually become *negative* when $\lambda = 2\pi/k$ approaches a typical molecular dimension (Chandra, A., and Bagchi, B., *J. Chem. Phys.* **91**, 3056 (1989); ibid. **90**, 1832 (1989)). Note that a negative value does not violate any stability conditions: the Kramers–Kronig relations (discussed in Appendix 11A) only forbid a dielectric constant in the range between zero and unity (Dolgov, O. V., Kirzhnits, D. A., and Maksimov, E. G., *Rev. Mod. Phys.* **53**, 81 (1981)).
258. This is the case even at optical frequencies. For yellow light, $k = 2\pi/\lambda = 10^{-3}$ Å^{-1}, and if $\sigma = 5$ Å, $k\sigma = 0.005$. The response of a rigid polar fluid to an optical field is of course not described by the formulae in this Appendix.
259. Neither derivation appears entirely satisfactory. The Nienhuis–Deutch treatment has been criticized by both Ramshaw[17] and Høye and Stell,[18] while that of Høye and Stell, which is based on a consideration of the interaction of two test dipoles in a dielectric fluid, suffers from a certain sleight-of-hand.
260. See e.g. Belinfante, F. J., *Physica* **12**, 1 (1946); Gagnon, R. J., *Am. J. Phys.* **38**, 879 (1970); Hannay, J. H., *Eur. J. Phys.* **4**, 141 (1983).
261. Gray, C. G., Karl, G., and Novikov, V. A., *Am. J. Phys.* **77**, 807 (2009).
262. Goldman, S., *J. Phys. Chem.* **83**, 3033 (1979).
263. Hernando, J. A., and Gamba, Z., *J. Chem. Phys.* **97**, 5142 (1992).
264. Different techniques for evaluating I_{332} (and other $I_{nn'l}$) are discussed in ref. 40. These include Fourier transform convolution methods and the hat function method (Høye, J. S., and Stell, G., *J. Chem. Phys.* **63**, 5342 (1975)). A substantial increase in accuracy can be obtained if the low-density limits of the integrals are subtracted off and integrated analytically.[192]
265. Anta, J. A., Lomba, E., Martín, C., Lombardero, M., and Lado, F., *Mol. Phys.* **84**, 743 (1995).
266. Lado, F., Lomba, E., and Lombardero, M., *J. Chem. Phys.* **103**, 481 (1995).
267. Kinoshita, M., and Harada, M., *Mol. Phys.* **74**, 443 (1991).
268. Richardi, J., Fries, P. H., and Krienke, H., *Mol. Phys.* **96**, 1411 (1999).
269. Joslin, C. G., *Mol. Phys.* **49**, 129 (1983).
270. Developments in the interaction site model to 1990 are reviewed by Monson, P. A., and Morriss, G. P., *Adv. Chem. Phys.* **77**, 451 (1990).
271. Chandler, D., *Faraday Disc.* **66**, 74 (1978).
272. Sullivan, D. E., and Gray, C. G., *Mol. Phys.* **42**, 443 (1981).
273. Cummings, P. T., and Stell, G., *Mol. Phys.* **46**, 383 (1982).
274. Cummings, P. T., and Sullivan, D. E., *Mol. Phys.* **46**, 665 (1982). The $c_{\alpha\beta}(r)$ required are quite unphysical; e.g. for triatomics, $c_{\alpha\beta}(r) \sim r$ as $r \to \infty$.
275. Raineri, F. O., and Stell, G., *J. Phys. Chem.* **B105**, 11880 (2001).
276. Chandler, D., Joslin, C. G., and Deutch, J. M., *Mol. Phys.* **47**, 871 (1982).
277. Lue, L., and Blankschtein, D., *J. Chem. Phys.* **102**, 5460 (1995).
278. These constraints can be derived using Kirkwood–Buff theory, discussed in Chapter 7.
279. Rossky, P. J., and Chiles, R. A., *Mol. Phys.* **51**, 661 (1984).
280. Kalyuzhnyi, Yu. V., and Cummings, P. T., *J. Chem. Phys.* **104**, 3325 (1996).
281. Wertheim, M. S., *J. Stat. Phys.* **35**, 19, 35 (1984).

11

SPECTROSCOPIC PROPERTIES

The study of light has resulted in achievements of insight, imagination, and ingenuity unsurpassed in any field of mental activity; it illustrates, too, better than any other branch of physics, the vicissitudes of theories.

J. J. Thomson (1925)

Life is but light in many-hued reflection.

Goethe, *Faust* (1832)

In this chapter we discuss phenomena relating to the interaction of a molecular system with a weak external time-dependent perturbation.[1] This is the realm of spectroscopy. We focus on three kinds of spectra: in § 11.1, we study those resulting from the absorption and emission of infrared light; in § 11.2, those associated with the scattering of light; and in § 11.3, those associated with the scattering of neutrons.

In the linear-response regime[2] each spectrum can be described as the spectral density $I(\omega)$—or Fourier transform—of a time correlation function[3–16] $C(t) = \langle A(t)A(0)\rangle$ of some dynamical variable $A(t)$, i.e.[17]

$$I(\omega) = \frac{1}{2\pi}\int_{-\infty}^{+\infty} \mathrm{d}t\, \exp(i\omega t)C(t). \tag{11.1}$$

For absorption, light scattering, and neutron scattering, A is the system dipole moment, polarizability, and nuclear density, respectively. Mathematically, this relation is an example of the Wiener–Khinchin theorem[18] for random variables. Physically, it is an example of the *fluctuation–dissipation* (or *fluctuation–response*) relation[20,21] between a system linear response function (such as the absorption coefficient) and the spontaneous equilibrium fluctuations (such as the dipole correlation function $\langle \boldsymbol{\mu}(t)\cdot\boldsymbol{\mu}(0)\rangle$).

The calculation of the spectral density itself is often a complicated dynamical problem.[22] Consider, however, the spectral moments,[36] defined by

$$M_n = \int_{-\infty}^{+\infty} \mathrm{d}\omega\, \omega^n I(\omega), \qquad n = 0, 1, 2 \ldots. \tag{11.2}$$

Now, the inverse transform of (11.1) is

$$C(t) = \int_{-\infty}^{+\infty} \mathrm{d}\omega\, \exp(-i\omega t)I(\omega), \tag{11.3}$$

and on expanding the exponential we find that

$$C(t) = \sum_{n=0}^{\infty} \frac{M_n}{n!}(-it)^n. \tag{11.4}$$

Thus the moments are essentially the Taylor series coefficients in the expansion of $C(t)$ in powers of t. The precise relation is

$$M_n = i^n C^{(n)}(0), \tag{11.5}$$

where $C^{(n)}(t) = (\mathrm{d}/\mathrm{d}t)^n C(t)$. In particular,

$$\begin{aligned} M_0 &= C(0) = \langle A^2 \rangle, \\ M_1 &= i\dot{C}(0) = i\langle \dot{A}A \rangle, \\ M_2 &= -\ddot{C}(0) = -\langle \ddot{A}A \rangle, \end{aligned} \tag{11.6}$$

etc., where $A \equiv A(0)$, $\dot{A} \equiv \dot{A}(0) \equiv (\mathrm{d}A/\mathrm{d}t)_{t=0}$, $\ddot{A} \equiv \ddot{A}(0) \equiv (\mathrm{d}^2A/\mathrm{d}t^2)_{t=0}$. Thus the moments are *equilibrium* properties, and are amenable to calculation using the techniques described in Vol. 1. For this reason they constitute the main focus of this chapter. We will restrict our study to isotropic systems.[47]

Spectral moments have several uses, including (i) the construction of approximate spectral densities and time correlation functions[37,38,48,49] (see e.g. § 11.3.6), (ii) as checks on both approximate theories[35,50,51] and computer simulations,[52] (iii) supplying parameters required as input for some theories (e.g. mean square forces and torques), and (iv) the estimation of inelasticity corrections to neutron scattering cross-sections (11.3.7).

11.1 Infrared spectra

11.1.1 Macroscopic relations for absorption

Consider a monochromatic light beam propagating along the z direction, of frequency ω and wavenumber k. The electric field a distance z within the sample satisfies

$$\mathbf{E} = \mathbf{E}_0 \exp i(kz - \omega t). \tag{11.7}$$

Here $\omega/k = v = c/n$, where v and c denote the phase velocity and speed of light *in vacuo*, respectively, and $n = n' + in''$ is the complex refractive index. Since the intensity $\mathcal{I}$ of the beam (energy per unit time per unit area) is proportional to $|E|^2$, we have

$$\mathcal{I} = \mathcal{I}_0 \exp(-2\omega n'' z/c), \tag{11.8}$$

where $\mathcal{I}_0$ is the incident intensity. If we now write

$$A(\omega) = 2\omega n''(\omega)/c \tag{11.9}$$

then (11.8) becomes

$$\mathcal{I}(\omega) = \mathcal{I}_0(\omega) \exp(-A(\omega)z). \tag{11.10}$$

Equation (11.10) is Lambert's law[53] for absorption. $A(\omega)$ is termed the absorption coefficient, and is typically assigned units of cm^{-1}. It may also be expressed in terms of the imaginary part of the complex dielectric constant $\epsilon(\omega) \equiv \epsilon'(\omega) + i\epsilon''(\omega) = n(\omega)^2$; we find

$$A(\omega) = \frac{\omega}{c} \frac{\epsilon''(\omega)}{n'(\omega)}. \tag{11.11}$$

11.1.2 The absorption coefficient of a dilute gas

A quantum-mechanical expression[19,40,54] for $A(\omega)$ may be derived in various ways, for example using quantum linear response theory.[19,40] Here we present a simple derivation[7,15] using time-dependent perturbation theory. Consider a dilute system in equilibrium perturbed by an incident light beam of frequency ω (in 11.1.3 we will consider the absorption of light by a dense medium). The average net energy absorbed per unit time from the radiation field (the net effect of both upward and downward transitions) is

$$P_A = \sum_{mn} P_m W_{mn} \hbar\omega_{nm}, \tag{11.12}$$

where $P_m = \exp(-\beta E_m)/Q$ is the probability of finding the system initially in state $|m\rangle$ with energy E_m, W_{mn} is the probability per unit time that the system will make a transition from $|m\rangle$ to final state $|n\rangle$, and $\hbar\omega_{nm} = E_n - E_m$ is the energy absorbed in this process. The transition rate W_{mn} is given by the Fermi golden rule of quantum time-dependent perturbation theory. In the dipole approximation[55]

$$W_{mn} = \frac{\pi E_0^2}{2\hbar^2} |\langle n|\boldsymbol{\mu} \cdot \mathbf{n}|m\rangle|^2 [\delta(\omega - \omega_{nm}) + \delta(\omega + \omega_{nm})], \tag{11.13}$$

where E_0 and $\mathbf{n}$ are the amplitude and polarization of the electric field of the light wave, respectively. For an isotropic fluid we can replace $|\langle n|\boldsymbol{\mu} \cdot \mathbf{n}|m\rangle|^2$ by $\frac{1}{3}|\boldsymbol{\mu}_{nm}|^2 \equiv \frac{1}{3}|\langle n|\boldsymbol{\mu}|m\rangle|^2$. Substituting (11.13) into (11.12) then yields

$$P_A = \frac{\pi E_0^2}{6\hbar} \sum_{mn} P_m |\boldsymbol{\mu}_{nm}|^2 \omega_{nm} [\delta(\omega - \omega_{nm}) + \delta(\omega + \omega_{nm})]. \tag{11.14}$$

For thin samples, or weak absorption, we have from (11.10) $\mathcal{I} \approx \mathcal{I}_0(1 - Az)$, so that the decrease in intensity in travelling a distance z is $\mathcal{I}_0 Az$. This is the power lost per unit area due to absorption. The total power loss is thus $P_A = \mathcal{I}_0 AzS = \mathcal{I}_0 AV$, where S is the cross-sectional area of the sample perpendicular to the beam, and V is the volume irradiated. Now $\mathcal{I}_0 = (c/8\pi)E_0^2$, and hence

$$A(\omega) = \frac{4\pi^2}{3\hbar c V} \sum_{mn} P_m |\boldsymbol{\mu}_{nm}|^2 \omega_{nm} [\delta(\omega - \omega_{nm}) + \delta(\omega + \omega_{nm})]. \tag{11.15}$$

We see from (11.15) that $A(\omega)$ is an *even* function[56] of ω, i.e. $A(-\omega) = A(\omega)$.

An alternative, and somewhat simpler, form of this result is obtained by interchanging the dummy indices m and n in the term containing $\delta(\omega + \omega_{nm})$ and noting that $P_n = P_m \exp(-\hbar\omega_{nm})$. This yields

$$A(\omega) = \frac{4\pi^2}{3\hbar cV} \sum_{mn} (P_m - P_n)|\boldsymbol{\mu}_{nm}|^2 \omega_{nm} \delta(\omega - \omega_{nm})$$

$$= \frac{4\pi^2 \omega}{3\hbar cV} [1 - \exp(-\beta\hbar\omega)] \sum_{mn} P_m |\boldsymbol{\mu}_{nm}|^2 \delta(\omega - \omega_{nm}). \qquad (11.16)$$

If we define the spectral density $I(\omega)$ as

$$I(\omega) = \sum_{mn} P_m |\boldsymbol{\mu}_{nm}|^2 \delta(\omega - \omega_{nm}) \qquad (11.17)$$

then we can write

$$A(\omega) = \kappa\omega[1 - \exp(-\beta\hbar\omega)] I(\omega), \qquad (11.18)$$

where $\kappa = 4\pi^2/3\hbar cV$.

We can write (11.18) in correlation function form as follows. Introducing the Fourier representation of the δ-function (cf. (B.72))

$$\delta(\omega - \omega_{nm}) = \frac{1}{2\pi} \int_{-\infty}^{+\infty} \mathrm{d}t \, \exp[i(\omega - \omega_{nm})t], \qquad (11.19)$$

we write the spectral density $I(\omega)$ of (11.17) as

$$I(\omega) = \frac{1}{2\pi} \int_{-\infty}^{+\infty} \mathrm{d}t \, \exp(i\omega t) C(t), \qquad (11.20)$$

where

$$C(t) = \sum_{mn} P_m |\boldsymbol{\mu}_{nm}|^2 \exp(-i\omega_{nm} t). \qquad (11.21)$$

To see that $C(t)$ is a correlation function, we write it as

$$C(t) = \sum_{mn} \langle n|\boldsymbol{\mu} P_m|m\rangle \cdot \langle m| \exp(iE_m t/\hbar) \boldsymbol{\mu} \exp(-iE_n t/\hbar)|n\rangle. \qquad (11.22)$$

Suppose $|m\rangle$ and $|n\rangle$ are eigenstates of the Hamiltonian H_0. Then

$$P|m\rangle = P_m|m\rangle,$$

$$\langle m| \exp(iH_0 t/\hbar) = \langle m| \exp(iE_m t/\hbar),$$

$$\exp(-iH_0 t/\hbar)|n\rangle = \exp(-iE_n t/\hbar)|n\rangle, \qquad (11.23)$$

where $P = \exp(-\beta H_0)/Q$ is the canonical density operator (see (B.59)). Hence

$$C(t) = \sum_{mn} \langle n|\boldsymbol{\mu} P|m\rangle \cdot \langle m|\boldsymbol{\mu}(t)|n\rangle, \qquad (11.24)$$

where $\boldsymbol{\mu}(t) = \exp(iH_0t/\hbar)\boldsymbol{\mu}\exp(-iH_0t/\hbar)$ is the time-dependent Heisenberg operator[57] corresponding to the time-independent Schrödinger operator $\boldsymbol{\mu} \equiv \boldsymbol{\mu}(0)$. Using the completeness relation (B.53), i.e. $\sum_m |m\rangle\langle m| = 1$, and the definition (B.61) of the trace, we thus have

$$C(t) = \mathrm{Tr}\,[\boldsymbol{\mu}P \cdot \boldsymbol{\mu}(t)] = \mathrm{Tr}\,[\boldsymbol{\mu}(t) \cdot \boldsymbol{\mu}P] \equiv \langle \boldsymbol{\mu}(t) \cdot \boldsymbol{\mu}(0)\rangle. \tag{11.25}$$

We note that $I(\omega)$ satisfies the detailed balance condition

$$I(-\omega) = \exp(-\beta\hbar\omega)I(\omega). \tag{11.26}$$

$C(t)$ satisfies

$$C(t)^* = C(-t) = C(t - i\beta\hbar) = \langle \boldsymbol{\mu}(0) \cdot \boldsymbol{\mu}(t)\rangle. \tag{11.27}$$

The first equality follows from the inverse Fourier transform relation

$$C(t) = \int_{-\infty}^{+\infty} \mathrm{d}t\, \exp(-i\omega t)I(\omega) \tag{11.28}$$

together with the reality of $I(\omega)$, which is apparent from (11.17). The second equality follows from (11.26) and (11.28), while the third equality follows from the stationarity of $C(t)$.

In the classical limit $\hbar \to 0$, Eqs. (11.26) and (11.27) establish that $I(\omega)$ is an even function of frequency, and that $C(t)$ is real, and an even function of time.

It sometimes proves useful to introduce the symmetrized and antisymmetrized densities

$$I_\pm(\omega) = \frac{1}{2}[I(\omega) \pm I(-\omega)] = \frac{1}{2}I(\omega)[1 \pm \exp(-\beta\hbar\omega)], \tag{11.29}$$

which satisfy

$$I_+(\omega)/I_-(\omega) = \coth(\beta\hbar\omega/2). \tag{11.30}$$

We can write (11.18) in terms of these functions as

$$A(\omega) = 2\kappa\omega I_-(\omega) = 2\kappa\omega \tanh\frac{1}{2}\beta\hbar\omega I_+(\omega). \tag{11.31}$$

From (11.27) and (11.29) we have

$$I_\pm(\omega) = \frac{1}{2\pi}\int_{-\infty}^{+\infty} \mathrm{d}t\, \exp(i\omega t)C_\pm(t), \tag{11.32}$$

where

$$C_\pm(t) = \frac{1}{2}[\langle \boldsymbol{\mu}(t) \cdot \boldsymbol{\mu}(0)\rangle \pm \langle \boldsymbol{\mu}(0) \cdot \boldsymbol{\mu}(t)\rangle]. \tag{11.33}$$

11.1.3 *The absorption coefficient for dense gases and liquids*

In the previous section we derived expressions for $A(\omega)$ assuming that every molecule experienced the external electric field $\mathbf{E}_0(t)$. In fact, each molecule experiences a different, fluctuating, field as a result of polarization of its neighbours by the light wave, and so our results are strictly valid only for dilute systems. Here we extend our analysis to dense gases and liquids. Our treatment is based on the concept of the Lorentz local field, which is the mean field experienced by a molecule (see Appendix 10B). It allows for dispersion in the dielectric constant and refractive index, but assumes that both are real.

In a dense medium the local field $\mathbf{E}_L(t) = \mathbf{E}_L \exp(-i\omega t)$ differs from both the applied field $\mathbf{E}_0(t) = \mathbf{E}_0 \exp(-i\omega t)$ and the macroscopic, Maxwell field $\mathbf{E}(t) = \mathbf{E} \exp(-i\omega t)$. From the macroscopic boundary conditions, the Maxwell field, i.e. the mean field at an arbitrary point in the system, is given by[58]

$$\mathbf{E} = \frac{1}{n'(\omega)^{1/2}} \mathbf{E}_0, \tag{11.34}$$

where $n'(\omega)$ is the refractive index at the frequency of the radiation. In turn $\mathbf{E}_L$ is related to $\mathbf{E}$ by

$$\mathbf{E}_L = \frac{\epsilon'(\omega) + 2}{3} \mathbf{E}, \tag{11.35}$$

where $\epsilon'(\omega) = n'(\omega)^2$ is the dielectric constant (a simple derivation of this result is given in § 10.4). Since the absorbed power is proportional to E_L^2, the absorption coefficient should be multiplied by a correction factor $f_A = (E_L/E_0)^2 = (E_L/E)^2(E/E_0)^2$. By the above arguments,

$$f_A(\omega) = \frac{1}{n'(\omega)} \left(\frac{n'(\omega)^2 + 2}{3} \right)^2. \tag{11.36}$$

If we assume the main contribution to the refractive index comes from the polarizability, then $n'(\omega) \approx n_\infty$, where $n_\infty \equiv n(\omega_\infty)$ is the high-frequency refractive index. In this case

$$f_A \approx \frac{1}{n_\infty} \left(\frac{n_\infty^2 + 2}{3} \right)^2, \tag{11.37}$$

which is often referred to as the Polo–Wilson correction factor.[59]

The effects predicted by (11.37) can be quite large for liquids, ranging from 10–15% for small molecules such as as Kr or N_2 to 30–40% for larger (and more polarizable) species such as CS_2 and C_6H_6.

Solvent effects on infrared lineshapes are discussed in more detail by Gray and Joslin.[60]

11.1.4 *Allowed and induced spectra*

In § 11.1.2 we showed that the absorption coefficient $A(\omega)$ is related to the frequency spectrum of the dipole–dipole correlation function $C(t) = \langle \boldsymbol{\mu}(t) \cdot \boldsymbol{\mu}(0) \rangle$.

The quantity $\boldsymbol{\mu}$ is the total dipole moment of the system, including polarization or collision-induced effects. It can be decomposed into one-, two-, three-body, etc., contributions:

$$\boldsymbol{\mu} = \sum_i \boldsymbol{\mu}_i + \sum_{i<j} \boldsymbol{\mu}_{ij} + \sum_{i<j<k} \boldsymbol{\mu}_{ijk} + \ldots. \tag{11.38}$$

Here $\boldsymbol{\mu}_i$ is the permanent dipole moment of molecule i, $\boldsymbol{\mu}_{ij}$ is the additional (collision-induced) moment of the ij pair resulting from mutual polarization, and so on. For *polar* molecules such as HCl, all the terms in (11.38) contribute to some extent, due in particular to long-range dipole–induced-dipole processes. Normally, however, the single-molecule spectrum dominates. For *nonpolar* molecules such as rare-gas mixtures, N_2, or CH_4, pair contributions to $\boldsymbol{\mu}$ are the most important. Only in a relatively few cases is it necessary to consider higher order effects.

We turn now to a consideration of allowed spectra, originating from fluctuations in the permanent dipoles, or, in the case of vibrational bands, their derivatives with respect to normal mode coordinates. Induced spectra are discussed in § 11.1.8.

11.1.5 Allowed spectral lineshapes [61]

Dilute gases

Figures 11.1 and 11.2 illustrate schematically the transitions of the pure rotation and vibration–rotation bands of linear molecules. If we ignore small anharmonic terms in the expansion of the intramolecular potential and molecular dipole moment, allowed dipole transitions are governed by the selection rules $\Delta v = 0, \pm 1$ and[65] $\Delta J = 0, \pm 1$, where v and J are the vibrational and rotational quantum numbers, respectively. Transitions for which $\Delta v = 0$ are pure rotational bands, and occur in the microwave or far-infrared region of the spectrum, typically betwen 1 and 500 cm^{-1}. Transitions with $\Delta v = \pm 1$ are the fundamental vibration–rotation bands, and occur in the (mid)-infrared region of the spectrum, centred on the vibrational frequency $\omega_v \sim 200$–$3000\ \text{cm}^{-1}$. Rotational transitions for which $\Delta J = -1, 0, 1$, are traditionally referred to as P-, Q-, and R-branch lines, respectively. For parallel modes of molecules possessing no orbital or spin angular momentum, the Q branch is absent.[65]

For pure rotational bands there is a single branch, since $\Delta J = 0$ is of course unobservable, and absorptive R(J) transitions, for which $J \to J + 1$, and emissive P($J + 1$) transitions, for which $J + 1 \to J$, coincide. Because the separation of adjacent rotational levels is often much smaller than kT, many levels are typically populated; the resulting cancellation then causes a substantial reduction in the net intensity of absorption.

In the vibration–rotation spectrum (Fig. 11.2), R-branch lines at $\omega' = \omega - \omega_v$ are more intense than the corresponding P-branch lines at $-\omega'$; in the absence of rovibrational coupling or vibrational perturbations, we have the relation[66]

$$I(-\omega') = \exp(-\beta\hbar\omega')I(\omega'). \tag{11.39}$$

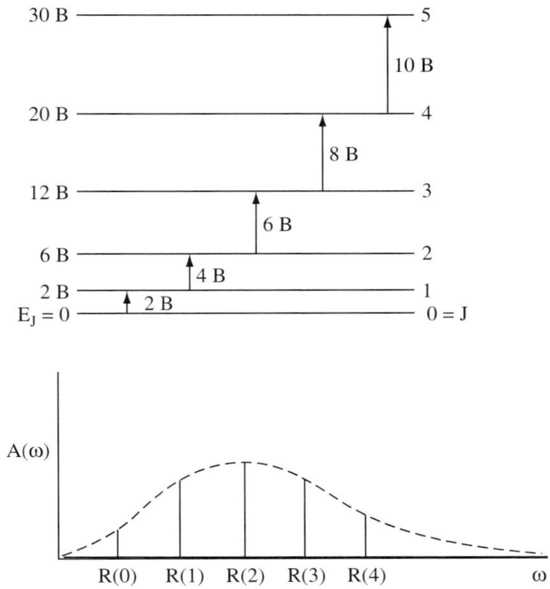

FIG. 11.1 Schematic of the first few lines of the isolated molecule far-infrared pure rotational absorption spectrum due to the R-branch transitions $J \to J+1$. For a rigid linear rotator the energy levels are given by $E_J = BJ(J+1)$, where $B = \hbar^2/2I$. The frequencies are hence given by $\omega_{J\to J+1} = 2B/\hbar\,(J+1)$ and are equally spaced. The idealized lines (δ-functions) have been chopped off at heights proportional to their integrated intensities. The classical envelope is shown by the dashed line. The characteristic 'wing shape' is due mainly to the Boltzmann population factor $P_J = (2J+1)\exp(-\beta E_J)/Q$.

Figure 11.3 shows the vibrational spectrum of the linear molecule HCN, which displays both a perpendicular mode, centred at 2800 cm^{-1}, and a parallel mode, centred at 4500 cm^{-1}. Note the Q-branch in the perpendicular band.

Spherical top molecules naturally show no pure rotational spectra; their vibrational bands display the simple PQR structure characteristic of a perpendicular band of a linear molecule. Other nonlinear molecules exhibit more complicated spectra, reflecting the more complex structure of their rotational energy levels. For a symmetric top such as NH_3, in addition to the selection rules on the total angular momentum quantum number J ($\Delta J = 0, \pm 1$, as for linear molecules), we also have a selection rule on the quantum number K for the component of angular momentum along the symmetry axis. This is $\Delta K = 0$, reflecting the fact that a change in the rotational motion of the molecule about this axis does not modify the dipole moment. Under low or medium resolution, therefore, the spectrum retains a simple PQR appearance. At higher resolution, however, centrifugal distortion causes each line to display K-structure.[68] The spectra of asymmetric tops are more complicated still.[68]

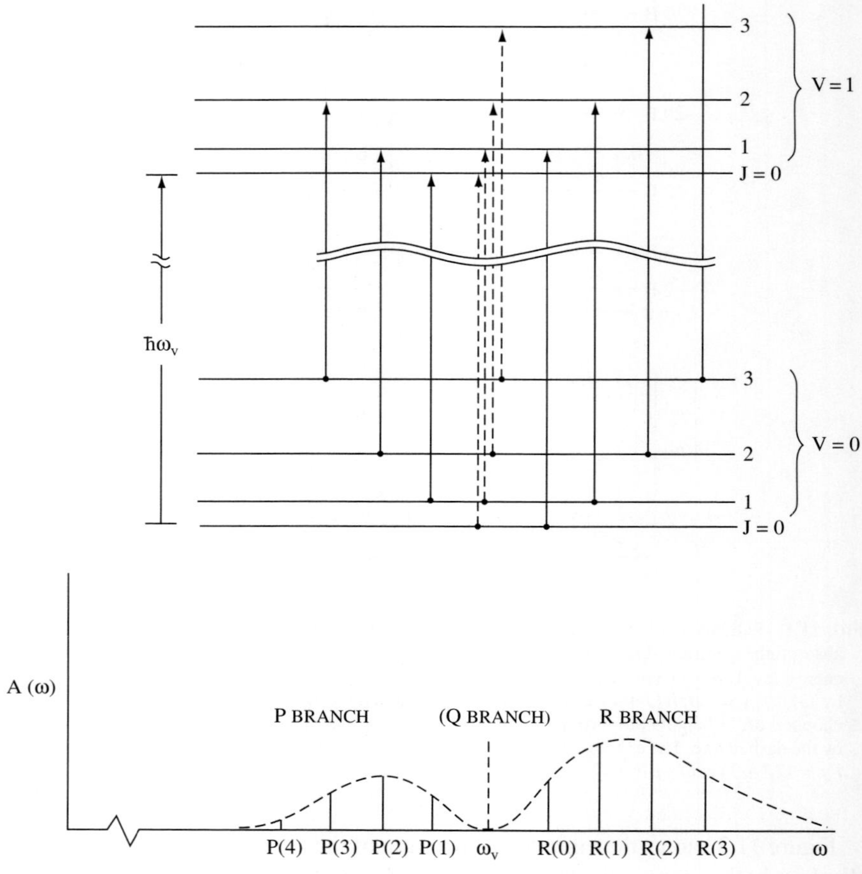

FIG. 11.2 Schematic of the isolated molecule infrared fundamental ($v = 0 \rightarrow 1$) vibration–rotation absorption spectrum. Q(J) lines actually have slightly different frequencies because of vibration–rotation coupling. The Q branch is absent for parallel modes of linear molecules. Other details as described in the legend to Fig. 11.1.

Dense gases and liquids

At extremely low densities, where the molecules absorb essentially independently, individual lines within a rotational or vibrational band are Gaussian, as a result of Doppler broadening associated with random thermal motion. Linewidth is proportional to the square root of the temperature. As the pressure increases, binary collisions occur, and the profiles assume a Lorentzian form, with a half-width approximately equal to the collision frequency, and therefore proportional to the density. At still higher pressures the lines begin to shift, and overlap, and to assume more complex shapes.[69] In dense gases and in liquids, the line structure

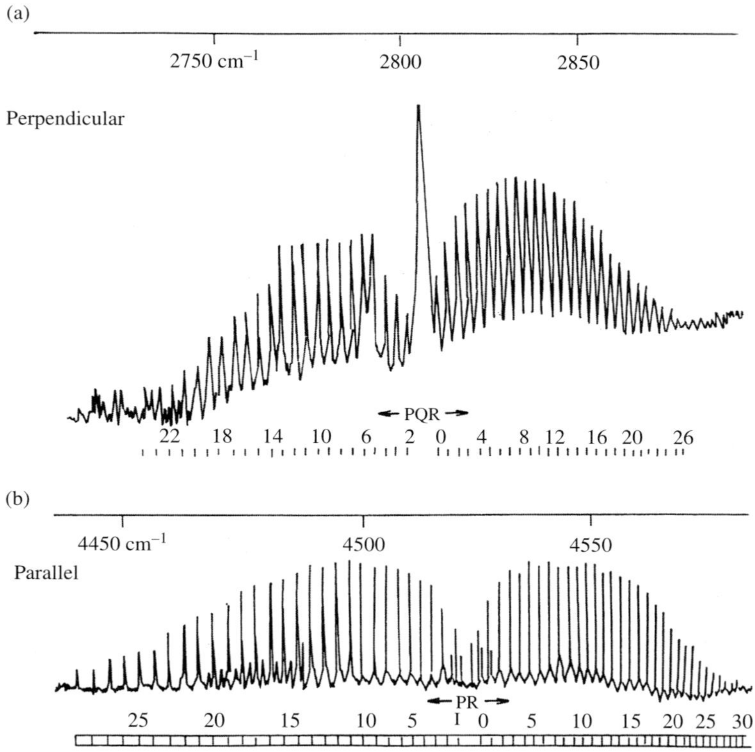

FIG. 11.3 Infrared vibration–rotation spectra of (a) perpendicular, and (b) parallel, modes of HCN. Reprinted with permission from ref. 67. Copyright 1978, Prentice-Hall.

is usually obliterated altogether. Figure 11.4 illustrates the behaviour described above for a dilute solution of HCl in argon as the rare-gas density increases.

In certain instances, increases in the density actually narrow the bandwidth. This is the phenomenon of motional (or collisional) narrowing.[72,73] It occurs when molecular rotation is strongly hindered, and is replaced by a librational, or rocking, motion. The P- and R-rotational wings then collapse; most of the intensity appears to reside in the Q-branch, although in fact there may be significant absorption far from the band centre, which is not observable in practice. Theoretical analysis of such effects requires consideration of vibrational and rotational relaxation.[74]

In dense gases and liquids, the calculation of infrared band shapes is a complicated problem in nonequilibrium statistical mechanics.[54,72,76] Quantitative analysis in condensed systems generally requires computer simulation.[77] Such computations are very resource-intensive. In this chapter, we shall mostly be concerned with an analysis of the band frequency *moments*, which in contrast is

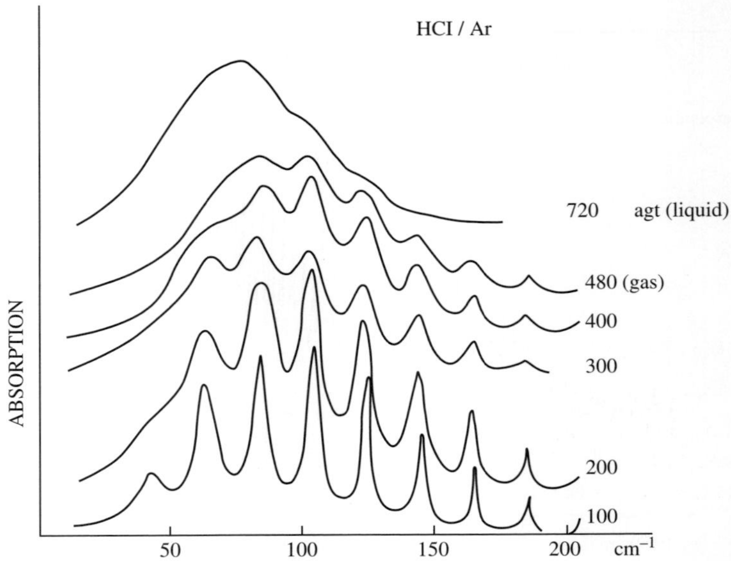

FIG. 11.4 Far-infrared rotational spectrum of dilute HCl in Ar at various densities (in amagat[70]). (From ref. 71.)

a relatively straightforward exercise in equilibrium statistical mechanics. As we shall see, such studies can yield very useful information on molecular properties and interactions.

Correlation functions for infrared and far-infrared spectra

Consider for simplicity a fluid of diatomic molecules. We let $q_i = r_i - r_e$, $i = 1, 2, \ldots, N$, where r_e is the equilibrium value of the internuclear separation r_i. The dipole moment $\boldsymbol{\mu}_i(q_i)$ of molecule i can be expanded in powers of q_i,

$$\boldsymbol{\mu}_i = \boldsymbol{\mu}_i^0 + \boldsymbol{\mu}_i' q_i + \ldots, \tag{11.40}$$

where $\boldsymbol{\mu}_i^0$ is the dipole at the equilibrium position, and $\boldsymbol{\mu}_i'$ is the derivative of the dipole with respect to q_i evaluated at the equilibrium position. As vectors, $\boldsymbol{\mu}_i^0$ and $\boldsymbol{\mu}_i'$ depend on the molecular orientation. The two terms on the right of (11.40) are responsible for the far-infrared (FIR) and fundamental infrared (IR) spectra, respectively. Higher terms involving q_i^2, etc., generate IR overtone spectra ($v = 0 \to 2$, etc.), but may be neglected for our purposes.[78]

Using (11.40) we can write the dipole–dipole correlation function

$$C(t) = \sum_{ij} \langle \boldsymbol{\mu}_i(t) \cdot \boldsymbol{\mu}_j(0) \rangle \tag{11.41}$$

as

$$C(t) \approx C^{\text{FIR}}(t) + C^{\text{IR}}(t), \tag{11.42}$$

where

$$C^{\text{FIR}}(t) = \sum_{ij} \left\langle \boldsymbol{\mu}_i^0(t) \cdot \boldsymbol{\mu}_j^0(0) \right\rangle \tag{11.43}$$

and

$$C^{\text{IR}}(t) = \sum_{ij} \left\langle \boldsymbol{\mu}_i'(t) q_i(t) \cdot \boldsymbol{\mu}_j'(0) q_j(0) \right\rangle \tag{11.44}$$

denote the FIR and IR correlation functions, respectively. We have neglected cross terms such as $\sum_{ij} \left\langle \boldsymbol{\mu}_i^0(t) \cdot \boldsymbol{\mu}_j'(0) q_j(0) \right\rangle$, assuming that intra- and intermolecular vibration–rotation correlation can be ignored.[79]

For the FIR spectrum, the dipole–dipole correlation function is a sum of one- and two-body contributions, or *self* and *distinct* parts:

$$C^{\text{FIR}}(t) = \sum_{i} \left\langle \boldsymbol{\mu}_i^0(t) \cdot \boldsymbol{\mu}_i^0(0) \right\rangle + \sum_{i \neq j} \left\langle \boldsymbol{\mu}_i^0(t) \cdot \boldsymbol{\mu}_j^0(0) \right\rangle. \tag{11.45}$$

For the IR spectrum, to a good approximation, only self terms contribute. To see this, we assume that the vibrational phases of different molecules are unrelated.[82] There is then no intermolecular vibrational correlation and $\langle q_i(t) q_j(0) \rangle = 0$ for $i \neq j$. $C^{\text{IR}}(t)$ can then be written as a product of rotational and vibrational terms, i.e.

$$C^{\text{IR}}(t) = C_r(t) C_v(t), \tag{11.46}$$

where

$$C_r(t) = N \langle \boldsymbol{\mu}_1'(t) \cdot \boldsymbol{\mu}_1'(0) \rangle \tag{11.47}$$

and

$$C_v(t) = \langle q_1(t) q_1(0) \rangle. \tag{11.48}$$

Correspondingly, the spectral density $I^{\text{IR}}(\omega)$ is a convolution of independent rotational, $I_r(\omega)$, and vibrational, $I_v(\omega)$, spectra. The IR moments also simplify, as discussed further in § 11.1.6.

11.1.6 Moments of allowed IR and FIR spectra

The spectral moments are defined as

$$M_n = \int_{-\infty}^{+\infty} \mathrm{d}\omega\, \omega^n I(\omega), \tag{11.49}$$

where $I(\omega)$ is the Fourier transform (11.1) of the dipole–dipole correlation function (11.41). Alternative expressions can be derived, and sometimes prove useful. In terms of the symmetrized and antisymmetrized densities of (11.29), we have

$$M_{2n} = \int_{-\infty}^{+\infty} \mathrm{d}\omega\, \omega^{2n} I_+(\omega),$$

$$M_{2n+1} = \int_{-\infty}^{+\infty} \mathrm{d}\omega\, \omega^{2n+1} I_-(\omega). \tag{11.50}$$

Using (11.30) and noting that $\coth x$ and $\tanh x$ are odd functions of x, it therefore follows that

$$M_{2n} = \int_{-\infty}^{+\infty} \mathrm{d}\omega\, \omega^{2n} \coth \frac{1}{2}\beta\hbar\omega\, I(\omega),$$

$$M_{2n+1} = \int_{-\infty}^{+\infty} \mathrm{d}\omega\, \omega^{2n+1} \tanh \frac{1}{2}\beta\hbar\omega\, I(\omega). \tag{11.51}$$

Thus the even moments can be calculated with or without inclusion of the factor $\coth \frac{1}{2}\beta\hbar\omega$, and the odd moments with or without the factor $\tanh \frac{1}{2}\beta\hbar\omega$.

Equation (11.51) implies that the even moments can be expressed in terms of the odd moments, and vice versa. Using the expansions[85] $\coth x = 1/x + x/3 - x^3/45 + \ldots$ and $\tanh x = x - x^3/3 + 2x^5/15 + \ldots$, we have

$$M_{2n} = \frac{2}{\beta\hbar} M_{2n-1} + \frac{1}{3}\frac{\beta\hbar}{2} M_{2n+1} - \frac{1}{45}\left(\frac{\beta\hbar}{2}\right)^3 M_{2n+3} + \ldots,$$

$$M_{2n+1} = \frac{\beta\hbar}{2} M_{2n+2} - \frac{1}{3}\left(\frac{\beta\hbar}{2}\right)^3 M_{2n+4} + \frac{2}{15}\left(\frac{\beta\hbar}{2}\right)^5 M_{2n+6} + \ldots. \tag{11.52}$$

In the semiclassical limit $\hbar \to 0$, we have the simple relation

$$M^{\mathrm{scl}}_{2n-1} = \frac{\beta\hbar}{2} M^{\mathrm{cl}}_{2n}. \tag{11.53}$$

Using (11.31), we can also write

$$M_{2n} = \int_0^{\infty} \mathrm{d}\omega\, \kappa^{-1} \omega^{2n-1} A(\omega) \coth \frac{1}{2}\beta\hbar\omega,$$

$$M_{2n+1} = \int_0^{\infty} \mathrm{d}\omega\, \kappa^{-1} \omega^{2n} A(\omega), \tag{11.54}$$

since $A(\omega)$ is an even function of ω (§ 11.1.2). Here $\kappa = f_A 4\pi^2/3\hbar c V$, where f_A is the local-field factor discussed in § 11.1.3. To the extent that we can neglect the frequency-dependence of f_A, we see that the *odd* moments of $I(\omega)$ are proportional to the *even* moments of $A(\omega)$.

Since higher moments weight the spectrum with higher powers of ω, they accentuate the high-frequency wings. It is difficult to measure intensities acurately at high frequencies for various reasons: the absorption is weak, there may be overlapping vibrational bands, and it becomes difficult to separate allowed and induced spectra. We shall therefore only consider moments up to M_4.

FIR moments

Molecular expressions for the moments can be derived from the zero-time derivatives of the permanent dipole–dipole correlation function (11.43) using the general relation (11.5). Thus

$$\begin{aligned}
M_0^{\mathrm{FIR}} &= \langle \mu^{0\,2} \rangle, \\
M_1^{\mathrm{FIR}} &= i\langle \dot{\boldsymbol{\mu}}^0 \cdot \boldsymbol{\mu}^0 \rangle = -i\langle \boldsymbol{\mu}^0 \cdot \dot{\boldsymbol{\mu}}^0 \rangle, \\
M_2^{\mathrm{FIR}} &= -\langle \ddot{\boldsymbol{\mu}}^0 \cdot \boldsymbol{\mu}^0 \rangle = \langle \dot{\mu}^{0\,2} \rangle, \\
M_3^{\mathrm{FIR}} &= -i\langle \dddot{\boldsymbol{\mu}}^0 \cdot \boldsymbol{\mu}^0 \rangle = i\langle \ddot{\boldsymbol{\mu}}^0 \cdot \dot{\boldsymbol{\mu}}^0 \rangle, \\
M_4^{\mathrm{FIR}} &= \langle \ddddot{\boldsymbol{\mu}}^0 \cdot \boldsymbol{\mu}^0 \rangle = \langle \ddot{\mu}^{0\,2} \rangle,
\end{aligned} \tag{11.55}$$

etc., where $\boldsymbol{\mu}^0 = \sum_i \boldsymbol{\mu}_i^0$ is the permanent dipole moment of the system. The second forms follow from the equality $\langle \dot{A} B \rangle = -\langle A \dot{B} \rangle$ (Appendix E). We note that, like the FIR correlation function itself, the moments contain both self and distinct terms.

IR moments

If it is assumed that vibration and rotation decouple, then as discussed in § 11.1.5 the IR correlation function factorizes into a product of rotational and vibrational terms. Differentiating (11.46) we find that

$$M_n^{\mathrm{IR}} = \sum_{m=0}^{n} \binom{n}{m} M_{\mathrm{r}m} M_{\mathrm{v}\,n-m}, \tag{11.56}$$

where $M_{\mathrm{r}n}$ and $M_{\mathrm{v}n}$ denote the rotational and vibrational moments, respectively. Our principal concern in this book is with the rotational moments; for a discussion of vibrational spectra see refs. 30 and 33. The vibrational contribution to the moments can be minimized by taking moments about a *shifted* band origin

$$\omega_{\mathrm{v}}' \equiv \frac{M_{\mathrm{v}1}}{M_{\mathrm{v}0}} = M_{\mathrm{v}0}^{-1} \int_{-\infty}^{+\infty} \mathrm{d}\omega\, \omega I_{\mathrm{v}}(\omega). \tag{11.57}$$

In dense media ω_{v}' differs slightly from ω_{v}, the free molecule vibrational frequency.[86] Experimentally, ω_{v}' can be determined using the relation $\omega_{\mathrm{v}}' = \bar{\omega} - M_{\mathrm{r}1}/M_{\mathrm{r}0}$ (which follows from (11.56)), where $\bar{\omega} \equiv M_1/M_0$ is the band mean frequency, and the rotational zeroth and first moments are discussed below.

The shifted moments are defined by

$$M_n^{\mathrm{IR}\,\prime} = \int_{-\infty}^{+\infty} \mathrm{d}\omega\, \left(\omega - \omega_{\mathrm{v}}'\right)^n I(\omega) \tag{11.58}$$

and are related to the unshifted moments by

$$M_n^{\mathrm{IR}\,\prime} = \sum_{m=0}^{n} (-)^{n-m} \binom{n}{m} M_m \, \omega_{\mathrm{v}}^{\prime\, n-m}. \tag{11.59}$$

Using (11.56) and (11.59), we have

$$M_n^{\mathrm{IR}\,\prime} = \sum_{m=0}^{n} \sum_{l=0}^{m} (-)^{n-m} \binom{n}{m} \binom{m}{l} M_{\mathrm{r}l} M_{\mathrm{v}\,m-l} \, \omega_{\mathrm{v}}^{\prime\, n-m}. \tag{11.60}$$

We can resum (11.60) into a single series,

$$M_n^{\mathrm{IR}\,\prime} = \sum_{m=0}^{n} \binom{n}{m} M_{\mathrm{r}m} M'_{\mathrm{v}\,n-m}, \tag{11.61}$$

where

$$\begin{aligned} M'_{\mathrm{v}n} &= \int_{-\infty}^{+\infty} \mathrm{d}\omega \, \left(\omega - \omega'_{\mathrm{v}}\right)^n I_{\mathrm{v}}(\omega) \\ &= \sum_{m=0}^{n} (-)^{n-m} \binom{n}{m} M_{\mathrm{v}m} \, \omega_{\mathrm{v}}^{\prime\, n-m} \end{aligned} \tag{11.62}$$

denotes a shifted *vibrational* moment.

Normally the vibrational linewidth is small, so that $M'_{\mathrm{v}n}/M_{\mathrm{r}n}$ decreases strongly with n. Therefore (11.60) becomes $M_n^{\mathrm{IR}\,\prime} \approx M_{\mathrm{r}n} M_{\mathrm{v}0}$. We therefore have

$$M_n^{\mathrm{IR}\,\prime} \approx M_{\mathrm{r}n} M_{\mathrm{v}0} = M_{\mathrm{r}n} \left\langle q_1^2 \right\rangle. \tag{11.63}$$

The rotational moments may be computed using (11.47). We see that the shifted IR moments are given by

$$\begin{aligned} M_0^{\mathrm{IR}\,\prime} &= N \left\langle \mu_1^{\prime 2} \right\rangle \left\langle q_1^2 \right\rangle, \\ M_1^{\mathrm{IR}\,\prime} &= N i \langle \dot{\boldsymbol{\mu}}'_1 \cdot \boldsymbol{\mu}'_1 \rangle \left\langle q_1^2 \right\rangle, \\ M_2^{\mathrm{IR}\,\prime} &= N \left\langle \dot{\mu}_1^{\prime 2} \right\rangle \left\langle q_1^2 \right\rangle, \\ M_3^{\mathrm{IR}\,\prime} &= -N i \left\langle \dddot{\boldsymbol{\mu}}'_1 \cdot \boldsymbol{\mu}'_1 \right\rangle \left\langle q_1^2 \right\rangle, \\ M_4^{\mathrm{IR}\,\prime} &= N \left\langle \ddot{\mu}_1^{\prime 2} \right\rangle \left\langle q_1^2 \right\rangle, \end{aligned} \tag{11.64}$$

etc., and contain only self contributions. Therefore the calculation of the IR moments is significantly simpler than that of the FIR moments.

Classical calculation of the moments

In this section we derive classical expressions for the even FIR and (shifted) IR spectral moments. Classically the odd moments vanish identically, since then

$I(\omega)$ is an even function of ω. Quantum expressions for the moments are given in § 11.1.7.

Note that the expressions given here neglect local field effects (§ 11.1.3). These may be approximately accounted for by multiplying by the factor f_A of (11.37).

Evaluation of M_0 From (11.55) we have at once

$$M_0^{\text{FIR}} = N\mu_1^{0\,2} g_1, \tag{11.65}$$

where

$$g_1 = 1 + \rho \int \mathbf{dr}_{12} \, \langle g(\mathbf{r}\omega_1\omega_2)\mathbf{u}_1 \cdot \mathbf{u}_2 \rangle_{\omega_1\omega_2} \tag{11.66}$$

is the Kirkwood orientational correlation parameter (see Chapter 10). Here $\mathbf{u}_i$ denotes a unit vector in the direction of $\boldsymbol{\mu}_i^0$.

For the IR band we have instead

$$M_0^{\text{IR}} = Nm_1^2, \tag{11.67}$$

where $m_1^2 \equiv \mu_1'^2 \langle q_1^2 \rangle$. The zeroth IR moment thus has a very simple form, and is independent of the state conditions and intermolecular forces.

Evaluation of M_2 The second FIR moment is given by (11.55) as $M_2^{\text{FIR}} = \langle \dot{\mu}^{0\,2} \rangle$. We shall evaluate this explicitly for nonlinear molecules. The result for linear molecules is obtained by a similar calculation, and is stated below.

To evaluate $\langle \dot{\mu}^{0\,2} \rangle$ we use the fundamental rigid-body kinematic relation[87]

$$\dot{\boldsymbol{\mu}}_i^0 = \boldsymbol{\Omega}_i \times \boldsymbol{\mu}_i^0, \tag{11.68}$$

where $\boldsymbol{\Omega}_i$ is the angular velocity of molecule i. Since $\dot{\boldsymbol{\mu}}^0 = \sum_i \dot{\boldsymbol{\mu}}_i^0$, $\langle \dot{\mu}^{0\,2} \rangle$ will contain both self ($i = j$) and distinct ($i \neq j$) terms. However, the distinct terms involve the averages $\langle \boldsymbol{\Omega}_i \boldsymbol{\Omega}_j \rangle$, which vanish because the angular velocities of different molecules are uncorrelated. Therefore

$$\begin{aligned} M_2^{\text{FIR}} &= N\left\langle \dot{\mu}_1{}^{0\,2} \right\rangle \\ &= N\left\langle \Omega_1^2 \mu_1^{0\,2} - \boldsymbol{\Omega}_1\boldsymbol{\Omega}_1 : \boldsymbol{\mu}_1^0 \boldsymbol{\mu}_1^0 \right\rangle \\ &= N\left(\left\langle \Omega_1^2 \right\rangle^{\omega_1} \mu_1^{0\,2} - \langle \boldsymbol{\Omega}_1\boldsymbol{\Omega}_1 \rangle^{\omega_1} : \boldsymbol{\mu}_1^0 \boldsymbol{\mu}_1^0 \right), \end{aligned} \tag{11.69}$$

where $\langle \ldots \rangle^{\omega_1}$ denotes an average taken at fixed orientation ω_1. Recalling the equipartition theorem (3.73) for the average rotational kinetic energy, we have

$$\frac{1}{2}\mathbf{I}_1 : \langle \boldsymbol{\Omega}_1\boldsymbol{\Omega}_1 \rangle^{\omega_1} = \frac{3}{2}kT, \tag{11.70}$$

where $\mathbf{I}_1$ is the molecular inertia tensor. This implies

$$\langle \boldsymbol{\Omega}_1\boldsymbol{\Omega}_1 \rangle^{\omega_1} = 2kT\mathbf{B}_1, \tag{11.71}$$

where we have introduced the rotational tensor $\mathbf{B}_1 \equiv \frac{1}{2}\mathbf{I}_1^{-1}$. Therefore

$$M_2^{\text{FIR}} = 2NkT\mu_1^{0\,2}(\text{Tr}\,\mathbf{B}_1 - \mathbf{u}_1\mathbf{u}_1 : \mathbf{B}_1), \tag{11.72}$$

where $\text{Tr}\,\mathbf{B}_1 \equiv B_{1\alpha\alpha}$ denotes the trace of $\mathbf{B}_1$.

The second IR moment is similarly given by

$$M_2^{\text{IR}} = 2NkTm_1^2(\text{Tr}\,\mathbf{B}_1 - \mathbf{u}_1\mathbf{u}_1 : \mathbf{B}_1). \tag{11.73}$$

Thus the second moments of IR and FIR bands depend only on the temperature, and are independent of the molecular environment. As we shall see in § 11.1.7, this is no longer true quantum mechanically.

We define *normalized* moments for the IR spectrum as

$$\hat{M}_n^{\text{IR}} \equiv M_n^{\text{IR}}/M_0^{\text{IR}}. \tag{11.74}$$

For FIR spectra it proves more convenient to normalize relative to the self part of M_0^{FIR}. Thus

$$\hat{M}_n^{\text{FIR}} \equiv M_n^{\text{FIR}}/N\mu_1^{0\,2}. \tag{11.75}$$

Then for *either* IR or FIR spectra, we have

$$\hat{M}_2 = 2kT(\text{Tr}\,\mathbf{B}_1 - \mathbf{u}_1\mathbf{u}_1 : \mathbf{B}_1). \tag{11.76}$$

For molecules with symmetry, the above relations simplify. Consider for example H_2O, where $\mu_{1z}^0 = \mu_1^0$, $\mu_{1x}^0 = \mu_{1y}^0 = 0$. For the FIR spectrum, or equivalently for the vibrational modes ν_1 and ν_2, where $\mu'_{1z} \neq 0$, but $\mu'_{1x} = \mu'_{1y} = 0$, we therefore have

$$\hat{M}_2^z = kT\left(\frac{1}{I_{xx}} + \frac{1}{I_{yy}}\right). \tag{11.77}$$

For the ν_3 mode, on the other hand, where $\mu'_{1x} \neq 0$, we find that

$$\hat{M}_2^x = kT\left(\frac{1}{I_{yy}} + \frac{1}{I_{zz}}\right). \tag{11.78}$$

For symmetric top molecules, we find for the FIR band and for parallel vibrational bands

$$\hat{M}_2^{\parallel} = \frac{2kT}{I_\perp}, \tag{11.79}$$

while for perpendicular IR modes, the corresponding result is

$$\hat{M}_2^{\perp} = kT\left(\frac{1}{I_\parallel} + \frac{1}{I_\perp}\right), \tag{11.80}$$

where $I_\parallel$ and $I_\perp$ denote the components of $\mathbf{I}_1$ parallel and perpendicular to the symmetry axis.

For spherical top molecules, where $\mathbf{I}_1 = I\mathbf{1}$, these results simplify further to

$$\hat{M}_2 = \frac{2kT}{I}. \tag{11.81}$$

However, this relation is relevant only for IR bands, since spherical tops do not possess a permanent dipole moment.

For linear molecules we find similarly for the normalized second IR moment

$$\hat{M}_2^{\parallel} = \frac{2kT}{I}, \tag{11.82}$$

where I is the moment of inertia. Equation (11.82) also holds for parallel IR modes. For perpendicular modes we have

$$\hat{M}_2^{\perp} = \frac{kT}{I}. \tag{11.83}$$

There have been several experimental studies of second moments, all of which verify the predicted independence on the molecular environment. For example, Table 11.1 shows second moments of the ν_3 vibrational band of CH_4 dissolved in various solvents. Although the band shapes differ significantly, the moments agree with each other, and with theory, within the experimental uncertainty of about 10%. Similarly, the second moments of the parallel and perpendicular IR bands of OCS/X gas mixtures (X = Ne, Ar, Kr, N_2, CH_4) agree well[89] with the predictions of (11.82) and (11.83) except at high densities, where collisional contributions become important.

Evaluation of M_4 The fourth moment of the FIR spectrum is given from (11.55) as $M_4^{\mathrm{FIR}} = \langle \ddot{\mu}^{0\,2} \rangle$. The normalized fourth moment is therefore

$$\hat{M}_4^{\mathrm{FIR}} = \left\langle \ddot{u}_1^2 \right\rangle + (N-1)\langle \ddot{\mathbf{u}}_1 \cdot \ddot{\mathbf{u}}_2 \rangle, \tag{11.84}$$

and contains both self and distinct contributions.

We will first calculate the self terms; for simplicity we consider linear molecules. Differentiating the kinematic relation $\dot{\mathbf{u}} = \boldsymbol{\Omega} \times \mathbf{u}$, and noting that $\boldsymbol{\Omega}$ is perpendicular to $\mathbf{u}$, we find that

Table 11.1

Normalized second moments of the ν_3 band of CH_4 in various solvents

		$\hat{M}_2/(10^{26}\ \mathrm{s}^{-2})$	
System	T(K)	Experiment	Theory
CH_4 (pure)	117	0.63	0.61
CH_4/Ar	111	0.58	0.58
CH_4/Kr	135	0.66	0.73
CH_4/Xe	163	0.77	0.84

From ref. 88.

$$\ddot{\mathbf{u}} = \dot{\mathbf{\Omega}} \times \mathbf{u} + \mathbf{\Omega} \times (\mathbf{\Omega} \times \mathbf{u})$$
$$= I^{-1}\boldsymbol{\tau} \times \mathbf{u} - \Omega^2\mathbf{u}, \tag{11.85}$$

where $\boldsymbol{\tau} = I\dot{\mathbf{\Omega}}$ is the torque experienced by a molecule (relative to its centre of mass). Since $\boldsymbol{\tau}$ is also perpendicular to $\mathbf{u}$, it follows that

$$\langle \ddot{u}^2 \rangle = \frac{\langle \tau^2 \rangle}{I^2} + \langle \Omega^4 \rangle. \tag{11.86}$$

Now $\Omega = J/I$, where J is the angular momentum. Evaluating the momentum average using the distribution function (3.57), we therefore have

$$\hat{M}_{s4}^{\mathrm{FIR}} = \frac{\langle \tau^2 \rangle}{I^2} + 8\left(\frac{kT}{I}\right)^2. \tag{11.87}$$

This relation determines the self part of the fourth moment of the FIR spectrum, or equivalently the *total* fourth moment of a parallel IR band. The fourth moment depends on the mean square torque $\langle \tau^2 \rangle$ acting on a molecule, and so reflects rotational hindering. Equation (11.87) provides a useful check of approximate theories or computer simulations of the correlation function $\langle \mathbf{u}_1(0) \cdot \mathbf{u}_1(t) \rangle$.

For *perpendicular* IR bands of linear molecules, we can show that $\hat{M}_4^{\perp} = \frac{1}{2}\hat{M}_{s4}^{\mathrm{FIR}}$.

Gordon[90] has calculated $\hat{M}_4$ for an arbitrary asymmetric top. His expression is rather cumbersome, however, so we only give the results for some special cases. If $\mathbf{u}$ lies along the z principal axis of $\mathbf{I}$, as for example in the FIR spectrum or ν_1 or ν_2 IR bands of H_2O, we have

$$\hat{M}_{s4}^{\mathrm{FIR}} = \frac{\langle \tau_x^2 \rangle}{I_{xx}^2} + \frac{\langle \tau_y^2 \rangle}{I_{yy}^2} + \frac{(kT)^2}{I_{xx}^2 I_{yy}^2 I_{zz}}[8I_{xx}I_{yy}I_{zz}$$
$$+ (I_{xx} + I_{yy})I_{zz}^2 + (I_{xx} - I_{yy})^2(I_{xx} + I_{yy} + I_{zz})]. \tag{11.88}$$

For FIR spectra or parallel bands of symmetric tops, this simplifies to

$$\hat{M}_{s4}^{\mathrm{FIR}} = \frac{\langle \tau_\perp^2 \rangle}{I_\perp^2} + 2\left(\frac{kT}{I_\perp}\right)^2\left(4 + \frac{I_\parallel}{I_\perp}\right), \tag{11.89}$$

where $\langle \tau_\perp^2 \rangle$ is the mean square torque perpendicular to the symmetry axis. For spherical tops, where $\langle \tau_\perp^2 \rangle = 2/3\langle \tau^2 \rangle$, this further simplifies to

$$\hat{M}_{s4}^{\mathrm{FIR}} = \frac{2\langle \tau^2 \rangle}{3I^2} + 10\left(\frac{kT}{I}\right)^2. \tag{11.90}$$

We next consider the distinct terms in (11.84). Using (11.85), we find that

$$\langle \ddot{\mathbf{u}}_1 \cdot \ddot{\mathbf{u}}_2 \rangle = I^{-2}\Big[\langle (\boldsymbol{\tau}_1 \times \mathbf{u}_1) \cdot (\boldsymbol{\tau}_2 \times \mathbf{u}_2) \rangle$$
$$-2kT\langle \mathbf{u}_1 \cdot (\boldsymbol{\tau}_2 \times \mathbf{u}_2) + \mathbf{u}_2 \cdot (\boldsymbol{\tau}_1 \times \mathbf{u}_1) \rangle + (2kT)^2\langle \mathbf{u}_1 \cdot \mathbf{u}_2 \rangle\Big]. \tag{11.91}$$

The distinct part of the normalized FIR fourth moment is therefore

$$\hat{M}_{d4}^{\mathrm{FIR}} = \rho I^{-2} \int d\mathbf{r} \Big\langle g(\mathbf{r}\omega_1\omega_2)\Big[(\boldsymbol{\tau}_1 \times \mathbf{u}_1)\cdot(\boldsymbol{\tau}_2 \times \mathbf{u}_2) - 4kT\mathbf{u}_1\cdot(\boldsymbol{\tau}_2 \times \mathbf{u}_2) + (2kT)^2\mathbf{u}_1\cdot\mathbf{u}_2\Big]\Big\rangle_{\omega_1\omega_2}, \tag{11.92}$$

where $g(\mathbf{r}\omega_1\omega_2)$ is the pair correlation function. Note that the presence of these terms implies that $\hat{M}_4^{\mathrm{FIR}} \neq \hat{M}_4^{\mathrm{IR}}$.

A number of workers have attempted to extract fourth moments, and mean square torques, from measured IR and FIR bands.[88–95] For example, Dreyfus *et al.*[89] measured $\hat{M}_4$ for dilute OCS–rare-gas mixtures, for both the ν_1 (parallel) and ν_2 (perpendicular) IR bands. They found that the expected relation $\hat{M}_4^{\perp} = \frac{1}{2}\hat{M}_4^{\parallel}$ was approximately obeyed, and that small deviations could be accounted for by considering vibration–rotation interactions. They determined $\langle\tau^2\rangle$ for both bands, and showed that the results agreed within experimental error (Fig. 11.5). For liquid CH_2Cl_2 (point group C_{2v}), Rothschild[93] estimated the torque about all three principal axes by applying (11.88) to a number of different IR bands: the results demonstrated that rotational hindering was somewhat anisotropic.

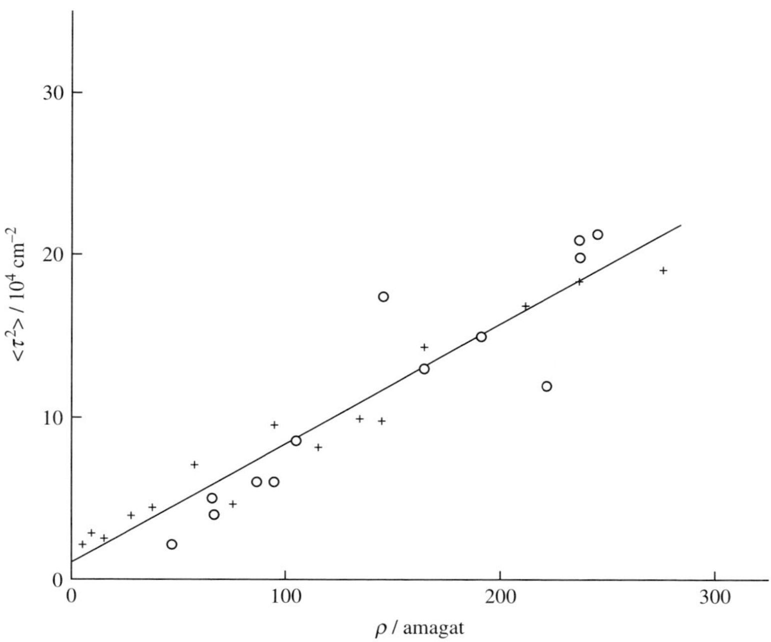

FIG. 11.5 Variation of mean square torque on a OCS molecule with density of the solvent gas Ar, as determined from the moments of the ν_1 and ν_2 bands, and corrected for vibration–rotation interaction. (From ref. 89.)

Such anisotropy has also been observed for the symmetric top CHD_3 in various solvents.[94]

Dreyfus *et al.*[95] measured the temperature dependence of $\langle\tau^2\rangle$ for the ν_3 vibration of N_2O in Ar, and also performed calculations using an anisotropic potential based on the electron–gas model supplemented by long-range dispersion terms. They showed that the latter did not significantly affect the computed torque, which is dominated by the short-range repulsive elements of the potential. Similar results were found for solutions of CO_2 and C_2D_2 in the rare gases.[96] However, in strongly multipolar solvents, long-range electrostatic terms (dipole–quadrupole, quadrupole–quadrupole, etc.) can significantly affect $\langle\tau^2\rangle$: for example, the mean square torque acting on a molecule of N_2O is 50% larger for N_2O/CO_2 mixtures than for N_2O/Xe under the same state conditions.[97]

Evaluation of M_6 The sixth moment of the FIR spectrum has been computed for asymmetric tops by Kalmykov and Titov.[98] The expressions are much too complicated to reproduce here. The moment depends on both the mean square molecular torque and averages over its angular gradient such as $\langle\boldsymbol{\nabla}_\omega\boldsymbol{\tau}\rangle$, $\langle\boldsymbol{\tau}\boldsymbol{\nabla}_\omega\boldsymbol{\tau}\rangle$, and $(\langle\boldsymbol{\nabla}_\omega\boldsymbol{\tau}\rangle^2)$, and could in principle yield information on these quantities.

11.1.7 Quantum expressions for the moments

The expressions for the moments given in (11.55) and (11.64) remain valid in a quantum treatment. When we come to evaluate these formulae quantum mechanically, we find that the classical results are modified by two kinds of quantum corrections, which we refer to as statistical and dynamical. Statistical corrections arise from the modification of the distribution functions for coordinates and velocities; dynamical corrections arise from the non-commutativity of the operators for position and momentum. As we shall see, quantum corrections to the even moments are of order $\hbar^2$, $\hbar^4$, etc., while away from the classical limit, the odd moments are nonzero, and contain terms of order $\hbar$, $\hbar^3$, etc.

Evaluation of M_0 For the zeroth moment, the only corrections are statistical. We consider first the FIR spectrum. Since $\boldsymbol{\mu}^0$ is a function only of the coordinates, and not momenta, the quantity $M_0^{\text{FIR}} = \langle\mu^{0\,2}\rangle$ can be calculated from the quantum configurational distribution function $P\left(\mathbf{r}^N\omega^N\right)$. This quantity can be expanded in powers of $\hbar^2$ according to the Wigner–Kirkwood expansion (Appendix 3D)

$$P_{\text{qu}}\left(\mathbf{r}^N\omega^N\right) = P_{\text{cl}}\left(\mathbf{r}^N\omega^N\right)\left(1+\hbar^2 P_2\left(\mathbf{r}^N\omega^N\right)+\ldots\right). \tag{11.93}$$

Here $P_{\text{cl}}\left(\mathbf{r}^N\omega^N\right)$ is the classical N-body distribution function. For linear molecules the leading correction is given explicitly by

$$P_2 = -\frac{1}{12}\beta^2\sum_{i=1}^{N}\left(\frac{1}{m}\left[\nabla_i^2\mathcal{U}-\frac{1}{2}\beta(\boldsymbol{\nabla}_i\mathcal{U})^2\right]+\frac{1}{I}\left[\nabla_{\omega_i}^2\mathcal{U}-\frac{1}{2}\beta(\boldsymbol{\nabla}_{\omega_i}\mathcal{U})^2\right]\right), \tag{11.94}$$

where m and I are the molecular mass and moment of inertia, $\mathcal{U}\left(\mathbf{r}^N\omega^N\right)$ is the intermolecular potential energy, and ∇^2 and ∇_ω^2 are the linear and angular Laplacians (Appendix A).

Clearly M_0^{FIR} is given by expressions similar to (11.65) and (11.66), but with the classical pair correlation function $g_{\mathrm{cl}}(12)$ replaced by the quantum function $g_{\mathrm{qu}}(12)$. An expansion for $g_{\mathrm{qu}}(12)$ can be derived from (11.93) and (11.94). For a dilute gas of linear molecules, we find that

$$g(12) = g_{\mathrm{cl}}(12)\left\{1 - \frac{1}{6}\beta^2\hbar^2\left(\frac{1}{m}\left[\nabla^2 u(12) - \frac{1}{2}\beta(\nabla u(12))^2\right]\right.\right.$$
$$\left.\left.+\frac{1}{I}\left[\nabla_\omega^2 u(12) - \frac{1}{2}\beta(\nabla_\omega u(12))^2\right]\right) + O(\hbar^4)\right\}. \tag{11.95}$$

For dense liquids the result is rather complicated, and involves many-body classical correlation functions $g_{\mathrm{cl}}(1\ldots h)$.

The zeroth IR moment has only self terms and is easily calculated: we find $M_0^{\mathrm{IR}} = N\mu_1'^2\langle q_1^2\rangle$. If no excited vibrational states are populated, $\langle q_1^2\rangle$ is just a ground state expectation value; for a harmonic oscillator of frequency ω_v, $\langle q_1^2\rangle = \hbar/2m\omega_v$, where m is the reduced mass.

Evaluation of M_1 From (11.55) we see that odd moments involve correlations between coordinates and velocities. These vanish in the classical limit, but yield terms of order $O(\hbar)$ in a quantum treatment.

To evaluate M_1^{FIR}, we use the quantum Heisenberg equation of motion (see Appendix E.1.2)

$$i\hbar\dot{A} = [A, H], \tag{11.96}$$

where H is the system Hamiltonian and $[A, B] = AB - BA$ is the commutator of the operators A and B. From (11.55) and (11.96) we have

$$M_1^{\mathrm{FIR}} = \frac{1}{\hbar}\langle\boldsymbol{\mu}^0\cdot[H, \boldsymbol{\mu}^0]\rangle. \tag{11.97}$$

We can also write this as a double commutator[99]

$$M_1^{\mathrm{FIR}} = \frac{1}{2\hbar}\langle[\boldsymbol{\mu}^0\cdot, [H, \boldsymbol{\mu}^0]]\rangle. \tag{11.98}$$

Since $H = K + \mathcal{U}$, and $\mathcal{U}$ commutes with $\boldsymbol{\mu}^0$ (both are functions only of coordinates), we can replace H in (11.98) by the total kinetic energy K,

$$M_1^{\mathrm{FIR}} = \frac{1}{2\hbar}\langle[\boldsymbol{\mu}^0\cdot, [K, \boldsymbol{\mu}^0]]\rangle. \tag{11.99}$$

Only the rotational part K_{r} will contribute to $[K, \boldsymbol{\mu}^0]$, since $\boldsymbol{\mu}^0$ does not depend on the coordinates of the molecular centres of mass. Writing $K_{\mathrm{r}} = \sum_i K_{\mathrm{r}i}$ and $\boldsymbol{\mu}^0 = \sum_i \boldsymbol{\mu}_i^0$, we get no cross terms from $[K, \boldsymbol{\mu}^0]$, since $K_{\mathrm{r}i}$ commutes with $\boldsymbol{\mu}_j^0$ for $i \neq j$.

The second commutator will not generate any cross terms either, since $\left[K_{\mathrm{r}i}, \boldsymbol{\mu}_i^0\right] \sim \mathbf{J}_i$, where $\mathbf{J}_i$ is the angular momentum of molecule i (cf. (11.103) below), and $\mathbf{J}_i$ also commutes with $\boldsymbol{\mu}_j^0$ for $i \neq j$. Therefore

$$M_1^{\mathrm{FIR}} = \frac{N}{2\hbar}\left\langle\left[\boldsymbol{\mu}_1^0\cdot, \left[K_{\mathrm{r}1}, \boldsymbol{\mu}_1^0\right]\right]\right\rangle. \tag{11.100}$$

Consider first linear molecules, for which $K_{\mathrm{r}1} = J_1^2/2I$. Then

$$M_1^{\mathrm{FIR}} = \frac{N}{4\hbar I}\left\langle\left[\mu_{1\beta}^0, \left[J_{1\alpha}^2, \mu_{1\beta}^0\right]\right]\right\rangle. \tag{11.101}$$

To evaluate the commutators in (11.101), we use the fundamental result[100]

$$[J_\alpha, \mu_\beta] = i\hbar\epsilon_{\alpha\beta\gamma}\mu_\gamma \tag{11.102}$$

where $\epsilon_{\alpha\beta\gamma}$ is the Levi-Civita tensor (see p. 520), $\mu_\alpha \equiv \mu_{1\alpha}^0$, and $J_\alpha \equiv J_{1\alpha}$. It then follows that

$$\left[J_\alpha^2, \mu_\beta\right] = i\hbar\epsilon_{\alpha\beta\gamma}\left(J_\alpha\mu_\gamma + \mu_\gamma J_\alpha\right), \tag{11.103}$$

and hence

$$\left[\mu_\beta, \left[J_\alpha^2, \mu_\beta\right]\right] = 2i\hbar\epsilon_{\alpha\beta\gamma}\mu_\beta[J_\alpha, \mu_\gamma] = 4\hbar^2\mu^2. \tag{11.104}$$

Therefore M_1^{FIR} is given by

$$M_1^{\mathrm{FIR}} = \frac{N\hbar\mu_1^{0\,2}}{I}. \tag{11.105}$$

For nonlinear molecules $K_{r1} = \mathbf{B}_1 : \mathbf{J}_1\mathbf{J}_1$, where $\mathbf{B}_1 = \frac{1}{2}\mathbf{I}_1^{-1}$. A parallel calculation yields

$$M_1^{\mathrm{FIR}} = N\hbar\mu_1^{0\,2}\left(\mathrm{Tr}\,\mathbf{B}_1 - \mathbf{u}_1\mathbf{u}_1 : \mathbf{B}_1\right). \tag{11.106}$$

It is noteworthy that while one cannot derive closed form expressions for the rotational energy levels and wavefunctions of an isolated asymmetric top, let alone a *liquid* of such molecules, a simple rigorous expression exists for the first spectral moment.

Equations (11.105) and (11.106) can also be derived using the Kramers–Kronig relations relating the real and imaginary parts of the dielectric constant; see Appendix 11A.

Evaluation of M_2 Using (11.55) and (11.96), we have

$$M_2^{\mathrm{FIR}} = -\frac{1}{\hbar^2}\langle[H, \boldsymbol{\mu}^0]^2\rangle. \tag{11.107}$$

H can be replaced by K, since $\mathcal{U}$ and $\boldsymbol{\mu}^0$ commute. However M_2^{FIR} will contain cross terms, since, as we shall see below, $\left\langle\left[K_i, \boldsymbol{\mu}_i^0\right]\cdot\left[K_j, \boldsymbol{\mu}_j^0\right]\right\rangle$ does not vanish.

We first consider the self ($i = j$) terms. For simplicity we restrict ourselves to linear molecules; for nonlinear molecules see ref. 90. We find that

$$M_{s2}^{\text{FIR}} = -\frac{N}{(2\hbar I)^2}\left\langle\left[J_1^2, \boldsymbol{\mu}_1^0\right]^2\right\rangle. \tag{11.108}$$

Using (11.102), we can write (11.103) in the two equivalent forms

$$\begin{aligned}\left[J_\alpha^2, \mu_\beta\right] &= 2i\hbar\epsilon_{\alpha\beta\gamma}J_\alpha\mu_\gamma - 2\hbar^2\mu_\beta \\ &= 2i\hbar\epsilon_{\alpha\beta\gamma}\mu_\gamma J_\alpha + 2\hbar^2\mu_\beta.\end{aligned} \tag{11.109}$$

Therefore

$$\begin{aligned}[J^2, \boldsymbol{\mu}]^2 &= \left(2i\hbar\epsilon_{\alpha\beta\gamma}J_\alpha\mu_\gamma - 2\hbar^2\mu_\beta\right)\left(2i\hbar\epsilon_{\alpha'\beta\gamma'}\mu_{\gamma'}J_{\alpha'} + 2\hbar^2\mu_\beta\right) \\ &= -4\hbar^2\epsilon_{\alpha\beta\gamma}\epsilon_{\alpha'\beta\gamma'}J_\alpha\mu_\gamma\mu_{\gamma'}J_{\alpha'} - 4\hbar^4\mu^2,\end{aligned} \tag{11.110}$$

since $\epsilon_{\alpha\beta\gamma}\mu_\beta\mu_\gamma = 0$. Using (B.22) to reduce the product $\epsilon_{\alpha\beta\gamma}\epsilon_{\alpha'\beta\gamma'}$, and noting that, for a linear molecule, $\mathbf{J}\cdot\boldsymbol{\mu} = 0$, we then find that

$$[J^2, \boldsymbol{\mu}]^2 = -4\hbar^2 J^2\mu^2 - 4\hbar^4\mu^2. \tag{11.111}$$

Thus

$$M_{s2}^{\text{FIR}} = N\mu_1^{0\,2}\left(\frac{2}{I}\langle K_{r1}\rangle + \frac{\hbar^2}{I^2}\right), \tag{11.112}$$

where $\langle K_{r1}\rangle = \langle J_1^2/2I\rangle$ is the average molecular rotational kinetic energy. The self part of the normalized FIR second moment, or equivalently the entire normalized second moment of a parallel IR band, is therefore

$$\hat{M}_{s2}^{\text{FIR}} = \frac{2}{I}\langle K_{r1}\rangle + \frac{\hbar^2}{I^2}. \tag{11.113}$$

A comparison of (11.113) with the classical result (11.79) shows that there are both statistical and dynamical quantum corrections to the second moment, the former resulting from the fact that $\langle K_{r1}\rangle \neq kT$, and the latter arising from the noncommutativity of the operators K_{r1} and $\boldsymbol{\mu}_1^0$. In the semiclassical limit, where $\hbar$ is small, we may generate these quantum corrections explicitly by expanding $\langle K_{r1}\rangle$ in powers of $\hbar^2$. Using (3.75) we find that

$$\hat{M}_{s2}^{\text{FIR}} = \hat{M}_2^{\text{cl}}\left(1 + \frac{\hbar^2\beta}{3I} + \frac{\hbar^2\beta^3}{24I}\langle\tau^2\rangle_{\text{cl}} + O(\hbar^4)\right), \tag{11.114}$$

where $\langle\tau^2\rangle_{\text{cl}}$ is the classical mean square torque and $\hat{M}_2^{\text{cl}} = 2kT/I$. Quantum corrections are thus expected to be appreciable when the torques are large and the moment of inertia is small (e.g. HCl, HF).

For a linear molecule the distinct contributions ($i \neq j$) to M_2^{FIR} are

$$M_{d2}^{\mathrm{FIR}} = -\frac{N(N-1)}{(2\hbar I)^2}\left\langle \left[J_1^2, \boldsymbol{\mu}_1\right] \cdot \left[J_2^2, \boldsymbol{\mu}_2\right]\right\rangle, \tag{11.115}$$

where from (11.109)

$$\left\langle \left[J_1^2, \boldsymbol{\mu}_1\right] \cdot \left[J_2^2, \boldsymbol{\mu}_2\right]\right\rangle = -4\hbar^2 \langle (\mathbf{J}_1 \cdot \mathbf{J}_2)(\boldsymbol{\mu}_1 \cdot \boldsymbol{\mu}_2) - \mathbf{J}_1 \cdot (\boldsymbol{\mu}_1 \cdot \mathbf{J}_2)\boldsymbol{\mu}_2 \rangle$$
$$-4i\hbar^3 \langle (\mathbf{J}_1 - \mathbf{J}_2) \cdot (\boldsymbol{\mu}_1 \times \boldsymbol{\mu}_2) \rangle + 4\hbar^4 \langle \boldsymbol{\mu}_1 \cdot \boldsymbol{\mu}_2 \rangle. \tag{11.116}$$

We note that, in contrast to the classical moment, the quantum M_2 depends on the nature of the intermolecular forces as well as the temperature.

Evaluation of M_3 and M_4 For parallel IR bands of linear molecules, the normalized third and fourth moments have been calculated by Gordon.[90] The derivations are too lengthy to reproduce here, but the results are

$$\hat{M}_3^{\parallel} = \frac{4\hbar}{I^2}\langle K_r \rangle + \frac{\hbar}{2I^2}\left\langle \nabla_\omega^2 \mathcal{U} \right\rangle + \frac{\hbar^3}{I^3},$$
$$\hat{M}_4^{\parallel} = \frac{4}{I^2}\left\langle K_r^2 \right\rangle + \frac{1}{I^2}\langle \tau^2 \rangle + \frac{6\hbar^2}{I^3}\langle K_r \rangle + \frac{\hbar^2}{I^3}\left\langle \nabla_\omega^2 \mathcal{U} \right\rangle + \frac{\hbar^4}{I^4}, \tag{11.117}$$

where ∇_ω^2 is the angular Laplacian of (A.20).

11.1.8 Induced spectra

Collision-induced absorption (CIA) was first detected experimentally by Crawford *et al.*[101] in 1949. It is associated with the two- and three-body terms, $\boldsymbol{\mu}_{ij}$, $\boldsymbol{\mu}_{ijk}$, etc., in (11.38). These are of particular importance when the permanent dipole $\boldsymbol{\mu}_i$ vanishes, so that there is no allowed spectrum. The pair contributions then invariably dominate, unless they too are forbidden by symmetry, as in the pure rare gases; indeed, there is very little *direct* experimental evidence for three-body dipoles, with the exception of recent obervations of triple–dipole transitions in the second overtone of hydrogen,[102] and, possibly, some rather old measurements of the intensity of the H_2 fundamental band.[103]

Collisional dipoles are typically of rather short range. Consider, for example, an overlap dipole resulting from the distortion of the electron clouds of two interacting molecules. This is roughly exponential, $\boldsymbol{\mu}_{12} \approx \boldsymbol{\mu}_0 \exp(-(r_{12} - r_{\mathrm{m}})/\rho)$, where $\rho \sim 0.1 r_{\mathrm{m}}$, and r_{m} denotes the minimum in the pair potential energy. This implies that $\boldsymbol{\mu}_{12}$ will vary rapidly with time; we may expect $\dot{\mu}_{12}/\mu_{12} \sim \bar{v}/\rho$, where $\bar{v}$ is a typical molecular velocity. The fluctuating dipole will absorb energy over a frequency range $\Delta\omega \sim \bar{v}/\rho$, typically 10–100 cm^{-1} at room temperature. Rather similar linewidths are expected for multipolar transitions. Generally, therefore, the rotational structure of CIA bands will not be resolved, even in the gas phase. Induced spectra are usually broad, featureless bands, as shown

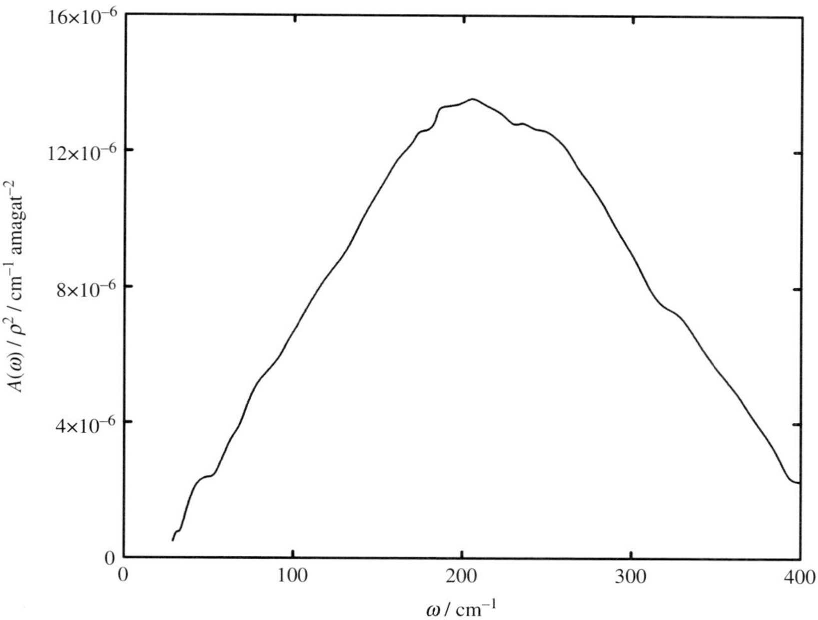

FIG. 11.6 CIA spectrum of gaseous CH_4 at 179 K. (From ref. 105.)

in Fig. 11.6 for methane. Only at very low temperatures may structure become visible, associated with the formation of bound and pseudobound dimer states.[106]

Collision-induced absorption often contributes significantly to the maintenance of radiative equilibrium in a hot gas. For this reason, it has practical consequences for the study of planetary[107] and stellar[108] atmospheres. As a result, the mechanisms contributing to CIA have often been studied in great detail[109] (although, surprisingly, there have as yet been very few *ab initio* calculations[110]). For example, Li *et al.*[111] have identified several different processes relevant to absorption by CH_4-N_2 mixtures, which are believed to be responsible for the atmospheric opacity of Saturn's moon Titan between 200 and 600 cm^{-1}. These include the induction of dipoles on CH_4 by the quadrupolar and hexadecapolar fields, plus their first and second spatial derivatives, of N_2; induction of dipoles on N_2 by the octopolar and hexadecapolar fields, plus their second derivatives, of CH_4; dispersion dipoles, resulting from intermolecular electronic correlation, and related to the hyperpolarizabilities of N_2 and CH_4; and short-range overlap effects.

The field of induced absorption is a very large one; more than a thousand papers dealing with all aspects of the phenomenon have appeared.[112] In this section we will not attempt to be comprehensive, nor will we attempt to identify and include every possible induction mechanism. Our aim is rather to illustrate the underlying physical principles. We will therefore confine ourselves to a few representative examples, concentrating on the calculation of induced spectral moments, and considering only the dominant induction processes. We refer the reader who is interested in a more complete review of the subject matter to the reviews by Tabisz and Neuman,[104] Birnbaum,[113] van Kranendonk,[34] and Frommhold.[114]

Pure rare gases

When two rare gas atoms interact, they are mutually polarized by dispersion and overlap forces. For identical atoms, the resulting atomic multipole moments have the form

$$\boldsymbol{\mu}_1 = -\boldsymbol{\mu}_2 = \mu(r_{12})\,\hat{\mathbf{r}}_{12}, \tag{11.118}$$

$$\mathbf{Q}_1 = \mathbf{Q}_2 = Q(r_{12})\,\frac{1}{2}(3\hat{\mathbf{r}}_{12}\hat{\mathbf{r}}_{12} - \mathbf{1}), \tag{11.119}$$

etc., where $\mathbf{r}_{12} = \mathbf{r}_1 - \mathbf{r}_2$.

The interactions of two nonoverlapping atoms have been studied using nonlocal response theory.[115,116] Unlike traditional *ab initio* methods, which generate only molecular properties, this theory directly yields expressions for the atomic multipole moments. As a result, it provides a much more accurate description of the resulting charge distribution. To leading order in r_{12}^{-1}, and neglecting damping effects,[109,117] the dispersion contributions to the moments of (11.118) and (11.119) are[115]

$$\begin{aligned} \mu_{\text{dis}}(r_{12}) &= -\frac{9\hbar}{\pi} r_{12}^{-7} \int_0^\infty d\omega\,\alpha(i\omega) B(-i\omega, i\omega), \\ Q_{\text{dis}}(r_{12}) &= \frac{3\hbar}{2\pi} r_{12}^{-6} \int_0^\infty d\omega\,\alpha(i\omega) B(i\omega, 0), \end{aligned} \tag{11.120}$$

where $\alpha(i\omega)$ and $B(i\omega_1, i\omega_2)$ are the imaginary-frequency dynamic polarizability and dipole–dipole–quadrupole hyperpolarizability, respectively. *Ab initio* results for these quantities are available in some cases.[118] Otherwise, approximate estimates of the integrals in (11.120) may be obtained using the expression for the dispersion coefficient C_6 (cf. § 2.6),

$$C_6 = \frac{3\hbar}{\pi} \int_0^\infty d\omega\,\alpha(i\omega)\alpha(i\omega). \tag{11.121}$$

Thus we can write, formally,

$$\begin{aligned} \int_0^\infty \mathrm{d}\omega\,\alpha(i\omega)B(-i\omega, i\omega) \Big/ \int_0^\infty \mathrm{d}\omega\,\alpha(i\omega)\alpha(i\omega) &= \frac{2B}{3\alpha}\sigma_{\text{D}}, \\ \int_0^\infty \mathrm{d}\omega\,\alpha(i\omega)B(i\omega, 0) \Big/ \int_0^\infty \mathrm{d}\omega\,\alpha(i\omega)\alpha(i\omega) &= \frac{2B}{3\alpha}\sigma_{\text{Q}}, \end{aligned} \tag{11.122}$$

where α and B are the static (zero-frequency) polarizability and hyperpolarizability, and σ_{D} and σ_{Q} are dimensionless numbers, presumably rather close to unity. Combining these results, we find that

$$\mu_{\rm dis}(r_{12}) = \sigma_{\rm D}\frac{\mu_7}{r_{12}^7}, \qquad \mu_7 = -\frac{2BC_6}{\alpha}, \tag{11.123}$$

and

$$Q_{\rm dis}(r_{12}) = \sigma_{\rm Q}\frac{Q_6}{r_{12}^6}, \qquad Q_6 = \frac{BC_6}{3\alpha}. \tag{11.124}$$

Approximations developed by Hunt and co-workers[111] suggest that $\sigma_{\rm D}$ lies somewhere in the range 1 to $\frac{5}{4}$, and $\sigma_{\rm Q}/\sigma_{\rm D} \approx \frac{27}{32}$. Since $B < 0$, the dispersion dipole is positive, reflecting the fact that dispersion forces pull electron density into the space between the two atoms. For argon, $\mu_7 = 1907$ au and $Q_6 = 318$ au,[119] while $\sigma_{\rm D} \approx 1.15$.[116]

At short range, overlap of the atomic charge clouds will also polarize the atomic charge distributions. Since electron exchange is not within the compass of nonlocal response theory, such effects must be treated in a more or less ad hoc manner. Jansen[120] has discussed a simple model in which a single electron pair is exchanged between Gaussian orbitals; this yields

$$\mu_{\rm ov}(r_{12}) = -\frac{1}{4}er_{12}\exp\left(-\beta^2 r_{12}^2/2\right)/\left[1-\exp\left(-\beta^2 r_{12}^2/2\right)\right], \tag{11.125}$$

although in practice the dependence on r_{12} is more likely to be exponential. The negative sign reflects the fact that overlap forces reduce the electron density between the atoms. For argon, SCF calculations[121] suggest that (11.125) gives a good representation of the overlap dipole over the range $0.8\,r_{\rm m}$–$1.1\,r_m$, where $r_{\rm m}$ is the minimum in the pair potential, provided we take $\beta = 0.495\,\text{bohr}^{-1}$.

Because dispersion and overlap forces operate on rather different length scales, we may reasonably assume their contribution is additive. The total atomic dipole is then just the sum of (11.123) and (11.125). For argon, $r_{\rm m} = 7.100$ bohr; at this separation, $\mu_{\rm dis} = 0.0024$ au and $\mu_{\rm ov} = -0.0037$ au. Thus the dispersion and overlap dipoles cancel for distances somewhat greater than $r_{\rm m}$. For the lighter rare gases, helium and neon, we expect that overlap effects will be relatively more important, while for the heavier gases, krypton and xenon, dispersion effects presumably dominate.

For the pair considered as a whole, the lowest nonzero moment is the quadrupole, of magnitude

$$\begin{aligned} Q_{12} &= 2r_{12}\,\mu(r_{12}) + 2Q(r_{12}) \\ &= \frac{Q_6^{\rm pair}}{r_{12}^6} - \frac{1}{2}er_{12}^2\exp\left(-\beta^2 r_{12}^2/2\right)/\left[1-\exp\left(-\beta^2 r_{12}^2/2\right)\right], \end{aligned} \tag{11.126}$$

where $Q_6^{\rm pair} = -\sigma_{\rm eff}(10BC_6/3\alpha)$, and $\sigma_{\rm eff} = (6\sigma_{\rm D} - \sigma_{\rm Q})/5 \approx (33/32)\sigma_D$. For argon, $Q_6^{\rm pair} = 3760$ au.

Because the pair dipole $\boldsymbol{\mu}_{12}$ vanishes, there is no pair absorption spectrum. However, each polarized atom of the pair generates an electric field, and because these fields do *not* cancel exactly, they will induce a dipole in a neighbouring *third* atom.

We can estimate the three-body dipole $\boldsymbol{\mu}_{123}$ quite easily using a DID (dipole-induced-dipole) model. From (2.63) and (C.6), the dipoles and quadrupoles of (11.118) and (11.119) will induce a further dipole in a neighbouring atom 3 given by

$$\begin{aligned}\boldsymbol{\mu}_3 &= \alpha \mathbf{T}^{(2)}(\mathbf{r}_{31}) \cdot \boldsymbol{\mu}_1 + \alpha \mathbf{T}^{(2)}(\mathbf{r}_{23}) \cdot \boldsymbol{\mu}_2 \\ &\quad - \frac{1}{3}\alpha \mathbf{T}^{(3)}(\mathbf{r}_{31}) : \mathbf{Q}_1 + \frac{1}{3}\alpha \mathbf{T}^{(3)}(\mathbf{r}_{23}) : \mathbf{Q}_2,\end{aligned} \tag{11.127}$$

where $\mathbf{T}^{(n)}(\mathbf{r}) = \nabla^n(1/r)$. Counting also the dipoles induced in atom 1 by the pair 23 and in atom 2 by the pair 31, we find that

$$\boldsymbol{\mu}_{123} = \sum_{ijk} \mathcal{P}_{ijk}\left(1 - \mathcal{P}_{ij}\right) \mu_{ij,k}\, \hat{\mathbf{r}}_{ij}, \tag{11.128}$$

where $\mathcal{P}_{ijk}$ is the permutation operator for cyclic interchange of i, j and k, $\mathcal{P}_{ij}$ is the permutation operator for interchange of i and j, and

$$\begin{aligned}\mu_{ij,k} &= \alpha \left(-3\cos\theta_j \frac{\mu(r_{jk})}{r_{ij}^3} + \frac{\mu(r_{ij})}{r_{jk}^3} \right) \\ &\quad + \frac{3}{2}\alpha \left((5\cos^2\theta_j - 1)\frac{Q(r_{jk})}{r_{ij}^4} - 2\cos\theta_j \frac{Q(r_{ij})}{r_{jk}^4} \right),\end{aligned} \tag{11.129}$$

where θ_i denotes the angle at atom i in the triangle formed by i, j, and k. The dipolar and quadrupolar terms are of the same order, since the two opposing atomic dipoles of a pair generate a far field qualitatively similar to that of the quadrupoles, varying as the inverse fourth power of the distance to a third atom. In principle, higher order multipole and overlap interactions could also be considered, but these are all of relatively short range, and probably unimportant. Note that there is *no* three-body dispersion dipole for the interaction of identical atoms.[116]

Modulation of the three-body dipole of (11.128) and (11.129) through atomic motion can, in principle, lead to collision-induced absorption. Rare-gas trimers are very floppy molecules, with an energy minimum in the D_{3h} (equilateral triangle) conformation; by symmetry, $\boldsymbol{\mu}_{123}$ vanishes here, but there are infrared-active modes. Cooper *et al.*[121] estimate that the transition dipole for the lowest allowed IR transition in Ar_3 is about 7×10^{-5} D, which is probably too weak to observe. However, Dullweber and Wales[122] have calculated oscillator strengths for clusters of up to 110 argon atoms, and their results suggest that, for some of these, absorption in the far infrared at frequencies between about 50 and 130 cm^{-1} may be detectable. For krypton at the triple point, the maximum absorption has been estimated[123] to be $A(\omega) = 0.0015\,cm^{-1}$ at $40\,cm^{-1}$; again, this is close to the current limits of detectability. None of these authors accounted properly for dispersion effects in their calculations.[124] It remains an open question, therefore, whether three-body CIA will ever be detected in the rare gases; to date, no observations have been reported.

Nitrogen

For N_2 the most important induction mechanism involves mutual polarization of an interacting pair of molecules by the electric fields generated by their quadrupole moments. From (2.63) and (2.186), this gives rise to a pair dipole moment

$$\boldsymbol{\mu}_{ij} = \frac{1}{3}\boldsymbol{\alpha}_j \cdot \mathbf{T}^{(3)}(\mathbf{r}_{ij}) : \mathbf{Q}_i - \frac{1}{3}\boldsymbol{\alpha}_i \cdot \mathbf{T}^{(3)}(\mathbf{r}_{ij}) : \mathbf{Q}_j, \tag{11.130}$$

where $\mathbf{T}^{(3)}(\mathbf{r})$ is defined through (2.53) and $\mathbf{r}_{ij} = \mathbf{r}_i - \mathbf{r}_j$ is the intermolecular separation. These dipoles absorb radiation at the frequencies at which they are modulated. This modulation may occur through molecular translation, rotation, or vibration. Vibrational modulation will lead to absorption near the fundamental frequency $\omega_{\text{vib}} = 2350\,\text{cm}^{-1}$; we will neglect this for the time being. Rotational modulation is caused by changes in both the quadrupole moments $\mathbf{Q}$ and the polarizabilities $\boldsymbol{\alpha}$. However, for nitrogen $\boldsymbol{\alpha}$ is almost isotropic, so we will ignore the latter mechanism here. For a linear molecule rotating freely about a fixed axis, fluctuations in $\mathbf{Q}$ occur at zero frequency and at twice the rotational frequency, causing transitions $\Delta J = 0, \pm 2$. Thus the rotational spectrum consists of a Q-branch and (degenerate) O- and S-branches. Because of the translational modulation of $\boldsymbol{\mu}_{ij}$, however, each line of this spectrum acquires a natural width of about $50\,\text{cm}^{-1}$, well in excess of the line spacing $4B = 8\,\text{cm}^{-1}$. Consequently, the individual transitions are not resolved.

From (11.38) and (11.130), the dipole-dipole correlation function is

$$C(t) = \frac{1}{9}\alpha^2 \sum_{ijkl} \langle \mathbf{Q}_i(t) : \mathbf{T}^{(3)}(\mathbf{r}_{ij}(t)) \cdot \mathbf{T}^{(3)}(\mathbf{r}_{kl}(0)) : \mathbf{Q}_k(0) \rangle. \tag{11.131}$$

Thus in general there are two-, three-, and four-body contributions to the spectrum. However, if we ignore the relatively small anisotropy in the N_2 intermolecular potential, we can perform the averages of the $\mathbf{T}^{(3)}$ tensors (which depend only on the intermolecular vectors) and of the $\mathbf{Q}$ tensors (which depend only on the molecular orientations) separately. This implies that $C(t)$ is the product of translational and rotational correlation functions, i.e.

$$C(t) = C_{\text{tr}}(t) C_{\text{rot}}(t). \tag{11.132}$$

The translational function now contains only two- and three-body terms, since $\langle \mathbf{Q}_i \mathbf{Q}_k \rangle_{\omega_i \omega_k} = 0$ for $i \neq k$. Explicitly, we have[125]

$$C_{\text{tr}}(t) = \frac{1}{30}\alpha^2 Q^2 \sum_{ijk} \langle \mathbf{T}^{(3)}(\mathbf{r}_{ij}(t)) \,\vdots\, \mathbf{T}^{(3)}(\mathbf{r}_{kj}(0)) \rangle, \tag{11.133}$$

while the single-molecule rotational function is (classically)[125]

$$C_{\text{rot}}(t) = \langle P_2(\mathbf{u}_1(t) \cdot \mathbf{u}_1(0)) \rangle = \frac{1}{4} + \frac{3}{4} M\left(1, \frac{1}{2}, -\frac{2kT}{I}t^2\right), \tag{11.134}$$

where $M(a, b, z)$ denotes a confluent hypergeometric function.[126]

Differentiating (11.132), we find that the spectral moments are given by

$$M_n = \sum_{m=0}^{n} \binom{n}{m} M_{tm} M_{r\,n-m}, \tag{11.135}$$

where, from (11.134), the rotational moments are given by

$$\begin{aligned} M_{r0} &= 1, \\ M_{r2n} &= \frac{3}{4} n! \left(\frac{8kT}{I} \right)^n, \qquad n \geq 1. \end{aligned} \tag{11.136}$$

The problem of computing the induced moments therefore reduces to that of calculating the translational moments

$$M_{t2n} = \frac{1}{30} \alpha^2 Q^2 \sum_{ijk} \left\langle \mathbf{T}_n^{(3)}(\mathbf{r}_{ij}) \vdots \mathbf{T}_n^{(3)}(\mathbf{r}_{kj}) \right\rangle, \tag{11.137}$$

where we have defined

$$\mathbf{T}_n^{(3)}(\mathbf{r}_{ij}) = \left(\frac{\mathrm{d}^n}{\mathrm{d}t^n} \mathbf{T}^{(3)}(\mathbf{r}_{ij}(t)) \right)_{t=0}. \tag{11.138}$$

To evaluate the time derivatives in (11.138), we write

$$\frac{\mathrm{d}}{\mathrm{d}t} = \sum_{k=1}^{N} (\mathbf{v}_k \cdot \nabla_k + \mathbf{a}_k \cdot \nabla_{\mathbf{v}_k}), \tag{11.139}$$

where $\nabla_k \equiv \partial/\partial \mathbf{r}_k$ and $\nabla_{\mathbf{v}_k} \equiv \partial/\partial \mathbf{v}_k$, and $\mathbf{v}_k = \mathrm{d}\mathbf{r}_k/\mathrm{d}t$ and $\mathbf{a}_k = \mathrm{d}\mathbf{v}_k/\mathrm{d}t$ are the velocity and acceleration of molecule k. Since[127] $\nabla_k \mathbf{T}^{(n)}(\mathbf{r}_{ij}) = (\delta_{ik} - \delta_{jk})\mathbf{T}^{(n+1)}(\mathbf{r}_{ij})$, it follows that

$$\begin{aligned} \mathbf{T}_0^{(3)}(\mathbf{r}_{ij}) &= \mathbf{T}^{(3)}(\mathbf{r}_{ij}), \\ \mathbf{T}_1^{(3)}(\mathbf{r}_{ij}) &= (\mathbf{v}_i - \mathbf{v}_j) \cdot \mathbf{T}^{(4)}(\mathbf{r}_{ij}), \\ \mathbf{T}_2^{(3)}(\mathbf{r}_{ij}) &= (\mathbf{v}_i - \mathbf{v}_j)^2 : \mathbf{T}^{(5)}(\mathbf{r}_{ij}) + (\mathbf{a}_i - \mathbf{a}_j) \cdot \mathbf{T}^{(4)}(\mathbf{r}_{ij}), \end{aligned} \tag{11.140}$$

etc. Since the acceleration $\mathbf{a}_i$ is itself a two-body term, we see that in a liquid M_{t4} and higher moments contain contributions arising from the interactions of four or more molecules. In practice these terms are very difficult to evaluate, requiring integrations over four-particle, or higher, correlation functions. However, M_{t0} and M_{t2} contain only two- and three-body terms and are easily calculated. Noting that $\langle \mathbf{v}_i \mathbf{v}_j \rangle = (kT/M)\mathbf{1}\delta_{ij}$ (where M is the molecular mass), and using the result[127]

$$\mathbf{T}^{(n)}(\mathbf{r}_{ij}) \bullet \mathbf{T}^{(n)}(\mathbf{r}_{kj}) = \frac{(2n)!}{2^n (r_{ij} r_{kj})^{n+1}} P_n(\cos\theta_{ijk}), \tag{11.141}$$

where the symbol $\bullet$ denotes a full tensor contraction, and θ_{ijk} is the angle at j in the triangle formed by molecules i, j, and k, we find that

$$M_{t0} = 3\alpha^2 Q^2 \left(N(N-1)\left\langle \frac{1}{r_{12}^8} \right\rangle + N(N-1)(N-2)\left\langle \frac{P_3(\cos\theta_{123})}{r_{12}^4 r_{32}^4} \right\rangle \right),$$

$$M_{t2} = 84\alpha^2 Q^2 \frac{kT}{M}\left(2N(N-1)\left\langle \frac{1}{r_{12}^{10}} \right\rangle + N(N-1)(N-2)\left\langle \frac{P_4(\cos\theta_{123})}{r_{12}^5 r_{32}^5} \right\rangle \right). \tag{11.142}$$

In terms of the liquid pair and triplet correlation functions, we can write

$$M_{t0} = 3\alpha^2 Q^2 \rho^2 V \left(\int d\mathbf{r}_{12}\, g(r_{12}) r_{12}^{-8} + \rho \int d\mathbf{r}_{12} d\mathbf{r}_{32}\, g(r_{12}r_{13}r_{32}) P_3(\cos\theta_{123})(r_{12}r_{32})^{-4} \right),$$

$$M_{t2} = 84\alpha^2 Q^2 \frac{kT}{M}\rho^2 V \left(2\int d\mathbf{r}_{12}\, g(r_{12}) r_{12}^{-10} + \rho \int d\mathbf{r}_{12} d\mathbf{r}_{32}\, g(r_{12}r_{13}r_{32}) P_4(\cos\theta_{123})(r_{12}r_{32})^{-5} \right). \tag{11.143}$$

Assuming a Lennard-Jones (LJ) potential with $\epsilon = 103\,\mathrm{K}$ and $\sigma = 3.615\,\text{Å}$, and taking $\alpha = 1.74\,\text{Å}^3$ and $Q = -1.5 \times 10^{-26}$ esu, Joslin *et al.*[127] calculated M_{t0} and M_{t2} in the superposition approximation. They found for the two- and three-body contributions at 76 K and 1 atm

$$M_{t0}/V[2B] = 4.6 \times 10^{-24}\,\mathrm{J}, \qquad M_{t0}/V[3B] = -4.0 \times 10^{-24}\,\mathrm{J},$$

$$M_{t2}/V[2B] = 36\,\mathrm{J\,s^{-2}}, \qquad M_{t2}/V[3B] = -12\,\mathrm{J\,s^{-2}}. \tag{11.144}$$

Note the substantial cancellation of two- and three-body terms in M_{t0}, which makes the computed moment rather sensitive to the approximations employed in its calculation. Such cancellation effects are widespread[128] throughout CIA and its light-scattering analogue CILS (§ 11.2.7).

The absolute values of the moments of (11.144) are rather too small,[127] even after allowance is made for local-field effects (§ 11.1.3). Agreement with experiment can be improved in a number of ways, including (a) inclusion of a wider range of induction mechanisms, e.g. hexadecapolar and overlap induction; (b) consideration of quantum effects; (c) inclusion of the anisotropic component of the polarizability, which has been shown to dramatically reduce cancellation effects;[129] (d) use of a more realistic intermolecular potential; (e) use of a more accurate expression for $g(r_{12}r_{13}r_{32})$; and (f) inclusion of many-body effects through ternary, etc. dipoles.

For a dilute gas the moments of (11.143) reduce to

$$M_{t0} = 3\alpha^2 Q^2 \rho^2 V \int d\mathbf{r}_{12} \exp[-\beta u(r_{12})] r_{12}^{-8},$$

$$M_{t2} = 168\alpha^2 Q^2 \frac{kT}{M} \rho^2 V \int d\mathbf{r}_{12} \exp[-\beta u(r_{12})] r_{12}^{-10}. \qquad (11.145)$$

Joslin *et al.*[125] have also computed the fourth and sixth moments for a gas. They find that

$$M_{t4} = 96\alpha^2 Q^2 \left(\frac{kT}{M}\right)^2 \rho^2 V \int d\mathbf{r}_{12} \exp[-\beta u(r_{12})]$$
$$\times \left(315 r_{12}^{-12} + 2\left(\beta u'\right)^2 r_{12}^{-10}\right),$$

$$M_{t6} = 384\alpha^2 Q^2 \left(\frac{kT}{M}\right)^3 \rho^2 V \int d\mathbf{r}_{12} \exp[-\beta u(r_{12})]$$
$$\times \left(31185 r_{12}^{-14} + 372\left(\beta u'\right)^2 r_{12}^{-12} - 30\beta u' \beta u'' r_{12}^{-11} + \left(\beta u''\right)^2 r_{12}^{-10}\right), \qquad (11.146)$$

where $u' \equiv du/dr$ and $u'' \equiv d^2u/dr^2$. Table 11.2 shows the values of $M_{t0}/\rho^2 V$, $\hat{M}_{t2}, \hat{M}_{t4}$, and $\hat{M}_{t6}$ for N_2 gas calculated at two different temperatures. (Reduced moments are defined by $\hat{M}_{tn} = M_{tn}/M_{t0}$.) We note that, to a very good approximation, $M_{t2n} \propto T^n$. This is typical of induced spectral moments of gases, and reflects the weak temperature dependence of the integrals in (11.145) and (11.146).

Dagg *et al.*[130] have estimated the quadrupole moment of nitrogen by comparing the theoretical expression for M_{t0} with experiment. The value obtained, $|Q| = 1.4 \pm 0.2 \times 10^{-26}$ esu, is consistent with other, more direct estimates (cf. Table D.3). It should be stressed, however, that the quantity directly measured in a CIA absolute intensity study is not Q per se, but rather (assuming there are no other sources of intensity) Q^2/σ^5, where σ measures the range of the intermolecular potential; this is easily seen from (11.145). Thus, whereas the estimated uncertainty in $|Q|$ is about 15%, this is principally due to the uncertainty in σ: the uncertainty in $(Q^2/\sigma^5)^{1/2}$ is only about 2%.

Table 11.2

Translational moments of the CIA spectrum of nitrogen gas

	343 K	228 K
$M_{t0}/\rho^2 V\,(\mathrm{J\,m^6})$	1.06×10^{-80}	1.07×10^{-80}
$\hat{M}_{t2}\,(\mathrm{s^{-2}})$	0.346×10^{-26}	0.223×10^{-26}
$\hat{M}_{t4}\,(\mathrm{s^{-4}})$	1.02×10^{-52}	0.433×10^{-52}
$\hat{M}_{t6}\,(\mathrm{s^{-6}})$	13.1×10^{-78}	3.68×10^{-78}

From ref. 125.

CIA in the fundamental band arises through vibrational modulation of both the molecular polarizability $\boldsymbol{\alpha}$ and quadrupole moment $\mathbf{Q}$ in (11.130). Moments up to M_{t6} have been calculated by Joslin.[131]

Chlorine–argon mixtures

Chlorine, Cl_2, possesses a sizeable hexadecapole moment ($\boldsymbol{\Phi}$) in addition to its quadrupole (cf. Table D.3), and both moments contribute to CIA. The Cl_2-Cl_2 potential is too anisotropic to permit separation of translational and rotational motions as was done for N_2. However, the interaction of Cl_2 with a rare-gas atom such as Ar is much more spherical. For a Cl_2-Ar mixture, the dipole–dipole correlation function is therefore approximately given by a sum of quadrupolar and hexadecapolar product functions, viz.

$$C(t) = \sum_{l=2,4} C_{\mathrm{tr}}^{(l)}(t) C_{\mathrm{rot}}^{(l)}(t). \tag{11.147}$$

For a dilute mixture we have ($A = Cl_2$, $B = \mathrm{Ar}$)

$$C_{\mathrm{tr}}^{(l)}(t) = N_A N_B \frac{2^l}{(2l+1)!} \alpha_B^2 Q_{Al}^2 \langle \mathbf{T}^{(l+1)}(\mathbf{r}_{12}(t)) \vdots \mathbf{T}^{(l+1)}(\mathbf{r}_{12}(0)) \rangle, \tag{11.148}$$

where $Q_{A2} \equiv Q_A$, $Q_{A4} \equiv \Phi_A$, and[132]

$$\begin{aligned} C_{\mathrm{rot}}^{(l)}(t) &= \langle P_{l+1}(\mathbf{u}_A(t) \cdot \mathbf{u}_A(0)) \rangle \\ &= 2^{-2l} \sum_m \binom{l+m}{\frac{1}{2}l + \frac{1}{2}m} \binom{l-m}{\frac{1}{2}l - \frac{1}{2}m} M\left(1, \frac{1}{2}, -\frac{kT}{2I_A} m^2 t^2\right), \end{aligned} \tag{11.149}$$

where $m = -l, -l+2, \ldots, l$.

The spectral moments are given by

$$M_n = \sum_{l=2,4} M_{t0}^{(l)} \sum_{m=0}^{n} \binom{n}{m} \hat{M}_{tm}^{(l)} M_{r\,n-m}^{(l)}, \tag{11.150}$$

where

$$M_{r\,2n}^{(l)} = n! \left(\frac{2kT}{I}\right)^n 2^{-2l} \sum_m \binom{l+m}{\frac{1}{2}l + \frac{1}{2}m} \binom{l-m}{\frac{1}{2}l - \frac{1}{2}m} m^{2n}, \tag{11.151}$$

$$M_{t0}^{(l)} = (l+1) \alpha_B^2 Q_{Al}^2 \rho_A \rho_B V \int d\mathbf{r}_{12} \exp[-\beta u_{AB}(r_{12})] r_{12}^{-(2l+4)}, \tag{11.152}$$

and $\hat{M}_{tn}^{(l)}$ is identical in form to the corresponding reduced moment for a pure fluid; expressions for $\hat{M}_{t2}^{(l)}$–$\hat{M}_{t6}^{(l)}$ can be found in ref. 133.

At 298 K, assuming a LJ interaction with $\epsilon_{AB} = 195\,\mathrm{K}$ and $\sigma_{AB} = 3.80\,\text{Å}$, and taking $Q_A = 3.23 \times 10^{-26}$ esu, $\Phi_A = 31.4 \times 10^{-42}$ esu, and $\alpha_B = 1.642\,\text{Å}^3$, we find that $M_{t0}^{(l)}/\rho_A \rho_B V = 4.75 \times 10^{-80}\,\mathrm{J\,m^6}$ and $2.29 \times 10^{-80}\,\mathrm{J\,m^6}$ for $l = 2$

and 4, respectively. Thus approximately one-third of the absorption intensity derives from hexadecapolar induction.[134]

Methane

For methane, CH_4, and other tetrahedral species (CCl_4, CF_4, etc.), the dominant induction mechanisms in the FIR involve polarization by the permanent octopole and hexadecapole moments. Therefore

$$\boldsymbol{\mu}_{ij}^{\text{mult}} = \frac{1}{15}\alpha \mathbf{T}^{(4)}(\mathbf{r}_{ij}) : \boldsymbol{\Omega}_i + \frac{1}{105}\alpha \mathbf{T}^{(5)}(\mathbf{r}_{ij}) : \boldsymbol{\Phi}_i + \frac{1}{15}\alpha \mathbf{T}^{(4)}(\mathbf{r}_{ij}) : \boldsymbol{\Omega}_j - \frac{1}{105}\alpha \mathbf{T}^{(5)}(\mathbf{r}_{ij}) : \boldsymbol{\Phi}_j, \tag{11.153}$$

where α is the isotropic molecular polarizability, and $\boldsymbol{\Omega}$ and $\boldsymbol{\Phi}$ denote the molecular octopole and hexadecapole tensors. The rotational selection rules for the octopole- and hexadecapole-induced spectra are $\Delta J = 0, \pm 1, \pm 2, \pm 3$ and $\Delta J = 0, \pm 1, \pm 2, \pm 3, \pm 4$, respectively. Because of the rapid translational modulation of $\boldsymbol{\mu}_{ij}$, each rotational line is broadened to the extent of $50\,\text{cm}^{-1}$ (octopole spectrum) or $65\,\text{cm}^{-1}$ (hexadecapole spectrum) at 300 K, and since $B \approx 5\,\text{cm}^{-1}$, this results in extensive intra- and interbranch overlap. Consequently individual transitions are not resolved, and at room temperature classical mechanics suffices for the calculation of the induced absorption spectrum.

Overlap induction makes a small, though non-negligible, contribution to the CH_4 CIA spectrum.[135] If we assume that the overlap dipole possesses the same symmetry as the octopolar and hexadecapolar terms of (11.153), then we can write

$$\boldsymbol{\mu}_{ij}^{\text{ov}} = f_3\mu_3(r_{ij})\hat{\mathbf{T}}^{(4)}(\hat{\mathbf{r}}_{ij}) : \mathbf{O}_i + f_4\mu_4(r)\hat{\mathbf{T}}^{(5)}(\hat{\mathbf{r}}_{ij}) : \mathbf{P}_i + f_3\mu_3(r_{ij})\hat{\mathbf{T}}^{(4)}(\hat{\mathbf{r}}_{ij}) : \mathbf{O}_j - f_4\mu_4(r)\hat{\mathbf{T}}^{(5)}(\hat{\mathbf{r}}_{ij}) : \mathbf{P}_j. \tag{11.154}$$

Here $\mu_l(r) = \lambda_l \exp[-(r-\sigma)/\rho_l]$, where σ is the zero of the CH_4-CH_4 potential, and λ_l and ρ_l measure the strength and range of $\mu_l(r)$, respectively. We have defined $\hat{\mathbf{T}}^{(l)}(\hat{\mathbf{r}}) \equiv r^{l+1}\mathbf{T}^{(l)}(\mathbf{r})$, and $\mathbf{O}$ and $\mathbf{P}$ are tensors normalized such that $\mathbf{O}\dot{:}\mathbf{O} = 6$ and $\mathbf{P}::\mathbf{P} = 15/2$. If we take $f_3 = (4320)^{-1/2}$ and $f_4 = (189000)^{-1/2}$, the above normalization is such as to make $\left\langle \mu_{ij}^{\text{ov}\,2} \right\rangle_{\omega_i\omega_j\omega} = \sum_l \mu_l(r_{ij})^2$.

Since the CH_4-CH_4 intermolecular potential is almost isotropic, the dynamical processes of relative translation and rotation of an interacting pair of molecules occur independently. The dipole–dipole correlation function is therefore given by

$$C(t) = \sum_{l=3,4} C_{\text{tr}}^{(l)}(t) C_{\text{rot}}^{(l)}(t), \tag{11.155}$$

with moments

$$M_n = \sum_{l=3,4} \sum_{m=0}^{n} \binom{n}{m} M_{tm}^{(l)} M_{rn-m}^{(l)}. \tag{11.156}$$

If $\Omega(t)$ measures the molecular orientation at time t, then the rotational correlation functions are

$$C_{\text{rot}}^{(l)}(t) = (2l+1)^{-1}\left\langle \sum_{mn} D_{mn}^{l}(\Omega(t))^{*} D_{mn}^{l}(\Omega(0)) \right\rangle, \tag{11.157}$$

where $D_{mn}^{l}(\Omega)$ denotes a generalized spherical harmonic (Appendix A.2). Classically, therefore,[133]

$$C_{\text{rot}}^{(l)}(t) = (2l+1)^{-1} \sum_{m=-l}^{l} \left(1 - m^2\omega_0^2 t^2\right) \exp\left(-\frac{1}{2} m^2 \omega_0^2 t^2\right), \tag{11.158}$$

where $\omega_0 = (kT/I)^{1/2}$. The rotational moments are therefore[133]

$$M_{r2n}^{(l)} = \frac{2(2n-1)!!}{2l+1} B_{2n+1}(l+1)\omega_0^{2n}, \tag{11.159}$$

where $B_n(x)$ denotes a Bernoulli polynomial.[136] As special cases of (11.159), we have $M_{r0}^{(l)} = 1$ and $M_{r2}^{(l)} = l(l+1)\omega_0^2$.

Quantum-mechanically we find for the rotational spectrum[135] that

$$I_{\text{rot}}^{(l)}(\omega) = (2l+1)^{-1} \sum_{J=0}^{\infty} \sum_{J'=|J-l|}^{J+l} P_J(2J+1)(2J'+1)\delta(\omega - \omega_{J'J}), \tag{11.160}$$

where $\omega_{J'J} = BJ'(J'+1) - BJ(J+1)$, with $B = \hbar/2I$ the rotational constant, and $P_J = \exp[-\beta B\hbar J(J+1)]/\sum_J (2J+1)^2 \exp[-\beta B\hbar J(J+1)]$. The moments of (11.160) are

$$M_{rn}^{(l)} = (2l+1)^{-1} \sum_{J=0}^{\infty} \sum_{J'=|J-l|}^{J+l} P_J(2J+1)(2J'+1)\omega_{J'J}^{n}, \tag{11.161}$$

so that

$$\begin{aligned} M_{r0}^{(l)} &= 1, \\ M_{r1}^{(l)} &= Bl(l+1), \\ M_{r2}^{(l)} &= l(l+1)\omega_0^2\left(1 + \frac{1}{12}(3l^2 + 3l - 1)\beta^2\hbar^2\omega_0^2\right), \end{aligned} \tag{11.162}$$

etc. Note that the quantum corrections are very substantial below 200 K because of the small moment of inertia.[135]

The translational correlation functions are

$$C_{\text{tr}}^{(l)} = \frac{2^l}{(2l+1)!}\alpha^2 Q_l^2 \sum_{ijk} \langle \mathbf{T}^{(l+1)\prime}(\mathbf{r}_{ij}(t)) \bullet \mathbf{T}^{(l+1)\prime}(\mathbf{r}_{kj}(0)) \rangle, \tag{11.163}$$

where $\mathbf{T}^{(l+1)\prime}(\mathbf{r}) = F_l(r)\mathbf{T}^{(l+1)}(\mathbf{r})$, $F_l(r) = 1 + \beta_l(r)$, and

$$\beta_l(r) = \frac{\lambda_l \exp[-(r-\sigma)/\rho_l]}{[2(l+1)]^{1/2}\alpha Q_l r^{-(l+2)}} \tag{11.164}$$

measures the fractional contribution to lth-order induction made by the overlap mechanism, $Q_l^2 \equiv 4\pi/(2l+1)\sum_m |Q_{lm}|^2$, where Q_{lm} is the spherical multipole moment component defined by (2.74); we note that $Q_3 = (12/5)^{1/2}\Omega$ and $Q_4 = (12/7)^{1/2}\Phi$, where Ω and Φ are defined by (2.105) and (2.106).

Using the relation

$$\frac{\mathrm{d}}{\mathrm{d}t}\mathbf{T}^{(l+1)\prime}(\mathbf{r}) = \Big(\beta_l'(r)\hat{\mathbf{r}}\mathbf{T}^{(l+1)}(\mathbf{r}) + F_l(r)\mathbf{T}^{(l+2)}(\mathbf{r})\Big)\cdot\mathbf{v}, \tag{11.165}$$

and noting that[135]

$$\begin{aligned}
\hat{\mathbf{T}}^{(n)}(\hat{\mathbf{r}}_{ij}) \bullet \hat{\mathbf{T}}^{(n)}(\hat{\mathbf{r}}_{kj}) &= \frac{(2n)!}{2^n}P_n(\cos\theta_{ijk}),\\
\hat{\mathbf{T}}^{(n)}(\hat{\mathbf{r}}_{ij}) \bullet [\hat{\mathbf{T}}^{(n-1)}(\hat{\mathbf{r}}_{kj})\hat{\mathbf{r}}_{kj}] &= -\frac{(2n)!}{2^n(2n-1)}P_n(\cos\theta_{ijk}),\\
[\hat{\mathbf{r}}_{ij}\hat{\mathbf{T}}^{(n-1)}(\hat{\mathbf{r}}_{ij})] \bullet [\hat{\mathbf{T}}^{(n-1)}(\hat{\mathbf{r}}_{kj})\hat{\mathbf{r}}_{kj}]&\\
= (2n-2)!\frac{nP_n(\cos\theta_{ijk}) + (n-1)P_{n-2}(\cos\theta_{ijk})}{2^{n-1}(2n-1)}&,
\end{aligned} \tag{11.166}$$

we find for the first two nonvanishing translational moments[135] that

$$\begin{aligned}
M_{t0}^{(l)} &= (l+1)\alpha^2 Q_l^2\rho^2 V\Bigg(\int \mathrm{d}\mathbf{r}_{12}\, g(r_{12})\frac{F_l(r_{12})^2}{r_{12}^{2l+4}}\\
&\quad + \rho\int \mathrm{d}\mathbf{r}_{12}\mathrm{d}\mathbf{r}_{32}\, g(r_{12}r_{13}r_{32})P_{l+1}(\cos\theta_{123})\frac{F_l(r_{12})F_l(r_{32})}{(r_{12}r_{32})^{l+2}}\Bigg),
\end{aligned} \tag{11.167}$$

$$\begin{aligned}
M_{t2}^{(l)} &= (l+1)(l+2)(2l+3)\alpha^2 Q_l^2\frac{kT}{M}\rho^2 V\Bigg(2\int \mathrm{d}\mathbf{r}_{12}\, g(r_{12})\\
&\quad\times\left[\frac{F_l(r_{12})^2}{r_{12}^{2l+6}} - \frac{2}{2l+3}\frac{r_{12}\beta_l'(r_{12})F_l(r_{12})}{r_{12}^{2l+6}} + \frac{2}{(2l+3)(2l+4)}\frac{\left[r_{12}\beta_l'(r_{12})\right]^2}{r_{12}^{2l+6}}\right]\\
&\quad + \rho\int \mathrm{d}\mathbf{r}_{12}\mathrm{d}\mathbf{r}_{32}\, g(r_{12}r_{13}r_{32})\\
&\quad\times\Bigg[P_{l+2}(\cos\theta_{123})\frac{F_l(r_{12})F_l(r_{32})}{(r_{12}r_{32})^{l+3}} - \frac{2}{2l+3}P_{l+2}(\cos\theta_{123})\frac{r_{12}\beta_l'(r_{12})F_l(r_{32})}{(r_{12}r_{32})^{l+3}}\\
&\quad + \frac{2}{(2l+3)(2l+4)}\frac{[(l+2)P_{l+2}(\cos\theta_{123}) + (l+1)P_l(\cos\theta_{123})]}{2l+3}\\
&\quad\times\frac{r_{12}\beta_l'(r_{12})r_{32}\beta_l'(r_{32})}{(r_{12}r_{32})^{l+3}}\Bigg]\Bigg).
\end{aligned} \tag{11.168}$$

Expressions for the multipolar contributions to $M_{t4}^{(l)}$ and $M_{t6}^{(l)}$ have also been derived for a low-density gas.[133]

M_{t0} and M_{t2} have been calculated[135] in the superposition approximation for a LJ model of CH_4 with $\epsilon = 193$ K and $\sigma = 3.655$ Å, taking $\alpha = 2.59$ Å^3, $Q_3 = 1.85 \times 10^{-34}$ esu, $Q_4 = -3.09 \times 10^{-42}$ esu, $\lambda_3 = 0.0018$ D, $\lambda_4 = 0.00065$ D, and $\rho_3 = \rho_4 = 0.11\sigma$. (The multipole moments are from *ab initio* calculations; the overlap parameters were obtained by fitting to the gas-phase spectrum.) The results are shown in Table 11.3 for $T = 122$ K and $\rho = 25.3$ mol dm^{-3}. Octopolar induction accounts for roughly 60% of $M_0 = M_{t0}$, hexadecapolar induction for 30%; overlap effects contribute the remainder. Note the significant cancellation of two- and three-body terms in $M_{t0}^{(3)}$; this reduces the relative importance of the octopolar mechanism, which accounts for almost 90% of the intensity in the gas phase.[133] Such cancellation apparently only occurs for $l = 2$ (cf. N_2 above) and, to a lesser extent, $l = 3$; the three-body term in $M_{t0}^{(4)}$ is positive. Note also that three-body contributions to $M_{t2}^{(l)}$ are positive for $l \geq 3$.

Table 11.4 shows a comparison between computed and experimental moments. The quantum formulae (11.162) were used to compute the rotational moments, and corrections were also applied to the translational moments.[135] Local field effects (§ 11.1.3) have also been included. Given the approximate nature of the superposition approximation, the agreement is probably as good as could be expected.

Tipping and co-workers[137] have studied CIA in the ν_2 fundamental band of methane perturbed by argon. During the ν_2 vibration, the molecular symmetry of

Table 11.3

Translational moments of the CIA spectrum of liquid methane at 122 K.

	Multipolar	Multipolar plus overlap
$M_{t0}^{(3)}/V[2B]$	2.19×10^{-24}	2.35×10^{-24}
$M_{t0}^{(3)}/V[3B]$	-1.29×10^{-24}	-1.36×10^{-24}
$M_{t2}^{(3)}/V[2B]$	76.5	85.9
$M_{t2}^{(3)}/V[3B]$	8.3	9.6
$M_{t0}^{(4)}/V[2B]$	0.33×10^{-24}	0.36×10^{-24}
$M_{t0}^{(4)}/V[3B]$	0.07×10^{-24}	0.08×10^{-24}
$M_{t2}^{(4)}/V[2B]$	17.7	19.5
$M_{t2}^{(4)}/V[3B]$	4.1	4.5
M_0/V	1.30×10^{-24}	1.43×10^{-24}

Units are J (zeroth moments) and Js^{-2} (second moments).
From ref. 135.

Table 11.4

Moments of the CIA spectrum of liquid methane at 122 K

	Calculated	Observed
M_0/V	1.62×10^{-24}	1.76×10^{-24}
M_1/V	2.74×10^{9}	3.05×10^{9}
M_2/V	1350	1610

Units of M_n are Js^{-n}.
From ref. 135.

methane changes from T_d to D_2, and as a result the molecule develops a quadrupole moment. This moment can induce a dipole on an argon atom, leading to absorption in the IR centred around $\nu_2 = 1530\,\text{cm}^{-1}$. The observed band second moment is consistent with *ab initio* estimates of the transition quadrupole. Further studies[138] on mixtures with hydrogen, nitrogen, and carbon dioxide showed that an alternative induction mechanism involving the permanent quadrupole of the perturbing gas was relatively less important as a source of IR intensity.

Carbon dioxide

The atmosphere of Venus is largely composed of carbon dioxide. Absorption of infrared radiation by CO_2 is primarily responsible for the 'greenhouse effect' that raises surface temperatures to 700 K. Accurate prediction of the CIA spectrum is necessary for a detailed analysis of the radiative transfer processes that govern the atmospheric temperature profile. As a first step in this direction, Gruszka and Borysow[139] have calculated the zeroth and second spectral moments; their calculations emphasize the sensitivity of these moments to anisotropic terms in the intermolecular potential.

The spherical components of the induced dipole $\boldsymbol{\mu}_{12}$ are given by

$$\mu_\nu(r\omega_1\omega_2\omega) = \left(\frac{64\pi^3}{3}\right)^{1/2} \sum_{l_1l_2lL} A_{l_1l_2lL}(r)\psi^{1\nu}_{l_1l_2lL}(\omega_1\omega_2\omega), \qquad \nu = 0, \pm 1, \tag{11.169}$$

where the angular basis function $\psi^{1\nu}_{l_1l_2lL}$ is defined in (A.196). The normalization in (11.169) ensures that

$$\left\langle \mu_{12}^2 \right\rangle_{\omega_1\omega_2\omega} = \left\langle \sum_\nu |\mu_\nu(r\omega_1\omega_2\omega)|^2 \right\rangle_{\omega_1\omega_2\omega} = \sum_{l_1l_2lL} A^2_{l_1l_2lL}(r). \tag{11.170}$$

Parity considerations dictate that only even values of l_1 and l_2 and odd values of L may occur, and (A.131) further restricts the sum over l to $l = L$ for $l_1 = l_2$ and $l = L \pm 1$ for $l_1 \neq l_2$. The leading terms in (11.169) are those for which $\{l_1l_2lL\} = \{2023\}$ and $\{0223\}$ (representing first-order quadrupolar induction *via* the isotropic component of the polarizability) and {4045} and {0445} (representing first-order hexadecapolar induction). For these terms

$$A_{l0l\,l+1} = -A_{0ll\,l+1} = \frac{(l+1)^{1/2}\alpha Q_l}{r^{l+2}}. \qquad (11.171)$$

Induction is also possible *via* anisotropy in the polarizability. The only sizeable component is the quadrupolar term {2233}, for which $A_{2233} = (2/15)^{1/2}\gamma Q_2/r^4$.

The terms $\{l0l\,l \pm 1\}$ and $\{0ll\,l \pm 1\}$ arise in second order, through induction of a dipole on molecule 1 by the first-order dipole induced on molecule 2 by the permanent moments of molecule 1 (and vice versa). Here

$$A_{l0lL} = (-)^l[30(l+1)(2l+3)(2L+1)]^{1/2} \times \begin{pmatrix} L & l+1 & 2 \\ 0 & 0 & 0 \end{pmatrix} \begin{Bmatrix} L & 1 & l \\ 1 & l+1 & 2 \end{Bmatrix} \frac{\alpha^2 Q_l}{r^{l+5}} \qquad (11.172)$$

(and $A_{0llL} = -A_{l0lL}$), where $(\cdots)$ and $\{\cdots\}$ denote $3j$ and $6j$ symbols, respectively. The most important of these terms are {2023} and {0223}.

If the density is not too high, the zeroth moment is given by

$$M_0 = \rho^2 V \int d\mathbf{r}_{12} \left\langle \exp[-\beta u(r\omega_1\omega_2\omega)] \left| \sum_{\nu} \mu_\nu(r\omega_1\omega_2\omega) \right|^2 \right\rangle_{\omega_1\omega_2}, \qquad (11.173)$$

and the second by

$$M_2 = \rho^2 V \int d\mathbf{r}_{12} \left\langle \exp[-\beta u(r\omega_1\omega_2\omega)] \times \left(\frac{2kT}{M} \sum_{\mu\nu} |\nabla_\mu \mu_\nu|^2 + \frac{kT}{I} \sum_{\mu\nu} \left(|\nabla_{\omega_1\mu}\mu_\nu|^2 + |\nabla_{\omega_2\mu}\mu_\nu|^2 \right) \right) \right\rangle_{\omega_1\omega_2}, \qquad (11.174)$$

where M is the molecular mass, and ∇_μ and $\nabla_{\omega_i\mu}$ denote spherical components of $\boldsymbol{\nabla} \equiv \partial/\partial\mathbf{r}_{12}$ and the angular gradient operator $\boldsymbol{\nabla}_{\omega_i}$ for molecule i, respectively. To evaluate the terms $\nabla_\mu\mu_\nu$ and $\nabla_{\omega_i\mu}\mu_\nu \equiv il_{i\mu}\mu_\nu$ in (11.174), we may use the expressions for $\nabla_\mu[f(r)Y_{lm}(\omega)]$ and $l_\mu Y_{lm}(\omega)$ given by Brink and Satchler.[140] The resulting formulae are too complicated to reproduce here; see ref. 139.

Gruszka and Borysow have computed M_0 and M_2 classically by four-dimensional quadrature for several different potential models of varying degrees of sophistication: isotropic LJ, LJ + point quadrupole, site–site LJ + point quadrupole, and site–site exp-6-1. Their results demonstrate the importance of considering anisotropy in the potential when dealing with a molecule as nonspherical as CO_2, and show that perturbational treatments are inadequate. However, none of the models studied could adequately reproduce the experimental results over the temperature range 230–350 K unless a rather substantial anisotropic overlap term, with an unrealistically large range, was added to the {2023} and {0223} contributions to $\boldsymbol{\mu}_{12}$.

Borysow and Moraldi[141] have performed similar calculations for CO_2-Ar mixtures.

11.2 Raman spectra

In this section we discuss isotropic and anisotropic Rayleigh and Raman scattering.[142] Our ultimate concern is the moments of the scattering spectra, which are equilibrium properties of the system. To set these in context, however, and to explain the meanings of various terms that arise, it is first necessary to broaden the discussion somewhat and consider the nonequilibrium processes that govern the scattering of light.

We therefore begin with a quantum–mechanical derivation of the cross-section for the scattering of light by a molecular fluid, and show how this can be related to the Fourier transform of the polarizability correlation function for the system. Our analysis includes an approximate treatment of local field effects, which affect scattered intensities in dense gases and liquids. Factors affecting the lineshape in dilute and dense atomic and molecular systems are considered. Finally, we discuss the calculation of the spectral moments, and show how experimental observations of these quantities can yield information on molecular properties and interactions.

11.2.1 Scattering cross-section

Consider a system irradiated by a monochromatic beam of light of wavevector $\mathbf{k}_0$, polarization $\mathbf{n}_0$, frequency ω_0, and intensity $\mathcal{I}(\omega_0)$. Scattering of light of wavevector $\mathbf{k}_S$, polarization $\mathbf{n}_S$, at frequencies between ω_S and $\omega_S + \mathrm{d}\omega_S$, into solid angle $\mathrm{d}\Omega$ causes energy removal from the beam at a rate given by

$$P(\omega_S) = \frac{\mathrm{d}^2\sigma}{\mathrm{d}\Omega \mathrm{d}\omega_S} \mathrm{d}\Omega \mathrm{d}\omega_S \mathcal{I}(\omega_0), \tag{11.175}$$

where $\mathrm{d}^2\sigma/\mathrm{d}\Omega\mathrm{d}\omega_S$ is the spectral differential cross-section.

To derive an expression for the cross-section in terms of molecular properties, it is convenient first to consider scattering by a single molecule; then by a dilute gas; and finally by a dense gas or liquid, where local field effects come into play.

Single molecule scattering[143]

Far from resonance,[144] and ignoring multiple and stimulated[147] scattering, the molecular scattering cross-section in the dipole approximation is[148]

$$\frac{\mathrm{d}^2\sigma}{\mathrm{d}\Omega d\omega_S} = k_S^4 \sum_{mn} P_m \left|\mathbf{n}_0 \cdot \boldsymbol{\alpha}_{mn}(\omega_0) \cdot \mathbf{n}_S\right|^2 \delta(\omega_{nm} - \omega). \tag{11.176}$$

Here $P_m = \exp(-\beta\hbar\omega_m)/Q$, where $Q = \sum_m \exp(-\beta\hbar\omega_m)$, is the probability of finding the molecule initially in state m, with energy $\hbar\omega_m$; $\hbar\omega = \hbar(\omega_0 - \omega_S)$ is the net molecular energy gain after scattering, when the molecule occupies state n, with energy $\hbar\omega_n$; and $\omega_{nm} = \omega_n - \omega_m$. The quantity

$$\boldsymbol{\alpha}_{mn}(\omega_0) = \sum_l{}'' \left(\frac{\langle m|\boldsymbol{\mu}|l\rangle\langle l|\boldsymbol{\mu}|n\rangle}{\hbar(\omega_{lm} - \omega_0)} + \frac{\langle n|\boldsymbol{\mu}|l\rangle^*\langle l|\boldsymbol{\mu}|m\rangle^*}{\hbar(\omega_{ln} + \omega_0)} \right) \tag{11.177}$$

is the complex transition polarizability; here $\boldsymbol{\mu}$ is the Schrödinger dipole moment operator, which is a function of both the electronic and nuclear coordinates. The double prime in (11.177) indicates that both m and n are excluded[149] from the sum over l.

These results simplify when electronic and nuclear (vibrational and rotational) degrees of freedom uncouple. We assume that the unperturbed molecule is in its ground electronic state $|0\rangle$, and that no electronic transitions are induced by the radiation. Then in the Born–Oppenheimer approximation $|m\rangle = |0\rangle|\mu\rangle$, $|n\rangle = |0\rangle|\nu\rangle$, and $|l\rangle = |L\rangle|\lambda\rangle$, where $|\mu\rangle$, $|\nu\rangle$, and $|\lambda\rangle$ denote nuclear eigenstates, and $|L\rangle$ signifies a virtual electronic state that the molecule assumes for the duration of the scattering process. Similarly $\omega_m = \omega_\mu$, $\omega_n = \omega_\nu$, and $\omega_l = \omega_L + \omega_\lambda$. The scattering cross-section for transitions within the ground electronic manifold can thus be written as

$$\frac{\mathrm{d}^2\sigma}{\mathrm{d}\Omega d\omega_\mathrm{S}} = k_\mathrm{S}^4 \sum_{\mu\nu} P_\mu \left|\mathbf{n}_0 \cdot \boldsymbol{\alpha}_{\mu\nu}(\omega_0) \cdot \mathbf{n}_S\right|^2 \delta(\omega_{\nu\mu} - \omega), \tag{11.178}$$

where $P_\mu = \exp(-\beta\hbar\omega_\mu)/\sum_\mu \exp(-\beta\hbar\omega_\mu)$, and

$$\boldsymbol{\alpha}_{\mu\nu}(\omega_0) = \boldsymbol{\alpha}_{\mu\nu}^{\mathrm{elec}}(\omega_0) + \boldsymbol{\alpha}_{\mu\nu}^{\mathrm{nuc}}(\omega_0), \tag{11.179}$$

where

$$\boldsymbol{\alpha}_{\mu\nu}^{\mathrm{elec}}(\omega_0) = \sum_{L\neq 0,\lambda} \left(\frac{\langle\mu|\langle 0|\boldsymbol{\mu}|L\rangle|\lambda\rangle\langle\lambda|\langle L|\boldsymbol{\mu}|0\rangle|\nu\rangle}{\hbar(\omega_L + \omega_{\lambda\mu} - \omega_0)} + \frac{\langle\mu|\langle 0|\boldsymbol{\mu}|L\rangle|\lambda\rangle^*\langle\lambda|\langle L|\boldsymbol{\mu}|0\rangle|\nu\rangle^*}{\hbar(\omega_L + \omega_{\lambda\nu} + \omega_0)} \right) \tag{11.180}$$

and

$$\boldsymbol{\alpha}_{\mu\nu}^{\mathrm{nuc}}(\omega_0) = \sum_{\lambda\neq \mu,\nu} \left(\frac{\langle\mu|\langle 0|\boldsymbol{\mu}|0\rangle|\lambda\rangle\langle\lambda|\langle 0|\boldsymbol{\mu}|0\rangle|\nu\rangle}{\hbar(\omega_{\lambda\mu} - \omega_0)} + \frac{\langle\mu|\langle 0|\boldsymbol{\mu}|0\rangle|\lambda\rangle^*\langle\lambda|\langle 0|\boldsymbol{\mu}|0\rangle|\nu\rangle^*}{\hbar(\omega_{\lambda\nu} + \omega_0)} \right) \tag{11.181}$$

represent contributions to $\boldsymbol{\alpha}_{\mu\nu}(\omega_0)$ resulting from electronic and nuclear excitations, respectively.

Typically electronic energies greatly exceed vibrational and rotational energies. Away from resonance, therefore, $|\omega_L \pm \omega_0| \gg \omega_{\lambda\mu}, \omega_{\lambda\nu}$. If we neglect the small nuclear frequencies in the denominators of (11.180), we can perform the sum over nuclear states using the completeness relation $\sum_\lambda |\lambda\rangle\langle\lambda| = 1$. If we also write $\boldsymbol{\mu} = \boldsymbol{\mu}_{\mathrm{elec}} + \boldsymbol{\mu}_{\mathrm{nuc}}$, where $\boldsymbol{\mu}_{\mathrm{elec}}$ and $\boldsymbol{\mu}_{\mathrm{nuc}}$ are the electronic and nuclear dipole moment operators, then since $\langle 0|\boldsymbol{\mu}_{\text{nuc}}|L\rangle = \boldsymbol{\mu}_{\text{nuc}}\delta_{L0}$, the electronic contribution can be written as

$$\boldsymbol{\alpha}^{\text{elec}}_{\mu\nu}(\omega_0) = \langle\mu|\boldsymbol{\alpha}^{\text{elec}}(\omega_0)|\nu\rangle, \tag{11.182}$$

where (cf. (C.20))

$$\boldsymbol{\alpha}^{\text{elec}}(\omega_0) = \sum_{L\neq 0}\left(\frac{\langle 0|\boldsymbol{\mu}_{\text{elec}}|L\rangle\langle L|\boldsymbol{\mu}_{\text{elec}}|0\rangle}{\hbar(\omega_L-\omega_0)} + \frac{\langle 0|\boldsymbol{\mu}_{\text{elec}}|L\rangle^*\langle L|\boldsymbol{\mu}_{\text{elec}}|0\rangle^*}{\hbar(\omega_L+\omega_0)}\right) \tag{11.183}$$

is the ground state electronic polarizability at the incident frequency.

For the nuclear contribution

$$\boldsymbol{\alpha}^{\text{nuc}}_{\mu\nu}(\omega_0) = \sum_{\lambda\neq\mu,\nu}\left(\frac{\langle\mu|\bar{\boldsymbol{\mu}}|\lambda\rangle\langle\lambda|\bar{\boldsymbol{\mu}}|\nu\rangle}{\hbar(\omega_{\lambda\mu}-\omega_0)} + \frac{\langle\mu|\bar{\boldsymbol{\mu}}|\lambda\rangle^*\langle\lambda|\bar{\boldsymbol{\mu}}|\nu\rangle^*}{\hbar(\omega_{\lambda\nu}+\omega_0)}\right), \tag{11.184}$$

where $\bar{\boldsymbol{\mu}} = \langle 0|\boldsymbol{\mu}_{\text{elec}}|0\rangle + \boldsymbol{\mu}_{nuc}$ is the dipole moment of the ground electronic state. We note that $\boldsymbol{\alpha}^{\text{nuc}}_{\mu\mu}(\omega_0)$ is the nuclear polarizability of the molecule in rovibrational level μ of the ground electronic state. At optical frequencies the nuclear polarizability is extremely small, and $\boldsymbol{\alpha}^{\text{nuc}}_{\mu\nu}(\omega_0) \ll \boldsymbol{\alpha}^{\text{elec}}_{\mu\nu}(\omega_0)$.

For the scattering of visible light, therefore, we may take

$$\frac{\mathrm{d}^2\sigma}{\mathrm{d}\Omega d\omega_{\text{S}}} = k_{\text{S}}^4\sum_{\mu\nu}P_\mu|\langle\mu|\mathbf{n}_0\cdot\boldsymbol{\alpha}^{\text{elec}}(\omega_0)\cdot\mathbf{n}_{\text{S}}|\nu\rangle|^2\delta(\omega_{\nu\mu}-\omega). \tag{11.185}$$

Dilute gases

In a dilute gas, by an extension of the above, the total scattering cross-section at optical frequencies is given by

$$\frac{\mathrm{d}^2\sigma}{\mathrm{d}\Omega d\omega_{\text{S}}} = k_{\text{S}}^4\sum_{\mu\nu}P_\mu|\langle\mu|\mathbf{n}_0\cdot\boldsymbol{\pi}^{\text{elec}}(\omega_0)\cdot\mathbf{n}_{\text{S}}|\nu\rangle|^2\delta(\omega_{\nu\mu}-\omega), \tag{11.186}$$

where μ and ν denote collective nuclear eigenstates of the system, and

$$\boldsymbol{\pi}^{\text{elec}}(\omega_0) = \sum_i \boldsymbol{\pi}_i \tag{11.187}$$

is the frequency-dependent N-body electronic polarizability, with $\boldsymbol{\pi}_i \equiv \boldsymbol{\pi}_i^{\text{elec}}(\omega_0)$ the electronic polarizability of molecule i at the frequency of the incident light.

Note that $\boldsymbol{\pi}_i$ is a function, in principle, of the nuclear coordinates of *all* N molecules. It differs from the polarizability of an isolated molecule: it is modified by pair, triplet, and other many-body interactions. It has the cluster expansion

$$\boldsymbol{\pi}_i = \boldsymbol{\alpha}_i + \sum_{j>i}\Delta\boldsymbol{\alpha}_{ij} + \sum_{k>j>i}\Delta\boldsymbol{\alpha}_{ijk} + \ldots, \tag{11.188}$$

where $\boldsymbol{\alpha}_i$ is the permanent polarizability of molecule i, $\Delta\boldsymbol{\alpha}_{ij}$ is the additional polarizability of the ij pair resulting from mutual polarization, $\Delta\boldsymbol{\alpha}_{ijk}$ is the irreducible triplet polarizability of *ijk*, and so on. For polar molecules, in particular, many-body terms contribute significantly to $\boldsymbol{\pi}_i$, because of long-range dipole–induced-dipole processes. The single-molecule term yields allowed Rayleigh and Raman spectra; other terms are responsible for the phenomenon of collision-induced light scattering (§ 11.2.7).

Dense fluids—local field effects

In dense fluids local field effects must be considered—that is, we must recognize that the electric fields responsible for the scattering are modified as a result of molecular interactions. We may develop an approximate theory of these effects based on a picture of scattering as a two-step process, in which absorption of energy from the incident beam is followed by spontaneous emission at a slightly different frequency. On this basis, we expect that the local field factor for light scattering can be written as the product

$$f_S(\omega_0, \omega_S) = f_A(\omega_0) f_E(\omega_S), \tag{11.189}$$

where the separate terms on the right represent local field factors for absorption and emission.

In § 11.1.3 we derived an expression for $f_A(\omega_0)$, Eq. (11.36). Note that this factor applies only to the absorption and *stimulated* emission of radiation. For the closely related phenomenon of *spontaneous* emission, in which an excited molecular state decays by virtue of coupling to quantum fluctuations in the radiation density, a slightly different result is obtained, as we now show.

We will assume that both the dielectric constant $\epsilon(\omega)$ and refractive index $n(\omega)$ are real.[150] Consider a transition between two molecular states l and m such that $\omega_{lm} = \omega_S$. The rate of spontaneous emission *in vacuo* is given by

$$\Gamma(\omega_S) = \frac{4}{3\pi\hbar} k_S^3 \left|\langle l|\boldsymbol{\alpha} \cdot \boldsymbol{\mathcal{E}}(\omega_S)|m\rangle\right|^2, \tag{11.190}$$

where $\boldsymbol{\mathcal{E}}(\omega_S)$ is the component of the fluctuating zero-point field of the vacuum at frequency ω_S, and $k_S = \omega_S/c$ is the wavevector of the emitted photon.

In a dense medium the amplitude of the fluctuations is modified. The field energy density at the transition frequency is $\epsilon'(\omega_S)\mathcal{E}(\omega_S)^2/8\pi$, where $\mathcal{E}(\omega_S)$ is the mean (macroscopic) field, and $\epsilon'(\omega) = n'(\omega)^2$ is the dielectric constant. Since this density is unchanged from its free-space form, it follows that $\mathcal{E}(\omega_S)$ is reduced by a factor of $n'(\omega_S)$. For spontaneous emission, of course, the relevant field is not $\mathcal{E}(\omega_S)$, but the local field $\mathcal{E}_L(\omega_S)$, which is the mean field experienced *by a molecule*. The two fields are related by an equation of the form (11.35); and therefore the amplitude of the fluctuations in $\mathcal{E}_L(\omega_S)$ will increase by a factor $(n'(\omega_S)^2 + 2)/3n'(\omega_S)$. The transition rate increases by the square of this amount.

In (11.190), the factor of k_S^3 represents the density of states for fluctuations at the emission frequency. In a dense medium, $k_S = n\omega_S/c$, and so dispersion will further increase the transition rate by a factor of $n'(\omega_S)^3$.

These arguments suggest a local field factor for spontaneous emission of

$$f_{\mathrm{E}}(\omega_{\mathrm{S}}) = n'(\omega_{\mathrm{S}})\left(\frac{n'(\omega_{\mathrm{S}})^2+2}{3}\right)^2. \tag{11.191}$$

Substituting our results for $f_{\mathrm{A}}(\omega_0)$ and $f_{\mathrm{E}}(\omega_{\mathrm{S}})$ into (11.189), we find that the local field factor for scattering is

$$f_{\mathrm{S}}(\omega_0,\omega_{\mathrm{S}}) = \frac{n'(\omega_{\mathrm{S}})}{n'(\omega_0)}\left(\frac{n'(\omega_0)^2+2}{3}\right)^2\left(\frac{n'(\omega_{\mathrm{S}})^2+2}{3}\right)^2. \tag{11.192}$$

At optical frequencies dispersion in the refractive index is often quite small, and we can approxinate $n'(\omega_0) = n'(\omega_{\mathrm{S}}) = n_\infty$, the high-frequency refractivity. In this case (11.192) simplifies to

$$f_{\mathrm{S}} = \left(\frac{n_\infty^2+2}{3}\right)^4. \tag{11.193}$$

In liquids local field effects can therefore be significant, increasing the intensity of light scattering by two to six times, a much greater effect than that found for infrared absorption (§ 11.1.3). Experimental measurements confirm that in most cases (11.193) is accurate to within 10–20%.[73]

11.2.2 Light scattering and polarizability correlation functions

Just as infrared absorption may be considered to arise from fluctuating molecular dipole moments, so light scattering may be considered to originate from polarizability fluctuations. In mathematical terms, the scattering cross-section is proportional to the spectral density of the time correlation function for the system polarizability.

To establish this result we may start from (11.186), suitably modified for local field effects. We introduce the Heisenberg representation of the polarizability,

$$\boldsymbol{\pi}(t) = \exp(iH_{\mathrm{nuc}}t/\hbar)\boldsymbol{\pi}^{\mathrm{elec}}(\omega_0)\exp(-iH_{\mathrm{nuc}}t/\hbar) \tag{11.194}$$

(implicitly dependent on the frequency), where H_{nuc} is the nuclear Hamiltonian (a function in general of both intramolecular and intermolecular coordinates), its correlation function

$$C(t) = \langle(\mathbf{n}_0\cdot\boldsymbol{\pi}(t)\cdot\mathbf{n}_S)(\mathbf{n}_0\cdot\boldsymbol{\pi}(0)\cdot\mathbf{n}_S)\rangle \tag{11.195}$$

and the associated Fourier transform

$$I(\omega) = \frac{1}{2\pi}\int_{-\infty}^{+\infty}\mathrm{d}t\,\exp(i\omega t)C(t). \tag{11.196}$$

Following the procedure outlined in § 11.1.2 we can then show that

$$\frac{\mathrm{d}^2\sigma}{\mathrm{d}\Omega d\omega_S} = f_{\mathrm{S}}k_{\mathrm{S}}^4 I(\omega), \tag{11.197}$$

which is the desired result.

Polarized and depolarized scattering

We now turn our attention to the correlation function of (11.195). The polarizability is a second-rank tensor; we will assume it is symmetric, which is the normal situation.[151] It may therefore be split into isotropic and traceless anisotropic parts according to

$$\boldsymbol{\pi}(t) = \boldsymbol{\pi}^{(0)}(t) + \boldsymbol{\pi}^{(2)}(t), \tag{11.198}$$

where $\boldsymbol{\pi}^{(0)} = \pi^{(0)}\mathbf{1}$ and $\boldsymbol{\pi}^{(2)} : \mathbf{1} = \langle \boldsymbol{\pi}^{(2)} \rangle = 0$. Here $\boldsymbol{\pi}^{(l)}$ denotes the irreducible component of rank l (cf. Appendix A.4.3). We therefore have

$$C(t) = (\mathbf{n}_0 \cdot \mathbf{n}_S)^2 \langle \pi^{(0)}(t)\pi^{(0)}(0)\rangle + \mathbf{n}_0\mathbf{n}_S : \langle \boldsymbol{\pi}^{(2)}(t)\boldsymbol{\pi}^{(2)}(0)\rangle : \mathbf{n}_0\mathbf{n}_S. \tag{11.199}$$

The ensemble average $\langle \boldsymbol{\pi}^{(2)}(t)\boldsymbol{\pi}^{(2)}(0)\rangle$, a fourth-rank isotropic tensor, may be evaluated using the procedure described in Appendix A.4.2. Introducing the invariant functions

$$\begin{aligned} C^{(0)}(t) &= \langle \pi^{(0)}(t)\pi^{(0)}(0)\rangle, \\ C^{(2)}(t) &= \langle \boldsymbol{\pi}^{(2)}(t) : \boldsymbol{\pi}^{(2)}(0)\rangle, \end{aligned} \tag{11.200}$$

we find that

$$C(t) = (\mathbf{n}_0 \cdot \mathbf{n}_S)^2 C^{(0)}(t) + \frac{1}{10}\left(1 + \frac{1}{3}(\mathbf{n}_0 \cdot \mathbf{n}_S)^2\right) C^{(2)}(t). \tag{11.201}$$

The frequency spectra corresponding to (11.200) are

$$I^{(0)}(\omega) = \frac{1}{2\pi}\int_{-\infty}^{+\infty} \mathrm{d}t\, \exp(i\omega t)\langle \pi^{(0)}(t)\pi^{(0)}(0)\rangle \tag{11.202}$$

and

$$I^{(2)}(\omega) = \frac{1}{2\pi}\int_{-\infty}^{+\infty} \mathrm{d}t\, \exp(i\omega t)\langle \boldsymbol{\pi}^{(2)}(t) : \boldsymbol{\pi}^{(2)}(0)\rangle \tag{11.203}$$

and satisfy

$$I(\omega) = (\mathbf{n}_0 \cdot \mathbf{n}_S)^2 I^{(0)}(\omega) + \frac{1}{10}\left(1 + \frac{1}{3}(\mathbf{n}_0 \cdot \mathbf{n}_S)^2\right) I^{(2)}(\omega). \tag{11.204}$$

Note that $I^{(0)}(\omega)$ is associated with fluctuations in the isotropic part of the system polarizability, while $I^{(2)}(\omega)$ is associated with fluctuations in the anisotropic part.

It is conventional, particularly in collision-induced light scattering, to refer to $I^{(0)}(\omega)$ and $I^{(2)}(\omega)$ as the *polarized* and *depolarized* spectra, respectively. Unfortunately this is somewhat misleading; better names would undoubtedly be *isotropic* and *anisotropic* spectra. The difficulty arises because, while the depolarized component of the scattered light (that is, light scattered with a polarization differing from that of the incident beam) arises solely from the anisotropic part of the polarizability, it is *not* true that the polarized component (having the same polarization as the incident beam) is completely determined by the isotropic part: as is apparent from (11.204), there is an additional contribution from the anisotropic term.

Clearly, polarized (isotropic) Rayleigh spectra[142] will be quasielastic, since inelasticity is introduced only through translational fluctuations; and polarized Raman spectra, caused by vibrational fluctuations, will show only Q-branches, with no accompanying rotational features unless rovibrational coupling is also present. Depolarized (anisotropic) spectra, on the other hand, will exhibit rotational fine structure associated with the rotational modulation of the anisotropic component of the polarizability.

11.2.3 Polarized scattering

Atomic fluids

For a dilute monatomic fluid such as argon, polarized scattering is the only possible mechanism. This scattering is associated with translational fluctuations in the isotropic part of the polarizability. This is outside the scope of the treatment in this section—although the generalization, involving consideration of the spatial dependence of the scattered and scattering fields, is straightforward. Although we shall not prove it here, the cross-section is given by

$$\frac{\mathrm{d}^2\sigma}{\mathrm{d}\Omega\mathrm{d}\omega_\mathrm{S}} = (\mathbf{n}_0 \cdot \mathbf{n}_S)^2 N\alpha_1^2 k_S^4 \frac{1}{\pi^{1/2}\bar{\omega}(k)} \exp\left(-\frac{\omega^2}{\bar{\omega}(k)^2}\right), \tag{11.205}$$

where $k = |\mathbf{k}_0 - \mathbf{k}_\mathrm{S}|$ is the scattering wavevector, $\bar{\omega}(k) = (2/\beta m)^{1/2}k$, and m is the atomic mass. The spectrum is therefore Gaussian, with a width of $\bar{\omega}(k)$. For Ar at 25°C, assuming a typical value of $k = 0.02\,\mathrm{nm}^{-1}$, this is approximately $0.035\,\mathrm{cm}^{-1}$, a typical Doppler width.[152]

In dense atomic fluids the approximations used to derive this result break down. As the density is increased at fixed temperature, collective translational fluctuations become important, and cause the single Gaussian of (11.205) to split into three separate peaks.[153] The central, or *Rayleigh*, peak represents quasielastic scattering caused by nonpropagating temperature fluctuations (thermal diffusion), while two side, or *Brillouin*, peaks represent inelastic scattering caused by propagating density fluctuations (sound waves). The latter are separated from the former by $\pm c_\mathrm{S}k$, where c_S is the adiabatic speed of sound. The entire spectrum normally lies within $0.1\,\mathrm{cm}^{-1}$ or less of the exciting line. In the hydrodynamical regime characterized by the limits $k\bar{l} \ll 1$ (where $\bar{l}$ is the mean free path or mean interparticle spacing) and $t \gg \bar{t}$ (where $\bar{t}$ is the mean time between collisions), the calculation of $\mathrm{d}^2\sigma/\mathrm{d}\Omega d\omega_\mathrm{S}$ is straightforwardly accomplished using the techniques of linear hydrodynamics.[155] Outside these limits the calculation is considerably more complex.[153]

Molecular fluids

Polarized Raman spectra of molecular fluids[156] are more complicated than those of atomic fluids because the vibrational molecular motion modulates α directly, and the rotational motion modulates it indirectly through

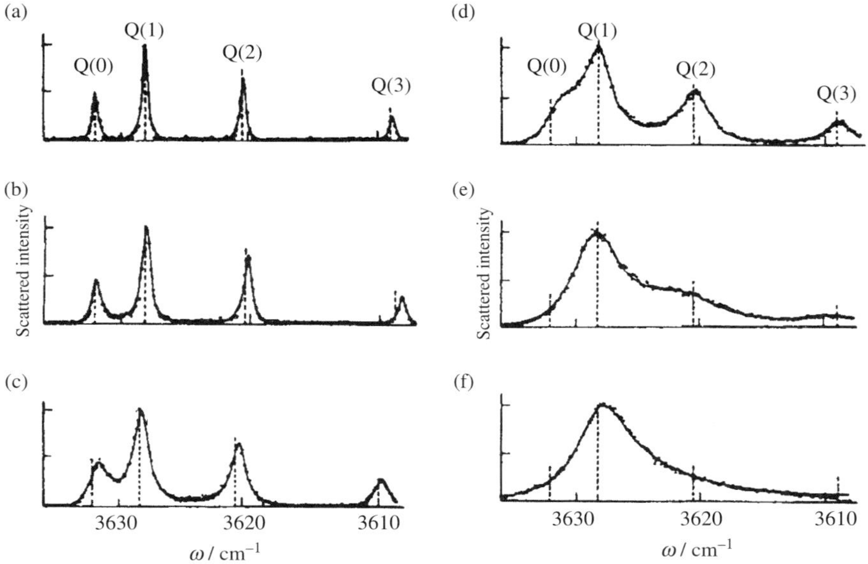

FIG. 11.7 Room-temperature polarized Raman spectrum (Stokes branch) of HD at several densities. Dashed lines indicate the positions of the gas-phase Q-branch lines. (a) 9 amagat; (b) 19 amagat; (c) 49 amagat; (d) 103 amagat; (e) 250 amagat; (f) 490 amagat. Redrawn from ref. 156.

vibration–rotation coupling. Figure 11.7 shows the fundamental ($\Delta v = 1$) Raman spectrum of HD over a range of densities. For a dilute gas, the angular momentum quantum number J is a good quantum number. Because α is a scalar, the rotational selection rule is $\Delta J = 0$, and this leads to scattering at frequencies $\omega = \omega_v + (B' - B)J(J+1)$, where B and B' are the rotational constants for the ground and excited vibrational states. Vibration–rotation coupling in the excited state increases the molecular moment of inertia relative to the ground state, so that $B' < B$. We therefore obtain a sequence of lines at ω_v, $\omega_v - 2(B - B')$, $\omega_v - 6(B - B')$, $\omega_v - 12(B - B')$, etc., labelled $Q(0)$, $Q(1)$, $Q(2)$, and $Q(3)$ in Fig. 11.7(a). Near atmospheric pressure the lineshape is approximately Lorentzian (cf. § 11.1.5). As the pressure increases to about 20 amagat,[70] the lines begin to shift and overlap. At low densities the shifts are to the red, because neighbouring molecules act primarily to weaken the intramolecular bonding through attractive forces. For pressures above 100 amagat, these shifts have disappeared, as short-range repulsive forces also come into play, strengthening the HD bond. At densities of a few hundred amagat, the lines have broadened so much that all structure has disappeared and we are left with a broad, featureless, asymmetric band. Line-broadening mechanisms are discussed in more detail by Clouter.[156]

In the hydrodynamic regime Mountain[13,157,158] has calculated polarized Rayleigh lineshapes for dense fluids by generalizing the hydrodynamical equations to incorporate a frequency-dependent bulk viscosity associated with fluctuations in the internal molecular energy. Since the timescale of these fluctuations is about 10 ps, the spectrum contains an extra diffusive mode with a width of $1/\tau \sim 0.1\,\text{ps}^{-1}$. This is usually much larger than the width of the Rayleigh line, and often overlaps the Brillouin lines.

11.2.4 Depolarized scattering

Dilute gases

For simplicity we will first consider scattering by a gas of diatomic molecules with no spin or orbital angular momentum (e.g. H_2, N_2). We will assume separability of vibration and rotation. The depolarized spectrum is

$$I^{(2)}(\omega) = \sum_{\mu\nu}\sum_{\alpha\beta} P_\mu \left|\left\langle \mu \left| \alpha^{(2)}_{\alpha\beta} \right| \nu \right\rangle\right|^2 \delta(\omega_{\nu\mu} - \omega), \tag{11.206}$$

where $|\mu\rangle = |vJM\rangle$, $|\nu\rangle = |v'J'M'\rangle$ are the rovibrational eigenfunctions. Now from (A.230), we have

$$\sum_{\alpha\beta} \alpha^{(2)\,2}_{\alpha\beta} = \sum_m |\alpha_{2m}|^2, \tag{11.207}$$

where the spherical polarizability tensor $\alpha_{2m} = \frac{1}{3}\gamma T_{2m}$, and T_{2m} is defined by (A.253). We can therefore write

$$I^{(2)}(\omega) = \sum_{vv'} I_{vv'}(\omega), \tag{11.208}$$

where

$$I_{vv'}(\omega) = \frac{8\pi}{15}|\langle v|\gamma|v'\rangle|^2 \sum_{JJ'}\sum_{MmM'} P_{vJ}|\langle JM|Y_{2m}|J'M'\rangle|^2 \delta(\omega_{v'J',vJ} - \omega). \tag{11.209}$$

Using (11.432) this is

$$I_{vv'}(\omega) = \frac{2}{3}|\langle v|\gamma|v'\rangle|^2 \sum_{JJ'} P_{vJ}(2J+1)(2J'+1)\begin{pmatrix} J & 2 & J' \\ 0 & 0 & 0 \end{pmatrix}^2 \delta(\omega_{v'J',vJ} - \omega), \tag{11.210}$$

or equivalently

$$I_{vv'}(\omega) = |\langle v|\gamma|v'\rangle|^2 \sum_{JJ'} P_{vJ}\left[\frac{J(J-1)}{2J-1}\delta_{J',J-2} + \frac{2J(J+1)(2J+1)}{3(2J-1)(2J+3)}\delta_{J',J} + \frac{(J+1)(J+2)}{2J+3}\delta_{J',J+2}\right]\delta(\omega_{v'J',vJ} - \omega). \tag{11.211}$$

Neglecting centrifugal distortion, and assuming that the vibrational motion is harmonic with frequency ω_v, $P_{vJ} = \exp(-v\beta\hbar\omega_v)\exp[-BJ(J+1)]/Q$, where B

is the rotational constant and Q is the partition function. Assuming $\omega_v \gg kT$, only $v = 0$ contributes significantly to the sum in (11.208).

By analogy with (11.40), we have

$$\gamma = \gamma^0 + \gamma' q + \ldots, \qquad (11.212)$$

where $q = r - r_e$ is the vibrational coordinate. If we neglect terms of order q^2 and above, $|\langle 0|\gamma|0\rangle|^2 = \gamma^{0\,2}$, $|\langle 0|\gamma|1\rangle|^2 = \gamma'^2 |\langle 0|q|1\rangle|^2 = \gamma'^2 \langle 0|q^2|0\rangle = \gamma'^2 \hbar/2\mu\omega_v$ (where μ is the oscillator reduced mass), and $|\langle 0|\gamma|v'\rangle|^2 = 0$ for $v' \geq 2$. We therefore have the vibrational selection rule $\Delta v = 0, \pm 1$. Thus the spectrum consists of a pure rotational band (the Rayleigh spectrum) and a fundamental vibrational band (the Raman spectrum).

From (11.211) we have the rotational selection rule $\Delta J \equiv J' - J = -2, 0, 2$, corresponding to O-, Q-, and S-branches of the spectrum, respectively. We note that the rotational selection rule differs from that for infrared absorption (§ 11.1.5). Thus, whereas in the IR spectrum the line spacing is $2B$, in the Raman it is $4B$ (Fig. 11.8). Another important difference between Raman and IR absorption

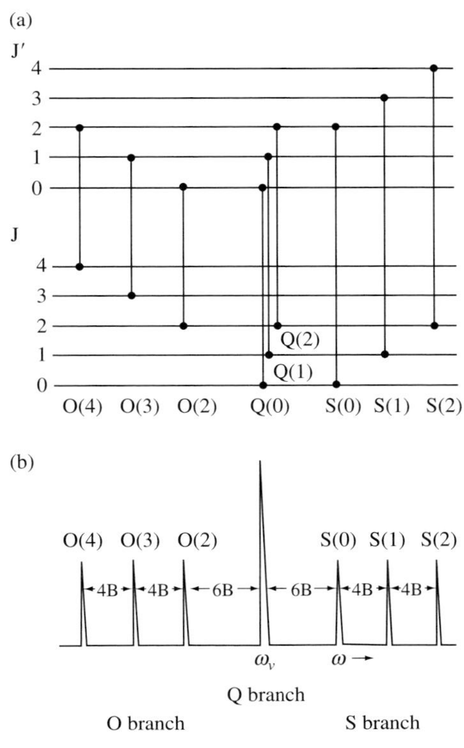

FIG. 11.8 (a) Energy levels and allowed transitions of the fundamental ($\Delta v = 1$) Raman spectrum of a linear molecule. (b) Schematic spectrum.

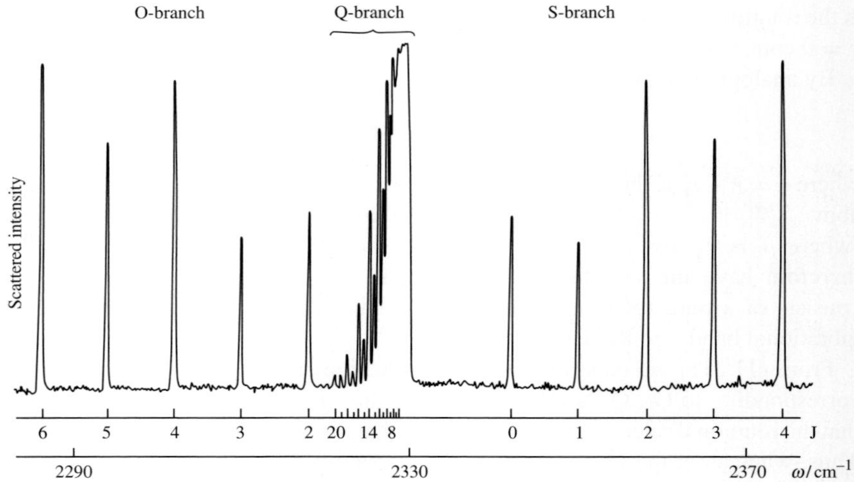

FIG. 11.9 Fundamental depolarized Raman spectrum (Stokes branch) of N_2 gas. Reprinted with permission from ref. 68. Copyright 1995, Oxford University Press.

spectra is that the former are displayed by *all* diatomic molecules, and not just those with permanent dipole moments. Thus light scattering studies may be used to measure molecular constants unobtainable from infrared spectroscopy, e.g. B for homonuclear diatomics.

The transitions $\Delta v = +1$ and $\Delta v = -1$ give rise to the *Stokes* and *anti-Stokes* branches of the fundamental band, respectively. Detailed balance considerations (cf. (11.26)) dictate that the low-frequency anti-Stokes branch is much weaker than the high-frequency Stokes branch.

The Raman spectrum of N_2 (Stokes branch) is shown in Fig. 11.9. The intensity alternation in the O- and S-branches is caused by the effects of nuclear spin; the splitting in the Q-branch is caused by vibration–rotation coupling.

For diatomic molecules possessing spin or orbital angular momentum, e.g. NO ($^2\Pi$) or O_2 $\left({}^3\Sigma_g^-\right)$, transitions with $\Delta J = \pm 1$ are also permitted.[73]

For polyatomic molecules the spectra are considerably more complicated.[73] We consider first the Rayleigh, or pure rotational, spectrum. Naturally this is absent for a (rigid) spherical top, since the polarizability is purely isotropic. For linear polyatomic molecules, it is the same as for a diatomic.

For a symmetric top, the selection rules are

$$\Delta J = 0, \pm 1, \pm 2, \qquad \Delta K = 0, \tag{11.213}$$

where K is the component of angular momentum along the symmetry axis. Here $\Delta J = \pm 1$ is forbidden if $K = 0$. Thus there are O-, P-, Q-, R-, and S-branches. The selection rules (11.213) follow from a consideration of the matrix elements

$$\langle JKM|\alpha_{2m}|J'K'M'\rangle \propto \int \mathrm{d}\Omega\, D^{J}_{MK}(\Omega)^{*} D^{J}_{m0}(\Omega) D^{J}_{M'K'}(\Omega)$$
$$\propto \begin{pmatrix} J & 2 & J' \\ \underline{M} & m & M' \end{pmatrix} \begin{pmatrix} J & 2 & J' \\ \underline{K} & 0 & K' \end{pmatrix}, \tag{11.214}$$

where the second line follows from (A.94) and (A.139) and $\underline{m} \equiv -m$. From the triangle rule (A.131) we thus have $|\Delta J| \equiv |J' - J| \leq 2$, while from (A.140) $\Delta K \equiv K' - K = 0$. Notice that the parity selection rule (A.155), which forbids the transition $\Delta J = \pm 1$ for *linear* molecules, applies only when $K = 0$, so that P- and R-branches appear for $K \neq 0$. The line spacing is therefore $2B$, as in the infrared absorption spectrum. Since $\omega_{J'K,JK} = B[J'(J' + 1) - J(J + 1)]$, we see that P- and R-branch lines originating in levels of even J coincide with the lines of the O- and S- branches. This yields an alternation in the intensities of successive lines unconnected with nuclear spin.

For an asymmetric top we also have the selection rules $\Delta J = 0, \pm 1, \pm 2$. Within each branch the spectrum is, however, very complicated.[73]

For the Raman, or vibration–rotation, spectrum of polyatomic molecules, we have the vibrational selection rule $\Delta v = \pm 1$ as before. For parallel bands of linear polyatomics, the rotational selection rule is $\Delta J = 0, \pm 2$; for perpendicular bands $\Delta J = \pm 1$ is also allowed. For spherical, symmetric, and asymmetric tops we also have the selection rule $\Delta J = 0, \pm 1, \pm 2$, although for totally symmetric vibrations of spherical tops there is no depolarized scattering since $\boldsymbol{\alpha}^{(2)\prime} = 0$. For symmetric tops we have the additional selection rule[73] $\Delta K = 0, \pm 1, \pm 2$, although not all these sub-branches may appear, depending on the value of $\boldsymbol{\alpha}^{(2)\prime}$. The spectra of asymmetric tops are discussed by Hills *et al.*[159]

Dense gases and liquids

Figure 11.10 shows the depolarized Raman spectrum of N_2 over a range of densities from 2 to 82 amagat. For pressures near atmospheric each line of the spectrum is resolved; the lineshape is approximately Lorentzian, with a linewidth proportional to the density. Above about 20 amagat, the lines begin to overlap, and at the highest pressures the spectrum is a featureless broad band.

A detailed treatment of the factors affecting scattering lineshapes in dense media is outside the scope of this chapter. The subject has been reviewed by Weber,[73] Kivelson and Madden,[32] Gelbart,[161] and Bauer *et al.*[162] More recently, instantaneous normal mode analysis[163] has been applied to the calculation of polarizability correlation functions[164,165] and Rayleigh and Raman lineshapes.[166] The theory is based on the observation that liquids are solid-like at short times, and correctly reproduces moments through M_4. It is most likely to be useful in the supercooled regime.

11.2.5 Allowed spectral moments

In this section we derive classical expressions for the even moments of allowed Rayleigh and Raman spectra (classically, the odd moments vanish). Quantum expressions for the moments are given in § 11.2.6.

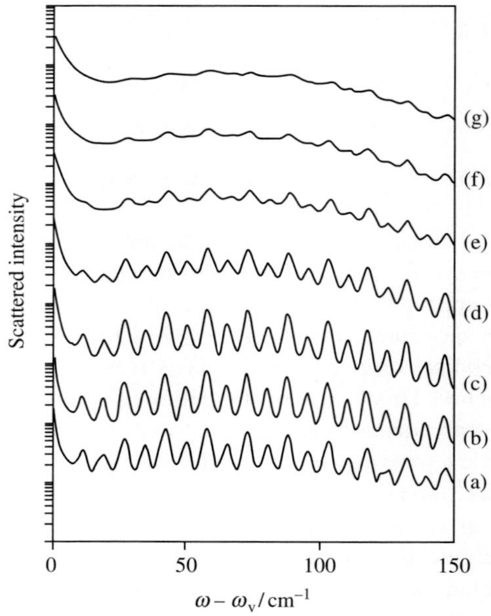

FIG. 11.10 Fundamental depolarized Raman spectrum (Stokes S-branch) of N_2 at 323 K as a function of density. (a) 2 amagat; (b) 4 amagat; (c) 8 amagat; (d) 21 amagat; (e) 40 amagat; (f) 62 amagat; (g) 82 amagat. (From ref. 160.)

The spectral moments can be obtained from the short-time behaviour of the correlation functions of (11.200) by making the substitutions

$$\pi^{(0)}(t) \to \alpha(t) = \sum_i \alpha_i(t),$$

$$\boldsymbol{\pi}^{(2)}(t) \to \boldsymbol{\alpha}^{(2)}(t) = \sum_i \boldsymbol{\alpha}_i^{(2)}(t), \tag{11.215}$$

which follow from (11.188). The isotropic and anisotropic components of the permanent molecular polarizability $\boldsymbol{\alpha}_i(t)$ can be expanded in the vibrational normal coordinates $q_{i\lambda}$ according to

$$\alpha_i(t) = \alpha_i^0 + \sum_\lambda \frac{\partial \alpha_i}{\partial q_{i\lambda}} q_{i\lambda}(t) + \ldots,$$

$$\boldsymbol{\alpha}_i^{(2)}(t) = \boldsymbol{\alpha}_i^{0\,(2)}(t) + \sum_\lambda \frac{\partial \boldsymbol{\alpha}_i^{(2)}(t)}{\partial q_{i\lambda}} q_{i\lambda}(t) + \ldots. \tag{11.216}$$

Rayleigh spectra are associated with the leading terms in (11.216), and Raman spectra with the terms linear in q.

From (11.5), it follows that the Rayleigh spectral moments are given by

$$\begin{aligned}
M_0^{(0)} &= \langle \alpha^0 \alpha^0 \rangle, & M_0^{(2)} &= \langle \boldsymbol{\alpha}^{0\,(2)} : \boldsymbol{\alpha}^{0\,(2)} \rangle, \\
M_1^{(0)} &= i \langle \dot{\alpha}^0 \alpha^0 \rangle, & M_1^{(2)} &= i \langle \dot{\boldsymbol{\alpha}}^{0\,(2)} : \boldsymbol{\alpha}^{0\,(2)} \rangle, \\
M_2^{(0)} &= \langle \dot{\alpha}^0 \dot{\alpha}^0 \rangle, & M_2^{(2)} &= \langle \dot{\boldsymbol{\alpha}}^{0\,(2)} : \dot{\boldsymbol{\alpha}}^{0\,(2)} \rangle, \\
M_3^{(0)} &= i \langle \ddot{\alpha}^0 \dot{\alpha}^0 \rangle, & M_3^{(2)} &= i \langle \ddot{\boldsymbol{\alpha}}^{0\,(2)} : \dot{\boldsymbol{\alpha}}^{0\,(2)} \rangle, \\
M_4^{(0)} &= \langle \ddot{\alpha}^0 \ddot{\alpha}^0 \rangle, & M_4^{(2)} &= \langle \ddot{\boldsymbol{\alpha}}^{0\,(2)} : \ddot{\boldsymbol{\alpha}}^{0\,(2)} \rangle,
\end{aligned} \tag{11.217}$$

etc., where $\alpha^0 = \sum_i \alpha_i^0$ and $\boldsymbol{\alpha}^{0\,(2)} = \sum_i \boldsymbol{\alpha}_i^{0\,(2)}$. Note that the Rayleigh moments will contain both self (one-body) and distinct (two-body) terms. We will normalize these moments with respect to the self parts of $M_0^{(0)}$ and $M_0^{(2)}$. Thus we define

$$\hat{M}_n^{(0)} = M_n^{(0)} / N\alpha_1^{0\,2}, \qquad \hat{M}_n^{(2)} = M_n^{(2)} / N\beta_1^{0\,2}, \tag{11.218}$$

where $\beta_1^{0\,2} \equiv 2\gamma_1^{0\,2}/3 = \boldsymbol{\alpha}_1^{0\,(2)} : \boldsymbol{\alpha}_1^{0\,(2)}$.

If vibration–rotation coupling can be ignored, we may define shifted Raman spectral moments, calculated with respect to $\omega = \omega_{v\lambda}$ as origin, which are to a good approximation the products of rotational and vibrational moments (see § 11.1.6). Provided the vibrational motion of different molecules is uncorrelated, these shifted moments contain only self terms, and are given by

$$\left[M_n^{(0)}\right]_\lambda' = \left[M_{rn}^{(0)}\right]_\lambda \left\langle q_{1\lambda}^2 \right\rangle, \qquad \left[M_n^{(2)}\right]_\lambda' = \left[M_{rn}^{(2)}\right]_\lambda \left\langle q_{1\lambda}^2 \right\rangle, \tag{11.219}$$

where $\left[M_{rn}^{(0)}\right]_\lambda$ and $\left[M_{rn}^{(2)}\right]_\lambda$ are moments of the single-body rotational correlation functions

$$\begin{aligned}
\left[C_r^{(0)}(t)\right]_\lambda &= N \left\langle \frac{\partial \alpha_1(t)}{\partial q_{1\lambda}} \frac{\partial \alpha_1(0)}{\partial q_{1\lambda}} \right\rangle, \\
\left[C_r^{(2)}(t)\right]_\lambda &= N \left\langle \left(\frac{\partial \boldsymbol{\alpha}_1^{(2)}(t)}{\partial q_{1\lambda}} \right) : \left(\frac{\partial \boldsymbol{\alpha}_1^{(2)}(0)}{\partial q_{1\lambda}} \right) \right\rangle.
\end{aligned} \tag{11.220}$$

If we define normalized vibrational moments by $\left[\hat{M}_n^{(0)}\right]_\lambda' \equiv \left[M_n^{(0)}\right]_\lambda' / \left[M_0^{(0)}\right]_\lambda$ and $\left[\hat{M}_n^{(2)}\right]_\lambda' \equiv \left[M_n^{(2)}\right]_\lambda' / \left[M_0^{(2)}\right]_\lambda$, it follows that

$$\left[\hat{M}_n^{(0)}\right]_\lambda' = \hat{M}_{sn}^{(0)}, \qquad \left[\hat{M}_n^{(2)}\right]_\lambda' = \hat{M}_{sn}^{(2)}, \tag{11.221}$$

where $\hat{M}_{sn}^{(0)}$ and $\hat{M}_{sn}^{(2)}$ are the self parts of the normalized Rayleigh moments.

Polarized spectrum

From the discussion in § 11.2.3, polarized scattering depends strongly on the scattering wavevector. For an atomic fluid, the width $\bar{\omega}(k)$ is proportional to k, and is typically $0.1\,\mathrm{cm}^{-1}$ or less. The spectrum is therefore quasielastic, and the normalized moments $\hat{M}_n^{(0)} \sim \bar{\omega}(k)^n$, $n \geq 1$, are essentially zero. For molecular fluids there are additional contributions from internal relaxation, but these are also very small. Polarized moments yield little information about molecular properties.

For the zeroth moment of the Rayleigh spectrum, one can show that

$$M_0^{(0)} = N\alpha_1^{0\,2}\beta^{-1}\rho\chi, \tag{11.222}$$

where χ is the isothermal compressibility. This result was first derived by Einstein.[167] At the critical point of the fluid, $\chi \to \infty$; therefore $M_0^{(0)}$ diverges, and the fluid scatters so much light it becomes opalescent.[168]

The zeroth Raman moment contains only self terms, and is given by the very simple expression

$$\left[M_0^{(0)}\right]_\lambda = Na_{1\lambda}^2, \tag{11.223}$$

where

$$a_{1\lambda}^2 = \left(\frac{\partial\alpha_1}{\partial q_{1\lambda}}\right)^2 \left\langle q_{1\lambda}^2\right\rangle. \tag{11.224}$$

Note that the Raman intensity remains finite at the critical point.

Depolarized spectrum

Calculation of the moments of the depolarized light scattering spectrum is considerably more tedious than that of the corresponding infrared spectral moments, because one must compute time derivatives of a second-rank tensor. Gordon has given results valid for arbitrary polyatomic molecules.[169] Here we derive expressions only for *linear* molecules, for which (cf. (C.27))

$$\boldsymbol{\alpha}_i^{0\,(2)} = \frac{1}{3}\gamma_1^0(3\mathbf{u}_i\mathbf{u}_i - \mathbf{1}), \tag{11.225}$$

where $\mathbf{u}_i$ is a unit vector lying along the internuclear axis of molecule i, and $\gamma_1^0 = \alpha_{1\parallel}^0 - \alpha_{1\perp}^0$. The first and second time derivatives are

$$\begin{aligned}\dot{\boldsymbol{\alpha}}_i^{0\,(2)} &= \gamma_1^0\,(\dot{\mathbf{u}}_i\mathbf{u}_i + \mathbf{u}_i\dot{\mathbf{u}}_i),\\ \ddot{\boldsymbol{\alpha}}_i^{0\,(2)} &= \gamma_1^0\,(\ddot{\mathbf{u}}_i\mathbf{u}_i + 2\dot{\mathbf{u}}_i\dot{\mathbf{u}}_i + \mathbf{u}_i\ddot{\mathbf{u}}_i),\end{aligned} \tag{11.226}$$

where (cf. § 11.1.6)

$$\dot{\mathbf{u}}_i = \boldsymbol{\Omega}_i \times \mathbf{u}_i, \qquad \ddot{\mathbf{u}}_i = I^{-1}\boldsymbol{\tau}_i \times \mathbf{u}_i - \Omega_i^2\mathbf{u}_i. \tag{11.227}$$

Here I, Ω_i, and $\boldsymbol{\tau}_i$ denote the moment of inertia and angular velocity of, and torque acting upon, molecule i.

Evaluation of $M_0^{(2)}$ From (11.217), the zeroth moment of the Rayleigh spectrum is

$$M_0^{\text{Ray}} = N\beta_1^{0\,2} g_2, \tag{11.228}$$

where the orientational correlation parameter g_2 is given by

$$g_2 = 1 + \rho \int d\mathbf{r}_{12} \, \langle g(\mathbf{r}\omega_1\omega_2)\mathbf{U}_1 : \mathbf{U}_2 \rangle_{\omega_1\omega_2}, \tag{11.229}$$

and $\mathbf{U}_i \equiv \boldsymbol{\alpha}_i^{0\,(2)}/\beta_1^0$. For a linear molecule, using (11.225), this becomes

$$g_2 = 1 + \rho \int d\mathbf{r}_{12} \, \langle g(\mathbf{r}\omega_1\omega_2) P_2(\mathbf{u}_1 \cdot \mathbf{u}_2) \rangle_{\omega_1\omega_2}. \tag{11.230}$$

Thus the zeroth moment yields information on intermolecular correlations that complements that obtainable from the FIR spectrum (cf. (11.65), (11.66)). Note that $g_2 < 1$, as the distinct contribution to M_0^{Ray}, is negative; while this is usually small, it can become significant at liquid densities.[170]

For nonpolar molecules, M_0^{Ray} is simply related to the zero-frequency Kerr constant (§ 10.8):

$$K(0) = \frac{2\pi\beta}{15V} M_0^{\text{Ray}}. \tag{11.231}$$

Deviations from this relation may reflect the operation of collision-induced effects.

For the Raman spectrum, where there are only self terms, we have

$$M_0^{\text{Ram}} = N\beta_1^{0\,2}. \tag{11.232}$$

Evaluation of $M_2^{(2)}$ From (11.217), (11.226), and (11.227), the self terms of the second Rayleigh depolarized moment are

$$M_2^{\text{Ray}} = 2N\gamma_1^{0\,2} \langle (\boldsymbol{\Omega} \times \mathbf{u})^2 \rangle = N\gamma_1^{0\,2} \frac{4kT}{I}. \tag{11.233}$$

There are no distinct terms since $\langle \boldsymbol{\Omega}_i \boldsymbol{\Omega}_j \rangle = \langle \boldsymbol{\Omega}_i \rangle \langle \boldsymbol{\Omega}_j \rangle = 0$. The normalized second moment is therefore

$$\hat{M}_2^{\text{Ray}} = \frac{6kT}{I}. \tag{11.234}$$

For parallel and perpendicular Raman bands we have

$$\hat{M}_2^{\parallel} = \frac{6kT}{I}, \qquad \hat{M}_2^{\perp} = \frac{3kT}{I}. \tag{11.235}$$

For an asymmetric top, Gordon[169] has shown that

$$\hat{M}_2 = kT(8\mathrm{Tr}\,\mathbf{B}_1 - 12\mathrm{Tr}\,\mathbf{U}_1 \cdot \mathbf{B}_1 \cdot \mathbf{U}_1), \tag{11.236}$$

where the rotational tensor $\mathbf{B}_1 \equiv \frac{1}{2}\mathbf{I}_1^{-1}$, and $\mathbf{I}_1$ is the inertia tensor. For Rayleigh spectra $\mathbf{U}_i \equiv \boldsymbol{\alpha}_i^{0\,(2)}/\beta_1^0$, while for the Raman band λ, $\mathbf{U}_i \equiv \boldsymbol{\alpha}_{i\lambda}^{(2)}/\beta_{1\lambda}$, where $\boldsymbol{\alpha}_{i\lambda}^{(2)} = \partial\boldsymbol{\alpha}_i^{(2)}/\partial q_{i\lambda}$ and $\beta_{1\lambda}^2 = \boldsymbol{\alpha}_{1\lambda}^{(2)} : \boldsymbol{\alpha}_{1\lambda}^{(2)}$.

For a symmetric top (11.236) reduces to

$$\hat{M}_2 = 6kT(\mathrm{Tr}\,\mathbf{B}_1 - \mathbf{u}_1\mathbf{u}_1 : \mathbf{B}_1). \tag{11.237}$$

For the Rayleigh band, or for a parallel Raman band, therefore,

$$\hat{M}_2^{\parallel} = \frac{6kT}{I_\perp}, \tag{11.238}$$

whereas for perpendicular Raman bands

$$\hat{M}_2^{\perp} = 3kT\left(\frac{1}{I_\parallel} + \frac{1}{I_\perp}\right). \tag{11.239}$$

Comparison of (11.238) and (11.239) with (11.79) and (11.80) shows that for symmetric tops the normalized second moment of the depolarized spectrum is three times the normalized second moment of the corresponding infrared band. The same relation obviously holds for linear molecules and for spherical tops.

Spherical tops have no depolarized Rayleigh spectrum since $\boldsymbol{\alpha}_1^{0\,(2)} = 0$. For the Raman spectrum

$$\hat{M}_2^{\mathrm{Ram}} = \frac{6kT}{I}. \tag{11.240}$$

Classically, the second Rayleigh and Raman moments depend only on the temperature, and not the molecular environment. As shown in § 11.2.6, this is no longer true quantum mechanically.

Experimental second moments of Rayleigh and Raman spectra of dense fluids are generally somewhat larger than predicted by the above equations. This reflects, primarily, the operation of collision-induced effects (§ 11.2.7). For liquid N_2 and O_2 the differences in M_2^{Ram} are on the order of 5%, while for CO_2 the difference is about 20%.[171] For M_2^{Ray} the differences appear to be larger.[172]

Evaluation of $M_4^{(2)}$ In general, we may expect the fourth moment of the depolarized Rayleigh spectrum to contain both self and distinct terms, i.e.

$$\hat{M}_4^{\mathrm{Ray}} = \hat{M}_{s4}^{\mathrm{Ray}} + \hat{M}_{d4}^{\mathrm{Ray}}. \tag{11.241}$$

For a linear molecule

$$\hat{M}_{s4}^{\text{Ray}} = 3\left\langle \ddot{u}_1^2 + (\ddot{\mathbf{u}}_1 \cdot \mathbf{u}_1)^2 + 2\dot{u}_1^4 \right\rangle \tag{11.242}$$

and

$$\begin{aligned}\hat{M}_{d4}^{\text{Ray}} &= 3(N-1)\langle(\ddot{\mathbf{u}}_1 \cdot \ddot{\mathbf{u}}_2)(\mathbf{u}_1 \cdot \mathbf{u}_2) + (\ddot{\mathbf{u}}_1 \cdot \mathbf{u}_2)(\mathbf{u}_1 \cdot \ddot{\mathbf{u}}_2) \\ &\quad + 2(\ddot{\mathbf{u}}_1 \cdot \dot{\mathbf{u}}_2)(\mathbf{u}_1 \cdot \dot{\mathbf{u}}_2) + 2(\dot{\mathbf{u}}_1 \cdot \ddot{\mathbf{u}}_2)(\dot{\mathbf{u}}_1 \cdot \mathbf{u}_2) + 2(\dot{\mathbf{u}}_1 \cdot \dot{\mathbf{u}}_2)^2\rangle.\end{aligned} \tag{11.243}$$

We will consider the distinct terms first. Using (E.5), and noting that $\langle \boldsymbol{\Omega}_i \rangle = 0$, we can easily show that

$$\langle(\ddot{\mathbf{u}}_1 \cdot \ddot{\mathbf{u}}_2)(\mathbf{u}_1 \cdot \mathbf{u}_2)\rangle = \langle(\ddot{\mathbf{u}}_1 \cdot \mathbf{u}_2)(\mathbf{u}_1 \cdot \ddot{\mathbf{u}}_2)\rangle = -\langle(\ddot{\mathbf{u}}_1 \cdot \dot{\mathbf{u}}_2)(\mathbf{u}_1 \cdot \dot{\mathbf{u}}_2)\rangle = \langle(\dot{\mathbf{u}}_1 \cdot \dot{\mathbf{u}}_2)^2\rangle. \tag{11.244}$$

Therefore $\hat{M}_{d4}^{\text{Ray}} = 0$, and the fourth Rayleigh moment contains *only* self terms; this is not generally appreciated.[84,171,173,174]

From (11.227), $\dot{u}_1^2 = -\ddot{\mathbf{u}}_1 \cdot \mathbf{u}_1 = \Omega_1^2$. Hence from (11.242)

$$\hat{M}_4^{\text{Ray}} = 3\left(\left\langle \ddot{u}_1^2 \right\rangle + 3\left\langle \Omega_1^4 \right\rangle\right), \tag{11.245}$$

and using (11.86) and the fact that $\langle \Omega^4 \rangle = 8(kT/I)^2$, we have

$$\hat{M}_4^{\text{Ray}} = 3\frac{\left\langle \tau_1^2 \right\rangle}{I^2} + 96\left(\frac{kT}{I}\right)^2. \tag{11.246}$$

Thus this moment depends on the mean square torque $\left\langle \tau_1^2 \right\rangle$ acting on a molecule. However, collision-induced effects typically increase the fourth moments of dense fluids by 10–40%.[171] In view of this, the prospects of extracting accurate mean square torques from experimental moments seem rather bleak.[175]

Equation (11.246) also determines the reduced fourth moment of a parallel Raman band.

Gordon[169] has calculated the fourth moment for an asymmetric top. The result is

$$\begin{aligned}\hat{M}_{s4}^{\text{Ray}} \equiv \hat{M}_4^{\text{Ram}} &= 4(kT)^2 \Big[5\mathbf{B}_1 : \mathbf{B}_1 + 27(\operatorname{Tr}\mathbf{B}_1)^2 \\ &\quad - 72\operatorname{Tr}\mathbf{U}_1 \cdot \mathbf{B}_1 \cdot \mathbf{U}_1 \operatorname{Tr}\mathbf{B}_1 + 6\operatorname{Tr}\mathbf{U}_1 \cdot \mathbf{B}_1 \cdot \mathbf{B}_1 \cdot \mathbf{U}_1 \\ &\quad + 18\operatorname{Tr}\mathbf{U}_1 \cdot \mathbf{B}_1 \cdot \mathbf{U}_1 \cdot \mathbf{B}_1 + 4\mathbf{B}_1 : \mathbf{B}_1 \operatorname{Tr}\mathbf{B}_1^{-1} \operatorname{Tr}\mathbf{B}_1 \\ &\quad - 6\operatorname{Tr}\mathbf{U}_1 \cdot \mathbf{B}_1 \cdot \mathbf{B}_1 \cdot \mathbf{U}_1 \operatorname{Tr}\mathbf{B}_1^{-1} \operatorname{Tr}\mathbf{B}_1 \\ &\quad - 4\operatorname{Tr}\mathbf{B}_1 \cdot \mathbf{B}_1 \cdot \mathbf{B}_1 \operatorname{Tr}\mathbf{B}_1^{-1} + 6\operatorname{Tr}\mathbf{U}_1 \cdot \mathbf{B}_1 \cdot \mathbf{B}_1 \cdot \mathbf{B}_1 \cdot \mathbf{U}_1 \operatorname{Tr}\mathbf{B}_1^{-1} \Big] \\ &\quad + 16\langle \boldsymbol{\tau}_1 \cdot \mathbf{B}_1 \cdot \mathbf{B}_1 \cdot \boldsymbol{\tau}_1 \rangle - 24\langle \boldsymbol{\tau}_1 \cdot \mathbf{B}_1 \cdot \mathbf{U}_1 \cdot \mathbf{U}_1 \cdot \mathbf{B}_1 \cdot \boldsymbol{\tau}_1 \rangle\end{aligned} \tag{11.247}$$

and

$$\hat{M}_{d4}^{\text{Ray}} = -8(N-1)(kT)^2\,[6(\mathbf{U}_1 : \mathbf{U}_2)(\mathbf{B}_1 : \mathbf{B}_2) + 2(\mathbf{U}_1 : \mathbf{U}_2)\,\mathrm{Tr}\,\mathbf{B}_1\,\mathrm{Tr}\,\mathbf{B}_2$$
$$+ 6(\mathbf{U}_1 : \mathbf{B}_2)(\mathbf{B}_1 : \mathbf{U}_2) - 12\mathrm{Tr}\,(\mathbf{U}_1 \cdot \mathbf{B}_2 \cdot \mathbf{B}_1 \cdot \mathbf{U}_2)$$
$$- 3\mathrm{Tr}\,(\mathbf{U}_1 \cdot \mathbf{B}_1 \cdot \mathbf{B}_2 \cdot \mathbf{U}_2) - 3\mathrm{Tr}\,(\mathbf{U}_1 \cdot \mathbf{B}_1 \cdot \mathbf{U}_2 \cdot \mathbf{B}_2)]$$
$$+ (N-1)\,[16\langle(\mathbf{U}_1 : \mathbf{U}_2)(\boldsymbol{\tau}_1 \cdot \mathbf{B}_1 \cdot \mathbf{B}_2 \cdot \boldsymbol{\tau}_2)\rangle$$
$$- 16\langle\boldsymbol{\tau}_1 \cdot \mathbf{B}_1 \cdot \mathbf{U}_2 \cdot \mathbf{U}_1 \cdot \mathbf{B}_2 \cdot \boldsymbol{\tau}_2\rangle - 8\langle\boldsymbol{\tau}_1 \cdot \mathbf{B}_1 \cdot \mathbf{U}_1 \cdot \mathbf{U}_2 \cdot \mathbf{B}_2 \cdot \boldsymbol{\tau}_2\rangle]. \quad (11.248)$$

For a symmetric top, we have

$$\mathbf{U}_i = 6^{-1/2}(3\mathbf{u}_i\mathbf{u}_i - \mathbf{1}),$$
$$\mathbf{B}_i = B_0\mathbf{1} + \frac{1}{3}B_2(3\mathbf{u}_i\mathbf{u}_i - \mathbf{1}), \quad (11.249)$$

where $B_0 = 1/3(B_{\parallel} + 2B_{\perp})$ and $B_2 = B_{\parallel} - B_{\perp}$, with $B_{\parallel} = 1/2I_{\parallel}$, $B_{\perp} = 1/2I_{\perp}$. Therefore

$$\hat{M}_{s4}^{\text{Ray}} \equiv \hat{M}_4^{\text{Ram}} = 3\frac{\langle\tau_{\perp}^2\rangle}{I^2} + 96\left(\frac{kT}{I_{\perp}}\right)^2\left(1 + \frac{I_{\parallel}}{16I_{\perp}}\right), \quad (11.250)$$

and

$$\hat{M}_{d4}^{\text{Ray}} = -36\left(\frac{kT}{I_{\perp}}\right)^2 (N-1)\langle P_2(\mathbf{u}_1 \cdot \mathbf{u}_2)\rangle + \text{torque terms as in (11.248)}. \quad (11.251)$$

Evaluation of $M_6^{(2)}$ Kalmykov and Titov[98] have given an expression for the sixth moment for molecules of arbitrary shape. Their results are too complicated to reproduce here. Experimental studies of $M_6^{(2)}$ could, in principle, yield information on molecular averages such as $\langle\nabla_\omega\boldsymbol{\tau}\rangle$, $\langle\boldsymbol{\tau}\nabla_\omega\boldsymbol{\tau}\rangle$, and $(\langle\nabla_\omega\boldsymbol{\tau})^2\rangle$, although in practice it would be very hard to separate collision-induced contributions to the spectrum.

11.2.6 *Quantum expressions for the allowed moments*

As discussed in § 11.1.7 for infrared absorption, there are two kinds of quantum corrections to the moments of polarized and depolarized light scattering spectra. Statistical corrections arise from the modification of the distribution functions for position and momentum; dynamical corrections arise from the noncommutativity of the operators for position and momentum. Otherwise the formulae of (11.217)–(11.221) remain valid, provided we interpret the time derivatives $\dot{\alpha}^0$ and $\dot{\boldsymbol{\alpha}}^{0\,(2)}$ according to (11.96).

Evaluation of M_0 The zeroth Rayleigh moments have only statistical corrections. Thus M_0^{Ray} is given by expressions similar to (11.222) and (11.228), but now χ and g_2 must be evaluated quantum-mechanically. For linear molecules the quantum g_2 is again given by (11.230), provided the quantum $g(\mathbf{r}\omega_1\omega_2)$ is used in place of its classical analogue.

The Raman zeroth spectral moments contain only self terms, and are unchanged from their classical values.

Evaluation of M_1 As discussed in § 11.2.5, M_1 and higher moments of the polarized spectrum, associated with fluctuations in the isotropic component of the polarizability, are essentially zero.

The first depolarized Rayleigh moment is given by

$$M_1^{\text{Ray}} = \frac{1}{\hbar}\langle \boldsymbol{\alpha}^{0\,(2)} : [H, \boldsymbol{\alpha}^{0\,(2)}]\rangle, \tag{11.252}$$

where H is the system Hamiltonian. Following the arguments used in § 11.1.7, this can be written as the double commutator

$$M_1^{\text{Ray}} = \frac{N}{2\hbar}\left\langle\left[\boldsymbol{\alpha}_1^{0\,(2)} :, \left[K_{r1}, \boldsymbol{\alpha}_1^{0\,(2)}\right]\right]\right\rangle, \tag{11.253}$$

where K_{r1} is the rotational kinetic energy of molecule 1.

For linear molecules, $K_{r1} = J_1^2/2I$, and $\boldsymbol{\alpha}_1^{0\,(2)}$ is given by (11.225). Therefore

$$M_1^{\text{Ray}} = \frac{N\gamma_1^{0\,2}}{4\hbar I}\left\langle\left[\mathbf{u}_1\mathbf{u}_1 :, \left[J_1^2, \mathbf{u}_1\mathbf{u}_1\right]\right]\right\rangle. \tag{11.254}$$

Now

$$\begin{aligned}
[J^2, u_\alpha u_\beta] &= [[J^2, u_\alpha], u_\beta] + u_\alpha[J^2, u_\beta] + u_\beta[J^2, u_\alpha] \\
&= 2i\hbar(\epsilon_{\alpha\gamma\delta}u_\beta u_\gamma J_\delta + \epsilon_{\beta\gamma\delta}u_\alpha u_\gamma J_\delta) + 2\hbar^2(3u_\alpha u_\beta - \delta_{\alpha\beta}) \\
&= 2i\hbar(\epsilon_{\alpha\gamma\delta}J_\delta u_\beta u_\gamma + \epsilon_{\beta\gamma\delta}J_\delta u_\alpha u_\gamma) - 2\hbar^2(3u_\alpha u_\beta - \delta_{\alpha\beta}),
\end{aligned} \tag{11.255}$$

where the second and third equalities follow from (11.102) and (11.103). The second form proves useful in the calculation of M_2 (see below). Therefore

$$[u_\alpha u_\beta, [J^2, u_\alpha u_\beta]] = -2i\hbar(\epsilon_{\alpha\gamma\delta}u_\beta u_\gamma + \epsilon_{\beta\gamma\delta}u_\alpha u_\gamma)[J_\delta, u_\alpha u_\beta] = 8\hbar^2, \tag{11.256}$$

and thus

$$M_1^{\text{Ray}} = \frac{2N\hbar\gamma_1^{0\,2}}{I}. \tag{11.257}$$

The normalized first moments for parallel and perpendicular Raman bands are

$$\hat{M}_1^{\parallel} \equiv \hat{M}_1^{\text{Ray}} = \frac{3\hbar}{I}, \qquad \hat{M}_1^{\perp} = \frac{3\hbar}{2I}. \tag{11.258}$$

For asymmetric tops a parallel calculation shows[169]

$$M_1^{\text{Ray}} = 2N\hbar\beta_1^{0\,2}\,(2\,\text{Tr}\,\mathbf{B}_1 - 3\,\text{Tr}\,\mathbf{U}_1\cdot\mathbf{B}_1\cdot\mathbf{U}_1). \tag{11.259}$$

For a symmetric top (11.259) reduces to

$$M_1^{\text{Ray}} = 2N\hbar\gamma_1^{0\,2}(\text{Tr}\,\mathbf{B}_1 - \mathbf{u}_1\mathbf{u}_1 : \mathbf{B}_1). \tag{11.260}$$

For the Rayleigh band, or for a parallel Raman band, therefore,

$$\hat{M}_1^{\parallel} = \frac{3\hbar}{I_\perp}, \tag{11.261}$$

whereas for perpendicular Raman bands

$$\hat{M}_1^{\perp} = \frac{3\hbar}{2}\left(\frac{1}{I_\parallel} + \frac{1}{I_\perp}\right). \tag{11.262}$$

Thus the normalized first moment of the depolarized spectrum is three times the value for the corresponding infrared band (§ 11.1.7).

Evaluation of M_2 Substituting (11.96) into (11.217), we find that

$$M_2^{\text{Ray}} = -\frac{1}{\hbar^2}\langle[H, \boldsymbol{\alpha}^{0\,(2)}] : [H, \boldsymbol{\alpha}^{0\,(2)}]\rangle. \tag{11.263}$$

We can replace H by the rotational kinetic energy $K_r = \sum_i K_{ri}$. Note that M_2^{Ray} will contain cross terms, since $\left\langle\left[K_{ri}, \boldsymbol{\alpha}_i^{0\,(2)}\right] : \left[K_{rj}, \boldsymbol{\alpha}_j^{0\,(2)}\right]\right\rangle$ does not vanish for $i \neq j$.

For linear molecules the self terms are given by

$$M_{s2}^{\text{Ray}} = -\frac{N\gamma_1^{0\,2}}{(2\hbar I)^2}\left\langle\left[J_1^2, \mathbf{u}_1\mathbf{u}_1\right] : \left[J_1^2, \mathbf{u}_1\mathbf{u}_1\right]\right\rangle, \tag{11.264}$$

and the distinct terms by

$$M_{d2}^{\text{Ray}} = -\frac{N(N-1)\gamma_1^{0\,2}}{(2\hbar I)^2}\left\langle\left[J_1^2, \mathbf{u}_1\mathbf{u}_1\right] : \left[J_2^2, \mathbf{u}_2\mathbf{u}_2\right]\right\rangle. \tag{11.265}$$

We will examine the self terms in more detail. From (11.255),

$$[J^2, \mathbf{uu}] : [J^2, \mathbf{uu}] = \Big(2i\hbar(\epsilon_{\alpha\gamma\delta}J_\delta u_\beta u_\gamma + \epsilon_{\beta\gamma\delta}J_\delta u_\alpha u_\gamma) - 2\hbar^2(3u_\alpha u_\beta - \delta_{\alpha\beta})\Big)$$
$$\times \Big(2i\hbar(\epsilon_{\alpha\gamma'\delta'}u_\beta u_{\gamma'}J_{\delta'} + \epsilon_{\beta\gamma'\delta'}u_\alpha u_{\gamma'}J_{\delta'}) + 2\hbar^2(3u_\alpha u_\beta - \delta_{\alpha\beta})\Big). \tag{11.266}$$

We can reduce the products of antisymmetric tensors using(A.202). Noting that $\epsilon_{\alpha\beta\gamma}u_\beta u_\gamma = 0$ and $\mathbf{J}\cdot\mathbf{u} = 0$, we then find that

$$[J^2, \mathbf{uu}] : [J^2, \mathbf{uu}] = -8\hbar^2 J^2 - 24\hbar^4. \tag{11.267}$$

Therefore

$$M_{s2}^{\text{Ray}} = N\gamma_1^{0\,2}\left(4\frac{\langle K_{r1}\rangle}{I} + 6\frac{\hbar^2}{I^2}\right). \tag{11.268}$$

The self part of the normalized Rayleigh second moment, or equivalently the entire normalized second moment of a parallel Raman band, is therefore

$$\hat{M}_{s2}^{\text{Ray}} = 6\frac{\langle K_{r1}\rangle}{I} + 9\frac{\hbar^2}{I^2}. \tag{11.269}$$

A comparison of (11.269) with the classical result (11.234) shows that there are both statistical and dynamical quantum corrections to the second moment. In the semiclassical limit we may generate these corrections explicitly by expanding $\langle K_{r1}\rangle$ in powers of $\hbar^2$. Using (3.75) we find that

$$\hat{M}_{s2}^{\text{Ray}} = \hat{M}_2^{\text{cl}}\left(1 + \frac{4\hbar^2\beta}{3I} + \frac{\hbar^2\beta^3}{24I}\langle\tau^2\rangle_{\text{cl}} + O(\hbar^4)\right), \tag{11.270}$$

where $\langle\tau^2\rangle_{\text{cl}}$ is the classical mean square torque and $\hat{M}_2^{\text{cl}} = 6kT/I$. Thus quantum corrections will be important when the torques are large and the moment of inertia is small.

Gordon[169] has derived quantum-mechanical expressions for the second moments of asymmetric tops. The general results are rather complicated. For symmetric tops, the self part of the Rayleigh moment, or equivalently the entire moment of a parallel Raman band, is given by

$$\hat{M}_{s2}^{\text{Ray}} = 6\frac{\langle K_{\text{r}1\perp}\rangle}{I_\perp} + 9\frac{\hbar^2}{I_\perp^2}, \tag{11.271}$$

where $K_{\text{r}1\perp}$ denotes that part of the molecular kinetic energy associated with rotations perpendicular to the symmetry axis. Equation (11.271) may be compared with the classical result (11.238).

In general, quantum corrections to M_2 may more easily be derived using (11.52). Thus

$$M_2 = \frac{2}{\beta\hbar}M_1 + \frac{1}{3}\left(\frac{\beta\hbar}{2}\right)^2 M_4 + O(\hbar^4), \tag{11.272}$$

and since $M_1 \propto \hbar$, it follows that

$$M_2 = M_2^{\text{cl}}\left(1 + \frac{1}{3}\left(\frac{\beta\hbar}{2}\right)^2\frac{\hat{M}_4^{\text{cl}}}{\hat{M}_2^{\text{cl}}} + O(\hbar^4)\right), \tag{11.273}$$

where $\hat{M}_4^{\text{cl}}$ is given by (11.247) and (11.248).

We note that, in contrast to the classical moment, the quantum M_2 depends on the nature of the intermolecular forces as well as the temperature.

Higher moments No quantum calculations of higher moments have yet been attempted. However, M_3 can be estimated using (11.52). This yields

$$M_3 = \frac{\beta\hbar}{2} M_4^{\rm cl}(1 + O(\hbar^2)). \tag{11.274}$$

11.2.7 *Induced spectra*

The phenomenon of collision-induced light scattering, or CILS, is associated with the presence of the terms $\Delta\boldsymbol{\alpha}_{ij}$, $\Delta\boldsymbol{\alpha}_{ijk}$, etc., in the expansion of the effective molecular polarizability $\boldsymbol{\pi}$, Eq. (11.188). In condensed phases collisional polarizabilities affect Rayleigh and Raman scattering significantly. Indeed, in polar liquids CILS is often as strong as, or in some cases even stronger than, the allowed scattering. For example, simulation of the depolarized Rayleigh spectrum of water shows that the induced intensity is some 30 × greater than the allowed intensity.[176] As discussed in § 11.2.5, second moments of the Raman spectra of nonpolar liquids are typically 5–20% larger than expected on the basis of purely allowed contributions,[171] while for Rayleigh spectra even larger discrepancies are found.[172] Similarly, the molecular torques extracted from fourth moments often do not agree particularly well with those derived from IR studies, and this too reflects the operation of collisional processes.[171]

Surprisingly, the first *purely* collisional spectra were not recorded until 1968, when McTague and Birnbaum[177] and Thibeau *et al.*[178] observed depolarized scattering from compressed argon and krypton. The first polarized spectrum was measured even later, in 1979, when Proffitt and Frommhold[179] observed scattering from helium. Since then, more than eight hundred publications have appeared[180] documenting induced scattering by gases[34,104,113,181–183] and liquids.[34,104,113,181]

In almost all cases, collision-induced spectra derive from the pair polarizability $\Delta\boldsymbol{\alpha}_{ij}$. Estimates of the triplet polarizability $\Delta\boldsymbol{\alpha}_{ijk}$ beyond the DID level[184] are available only for a very few systems;[185–187] three-body CILS is usually weak, and has only recently been observed.[188,189]

In this section we discuss the contributions made by the collisional terms to light scattering spectra. For reasons of space, we will largely restrict the discussion to depolarized spectra, which are generally much stronger than their polarized counterparts, and significantly easier to measure.[183] As always, the emphasis is on the spectral moments, which can be calculated within the framework of equilibrium statistical mechanics. Except at very low temperatures, these may be evaluated classically, quantum corrections generally being small.[190,191] Moments can be used to construct approximate spectral lineshapes: see, e.g., refs. 192 and 193 for an application to the polarized and depolarized interaction-induced spectra of N_2 gas.

Using (11.200) and (11.188), and neglecting three-body and higher contributions to $\boldsymbol{\pi}$, the correlation function for depolarized scattering can be written as the sum of allowed, cross, and pure induced terms:

$$C^{(2)}(t) = C^{(2)}_{\rm all}(t) + C^{(2)}_{\rm cross}(t) + C^{(2)}_{\rm ind}(t), \tag{11.275}$$

where

$$C_{\text{all}}^{(2)}(t) = \sum_{ij} \left\langle \boldsymbol{\alpha}_i^{(2)}(t) : \boldsymbol{\alpha}_j^{(2)}(0) \right\rangle,$$

$$C_{\text{cross}}^{(2)}(t) = \frac{1}{2} \sum_{ijk} \left\langle \boldsymbol{\alpha}_i^{(2)}(t) : \Delta\boldsymbol{\alpha}_{jk}^{(2)}(0) + \Delta\boldsymbol{\alpha}_{jk}^{(2)}(t) : \boldsymbol{\alpha}_i^{(2)}(0) \right\rangle,$$

$$C_{\text{ind}}^{(2)}(t) = \frac{1}{4} \sum_{ijkl} \left\langle \Delta\boldsymbol{\alpha}_{ik}^{(2)}(t) : \Delta\boldsymbol{\alpha}_{jl}^{(2)}(0) \right\rangle. \tag{11.276}$$

Rovibrational (Raman) as well as rotational (Rayleigh) induced bands are observed. To simplify our treatment of Raman spectra we will consider a fluid of diatomic molecules. The expansion of the permanent molecular polarizability $\boldsymbol{\alpha}_i$ in terms of the vibrational coordinate $q_i = r_i - r_e$ is given by (11.216); the collisional polarizability $\Delta\boldsymbol{\alpha}_{ij}$ may similarly be expanded about the equilibrium bond lengths of molecules i and j as

$$\Delta\boldsymbol{\alpha}_{ij} = \sum_{m,n=0}^{\infty} \frac{\Delta\boldsymbol{\alpha}_{ij}^{mn}}{m!n!} q_i^m q_j^n, \tag{11.277}$$

where $\Delta\boldsymbol{\alpha}_{ij}^{mn} \equiv \partial^{m+n} \Delta\boldsymbol{\alpha}_{ij} / \partial q_i^m \partial q_j^n$. Rayleigh scattering is associated with the term $m = n = 0$, while scattering in the fundamental Raman band is associated with the terms $m = 0$, $n = 1$ and $m = 1$, $n = 0$.

Rayleigh spectrum

The cross-correlation function is a sum of two- and three-body contributions,

$$\begin{aligned} C_{\text{cross}}^{(2)}(t) = {} & N(N-1) \left\langle \boldsymbol{\alpha}_1^{0\,(2)}(t) : \Delta\boldsymbol{\alpha}_{12}^{00\,(2)}(0) + \Delta\boldsymbol{\alpha}_{12}^{00\,(2)}(t) : \boldsymbol{\alpha}_1^{0\,(2)}(0) \right\rangle \\ & + \frac{1}{2} N(N-1)(N-2) \\ & \times \left\langle \boldsymbol{\alpha}_1^{0\,(2)}(t) : \Delta\boldsymbol{\alpha}_{23}^{00\,(2)}(0) + \Delta\boldsymbol{\alpha}_{23}^{00\,(2)}(t) : \boldsymbol{\alpha}_1^{0\,(2)}(0) \right\rangle, \end{aligned} \tag{11.278}$$

while the pure induced correlation function also contains four-body terms,

$$\begin{aligned} C_{\text{ind}}^{(2)}(t) = {} & \frac{1}{2} N(N-1) \left\langle \Delta\boldsymbol{\alpha}_{12}^{00\,(2)}(t) : \Delta\boldsymbol{\alpha}_{12}^{00\,(2)}(0) \right\rangle \\ & + N(N-1)(N-2) \left\langle \Delta\boldsymbol{\alpha}_{12}^{00\,(2)}(t) : \Delta\boldsymbol{\alpha}_{13}^{00\,(2)}(0) \right\rangle \\ & + \frac{1}{4} N(N-1)(N-2)(N-3) \left\langle \Delta\boldsymbol{\alpha}_{12}^{00\,(2)}(t) : \Delta\boldsymbol{\alpha}_{34}^{00\,(2)}(0) \right\rangle, \end{aligned} \tag{11.279}$$

limiting our ability to compute moments of Rayleigh CILS spectra for liquids, except by direct simulation.

Raman spectrum

Consider now Raman scattering in the fundamental band. If we assume that the vibrational motion decouples adiabatically from the rotational motion, the correlation functions can be separated into vibrational and rotational terms, $C(t) = C_r(t)C_v(t)$. Assuming the vibrational phases of neighbouring molecules are uncorrelated, we have $C_v(t) = \langle q_1(t)q_1(0)\rangle$.

For the rotational cross-correlation function, there are only two-body contributions:

$$C^{(2)}_{\text{r cross}}(t) = N(N-1)\left\langle \boldsymbol{\alpha}'_1(t) : \Delta\boldsymbol{\alpha}^{10\,(2)}_{12}(0) + \Delta\boldsymbol{\alpha}^{10\,(2)}_{12}(t) : \boldsymbol{\alpha}'_1(0)\right\rangle. \tag{11.280}$$

The rotational correlation function for pure induced scattering contains two- and three-body terms:

$$\begin{aligned} C^{(2)}_{\text{r ind}}(t) &= N(N-1)\left\langle \Delta\boldsymbol{\alpha}^{10\,(2)}_{12}(t) : \Delta\boldsymbol{\alpha}^{10\,(2)}_{12}(0)\right\rangle \\ &\quad + N(N-1)(N-2)\left\langle \Delta\boldsymbol{\alpha}^{10\,(2)}_{12}(t) : \Delta\boldsymbol{\alpha}^{10\,(2)}_{13}(0)\right\rangle. \end{aligned} \tag{11.281}$$

Since there is no four-body contribution, this function is much more easily evaluated than its Rayleigh analogue.

Atomic fluids

The collisional polarizability of a pair of atoms separated by a vector $\mathbf{r}$ has the form

$$\Delta\boldsymbol{\alpha}(\mathbf{r}) = \Delta\alpha(r)\mathbf{1} + \frac{1}{3}\Delta\gamma(r)(3\hat{\mathbf{r}}\hat{\mathbf{r}} - \mathbf{1}), \tag{11.282}$$

where $\Delta\alpha$ and $\Delta\gamma$ determine the trace and anisotropy of $\Delta\boldsymbol{\alpha}$, respectively. Various methods have been used to calculate these quantities. The simplest treatment is based on the dipole–induced-dipole (DID) model (§ 10.4). However, this is a purely classical formalism, and ignores both short-range quantum-mechanical exchange and hyperpolarization effects. The latter are associated with the non-linear mutual polarization of the interacting atoms, and significantly affect the trace.[194] Only *ab initio* methods are currently able to compute the pair polarizability accurately at small separations, where the atomic charge clouds overlap. Moszynski *et al.*[195] have performed symmetry-adapted perturbation theory (SAPT)[196] calculations for He, Dacre[197] has performed CISD calculations for He and Ne, Joslin *et al.*[198] have performed second-order Møller-Plesset calculations for Ar, and most recently Hättig *et al.*[199] have performed large-basis set CCSD calculations for He and Ar. Coupled-cluster calculations have also been reported for the mixed pairs He-Ne, He-Ar, and Ne-Ar.[110] For the heavier rare gases, only Hartree-Fock estimates are available.[200]

For the depolarized scattering we will define reduced moments according to

$$M_n^{(2)} = \frac{1}{2}\rho^2 V \left(\frac{2kT}{m}\right)^{n/2} m_n^{(2)}, \tag{11.283}$$

where m is the atomic mass. For a dilute gas we then find,[198] on substituting (11.282) into (11.279), that

$$\begin{aligned}
m_0^{(2)} &= \frac{2}{3}\int d\mathbf{r}\, \exp[-\beta u(r)]\, \Delta\gamma^2, \\
m_2^{(2)} &= \frac{2}{3}\int d\mathbf{r}\, \exp[-\beta u(r)] \left(\Delta\gamma'^2 + 6\frac{\Delta\gamma^2}{r^2}\right), \\
m_4^{(2)} &= \frac{2}{3}\int d\mathbf{r}\, \exp[-\beta u(r)] \left(3\Delta\gamma''^2 + 4\frac{\Delta\gamma'\Delta\gamma''}{r} - 12\frac{\Delta\gamma\Delta\gamma''}{r^2}\right. \\
&\quad +32\frac{\Delta\gamma'^2}{r^2} - 96\frac{\Delta\gamma\Delta\gamma'}{r^3} + 120\frac{\Delta\gamma^2}{r^4} \\
&\quad \left. -2\beta u'\Delta\gamma'\left(\Delta\gamma'' + 2\frac{\Delta\gamma'}{r} - 6\frac{\Delta\gamma}{r^2}\right) + \beta^2 u'^2 \Delta\gamma'^2\right),
\end{aligned} \tag{11.284}$$

where $u(r)$ is the interatomic potential, and a prime denotes differentiation with respect to the interatomic separation.

Recently, Guillot-Noël *et al.*[201] have performed very accurate measurements of scattering by He over more than four decades of intensity. Their results show that the zeroth and second spectral moments are accurately described by the *ab initio* pair polarizability anisotropies of both Moszynski *et al.*[195] and Dacre.[197] The far wings of the spectrum were shown to result almost exclusively from short-range overlap effects, with negligible contributions from induction and dispersion.

Joslin *et al.*[198] have computed the moments of (11.284) for a dilute gas of argon atoms using their *ab initio* static pair polarizability. (Strictly speaking, $\Delta\gamma$ should be evaluated at the frequency of the incident light.) The calculated zeroth moment was about 15–20% too low at all temperatures; the second and fourth moments were also too low, although the predicted temperature dependence was close to what is observed. There is some evidence that at least some of the experimental data may be inaccurate.[202]

Barocchi *et al.*[203] have studied scattering from mercury vapour over a range of temperatures. Second-order DID contributions were found to be significantly larger than for the rare gases, reflecting the larger atomic polarizability of Hg. To bring theory and experiment into quantitative agreement, it was also necessary to postulate an additional sizeable positive contribution to $\Delta\gamma$ at intermediate r. This may be associated with a specific electronic pair interaction between the open-shell ($6s^2$) Hg atoms; overlap contributions in closed-shell systems would be expected to be negative.

For a dense fluid the zeroth and second depolarized moments are

$$M_0^{(2)} = 6\alpha_1^4 V \Bigg(2\rho^2 \int d\mathbf{r}_{12}\, g(r_{12}) \frac{\kappa(r_{12})^2}{r_{12}^6}$$

$$+4\rho^3 \int d\mathbf{r}_{12} d\mathbf{r}_{32}\, g(r_{12}r_{13}r_{32}) P_2(\cos\theta_{123}) \frac{\kappa(r_{12})\kappa(r_{32})}{(r_{12}r_{32})^3}$$

$$+\rho^4 \int d\mathbf{r}_{12} d\mathbf{r}_{32} d\mathbf{r}_{14}\, [g(r_{12}r_{13}r_{14}r_{23}r_{24}r_{34}) - g(r_{13})g(r_{24})]$$

$$\times P_2(\cos\theta_{13,24}) \frac{\kappa(r_{13})\kappa(r_{24})}{(r_{13}r_{24})^3} \Bigg) \qquad (11.285)$$

and

$$M_2^{(2)} = 2\frac{2kT}{M}\alpha_1^4 V \Bigg(\rho^2 \int d\mathbf{r}_{12}\, g(r_{12}) \Bigg[90\frac{\kappa(r_{12})^2}{r_{12}^8}$$

$$- 36\frac{\kappa(r_{12})\kappa'(r_{12})}{r_{12}^7} + 6\frac{\kappa'(r_{12})^2}{r_{12}^6} \Bigg]$$

$$+\rho^3 \int d\mathbf{r}_{12} d\mathbf{r}_{32}\, g(r_{12}r_{13}r_{32}) \Bigg[90 P_3(\cos\theta_{123}) \frac{\kappa(r_{12})\kappa(r_{32})}{(r_{12}r_{32})^4}$$

$$-36 P_3(\cos\theta_{123}) \frac{\kappa(r_{12})\kappa'(r_{32})}{r_{12}^4 r_{32}^3}$$

$$+ \frac{6}{5}[3P_3(\cos\theta_{123}) + 2P_1(\cos\theta_{123})] \frac{\kappa'(r_{12})\kappa'(r_{32})}{(r_{12}r_{32})^3} \Bigg] \Bigg), \qquad (11.286)$$

where $\cos\theta_{123} = \hat{\mathbf{r}}_{12} \cdot \hat{\mathbf{r}}_{32}$ and $\cos\theta_{13,24} = \hat{\mathbf{r}}_{13} \cdot \hat{\mathbf{r}}_{24}$. Here the induced polarizability appears via the dimensionless factor $\kappa(r) = \Delta\gamma(r)/\left(6\alpha_1^2/r^3\right)$. In the first-order DID approximation, $\kappa = 1$. If accurate *ab initio* estimates of $\Delta\gamma$ are not available, we may include the effects of other induction mechanisms approximately using the semi-empirical Meinander–Tabisz–Zoppi model,[204] which sets

$$\kappa(r) = 1 + \left(\frac{6\alpha_1^2}{r^3}\right)^{-1} \left(\frac{A_6}{r^6} - \lambda \exp[-(r-\sigma)/\rho]\right), \qquad (11.287)$$

where λ and ρ measure the strength and range of overlap interactions, σ is the range of the pair potential, and A_6 is a composite factor representing the combined effects of second-order DID and dispersion interactions. The derivation of (11.285) and (11.286) follows that given for the moments of the CIA spectrum in § 11.1.8.

Ladanyi *et al.*[205] have simulated $M_0^{(2)}$, $M_2^{(2)}$, and $M_4^{(2)}$ for argon and krypton, and have studied their temperature and density dependence. The moments turn out to be insensitive to the details of the potential, depending only on its range, and scaling approximately as $M_{2n}^{(2)} \propto \left(\alpha_1^4/\sigma^6\right)(kT/m\sigma^2)^n$. As the density increases, the zeroth and second moments both show maxima; however, as

Table 11.5

Moments of the depolarized CILS spectrum for liquid argon at 130 K and 615 amagat

	Calculated	Observed
$\hat{M}_2^{(2)}/(\mathrm{cm}^{-2})$	524	528
$\hat{M}_4^{(2)}/(\mathrm{cm}^{-4})$	1.97×10^6	1.96×10^6

From ref. 207.

the moment *ratios* $M_2^{(2)}/M_0^{(2)}$ and $M_4^{(2)}/M_2^{(2)}$ both increase monotonically, the scattering spectrum broadens uniformly, as observed experimentally. At liquid-like densities use of the superposition approximation[206] was shown to lead to serious overestimates of $M_0^{(2)}$, reflecting the very close cancellation of the two-, three-, and four-body terms in (11.285); for $M_2^{(2)}$ cancellation is less complete, and the superposition approximation does better. Non-DID effects, as represented by (11.287), were found to be important, yielding moments significantly smaller than predicted by first-order DID. These differences were particularly pronounced at high temperatures and densities, and increased as the order of the moment increased; this is because smaller interatomic separations were then probed.

Teboul and le Duff[207] have simulated light scattering from liquid argon at 130 K and 615 amagat using Joslin's $\Delta\boldsymbol{\alpha}$. They find excellent agreement with experiment over the whole frequency range for which data are available. Table 11.5 shows results for the reduced second and fourth moments.

Influence of the triplet polarizability

There are some indications that the three-body irreducible polarizability $\Delta\boldsymbol{\alpha}_{123}$ may contribute to light scattering in moderately dense rare gas systems. Table 11.6 shows the third virial coefficients of the moments of the room-temperature depolarized spectra of neon, argon, and krypton. These coefficients are defined through the density expansion

Table 11.6

Third virial coeficients of the moments of the depolarized CILS spectrum for Ne, Ar and Kr at 296 K

	$C_0^{(2)}$		$C_1^{(2)}$		$C_2^{(2)}$	
	Expt	Calc	Expt	Calc	Expt	Calc
Ne	6.5 ± 1.3	6.5	1.12 ± 0.46	0.48	87 ± 37	37
Ar	2530 ± 160	2510	—	—	12300 ± 3200	10400
Kr	10400 ± 900	13900	—	—	22900 ± 1600	31600

Units of $C_n^{(2)}$ are $\mathrm{Å}^{12} ps^{-n}$
From ref. 188.

$$M_n^{(2)} = B_n^{(2)}\rho^2 + C_n^{(2)}\rho^3 + \dots. \tag{11.288}$$

In the table, experimental values for $C_n^{(2)}$ for $n = 0, 1, 2$ are compared with theoretical estimates calculated in the pair approximation, using highly accurate pair potentials and polarizabilities $\Delta\alpha_{12}^{(2)}$ obtained by fitting to the two-body spectra.[188] For Ne, quantum corrections at the level of the first Wigner–Kirkwood approximation have been applied.[189]

The agreement for Ar is good, and for this gas irreducible three-body effects are either absent or below the limits of detectability. For other two gases, however, there are definite differences. For Ne, $C_0^{(2)}$ is given accurately, but $C_1^{(2)}$ and $C_2^{(2)}$ are underestimated. For Kr, on the other hand, the pair approximation *overestimates* the moments. These results have been interpreted[188,189] as suggesting the presence of two distinct contributions to $\Delta\boldsymbol{\alpha}_{123}$ of opposite sign, one operating at short range (presumably an overlap or exchange effect), the other at long range, primarily first-order DID.[184,185] For Ne the former appears to be dominant, while for Kr, with its much larger polarizability, the latter is more important. For Ar it seems that the effects approximately cancel. This hypothesis remains to be verified, however; no calculations of the irreducible three-body contributions to the virial coefficients have yet appeared, although estimates of both the triplet potential and $\Delta\boldsymbol{\alpha}_{123}$ are available.[184,185,187]

It might be imagined that the *polarized* spectra offer a better test for the presence of three-body effects, since long-range contributions to the trace $\Delta\alpha_{123}$ are on the same order, α_1^3/r^6, as the leading DID term in $\Delta\alpha_{12}$. Unfortunately, polarized scattering by the rare gases is so weak that it has not so far been possible to determine $C_n^{(0)}$ with the accuracy required.

Molecular fluids: the pair polarizability

Accurate *ab initio* results for molecular collisional polarizabilities are available only for a very limited number of cases. Short-range overlap contributions, including correlation effects, have been computed for only one molecular pair, He-CH_4.[208] Fortunately, most scattered intensity derives from long-range interactions beyond the van der Waals minimum, and for such intermolecular separations, $\Delta\boldsymbol{\alpha}_{12}$ is well approximated as a sum of induction and dispersion terms.[109]

The form of the long-range collisional polarizability of a pair of centrosymmetric linear molecules (e.g. H_2, N_2, CO_2) has been given by Hunt and co-workers.[109] The induction contribution is

$$\begin{aligned}\Delta\alpha_{\alpha\beta}^{\text{ind}}(12) = (1+\mathcal{P}_{12})\Big(&\alpha_{\alpha\gamma}(1)T_{\gamma\delta}^{(2)}(\mathbf{r})\alpha_{\delta\beta}(2)\\ &+\alpha_{\alpha\gamma}(1)T_{\gamma\delta}^{(2)}(\mathbf{r})\alpha_{\delta\epsilon}(2)T_{\epsilon\zeta}^{(2)}(\mathbf{r})\alpha_{\zeta\beta}(1)+\dots\\ &+\frac{1}{15}\alpha_{\alpha\gamma}(1)T_{\gamma\delta\epsilon\zeta}^{(4)}(\mathbf{r})E_{\beta,\delta\epsilon\zeta}(2)+\frac{1}{15}E_{\alpha,\gamma\delta\epsilon}(1)T_{\gamma\delta\epsilon\zeta}^{(4)}(\mathbf{r})\alpha_{\zeta\beta}(2)\\ &-\frac{1}{9}B_{\alpha\beta,\gamma\delta}(1)T_{\gamma\delta\epsilon\zeta}^{(4)}(\mathbf{r})Q_{\epsilon\zeta}(2)+\dots\Big),\end{aligned} \tag{11.289}$$

where $\mathcal{P}_{12}$ is the permutation operator for molecules 1 and 2, and $\mathbf{r} = \mathbf{r}_2 - \mathbf{r}_1$. The first two terms represent first- and second-order DID contributions. The third and fourth terms, which involve the dipole–octopole polarizability $\mathbf{E}$, represent octopole–induced-dipole and nonuniform dipole–induced-dipole interactions. The final term involves the dipole–dipole–quadrupole hyperpolarizability $\mathbf{B}$, and reflects hyperpolarization effects associated with the field of the molecular quadrupole moment $\mathbf{Q}$.

It has been suggested[172,209] that a distributed polarizability model may provide a better description of the induced polarizability, especially for highly elongated species such as CO_2. However, calculations have shown that this approach is not obviously superior, and does not necessarily produce moments in better agreement with experiment.[210]

The leading dispersion term is given by

$$\Delta\alpha^{\text{disp}}_{\alpha\beta}(12) = \frac{\hbar}{2\pi}(1 + \mathcal{P}_{12})T^{(2)}_{\gamma\epsilon}(\mathbf{r})T^{(2)}_{\delta\zeta}(\mathbf{r}) \times \int_0^\infty d\omega\, \gamma_{\gamma\delta\alpha\beta}(i\omega, 0, 0)(1)\alpha_{\epsilon\zeta}(i\omega)(2), \tag{11.290}$$

where $\boldsymbol{\alpha}(i\omega)$ is the polarizability at the imaginary frequency $i\omega$ (cf. Appendix C.3), and $\boldsymbol{\gamma}(i\omega, 0, 0)$ is an imaginary-frequency second hyperpolarizability. Dispersion contributions of order r^{-8} have also been derived.[117]

Equations (11.289) and (11.290) are complete to order r^{-6}. In practice the first-order DID term always gives the dominant contribution to $\Delta\boldsymbol{\alpha}$: for example, for T-shaped pairs of H_2 or N_2 molecules, at distances just beyond the van der Waals minimum, the first-order DID term in the anisotropic part of the collisional polarizability differs from the long-range total by 5% or less.

The spherical tensor components[211] of $\Delta\boldsymbol{\alpha}_{12}$ can be written as

$$\Delta\alpha_{JK}(r\omega_1\omega_2\omega) = \left(\frac{64\pi^3}{2J+1}\right)^{1/2} \sum_{l_1 l_2 l L} B^{(J)}_{l_1 l_2 l L}(r)\psi^{JK}_{l_1 l_2 l L}(\omega_1\omega_2\omega), \tag{11.291}$$

where $J = 0$, $K = 0$ for the isotropic, and $J = 2$, $-2 \le K \le 2$ for the anisotropic, components. The angular basis function $\psi^{JK}_{l_1 l_2 l L}$ is defined by

$$\psi^{JK}_{l_1 l_2 l L} = \sum_{m_1 m_2 m M} C(l_1 l_2 l; m_1 m_2 m)C(lLJ; mMK) \times Y_{l_1 m_1}(\omega_1)Y_{l_2 m_2}(\omega_2)Y_{LM}(\omega). \tag{11.292}$$

The normalization in (11.291) ensures that

$$\left\langle \Delta\alpha^2_{12} \right\rangle_{\omega_1\omega_2\omega} = \frac{1}{3}\langle \Delta\alpha_{00}(r\omega_1\omega_2\omega)^2 \rangle_{\omega_1\omega_2\omega} = \frac{1}{3}\sum_{l_1 l_2 l L} B^{(0)\,2}_{l_1 l_2 l L}(r) \tag{11.293}$$

and

$$\left\langle \Delta\boldsymbol{\alpha}_{12}^{(2)} : \Delta\boldsymbol{\alpha}_{12}^{(2)} \right\rangle_{\omega_1\omega_2\omega} = \left\langle \sum_K |\Delta\alpha_{2K}(r\omega_1\omega_2\omega)|^2 \right\rangle_{\omega_1\omega_2\omega} = \sum_{l_1l_2lL} B_{l_1l_2lL}^{(2)\,2}(r). \tag{11.294}$$

The coefficients $B_{l_1l_2lL}^{(J)}$ have been determined[212] for each of the mechanisms in (11.289) and (11.290). For the leading (first-order DID) terms, we have explicitly

$$B_{0222}^{(0)} = \frac{4}{(15)^{1/2}}\alpha_1\gamma_2 r^{-3}, \qquad B_{2022}^{(0)} = \frac{4}{(15)^{1/2}}\gamma_1\alpha_2 r^{-3}, \tag{11.295}$$

$$B_{2222}^{(0)} = -\left(\frac{56}{675}\right)^{1/2}\gamma_1\gamma_2 r^{-3}, \tag{11.296}$$

and

$$B_{0002}^{(2)} = (24)^{1/2}\alpha_1\alpha_2 r^{-3}, \tag{11.297}$$

$$B_{0222}^{(2)} = -\left(\frac{28}{15}\right)^{1/2}\alpha_1\gamma_2 r^{-3}, \qquad B_{2022}^{(2)} = -\left(\frac{28}{15}\right)^{1/2}\gamma_1\alpha_2 r^{-3}, \tag{11.298}$$

$$B_{2202}^{(2)} = \left(\frac{8}{3375}\right)^{1/2}\gamma_1\gamma_2 r^{-3}, \tag{11.299}$$

$$B_{2222}^{(2)} = -\left(\frac{8}{675}\right)^{1/2}\gamma_1\gamma_2 r^{-3}, \tag{11.300}$$

$$B_{2242}^{(2)} = \left(\frac{96}{125}\right)^{1/2}\gamma_1\gamma_2 r^{-3}. \tag{11.301}$$

The leading contribution to $B_{0000}^{(0)}$ is the second-order DID term

$$B_{0000}^{(0)} = \frac{2}{3^{1/2}}\left(3\alpha_1^2\alpha_2 + 3\alpha_1\alpha_2^2 + \frac{2}{3}\gamma_1^2\alpha_2 + \frac{2}{3}\alpha_1\gamma_2^2\right)r^{-6}, \tag{11.302}$$

while those to $B_{0220}^{(2)}$ and $B_{2020}^{(2)}$ (contributors to the cross depolarized spectrum) are

$$B_{0220}^{(2)} = \left(\frac{2}{3}\right)^{1/2}\left(4\alpha_1\alpha_2\gamma_2 + \frac{1}{5}\alpha_1^2\gamma_2 + \frac{2}{3}\alpha_1\gamma_2^2 + \frac{1}{225}\gamma_1^2\gamma_2\right)r^{-6},$$

$$B_{2020}^{(2)} = \left(\frac{2}{3}\right)^{1/2}\left(4\gamma_1\alpha_1\alpha_2 + \frac{1}{5}\gamma_1\alpha_2^2 + \frac{2}{3}\gamma_1^2\alpha_2 + \frac{1}{225}\gamma_1\gamma_2^2\right)r^{-6}. \tag{11.303}$$

In each case there is also a sizeable dispersion contribution.[109]

Molecular fluids: spectral moments

We next consider the calculation of the moments of the depolarized spectrum. For reasons of space, we will mostly treat Rayleigh spectra. With appropriate modifications, Raman moments may be calculated in a similar fashion; note, however, that for some vibrational modes, e.g. the ν_2 (π_u) doubly degenerate bending mode of CO_2 and CS_2, the derivative $(\partial \boldsymbol{\alpha}_1/\partial q)_0$ vanishes. The dipole–induced-dipole mechanism, which dominates Rayleigh spectra, is then no longer the leading source of spectral intensity; its place is taken by other mechanisms, chief among which[213,214] is the quadrupole–induced-dipole mechanism, involving (the derivative of) the quadrupole polarizability **A** (Appendix C). We will concentrate primarily on linear centrosymmetric species, such as N_2 and CS_2. Results for other symmetries are briefly considered at the end of this section.

Zeroth moment

Cross moment We begin by noting that the spherical tensor representation of the anisotropic component of the permanent polarizability $\boldsymbol{\alpha}_1^{(2)}$ is (cf. Appendix A of Vol. 1)

$$\alpha_{2K}(1) = \left(\frac{8\pi}{15}\right)^{1/2} \gamma_1 Y_{2K}(\omega_1). \tag{11.304}$$

If we assume an isotropic potential, it then follows that

$$M_0^{(2)} = 2\left(\frac{2}{3}\right)^{1/2} \gamma_1 \rho^2 V \int \mathrm{d}\mathbf{r}\, g(r) B_{0220}^{(2)}(r). \tag{11.305}$$

Note the absence of three-body terms. In the second-order DID approximation, (11.305) becomes

$$M_0^{(2)} = 4\gamma_1^2 \left(\frac{7}{5}\alpha_1^2 + \frac{2}{9}\alpha_1\gamma_1 + \frac{1}{675}\gamma_1^2\right) \rho^2 V \int \mathrm{d}\mathbf{r}\, g(r) r^{-6}, \tag{11.306}$$

which is positive definite.

For a more general, anisotropic, potential, there will be a nonvanishing first-order DID contribution to $M_0^{(2)}$ that can easily outweigh (11.306). The cross moment can then be *negative* (this is, in fact, the normal situation). Two- and three-body contributions to $M_0^{(2)}$ have been simulated for both N_2 (both Rayleigh and Raman)[215] and for the strongly anisotropic species CS_2 (Rayleigh only).[216] In the Rayleigh moment, the two-body term dominates, is negative, and becomes more so as the temperature is lowered and the density rises; the three-body term is positive for N_2, but can be either positive or negative for CS_2, depending on the state condition. The N_2 Raman moment contains only a two-body term, which is negative, and which also increases in magnitude as ρ increases.

At high densities the cross contribution to the zeroth moment generally dominates the pure induced contribution, because of substantial cancellation in the latter (see below).

Pure induced moment The depolarized zeroth moment is

$$M_0^{(2)} = 4\left(3\alpha_1^4 + \frac{7}{15}\alpha_1^2\gamma_1^2 + \frac{22}{225}\gamma_1^4\right)\rho^2 V \int \mathrm{d}\mathbf{r}\, g(r) r^{-6}$$
$$+4\left(6\alpha_1^4 + \frac{7}{15}\alpha_1^2\gamma_1^2\right)\rho^3 V \int \mathrm{d}\mathbf{r}_{12}\mathrm{d}\mathbf{r}_{32}\, g(r_{12}r_{13}r_{32}) P_2(\cos\theta_{123})(r_{12}r_{32})^{-3}$$
$$+6\alpha_1^4\rho^4 V \int \mathrm{d}\mathbf{r}_{12}\mathrm{d}\mathbf{r}_{32}\mathrm{d}\mathbf{r}_{14}$$
$$\times\left[g(r_{12}r_{13}r_{14}r_{23}r_{24}r_{34}) - g(r_{13})g(r_{24})\right] P_2(\cos\theta_{13,24})(r_{13}r_{24})^{-3}. \quad (11.307)$$

Mueller *et al.*[215] have simulated the Rayleigh and Raman spectra of N_2 at 323 K for a range of densities, $0.5 < \rho/\rho_c < 2.5$, where ρ_c is the critical density. Using a three-site 12-6-1 potential, and assuming a first-order DID mechanism, they found good agreement with experiment over the whole density range. Figure 11.11 shows the behaviour of the dimensionless quantity $M_0^{\mathrm{ind}*} = M_0^{(2)}/\left(2/3N\gamma_1^2\right)$ as a function of density. Note the significant cancellation that occurs between the positive two- and four-body contributions and the negative three-body contribution in the Rayleigh spectrum, and between the positive two-body and negative three-body contributions in the Raman spectrum. While each of these terms is separately quite large, and increases rapidly with density, the net moments always remain small. This behaviour is analogous to that observed for the rare gases (see above), and reflects the relative insensitivity of $M_0^{(2)}$ to anisotropy in $\boldsymbol{\alpha}_1$, as well as the near sphericity of the nitrogen intermolecular potential. As a result of this cancellation, the cross term dominates the pure induced term for densities higher than the critical density.[215]

Stassen and Steele[216] have performed a detailed analysis of the various contributions to the zeroth moment of the Rayleigh spectrum for liquid CS_2. In a comprehensive study covering various state points, they have simulated the two-, three-, and four-body terms in $M_0^{(2)}$ in the first-order DID approximation. They have further subdivided these into contributions arising from the isotropic and anisotropic parts of $\boldsymbol{\alpha}_1$. As for more nearly spherical species, two- and four-body contributions are positive. Three-body contributions are predominantly negative, although the anisotropy in $\boldsymbol{\alpha}_1$ acts to reduce their absolute magnitude. However, the cancellation effect is much less complete. As the density increases and the temperature decreases, the pure induced moment increases, as does the allowed term; however, the effect on the total zeroth moment is more than offset by a concomitant increase in the magnitude of the negative cross term.

In a simulation study of $M_0^{(2)}$ for N_2, O_2, and CO_2, Barreau *et al.*[217] have shown that the effect of anisotropy in the molecular polarizability is felt most strongly at high densities and for more elongated molecules.

In most cases second-order DID contributions can be neglected.[218,219] For highly polarizable species such as CS_2, however, they may substantially reduce

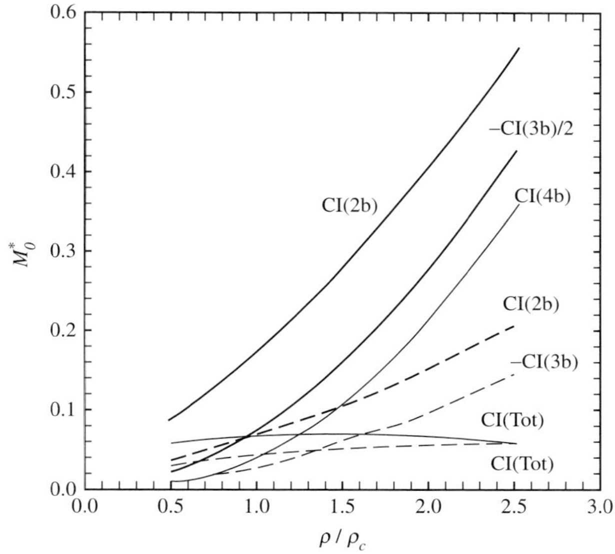

FIG. 11.11 Two-, three-, and four-body contributions to the dimensionless zeroth moments $M_0^{\text{ind}\,*}$ of the induced depolarized Rayleigh (solid lines) and Raman (dashed lines) light scattering spectra of N_2 at 323 K. Here ρ/ρ_c is the ratio of the density to the critical density. (From ref. 215.)

the pure induced moment. Since, as discussed above, they also increase the cross moment, the *net* effect on the spectral intensity is even here often quite small.[220]

Total zeroth moment Figure 11.12 shows the total reduced zeroth moments of the depolarized Rayleigh and Raman spectra of ethane, C_2H_6, simulated by Keller *et al.*[221] as a two-centre LJ diatomic, at 336 K and densities varying from 0.7 to 2.5ρ_c. We have defined

$$M_0^* = M_0^{(2)}/M_0^{(2)\,\text{self}} = 1 + m_0^{\text{dis}} + m_0^{\text{cross}} + m_0^{\text{ind}}, \tag{11.308}$$

where $m_0^{\text{dis}} = M_0^{(2)\,\text{dis}}/M_0^{(2)\,\text{self}}$, $m_0^{\text{cross}} = M_0^{(2)\,\text{cross}}/M_0^{(2)\,\text{self}}$, and $m_0^{\text{ind}} = M_0^{(2)\,\text{ind}}/M_0^{(2)\,\text{self}}$ represent the contributions of distinct, cross, and pure induced terms relative to that of the self term (there are of course no distinct terms for Raman scattering).

For the Rayleigh moment, the distinct term is always very small. For densities near critical, the total moment may be as much as 2.5 times larger than the allowed contribution, but this reflects mostly the very small magnitude of the latter, associated with the weak anisotropy in $\boldsymbol{\alpha}_1$; for other species, e.g. N_2, the ratio is much smaller.[222] The increase in M_0^* at low ρ is associated with the dominance of the positive two-body contribution to m_0^{ind}. Above $\rho \approx \rho_c$, however, cancellation in this term, and the continued growth of the negative m_0^{cross}, combine to reduce M_0^*.

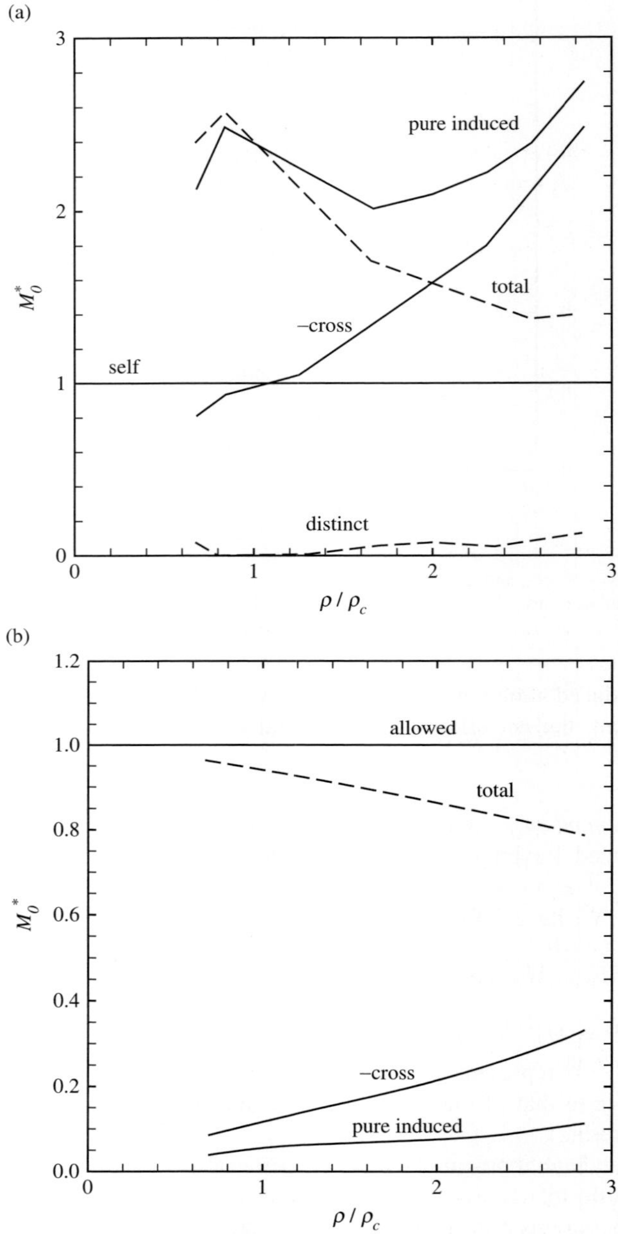

FIG. 11.12 Allowed, cross, and pure induced contributions to the reduced total zeroth moments M_0^* of the depolarized (a) Rayleigh and (b) Raman spectra of C_2H_6 at 336 K. (From ref. 221.)

For the Rayleigh spectrum, the relative importance of the negative cross and positive pure induced contributions is determined by the ratio γ_1^*/α_1^{*2}, where $\alpha_1^* = \alpha_1/\sigma^3$ and $\gamma_1^* = \gamma_1/\sigma^3$, and σ is the range of the intermolecular potential. For the Raman spectrum the corresponding ratio is $\gamma_1'^*/\alpha_1'^*\alpha_1^*$, where $\alpha_1'^* = \alpha_1'/\sigma^3$ and $\gamma_1'^* = \gamma_1'/\sigma^3$. For ethane, these ratios are approximately[221] 2 and 20; this explains why the cross term can be relatively more important than the pure induced term in Raman scattering, even though the opposite situation holds for Rayleigh scattering. The Raman moment is in fact almost entirely allowed, except at the highest densities, and this also reflects the much larger anisotropy in $\boldsymbol{\alpha}_1'$ than in $\boldsymbol{\alpha}_1$.[223]

Second moment

Cross moment For the depolarized spectrum there are only two-body contributions:

$$M_2^{(2)} = 2N(N-1)\left\langle \dot{\boldsymbol{\alpha}}_1^{(2)} : \Delta\dot{\boldsymbol{\alpha}}_{12}^{(2)} \right\rangle. \tag{11.309}$$

If we neglect anisotropy in the intermolecular potential, only the term in (11.291) involving $\psi_{2020}^{2K} = Y_{2K}(\omega_1)/4\pi$ contributes, and

$$M_2^{(2)} = \frac{8\pi}{5}\left(\frac{2}{3}\right)^{1/2} \gamma_1 \sum_K \langle |\dot{Y}_{2K}(\omega_1)|^2\rangle_{\omega_1} \rho^2 V \int \mathrm{d}\mathbf{r}\, g(r) B_{2020}^{(2)}(r). \tag{11.310}$$

To evaluate the time derivatives in (11.310) we note that

$$\begin{aligned}\dot{Y}_{lm}(\omega_1) &= \sum_\mu (-)^\mu \Omega_{1\mu} \nabla_{\omega_1 \bar{\mu}} Y_{lm}(\omega_1) \\ &= i[l(l+1)]^{1/2} \sum_{m_1 m_2} C(l1l; m_1 m_2 m) Y_{lm_1}(\omega_1)\Omega_{1m_2},\end{aligned} \tag{11.311}$$

where the second line follows from (A.18b) and (A.26), and Ω_{1m} denotes a spherical component of the molecular angular velocity. Noting that $\left\langle \Omega_{1m_1}\Omega_{1m_2}^* \right\rangle = (kT/I)\delta_{m_1 m_2}$, we now find that

$$\begin{aligned}M_2^{(2)} &= 12\left(\frac{2}{3}\right)^{1/2} \gamma_1 \frac{kT}{I}\rho^2 V \int \mathrm{d}\mathbf{r}\, g(r) B_{0220}^{(2)}(r) \\ &= 24\gamma_1^2\left(\frac{7}{5}\alpha_1^2 + \frac{2}{9}\alpha_1\gamma_1 + \frac{1}{675}\gamma_1^2\right)\frac{kT}{I}\rho^2 V \int \mathrm{d}\mathbf{r}\, g(r) r^{-6},\end{aligned} \tag{11.312}$$

where the second equality follows in the second-order DID approximation using (11.303). Note that (11.312) is positive definite. In practice, anisotropy in the potential may introduce nonvanishing first-order DID contributions, leading to a negative second moment; this is the case even for the weakly nonspherical N_2.[222,224]

For a quasispherical molecule such as H_2, (11.305) and (11.312) imply

$$\hat{M}_2^{(2)} \equiv M_2^{(2)}/M_0^{(2)} = 6kT/I, \tag{11.313}$$

exactly as for allowed depolarized scattering, cf. (11.234).

Pure induced moment A first-order DID approximation yields the dominant terms.[225] Neglecting anisotropy in the intermolecular potential, we find that

$$\begin{aligned} M_2^{(2)} &= 4\frac{kT}{M}\left(\alpha_1^4 + \frac{7}{45}\alpha_1^2\gamma_1^2 + \frac{22}{675}\gamma_1^4\right) N(N-1)\langle \mathbf{T}^{(3)}(\mathbf{r}_{12})\vdots\mathbf{T}^{(3)}(\mathbf{r}_{12})\rangle \\ &+4\frac{kT}{M}\alpha_1^2\left(\alpha_1^2 + \frac{7}{90}\gamma_1^2\right) N(N-1)(N-2)\langle \mathbf{T}^{(3)}(\mathbf{r}_{12})\vdots\mathbf{T}^{(3)}(\mathbf{r}_{13})\rangle \\ &+\frac{28}{15}\frac{kT}{I}\gamma_1^2\left(\alpha_1^2 + \frac{44}{105}\gamma_1^2\right) N(N-1)\langle \mathbf{T}^{(2)}(\mathbf{r}_{12}) : \mathbf{T}^{(2)}(\mathbf{r}_{12})\rangle \\ &+\frac{28}{15}\frac{kT}{I}\alpha_1^2\gamma_1^2 N(N-1)(N-2)\langle \mathbf{T}^{(2)}(\mathbf{r}_{12}) : \mathbf{T}^{(2)}(\mathbf{r}_{13})\rangle. \end{aligned} \tag{11.314}$$

$M_2^{(2)}$ is apparently rather sensitive to the details of the intermolecular potential.[210] Note that the two-body contributions are positive, and the three-body negative; however, the cancellation is less complete than that for the zeroth moment.[224]

Total depolarized second moment From (11.275), the total second moment of a depolarized Rayleigh or parallel Raman band is the sum of the allowed, cross, and pure induced contributions. Relative to the contribution from the allowed spectrum this is

$$M_2^* \equiv M_2^{(2)}/M_2^{(2)\,\text{all}} = 1 + m_2^{\text{cross}} + m_2^{\text{ind}}, \tag{11.315}$$

where, from (11.233), $M_2^{(2)\text{all}} = N\gamma_1^{0\,2}(4kT/I)$ (Rayleigh) or $N\gamma_1'^2$ $(4kT/I)$ (Raman), and we have defined $m_2^{\text{cross}} = M_2^{(2)\,\text{cross}}/M_2^{(2)\,\text{all}}$ and $m_2^{\text{ind}} = M_2^{(2)\,\text{ind}}/M_2^{(2)\,\text{all}}$. At low densities $M_2^* \to 1$. For many fluids, including nitrogen,[226,227] oxygen,[172] and carbon dioxide,[172,209] M_2^* is found to be almost independent of temperature. Table 11.7 shows the cross and pure induced contributions to M_2^* for simulated first-order DID Rayleigh scattering by N_2 at 323 K over a range of densities. We have separated the pure induced moment into two- and three-body terms.[222,224] Note how closely M_2^* approaches unity; the separate induced terms never exceed several percent, and the cancellation among these is almost complete. Similar results are found for the Raman spectrum.[222]

The width of a spectral band is determined by the reduced second moment $\hat{M}_2^{(2)} = M_2^{(2)}/M_0^{(2)}$. Relative to its value for a low-density gas, this is $M_2^{**} \equiv \hat{M}_2^{(2)}/(6kT/I)$, which can be expressed as

Table 11.7

Contributions to the second moment of the depolarized Rayleigh spectrum of N_2 *at* 323 K

ρ/ρ_c	m_2^{cross}	$m_2^{\text{ind}}[2B]$	$m_2^{\text{ind}}[3B]$	M_2^*
0.48	−0.006	0.010	−0.001	1.003
0.80	−0.009	0.018	−0.004	1.005
1.45	−0.017	0.035	−0.015	1.003
2.09	−0.026	0.057	−0.032	0.999
2.54	−0.032	0.076	−0.049	0.995

From ref. 224.

$$M_2^{**} = \frac{M_2^*}{M_0^*}, \tag{11.316}$$

with M_0^* defined by (11.308). From the above remarks, it is clear that the observed variation of M_2^{**} with density[222,223] is mostly attributable to changes in M_0^*. Now, the Rayleigh M_0^* typically reaches a maximum for densities near the critical density, as shown in Fig. 11.12(a), and thus M_2^{**} typically goes through a *minimum* here. For Raman spectra, on the other hand, where M_0^* usually shows a monotonic decrease with ρ (Fig. 11.12(b)), M_2^{**} shows a corresponding monotonic increase.

Figure 11.13 shows M_2^{**} for ethane, where in fact there are nonvanishing induced contributions to the Rayleigh M_2^* as well as to M_0^*.[221]

Higher moments

Higher moments have been computed for a dilute gas, assuming that the intermolecular potential is isotropic.[193,228] The first-order DID approximation becomes less satisfactory as the order of the moment increases, and smaller intermolecular separations are probed. This is particularly true in the case of polarized scattering. The principal long-range corrections often come from terms involving the tensor **E** (cf. (11.289)), and at short range overlap contributions may also be significant.

Collisional effects on the fourth moments of liquids have been investigated by Frenkel and McTague.[171]

Other molecular symmetries

Tetrahedral molecules Rayleigh moments in the DID approximation can be derived from the corresponding expressions for linear molecules, given above, by setting $\gamma_1 = 0$. It may be noted that there is no cross depolarized scattering, and no rotational contribution to the second moment of the pure induced depolarized spectrum.

Based on fits to the experimental $M_0^{(2)}$, $M_2^{(2)}$, and $M_4^{(2)}$ at several temperatures, Meinander *et al.*[229] first described the anisotropic part of the pair polarizability of

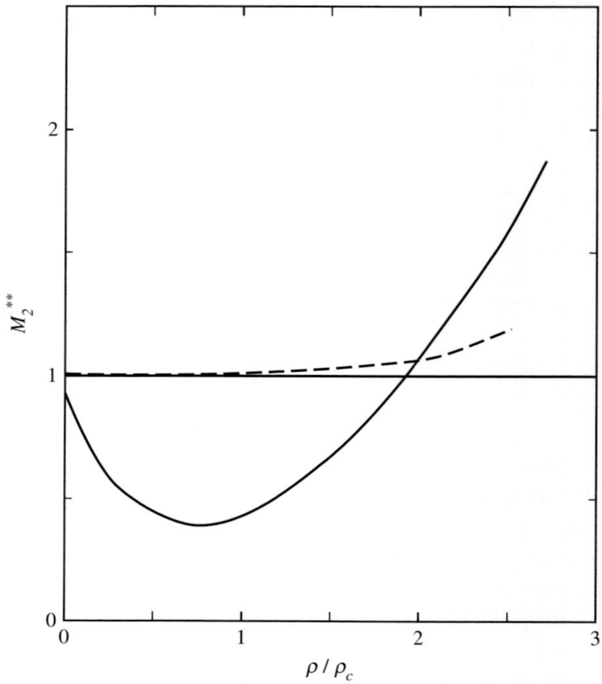

FIG. 11.13 The reduced second moments M_2^{**} of the depolarized Rayleigh (solid line) and Raman (dashed line) spectra of C_2H_6 at 336 K. (From ref. 223.)

CH_4 as a sum of positive long-range DID terms, corrected for dispersion at second order, and a negative short-range (exponential) overlap term. This is similar to the Meinander–Tabisz–Zoppi model[204] for atomic fluids, Eq. (11.287). Using this model, El-Sheikh *et al.*[230] estimated that overlap effects reduce $M_0^{(2)}$ and $M_2^{(2)}$ for liquid CF_4 by 7 and 11%, respectively. More recently, Elliasmine *et al.*[231] have emphasized the importance of additional long-range induction mechanisms involving the dipole–quadrupole polarizability **A** and the dipole–octopole polarizability **E**. By adding such terms to a DID model, one can largely explain the observed Rayleigh spectra of CF_4 gas. Computed lineshapes are consistent with *ab initio* estimates of the polarizabilities. Meinander *et al.*[232] find similar results for CH_4.

For molecules for which no theoretical estimates are available, information on **A** and **E**, and their derivatives, could in principle be obtained by fitting to the spectral moments.

Barocchi *et al.*[191] have calculated second-order quantum corrections for gaseous CH_4, and find them to be negligible for $M_0^{(2)}$ and $M_2^{(2)}$, but significant for $M_4^{(2)}$ below 200 K.

Polar molecules For polar molecules, including noncentrosymmetric linear species such as HCl, and molecules of lower symmetry such as H_2O and H_2S, an additional polarizability induction mechanism is possible, involving the dipole moment $\boldsymbol{\mu}$ and the first hyperpolarizability $\boldsymbol{\beta}$. In practice, however, this can usually be neglected as a source of intensity.[233]

11.2.8 Scattering geometry, depolarization ratios, and total intensities

In this section we consider some of the more practical aspects related to the scattering of light. These include (i) the angular distribution of the radiation, (ii) its degree of depolarization, and (iii) its total intensity.

In § 11.2.2 we introduced the frequency spectra for polarized and depolarized scattering, Eqs. (11.202) and (11.203). If we now define

$$I_{\parallel}(\omega) = I^{(0)}(\omega) + \frac{2}{15} I^{(2)}(\omega),$$
$$I_{\perp}(\omega) = \frac{1}{10} I^{(2)}(\omega), \tag{11.317}$$

then using (11.204) we can write the spectrum of scattered light as

$$I(\omega) = (\mathbf{n}_0 \cdot \mathbf{n}_S)^2 I_{\parallel}(\omega) + (\mathbf{n}_0 \times \mathbf{n}_S)^2 I_{\perp}(\omega). \tag{11.318}$$

We see that $I_{\parallel}(\omega)$ determines the intensity of scattered light having the same polarization as the incident beam, while $I_{\perp}(\omega)$ determines the intensity polarized at 90° to this. In a typical light-scattering experiment one observes the scattering at two mutually perpendicular scattered polarizations, $\mathbf{n}_S^{(1)}$ and $\mathbf{n}_S^{(2)}$, say. This enables measurement of both $I_{\parallel}(\omega)$ and $I_{\perp}(\omega)$, and allows separation of the polarized and depolarized spectra. For example, one may choose $\mathbf{n}_S^{(1)}$ to lie in the plane defined by $\mathbf{n}_0$ and $\mathbf{k}_S$, so that

$$\mathbf{n}_S^{(1)} = \frac{\mathbf{n}_0 - (\mathbf{n}_0 \cdot \hat{\mathbf{k}}_S)\hat{\mathbf{k}}_S}{|\mathbf{n}_0 \times \hat{\mathbf{k}}_S|}, \qquad \mathbf{n}_S^{(2)} = \frac{\mathbf{n}_0 \times \hat{\mathbf{k}}_S}{|\mathbf{n}_0 \times \hat{\mathbf{k}}_S|}. \tag{11.319}$$

The corresponding spectra are then

$$I_1(\omega) = (\mathbf{n}_0 \times \hat{\mathbf{k}}_S)^2 I_{\parallel}(\omega) + (\mathbf{n}_0 \cdot \hat{\mathbf{k}}_S)^2 I_{\perp}(\omega),$$
$$I_2(\omega) = I_{\perp}(\omega). \tag{11.320}$$

Since $\mathbf{n}_S^{(2)}$ is orthogonal to $\mathbf{n}_0$, the appearance of radiation with this polarization represents a *depolarization* of the incident radiation.

Depolarization ratio

It is conventional to define the *depolarization ratio*, ϱ, as the ratio of integrated intensities for the two polarization directions, i.e.

$$\varrho = \frac{(\mathrm{d}\sigma/\mathrm{d}\Omega)_2}{(\mathrm{d}\sigma/\mathrm{d}\Omega)_1} = \frac{\int_{-\infty}^{+\infty} \mathrm{d}\omega\, I_2(\omega)}{\int_{-\infty}^{+\infty} \mathrm{d}\omega\, I_1(\omega)}$$

$$= \frac{M_{\perp 0}}{(\mathbf{n}_0 \times \hat{\mathbf{k}}_S)^2 M_{\parallel 0} + (\mathbf{n}_0 \cdot \hat{\mathbf{k}}_S)^2 M_{\perp 0}}, \tag{11.321}$$

where $M_{\parallel 0}$ and $M_{\perp 0}$ denote zeroth moments of $I_{\parallel}(\omega)$ and $I_{\perp}(\omega)$, respectively.

ϱ is clearly a function of the scattering geometry. We distinguish three cases:

V geometry For incident light polarized perpendicular to the scattering plane, the so-called *V* geometry,[234] $\mathbf{n}_0 \cdot \hat{\mathbf{k}}_S = 0$. Therefore

$$\varrho_V = \frac{M_{\perp 0}}{M_{\parallel 0}}, \tag{11.322}$$

independent of the direction of detection. From (11.317),

$$\varrho_V = \frac{\frac{1}{10} M_0^{(2)}}{M_0^{(0)} + \frac{2}{15} M_0^{(2)}}, \tag{11.323}$$

and we see that ϱ_V is bounded,

$$0 \leq \varrho_V \leq \frac{3}{4}. \tag{11.324}$$

Violations of this inequality have been observed in cases where the polarizability contains an antisymmetric component, or near resonance.[146]

H geometry For incident light polarized in the scattering plane (*H* geometry), $(\mathbf{n}_0 \cdot \hat{\mathbf{k}}_S)^2 = (\hat{\mathbf{k}}_0 \times \hat{\mathbf{k}}_S)^2$. Therefore

$$\varrho_H = \frac{M_{\perp 0}}{M_{\parallel 0} \cos^2 \theta + M_{\perp 0} \sin^2 \theta}, \tag{11.325}$$

where θ denotes the angle between incident and scattered light. Thus ϱ_H depends on the direction of observation. For forward or backward scattering, $\varrho_H = \varrho_V$; while for 90° scattering, $\varrho_H = 1$, meaning that the scattered light is naturally polarized.

N geometry For naturally polarized (that is, unpolarized) incident light (*N* geometry), we have

$$\langle (\mathbf{n}_0 \cdot \hat{\mathbf{k}}_S)^2 \rangle = \frac{1}{2} (\hat{\mathbf{k}}_0 \times \hat{\mathbf{k}}_S)^2. \tag{11.326}$$

The depolarization ratio is therefore given by

$$\varrho_N^{-1} = \frac{1}{2} \left(\varrho_V^{-1} + \varrho_H^{-1} \right). \tag{11.327}$$

For forward or backward scattering, $\varrho_N = \varrho_V$. For 90° scattering

$$\varrho_N(90^\circ) = \frac{\frac{1}{5}M_0^{(2)}}{M_0^{(0)} + \frac{7}{30}M_0^{(2)}}, \tag{11.328}$$

which satisfies $0 \leq \varrho_N(90^\circ) \leq \frac{6}{7}$.

Virial expansion of ϱ_V

In a dilute gas, we can expand ϱ_V in powers of the density. This yields

$$\varrho_V = A_\varrho + B_\varrho \rho + \ldots, \tag{11.329}$$

where A_ϱ, B_ϱ, ... are the first, second, ... virial coefficients for the depolarization of scattered light. Studies of these coefficients can provide information on molecular polarizabilities and interactions.[235]

Expressions for the virial coefficients may be obtained from the density expansions of the moments $M_0^{(0)}$ and $M_0^{(2)}$. For Rayleigh scattering, we can write, using (11.222) and (11.228),

$$M_0^{(0)} = N\alpha_1^{0\,2}\left(1 + \mathcal{B}^{(0)}\rho + \ldots\right) \tag{11.330}$$

and

$$M_0^{(2)} = N\beta_1^{0\,2}\left(1 + \mathcal{B}^{(2)}\rho + \ldots\right). \tag{11.331}$$

In (11.330),

$$\mathcal{B}^{(0)} = -2B_p + \mathcal{B}^{(0)}_{\text{ind}}, \tag{11.332}$$

where

$$B_{\text{p}} = -\frac{1}{2}\int d(12)\,[\exp(-\beta u_{12}) - 1] \tag{11.333}$$

is the second pressure virial coefficient, and

$$\mathcal{B}^{(0)}_{\text{ind}} = \alpha_1^{0\,-2}\left\langle \int \mathbf{dr}_{12}\, \exp(-\beta u_{12})\, \frac{1}{2}\left(\alpha_{12}^{(0)2} - \alpha_{12}^{0\,(0)2}\right)\right\rangle_{\omega_1\omega_2} \tag{11.334}$$

measures fluctuations in the trace of the total pair polarizability $\boldsymbol{\alpha}_{12} = \boldsymbol{\alpha}^0_{12} + \Delta\boldsymbol{\alpha}_{12}$. Similarly, in (11.331),

$$\mathcal{B}^{(2)} = \mathcal{B}^{(2)}_{\text{or}} + \mathcal{B}^{(2)}_{\text{ind}} \tag{11.335}$$

is the sum of an orientational term,

$$\mathcal{B}^{(2)}_{\text{or}} = \beta_1^{0\,-2}\left\langle \int \mathbf{dr}_{12}\,[\exp(-\beta u_{12}) - 1]\,\boldsymbol{\alpha}_1^{0\,(2)} : \boldsymbol{\alpha}_2^{0\,(2)}\right\rangle_{\omega_1\omega_2}, \tag{11.336}$$

plus a term measuring fluctuations in the anisotropic part of $\boldsymbol{\alpha}_{12}$,

$$\mathcal{B}_{\text{ind}}^{(2)} = \beta_1^{0\,-2} \left\langle \int d\mathbf{r}_{12} \exp(-\beta u_{12}) \frac{1}{2} \left(\boldsymbol{\alpha}_{12}^{(2)} : \boldsymbol{\alpha}_{12}^{(2)} - \boldsymbol{\alpha}_{12}^{0\,(2)} : \boldsymbol{\alpha}_{12}^{0\,(2)} \right) \right\rangle_{\omega_1 \omega_2}. \tag{11.337}$$

Measurements of A_ϱ provide information on the molecular polarizability tensor (Appendix D.2). Combining (11.323), (11.330), and (11.331), we find that

$$A_\varrho = \frac{3\beta_1^{0\,2}}{30\alpha_1^{0\,2} + 4\beta_1^{0\,2}}. \tag{11.338}$$

The relative magnitudes of β_1^0 and α_1^0 are constrained by the requirement that the principal components of the polarizability tensor be positive, and thus the upper limit of (11.324) can never be reached. Even for strongly anisotropic molecules, A_ϱ is generally quite small, and is typically in the range 0.01–0.1. Thus Rayleigh scattering from gases tends to be strongly polarized. For Raman scattering, where an analogous result to (11.338) holds but with α_1^0 replaced by $\alpha_1^{0\prime}$ and β_1^0 by $\beta_1^{0\prime}$, much larger degrees of depolarization may be recorded, since the derivatives may be of either sign.

The second virial coefficient, B_ϱ, is similarly found to be

$$B_\varrho = A_\varrho \left(1 - \frac{4}{3} A_\varrho\right) \left(\mathcal{B}^{(2)} - \mathcal{B}^{(0)}\right). \tag{11.339}$$

Graham and co-workers[236] have measured B_ϱ for several molecular gases including ethylene, sulphur dioxide, and dimethyl ether, and have investigated the relative importance of the various contributions to the coefficients $\mathcal{B}^{(0)}$ and $\mathcal{B}^{(2)}$. These are found to be in the order (without regard to sign)

$$2B_p > \mathcal{B}_{\text{ind}}^{(2)} \gg \mathcal{B}_{\text{or}}^{(2)} \gg \mathcal{B}_{\text{ind}}^{(0)}. \tag{11.340}$$

Additionally, the DID mechanism was shown to give a good description of $\mathcal{B}_{\text{ind}}^{(2)}$, with terms of order $\boldsymbol{\alpha}^2$ (the lowest order pure induced term) and $\boldsymbol{\alpha}$ (the lowest order cross term) dominating. For the polar molecules SO_2 and $(CH_3)_2O$, terms involving the quadrupole polarizability $\mathbf{A}$ contributed on the order of a few percent. Using an anisotropic intermolecular potential whose parameters were fitted to B_p, the experimental B_ϱ could be reproduced to within experimental error.

Angular dependence of the scattering of naturally polarized light

The spectral density for scattering of naturally polarized light is, from (11.320),

$$\begin{aligned} I_{1\oplus 2}(\omega) &= \langle I_1(\omega) + I_2(\omega) \rangle \\ &= I_\parallel(\omega) \langle (\mathbf{n}_0 \times \hat{\mathbf{k}}_S)^2 \rangle + I_\perp(\omega) \langle (\mathbf{n}_0 \cdot \hat{\mathbf{k}}_S)^2 + 1 \rangle, \end{aligned} \tag{11.341}$$

where $\langle \ldots \rangle$ denotes an average over all possible directions of $\mathbf{n}_0$. Using (11.326), it follows that

$$I_{1\oplus 2}(\omega) = I_0(\omega) + I_2(\omega)P_2(\cos\theta), \tag{11.342}$$

where $P_2(x)$ denotes a second-order Legendre polynomial, and

$$\begin{aligned} I_0(\omega) &= \frac{2}{3}\left[I_{\parallel}(\omega) + 2I_{\perp}(\omega)\right], \\ I_2(\omega) &= \frac{1}{3}\left[I_{\parallel}(\omega) - I_{\perp}(\omega)\right] \end{aligned} \tag{11.343}$$

determine the magnitudes of the isotropic and anisotropic components of the scattering. (Note that $I_0(\omega)$ and $I_2(\omega)$ differ from the quantities $I^{(0)}(\omega)$ and $I^{(2)}(\omega)$, which appear in Eq. (11.317) at the beginning of this subsection.) It is apparent that the scattering is strongly anisotropic even for a system of isotropically polarizable molecules. In this case the intensity $\propto 1 + \cos^2\theta$, so that twice as much radiation is scattered along the direction of the incident beam as at right angles.

Total scattering cross-section

The total cross-sections σ_1 and σ_2 for the scattering of linearly polarized light are obtained by integrating the spectra $I_1(\omega)$ and $I_2(\omega)$ of (11.320) over all frequencies and solid angles. That is,

$$\begin{aligned} \sigma_1 &= f_{\mathrm{S}} k_{\mathrm{S}}^4 \int \mathrm{d}\Omega \int \mathrm{d}\omega \, \Big((\mathbf{n}_0 \times \hat{\mathbf{k}}_{\mathrm{S}})^2 I_{\parallel}(\omega) + (\mathbf{n}_0 \cdot \hat{\mathbf{k}}_{\mathrm{S}})^2 I_{\perp}(\omega) \Big), \\ \sigma_2 &= f_{\mathrm{S}} k_{\mathrm{S}}^4 \int \mathrm{d}\Omega \int \mathrm{d}\omega \, I_{\perp}(\omega). \end{aligned} \tag{11.344}$$

Since $\int \mathrm{d}\Omega \, (\mathbf{n}_0 \cdot \hat{\mathbf{k}}_S)^2 = 4\pi/3$ and $\int \mathrm{d}\Omega \, (\mathbf{n}_0 \times \hat{\mathbf{k}}_{\mathrm{S}})^2 = 8\pi/3$, it follows that

$$\begin{aligned} \sigma_1 &= \frac{4\pi}{3} f_S k_S^4 \left(2M_{\parallel 0} + M_{\perp 0}\right), \\ \sigma_2 &= 4\pi f_S k_S^4 M_{\perp 0}. \end{aligned} \tag{11.345}$$

The total cross-section without regard to polarization is

$$\begin{aligned} \sigma_{1\oplus 2} &= \sigma_1 + \sigma_2 \\ &= \frac{8\pi}{3} f_{\mathrm{S}} k_{\mathrm{S}}^4 (M_{\parallel 0} + 2M_{\perp 0}) \\ &= \frac{8\pi}{9} f_{\mathrm{S}} k_{\mathrm{S}}^4 \langle \boldsymbol{\pi} : \boldsymbol{\pi} \rangle. \end{aligned} \tag{11.346}$$

Note that $\sigma_{1\oplus 2} = P/\mathcal{I}(\omega_0)$, where $P = \sum_{\mathbf{n}_{\mathrm{S}}} \int \mathrm{d}\Omega \int d\omega_{\mathrm{S}} \, P(\omega_{\mathrm{S}})$ is the total scattered power. This is an extensive quantity, proportional to the size of the system.

In a low-density gas, each molecule scatters independently, and the total scattering cross-section per molecule is therefore

$$\sigma \equiv \sigma_{1\oplus 2}/N = \frac{8\pi}{9} k_S^4 \boldsymbol{\alpha}_1 : \boldsymbol{\alpha}_1, \tag{11.347}$$

where $\boldsymbol{\alpha}_1$ is the molecular polarizability tensor. To investigate the magnitude of this term we may consider Rayleigh and Raman scattering by carbon dioxide at the argon blue line of 488 nm. For Rayleigh scattering,[237] $\boldsymbol{\alpha}_1 : \boldsymbol{\alpha}_1 = 24.4 \times 10^{-48}\,\text{cm}^6$. For Raman scattering by the ν_1 Σ_g^+ (symmetric stretch) mode,[238] $\boldsymbol{\alpha}_1 : \boldsymbol{\alpha}_1 = 0.0132 \times 10^{-48}\,\text{cm}^6$. Therefore

$$\sigma^{\text{Ray}} = 1.87 \times 10^{-26}\,\text{cm}^2,$$

$$\sigma^{\text{Ram}} = 1.01 \times 10^{-29}\,\text{cm}^2.$$

Such cross-sections are typical of the values for many small molecules; note that Raman scattering is typically three or more orders of magnitude weaker than Rayleigh scattering.

11.3 Neutron scattering

11.3.1 Introduction

The differential cross-section for the inelastic scattering of neutrons by a molecular fluid is[14,239–241]

$$\frac{d^2\sigma}{d\Omega d\omega_S} = \frac{k_S}{k_0} \sum_{mn} P_m \langle m| \sum_{\lambda\alpha} b_\alpha \exp(-i\mathbf{k}\cdot\mathbf{r}_{\lambda\alpha})|n\rangle \times \sum_{\lambda'\beta} b_\beta \langle n| \exp(i\mathbf{k}\cdot\mathbf{r}_{\lambda'\beta})|m\rangle \delta(\omega - \omega_{nm}). \tag{11.348}$$

Here $\mathbf{r}_{\lambda\alpha}$ denotes the position of nucleus α in molecule λ, and b_α is its scattering length. $\mathbf{k}_0$ and $\mathbf{k}_S$ are the wavectors of incident and scattered neutrons; $\mathbf{k} = \mathbf{k}_0 - \mathbf{k}_S$. $\hbar\omega$ and $\hbar\mathbf{k}$ denote the energy and momentum, respectively, transferred to the system by a scattering event. $|m\rangle$ and $|n\rangle$ are state vectors describing the initial and final states of the system, with energies ω_m and ω_n, respectively, and $\omega_{nm} = \omega_n - \omega_m$. P_m is the probability of finding the system initially in state m. Conservation of energy in the scattering process requires $\omega = \omega_{nm} = \hbar/2m_N\left(k_0^2 - k_S^2\right)$, where m_N is the neutron mass.

By a procedure similar to that described for IR and Raman spectra, we may cast (11.348) in the form[242]

$$\frac{d^2\sigma}{d\Omega d\omega_S} = \frac{k_S}{k_0}\frac{1}{2\pi}\int_{-\infty}^{+\infty} dt \exp(-i\omega t) \times \sum_{\lambda\lambda'}\sum_{\alpha\beta} \overline{b_\alpha b_\beta} \left\langle \exp[-i\mathbf{k}\cdot\mathbf{r}_{\lambda\alpha}(0)] \exp[i\mathbf{k}\cdot\mathbf{r}_{\lambda'\beta}(t)]\right\rangle. \tag{11.349}$$

Here $\mathbf{r}_{\lambda\alpha}(t) = \exp(iHt/\hbar)\mathbf{r}_{\lambda\alpha}\exp(-iHt/\hbar)$ denotes the position of nucleus α of molecule λ at time t (H is the system Hamiltonian). $\overline{(\cdots)}$ denotes an average over nuclear spin states, while $\langle\cdots\rangle$ denotes an average over translational, rotational, and vibrational molecular degrees of freedom in the grand canonical ensemble. In what follows we shall restrict our discussion to an isotropic fluid, such as a bulk liquid or gas that is not subjected to any external fields. The right-hand side of (11.349) is then a function only of the magnitude of $\mathbf{k}$, and not its direction.

Assuming the spin states of neighbouring nuclei are uncorrelated, the scattering cross-section can be split into coherent and incoherent parts according to

$$\frac{\mathrm{d}^2\sigma}{\mathrm{d}\Omega d\omega_{\mathrm{S}}} = \left(\frac{\mathrm{d}^2\sigma}{\mathrm{d}\Omega d\omega_{\mathrm{S}}}\right)_{\mathrm{inc}} + \left(\frac{\mathrm{d}^2\sigma}{\mathrm{d}\Omega d\omega_{\mathrm{S}}}\right)_{\mathrm{coh}}, \tag{11.350}$$

where

$$\left(\frac{\mathrm{d}^2\sigma}{\mathrm{d}\Omega d\omega_{\mathrm{S}}}\right)_{\mathrm{inc}} = \frac{k_{\mathrm{S}}}{k_0}N\sum_\alpha b_\alpha^{\mathrm{inc}\,2}S_{\mathrm{self}}(k,\omega) \tag{11.351}$$

and

$$\left(\frac{\mathrm{d}^2\sigma}{\mathrm{d}\Omega d\omega_{\mathrm{S}}}\right)_{\mathrm{coh}} = \frac{k_{\mathrm{S}}}{k_0}N\left(\sum_\alpha \bar{b}_\alpha\right)^2 S(k,\omega). \tag{11.352}$$

Here N is the mean number of molecules in the fluid. $b_\alpha^{\mathrm{inc}} = \left(\overline{b_\alpha^2} - \bar{b}_\alpha^2\right)^{1/2}$ is the incoherent scattering length for nucleus α. $S_{\mathrm{self}}(k,\omega)$ and $S(k,\omega)$ are the dynamic self structure and structure factors, respectively, and are given by

$$S_{\mathrm{self}}(k,\omega) = \frac{1}{2\pi}\int_{-\infty}^{+\infty}\mathrm{d}t\,\exp(-i\omega t)F_{\mathrm{self}}(k,t) \tag{11.353}$$

and

$$S(k,\omega) = \frac{1}{2\pi}\int_{-\infty}^{+\infty}\mathrm{d}t\,\exp(-i\omega t)F(k,t), \tag{11.354}$$

where we have introduced the intermediate scattering functions $F_{\mathrm{self}}(k,t)$ and $F(k,t)$, defined by

$$F_{\mathrm{self}}(k,t) = N^{-1}\left(\sum_\alpha b_\alpha^{\mathrm{inc}\,2}\right)^{-1}\sum_\alpha b_\alpha^{\mathrm{inc}\,2} \times \sum_\lambda \langle\exp[-i\mathbf{k}\cdot\mathbf{r}_{\lambda\alpha}(0)]\exp[i\mathbf{k}\cdot\mathbf{r}_{\lambda\alpha}(t)]\rangle \tag{11.355}$$

and

$$F(k,t) = N^{-1}\left(\sum_{\alpha} \bar{b}_\alpha\right)^{-2} \sum_{\alpha\beta} \bar{b}_\alpha \bar{b}_\beta \times \sum_{\lambda\lambda'} \left\langle \exp[-i\mathbf{k}\cdot\mathbf{r}_{\lambda\alpha}(0)] \exp[i\mathbf{k}\cdot\mathbf{r}_{\lambda'\beta}(t)]\right\rangle. \quad (11.356)$$

The relations converse to (11.353) and (11.354) are

$$F_{\text{self}}(k,t) = \int_{-\infty}^{+\infty} \mathrm{d}t\, \exp(i\omega t) S_{\text{self}}(k,\omega),$$
$$F(k,t) = \int_{-\infty}^{+\infty} \mathrm{d}t\, \exp(i\omega t) S(k,\omega). \quad (11.357)$$

Equation (11.348) shows that $S_{\text{self}}(k,\omega)$ and $S(k,\omega)$ are real, and implies the detailed balance conditions

$$S_{\text{self}}(k,-\omega) = \exp(-\beta\hbar\omega) S_{\text{self}}(k,\omega),$$
$$S(k,-\omega) = \exp(-\beta\hbar\omega) S(k,\omega). \quad (11.358)$$

These results can be used to establish a relation between the real and imaginary parts of $F(k,t)$ (or $F_{\text{self}}(k,t)$). We first introduce the symmetrized and antisymmetrized functions

$$S_\pm(k,\omega) = \frac{1}{2}(S(k,\omega) \pm S(k,-\omega)) = \frac{1}{2} S(k,\omega)(1 \pm \exp(-\beta\hbar\omega)), \quad (11.359)$$

which are related by

$$S_+(k,\omega)/S_-(k,\omega) = \coth(\beta\hbar\omega/2). \quad (11.360)$$

In terms of these functions we have

$$\Re F(k,t) = \int_{-\infty}^{+\infty} \mathrm{d}\omega\, \exp(i\omega t) S_+(k,\omega),$$
$$\Im F(k,t) = -i\int_{-\infty}^{+\infty} \mathrm{d}\omega\, \exp(i\omega t) S_-(k,\omega). \quad (11.361)$$

Substituting (11.360) in (11.361), and using the relations $\tan ix = i\tanh x$, $\cot ix = -i\coth x$, we find that

$$\Re F(k,t) = \int_{-\infty}^{+\infty} \mathrm{d}\omega\, \coth\left(\frac{1}{2}\beta\hbar\omega\right) \exp(i\omega t) S_-(k,\omega)$$
$$= -\cot\left(\frac{1}{2}\beta\hbar\frac{d}{dt}\right) \Im F(k,t), \quad (11.362)$$

and

$$\Im F(k, t) = -i \int_{-\infty}^{+\infty} d\omega \tanh\left(\frac{1}{2}\beta\hbar\omega\right) \exp(i\omega t) S_{+}(k, \omega)$$
$$= -\tan\left(\frac{1}{2}\beta\hbar\frac{d}{dt}\right) \Re F(k, t). \tag{11.363}$$

Similar relations hold for $F_{\text{self}}(k, t)$.

We also note the relations

$$F_{\text{self}}(k, t)^* = F_{\text{self}}(k, -t) = F_{\text{self}}(k, t + i\beta\hbar),$$
$$F(k, t)^* = F(k, -t) = F(k, t + i\beta\hbar). \tag{11.364}$$

The first equality follows from the fact that $S(k, \omega)$ is real, the second from (11.358).

For rigid molecules the intermediate scattering functions functions may be expressed[14,239] as spatial Fourier transforms of the van Hove correlation function[243]

$$G(\mathbf{r}\omega_1\omega_2 t) = \rho^{-2}\langle\rho(\mathbf{r}_1\omega_1 0)\rho(\mathbf{r}_2\omega_2 t)\rangle, \tag{11.365}$$

where $\mathbf{r} = \mathbf{r}_2 - \mathbf{r}_1$, and

$$\rho(\mathbf{r}_1\omega_1 t) = \sum_{\lambda} \delta[\mathbf{r}_1 - \mathbf{r}_\lambda(t)]\delta[\omega_1 - \omega_\lambda(t)] \tag{11.366}$$

is the local number density for molecules with orientation ω_1 at position $\mathbf{r}_1$ at time t; we note that $\langle\rho(\mathbf{r}_1\omega_1 t)\rangle = \rho$. $G(\mathbf{r}\omega_1\omega_2 t)$ determines the probability that a molecule will be at position $\mathbf{r}_2$ in orientation ω_2 at time t, given that a molecule (possibly the same one) was at position $\mathbf{r}_1$ in orientation ω_1 at time zero. We can split this function into self and distinct parts as follows:

$$G(\mathbf{r}\omega_1\omega_2 t) = G_{\text{self}}(\mathbf{r}\omega_1\omega_2 t) + G_{\text{dis}}(\mathbf{r}\omega_1\omega_2 t), \tag{11.367}$$

where

$$G_{\text{self}}(\mathbf{r}\omega_1\omega_2 t)$$
$$= \sum_{\lambda} \langle\delta[\mathbf{r}_1 - \mathbf{r}_\lambda(0)]\delta[\mathbf{r}_2 - \mathbf{r}_\lambda(t)]\delta[\omega_1 - \omega_\lambda(0)]\delta[\omega_2 - \omega_\lambda(t)]\rangle \tag{11.368}$$

and

$$G_{\text{dis}}(\mathbf{r}\omega_1\omega_2 t)$$
$$= \sum_{\lambda \neq \lambda'} \langle\delta[\mathbf{r}_1 - \mathbf{r}_\lambda(0)]\delta[\mathbf{r}_2 - \mathbf{r}_{\lambda'}(t)]\delta[\omega_1 - \omega_\lambda(0)]\delta[\omega_2 - \omega_{\lambda'}(t)]\rangle\,. \tag{11.369}$$

$G_{\text{dis}}(\mathbf{r}\omega_1\omega_2 t)$ is the time-dependent generalization of the pair correlation function $g(\mathbf{r}\omega_1\omega_2)$ of (3.108). We note that[239,244]

$$G_{\text{self}}(\mathbf{r}\omega_1\omega_2 0) = \rho^{-1}\delta(\mathbf{r})\delta(\omega_1 - \omega_2), \qquad G_{\text{self}}(\mathbf{r}\omega_1\omega_2\infty) = N^{-1},$$
$$G_{\text{dis}}(\mathbf{r}\omega_1\omega_2 0) = g(\mathbf{r}\omega_1\omega_2), \qquad G_{\text{dis}}(\mathbf{r}\omega_1\omega_2\infty) = 1. \tag{11.370}$$

The infinite-time limits follow from the fact that as $t \to \infty$, molecules at $(\mathbf{r}_1\omega_1 0)$ and $(\mathbf{r}_2\omega_2 t)$ become uncorrelated. For similar reasons, for all t we have

$$\lim_{r\to\infty} G_{\text{self}}(\mathbf{r}\omega_1\omega_2 t) = N^{-1}, \qquad \lim_{r\to\infty} G_{\text{dis}}(\mathbf{r}\omega_1\omega_2 t) = 1. \tag{11.371}$$

For molecules that are only weakly anisotropic (e.g. nitrogen, methane), the self part of the van Hove correlation function approximately factors into translational and rotational parts,

$$G_{\text{self}}(\mathbf{r}\omega_1\omega_2 t) \approx G_{\text{self}}(r,t)G(\omega_1\omega_2 t), \tag{11.372}$$

while for the distinct part we have

$$G_{\text{dis}}(\mathbf{r}\omega_1\omega_2 t) \approx G_{\text{dis}}(r,t). \tag{11.373}$$

Here $G_{\text{self}}(r,t)$ and $G_{\text{dis}}(r,t)$ are the self and distinct centres correlation functions, while $G(\omega_1\omega_2 t)$ is the rotational van Hove function describing single-molecule reorientational dynamics. Expressions for $G(\omega_1\omega_2 t)$ for linear, spherical, and symmetric top molecules have been derived.[245]

If we also define the form factors

$$F_\alpha(\mathbf{k}\omega_1) = \exp(i\mathbf{k}\cdot\mathbf{d}_\alpha(\omega_1)),$$
$$F(\mathbf{k}\omega_1) = \sum_\alpha \bar{b}_\alpha F_\alpha(\mathbf{k}\omega_1) \Big/ \sum_\alpha \bar{b}_\alpha, \tag{11.374}$$

where $\mathbf{d}_\alpha(\omega_1)$ is the vector specifying the position of nucleus α within a molecule, we can now write the self scattering function as

$$F_{\text{self}}(k,t) = \rho\left(\sum_\alpha b_\alpha^{\text{inc}\,2}\right)^{-1} \sum_\alpha b_\alpha^{\text{inc}\,2}$$
$$\times \int d\mathbf{r}\, \exp(i\mathbf{k}\cdot\mathbf{r}) \left\langle G_{\text{self}}(\mathbf{r}\omega_1\omega_2 t)F_\alpha(\mathbf{k}\omega_1)^* F_\alpha(\mathbf{k}\omega_2)\right\rangle_{\omega_1\omega_2}, \tag{11.375}$$

and the scattering function as a sum of intramolecular and intermolecular terms,

$$F(k,t) = F_{\text{intra}}(k,t) + F_{\text{inter}}(k,t), \tag{11.376}$$

where

$$F_{\text{intra}}(k,t) = \rho \int d\mathbf{r}\, \exp(i\mathbf{k}\cdot\mathbf{r}) \left\langle G_{\text{self}}(\mathbf{r}\omega_1\omega_2 t)F(\mathbf{k}\omega_1)^* F(\mathbf{k}\omega_2)\right\rangle_{\omega_1\omega_2} \tag{11.377}$$

and

$$F_{\text{inter}}(k,t) = \rho \int d\mathbf{r}\, \exp(i\mathbf{k}\cdot\mathbf{r}) \left\langle G_{\text{dis}}(\mathbf{r}\omega_1\omega_2 t)F(\mathbf{k}\omega_1)^* F(\mathbf{k}\omega_2)\right\rangle_{\omega_1\omega_2}. \quad (11.378)$$

The intermolecular scattering function can be written as

$$F_{\text{inter}}(k,t) = F_{\text{fs}}(k) + \rho \int d\mathbf{r}\, \exp(i\mathbf{k}\cdot\mathbf{r}) \left\langle (G_{\text{dis}}(\mathbf{r}\omega_1\omega_2 t) - 1)F(\mathbf{k}\omega_1)^* F(\mathbf{k}\omega_2)\right\rangle_{\omega_1\omega_2}, \quad (11.379)$$

where $F_{\text{fs}}(k) = \rho \int d\mathbf{r}\, \exp(i\mathbf{k}\cdot\mathbf{r}) = (2\pi)^3 \rho\delta(\mathbf{k})$ is the 'forward scattering', which vanishes except along the direction of the incident neutron beam, and may be ignored for most purposes. (We have used the fact that $F(\mathbf{k},\omega) = 1$ for $\mathbf{k} = 0$, cf. (11.314) above.) From (11.371), the remaining term on the right of (11.379) is absolutely convergent.

The above relations may be combined to express $S_{\text{self}}(k,\omega)$ and $S(k,\omega)$ directly in terms of $G_{\text{self}}(\mathbf{r}\omega_1\omega_2 t)$ and $G_{\text{dis}}(\mathbf{r}\omega_1\omega_2 t)$. We find that

$$S_{\text{self}}(k,\omega) = \frac{1}{2\pi}\rho \left(\sum_\alpha b_\alpha^{\text{inc}\,2}\right)^{-1} \sum_\alpha b_\alpha^{\text{inc}\,2} \times \iint d\mathbf{r}\, dt\, \exp[i(\mathbf{k}\cdot\mathbf{r} - \omega t)] \left\langle G_{\text{self}}(\mathbf{r}\omega_1\omega_2 t)F_\alpha(\mathbf{k}\omega_1)^* F_\alpha(\mathbf{k}\omega_2)\right\rangle_{\omega_1\omega_2}, \quad (11.380)$$

and

$$S(k,\omega) = S_{\text{intra}}(k,\omega) + S_{\text{inter}}(k,\omega), \quad (11.381)$$

where

$$S_{\text{intra}}(k,\omega) = \frac{1}{2\pi}\rho \iint d\mathbf{r}\, dt\, \exp[i(\mathbf{k}\cdot\mathbf{r} - \omega t)] \times \left\langle G_{\text{self}}(\mathbf{r}\omega_1\omega_2 t)F(\mathbf{k}\omega_1)^* F(\mathbf{k}\omega_2)\right\rangle_{\omega_1\omega_2} \quad (11.382)$$

and

$$S_{\text{inter}}(k,\omega) = \frac{1}{2\pi}\rho \iint d\mathbf{r}\, dt\, \exp[i(\mathbf{k}\cdot\mathbf{r} - \omega t)] \times \left\langle [G_{\text{dis}}(\mathbf{r}\omega_1\omega_2 t) - 1]F(\mathbf{k}\omega_1)^* F(\mathbf{k}\omega_2)\right\rangle_{\omega_1\omega_2}. \quad (11.383)$$

Thus the dynamic structure factor is a space-time transform of the van Hove correlation function, and represents the decomposition of molecular translational and rotational motions into travelling waves. Essentially, $S(k,\omega)$ measures the amplitude of those waves that propagate with wavevector k and frequency ω. Note that (11.380), (11.382) and (11.383) are not, in general, invertible—that is, knowledge of $S(k,\omega)$ is not sufficient to determine $G(\mathbf{r}\omega_1\omega_2 t)$ unlike atomic fluids (see below).

It is sometimes useful to introduce the concept of *partial* dynamic structure factors. From (11.354) and (11.356), we can write

$$S(k,\omega) = \left(\sum_{\alpha} \bar{b}_{\alpha}\right)^{-2} \sum_{\alpha\beta} \bar{b}_{\alpha}\bar{b}_{\beta}S_{\alpha\beta}(k,\omega), \tag{11.384}$$

where

$$S_{\alpha\beta}(k,\omega) = \frac{1}{2\pi}\int_{-\infty}^{+\infty} \mathrm{d}t\ \exp(-i\omega t)F_{\alpha\beta}(k,t) \tag{11.385}$$

and we have defined

$$\begin{aligned} F_{\alpha\beta}(k,t) &= N^{-1}\sum_{\lambda\lambda'}\langle \exp[-i\mathbf{k}\cdot\mathbf{r}_{\lambda\alpha}(0)]\exp[i\mathbf{k}\cdot\mathbf{r}_{\lambda'\beta}(t)]\rangle \\ &= \rho\int \mathrm{d}\mathbf{r}\ \exp(i\mathbf{k}\cdot\mathbf{r})G_{\alpha\beta}(r,t). \end{aligned} \tag{11.386}$$

Here the partial van Hove correlation function is given by

$$G_{\alpha\beta}(r,t) = \rho^{-2}\sum_{\lambda\lambda'}\left\langle \delta[\mathbf{r}_1 - \mathbf{r}_{\lambda\alpha}(0)]\delta[\mathbf{r}_2 - \mathbf{r}_{\lambda'\beta}(t)]\right\rangle_{\omega_1\omega_2}, \tag{11.387}$$

and is an obvious generalization of the site–site correlation function $g_{\alpha\beta}(r)$ of § 3.1.6.

Special cases

In the remainder of this section we consider some special cases in which the above expressions simplify.[239]

Atomic fluids For monatomic fluids such as argon, we have

$$\begin{aligned} S_{\mathrm{self}}(k,\omega) &= \frac{1}{2\pi}\rho\iint \mathrm{d}\mathbf{r}\,\mathrm{d}t\ \exp[i(\mathbf{k}\cdot\mathbf{r}-\omega t)]G_{\mathrm{self}}(r,t), \\ S(k,\omega) &= \frac{1}{2\pi}\rho\iint \mathrm{d}\mathbf{r}\,\mathrm{d}t\ \exp[i(\mathbf{k}\cdot\mathbf{r}-\omega t)][G(r,t)-1], \end{aligned} \tag{11.388}$$

which can be inverted to yield

$$\begin{aligned} G_{\mathrm{self}}(r,t) &= \left(\frac{1}{2\pi}\right)^3\rho^{-1}\iint \mathrm{d}\mathbf{k}\,\mathrm{d}\omega\ \exp[-i(\mathbf{k}\cdot\mathbf{r}-\omega t)]S_{\mathrm{self}}(k,\omega), \\ G(r,t)-1 &= \left(\frac{1}{2\pi}\right)^3\rho^{-1}\iint \mathrm{d}\mathbf{k}\,\mathrm{d}\omega\ \exp[-i(\mathbf{k}\cdot\mathbf{r}-\omega t)]S(k,\omega). \end{aligned} \tag{11.389}$$

Measurement of the coherent and incoherent scattering cross-sections over a range of momentum and energy transfers could therefore serve, in principle, to determine the time-dependent atomic correlation functions $G(r,t)$ and $G_{\mathrm{self}}(r,t)$.

Short times Setting $t = 0$ in (11.375)–(11.378), and using (11.370), we find that

$$F_{\text{self}}(k, 0) = 1,$$
$$F_{\text{intra}}(k, 0) = S_{\text{intra}}(k),$$
$$F_{\text{inter}}(k, 0) = S_{\text{inter}}(k), \tag{11.390}$$

where

$$S_{\text{intra}}(k) = \left\langle |F(\mathbf{k}\omega_1)|^2 \right\rangle_{\omega_1},$$
$$S_{\text{inter}}(k) = \rho \int d\mathbf{r}\ \exp(i\mathbf{k}\cdot\mathbf{r}) \left\langle g(\mathbf{r}\omega_1\omega_2) F(\mathbf{k}\omega_1)^* F(\mathbf{k}\omega_2) \right\rangle_{\omega_1\omega_2}, \tag{11.391}$$

are the intramolecular and intermolecular static structure factors discussed in Chapter 9.

Long times In the limit $t \to \infty$, again using (11.370), we find that

$$F_{\text{self}}(k, \infty) = F_{\text{intra}}(k, \infty) = F_{\text{inter}}(k, \infty) = 0. \tag{11.392}$$

Small k In the limit $k \to 0$, we have from (11.371)

$$F_{\text{self}}(0, t) = 1,$$
$$F_{\text{intra}}(0, t) = 1,$$
$$F_{\text{inter}}(0, t) = \rho \int d\mathbf{r}\, [g(\mathbf{r}\omega_1\omega_2) - 1] = \rho kT\chi - 1, \tag{11.393}$$

where χ is the isothermal compressibility of (3.113). Strictly speaking, the result for $F_{\text{inter}}(0, t)$ applies for arbitrarily small $k > 0$, since it neglects the extensive forward scattering contribution. For the zero-wavevector structure factors we therefore have

$$S_{\text{self}}(0, \omega) = \delta(\omega),$$
$$S(0, \omega) = \rho kT\chi\ \delta(\omega), \tag{11.394}$$

and the scattering is purely elastic.

Large k We next consider the limit $k \to \infty$. From (11.378), intermolecular contributions to the scattering can only be significant for $k \lesssim \sigma^{-1}$, where σ is the closest distance of molecular approach. Therefore

$$F_{\text{inter}}(\infty, t) = 0, \qquad S_{\text{inter}}(\infty, \omega) = 0. \tag{11.395}$$

For the self scattering and intramolecular terms, there will be contributions to (11.375) and (11.377) for $k \lesssim (\bar{v}t)^{-1}$, where $\bar{v}$ is the mean molecular speed.

Therefore $F_{\text{self}}(\infty, t)$ and $F_{\text{intra}}(\infty, t)$ vanish except at zero time. From the zero-time behaviour of $G_{\text{self}}(\mathbf{r}\omega_1\omega_2 t)$ as given by (11.370), we have

$$F_{\text{self}}(\infty, 0) = 1,$$
$$F_{\text{intra}}(\infty, 0) = \sum_\alpha \bar{b}_\alpha^2 \Big/ \left(\sum_\alpha \bar{b}_\alpha\right)^2. \tag{11.396}$$

In this limit $S_{\text{self}}(k, \omega)$ and $S_{\text{intra}}(k, \omega)$ extend up to frequencies of about $k\bar{v} \sim k(k_B T/m)^{1/2}$, where m is the molecular mass. Thus the inelasticity increases proportionally to k.

Free rotation In the free-rotation limit the intermediate scattering functions factorize into translational and rotational parts. Substituting (11.372) and (11.373) into (11.375)–(11.378), we find that

$$F_{\text{self}}(k, t) = F_{\text{self}}^{\text{trans}}(k, t) F_{\text{self}}^{\text{rot}}(k, t),$$
$$F_{\text{intra}}(k, t) = F_{\text{self}}^{\text{trans}}(k, t) F_{\text{intra}}^{\text{rot}}(k, t),$$
$$F_{\text{inter}}(k, t) = F_{\text{dis}}^{\text{trans}}(k, t) F_{\text{inter}}^{\text{rot}}(k, t), \tag{11.397}$$

where

$$F_{\text{self}}^{\text{trans}}(k, t) = \rho \int \mathrm{d}\mathbf{r}\, \exp(i\mathbf{k}\cdot\mathbf{r}) G_{\text{self}}(r, t),$$
$$F_{\text{dis}}^{\text{trans}}(k, t) = \rho \int \mathrm{d}\mathbf{r}\, \exp(i\mathbf{k}\cdot\mathbf{r}) G_{\text{dis}}(r, t), \tag{11.398}$$

and

$$F_{\text{self}}^{\text{rot}}(k, t) = \left(\sum_\alpha b_\alpha^{\text{inc}\,2}\right)^{-1} \sum_\alpha b_\alpha^{\text{inc}\,2} \langle G(\omega_1\omega_2 t) F_\alpha(\mathbf{k}\omega_1)^* F_\alpha(\mathbf{k}\omega_2)\rangle_{\omega_1\omega_2},$$
$$F_{\text{intra}}^{\text{rot}}(k, t) = \langle G(\omega_1\omega_2 t) F(\mathbf{k}\omega_1)^* F(\mathbf{k}\omega_2)\rangle_{\omega_1\omega_2},$$
$$F_{\text{inter}}^{\text{rot}}(k, t) = |\langle F(\mathbf{k}\omega_1)\rangle|^2. \tag{11.399}$$

The associated dynamic structure factors are then given by the convolutions

$$S_{\text{self}}(k, \omega) = \int_{-\infty}^{+\infty} \mathrm{d}\omega'\, S_{\text{self}}^{\text{trans}}(k, \omega) S_{\text{self}}^{\text{rot}}(k, \omega - \omega'),$$
$$S_{\text{intra}}(k, \omega) = \int_{-\infty}^{+\infty} \mathrm{d}\omega'\, S_{\text{self}}^{\text{trans}}(k, \omega) S_{\text{intra}}^{\text{rot}}(k, \omega - \omega'),$$
$$S_{\text{inter}}(k, \omega) = \int_{-\infty}^{+\infty} \mathrm{d}\omega'\, S_{\text{dis}}^{\text{trans}}(k, \omega') S_{\text{inter}}^{\text{rot}}(k, \omega - \omega'). \tag{11.400}$$

11.3.2 Dynamic structure factor for gases

We now move on to a discussion of the form of the dynamic structure and self structure factors for dilute atomic and molecular gases. We will show how $S_{\mathrm{self}}(k,\omega)$ and $S(k,\omega)$ may be calculated both classically and quantum-mechanically. At the end of this section we briefly discuss the density dependence of these quantities.

Atomic gases

Classically, the van Hove correlation function for a dilute atomic gas is[246]

$$\rho G^{\mathrm{cl}}(r,t) = \langle \delta(\mathbf{r}-\mathbf{v}t)\rangle_p, \tag{11.401}$$

where $\langle\cdots\rangle_p$ denotes an average over the momentum $\mathbf{p} = M\mathbf{v}$, with M the atomic mass. Using the results of § 3.1.2 we therefore have

$$\rho G^{\mathrm{cl}}(r,t) = \left(\frac{\beta M}{2\pi t^2}\right)^{3/2} \exp\left(-\frac{\beta M r^2}{2t^2}\right). \tag{11.402}$$

Thus, from (11.398), (11.402) and (B.89), we find

$$\begin{aligned} F^{\mathrm{cl}}(k,t) &= 4\pi\left(\frac{\beta M}{2\pi t^2}\right)^{3/2}\int_0^\infty \mathrm{d}r\, r^2 j_0(kr)\exp\left(-\frac{\beta M r^2}{2t^2}\right) \\ &= \exp\left(-\frac{k^2t^2}{2\beta M}\right). \end{aligned} \tag{11.403}$$

The corresponding dynamic structure factor is therefore Gaussian,

$$S^{\mathrm{cl}}(k,\omega) = \frac{1}{\pi^{1/2}\bar{\omega}(k)}\exp\left(-\frac{\omega^2}{\bar{\omega}(k)^2}\right), \tag{11.404}$$

with linewidth $\bar{\omega}(k) = (2/\beta M)^{1/2}k$.

In a quantum treatment, a convenient starting point is

$$F^{\mathrm{qu}}(k,t) = \langle \exp(-i\mathbf{k}\cdot\mathbf{r}(0))\exp(i\mathbf{k}\cdot\mathbf{r}(t))\rangle\,. \tag{11.405}$$

We note that the operators $\mathbf{r}(0)$ and $\mathbf{r}(t)$ do not commute. For a free particle $\mathbf{r}(t) = \mathbf{r}(0) + \mathbf{p}(0)t/M$, and therefore $[r_\alpha(0), r_\beta(t)] = t/M[r_\alpha, p_\beta] = i\hbar t/M\delta_{\alpha\beta}$; thus $[\mathbf{k}\cdot\mathbf{r}(0), \mathbf{k}\cdot\mathbf{r}(t)] = i\hbar k^2 t/M$. Now for two operators A and B whose commutator is a C-number,[247]

$$\exp(A)\exp(B) = \exp\left(A + B + \frac{1}{2}[A,B]\right). \tag{11.406}$$

Thus

$$F^{\mathrm{qu}}(k,t) = \left\langle \exp\left(\frac{i(\mathbf{k}\cdot\mathbf{p})t}{M} + \frac{i\hbar k^2 t}{2M}\right)\right\rangle. \tag{11.407}$$

The average over momenta may now be performed classically, since the translational motion of a free particle is not quantized. This yields

$$F^{\mathrm{qu}}(k,t) = \exp\left(-\frac{k^2t^2}{2\beta M} + \frac{i\hbar k^2 t}{2M}\right)$$
$$= F^{\mathrm{cl}}(k,t)\exp\left(\frac{i\hbar k^2 t}{2M}\right). \tag{11.408}$$

Introducing the complex or Egelstaff time τ, defined by $\tau^2 = t(t - i\beta\hbar)$, this is

$$F^{\mathrm{qu}}(k,t) = \exp\left(-\frac{k^2\tau^2}{2\beta M}\right). \tag{11.409}$$

The quantum van Hove correlation function is similarly given by an expression of the form (11.402), but with t replaced by τ. We see that quantum effects render both $F(k,t)$ and $G(r,t)$ complex for times $t \gtrsim 2M/\hbar k^2$. However, these functions become very small for $t \gtrsim (2\beta M)^{1/2}/k$. Therefore quantum effects are only important in the range $2M/\hbar k^2 \lesssim t \lesssim (2\beta M)^{1/2}/k$, a condition that can only be satisfied provided $k \gtrsim k_{\max}$, where $k_{\max} = 2\pi/\Lambda_t$ and $\Lambda_t = (2\pi\beta\hbar^2/M)^{1/2}$ is the de Broglie wavelength (§ 1.2.2).

The dynamic structure factor corresponding to (11.408) is

$$S^{\mathrm{qu}}(k,\omega) = \frac{1}{\pi^{1/2}\bar{\omega}(k)}\exp\left(-\frac{(\hbar k^2/2M - \omega)^2}{\bar{\omega}(k)^2}\right). \tag{11.410}$$

Considered in frequency space, the quantum structure factor is therefore a Gaussian with the same width as its classical counterpart, but shifted by $\Delta\omega = \hbar k^2/2M$. Physically, this shift may be considered to arise from the recoil of the atom when it scatters a neutron: transfer of momentum $\hbar\mathbf{k}$ to the atom skews the underlying Maxwellian velocity distribution. If we write

$$S^{\mathrm{qu}}(k,\omega) = S^{\mathrm{cl}}(k,\omega)\exp\left(\frac{1}{2}\beta\hbar\omega - \frac{\beta\hbar^2k^2}{8M}\right) \tag{11.411}$$

we see that quantum effects are significant only for large energy or momentum transfers, i.e. when $|\omega| \gtrsim \omega_{\max}$, where $\omega_{\max} = k_B T/\hbar$, or $k \gtrsim k_{\max}$. For argon at room temperature, $\omega_{\max} = 40\,\mathrm{ps}^{-1}$ and $k_{\max} = 40\,\text{Å}^{-1}$.

In moderately dense systems, two- and three-body interactions become important. Provided the density is not too high, $S(k,\omega)/S(k)$ may be expanded as a virial series,

$$\frac{S(k,\omega)}{S(k)} = \sum_{n\geq 1} S^{(n)}(k,\omega)\rho^{n-1}, \tag{11.412}$$

where $S^{(n)}(k, \omega)$ is the nth virial coefficient for coherent scattering, describing n-body dynamics in the fluid. Moraldi *et al.*[248] have given formal expressions for the second and third coefficients for an atomic fluid.

Miyazaki and de Schepper[249] have computed $S^{(2)}(k, \omega)$ for LJ argon using generalized Enskog theory. They find that

$$S^{(2)}(k, \omega) = \frac{1}{\pi}\Re\left\langle \frac{1}{i\omega + i\mathbf{k}\cdot\mathbf{v}} \Lambda_E(\mathbf{k}, \omega) \frac{1}{i\omega + i\mathbf{k}\cdot\mathbf{v}} \right\rangle_{\mathbf{v}}, \tag{11.413}$$

where $\langle \cdots \rangle_{\mathbf{v}} = \int d\mathbf{v} f(v)(\cdots)$ denotes an average over the Maxwell velocity distribution, and $\Lambda_{\mathrm{E}}(\mathbf{k}, \omega)$ is the Enskog operator in the pair approximation, a k- and ω-dependent generalization of the well-known Boltzmann operator Λ_{B}. $\Lambda_{\mathrm{E}}(\mathbf{k}, \omega)$ is given explicitly as a sum of mean-field and collision terms; the latter can be evaluated numerically by integrating the equations of motion of a colliding pair of atoms. Miyazaki and de Schepper found that (11.413) gave results in good overall agreement with the experimental measurements of Verkerk *et al.*,[250] whereas the Boltzmann equation failed for k values on the order of the inverse interaction length σ^{-1}.

Molecular gases—classical treatment

For simplicity we will consider a gas of homonuclear diatomic molecules.[251] An analogous treatment of spherical tops has been given by Sears.[252] Since molecular vibration cannot be adequately described within the framework of classical mechanics, we will assume the molecules are rigid.

In a dilute gas, the translational and rotational molecular motions decouple. The problem therefore reduces to that of computing the translational and rotational functions defined in (11.397)–(11.400). The translational structure factor $S_{\mathrm{self}}^{\mathrm{trans}}(k, t)$ is identical to $S(k, \omega)$ for an atomic gas, (11.404). It remains to calculate the rotational functions.

For a homonuclear diatomic of bond length $2d$, the intermediate rotational self scattering function is given from (11.399) by

$$F_{\mathrm{self}}^{\mathrm{rot}}(k, t) = \frac{1}{2}\langle G(\omega_1\omega_2 t)[F_a(\mathbf{k}\omega_1)^* F_a(\mathbf{k}\omega_2) + F_b(\mathbf{k}\omega_1)^* F_b(\mathbf{k}\omega_2)]\rangle, \tag{11.414}$$

where a, b label the two nuclei and $F_a(\mathbf{k}, \omega_1) = F_b(\mathbf{k}, \omega_1)^* = \exp(i\mathbf{k}\cdot\mathbf{d}_1)$, wth $\mathbf{d}_1 \equiv \mathbf{d}_a(\omega_1)$. Hence

$$\begin{aligned} F_{\mathrm{self}}^{\mathrm{rot}}(k, t) = \langle G(\omega_1\omega_2 t)[&\cos(\mathbf{k}\cdot\mathbf{d}_1)\cos(\mathbf{k}\cdot\mathbf{d}_2) \\ &+ \sin(\mathbf{k}\cdot\mathbf{d}_1)\sin(\mathbf{k}\cdot\mathbf{d}_2)]\rangle_{\omega_1\omega_2}. \end{aligned} \tag{11.415}$$

Similarly, the intramolecular function is given by

$$F^{\text{rot}}_{\text{intra}}(k,t) = \frac{1}{4}\langle G(\omega_1\omega_2 t)[F_a(\mathbf{k}\omega_1)^* F_a(\mathbf{k}\omega_2) + F_a(\mathbf{k}\omega_1)^* F_b(\mathbf{k}\omega_2)$$
$$+ F_b(\mathbf{k}\omega_1)^* F_a(\mathbf{k}\omega_2) + F_b(\mathbf{k}\omega_1)^* F_b(\mathbf{k}\omega_2)]\rangle$$
$$= \langle G(\omega_1\omega_2 t)\cos(\mathbf{k}\cdot\mathbf{d}_1)\cos(\mathbf{k}\cdot\mathbf{d}_2)\rangle_{\omega_1\omega_2}\,. \tag{11.416}$$

Since these functions depend only on the magnitude of **k** and not its direction, we can average them over all orientations ω_k. We first expand $\cos(\mathbf{k}\cdot\mathbf{d}) = \Re\exp(i\mathbf{k}\cdot\mathbf{d})$ and $\sin(\mathbf{k}\cdot\mathbf{d}) = \Im\exp(i\mathbf{k}\cdot\mathbf{d})$ in spherical harmonics using (B.89). Using also (A.27) and (A.33), we find that

$$\langle\cos(\mathbf{k}\cdot\mathbf{d}_1)\cos(\mathbf{k}\cdot\mathbf{d}_2)\rangle_{\omega_k} = \sum_{l\,\text{even}}(2l+1)j_l(kd)^2 P_l(\cos\Omega),$$
$$\langle\sin(\mathbf{k}\cdot\mathbf{d}_1)\sin(\mathbf{k}\cdot\mathbf{d}_2)\rangle_{\omega_k} = \sum_{l\,\text{odd}}(2l+1)j_l(kd)^2 P_l(\cos\Omega), \tag{11.417}$$

where $j_l(x)$ denotes a spherical Bessel function and $P_l(x)$ a Legendre polynomial, and $\cos\Omega = \mathbf{d}_1\cdot\mathbf{d}_2$. Note that Ω measures the angle of rotation about the axis defining the angular momentum vector **J**.

For a freely rotating linear molecule,[245]

$$G(r\omega_1\omega_2 t) = \left\langle\sum_l(2l+1)P_l(\cos\Omega)P_l(\cos\dot{\Omega}t)\right\rangle_J, \tag{11.418}$$

where $\dot{\Omega} = J/I$ is the angular velocity, with I the moment of inertia. From § 3.1.2, the average $\langle\cdots\rangle_J \equiv (\beta/I)\int_0^\infty \mathrm{d}J\,J\exp(-\beta J^2/2I)(\ldots)$.

Using (11.415)–(11.416), and noting that $\langle P_l(\cos\Omega)P_{l'}(\cos\Omega)\rangle_{\omega_1\omega_2} = (2l+1)^{-1}\delta_{ll'}$, we find that[253]

$$F^{\text{rot}}_{\text{self}}(k,t) = \sum_l(2l+1)j_l(kd)^2\langle P_l(\cos\dot{\Omega}t)\rangle_J,$$
$$F^{\text{rot}}_{\text{intra}}(k,t) = \sum_{l\,\text{even}}(2l+1)j_l(kd)^2\langle P_l(\cos\dot{\Omega}t)\rangle_J. \tag{11.419}$$

The rotational structure factors are correspondingly

$$S^{\text{rot}}_{\text{self}}(k,\omega) = \sum_l(2l+1)j_l(kd)^2 S^{\text{rot}}_l(k,\omega),$$
$$S^{\text{rot}}_{\text{intra}}(k,\omega) = \sum_{l\,\text{even}}(2l+1)j_l(kd)^2 S^{\text{rot}}_l(k,\omega),$$
$$\tag{11.420}$$

where $S_l^{\mathrm{rot}}(k,\omega) = \langle s_l^{\mathrm{rot}}(k,\omega)\rangle_J$ and

$$s_l^{\mathrm{rot}}(k,\omega) = \frac{1}{2\pi}\int_{-\infty}^{+\infty} \mathrm{d}t\ \exp(-i\omega t) P_l(\cos\dot{\Omega}t). \tag{11.421}$$

Using the expansion[136]

$$P_l(\cos x) = \sum_m a_{lm} \cos mx,$$

$$a_{lm} = \frac{(l-m-1)!!(l+m-1)!!}{(l-m)!!(l+m)!!}, \tag{11.422}$$

where the sum over m runs over the values $-l, -l+2, \ldots l$, we find that

$$s_l^{\mathrm{rot}}(k,\omega) = \frac{1}{2}\sum_m a_{lm}[\delta(\omega - m\dot{\Omega}) + \delta(\omega + m\dot{\Omega})], \tag{11.423}$$

and therefore

$$S_l^{\mathrm{rot}}(k,\omega) = \frac{1}{2}\left(1+(-)^l\right)\left(\frac{(l-1)!!}{(l+1)!!}\right)^2 \delta(\omega) + \frac{1}{2}\beta I|\omega| \sum_{m\neq 0} \frac{a_{lm}}{m^2} \exp\left(-\frac{\beta I\omega^2}{2m^2}\right). \tag{11.424}$$

The convolution of (11.424) with the translational function (11.404) is elementary. Noting that $I = Md^2$, we find for the total structure factors

$$S_{\mathrm{self}}(k,\omega) = \sum_l (2l+1) j_l(kd)^2 S_l(k,\omega),$$

$$S(k,\omega) = \sum_{l\,\mathrm{even}} (2l+1) j_l(kd)^2 S_l(k,\omega), \tag{11.425}$$

where

$$S_l(k,\omega) = \frac{1}{\pi^{1/2}\bar{\omega}(k)} \exp\left(-\frac{\omega^2}{\bar{\omega}(k)^2}\right) \times \sum_m a_{lm} \frac{k^2d^2}{k^2d^2+m^2}\left(1+\pi^{1/2}\phi_m \exp\left(\phi_m^2\right) \mathrm{erf}\phi_m\right), \tag{11.426}$$

where erf$x$ denotes the error function and

$$\phi_m = \frac{|m\omega|}{(k^2d^2+m^2)^{1/2}\bar{\omega}(k)}. \tag{11.427}$$

Molecular gases—quantum treatment

The quantum-mechanical calculation of the structure factors is most easily performed in the Schrödinger representation. We will consider only $S(k,\omega)$ here.

From (11.348) the internal (vibration–rotation) contribution to this function is

$$S_{\text{int}}(k,\omega) = \sum_{mn} P_m |\langle m| \cos(\mathbf{k}\cdot\mathbf{d})|n\rangle|^2 \delta(\omega - \omega_{nm}), \tag{11.428}$$

where m and n label internal nuclear states of the molecule. Using the Rayleigh expansion (B.89), and averaging over orientations of $\mathbf{k}$, we find that

$$S_{\text{int}}(k,\omega) = 4\pi \sum_{l\,\text{even}} \sum_{\lambda} \sum_{mn} P_m |\langle m| j_l(kd) Y_{l\lambda}(\theta_d\phi_d)|n\rangle|^2 \delta(\omega - \omega_{nm}). \tag{11.429}$$

In what follows we will neglect the effects of nuclear spin[240,251] and of centrifugal distortion.[254] Vibration and rotation then decouple, and $|m\rangle = |\nu\rangle|JM\rangle$, where ν labels the vibrational state (with energy $\hbar\omega_\nu$), and J and M are the usual rotational quantum numbers. In the coordinate representation, the rotational eigenfunctions are the spherical harmonics $\langle\theta_d\phi_d|JM\rangle = Y_{JM}(\theta_d\phi_d)$; the rotational eigenvalues are $\hbar\omega_J = \hbar BJ(J+1)$, where $B = \hbar/2I$ is the rotational constant. Therefore

$$\begin{aligned} S_{\text{int}}(k,\omega) = 4\pi &\sum_{l\,\text{even}} \sum_{\nu\nu'} P_\nu |\langle\nu| j_l(kd)|\nu'\rangle|^2 \\ &\times \sum_{JJ'} \sum_{M\lambda M'} P_J |\langle JM| Y_{l\lambda}|J'M'\rangle|^2 \delta(\omega - \omega_{\nu'J',\nu J}), \end{aligned} \tag{11.430}$$

where

$$\begin{aligned} P_\nu &= \exp(-\beta\hbar\omega_\nu) \left(\sum_\nu \exp(-\beta\hbar\omega_\nu) \right)^{-1}, \\ P_J &= \exp(-\beta\hbar BJ(J+1)) \left(\sum_J (2J+1) \exp(-\beta\hbar BJ(J+1)) \right)^{-1}. \end{aligned} \tag{11.431}$$

Now from (A.40), (A.141), and (A.139) we have

$$\sum_{M\lambda M'} |\langle JM| Y_{l\lambda}|J'M'\rangle|^2 = \frac{(2J+1)(2l+1)(2J'+1)}{4\pi} \begin{pmatrix} J & l & J' \\ 0 & 0 & 0 \end{pmatrix}^2. \tag{11.432}$$

We can therefore write

$$S_{\text{int}}(k,\omega) = \sum_{l\,\text{even}} \sum_{\nu\nu'} P_\nu (2l+1) |\langle\nu| j_l(kd)|\nu'\rangle|^2 S_l^{\text{rot}}(\omega - \omega_{\nu'\nu}), \tag{11.433}$$

where we have defined the rotational function

$$S_l^{\text{rot}}(\omega) = \sum_{JJ'} P_J (2J+1)(2J'+1) \begin{pmatrix} J & l & J' \\ 0 & 0 & 0 \end{pmatrix}^2 \delta(\omega - \omega_{J'J}). \tag{11.434}$$

To compute $S(k, \omega)$ we convolve $S_{\rm int}(k, \omega)$ with the quantum translational function of (11.410). This yields

$$S(k,\omega) = \sum_{l\,\text{even}} \sum_{\nu\nu'} P_\nu (2l+1) |\langle \nu | j_l(kd) | \nu' \rangle|^2 S_l(k, \omega - \omega_{\nu'\nu}), \qquad (11.435)$$

where

$$S_l(k,\omega) = \frac{1}{\pi^{1/2}\bar{\omega}(k)} \sum_{JJ'} P_J (2J+1)(2J'+1) \begin{pmatrix} J & l & J' \\ 0 & 0 & 0 \end{pmatrix}^2 \times \exp\left(-\frac{(\hbar k^2/2M + \omega_{J'J} - \omega)^2}{\bar{\omega}(k)^2} \right) \qquad (11.436)$$

is the quantum analogue of (11.426).

From (11.435) we see that $S(k, \omega)$ consists of a series of vibrational bands. For many molecules, e.g. H_2, N_2, Cl_2, only the vibrational ground state $\nu = 0$ is appreciably populated at or below room temperature. $S(k, \omega)$ therefore contains a pure rotational band corresponding to $\nu' = 0$ as well as fundamental ($\nu' = 1$) and overtone ($\nu' = 2, 3 \ldots$) vibrational bands.

Figure 11.14(a) shows the pure rotational band for $k\bar{d} = 2$ (where $2\bar{d}$ is the mean bond length) at temperatures below, equal to, and above the rotational temperature $\theta_{\rm r} = \hbar B/k_{\rm B}$ (§ 1.2.2). The vibrational matrix elements were calculated using parameters appropriate to H_2; for this gas the temperatures correspond to 18, 88, and 438 K and the wavector to 5 Å^{-1}, while one unit of reduced frequency is $B = 61$ cm^{-1}. In the low-temperature spectrum, the strong peaks at $\omega/B = 4$ and 10 are associated with the terms $l = 0$ and 2 in the sum (11.435), respectively, while the much weaker peak at $\omega/B = 24$ (barely visible in the figure) is associated with $l = 4$. These peaks correspond to transitions from the ground rotational state $J = 0$ to excited rotational levels $J' = 0, 2, \ldots, l$.

For small $k\bar{d}$, the fundamental vibrational band is a factor $\sim \gamma^2 (k\bar{d})^4 \ll 1$ weaker than the rotational band, where $\gamma = (2\alpha\bar{d})^{-1}$, with $\alpha = (\mu\omega_0/\hbar)^{1/2}$ and $\mu = M/4$ the oscillator reduced mass. Harmonics are weaker still. Figure 11.14(b) shows the fundamental band of H_2 for $k\bar{d} = 2$; here $\gamma = 0.166$. Note that because the l-dependence of the matrix elements $\langle 0 | j_l(kd) | 0 \rangle$ and $\langle 0 | j_l(kd) | 1 \rangle$ is different, the fundamental band is not simply a shifted copy of the rotational band, but displays an altered profile.

For large $k\bar{d}$, the vibrational bands attain a significant fraction of the intensity of the rotational band. The H_2 spectrum for $k\bar{d} = 5$ ($k = 13$ Å^{-1}) is shown in Fig. 11.15(a). At low temperatures the fundamental band is that portion of the spectrum lying above $\omega/B = 90$; at high temperatures it overlaps the rotational band. In the low-temperature rotational band the three peaks visible are associated with $l = 2$, 4 (the strongest), and 6; $l = 0$ transitions are hidden underneath the $l = 2$ peak. In the vibrational band the strongest peak is associated with $l = 2$. Figure 11.15(b) shows the spectrum for $k\bar{d} = 10$ ($k = 27$ Å^{-1}); here the

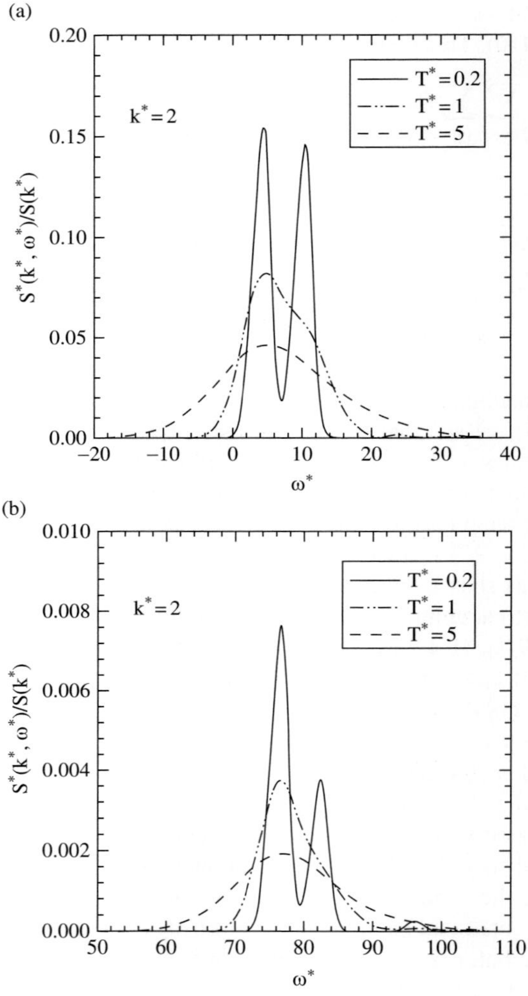

FIG. 11.14 $S(k, \omega)$ for H_2 gas at $k^* = 2$. (a) Rotational band; (b) fundamental band. Reduced units are used such that $k^* = k\bar{d}$, $\omega^* = \omega/B$, $S^*(k^*, \omega^*) = BS(k, \omega)$ and $T^* = T/\theta_r$.

vibrational and rotational bands have completely merged even at low temperatures. There are two strong peaks: that at lower frequency is primarily associated with $l = 8$ pure rotational and $l = 6$ fundamental transitions, while that at higher frequency derives from $l = 10$ rotational and $l = 8$ fundamental transitions. The two smaller peaks at lower frequencies are $l = 4$ and 6 rotational transitions, while the higher frequency peaks are mixtures of rotational band transitions with increasing amounts, as ω/B increases, of vibrational fundamental and overtone (primarily $\nu' = 2$ and 3) transitions.

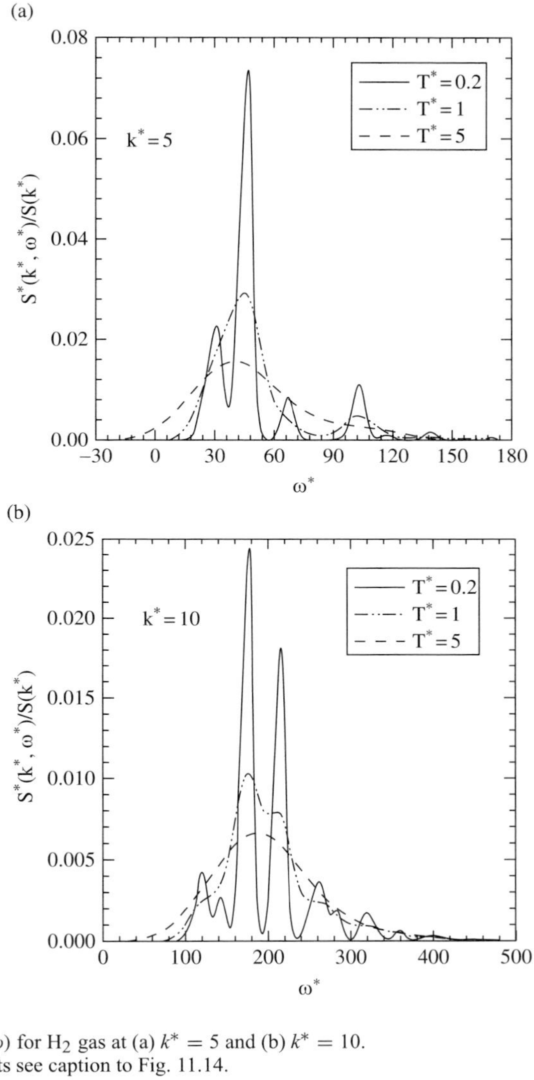

FIG. 11.15 $S(k, \omega)$ for H_2 gas at (a) $k^* = 5$ and (b) $k^* = 10$. For reduced units see caption to Fig. 11.14.

In contrast to infrared (§ 11.1) and Raman spectra (§ 11.2), we see that the band fine structure is only resolved for temperatures below about the rotational temperature. To understand this, note that the structure will be suppressed provided the translational linewidth $\bar{\omega}(k) = 2Bk\bar{d}(T/\theta_r)^{1/2}$ exceeds the line spacing $2Bl_{\max}$, where $l_{\max}$ is the maximum value of l, for a given k, contributing significantly to the sum (11.435). For $k\bar{d} \gtrsim 1$, $l_{\max} \approx k\bar{d}$ (for $k\bar{d} \lesssim 1$, only the quasielastic $l = 0$ term is important). Consequently, the structure disappears for $T/\theta_r \gtrsim 1$. Above the rotational temperature, quantum corrections to the lineshape are essentially translational in nature, originating in detailed balance and recoil effects.

The static structure factor $S(k) = \int d\omega S(k,\omega)$ shows clearly the influence of molecular vibration at large k. From (11.435) we find that

$$S(k) = \sum_{l\,\text{even}} \sum_{\nu\nu'} P_\nu (2l+1) |\langle \nu | j_l(kd) | \nu' \rangle|^2 \tag{11.437}$$

since $\int d\omega S_l(\omega) = 1$. Using closure and noting the identity[126]

$$\sum_{l\,\text{even}} (2l+1) j_l(x)^2 = \frac{1}{2}(1 + j_0(2x)), \tag{11.438}$$

we find that

$$S(k) = \frac{1}{2} \sum_{\nu} P_\nu \langle \nu | 1 + j_0(2kd) | \nu \rangle. \tag{11.439}$$

If we assume that only the ground vibrational state is occupied, that the vibrations are harmonic and of small amplitude $\Delta = 1/(\sqrt{2}\alpha)$, then we have approximately

$$S(k) = \frac{1}{2}\left(1 + \exp\left(-\frac{1}{2}k^2\Delta^2\right) j_0(2kd)\right), \tag{11.440}$$

where $\exp(-k^2\Delta^2/2)$ is the Debye–Waller factor.[240] For $k \gtrsim \Delta^{-1}$, we see that the oscillations in $S(k)$ are damped by molecular vibrations. Physically, this reflects decorrelation of the phase of scattering by the two nuclei of the diatomic.

We note, however, that the dependence of $S(k,\omega)$ on the vibrational amplitude is much more complex than that of its integral $S(k)$, and depends on the frequency as well as the wavevector—in contrast to the situation encountered for vibration of atoms on lattice sites in a solid.[240] As we shall see below, it is such as to render the first frequency moment $\int d\omega\, \omega S(k,\omega)$ a constant, dependent only on k and the molecular mass M.

For molecular fluids there appear as yet to have been no calculations of the virial coefficients $S^{(n)}(k,\omega)$, $n > 1$, representing n-body corrections to the structure factor for a gas (cf. (11.412)).

11.3.3 Dynamic structure factor for liquids

In a dilute gas, the individual molecules scatter essentially independently. As the density increases, however, molecular motions become correlated over a distance that is roughly the mean free path $\bar{l}$. If the scattering wavevector is sufficiently large that $k\bar{l} \gg 1$, these correlations are unimportant, and $S(k,\omega)$ resembles that for an ideal gas. In contrast, in the opposite—hydrodynamic—limit $k\bar{l} \ll 1$, the scattering is a collective property of the system. It may be considered to arise from density fluctuations, which are subject to the laws of thermodynamics and hydrodynamics,[255] and may be calculated from purely macroscopic quantities such as specific heats and transport coefficients. In practice, this limit is rarely realized in neutron scattering.[256] In dense fluids or liquids one is normally in the intermediate regime where $k\bar{l} \sim 1$; here molecular collisions cannot be ignored, but are insufficient to maintain local thermodynamic equilibrium. The calculation of $S(k, \omega)$ is then considerably more complicated.

The hydrodynamic limit

We will begin our discussion by considering coherent scattering in the hydrodynamic limit. This is a useful starting point because most modern theories of $S(k, \omega)$ build upon generalizations of hydrodynamic concepts by introducing frequency- and wavevector-dependent transport coefficients.

Consider first a classical atomic fluid[154,157,158,244,257,258] such as argon. Such a fluid can be treated as a continuous medium, characterized by thermodynamic variables such as the number, energy and entropy densities, and pressure. Calculation of $S(k, \omega)$ requires computation of the spectrum of the density–density correlation function $\langle \delta\rho(k, 0)^* \delta\rho(k, t) \rangle$, where $\delta\rho(k, t) = \int d\mathbf{r} \exp[i\mathbf{k} \cdot \mathbf{r(t)}](\rho(\mathbf{r}, t) - \rho)$ is the Fourier component of the number density fluctuation at wavevector k. There are five hydrodynamic modes, or fluctuations whose lifetime approaches infinity in the limit $k \to 0$, corresponding to the conserved densities for number, momentum, and energy. Supplementing the conservation equations for these quantities are the constitutive relations expressing their fluxes in terms of the local fluid velocity, velocity and temperature gradients, and the transport coefficients for momentum (the bulk and shear viscosities η_V and η_S) and energy (the thermal conductivity λ). Assuming small deparatures from equilibrium, we can derive a set of five linear equations relating fluctuations in seven quantities: density, pressure, entropy, temperature, and three components of velocity. Two of the four scalar variables can be eliminated assuming local thermodynamic equilibrium, and two of the velocity equations can be decoupled, since they correspond to transverse shear modes that do not affect the density fluctuations. This leaves three coupled linear partial differential equations relating fluctuations in the density, temperature, and the divergence of the velocity, which can be solved by Fourier–Laplace transformation to (k, ω) space and subsequent matrix inversion.

The resulting intermediate scattering function can be written as

$$F(k, t) = S(k) \left[\left(1 - \gamma^{-1}\right) \exp(-D_T k^2 |t|) + \gamma^{-1} \exp(-\Gamma k^2 |t|) \cos c_S kt \right], \tag{11.441}$$

where $\gamma = C_p / C_V$ is the specific heat ratio, $c_S = [(\partial p / \partial \rho)_S / M]^{1/2}$ is the adiabatic speed of sound, $D_T = \lambda / M\rho C_p$ is the thermal diffusion coefficient, and $\Gamma = \frac{1}{2}[(\gamma - 1)D_T + D_L]$ is the sound attenuation coefficient.[259] Here $D_L = \eta_L / M\rho$, where $\eta_L = \eta_V + \frac{4}{3}\eta_S$ is the longitudinal viscosity. The associated dynamic structure factor is

$$S(k,\omega) = S(k)\frac{1}{\pi}\left[\left(1-\gamma^{-1}\right)\frac{D_T k^2}{\omega^2 + (D_T k^2)^2}\right.$$
$$\left. + \frac{1}{2}\gamma^{-1}\frac{\Gamma k^2}{(\omega - c_S k)^2 + (\Gamma k^2)^2} + \frac{1}{2}\gamma^{-1}\frac{\Gamma k^2}{(\omega + c_S k)^2 + (\Gamma k^2)^2}\right], \quad (11.442)$$

and is a sum of three Lorentzians, representing scattering by different modes of the fluid. The first term, the quasielastic *Rayleigh line*, can be considered to arise from scattering caused by entropy, or heat, fluctuations, which are purely diffusive modes. The next two terms are the inelastic *Brillouin doublet*. These lines are shifted by $\pm c_S k$, and are caused by collisions in which the neutron exchanges energy $\pm c_S k$ with the phonons, or sound modes; in other words, Brillouin scattering is caused by propagating pressure fluctuations. The linewidths—$D_T k^2$ and Γk^2 for the Rayleigh and Brillouin lines, respectively—are just the inverse mode lifetimes.

For molecular liquids, the calculation of $S(k,\omega)$ is more complicated. Internal degrees of freedom may couple to the translational motion, and provide additional relaxation mechanisms for the density fluctuations. A simple theory of the scattering was proposed by Mountain,[13,157,158] based on Zwanzig 's[260] observation that the bulk viscosity contains a frequency-dependent part arising from fluctuations in the energy contained in internal degrees of freedom. Typically these occur on a timescale of about 10 ps. If we substitute this frequency-dependent viscosity into the hydrodynamic equations, we find that $S(k,\omega)$ contains an extra diffusive mode with a width of roughly 0.1 ps^{-1}; this is often much larger than the Rayleigh linewidth $D_T k^2$. This new *relaxation* mode is nonhydrodynamic inasmuch as its lifetime does not become infinite as $k \to 0$. We also note that the Brillouin lines are shifted to slightly higher frequencies because of sound dispersion caused by the frequency-dependent response of the fluid.

The intermediate regime

Once we leave the hydrodynamic limit, and the range of spatial correlations becomes comparable to k^{-1}, the approach described above breaks down. To illustrate the concepts involved, we consider a classical hard sphere fluid. For values of $k\bar{l}$ that do not significantly exceed unity, approximate solutions for $S(k,\omega)$ may be obtained from generalized Enskog theory.[261] Here one introduces an operator L_E, which is an effective one-particle replacement for the N-particle Liouvillian $i\mathcal{L}$. The intermediate scattering function is then

$$F(k,t) = S(k)\,\langle \exp(L_E(k)t)\rangle\,, \qquad t \geq 0, \quad (11.443)$$

where the angular brackets denote a velocity average over the normalized Maxwellian distribution $f(v) = (\beta M/2\pi)^{3/2}\exp(-\beta M v^2/2)$, i.e. $\langle\cdots\rangle = \int d\mathbf{v} f(v)(\cdots)$. Correspondingly, the dynamic structure factor becomes

$$S(k,\omega) = S(k)\frac{1}{\pi}\Re\left\langle\frac{1}{i\omega - L_E(k)}\right\rangle. \quad (11.444)$$

We seek an expansion of $F(k, t)$ and $S(k, \omega)$ in the eigenmodes of L_E. This yields

$$F(k, t) = S(k) \sum_{n=-\infty}^{\infty} A_n(k) \exp(-z_n(k)t), \qquad t \geq 0, \tag{11.445}$$

and

$$S(k, \omega) = S(k) \frac{1}{\pi} \Re \sum_{n=-\infty}^{\infty} \frac{A_n(k)}{i\omega + z_n(k)}, \tag{11.446}$$

where $z_n(k) = \Gamma_n(k) + i\omega_n(k)$ denotes a complex eigenvalue of $-L_E$, and $A_n(k)$ is related to the corresponding eigenfunction. These quantities can be determined by diagonalization of L_E in a suitable basis.[262] Modes with $\omega_n \neq 0$ (propagating modes) occur in conjugate pairs with $z_{-n} = z_n^*$ and $A_{-n} = A_n^*$.

The three lowest modes (i.e. those with the smallest $\Gamma_n(k)$) turn out to be direct extensions of the hydrodynamic heat and sound modes. For these modes $z_n(k) \to 0$ as $k \to 0$, while the amplitudes $A_n(k)$ remain finite. For all other modes $z_n(k)$ approaches a finite, positive, real value in the limit of small k, while $A_n(k) \to 0$. These so-called *kinetic* modes are important outside the hydrodynamic regime, especially at large frequencies.

$S(k, \omega)$ has been computed using this approach for a wide range of wavevectors and densities.[262] For $k\bar{l} < 0.1$, the results are indistinguishable from those obtained from hydrodynamics. For $0.1 < k\bar{l} < 1$, at densities $\rho\sigma^3 < 0.15$, it is sufficient to consider the finite-k extensions of the three hydrodynamic modes, while for $\rho\sigma^3 > 0.5$, a three-Lorentzian description is possible using 'effective' modes.[262] At intermediate densities, two additional effective modes are required. In the range $1 < k\bar{l} < 3$, a proper description of $S(k, \omega)$ requires many modes; a simplified description in terms of effective modes is not possible. For yet larger values of $k\bar{l}$, (11.446) does not yield a suitable representation of the structure factor, although this may still be computed directly from (11.444).

The behaviour of the modes as a function of k is especially interesting at high densities. The Brillouin doublet is predicted to survive for $k\sigma < 0.5$, beyond which it has broadened so much as to be unobservable. For $k \approx k_G$, where $k_G = 2\pi/\sigma$, the heat mode eigenvalue z_0 softens appreciably, and for larger values of k oscillates around the diffusive mode of the single-particle Enskog operator—that is, $\Gamma_0/k^2 \to D$, the coefficient of self diffusion. The minimum of Γ_0 at $k = k_G$ is associated with the phenomenon of de Gennes narrowing.[46,154] Physically, the increased lifetime of density fluctuations in the vicinity of the de Gennes wavevector reflects the ease with which the atomic interactions can support oscillations at wavelengths on the order of the atomic spacing. For the same reason, $S(k)$ maximizes near $k = k_G$.[257] There are also propagation gaps in the sound modes in the vicinity of $k = k_G$. In other words, $\omega_S(k) = 0$, and the modes become diffusive, so that the fluid cannot transmit sound waves at these wavelengths. This disappearance and reappearance of propagating modes can be traced to a competition between elastic and dissipative response in the fluid.[262] The phenomenon has been confirmed both by molecular dynamics (MD) simulation of hard sphere[263] and LJ[264] argon and by neutron scattering studies of the liquid.[265]

Calculation of $S(k, \omega)$ is naturally more difficult when the potential is smooth.[262,266] For simple atomic fluids, provided k is not too large, one may use generalized hydrodynamics.[154,157] In essence, one neglects the kinetic modes, and assumes that the equations of linearized hydrodynamics apply, but with the transport coefficients replaced by k- and ω-dependent generalizations. These generalizations are essentially ad hoc. For example, in the example discussed in § 11.3.6, one effectively assumes that

$$D_{\mathrm{T}}(k,\omega) = D_{\mathrm{T}}\,(1 + \gamma D_{T} k/\bar{v})^{-1}\,,$$
$$D_{\mathrm{L}}(k,\omega) = D_{\mathrm{L}}\left(\left[1 + (k\bar{v}\tau(k))^2\right]^{1/2} + i\omega\tau(k)\right)^{-1}, \tag{11.447}$$

where $\tau(k) = D_{\mathrm{L}} k^2/(\hat{M}_4/\hat{M}_2 - \gamma \hat{M}_2)$, where the $\hat{M}_n$ are reduced moments.

In order to dispense with phenomenological approximations, Götze and Lücke[267] have proposed that $S(k, \omega)$ be calculated using mode-coupling theory.[154,157] This theory is, like generalized hydrodynamics, most naturally constructed within the Mori formalism.[268–270] The physical idea underlying the theory is that fluctuations in fast (unconserved) variables couple the slow modes of the fluid, corresponding to the quasi-conserved set of variables ξ_n, into products $\xi_n\xi_m$, $\xi_n\xi_m\xi_l$, etc. The dominant process involves decay into double excitations. By computing the decay rates one obtains *microscopic* expressions for the generalized transport coefficients.

Using mode-coupling theory one generates an $S(k, \omega)$ that possesses the correct zeroth, second, and fourth moments, and the correct free-particle and hydrodynamic limits. Long-time anomalies in associated correlation functions are also correctly predicted. Mode-coupling calculations on simple atomic fluids[271] show reasonably good agreement with MD simulations and experiment[257] over a wide range of k and ω. Because of its complexity, however, it has yet to be applied to molecular fluids.

For molecular fluids, the only practical route to $S(k, \omega)$ appears to be computer simulation. In a traditional MD experiment, a lower limit to the range of available wavectors is set by the length L of the simulation cell: periodicity demands that k cannot be smaller than $2\pi/L$. Since L increases only as the cube root of the number of molecules, simulations at small k require very large systems, and are probably computationally prohibitive for $k < 0.2\,\text{Å}^{-1}$. An alternative procedure, which is applicable for arbitrary k, but is limited to short simulation runs, has been described by Salacuse and Egelstaff.[272]

Simulations have been performed on several molecular fluids including nitrogen,[273] carbon tetrachloride,[274] methylene dichloride,[275] water,[276–283] methanol,[284] hydrogen sulphide,[285] butane,[286] and benzene.[287]

Simulations of $S(k, \omega)$ clearly offer a more sensitive test of the intermolecular potential than those of $S(k)$ (Chapter 9). Backx *et al.*[273] have simulated $S(k, \omega)$ for dense nitrogen gas, using a potential consisting of site–site LJ interactions plus a point quadrupole term. Comparison with experiment reveals some clear discrepancies, which are shown in Fig. 11.16. For $k < 1\,\text{Å}^{-1}$, the simulated data are too low for $\omega < 3\,\text{ps}^{-1}$. For larger k, the agreement is good for $\omega < 10\,\text{ps}^{-1}$. This suggests the potential describes the short-range features of intermolecular collisions well, but fails adequately to describe long-range effects: over the range 5–15 Å some modification that will effectively slow the dynamic behaviour is needed, thereby providing an enhancement in $S(k, \omega)$ at small frequencies. For $5 < \omega < 8\,\text{ps}^{-1}$ there is good agreement for all k, while for larger frequencies the shapes of the curves are similar, but the simulated structure factor is too small. This may mean that the potential should also be modified to include high-frequency processes occurring on a timescale of 0.1 ps or less, possibly by changing its steepness.

In water $S(k, \omega)$ is also found to be sensitive to the details of the intermolecular potential.[277] The effect of three-body interactions has been studied.[278] In the hydrodynamic regime these act to reduce the thermal diffusivity, significantly narrowing the Rayleigh line; at the same time the viscosity increases, broadening the Brillouin doublet. Observable effects persist to k values corresponding to the length scale of nearest-neighbour interactions.

Simulations on water indicate the presence of two prominent collective modes at intermediate wavevectors.[276] Figure 11.17 shows partial structure factors for O-O and H-H correlations. The mode eigenvalues (or more precisely, their imaginary parts) approximately coincide with the peaks in the structure factors. The low-frequency acoustic mode ω_1 is visible in both $S_{\text{OO}}(k, \omega)$ and $S_{\text{HH}}(k, \omega)$, although at high k its contribution to the latter is overdamped. For small k the propagation speed ω_1/k is about $1500\,\text{m s}^{-1}$, which identifies it as an extension of the hydrodynamic sound mode. The high-frequency mode ω_2 is visible only in the spectrum of H-atom fluctuations. This is an optical mode, with a frequency $\omega_2 \approx 150\,\text{ps}^{-1}$, which is almost independent of k. Neutron scattering studies[288] and computer simulations[284] indicate that such modes may also exist in other hydrogen-bonded liquids; they are believed to be related to rapid H-atom librational motions, although their exact nature is a subject of debate.[279,280] Coupling of the acoustic mode with a further, low-energy, optical mode, possibly associated with the relative motion of nearest neighbours, is believed to be responsible for the phenomenon of 'fast sound'.[289] Balucani *et al.*[282] and Petrillo *et al.*[289] have summarised recent theoretical and experimental developments.

In hydrocarbons the scattering is almost purely incoherent. Comparison of the calculated self structure factor with experiment has been used to evaluate models for the vibrational force field.[286,287] Figure 11.18 shows a comparison of the simulated and measured $S_{\text{self}}(k, \omega)$ for n-butane. The most notable discrepancy occurs in the vicinity of $1100\,\text{cm}^{-1}$, where a distinct peak is predicted

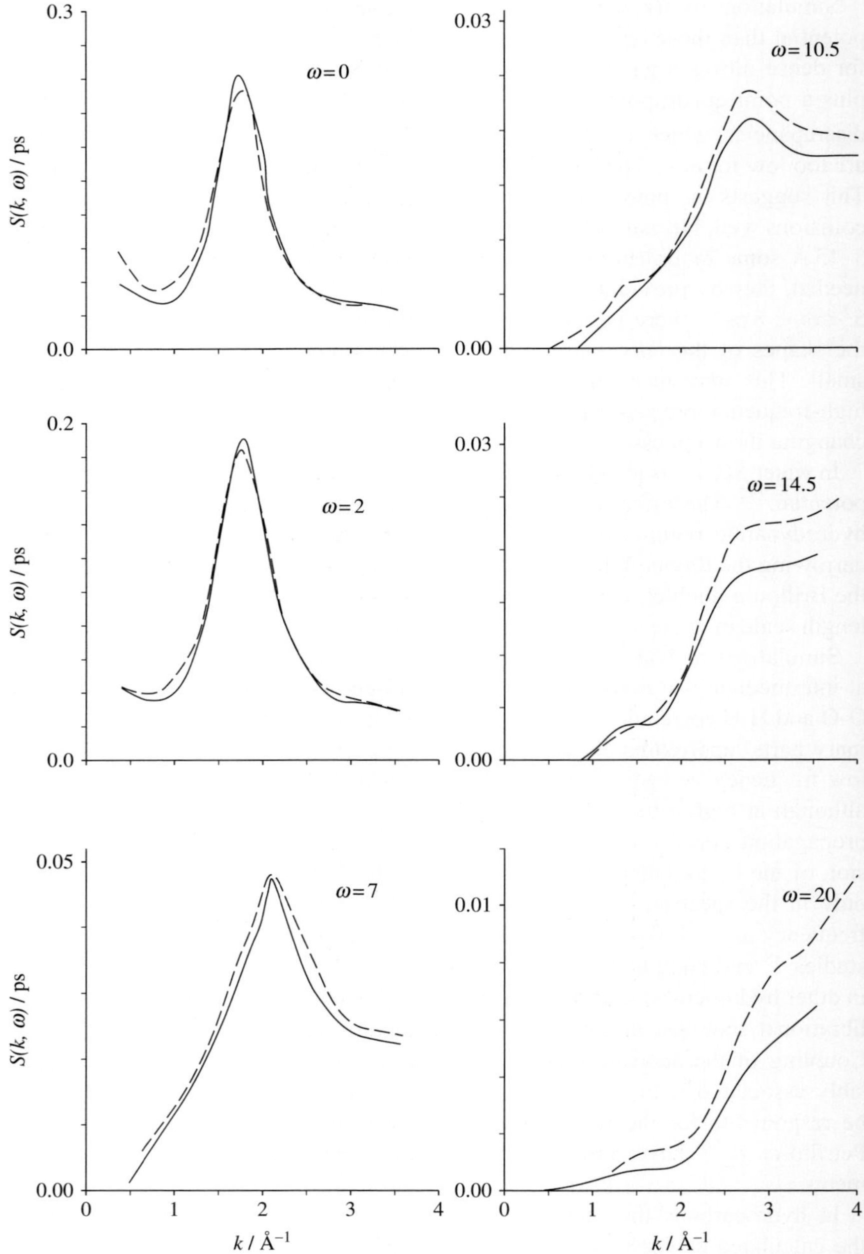

FIG. 11.16 $S(k, \omega)$ for nitrogen gas at $\rho = 0.0121$ Å^{-3} and $T = 297$ K, with ω in units of ps^{-1}. Comparison of MD simulation (solid line) with experiment (broken line). (From ref. 273.)

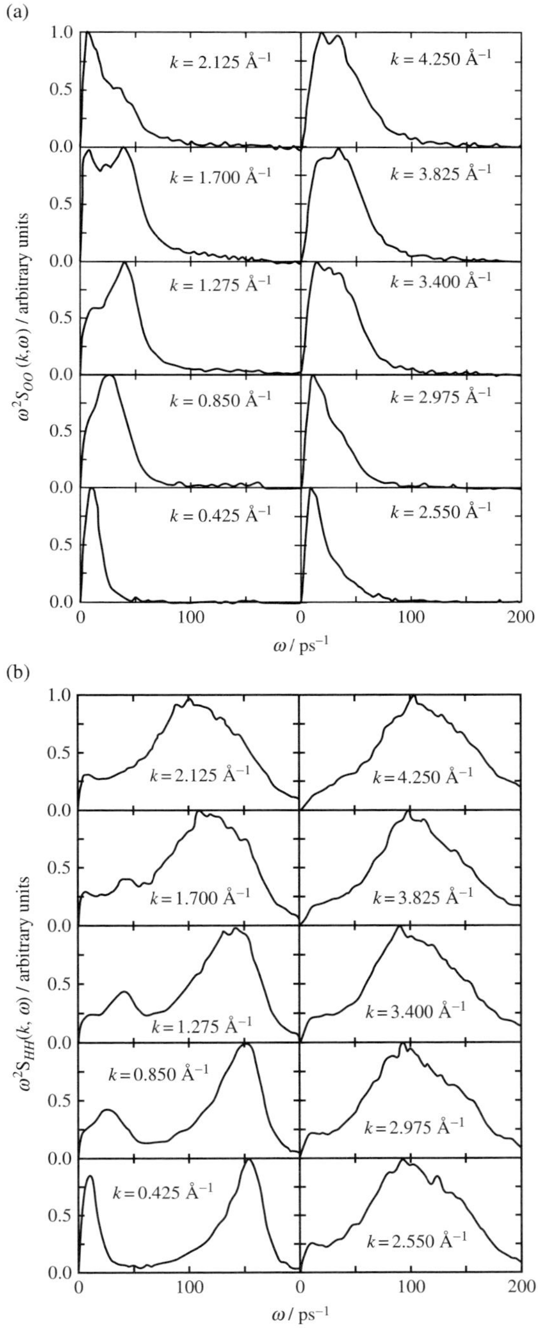

FIG. 11.17 Simulated partial dynamic structure factors for (a) O-O and (b) H-H correlations in TIP4P water at $\rho = 1\ \mathrm{g\,cm^{-3}}$ and $T = 310\ \mathrm{K}$. Reprinted with permission from ref. 276. Copyright 1989, American Physical Society.

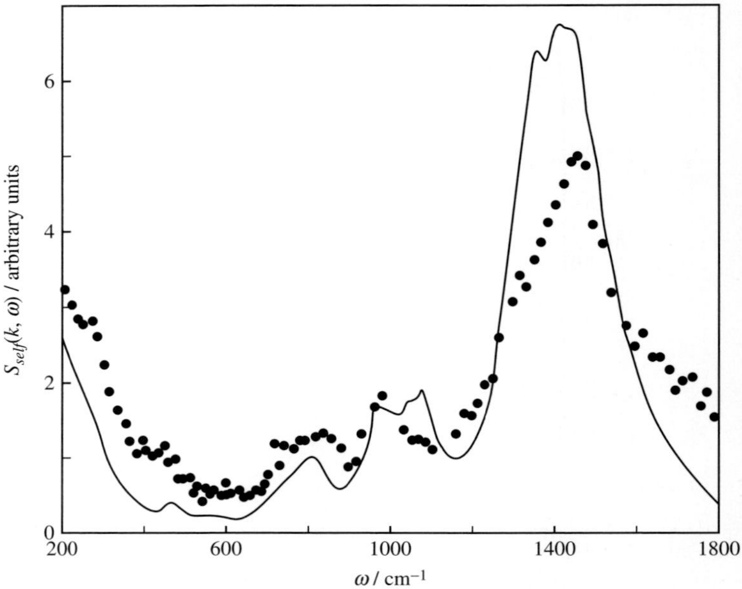

FIG. 11.18 $S_{\text{self}}(k, \omega)$ of liquid n-butane at 140 K. Comparison of MD simulation (solid line) with neutron scattering data (points). The data were recorded at a constant scattering angle of $\theta = 5.25^{\circ}$, so that k varies with ω. (From ref. 286.)

but not observed, indicating that the C-C-C bending motion is not properly described.[286]

The transition to large k

It is interesting to see how the transition to a Gaussian lineshape is achieved as the wavevector k increases. We will give a very simplified treatment of collisional effects, approximating the Enskog operator by $L_{\text{E}} = i\mathbf{k} \cdot \mathbf{v} - \gamma$, where $\gamma > 0$, and is on the order of the inverse of the mean time between collisions, $\bar{t} = \bar{l}/\bar{v}$. The presence of a negative real component in L_{E} reflects collisional damping at large times, and ensures that $F(k, t) \to 0$ in the limit $t \to \infty$. Equation (11.444) then implies

$$S(k, \omega) = \frac{1}{\pi} \Im \int \mathrm{d}\mathbf{v} \, \frac{f(v)}{\mathbf{k} \cdot \mathbf{v} - \omega - i\gamma}. \tag{11.448}$$

Integrating over the components of $\mathbf{v} \perp \mathbf{k}$, and setting $\zeta = (\omega + i\gamma)/\bar{\omega}(k)$, where $\bar{\omega} = (2/\beta M)^{1/2} k$, we find that

$$S(k, \omega) = \frac{1}{\pi \bar{\omega}(k)} \Im[Z(\zeta)], \tag{11.449}$$

where

$$Z(\zeta) = \frac{1}{\pi^{1/2}} \int_{-\infty}^{+\infty} \mathrm{d}t \, \frac{\exp(-t^2)}{t-\zeta}$$
$$= -2\zeta \sum_{n=0}^{\infty} \frac{(-2\zeta^2)^n}{(2n+1)!!} + i\pi^{1/2} \exp(-\zeta^2), \qquad \Im(\zeta) > 0, \qquad (11.450)$$

is the plasma dispersion function.[290] In the limit $k\bar{l} \gg 1$, we have $\gamma \ll \bar{\omega}(k)$, and it follows that

$$S(k,\omega) = \frac{1}{\pi^{1/2}\bar{\omega}(k)} \exp\left(-\frac{\omega^2}{\bar{\omega}(k)^2}\right). \qquad (11.451)$$

Thus the structure factor is Gaussian, and is in fact just that for a dilute gas, as expected (cf. (11.404)).

11.3.4 Quantum corrections to $S(k,\omega)$

At high frequencies or momenta, quantum corrections to $S(k,\omega)$ become important. Various schemes have been proposed to incorporate these corrections into calculated classical structure factors to enable a direct comparison with experiment. One very commonly used approximation is

$$S^{\mathrm{qu}}(k,\omega) \approx S^{\mathrm{cl}}(k,\omega) \exp\left(\frac{1}{2}\beta\hbar\omega - \frac{\beta\hbar^2 k^2}{8M}\right), \qquad (11.452)$$

which is correct to order $\hbar$. The 'recoil' correction $\exp(-\beta\hbar^2k^2/8M)$ is not the only correction of order $\hbar^2$, but is included because of its importance at high k. The quantum structure factor of (11.452) satisfies the detailed balance relation (11.358), possesses the correct first frequency moment to order $\hbar^2$ (since $M_1 = (\beta\hbar/2)M_2 + O(\hbar^3)$, cf. (11.52)), and reduces to the exact result for an ideal atomic gas, Eq. (11.411).

The equivalent approximation for the quantum intermediate scattering function is

$$F^{\mathrm{qu}}(k,t) \approx F^{\mathrm{cl}}\left(k, t - \frac{1}{2}i\beta\hbar\right) \exp\left(-\frac{\beta\hbar^2k^2}{8M}\right). \qquad (11.453)$$

One sometimes sees instead[257]

$$F^{\mathrm{qu}}(k,t) \approx F^{\mathrm{cl}}(k,\tau), \qquad (11.454)$$

where $\tau = [t(t - i\beta\hbar)]^{1/2}$ is the complex time introduced in (11.409). For a dilute atomic gas, Eqs. (11.453) and (11.454) are equivalent.

In liquids the remaining $O(\hbar^2)$ corrections are typically on the order of a few percent; since the accuracy of measurement is about 1%, it seems worthwhile to attempt to account for them. Sears[291] has developed a theory of quantum corrections for neutron scattering in near-classical liquids that considers not only detailed balance and recoil corrections, but also includes quantum effects in the structure and dynamics of the liquid. Since the method is based on an expansion of the frequency moments $M_n = \int d\omega\, \omega^n S(k,\omega)$ in powers of $\hbar^2$, it is valid only for small $\hbar k$.

Kneller[292] and Bonella *et al.*[293] have shown how quantum diffraction effects may be incorporated into a Feynman path-integral description of the scattering. The theory omits quantum symmetry effects, but these are generally very small, as discussed in § 1.2.2. In principle, this method enables direct computation of the quantum $S(k,\omega)$ by classical MD simulation, and could be especially useful at large k, where other semiclassical approximations break down. In practice, however, its numerical implementation requires the computation of the Maslov index for each contributing trajectory, and there is at present no satisfactory way to achieve this.[293]

11.3.5 Spectral moments

The frequency moments of the dynamic structure and self structure factors are defined as

$$M_n^{\text{self}} = \int_{-\infty}^{+\infty} d\omega\, \omega^n S_{\text{self}}(k,\omega),$$

$$M_n = \int_{-\infty}^{+\infty} d\omega\, \omega^n S(k,\omega), \tag{11.455}$$

and are functions of the wavevector k, that is, $M_n^{\text{self}} = M_n^{\text{self}}(k)$, $M_n = M_n(k)$. These moments have many uses, including (i) the construction of approximate theories for $S_{\text{self}}(k,\omega)$ and $S(k,\omega)$ (§ 11.3.6), (ii) the extraction of mean square forces and torques from experimental data, (iii) as checks on molecular dynamics simulation data, and (iv) the estimation of inelasticity corrections to the integrated scattering cross-section (§ 11.3.7). From (11.357), we have

$$M_n^{\text{self}} = (-i)^n \left(\frac{d^n F_{\text{self}}(k,t)}{dt^n} \right)_{t=0},$$

$$M_n = (-i)^n \left(\frac{d^n F(k,t)}{dt^n} \right)_{t=0}. \tag{11.456}$$

In this section we will discuss the calculation of M_n^{self} and M_n. We will initially assume that the moments may be computed within the framework of classical dynamics, and that the intermolecular potentials are smooth. In this case $S_{\text{self}}(k,\omega)$ and $S(k,\omega)$ are even functions of frequency and all odd moments vanish.

It follows immediately from (11.390) that the zeroth moments are given by

$$M_0^{\text{self}} = 1,$$

$$M_0 = S(k). \tag{11.457}$$

The static structure factor $S(k)$ is discussed at length in Chapter 9.

Atomic fluids

For an atomic fluid the intermediate self scattering function of (11.355) becomes

$$F_{\text{self}}(k,t) = \langle A_1(0)^* A_1(t) \rangle, \tag{11.458}$$

where $A_\lambda(t) = \exp(i\mathbf{k}\cdot\mathbf{r}_\lambda(t))$. Consequently

$$M_{2n}^{\text{self}} = (-)^n \left\langle A_1^{(0)*} A_1^{(2n)} \right\rangle = \left\langle A_1^{(n)*} A_1^{(n)} \right\rangle, \tag{11.459}$$

where $A_\lambda^{(n)} = [(\mathrm{d}^n/\mathrm{d}t^n)A_\lambda(t)]_{t=0}$. Noting that $\mathrm{d}\mathbf{r}_\lambda(t)/\mathrm{d}t = \mathbf{v}_\lambda(t)$ and $\mathrm{d}\mathbf{v}_\lambda(t)/\mathrm{d}t = \mathbf{F}_\lambda(t)/M$, where M, $\mathbf{v}_\lambda$, and $\mathbf{F}_\lambda$ denote the mass and velocity of atom λ and the force acting on it, respectively, we have

$$\begin{aligned} A_\lambda^{(1)} &= i\mathbf{k}\cdot\mathbf{v}_\lambda \exp(i\mathbf{k}\cdot\mathbf{r}_\lambda), \\ A_\lambda^{(2)} &= \Big(-(\mathbf{k}\cdot\mathbf{v}_\lambda)^2 + i(\mathbf{k}\cdot\mathbf{F}_\lambda)/M\Big)\exp(i\mathbf{k}\cdot\mathbf{r}_\lambda), \end{aligned} \tag{11.460}$$

etc. Therefore

$$\begin{aligned} M_2^{\text{self}} &= \left\langle (\mathbf{k}\cdot\mathbf{v}_1)^2 \right\rangle, \\ M_4^{\text{self}} &= \left\langle (\mathbf{k}\cdot\mathbf{v}_1)^4 \right\rangle + M^{-2}\left\langle (\mathbf{k}\cdot\mathbf{F}_1)^2 \right\rangle. \end{aligned} \tag{11.461}$$

The averages in these equations are over all atomic positions and velocities. Since $S_{\text{self}}(k,\omega)$ depends only on the magnitude of $\mathbf{k}$ and not its direction, the same must be true of its moments. Thus we may also average (11.461) over all possible directions of $\mathbf{k}$. The average of a product of $2n$ Cartesian components of $\hat{\mathbf{k}} = \mathbf{k}/k$ is

$$\langle \hat{k}_\alpha \hat{k}_\beta \hat{k}_\gamma \hat{k}_\delta \ldots \hat{k}_\mu \hat{k}_\nu \rangle = \frac{1}{(2n+1)!!}\mathcal{P}\,\delta_{\alpha\beta}\delta_{\gamma\delta}\ldots\delta_{\mu\nu}, \tag{11.462}$$

where $\mathcal{P}$ denotes a sum over the $(2n-1)!!$ possible pairwise combinations of the indices $\alpha, \beta \ldots \nu$. From § 3.1.2, $\langle v^{2n} \rangle = (2n+1)!!(\beta M)^{-n}$. Therefore

$$\begin{aligned} M_2^{\text{self}} &= (\beta M)^{-1}k^2, \\ M_4^{\text{self}} &= (\beta M)^{-2}\left(3k^4 + \frac{1}{3}k^2\left\langle (\beta F_1)^2 \right\rangle\right), \end{aligned} \tag{11.463}$$

where the remaining average is configurational. We see that the second moment is independent of the nature of the interatomic interactions, while the fourth moment depends on the mean square force acting on an atom of the fluid. From (E.31), $\langle (\beta F_1)^2 \rangle = \beta \left\langle \nabla_1^2 \mathcal{U} \right\rangle$, where $\mathcal{U}$ is the N-body potential energy. Assuming $\mathcal{U} = \sum_{\lambda<\lambda'} u(r_{\lambda\lambda'})$, we therefore have

$$M_4^{\text{self}} = (\beta M)^{-2}\left[3k^4 + k^2\frac{4\pi}{3}\beta\rho \int_0^\infty \mathrm{d}r\, r^2\, g(r)\left(u'' + 2\frac{u'}{r}\right)\right], \tag{11.464}$$

where $u' = \mathrm{d}u/\mathrm{d}r$, $u'' = \mathrm{d}^2u/\mathrm{d}r^2$, and $g(r)$ is the pair correlation function. This form is more convenient for computation than (11.463), since it does not require knowledge of the higher order correlation functions $g(123)$ and $g(1234)$.

Sears[294] has given expressions for higher order moments, M_6^{self}, M_8^{self}, etc.

From (11.356), the intermediate scattering function is

$$F(k,t) = N^{-1}\sum_{\lambda\lambda'}\langle \exp(i\mathbf{k}\cdot\mathbf{r}_\lambda(0))\exp(-i\mathbf{k}\cdot\mathbf{r}_{\lambda'}(t))\rangle\,, \tag{11.465}$$

with moments

$$M_{2n} = N^{-1}\sum_{\lambda\lambda'}\left\langle A_\lambda^{(n)*}A_{\lambda'}^{(n)}\right\rangle. \tag{11.466}$$

The zeroth moment is

$$\begin{aligned} S(k) &= 1 + \sum_{\lambda\neq\lambda'}\langle \exp(i\mathbf{k}\cdot\mathbf{r}_{\lambda\lambda'})\rangle \\ &= 1 + 4\pi\rho\int_0^\infty \mathrm{d}r\, r^2\, h(r) j_0(kr), \end{aligned} \tag{11.467}$$

where we have used (B.89) to expand the exponential as a sum of spherical waves.

Since the velocities of different molecules are uncorrelated and $\langle \mathbf{v}_\lambda\rangle = 0$, the second moment is

$$\begin{aligned} M_2 &= N^{-1}\sum_{\lambda\lambda'}\langle(\mathbf{k}\cdot\mathbf{v}_\lambda)(\mathbf{k}\cdot\mathbf{v}_{\lambda'})\rangle \exp(i\mathbf{k}\cdot\mathbf{r}_{\lambda\lambda'}) \\ &= (\beta M)^{-1}k^2, \end{aligned} \tag{11.468}$$

and is therefore identical to M_2^{self}.

The calculation of the fourth moment is rather more complicated. We have

$$\begin{aligned} M_4 = N^{-1}&\sum_{\lambda\lambda'}\Big\langle\Big(-(\mathbf{k}\cdot\mathbf{v}_\lambda)^2 + i(\mathbf{k}\cdot\mathbf{F}_\lambda)/M\Big) \\ &\times\ \Big(-(\mathbf{k}\cdot\mathbf{v}_{\lambda'})^2 - i(\mathbf{k}\cdot\mathbf{F}_{\lambda'})/M\Big)\exp(i\mathbf{k}\cdot\mathbf{r}_{\lambda\lambda'})\Big\rangle \\ &= 3(\beta M)^{-2}k^4 + (\beta M)^{-2}k^4 N^{-1}\sum_{\lambda\neq\lambda'}\langle\exp(i\mathbf{k}\cdot\mathbf{r}_{\lambda\lambda'})\rangle \\ &+ i(\beta M^2)^{-1}k^2 N^{-1}\sum_{\lambda\neq\lambda'}\langle(-\mathbf{k}\cdot\mathbf{F}_\lambda + \mathbf{k}\cdot\mathbf{F}_{\lambda'})\exp(i\mathbf{k}\cdot\mathbf{r}_{\lambda\lambda'})\rangle \\ &+ M^{-2}N^{-1}\sum_{\lambda\lambda'}\langle(\mathbf{k}\cdot\mathbf{F}_\lambda)(\mathbf{k}\cdot\mathbf{F}_{\lambda'})\exp(i\mathbf{k}\cdot\mathbf{r}_{\lambda\lambda'})\rangle\,. \end{aligned} \tag{11.469}$$

To simplify (11.469) we use the hypervirial theorem (Appendix E). Consider the function $B = \sum_{\lambda\lambda'} (\mathbf{k} \cdot \mathbf{p}_\lambda)(\mathbf{k} \cdot \mathbf{F}_{\lambda'}) \exp(i\mathbf{k} \cdot \mathbf{r}_{\lambda\lambda'})$. Since $\langle \dot{B} \rangle = 0$, we have

$$\langle \sum_{\lambda\lambda'} \big[(\mathbf{k} \cdot \dot{\mathbf{p}}_\lambda)(\mathbf{k} \cdot \mathbf{F}_{\lambda'}) + (\mathbf{k} \cdot \mathbf{p}_\lambda)(\mathbf{k} \cdot \dot{\mathbf{F}}_{\lambda'})$$
$$+ \; i(\mathbf{k} \cdot [\mathbf{v}_\lambda - \mathbf{v}_{\lambda'}])(\mathbf{k} \cdot \mathbf{p}_\lambda)(\mathbf{k} \cdot \mathbf{F}_{\lambda'}) \big] \exp(i\mathbf{k} \cdot \mathbf{r}_{\lambda\lambda'}) \rangle = 0. \tag{11.470}$$

Now $\dot{\mathbf{p}}_\lambda = \mathbf{F}_\lambda$ and $\dot{\mathbf{F}}_\lambda = \sum_{\lambda'} (\mathbf{v}_{\lambda'} \cdot \nabla_{\lambda'}) \mathbf{F}_\lambda$. Substituting these results into (11.470) and noting that $\langle v_{\lambda\alpha} v_{\lambda'\beta} \rangle = (\beta M)^{-1} \delta_{\lambda\lambda'} \delta_{\alpha\beta}$, $\mathbf{F}_\lambda = -\boldsymbol{\nabla}_\lambda \mathcal{U}$, we find that

$$\sum_{\lambda\lambda'} \langle (\mathbf{k} \cdot \mathbf{F}_\lambda)(\mathbf{k} \cdot \mathbf{F}_{\lambda'}) \exp(i\mathbf{k} \cdot \mathbf{r}_{\lambda\lambda'}) \rangle + i\beta^{-1} k^2 \sum_{\lambda \neq \lambda'} \langle \mathbf{k} \cdot \mathbf{F}_{\lambda'} \exp(i\mathbf{k} \cdot \mathbf{r}_{\lambda\lambda'}) \rangle$$
$$= \beta^{-1} \mathbf{kk} : \sum_{\lambda\lambda'} \langle \exp(i\mathbf{k} \cdot \mathbf{r}_{\lambda\lambda'}) \boldsymbol{\nabla}_\lambda \boldsymbol{\nabla}_{\lambda'} \mathcal{U} \rangle. \tag{11.471}$$

Consider next the function $C = \sum_{\lambda \neq \lambda'} \mathbf{k} \cdot \mathbf{p}_\lambda \exp(i\mathbf{k} \cdot \mathbf{r}_{\lambda\lambda'})$. Setting $\langle \dot{C} \rangle = 0$, we find similarly that

$$\sum_{\lambda \neq \lambda'} \langle \mathbf{k} \cdot \mathbf{F}_\lambda \exp(i\mathbf{k} \cdot \mathbf{r}_{\lambda\lambda'}) \rangle + i\beta^{-1} k^2 \sum_{\lambda \neq \lambda'} \langle \exp(i\mathbf{k} \cdot \mathbf{r}_{\lambda\lambda'}) \rangle = 0. \tag{11.472}$$

Using (11.471) and (11.472) in (11.469), we obtain the simple result

$$M_4 = (\beta M)^{-2} \left(3k^4 + \beta N^{-1} \mathbf{kk} : \sum_{\lambda\lambda'} \langle \exp(i\mathbf{k} \cdot \mathbf{r}_{\lambda\lambda'}) \boldsymbol{\nabla}_\lambda \boldsymbol{\nabla}_{\lambda'} \mathcal{U} \rangle \right). \tag{11.473}$$

We now note that $\boldsymbol{\nabla}_\lambda \boldsymbol{\nabla}_\lambda \mathcal{U} = \sum_{\lambda'} \boldsymbol{\nabla}_{\lambda\lambda'} \boldsymbol{\nabla}_{\lambda\lambda'} u(r_{\lambda\lambda'})$ in an obvious notation, while for $\lambda \neq \lambda'$, $\boldsymbol{\nabla}_\lambda \boldsymbol{\nabla}_{\lambda'} \mathcal{U} = -\boldsymbol{\nabla}_{\lambda\lambda'} \boldsymbol{\nabla}_{\lambda\lambda'} u(r_{\lambda\lambda'})$. We can therefore write (11.473) as

$$M_4 = (\beta M)^{-2} \left(3k^4 + k^2 \beta\rho \int d\mathbf{r}\, g(r)(1 - \cos kz) \frac{\partial^2 u}{\partial z^2} \right), \tag{11.474}$$

where z is the component of $\mathbf{r}$ in the direction of $\mathbf{k}$; that is, $kz = \mathbf{k} \cdot \mathbf{r} = kr \cos\theta$, where θ is the angle between $\mathbf{k}$ and $\mathbf{r}$. Since

$$\frac{\partial^2 u}{\partial z^2} = \frac{1}{3} \left(u'' + 2\frac{u'}{r} \right) + \frac{2}{3} \left(u'' - \frac{u'}{r} \right) P_2(\cos\theta), \tag{11.475}$$

it follows using (B.90) and (A.9b) that

$$M_4 = (\beta M)^{-2} \left(3k^4 + k^2 \frac{4\pi}{3} \beta\rho \int_0^\infty dr\, r^2\, g(r) \right.$$
$$\left. \times \left[[1 - j_0(kr)] \left(u'' + 2\frac{u'}{r} \right) + 2j_2(kr) \left(u'' - \frac{u'}{r} \right) \right] \right). \tag{11.476}$$

If we define $\mu_4 = M_4 - 3M_2^2$, $\mu_4^{\text{self}} = M_4^{\text{self}} - 3M_2^{\text{self}2}$, then from (11.464) and (11.476) we see that for small k

$$\mu_4^{\text{self}} = O(k^2), \qquad \mu_4 = O(k^4), \tag{11.477}$$

while for large k

$$\mu_4^{\text{self}} \sim k^2, \qquad \mu_4 \sim k^2 \tag{11.478}$$

and

$$\mu_4 - \mu_4^{\text{self}} \to 0. \tag{11.479}$$

The above results for M_4 and M_4^{self} apply only when the atoms of the fluid interact with a continuous, or soft, potential; for hard potentials, different results are obtained. For example, for a fluid of hard spheres, the discontinuity in the potential generates an ω^{-4} tail in $S_{\text{self}}(k, \omega)$, and as a result the fourth and higher moments of this function are infinite.[266]

Molecular fluids

The calculation of the moments for a molecular fluid is, naturally, considerably more complicated. We begin by discussing the simplest possible case, that of a fluid of homonuclear diatomics, and thereafter we briefly review results for molecules of other symmetries. For simplicity, we will ignore the effects of molecular vibration.

Homonuclear diatomics For a homonuclear diatomic, the self structure factor is, from (11.355),

$$F_{\text{self}}(k, t) = \frac{1}{2}\left\langle C_1(0)^* C_1(t) + D_1(0)^* D_1(t)\right\rangle, \tag{11.480}$$

where we have defined $C_\lambda(t) = A_\lambda(t)B_\lambda(t)$, $D_\lambda(t) = A_\lambda(t)B_\lambda(t)^*$, and $A_\lambda(t) = \exp(i\mathbf{k} \cdot \mathbf{r}_\lambda(t))$, $B_\lambda(t) = \exp(i\mathbf{k} \cdot \mathbf{d}_\lambda(t))$, where $\mathbf{r}_\lambda$ denotes the position of the centre of mass of molecule λ, and $\mathbf{d}_\lambda$ is one-half the interatomic vector. The moments of $F_{\text{self}}(k, t)$ are therefore given by

$$M_{2n}^{\text{self}} = \frac{1}{2}\left\langle C_1^{(n)*} C_1^{(n)} + D_1^{(n)*} D_1^{(n)}\right\rangle, \tag{11.481}$$

where $C_\lambda^{(1)} = A_\lambda^{(1)} B_\lambda^{(0)} + A_\lambda^{(0)} B_\lambda^{(1)}$, $C_\lambda^{(2)} = A_\lambda^{(2)} B_\lambda^{(0)} + 2A_\lambda^{(1)} B_\lambda^{(1)} + A_\lambda^{(0)} B_\lambda^{(2)}$, etc. Here $A_\lambda^{(1)}$ and $A_\lambda^{(2)}$ are given by (11.460), while

$$\begin{aligned} B_\lambda^{(1)} &= i\mathbf{k} \cdot (\boldsymbol{\Omega}_\lambda \times \mathbf{d}_\lambda) \exp(i\mathbf{k} \cdot \mathbf{d}_\lambda), \\ B_\lambda^{(2)} &= \left(-[\mathbf{k} \cdot (\boldsymbol{\Omega}_\lambda \times \mathbf{d}_\lambda)]^2 + i\mathbf{k} \cdot \left[(\boldsymbol{\tau}_\lambda \times \mathbf{d})/I - \Omega_\lambda^2 \mathbf{d}_\lambda\right]\right) \exp(i\mathbf{k} \cdot \mathbf{d}_\lambda), \end{aligned} \tag{11.482}$$

where $\boldsymbol{\tau}_\lambda$ is the torque acting on molecule λ. In deriving (11.482), we have used the relations $\dot{\mathbf{d}}_\lambda = \boldsymbol{\Omega}_\lambda \times \mathbf{d}_\lambda$, $\dot{\boldsymbol{\Omega}}_\lambda = \boldsymbol{\tau}_\lambda/I$, and $\boldsymbol{\Omega}_\lambda \cdot \mathbf{d}_\lambda = 0$. Thus

$$M_2^{\text{self}} = \left\langle (\mathbf{k}\cdot\mathbf{v}_1)^2 + [\mathbf{k}\cdot(\boldsymbol{\Omega}_1 \times \mathbf{d}_1)]^2 \right\rangle, \tag{11.483}$$

and since $\langle v_{1\alpha} v_{1\beta}\rangle = (\beta M)^{-1}\delta_{\alpha\beta}$, $\langle \Omega_{1\alpha}\Omega_{1\beta}\rangle = (\beta I)^{-1}\delta_{\alpha\beta}$, we have

$$M_2^{\text{self}} = (\beta M)^{-1}k^2 + (\beta I)^{-1}\left(k^2 d^2 - \langle \mathbf{k}\cdot\mathbf{d}_1\rangle^2\right). \tag{11.484}$$

Averaging over all possible directions of $\mathbf{k}$, or equivalently of $\mathbf{d}_1$, and noting that $I = Md^2$, we find that

$$M_2^{\text{self}} = \frac{5}{3}(\beta M)^{-1}k^2. \tag{11.485}$$

Similarly

$$\begin{aligned} M_4^{\text{self}} &= \frac{1}{2}\left\langle \Big| -[\mathbf{k}\cdot(\mathbf{v}_1 + \boldsymbol{\Omega}_1 \times \mathbf{d}_1)]^2 \right. \\ &\quad \left. + i\,\mathbf{k}\cdot\left(\mathbf{F}_1/M + (\boldsymbol{\tau}_1 \times \mathbf{d}_1)/I - \Omega_1^2\mathbf{d}_1\right)\Big|^2 \right\rangle + \frac{1}{2}\,\langle \mathbf{d}_1 \to -\mathbf{d}_1\rangle \\ &= \left\langle (\mathbf{k}\cdot\mathbf{v}_1)^4 + 6(\mathbf{k}\cdot\mathbf{v}_1)^2[\mathbf{k}\cdot(\boldsymbol{\Omega}_1 \times \mathbf{d}_1)]^2 + [\mathbf{k}\cdot(\boldsymbol{\Omega}_1 \times \mathbf{d}_1)]^4 \right. \\ &\quad \left. + [(\mathbf{k}\cdot\mathbf{F}_1)/M]^2 + [\mathbf{k}\cdot(\boldsymbol{\tau}_1 \times \mathbf{d}_1)/I]^2 + (\mathbf{k}\cdot\mathbf{d}_1)^2\Omega_1^4 \right\rangle. \end{aligned} \tag{11.486}$$

To simplify this result, we note that

$$\begin{aligned} \left\langle [\mathbf{k}\cdot(\boldsymbol{\Omega}_1 \times \mathbf{d}_1)]^{2n} \right\rangle &= \frac{1}{2n+1}\left\langle (\boldsymbol{\Omega}_1 \times \mathbf{d}_1)^{2n} \right\rangle \\ &= \frac{1}{2n+1}\mathrm{d}^{2n}\left\langle \Omega_1^{2n} \right\rangle = \frac{(2n)!!}{2n+1}(\beta M)^{-n}, \end{aligned} \tag{11.487}$$

since $\boldsymbol{\Omega}_1 \cdot \mathbf{d}_1 = 0$ and $\langle \Omega^{2n}\rangle = (2n)!!(\beta I)^{-n}$. We also have $\langle [\mathbf{k}\cdot(\boldsymbol{\tau}_1 \times \mathbf{d}_1)]^2\rangle = (1/3)d^2\langle \tau^2\rangle$, since $\boldsymbol{\tau}_1 \cdot \mathbf{d}_1 = 0$. Then

$$M_4^{\text{self}} = (\beta M)^{-2}\left(\frac{43}{5}k^4 + \frac{8}{3}k^2 d^{-2} + \frac{1}{3}k^2\beta^2\left\langle F_1^2\right\rangle + \frac{1}{3}k^2 d^{-2}\beta^2\left\langle \tau_1^2\right\rangle\right). \tag{11.488}$$

Figure 11.19 shows a plot of M_4^{self} for a fluid of quadrupolar LJ diatomics. The mean square force and torque were computed by Monte Carlo simulation.[295,296] Quadrupolar effects are seen to be significant at small and intermediate $k^* = k\sigma$.

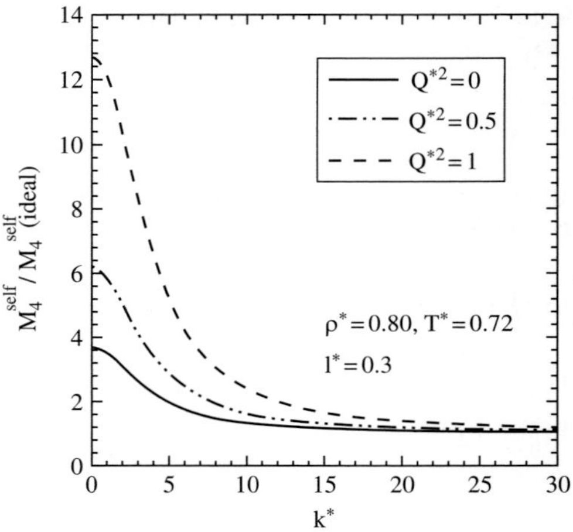

FIG. 11.19 The fourth moment of the dynamic self structure factor, shown relative to its ideal gas value, for a fluid of quadrupolar LJ diatomics of bond length $l^* = 2d/\sigma = 0.3$, as a function of the wavevector. All units are reduced relative to the atomic Lennard-Jones parameters ϵ and σ.

Using the hypervirial theorems (E.31) and (E.32), our expression for M_4^{self} can also be written as

$$M_4^{\text{self}} = (\beta M)^{-2} \left(\frac{43}{5} k^4 + \frac{8}{3} k^2 d^{-2} + \frac{1}{3} k^2 \beta\rho \int d\mathbf{r}\, g(\mathbf{r}\omega_1\omega_2) \left(\nabla_1^2 + d^{-2} \nabla_{\omega 1}^2 \right) u(\mathbf{r}\omega_1\omega_2) \right), \tag{11.489}$$

where ∇^2 and ∇_ω^2 are the Laplacian and angular Laplacian (Vol. 1, p. 448) operators, respectively.

The intermediate scattering function is given by

$$F(k,t) = N^{-1} \sum_{\lambda\lambda'} \left\langle E_\lambda(0)^* E_{\lambda'}(t) \right\rangle, \tag{11.490}$$

where $E_\lambda(t) = \exp(i\mathbf{k} \cdot \mathbf{r}_\lambda(t)) \cos(\mathbf{k} \cdot \mathbf{d}_\lambda(t))$. The moments are therefore

$$M_{2n} = N^{-1} \sum_{\lambda\lambda'} \left\langle E_\lambda^{(n)*} E_{\lambda'}^{(n)} \right\rangle, \tag{11.491}$$

so that

$$M_0 = N^{-1} \sum_{\lambda\lambda'} \langle \cos(\mathbf{k} \cdot \mathbf{d}_\lambda) \cos(\mathbf{k} \cdot \mathbf{d}_{\lambda'}) \rangle, \tag{11.492}$$

$$M_2 = \left\langle |i\mathbf{k} \cdot \mathbf{v}_1 \cos(\mathbf{k} \cdot \mathbf{d}_1) - \mathbf{k} \cdot (\mathbf{\Omega}_1 \times \mathbf{d}_1) \sin(\mathbf{k} \cdot \mathbf{d}_1)|^2 \right\rangle, \tag{11.493}$$

etc. Substituting (11.417) in (11.492), we find for the zeroth moment that

$$S(k) = \frac{1}{2}(1 + j_0(2kd)) + \sum_{l\,\text{even}} (2l+1) j_l(kd)^2 G_l, \tag{11.494}$$

where G_l is the angular correlation parameter discussed in Chapter 10.[297]

The second moment contains only self terms, and can be written as

$$\begin{aligned} M_2 &= \left\langle (\mathbf{k}\cdot\mathbf{v}_1)^2 \cos^2(\mathbf{k}\cdot\mathbf{d}_1) + [\mathbf{k}\cdot(\boldsymbol{\Omega}_1 \times \mathbf{d}_1)]^2 \sin^2(\mathbf{k}\cdot\mathbf{d}_1)\right\rangle \\ &= (\beta M)^{-1} k^2 \left\langle \cos^2(\mathbf{k}\cdot\mathbf{d}_1)\right\rangle \\ &\quad + (\beta I)^{-1} \left\langle \left(k^2 d^2 - (\mathbf{k}\cdot\mathbf{d}_1)^2\right) \sin^2(\mathbf{k}\cdot\mathbf{d}_1)\right\rangle \\ &= \frac{1}{6}(\beta M)^{-1} k^2 \,(5 + j_0(kl) - 2 j_2(kl)). \end{aligned} \tag{11.495}$$

Other molecular geometries Chapter 9 contains a detailed discussion of the form of the zeroth moment $M_0 = S(k)$ of the dynamic structure factor for molecules of several different symmetries.

For the second moments of $S(k,\omega)$ and $S_{\text{self}}(k,\omega)$, it can be shown that[298]

$$\begin{aligned} M_2^{\text{self}} &= \left(\sum_\alpha b_\alpha^{\text{inc}\,2}\right)^{-1} (\beta M)^{-1} k^2 \sum_\alpha b_\alpha^{\text{inc}\,2} A_{\alpha\alpha}, \\ M_2 &= \left(\sum_\alpha \bar{b}_\alpha\right)^{-2} (\beta M)^{-1} \sum_{\alpha\beta} \bar{b}_\alpha \bar{b}_\beta [A_{\alpha\beta} j_0(kd_{\alpha\beta}) + B_{\alpha\beta} j_2(kd_{\alpha\beta})], \end{aligned} \tag{11.496}$$

where

$$\begin{aligned} A_{\alpha\beta} &= 1 + \frac{1}{3} M \left(\mathbf{d}_\alpha \cdot \mathbf{d}_\beta \,\mathrm{Tr}\,\mathbf{I}^{-1} - \mathbf{d}_\alpha \cdot \mathbf{I}^{-1} \cdot \mathbf{d}_\beta\right), \\ B_{\alpha\beta} &= A_{\alpha\beta} - 1 - M(\mathbf{d}_\alpha \times \hat{\mathbf{d}}_{\alpha\beta}) \cdot \mathbf{I}^{-1} \cdot (\mathbf{d}_\beta \times \hat{\mathbf{d}}_{\alpha\beta}), \end{aligned} \tag{11.497}$$

and $\mathbf{I}^{-1}$ is the inverse of the molecular inertia tensor $\mathbf{I} = \sum_\alpha M_\alpha \mathbf{d}_\alpha \mathbf{d}_\alpha$.

M_4 for water has been computed[278] by evaluating $\int d\omega\, \omega^4 S(k,\omega)$ directly using a simulated $S(k,\omega)$. Three-body forces are found to have a significant effect for k values that probe the scale of nearest-neighbour interactions. The dependence on k is shown in Fig. 11.20.

Quantum expressions for the moments

Quantum expressions for the moments of $S_{\text{self}}(k,\omega)$ and $S(k,\omega)$ can be derived in the same way as was done in § 11.1.7 for infrared spectra. Away from the classical limit, the odd moments M_{2n+1}^{self} and M_{2n+1} are nonzero, and are proportional to

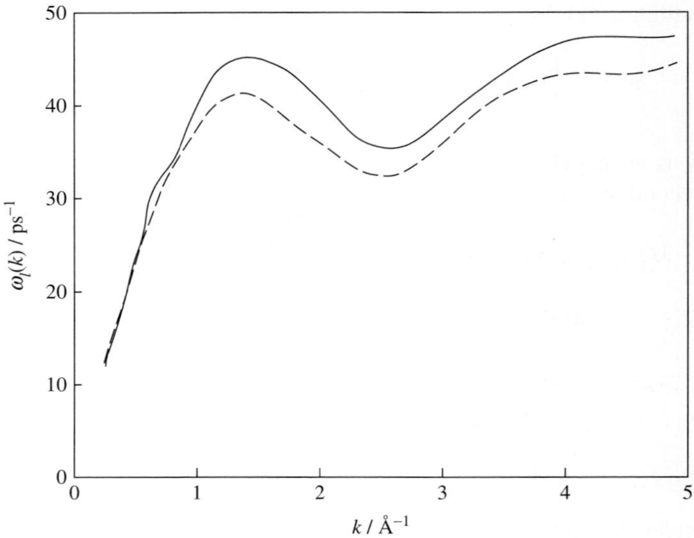

FIG. 11.20 $\omega_l(k) = (M_4/M_2)^{1/2}$ from simulations of liquid water at 300 K. Dashed line, two-body forces only; full line, two- plus three-body forces. (From ref. 278.)

$\hbar$ at high temperatures, while the even moments M_{2n}^{self} and M_{2n} have corrections of order $\hbar^2$, $\hbar^4$, etc.

Zeroth moment For the zeroth moment $M_0 = S(k)$, the only corrections are statistical, and may be calculated from the Wigner–Kirkwood expansion of the quantum configurational distribution function (Appendix 3D). Quantum effects on $S(k)$ are normally only important for molecules containing light atoms, such as hydrogen and water; see Chapter 9 for details.

First moments The moments M_1^{self} and M_1 can be computed exactly. We first consider atomic fluids. Using (11.355) and (11.456), we have

$$M_1^{\text{self}} = -i\langle A_1^* \dot{A}_1 \rangle, \tag{11.498}$$

where $A_\lambda = \exp(i\mathbf{k} \cdot \mathbf{r}_\lambda)$. From (11.96) $i\hbar\dot{A}_1 = [A_1, H] = [A_1, K_1]$, where $K_1 = p_1^2/2M$ is the kinetic energy of atom 1, and $\mathbf{p}_1 = -i\hbar\boldsymbol{\nabla}_1$ is its momentum. Now

$$\left[A_1, p_{1\mu}^2\right] = A_1 p_{1\mu} p_{1\mu} - p_{1\mu} p_{1\mu} A_1 = [A_1, p_{1\mu}] p_{1\mu} + p_{1\mu}[A_1, p_{1\mu}], \tag{11.499}$$

and since $[A_1, p_{1\mu}] = -\hbar k_\mu A_1$,

$$\begin{aligned}\left[A_1, p_{1\mu}^2\right] &= -\hbar k_\mu \left(A_1 p_{1\mu} + p_{1\mu} A_1\right) = -\hbar k_\mu \left(2A_1 p_{1\mu} - [A_1, p_{1\mu}]\right) \\ &= -A_1 \left(2\hbar\mathbf{k} \cdot \mathbf{p}_1 + \hbar^2 k^2\right).\end{aligned} \tag{11.500}$$

Therefore

$$\dot{A}_1 = iM^{-1}A_1\left(\mathbf{k}\cdot\mathbf{p}_1 + \frac{1}{2}\hbar k^2\right), \tag{11.501}$$

and since $\langle\mathbf{p}_1\rangle = 0$, it follows that

$$M_1^{\text{self}} = \frac{\hbar k^2}{2M}. \tag{11.502}$$

The calculation of M_1 is somewhat trickier.[45] From (11.356) and (11.456) we have

$$M_1 = -iN^{-1}\sum_{\lambda\lambda'}\langle A_\lambda^*\dot{A}_{\lambda'}\rangle, \tag{11.503}$$

and using (11.501) this becomes

$$M_1 = N^{-1}M^{-1}\sum_{\lambda\lambda'}\left\langle A_\lambda^*A_{\lambda'}\left(\mathbf{k}\cdot\mathbf{p}_{\lambda'} + \frac{1}{2}\hbar k^2\right)\right\rangle. \tag{11.504}$$

This may be decomposed into self and distinct terms; the former are given by (11.502), while the latter are

$$M_1^{\text{dis}} = N^{-1}M^{-1}\sum_{\lambda\neq\lambda'}\left\langle A_\lambda^*A_{\lambda'}\left(\mathbf{k}\cdot\mathbf{p}_{\lambda'} + \frac{1}{2}\hbar k^2\right)\right\rangle. \tag{11.505}$$

We note that the first term on the right of (11.505) is *not* zero, since quantum-mechanically $\mathbf{p}_{\lambda'}$ and $\mathbf{r}_{\lambda'}$ do not commute, so that the averaging over coordinates and momenta cannot be carried out separately. We will now show that, in fact, $M_1^{\text{dis}} = 0$. To this end we may use a quantum hypervirial theorem (Appendix E.2.2). Consider the time dependence of the function $\langle A_\lambda(t)^*A_{\lambda'}(t)\rangle$. From stationarity, this function is independent of time; hence

$$\frac{\mathrm{d}}{\mathrm{d}t}\langle A_\lambda(t)^*A_{\lambda'}(t)\rangle = \left\langle\dot{A}_\lambda^*A_{\lambda'}\right\rangle + \langle A_\lambda^*\dot{A}_{\lambda'}\rangle = 0. \tag{11.506}$$

Using (11.501), it follows that

$$\left\langle A_\lambda^*\left(-\mathbf{k}\cdot\mathbf{p}_\lambda + \frac{1}{2}\hbar k^2\right)A_{\lambda'}\right\rangle + \left\langle A_\lambda^*A_{\lambda'}\left(\mathbf{k}\cdot\mathbf{p}_\lambda + \frac{1}{2}\hbar k^2\right)\right\rangle = 0, \tag{11.507}$$

and since $\left[\mathbf{p}_\lambda, A_\lambda'\right] = 0$ for $\lambda\neq\lambda'$, we have

$$\left\langle A_\lambda^*A_{\lambda'}\left(\mathbf{k}\cdot[\mathbf{p}_{\lambda'} - \mathbf{p}_\lambda] + \hbar k^2\right)\right\rangle = 0, \qquad \lambda\neq\lambda'. \tag{11.508}$$

Using (11.505) and (11.508) we can write

$$M_1^{\text{dis}} = N^{-1}M^{-1}\frac{1}{2}\sum_{\lambda\neq\lambda'}\langle\exp(i\mathbf{k}\cdot[\mathbf{r}_{\lambda'} - \mathbf{r}_\lambda])\,\mathbf{k}\cdot[\mathbf{p}_{\lambda'} + \mathbf{p}_\lambda]\rangle\,. \tag{11.509}$$

Now $\langle f(\mathbf{r}_{\lambda'} - \mathbf{r}_\lambda)(\mathbf{p}_{\lambda'} + \mathbf{p}_\lambda)\rangle = \langle f(\mathbf{r}'_\lambda - \mathbf{r}_\lambda)\rangle \langle \mathbf{p}_{\lambda'} + \mathbf{p}_\lambda\rangle$, for an arbitrary function $f(\mathbf{r}_{\lambda'} - \mathbf{r}_\lambda)$, since $[\mathbf{p}_{\lambda'} + \mathbf{p}_\lambda, \mathbf{r}_{\lambda'} - \mathbf{r}_\lambda] = 0$. Therefore

$$M_1^{\text{dis}} = 0 \tag{11.510}$$

and we obtain finally the very simple result

$$M_1 = M_1^{\text{self}} = \frac{\hbar k^2}{2M}. \tag{11.511}$$

We will next consider a rigid homonuclear diatomic. From (11.480),

$$M_1^{\text{self}} = -\frac{1}{2}i\langle C_1^*\dot{C}_1 + D_1^*\dot{D}_1\rangle. \tag{11.512}$$

Since $[A_1, B_1] = 0$, we have

$$M_1^{\text{self}} = M_{1\,\text{trans}}^{\text{self}} + M_{1\,\text{rot}}^{\text{self}}, \tag{11.513}$$

where, from (11.502), $M_{1\,\text{trans}}^{\text{self}} = \hbar k^2/2M$, and

$$M_{1\,\text{rot}}^{\text{self}} = -\frac{1}{2}i\langle B_1^*\dot{B}_1 + B_1\dot{B}_1^*\rangle. \tag{11.514}$$

Since $i\hbar\dot{B}_1 = [B_1, H]$, it follows that

$$M_{1\,\text{rot}}^{\text{self}} = -\frac{1}{2\hbar}\langle[B_1^*, [B_1, H]]\rangle = -\frac{1}{4\hbar I}\left\langle\left[B_1^*, \left[B_1, J_1^2\right]\right]\right\rangle, \tag{11.515}$$

where $\mathbf{J}_1$ is the angular momentum of molecule 1. Now $\left[B_1, J_1^2\right] = x_{1\mu}J_{1\mu} + J_{1\mu}x_{1\mu}$, where $x_{1\mu} \equiv [B_1, J_{1\mu}]$ is obviously a function of $\mathbf{d}_1$, but not of $\mathbf{J}_1$. Hence $\left[B_1, J_1^2\right] = 2x_{1\mu}J_{1\mu} - x_2$, where $x_2 \equiv [x_{1\mu}, J_{1\mu}]$ is also a function of $\mathbf{d}_1$ only. Consequently $\left[B_1^*, \left[B_1, J_1^2\right]\right] = 2[B_1, J_{1\mu}]\left[B_1^*, J_{1\mu}\right]$, and therefore

$$M_{1\,\text{rot}}^{\text{self}} = -\frac{1}{2\hbar I}\langle[B_1, \mathbf{J}_1]\cdot\left[B_1^*, \mathbf{J}_1\right]\rangle. \tag{11.516}$$

Since $\mathbf{J}_1 = -i\hbar\sum_\alpha \mathbf{d}_{1\alpha}\times\boldsymbol{\nabla}_{1\alpha}$, we have $[B_1, \mathbf{J}_1] = \hbar(\mathbf{k}\times\mathbf{d})B_1$. Hence

$$M_{1\,\text{rot}}^{\text{self}} = \frac{\hbar}{2I}\left\langle(\mathbf{k}\times\mathbf{d})^2\right\rangle = \frac{\hbar k^2}{3M}. \tag{11.517}$$

Finally, therefore,

$$M_1^{\text{self}} = \frac{5\hbar k^2}{6M}. \tag{11.518}$$

Similarly, we can show from (11.490) that $M_1 = M_{1\,\text{trans}} + M_{1\,\text{rot}}$, where

$$M_{1\,\text{trans}} = \frac{\hbar k^2}{2M}\left\langle\cos^2(\mathbf{k}\cdot\mathbf{d})\right\rangle = \frac{\hbar k^2}{4M}(1 + j_0(2kd)), \tag{11.519}$$

and

$$M_{1\,\text{rot}} = -\frac{1}{2\hbar I}\left\langle[\cos(\mathbf{k}\cdot\mathbf{d}), \mathbf{J}_1]^2\right\rangle. \tag{11.520}$$

Since $[\cos(\mathbf{k}\cdot\mathbf{d}), \mathbf{J}_1] = i\hbar(\mathbf{k}\times\mathbf{d})\sin(\mathbf{k}\cdot\mathbf{d})$, we have

$$M_{1\,\mathrm{rot}} = \frac{\hbar}{2I}\left\langle(\mathbf{k}\times\mathbf{d})^2\sin^2(\mathbf{k}\cdot\mathbf{d})\right\rangle = \frac{\hbar k^2}{6M}(1 - j_0(2kd) - j_2(2kd)). \quad (11.521)$$

Thus

$$M_1 = \frac{\hbar k^2}{12M}(5 + j_0(2kd) - 2j_2(2kd)). \quad (11.522)$$

For rigid molecules of more general symmetry the first moments can be shown to be

$$M_1^{\mathrm{self}} = \left(\sum_\alpha b_\alpha^{\mathrm{inc}\,2}\right)^{-1}\frac{\hbar k^2}{2M}\sum_\alpha b_\alpha^{\mathrm{inc}\,2}A_{\alpha\alpha},$$

$$M_1 = \left(\sum_\alpha \bar{b}_\alpha\right)^{-2}\frac{\hbar k^2}{2M}\sum_{\alpha\beta}\bar{b}_\alpha\bar{b}_\beta[A_{\alpha\beta}j_0(kd_{\alpha\beta}) + B_{\alpha\beta}j_2(kd_{\alpha\beta})], \quad (11.523)$$

where $A_{\alpha\beta}$ and $B_{\alpha\beta}$ are given by (11.497).

When the effects of molecular vibration are included, *completely different expressions for the moments result.* Thus, consider that the Hamiltonian for a system of nonrigid molecules may be written quite generally as

$$H = \sum_\alpha \frac{\mathbf{p}_\alpha^2}{2M_\alpha} + \mathcal{U}(\mathbf{r}_1, \mathbf{r}_2 \ldots \mathbf{r}_N), \quad (11.524)$$

where the sum over α runs over all atoms in the system without regard to molecule, and the potential energy $\mathcal{U}(\mathbf{r}_1, \mathbf{r}_2 \ldots \mathbf{r}_N)$ includes intramolecular as well as intermolecular terms. If we now write

$$M_1^{\mathrm{self}} = -i\left(\sum_\alpha b_\alpha^{\mathrm{inc}\,2}\right)^{-1}\sum_\alpha b_\alpha^{\mathrm{inc}\,2}\left\langle A_\alpha^*\dot{A}_\alpha\right\rangle, \quad (11.525)$$

where $i\hbar\dot{A}_\alpha = [A_\alpha, H] = \left[A_\alpha, p_\alpha^2\right]/2M_\alpha$, then by an analysis paralleling that of the atomic case, we see that

$$M_1^{\mathrm{self}} = \left(\sum_\alpha b_\alpha^{\mathrm{inc}\,2}\right)^{-1}\frac{\hbar k^2}{2}\sum_\alpha b_\alpha^{\mathrm{inc}\,2}M_\alpha^{-1}. \quad (11.526)$$

Similarly, we find that

$$M_1 = \left(\sum_\alpha b_\alpha\right)^{-2}\frac{\hbar k^2}{2}\sum_\alpha \left(\bar{b}_\alpha\right)^2 M_\alpha^{-1}. \quad (11.527)$$

For a homonuclear diatomic, we therefore find that

$$M_1^{\mathrm{self}} = \frac{\hbar k^2}{M}, \qquad M_1 = \frac{\hbar k^2}{2M}, \quad (11.528)$$

in place of (11.518) and (11.522). The differences are substantial. Molecular non-rigidity increases M_1^{self} by 20% for all k; it completely eliminates the oscillations in M_1, and increases it by over 40% for some values of k. These results are independent of the amplitude of the vibration.

To understand these differences, we will consider the particular case of a dilute gas of vibrating diatomic molecules, for which $S(k, \omega)$ is given explicitly by (11.435). The moments of this function are

$$M_n = \sum_{l\,\text{even}} \sum_{\nu\nu'} P_\nu (2l+1) |\langle \nu | j_l(kd) | \nu' \rangle|^2 M^{(l)}_{\nu\nu' n}, \tag{11.529}$$

where

$$M^{(l)}_{\nu\nu' n} = \int_{-\infty}^{+\infty} \mathrm{d}\omega\, \omega^n S_l(\omega - \omega_{\nu'\nu}) = \int_{-\infty}^{+\infty} \mathrm{d}\omega\, (\omega + \omega_{\nu'\nu})^n S_l(\omega), \tag{11.530}$$

and $S_l(k, \omega)$ is given by (11.436). We have already seen that $M_0 \equiv S(k)$ is changed only slightly by the inclusion of vibrational effects (see (11.440)). For $n \geq 1$, however, the appearance of factors of $\omega_{\nu'\nu}$ in (11.530) produces substantial changes in the moments. For $n = 1$ we have

$$M^{(l)}_{\nu\nu' 1} = M^{(l)}_1 + \omega_{\nu'\nu} M^{(l)}_0, \tag{11.531}$$

where

$$M^{(l)}_n = \int_{-\infty}^{+\infty} \mathrm{d}\omega\, \omega^n S_l(\omega). \tag{11.532}$$

We can easily show that $M^{(l)}_0 = 1$ and $M^{(l)}_1 = \hbar k^2/2M + Bl(l+1)$, where $B = \hbar/2I$ is the rotational constant. Hence

$$M_1 = \sum_{l\,\text{even}} \sum_{\nu\nu'} P_\nu (2l+1) |\langle \nu | j_l(kd) | \nu' \rangle|^2 \left(\frac{\hbar k^2}{2M} + Bl(l+1) + \omega_{\nu'\nu} \right). \tag{11.533}$$

Using (B.53) the first and second terms on the right can be written as

$$M_1^{\text{tr}\oplus\text{rot}} = \sum_\nu P_\nu \left(\frac{\hbar k^2}{2M} \langle \nu | T(kd) | \nu \rangle + B \langle \nu | R(kd) | \nu \rangle \right), \tag{11.534}$$

where

$$T(x) \equiv \sum_{l\,\text{even}} (2l+1) j_l(x)^2 = \frac{1}{2}[1 + j_0(2x)],$$

$$R(x) \equiv \sum_{l\,\text{even}} (2l+1) l(l+1) j_l(x)^2 = \frac{1}{3} x^2 [1 - j_0(2x) - j_2(2x)]. \tag{11.535}$$

If we neglect the dependence of d on the vibrational state of the molecule, we therefore have

$$M_1^{\mathrm{tr}\oplus\mathrm{rot}} = \frac{\hbar k^2}{12M}[5 + j_0(2kd) - 2j_2(2kd)]. \tag{11.536}$$

This is the contribution to M_1 from the rotational band of the spectrum ($\nu = \nu'$), and is exactly Eq. (11.522) for a rigid molecule.

The third term on the right of (11.533) is the contribution to M_1 from the vibrational bands of the spectrum ($\nu \neq \nu'$). It can be written as

$$M_1^{\mathrm{vib}} = \frac{1}{\hbar}\sum_{l\,\mathrm{even}}\sum_{\nu} P_\nu(2l+1)\langle\nu|j_l(kd)[H, j_l(kd)]|\nu\rangle, \tag{11.537}$$

and using arguments similar to those following (11.515), we can show that

$$\begin{aligned} M_1^{\mathrm{vib}} &= -\frac{2}{M\hbar}\sum_{\nu} P_\nu \sum_{l\,\mathrm{even}}(2l+1)\left\langle\nu\left|[j_l(kd), p]^2\right|\nu\right\rangle \\ &= \frac{\hbar k^2}{2M}\sum_{l\,\mathrm{even}}\sum_{\nu} P_\nu(2l+1)\left\langle\nu\left|j_l'(kd)^2\right|\nu\right\rangle. \end{aligned} \tag{11.538}$$

Since[126] $(2l+1)j_l'(x) = lj_{l-1}(x) - (l+1)j_{l+1}(x)$, we have

$$M_1^{\mathrm{vib}} = \frac{\hbar k^2}{2M}\sum_{\nu} P_\nu\langle\nu|V(kd)|\nu\rangle, \tag{11.539}$$

where

$$\begin{aligned} V(x) &= \sum_{l\,\mathrm{even}}(2l+1)^{-1}[lj_{l-1}(x) - (l+1)j_{l+1}(x)]^2 \\ &= \frac{1}{6}[1 - j_0(2x) + 2j_2(2x)]. \end{aligned} \tag{11.540}$$

Neglecting the dependence of d on ν, we therefore find that

$$M_1^{\mathrm{vib}} = \frac{\hbar k^2}{12M}[1 - j_0(2x) + 2j_2(2x)]. \tag{11.541}$$

The sum of (11.536) and (11.541) is thus

$$M_1 = M_1^{\mathrm{tr}\oplus\mathrm{rot}} + M_1^{\mathrm{vib}} = \frac{\hbar k^2}{2M}, \tag{11.542}$$

in agreement with (11.528).

The above analysis shows that neglect of molecular nonrigidity leads to serious errors in the *formal* expression for M_1, and the same will obviously be true for the higher moments also. However, (11.435) shows that the vibrational contribution to these moments comes from frequencies near $\omega_{\nu'\nu} = \omega_{\nu'} - \omega_\nu$. In most cases, vibrational frequencies are so high that experimentally measured moments are better approximated by the rigid molecule expressions (11.523) than by their nonrigid counterparts.

Higher moments Fredrikze[299] has calculated the lowest order quantum corrections for moments up to M_4 for a monatomic fluid mixture. More generally, quantum corrections to the higher moments can be derived by the indirect method discussed in § 11.1.7. In terms of the symmetrized and antisymmetrized functions introduced in (11.359), we have

$$M_{2n} = \int_{-\infty}^{+\infty} d\omega\, \omega^{2n} S_+(k,\omega),$$

$$M_{2n+1} = \int_{-\infty}^{+\infty} d\omega\, \omega^{2n+1} S_-(k,\omega). \tag{11.543}$$

Using (11.360), and noting that $\coth x$ and $\tanh x$ are odd functions of x, it follows that

$$M_{2n} = \int_{-\infty}^{+\infty} d\omega\, \omega^{2n} \coth \frac{1}{2}\beta\hbar\omega\, S(k,\omega),$$

$$M_{2n+1} = \int_{-\infty}^{+\infty} d\omega\, \omega^{2n+1} \tanh \frac{1}{2}\beta\hbar\omega\, S(k,\omega). \tag{11.544}$$

The odd and even moments are therefore related by (11.52). These relations can also be derived using Eqs. (11.362) and (11.363) relating the real and imaginary parts of $F(k,t)$.

The series for M_2 is therefore

$$M_2 = \frac{2}{\beta\hbar} M_1 + \frac{1}{3}\left(\frac{\beta\hbar}{2}\right)^2 M_4 + O(\hbar^4), \tag{11.545}$$

and since M_1 is strictly proportional to $\hbar$, it follows that

$$M_2 = M_2^{\mathrm{cl}} + \frac{1}{3}\left(\frac{\beta\hbar}{2}\right)^2 M_4^{\mathrm{cl}} + O(\hbar^4). \tag{11.546}$$

The $O(\hbar^2)$ quantum correction to the second moment is thus obtainable if M_4^{cl} is available from theory or experiment. Similarly, M_3 may be computed to lowest order using

$$M_3 \approx \frac{\beta\hbar}{2} M_4^{\mathrm{cl}}. \tag{11.547}$$

11.3.6 Construction of approximate theories for $S(k,\omega)$ using spectral moments

The computation of spectral moments, which is accomplished within the confines of equilibrium statistical mechanics, offers a simple route to the approximate calculation of the dynamical quantities $S(k,\omega)$ and $S_{\mathrm{self}}(k,\omega)$.

There are in principle a variety of methods by which to translate knowledge of the M_n into an estimate of $S(k, \omega)$. One possibility is to construct the cumulant expansion of the intermediate scattering function. The cumulants ρ_n of $F(k, t)$ are defined through the equation

$$F(k,t) = F(k,0) \exp\left(\sum_{n=1}^{\infty} \rho_n \frac{(it)^n}{n!}\right). \tag{11.548}$$

We recall that the moment expansion can be written

$$F(k,t) = F(k,0) \sum_{n=0}^{\infty} \mu_n \frac{(it)^n}{n!}, \tag{11.549}$$

where $\mu_n = M_n/M_0$. It is obvious that ρ_n can be expressed in terms of the $\mu_{n'}$ for $n' \leq n$ (and vice versa). The relation for the first few n is

$$\begin{aligned} \rho_1 &= \mu_1, \qquad & \rho_2 &= \bar{\mu}_2, \\ \rho_3 &= \bar{\mu}_3, \qquad & \rho_4 &= \bar{\mu}_4 - 3\bar{\mu}_2^2, \end{aligned} \tag{11.550}$$

where

$$\begin{aligned} \bar{\mu}_n &= M_0^{-1} \int_{-\infty}^{+\infty} \mathrm{d}\omega\, (\omega - \mu_1)^n S(k,\omega) \\ &= \sum_{m=0}^{n} (-)^{n-m} \binom{n}{m} \mu_m \mu_1^{n-m} \end{aligned} \tag{11.551}$$

is the nth. shifted reduced moment.

Equation (11.548) is formally exact. However, since only a few moments are generally available, the cumulant expansion must be truncated. For a classical fluid, at the lowest nontrivial level the prediction for $F(k, t)$ is

$$F(k,t) = \exp\left(-\frac{1}{2}\mu_2 t^2\right), \tag{11.552}$$

and therefore $S(k, \omega)$ is Gaussian,

$$S(k,\omega) = M_0 \left(\frac{1}{2\pi\mu_2}\right)^{1/2} \exp\left(-\frac{\omega^2}{2\mu_2}\right). \tag{11.553}$$

This result is obviously unsatisfactory at small k, where $S(k, \omega)$ shows observable maxima at nonzero ω.

At the next highest level, when we incorporate the fourth moment into the calculation, the cumulant expansion yields

$$F(k,t) = F(k,0) \exp\left(-\frac{1}{2}\rho_2 t^2 + \frac{1}{24}\rho_4 t^4\right), \tag{11.554}$$

which will yield sensible results only if $\rho_4 < 0$, so that $F(k,t)$ is bounded as $t \to \infty$. Unfortunately, this is not generally true. Consider for example the moments of the self scattering function of an atomic fluid. From (11.550) and (11.463), $\rho_4^{\text{self}} = k^2 \langle F_1^2 \rangle / 3M^2 > 0$. Obviously the cumulant expansion fails here.

A more promising route to the calculation of $S(k,\omega)$ is based on Mori theory.[268–270] We will consider only the classical case. Mori defines a sequence of *memory functions* $K_0(t) \equiv F(k,t)$, $K_1(t)$, $K_2(t), \ldots$, etc., such that $K_{n+1}(t)$ determines the causal time evolution of $K_n(t)$ according to

$$\frac{\partial K_n}{\partial t} = -\int_0^t \mathrm{d}t'\, K_n(t-t')K_{n+1}(t'), \qquad n = 0, 1, 2, \ldots. \tag{11.555}$$

Taking the Laplace transform of each equation of this infinite hierarchy yields

$$\tilde{K}_n(s) = \frac{K_n(0)}{s + \tilde{K}_{n+1}(s)}, \tag{11.556}$$

where

$$\tilde{K}_n(s) = \int_0^\infty \mathrm{d}t\, \exp(-st) K_n(t). \tag{11.557}$$

When $s = i\omega$, $\tilde{K}_0(s)$ is related to $S(k,\omega)$: actually $S(k,\omega) = \pi^{-1}\Re\tilde{K}_0(i\omega)$. We thus arrive at the following continued-fraction representation of the dynamic structure factor:

$$S(k,\omega) = \frac{1}{\pi}\Re\frac{K_0(0)}{i\omega+}\,\frac{K_1(0)}{i\omega+}\,\frac{K_2(0)}{i\omega+}\cdots\frac{K_n(0)}{i\omega + \tilde{K}_{n+1}(i\omega)}. \tag{11.558}$$

The parameters $K_n(0)$ in (11.558) can be related to the spectral moments of $S(k,\omega)$. For small values of n, this relationship is easily established using the small-time expansion of the $K_n(t)$,

$$K_n(t) = \sum_{m=0}^\infty \frac{(-)^m}{(2m)!} M_{2m}^{(n)} t^{2m}. \tag{11.559}$$

Substituting (11.559) in (11.555) gives

$$\begin{aligned}
M_2^{(n)} &= M_0^{(n)} M_0^{(n+1)}, \\
M_4^{(n)} &= M_2^{(n)} M_0^{(n+1)} + M_0^{(n)} M_2^{(n+1)}, \\
M_6^{(n)} &= M_4^{(n)} M_0^{(n+1)} + M_2^{(n)} M_2^{(n+1)} + M_0^{(n)} M_4^{(n+1)},
\end{aligned} \tag{11.560}$$

etc., which can be rearranged to yield

$$\begin{aligned}
M_0^{(n+1)} &= \hat{M}_2^{(n)}, \\
\hat{M}_2^{(n+1)} &= \hat{M}_4^{(n)} / \hat{M}_2^{(n)} - \hat{M}_2^{(n)}, \\
\hat{M}_4^{(n+1)} &= \hat{M}_6^{(n)} / \hat{M}_2 - 2\hat{M}_4^{(n)} + \hat{M}_2^{(n)\,2},
\end{aligned} \tag{11.561}$$

etc. Solving (11.561) recursively, and noting that $M_0^{(n)} = K_n(0)$, we find that

$$\begin{aligned} K_0(0) &= M_0, \\ K_1(0) &= \hat{M}_2, \\ K_2(0) &= \hat{M}_4/\hat{M}_2 - \hat{M}_2, \\ K_3(0) &= \frac{\hat{M}_6/\hat{M}_2 - (\hat{M}_4/\hat{M}_2)^2}{\hat{M}_4/\hat{M}_2 - \hat{M}_2}, \end{aligned} \tag{11.562}$$

etc. The general relationship has been given by Dupuis.[300] To determine $K_n(0)$ we must know M_{2n} and all the lower even moments.

Setting $n = 0$ in (11.558), and noting that $M_0 = S(k)$, we now obtain

$$S(k,\omega)/S(k) = \frac{1}{\pi}\Re\left(i\omega + \tilde{K}_1(i\omega)\right)^{-1}, \tag{11.563}$$

while for $n = 1$

$$S(k,\omega)/S(k) = \frac{1}{\pi}\Re\left(i\omega + \frac{\hat{M}_2}{i\omega + \tilde{K}_2(i\omega)}\right)^{-1}. \tag{11.564}$$

These results, and their analogues for higher values of n, are formally exact. In addition to a knowledge of the moments through order $2n$, however, we see that the nth level of Mori theory also requires a closure, in the form of a relation for $\tilde{K}_{n+1}(i\omega)$, the transform of the $(n+1)$th order memory function.

We now show how such a closure may be provided. For simplicity we will consider the case of an atomic fluid such as argon. Consider first the hydrodynamic limit $k \to 0$. Here $S(k,\omega)$ is well represented by the Rayleigh–Brillouin lineshape (11.442), with a central peak at $\omega = 0$, and two side peaks at $\omega = \pm c_S k$. The salient features of this spectrum are captured by an equation of the form (11.564) provided we take $\hat{M}_2 = \gamma^{-1}c_S^2k^2$ and[240]

$$\tilde{K}_2(i\omega) = \frac{\left(1-\gamma^{-1}\right)c_S^2k^2}{i\omega + \gamma D_T k^2} + D_L k^2. \tag{11.565}$$

If we extend this closure by introducing k-dependence into the speed of sound c_S and the transport coefficients D_T and D_L, we can represent $S(k,\omega)$ over a much wider range of k and ω. Various generalizations have been proposed.[154,157,240,244] Sears[301] suggests we set

$$\tilde{K}_2(i\omega) = \frac{A(k)}{i\omega + \alpha(k)} + \frac{B(k)}{i\omega + \beta(k)}. \tag{11.566}$$

Comparing (11.565) and (11.566) in the hydrodynamic limit, we find that

$$A(0) = \left(1 - \gamma^{-1}\right) c_{\mathrm{S}}^2 k^2, \tag{11.567}$$

$$\alpha(0) = \gamma D_{\mathrm{T}} k^2, \tag{11.568}$$

$$\frac{B(0)}{\beta(0)} = D_{\mathrm{L}} k^2. \tag{11.569}$$

In view of (11.567) it is natural to take

$$A(k) = \left(1 - \gamma^{-1}\right) c_{\mathrm{S}}(k)^2 k^2, \tag{11.570}$$

where $c_S(k) = (\gamma/\beta M S(k))^{1/2}$ is the k-dependent adiabatic speed of sound; this preserves the relation $K_1(0) = \hat{M}_2$ for all k. Next, we note that the sum rule

$$\frac{1}{\pi} \int_{-\infty}^{+\infty} \mathrm{d}\omega \, \Re \tilde{K}_2(i\omega) = K_2(0) \tag{11.571}$$

provides a relation between $A(k)$ and $B(k)$. From (11.562), we have

$$A(k) + B(k) = \hat{M}_4/\hat{M}_2 - \hat{M}_2, \tag{11.572}$$

and therefore

$$B(k) = \hat{M}_4/\hat{M}_2 - \gamma \hat{M}_2. \tag{11.573}$$

All that now remains is to fix the values of $\alpha(k)$ and $\beta(k)$ away from the hydrodynamic limit. To this end we may consider the limit $k \to \infty$, in which the dynamic structure factor approaches that of an ideal gas, (11.404). Although we cannot duplicate a Gaussian lineshape with a closure of the form (11.566), by matching intensities at $\omega = 0$ we find that

$$\frac{A(\infty)}{\alpha(\infty)} + \frac{B(\infty)}{\beta(\infty)} = \frac{2\hat{M}_2}{M_0 \bar{v} k}, \tag{11.574}$$

where $\bar{v} = (8/\pi \beta M)^{1/2}$ is the mean atomic speed. Assuming $\alpha(\infty) = \beta(\infty)$, then using (11.572), together with the limiting values $M_0 \to 1, M_2 \to k^2/\beta M, M_4 \to 3k^4/(\beta M)^2$ (cf. (11.467), (11.468), and (11.476)), we find that

$$\alpha(\infty) = \beta(\infty) = \bar{v} k. \tag{11.575}$$

Equations (11.568), (11.569), (11.573), and (11.575) suggest that it would be reasonable to take

$$\alpha(k) = \frac{\bar{v} \gamma D_{\mathrm{T}} k^2}{\bar{v} + \gamma D_{\mathrm{T}} k} \tag{11.576}$$

and

$$\beta(k)^2 = \left(\frac{\hat{M}_4/\hat{M}_2 - \gamma \hat{M}_2}{D_{\mathrm{L}} k^2}\right)^2 + \bar{v}^2 k^2. \tag{11.577}$$

Figure 11.21 shows the resulting $S(k,\omega)$ for liquid argon at 94 K, computed using the experimental values of the transport coefficients.[301] For small k, the coherent scattering function has the three-peak shape characteristic of the hydrodynamic limit. The Brillouin lines broaden with increasing k, and are not visible above 0.1 Å^{-1}. For larger wavectors, $S(k,\omega)$ is characterized by a single sharp peak that is the extension of the Rayleigh line. For $\omega^2 < \hat{M}_2$, the lineshape is approximately Lorentzian; for larger frequencies it decays more rapidly.

11.3.7 Inelasticity corrections to the integrated cross-section

The integrated cross-section $(\mathrm{d}\sigma/\mathrm{d}\Omega)_{\mathrm{coh}}$ determines the total scattered (coherent) neutron flux for a particular incident energy ω_0 at an observation angle $\Omega = (\theta, \phi)$, irrespective of the energy ω lost in the scattering process. It is related to the differential cross-section $(\mathrm{d}^2\sigma/\mathrm{d}\Omega\mathrm{d}\omega_\mathrm{S})_{\mathrm{coh}}$ by

$$\left(\frac{\mathrm{d}\sigma}{\mathrm{d}\Omega}\right)_{\mathrm{coh}} = \int \mathrm{d}\omega_\Omega \left(\frac{\mathrm{d}^2\sigma}{\mathrm{d}\Omega\mathrm{d}\omega_\mathrm{S}}\right)_{\mathrm{coh}}, \tag{11.578}$$

where the subscript Ω indicates that the frequency integration is to be performed at fixed solid angle. Using (11.352), we can write (11.578) as

$$\left(\frac{\mathrm{d}\sigma}{\mathrm{d}\Omega}\right)_{\mathrm{coh}} = N\left(\sum_\alpha \bar{b}_\alpha\right)^2 \int \mathrm{d}\omega_\Omega \frac{k_\mathrm{S}}{k_0} S(k,\omega). \tag{11.579}$$

In this equation, both k_S and k are functions of ω at fixed ω_0 (or k_0) and Ω. If this dependence is ignored, we find that

$$\left(\frac{\mathrm{d}\sigma}{\mathrm{d}\Omega}\right)_{\mathrm{coh}} \approx N\left(\sum_\alpha \bar{b}_\alpha\right)^2 S(k), \tag{11.580}$$

which yields an approximate relation between the measured integrated cross-section and the static structure factor $S(k) = \int \mathrm{d}\omega\, S(k,\omega)$. Equation (11.580) is referred to as the *static approximation*, since it is the result that would be obtained rigorously in the limit of infinite molecular mass. It is not, however, very accurate, particularly for scattering by light molecules. Using the expressions for the spectral moments derived in § 11.3.5, we can derive inelasticity corrections[45] to $S(k)$, as we now show.

The equations of conservation of energy and momentum yield relations between k_0, k_S, k, and ω. We have

$$\frac{k_\mathrm{S}}{k_0} = \left(1 - \frac{\omega}{\omega_0}\right)^{1/2} = 1 - \frac{1}{2}\frac{\omega}{\omega_0} - \frac{1}{8}\left(\frac{\omega}{\omega_0}\right)^2 + \ldots \tag{11.581}$$

and

$$k^2 = k_0^2 + k_\mathrm{S}^2 - 2k_0k_S\cos\theta. \tag{11.582}$$

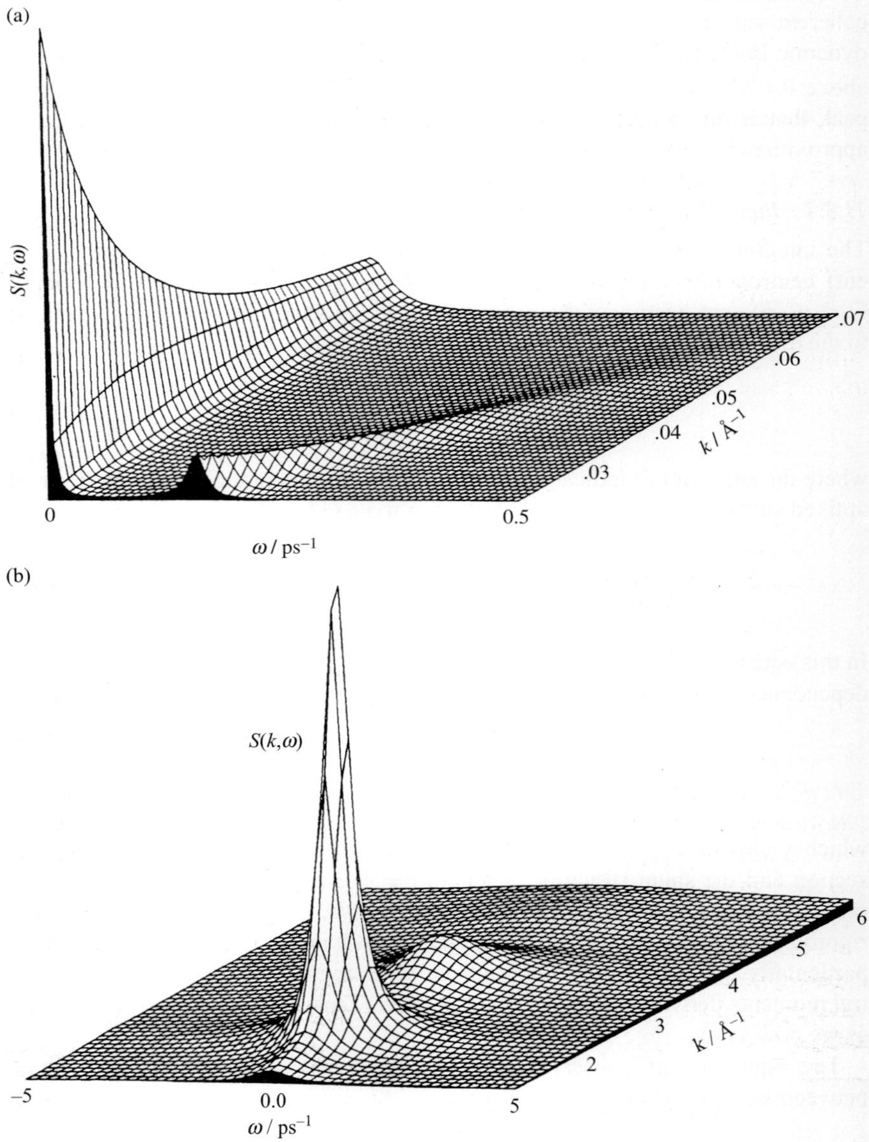

FIG. 11.21 $S(k, \omega)$ for liquid argon at 94 K from Mori theory. (a) Small k values appropriate to Brillouin light scattering; (b) larger k values typical of most neutron scattering studies. Reprinted with permission from ref. 301. Copyright 1970, NRC Research Press.

Let us introduce the wavevector $k_{\text{el}} = 2k_0 \sin \frac{1}{2}\theta$, which determines the momentum transfer in an elastic collision. From (11.581) and (11.582),

$$\left(\frac{k}{k_{\text{el}}}\right)^2 = 1 - \frac{1}{2}\frac{\omega}{\omega_0} + \frac{1}{8}\left(\frac{\omega}{\omega_0}\right)^2 \frac{\cos\theta}{1-\cos\theta} + \ldots. \tag{11.583}$$

Next, we perform a Taylor expansion of $S(k, \omega)$ about $k = k_{\text{el}}$. Using (11.579),

$$\left(\frac{\mathrm{d}\sigma}{\mathrm{d}\Omega}\right)_{\text{coh}} = N\left(\sum_\alpha \bar{b}_\alpha\right)^2 \int \mathrm{d}\omega_\Omega \frac{k_{\text{S}}}{k_0}$$
$$\times \left(S(k_{\text{el}}, \omega) + \left(k^2 - k_{\text{el}}^2\right)\frac{\partial}{\partial k^2}S + + \frac{1}{2}\left(k^2 - k_{\text{el}}^2\right)^2 \left(\frac{\partial}{\partial k^2}\right)^2 S + \ldots\right). \tag{11.584}$$

Substituting for k_{S} and k using (11.581) and (11.583) and integrating, we then find, in place of (11.580),

$$\left(\frac{\mathrm{d}\sigma}{\mathrm{d}\Omega}\right)_{\text{coh}} = N\left(\sum_\alpha \bar{b}_\alpha\right)^2 S'(k_{\text{el}}), \tag{11.585}$$

where

$$\begin{aligned} S'(k_{\text{el}}) = S(k_{\text{el}}) &- \frac{1}{2}m_1 - \frac{1}{8}m_2 + \ldots \\ &- \frac{1}{2}k_{\text{el}}^2 \frac{\partial}{\partial k^2}m_1 + \frac{1}{8}\left(k_{\text{el}}^2 + 2k_0^2\right)\frac{\partial}{\partial k^2}m_2 + \ldots \\ &+ \frac{1}{8}k_{\text{el}}^4\left(\frac{\partial}{\partial k^2}\right)^2 m_2 + \ldots. \end{aligned} \tag{11.586}$$

Here $m_n = M_n/\omega_0^n$ denotes a dimensionless spectral moment of $S(k, \omega)$.

Normally the measured cross-section includes contributions only up to a certain cutoff frequency $\omega_{\max}$; obviously this cannot exceed the incident neutron energy. If $\omega_{\max} < \omega_{\text{vib}}$, where ω_{vib} is the frequency of a molecular normal mode, the appropriate moments to use in (11.586) are those in which that mode is frozen. For a completely rigid molecule, M_1 is given by (11.523), and M_2 (classically) by (11.496).

For a classical homonuclear diatomic of mass M, assuming $\omega_{\max} > \omega_{\text{vib}}$, (11.586) becomes

$$S'(k_{\text{el}}) = S(k_{\text{el}}) - \frac{m_N}{M}\left[\left(\frac{k_{\text{el}}}{k_0}\right)^2 + \frac{1}{2}\frac{k_{\text{B}}T}{\hbar\omega_0} + \ldots\right], \tag{11.587}$$

where m_{N} is the neutron mass. The corrections are therefore small for small-angle scattering of epithermal neutrons by heavy molecules. For the scattering of thermal neutrons by light molecules, one may have to include additional terms involving the third and higher moments.

Appendix 11A Kramers–Kronig relations

The real and imaginary parts of the linear response functions of a stationary causal system are mutual Hilbert transforms. Consider, for example, the complex susceptibility $\chi(\omega) = \chi'(\omega) + i\chi''(\omega)$. We can show that

$$\chi'(\omega) = \frac{2}{\pi} P \int_0^\infty \mathrm{d}\omega' \, \frac{\omega' \chi''(\omega')}{\omega'^2 - \omega^2},$$

$$\chi''(\omega) = -\frac{2}{\pi} P \int_0^\infty \mathrm{d}\omega' \, \frac{\omega \chi'(\omega')}{\omega'^2 - \omega^2}, \tag{11A.1}$$

where P denotes the principal value of the integral. Such relations were first investigated by Kramers[302] and Kronig[303] in connection with the complex index of refraction, and are called Kramers–Kronig, or dispersion, relations. Their importance stems from the fact that knowledge of one of χ' or χ'' as a function of frequency enables calculation of the other. In this Appendix, we give a derivation of (11A.1), and discuss some of its consequences.

We begin our discussion with the equation relating the polarization $\mathbf{P}(t)$ to the macroscopic electric field $\mathbf{E}(t')$. For a stationary causal system in the linear response regime,

$$\mathbf{P}(t) = \int_{-\infty}^{t} \mathrm{d}t' \, \chi(t - t')\mathbf{E}(t'). \tag{11A.2}$$

Linearity is expressed by the fact that this equation contains only linear powers of $\mathbf{P}$ and $\mathbf{E}$. Stationarity is expressed by the fact that the response function χ depends only on the difference, $t - t'$, of the times between application of cause, $\mathbf{E}(t')$, and measurement of effect, $\mathbf{P}(t)$. Causality is expressed by the fact that the upper limit of the integral is t, so that $\mathbf{P}(t)$ depends only on $\mathbf{E}(t')$ at earlier times. Note that since both $\mathbf{P}(t)$ and $\mathbf{E}(t')$ are real functions, $\chi(t - t')$ is *also* real.

Changing integration variable to the time-delay parameter $\tau = t - t'$, we can write (11A.2) as

$$\mathbf{P}(t) = \int_0^\infty \mathrm{d}\tau \, \chi(\tau)\mathbf{E}(t - \tau). \tag{11A.3}$$

Fourier–Laplace transformation of (11A.3) then yields the frequency domain relation $\mathbf{P}(\omega) = \chi(\omega)\mathbf{E}(\omega)$, where

$$\chi(\omega) = \int_0^\infty \mathrm{d}\tau \, \exp(i\omega\tau)\chi(\tau). \tag{11A.4}$$

In this equation ω is in general complex. Since $\chi(\tau)$ is real, we must have

$$\chi(-\omega) = \chi^*(\omega^*). \tag{11A.5}$$

Equation (11A.5) implies that, for *real* ω,

$$\chi'(-\omega) = \chi'(\omega), \qquad \chi''(-\omega) = -\chi''(\omega), \tag{11A.6}$$

so that the dispersive and absorptive parts of $\chi(\omega)$ are even and odd functions of the frequency, respectively.

It is apparent from (11A.4) that, provided $\chi(\tau)$ is everywhere finite, $\chi(\omega)$ is analytic in the upper half plane, i.e. for $\Im\omega > 0$, since the integral and its derivative are then finite. We can therefore apply Cauchy's integral formula,[304] yielding

$$\chi(\omega) = \frac{1}{2\pi i}\int_{-\infty}^{+\infty} \mathrm{d}\omega' \, \frac{\chi(\omega')}{\omega' - \omega}, \qquad \Im\omega > 0. \tag{11A.7}$$

Here the integration is taken along the real axis, and we have assumed that $\chi(\omega)$ vanishes sufficiently rapidly for $\omega \to \infty$ to eliminate any contribution from the large semicircle needed to close the contour integral. Taking the limit $\Im\omega \to 0$, and using (B.73), we then obtain

$$\chi(\omega) = \frac{1}{\pi i} P \int_{-\infty}^{+\infty} \mathrm{d}\omega' \, \frac{\chi(\omega')}{\omega' - \omega}, \tag{11A.8}$$

or equivalently

$$\chi'(\omega) = \frac{1}{\pi} P \int_{-\infty}^{+\infty} \mathrm{d}\omega' \, \frac{\chi''(\omega')}{\omega' - \omega},$$

$$\chi''(\omega) = -\frac{1}{\pi} P \int_{-\infty}^{+\infty} \mathrm{d}\omega' \, \frac{\chi'(\omega')}{\omega' - \omega}. \tag{11A.9}$$

These relations may be written in the form (11A.1) using the symmetry properties (11A.6).

We stress that because the Kramers–Kronig relations are based on the general principles of linearity,[305] stationarity, and causality, they are *exact*. In particular, they are valid quantum mechanically as well as classically.

Since

$$\chi(\omega) = \frac{\epsilon(\omega) - 1}{4\pi}, \tag{11A.10}$$

the real and imaginary parts of the complex dielectric constant satisfy similar relations. Thus

$$\epsilon'(\omega) - 1 = \frac{2}{\pi} P \int_{0}^{\infty} \mathrm{d}\omega' \, \frac{\omega'\epsilon''(\omega')}{\omega'^2 - \omega^2}. \tag{11A.11}$$

It is also possible to derive Kramers–Kronig relations for the complex refractive index $n(\omega) = \epsilon(\omega)^{1/2}$, even though $n(\omega)$ does not itself express a linear causal relationship between two physical quantities. One finds that[306–309]

$$n'(\omega) - 1 = \frac{2}{\pi} P \int_{0}^{\infty} \mathrm{d}\omega' \, \frac{\omega' n''(\omega')}{\omega'^2 - \omega^2}. \tag{11A.12}$$

Using the relation (11.9) between $n''(\omega)$ and $A(\omega)$, this may be written

$$n'(\omega) - 1 = \frac{c}{\pi} P \int_0^\infty d\omega' \frac{A(\omega')}{\omega'^2 - \omega^2}, \tag{11A.13}$$

which gives a direct relation between dispersion and absorption.

As one consequence of (11A.13), consider the integral

$$\int_0^\infty d\omega\, A(\omega)(n'(\omega) - 1) = \frac{c}{\pi} P \int_0^\infty d\omega \int_0^\infty d\omega' \frac{A(\omega)A(\omega')}{\omega'^2 - \omega^2}. \tag{11A.14}$$

The double integral on the right vanishes, since the integrand is antisymmetric in ω and ω'. Therefore

$$\int_0^\infty d\omega\, A(\omega)(n'(\omega) - 1) = 0, \tag{11A.15}$$

so that $n'(\omega) - 1$ and $A(\omega)$ are orthogonal functions considered in frequency space.

The relation between the dielectric constant and infrared absorption

Consider (11A.11) in the limit $\omega \to 0$. Writing $\epsilon'(0) = \epsilon$, the static dielectric constant, we have

$$\epsilon - 1 = \frac{2}{\pi} \int_0^\infty d\omega \frac{\epsilon''(\omega)}{\omega}, \tag{11A.16}$$

and using (11.11) it follows that

$$\epsilon = 1 + \frac{2c}{\pi} \int_0^\infty d\omega\, A(\omega) n'(\omega) \omega^{-2}, \tag{11A.17}$$

which consitutes a sum rule for the absorption coefficient. Since $n'(\omega)$ and $A(\omega)$ are positive definite, (11A.17) shows that $\epsilon > 1$.[310] A different relation between ϵ and $A(\omega)$ follows from (11A.12). We find that

$$\epsilon = \left(1 + \frac{c}{\pi} \int_0^\infty d\omega\, A(\omega)\omega^{-2}\right)^2. \tag{11A.18}$$

Comparing (11A.17) and (11A.18), we find that

$$\int_0^\infty d\omega\, A(\omega)(n'(\omega) - 1)\omega^{-2} = \frac{c}{2\pi} \left(\int_0^\infty d\omega\, A(\omega)\omega^{-2}\right)^2, \tag{11A.19}$$

which complements (11A.15).

The relations (11A.17) and (11A.18) are exact. In practice, an approximation is often used in which one restricts the integration to the far-infrared rotational absorption band. We then have

$$\epsilon - \epsilon_\infty \approx \frac{2c}{\pi} \int_0^{\omega_\infty} d\omega \, A(\omega) n'(\omega) \omega^{-2},$$

$$\epsilon^{1/2} - n_\infty \approx \frac{c}{\pi} \int_0^{\omega_\infty} d\omega \, A(\omega) \omega^{-2}, \qquad (11A.20)$$

where $\epsilon_\infty \equiv \epsilon'(\omega_\infty)$ and $n'_\infty \equiv n'(\omega_\infty)$ are the dielectric constant and refractive index, respectively, at a frequency ω_∞ lying above molecular translational and rotational frequencies, but below any vibrational and electronic absorption bands. From an experimental point of view, these relations provide a check on the consistency of measurements of ϵ, ϵ_∞ and $A(\omega)$; the second form is preferred because $n'(\omega)$ is not generally available across absorption bands.[311] It is now recognized that discrepancies once attributed[312] (in the case of C_2H_4) to vibrational difference bands, or (in the case of SF_6) to undetected collision-induced absorption bands, actually reflect experimental error in the measurement of the high-frequency refractive index.[313,314]

Quantum expresssions for the dielectric constant

Kramers–Kronig relations may also be used to derive a quantum-mechanical fluctuation formula for the dielectric constant. For this purpose it is necessary to consider local field effects related to the absorption of radiation (cf. § 11.1.3). Joslin[315] has shown that when these are properly taken into account, one obtains the simple relation

$$\frac{\epsilon - 1}{\epsilon + 2} = \frac{8\pi}{9\hbar V} M_{-1}, \qquad (11A.21)$$

where

$$M_{-1} = \int d\omega \, \omega^{-1} I(\omega) \qquad (11A.22)$$

is the first negative moment of the infrared absorption spectrum. While moments of negative index cannot be computed from the short-time behaviour of $C(t)$, an expression for M_{-1} can be obtained directly from (11.17). We find that

$$M_{-1} = \sum_{mn} P_m \frac{|\langle m|\boldsymbol{\mu}|n\rangle|^2}{\omega_n - \omega_m}, \qquad (11A.23)$$

where $|m\rangle$ and $|n\rangle$ are eigenstates of the system with eigenvalues $\hbar\omega_m$ and $\hbar\omega_n$. From (C.15), this is clearly related to the polarizability of the system, and we can write

$$\frac{\epsilon - 1}{\epsilon + 2} = \frac{4\pi}{3V} \langle \alpha \rangle, \qquad (11A.24)$$

where

$$\langle \alpha \rangle = \frac{2}{3} \sum_{mn} P_m \frac{|\langle m|\boldsymbol{\mu}|n\rangle|^2}{\hbar(\omega_n - \omega_m)} \qquad (11A.25)$$

is a weighted sum of states, with the principal contribution originating in the ground electronic manifold. In the Born–Oppenheimer approximation, this can be decomposed into a sum of nuclear and electronic terms, much as was done for Raman scattering in § 11.2. The electronic terms are associated with virtual transitions outside the electronic ground state, and can be related to the molecular optical polarizabilities. The nuclear terms represent transitions within the ground state. They can themselves be decomposed into rototranslational and vibrational terms; the latter are associated with contributions to the polarizability arising from molecular nonrigidity, while the former describe dipole moment fluctuations caused by the rotation and relative translation of the molecules of the system.[315]

To obtain a semiclassical expansion of the quantum ϵ in powers of $\hbar$, we may use (11.52). This yields

$$M_{-1} = \frac{\beta\hbar}{2}\left(M_0 - \frac{1}{3}\left(\frac{\beta\hbar}{2}\right)^2 M_2 + \frac{2}{15}\left(\frac{\beta\hbar}{2}\right)^4 M_4 + O(\hbar^6)\right). \quad (11A.26)$$

Note that in this expression M_0 is the *quantum* zeroth moment, which may be related to its classical counterpart using the Wigner–Kirkwood expansion (11.93).

As an example, consider a dilute gas of rigid asymmetric tops. Here $M_0^{\mathrm{qu}} = M_0^{\mathrm{cl}} = N\mu_1^2$, since μ_1^2 is a constant. To lowest order in $\hbar^2$, from (11.53) and (11.106), $M_2 = \left(2N\mu_1^2/\beta\right)(\mathrm{Tr}\,\mathbf{B}_1 - \mathbf{u}_1\mathbf{u}_1 : \mathbf{B}_1)$, where $\mathbf{B}_1 = \frac{1}{2}\mathbf{I}_1^{-1}$, $\mathbf{u}_1$ is a unit vector in the direction of the molecular dipole moment, and $\mathbf{I}_1$ is the molecular inertia tensor. Thus the leading quantum correction to the first dielectric virial coefficient $A_\epsilon = \lim_{\rho\to 0}\rho^{-1}(\epsilon - 1)/(\epsilon + 2)$ is

$$\Delta A_\epsilon = -\frac{2\pi}{27}\beta^2\hbar^2\mu_1^2\,(\mathrm{Tr}\,\mathbf{B}_1 - \mathbf{u}_1\mathbf{u}_1 : \mathbf{B}_1), \quad (11A.27)$$

in agreement with the result given in § 10.6.

The f-sum rule

We next consider (11A.11) in the limit $\omega \to \infty$:

$$\epsilon'(\omega) - 1 \to -\frac{2}{\pi}\frac{1}{\omega^2}\int_0^\infty \mathrm{d}\omega'\,\omega'\epsilon''(\omega'). \quad (11A.28)$$

Using (11.11), (11A.10), and (11A.15), we can rewrite this as

$$\int_0^\infty \mathrm{d}\omega\, A(\omega) = -\frac{2\pi^2}{c}\lim_{\omega\to\infty}[\omega^2\chi'(\omega)]. \quad (11A.29)$$

The limit will be finite provided $\chi'(\omega)$ is $O(\omega^{-2})$ for large ω. This constitutes a sum rule for the integrated intensity; for historical reasons, it is referred to as the *f*-sum rule.[54]

The right-hand side of (11A.29) can often be evaluated quite simply.[316] Consider the following classical argument. We suppose a system of linear, dipolar molecules is subjected to a weak electric field $\mathbf{E} = E_0\mathbf{e}\exp(-i\omega t)$. For sufficiently large ω, intermolecular forces cannot affect the molecular motion. The rotational equation of motion is therefore $\ddot{\boldsymbol{\theta}} = \boldsymbol{\tau}/I$, where from (2.142) $\boldsymbol{\tau} = \boldsymbol{\mu} \times \mathbf{E}$. This causes a change in $\boldsymbol{\mu}$ given by

$$\Delta\boldsymbol{\mu} = \Delta\boldsymbol{\theta} \times \boldsymbol{\mu} = (I\omega^2)^{-1}\boldsymbol{\mu} \times (\boldsymbol{\mu} \times \mathbf{E}). \tag{11A.30}$$

The induced polarization $\mathbf{P} = \rho\langle \mathbf{e} \cdot \Delta\boldsymbol{\mu}\rangle$, where ρ is the number density. Therefore $P = \chi' E$, where the susceptibility

$$\chi'(\omega) = \frac{\rho}{I\omega^2}\langle \mathbf{e} \cdot \boldsymbol{\mu} \times (\boldsymbol{\mu} \times \mathbf{e})\rangle = -\frac{2\rho\mu^2}{3I\omega^2}. \tag{11A.31}$$

We see that, indeed, $\chi'(\omega) \sim \omega^{-2}$ at large ω. Combining (11A.29) and (11A.31), we find that

$$\int_0^\infty \mathrm{d}\omega\, A(\omega) = \frac{4\pi^2\rho\mu^2}{3cI}. \tag{11A.32}$$

The integrated intensity is seen to be independent both of the nature of the intermolecular forces and of the temperature. Note that (11A.32) also follows as a consequence of (11.54) and (11.105).

References and notes

1. In this chapter equations are written using the esu/Gaussian system. For conversion to the SI system of units, see Appendix D.3.
2. Nonlinear-response spectra may be expressed as the transforms of many-time correlation functions.
3. The use of time correlation function methods in spectroscopy seems to have originated in the treatment of collisional broadening of absorption and emission lines of low density gases;[4–9] for a paedagogical discussion, see ref. 9. Later the method was applied to many other kinds of spectra, e.g. light scattering (Rayleigh,[10] Raman,[11,12] and Brillouin[13]), neutron,[12,14] X-ray,[12] and electron[12] scattering, and magnetic resonance[7,15] and Mössbauer[16] spectra.

4. Lenz, W. *Z. Physik*. **80**, 423 (1933).
5. Lindholm, E. *Arkiv Mat. Astr. Fys*. **28**, 1 (1941); ibid. **32**, 1 (1945).
6. Foley, H. M. *Phys. Rev*. **69**, 616 (1946).
7. Anderson, P. W. *Phys. Rev*. **76**, 647 (1949); ibid. **86**, 809; *J. Phys. Soc. Japan* **9**, 316 (1954).
8. Baranger, M. *Phys. Rev*. **111**, 481, 494 (1958); ibid. **112**, 855 (1958).
9. Gray, C. G. *Am. J. Phys*. **40**, 491 (1972).
10. Komarov, L. I., and Fisher, I. Z. *Sovi. Phys. JETP* **16**, 1358 (1963); Pecora, R., *J. Chem. Phys.* **40**, 1604 (1964).
11. Cowley, R. A. *Proc. Phys. Soc. (London)* **84**, 281 (1964).
12. Sjölander, A. in *Phonons and Phonon Interactions*, ed. T. A. Bak, Benjamin, New York (1964), p. 76.
13. Mountain, R. D. *J. Res. Natl. Bur. Std*. **A70**, 207 (1966); *Rev. Mod. Phys*. **38**, 205 (1966).
14. Van Hove, L. *Phys. Rev*. **95**, 249 (1954).
15. Kubo, R., and Tomita, K., *J. Phys. Soc. Japan* **9**, 888 (1954), and refs. therein.
16. Singwi, K. S. in ref. 12, p. 167.

17. We define the (ω, t) Fourier transform as in (11.1). Some authors replace i by $-i$, which leads to the correlation function $\langle A(0)A(t)\rangle$ instead of $\langle A(t)A(0)\rangle$ (the difference is significant in the quantum case). Our definition leads naturally to the definition of the complex susceptibility as $\chi = \chi' + i\chi''$, whereas the other convention leads to $\chi = \chi' - i\chi''$.
18. For a modern discussion of this theorem see, e.g., ref. 19, p. 553.
19. McQuarrie, D. A. *Statistical Mechanics*, Harper and Row, New York (1976).
20. Besides the spectroscopic examples, there are many other early derivations of the fluctuation-dissipation theorem, e.g. in thermodynamics, dielectric theory, transport theory, and electrical noise theory, by Gibbs, Einstein, Onsager, Kirkwood, Nyquist and others; see, e.g., Takahasi, H. *J. Phys. Soc. Japan* **7**, 439 (1952). The theorem was formalized by Callen, H. B., and Welton, T. A. *Phys. Rev.* **83**, 34 (1951).
21. Kubo, R. *Rep. Prog. Phys.* **29**, 255 (1966), and refs. therein.
22. See, e.g., refs. 19, 23–35.
23. Clarke, J. H. R. *Adv. Infrared Raman Spec.* **4**, 109 (1978).
24. Yarwood, J., and Arndt, R. in *Molecular Association*, Vol. 2, ed. R. Foster, Academic Press, London (1979), p. 267; Yarwood, J. *Ann. Rep. Prog. Chem.* **76**, 100 (1979).
25. Bailey, R. T. Specialist periodical reports of the Chemical Society, Molecular spectroscopy **2**, 173 (1974).
26. French, M. J. Specialist periodical reports of the Chemical Society, Molecular spectroscopy **1**, 229 (1973).
27. Griffiths, J. E. *Vibrational Spectra and Structure* **6**, ed. J. Durig, Elsevier, Amsterdam (1977), p. 273.
28. Perrot, M., and Lascombe, J. ref. 29, p. 49.
29. Buckingham, A. D., Lippert, E., and Bratos, S. (Eds), *Organic Liquids*, Wiley, Chichester (1978).
30. Bratos, S., and Pick, R. M. (eds.), *Vibrational Spectroscopy of Molecular Liquids and Solids*, Plenum, New York (1980).
31. Marabella, L. J. *Appl. Spectr. Rev.* **7**, 313 (1973).
32. Kivelson, D., and Madden, P. A. *Ann. Rev. Phys. Chem.* **31**, 523 (1980).
33. Oxtoby, D. W. *Adv. Chem. Phys.* **40**, 1 (1979); *Ann. Rev. Phys. Chem.* **32**, 77 (1981).
34. Van Kranendonk, J. (ed.), *Intermolecular Spectroscopy and Dynamical Properties of Dense Systems*, North-Holland, Amsterdam (1980).
35. Steele, W. A., ref. 34, p. 325; ref. 30, p. 61; *Adv. Chem. Phys.* **34**, 1 (1976).
36. The method of moments has a long history in both mathematics[37–39] and physics.[7,15,40–46]
37. Akhiezer, N. I. *The Classical Moment Problem*, Oliver and Boyd, Edinburgh (1965).
38. Shohat, J. A., and Tamarkin, J. D. *The Problem of Moments*, 2nd. edn, Am. Math. Soc., Providence (1950).
39. David, F. N. *Probability Theory for Statistical Methods*, Cambridge University, Cambridge (1951).
40. Kubo, R. *J. Phys. Soc. Japan* **12**, 570 (1957); *Lectures in Theoretical Physics*, ed. W. Brittin, Vol. 1, Interscience, New York (1959), p. 120.
41. Van Vleck, J. H. *Phys. Rev.* **74**, 1168 (1948). See also Abragam, A. *The Principles of Nuclear Magnetism*, Oxford University Press, Oxford (1961), p. 97.
42. Maradudin, A. A., Montroll, E. W., Weiss, G. H., and Apatova, I. P. *Theory of Lattice Dynamics in the Harmonic Approximation*, 2nd. edn, Academic Press, New York (1971), and refs. therein.
43. Vinti, J. P. *Phys. Rev.* **41**, 432 (1932); Bethe, H. A., and Salpeter, E. E., *Quantum Mechanics of One- and Two-Electron Atoms*, Plenum, New York (1977), p. 255, and refs. therein.
44. Lax, M. *J. Chem. Phys.* **20**, 1752 (1952).

45. Placzek, G. *Phys. Rev.* **86**, 377 (1952).
46. De Gennes, P. G. *Physica* **25**, 825 (1959).
47. See Pasini, P., and Zannoni, C. *Mol. Phys.* **52**, 749 (1984), for a discussion of moments of uniaxially oriented systems.
48. Paul, R. *Mol. Phys.* **41**, 1 (1980), and refs. therein.
49. Joslin, C. G., and Gray, C. G. *Mol. Phys.* **58**, 789 (1986).
50. Wilde, R. E. *J. Chem. Phys.* **74**, 4334 (1981).
51. Roney, P. L. *J. Chem. Phys.* **102**, 4757 (1995).
52. Gruszka, M., and Borysow, A. *Mol. Phys.* **93**, 1007 (1998).
53. This law was first proposed by P. Bouguer in 1729, and later by J. H. Lambert in 1760. It is often referred to, rather loosely, as the Beer–Lambert law. In fact, Beer's contribution (in 1851) was to suggest that $A(\omega)$ was additionally proportional to the density—which is not, in general, true.
54. Van Vleck, J. H., and Huber, D. L. *Rev. Mod. Phys.* **49**, 939 (1977); Huber, D. L., and van Vleck, J. H. *Rev. Mod. Phys.* **38**, 187 (1966).
55. Ref. 19, Appendix F.
56. Operationally $A(\omega)$ is defined through Lambert's law (11.10) only for positive ω. An extension to negative frequencies is, however, useful in the formal development of the theory.
57. Cohen-Tannoudji, C., Diu, B., and Laloë, F. *Quantum Mechanics*, Vol. 1, Wiley (1977), p. 312 *et seq.*
58. To derive (11.34) we note that the energy flux, or Poynting vector, is continuous across the boundary of the system (this neglects reflected radiation). The outside and inside fluxes are given by $\rho_0 v_0$ and ρv, respectively, where ρ is the energy density of radiation and v the velocity of the light wave. Since $\rho \propto DE = \epsilon' E^2$, $\rho_0 \propto E_0^2$, and $v = c/n'$, $v_0 = c$, we get $(E/E_0)^2 = n'/\epsilon' = 1/n'$.
59. Polo, S. R., and Wilson, M. K. *J. Chem. Phys.* **23**, 2376 (1955). See also Chako, N. Q. *J. Chem. Phys.* **2**, 644 (1934).
60. Gray, C. G., and Joslin, C. G. *Optical, Electric and Magnetic Properties of Molecules: A Review of the Papers of A. D. Buckingham*, ed. D. C. Clary and B. J. Orr, Elsevier, Amsterdam (1997). Buckingham's theory of solvent shifts of vibrational fundamental and overtone bands is reviewed here. An extension, with applications to liquid N_2 and N_2/Ar, is given by Alessi, N., Tolokh, I. S., Goldman, S., and Gray, C. G. *Mol. Phys.* **103**, 2381 (2005).
61. For reviews of experiments in this field, see refs. 23–27, 31, 62–64.
62. Bulanin, M. O. *J. Mol. Struct.* **19**, 59 (1973).
63. Evans, M. Evans, G., and Davies, R. *Adv. Chem. Phys.* **44**, 255 (1980).
64. Vincent-Geisse, J., ref. 30, p. 117; Vincent-Geisse, J. and Soussen-Jacob, J., ref. 29, p. 71.
65. For linear molecules the radiative dipole matrix element $\langle J'm'|\mu_\mu|Jm\rangle$ (where $\mu_\mu \propto Y_{1\mu}$—see (A.165)) can be evaluated using (A.40), and is proportional to $C(J1J';m\mu m')C(J1J';000)$, where C is a Clebsch–Gordan (CG) coefficient. The CG coefficients obey the triangle selection rule $\Delta(J1J')$ (see (A.131)), or $J' = J+1, J, J-1$. These selection rules are based on the rotational transformation laws of $|Jm\rangle$ and $\boldsymbol{\mu}$. The parity selection rule $J' \neq J$ is contained in the CG coefficient $C(J1J';000)$ (see (A.155)), and is based on the parity properties of $|Jm\rangle$ and $\boldsymbol{\mu}$. For nonlinear molecules the selection rules derive from a consideration of the matrix elements $\left\langle J'm'n'|D^{1*}_{\mu\nu}|Jmn\right\rangle$, where the wavefunction $\langle\Omega|Jmn\rangle \propto D^{J}_{mn}(\Omega)^*$. These can be evaluated using (A.94). The parity selection rule is absent, so that $J \to J$ transitions are allowed.

66. This relation may be derived from (11.17) by assuming that vibrational and rotational motions are separable. This implies that the spectrum may be deconvoluted into independent vibrational and rotational parts, each satisfying separate detailed balance criteria. Any observed deviations from (11.39) indicate the presence of vibrational or vibration–rotation perturbations.
67. Flygare, W. H. *Molecular Structure and Dynamics*, Prentice-Hall, Englewood Cliffs (1978), p. 226.
68. Bernath, P. F. *Spectra of Atoms and Molecules*, Oxford University Press, Oxford (1995).
69. There is an extensive literature on the theory of overlapping spectral lines. See, e.g. Gordon, R. G., and McGinnis, R. P. *J. Chem. Phys.* **55**, 4898 (1971); Koszykowski, M. L., and Marcus, R. A. *J. Chem. Phys.* **68**, 1216 (1978); Lévi, G., Marsault, J.-P., Marsault-Herail, F., and McClung, R. E. D. *J. Chem. Phys.* **63**, 3543 (1975); Van-Thanh, N., Bouanich, J. P. and Rossi, I. *Mol. Phys.* **40**, 869 (1980); Leicknam, J. C., Guissani, Y., and Bratos, S. *J. Chem. Phys.* **68**, 3380 (1978); Boulet, C., Boissoles, J., and Robert, D. *J. Chem. Phys.* **89**, 625 (1988); Bratos, S., Leicknam, J. C., and Guissani, Y. *J. Mol. Struct.* **47**, 15 (1978); Roney, P. L. *J. Chem. Phys.* **101**, 1037, 1050 (1994); and ref. 8.
70. The density in amagat units is the ratio of the number density to that of an ideal gas at STP (i.e. 0° C and 1 atmosphere), viz. 2.687×10^{19} molecules cm^{-3}.
71. Van der Elsken, J., Wegdam, G. H., Frenkel, D., and van Aalst, R. M. in *Molecular Spectroscopy of Dense Phases*, ed. M. Grosmann, S. G. Elkomoss, and J. Ringeissen, Elsevier, Amsterdam (1976), p. 529; for lower densities, see van Kreveld, M. E., van Aalst, R. M., and van der Elsken, J. *J. Chem. Phys.* **55**, 2853 (1971).
72. Burshtein, A. I., and Temkin, S. I. *Spectroscopy of Molecular Rotation in Gases and Liquids*, Cambridge University Press, Cambridge (1994).
73. Weber, A. (ed.), *Raman Spectroscopy of Gases and Liquids*, Springer-Verlag, Berlin (1979).
74. See refs. 32, 33, 35 and 75.
75. Vaughan, W. E. *Ann. Rev. Phys. Chem*. **30**, 103 (1979); Chesnoy, J., and Gale, G. M. *Adv. Chem. Phys.* **70**, Pt. 2, 297 (1988); Westlund, P.-O., and Lynden-Bell, R. M. *Mol. Phys.* **60**, 1189 (1987).
76. Breene, R. G., Jr. *Theories of Spectral Line Shape*, Wiley, New York (1981).
77. See e.g. ref. 52, which describes the simulation of the FIR CIA spectrum of carbon dioxide.
78. As a rule the intensities of overtone (harmonics $2\omega_v$ etc. of the fundamental frequency ω_v) and combination bands (sum and difference frequencies $\omega_v \pm \omega_{v'}$, $\omega_v \pm 2\omega_{v'}$ etc.) are determined mainly by 'mechanical' anharmonicity (anharmonic terms in the intramolecular vibrational potential), with smaller contributions from 'electrical' anharmonicity (q^2 etc. terms in (11.40)). See for example Lehman, K. L., and Smith, A. M. *J. Chem. Phys.* **93**, 6140 (1990). For a counterexample in which electrical anharmonicity contributions dominate, see Joslin, C. G., Gray, C. G., Goddard, J. D., Goldman, S., Yang, J., and Poll, J. D., *Chem. Phys. Lett.* **213**, 377 (1993).
79. Intramolecular vibration–rotation or *Coriolis* effects can occasionally be appreciable, particularly for degenerate modes.[68] Intermolecular coupling may also be important. Thus, suppose the molecular interaction is such as to change the bond length (a vibrational perturbation). This will affect the moment of inertia, and hence the rotational motion. In addition, the anisotropic intermolecular force parameters (dipole moments, quadrupole moments, polarizabilities, shape factors, etc.) depend on the bond lengths and bond angles, so that the torque experienced by a molecule depends on the vibrational state of its neighbours. Conversely, as a molecule rotates,

vibration–rotation coupling leads to perturbations of its vibrational motion. Effects on correlation functions have been investigated both theoretically[35,80,81] and experimentally.[23,24,27,81]

80. Müller, K., and Kneubühl, F. *Chem. Phys.* **8**, 468 (1975); Bratos, S., and Chestier, J. P. *Phys. Rev.* **A9**, 2136 (1974); Cattani, M., Van-Thanh, N., and Rossi, I. *Can. J. Phys.* **52**, 2313 (1974); Dreyfus, C., Breuillard, C., Nguyen-Tan, T., Grosjean, A., and Vincent-Geisse, J. *J. Mol. Struct.* **47**, 41 (1978); Van Woerkom, P. C. M., de Bleyser, J., de Zwart, M., and Leyte, J. C. *Chem. Phys.* **4**, 236 (1974); Lynden-Bell, R. M. *Mol. Phys.* **36**, 1529 (1978); Berens, P. H., and Wilson, K. R. *J. Chem. Phys.* **74**, 4872 (1981).
81. Goldberg, H. S., and Pershan, P. S. *J. Chem. Phys.* **58**, 3816 (1973).
82. Of course in condensed media one knows there must be *some* correlation between neighbouring molecules.[24,33] In pure solids, for example, vibration–vibration coupling is responsible for the vibrational exciton splittings.[83] From the observed exciton splitting one can estimate the strength of the vibration–vibration interaction and its importance for liquid vibrational band widths; it is generally found to be small.[84]
83. Van Kranendonk, J., and Karl, G. *Rev. Mod. Phys.* **40**, 531 (1968); Thiéry, M. M., and Chandrasekharan, V. *J. Chem. Phys.* **67**, 3659 (1977) and refs. therein; Decius, J. C., and Hexter, R. M. *Molecular Vibrations in Crystals*, McGraw-Hill, New York (1977), p. 246.
84. Hanson, F. E., and McTague, J. P. *J. Chem. Phys.* **72**, 1733 (1980).
85. The expansions for $\coth x$ and $\tanh x$ converge only for $|x| < \pi/2$ (i.e. $\hbar\omega < \pi kT$), so that (11.51) are in general asymptotic expansions.
86. In liquids, the mean vibrational frequency is shifted from its value in the gas as a result of molecular interactions (see Tables 1.2 and 1.3). Vibrational lines are also broadened by intermolecular interactions, as a result of both frequency modulation (dephasing) and amplitude modulation (relaxation).
87. Goldstein, H. *Classical Mechanics*, 2nd. edn, Addison-Wesley, Reading (1980), p. 174.
88. Cabana, A., Bardoux, R., and Chamberland, A. *Can. J. Chem.* **47**, 2915 (1969).
89. Dreyfus, C., Dayan, E., and Vincent-Geisse, J. *Mol. Phys.* **30**, 1453 (1975); Dreyfus, C., Berreby, L., Dayan, E., and Vincent-Geisse, J. *J. Chem. Phys.* **68**, 2630 (1978); Dreyfus, C., Berreby, L., and Dayan, E. *Chem. Phys. Lett.* **79**, 476 (1981).
90. Gordon, R. G. *J. Chem. Phys.* **38**, 1724 (1963); ibid. **39**, 2788 (1963); **41**, 1819 (1964).
91. Armstrong, R. L., Blumenfeld, S. M., and Gray, C. G. *Can. J. Phys.* **46**, 1331 (1968).
92. Armstrong, R. L., and Welsh, H. L. *Can. J. Phys.* **43**, 547 (1965).
93. Rothschild, W. G. *J. Chem. Phys.* **53**, 990 (1970).
94. Marsault, J.-P., Marsault-Herail, F., and Lévi, G. *Mol. Phys.* **33**, 735 (1977).
95. Dreyfus, C., and Cartigny, J. *J. Chem. Phys.* **80**, 5388 (1984); Dreyfus, C., Balou, D., and Brigot-Dutartre, N. *J. Chem. Phys.* **80**, 5393 (1984).
96. Berreby, L., and Dayan, E. *Mol. Phys.* **48**, 581 (1983).
97. Berreby, L., and Dayan, E. *Chem. Phys. Lett.* **243**, 85 (1995).
98. Kalmykov, Y. P., and Titov, S. V. *Mol. Phys.* **98**, 1907 (2000).
99. Expand the double commutator to $\langle \boldsymbol{\mu} \cdot H\boldsymbol{\mu} - \mu^2 H - H\mu^2 + \boldsymbol{\mu} \cdot H\boldsymbol{\mu} \rangle$, and replace the term $\langle H\mu^2 \rangle$ by $\langle \mu^2 H \rangle$, which follows from the cyclic invariance of the trace $\mathrm{Tr}\, H\mu^2 P$, and the fact that H commutes with $P = \exp(-\beta H)/Q$.
100. This relation applies not just to $\boldsymbol{\mu}$ but to any vector function of the coordinates alone. See ref. 57, p. 705; or Messiah, A. *Quantum Mechanics*, Vol. 2, North-Holland, Amsterdam (1961), p. 532.
101. Crawford, M. F., Welsh, H. L., and Locke, J. L. *Phys. Rev.* **75**, 1607 (1949).

102. Reddy, S. P., Xiang, F., and Varghese, G. *Phys. Rev. Lett.* **74**, 367 (1995).
103. Moraldi, M., and Frommhold, L. *Phys. Rev.* **A49**, 4508 (1994). See also ref. 104, p. 41.
104. Tabisz, G. C., and Neuman, M. N. (eds), *Collision- and Interaction-Induced Spectroscopy*, NATO ASI Series C **452**, Kluwer Academic, Dordrecht (1995).
105. Dagg, I. R., Anderson, A., Yan, S., Smith, W., Joslin, C. G., and Read, L. A. A. *Can. J. Phys.* **64**, 763 (1986).
106. Wishnow, E. H., Gush, H. P., and Ozier, I. *J. Chem. Phys.* **104**, 3511 (1996); McKellar, A. R. W. *J. Chem. Phys.* **88**, 4190 (1988).
107. Lorenz, R. D., McKay, C. P., and Lunine, J. I. *Science* **275**, 642 (1997).
108. Borysow, A. *Astron. Astrophys.* **390**, 779 (2002).
109. Hunt, K. L. C., and Li, X., ref. 104, p. 61.
110. For calculations on mixed rare-gas pairs see Cacheiro, J. L., Fernández, B., Marchesan, D., Coriani, S., Hättig, C., and Rizzo, A. *Mol. Phys.* **102**, 101 (2004).
111. Li, X., Champagne, M. H., and Hunt, K. L. C. *J. Chem. Phys.* **109**, 8416 (1998).
112. For complete bibliographies up to 1990 see Hunt, J. L., and Poll, J. D. *Mol. Phys.* **59**, 163 (1986), and also *A First Update for a Second Bibliography on Collision-Induced Absorption* (1990), available from J. L. H. at the University of Guelph, Guelph, Ontario, Canada.
113. Birnbaum, G. (ed.). *Phenomena Induced by Intermolecular Interactions*, NATO ASI Series B **127**, Plenum, New York (1985).
114. Frommhold, L. *Collision-Induced Absorption in Gases*, Cambridge University Press, Cambridge (1993).
115. Li, X., and Hunt, K. L. C. *J. Chem. Phys.* **105**, 4076 (1996).
116. Li, X., and Hunt, K. L. C. *J. Chem. Phys.* **107**, 4133 (1997).
117. Hunt, K. L. C., and Bohr, J. E. *J. Chem. Phys.* **84**, 6141 (1986).
118. Fowler, P. W. *Chem. Phys.* **143**, 447 (1990).
119. This assumes $\alpha = 11.08$ au, $B = -164.3$ au, $C_6 = 64.30$ au; see ref. 115.
120. Jansen, L. *Phys. Rev.* **125**, 1798 (1962).
121. Cooper, A. R., Jain, S., and Hutson, J. M. *J. Chem. Phys.* **98**, 2160 (1993).
122. Dullweber, A., and Wales, D. J. *Mol. Phys.* **94**, 651 (1998).
123. Guillot, B. *J. Chem. Phys.* **91**, 3456 (1989).
124. Cooper *et al.* ignore dispersion altogether, while Dullweber and Wales and Guillot significantly underestimate its magnitude.
125. Joslin, C. G., Gray, C. G., and Gburski, Z. *Mol. Phys.* **53**, 203 (1984); Joslin, C. G., and Gray, C. G. *Chem. Phys. Lett.* **107**, 249 (1984).
126. Abramowitz, M., and Stegun, I. A. *Handbook of Mathematical Functions*, National Bureau of Standards (1964).
127. Joslin, C. G., Singh, S., and Gray, C. G. *Can. J. Phys.* **63**, 76 (1985).
128. Birnbaum, G., and Guillot, B., ref. 104, p. 1.
129. Stassen, H., and Steele, W. A. *Mol. Phys.* **89**, 1603 (1996).
130. Dagg, I. R., Anderson, A., Yan, S., Smith, W., and Read, L. A. A. *Can. J. Phys.* **63**, 625 (1985).
131. Joslin, C. G. *Can. J. Phys.* **65**, 1629 (1987).
132. Dagg, I. R., Anderson, A., Smith, W., Missio, M., Joslin, C. G., and Read, L. A. A. *Can. J. Phys.* **66**, 453 (1988).
133. Joslin, C. G., Gray, C. G., and Singh, S. *Mol. Phys.* **54**, 1469 (1985).
134. Dagg, I. R., Anderson, A., Mooney, M. C., Joslin, C. G., Smith, W., and Read, L. A. A. *Can. J. Phys.* **68**, 121 (1990).
135. Birnbaum, G., Buechele, A., Jiang, T., Joslin, C. G., and Goldman, S. *Mol. Phys.* **93**, 573 (1998).

136. Gradshteyn, I. S., and Ryzhik, I. M. *Table of Integrals, Series, and Products*, Academic Press, San Diego (1980).
137. Tipping, R. H., Brown, A., Ma, Q., Hartmann, J. M., Boulet, C., and Liévin, J. *J. Chem. Phys.* **115**, 8852 (2001).
138. Hartmann, J. M., Brodbeck, C., Flaud, P.-M., Tipping, R. H., Brown, A., Ma, Q., and Liévin, J. *J. Chem. Phys.* **116**, 123 (2002).
139. Gruszka, M., and Borysow, A. *Mol. Phys.* **88**, 1173 (1996); *J. Chem. Phys.* **101**, 3573 (1994).
140. Brink, D. M., and Satchler, G. R. *Angular Momentum*, 3rd. edn, Oxford University Press, Oxford (1993), p. 150.
141. Borysow, A., and Moraldi, M. *J. Chem. Phys.* **99**, 8424 (1993); *Phys. Rev. Lett.* **68**, 3686 (1992).
142. The terms *Rayleigh* and *Raman* have different connotations in chemistry and physics. To a chemist depolarized Rayleigh scattering means the Q-branch of the pure rotational spectrum; to a physicist it means the entire rotational spectrum, O-, Q-, and S-branches. We adopt the physics terminology here.
143. A quantum-mechanical calculation (using the old quantum theory) of the scattering cross-section was first given by Kramers and Heisenberg using second-order time-dependent perturbation theory. See Kramers, H. A., and Heisenberg, W. *Z. Phys.* **31**, 681 (1925).
144. For a discussion of resonance Raman scattering see ref. 145, p. 145, and ref. 146.
145. Koningstein, J. A., *Introduction to the Theory of the Raman Effect*, Reidel, Dordrecht (1972).
146. Hamaguchi, H. *Adv. Infrared Raman Spec.* **12**, 273 (1985).
147. For a discussion of stimulated Raman scattering see ref. 145, p. 137.
148. Loudon, R. *The Quantum Theory of light*, 2nd. edn, Oxford (1983), p. 311 *et seq.*; Placzek, G. *The Rayleigh and Raman Scattering*, UCRL-Trans-526(L), US Atomic Energy Commission, Oak Ridge (1962).
149. Tang, A., and Albrecht, A. C. in *Raman Spectroscopy*, Vol. 2 ed. H. A. Szymanski, Plenum, New York (1970).
150. Away from resonance, this approximation is a good one. For absorbing dielectric media, where $\epsilon(\omega)$ and $n(\omega)$ are both complex, treatment of local field effects is more contentious. The results probably depend on the nature of the model. See e.g. Barnett, S. M., Huttner, B., and Loudon, R. *Phys. Rev. Lett.* **68**, 3698 (1992); Barnett, S. M., Huttner, B., Loudon, R., and Matloob, R. *J. Phys.* **B29**, 3763 (1996); Scheel, S., Knöll, L., Welsch, D.-G., and Barnett, S. M. *Phys. Rev.* **A60**, 1590 (1999); Fleischhauer, M. *Phys. Rev.* **A60**, 2534 (1999).
151. See Appendix C.2.1 of Vol. 1, p. 544. For an analogous treatment in the more general case when $\boldsymbol{\alpha}$ contains an imaginary antisymmetric component, see Landau, L. D., Lifshitz, E. M., and Pitaevskiĭ, L. P. *Electrodynamics of Continuous Media*, 2nd. edn, Pergamon, Oxford (1984), p. 416 *et seq.*
152. Translational fluctuations also broaden molecular depolarized spectra to a similar extent. Except at very low densities, however, Doppler broadening is much less important than pressure broadening; thus lineshapes in gases are usually Lorentzian. See e.g. ref. 76.
153. See e.g. ref. 154, p. 90 *et seq.*, and refs. therein, especially Clark, N. A. *Phys. Rev.* **A12**, 232 (1975).
154. Balucani, U., and Zoppi, M. *Dynamics of the Liquid State*, Oxford University Press, Oxford (1994).
155. See e.g. ref. 158, Chapter 10, p. 223.
156. Clouter, M. J. *Ann. Rev. Phys. Chem.* **39**, 69 (1988).
157. Boon, J. P., and Yip, S. *Molecular Hydrodynamics*, McGraw-Hill, New York (1980).

158. Berne, B. J., and Pecora, R. *Dynamic Light Scattering*, Wiley, New York (1976).
159. Hills, G. W., Foster, R. B., and Jones, W. J. *Mol. Phys.* **33**, 1571 (1977).
160. Reh, M., Röder, M., and Versmold, H. *Mol. Phys.* **80**, 407 (1993).
161. Gelbart, W. M. *Adv. Chem. Phys.* **26**, 1 (1974).
162. Bauer, D. R., Brauman, J. I., and Pecora, R. *Ann. Rev. Phys. Chem.* **27**, 443 (1976).
163. Keyes, T. *J. Phys. Chem.* **A101**, 2921 (1997).
164. Keyes, T. *J. Chem. Phys.* **104**, 9349 (1996).
165. Ladanyi, B. M., and Klein, S. *J. Chem. Phys.* **105**, 1552 (1996).
166. Keyes, T. *J. Chem. Phys.* **106**, 46 (1997).
167. Einstein, A. *Ann. Physik.* **33**, 1275 (1910).
168. For a fuller discussion of this subject see ref. 158, p. 257 *et seq.*
169. Gordon, R. G. *J. Chem. Phys.* **40**, 1973 (1964).
170. Steele, W. A. *J. Mol. Liq.* **48**, 321 (1991).
171. Frenkel, D., and McTague, J. P. *J. Chem. Phys.* **72**, 2801 (1980).
172. Ladanyi, B. M. *J. Chem. Phys.* **78**, 2189 (1983).
173. De Santis, A., Frattini, R., and Sampoli, M. *Chem. Phys. Lett.* **114**, 71 (1985).
174. Geiger, L. C., Ladanyi, B. M., and Chapin, M. E. *J. Chem. Phys.* **93**, 4533 (1990).
175. De Santis, A., Sampoli, M., and Vallauri, R. *Mol. Phys.* **53**, 695 (1984).
176. Ricci, M. A., Ruocco, G., and Sampoli, M. *Mol. Phys.* **67**, 19 (1989).
177. McTague, J. P., and Birnbaum, G. *Phys. Rev. Lett.* **21**, 661 (1968).
178. Thibeau, M., Oksengorn, B., and Vodar, B. *J. Phys.* **29**, 287 (1968).
179. Proffitt, M. H., and Frommhold, L. *Phys. Rev. Lett.* **42**, 1473 (1979).
180. Borysow, A., and Frommhold, L. *Adv. Chem. Phys.* **75**, 439 (1989); Ulivi, L., and Frommhold, L. *A Second Bibliography on Collision-Induced Light Scattering* (1993), available from the authors.
181. Madden, P. A. in *Spectroscopy and Relaxation of Molecular Liquids*, ed. D. Steele and J. Yarwood, Elsevier, Amsterdam (1991), p. 124.
182. Birnbaum, G., Guillot, B., and Bratos, S. *Adv. Chem. Phys.* **51**, 49 (1982).
183. Frommhold, L. *Adv. Chem. Phys.* **46**, 1 (1981).
184. Buckingham, A. D., and Hands, I. D. *Chem. Phys. Lett.* **185**, 544 (1991).
185. Champagne, M. H., Li, X., and Hunt, K. L. C. *J. Chem. Phys.* **112**, 1893 (2000).
186. Pérez, J. J., Clarke, J. H. R., and Hinchliffe, A. *Chem. Phys. Lett.* **104**, 583 (1984).
187. Aleman, C., Pérez, J. J., and Hinchliffe, A. *Int. J. Mass Spectrom. Ion Processes* **122**, 331 (1992).
188. Bafile, U., Pestelli, S., Ulivi, L., and Zoppi, M. ref. 104, p. 31.
189. Pestelli, S., Bafile, U., Ulivi, L., and Zoppi, M. *Phys. Rev.* **A49**, 4602 (1994).
190. Quantum corrections to CILS moments have been considered by various authors. See, for example, Bafile, U., Ulivi, L., Zoppi, M., and Pestelli, S. *Mol. Phys.* **79**, 179 (1993); Bafile, U., Ulivi, L., Zoppi, M., Moraldi, M., and Frommhold, L. *Phys. Rev.* **A44**, 4450 (1991); Moraldi, M., Celli, M., and Barocchi, F. *Phys. Rev.* **A40**, 1116 (1989); and ref. 191.
191. Barocchi, F., Guasti, A., Zoppi, M., El-Sheikh, S. M., Tabisz, G. C., and Meinander, N. *Phys. Rev.* **A39**, 4537 (1989).
192. Bancewicz, T., Teboul, V., and le Duff, Y. *Mol. Phys.* **81**, 1353 (1994).
193. Le Duff, Y., Bancewicz, T., and Głaz, W., ref. 104, p. 423.
194. Buckingham, A. D., and Clarke, K. L. *Chem. Phys. Lett.* **57**, 321 (1978).
195. Moszynski, R., Heijmen, T. G. A., Wormer, P. E. S., and van der Avoird, A. *J. Chem. Phys.* **104**, 6997 (1996).
196. For a recent review of SAPT see Jeziorski, B., Moszynski, R., and Szalewicz, K. *Chem. Rev.* **94**, 1887 (1994). There are several attractive features of this method, not least its ability to unambiguously ascribe contributions to *ab initio* pair properties such as $\Delta\boldsymbol{\alpha}$ to the operation of induction, dispersion and overlap mechanisms.

197. Dacre, P. D. *Mol. Phys.* **45**, 17 (1982); *Can. J. Phys.* **60**, 963 (1982); ibid. **59**, 1439 (1981).
198. Joslin, C. G., Goddard, J. D., and Goldman, S. *Mol. Phys.* **89**, 791 (1996).
199. Hättig, C., Larsen, H., Olsen, J., Jørgensen, P., Koch, H., Fernández, B., and Rizzo, A. *J. Chem. Phys.* **111**, 10099 (1999).
200. Dacre, P. D. *Mol. Phys.* **47**, 193 (1982).
201. Guillot-Noël, C., Chrysos, M., le Duff, Y., and Rachet F. *J. Phys.* **B33**, 569 (2000).
202. Koch, H., Hättig, C., Larsen, H., Olsen, J., Jørgensen, P., Fernández, B., and Rizzo, A. *J. Chem. Phys.* **111**, 10108 (1999).
203. Barocchi, F., Hensel, F., and Sampoli, M. *Chem. Phys. Lett.* **232**, 445 (1995).
204. Meinander, N., Tabisz, G. C., and Zoppi, M. *J. Chem. Phys.* **84**, 3005 (1986). For Ar the MTZ model agrees quite well with the *ab initio* calculations of Joslin *et al.*[198] except at small r; this reflects the fact that MTZ chose α_1 rather too large.
205. Ladanyi, B. M., Barreau, A., and Dumon, B. *Mol. Phys.* **77**, 735 (1992).
206. The superposition approximation for the four-body distribution function sets this equal to the product of six pair functions. Improved approximations are discussed by Pizio, O. A. *Mol. Phys.* **77**, 1011 (1992).
207. Teboul, V., and le Duff, Y. *J. Chem. Phys.* **107**, 10415 (1997).
208. Woliński, K., and Sadlej, A.J. *Mol. Phys.* **75**, 221 (1992). For this system overlap terms probably affect the Raman scattering to a similar degree as the induction terms involving **E**.
209. De Santis, A., and Sampoli, M. *Mol. Phys.* **53**, 717 (1984).
210. De Santis, A., Frattini, R., Sampoli, M., and Vallauri, R. *Europhys. Lett.* **2**, 17 (1986).
211. Note that we have defined $\Delta\alpha_{00} = 3^{1/2}\Delta\alpha$. The relation between $\Delta\alpha_{2K}$ and $\Delta\alpha^{(2)}_{\alpha\beta}$ is as given in (A.231c).
212. See ref. 109, p. 66. Hunt's coefficient $A_{Jl_1l_2lL} = (3/(2J+1))^{1/2}B^{(J)}_{l_1l_2lL}$.
213. Amos, R. D., Buckingham, A. D., and Williams, J. H. *Mol. Phys.* **39**, 1519 (1980).
214. Song, X., Frattini, R., and Jonas, J. *J. Chem. Phys.* **89**, 156 (1988).
215. Mueller, A., Steele, W. A., and Versmold, H. *J. Chem. Phys.* **99**, 4993 (1993).
216. Stassen, H., and Steele, W. A. *J. Chem. Phys.* **103**, 4408 (1995).
217. Barreau, A., Chave, A., Dumon, B., Thibeau, M., and Ladanyi, B. M. *Mol. Phys.* **67**, 1241 (1989).
218. Bancewicz, T. *Mol. Phys.* **50**, 173 (1983).
219. Ladanyi, B. M. *Chem. Phys. Lett.* **121**, 351 (1985).
220. Geiger, L. C., and Ladanyi, B. M. *J. Chem. Phys.* **89**, 6588 (1988); ibid. **87**, 191 (1987); **91**, 2764 (1989).
221. Keller, M., Mueller, A., Versmold, H., and Steele, W. A. *J. Chem. Phys.* **103**, 8854 (1995).
222. Mueller, A., Reh, M., Röder, M., Steele, W., and Versmold, H. *Mol. Phys.* **85**, 233 (1995).
223. Keller, M., Mueller, A., Reh, H., Röder, M., Steele, W. A., and Versmold, H. ref. 104, p. 87.
224. Mueller, A., Versmold, H. and Steele, W. A., ref. 104, p. 107.
225. Other contributions, operating at shorter range, tend to affect the spectra more at high frequencies (see, e.g. ref. 193). Therefore second moments evaluated in the DID approximation will be somewhat less accurate than zeroth moments. When second-order DID, dispersion, and overlap contributions are included, agreement with experiment is generally improved, especially for polarized spectra. See, e.g. Bafile, U., Ulivi, L., Zoppi, M., Barocchi, F., Moraldi, M. and Borysow, A. *Phys. Rev.* **A42**, 6916 (1990); Brown, M. S., Wang, S.-K., and Frommhold, L. *Phys, Rev.* **A40**, 2276 (1989); Brown, M. S., and Frommhold, L. *Mol. Phys.* **66**, 527 (1989).

226. De Santis, A., Moretti, E., and Sampoli, M. *Mol. Phys.* **46**, 1271 (1982).
227. De Santis, A., and Sampoli, M. *Mol. Phys.* **51**, 97 (1984).
228. Bancewicz, T. *Chem. Phys. Lett.* **213**, 363 (1993).
229. Meinander, N., Penner, A. R., Bafile, U., Barocchi, F., Zoppi, M., Shelton, D. P., and Tabisz, G. C. *Mol. Phys.* **54**, 493 (1985).
230. El-Sheikh, S. M., Tabisz, G. C., and Ulivi, L. *Mol. Phys.* **72**, 345 (1991).
231. Elliasmine, A., Godet, J.-L., le Duff, Y., and Bancewicz, T. *Mol. Phys.* **90**, 147 (1997); *Phys. Rev.* **A55**, 4230 (1997).
232. Meinander, N., Tabisz, G. C., Barocchi, F., and Zoppi, M. *Mol. Phys.* **89**, 521 (1996).
233. De Lorenzi, A., de Santis, A., Frattini, R., and Sampoli, M. *Phys. Rev.* **A33**, 3900 (1986).
234. V (vertical), H (horizontal), and N (natural, or unpolarized) refer to the direction of polarization of the incident light relative to the scattering plane, which is defined by the directions of the incident and scattered beams.
235. Buckingham, A. D., and Stephen, M. J. *Trans. Faraday Soc.* **53**, 884 (1957); Gray, C. G., and Ralph, H. I. *Phys. Lett.* **A33**, 165 (1970); Ralph, H. I., and Gray, C. G., *Mol. Phys.* **27**, 1683 (1974).
236. Couling, V. W., and Graham, C. *Mol. Phys.* **87**, 779 (1996); Couling, V. W., Graham, C., and McKenzie, J. M. *Mol. Phys.* **96**, 921 (1999); Couling, V. W., and Nhlebela, R. V. *Phys. Chem. Chem. Phys.* **3**, 4551 (2001).
237. Bogaard, M. P., Buckingham, A. D., Pierens, R. K., and White, A. H. *J. Chem. Soc. Faraday* I **74**, 3008 (1978).
238. Yoshino, T., and Bernstein, H. J. *J. Mol. Spectr.* **2**, 213 (1958).
239. Egelstaff, P. A., Gray, C. G., Gubbins, K. E., and Mo, K. C. *J. Stat. Phys.* **13**, 315 (1975).
240. Lovesey, S. W. *Theory of Neutron Scattering from Condensed Matter*, Vol. 1, Oxford University Press, Oxford (1984).
241. Golub, R., Richardson, D. J., and Lamoreaux, S. K. *Ultra-Cold Neutrons*, Adam Hilger, Bristol (1991).
242. The factor $\exp(-i\omega t)$ is conventional in inelastic neutron scattering. We have adhered to this usage, which differs from that adopted in our description of infrared absorption and Raman scattering, to avoid confusion.
243. Our definition of the van Hove correlation function differs from that of ref. 14 by a factor of ρ^{-1}.
244. March, N. H., and Tosi, M. P. *Atomic Dynamics in Liquids*, Macmillan, New York (1976).
245. Steele, W. A. in *Spectroscopy and Relaxation of Molecular Liquids*, ed. D. Steele and J. Yarwood, Elsevier (1991), p. 63.
246. For a dilute atomic gas $G(r,t) = G_{\text{self}}(r,t)$, $F(k,t) = F_{\text{self}}(k,t)$ and $S(k,\omega) = S_{\text{self}}(k,\omega)$.
247. Galindo, A., and Pascual, P. *Quantum Mechanics I*, Springer-Verlag (1990), p. 81.
248. Moraldi, M., Celli, M., and Barocchi, F. *Phys. Rev.* **A46**, 7561 (1992).
249. Miyazaki, K., and de Schepper, I. M. *Phys. Rev. Lett.* **87** 214502 (2001); *Phys. Rev.* **E63**, 060201 (2001).
250. Verkerk, P., Bafile, U., Barocchi, F., de Graaf, L. A., Suck, J.-B., and Mutka, H. *Phys. Rev. Lett.* **67**, 1262 (1991).
251. Sears, V. F. *Can. J. Phys.* **44**, 1279 (1966).
252. Sears, V. F. *Can. J. Phys.* **45**, 237 (1967).
253. The averages over angular momentum can be performed explicitly, if desired, to yield expressions for the intermediate scattering functions in terms of confluent hypergeometric functions. See St. Pierre, A. G., and Steele, W. A. *Phys. Rev.* **184**, 172 (1969).

254. This is valid provided $(\omega_{\mathrm{rot}}/\omega_{\mathrm{vib}})^2 \ll 1$, where ω_{rot} and ω_{vib} are molecular rotational and vibrational frequencies; typically $\omega_{\mathrm{rot}}/\omega_{\mathrm{vib}} \approx 0.01$.
255. Landau, L. D., and Lifshitz, E. M. *Fluid Mechanics*, 2nd. edn, Pergamon, Oxford (1987).
256. See however Bafile, U., Verkerk, P., Barocchi, F., de Graaf, L. A., Suck, J.-B., and Mutka, H. *Phys. Rev. Lett.* **65**, 2394 (1990), and Youden, J., Egelstaff, P. A., Mutka, J., and Suck, J.-B. *J. Phys. Cond. Matter* **4**, 8945 (1992), who have studied the transition between the hydrodynamic and kinetic regimes in gases at intermediate densities using small-angle neutron scattering.
257. Egelstaff, P. A., *An Introduction to the Liquid State*, 2nd edn, Oxford (1992).
258. Hansen, J. P., and McDonald, I. R. *Theory of Simple Liquids*, 2nd. edn, Academic Press, San Diego (1986).
259. The intensity of a sound wave travelling through a medium decreases exponentially with distance x travelled. We have $I = I_0 \exp(-\alpha x)$, where the absorption coefficient $\alpha = 2\Gamma k^2/c_S$.[255]
260. Zwanzig, R. *J. Chem. Phys.* **43**, 714 (1965).
261. De Schepper, J.M., and Cohen, E. G. D. *J. Stat. Phys.* **27**, 223 (1982). Note that this is a mean-field theory, which treats binary collisions as dynamically uncorrelated events. Correlation effects have been treated by Leutheusser, E. *J. Phys.* **C15**, 2801, 2827 (1982), using a mode-coupling approximation.
262. Kamgar-Parsi, B., Cohen, E. G. D., and de Schepper, I. M. *Phys. Rev.* **A35**, 4781 (1987); Cohen, E. G. D., de Schepper, I. M., and Zuilhof, M. J. *Physica* **B127**, 282 (1984); De Schepper, I. M., Cohen, E. G. D., and Zuilhof, M. J. *Phys. Lett.* **A101**, 399 (1984); Zuilhof, M. J., Cohen, E. G. D., and de Schepper, I. M. *Phys. Lett.* **A103**, 120 (1984).
263. Bruin, C., Michels, J. P. J., van Rijs, J. C., de Graaf, L. A., and de Schepper, I. M. *Phys. Lett.* **A110**, 40 (1985).
264. De Schepper, I. M., van Rijs, J. C., van Well, A. A., Verkerk, P., de Graaf, L. A., and Bruin, C. *Phys. Rev.* **A29**, 1602 (1984).
265. De Schepper, I. M., Verkerk, P., van Well, A. A., de Graaf, L. A., and Bruin, C. *Phys. Rev. Lett.* **50**, 974 (1983).
266. De Schepper, I. M., Ernst, M. H., and Cohen, E. G. D. *J. Stat. Phys.* **25**, 321 (1981).
267. Götze, W., and Lücke, M. *Phys. Rev.* **A11**, 2173 (1975); Bosse, J., Götze, W., and Lücke, M. *Phys. Rev.* **A17**, 434 (1978).
268. Berne, B. J., and Harp, G. D. *Adv. Chem. Phys.* **17**, 63 (1970).
269. Mori, H. *Prog. Theor. Phys.* **33**, 423 (1965); ibid. **34**, 399 (1965).
270. Evans, M. W., Grigolini, P., and Parravicini, G. P. (eds), *Adv. Chem. Phys.* **62**, 81 (1985).
271. Bosse, J., Götze, W., and Lücke, M. *Phys. Rev.* **A17**, 447 (1978).
272. Salacuse, J.J., and Egelstaff, P. A. *Phys. Rev.* **E64**, 051201 (2001).
273. Backx, P., Lonngi, P. A., and Egelstaff, P. A. *J. Phys. Cond. Matter* **5**, 4265 (1993).
274. García-Hernández, M., Martínez, J. L., Bermejo, F. J., Chahid, A., and Enciso, E. *J. Chem. Phys.* **96**, 8477 (1992); Chahid, A., Bermejo, F. J., Enciso, E., García-Hernández, M., and Martínez, J. L. *J. Phys. Cond. Matter* **4**, 1213 (1992).
275. Kneller, G. R., and Geiger, A. *Mol. Phys.* **70**, 465 (1990).
276. Ricci, M. A., Rocca, D., Ruocco, G., and Vallauri, R. *Phys. Rev.* **A40**, 7226 (1989), *Phys. Rev. Lett.* **61**, 1958 (1988); Ricci, M. A., Ruocco, G., and Vallauri, R. *Helv. Phys. Acta* **62**, 676 (1989).
277. Impey, R. W., Madden, P. A., and McDonald, I. R. *Mol. Phys.* **46**, 513 (1982).
278. Wojcik, M., and Clementi, E. *J. Chem. Phys.* **85**, 6085 (1986).
279. Bermejo, F. J., Alvarez, M., Bennington, S. M., and Vallauri, R. *Phys. Rev.* **E51**, 2250 (1995).

280. Sciortino, F., and Sastry, S. *J. Chem. Phys.* **100**, 3881 (1994).
281. Saito, S., and Ohmine, I. *J. Chem. Phys.* **102**, 3566 (1995).
282. Balucani, U., Brodholt, J. P., and Vallauri, R. *J. Phys. Cond. Matter* **8**, 9269 (1996).
283. Sampoli, M., Ruocco, G., and Sette, F. *Phys. Rev. Lett.* **79**, 1678 (1997).
284. Alonso, J., Bermejo, F. J., García-Hernández, M., Martínez, J. L., Howells, W. S., and Criado, A. *Phys. Lett.* **A172**, 177 (1992); *J. Chem. Phys.* **96**, 7696 (1992).
285. Ricci, M. A., Ruocco, G., Rocca, D., and Vallauri, R. *J. Chem. Phys.* **93**, 9012 (1990).
286. Ullo, J. J., and Yip, S. *J. Chem. Phys.* **85**, 4056 (1986).
287. Anderson, J., Ullo, J. J., and Yip, S. *J. Chem. Phys.* **86**, 4078 (1987).
288. García-Hernández, M., and Enciso, E. *J. Phys. Cond. Matter* **2**, 6659 (1990).
289. Petrillo, C., Sacchetti, F., Dorner, B., and Suck, J.-B. *Phys. Rev.* **E62**, 3611 (2000).
290. Swanson, D. G. *Plasma Waves*, Academic Press, San Diego (1989).
291. Sears, V. F. *Phys. Rev.* **A31**, 2525 (1985).
292. Kneller, G. R. *Mol. Phys.* **83**, 63 (1994).
293. Bonella, S., Ciccotti, G., and Coker, D. F. *Mol. Phys.* **89**, 1203 (1996).
294. Sears, V. F. *Phys. Rev.* **A5**, 452 (1972).
295. Gray, C. G., Wang, S. S., and Gubbins, K. E. *Chem. Phys. Lett.* **26**, 610 (1974).
296. Twu, C. H., Gray, C. G., and Gubbins, K. E. *Mol. Phys.* **27**, 1601 (1974).
297. Note that $G_0 = \rho kT\chi - 1$, where χ is the isothermal compressibility; cf. (3.113).
298. Blum, L., and Narten, A. H. *J. Chem. Phys.* **64**, 2804 (1976).
299. Fredrikze, H. *Mol. Phys.* **48**, 903 (1983).
300. Dupuis, M. *Prog. Theor. Phys.* **35**, 752 (1966).
301. Sears, V. F. *Can. J. Phys.* **48**, 616 (1970); ibid. **47**, 199 (1969).
302. Kramers, H. A. *Atti Congr. Int. Fisici, Como* **2**, 545 (1927); *Collected Scientific Papers*, North-Holland, Amsterdam (1956), p. 333.
303. Kronig, R. de L. *J. Optical Soc. Am.* **12**, 547 (1926).
304. Marsden, J. E. *Basic Complex Analysis*, Freeman, San Francisco (1973).
305. The Kramers–Kronig relations can be extended to nonlinear response functions. See, e.g. Flytzanis, C. in *Quantum Electronics: A Treatise*, Vol. 1, ed. H. Rabin and C. L. Tang, Academic Press, New York (1975), p. 9.
306. Newton, R. G., *Scattering Theory of Waves and Particles*, 2nd. edn, Springer-Verlag, New York (1982), p. 103 *et seq.*
307. Byron, F. W., Jr., and Fuller, R. W. *Mathematics of Classical and Quantum Physics*, Vol. 2, Addison-Wesley, Reading (1970), p. 344 *et seq.*
308. Nussenzveig, H. M. *Causality and Dispersion Relations*, Academic Press, New York (1972).
309. Cardona, M. in *Optical Properties of Solids*, ed. S. Nudelman and S. S. Mitra, Plenum Press, New York (1969), Chapter 6, p. 137.
310. See also Dolgov, O. V., Kirzhnits, D. A., and Maksimov, E. G. Rev. Mod. Phys. **53**, 81 (1981).
311. Adjouri, C., Balu, N., Obriot, J., and Bose, T. K. *J. Chem. Phys.* **106**, 7491 (1997).
312. Birnbaum, G., and Bose, T. K. *J. Chem. Phys.* **71**, 17 (1979).
313. Bose, T. K., Boudjarane, K., Huot, J., and St-Arnaud, J. M. *J. Chem. Phys.* **89**, 7435 (1988).
314. St-Arnaud, J. M., and Bose, T. K. *J. Chem. Phys.* **71**, 4951 (1979).
315. Joslin, C. G., unpublished.
316. If the absorbing particles are point charges q with mass m (e.g. electrons in an atom), the RHS is proportional to q^2/m, and we have the well-known Thomas–Reiche–Kuhn sum rule from atomic physics. See, e.g. ref. 54 and Bethe, H. A., and Jackiw, R. W. *Intermediate Quantum Mechanics*, 2nd. edn, Benjamin, New York (1968), p. 215.

INDEX

This is a combined index for Volume 1: Fundamentals and Volume 2: Applications.